Handbook of Applied Mycology

Volume 4: Fungal Biotechnology

Handbook of Applied Mycology

Series Editor
Dilip K. Arora

Banaras Hindu University
Varanasi, India

Volume 1 Soil and Plants, edited by *Dilip K. Arora, Bharat Rai, K. G. Mukerji*, and *Guy R. Knudsen*

Volume 2 Humans, Animals, and Insects, edited by *Dilip K. Arora, Libero Ajello*, and *K. G. Mukerji*

Volume 3 Foods and Feeds, edited by *Dilip K. Arora, K. G. Mukerji*, and *Elmer H. Marth*

Volume 4 Fungal Biotechnology, edited by *Dilip K. Arora, Richard P. Elander, and K. G. Mukerji*

Volume 5 Mycotoxins in Ecological Systems, edited by *Deepak Bhatnagar, Eivind B. Lillehoj, and Dilip K. Arora*

Handbook of Applied Mycology
Volume 4: Fungal Biotechnology

edited by

Dilip K. Arora

Banaras Hindu University
Varanasi, India

Richard P. Elander

Bristol-Myers Squibb Company
Syracuse, New York

K. G. Mukerji

University of Delhi
Delhi, India

MARCEL DEKKER, INC. NEW YORK • BASEL

ISBN 0-8247-8501-0

This book is printed on acid-free paper.

MARCEL DEKKER, INC.
270 Madison Avenue, New York, New York 10016

Current printing (last digit):
10 9 8 7 6 5

PRINTED IN THE UNITED STATES OF AMERICA

HANDBOOK INTRODUCTION

Applied mycology is one of the most stimulating and rapidly evolving areas of the biological sciences. It encompasses many facets of agricultural, industrial, pharmacological, medical, and food sciences, all of which include organisms or processes that are subject to biotechnological manipulation. However, published information on the practical significance of fungi is fragmented in many specialized publications. There is an obvious need for a comprehensive treatment of the basic principles, methods, and applications of mycology as an applied, integrated, and multidisciplinary subject. The potential usefulness of such a handbook is suggested by the practical significance of mycological research, by the uniqueness and increasing coherence of the field's conceptual directions, and by the international scope of the subject. These five volumes of the *Handbook of Applied Mycology* attempt to fulfill this need. These volumes present and collate the major aspects of applied mycology and are designed as a modern standard reference for the work in this field.

In developing these volumes we decided that the treatment of the subject should be comprehensive but concise enough to enable completion within a set of five volumes published closely together rather than a continuous encyclopedic series taking several years to complete. However, this does not preclude the possibility of publishing supplementary volumes, as appropriate, to communicate new information. Further, we wanted to develop a treatise that would offer a detailed overview of applied aspects of mycology in terms of its theoretical, methodological, and empirical contributions. By emphasizing the interdisciplinary aspects of the subject, we hope that the handbook will highlight recent prospective linkages among diverse research paradigms.

Since the information presented in these volumes is expected to serve students as well as veterans in the field, both basic and advanced specialized materials are included. In addition to addressing researchers in mycology, the handbook will prove useful to a broad spectrum of professional groups, including biotechnologists, medical scientists, molecular biologists, food scientists, pharmacologists, soil microbiologists, microbial ecologists, and plant pathologists. Each chapter is intended to provide a balanced view of the current state of knowledge within an expanding field of interest. The selection of the chapters and overall organization of the book were guided by several insightful colleagues from around the world. These volumes were authored by a large team of international experts. Every chapter within the five volumes has been edited by a distinguished group of scientists. As a result, nine editors and over 200 authors representing 20 countries have contributed to this project.

Because of the multidisciplinary nature of the subject, judgment had to be made about the relative importance and significance of the hundreds of potential topics related to applied mycology. Also, a balance had to be struck between comprehensive coverage and intellectual appeal of the field. Readers should find this work useful for in-depth information and as a route to additional information contained in its critical discussions and wide range of references provided. Each of the five volumes is intended to be self-contained. Therefore, some degree of duplication of materials, especially basic principles, is inevitable.

No work of this magnitude can be accomplished without the support and contributions of many individuals. The planning and eventual production of this handbook were a collaborative effort. I am deeply indebted to the group of coeditors and many colleagues who have assisted me throughout all stages of the project. I also appreciate the dedication and hard work of the chapter authors, who cooperated persistently in the production of these volumes. I acknowledge with gratitude the support of M. V. Wiese, N. W. Schaad, S. Stotzsky, J. S. Singh, Bharat Rai, R. S. Upadhyay, V. S. Jaiswal, B. Thompson, Jon Olson, A. K. Pandey, Rejeev Gaur, and Shusma Gupta, whose interest and cooperation have been invaluable to me. I am grateful to the University of Idaho for providing assistance in the form of staff support and facilities, and the Banaras Hindu University, India, for granting leave during the period I edited these volumes. I appreciate the encouraging support provided by Ms. Sandra Beberman of Marcel Dekker, Inc., at all stages of production of this book. My gratitude is expressed to my teacher, J. L. Lockwood, and to my parents, whose example has always been my guiding principle. Finally, to Meenakshi, Sidhartha, and Gautam for their patience, encouragement, and willingness to forgo other interests so that the handbook might be completed, I am grateful.

Dilip K. Arora

PREFACE

Fungal biotechnology can be defined as the application of fungal organisms or their subcellular components along with associated process technology applied to manufacturing service industries and environmental management. Successful application of biotechnology processes associated with fungi is achieved through the combined integration of a number of scientific disciplines and technologies including genetics, molecular biology, biochemistry, and chemical and process engineering.

Biotechnology processes using fungi normally involve the production of cells or mycelial biomass for the purpose of catalyzing desired chemical transformations. The transformations can further be divided into the production of useful end products (steroid conversions; antibiotics; and biochemical products from intermediary metabolism such as amino acids, alcohols, purines and pyrimidines, lipids, polysaccharides, and other structural cell wall components). Other transformations involve the decomposition of biological and chemical materials (bioremediation of chemical and biological wastes, sewage disposal, hydrocarbon degradation, etc.).

The practical use of newer genetic techniques such as protoplast fusion and genetic engineering has now created a new dimension for fungal biotechnology. The new genetic technologies, the use of immobilized enzymes, and powerful computerized and robotic process technologies now available for fungal fermentation processes will culminate in continued expansion of the fungal technology industry.

The industrial use of fungi is now at the forefront of biotechnology along with the use of other microbial and animal cells. Recent advances in the understanding of the molecular biology of fungi will further increase the growth of fungal technology in both fundamental and applied aspects. The major use of fungi continues to be associated with their ability to produce economically large amounts of a variety of small molecules and excreted enzymes. The present contributions of biotechnology to the current world economy are estimated to be in excess of $15 billion. Two chapters in this volume deal with β-lactam antibiotics produced commercially by strains of *Penicillium* and *Cephalosporium*. In 1990, the worldwide sales of penicillins and cephalosporins were estimated to be in excess of $8 billion. Thus, two species of fungi contribute more than one-half of the world's biotechnology market! Other currently important marketed products include cellulolytic enzymes produced by engineered strains of *Trichoderma*, the use of cyclosporin antibiotics for human organ transfer produced by strains of *Tolypocladium*, and the use of important cholesterol-lowering compactin molecules produced by several species of *Penicillium* and *Aspergillus*. The worldwide sales of these latter fungal secondary metabolites are estimated to be in excess of $1 billion.

Despite the abundance of new biotechnology literature, there remains a scarcity of compiled literature related strictly to fungal biotechnology. It is our intention that this volume, the fourth in the *Handbook of Applied Mycology* series, will partially fill this void.

The major purpose of the volume is to highlight important commercial aspects of fungi to the rapidly expanding biotechnology industry. With this in mind, we assembled a list of contributors who are international experts from both industry and academia. The contents are important to biotechnology business analysts as well as to biotechnologists.

The 34 chapters are divided into major categories of molecular technologies; commercial applications; fungal decomposition of biological and chemical wastes; and culture collections, aspects of legal protection for biotechnology inventions, and biosafety.

It is impossible to cover all aspects of fungal biotechnology in a single volume. Indeed, most of the topics related to food biotechnology have been covered in Volume 3 of the *Handbook of Applied Mycology: Foods and Feeds* (Marcel Dekker, Inc., 1991). In the selection process, we have intentionally overlooked aspects of fungal biotechnology dealing with baker's and brewer's yeast mass cell production, steroid conversions, etc., as these areas are usually well covered in standard industrial microbiology and biotechnology texts. We feel the chapters devoted to the potential use of mycorrhizal fungi, fungal polysaccharides, phytohormone production by fungi, use of thermophilic fungi in agriculture, fungal degradation of industrial wastes and wastewater, fungal degradation of lignin, etc., represent areas of fungal biotechnology that currently receive little attention in the published biotechnology literature. It is our opinion that these less publicized subject categories represent important future fungal biotechnology applications.

We are indebted to many authors for their up-to-date discussions on various topics. We thank Ms. Sandra Beberman and Mr. Andrew Berin for their dedicated assistance in editorial structuring.

Dilip K. Arora
Richard P. Elander
K. G. Mukerji

CONTENTS

Handbook Introduction iii
Preface v
Contributors xi

Part I: MOLECULAR TECHNOLOGIES

1. Fungal Biotechnology: An Overview 1
Richard P. Elander and David A. Lowe

2. Molecular Methods for Identification and Taxonomy of Filamentous Fungi 35
Maren A. Klich and Edward J. Mullaney

3. Isozyme Analysis in Fungal Taxonomy and Molecular Genetics 57
J. A. Micales, M. R. Bonde, and G. L. Peterson

4. The Application of Molecular Genetic Methods to Filamentous Fungi 81
Alan Upshall

5. Protoplast Fusion in Fungi 101
Michael E. Gadau and A. J. Lingg

6. Isolation of Fungal Protoplasts 129
Teruyoshi Hashiba

7. Transformation and Gene Manipulation in Filamentous Fungi: An Overview 151
Theo Goosen, Cees J. Bos, and Henk van den Broek

8. The Application of Genetic Engineering to Strain Improvement of β-Lactam-Producing Filamentous Fungi 197
Shu-Jen D. Chiang and Richard P. Elander

9. The Fungal Mitochondrial Genome—A Broader Perspective 213
Michael E. S. Hudspeth

10. The Structure and Function of Plasmids in Filamentous Fungi 243
Tim Rubidge

Part II: COMMERCIAL APPLICATIONS

11. The Biochemistry of Some Fungal Polysaccharides with Industrial Potential 259
Brian J. Catley

12. Chitin Biosynthesis in Fungi 281
Jose Ruiz-Herrera, Rafael Sentandreu Ramon, and J. Pedro Martinez

13. Production and Biosynthesis of Amino Acids by Fungi 313
Eva M. Kubicek-Pranz and Christian P. Kubicek

14. Organic Acids of Fungi 357
Ramunas Bigelis and Dilip K. Arora

15. Potential for Fungal Lipids in Biotechnology 377
John D. Weete and Shailesh Gandhi

16. Carotenoid Biosynthesis and Its Regulation in Fungi 401
Peter M. Bramley and Alan Mackenzie

17. Nucleic Acids and Nucleotides 445
David H. Griffin

18. Biosynthesis and Physiology of Ergot Alkaloids 475
Helena Sočič and Vladimira Gaberc-Porekar

19. Production of Penicillins and Cephalosporins by Fungi 517
Abdel-Halim M. M. El-Sayed

20. Fungal Plant Growth Regulators 565
B. K. Lonsane and P. K. R. Kumar

21. Production of Ethanol by Yeast 603
Tahía Benítez, Enrique Martínez-Force, Juan Jiménez, and Andrés Aguilera

22. Fungal Enzymes 681
David A. Lowe

23. The Role of Thermophilic Fungi in Agriculture 707
H. S. Shekhar Sharma and B. N. Johri

24. Biotechnological Potential of Thermophilic Fungi 729
T. Satyanarayana, B. N. Johri, and Joachim Klein

25. Fungal Degradation of Lignin: Biotechnological Applications 763
K. Boominathan and C. Adinarayana Reddy

26. VA Mycorrhizal Fungi: Challenges for Commercialization 823
Tim Wood

27. Theoretical Considerations in the Commercial Utilization of Mycorrhizal Fungi 849
R. Whetten and Anne J. Anderson

Part III: DECOMPOSITION OF BIOLOGICAL AND CHEMICAL WASTE

28. Metabolism of Foreign Compounds by Fungi 881
H. G. Raj, Manju Saxena, Abdelamir Allameh, and K. G. Mukerji

29. Hydrocarbon-Degrading Yeasts and Filamentous Fungi of Biotechnological Importance 905
Nicholas D. Lindley

30. Fungal Degradation of Industrial Wastes and Wastewater 931
Ratna Nandan and S. Raisuddin

31. Use of Fungi for Bioconversion of Distillery Waste 963
Jožefa Friedrich, Aleska Cimerman, and Anton Perdih

Part IV: CULTURE COLLECTIONS, LEGAL ASPECTS, AND BIOSAFETY

32. Current Status of Fungal Culture Collections and Their Role in Biotechnology 993
Shung-Chang Jong and Jeannette M. Birmingham

33. Legal Protection for Biotechnology Inventions 1025
Roman Saliwanchik and David R. Saliwanchik

34 Biohazard and Biosafety in Biotechnological Industry 1055
Eitan Israeli, Janina Gitelman, and Daniel F. Liberman

Index 1091

CONTRIBUTORS

Andrés Aguilera* Department of Biochemistry, New York University, New York, New York

Abdelamir Allameh† Biochemistry Department, Vallabhbhai Patel Chest Institute, University of Delhi, Delhi, India

Anne J. Anderson Department of Biology, Utah State University, Logan, Utah

Dilip K. Arora Centre of Advanced Study in Botany, Banaras Hindu University, Varanasi, India

Tahía Benítez Department of Genetics, Faculty of Biology, University of Seville, Seville, Spain

Ramunas Bigelis Biotechnology Department, Technology Division, Amoco Research Center, Amoco Technology Company, Naperville, Illinois

Jeannette M. Birmingham Mycology and Botany Department, American Type Culture Collection, Rockville, Maryland

M. R. Bonde Foreign Disease—Weed Science Research Unit, Agricultural Research Service, United States Department of Agriculture, Frederick, Maryland

K. Boominathan Department of Microbiology and Public Health, Michigan State University, East Lansing, Michigan

Cees J. Bos Department of Genetics, Wageningen Agricultural University, Wageningen, The Netherlands

---

Present affiliations:
*Department of Genetics, University of Seville, Seville, Spain.
†Department of Biochemistry, Faculty of Medical Sciences, Tarbiat Modaress University, Tehran, Iran.

Peter M. Bramley Department of Biochemistry, Royal Holloway and Bedford New College, Egham, Surrey, England

Brian J. Catley Department of Biological Sciences, Heriot-Watt University, Edinburgh, Scotland

Shu-Jen D. Chiang Genetic Engineering Department, Industrial Division, Bristol-Myers Squibb Company, Syracuse, New York

Aleska Cimerman Department of Biotechnology and Industrial Mycology, Boris Kidrič Institute of Chemistry, Ljubljana, Slovenia, Yugoslavia

Richard P. Elander Biotechnology Administration Department, Industrial Division, Bristol-Myers Squibb Company, Syracuse, New York

Abdel-Halim M. M. El-Sayed* Department of Microbiology, Faculty of Pharmacy, Zagazig University, Zagazig, Egypt

Jožefa Friedrich Department of Biotechnology and Industrial Mycology, Boris Kidrič Institute of Chemistry, Ljubljana, Slovenia, Yugoslavia

Vladimira Gaberc-Porekar Department of Biosynthesis and Biotransformations of Biologically Active Compounds, Boris Kidrič Institute of Chemistry, Ljubljana, Slovenia, Yugoslavia

Michael E. Gadau† General Medical Research, Seattle Veterans Administration Medical Center, Seattle, Washington

Shailesh Gandhi Department of Botany and Microbiology, Auburn University, Auburn, Alabama

Janina Gitelman Department of Virology, Israel Institute for Biological Research, Ness-Ziona, Israel

Theo Goosen Department of Genetics, Wageningen Agricultural University, Wageningen, The Netherlands

David H. Griffin Faculty of Environmental and Forest Biology, College of Environmental Science and Forestry, State University of New York, Syracuse, New York

Teruyoshi Hashiba Department of Plant Pathology, Faculty of Agriculture, Tohoku University, Sendai, Japan

Michael E. S. Hudspeth Department of Biological Sciences, Northern Illinois University, DeKalb, Illinois

Eitan Israeli Department of Virology, Israel Institute for Biological Research, Ness-Ziona, Israel

---

Present affiliations:
*SymBiotech, Inc., Wallingford, Connecticut.
†ICOS Corporation, Bothell, Washington.

Juan Jiménez* Department of Biochemistry, University of Oxford, Oxford, England

B. N. Johri Department of Microbiology, G. B. Pant University of Agriculture and Technology, Pantnagar, India

Shung-Chang Jong Mycology and Botany Department, American Type Culture Collection, Rockville, Maryland

Joachim Klein Gesellschaft für Biotechnologische Forschung mbH, Braunschweig, Germany

Maren A. Klich Southern Regional Research Center, Agricultural Research Service, United States Department of Agriculture, New Orleans, Louisiana

Christian P. Kubicek Institute of Biochemical Technology and Microbiology, University of Technology, Vienna, Austria

Eva M. Kubicek-Pranz Institute of Biochemical Technology and Microbiology, University of Technology, Vienna, Austria

P. K. R. Kumar† Fermentation Technology and Bioengineering Discipline, Central Food Technological Research Institute, Mysore, India

Daniel F. Liberman Environmental Medical Service, Massachusetts Institute of Technology, Cambridge, Massachusetts

Nicholas D. Lindley Department of Biochemical Engineering, Institut National des Sciences Appliquées, Toulouse, France

A. J. Lingg‡ Department of Bacteriology and Biochemistry, University of Idaho, Moscow, Idaho

B. K. Lonsane Fermentation Technology and Bioengineering Discipline, Central Food Technological Research Institute, Mysore, India

David A. Lowe Department of Fermentation Development, Industrial Division, Bristol-Myers Squibb Company, Syracuse, New York

Alan Mackenzie§ Department of Biochemistry, Royal Holloway and Bedford New College, Egham, Surrey, England

J. Pedro Martinez Departamento de Microbiologia, Facultat de Farmacia, Universitat de Valencia, Valencia, Spain

---

Present affiliations:
*Department of Genetics, Faculty of Sciences, University of Malaga, Malaga, Spain.
†Institute für Technische Chimie, University of Hannover, Hannover, Germany.
‡College of Agriculture.
§Department of Biology, Eton College, Eton, Berkshire, England.

Enrique Martínez-Force Department of Genetics, Faculty of Biology, University of Seville, Seville, Spain

J. A. Micales Forest Service, Forest Products Laboratory, United States Department of Agriculture, Madison, Wisconsin

K. G. Mukerji Department of Botany, University of Delhi, Delhi, India

Edward J. Mullaney Southern Regional Research Center, Agricultural Research Service, United States Department of Agriculture, New Orleans, Louisiana

Ratna Nandan Environmental Microbiology Division, Industrial Toxicology Research Centre, Lucknow, India

Anton Perdih Department of Chemistry and Chemical Technology, University of Ljubljana, Ljubljana, Slovenia, Yugoslavia

G. L. Peterson Foreign Disease—Weed Science Research Unit, Agricultural Research Service, United States Department of Agriculture, Frederick, Maryland

S. Raisuddin Preventive Technology Division, Industrial Toxicology Research Centre, Lucknow, India

H. G. Raj Department of Biochemistry, Vallabhbhai Patel Chest Institute, University of Delhi, Delhi, India

Rafael Sentandreu Ramon Departamento de Microbiologia, Facultat de Farmacia, Universitat de Valencia, Valencia, Spain

C. Adinarayana Reddy Department of Microbiology and Public Health, Michigan State University, East Lansing, Michigan

Tim Rubidge Chemical Defence Establishment, Salisbury, Wiltshire, England

Jose Ruiz-Herrera* Departamento de Genética y Biología Molecular, Centro de Investigación y de Estudios Avanzados, Mexico City, Mexico

David R. Saliwanchik Saliwanchik & Saliwanchik, Biotechnology Patent Law, Gainesville, Florida

Roman Saliwanchik Saliwanchik & Saliwanchik, Biotechnology Patent Law, Gainesville, Florida

T. Satyanarayana Department of Microbiology, University of Delhi, Delhi, India

---

**Present affiliation:* Instituto de Investigación en Biología Experimental, Universidad de Guanajuato, Guanajuato, Mexico.

Manju Saxena* Department of Botany, Mycology Laboratory, University of Delhi, Delhi, India

H. S. Shekhar Sharma Plant Pathology Research Division, Department of Agriculture for Northern Ireland, Belfast, Northern Ireland

Helena Sočič Department of Biosynthesis and Biotransformation of Biologically Active Compounds, Boris Kidrič Institute of Chemistry, Ljubljana, Slovenia, Yugoslavia

Alan Upshall ZymoGenetics Incorporated, Seattle, Washington

Henk van den Broek Department of Genetics, Wageningen Agricultural University, Wageningen, The Netherlands

John D. Weete Department of Botany and Microbiology, Auburn University, Auburn, Alabama

R. Whetten Department of Biology, Utah State University, Logan, Utah

Tim Wood NPI, Salt Lake City, Utah

**Present affiliation*: Department of Human Biological Chemistry and Genetics, University of Texas Medical Branch, Galveston, Texas.

Part I

MOLECULAR TECHNOLOGIES

1

FUNGAL BIOTECHNOLOGY: AN OVERVIEW

RICHARD P. ELANDER and DAVID A. LOWE *Industrial Division, Bristol-Myers Squibb Company, Syracuse, New York*

I. HISTORY AND INTRODUCTION

Man has unknowingly exploited the properties of fungi for millenia. The fermentation of grape juice and the use of fermented foods was first recorded in Sumeria and Babylon in the sixth century B.C., and since that time man has successfully used yeasts and fungi to modify less palatable food sources. Fungi have been cultivated for food in the Orient for over 2000 years, hieroglyphics of mushrooms have been found in Egyptian tombs, and many different mushroom cults have developed in Central America, Mexico, and Siberia. Although man's first encounters with fungi were beneficial, his first scientific investigations on fungi were concerned with their devastating destructive effects on plant crops. Vast inroads in the systematic knowledge of fungi were gained from studies on the pathogens of the potato (*Phytophthora infestans*), wheat rust (*Puccinia graminis*), wheat bunt (*Tilletia tritici*), corn smut (*Ustilago zeae*), vine and hop mildew (*Erysiphe* species), and coffee rust (*Hemileia vastatrix*).

The industrial use of fungi is now at the forefront of biotechnology along with the use of bacteria and animal cells. Recent advances in the understanding of the molecular biology of fungi will further increase the growth of fungal technology in both fundamental and applied aspects. The main use of fungi at present is associated with their ability to produce economically large amounts of a variety of small molecules and excreted enzymes.

The present contribution of biotechnology to the current world economy has been estimated to be in excess of $15 billion [1]. It is the intent of this chapter to highlight some of the major industrial applications of fungi with the following major objectives:

1. To describe both traditional and current genetic methodologies which have been successful in the commercial exploitation of industrial fungi
2. To describe bioprocess development and fermentation scale-up of a variety of industrial processes based on modern genetic and bioengineering technologies
3. To identify particular areas of biotechnology where molecular biology methods can be applied to develop and further exploit filamentous fungi

4. To describe new advances in modern genetic cloning technology that can be used for the synthesis of improved, novel fungal metabolites

II. DIVERSITY OF FUNGAL FERMENTATION PRODUCTS

Taxonomically, filamentous fungi comprise a heterogeneous collection of microorganisms [2]. Members of this group occupy a wide range of micro- and macrohabitats, from the heterotrophic saprophytic through symbiotic associations, as in mycorrhizae and lichens, to parasites.

The major factor associated with the biological success of filamentous fungi is their metabolic and biochemical diversity. This has allowed adaption to and exploitation of unique environments inhospitable to other organisms. A practical example can be seen in the decomposition of compost where the initial attack is by bacteria which utilize the more readily available nutrients followed by the thermophilic streptomycetes. The major decomposition is, however, carried out by fungi due to their diverse ability to utilize substrates that are recalcitrant to prior metabolism. Classically, this is displayed by the extreme specificity of the many fungal pathogens and mycorrhizal associations, and the uniqueness of the various antibiotics and other secondary metabolites produced by different saprophytic species. Such metabolic diversity has long been identified and well documented although it is not unique to fungi [3]. However, commercial exploitation by the industry has concentrated on a small number of metabolites from an equally small number of species (see Table 1). These limitations are the consequence of several factors. Obviously, the need for a compound must exist at a level which makes profitable production possible. Another, and probably a more limiting, factor has been the nature of the species producing the metabolite. All the species exploited by industry are able to grow vigorously in a defined, inexpensive, artificial culture and are capable of empirical genetic manipulation. There are many fungal species that do not meet these criteria, and the gene pools of these have been unavailable to researchers. The ability to clone genes from one host and express them in another opens a whole new avenue of exploitation and potential commercialization.

III. FUNGAL METABOLISM

A. Primary Metabolism

Primary metabolism is essentially the utilization of nutrients for the production of small molecules involved in the generation of energy and the building blocks for cell structure, reproduction, and physiological function. Fungi in their natural environment tend to colonize the surface of liquids and solids, and in doing so expose their hyphae to the surrounding air and can rapidly exhaust local nutrients. This has led to the development of efficient translocation mechanisms to fully utilize available substrates. Oxygen is readily available from the gaseous phase. Nutrients can be utilized directly or indirectly by the excretion of extracellular hydrolytic enzymes, e.g., amylases and proteases. On solid media this enables the hyphae to penetrate fully into the supporting material. In liquid culture all nutrients are available by diffusion, and homogeneous environments can easily be obtained by agitation. In this manner, the fungal organism can be presented with a wide variety of soluble and insoluble materials for utilization as nutrient sources.

TABLE 1. Some Examples of Fungi and Fungal Products

Organism	Product
	Antibiotics
Cephalosporium acremonium	Cephalosporin C
Emericellopsis sp.	Penicillin N
Fusidium coccineum	Fusidic acid
Paecilomyces varioti	Variotin
Penicillium chrysogenum	Penicillins G and V
Penicillium patulum	Griseofulvin
Tolypocladium inflatum	Cyclosporin
	Metabolites Other Than Antibiotics
Ashbya gossypii	Riboflavin
Aspergillus niger	Citric acid
Aspergillus oryzae	Kojic acid
Aspergillus terreus	Itaconic acid
Ceratocystis virescens	Flavors and fragrances
Claviceps purpurea	Alkaloids
Fusarium moniliforme	Gibberellins
Pencillium stoloniferum	Mycophenolic acid
Phycomyces blakesleanus	β-Carotene
Rhizopus sp.	Fumaric acid
	Enzymes
Aspergillus sp.	Glucoamylase
Aspergillus sp.	Lipase
Aspergillus sp.	Pectinase
Aspergillus sp.	Pentosanase
Aspergillus niger	β-Glucanase
Aspergillus niger	Glucose oxidase
Aspergillus niger	Lactase
Aspergillus oryzae	α-Amylase
Aspergillus oryzae	Protease
Fusarium spp.	Penicillin acylase
Mucor sp.	Rennet
Penicillium sp.	Dextranase
Trichoderma reesei	Cellulase
	Miscellaneous Compounds
Aspergillus sp.	Polyols
Aureobasidium pullulans	Pullulan
Fusarium sp.	Microbial protein
Metarrhizium anisopliae	Microbial insecticides
Neurospora crassa	Insulin
Pennicillium chrysogenum	Double-stranded RNA
Penicillium roquefortii	Cheese manufacture
Rhizopus nigricans	Steroid conversions
Saccharomyces sp.	Ethanol
Saccharomyces cerevisiae	Interferons
Saccharomyces cerevisiae	Vaccines
Schizophyllum commune	Schizophyllan
Trigonopsis sp.	Glycerol

1. Carbon Metabolism

Most fungi can utilize a broad range of carbohydrates, fatty acids, organic acids, and amino acids as a source of carbon for growth and for energy. First preference is usually given to the soluble monomer carbohydrates and simple fatty acid and organic acids, as these can be utilized without the synthesis and excretion of hydrolytic enzymes. Disaccharides, polysaccharides, and lipids are usually broken down by extracellular or cell surface enzymes which are under the control of glucose repression and induction by that particular substrate. Some sugars are utilized slowly either due to a slow active uptake rate, as is the case with galactose, or by the production of limited quantities of extracellular enzyme, as is the case with lactose metabolism. The utilization of insoluble substrates such as starch and cellulose is usually accomplished by a sequential action of several extracellular enzymes with somewhat broad substrate specificities that facilitate the eventual total hydrolysis of the solid.

In terms of catabolism, fungi have similar glycolytic and tricarboxylic acid cycles as well as oxidative phosphorylation mechanisms that are to be found in all living organisms. As will be discussed later, the tricarboxylic acid cycle can provide a varied source of organic acids as primary excretion products.

2. Nitrogen Metabolism

Since most fungi are saprophytes, they can utilize a wide range of nitrogenous compounds. As is the case with carbon utilization, the smallest units tend to be utilized first as they do not require the production of extra enzymes. Ammonia can be utilized by all fungi, and most fungi can use nitrate. Proteins and amino acids are excellent sources of nitrogen, which can also provide a source of carbon. Many fungi can grow on a single amino acid as a sole source of nitrogen with asparagine and glutamine being the best examples. Amino acids can be utilized directly or by initial deamination to release ammonia. The majority of amino acids are synthesized from the involvement of ammonia fixation with glutamate/glutamine and by further transamination.

The nature and level of various nitrogen sources in the environment appear to be important in controlling fungal development. For example, nitrogen exhaustion can stimulate conidiation in some fungi; in others it can lead to autolysis.

Primary metabolism, particularly of amino acids, is important in generating the building units for secondary metabolism, and it is these metabolic pathways that have been successfully altered by mutation that can lead to the initial overproduction of certain primary building units for eventual polymerization into secondary metabolites.

B. Secondary Metabolism

Fungi can produce a wide variety of secondary metabolites which are usually characterized by their complex structures, most being the polymerization of simple building units such as acetate, sugars, amino acids, nucleosides, or combinations of these. The reasons for their production are uncertain, but they are all a result of unbalanced or uncontrolled metabolism usually occurring after the growth phase has stopped. It is perhaps the adaptability of fungi for the rapid utilization of varied substrates that has led to the sudden shortage or limitation of one particular nutrient, be it carbon or nitrogen, which in turn creates an unbalanced environment suitable for secondary metabolite production. The production and importance of secondary metabolites will be discussed in detail in succeeding chapters.

IV. FUNGAL GENETICS AND STRAIN IMPROVEMENT

In contrast to the viruses, bacteria, and streptomycetes, the yeasts and filamentous fungi are eukaryotic in that they possess true delimited nuclei with discernible chromosomes. Thus, with true sexual meiotic mechanisms of crossing over and classic Mendelian genetic ratios, the fungi can be compared to higher eukaryotic plants and animals [4].

A. Life Cycles and Sexuality in Filamentous Fungi

True diploid cells are produced in the life cycles of most fungi. Unlike the partial diploids of bacteria, the homologous chromosome pairs are made up of complete chromosomes. The fungi as a group exhibit a great variety of life cycles that may be categorized into three major types: haploid, diploid, and dikaryon. The dikaryon results from delayed fusion of nuclei following anastomosis of two haploid cells. After cytoplasmic fusion, the binucleate dikaryotic cells divide mitotically to produce two daughter cells, each with the same dikaryotic nuclei. The dikaryon is found as a transient phase in the sac fungi (Ascomycetes) but may represent the dormant phase in many mushroom and toadstool fungi (Basidiomycetes).

Neurospora crassa is an ascomycetous fungus with a haploid vegetative phase and has been studied extensively as a genetic model. It is heterothallic and requires two mating types to carry out its complete life cycle. The vegetative mycelium is haploid and can be propagated almost indefinitely by serial transfers of hyphal fragments or asexual spores (conidia) to fresh nutrient media. *N. crassa* produces two types of spores—macroconidia and microconidia.

Of the two mating types, A or a, either will produce protoperithecia on appropriate media. After gametangia are produced, the sexual cycle is initiated by fertilization with spermatia or mycelium of the mating type opposite to that of the protoperithecium.

After the nucleus has migrated through a specialized structure known as a trichogyne into the protoperithecium, a dikaryon is formed which is reproduced mitotically to form a large number of dikaryotic cells which become zygotes upon fusion of pairs of the dikaryotic cells. Each fused nucleus undergoes meiosis (reduction division) to form four haploid nuclei which, in turn, undergo two successive mitotic divisions to form eight ascospores which are contained linearly in an ascus sac.

A perithecium may contain 2-300 asci, each derived from a fused nucleus, and when the perithecia mature the spores are forcibly discharged from the sac. The ascospores generally require a heat shock which induces germination.

The common baker's yeast, *Saccharomyces cerevisiae*, is a heterothallic ascomycetous budding fungus that has been used for many important genetic studies. Many yeasts are thought to be degenerate sac fungi which exist vegetatively as budding unicells in which the vegetative cell also acts as the sexual cell (ascus). *Saccharomyces* strains reproduce asexually by a process called budding, which is a mitotic division in which equal chromosomal but unequal cytoplasmic division occurs. As in *Neurospora* species, the diploid zygote of a common yeast undergoes meiosis within a closed sac; hence, all products of single meiotic division are kept together. Polyploidy is also common in yeast and arises as the result of a fusion of a haploid and a diploid or of diploid cells. Genetic analyses indicate that ploidy levels of up to the tetraploid (4N) condition are possible.

Another important ascomycetous fungus which displays a variety of genetic mechanisms is *Aspergillus nidulans*. Like species of *Neurospora*, its vegetative mechanisms convey on it certain advantages as a tool in genetic investigations. In addition to its regular asexual and sexual cycles, *A. nidulans* possesses unique heterokaryotic and parasexual mechanisms.

The normal asexual cycle consists of the germination of uninucleate conidia which produce typical haploid vegetative mycelia, a portion of which is differentiated into specialized conidial heads. The mature conidia separate and germinate, thereby repeating the cycle.

In contrast to neurosporae, which are usually heterothallic and have two mating types, *A. nidulans* is homothallic and possesses no mating types. Hence, a conidium can fertilize an ascocarp initial. In contrast to *Neurospora*, the ascocarp is a closed, thin-walled globose structure called a cleistothecium. Each cleistothecium contains numerous globose asci containing eight ascospores which are not arranged in a definite order.

B. Parasexuality

Although *A. nidulans* is generally considered to be an organism with a haploid vegetative phase, it can diploidize. The ability to form rare diploid spores conveys on it certain advantages not found in species of *Neurospora*. The unusual diploid phase in the asexual cycle of the fungus is a result of the fusion of somatic vegetative nuclei which are in the heterozygous condition. Unusual mitotic crossing over occurs at the four-strand stage, segregation is random and mitotic, and the expected reciprocal products are found. The somatic recombination cycle has been termed parasexuality by Pontecorvo and his associates.

Parasexuality has been discovered in a number of industrially important fungi, including the penicillin fungus *Penicillium chrysogenum*. The unusual cycle makes it possible to test the effects of genes in the heterozygous condition and has been used to map genes in *A. nidulans*. With respect to recombination efficiency, however, somatic recombination is 0.2% as effective as sexual recombination in *A. nidulans* [5]. A comparison of these two important recombinational mechanisms is presented in Table 2.

Filamentous microorganisms also display another unusual method for genetic variation. Many fungi show hyphal anastomosis when they are actively growing in culture. Fusion results in a mixing of both the cytoplasm and the nuclei of the fused hyphae. If the hyphae represent genetically different strains (biochemically deficient, drug-resistant, or conidial color mutants), the fusion may result in the production of a strain which will have nuclei of both strains. Such a strain is called a heterokaryon. If each parental strain has a different nutritional requirement due to gene mutation, the heterokaryon will have the wild-type allele of each and will grow on a minimal medium (prototrophic). Frequently, heterokaryotic growth is better than that obtained from homokaryotic strains [6].

Heterokaryosis is of great significance in organisms in which it is found. Many fungi which are capable of forming heterokaryons are probably heterokaryotic in their natural habitats. Heterokaryosis is also an important consideration in the cultivation and maintenance of important, higher yielding

TABLE 2. Comparison of the Sexual and Parasexual Cycles in Fungi

Step	Sexual cycle	Parasexual cycle
1	Hyphal anastomosis (plasmogamy)	Hyphal anastomosis (plasmogamy)
2	Nuclear exchange to form heterokaryon	Nuclear exchange to form heterokaryon
3	Nuclear fusion (karyogamy) in specialized sexual structures to yield "selfed" or "hybrid" ascocarps	Rare nuclear fusion in vegetative cells, heterozygotes, recognized by color or nutrition. Homozygotes, if formed, are not detected
4	Diploidization: zygote persists for one nuclear generation	Diploidization: zygote may persist through many mitotic divisions
5	Recombination at meiosis; crossing over at four-strand stage, in all chromosome pairs, random assortment of each chromosome pair followed by reduction to haploid state	Recombination by rare accident of mitosis. (1) Mitotic crossing over at four-strand stage and usually only one exchange in a single chromosome arm; (2) haploidization: independent of crossing over and random assortment of each chromosome pair
6	Products of meiosis readily recognized and isolated	Recombinants occur among vegetative cells and recognized only by use of suitable markers and selection techniques

Source: From Elander and Espenshade, 1976.

strains of industrially important microorganisms. Mutation to lower levels of productivity often occurs, which results in "rundown" and lower production yields of valuable production strains. Such strains have to be repurified by single-spore selection.

Heterokaryosis is most significant genetically since it allows for the establishment of balanced genetic systems which have a selective advantage over homokaryotic strains. Heterokaryosis may be considered a process midway between syntrophism and heterozygosity in a diploid organism. Syntrophism is a relationship between two different strains of an organism resembling symbiosis, which enables both strains to survive in a particular environment whereas neither would survive alone.

The discovery of diploid nuclei in *aspergilli* and *penicillia* has resulted in a more profound understanding of recombination potential among fungal organisms. In order to detect rare heterozygous diploid nuclei, an environment which selects for a unique phenotype from an extremely large population of parental strains is necessary. A method commonly employed to demonstrate parasexual recombination is to propagate complementary biochemically deficient strains on a nonselective medium.

Hyphal segments of copropagated mycelium are then plated on a selective medium, a condition which selects for heterokaryotic mycelium. Following anastomosis and fusion of the haploid mutant strains, the conidia of the heterokaryotic mycelium are then examined for the parental types employed for the initial mating. If parental types are found, the mycelium is heterokaryotic. Rare heterozygous conidia may also be found although their frequency of occurrence is one in a million. The spores are characteristically larger than their haploid parents, they may be more resistant to ultraviolet radiation than haploid cultures, their desoxyribonucleic acid content is doubled, their conidial color is usually of the wild phenotype, and they are prototrophic with respect to nutritional requirements. The heterozygous diploid nuclei may give rise to recombinants, either by vegetative haploidization (via random assortment of chromosomes) or by mitotic crossing over [7].

Elander [8] described a highly stable diploid strain of *P. chrysogenum* along with a spontaneous segregant that produced nearly 25% more antibiotic than its parent, which itself yielded better than the production haploid strain. Ball [9] also described recombinant segregants from diploids with improved yields.

C. Mutation and Selection

Historically, random mutation and selection have been the most successful strategies used for the manipulation and screening of fungi to generate strains that produce large amounts of commercially important products. These methods, although successful in most cases, are based on empirical experience and have, in general, a moderate probability of success. They are "hit-or-miss" methods that require brute force effort, persistence, and skill in the art of microbiology. These older technologies were developed at a time when knowledge of the genetics and biochemistry of fungal cultures was very limited.

These classical methods have in recent years been replaced by less empirical, directed selection techniques such as:

1. Direct colony selection following overlay bioassay
2. Mutants resistant to toxic precursors or to the toxicity of end products
3. Mutants resistant to analogs of intermediary metabolites such as amino acids, carbohydrates, and organic acids
4. Mutants resistant to metallic ions that complex antibiotic molecules or their biosynthetic intermediates
5. Use of auxotrophic strains deficient for biosynthetic intermediates followed by prototrophic reversion
6. Selection for improved strains after protoplast fusion
7. Use of antibiotic selection for the isolation of multiple plasmid copy strains

Examples of these techniques can be selected from studies on the industrial strain selection programs with *Penicillium chrysogenum* and *Acremonium chrysogenum* (also known as *Cephalosporium acremonium*), the cultures that produce β-lactam antibiotics. Here aminoadipic acid, cysteine, valine, and methionine have been shown to be either direct biosynthetic precursors or stimulants of the antibiotics penicillin and cephalosporin. Mutants of *Penicillium* and *Cephalosporium*, selected by their resistance to analogs of these and other related amino acids, have been examined for their β-lactam antibiotic productivity. In comparison with random selection, mutants resistant to certain methionine or lysine analogs yielded a much higher frequency of superior strains. Although methionine is not a precursor for cephalosporin

C biosynthesis, mutants resistant to methionine analogs such as methionine, trifluoromethionine, or selenomethionine have been selected in the expectation that they would overproduce methionine, thus alleviating the need for added methionine [10].

D. Protoplast Fusion

Cell and protoplast fusion represent simple approaches to gene transfer in microbial and fungal organisms. Microbial protoplasts have important value in genetic manipulations because of the relative ease with which they can be prepared and their capability for undergoing reversion to the normal cellular form [11]. Protoplast fusion also offers an approach to bypass the natural barriers to genetic exchange in nonfertile microorganisms or distantly related microorganisms. In addition, it offers an opportunity for recombinational breeding and genetic exchange in a similar manner to sexual breeding.

Polyethylene glycol (PEG)-induced fusion of protoplasts has been used effectively to force cells of differing genotypes to form transient hybrids or diploids, thus allowing chromosomes to rearrange and DNA sequences to recombine. In this manner a high frequency of genetic exchange can occur between more than two parents [12]. This advantage is applicable to industrial strain improvement programs.

Species of *Acremonium chrysogenum* that overproduce cephalosporin C have been obtained by protoplast fusion [13]. The value of protoplast fusion in developing hybrid drugs is enhanced by the possibility of interspecific genetic recombination. Interspecific hybrid formation via protoplast fusion was achieved between species of *Aspergillus* and *Penicillium* [14]. Fusions between *P. chrysogenum* and *A. chrysogenum* auxotrophs were reported by Peberdy [11] which resulted in prototroph hybrids with some intermediary morphological forms suggesting gene recombination and nutritional complementation. Stable isolates were obtained that had *Acremonium* morphology but differed in their response to phenoxyacetic acid as compared to the *Acremonium* parent in shake-flask fermentations. Protoplast fusion technology has also been applied for the development of fast-growing, low-*p*-hydroxypenicillin V-producing strains of *P. chrysogenum* [15].

E. Molecular Genetics

The molecular biology of fungi is being successfully developed and refined to the degree that commercial exploitation is now a reality [16,17]. In many cases the properties of fungi have allowed the molecular geneticists to expand their horizons, which had become somewhat limiting with working with prokaryotes. For example, fungi are known for their ability to excrete large quantities of extracellular proteins, many of which are glycosylated. In addition, fungi being eukaryotes are able to process genetic material with both introns and exons.

1. Genetic Transformation Strategies

An efficient transformation system is required for the shuttling of rDNA between fungal cultures. Transformations have been reported in at least six different genera of fungi to date (Table 3). This list will grow as the various techniques become common. Due to their extensive use and recorded genetic history, *A. nidulans* and *N. crassa* have been studied in detail. Transformation techniques developed with these two cultures

TABLE 3. Transformation in Filamentous Fungal Species

Organism	Cloned gene/product
Acremonium chrysogenum	Cyclase, expandase
Aspergillus nidulans	Orotidine-5-phosphate decarboxylase, alcA, alcR, argB, amdS, trpC, oligomycin-resistant ATP synthetase
Aspergillus niger	Glucoamylase, protease, β-glucosidase, amdS
Aspergillus awamori	Glucoamylase, protease, β-glucosidase,
Aspergillus terreus	amdS
Mucor circinelloides	Lipase, acid protease, amylase
Neurospora crassa	Glutamate dehydrogenase
Penicillium chrysogenum	Acyltransferase, cyclase
Podospora anserina	Mitochondrial DNA

have been successfully used and modified for studies on industrial cultures [17,18].

Currently, fungal transformations are carried out with the protoplast form as this naked membrane-coated state is able to accept the various forms of DNA. Two basic selection pressures have been used to facilitate these processes, namely, the use of nutritional requirements such as auxotrophic markers or acetamide utilization, and antibiotic resistance. Genes have been cloned from plasmid [19] or cosmid libraries [20] and by isolating cDNA clones from induced cultures [21-23].

From published evidence it appears that fungi have a greater potential for expressing genes from a wide variety of hosts compared to the use of prokaryotes and even yeasts. For example, *Aspergillus* can recognize transcriptional control and secretion signals from a variety of sources and can process out introns from other fungi. Transformations of industrial fungal cultures have been reported by several authors [17,24,25], e.g., the expression of rennin protease from *Mucor miehei* in *A. nidulans* at Genencor, and yield increases of glucoamylase in *A. niger* at Panlabs.

Of even greater interest is the recognized ability of fungi to express genes of higher eukaryotic origin that cannot be expressed in yeast or prokaryotes. For example, bovine chymosin (rennin) and its inactive precursor prochymosin can be produced in *A. nidulans* by the fusion of the rennin genes with the promoter and terminator of the glucoamylase gene from *A. niger* [24]. In a similar manner, using the alcohol dehydrogenase promoter from *A. nidulans*, Allelix obtained expression of interferon, and tissue plasminogen activator was expressed at ZymoGenetics [17,24]. Gene manipulation in fungi also has the potential to improve the biosynthetic ability of the host for secondary metabolite production. *A. chrysogenum* was recently shown to be transformable, and the isopenicillin N synthetase (cyclase) gene has been isolated from this organism, inserted, and expressed in *E. coli*, then reintroduced back into the original culture [26,27].

2. Plasmids and Cloning Vectors

One of the major problems in applying recombinant DNA technology to model fungal systems has been the lack of vectors suitable for the development of efficient cloning systems, i.e., plasmids that can replicate autonomously in *Neurospora*, *Aspergillus*, and *E. coli*. Of particular interest is the development of shuttle vectors in order to carry out gene manipulations such as the substitution of modified cloned sequences for resident chromosomal sequences. Stohl and Lambowitz [28] constructed useful shuttle vectors, pALS-1, which consist of a mitochondrial plasmid, pBR325, and the qa-2+ gene of *Neurospora*, and pJP102 which contains an autonomously replicating sequence (ARS). The vector made possible specific gene disruption using linear fragments carrying the am+ sequence disrupted by the selectable marker qa-2+. Weaknesses of the current *Neurospora* system are inherent instability of the plasmids and a low transformation frequency.

Transformation systems and cloning vectors for *Aspergillus* are also available [20,29]. In contrast to the *Neurospora*, the *Aspergillus* system is characterized by an increased transformation efficiency using an oligomycin resistance gene as the selectable marker [29]. A vector carrying a chromosomal sequence was used to isolate the isocitrate lyase and fawn genes by mutant complementation.

Direct cloning by transformation can be achieved even with moderate transformation frequencies (1000 transformants/μg DNA) now routinely obtained in most laboratories with *A. nidulans*. Multimarked strains are transformed and then screened for complementation of individual nonelected markers. It appears that with the current transformation frequencies reported for *A. nidulans*, any structural or regulatory, metabolic, or developmental gene can be cloned in this organism.

An efficient transformation system and cosmid-cloning vector was recently reported for the cephalosporin C fungus *A. chrysogenum* [26]. The isopenicillin N synthetase (cyclase) gene was isolated with a 20-mer probe using hygromycin phosphate transferase gene (HPT) as the selectable marker. The final cyclase gene-cloning vector consisted of an HPT gene cloned from *E. coli* placed under control of the phosphoglycerokinase promoter from *S. cerevisiae*, a fragment of *Acremonium* mitochondrial DNA which functioned as an ARS sequence in yeast, and pBR322 DNA including an origin of replication for easy isolation in *E. coli*. Expression of the cyclase gene was magnified to the point that nearly 20% of the *E. coli* protein was the cyclase enzyme.

When recloned back into *Acremonium*, there was an interesting gene dosage effect with concomitant increase in cephalosporin synthesis. The fungal gene was also expressed in cephamycin-producing strains of *Streptomyces clavuligerus* [26].

V. BIOPROCESS TECHNOLOGY

The development of a typical biotechnology process (Fig. 1) includes three steps: development on a laboratory scale, scale-up to a pilot plant vessel which closely resembles production configurations, and final adjustment and run of the process on an industrial scale.

A. Laboratory Fermentations

The goals of laboratory research are to develop (a) culture preservation procedures, (b) culture and propagation on suitable media, (c) nutrient

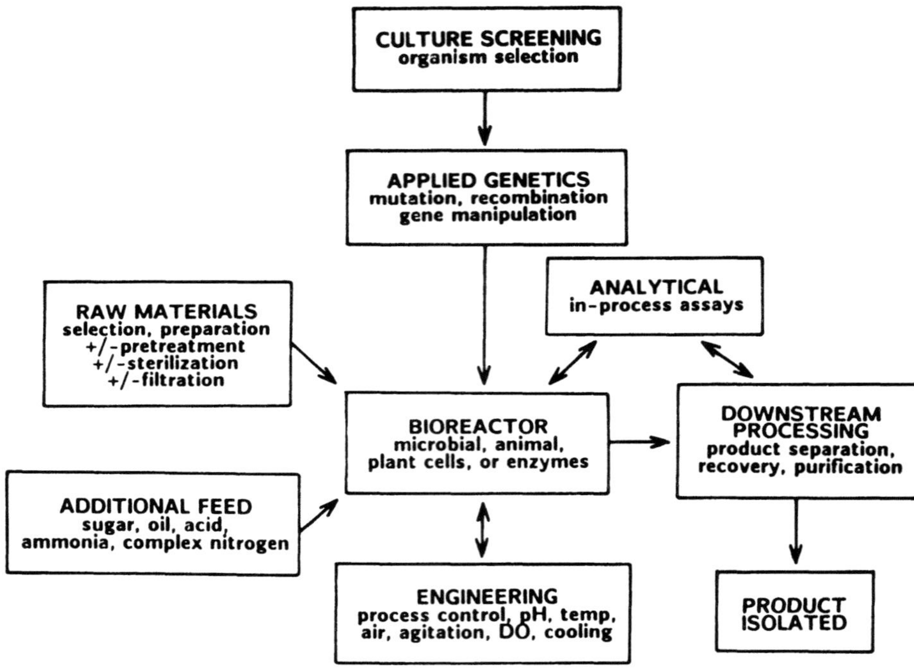

FIGURE 1. Schematic overview of a biotechnological process.

medium of suitable composition, (d) techniques of laboratory scale culture, and (e) small samples for initial extraction and product recovery studies.

The medium for propagation of the strain has to meet several criteria. It has to ensure good growth and/or sporulation, and at the same time assure satisfactory genetic stability of the strain, the latter being very important with highly mutated strains. Both the strain and its culture conditions, i.e., the composition of the nutrient medium, incubation temperature, time of incubation, etc., should be conceived as forming an integral system which can be severely disturbed by changing only some of the conditions. The appropriate preservation techniques are to be determined by studying the stability of the production strain preserved by different techniques in long-term experiments. Each type of stock culture should be prepared in a sufficient amount even when the preserved strain is not yet optimal for production. This procedure ensures the supply of a uniform inoculum material for sufficient periods of time; the stability of the strain and the efficiency of preservation must be examined in long-term studies over several years.

The next step is a detailed study of the composition of the medium which provides for the optimum quantitative ratios of individual components and ensures the maintenance of their optimum proportions throughout the process. Apart from productivity, a number of additional factors influence suitability of a medium for industrial purposes. These include the availability and price of individual raw materials, their sterilizability, requirements for an economical isolation procedure, and final purification of the product. Cost of raw materials is initially not important during early development of a process or where the product is of very high value; however, in the case of established fermentations, low-cost raw materials can contribute significantly to the final overall cost of the product.

Aeration conditions are an important aspect of laboratory studies. An optimum aeration regimen for laboratory cultures can be attained by using shakers with different types of motions, i.e., rotary or reciprocal, altering the speed of agitation, changing the shape and type of closure of the flasks and the volume of the nutrient medium. Other optimization factors include techniques of inoculation with spores or vegetative cells, the size and age of inoculum, the length of time necessary to attain maximum yield, and variations of temperature and pH.

The above scheme forms a basis for the development of a tentative laboratory process. When this is set up successfully, the factors exhibiting a crucial effect on production yields are studied in detail. This cyclic sequence of research operations is widely used for gradually stepping up the production of a given substance to an economically desirable level and, consequently, for determining optimum conditions for industrial production, and to provide a basis for the scale-up of the process to the pilot plant level.

In the case of well-established fermentation processes, the role of the laboratory is directed toward the evaluation of new, improved mutants generated in the strain improvement programs. Here raw material composition and fermentation conditions can be altered to realize the full potential of new cultures. In addition, new sources and different types of raw materials have to be evaluated in order to take advantage of the economic changes that occur in agriculturally based raw materials due to changes in seasonal weather conditions, political price supports, and new advances in process technologies.

B. Pilot Plant and Production Scale Fermentations

Initial stages in pilot plant development are directed toward obtaining reproducible results with the main emphasis on the standard course of the microbial process following changes in growth, pH, consumption of nutrients, production of metabolites, and in the physiological state of the microorganism as would be expected from laboratory results. On attaining reproducible results, the search for individual optimum conditions can begin.

The first factors to be explored are usually aeration and agitation in order to maximize mass transfer of both nutrients and oxygen to the culture. Such studies often lead to a more active fermentation which necessitates alterations in the ratio of carbon and nitrogen. Controlled growth conditions can conveniently be attained in stirred-tank fermentations by the continual feeding of carbon sources such as sucrose, molasses, corn syrups, or soybean oil. Such feeding of nutrients, together with the ability to control pH by acid or caustic addition, can produce dramatic increases in productivity. In a similar manner, ammonia gas can be mixed with the ingoing air to control pH and add inorganic nitrogen to the fermentation. Vigorous aerated fermentations are often subject to excessive foaming which, if left uncorrected, can lead to excessive loss of broth. This can be controlled by the use of antifoam agents such as silicones and polypropylene glycol. Antifoams also influence the dispersion of dissolved gases to the culture and the overall respiration rate of the culture. Judicial use of antifoam is necessary, as excess can have detrimental effects on extraction. The culture fluid from pilot plant fermentors can be used for a tentative isolation of metabolites and preliminary tests of their quality.

Inoculum and seed buildup have to be developed at the pilot stage. The propagation of a sufficient amount of inoculum is usually achieved by two- or three-stage culture in which the volume of the medium is increased in each subsequent stage. The inoculum produced in the final stage should have sufficient quality to ensure the highest yield in the shortest possible time. Modifications in the seed medium can also lead to improved culture quality and greater potential for higher productivity in the final stage.

Throughout the seed buildup and production stages, extreme caution has to be taken to maintain aseptic conditions to avoid contamination. This is particularly important during the transfer of seed broth to the production vessel. Samples should be taken as often as possible to check for the presence of microorganisms, particularly bacteria, due to their fast growth rates as compared to fungal cultures. In some antibiotic fermentations, the conditions can be self-sterilizing with regard to bacteria, but even in these cases yeast contaminations can occur.

Temperature and pH levels can be optimized at the pilot stage, although fine tuning and the establishment of improved profiles should be left until established consistent results have been obtained at the production stage.

C. Scale-up of Fermentations

The scale-up of culture procedures from the laboratory flask level to pilot plant vessel and, eventually, production fermentors involves many problems, one of which is the decrease in yields observed in fermentors as compared with the laboratory process. The main reason is the difference in conditions: aeration and agitation techniques, sterilization of the medium, raw materials, inoculation, presence of antifoam, and type of construction materials.

To be able to reproduce laboratory results in production scale fermentors in the shortest possible time, the reason for the differences in yields must be established. Parallel trials are therefore carried out in flasks on shakers and in fermentors, with a close check kept on the differences between the two runs. A prerequisite of a successful parallel experiment is satisfactory reproducibility of results in the two systems using, as nearly as possible, the same medium components. However, in some cases, e.g., where fed fermentations are used, shake-flask conditions cannot be repeated as a readily used carbon source like glucose cannot be batched in to a shake-flask medium. If parallel shake-flask experiments cannot be performed, more indirect comparisons can be studied such as measuring the utilization patterns of raw materials, i.e., ammonia, soluble protein, and amino acids, and complex carbohydrates. Cell mass, viscosity, and mycelial morphology are other important factors. More indirect measurements can include analysis of the levels of both extracellular and intracellular proteins.

In the case of mycelial cultures growing on insoluble complex raw materials, the direct measurement of cell mass can be difficult. Most measurements are indirect and are a compromise of ease of use and interpretation of results. Common techniques include centrifugation to yield a sedimentation value, dilution and settling under gravity to yield arbitrary patterns, DNA and RNA determination, estimation of unique fungal components such as ergosterol, viscosity measurement, measurement of rate of filtration, and counting of different morphological forms. All of these techniques can be used in both shake-flask and stirred-tank samples. Unique to stirred tanks is the ability to measure the concentration of gases in the spent air. Here the measurement of carbon dioxide is very useful as CO_2 is a direct consequence of the metabolism of carbon. Analysis of the rate of production and the accumulated total can be useful in following the progress of a fermentation. Unfortunately, carbon dioxide liberation cannot be measured routinely in shake flasks.

D. Production Fermentation

Typical production vessels range from 30,000 to 200,000 liters. Agitation is provided by turbine impellers at a power input of 1-4 W/liter. Volumetric oxygen uptake rates of a rapidly growing culture can attain 0.4-0.8 mmol/liter per minute, and it is necessary to supply air at rates of 0.5-1.0 v/v per minute. Compressed air is normally sterilized by using depth filters or membrane filters. The majority of the carbohydrate is supplied by feeding glucose, sucrose, corn syrup, or a similar carbon source throughout the fermentation at predetermined rates to provide a limited but sustained growth during the early stages of the fermentation, and restricted growth during the latter stage to allow for maximum production formation. However, in some fermentations it is still convenient or economical to batch in all the raw material ingredients in the starting medium. Oil, such as soybean oil, methyl oleate, or lard oil, can be fed as an alternate high-energy carbon source. Complex nitrogen is an important factor in all fermentations. Typical sources include cotton seed flour; soybean flour or meal; corn steep liquor; brewer's yeast; distiller's solubles; rape, linseed, or similar oil seeds; fish meal; and dried blood. Where sugar feeding is used, volume expansion can be very high and miniharvest schedules can be used to obtain high productivity rates.

The fermentation environment inside the tanks is controlled either by using in situ sensors or by sampling the broth and manually measuring the necessary parameters. In situ, on-line analysis can easily be applied to pH, dissolved oxygen, and exit gas CO_2 by the use of the appropriate probes and analytical instruments.

Dissolved oxygen can be used to regulate the air flow in order to use the minimumamount of air necessary to maintain predetermined dissolved oxygen levels. Exit gas analysis, by infrared for CO_2 or mass spectrometry for CO_2, O_2, and N_2, can provide metabolic information to regulate the feed rates. Periodic broth sampling is always necessary to determine concentrations of product, precursors, phosphate, ammonia, aminonitrogen, and carbohydrates. Such values can be used to regulate the addition of precursor and other required nutrients, and in addition, the recorded information can provide detailed fermentation patterns for retrospective analysis.

With very vigorous fermentations, some consideration should be given to the removal of heat from the fermentation vessel. Heat generation is autocatalytic in that raised temperatures increase metabolic activity which in turn produces more heat. This heat has to be removed by jacketed cooling in small tanks and by both jacketed and internal coils in larger tanks. In most cases, the use of mechanically chilled water is sufficient to remove all metabolic heat, provided that adequate heat transfer surfaces are available.

E. Downstream Processing

The extraction and purification of fermentation products may be difficult and costly. Practically, the high-quality product should be obtained as quickly as possible at an efficient recovery rate using minimum plant investment operated at minimal costs. Unfortunately, recovery costs of microbial products may vary from as low as 20% to as high as 60% of the total manufacturing costs [30]. However, in general, most of the labor costs occur in the extraction processes, and thus extraction efficiency and scheduling have the highest priority in the whole production train. It is often more cost-effective to harvest a fermentation at a suboptimal time than to create a gap in an extraction run.

At harvest time the specific product will be present at a varying concentration (100 μg to 50 mg/ml) in an aqueous suspension that contains intact microorganisms, cell fragments, soluble and insoluble medium components, and other metabolic products. The product may also be intracellular, heat-labile, and easily broken down by contaminating microorganisms. All of these factors tend to increase the difficulties of product recovery. To ensure good recovery or purification, speed of operation may be necessary with labile products. The processing equipment must therefore be of the correct type and correct size to ensure that the harvested broth can be processed within a satisfactory time limit.

The choice of recovery process can be based on the following:

1. Intracellular or extracellular location of the product
2. Concentration of the product in the fermentation broth
3. Physical and chemical properties of the desired product as an aid to selecting separation procedures
4. Intended use of the product
5. Minimal acceptable standard of purity
6. Impurities in the fermentor broth
7. Marketable price for the product
8. Available facilities

The first stage for the recovery of an extracellular product is the removal of large solid particles and microbial cells by centrifugation or filtration. Filtration can be achieved either by the use of filter aids or by ultra- or micromembrane filtration. With the latter, diafiltration is possible. In the next stage, the cell-free solution is fractionated or extracted using absorption or ion exchange chromatography, liquid-liquid solvent extraction, acid/organic solvent, or precipitation. Finally, the product-containing fraction is purified by fractional precipitation, chromatography, crystallization, etc. In some cases the first two stages can be combined as a whole-broth extraction, e.g., using solvent extraction or ion exchange resins with whole broths.

It may be possible to modify the handling characteristics of the broth so that it can be handled faster with simpler equipment, making use of a number of techniques:

1. Selection of a microorganism which does not produce pigments or undesirable metabolites
2. Modification of the fermentation conditions to reduce the production of undesirable metabolites
3. Precise timing of harvesting

4. pH control after harvesting
5. Temperature treatment after harvesting
6. Addition of flocculating agents
7. Use of enzymes to attack cell walls
8. Early termination of feeds such as oil
9. Limited use of antifoam agents

The above list emphasizes the close association of fermentation and product recovery as integral parts of the overall process.

Future bioproducts derived from fungal fermentations will be proteins and enzymes, and many of the new developments in fermentation technology and genetic engineering are already having considerable impact on downstream processing. Most of the important industrial fungal enzymes (amylases, proteases, penicillin acylases, etc.) are actively secreted by the producer fungus, thus making the downstream processing much easier. It may well be possible to genetically engineer the excretion of normally intracellular proteins, making their purification less complicated [31]. Similarly, it may be possible to selectively degrade contaminating proteins and nucleic acids by controlled genetic activity. Finally, it may be possible to transfer the genes of product formation to other organisms with more suitable downstream processing characteristics [32,33].

VI. REPRESENTATIVE FUNGAL FERMENTATIONS

A. Enzymes

Enzymes have been produced from fungi on a regular basis from the end of the nineteenth century with the most noted example being amylase activity in Taka-diastase preparations from *A. oryzae*. A wide variety of enzymes are produced from fungal cultures (see Table 1) due to the ease and cost effectiveness of their production, absence of seasonal or political effects, and ease of increased productivity due to genetic strain improvement.

Although submerged agitated fermentations are the commonest means of fungal enzyme production, surface culture is still the best method for some processes due to its relatively simple setup and the high concentrations of extracellular enzymes produced initially, which makes subsequent recovery easy. Amyloglucosidase from *Rhizopus* and the acid protease from *A. niger* are examples of surface-produced enzymes [34 35].

Enzyme recovery depends to a great extent on the final use of the catalyst. Crude preparations can be made by a direct spray drying or lyophilization of the culture supernatant or even from the whole broth. For use in immobilized systems or diagnostic purposes, considerable effort may be spent in purification, which can include salt or solvent precipitation, ion exchange or affinity chromatography, or electrophoresis.

Immobilized enzymes have many important industrial uses as they have replaced many chemical reactions due to the substantial cost savings that have resulted from their use (Table 4).

1. Starch-Related Hydrolyzing Enzymes

There are two important fungal enzymes involved in starch degradation. α-Amylase, which attacks the internal 1,4-glycosidic linkages in amylose and amylopectin, is obtained from *A. niger* and *A. oryzae*. Amyloglucosidase or glucoamylase hydrolyzes the terminal 1,4-glucan links to release glucose

TABLE 4. Advantages in the Use of Immobilized Enzymes

Properties	Uses/examples
High substrate selectivity	Isomeric, regio- or enantiomer specificity
High substrate affinity	High conversion yields as enzyme operates well at low substrate concentrations, minimum of substrate in final product
High substrate tolerance	High conversion yields at high substrate concentrations; high product concentration aids product recovery
Low product affinity	High conversion yields due to low enzyme activity inhibition even at high product concentration
Ambient conditions	Operations in noncorrosive, near-neutral pH, ambient temperatures, safe operating conditions less operator stress
Aqueous conditions	Elimination of sovents and reactive chemicals
Good environmental impact	No hazardous wastes generated, limited disposal problems
Insoluble enzyme	Enzyme catalyst reused many times, cost per run minimized; product easily removed from enzyme, product not contaminated with enzyme

and is obtained from a variety of *Aspergillus* and *Rhizopus* species. In combination, these enzymes are used to supplement the natural starch-degrading enzymes in wheat flour and other confectionaries where their action releases glucose, which is in turn used by yeast to raise the dough. These enzymes have also found use as supplements in beer making. Their most current use is in the manufacture of glucose syrups, considerable proportions of which are further converted to the high-fructose sucrose replacements.

2. Lactase or β-Galactosidase

This enzyme is produced by *A. niger* and *A. oryzae* and hydrolyzes the α-1,4-glycosidic linkage in lactose to glucose and galactose, thus giving a sweeter and more soluble product. Milk treated by this enzyme reduces the lactose intolerance found in many people.

3. Invertase

Although yeast is the most common source of this enzyme, invertase can be obtained from *Aspergillus* species. It is used in the confectionary industry to hydrolyze sucrose to glucose and fructose, products that are sweeter and more soluble and present improved texture to the product.

4. Pectinolytic Enzymes

Pectins are composed of α-1,4-glycosidic linkages of galacturonic acids. Both exo- and endopectinases exist, and common sources include *Aspergillus*,

Botrytis, *Mucor*, and *Penicillium*. Pectinases aid the extraction and clarification of fruit juices and some wines.

5. Cellulases

Cellulases hydrolyze the β-1,4-glycosidic linkages of cellulose, basically releasing glucose. This process has a great potential for the production of glucose from readily available raw materials; however, enzyme activities are low, and the commercialization of this type of process has not yet been realized. Cellulases are produced from a variety of *Penicillium*, *Aspergillus*, *Fusarium*, and *Chrysosporium* species; however, the most active come from *Trichoderma viride*. Cellulases have uses in cattle feed supplements, digestive aids, food modification, and paper-making processes.

6. Glucose Oxidase

Glucose oxidase is produced by *A. niger* and *P. notatum*. It is used as an analytical tool for the measurement of glucose either in test strips or in enzyme electrodes. It is also used to remove oxygen from some food preparations such as fruit juices and mayonnaise.

7. Protease

Although most detergent alkaline proteases are of bacterial origin, the acid proteases are found in the fungi, e.g., *A. saitoi*, *A. niger*, *A. oryzae*, *Trametes sanguineara*, and *Mucor pusillus*. Proteases are used to tenderize meat, remove bitter flavors, replace rennin in cheese making, reduce the elasticity of gluten proteins in bread making, and in the chill proofing of beer.

8. Enzymes Associated with β-Lactam Antibiotic Manufacture

Immobilized fungal enzymes from *Bovista* and *Fusarium* [37,37] have now been established as important manufacturing tools in the large-scale production of β-lactam antibiotics [38]. Immobilized penicillin acylase enzymes currently account for more than 80% of the 6-aminopenicillanic acid (6-APA) produced and can yield up to 1000 kg 6-APA per kg enzyme activity used. Similar enzyme systems can be used to yield 7-aminodesacetoxycephalosporanic acid (7-ADCA) from ring-expanded penicillins. 7-Aminocephalosporanic acid (7-ACA) appears to be more difficult to obtain by enzymic means, but industrially useful enzymes capable of hydrolyzing the fermented cephalosporin C to 7-ACA have been reported in Japan [39].

Enzyme activities have been patented that yield 7-ACA from N-acylated cephalosporin C and from a *Trigonopsis*-catalyzed deaminated product, glutaryl-7-ACA [40]. The ester group on cephalosporin C can also be hydrolyzed by enzymic means to yield the desacetyl derivatives, useful intermediates in the synthesis of certain injectable cephalosporins [41].

B. Food Products

Fungi and their associated products are consumed by animals and humans in a great variety of forms. They can provide nutrient directly in the form of mushrooms, as in the form of processed biomass food supplements, or as flavorings and seasonings as can be found in the many types of Oriental sauces.

Mushrooms in their many forms are cultivated worldwide, and their annual consumption is in excess of 350,000 metric tons. These edible fungi are propagated on a variety of plant nutrients unique to each species. The common mushroom *Agaricus bisporus* is the most widely cultivated, using composted straw as its main nutrient. For further information on the culture of this fungus, see the review by Hayes and Nair [42].

The Chinese Padi-straw mushroom *Volvariella volvaceae* is cultivated on rice straw, and the Shiitake mushroom *Lentinus edodes* is cultivated on hardwoods such as oak and chestnut. The oyster mushroom *Pleurotus ostreatus* is grown historically on beech trunks; however, it can be cultivated commercially on a variety of other complex nutrients such as crushed corn cobs.

Single-cell protein has been in and out of fashion over the last 20 years, depending on the cost and availability of raw materials such as glucose and oil and of competing products such as fish meal, whey, and soybean protein. Two fungal systems of note have been developed. Rank Hovis McDougal in England have cultivated *Fusarium graminearum* to produce a textured product called Mycoprotein, which has been approved for human consumption. A company in Finland has cultivated *Paecilomyces varioti* on sulfite waste liquors at a level of 10,000 tons/year to reduce the biological oxygen demand (BOD) by 80% and provide a salable animal nutrient.

Oriental cuisine has utilized a variety of products and flavorings derived of fungal origin. The Koji process is common to the manufacture of both soya sauce (Shoyu) and miso. Our present knowledge of these processes is due to the pioneering effort of Hesseltine [43]. Koji fermentation involves the solid cultivation of *A. oryzae* on mixtures of soybeans and wheat for soy sauce, or on mixtures of soybeans with rice or barley for miso. The fermentations are well controlled in terms of temperature and water content to minimize the contamination with other opportunistic fungi, although infestations of *Rhizopus nigricans* can occur. These fermented products are subsequently mixed with salt and allowed to undergo an anaerobic fermentation involving *Saccharomyces rouxii*.

Chinese cheese (sufu) is produced from the curd of soy milk (tofu) by fermentation with a variety of fungi, including *Actinomucor elegans* and *Mucor* species.

Fermented foods also have an important dietary role in Indonesian foods. Tempeh is fermented from soybeans with *Rhizopus oligosporus*, and ontjom is fermented from peanut meal with *Neurospora sitophila*. In both of these cases, as also occurs in other Chinese preparations, the fungal fermentation renders the basic food more palatable in that it is converted to a more digestible form; in addition, it can provide a more balanced nutrient and a source of vitamins, which is often necessary in areas where dietary intake is restricted to plant products.

C. Organic Acids

Certain fungal cultures when grown under specific conditions can be prolific producers of organic acids. These primary metabolites are derived mainly from the Krebs cycle or from glucose metabolism. See Chapter 14 for a more extensive review on this subject.

1. Citric Acid

Citric acid is the most important organic acid produced by industrial fermentation and is widely used in the food and pharmaceutical industries.

It serves as an acidulant in soft drinks; a buffer in jellies, jams, and desserts; an antioxidant of fats; an emulsifying and stabilizing agent; a detergent; a blood anticoagulant; a metal complexing and cleaning agent; and a component of effervescent tablets and powders. Other industrial uses of citric acid involve electroplating processes, plasticizers, water conditioners, and secondary oil recovery.

Citric acid production is not uncommon among fungi, especially the penicillia and aspergilli. Commercially, it has been produced Koji style, in shallow pans, or by submerged culture in yields of over 60% [44]. Present day fermentations with selected strains of *A. niger*, however, employ submerged cultures which substantially reduce fermentation times, utilize cheap raw carbohydrate-based media, and yield up to 90% (w/w) citric acid from sucrose [45]. The removal of manganese and iron is required for optimum production.

Selective breeding after chemical mutagenesis has generated *A. niger* strains that produce 70% more citric acid, but only one-half the amount of oxalic acid, an undesirable byproduct [46]. It is surprising, however, that after many decades of research no selection procedures for *A. niger* or genetic loci have been described that involve specific steps of citric acid metabolism in this organism. More recent approaches with morphological, auxotrophic, and Benlate-resistant mutants and their revertants may eventually allow association of genetic loci with citric acid excretion phenotypes [47].

2. Itaconic Acid

Itaconic acid, a substituted methylacrylic acid formed from *cis*-aconitic acid by various fungi, is marketed as a refined product or as esters. It is used in the manufacture of plastics such as resin coatings and synthetic detergents as it can alter the properties of polymers and enhance dyeing characteristics. *Aspergillus terreus* is used for the industrial fermentation of itaconic acid, and this process was recently reviewed [45,47].

3. Gluconic Acid

About 45,000 metric tons of gluconic acid, an oxidation product of glucose, are manufactured annually worldwide, mainly as sodium gluconate, a versatile sequestering agent—particularly for calcium in detergents—or as the calcium salt, a pharmaceutical used in treating calcium deficiencies. It is also made as the free acid, a noncorrosive cleaning agent, or β-gluconolactone, a slow-acting acidulant.

The first large-scale production of gluconic acid used *Penicillium purpurogenum* var. *rubri-sclerotium* and *Penicillium chrysogenum* in surface culture. *Aspergillus niger* is now the organism of choice and is utilized in submerged fermentations. A recent review on strain improvement reports that present strains function at near-maximum theoretical efficiency [48].

D. Vitamins

Fungi are quite capable of producing a variety of vitamins including biotin, folic acid, pantothenic acid, pyridoxine, riboflavin, thiamine, and cyanocobalamin. However, only the production of riboflavin has been commercialized, although fungi can be used as useful sources of vitamins A and D. A more detailed description of microbiological vitamin production can be found in the literature [50,51].

1. Riboflavin

The annual production of riboflavin is greater than 1250 tons, and the microbial production by *Ashbya gossypii* competes well with the chemical synthesis. A number of bacteria and fungi can produce riboflavin, but *Eremothecium ashbyii* and *A. gossypii* are the most important since these fungi can make up to 2.5 and 7 g/liter, respectively. Details regarding the processes and pathways for microbial production of this vitamin have been discussed elsewhere [51].

Hyperproducing mutants of *A. gossypii* have internal concentrations of precursor flavin nucleotides similar to those in wild-type strains; however, intracellular riboflavin levels are higher. Flavinogenic strains have a correlation between sporulation and productivity. Superior strains sporulated well, while mutants incapable of overproducing riboflavin do not sporulate at all, suggesting that common factors may play a role in regulating both processes. Early studies employed sodium dithionate to eliminate low-yielding mutants and recommended frequent reisolation of flavinogenic strains and immediate preservation of useful strains due to the inherent instability of the riboflavin-producing mutants.

Extensive studies on fermentation processes have complemented genetic manipulation of *A. gossypii*. Production by flavinogenic mutants is enhanced by a number of compounds, among them purines and surface-active agents [52].

2. Vitamin A

Carotenoids are precursors of vitamin A when ingested by animals and are added to a variety of foods and feed as a coloring agent. Though such additives are synthetic, much interest has focused on *Blakeslea trispora*, a heterothallic fungus of the Mucorales group which is extremely rich in mycelial β-carotene, as a potential source of carotenoids. The yield of β- is as much as 1 g/liter of medium or 20 mg/g of dry mycelia when both sexual forms are grown together. *Phycomyces blakesleeanus*, a fungus amenable to genetic manipulation, also produces substantial amounts of β-carotene. Recently, superproducing strains of *P. blakesleeanus* have been isolated which synthesize 25 mg of β-carotene per gram dry weight of cells [53].

3. Vitamin D

Sterols, which are secondary metabolites but essential for cellular functions, are widely distributed among the fungi. Ergosterol, a commercially important lipid, can be extracted from dehydrated yeast, irradiated with ultraviolet light, and converted to vitamin D. Efficient ergosterol-producing yeast strains have been isolated by screening natural isolates, hybridization and polyploidization, and mutagenesis followed by enrichment with echinocandin and nystatin [54].

VII. TRANSFORMATIONS

The use of fungal cultures, particularly fungal spores, in transformation reactions has been well documented [54,55]. Of greatest importance commercially is the transformation of steroids. It was the discovery in 1952 of the 11α-hydroxylation of progesterone by *Rhizopus* that gave the impetus for the industrial manufacture of a whole range of adrenocortical hormones. Today's female hormones (estrogens) and male hormones (androgens) are derived from the basic steroid diosgenin isolated from the Mexican yam. Fungal transformation reactions are good examples of regiospecific enzyme reactions in that only selective regions of the molecule are reacted on, often with the formation of stereospecific isomers. With steroids, such structural changes can have profound therapeutic effects. These reactions cannot be performed by chemical means. The most important reactions include the 11α hydroxylation by *R. nigricans* the 11β hydroxylation by *Cunninghamella blakesleeana* and *Curvularia lunata*, the C-1 dehydrogenation by *Fusarium* species, and lactonization by *Cylindrocarpon radiciola* (see Fig. 2). Other hydroxylations at other specific sites on the steroid molecule, double-bond isomerization, and other oxidations and reductions can be performed by a variety of fungal cultures [54,55].

VIII. FUNGAL β-LACTAM ANTIBIOTIC FERMENTATION: PENICILLINS AND CEPHALOSPORINS

A. Biology of the Producer Organisms

The initial discovery of Fleming's contaminating *Penicillium* fungus and its remarkable antibacterial properties of penicillin now dates back more than 50 years. Today's culture technology started during the second world war as a combined effort between England and America. Exploitation of

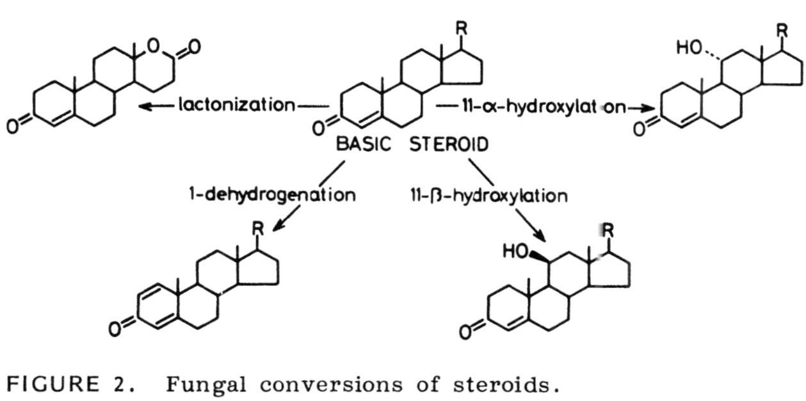

FIGURE 2. Fungal conversions of steroids.

the second major class of β-lactam substances from fungi began with the discovery of the cephalosporin-producing organism *Acremonium chrysogenum* by Brotzu in Sardinia some 30 years ago and the chemical isolation of cephalosporin C as a contaminant in a crude preparation of penicillin N by Newton and Abraham. Cephalosporin C was found to be resistant to enzymatic hydrolysis and more active than the natural penicillins against gram-negative bacteria. These characteristics attracted the attention of clinicians because of the increasing incidence of bacterial resistance to penicillins by penicillinase-producing organisms. A major impetus for the commercial exploitation of cephalosporin C was the discovery of a practical method for chemically splitting cephalosporin C to 7-ACA. This stimulated the development of the semisynthetic cephalosporins which was analogous historically to the rapid progress in chemotherapy induced by the discovery of 6-APA nucleus and its semisynthetic derivatives.

In the past 40 years, penicillins and cephalosporins have been developed from unstable, low-purity laboratory extracts into multiproduct, multinational, multibillion dollar industries. Today there is sufficient fermentation and production capacity worldwide to provide each individual on this planet with sufficient penicillin for one therapeutic treatment each year.

The market for penicillins and cephalosporins has reached maturity in the developed countries with worldwide sales in excess of 5 billion (U.S.) in 1983 [1]. Future expansion will depend on penetration into developing countries, which at the moment is restricted by currency exchange and low per capita income. With patent expiration and the increase in the availability of high-producing mutant strains and relevant manufacturing technology, developing countries should be able to establish their own processes.

B. Strain Improvement

Pharmaceutical companies have carried out extensive mutation and selection programs to screen for mutant cultures of *P. chrysogenum* and *A. chrysogenum* with higher productivities. These independent programs have created a wide variety of subspecies, each unique in its ability to perform optimally in its specific manufacturing environment. Although strain improvement programs are considered proprietary information, some details have been made available in the reviews [16,56]. Panlabs has had nonexclusive penicillin and cephalosporin strain development programs in operation for many years. Emphasis in strain improvement has been directed toward titer increase, with secondary emphasis on mycelial vigor and sporulation and germination efficiency. The latter properties are important for strain preservation and inoculum propagation. Today's cultures are capable of producing in excess of 30 mg/ml penicillin G [56].

C. Strain Maintenance

Highly mutated industrial strains of *P. chrysogenum* and *A. chrysogenum* have inherent genetic instabilities, and it is of prime importance to maintain both primary stock cultures and working cultures under conditions where phenotypic changes are minimized. This normally entails storage at the temperature of the gaseous (-167°C) or liquid (-196°C) phase of nitrogen. Long-term preservation procedures used at Bristol-Myers Squibb have been reported [10]. Squibb reported the storage of *P. chrysogenum* for 3 years in liquid nitrogen with no loss in conidial viability and minimal loss in productivity [57].

D. Biosynthesis and Regulation

1. Penicillin

The β-lactam thiazolidine ring of penicillin is constructed from L-cysteine and L-valine. The molecule is formed from a tripeptide of α-aminoadipate, cysteine, and valine which cyclized to form isopenicillin N. Benzylpenicillin is produced in the exchange of L-α-AAA with activated phenylacetic acid. 6-APA, which is not an intermediary product of biosynthesis, is excreted in the absence of a side chain precursor.

Several regulatory mechanisms are known in penicillin biosynthesis. The amino acid lysine is synthesized from a pathway that involves L-α-aminoadipic acid, so that penicillin and lysine share a common branched biosynthetic pathway. Lysine inhibits penicillin synthesis because it is a feedback inhibitor of homocitrate synthase, an enzyme involved in L-α-AAA synthesis. If L-α-AAA is deficient, penicillin cannot be synthesized, but addition of small amounts of L-α-AAA stimulates penicillin synthesis. The inhibition by lysine of homocitrate synthase is less than the inhibition of penicillin formation, however, so that additional regulatory processes are postulated.

Penicillin biosynthesis is affected by phosphate concentration and also shows a catabolite repression by glucose. Originally penicillin fermentations were carried out with the slowly metabolizable sugar lactose. However, in today's fed-batch stirred-tank fermentations, glucose is used and its consumption by the culture controlled by the rate of addition.

2. Cephalosporin

The first stages of biosynthesis to isopenicillin N are similar to those in penicillin. Cyclization activity, i.e., the formation of penicillin N, was first discovered in a cell-free system of *A. chrysogenum*. In the next stage, penicillin N is produced by transformation of the L-α-AAA side chain into the D form by a racemase. By ring expansion of penicillin N, deacetoxycephalosporin C is produced and, after oxidation and acetylation, is converted to cephalosporin C. In contrast to penicillin, the D-α-AAA moiety cannot be changed with precursor feeding.

Cephalosporin production is affected by carbohydrate catabolite repression and phosphate regulation. The addition of lysine in low concentrations promotes cephalosporin production, whereas the inhibition caused by higher amounts of lysine can be overcome by the addition of DL-α-AAA. Methionine stimulates cephalosporin synthesis.

The cyclase enzyme is capable of generating new penicillins and cephalosporins when tripeptide analogs are added to cell-free systems [58]. The entire cyclase gene was recently isolated, sequenced, and cloned in bacteria and streptomycetes [26].

E. Assay Development

Microbiological bioassay can be used, particularly as a preliminary assay, but such techniques are labor-intensive and usually lack the precision of automated chemical assays. Automated assays such as Technicon have been used extensively over the past 20 years using hydroxylamine [59] or penicillanic acid detection [60]. Over the past 10 years the use of high-pressure liquid chromatography (HPLC) has become common, using reverse phase C18 columns and UV detection [61]. HPLC has the advantage of high accuracy and good precision, and can give excellent quantitative results for penicillin, precursor, and breakdown products. Analytical departments are now using automated or semiautomated systems to improve the productivity and accuracy of their measurements. Laboratory robots, such as the Zymate [62], are now capable of sample preparation, e.g., automated weighing, diluting, mixing, and filtering. Such machines even have their own automated injection station for HPLC analysis. The majority of HPLC manufacturers supply automated sample injection systems together with automated data acquisition, which can greatly facilitate sample throughput by running unattended overnight.

IX. OTHER SECONDARY METABOLITES

A. Griseofulvin

Griseofulvin is an effective systemic antifungal antibiotic used in the systemic infections of nails, hair, and skin. It is produced from a variety of *Penicillium* species, first being isolated from *P. griseofulvum*; however, it is now produced commercially by mutants of *P. patulum*. The antibiotic was first known as the "curling factor" due to its morphological effect on hyphae. It is also produced in *Nigrospora oryzae*, an organism quite distinct from the penicillia. Further details of the fermentation and chemistry of griseofulvin can be found in the review articles [63].

B. Gibberellic Acid

Gibberellins were isolated from *Gibberella fujikuroi* (perfect stage of *F. moniliforme*) after it was observed that *G. fujikuroi*-infected rice plants grew faster than normal. Gibberellins were subsequently identified as endogenous plant hormones, and thus the fungal production of gibberellins has had a significant commercial impact on agriculture with over 4 tons produced annually.

The effect of gibberellin depends on the plant species. It can be used to promote growth and aid in the propagation of plantlets from meristematic tissue. In other applications, it can control fruit drop (apples), increase fruit set (pears), and improve fruit yield (seedless grapes).

A description of mutants with enhanced yields of gibberellin phytohormones in *Fusarium moniliforme* has been given [64], and work with this fungus, attempting to exploit the parasexual cycle to improve the yield of gibberellic acid, has been reported with the claim that recombinants with improved yields can be obtained [65]. The chemistry and biological properties of the gibberellins were reviewed by Grove [63].

C. Ergot Alkaloids

Rye, if left to overwinter under wet conditions, can become infected with the ergot fungus *Claviceps purpurea*. The subsequent grazing of this pasture by animals or the ingestion of bread made from the infected rye have had severe poisonous effects on the victim. The pharmacological effects of these alkaloids vary from vasoconstruction and uterine contraction to adrenergic blocking and serotonin antagonism, due to the large variety of active molecules produced from the fungus. The alkaloids were first obtained directly from contaminated rye; however, this mode of production is not cost-effective. Initial fermentation attempts with *Claviceps purpurea* were not successful. However, Arcamone et al. [64] successfully developed *Claviceps paspali*, through media improvement and superior strain selection, to produce submerged fermentations of alkaloid at 5 mg/g. These fermentations, combined with chemical modification, enabled Farmitalia and Sandoz to market a variety of D-lysergic acid related compounds [65].

D. Other Fungal Toxins

Several fungi are known to produce powerful carcinogens but, unlike the ergots, they have not been developed into major fermentations as the demand for these products is not high. However, they have been studied in detail in order to assess their presence in a variety of human and animal food sources. The most prevalent cause of food contamination is *Aspergillus flavus*, which produces aflatoxins in plant products, such as peanuts and cereal seed, which cause liver tumors. Ochratoxin and citrinin, produced by some penicillia and aspergilli, cause kidney disease in hogs, and zearalenone, a product from *F. graminearum*, can cause estrogenism in hogs.

Basidiomycetes also produce a variety of toxic compounds, e.g., tricholomoic acid from *Tricholoma muscarium* is a fly killer; psilocin from *Psilocybe mexicana* is a hallucinogen; muscarine from *Amanita muscaria* causes respiratory failure; phallotoxins and amatoxins from *Amanita* species cause liver damage.

E. Antihypercholesterolemic Agents from Fungi

A number of filamentous fungi have been shown to produce compounds which interfere with cholesterol biosynthesis (Fig. 3). Caerulenin is a potent inhibitor of fatty acid biosynthesis produced by strains of *Cephalosporium caerulens* and is now widely used as an important biochemical reagent.

Lovastatin (mevalonin) produced by strains of *Aspergillus terreus* is a recent breakthrough drug for therapy for atherosclerosis and ischemic heart disease [66]. Lovastatin is a member of a group of compounds which inhibit hydroxymethylglutaryl coenzyme A(HMG-CoA) reductase which catalyzes a major rate-limiting step in the biosynthesis of cholesterol (Fig. 3). Other compounds related to lovastatin are the compactins, produced by strains of *Penicillium citrinum*, and the monacolins, produced by strains of *Monascus ruber*. Lovastatin has been marketed as a "wonder drug" for decreasing levels of low-density lipoproteins (LDLs) in patients suffering from hypercholesterolemia.

F. Fungal Insecticides and Herbicides

Natural infections by fungi play a major role in the control of many important economic pests. This phenomenon has been practically applied since the early 1900s when the introduction of species of the fungal genus *Aschersonia* resulted in the complete control of citrus white fly. There are more than 500 species of fungi capable of infecting insect hosts. In spite of this fact, the use of fungi to control insect pest populations on a commercial basis is still largely unrealized [67]. Those fungal insects which have been used on a large scale include the use of *Beauveria bassiana* conidia to control the Colorado potato beetle (*Leptinotarsa decemlineata*) and the European corn borer (*Ostrina nubialis*), the use of *Metarhizium*

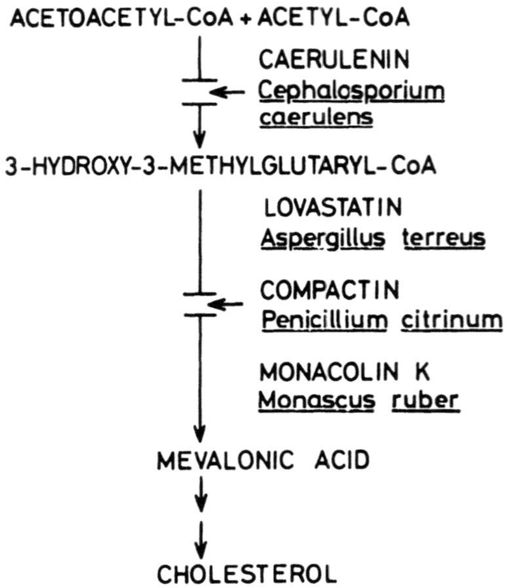

FIGURE 3. Fungal inhibitors of cholesterol biosynthesis.

anisopliae preparations to control a spittle bug (*Mahanarus posticata*) of sugarcane, and the use of *Verticillium lecanii* to control aphids and other insect pests of glasshouse crops. Species of the genus *Entomophordes* produce disease and therefore potential control of a wide variety of insect pests, but the use of these fungi for pest control remains largely unexplored. This stems primarily from difficulties in propagating species of *Entomophordes* in artificial media. Species of *Culcinomyces*, *Coelomomyces*, and *Lagenidium* can be effective in mosquito population control, and their utility is also affected by poor growth on artificial media.

The use of fungi to control weeds and other noxious plants represents a fertile future use for fungal biotechnology. Fungus-based herbicides (mycoherbicides) have advantages over toxic, often carcinogenic, chemical herbicides. Their development and raw material costs can be cheaper than herbicides manufactured by chemical synthesis. Their major disadvantage is that mycoherbicides usually affect only a single noxious plant in contrast to chemical herbicide and therefore are only limited to restricted markets. Mycogen, a new biotechnology company in San Diego, anticipates marketing a variety of mycoherbicides and expects a potential U.S. market in the 1990s of $30 million for the two mycoherbicides. *Alternaria cassiae* is aimed at the control of sicklepod, a weed that infests soybeans and peanuts in the southeastern United States. The plant pathogen *Fusarium lateritium* will be used to control the velvet leaf weed, which infests 20 million acres of soybean in the midwestern United States.

X. FUTURE PROSPECTS FOR INDUSTRIAL FUNGI

This chapter has examined the application of modern genetics and biotechnology to industrial fungal fermentations. The volume and variety of metabolites produced by such fermentations clearly indicate that in specific cases traditional genetic approaches have been quite successful in generating fungal strains that can efficiently produce marketable substances. The disadvantages of industrial organisms have been overcome by persistence and ingenuity, and some industrial fermentations may operate at near-maximal capability. Now industrial strain improvement programs are embracing newer genetic technology and seeking even better strains and processes as well as new products.

Tremendous progress has been made over the past 2 years in applying genetic engineering strategies to filamentous bacteria and fungi-producing antibiotics, the present largest market for biotechnology products. Gene amplification and cascade hybridization strategies offer exciting new strain improvement techniques for increasing the levels or modifying the regulation of the important β-lactam synthase enzymes (AC synthetase, cyclase, expandase, acyltransferase, etc.) in fungi producing penicillins and cephalosporins. Such programs can lead to the isolation and cloning of entire antibiotic biosynthetic pathways [68] in different microbial strains and to the synthesis of novel, more useful, "hybrid" antibiotics [69,70].

Can strain improvement programs for important metabolites be expected to match the recent dramatic successes in producing valuable proteins in genetically engineered yeast? A number of such proteins, many with medical applications, have been synthesized in *S. cerevisiae* using cloned mammalian or viral genes. These scarce substances will eventually be marketed as high-value specialty chemicals or will be used in the manufacture of important new therapeutic products. Examples of some of these substances are human leukocyte interferon D; human epidermal growth factor; urogastrone; human proinsulin; human preproparathyroid hormone; human interleukin-2; human blood factor VIII human α_1-antitrypsin; human β-endorphin; calf chymosin; bovine growth hormone; and hepatitis B virus surface antigen. This list will certainly grow as more genes important in medicine, agriculture, and industry are cloned and expressed in suitable hosts.

The advantages of using fungi as expression systems and the problems that need to be alleviated are common to all types of proteins. In outline, filamentous fungi, especially commercial strains, secrete large amounts of proteins. Exact levels are hard to assess, but they are obviously high and commercially rewarding. Cleavage of signal peptides to generate mature active protein is dependent on amino acid signals common to many species. Proteins are modified by glycosylation with mannose sugars attached to asparagine-linked *N*-acetyl glucosamine core molecules in a fashion similar to mammalian systems.

The requirements for mRNA processing are also akin to those present in higher eukaryotes. Intervening sequences present in the primary transcript carry consensus splice donor, splice acceptor, and internal splice recognition signals. Fungal systems thus offer all features necessary for mRNA transcription and translation and for the processing of both mRNA and protein. But these are not enough in themselves. The product has to be produced at high economic levels. In other systems, the approach to achieve this goal has been based on gene amplification using high-copy-number plasmid vectors. In filamentous fungi, the stable, and hence manipulatable, transformants are the result of the integration of the DNA into the chromosome. While several copies can be integrated, the very-high-copy-number characteristics of plasmid-based systems are rarely achieved. A key question is whether high-copy-number vectors are absolutely necessary.

A consideration must be given to the problem seen by many as the major stumbling block to the exploitation of filamentous fungi—proteases. *Aspergillus oryzae* is a major organism commercially used for the production of all types of protease. Copious amounts of these protein-degrading enzymes are produced, especially under derepressed conditions. There is much literature on the presence of proteases and some on their activity, but they are genetically undefined. It is possible that the industrial strains of fungi selected for high levels of enzyme secretion are deficient in specific proteases or in a number of them, except, of course, for *A. oryzae*. This is another strong reason for using the ready-prepared genotypes in any experiments testing heterologous protein expression. Undoubtedly, the protease problem will have to be addressed, but its magnitude has yet to be determined. An empirical, initial solution may be to ensure that the culture medium contains sufficient quantities of the agents known to repress protease production.

In conclusion, there is mounting evidence that filamentous fungi are emerging as candidates for strong hosts capable of expressing primary and secondary metabolite foreign proteins. In addition, rDNA methods are also being applied to improve the long-standing existing processes and to widen their use in order to expand the range of products of fungal origin currently being produced.

REFERENCES

1. Hacking, A. J. *Economic Aspects of Biotechnology*. Cambridge University Press, London, 1986.
2. Bennett, J. W. Molds, manufacturing and molecular genetics. In *Molecular Genetics of Filamentous Fungi* (W. E. Timberlake, ed.), Alan R. Liss, New York, 1986, pp. 345-366.
3. Turner, W. B. *Fungal Metabolites*. Academic Press, London, 1971.
4. Elander, R. P., and Demain, A. L. Genetics of microorganisms in relation to industrial requirements. In *Biotechnology: Microbial Fundamentals*, Vol. 1 (H. J. Rehm and G. Reed, eds.), Verlag-Chemie, Weinheim, 1981, pp. 237-277.
5. Elander, R. P., and Espenshade, M. A. The role of microbial genetics in industrial microbiology. In *Industrial Microbiology* (B. M. Miller and W. Litsky, eds.), McGraw-Hill, New York, 1976, pp. 192-256.
6. Elander, R. P. Mutations affecting antibiotic synthesis in fungi producing β-lactam antibiotics. In *Genetics of Industrial Microorganisms* (O. K. Sebek and A. I. Laskin, eds.), American Society for Microbiology, Washington, D.C., 1979, pp. 21-35.
7. Elander, R. P., and Chang, L. T. Microbial culture selection. In *Microbial Technology: Fermentation Technology*, Vol. 2 (H. J. Peppler and D. Perlman, eds.), Academic Press, New York, 1979, pp. 244-302.
8. Elander, R. P. Enhanced penicillin synthesis in mutant and recombinant strains of *Penicillium chrysogenum*. In *Induced Mutations and Their Utilization* (H. Stubbe, ed.), Akademie-Verlag, Berlin, 1967, pp. 403-423.
9. Ball, C. Genetic approaches to overproduction of β-lactam antibiotics in eukaryotes. In *Overproduction of Microbial Products* (V. Krumphanzl, B. Sikyta, and Z. Vanek, eds.), Academic Press, London, 1982, pp. 515-534.

10. Elander, R. P. Strain improvement and preservation of β-lactam producing microorganisms. In *Antibiotics Containing the β-Lactam Structure* (A. L. Demain and N. A. Solomon, eds.), Springer-Verlag, New York, 1983, pp. 97-146.
11. Peberdy, J. F. Protoplast fusion: a tool for genetic manipulation and breeding of industrial microorganisms. *Enz. Microb. Technol.*, *2*:23-39, 1980.
12. Hopwood, D. A. The genetic programming of industrial microorganisms. In *Industrial Microbiology and the Advent of Genetic Engineering*, W. H. Freeman, San Francisco, 1981, pp. 31-42.
13. Hamlyn, P. F., and Ball, C. Recombination studies with Cephalosporium acremonium. In *Genetics of Industrial Microorganisms* (O. K. Sebek and A. I. Laskin, eds.), American Society for Microbiology, Washington, D.C., 1979, pp. 185-191.
14. Ferenczy, L. Microbial protoplast fusion. In *Genetics as a Tool in Microbiology* (S. W. Glover and D. A. Hopwood, eds.), Cambridge University Press, London, 1981, pp. 1-34.
15. Chang, L. T., Terasaka, D. T., and Elander, R. P. Protoplast fusion in industrial fungi. *Dev. Ind. Microbiol.*, *23*:21-29, 1982.
16. Elander, R. P., and Vournakis, J. N. Genetic aspects of overproduction of antibiotics and other secondary metabolites. In *Overproduction of Microbial Metabolites* (Z. Vanek and Z. Hostalek, eds.), Butterworths, Boston, 1986, pp. 63-79.
17. Upshall, A. Filamentous fungi in biotechnology. *Biotechniques*, *4*:158-166, 1986.
18. Kelly, J. M., Hynes, M. J. Transformation of Aspergillus niger by the amdS gene of Aspergillus nidulans. *EMBO J.*, *4*:475-479, 1983.
19. Johnstone, Y. L., Hughes, S. G., and Clutterbuck, A. J. Cloning of an Aspergillus developmental gene by transformation. *EMBO J.*, 5:1307-1311, 1985.
20. Yelton, M. M., Timberlake, W. E., and van den Hondel, C. A. M. J. J. A cosmid selecting genes by complementation in Aspergillus nidulans: Selection of the developmentally regulated by A locus. *Proc. Natl. Acad. Sci.* USA, *82*:834-838, 1983.
21. Doy, C. H., Pateman, J. A., Olsen, J. E., Kane, H. J., and Creaser, E. H. Genomic clones of Aspergillus nidulans containing *alc*A, the structural gene for alcohol dehydrogenase and *alc*R, a regulatory gene for ethanol metabolism. *DNA*, *4*:105-114, 1985.
22. Hynes, M. J., Corrick, C. M., King, J. A. Isolation of genomic clones containing the amdS gene of Aspergillus nidulans and their use in the analysis of structural and regulatory mutations. *Mol. Cell. Biol.*, *3*:1430-1439, 1983.
23. Nunberg, J. H., Meade, J. H., Cole, G., Lawyer, F. C., McCabe, P., Schweikhardt, V., Tal, R., Wittman, V. P., Flatgaard, J. E., and Innes, M. A. Molecular cloning and characterization of the glucoamylase gene from Aspergillus awamori. *Mol. Cell. Biol.*, *11*:2306-2315, 1984.
24. Cullen, D., and Leong, S. Recent advances in the molecular genetics of industrial filamentous fungi. *Trends Biotechnol.*, *5*:285-288, 1986.
25. Hynes, M. J. Transformation of filamentous fungi. *Exp. Mycol.*, *10*:1-8, 1986.
26. Samson, S. M., Belagaje, R., Blankenship, D. T., Chapman, J. L., Perry, D., Skatrud, P. L., VanFrank, R. M., Abraham, E. P., Baldwin, J. E., Queener, S. W., and Ingolia, T. D. Isolation, sequence

determination and expression in Escherichia coli of the isopenicillin N synthetase gene of Cephalosporium acremonium. *Nature, 318*:191-194, 1985.
27. Penalva, M. A., Tourino, A., Patino, C., Sanchez, F., Fernandez Sousa, J. M., and Rubio, V. Studies on transformation of Cephalosporium acremonium. In *Molecular Genetics of Filamentous Fungi* (W. Timberlake, ed.), A. R. Liss, New York, 1986, pp. 59-68.
28. Stohl, L. L., Lambowitz, A. M. Construction of a shuttle vector for the filamentous fungus, Neurospora crassa. *Proc. Natl. Acad. Sci.*, USA, *80*:1058-1062, 1984.
29. Turner, G., Ballance, D. J., Ward, M., and Beri, R. K. Development of cloning vectors and a marker for gene replacement techniques in Aspergillus nidulans. In *Molecular Genetics of Filamentous Fungi* (W. E. Timberlake, ed.), A. R. Liss, New York, 1985, pp. 15-28.
30. Swartz, R. W. The use of economic analysis of penicillin G manufacturing costs in establishing priorities for fermentation process improvement. *Ann. Rep. Ferment. Processes, 3*:75-110, 1979.
31. Priest, F. G. *Extracellular Enzymes*. Van Nostrand Reinhold, Wokingham, 1984.
32. Fish, N. M., and Lilly, M. D. The interactions between fermentation and protein recovery. *Biotechnology, 2*:623-627, 1984.
33. Flemming, R. J., and Steinschneider, A. Fermentation economics in relation to genetic engineering. *Biotechnology, 3*:773-776, 1983.
34. Arima, K. Microbial enzyme production. In *Global Impacts of Applied Microbiology* (M. P. Starr, ed.), John Wiley and Sons, New York, 1964, pp. 277-294.
35. Underkofler, L. A. Development of a commercial enzyme process: Glucoamylase. In *Cellulases and Their Applications* (R. F. Gould, ed.), American Chemical Society, Washington, D.C., 1969, pp. 343-358.
36. Schneider, W. J., and Roehr, M. Purification and properties of penicillin acylase of Bovista plumbea. *Biochim. Biophys. Acta, 425*:177-185, 1976.
37. Lowe, D. A., Romancik, G., and Elander, R. P. Enzymatic hydrolysis of penicillin V to 6-aminopenicillanic acid by Fusarium oxysporum. *Biotechnol. Lett., 8*:151-156, 1986.
38. Lowe, D. A. Industrial importance of biotransformations of β-lactam antibiotics. *Dev. Ind. Microbiol., 26*:143-155, 1985.
39. Vandamme, E. J. Immobilized biocatalysts and antibiotic production: biochemical, genetical, and biotechnical aspects. In *Bioreactor Immobilized Enzymes and cells* (M. Moo-Young, ed.), Elsevier, New York, 1988, pp. 261-286.
40. Shibuya, Y., Matsumoto, K., and Fujii, T. Isolation and properties of 7-(4-carboxybutanamido) cephalosporanic acid acylase producing bacteria. *Agr. Biol. Chem., 45*:1561-1567, 1981.
41. Abbott, B. J. Preparation of pharmaceutical compounds by immobilized enzymes and cells. *Adv. Appl. Microbiol., 20*:203-257, 1976.
42. Hayes, W. A., and Nair, N. G. The cultivation of Agaricus bisporus and other edible mushrooms. In *The Filamentous Fungi*, Vol. 1 (J. E. Smith and D. R. Berry, ed.), John Wiley and Sons, New York, 1975, pp. 212-248.
43. Hesseltine, C. W. A millenium of fungi, food and fermentation. *Mycologia, 57*:1-148, 1965.
44. Lockwood, L. B. Organic acid production. In *The Filamentous Fungi*, Vol. 1 (J. E. Smith and D. R. Berry, eds.), John Wiley and Sons, New York, 1975, pp. 140-157.

45. Kubicek, C. P., and Rohr, M. Novel trends in physiology and technology of citric acid production. In *Overproduction of Microbial Products* (V. Krumphanzl, B. Sikyta, and Z. Vanek, eds.), Academic Press, New York, 1982, pp. 253-262.
46. Golubtsova, V. M., Shcherbakova, E. Y., Runkovshaya, L. Y., and Ermakova, V. P. Change in the ratio of citric and oxalic acids in Aspergillus niger under the influence of mutagenic factors. *Mikrobiologiya, 48*:1061-1065, 1979.
47. Kubicek, C. P., and Rohr, M. Citric acid fermentation. *Crit. Rev. Biotechnol., 3*:331-374, 1986.
48. Kundu, P. N., and Das, A. Calcium gluconate production by a nonconventional fermentation method. *Biotechnol. Lett., 4*:365-368, 1982.
49. Queener, S. W., Wilkerson, S. G., and Nash, C. H. Sterol content and titer of mycophenolic acid in polyene antibiotic resistant mutants of *Penicillium stoloniferum*. In *Overproduction of Microbial Products* (V. Krumphanzl, B. Sikyta, and Z. Vanek, eds.), Academic Press, London, 1982, pp. 535-548.
50. Yameda, K., Nakahara, T., and Fukui, S. Petroleum microbiology and vitamin production. In *Biochemical and Industrial Aspects of Fermentation* (K. Sakaguchi, T. Umura, and S. Konoshita, eds.), Kodansha, Tokyo, 1971, pp. 62-90.
51. Perlman, D. Microbial process for riboflavin production. In *Microbial Technology, Vol. I, Microbial Processes*, 2nd ed. (H. J. Peppler and D. Perlman, eds.), Academic Press, New York, 1979, pp. 521-527.
52. Lago, B. D., and Kaplan, L. Vitamin fermentations: B2 and B12. In *Advances in Biotechnology: Fermentation Products*, Vol. 3 (C. Vezina and K. Singh, eds.), Pergamon Press, New York, 1981, pp.241-257.
53. Murillo, F. J., Calderon, I. L., Lopez-Diaz, I., and Cerda-Olmedo, E. Carotene super-producing strains of Phycomyces. *Appl. Environ. Microbiol., 36*:639-642, 1983.
54. Vezina, C., and Singh, K. Transformation of organic compounds by fungal spores. In *The Filamentous Fungi*, Vol. 1 (J. E. Smith and D. R. Berry, eds.), John Wiley and Sons, New York, 1975, pp. 158-192.
55. Smith, L. L. Steroids. In *Biotechnology*, Vol. 6a (K. Kieslich, eds.), Verlag Chemie, Florida, 1984, pp. 31-78.
56. Lein, J. The Panlabs penicillin strain improvement program. In *Overproduction of Microbial Metabolites* (Z. Vanek and Z. Hostalek, eds.), Butterworths, Boston, 1986, pp. 105-139.
57. Fortney, K. L., and Thoma, R. W. Stabilization of culture productivity. *Dev. Ind. Microbiol., 18*:319-325, 1977.
58. Baldwin, J. W., Aalington, R. M., Derome, A. E., Ting, H-H., and Turner, N. J. Penicillin biosynthesis: Multiple pathways from a modified substrate. *J. Chem. Soc. Commun.*, 1984, pp. 1211-1214.
59. Boxer, G. E., and Everett, G. E. Colorimetric determination of benzyl penicillin. *Anal. Chem., 21*:670-675, 1949.
60. Bundgaard, H., and Ilver, K. A new spectrophotometric method for the determination of penicillins. *J. Pharm. Pharmacol., 24*:790-794, 1972.
61. Miller, R. D., and Neuss, N. High performance liquid chromatography of β-lactam antibiotics. In *Antibiotics Containing the β-Lactam Structure* (A. L. Demain and N. A. Solomon, eds.), Springer-Verlag, New York, 1983, pp. 301-308.

62. Zymark Corporation, Zymark Center, Hopkinton, Massachusetts.
63. Grove, J. F. Griseofulvin and some analogues. *Prog. Chem. Organic Natural Products, 22*:203-264, 1964.
64. Arcamone, F., Chain, E. B., Ferretti, A., Minghetti, A., Pennella, P., Tonolo, A., and Vero, L. Production of a new lysergic acid derivative in submerged culture by a strain of Claviceps paspali. *Proc. Roy. Soc. Ser. B, 155*:26-54, 1961.
65. Stadler, P. A., Guttman, S., Hauth, H., Huguenin, R. L., Sandrin, E. D., Wersin, G., and Hofmann, A. Die synthese der alkaloide der ergotoxin-gruppe. *Helv. Chim. Acta, 52*:1549-1564, 1969.
66. Buckland, B., Gbewonyo, K., Hallada, T., Kaplan, L., and Masurakar, P. Production of lovastatin an inhibitor of cholesterol accumulation in humans. *Abstr. Int. Conf. Biotech. Microb. Prod.*, San Diego, Soc. Indust. Microbiol., Arlington, Virginia, 1988, p. 10.
67. Miller, L. K., Lingg, J., and Bulla, Jr., L. A. Bacterial, viral and fungal insecticides. *Science, 219*:715-721, 1983.
68. Stanzak, K., Matsushima, R. H., Baltz, R. H., and Rao, R. N. Cloning and expression in Streptomyces lividans of clustered erythromycin biosynthesis genes from Streptomyces erythreus. *Biotechnology, 4*:229-232, 1986.
69. Omura, S., Ikeda, H., Malpartida, F., Keiser, H. M., and Hopwood, D. A. Production of new hybrid antibiotics, mederrhodins B, by a genetically engineered strain. *Antimicrob. Ag. Chemother., 29*:13-19, 1986.
70. Fayerman, J. T. New developments in gene cloning in antibiotic producing microorganisms. *Biotechnology, 4*:786-789, 1986.

2

MOLECULAR METHODS FOR IDENTIFICATION AND TAXONOMY OF FILAMENTOUS FUNGI

MAREN A. KLICH and EDWARD J. MULLANEY *Agricultural Research Service, United States Department of Agriculture, New Orleans, Louisiana*

I. INTRODUCTION

Molecular techniques have provided insights in many disciplines in the biological sciences, including systematics. Two of the most obvious applications of molecular methods in systematics are the elucidation of evolutionary relationships among the major groups of fungi and the delineation of species concepts. These techniques used in conjunction with standard methodologies may be especially valuable in solving problems in fungal systematics. Many fungi have little or no fossil record on which to base possible evolutionary relationships. Information on genetic similarities and differences at the molecular level could give us a powerful tool for assessing phylogenetic relationships. Standard taxonomic methodologies for determining taxa are based primarily on mating capabilities or morphological similarities. Such methods are not applicable to the common fungi that are not known to reproduce sexually. These fungi also often have few available morphological characteristics on which to base taxonomic decisions. Data from molecular studies could help to overcome these problems.

Many molecular methodologies were developed recently but initially were used mostly by researchers whose main interests were in molecular biology. Few mycologists with interests in evolutionary biology or systematics have devoted their time and money to the training and equipment needed to use these methods. In recent years, however, both the techniques and equipment necessary for molecular biological studies have changed, so that molecular methods can be readily used as tools in other disciplines. There are now several laboratory "cookbooks" that give very detailed protocols for many of the basic molecular techniques. They also contain other valuable information such as support protocols and lists of vendors that would enhance anyone's competence in the field [1-3]. Methods of isolating both DNA and RNA have been refined and improved over the years. Today methods are available for DNA isolation from many fungal species [e.g., 4-12]. Since many of these techniques require a minimal amount of specialized equipment, they can be utilized in most laboratories. Techniques for RNA isolation are also available [11,13-15]. However, the widespread occurrence and thermal stability of RNAses make the successful isolation of RNA more difficult than that of DNA.

Four general molecular approaches have been used in studies of evolutionary relationships and identification of filamentous fungi. The oldest technique, G + C molar percentage, involves determination of the relative frequency of the nucleotides guanine and cytosine in the genome and comparison of these values from different organisms. A second technique, DNA-DNA complementarity, involves "unzipping" DNA from two isolates or taxa, mixing the DNA from the two and measuring the rate or amount of reassociation of the DNA fragments between the two. In a third method, ribosomal RNA (rRNA) sequence comparison, the actual sequence of nucleotides in these relatively small nucleic acids (or portions thereof) is compared for homology. In the fourth method, generally referred to as restriction fragment length polymorphism, the nucleic acids are cut enzymatically and the resulting DNA fragments separated electrophoretically on gels. Different patterns of bands (polymorphisms) which result may sometimes be observed directly, but DNA from the gels is often transferred to a special membrane and exposed to radioactively labeled fragments of nucleic acid (probes) which will adhere (hybridize) to portions of DNA on the gel that have similar nucleotide sequences. Banding patterns of these radioactive sites are then observed on X-ray film. For clarity, we will refer to the procedures without hybridization with probes as restriction fragment length polymorphisms (RFLPs) and those with probes as restriction fragment length polymorphisms with hybridization (RFLPHs).

Assuming that phenotypic characters are reflections of the genotype of an organism, we would generally expect molecular taxonomic studies to support current phenotypically based taxonomic systems. When this is not the case, researchers must carefully examine the strengths and weaknesses of the molecular methods used as well as the rationale for the phenotypically based system before drawing any conclusions. No one molecular technique is appropriate for all taxonomic work. Convergent evolution does occur at both the phenotypic and molecular levels. Too much taxonomic importance may have been placed on a phenotypic character that is of little evolutionary importance. In short, there is much to be gained from integrating molecular methods in fungal systematics, but it must not be done with a blind eye to the limitations of either discipline.

Other reviews have been published on the molecular methods in the classification of fungi. Kurtzman [16,17] reviewed the G + C and reassociation studies, with emphasis on the yeasts. The nucleic acid work with the genus *Aspergillus* was reviewed by Croft [18]. Blanz and Unseld [19] reviewed the use of ribosomal RNA (rRNA) as a taxonomic tool in mycology. Taylor [20] reviewed the mitochondrial DNA (mtDNA) work in fungal evolutionary biology. In this chapter, we will limit our discussion to applications of molecular techniques used in systematics and identification of filamentous fungi. Emphasis will be placed on applications of recently developed methods. We will describe each method, give examples of its application in the study of filamentous fungi, and discuss its strong and weak points as a tool in evolutionary biology and identification of filamentous fungi.

II. G + C VALUES

The molar percentage of guanine plus cytosine (G + C) is a measure of DNA base composition. This molecular method is among the simplest, but also gives the least resolution to many taxonomic problems because there is a large overlap among unrelated species [17]. The technique is based on the fact that different species will have different relative amounts of the four different bases in their DNA. It provides no information about how these bases might be arranged. Taxa with the same G + C values are not necessarily related and there are also reports of intraspecific variation in G + C values [e.g., 21-23]. Many authors use this method in conjunction with DNA-DNA complementarity, with the relatively simple G + C method serving as a screen for the labor-intensive reassociation methods.

The principal methods used for determining the G + C values in filamentous fungi have been thermal denaturation and cesium chloride density gradient equilibrium centrifugation, with the latter method generally considered to be more accurate [16]. The thermal denaturation method takes advantage of the fact that heating native double-stranded DNA will denature it into single strands and that the higher the G + C content of the DNA, the higher the denaturation temperature. When a sample of DNA is gradually heated, its absorbance of DNA at 260 nm increases until all of the DNA is denatured. The temperature at the midpoint of this absorbance increase, the T_m, relates directly to the DNA base composition [24,25]. Measuring G + C content by cesium chloride density gradient equilibrium centrifugation is possible because the buoyant density of DNA increases linearly with its G + C content [26,27]. In studies involving a comparison between these two methods for determining fungal G + C values, differences of up to 3% in G + C values were reported for zoosporic fungi [28], less than 1% for agarics [29], and in a survey including representatives of all of the classes of fungi, the largest discrepancy between the two methods was 5% for a zygomycete [21].

The different types of DNA (nuclear, mitochondrial, etc.) in an organism will vary in their G + C content. For example, Dons et al. [30] determined that the nuclear G + C content of *Schizophyllum commune* was 57%, whereas the mtDNA G + C content was only 22%. The inherent differences in G + C content of the different DNAs have been utilized by Garber and Yoder [5] to develop a method of separating nuclear, mitochondrial, and ribosomal DNA by cesium chloride/bisbenzimide density gradient centrifugation.

Some of the earliest indications that G + C values could be useful in studying the systematics and phylogeny of filamentous fungi were in reports by Uryson and Belozersky [31] and Vanyushin et al. [32]. They analyzed two to four isolates from each of the major classes of fungi and found that the G + C contents of Ascomycetes and Deuteromycetes were close to 50% whereas G + C contents of Basidiomycetes were consistently greater than 50%. In a large survey of fungi from all the major groups, including yeasts, Storck and Alexopoulos [22] found the G + C values for fungi ranged from 26 to 70%. Intraspecific variation in G + C values was usually less than 10%. In general, the Zygomycetes were characterized by G + C content lower than 50%; the Homobasidiomycetes by G + C content greater than 50%, and the Ascomycetes were variable. Storck [21] reported that the Hemiascomycetes usually had G + C values below 45% whereas Euascomycetes G + C values were generally greater than 45%.

Ojha et al. [28] found that G + C values did not differentiate two genera in the Blastocladiales, *Allomyces* and *Blastocladiella*; both had values between 62 and 66%. They examined one isolate from the Saprolegniales, *Saprolegnia ferax*, and its G + C mole percentage was 55-57, lower than that for the Blastocladiales. In another study, G + C values obtained for several species of *Saprolegnia* were all between 58 and 59%, indicating that this is not a useful method for separating species of *Saprolegnia* [33]. However, *Achlya*, another member of the Saprolegniaceae, varied in G + C content from species to species, and this variation correlated with some of the intrageneric groups which had been established based on morphology [34].

The G + C values did not yield a means of separating species of the Ascomycete genus *Neurospora*; all G + C values ranged between 55 and 57% [35].

In examining the G + C content of *Geotrichum* species and other closely related taxa, Guêho [36] found that the values fell into three groups. These groups correlated well with biochemical and morphological data already in the literature and substantiated the work indicating that these genera represented both ascomycetous and basidiomycetous forms. The G + C content of 55 dermatophyte isolates (*Trichophyton*, *Microsporum*, and *Epidermophyton*) were all between 48.7 and 50.3% [37]. This method was not useful in distinguishing among these three closely related genera.

Horgen et al. [29] examined seven species in the Agaricales and found that the G + C base composition varied from 43.5 to 56%. Two strains of one species, *Agaricus brunnescens*, had quite different base compositions, one with 43.5%, the other with 47.3%. Dons et al. [30], working with another agaric, *Schizophyllum commune*, determined that the nuclear DNA had a G + C content of 57% whereas the mtDNA G + C content was only 22%.

In response to the speculation that *Ceratocystis* might be the perfect form of *Sporothrix schenckii*, Mendonça-Hagler et al. [38] examined four isolates of *S. schenckii* and 13 isolates of various *Ceratocystis* species. They found that the G + C values for *S. schenckii* were between 54.4 and 55.3%, and that only one *Ceratocystis* isolate, *C. minor*, had a G + C reading in this range. Moreover, other information did not substantiate the relatedness of these two species. *C. minor* did not hydrolyze starch, whereas *S. schenckii* did. The rhamnomannans of the two species were similar but not identical.

III. DNA COMPLEMENTARITY

The extent that denatured DNA from one isolate will hybridize to denatured DNA from another is a measure of the relatedness of the two genomes. This technique has been referred to in the literature as DNA-DNA hybridization, DNA nucleotide sequence homology, DNA complementarity, and DNA reassociation. To prevent confusion with other methodologies, we will use the terms reassociation and complementarity for this technique. It was among the first molecular techniques to be adapted and used as an aid in classification of organisms [39].

A variety of methods have been employed to measure DNA relatedness by this technique. Some require use of radioisotopes, with the level of radioactivity in the reassociated DNA being the indicator of the amount of complementarity. DNA reassociation can also be assayed by heating the test DNAs and spectrophotometrically measuring the extent of renaturation between the two DNAs as the temperature drops [40,41]. Kurtzman [16] reviewed the use of these techniques in fungi and detailed the procedures. The accuracy of this technique is improved by having highly purified nuclear DNA consisting of unique (nonrepeated) sequences. Repeated sequences from an isolate can reassociate with each other, thereby obscuring the true interisolate reassociation. Repeated DNA sequences may be removed from the sample by use of hydroxyapatite chromatography.

In most studies, the complementarity values are normalized so that the level of reassociation of the DNA of an isolate to its own DNA is set at 100%. Unless otherwise indicated, all of the reassociation values referred to in this chapter are normalized. Researchers unfamiliar with this normalization process may be misled into thinking that complementarity of DNA from one isolate to itself really is close to 100%. Actually, the reassociation value may be much lower. For instance, the percentage of labeled DNA actually binding to unlabeled DNA of the same isolate was 78.7% for *Agaricus brunnescens* [29] and 94% for *Neurospora crassa* [42]. Using the buoyant density method, actual self-complementarity for *Allomyces macrogynus* was 93% [28]. Normalization can also lead to complementarity values greater than 100% [43].

Ojha et al. [28] used ^{32}P to label in vivo the DNA of representative isolates of two genera of the Blastocladiales, *Allomyces* and *Blastocladiella*. Labeled DNA was allowed to reassociate with the unlabeled test DNA and hybridization was measured by thermal elution of duplexes [44]. They obtained somewhat different results in the normalized percent binding depending on which of the test pair was labeled. In the most extreme case, labeled DNA of *Allomyces arbuscula* to unlabeled *Saprolegnia ferax* gave a normalized binding of 48%, whereas labeled DNA of *S. ferax* with unlabeled *A. arbuscula* gave a normalized binding of 37%. The complementarity data did show that the two members of the Blastocladiales, *Allomyces* and *Blastocladiella*, had a greater degree of complementarity with one another than with the representative of the Saprolegniales, *S. ferax*. Another saprolegniaceous fungus, *Achlya radiosa*, had greater nucleotide sequence complementarity with *S. ferax* than with the blastocladiaceous fungi.

Using a spectrophotometric method, Ellis [40] examined the complementarity between members of the *Rhizopus arrhizus-Rhizopus oryzae* group. Several of the species in this group are difficult to distinguish from one another on morphological characteristics. *Rhizopus oryzae* × *R. arrhizus* reassociations were all above 95% and it was proposed that the two species were conspecific. *Amylomyces rouxii* reassociated with *R. oryzae* or *R. arrhizus* also gave complementarity values of at least 95%, and it was suggested that *A. rouxii* be reduced to varietal status in *R. arrhizus*. Although the *R. delemar* isolates and their varieties had lower complementarity values with *R. arrhizus* (down to 60%), it was suggested that *R. delemar* also be reduced to varietal status within *R. arrhizus*.

Reassociation experiments on *Fusarium* Sections *Liseola* and *Elegans* have added to our understanding of this complex genus [45-47]. The morphological complexity of this group is also reflected in the DNA. The reassociation work has shown that the presence of monophialides vs. polyphialides and chlamydospores vs. no chlamydospores are reliable characters for distinguishing species in the section *Liseola*; that several anamorphic (imperfect) species have the same teleomorph (perfect state); and that an association exists between complementarity values and mycotoxin production by some species.

Dutta [42] used in ^{32}P-labeled nonrepeated DNA and hydroxyapatite chromatography on heterothallic species of the ascomycete genus *Neurospora*. All *N. crassa* isolates had normalized complementarity values of 97-99%, even for isolates of different geographic origin. The same labeled *N. crassa* DNA reassociated to *N. intermedia* and *N. sitophila* DNA 90 and 84%, respectively. When *N. sitophila* was labeled and hybridized with the other two species, reassociation with *N. crassa* was 82% and with *N. intermedia* 87%. Based on these DNA reassociation data, it was possible to assign some unidentified isolates to species, thereby avoiding the time-consuming process of making genetic crosses between known and unknown isolates to identify species. In another study of *Neurospora* [35], heterothallic, homothallic, and pseudohomothallic species were compared in all possible combinations in DNA reassociation. The reassociation values depended to some degree on which species provided the reference DNA. Labeled *N. tetrasperma* (pseudohomothallic) hybridized to unlabeled *N. crassa* (heterothallic) gave a complementarity of 87%; the inverse gave a reading of 81%. The reassociation value for labeled *N. tetrasperma* × unlabeled *N. africana* (homothallic) was 64%; the inverse was 70%. The values for the *N. crassa*, *N. africana* reassociations were 64 and 59%. Several other species were also hybridized to these three species, and it was concluded that the heterothallic and pseudohomothallic species of this genus were more closely related to one another than to the homothallic species. However, no final conclusion on the phylogeny of *Neurospora* was possible from these data. In a study of the homothallic species of *Neurospora* [48], it was determined that the morphologically similar species, *N. lineolata*, *N. galapagosensis*, *N. africana*, and *N. dodgei*, had interspecific hybridization values of over 90% for the nonrepeated DNA. The morphologically distinct species *N. terricola*, however, when renatured with the other species had values of only 81-83% when renatured with the other species. Based on the greater diversity in DNA sequences in the heterothallic species, they proposed the evolutionary sequence of *Neurospora* species to be from homothallism to pseudohomothallism to heterothallism.

Rhizoctonia anastomosis groups have been studied by measuring DNA-DNA reassociation with ^{125}I in vitro-labeled DNA and S1 nuclease to hydrolyze unhybridized single-stranded DNA [10] and with spectrophotometry [49-56]. These researchers both demonstrated a high degree of reassociation for isolates within an anastomosis group and relatively low intergroup reassociation values. The values, however, were quite different. Kuninaga and Yokosawa [49-56] generally reported intragroup reassociation values of 90% or more whereas the intragroup reassociation values in the Vilgalys [10] study were as low as 57%. The differences in the methodologies could account for some of this variability. The observed differences in intragroup reassociation values could also reflect genetic changes associated with geographic diversity. The isolates in the Vilgalys [10] study were from geographically diverse sources, whereas all of the isolates in the Kuninaga and Yokosawa [49-56] studies were from Japan.

Horgen et al. [29] used nonrepeated (unique sequence) DNA labeled with ^{32}P by nick translation and hydroxyapatite chromatography on DNA of isolates of the Agaricales. They found intraspecific complementarity of 96.7%, intrageneric binding of 52.8%, and intergeneric binding consistently below 10%. The authors concluded that this method may be useful for determining relationships among closely related Basidiomycetes but would not be appropriate for studying distantly related Basidiomycetes. One such study of five intersterility groups in the mushroom *Collybia dryophila* demonstrated that intragroup reassociation values were 75% or above. Intergroup values ranged from 34 to 107%. They observed that intersterility barriers may occur independently of DNA sequence divergence and that intersterility groups in *C. dryophila* may have arisen primarily via an allopatric mode of evolution [43].

Mendonça-Hagler et al. [38] examined DNA complementarity of several isolates of *Sporothrix schenckii* by affixing purified DNA to a nitrocellulose filter and then using ^{125}I-labeled test DNA. Based on this technique, they found a binding range of 70-93% for the various isolates of this species.

A few authors suggested taxonomic changes based predominantly on the single character of DNA reassociation values. All of these were based on values from spectrophotometric measurements. Using nonreciprocal reassociation data from one isolate of each *Aspergillus flavus*, *A. oryzae*, *A. sojae*, and *A. parasiticus*, which yielded reassociation values between 69 and 100%, all of these species were formally reduced to subspecific status under *A. flavus* by Kurtzman et al. [57]. One new species, *Aspergillus nomius*, has been proposed for isolates formerly considered to be *A. flavus* whose only consistently distinct characteristics are the DNA reassociation values with *A. flavus* and low growth rates at 42°C [58]. Ellis [59] examined reassociation data from one isolate of each *Rhizopus microsporus*, *R. cohnii*, *R. pseudochinensis*, *R. pygmaeus*, *R. rhizopodiformis*, *R. oligosporus*, and three isolates of *R. chinensis*, including one of *R. chinensis* var. *liquifaciens*. Reassociation values were all above 70%, and the author proposed that all of these species should be considered as varieties of *R. microsporus*. This work supported the morphologically based work of Schipper and Stalpers [60], which also suggested that many species of *Rhizopus* should be reduced to varietal status.

IV. RNA

Because ribosomes serve the very basic function of translating mRNA into peptides and proteins, rRNA contains some of the most highly conserved nucleic acid sequences known. In addition to being highly conserved, the rRNA sequences and the DNA sequences coding for them (rDNA) are highly repetitive. Ribosomes consist of protein and RNA subunits of various sizes, which are referred to by their sedimentation rates (e.g., 5S, 18S, 25S). Because of the conservative nature of the RNA molecule, studies with RNA have proved to be most useful in taxonomic/phylogenetic comparisons above the species level [16,61].

Researchers have used several methods to evaluate RNA similarities. The earliest studies were simply comparisons of molecular weights of the RNA subunits [62]. Ribosomal RNA-DNA complementarity between taxa has also been used [63]. The actual rRNA sequences of the smaller subunits have been compared [64]. Since RNA sequence determines the secondary structure of the molecule, secondary structure has also been used to compare taxa [65].

An early study of fungal RNA in systematics was a comparison of molecular weights of the 18S and 25S subunits of taxa representing all of the major classes of fungi [62]. Results indicated that the molecular weights of the 18S rRNA were similar for all fungi (0.71-0.75 million) but that the 25S rRNA molecular weights were different for the major groups of fungi. Data were consistent with the concept that fungi are a polyphylogenetic group separate from the plant and animal kingdoms. The Myxomycetes appeared to be closer to animals than plants. Oomycetes seemed to have evolved independently from the rest of the fungi. Chytridiomycetes, Zygomycetes, Ascomycetes, and Basidiomycetes represent apparently related groups with the 25S rRNA molecular weights within 3% of one another, with the highest weights in the Chytridiomycetes.

In a variation of the DNA-DNA complementarity discussed in the preceding section, DNA-rRNA reassociations have also been studied using DNA from *Saccharomyces* sensu Lodder and 25S rRNA from a variety of other species of yeasts, filamentous fungi, and plants [63]. They found that *Saccharomyces* species were most closely related to one another, less related to Ascomycetes (or Deuteromycetes with Ascomycetes teleomorphs) and Basidiomycetes, and least related to Zygomycetes, spinach, and the alga *Chlamydomonas*.

The small 5S rRNA has been sequenced in many organisms and these data have been used as indicators of phylogenetic relationships [61,64]. The 5S rRNA sequences from a number of Basidiomycetes are known [13, 65-68]. Work by Walker and Doolittle [68] delineated two distinct sequence groupings of 5S rRNA in the Basidiomycetes which correlated exactly with the presence or absence of dolipores in the septal walls of the hyphae. They suggested that the nature of the septal wall may be a more reliable morphological means of separating Homobasidiomycetes from Heterobasidiomycetes than traditional methods. The 5S rRNA sequences of the smut fungi indicate that the smuts may be divided into two major groups: one which parasitizes dicots and one which parasitizes graminicolous plants [13,65,69]. However, in other basidiomycete groups, 5S rRNA sequence comparisons, the resolution of taxonomic relationships has been poor [66,69]. The use of 5S rRNA in a study to compare sequence homologies between the zygomycete *Phycomyces blakesleeanus* and several other organisms produced surprising results [70]. The percentages of sequence homologies between this lower fungal representative and the other organisms were (in order): the protozoan *Tetrahymena* (68%), human KB cells (67%), *Euglena* (66%), spinach (62%), with homologies to the representative higher fungi (*Aspergillus*, *Neurospora*, *Saccharomyces*, etc.) having an average of 58% sequence homology. Percentages of 5S rRNA sequence homologies among the higher fungal isolates ranged from 64 to 98%. Use of such data alone in developing a phylogenetic scheme could drive even the most ardent evolutionary biologist to consider the creationist point of view.

Obviously, the rRNA methodologies have limitations as tools in fungal systematics. Hasegawa et al. [64] noted that the 5S rRNA molecule is too small (only about 120 nucleotides) to be statistically reliable as an indicator of phylogenetic relationships. Reports of intraspecific heterogeneity in the highly conserved 5S rRNA sequences [71,72] have raised concern about precision of this technique. Hasegawa et al. [64] favored the larger subunit rRNA as being more statistically reliable for phylogenetic analysis. A new rapid rRNA sequencing technique using crude cellular lysates [73] will probably facilitate this research. Blanz and Unseld [19] reviewed the current work on rRNA as a taxonomic tool in mycology and acknowledge its limitations. They cite preliminary sequence data and secondary structure analyses for 18S and 28S rRNA, and suggest that these may yield possible means to refine the precision of rRNA techniques in verifying taxonomic relationships, but also suggest that as many characters as possible be used when developing phylogenetic schemes. Rothschild et al. [74] questioned the validity of rRNA sequence comparisons as phylogenetic tools and also wisely caution that taxonomic decisions should not be based solely on rRNA or any other single characteristic.

V. RESTRICTION ENZYMES AND PROBES

Gel electrophoresis and restriction endonucleases have provided the basis for another approach to the molecular systematics of filamentous fungi. Restriction endonucleases (restriction enzymes) cleave double-stranded DNA at specific base pair sequences called recognition sequences. For example, the restriction enzyme *SmaI* recognizes the sequence CCCGGG and cleaves it and its complement into CCC-GGG. Because they have been used in genetic engineering technology, many different restriction enzymes are commercially available. Once the DNA is broken into fragments by the restriction enzyme, these fragments may be separated by length using gel electrophoresis. The resulting pattern of DNA fragments may be observed by staining the gel with ethidium bromide and observing it under UV light. Different restriction enzymes cleave the DNA into different sizes of fragments and therefore yield different patterns on the gel after electrophoresis.

If the DNA molecules cleaved with a restriction enzyme are small and homogeneous like mtDNA and rDNA, discrete fragment patterns will be observed on the gel. If the DNA sample is from a heterogeneous population of large molecules such as chromosomal DNA, there will be so many different sized fragments produced that the gel will appear to have a continuous smear of DNA.

Regions of similar DNA sequences shared between isolates may be determined by Southern blot analysis [75]. The DNA fragments on the gel are denatured into single strands, and then transferred and affixed to nitrocellulose or some other suitable membrane. A previously isolated or synthesized fragment of DNA (the probe) is then radiolabeled. The probe and the membrane-bound DNA are brought together under conditions which favor hybridization of the probe to any DNA fragment with the same or a very similar DNA sequence. Excess probe is then removed by washing and regions of hybridization may be detected by placing the membrane in contact with X-ray film.

Differences in DNA restriction fragment lengths (RFLPs) and differences in restriction fragment hybridization patterns (RFLPHs) have been used in fungal systematics. The proportion of bands common to the DNA of two isolates in an RFLP or RFLPH analysis is a measure of similarity between the isolates. While this method cannot detect all genetic variation, it is simpler than either DNA sequencing or DNA complementarity and does correlate well with DNA sequence analysis [76].

Several factors can limit the accuracy of RFLP/RFLPH analyses [20,77, 78]. Mutational events which alter the length of DNA fragments will alter the fragment patterns on the gels. Any mutation at a restriction site will also alter the fragment patterns. Therefore, a deletion or insertion could yield a polymorphism which in reality has little significance. If a probe is not used, similarly sized nonhomologous DNA fragments will appear to be identical, giving the impression of greater similarity between isolates than actually exists. For phylogenetic analyses based on mathematical models, the data may be biased by the use of too few restriction enzymes. The frequency of the four nucleotides is not identical in an organism (or they all would have G + C values of 50%), and not all restriction sites have identical mutation rates. To minimize the effects of these factors, results of many different restriction enzyme digests should be used in phylogenetic studies [79,80].

Both RFLPs and RFLPHs have been used in studying genetic relatedness in fungi. For phylogenetic studies, mathematical models are available for the quantitative analysis of RFLP/RFLPH results. These models give an estimate of nucleotide sequence diversity based on the proportion of fragments shared by the different DNA samples [79]. However, as previously noted, length mutations will also alter these fragment patterns and need to be considered when using models based on nucleotide substitutions.

Mitochondrial DNA, like RNA, has appeal for evolutionary studies because it is small enough to be studied in its entirety. Generally, mitochondrial DNA is assumed to mutate at a faster rate than nuclear DNA making it more suitable for assessing differences between closely related species [79,81]. The work on mtDNA in fungal evolutionary biology was recently reviewed by Taylor [20]. Fungal mtDNAs are all circular but are extremely variable in size, ranging from 18.9 kb for *Torulopsis glabrata* [82] to 176 kb for *Agaricus bitorquis* [83]. These are between the sizes of most animal mtDNAs, which are quite small (±16 kb [81]), and plant mtDNAs, which are quite large (130-350 kb [84]). To use mtDNA in evolutionary studies, the means of inheritance of mtDNA must be considered. Among the fungi, maternal, paternal, recombinant, and biparental inheritance of mtDNA has been demonstrated (reviewed by Taylor [20]).

Preliminary work has indicated that mtDNA may be useful in oomycete taxonomy. Using 14 restriction endonucleases Förster and Kinscherf [77] quantitatively compared restriction patterns of mtDNA of several *Phytophthora* isolates. The intraspecific patterns for the six *P. cactorum* isolates were very similar to one another. The relatedness values calculated for the *P. megasperma* isolates indicated as much divergence within this species as there was between *P. megasperma* and the other species. These data support the contention that *P. megasperma* consists of several biological species. Species of another oomycete, *Pythium*, could also be separated from one another by restriction banding pattern differences [85]. Analysis of restriction digests of mtDNA from 21 isolates revealed a high frequency of interspecific variation in banding patterns. A low frequency of intraspecific polymorphisms, even the 10 *P. oligandrum* isolates from diverse locations, was observed.

The chytridiomycete *Allomyces* was also investigated by mtDNA restriction enzyme analysis. Grouping of species by RFLPHs was not possible because of the considerable intraspecific variation of the seven species in the study. Part of the intraspecific variation was attributable to the extreme variation in the mtDNA size of some of the species. However, although the four isolates of *A. arbuscula* had mtDNA of similar size, they had such different restriction patterns that they could not be grouped in any way [86]. Further study showed that while gene order appears to be similar among at least some of the *Allomyces* species, frequent spontaneous insertion mutations occur in the mitochondrial genome of *A. macrogynus* [87].

Analysis of RFLPHs has been useful in taxonomic studies of the genus *Fusarium* [88]. The species in this genus are morphologically variable and tend to lose taxonomically important characteristics when maintained in culture. Four probes were randomly generated from an isolate of *Fusarium oxysporum* f. sp. *dianthi*. These were hybridized to representatives of the eight species of *Fusarium* proposed by Snyder and Hansen [89]. Using different probes and different restriction enzymes, several of the taxa could be distinguished from the others. One combination of probes and restriction enzymes yielded identical patterns for the three *F. oxysporum* f. sp. *dianthi* isolates, even though one of these isolates had been in culture for 7 years and had degenerated morphologically. This method could be useful in distinguishing strain degeneration from strain contamination.

Kistler et al. [90] using Southern blotting examined rDNA and mtDNA from virulent strains of the three formae speciales of *Fusarium oxysporum* infecting crucifers. The rDNA restriction fragments were the same for all isolates. RFLPHs were present in the mtDNA preparations, and these corresponded with the formae speciales. Three different mitochondrial plasmidlike DNAs were also detected, and these also correlated with the three formae speciales.

Kozłowski and Stepień [91] compared the mtDNA of several species of *Aspergillus* by RFLPH analysis using plasmids containing cloned fragments of *A. nidulans* mtDNA as probes. This was done to determine if mtDNA data would correspond to the morphological data which formed the basis of the taxonomy of this genus. The two members of the section *Nidulantes* (*A. nidulans* and *A. echinulatus*) they examined had a high degree of homology to one another. The two members of the section *Flavi* (*A. oryzae* and *A. tamarii*) were almost identical to one another. Unexpectedly, *A. wentii*, a member of a different section, had an identical restriction pattern to that of *A. tamarii*. Even more unexpected was the very distant relationship in the mtDNA between *A. niger* and *A. awamori*, two species of the section *Nigri*. These species are so similar in morphology and growth rates that Al-Musallam [92] proposed that *A. awamori* be considered a variety of *A. niger*. The authors proposed that the limitations in RFLPH analyses (discussed above) were responsible for these apparent discrepancies.

In a review of the evolutionary biology of *Aspergillus*, Croft [18] reported no intraspecific variation in restriction maps of mtDNA. Restriction site polymorphisms in the mtDNA of different taxa in the section *Nidulantes* did occur. Most of these differences were attributed to the presence of optional introns in the mitochondrial genome. These introns may limit the value of RFLPH in defining evolutionary relationships even among closely related *Aspergillus* species. Introns are apparently not a problem in the nuclear DNA of *Aspergillus*. Southern blotting with random probes from nuclear DNA of an *A. nidulans* isolate proved to be very diagnostic for species within the section *Nidulantes*. For most of the probes, intraspecific polymorphisms were absent or few in number, and these minor differences were between sexual incompatibility groups within a species. Interspecific polymorphisms did exist between species in the section *Nidulantes*. However, none of the probes hybridized to *A. unguis*, even though it is classified as a member of that group. For fungi outside the section *Nidulantes*, the probes did not hybridize to *A. niger* DNA, but hybridized strongly to *A. terreus* DNA, indicating that this species may be more closely related to the section *Nidulantes* than its present classification would indicate. Strong hybridization between *A. terreus* DNA and a cloned *A. nidulans* developmental gene [93] adds further support to this hypothesis. The use of cloned developmental *A. nidulans* genes as probes has revealed that genomic DNA from several other species of *Aspergillus* and other hyphomycetes contain DNA sequences that hybridize to these genes [93,94]. There is therefore the potential of using these or other cloned *A. nidulans* genes to establish the degree of DNA relatedness between *A. nidulans* and other filamentous fungi. There is also the possibility of determining if there is a widespread distribution of similar genes in the Hyphomycetes.

If so, *A. nidulans* may be used as a model to facilitate developmental regulation studies in fungal species that are less characterized genetically.

RFLP technology has also proved to be useful in distinguishing two closely related, economically important species of *Aspergillus* [6]. About half of the isolates of *A. flavus* produce alfatoxin, a secondary metabolite which is a potent carcinogen. *A. oryzae* does not produce aflatoxin and is used in oriental food fermentations (Koji process). Morphologically, these two species are very difficult to distinguish. Total DNA of 11 isolates of *A. flavus* and seven isolates of *A. oryzae* which had been selected to represent a high degree of morphological variability and diverse geographic origin were digested with several restriction enzymes. One of these, *Sma*I, yielded RFLPs that were intraspecifically consistent and interspecifically distinct, indicating that this restriction enzyme could be used as a tool in distinguishing these two species. It was further determined that the source of the differences observed was the nuclear rather than the mitochondrial DNA. One *A. flavus* isolate had degenerated morphologically in culture but still had the restriction pattern of the species [6]. As with *Fusarium*, RFLPs can be used to determine if these *Aspergillus* cultures have changed over time or if they have been contaminated.

Intraspecific variation in *Neurospora* mtDNA has been studied extensively using RFLPH analyses with cloned fragments of mtDNA from a wild-type *N. crassa* as probes [95]. It was demonstrated that most of the variability in the mtDNA genome of *N. crassa* was a result of length mutations (insertions and deletions) and not nucleotide substitutions. Their data suggest that geographically distinct mtDNA populations exist within *N. crassa*. *Neurospora* mitochondrial plasmids have been used to study evolutionary relationships in this genus [96,97]. Plasmids from *N. intermedia* and *N. tetrasperma* were characterized by restriction endonucleases and RFLPH analysis using labeled plasmid DNA [96]. Taylor et al. [97] used a similar analysis to demonstrate that another mitochondrial plasmid from *N. crassa* is homologous to a plasmid for *N. intermedia* and *N. tetrasperma*. This suggested that plasmids could be transmitted independently of their host mitochondria. Natvig et al. [78] used randomly generated *Neurospora* nuclear DNA probes for an RFLPH analysis of 14 *Neurospora* isolates representing four species. The different species could be distinguished from one another by this technique. As with mtDNAs, nuclear DNAs of *N. tetrasperma* had geographically distinct restriction patterns. The results also supported earlier work with mtDNA suggesting that the four-spored *Neurospora* strains represent a monophyletic group. The authors suggested that both nuclear and mitochondrial DNA should be examined to develop a more precise phylogeny within this genus.

Recently, Taylor and Natvig [98], using numerical analysis of mtDNA length mutations in heterothallic and pseudohomothallic *Neurospora* species, concluded that *N. tetrasperma* diverged from the eight-spored species before the eight-spored species diverged from one another. They also concluded that *N. tetrasperma* and *N. sitophila* are monophyletic but *N. crassa* and *N. intermedia* are not.

Probes containing rDNA from *Neurospora crassa* were used in an RFLPH study of *Sclerotinia* taxonomy [99]. Not all isolates of this ascomycetous genus of plant pathogens produce all of the morphological features necessary for identification. The DNAs of 42 isolates representing seven species of *Sclerotinia* from a variety of host tissues and geographic origins were analyzed. The cloned *Neurospora* rDNA hybridized to analagous *Sclerotonia* rDNA. Identical restriction fragment patterns were obtained when rDNAs from two other fungi (*Armillaria mellea* and *Schizophyllum commune*) were used as probes, confirming that the bands did indeed represent rDNA. All of the *Sclerotinia* species had distinct, characteristic rDNA fragment patterns except *S. sclerotiorum* and *S. fircariae*, which had identical patterns. This observation supported other evidence that these two species are synonymous [100]. There was some intraspecific variation. One of the 21 *S. sclerotiorum* isolates had a banding pattern slightly different from the others which the authors hypothesized may have been due to the loss of a single *Eco*RI site. Three slightly different patterns were observed among the six *S. trifoliorum* isolates. These were not correlated with differences in host or geographic distribution. When DNAs of three of the *Sclerotinia* species were hybridized with a mitochondrial rRNA gene probe from *Neurospora crassa*, there were extensive polymorphisms between species and among the isolates of each species. The authors felt that because there was so much variability in mtDNA restriction patterns, nuclear rDNA was more useful in studies such as theirs involving more than one species of a genus. They also felt that molecular techniques should be used in combination with conventional taxonomic characters to further elucidate evolutionary relationships in this group.

Garber and Yoder [101] conducted an RFLPH study of the mtDNA of 23 isolates of the pathogenic filamentous ascomycete *Cochiobolus heterostrophus*. One of the resulting polymorphisms correlated with mating type, race, geographic origin, and a nuclear DNA polymorphism.

Studies of the mtDNA relatedness of the Agaricales have provided interesting information regarding the taxonomy of this order. Hintz et al. [83] examined four *Agaricus brunnescens* (*bisporus*) isolates and 10 *A. bitorquis* isolates using by comparing RFLPHs using labeled *A. brunnescens* mtDNA as a probe. The restriction patterns were identical for the four *A. brunnescens* isolates. The *A. bitorquis* isolates yielded high restriction pattern heterogeneity, and the authors tentatively assigned these isolates to four groups based on this heterogeneity. The groups did not represent geographic differences. Future work delineating the genomic organization of these groups should provide interesting information on the evolutionary biology of this species. In another study of *Agaricus* species, nuclear RFLPHs were used to identify probes that could serve as genetic markers for *Agaricus brunnescens*, to distinguish *A. brunnescens* from the closely related species *A. bitorquis* and to determine if homokaryotic isolates would have less complex hybridization patterns than heterokaryons of *A. brunnescens* [102]. Ten randomly generated nuclear recombinant clones from these two species were used to probe 11 *A. brunnescens* and seven *A. bitorquis* isolates. *A. brunnescens* isolates had a limited number of RFLPHs; four of the 11 isolates had identical patterns with all 10 probes. The homokaryotic strains of *A. brunnescens* had fewer bands than the heterokaryotic strains. Seven probes that would distinguish the two species were found. However, all but two also produced intraspecific polymorphisms. This information could be very useful in the commercial mushroom industry because crosses could be quickly and easily distinguished by their polymorphisms.

May and Taylor [103] studied the mtDNA of another agaric, *Coprinus cinereus*, and found that the mating system of this species yielded colonies composed of sectors differing in mtDNA. This indicates that mitochondria do not migrate along with nuclei during mating. Wu et al. [11] generated random probes from *C. cinereus* and used these to probe genomic DNA from 16 strains of *C. cinereus*. Nuclear DNA polymorphisms were observed, both as gains and losses of restriction sites and as frequent insertion/deletion polymorphisms.

Armillaria mellea contains eight biological species (reproductively isolated groups). To determine whether the observed mating group differences were reflected in genetic differences, the DNAs of two isolates of each of the eight biological species were hybridized with a variety of probes including random nuclear DNA fragments, mitochondrial DNA, and cloned *Schizophyllum commune* rDNA [104]. Generally, the fragment patterns within a biological species showed much less diversity than was present between biological species. The authors concluded that "the biological species of *A. mellea* are distinct entities among which genetic diversity has occurred."

Randomly selected clones from gene libraries proved useful as probes in testing the hypothesis that two commonly used lignin-degrading strains, *Phanerochaete chrysosporium* ATCC 34541 and *Sporotrichum pulverulentum* ATCC 24725, were teleomorph and anamorph, respectively, of the same fungus [105]. Autoradiographs showed similar patterns for these two for both mitochondrial and nuclear DNA. The G + C content from both the chromosomal and mitochondrial DNA were also identical for the two species. The authors concluded that these DNA data confirm the classification of these two as members of the same species. Two other *P. chrysosporium* isolates had restriction digest patterns that were different from those of the commonly used strain, demonstrating intraspecific polymorphism in this species.

To study potential evolutionary relationships among simple eukaryotes, Kwok et al. [15] used the hybridization melt procedure [106] and a plasmid containing a cloned restriction fragment coding for the 28S and 18S RNAs from the basidiomycete *Tyromyces unicolor*. In this method the sample DNA was bound to a nitrocellulose filter disk and hybridized with the radioactively labeled probe. After about 48 h the disks were eluted at progressive temperature increases of 5°C. Thermal dissociation profiles were generated by plotting cumulative percent total radioactivity vs. temperature. T_m was the temperature at which 50% of the radioactivity was eluted. This procedure was used with fungi (seven Basidiomycetes, five Ascomycetes, one Zygomycetes, one Oomycetes, and one Chytridiomycetes), algae (two Rhodophyta, one Chlorophyta, and one Euglenophyta), protozoans (two Flagellata and two Ciliata), and Mycetozoa (one Dictyostelia). Based on the relative hybridization data from this single probe, they concluded that there was no evidence to support the theory of a close evolutionary relationship between red algae and higher fungi and that the Chytridiomycetes were more closely related to the higher fungi than to any of the other major taxa.

VI. CONCLUSIONS

Molecular techniques are obviously very useful in fungal systematics. However, one of the major problems in using molecular methodologies is "aiming." How do we select the appropriate technique for the taxonomic level being examined? The fungi are an extremely diverse group and it is unlikely that any one current method would be appropriate for either distinguishing all taxa or developing phylogenetic schemes for all major classes. Even within the same family, results from one technique used successfully on one genus may not prove successful for another. For example, G + C values distinguished groups in the genus *Achlya* in the Saprolegniaceae but did not distinguish the different members of the genus *Saprolegnia* which is in the same family [34].

Powerful new molecular tools are being developed. The ability to sequence significant portions of fungal genomes will certainly aid systematic mycology. There is an increasing demand in such areas as medicine and forensics that the methods for molecular data acquisition be further streamlined. The need to use radioactive material in several of the procedures has inhibited their adoption in some laboratories. Nonradioactive methods of labeling nucleic acid probes have been used successfully in recent taxonomic studies [e.g., 97] and provide a means to widen acceptance of these techniques. Methods for the rapid sequencing of rRNA from crude cellular lysates [73] and methods for polymerase chain reaction amplification of DNA and RNA to facilitate sequencing analysis [107] may have tremendous impact on the future of fungal molecular genetics.

Molecular techniques have given us the ability to distinguish individual strains from one another. This is extremely useful for patented organisms and for ecological studies in which one needs to know the distribution of a given strain. However, to the taxonomist, this degree of precision raises the same questions that have been asked since classification began: how does one decide which characters are important at which taxonomic levels? Ultimately, taxonomy has always involved subjective decisions on the part of the taxonomist whether the system be based on standard taxonomic techniques or numerical approaches (even the numerical approaches require subjective decisions on program selection, character weighting, etc.). This inherent subjectivity will not be eliminated with the use of molecular techniques. They are powerful new tools for the study of taxonomy, but they are not a panacea; decisions such as "where to draw the line" between taxa will still need to be made. Nevertheless, as the molecular methods are integrated into fungal systematics, many uncertainties in the standard taxonomies certainly will be resolved.

REFERENCES

1. Ausubel, F. M., Brent, R., Kingston, R. E., Moore, D. D., Seidman, J. G., Smith, J. A., and Struhl, K. (eds.). *Current Protocols in Molecular Biology*. John Wiley and Sons, New York, 1987.
2. Davis, L. G., Dibner, M. D., and Battey, J. F. *Basic Methods in Molecular Biology*. Elsevier, New York, 1986, pp. 1-388.
3. Maniatis, T., Fritsch, E. F., and Sambrook, J. *Molecular Cloning*. Cold Spring Harbor Laboratory, New York, 1982.
4. Biel, S. W., and Parrish, F. W. Isolation of DNA from fungal mycelia and sclerotia without use of density gradient ultracentrifugation. *Anal. Biochem.*, *154*:21-25, 1986.
5. Garber, R. C., and Yoder, O. C. Isolation of DNA from filamentous fungi and separation into nuclear, mitochondrial, ribosomal and plasmid components. *Anal. Biochem.*, *135*:416-422, 1983.
6. Klich, M. A., and Mullaney, E. J. DNA restriction enzyme fragment polymorphism as a tool for rapid differentiation of Aspergillus flavus from Aspergillus oryzae. *Exp. Mycol.*, *11*:170-175, 1987.

7. Lee, S. B., Milgroom, M. G., and Taylor, J. W. A rapid, high yield miniprep method for isolation of total genomic DNA from fungi. *Fungal Genet. Newslett.*, *35*:23-24, 1988.
8. Sprecht, C. A., DiRusso, C. C., Novotny, C. P., and Ullrich, R. C. A method for extracting high-molecular-weight deoxyribonucleic acid from fungi. *Anal. Biochem.*, *119*:158-163, 1982.
9. Taylor, J. W., and Natvig, D. O. Isolation of fungal DNA. In *Zoosporic Fungi in Teaching and Research* (M. S. Fuller and A. Jaworski, eds.), Southeastern Publishing, Athens, Georgia, 1987, pp. 252-258.
10. Vilgalys, R. Genetic relatedness among anastomosis groups in Rhizoctonia as measured by DNA/DNA hybridization. *Phytopathology*, *78*:698-702, 1988.
11. Wu, M. M. J., Cassidy, J. R., and Pukkila, P. J. Polymorphisms in DNA of Coprinus cinereus. *Curr. Genet.*, *7*:385-392, 1983.
12. Yelton, M. A., Hamer, J. E., and Timberlake, W. E. Transformation of Aspergillus nidulans by using a trpC plasmid. *Proc. Natl. Acad. Sci. USA*, *81*:1470-1474, 1984.
13. Blanz, P. A., and Gottschalk, M. A comparison of 5S ribosomal RNA nucleotide sequences from smut fungi. *System. Appl. Microbiol.*, *5*:518-526, 1984.
14. Johnson, J. L. Genetic characterization. In *Manual for Methods for General Bacteriology* (P. Gerhardt, ed.), American Society for Microbiology, Washington, D.C., 1981, pp. 450-472.
15. Kwok, S., White, T. J., and Taylor, J. W. Evolutionary relationships between fungi, red algae, and other simple eucaryotes inferred from total DNA hybridizations to a cloned basidiomycete ribosomal DNA. *Exp. Mycol.*, *10*:196-204, 1986.
16. Kurtzman, C. P. Molecular taxonomy of the fungi. In *Gene Manipulations in Fungi* (J. W. Bennett and L. L. Lasure, eds.), Academic Press, Orlando, 1985, pp. 35-63.
17. Kurtzman, C. P. Impact of nucleic acid comparisons on the classification of fungi. *Proc. Indian Acad. Sci. (Plant Sci.)*, *97*:185-201, 1987.
18. Croft, J. H. Genetic variation and evolution in Aspergillus. In *Evolutionary Biology of the Fungi* (A. D. M. Rayner, C. M. Brasier, and D. Moore, eds.), Cambridge University Press, New York, 1986, pp. 311-323.
19. Blanz, P. A., and Unseld, M. Ribosomal RNA as a taxonomic tool in mycology. *Stud. Mycol.*, *30*:247-258, 1987.
20. Taylor, J. W. Fungal evolutionary biology and mitochondrial DNA. *Exp. Mycol.*, *10*:259-269, 1986.
21. Storck, R. Nucleotide composition of nucleic acids of fungi. II. Deoxyribonucleic acids. *J. Bacteriol.*, *91*:227-230, 1966.
22. Storck, R., and Alexopoulos, C. J. Deoxyribonucleic acid of fungi. *Bact. Rev.*, *24*:126-154, 1970.
23. Villa, V. D., and Storck, R. Nucleotide composition of nuclear and mitochondrial deoxyribonucleic acid of fungi. *J. Bacteriol.*, *96*:184-190, 1968.
24. Mandel, M., and Marmur, J. Use of ultraviolet absorbance-temperature profile for determining the guanine plus cytosine content of DNA. In *Methods in Enzymology, Vol. 12, Nucleic Acids, Part B* (L. Grossman and K. Moldave, eds.), Academic Press, New York, 1968, pp. 195-206.
25. Marmur, J., and Doty, P. Determination of the base composition of deoxyribonucleic acid from its thermal denaturation temperature. *J. Mol. Biol.*, *5*:109-118, 1962.

26. Mandel, M., Schildkraut, C. L., and Marmur, J. Use of CsCl density gradient analysis for determining the guanine plus cytosine content of DNA. In *Methods in Enzymology, Vol. 12, Nucleic Acids, Part B* (L. Grossman and K. Moldave, eds.), Academic Press, New York, 1968, pp. 184-195.
27. Schildkraut, C. L., Marmur, J., and Doty, P. Determination of the base composition of deoxyribonucleic acid from its buoyant density in CsCl. *J. Mol. Biol.*, *4*:430-443, 1962.
28. Ojha, M., Dutta, S. K., and Turian, G. DNA nucleotide sequence homologies between some zoosporic fungi. *Mol. Gen. Genet.*, *136*:151-165, 1975.
29. Horgen, P. A., Arthur, R., Davy, O., Moum, A., Herr, F., Straus, N., and Anderson, J. The nucleotide sequence homologies of unique DNAs of some cultivated and wild mushrooms. *Can. J. Microbiol.*, *30*:587-593, 1984.
30. Dons, J. J. M., DeVries, O. M. H., and Wessels, J. G. H. Characterization of the genome of the basidiomycete Schizophyllum commune. *Biochim. Biophys. Acta*, *563*:100-112, 1979.
31. Uryson, S. O., Belozersky, A. N. Nucleotide composition of deoxyribonucleic and ribonucleic acids of certain fungi. *Dokl. Akad. Nauk. SSSR*, *133*:708-710, 1960.
32. Vanyushin, B. F., Belozersky, A. N., and Bogdanova, S. L. A comparative study of the nucleotide composition of ribonucleic acids in some fungi and myxomycetes. *Dokl. Akad. Nauk. SSSR*, *134*:1222-1225, 1960.
33. Neish, G. A., and Green, B. R. Nuclear and satellite DNA base composition and the taxonomy of Saprolegnia (Oomycetes). *J. Gen. Microbiol.*, *96*:215-219, 1976.
34. Green, B. R., and Dick, M. W. DNA base composition and the taxonomy of the Oomycetes. *Can. J. Microbiol.*, *18*:963-968, 1972.
35. Dutta, S. K., Sheikh, I., Choppala, J., Aulakh, G. S., and Nelson, W. H. DNA homologies among homothallic, pseudo-homothallic and heterothallic species of Neurospora. *Mol. Gen. Genet.*, *147*:325-330, 1976.
36. Guého, E. Deoxyribonucleic acid base composition and taxonomy in the genus Geotrichum link. *Antonie van Leeuwenhoek*, *45*:199-210, 1979.
37. Davison, F. D., MacKenzie, D. W. R., and Owen, R. J. Deoxyribonucleic acid base compositions of dermatophytes. *J. Gen. Microbiol.*, *118*:465-470, 1980.
38. Mendonça-Hagler, L. C., Travassos, L. R., Lloyd, K. O., and Phaff, H. J. Deoxyribonucleic acid base composition and hybridization studies on the human pathogen Sporothrix schenckii and Ceratocystis species. *Infect. Immun.*, *9*:934-938, 1974.
39. Hoyer, B. H., McCarthy, B. J., and Bolton, E. T. A molecular approach in the systematics of higher organisms. *Science*, *144*:959-967, 1964.
40. Ellis, J. J. Species and varieties in the Rhizopus arrhizus-Rhizopus oryzae group as indicated by their DNA complementarity. *Mycologia*, *77*:243-247, 1985.
41. Seidler, R. J., and Mandel, M. Quantitative aspects of deoxyribonucleic acid renaturation: base composition, state of chromosome replication and polynucleotide homologies. *J. Bacteriol.*, *106*:608-614, 1971.

42. Dutta, S. K. DNA homologies among heterothallic species of Neurospora. *Mycologia, 68*:388-401, 1976.
43. Vilgalys, R. J., and Johnson, J. L. Extensive genetic divergence associated with speciation in filamentous fungi. *Proc. Natl. Acad. Sci. USA, 84*:2355-2358, 1987.
44. Dutta, S. K., and Ojha, M. Relatedness between major taxonomic groups of fungi based on measurements of DNA nucleotide sequence homology. *Mol. Gen. Genet., 114*:232-240, 1972.
45. Ellis, J. J. Section Liseola of Fusarium. *Mycologia, 80*:255-258, 1988.
46. Ellis, J. J. Lesser known species in section Liseola of Fusarium and near relatives in the Fusarium oxysporum group. *Mycologia, 80*:734-738, 1988.
47. Ellis, J. J. An alignment of toxigenic Gibberella strains having anamorphs in the section Liseola of Fusarium. *Mycologia 81* (in press).
48. Williams, N. P., Mukhopadhyay, D., and Dutta, S. K. Homologies of Neurospora homothallic species using repeated and nonrepeated DNA sequences. *Experientia, 37*:1157-1158, 1981.
49. Kuninaga, S., and Yokosawa, R. A comparison of DNA base compositions among anastomosis groups in Rhizoctonia solani Kühn. *Ann. Phytopathol. Soc. Jpn., 46*:150-158, 1980.
50. Kuninaga, S., and Yokosawa, R. DNA base sequence homology in Rhizoctonia solani Kühn I. Genetic relatedness within anastomosis group 1. *Ann. Phytopathol. Soc. Jpn., 48*:659-667, 1982.
51. Kuninaga, S., and Yokosawa, R. DNA base sequence homology in Rhizoctonia solani Kühn II. Genetic relatedness within anastomosis group 2. *Ann. Phytopathol. Soc. Jpn., 48*:668-673, 1982.
52. Kuninaga, S., and Yokosawa, R. DNA base sequence homology in Rhizoctonia solani Kühn III. Genetic relatedness within AG-3, AG-5, AG-7 and AG-BI. *Ann. Phytopathol. Soc. Jpn., 49*:647-652, 1983.
53. Kuninaga, S., and Yokosawa, R. DNA base sequence homology in Rhizoctonia solani Kühn IV. Genetic relatedness within AG-4. *Ann. Phytopath. Soc. Jpn., 50*:322-330, 1984.
54. Kuninaga, S., and Yokosawa, R. DNA base sequence homology in Rhizoctonia solani Kühn V. Genetic relatedness within AG-6. *Ann. Phytopathol. Soc. Jpn., 50*:346-352, 1984.
55. Kuninaga, S., and Yokosawa, R. DNA base sequence homology in Rhizoctonia solani Kühn VI. Genetic relatedness among seven anastomosis groups. *Ann. Phytopathol. Soc. Jpn., 51*:127-132, 1985.
56. Kuninaga, S., and Yokosawa, R. DNA base sequence homology in Rhizoctonia solani Kühn VII. Genetic relatedness between AG-BI and other anastomosis groups. *Ann. Phytopathol. Soc. Jpn., 51*:133-138, 1985.
57. Kurtzman, C. P., Smiley, M. J., Robnett, C. J., and Wicklow, D. T. DNA relatedness among wild and domesticated species in the Aspergillus flavus group. *Mycologia, 78*:955-959, 1986.
58. Kurtzman, C. P., Horn, B. W., and Hesseltine, C. W. Aspergillus nomius, a new aflatoxin-producing species related to Aspergillus flavus and Aspergillus tamarii. *Antonie van Leeuwenhoek, 53*:147-158, 1987.
59. Ellis, J. J. Species and varieties in the Rhizopus microsporus group as indicated by their DNA complementarity. *Mycologia, 78*:508-510, 1986.
60. Schipper, M. A. A., and Stalpers, J. A. A revision of the genus Rhizopus. *Studies in Mycology*, No. 25. Baarn, Netherlands, Centraalbureau voor Schimmelcultures, 1984, pp. 1-34.

61. Hori, H., and Osawa, S. Origin and evolution of organisms as deduced from 5S ribosomal RNA sequences. *Mol. Biol. Evol.*, *4*:445-472, 1987.
62. Lovett, J. S., and Haselby, J. A. Molecular weights of the ribosomal ribonucleic acid of fungi. *Arch. Mikrobiol.*, *80*:191-204, 1971.
63. Bicknell, J. N., Douglas, H. C. Conservation of ribosomal RNA sequence in yeast, filamentous fungi, and plants. *Bacteriological Proceedings*, 69th Annual Meeting of the American Society for Microbiology, Florida, 1969, pp. 39 (Abstr.).
64. Hasegawa, M., Iida, Y., Yano, T., Takaiwa, F., and Iwabuchi, M. Phylogenetic relationships among eukaryotic kingdoms inferred from ribosomal RNA sequences. *J. Mol. Evol.*, *22*:32-38, 1985.
65. Blanz, P. A., and Gottschalk, M. Systematic position of Septobasidium, Graphiola and other basidiomycetes as deduced on the basis of their 5S ribosomal RNA nucleotide sequences. *System. Appl. Microbiol.*, *8*:121-127, 1986.
66. Gottschalk, M., and Blanz, P. A. Untersuchungen an 5S ribosomalen Ribonukleinsäuren als Beitrag zur Klärung von Systematik und Phylogenie der Basidiomyceten. *Z. Mykol.*, *51*:205-243, 1985.
67. Huysmans, E., Dams, E., Vandenberghe, A., and Wachter, R. De. The nucleotide sequences of the 5S rRNAs of four mushrooms and their use in studying the phylogenetic position of basidiomycetes among the eukaryotes. *Nucleic Acid Res.*, *11*:2871-2880, 1983.
68. Walker, W. F., and Doolittle, W. F. Redividing the Basidiomycetes on the basis of 5S rRNA sequences. *Nature*, *299*:723-724, 1982.
69. Blanz, P. A. 5S ribosomal RNA nucleotide sequences as a taxonomic tool in higher fungi. XIV International Botanical Congress, Berlin, 1987. *Int. Bot. Congr. Abstr.*, *17*:297, 1987.
70. Andersen, J., Andresini, W., and Delihas, N. On the phylogeny of Phycomyces blakesleeanus; nucleotide sequence of 5S ribosomal RNA. *J. Biol. Chem.*, *257*:9114-9118, 1982.
71. Bartoszewski, S., Borsuk, P., Kern, I., Bartnik, E. Microheterogeneity in Aspergillus nidulans 5S rRNA genes. *Curr. Genet.*, *11*:571-573, 1987.
72. Selker, E. U., Stevens, J. N., and Metzenberg, R. L. Heterogeneity of 5S RNA in fungal ribosomes. *Science*, *227*:1340-1343, 1985.
73. Lane, D. J., Pace, B., Olsen, G. J., Stahl, D. A., Sogin, M. L., and Pace, N. L. Rapid determination of 16S ribosomal RNA sequences for phylogenetic analyses. *Proc. Natl. Acad. Sci. USA*, *82*:6955-6959, 1985.
74. Rothschild, L. J., Ragan, M. A., Coleman, A. W., Heywood, P., and Gerbi, S. A. Are rRNA sequence comparisons the Rosetta Stone of phylogenetics? *Cell*, *47*:640, 1986.
75. Southern, E. M. Detection of specific sequences among DNA fragments separated by gel electrophoresis. *J. Mol. Biol.*, *98*:503-517, 1975.
76. Nei, M. Genetic polymorphism and the role of mutation in evolution. In *Evolution of Genes and Proteins* (M. Nei and R. K. Koehn, eds.), Sinauer Associates, Sunderland, Massachusetts, 1983, pp. 165-190.
77. Förster, H., Kinscherf, T. G. Estimation of relatedness between Phytophthora species by analysis of mitochondrial DNA. *Mycologia*, *80*:466-478, 1988.
78. Natvig, D. O., Jackson, D. A., Taylor, J. W. Random-fragment hybridization analysis of evolution in the genus Neurospora: the status of four-spored strains. *Evolution*, *41*:1003-1021, 1987.

79. Nei, M., and Li, W-H. Mathematical model for studying genetic variation in terms of restriction endonucleases. *Proc. Natl. Acad. Sci. USA*, *76*:5269-5273, 1979.
80. Nei, M., and Tajima, F. Maximum likelihood estimation of the number of nucleotide substitutions from restriction sites data. *Genetics*, *105*:207-217, 1983.
81. Brown, W. M., George, M., and Wilson, A. C. Rapid evolution of animal mitochondrial DNA. *Proc. Natl. Acad. Sci. USA*, *76*:1967-1971, 1979.
82. Clark-Walker, G. D., and Sriprakash, K. S. Sequence rearrangements between mitochondrial DNAs of Torulopsis glabrata and Kloeckera africana identified by hybridization with six polypeptide encoding regions from Saccharomyces cerevisiae mitochondrial DNA. *J. Mol. Biol.*, *151*:367-387, 1981.
83. Hintz, W. E., Mohan, M., Anderson, J. B., and Horgen, P. A. The mitochondrial DNAs of Agaricus: heterogeneity in A. bitorquis and homogeneity in A. brunnescens. *Curr. Genet.*, *9*:127-132, 1985.
84. Ward, B. L., Anderson, R. S., and Bendich, A. J. The mitochondrial genome is large and variable in a family of plants (Cucurbitaceae). *Cell*, *35*:793-803, 1981.
85. Martin, F. N., and Kistler, H. C. The use of mitochondrial DNA (mtDNA) restriction fragment length polymorphisms as a taxonomic aid for the genus Pythium. *Phytopathology*, *77*:1700 (Abstr.), 1987.
86. Borkhardt, B., Pedersen, M. B., and Olson, L. W. Mitochondrial DNA in the aquatic fungus Allomyces. *Curr. Genet.*, *12*:149-156, 1987.
87. Borkhardt, B., Brown, T. A., Thim, P., and Olson, L. W. The mitochondrial genome of the aquatic phycomycete Allomyces macrogynus: Physical mapping and mitochondrial DNA instability. *Curr. Genet.*, *13*:41-47, 1988.
88. Manicom, B. Q., Bar-Joseph, M., Rosner, A., Vigodsky-Haas, H., and Kotze, J. M. Potential applications of random DNA probes and restriction fragment length polymorphisms in the taxonomy of the fusaria. *Phytopathology*, *77*:669-672, 1987.
89. Snyder, W. C., and Hansen, H. N. The species concept in Fusarium with reference to the section Martiella. *Am. J. Bot.*, *27*:64-67, 1940.
90. Kistler, H. C., Bosland, P. W., Benny, U., Leong, S., and Williams, P. H. Relatedness of strains of Fusarium oxysporum from crucifers measured by examination of mitochondrial and ribosomal DNA. *Phytopathology*, *77*:1289-1293, 1987.
91. Kozłowski, M., and Stepień, P. P. Restriction enzyme analysis of mitochondrial DNA of members of the genus Aspergillus as an aid in taxonomy. *J. Gen. Microbiol.*, *128*:471-476, 1982.
92. Al-Musallam, A. Revision of the black Aspergillus species. Ph.D. thesis. University of Utrecht, Netherlands, 1980, pp. 1-92.
93. Mullaney, E. J., and Klich, M. A. Survey of representative species of Aspergillus for regions of DNA homology to Aspergillus nidulans developmental genes. *Appl. Microbiol. Biotechnol.*, *25*:476-479, 1987.
94. Mullaney, E. J., and Klich, M. A. Representative hyphomycetes have regions of DNA homologous to Aspergillus nidulans developmental genes. *Mycologia*, *80*:582-585, 1988.
95. Taylor, J. W., Smolich, B. D., and May, G. Evolution and mitochondrial DNA in Neurospora crassa. *Evolution*, *40*:716-739, 1986.
96. Natvig, D. O., May, G., and Taylor, J. W. Distribution and evolutionary significance of mitochondrial plasmids in Neurospora spp. *J. Bacteriol.*, *159*:288-293, 1984.

97. Taylor, J. W., Smolich, B. D., and May, G. An evolutionary comparison of homologous mitochondrial plasmid DNAs from three Neurospora species. *Mol. Gen. Genet.*, *201*:161-167, 1985.
98. Taylor, J. W., and Natvig, D. O. Mitochondrial DNA and evolution of heterothallic and pseudohomothallic Neurospora species. *Mycol. Res.* (in press).
99. Kohn, L. M., Petsche, D. M., Bailey, S. R., Novak, L. A., and Anderson, J. B. Restriction fragment length polymorphisms in nuclear and mitochondrial DNA of Sclerotinia species. *Phytopathology*, *78*:1047-1051, 1988.
100. Kohn, L. M. A monographic revision of the genus Sclerotinia. *Mycotaxon*, *9*:365-444, 1979.
101. Garber, R. C., and Yoder, O. C. Mitochondrial DNA of the filamentous ascomycete Cochliobolus heterostrophus. *Curr. Genet.*, *8*:621-628, 1984.
102. Castle, A. J., Horgen, P. A., Anderson, J. B. Restriction fragment length polymorphisms in the mushrooms Agaricus brunnescens and Agaricus bitorquis. *Appl. Environ. Microbiol.*, *53*:816-822, 1987.
103. May, G., and Taylor, J. W. Patterns of mating and mitochondrial DNA inheritance in the agaric basidiomycete Coprinus cinereus. *Genetics*, *118*:213-220, 1988.
104. Anderson, J. B., Petsche, D. M., and Smith, M. L. Restriction fragment polymorphisms in biological species of Armillaria mellea. *Mycologia* *79*:69-76, 1987.
105. Raeder, U., and Broda, P. Comparison of the lignin-degrading white rot fungi Phanerochaete chrysosporium and Sporotrichum pulverulentum at the DNA level. *Curr. Genet.*, *8*:499-506, 1984.
106. Kafatos, F. C., Jones, C. W., and Efstratiadis, A. Determination of nucleic acid sequence homologies and relative concentrations by a dot hybridization procedure. *Nucleic Acids Res.*, *7*:1541-1552, 1979.
107. Saiki, R. K., Gelfand, D. H., and Stoffell, S., Scharf, S. J., Higuchi, R., Horn, G. T., Mullis, K. B., and Erlich, H. A. Primer-directed enzymatic amplification of DNA with a thermostable DNA polymerase. *Science*, *239*:487-491, 1988.

3

ISOZYME ANALYSIS IN FUNGAL TAXONOMY AND MOLECULAR GENETICS

J. A. MICALES *Forest Service, Forest Products Laboratory, United States Department of Agriculture, Madison, Wisconsin*

M. R. BONDE and G. L. PETERSON *Agricultural Research Service, United States Department of Agriculture, Frederick, Maryland*

I. INTRODUCTION

Isozyme analysis is a powerful technique which has many applications in plant pathology and mycology. For years, its use was restricted to studies in fungal taxonomy, often with ambiguous results [1]. Concurrently, geneticists were using this procedure to examine the population genetics of fish [2-5], mammals [6,7], insects [8,9], nematodes [10], and plants [11]. Isozyme analysis is now used by mycologists to resolve taxonomic disputes, identify unknown fungal taxa, "fingerprint" patentable fungal lines, analyze the amount of genetic variability in a population, trace the origin of pathogens, follow the segregation of loci, and identify ploidy levels throughout the life cycle of an organism. This topic has been previously reviewed [12,13], but the many advances in recent years necessitate its revision.

Enzymes which are coded by different alleles or separate genetic loci frequently possess different electrophoretic mobilities. Such differences are due to variations in the amino acid content of the molecule, which in turn is dependent on the sequence of nucleotides in the DNA. Electrophoretic banding patterns are frequently predictable since they are dependent on the genetic and nuclear condition of the organism. Most mycologists and plant pathologists have not taken advantage of this relationship and restrict their interpretation of electrophoretic data to simple "band-counting" techniques [1,14]. A genetic interpretation of the data frequently provides much additional information about the genetics and taxonomy of a group of organisms. Genetic crosses may be necessary to confirm genetic interpretations, but certain banding patterns are readily recognizable from comparable studies in human and animal genetics [15].

II. APPLICATIONS OF ISOZYME ANALYSIS

A. Delineation of Fungal Taxa

Isozyme analysis is most frequently used to address taxonomic problems, particularly when morphological characteristics overlap or are plastic within a genus or species. The application of electrophoretic techniques to systematics has been reviewed [16]. A genetic interpretation of banding patterns may not be necessary to allow separation of fungal taxa; genetic similarities and distances may be calculated by simple band-counting procedures. Alternatively, the interpretation of banding patterns in terms of specific alleles allows the determination of ratios of alleles expressed in common among fungal isolates. These ratios are an excellent means of determining phylogenetic relationships among organisms. In studies of closely related species, the assumption is made that the same loci are present and are being compared among isolates, even though the organisms may be from different species and the number of chromosomes may not be the same.

Extensive electrophoretic studies have been made of fungi pathogenic to insects and mites [17-26], mycorrhizal fungi [27-29], as well as many plant pathogens and saprophytes (as discussed below).

Isozyme analysis has been applied at many different taxonomic levels. It generally is most successful at distinguishing species and has been used to make recommendations on the separation or combination of species. For example, Hodges et al. [30] and Micales et al. [31] used isozyme analysis to confirm the conspecificity of *Endothia eugeniae* (Nutman and Roberts) Reid and Booth and *Cryphonectria cubensis* (Bruner) Hodges, thus supporting studies on cultural characteristics and pathogenicity of these organisms [30]. Also, the close relationship of *Peronosclerospora sacchari* (T. Miyake) Shirai and Hara and *P. philippinensis* (Weston) C. G. Shaw, important pathogens of sugarcane and maize in the tropics, has been demonstrated by starch gel electrophoresis [32,33]. Differences in isozyme banding patterns have been used to separate species of *Agaricus* [34], *Aspergillus* [35], *Botrytis* [36], *Gaeumannomyces* [37], *Glomus* [27], *Penicillium* [38], *Phytophthora* [39,40], *Pleurotus* [41-43], *Puccinia* [44-46], *Rhizopogon* [28], and *Scytinostroma* [47], and intersterilily groups (biological species) of *Heterobasidion annosum* (Fr.) Bref. [48] and *Armillaria* [49,50].

The ability of electrophoresis to distinguish fungal species is dependent on the amount of genetic variation within the population. Organisms with high levels of genetic variability will express large levels of intraspecific variation, thus obscuring interspecific differences. The choice of enzyme systems also is important. Johnson [51] proposed, and many studies [52-55] support the concept, that different types of enzymes exhibit different degrees of variation due to the intensity of selection pressure. Regulatory enzymes involved with energy metabolism often display less variation than nonregulatory enzymes, such as esterases and phosphatases. Any study that utilizes only nonregulatory enzymes may detect disproportionately high levels of intraspecific variation and thus be unable to distinguish species. This problem can be avoided by screening for both regulatory and nonregulatory enzymes.

Electrophoresis also has been used to identify and distinguish subspecific taxa. Varieties and subspecies of *Phytophthora megasperma* Drechsler [39] and *Leptographium wageneri* (Kendrick) Wingfield [56,57] have been differentiated. Faris et al. [58] used isozyme patterns to detect two subgroups of *Phytophthora megasperma* var. *medicaginus* which also expressed diverse cultural characteristics and differed in pathogenicity. Different researchers were unable to detect subspecific and varietal differences in other fungi, including *Gaeumannomyces graminis* (Sacc.) Arx and Olivier [37] and *Scytinostroma galactinum* (Fr.) Donk [47]. Mixed results also have been obtained in differentiating races, formae speciales, and subgroups differing in host preference, pathogenicity and virulence. Such separations have been made for some fungi, including *Cryphonectria cubensis* [59], *Fusarium oxysporum* Schlect. emend. Snyd. and Hans. [60], *Puccinia graminis* Pers. [46], *Puccinia sorghi* Schwein. [44] and *Phyllotopsis nidulans* (Pers. ex Fr.) Sing. [= *Pleurotus nebrodensis* (Pers. ex Fr.) Kummer] (42). Burdon and Roelfs [61] were able to find such a correlation in asexual populations of *P. graminis* f. sp. *tritici*; this association did not occur in sexual populations which had a much higher degree of genetic variability. More commonly, host preferences, pathogenicity, and virulence are not correlated with differences in isozyme patterns, e.g., *Erysiphe graminis* DC ex Murat f. sp. *hordei* [62], *Pleurotus ostreatus* (Jacq. ex Fr.) Kummer [41], *Rhizopogon* [28], and *Rhynchosporium secalis* (Oud.) Davis [54].

The geographic source of an isolate sometimes can be determined by isozyme analysis. Such information can be used to determine the origin of introduced pathogens and to trace the development of epiphytotics. Local populations of *Morchella deliciosa* Fr., *Morchella esculenta* (L.) Pers. [63], and *Neurospora intermedia* Tai [64] were shown to be genetically distinct. Bonde et al. [65] were able to separate isolates of *Phakopsora pachyrhizi* Sydow, the causal agent of soybean rust, collected from the eastern and western hemispheres. The differences were so great that two species may be causing rust on soybeans. Isozyme patterns have also been used to trace independent introductions of *Puccinia recondita* Robs. ex Desm. to the United States [66] and *P. graminis* f. sp. *tritici* to Australia [67]. Another potential application of isozyme analysis is the tracking of organisms specifically released for biological control of either weed or plant pathogens.

B. Identification of Unknown Fungi

The capability of using isozyme analysis to distinguish taxa leads to its application in the identification of unknown fungi.

The importance of the correct identification of pathogens cannot be overstated. Regulatory agencies at the federal and state levels in the United States must correctly identify pathogens of regulatory significance. In some cases, only a few spores may be available for examination; isozyme analysis is a tool which can greatly facilitate identification. Scientists at Foreign Disease-Weed Science Research Unit (FD-WSRU) are routinely using isozyme techniques to examine stored grain, storage facilities, and transportation vehicles for teliospores of *Tilletia indica* Mitra, causal agent of Karnal bunt of wheat. This organism has not yet been found established in the United States but has been intercepted in shipments of grain from Mexico. It is easily confused with the endemic pathogen *Tilletia barclayana* (Bref.) Sacc. and Syd., which causes kernal smut of rice. These two organisms are difficult to differentiate by spore morphology but are easily distinguished by isozyme analysis [69].

Within any species, some genetic loci tend to have little or no variation among individuals. These loci are termed "monomorphic" and are defined as loci in which 99% or more of the individuals express the most common allele. It is the presence of these monomorphic loci which should be stressed when comparing unknown isolates to previously identified cultures in order to determine identity. The amount of genetic variation within fungi is extremely variable [68], but almost all species have some monomorphic loci which can be used for this purpose.

C. Genetic Variation

Isozyme patterns can be used as markers to estimate the amount of variation in a fungal population or species. They are more neutral than virulence markers since they are not usually exposed to the strong selection pressures of the host. As genetic markers, isozymes are useful for studying population structure, following epidemics, tracing origins of new pathogenic forms, and analyzing crosses [68].

The amount of polymorphism detected by isozyme analysis in recent studies of fungal species is presented in Table 1. One factor which must be considered in evaluating any estimate of genetic variability is the quantity of individuals and the number of loci examined. The best estimates of polymorphism are obtained when large numbers of isolates and enzyme systems are utilized. Results may vary among experiments if widely diverse sample sizes and enzyme systems are employed [56,57].

Levels of genetic variability appear to be quite diverse among fungi. Newton [68] correlated the amount of polymorphism in a species with the physiological specialization of the fungus. Obligate parasites and highly specialized pathogens, which have a relatively uniform substrate and environment, exhibited low levels of genetic variability. Facultative pathogens and saprophytes, which have a much wider host range and are exposed to more diverse substrates and environments, expressed much greater genetic variation. Recent studies (Table 1) do not entirely support this concept. *Uromyces appendiculatus* (Pers.) Unger, the causal agent of bean rust, is an obligate, autoecious, macrocyclic pathogen, yet 67% of its isozyme loci were polymorphic [70]. A worldwide sampling of *Puccinia graminis* f. sp. *tritici* revealed that 61% of its loci were polymorphic [71]. In contrast, *Fusarium oxysporum*, a pathogen with a broad host range, exhibited only 24% polymorphism [60]. The definition of species is essential to evaluations of genetic diversity. Leung and Williams [72] determined that *Pyricularia oryzae* Cavara collections from rice exhibited limited polymorphism (11%) compared to isolates from 34 other graminaceous hosts (55%). In such a case, traditional taxonomic criteria may not reflect the true genetic condition of the fungus, and more than one species may actually exist.

Other researchers have suggested that the amount of polymorphism corresponds to the degree of sexual reproduction in a population or species. Burdon et al. [73] determined that there was no genetic variability among *Puccinia graminis* f. sp. *tritici* collections in Australia, where the fungus is maintained asexually. Collections from the United States, which underwent sexual reproduction until the 1920s, when the barberry was eradicated, exhibited 38% polymorphism [61]. Low levels of isozyme diversity also were associated with populations of *Phakopsora pachyrhizi* [74] and *Puccinia striiformis* West. [46]; neither pathogen is known to have a functioning sexual cycle. In contrast, sexual populations of *Uromyces appendiculatus* displayed less genetic diversity than asexual populations [70]. The authors concluded that mutation and selection would lead to greater genetic divergence and higher levels of polymorphism in asexual populations since there is no exchange of genes. Lack of genetic variation may indicate that a pathogen has been introduced to an area only recently and that insufficient

TABLE 1. Percentage of Polymorphic Loci in Fungal Populations and Species

Species	Percent polymorphism	Ref.
Pleurotus ostreatus (Pers. ex Fr.) Sing.	93	132
Agaricus campestris Fr.	87	77
Uromyces appendiculatus (Pers.) Unger	67	70
Heterobasidion annosum (Fr.) Bref., intersterility group "spruce"	60	48
Puccinia graminis Pers. f. sp. *tritici* (worldwide)	61	71
Pyricularia oryzae Cavara (graminaceous hosts)	55	72
Lentinula edodes (Berk.) Pegler	55	88
Phytophthora infestans (Mont.) deBary	54	80
Tilletia indica Mitra	44,52	69,133
Agaricus brunnescens Peck	43	89
Cryphonectria cubensis (Bruner) Hodges	38	31
Puccinia graminis f. sp. *tritici* (United States)	38	61
Leptographium wageneri (Kendrick) Wingfield	30,48	56,57
Volvariella volvacea (Bull. ex Fr.) Sing.	29	83
Fusarium oxysporum Schlect. emend Snyd. and Hans.	24	60
Peronosclerospora sorghi (Weston and Uppal) C. G. Shaw	23	32
Phytophthora cinnamomi Rands	23	75
Heterobasidion annosum, intersterility group "pine"	20	48
Ustilago spinificis Ludw.	12	134
Pyricularia oryzae (rice isolates)	11	72
Ustilago bullata Berk.	11	76
Puccinia recondita Robs. ex Desm. f. sp. *tritici*	9	73
Puccinia striiformis West. f. sp. *tritici*	0	46
Puccinia striiformis f. sp. *hordei*	0	46
Puccinia graminis tritici (Australia)	0	73
Puccinia hordei Otth.	0	46

time has elapsed to allow for variation [75]. Low levels of variation also may indicate that a pathogen has developed from a limited number of introductions [72]. Clearly, genetic diversity can result from several different factors.

Isozyme analysis can be used to help interpret the amount of heterozygosity in a population. Here the assumption must be made that the isozyme loci tested are representative of the organism's genome; this may be incorrect. Heterozygous banding patterns are often readily recognizable (Section VI). The frequency of heterozygous loci appears to be quite variable in fungi. Bonde et al. [69] estimated 7.2% heterozygosity in *Tilletia indica*. This level of heterozygosity is comparable to 6.3% reported for humans [15]. Kirby [76] estimated a mean heterozygosity level of 4.8% in populations of *Ustilago bullata* Berk. in Australia and New Zealand, whereas Burdon et al. [73] obtained estimates of 33 and 15% for populations of *P. graminis* f. sp. *tritici* and *P. recondita* f. sp. *tritici*, respectively. May and Royse [77] estimated that 28% of the loci were heterozygous for *Agaricus campestris* Fr. The amount of heterozygosity may be quite different among closely related taxa. No heterozygous loci were detected for the wood decay fungus *Scytinostroma galactinum*. The heterozygosity of two subspecies of *Scytinostroma protrusum* (Burt) Nakas, varied from 12.5% (*S. protrusum* subsp. *protrusum* Nakas.) to 37.5% (*S. protrusum* subsp. *septentrionale* Nakas.) [47]. Two different intersterility groups of *Heterobasidion annosum* exhibited heterozygosity levels of 6 and 2% [48]. In most instances, estimates of heterozygosity should be considered conservative since electrophoresis may only detect a third of the heterozygosity that actually exists [78]. Likewise, overestimates of heterozygosity can occur for heterokaryotic species since different genetic nuclei may be present [75].

D. Identification of Crosses, Hybrids, and Genetic Lines

Isozyme banding patterns have been used as genetic markers to trace crosses and to identify hybrids in the field and the laboratory. Burdon et al. [79] used isozyme analysis to determine that a common Australian race of *P. graminis* f. sp. *tritici* originated as a somatic hybrid of other races. An additional study [45] with *P. graminis* demonstrated that collections made from the Australian grass *Agropyron scabrum* (Labill.) Beauv. originated as a somatic hybrid of *P. graminis* f. sp. *tritici* and *P. graminis* f. sp. *secalis*. Tooley et al. [80] studied the distribution of proteins coded by alleles at two polymorphic loci and determined that mating patterns of *Phytophthora infestans* (Mont.) deBary were random in Mexico, where the sexual stage of the pathogen exists. Almost every individual in the asexual population, including isolates from the United States, Canada, and Europe, but not Mexico, was heterozygous at these two loci, thus demonstrating the near absence of mating. The lack of recombination of mating types also occurred among *Phytophthora cinnamomi* Rands isolates in Australia [75,81]. In the laboratory, isozyme analysis has been used to analyze the progeny of crosses [82-85] and to prove that protoplast fusion has occurred between parental strains [86,87]. The variability of polymorphic loci can also be used to "fingerprint" patentable lines of edible mushrooms [84,88-90], biological control agents [91,92], and other commercially important fungi.

E. Segregation and Linkage of Loci

Careful analysis of the inheritance of isozyme patterns can reveal whether the loci which code for individual enzymes segregate independently or are linked. Genetic maps of the chromosomes also can be constructed. Royse et al. [93] traced the inheritance of six segregating loci in *Lentinula edodes* (Berk.) Pegler. Random segregation was observed for five of the loci; joint segregation with a putative lethal allele was proposed for the locus coding for aspartate aminotransferase. The loci coding for glucose phosphate isomerase and phosphoglucokinase were shown to be linked. Spear et al. [94] determined that the loci coding for two peptidases in *Agaricus brunnescens* Peck were linked; the frequency of crossing over was used to estimate the distance of the two loci from the centromere. Similar linkage studies have been conducted with *Agaricus campestris* [77], *Volvariella volvacea* (Bull. ex Fr.) Sing. [83], *P. graminis* f. sp. *tritrici* [71], and *Ustilago bullata* [95].

F. Determination of Ploidy Levels and Tracing Life Cycles

The pattern of isozyme bands can sometimes be used to determine the ploidy level of a fungal isolate; studies of the life cycle of the organism can thus be performed. Haploid and homozygous isolates should produce single-isozyme bands. Heterozygous dikaryotes, diploids, and organisms with higher ploidy levels typically produce more complicated banding patterns. Expected banding patterns are discussed in Section VI. Such interpretations should be confirmed by genetic crosses or cytological studies since secondary, artifactual bands can arise for a variety of reasons (Section VI.C), thus invalidating a genetic interpretation.

Atypical meiosis also has been detected by isozyme analysis. Bonde et al. [74] showed that basidiospores arising from germinated teliospores of *Tilletia indica* did not inherit alleles with equal frequency and that some basidiospores appeared to inherit both alleles. The authors proposed that some basidiospores may receive two haploid nuclei from the promycelium or that the spores are actually aneuploids. This interpretation has been supported by cytological evidence [96]. Atypical meiosis has also been demonstrated in homokaryotic lines of *Agaricus brunnescens* [94].

III. SELECTION OF FUNGAL TISSUES AND SAMPLE PREPARATION

Electrophoresis can be performed successfully with a wide variety of fungal tissues, including vegetative mycelia, spores, basidiocarps, and sclerotia. Care must be taken in the selection of fungal tissues since inconsistent results may be obtained when a mixture of different tissues is used. Such variation may be due to differential expression of alleles at different stages of development or from variations in the nuclear condition, as in the expression of monokaryotic and dikaryotic stages of an organism. Variations in catalase banding patterns have been reported for *Fusarium solani* (Mart.) Sacc. [97]. Similar differences in isozyme banding patterns have been observed among different portions of the basidiocarp in *Lentinula edodes* [98], *Coprinus cinerens* (Schaeff.: Fr.) [99], and *Agaricus brunnescens* [100]. Variations in isozyme patterns reported for *Schizophyllum commune* Fr. at different stages of the life cycle [101] have been disputed by other researchers [102,103].

Proteins can be extracted from spores, mycelia, or fruiting bodies which have been grown in culture or collected in the greenhouse or field. Uniform conditions of light, temperature, and nutrition should be used when sample material is grown in culture in order to minimize the chance for the differential expression of induced enzymes. Ideally, samples should be collected at a specific stage of the growth curve; this is usually difficult to coordinate, and most researchers harvest their material after a specified period of growth. Initial studies should be conducted to determine the amount of electrophoretic variation detected at different stages of the growth cycle. The amount of protein in the mycelium can also vary with the stage of growth; protein assays [104,105] should be conducted to determine the time of maximum protein production. Culture media should be used that allow rapid growth of the fungus; slowly growing cultures often possess low enzyme activity.

Proteins can be extracted from fungal tissues by a number of different procedures. Thin-walled structures, such as certain conidia and vegetative tissues, can be crushed with a glass rod after the sample has been frozen in liquid nitrogen. Lyophilization is frequently used with vegetative tissues and large fruiting bodies; this process removes the water from the sample and allows it to be crushed readily. The preparation of acetone powders can be used in lieu of lyophilization although we have obtained lesser quantities of enzymes with this technique. Large fungal structures, such as basidiocarps, ascocarps, or mycelial mats, can be disrupted with a variety of commercial homogenizers or simply crushed in a chilled mortar and pestle. Sample buffer should be added to frozen and freeze-dried samples before they thaw or rehydrate; nonfrozen samples should be disrupted in sample buffer. The amount of buffer which should be added will vary with the nature and size of the sample and is best determined empirically. There is a thin line between the addition of enough buffer to generate adequate sample size and diluting the sample below the level of detection. Proteins in a sample may precipitate or aggregate if they are in low concentrations, if insufficient sample buffer is added, or if the sample buffer has the wrong pH or ionic strength.

The denaturation of enzymes must be prevented during and after homogenization. Samples should be kept cold (below 4°C) during preparation and storage. Agents which inhibit proteases and stabilize enzymes such as phenylmethylsulfonyl fluoride (PMSF) and polyvinylpyrrolidone (PVP), can be added to the sample buffer to maintain enzyme activity. Sample preparation is the most critical step in isozyme analysis and other electrophoretic techniques. Poor resolution, faint staining of bands, and irregular banding patterns are frequently caused by the incorrect choice of sample buffer or improper extraction techniques. Troubleshooting techniques associated with extraction procedures have been described [13].

IV. ELECTROPHORETIC TECHNIQUES

A variety of electrophoretic procedures, including starch gel electrophoresis, polyacrylamide gel electrophoresis (PAGE), isoelectric focusing, and two-dimensional electrophoresis, have been used to separate isozymes. Each technique has advantages and disadvantages. The goal of this chapter is to acquaint the reader with the different procedures rather than to provide detailed protocols. References will be provided for more elaborate treatments of each technique.

A. Starch Gel Electrophoresis

Starch gel electrophoresis is one of the oldest electrophoretic procedures, and it is still one of the best for isozyme analysis. Most of the major studies in fungal genetics and taxonomy have been conducted with this technique. It is inexpensive, uses nontoxic chemicals, and can be used to screen rapidly for many different enzyme activities. The preparation of starch gels is something of an art and requires some practice in order to obtain reproducible electrophoretic patterns. Many different systems are available commercially, but most researchers build their own equipment quite inexpensively. Highly detailed descriptions of equipment and electrophoretic procedures are available [13,106-110]. A list of troubleshooting tips was assembled by Micales et al. [13].

Two types of experiments are usually conducted with starch gel electrophoresis. A "screening" run should be conducted early in a study with a limited number of isolates [8-10] and a large number of stains and buffer systems. This experiment will determine which enzymes are active in the taxon and which buffers provide optimal resolution and activity of each enzyme. Many different buffer systems have been used for starch gel electrophoresis. The taxonomy and genetics of large numbers of isolates can then be studied in subsequent "analytical" runs, using the enzyme systems determined to be best in the screening run. The arrangement of samples on the gel depends on the type of experiment. In screening runs at FD-WDRU, eight to 10 samples are repeated four times across the width of the gel. Dye markers are placed between each group to identify places where the gel will be cut following electrophoresis. A larger number of samples are used for analytical experiments. The sequence of samples should be repeated at least once in an analytical run if sufficient space is available; variations in migration in different portions of the gel can then be detected. This greatly facilitates interpretation of the results. It is advisable to repeat one or two samples at intervals on the gel in order to provide a standard for comparison. If specific comparisons are desired among isolates, those samples should be placed in close proximity to ensure accurate analysis.

Starch gels can be quite thick and must be adequately cooled. Specialized cooling plates are commercially available, and ice packs can also be used. This thickness dramatically increases the number of enzymes which can be analyzed at any one time. The gel can be sliced vertically between each group of samples, and thin horizontal slices can also be made with a commercial gel slicer or with nylon thread and plexiglas supports. Slicing techniques are illustrated in Micales et al. [13]. More than twenty different pieces can be easily obtained from a single gel. Each slice is then placed in a staining tray. Staining procedures are described in Section V.

B. Polyacrylamide Gel Electrophoresis

PAGE also is commonly used to separate proteins. It usually provides much better resolution than starch gel electrophoresis, and more complex banding patterns are usually encountered. There is a greater chance that secondary, "artifactual" bands will be detected due to the increased resolution. Unfortunately, fewer staining procedures are available for PAGE and only a small number of enzymes can be studied at one time. Detailed procedures for PAGE are presented by Hames [111] and Andrews [112]; the former contains a useful list of troubleshooting procedures.

C. Isoelectric Focusing

Isoelectric focusing has also been used to study the distribution of isozymes, primarily in taxonomic studies [37,40,49,50,113-117]. In isoelectric focusing, a pH gradient is established across the gel due to the inclusion of highly charged moieties, termed "ampholytes," in the gel. Proteins migrate to their isoelectric point, where they have a net charge of zero. Excellent resolution is obtained with isoelectric focusing, and more is learned about the physical properties of the enzymes with this technique. Unfortunately, proteins often precipitate at their isoelectric point and may lose activity. Isoelectric focusing cannot be used for the genetic interpretations in Section VI, but simple band-counting techniques can be substituted. Detailed protocols are presented by Andrews [112], An der Lan and Chrambach [118], and Righetti and Drysdale [119].

D. Two-Dimensional Electrophoresis

This technique has only been used rarely in taxonomic studies. Torp and Andersen [62] separated 174 proteins from *Erysiphe graminis* f. sp. *hordei*, the causal agent of barley powdery mildew. Howes et al. [120] used two-dimensional electrophoresis to separate races of *P. graminis* f. sp. *tritici*; they detected 290 separate polypeptides with this technique. As demonstrated by these two examples, two-dimensional electrophoresis has tremendous resolution, and quantification of the numerous protein spots can be a major problem. In two-dimensional electrophoresis, proteins are separated by two different parameters, typically isoelectric focusing in the first dimension and PAGE or SDS-PAGE in the second dimension. Proteins are usually inactivated by this procedure and thus are detected by general protein stains. Detailed descriptions on the theory and practice of two-dimensional gel electrophoresis are presented by Sinclair and Rickwood [121] and Andrews [112].

V. STAINING PROCEDURES

Specific activity stains are used to detect the location of specific enzymes after completion of electrophoresis. Detection of specific enzymes is possible because the appropriate substrates and cofactors required for activity are provided in the staining solution. The enzymatic reaction then forms a colored product, either through direct activity with a dye or by involving other enzymes in a series of reactions with the generation of a colored product as a final result. Fluorescent products can be detected with ultraviolet light; alternatively, nonfluorescent products can also be visualized as a negative stain by reacting the starch with a fluorescent compound. The biochemistry of such staining reactions has been discussed by Gabriel [122] and Vallejos [123].

Staining recipes for many different enzymes have been described [13,15,107,108,110,122-124]. At FD-WSRU, stain components are weighed into prelabeled styrofoam cups during an electrophoretic run. This can be done quite rapidly if stains with similar reagents are grouped together. One or two key enzymes and cofactors should be left out initially to prevent premature reactions. They can be added just before the staining solution is poured over the gel. Many stains are light-sensitive, so that development of the gels should be done in the dark. Reactions occur more rapidly if the staining reactions are performed in a 37°C incubator. Agar overlays can be used for enzymes with low activity or soluble products. This is done by incorporating the staining reagents and buffer into 6-8% molten agar and immediately pouring the mixture over the gel. A more detailed description is provided by Vallejos [123].

VI. GENETIC INTERPRETATION OF BANDING PATTERNS

Genetic interpretations of electrophoretic banding patterns can result in a better understanding of the nuclear condition and genetic makeup of an organism. A brief summary of the principles of genetic interpretation is provided; the reader is also referred to more detailed accounts [15,124].

Isozymes are defined as multiple molecular forms of a given enzyme. These forms usually have similar, if not identical, enzymatic properties. Isozymes can occur within a single individual or among different individuals of the same species or taxa. Isozymes also can be localized in specific tissues of an organism or be compartmentalized within different areas of the cell. Only those isozymes with amino acid compositions of different net charge, or those which result in large differences in the shape of the enzyme, will be differentiated by electrophoresis. This represents only one-third of all possible isozymes which may be present within a genetic system [78].

Detectable isozymes can arise from three different genetic and biochemical phenomena: (a) multiple allelism at a single locus; (b) multiple loci coding for a single enzyme; and (c) posttranslational processing and the formation of secondary isozymes [15]. All of these possibilities must be considered when trying to interpret electrophoretic data.

A. Multiple Alleles at a Single Locus

Variations in banding patterns frequently are caused by the expression of multiple alleles at a single locus, each allele coding for a structurally distinct polypeptide chain. Isozymes which are coded by alleles at a single locus are termed "allozymes." The actual electrophoretic banding pattern is dependent on the number of alleles present at the locus, which in turn is dependent on the nuclear condition (monokaryotic, dikaryotic), ploidy number (haploid, diploid, polyploid), and genetic makeup (homozygous, heterozygous) of the organism. Individuals which are haploid or homozygous will produce simple banding patterns due to the expression of a single allele. Organisms which are dikaryotic or diploid and heterozygous will produce more complex banding patterns due to the expression of two separate alleles.

Banding patterns are also dependent on the quaternary structure of the enzyme. Enzymes which are monomeric are composed of a single polypeptide chain; monomeric isomers of a heterozygote will appear as a mixture of the allozymes produced by the two corresponding homozygotes (Fig. 1). In contrast, multimeric enzymes consist of two or more polypeptide chains; heteromeric (hybrid) bands will be produced by heterozygous individuals as the result of the random combination of polypeptide chains which produce hybrid enzyme molecules. Heteromeric bands cannot occur unless different alleles (coding for the different polypeptide chains) are present in the

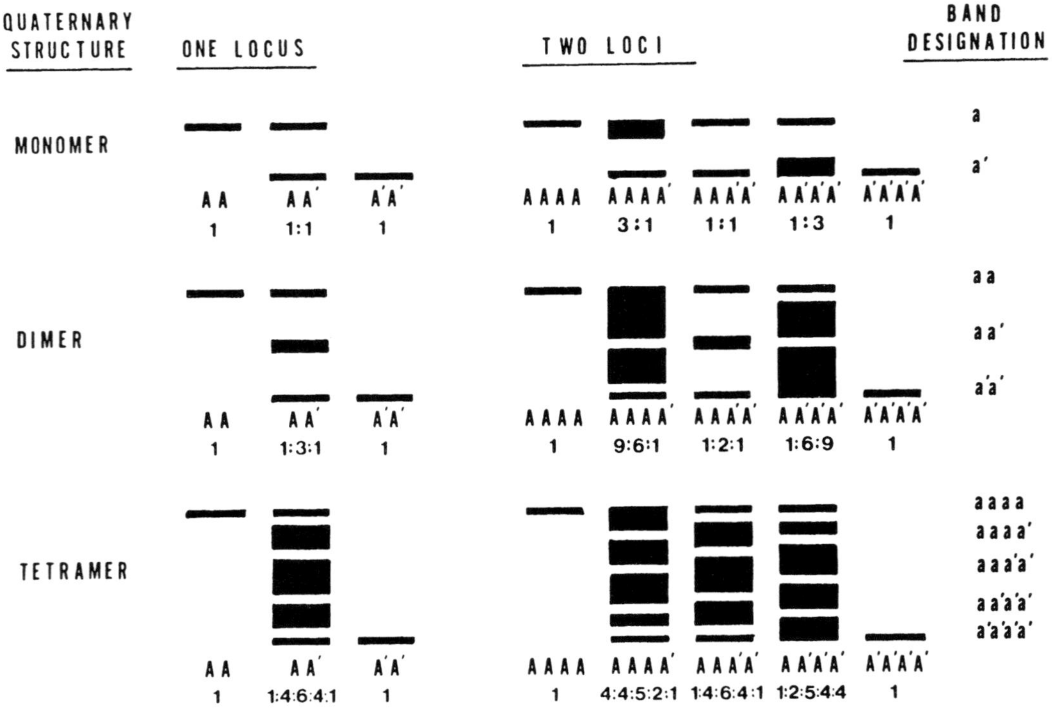

FIGURE 1. Predicted banding patterns for one locus with two alleles (A and A') and two segregating loci which share the same two alleles for monomeric, dimeric, and tetrameric enzymes of a diploid organism. Genotypes are listed beneath each banding pattern; alleles are designated by capital letters. The subunit composition of each protein band is shown on the right; lower case letters refer to subunit designations. The expected ratios of banding intensity for each phenotype is presented beneath the genotype. (Originally adapted from Ref. 3 and reproduced from Ref. 13.)

same individual. The expected banding patterns for monomeric, dimeric, and tetrameric enzymes coded by two alleles at a single locus are presented in Fig. 1. The frequency of occurrence of the different allozyme molecules, assuming random combinations of polypeptide chains, should follow Mendelian ratios. The ratio of isomers formed in heterozygotes will be 1:1 for monomers, 1:2:1 for dimers, 1:3:3:1 for trimers, and 1:4:6:4:1 for tetramers. In some cases, a particular polypeptide will not contribute equally to the activity of the enzyme due to a slow rate of synthesis, low stability, or a tendency to break down before it can be assembled into the final enzyme. A particular polypeptide chain can also reduce the effectiveness of an enzyme by decreasing its stability or reducing its catalytic ability, such enzymes would not be detected in the expected ratios of intensity on the electrophoretic gel. A complex series of banding patterns is produced when three or more alleles are present in a population [15]. Further examples of banding patterns of dimeric and tetrameric enzymes are diagrammed in Fig. 2.

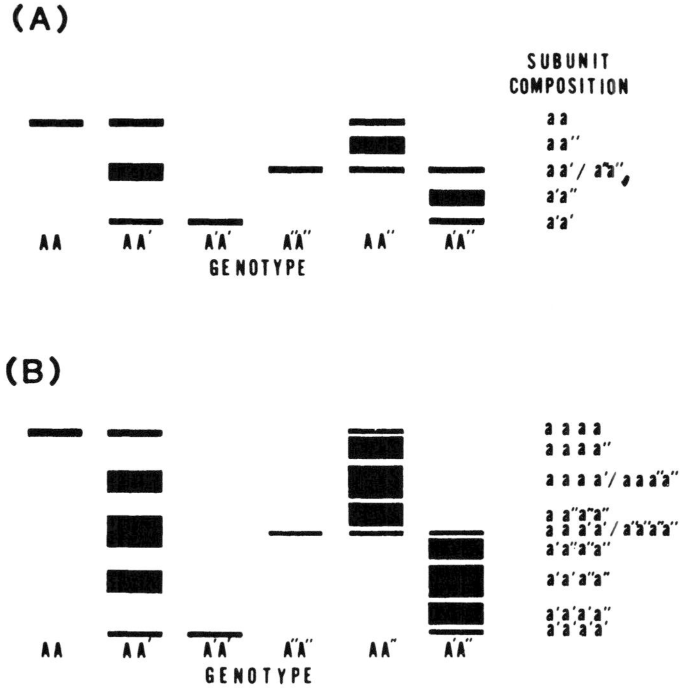

FIGURE 2. Predicted banding patterns of dimeric (A) and tetrameric (B) enzymes coded by a single locus and three electrophoretically distinct alleles (A, A', and A'') of a diploid organism. Genotypes are listed below each banding pattern. The subunit composition of each protein band is shown on the right; lower case letters refer to subunit designations. (Originally adapted from Ref. 3 and reproduced from Ref. 13.)

B. Multiple Loci

Multiple loci also can code for a structurally distinct series of isozymes; up to four different loci have been reported for some enzymes. Multiple loci are thought to originate from gene duplications and subsequent divergence by point mutations; they are frequently expressed in different tissues of an organism or compartmentalized in different areas of the cell. The distribution of loci is usually constant within a given species [15]. Isozymes coded by different loci are frequently detected in separate regions of the gel due to their differences in charge and conformation from that usually associated with multiple alleles at a single locus. Heteromeric bands can be formed by polypeptides coded by different loci, and banding patterns can thus become quite complex when a number of different loci and alleles are involved (Fig. 3). For this reason, it is often difficult to interpret banding patterns of enzymes which are not substrate-specific (such as esterases and peroxidases).

C. Secondary Isozymes

Banding sequences may not appear to follow expected genetic patterns due to posttranslational processing and other events which form secondary

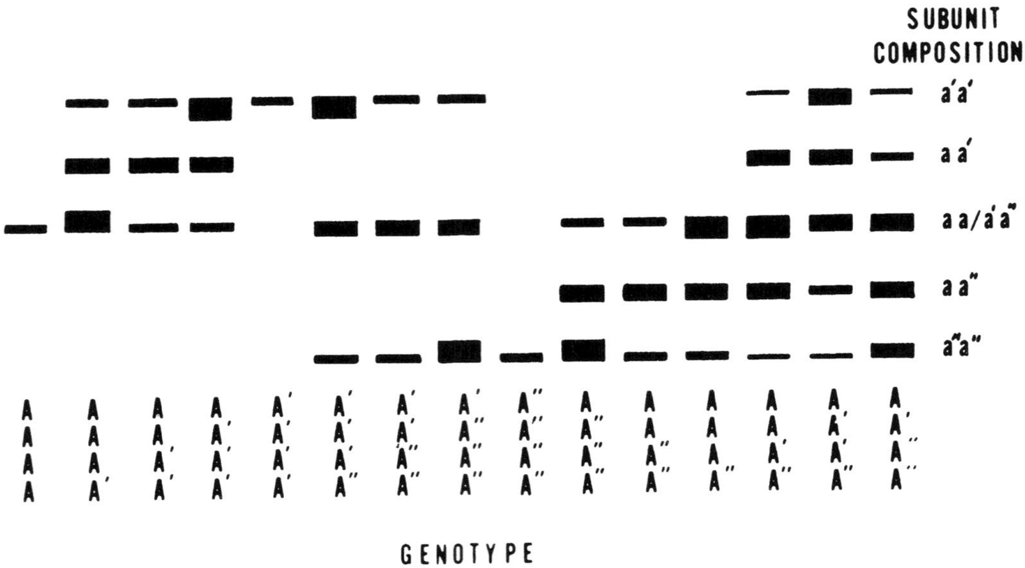

FIGURE 3. Predicted banding patterns of a dimeric enzyme coded by two loci which share the same three electrophoretically distinct alleles (A, A', and A''). Genotypes are listed below each banding pattern. The subunit composition of each protein band is shown on the right; lower case letters refer to subunit designations. (Originally adapted from Ref. 3 and reproduced from Ref. 13.)

isozymes. Common modifications include the deamidation of glutamine and asparagine residues, acetylations, oxidation of sulfhydryl groups, additions and removals of carbohydrate and phosphate moieties, cleavage of the enzyme by proteases, and aggregation or polymerization of the protein. The formation of such secondary isozymes is usually quite uniform within a species or group and can be recognized by the production of a series of bands for each allele [15].

Another process responsible for the generation of secondary isozymes is conformational isomerism. Some enzymes may have several stable configurations which vary in tertiary or quaternary structure; these will frequently have different electrophoretic mobilities. All configurations are usually isolated from a single preparation, and a series of electrophoretic bands will be observed for a single allele.

Enzymes which require cofactors, such as NAD and NADH, may vary in their electrophoretic mobility depending on the degree of saturation of the enzyme with the cofactor. The amount of cofactor in the staining solution must not be limiting or inconsistent results are likely to occur. It may be necessary to add the cofactor to the sample buffer and/or the gel buffer in order to maintain the stability of the enzyme during electrophoresis [15].

Artifact bands can also arise from the proteolysis of the sample during extraction and storage. Sample material can be preserved at -80°C or below, but it should not be kept for more than a year as activity will generally decrease even at ultracold temperatures. Some enzymes, such as those located in membranes, may be bound to other cell constituents and will not migrate freely in an electrophoretic field. Alternate methods of sample extraction and storage may be necessary. Stabilizing agents can be added to the sample to protect the enzymes from other cell constituents [15].

VII. GENIC NOMENCLATURE AND PRESENTATION OF RESULTS

Taxonomic and genetic information can be summarized by several different methods. The technique of May et al. [126], as further described by Bonde et al. [32], allows for direct comparison among experiments regardless of absolute migration distances as long as standard alleles are included in each test. Alleles at a single locus are described by the relative anodal or cathodal movement from the origin as compared to the movement of the homomeric protein products coded by the most frequently observed allele (which is designated 100). A dimeric heterozygote (or heterokaryon), which produces three bands on the electrophoretic gel, would be designated by two different alleles (e.g., 100/130). A diploid homozygous individual would be described by two identical alleles (e.g., 100/100 or 130/130). The frequency of the shared alleles can be used to calculate the coefficients of similarity or distance among organisms. Several different formulas can be used to determine these quantities [125,127,128]; commercial computer programs are available to do these calculations [129]. The final information can be summarized in the form of a dendrogram or by cluster analysis. This type of diagram can be constructed with a variety of clustering procedures [125,130,131] using commercial computer programs.

VIII. SUMMARY

Isozyme analysis is not a new technique, but only recently has it been used to help address important problems of fungal taxonomy and genetics. A virtual explosion of publications is anticipated in this area within the next few years. Objections raised by "classical" geneticists are subsiding as genetic interpretations of banding patterns are being confirmed by crossing experiments. Isozyme analysis will soon become a standard tool for plant pathologists and mycologists, and will help resolve many mysteries of the fungal kingdom.

REFERENCES

1. Shecter, Y. (ed.). Symposium on the use of electrophoresis in the taxonomy of algae and fungi. *Bull. Torrey Bot. Club, 100*:253-312, 1973.
2. Allendorf, F. W., Mitchell, N., Ryman, N., and Stahl, G. Isozyme loci in brown trout (Salmo trutta L.): detection and interpretation from population data. *Hereditas, 86*:179-190, 1977.
3. May, B. *Salmonid genome: Evolutionary restructuring following a tetraploid event*. Ph.D. thesis, The Pennsylvania State University, 1980.
4. May, B., Wright, J. E., and Stoneking, M. Joint segregation of biochemical loci in Salmonidae: results from experiments with Salvelinus and review of the literature on other species. *J. Fish. Res. Board Can., 36*:1114-1128, 1979.
5. Markert, C. L., and Faulhaber, I. Lactate dehydrogenase isozyme patterns of fish. *J. Exp. Zool., 159*:319-332, 1965.
6. Selander, R. K., Smith, M. H., Yang, S. Y., Johnson, W. E., and Gentry, J. B. IV. Biochemical polymorphism and systematics in the

genus Peromyscus. I. Variation in the old-field mouse (Peromyscus polionotus). *Univ. Texas Publ.*, *103*:49-90, 1971.
7. Wheeler, L. L. XVI. Inheritance of allozymes in subspecific F_1 hybrids of Mus musculus from Denmark. *Stud. Genet.*, *7*:319-325, 1972.
8. Yang, S. Y., Wheeler, L. L., and Bock, I. R. IX. Isozyme variations and phylogenetic relationships in the Drosophila bipectinata species complex. *Stud. Genet.*, *7*:213-227, 1972.
9. Steiner, W. W. M., and Joslyn, D. J. Electrophoretic techniques for the genetic study of mosquitoes. *Mosquito News*, *39*:35-53, 1979.
10. Huettel, R. N., Dickson, D. W., and Kaplan, D. T. Biochemical identification of the two races of Radopholus similis by starch gel electrophoresis. *J. Nematol.*, *15*:338-344, 1983.
11. Conkle, M. T. (technical coordinator). Proceedings of the symposium on isozymes of North American forest trees and forest insects. *Gen. Tech. Rep. PSW-48*, Pacific Southwest Forest and Range Exp. Stn., U.S. Forest Service, U.S. Dept. Agr., Berkeley, 1981.
12. Burdon, J. J., and Marshall, D. R. The use of isozymes in plant disease research. In *Isozymes in Plant Genetics and Breeding*, Vol. A (S. D. Tanksley and T. J. Orton, eds.), Elsevier, New York, 1983, pp. 401-412.
13. Micales, J. A., Bonde, M. R., and Peterson, G. L. The use of isozyme analysis in fungal taxonomy and genetics. *Mycotaxon*, *27*:405-449, 1986.
14. Shannon, M. C., Ballal, S. K., and Harris, J. W. Starch gel electrophoresis of enzymes from nine species of Polyporus. *Am. J. Bot.*, *60*:96-100, 1973.
15. Harris, H., and Hopkinson, D. A. *Handbook of Enzyme Electrophoresis in Human Genetics* [with supplements]. American Elsevier, New York, 1976.
16. Buth, D. G., The application of electrophoretic data in systematic studies. *Ann. Rev. Ecol. Systemat.*, *15*:501-522, 1984.
17. Boucias, D. G., McCoy, C. W., and Joslyn, D. J. Isozyme differentiation among 17 geographical isolates of Hirsutella thompsonii. *J. Invert. Pathol.*, *39*:329-337, 1982.
18. Joslyn, D. J., and Boucias, D. G. Isozyme differentiation among three pathotypes of the entomogenous fungus Nomuraea rileyi. *Can. J. Microbiol.*, *27*:364-366, 1981.
19. Maghrabi, H. A., and Kish, L. P. Isozyme characterization of Ascosphaerales associated with bees. I. Ascosphaera apis, Ascosphaera proliperda, and Ascosphaera aggregata. *Mycologia*, *77*:358-365, 1985.
20. Maghrabi, H. A., and Kish, L. P. Isozyme characterization of Ascosphaerales associated with bees. II. Ascosphaera major, Ascosphaera atra, and Ascosphaera asterophora. *Mycologia*, *77*:366-372, 1985.
21. Maghrabi, H. A., and Kish, L. P. Isozyme characterization of Ascosphaerales associated with bees. III. Bettsia alvei. *Mycologia*, *78*:676-677, 1986.
22. Maghrabi, H. A., and Kish, L. P. Isozyme characterization of Ascosphaerales associated with bees. IV. Analyses. *Mycologia*, *79*:519-523, 1987.
23. Maghrabi, H. A., and Kish, L. P. Morphological and isozyme survey of chalkbrood disease of the alfalfa leafcutting bee in the western United States. *Mycologia*, *79*:565-570, 1987.

24. Poprawski, T. J., Riba, G., Jones, W. A., and Aioun, A. Variation in isoesterase profiles of geographical populations of Beauveria bassiana (Deuteromycotina: Hyphomycetes) isolated from Sitona weevils (Coleoptera: Curculionidae). *Environ. Entomol.*, *17*:275-279, 1988.
25. Riba, G., Bouvier-Fourcade, I., and Caudal, A. Isoenzyme polymorphism in Metarhizium anisopliae (Deuteromycotina: Hyphomycetes) entomogenous fungi. *Mycopathologia*, *96*:161-169, 1986.
26. Riba, G., Soares, Jr., G. G., Samson, R. A., Onillon, J., and Caudal, A. Isoenzyme analysis of isolates of the entomogenous fungi Tolypocladium cylindrosporum and Tolypocladium extinguens (Deuteromycotina: Hyphomycetes). *J. Invert. Pathol.*, *48*:362-367, 1986.
27. Hepper, C. M., Sen, R., Azcon-Aguilar, C., and Grace, C. Variation in certain isozymes amongst different geographical isolates of the vesicular-arbuscular mycorrhizal fungi Glomus clarum, Glomus monosporum, and Glomus mosseae. *Soil Biol. Biochem.*, *20*:51-59, 1988.
28. Ho, I., and Trappe, J. M. Enzymes and growth substances of Rhizopogon species in relation to mycorrhizal hosts and infrageneric taxonomy. *Mycologia*, *79*:553-558, 1987.
29. Sen, R., and Hepper, C. M. Characterization of vesicular-arbuscular mycorrhizal fungi (Glomus spp.) by selective enzyme staining following polyacrylamide gel electrophoresis. *Soil Biol. Biochem.*, *18*:29-34, 1986.
30. Hodges, Jr., C. S., Alfenas, A. C., and Ferreira, F. A. The conspecificity of Cryphonectria cubensis and Endothia eugeniae. *Mycologia*, *78*:343-350, 1986.
31. Micales, J. A., Stipes, R. J., and Bonde, M. R. On the conspecificity of Endothia eugeniae and Cryphonectria cubensis. *Mycologia*, *79*:707-720, 1987.
32. Bonde, M. R., Peterson, G. L., Dowler, W. M., and May, B. Isozyme analysis to differentiate species of Peronosclerospora causing downy mildews of maize. *Phytopathology*, *74*:1278-1283, 1984.
33. Micales, J. A., Bonde, M. R., and Peterson, G. L. Isozyme analysis and aminopeptidase assays of the genus Peronosclerospora. *Phytopathology*, *78*:1396-1402, 1988.
34. Kerrigan, R. W., and Ross, I. K. Extracellular laccases: biochemical markers for Agaricus systematics. *Mycologia*, *80*:689-695, 1988.
35. Kurzeja, K. C., and Garber, E. D. A genetic study of electrophoretically variant extracellular amylolytic enzymes of wild-type strains of Aspergillus nidulans. *Can. J. Genet. Cytol.*, *15*:275-287, 1973.
36. Backhouse, D., Willetts, H. J., and Adam, P. Electrophoretic studies of Botrytis species. *Trans. Br. Mycol. Soc.*, *82*:625-630, 1984.
37. Abbott, L. K., and Holland, A. A. Electrophoretic patterns of soluble proteins and isoenzymes of Gaeumannonmyces graminis. *Aust. J. Bot.*, *23*:1-12, 1975.
38. Cruickshank, R. H., and Pitt, J. I. Identification of species in Penicillium subgenus penicillium by enzyme electrophoresis. *Mycologia*, *79*:614-620, 1987.
39. Bielenin, A., Jeffers, S. N., Wilcox, W. F., and Jones, A. L. Separation by protein electrophoresis of six species of Phytophthora associated with deciduous fruit crops. *Phytopathology*, *78*:1402-1408, 1988.
40. Erselius, L. J., and de Vallavieille, C. Variation in protein profiles of Phytophthora: comparison of six species. *Trans. Br. Mycol. Soc.*, *83*:463-472, 1984.

41. Blaich, R., and Esser, K. Function of enzymes in wood destroying fungi. II. Multiple forms of laccase in white rot fungi. *Arch. Microbiol.*, *103*:271-277, 1975.
42. Boisselier-Dubayle, M. C. Taxonomic significance of enzyme polymorphism among isolates of Pleurotus (Basidiomycetes) from umbellifers. *Trans. Br. Mycol. Soc.*, *81*:121-127, 1983.
43. May, B., and Royse, D. J. Interspecific allozyme variation within the fungal genus Pleurotus. *Trans. Br. Mycol. Soc.*, *90*:29-36, 1988.
44. Bergquist, R. R., and Pryor, A. J. Virulence and isozyme differences for establishing racial identity in rusts of maize. *Plant Dis.*, *68*:281-283, 1984.
45. Burdon, J. J., and Marshall, D. R. Isozyme variation between species and formae speciales of the genus Puccinia. *Can. J. Bot.*, *59*:2628-2634, 1981.
46. Newton, A. C., Caten, C. E., and Johnson, R. Variation for isozymes and double-stranded RNA among isolates of Puccinia striiformis and two other cereal rusts. *Plant Pathol.*, *34*:235-247, 1985.
47. Nakasone, K. K., and Micales, J. A. Scytinostroma galactinum species complex in the United States. *Mycologia*, *80*:546-559, 1988.
48. Otrosina, W. J., Chase, T. E., Cobb, Jr., F. W., and Taylor, J. W. Isozyme structure of Heterobasidion annosum isolates relating to intersterility groups. In *Proceedings of the Seventh International Conference on Root and Butt Rots of Forest Trees* (D. J. Morrison, ed.), Forestry Canada, Victoria, British Columbia, 1989, pp. 406-416.
49. Morrison, D. J., Thomson, A. J., Chu, D., Peet, F. G., and Sahota, T. S. Isozyme patterns of Armillaria intersterility groups occurring in British Columbia. *Can. J. Microbiol.*, *31*:651-653, 1985.
50. Morrison, D. J., Thomson, A. J., Chu, D., Peet, F. G., Sahota, T. S., and Rink, U. Characterization of Armillaria intersterility groups by isozyme patterns. In *Proceedings of the Sixth International Conference on Root and Butt Rots of Forest Trees* (G. A. Kile, ed.), Commonwealth Scientific and Industrial Research Organization, Melbourne, 1984, pp. 2-11.
51. Johnson, G. B. Enzyme polymorphism and metabolism. *Science*, *184*:28-37, 1974.
52. Brown, A. J. L., and Langley, C. H. Reevaluation of level of genic heterozygosity in natural populations of Drosophila melanogaster by two dimensional electrophoresis. *Proc. Natl. Acad. Sci. USA*, *76*:2381-2384, 1979.
53. Huettermann, A., Volger, C., Schorn, R., Ahnert, G., and Ganser, H. G. Studies on isoenzyme polymorphism in Fomes annosus. *Eur. J. For. Pathol.*, *9*:265-274, 1979.
54. Newman, P. Variation amongst isozymes of Rhynchosporium secalis. *Plant Pathol.*, *34*:329-337, 1985.
55. Racine, R. R., and Langley, C. H. Genetic heterozygosity in a natural population of Mus musculus assessed using two-dimensional electrophoresis. *Nature*, *283*:855-857, 1980.
56. Otrosina, W. J., and Cobbs, Jr., F. W. Analysis of allozymes of three distinct variants of Verticicladiella wageneri in western North America. *Phytopathology*, *77*:1360-1363, 1987.
57. Zambino, P. J., and Harrington, T. C. Isozyme variation within and among host-specialized varieties of Leptographium wageneri. *Mycologia*, *81*:122-133, 1989.

58. Faris, M. A., Sabo, F. E., and Cloutier, Y. Intraspecific variation in gel electrophoresis patterns of soluble mycelial proteins of Phytophthora megasperma isolated from alfalfa. *Can. J. Bot.*, *64*:262-265, 1986.
59. Alfenas, A. C., Jeng, R., and Hubbes, M. Isoenzyme and protein patterns of isolates of Cryphonectria cubensis differing in virulence. *Can. J. Bot.*, *62*:1756-1762, 1984.
60. Bosland, P. W., and Williams, P. H. An evaluation of Fusarium oxysporum from crucifers based on pathogenicity, isozyme polymorphism, vegetative compatibility, and geographic origin. *Can. J. Bot.*, *65*:2067-2073, 1987.
61. Burdon, J. J., and Roelfs, A. P. The effect of sexual and asexual reproduction on the isozyme structure of populations of Puccinia graminis. *Phytopathology*, *75*:1068-1073, 1985.
62. Torp, J., and Andersen, B. Two-dimensional electrophoresis of proteins from cultures of Erysiphe graminis f. sp. hordei. *Physiol. Plant Pathol.*, *21*:151-160, 1982.
63. Gessner, R. V., Romano, M. A., and Schultz, R. W. Allelic variation and segregation in Morchella deliciosa and M. esculenta. *Mycologia*, *79*:683-687, 1987.
64. Spieth, P. T. Population genetics of allozyme variation in Neurospora intermedia. *Genetics*, *80*:785-805, 1975.
65. Bonde, M. R., Peterson, G. L., and Dowler, W. M. A comparison of isozymes of Phakopsora pachyrhizi from the eastern hemisphere and the new world. *Phytopathology*, *78*:1491-1494, 1988.
66. Burdon, J. J., and Roelfs, A. P. Isozyme and virulence variation in asexually reproducing populations of Puccinia graminis and P. recondita on wheat. *Phytopathology*, *75*:907-913, 1985.
67. Burdon, J. J., Marshall, D. R., Luig, N. H., and Gow, D. J. S. Isozyme studies on the origin and evolution of Puccinia graminis f. sp. tritici in Australia. *Aust. J. Biol. Sci.*, *35*:231-238, 1982.
68. Newton, A. C. Markers in pathogen populations. In *Genetics and Plant Pathogenesis* (P. R. Day and G. J. Jellis, eds.), Blackwell Scientific, Oxford, 1987, pp. 187-194.
69. Bonde, M. R., Peterson, G. L., and Matsumoto, T. T. The use of isozymes to identify teliospores of Tilletia indica. *Phytopathology*, *79*:596-599, 1989.
70. Lu, T. H., and Groth, J. V. Isozyme detection and variation in Uromyces appendiculatus. *Can. J. Bot.*, *66*:885-890, 1987.
71. Burdon, J. J., Roelfs, A. P., and Brown, A. H. D. The genetic basis of isozyme variation in the wheat stem rust fungus (Puccinia graminis tritici). *Can. J. Genet. Cytol.*, *28*:171-175, 1986.
72. Leung, H., and Williams, P. H. Enzyme polymorphism and genetic differentiation among geographic isolates of the rice blast fungus. *Phytopathology*, *76*:778-783, 1986.
73. Burdon, J. J., Luig, N. H., and Marshall, D. R. Isozyme uniformity and virulence variation in Puccinia graminis f. sp. tritici and P. recondita f. sp. tritici in Australia. *Aust. J. Biol. Sci.*, *36*:403-410, 1983.
74. Bonde, M. R., Peterson, G. L., and Royer, M. H. Inheritance of isozymes in the smut pathogen Tilletia indica. *Phytopathology*, *78*:1276-1279, 1988.

75. Old, K. M., Moran, G. F., and Bell, J. C. Isozyme variability among isolates of Phytophthora cinnamomi from Australia and Papua New Guinea. *Can. J. Bot.*, *62*:2016-2022, 1984.
76. Kirby, G. C. Variation in Ustilago bullata on Agropyron scabrum in Australia and New Zealand. *Australasian Plant Pathol.*, *16*:10-12, 1987.
77. May, B., and Royse, D. J. Genetic variation and joint segregation of biochemical loci in the common meadow mushroom, Agaricus campestris. *Biochem. Genet.*, *20*:1165-1173, 1982.
78. Marshall, D. R., and Brown, A. H. D. The charge-state model of protein polymorphism in natural populations. *J. Mol. Ecol.*, *6*:149-163, 1975.
79. Burdon, J. J., Marshall, D. R., and Luig, N. H. Isozyme analysis indicates that a virulent cereal rust pathogen is a somatic hybrid. *Nature*, *293*:565-566, 1981.
80. Tooley, P. W., Fry, W. E., and Villarreal Gonzalez, M. J. Isozyme characterization of sexual and asexual Phytophthora infestans populations. *J. Hered.*, *76*:431-435, 1985.
81. Old, K. M., Dudzinski, M. J., and Bell, J. C. Isozyme variability in field populations of Phytophthora cinnamomi in Australia. *Aust. J. Bot.*, *36*:355-360, 1988.
82. May, B., and Royse, D. J. Confirmation of crosses between lines of Agaricus brunnescens by isozyme analysis. *Exp. Mycol.*, *6*:283-292, 1982.
83. Royse, D. J., Jodon, M. H., Antoun, G. G., and May, B. P. Confirmation of intraspecific crossing and single and joint segregation of biochemical loci of Volvariella volvacea. *Exp. Mycol.*, *11*:11-18, 1987.
84. Royse, D. J. and May, B. Identification of shiitake genotypes by multilocus enzyme electrophoresis: catalog of lines. *Biochem. Genet.*, *25*:705-716, 1987.
85. Shattock, R. C., Tooley, P. W., and Fry, W. E. Genetics of Phytophthora infestans: determination of recombination, segregation, and selfing by isozyme analysis. *Phytopathology*, *76*:410-413, 1986.
86. Anne, J. and Peberdy, J. F. Characterisation of inter-specific hybrids between Penicillium chrysogenum and P. roqueforti by iso-enzyme analysis. *Trans. Br. Mycol. Soc.*, *77*:401-408, 1981.
87. Toyomasu, T., Matsumoto, T., and Mori, K. I. Interspecific protoplast fusion between Pleurotus ostreatus and Pleurotus salmoneo-stramineus. *Agr. Biol. Chem.*, *50*:223-225, 1986.
88. Royse, D. J., Spear, M. C., and May, B. Cell line authentication and genetic relatedness of lines of the shiitake mushroom, Lentinus edodes. *J. Gen. Appl. Microbiol.*, *29*:205-216, 1983.
89. Royse, D. J., and May, B. Use of isozyme variation to identify genotypic classes of Agaricus brunnescens. *Mycologia*, *74*:93-102, 1982.
90. Royse, D. J., and May, B. Genetic relatedness and its application in selective breeding of Agaricus brunnescens. *Mycologia*, *74*:569-575, 1982.
91. Stasz, T. E., Weeden, N. F., and Harman, G. E. Methods of isozyme electrophoresis for Trichoderma and Gliocladium species. *Mycologia*, *80*:870-874, 1988.
92. Zamir, D. and Chet, I. Application of enzyme electrophoresis for the identification of isolates of Trichoderma harzianum. *Can. J. Microbiol.*, *31*:578-580, 1985.

93. Royse, D. J., Spear, M. C., and May, B. Single and joint segregation of marker loci in the shiitake mushroom, Lentinus edodes. *J. Gen. Appl. Microbiol.*, *29*:217-222, 1983.
94. Spear, M. C., Royse, D. J., and May, B. Atypical meiosis and joint segregation of biochemical loci in Agaricus brunnescens. *J. Hered.*, *74*:417-420, 1983.
95. Kirby, G. C., and Mulley, A. M. Inheritance of isozyme mobility differences between two races of the head smut of grasses, Ustilago gullata Berk. *Bot. Gaz.* (Chicago), *143*:243-248, 1982.
96. Goates, B. J., and Hoffmann, J. A. Nuclear behavior during teliospore germination and sporidial development in Tiletia caries, T. foetida, and T. controversa. *Can. J. Bot.*, *65*:512-517, 1986.
97. Hall, R. Proteins and catalase isoenzymes from Fusarium solani and their taxonomic significance. *Aust. J. Biol.*, *20*:419-428, 1967.
98. Okunishi, M., Yamada, K., and Komagata, K. Electrophoretic comparison of enzymes from basidiomycetes in different stages of development. *J. Gen. Appl. Microbiol.*, *25*:329-334, 1979.
99. Moore, D., and Jirjis, R. Electrophoretic studies of carpophore development in the basidiomycete Coprinus cinereus. *New Phytol.*, *87*:101-113, 1981.
100. Paranjpe, M. S., Chen, P. K., and Jong, S. C. Morphogenesis of Agaricus bisporus: changes in protein and enzyme activity. *Mycologia*, *71*:469-478, 1979.
101. Wang, C. S., and Raper, J. R. Isozyme patterns and sexual morphogenesis in Schizophyllum. *Proc. Natl. Acad. Sci. USA*, *66*:882-889, 1970.
102. Lilly, W. W., and Charvat, I. Activities and isozymes of acid phosphatase in Schizophyllum commune: a re-examination. *Mycologia*, *79*:314-319, 1987.
103. Ullrich, R. C. Isozyme patterns and cellular differentiation in Schizophyllum. *Mol. Gen. Genet.*, *156*:157-161, 1977.
104. Bradford, M. M. A rapid and sensitive method for the quantitation of microgram quantities of protein using the principle of protein-dye binding. *Anal. Biochem.*, *72*:248-254, 1976.
105. Lowry, O. H., Rosebrough, N. J., Farr, A. L., and Randall, R. J. Protein measurement with the folin phenol reagent. *J. Biol. Chem.*, *193*:265-275, 1951.
106. Cheliak, W. M., and Pitel, J. A. Techniques for starch gel electrophoresis of enzymes from forest trees. *Information Report PI-X-42*. Petawawa National Forestry Institute, Canadian Forest Service, 1984.
107. Conkle, M. T., Hodgskiss, P. D., Nunnally, L. B., and Hunter, S. C. Starch gel electrophoresis of conifer seeds: a laboratory manual. *Gen. Tech. Rep. PSW-64*, Pacific Southwest Forest and Range Exp. Stn., U.S. Forest Service, U.S. Dept. Agr., Berkeley, 1982.
108. Shaw, C. R., and Prasad, R. Starch gel electrophoresis of enzymes: a compilation of recipes. *Biochem. Genet.*, *4*:297-320, 1970.
109. Shields, C. R., Orton, T. J., and Stuber, C. W. An outline of general resource needs and procedures for the electrophoretic separation of active enzymes from plant tissue. In *Isozymes in Plant Genetics and Breeding*, Vol. A (S. D. Tanksley and T. J. Orton, eds.), Elsevier, New York, 1983, pp. 443-468.
110. Werth, C. R. Implementing an isozyme laboratory at a field station. *Va. J. Sci.*, *36*:53-76, 1985.

111. Hames, B. D. An Introduction to Polyacrylamide Gel Electrophoresis. In *Gel Electrophoresis of Proteins: A Practical Approach* (B. D. Hames and D. Rickwood, eds.), IRL Press, Washington, D.C., 1981, pp. 1-91.
112. Andrews, A. T. *Electrophoresis: Theory, Techniques, and Biochemical and Clinical Applications*. Clarendon Press, Oxford, 1986.
113. Adaskaveg, J. E., Stanghellini, M. E., Gilbertson, R. L., and Egen, N. B. Comparative protein studies of several Pythium species using isoelectric focusing. *Mycologia, 80*:665-672, 1988.
114. de Vallavieille, C., and Erselius, L. J. Variation in protein profiles of Phytophthora: survey of a composite population of three species on citrus. *Trans. Br. Mycol. Soc., 83*:473-479, 1984.
115. Magro, P., Di Lenna, P., Marciano, P., and Pallavicini, C. Variability of polygalacturonase and protein isoelectric focusing patterns of Botrytis cinerea isolates. *J. Gen. Microbiol., 120*:105-109, 1980.
116. Prillinger, H., and Molitoris, H. P. Genetic analysis in wood-decaying fungi. I. Genetic variation and evidence for allopatric speciation in Pleurotus ostreatus using phenoloxidase zymograms and morphological criteria. *Physiol. Plant., 46*:265-277, 1979.
117. Stenlid, J. Population structure of Heterobasidion annosum as determined by somatic incompatibility, sexual incompatibility, and isoenzyme patterns. *Can. J. Bot., 63*:2268-2273, 1985.
118. An der Lan, B., and Chrambach, A. Analytical and preparative gel electrofocusing. In *Gel Electrophoresis of Proteins: A Practical Approach* (B. D. Hames and D. Rickwood, eds.), IRL Press, Washington, D.C., 1981, pp. 157-187.
119. Righetti, P. G., and Drysdale, J. W. Isoelectric focusing. In *Laboratory Techniques in Biochemistry and Molecular Biology* (T. S. Work and E. Work, eds.), North-Holland, Amsterdam, 1976, pp. 341-587.
120. Howes, N. K., Kim, W. K., and Rohringer, R. Detergent-soluble polypeptides extracted from uredospores of four physiological races of Puccinia graminis f. sp. tritici. *Physiol. Plant Pathol., 21*:361-366, 1982.
121. Sinclair, J., and Rickwood, D. Two-dimensional gel electrophoresis. In *Gel Electrophoresis of Proteins: A Practical Approach* (B. D. Hames and D. Rickwood, eds.), IRL Press, Washington, D.C., 1981, pp. 189-218.
122. Gabriel, O. Locating enzymes on gels. In *Methods of Enzymology*, Vol. 2 (S. P. Colowick and N. O. Kaplan, eds.), Academic Press, New York, 1971, pp. 578-604.
123. Vallejos, C. E. Enzyme activity staining. In *Isozymes in Plant Genetics and Breeding*, Vol. A (S. D. Tanksley and T. J. Orton, eds.), Elsevier, New York, 1983, pp. 469-516.
124. Siciliano, M. J., and Shaw, C. R. Separation and visualization of enzymes on gels. In *Chromatographic and Electrophoretic Techniques, Vol. 2, Zone Electrophoresis* (I. Smith, ed.), A. W. Heinemann, London, 1976, pp. 185-209.
125. Ferguson, A. *Biochemical Systematics and Evolution*. John Wiley and Sons, New York, 1980.
126. May, B., Roberts, D. W., and Soper, R. S. Intraspecific genetic variability in laboratory strains of Entomophthora as determined by enzyme electrophoresis. *Exp. Mycol., 3*:289-297, 1979.
127. Nei, M. Estimation of average heterozygosity and genetic distance from a small number of individuals. *Genetics, 89*:583-590, 1978.

128. Rogers, J. S. IV. Measures of genetic similarity and genetic distance. *Stud. Genet.*, 7:145-153, 1972.
129. Swofford, D. L., and Selander, R. B. BIOSYS-I: a FORTRAN program for the comprehensive analysis of electrophoretic data in population genetics and systematics. *J. Hered.*, *72*:281-283, 1981.
130. Dunn, G., and Everitt, B. S. *An Introduction to Mathematical Taxonomy*. Cambridge University Press, New York, 1982.
131. Sneath, P. H. A., and Sokol, R. R. *Numerical Taxonomy*. W. H. Freeman, San Francisco, 1973.
132. Kulkarni, R. K., Kamerath, C. D., and Allred, K. L. Genetic diversity between isolates of Pleurotus ostreatus as revealed by isozyme analysis. *Proc. Int. Symp. Scientific and Technical Aspects of Cultivating Edible Fungi*. Pennsylvania State University, University Park, 1986, pp. 171-181.
133. Bonde, M. R., Peterson, G. L., Dowler, W. M., and May, B. Comparison of Tilletia indica isolates from India and Mexico by isozyme analysis. (Abstr.) *Phytopathology*, *75*:1309, 1985.
134. Andrews, R., Kirby, G. C., and Adams, M. The population biology of a smut fungus, Ustilago spinificis Ludw. II. Isozyme polymorphism and geographic differentiation. *Aust. J. Bot.*, *36*:347-353, 1988.

4

THE APPLICATION OF MOLECULAR GENETIC METHODS TO FILAMENTOUS FUNGI

ALAN UPSHALL *ZymoGenetics Incorporated, Seattle, Washington*

I. INTRODUCTION

The filamentous fungi are a heterogeneous group of eukaryotic microorganisms inhabiting a wide range of diverse habitats. The group includes, for example, the aquatic Phycomycetes, the ubiquitous soilborne saprophytes, the species-limited saprophytes such as those found metabolizing the lignified tissue in decaying trees, the obligate host-specific plant and animal pathogens, and the opportunistic pathogens capable of surviving as saprophytes in the absence of host tissue. Such a diversity of habitats and biological function not only gives these organisms a central role in ecology but has provided a rich resource for both the academic and the industrial scientists to investigate and exploit [1-6]. In genetic terms, perhaps the most famous contribution of academic fungal research was the formulation of the one gene-one enzyme hypothesis by the use of single-gene mutants of *Neurospora crassa* [7]. Another example is the elucidation of the details of parasexual genetics, especially the process of haploidization of diploid nuclei with concurrent random chromosome segregation in *Aspergillus nidulans* [7] which provided the experimental basis for the genetic analysis of somatic cell cultures of mammalian cells.

A comprehensive genetic understanding does not exist for other filamentous fungal species. Some of the reasons for this include the inability of interesting species to either grow or undergo sexual reproduction in artifical culture. This is especially true of pathogens. Often there is the added complication of unknown ploidy. In imperfect species, genetic analysis based on parasexual methods is cumbersome and limited in scope. A lack of genetic understanding is especially true for the limited number of species which are commercially exploited [9]. Most of these species were initially selected on the basis of their ability to perform a function better than other species [3-6]. Those which immediately come to mind are *Aspergillus oryzae* and *Aspergillus awamori* for the production of enzymes; *Penicillium chrysogenum* and *Cephalosporium acremonium* for the production of antibiotics; *Aspergillus niger* for the production of citric acid. Additional consideration must, however, be given to other commercial species such as the use of *Agaricus bisporus* as a food. Commercial mushrooms form the largest cash crop industry in America and there is a growing industry exploiting the more exotic mushrooms such as Shiitake. Filamentous fungi are also directly involved in food production such as in cheese (*Penicillium roquefortii*) and oriental sauce manufacture (*Aspergillus sojae*).

The methods by which industry has exploited these species does not rely on an understanding of their genetic systems. The standard method has been mutation, followed by a screening of the survivors for improvement of the trait [10]. This is often coupled with a modification of culture methods to more efficiently utilize the new improved genotype. This approach has been highly successful. Current yields of glucoamylase production from *A. niger* are reported to be 20 g/liter or more [11]. When we consider that these production runs, as well as those used for producing either primary or secondary metabolites, which we can assume to be equally productive, are carried out on a scale of thousands of liter batch fermentations, we can imagine the magnitude of production per annum.

Even though the quantity of annual production is enormously high, there is always the requirement to improve. It is important for companies to retain market share, not only in real terms by producing the product more cheaply, but by continually introducing new and more efficient products. Economy of production has become increasingly more difficult to achieve by the standard methods. The mutation-screening approach eventually saturates the genotype for titer-increasing combinations, and advances become progressively smaller and therefore more costly to introduce into scale-up. As the industrial species are generally imperfect, classical methods of genetic manipulation are not available. Parasexual genetic methods require selectable gene mutations, usually auxotrophic [8], but the introduction of such mutations into production strains usually lowers their productive capacity [12]. In addition, production strains are chromosomally aberrant, a genetic situation which gives low frequencies of recombinants from both the sexual and parasexual processes [13].

Improvement also can take the form of the introduction of a new version of a product and even new products. This is an area which has engendered considerable commercial excitement since it is one which can be directly approached by the use of recombinant DNA technology [14,15]. Many species have for a long time been known to produce useful metabolites but these species could not be exploited for various reasons. For example, they may have failed to adapt to large-scale economical production methods. Alternatively, they may concurrently produce other undesirable metabolites which would interact with the desired product, or which contaminate the product rendering it unacceptable for commercial purposes. Proteases are an example of the first type of contaminant whereas aflatoxins are typical of the latter.

Recombinant DNA (rDNA) methods allow the exploitation of an unlimited range of fungal species. There are many methods by which genes can be cloned. Manipulation of genomes by the introduction of cloned genes avoids the necessity to fuse two possibly incompatible genomes, which is required for a standard genetic or protoplast fusion approach [14,15]. These rDNA methods do not involve mutagens which, although efficient in developing new strains, introduce other genetic changes which limit the capacity of the new genotype for change [13]. The specific transfer of genes from one species to another by rDNA methods allows not only the academic comparisons of interspecies structural, functional, and regulatory relationships, but in a commercial situation also allows exploitation of the genotype of the current improved production strains to produce the new product [14]. Of course such a new product need not be of fungal origin. Success in these strategies would prove a great economic advantage since it would bypass the need for strain development. New products would therefore reach market earlier and with lower development costs. In recent months, reports describing the successful application of these strategies have been published and descriptions of these successes will be discussed in a later section. In this Introduction I have not considered one group of fungi which have great importance—the plant pathogens. Their negative impact on the agricultural economy is considerable. An understanding of their genetic structure and how they interact with the plant to create the specific diseased state is fundamental both to an understanding of the epidemiology of the disease and to attempts to control epidemics. Considerable research is now being carried out in many laboratories attempting to elucidate the genetic basis of the host-parasite interaction. Such research is methodologically no different from that being performed on all species which are being investigated by molecular means. The methods and strategies that will be described in the rest of this chapter can be applied to all species.

II. DNA-MEDIATED TRANSFORMATION

A. Methods

The basic transformation methods which have been applied to all filamentous fungi were developed some time ago for *N. crassa* [16] and *A. nidulans* [17]. In contrast to the yeast *Saccharomyces cerevisiae*, efficiencies of transformation in filamentous systems are low, generally of the order of about 100 transformants per microgram of DNA. Since the first papers, the frequency of reports of successful transformation of other fungal species has been rapidly increasing. So many species have now been transformed that it is superfluous to list them. It is clear that given an appropriate selection system any culturable fungus can be transformed.

Transformation involves the addition of DNA to cells which have been treated in such a way that they can take up the DNA and incorporate it into their genome. Most protocols require the generation of protoplasts by treatment with digestive enzymes, the most common of which is Novozym 234. While the procedure for making protoplasts is relatively simple, the level of competence and regenerative ability of the protoplast preparation is the most critical factor in the success of the transformation. Many laboratories, including my own, have invariably experienced periods when, for no apparent reason, transformation either becomes highly inefficient or fails. Typically it returns to normal at a later time, for equally inexplicable reasons. An outline generic protocol is shown in Table 1.

There are many variations on this general theme such as using the young hyphae of germinating spores as the protoplast source [18], or using either sorbitol [19] or potassium chloride [17] as the osmotic stabilant. However, because of the inherent variability of the system, it is difficult to conclusively prove that any one of the published methods is consistently better than another in terms of efficiency, usually expressed as the number of transformants per microgram of DNA. There have been several reports identifying various factors which purportedly increase the efficiency of transformation. These extend from manipulative procedures [20] to changes in the makeup of the transforming DNA. Both Ballance and Turner [21] and Johnstone [22] reported isolating DNA sequences from *A. nidulans*

TABLE 1. Steps Required for the Transformation of Filamentous Fungi

1. Overnight culture of mycelium
2. Treatment with cell wall-degrading enzymes in osmotically balanced buffer
3. Separation of protoplasts from cell debris by filtration or centrifugation
4. Thorough washing of protoplasts, generally with calcium chloride buffer
5. Mixing of DNA with protoplasts in either the presence or the absence of polyethylene glycol
6. Plating the mixture to regeneration medium and the imposition of selection to recover the transformants

which, when incorporated into the transforming plasmids, increased the frequency of recovery of transformants.

Generally, experience has shown that in developing a transformation system the major stumbling block is to achieve the first successful experiment. When this is achieved one can know that the selection system will work and subsequent efforts can be concentrated on optimization of the protocols. Consequently, most of the transformation procedures developed for new fungal species rely heavily on those which have been shown to function in other species. By using a successful, proven system, one can concentrate on generating protoplasts of sufficient competence.

B. Selection Strategies

There are now a wide variety of selective strategies to obtain transformants. The first systems developed, the complementation of mutations of the *qa-2* gene in *N. crassa* [16] and of the *pyr*G gene of *A. nidulans* [17], relied on the complementation of auxotrophy by wild-type allele on the transforming DNA. Auxotrophic complementation systems became more diverse with the cloning of additional genes and many different ones are now routinely used, e.g., *arg*B [23,24], *trp*C [19,25,26] in *Aspergillus* and related species, *am* in *N. crassa* [27], *leu* in *Mucor circinelloides* [28], *pyr* in *A. niger* [29,30], TRP 1 in *Coprinus cinereus* [31], and the *trp* gene in *Schizophyllum commune* [32]. In the academic laboratories, where mutant strains are readily available or easy to isolate, selection by complementation of auxotrophy is convenient and routine. With complete dominance of the wild-type allele, only a single copy of the transforming DNA is needed to achieve complementation. This single-copy transformation is important when attempting to clone genes by complementing mutant strains with plasmid library DNA. Nevertheless, there are many disadvantages to these methods. The integration of multiple plasmids into the genome (see below) creates difficulties when attempting to isolate and identify the complementing DNA by transformation of *E. coli* with DNA obtained from the transformed fungal isolate. Further, complementation of auxotrophies is not feasible for many fungal strains. It is often impractical to obtain a genotype carrying a mutant allele of the selected gene, and this is especially

TABLE 2. A Sample of Selection Systems Used for Transforming Filamentous Fungal Species[a]

Fungus	Selection/gene									
	argB	*trpC*	*pyrG*(4)	ben^r	hph	arom	gpdh	*amdS*	ura	leu
A. nidulans	22,24	19	17	b	39,41			34		
A. niger			29,30					34		
A. oryzae	95,96							94		
P. chrysogenum		26						35		
N. crassa				47		16	20			
C. heterostrophus		25			38			36		
Coprinus cinerea		31								
Colleotothrichum spp.				97	97			98		
Magnaporthe griseae	99									
Mucor circinelloides										28
Schizophyllum commune		32								
Ustilago maydis					94					
Fulvia fulvia					42					
Septoria nodorum					43					
Podospora anserina										53
Fusarium solani (pisi)					100					

[a]Numbers indicate reference.
[b]Upshall, unpublished.
Note: This list is not exhaustive. It is meant to display the range of fungi and selective systems that have been reported.

true of commercial isolates. These limitations have encouraged the development of "positive" selection systems. Although most of these are based on resistance to a toxin, the first system developed for filamentous fungi was based on the cloned *amd*S gene of *A. nidulans* [33]. This gene allows the fungus to utilize acetamide as a sole carbon source. *Aspergillus niger* does not have an equivalent gene and is therefore unable to grow on medium containing acetamide. This observation enabled the first transformation of an industrial species [34] and *amd*S selection was subsequently used to develop a transformation system for several other fungi [35,36]. Although this selection system has proven of value, the transformation frequency is generally low. Invariably, multiple copies of the plasmid are integrated into the genome and there is a direct relationship between the strength of growth of the transformants on the selective medium and the number of copies of the *amd*S gene that have been incorporated into the genome [34,37]. Those transformants that develop early and which show the most vigorous growth have a higher plasmid copy number. Subsequent to the development of the *amd*S system, other positive selections have been developed. Of note are those using the bacterial genes that confer resistance to the aminoglycoside antibiotics G418 and hygromycin B. To allow the expression of these genes in fungal hosts, Turgeon et al. [38] constructed a plasmid which contained a promoterless hygromycin resistance coding sequence (hmh) engineered to carry a *Bam*HI restriction enzyme site at its 5' end. Random 2-kb fragments of DNA from the plant pathogen *Cochliobolous heterostrophus* were cloned into this restriction site and the resultant library used to transform protoplasts of the same species. Transformants were deduced to have arisen from plasmids in which a promoter sequence present on one of the 2-kb fragments had ligated adjacent to, and allowed the expression of, the hmh gene. More specifically, Punt and his colleagues [39] constructed a fusion between the hmh coding sequence and the promoter of the *gpd*A (glucose phosphate dehydrogenase) gene isolated from *A. nidulans* [40]. This construction has been successfully used to transform several fungal species, including plant pathogens [41,43].

Other positive selection systems, which are more suitable for industrial fungi because they use only species homologous genes, are based on genetically dominant resistance to fungicides. Two such systems—one based on resistance to benomyl fungicide, the other on resistance to oligomycin—are in use. Dominant mutations conferring resistance to benomyl occur in the gene encoding β-tubulin (*ben*A in *A. nidulans*; Ref. 44), and for oligomycin resistance the mutations occur in the gene encoding subunit 9 of the ATP synthase (*oli*C in *A. nidulans*; Ref. 45). The alleles encoding resistance to benomyl have been isolated from *A. nidulans* [46], *A. niger* [15], and *N. crassa* (bml^r, [47]), and in each species an efficient transformation system has been developed. The benomyl resistance gene of *A. niger* has been used to transform *Penicillium chrysogenum* [15] and it is likely that these isolated genes could be used to initiate a transformation protocol for any fungal species that is sensitive to this fungicide. Similarly, the alleles encoding the oligomycin resistance subunit 9 components of ATP synthase have been cloned from *A. nidulans* [48], *P. chrysogenum* [49], and *A. niger* [50], and have been successfully used to develop homologous transformation systems. It is clear that these transformation systems, based on toxin resistance, should provide the starting position for any one needing to initiate molecular biological studies with a new fungal species. The successful use of the "fungal promoter-hygromycin resistance" plasmids and of the dominant benomyl-resistant alleles to transform a range of fungal species suggests that filamentous fungi are to a great degree promiscuous in their ability to express genes isolated from other fungal species, especially those genes which are not subject to rigorous and complicated regulation. The development of these dominant selectable systems carries an additional bonus with respect to their use in cotransformation experiments. The frequency of uptake of an unselected DNA molecule along with the selected plasmid has been shown to be often greater than 50% [43,51]. Thus, for general transformation purposes it is unnecessary to create sophisticated plasmid constructs combining both selectable and experimental molecules in a single unit.

C. Genetics of Transformants

Filamentous fungi share the common feature of integrative transformation, with the exception of *Mucor circinelloides* [28] and *Podospora anserina* [52] (which also displays integrative transformation when the homologous *ura*-5 gene is used for selection; Ref. 53). In transformation experiments the colonies recovered on selection plates can be either of two basic phenotypes: large, vigorously growing sporulating colonies (always in the minority) and small, weakly growing colonies most of which fail to continue to grow when propagated to fresh selective medium. This latter group has been termed abortive transformants [54]. Occasionally, these weak colonies release faster growing sectors which assume a stable transformed phenotype. Southern blot analysis of the stable transformant genotypes show that the donor DNA has integrated into the host chromosomal DNA, but not necessarily by recombination with homologous DNA. Depending on the selective system and the species, integration can be at a homologous location or elsewhere in the genome. Even when using homologous genes for selection, nonhomologous integration is very common in *N. crassa* [16] and in *C. cinerea* [31]. Alternatively, in *A. nidulans*, homologous integration is generally the rule [24,48] but in transformants carrying many copies of the donor plasmid there is usually simultaneous integration at both homologous and nonhomologous locations [21,34,38]. In *A. niger*, with selection for benomyl resistance integration was most frequent at sites other than the resident locus [15], but for *pyr*G-mediated transformation integration was at the homologous site [29,30].

In experiments which require that the location of chromosomal integration be controlled, targeting of the donor DNA is essential. To achieve such targeting, Hamer and Timberlake [55] created a mutant allele of the *arg*B gene of *A. nidulans* by inserting the CAT (chloramphenicol acetyltransferase) gene into the coding sequence. Upon integration into the genome of an *arg*B mutant strain, only those transformants in which a functional *arg*B gene is reconstituted, by recombination between the two *arg*B mutant alleles, can survive. A similar strategy has been used by van Gorcom et al. [56]. A mutant allele of the *A. niger pyr*G gene was created by removing a *Bam*HI restriction enzyme site present in the coding sequence. This mutant allele was then used to transform a *pyr*G mutant strain. Complementation can only be achieved by recombination with the homologous *pyr*G gene.

A feature of filamentous fungal transformation is the tandem incorporation of large numbers of transforming DNA molecules at the site of integration [24,34]. Whether these tandem repeats are the result of recombination between the DNA molecules prior to integration or the creation of localized regions of homology due to the first integration event is not known. Transformation involving plasmid molecules with no DNA homology to the host genome are generally inefficient, but even in these cases all stable transformants are the result of integration events. It is clear that there is no stringent requirement for homology between donor and recipient DNA for recombination to occur.

III. GENE CLONING

In principle, the methods used for the cloning of genes from filamentous fungi are no different from those used in all species. Realistically, the methods employed are dictated by convenience and prior information.

A. Complementation

The first attempts to clone fungal genes were by complementation of *E. coli* auxotrophs, a strategy which had limited success. Our current knowledge of the structure of fungal transcription and translation signals and of the intron/exon arrangement of most genes [57] provides the explanation for the limitations of these methods. Simply put, *E. coli* cellular components are not able to recognize filamentous fungal eukaryotic regulatory signals or remove introns from mRNA molecules. The same reasons are responsible for the general failure to isolate genes by complementing mutant yeast strains. Although some genes were isolated by these methods [25,58-60], these successes have proven to be the exception. To bypass these problems McKnight and McConaghy [61] developed a plasmid for the construction of libraries in which cDNA molecules are ligated between a yeast promoter and a yeast terminator. Transforming suitable yeast strains with these libraries selects for full-length functional cDNA molecules, which can then be used as probes for the isolation of the chromosomal gene. This method has proven very successful in the isolation of cDNA clones encoding alcohol dehydrogenases and other clones encoding triose phosphate isomerase from both *A. nidulans* [62,63] and *A. niger* (Upshall and McKnight, unpublished).

Isolation of genes by direct complementation of fungal mutants with genomic plasmid libraries has been successful in a number of cases [18, 64,65]. To recover the construct containing the complementing allele, either total genomic DNA extracted from the transformant can be used directly to transform *E. coli* or, alternatively, the extracted DNA can first be partially digested and relegated before the *E. coli* transformation. An alternative approach is to employ cosmid libraries which carry much larger segments of DNA. Although the efficiency of transformation with cosmid libraries is low, fewer colonies have to be screened. This technique was used by Yelton et al. [19] to isolate the *yA* gene from *A. nidulans* and by Weltring et al. to clone the pisatin demethylase gene from *Nectria haematococca* [66]. In each case, DNA from the fungal transformants containing the desired gene was directly packaged into phage λ and rescued into *E. coli*. A problem with cosmid clones is that they are large and can require considerable subcloning and restriction mapping to locate the specific complementing gene to a defined region of the DNA. Timberlake et al. [51] eliminated the requirement for continued subcloning and testing by cotransforming individual restriction digestion fragments with a selectable plasmid. An alternative approach to this cosmid recovery problem is to construct ordered cosmid libraries and to perform the initial transformation with pools containing a fraction of the total number. With this sib selection procedure [67], positive pools are divided into progressively smaller units until the desired cosmid clone is identified. A further advantage of ordered cosmid libraries is that they organize analysis of genome structure by chromosome walking.

B. Hybridization

DNA hybridization using a cloned gene from another species has been a popular method and, even in the absence of information on nucleotide homology, is always a suitable first attempt strategy. Many genes have been isolated this manner using either cloned yeast or filamentous fungal DNA as probes [40,49,50,68]. With highly conserved genes, probes made from higher eukaryotic species have been used successfully, notably the B-tubulin gene of chicken was used to isolate the *ben*A gene from *A. nidulans* [69]. Obviously, as more filamentous fungal genes become available, heterologous hybridization will be of even greater value in the isolation of genes, especially from the pathogenic species.

Hybridization with oligonucleotide probes is also becoming increasingly useful especially in situations where information exists on the structure of the protein. In many cases, such as with industrial species, proteins have often been purified and amino acid sequences determined. With this information, a family of oligonucleotides can be synthesized in which the DNA sequences reflect the possible codons that would translate into the known polypeptide. As an example, this procedure was recently used to obtain clones of the isopenicillin N synthase enzyme from *Cephalosporium acremonium* [70].

With genes that are subject to regulation and hence are expressed only during induced or derepressed conditions, hybridization of abundant mRNA or derived cDNAs with nucleic acids obtained from strains not expressing the gene has proved an effective tool [71,72]. Various protocols have been elaborated around these procedures. For example, Orr and Timberlake [73] used cascade hybridization to isolate conidiation-specific cDNA sequences. Hynes et al. [33] exploited differential hybridization using cDNAs taken from polypeptide-producing and non-polypeptide-producing cells in isolating the *amd*S gene.

In this section I have not presented an exhaustive list of the examples that can be found in the literature. As stated previously, just about any gene-cloning method that has been applied in molecular biology can be appropriate for filamentous fungi.

IV. STRAIN MANIPULATION

Classical genetics relies on mutations to generate a change in phenotype. By the conventional genetic methods of complementation, recombinational mapping, and reverse or suppressor mutation studies, it is possible to ascribe a function to the gene. With the isolation of a gene it becomes possible to perform reverse genetics, i.e., to make specific changes in the gene and to reintroduce this gene back into the genome. Often it is required that the mutant gene replace the resident allele so that the effect of the mutant can be specifically assessed [74]. A further use of cloned genes is to disrupt and destroy the function of a resident allele, i.e., generate a null mutation. In this way it can be determined whether the gene is essential for the normal functioning of the organism [75]. Such gene-disrupted strains can then be used as recipient hosts for transformation purposes. This method is valuable since it can generate mutations in genes which have no known mutant phenotype or function.

To employ this technique, homologous recombination between the resident gene and the donor DNA is essential. Two procedures, which have been called the one-step [74] and two-step [75] methods, are available. In the first, the resident allele is replaced directly by the donor DNA whereas in the second, the initial step is for the donor DNA to integrate in tandem with the resident allele and then to allow the selected transformant to undergo meiosis. In this division cycle, there is always a measure of instability of the duplicated DNA [24] and progeny which exhibit a mutant phenotype are then screened for ones in which the resident allele is replaced by the allele on the plasmid. Clearly, the two-step method can only be used in sexually reproducing fungi. Successful one-step and two-step gene replacement have been achieved in *A. nidulans* by Miller et al. for the *arg*B gene, which shows a high frequency of site-specific homologous integration, and for the *Spo*C1-C gene [76]. One-step gene replacement has also been reported for *A. niger* [15]. In both cases the frequency of replacement was not high and many transformants had to be screened to identify the correct genotype. Paietta and Marzluf [77] detailed attempts at gene replacements at the *am* locus of *N. crassa*, and although they were able to detect site-specific integration at *am*, they were unable to detect strains which showed a simple gene replacement phenotype.

V. INDUSTRIAL APPLICATIONS

As outlined in the Introduction, filamentous fungi have a long history of exploitation by the pharmaceutical and chemical industries. Recombinant DNA methods add more tools to the arsenal of strain improvement methods and consequently allow the exploitation of existing genotypes by modification of biosynthetic pathways, or the expression of genes encoding valuable metabolites from noncommercial species. In addition, the possibility of exploiting their enormous secretion capacity to commercially produce previously unavailable and commercially extremely valuable protein therapeutics (such as those derived from the cloning and expression of mammalian genes) provides a strong incentive for their evaluation. There are three excellent examples of the application of recombinant DNA methods in the exploitation of filamentous fungi for commercial gain.

A. Enzymes

A long-envisioned goal for the industrial enzyme industry has been to develop a lipase for use in washing powders. Novo Industri A/S was the first company to achieve this goal (Lipolase), in a spectacular manner. A cDNA clone encoding a lipase was isolated from a fungal species (unidentified) which was not suitable for commercial exploitation. This DNA was transformed into an improved strain of *Aspergillus* which was being used for the high-level production of a different enzyme. Excitingly, the transformed strain expressed the lipase at high, commercial levels and production of the enzyme became an instant priority for the company. The reports indicate that the time scale between the cloning of the gene and the first commercial fermentations was only 8 months, a remarkable achievement. Although the scientific details are not available, this example demonstrates the power of the new technology. The ability to clone the lipase was fundamental, but of course equally important was the ability to introduce the cloned DNA into the *Aspergillus* host. The fact that this strain was already used for manufacturing a different enzyme is also of great significance since the company is making use of a familiar strain. In essence, they have commercial expression of a new product without the expensive and laborious process of strain development. The ability to use ready-made hosts in this fashion is an economic windfall.

B. Secondary Metabolites

The second example involves the cloning of genes involved in the biosynthesis of the β-lactam antibiotics cephalosporin and penicillin. Both of these antibiotics are made from a common precursor (a tripeptide containing L-α-aminoadipic acid, L-cysteine, and L-valine, LLD-ACV), which is converted into isopenicillin N by isopenicillin N synthase or cyclase [78]. The gene encoding this enzyme was first cloned from *Cephalosporium* by using radiolabeled oligonucleotide probes which were made on the basis of amino acid sequence obtained from purified protein [70]. Subsequent to this cloning the IPS synthase genes of both *P. chrysogenum* [79] and *A. nidulans* [80] was cloned using the *Cephalosporium* gene as the probe. DNA and amino acid sequence comparisons showed these three genes to be remarkably conserved in structure with very few amino acid differences. There are no reports on the effect of the reintroduction of these genes back into a production strain genome to determine whether increased copy number would have any effect on the titer of antibiotic. However, successful experiments of this nature were recently reported for another of the genes in the cephalosporin pathway. Cantwell et al. [81] described the cloning and use of the *cef*EF gene from *C. acremonium*. This gene encodes an expandase/hydroxylase which converts penicillin N into deacetoxycephalosporin C, the first compound in the specific biosynthesis of one of the cephalosporin molecules. Following the integration of additional copies of this gene back into a production strain of *C. acremonium*, a transformant was recovered that produced, on average, a 47% increase in yield of the antibiotic. Extracts of the cells showed a greater activity of the enzyme and also accumulated less of the precursor molecule penicillin N. Chromosome blot analysis showed the insertion of a copy of *cef*EF gene into chromosome 3 in addition to the resident locus located on chromosome 2. This result is perhaps surprising since one would expect that gene duplication is a genetic change that would have been induced by routine mutagenesis and that a rate-limiting step of this direct importance to titer levels would not exist in current developed strains. The example graphically illustrates how new genotypes carrying multiple copies of genes which control rate-limiting steps can be precisely generated and also their value in the commercial environment. Such precision enables more rational, structured, and economic breeding programs.

C. Mammalian Proteins

Finally, there are the examples of using filamentous fungal species as expression systems for the production of mammalian therapeutic proteins.

TABLE 3. Mammalian Proteins Expressed from *Aspergillus nidulans*

Protein[a]	Approx. titer (mg/liter)	Ref.
Prochymosin	1	83
α-Interferon	1	84
t-PA	2.5	87,88
hGMCSF	0.5-1.0	88
AAT	1.5	88

[a]For abbreviations, see text.

Three groups have reported on the success of this approach and in each case the fungus used was *A. nidulans* and the product was secreted into the culture medium. To achieve controlled expression, each group used a promoter of fungal origin. For the expression of prochymosin, Cullen et al. [83] employed the transcriptional and protein secretion signals of the glucoamylase gene from *A. niger* which was cloned using probes derived from the known nucleotide sequence [71,72]. In *A. niger*, this gene is induced by starch and it shows the same regulatory control in *A. nidulans*. The approach taken by Gwynne et al. was to clone the ethanol regulated *alc*A (alcohol dehydrogenase) gene from *A. nidulans* and use the regulatory sequences of this gene in the expression of α-interferon [84]. This gene is carbon-repressed, and is activated in the absence of glucose and in the presence of ethanol or threonine, this induction being genetically regulated in a positive manner by the *alc*R gene [85]. Expression and secretion of α-interferon was regulated as expected, but initially there was a limit to the level of expression achieved in high-copy-number transformants. This limitation was overcome by introducing multiple copies of the *alc*R regulatory gene into the α-interferon-expressing transformant [86]. The interpretation was that in the transformant carrying only one copy of *alc*R, the multiplicity of copies of the *alc*A promoter effectively titrated the *alc*R regulatory protein product. Maximal expression was therefore controlled not by *alc*A but by the level of the *alc*R product. Overexpressing *alc*R provided enough regulatory molecules to allow continued transcription of the *alc*A-controlled α-interferon gene. Upshall et al. [87] used the promoter of either the constitutively expressed *tpi*A gene from *A. nidulans* [63], the *A. nidulans* ethanol-regulated but not glucose-repressed *alc*C gene [62], or the constitutively expressed *A. niger adh*A gene (unpublished) to control the expression and secretion of human tissue plasminogen activator from *A. nidulans*. Each of these genes was cloned by using cDNA libraries to complement yeast mutant strains as described earlier. In each case, active protein was recovered in the culture medium and highest levels were observed with the strongest promoter, *adh*A.

In addition to defining the possibility of using fungi to produce mammalian protein therapeutics, these experiments have contributed extensively to our basic knowledge of fungal systems. For example, in attempting to extend the expression studies to *A. niger*, it was found that the *acl*C gene of *A. nidulans* did not function in *A. niger*, while the *adh*A gene of *A. niger* functioned in *A. nidulans* (Upshall and McKnight, unpublished.) An interpretation is that *A. nidulans* and *A. niger* differ in their regulation of alcohol dehydrogenases, perhaps as a consequence of differences in their internal metabolisms, especially in relation to the glycolytic and ancillary pathways.

The major reason for using filamentous fungi to express mammalian proteins is to take advantage of their high secretion capability, a characteristic of many species. Different species secrete proteins of different types, e.g., the proteases, ligninases, cellulases, amylases. In each case, the primary translation product comprises not only the active protein molecule but also signals, usually at the N terminus, which direct the protein through the secretory pathway of the cell [88-90]. These signals are processed away from the molecule by specific cellular proteases. The structure, protease recognition sites, and the methods of processing of these fungal secretory signals have been found in some cases to be similar to those present in proteins secreted from mammalian cells. For example, the glucoamylase gene of *A. niger* and *A. awamori* has two processing steps. The first removes the hydrophobic N-terminal signal and the second cleaves after a dibasic lys-arg doublet to remove the remaining leader peptide [71,72]. These events are similar to those involved in the processing of mammalian t-PA [91]. The functional equivalence was identified in the work of Upshall et al. [87] who tested the ability of the natural t-PA molecule, i.e., with its own secretory and processing signals, to be recognized by the fungal system. Amino acid analysis of the N terminus of the secreted active protein indicated that *A. nidulans* had accurately processed and hence had recognized the mammalian signals. Clearly, certain aspects of secretion in the filamentous fungi are the same in higher organisms. This is especially significant when one realizes that t-PA could not be secreted to any great quantity by *S. cerevisiae*, even when yeast secretory signals were used to replace the natural t-PA signals [92,93].

VI. CONCLUSIONS

Molecular methods can now be applied to almost any filamentous fungal species that can be cultured in the laboratory. This has been made possible by the development of basic transformation protocols and selective systems based on dominant resistant genetic markers. This advance has opened new avenues to further both our understanding of the biology of these species and their exploitation.

REFERENCES

1. Turner, W. B. *Fungal Metabolites*. Academic Press, London, 1971.
2. Fincham, J. R. S., Day, P. R., and Radford, A. *Fungal Genetics*. Blackwell, London, 1979, pp. 1-636.
3. Smith, J. E., and Berry, D. R. *The Filamentous Fungi*, Vol. 1. Edward Arnold, London, 1975.
4. Smith, J. E., and Berry, D. R. *The Filamentous Fungi*, Vol. 2. Edward Arnold, London, 1976.
5. Smith, J. E., and Berry, D. R. *The Filamentous Fungi*, Vol. 3. Edward Arnold, London, 1978.

6. Smith, J. E., Berry, D. R., and Kristianson. *The Filamentous Fungi*, Vol. 4. Edward Arnold, London, 1983.
7. Beadle, G. W., and Tatum, E. L. Genetic control of biochemical reactions in Neurospora. *Proc. Natl. Acad. Sci. USA, 27*:409-499, 1941.
8. Pontecorvo, G., and Kafer, E. Genetic analysis based on mitotic recombination. *Adv. Genet., 9*:71-104, 1958.
9. Ball, C. Filamentous Fungi. In *Genetics and Breeding of Industrial Micro-organisms* (C. Ball, ed.), CRC Press, Boca Raton, 1983, pp. 159-188.
10. Rowlands, R. T. Industrial strain improvement: mutagenesis and random screening procedures. *Enz. Microb. Technol., 6*:3-10, 1984.
11. van Brunt, J. Fungi: the perfect hosts? *Biotechnology, 4*:1057-1062, 1986.
12. MacDonald, K. D., Hutchinson, J. M., and Gillett, W. A. Isolation of auxotrophs of Penicillium chrysogenum and their penicillin yields. *J. Gen. Microbiol., 33*:365-374, 1963.
13. MacDonald, K. D. The persistence of parental genome segregation in Penicillium chrysogenum after nitrogen mustard treatments. *Mutat. Res., 5*:302-305, 1968.
14. Upshall, A. Filamentous fungi in biotechnology. *Biotechniques, 4*:158-166, 1986.
15. Rambosek, J., and Leach, J. Recombinant DNA in filamentous fungi: progress and prospects. In *Critical Reviews in Biotechnology*, Vol. 6 (G. G. Stewart and I. Russell, eds.), CRC Press, Boca Raton, pp. 357-393.
16. Case, M. E., Schweizer, M., Kushner, S. R., and Giles, N. H. Efficient transformation of Neurospora crassa by utilizing hybrid plasmid DNA. *Proc. Natl. Acad. Sci. USA, 76*:5259-5363, 1979.
17. Ballance, D. J., Buxton, F. P., and Turner, G. Transformation of Aspergillus nidulans by the orotidine-5'-phosphate decarboxylase gene of Neurospora crassa. *Biochem. Biophys. Res. Commun., 112*:284-289, 1983.
18. Oakley, B. R., Rinehart, J. E., Mitchell, B. L., Oakley, C. E., Carmona, C., Gray, G. L., and May, G. S. Cloning, mapping and molecular analysis of the pyrG, orotidine-5'-phosphate decarboxylase gene of Aspergillus nidulans. *Gene, 61*:385-400, 1987.
19. Yelton, M. M., Hamer, J. E., and Timberlake, W. E. Transformation of Aspergillus nidulans by using a trpC plasmid. *Proc. Natl. Acad. Sci. USA, 81*:1470-1474, 1984.
20. Buxton, F. P., and Radford, A. The transformation of mycelial spheroplasts of Neurospora crassa and the attempted isolation of an autonomous replicator. *Mol. Gen. Genet., 196*:339-344, 1984.
21. Ballance, D. J. and Turner, G. Development of a high frequency transforming vector for Aspergillus nidulans. *Gene, 36*:321-331, 1985.
22. Johnstone, I. L. Transformation of Aspergillus nidulans. *Microbiol. Sci., 2*:307-311, 1985.
23. John, M. A., and Peberdy, J. F. Transformation of Aspergillus nidulans using the argB gene. *Enz. Microb. Technol., 6*:386-389, 1984.
24. Upshall, A. Genetic and molecular characterization of argB transformants of Aspergillus nidulans. *Curr. Genet., 10*:593-599, 1986.
25. Turgeon, B. G., MacRae, W. D., Garber, C. G., Fink, G. R., and Yoder, O. C. A cloned tryptophan-synthesis gene from the Ascomycete Cochliobolus heterostrophus functions in Escherichia coli, yeast and Aspergillus nidulans. *Gene, 42*:79-88, 1986.

26. Picknett, T. M., Saunders, G., Ford, P., and Holt, G. Development of a gene transfer system for Penicillium chrysogenum. *Curr. Genet.*, *12*:449-455, 1987.
27. Kinsey, J. A. and Rambosek, J. A. Transformation of Neurospora crassa with the cloned am (glutamate dehydrogenase) gene. *Mol. Cell. Biol.*, *4*:117-122, 1984.
28. van Heeswijk, R., and Roncero, M. I. G. High frequency transformation of Mucor with recombinant plasmid. *Carlsberg Res. Commun.*, *49*:691-702, 1984.
29. van Hartingsveldt, W., Mattrern, I. P., van Zeijl, C. M. J., Pouwels, P. H., and van den Hondel, C. A. M. J. J. Development of a homologous transformation system for Aspergillus niger based on the pyrG gene. *Mol. Gen. Genet.*, *206*:71-75, 1987.
30. Goosen, T., Bloemheuvel, G., Gysler, C., de Bie, D., van dan Broek, H. W. J., and Swart, K. Transformation of Aspergillus niger using the homologous orotidine-5'-phosphate decarboxylase gene. *Curr. Genet.*, *11*:499-503, 1987.
31. Binnenger, D. M., Skrzynia, C., Pukkila, P. J., and Casselton, L. A. DNA mediated transformation of the basidiomycete Coprinus cinereus. *EMBO J.*, *6*:835-840, 1987.
32. Munoz-Rivas, A., Specht, C. A., Drummond, B. J., Froeliger, E., Novotny, C. P., and Ullrich, R. C. Transformation of the basidiomycete, Schizophyllum commune. *Mol. Gen. Genet.*, *205*:103-106, 1986.
33. Hynes, M. J., Corrick, C. M., and King, J. A. Isolation of genomic clones containing the amdS gene of Aspergillus nidulans and their use in the analysis of structural and regulatory mutations. *Mol. Cell. Biol.*, *3*:1430-1439, 1983.
34. Kelly, J. M., and Hynes, M. J. Transformation of Aspergillus niger by the amdS gene of Aspergillus nidulans. *EMBO J.*, *4*:475-479, 1985.
35. Upshall, A., and Favreau, M. A., unpublished results.
36. Turgeon, B. G., Garber, R. C., and Yoder, O. C. Transformation of the fungal maize pathogen Cochliobolus heterostrophus using the Aspergillus nidulans amdS gene. *Mol. Gen. Genet.*, *201*:450-453, 1985.
37. Wernars, K., Goosen, T., Swart, K., and van den Broek, H. W. J. Genetic analysis of Aspergillus nidulans $AmdS^+$ transformants. *Mol. Gen. Genet.*, *205*:312-317, 1986.
38. Turgeon, B. G., Garber, R. C., and Yoder, O. C. Development of fungal transformation systems based on selection of sequences with promoter activity. *Mol. Cell. Biol.*, *7*:3297-3305, 1987.
39. Punt, P. J., Oliver, R. P., Dingemanse, M. A., Pouwels, P. H., and van den Hondel, C. A. M. J. J. Transformation of Aspergillus nidulans based on the hygromycin B resistance marker from Escherichia coli. *Gene*, *56*:117-124, 1987.
40. Punt, P. J., Dingemanse, M. A., Jacobs-Meijsing, B. J. M., Pouwels, P. H., and van den Hondel, C. A. M. J. J. Isolation and characterization of the glyceraldehyde-3-phosphate dehydrogenase gene of Aspergillus nidulans. *Gene*, *69*:49-57, 1988.
41. Cullen, D., Leong, S. A., Wilson, L. J., and Henner, D. J. Transformation of Aspergillus nidulans with the hygromycin-resistance gene, hph. *Gene*, *57*:21-26, 1987.
42. Oliver, R. P., Roberts, I. N., Harling, R., Kenyon, L., Punt, P. J., Dingemanse, M. A., and van den Hondel, C. A. M. J. J. Transformation of Fulvia fulva, a fungal pathogen of tomato, to hygromycin B resistance. *Curr. Genet.*, *12*:231-233, 1987.

43. Cooley, R. N., Shaw, R. K., Franklin, F. H. C., and Caten, C. E. Transformation of the phytopathogenic fungus Septoria nodorum to hygromycin resistance. *Curr. Genet.*, *13*:383-389, 1988.
44. Sheir-Neiss, G., Lai, M. H., and Morris, N. R. Identification of a gene for beta tubulin in Aspergillus nidulans. *Cell*, *15*:639-647.
45. Rowlands, R. T. R., and Turner, G. Nuclear and extranuclear inheritance of oligomycin resistance in Aspergillus nidulans. *Mol. Gen. Genet.*, *126*:201-216, 1973.
46. May, G. S. Gambino, J., Weatherbee, J. A., and Morris, N. R. Identification and functional analysis of beta tubulin genes by site specific integrative transformation in Aspergillus nidulans. *J. Cell Biol.*, *101*: 712-719, 1985.
47. Orbach, M. J., Porro, E. G., and Yanofsky, C. Cloning and characterization of the gene for B-tubulin from a benomyl resistant mutant of Neurospora crassa and its use as a dominant selectable marker. *Mol. Cell. Biol.*, *6*:2452-2461, 1986.
48. Ward, M., Wilkinson, B., and Turner, G. Transformation of Aspergillus nidulans with a cloned, oligomycin resistant ATP synthase subunit 9 gene. *Mol. Gen. Genet.*, *202*:265-270, 1986.
49. Bull, J. H., Smith, D. J., and Turner, G. Transformation of Penicillium chrysogenum with a dominant selectable marker. *Curr. Genet.*, *13*:377-382, 1988.
50. Ward, M., Wilson, L. J., Carmona, C. L., and Turner, G. The oliC3 gene of Aspergillus niger: isolation and use as a selectable marker for transformation. *Curr. Genet.*, *14*:37-42, 1988.
51. Timberlake, W. E., Boylan, M. T., Cooley, M. B., Mirabito, P. M., O'Hara, E. G., and Willet, C. E. Rapid identification of mutation complementing restriction fragments from Aspergillus nidulans cosmids. *Exp. Mycol.*, *9*:351-355, 1985.
52. Tudzynski, P., Stahl, U., and Esser, K. Transformation to senescence with plasmid-like DNA in the ascomycete Podospora anserina. *Curr. Genet.*, *2*:181-184, 1980.
53. Begueret, J., Razanamparany, V., Perrot, M., and Barreau, C. Cloning gene ura5 for the orotidine acid phosphorylase of the filamentous fungus Podospora anserina: transformation of protoplasts. *Gene*, *32*:487-492, 1984.
54. Tilburn, J., Scazzochio, C., Taylor, G. G., Zabicky-Zissman, J. H., Lockington, R. A., and Davis, R. W. Transformation by integration in Aspergillus nidulans. *Gene*, *26*:205-221, 1983.
55. Hamer, J. E. and Timberlake, W. E. Functional organization of the Aspergillus nidulans trpC promoter. *Mol. Cell. Biol.*, *7*:2352-2359, 1987.
56. van Gorcom, R. F. M., and van den Hondel, C. A. M. J. J. Expression analysis vectors for Aspergillus niger. *Nucleic Acid Res.*, *16*:9052, 1988.
57. Ballance, D. J. Sequences important for gene expression in filamentous fungi. *Yeast*, *2*:229-236, 1986.
58. Vapnek, K., Hautala, J. A., Jacobsen, J. W. Giles, N. H., and Kushner, S. R. Expression in Escherichia coli K-12 of the structural gene for catabolic dehydroquinase of Neurospora crassa. *Proc. Natl. Acad. Sci. USA*, *74*:3508-3512, 1977.
59. Berse, B., Dmochowska, A., Skrzypek, M., Weglenski, P., Bates, M. A., and Weiss, R. Cloning and characterization of the ornithine

carbamoyl transferase gene fromAspergillus nidulans. *Gene, 25*:109-117, 1983.

60. Friedlin, E. and Nuesch, J. Isolation of a selective marker from Cephalosporium acremonium by complementation of an auxotrophic mutant of E. coli. *Curr. Genet., 8*:271-276, 1984.
61. McKnight, G. L., and McConaughy, B. L. Selection of functional cDNAs by complementation in Yeast. *Proc. Natl. Acad. Sci. USA, 80*:4412-4416, 1983.
62. McKnight, G. L., Kato, H., Upshall, A., Parker, M. D., Saari, G., and O'Hara, P. J. Identification and molecular analysis of a third Aspergillus nidulans alcohol dehydrogenase gene. *EMBO J., 4*:2093-2099, 1985.
63. McKnight, G. L., O'Hara, P. J., and Parker, M. L. Nucleotide sequence of the triosephosphate isomerase gene from Aspergillus nidulans: implications for a differential loss of introns. *Cell, 46*:143-147, 1986.
64. Johnstone, I. L., Hughes, S. G., and Clutterbuck, A. J. Cloning an Aspergillus nidulans developmental gene by transformation. *EMBO J., 4*:1307-1311, 1985.
65. Ballance, D. J., and Turner, G. Gene cloning in Aspergillus nidulans: isolation of the isocitrate lyase gene (acuD). *Mol. Gen. Genet., 202*: 271-275, 1986.
66. Weltring, K-M., Turgeon, B. G., Yoder, O. C., and van Etten, H. D. Cloning a phytoalexin-detoxifying gene from the plant pathogenic fungus Nectria haematococca by expression in Aspergillus nidulans. *Gene, 68*:335-344, 1988.
67. Akins, R. A., and Lambowitz, A. M. General method for cloning Neurospora crassa nuclear genes by complementation of mutants. *Mol. Cell. Biol., 5*:2272-2278, 1985.
68. Goyon, C., Faugeron, G., and Rossignol, J-L. Molecular cloning and characterization of the met2 gene from Ascobolus immersus. *Gene, 63*:297-308, 1988.
69. May, G. S., Tsang, M.L.-S., Smith, H., Fidel, S., and Morris, N. R. Aspergillus nidulans beta tubulin genes are unusually divergent. *Gene, 55*:231-244, 1987.
70. Samson, S. M., Belagaje, R., Blankenship, D. T., Chapman, J. L., Perry, D., Skatrud, P. L., VanFrank, R. M., Abraham, E. P., Baldwin, J. E., Queener, S. W., and Ingolia, T. D. Isolation, sequence determination and expression in Escherichia coli of the isopenicillin N synthase gene from Cephalosporium acremonium. *Nature, 318*:191-194, 1985.
71. Nunberg, J. H., Meade, J. H., Cole, G., Lawyer, F. C., McCabe, P., Schweickart, V., Tal, R., Wittman, V. P., Flatgaard, J. E., and Innes, M. A. Molecular cloning and characterization of the glycoamylase gene of Aspergillus awamori. *Mol. Cell. Biol., 4*:2306-2315, 1984.
72. Boel, E., Hjort, I., Svennson, B., Norris, F., Norris, K. E., and Fiil, N. P. Glucoamylases G1 and G2 from Aspergillus niger are synthesized from two different but closely related mRNAs. *EMBO J., 3*:1097-1102, 1984.
73. Orr, W. C. and Timberlake, W. E. Clustering of spore specific genes in Aspergillus nidulans. *Proc. Natl. Acad. Sci. USA, 79*:5976-5980, 1982.
74. Rothstein, R. J. One step gene disruption in yeast. *Meth. Enzymol., 101*:202-211, 1983.

75. Scherer, S., and Davis, R. W. Replacement of chromosome segments with altered DNA sequences constructed in vitro. *Proc. Natl. Acad. Sci. USA*, *76*:4951-4955, 1979.
76. Miller, B. L., Miller, K. Y., and Timberlake, W. E. Direct and indirect gene replacements in Aspergillus nidulans. *Mol. Cell. Biol.*, *5*:1714-1721, 1985.
77. Paietta, J. V., and Marzluf, G. A. Gene disruption by transformation in Neurospora crassa. *Mol. Cell. Biol.*, *5*:1554-1559, 1985.
78. Queener, S. W., and Neuss, N. The biosynthesis of B-lactam antibiotics. In *The Chemistry and Biology of B-Lactam Antibiotics*, Vol. 3 (R. B. Morin and M. Gorman, eds.), Academic Press, New York, 1982, pp. 1-81.
79. Carr, L. G., Skatrud, P. L., Scheetz II, M. E., Queener, S. W., and Ingolia, T. D. Cloning and expression of the isopenicillin N synthase gene from Penicillium chrysogenum. *Gene*, *48*:257-266, 1986.
80. Ramon, D., Carramolino, L., Patino, C., Sanchez, F., and Penalva, M. A. Cloning and characterization of the isopenicillin N synthase gene mediating the formation of the B-lactam ring in Aspergillus nidulans. *Gene*, *57*:171-181, 1987.
81. Cantwell, C., Skatrud, P., Chapman, J., Dotzlaf, J., Yeh, W. K., Fisher, T., Samson, S., Ingolia, T. D., and Queener, S. W. Biosynthetic studies with a recombinant strain of Cephalosporium acremonium. *SIM News*, *38* (suppl):22, 1988.
82. Kupka, J., Shen, Y. Q., Wolfe, S., and Demain, A. L. Studies on the ring cyclization and ring expansion enzymes of beta lactam biosynthesis in Cephalosporium acremonium. *Can. J. Microbiol.*, *29*:488-496, 1983.
83. Cullen, D., Gray, G. L., Wilson, L. J., Hayenga, K. J., Lamsa, M. H., Rey, M. W., Norton, S., and Berka, R. M. Controlled expression and secretion of bovine chymosin in Aspergillus nidulans. *Biotechnology*, *5*:369-376, 1987.
84. Gwynne, D. I., Buxton, F. P., Williams, S. A., Garven, S., and Davis, R. W. Genetically engineered secretion of human interferon and a bacterial endoglucanase from Aspergillus nidulans. *Biotechnology*, *5*:713-719, 1987.
85. Pateman, J. A., Doy, C. H., Olsen, J. E., Norris, U., Creaser, E. H., and Hynes, M. Regulation of alcohol dehydrogenase (ADH) and aldehyde dehydrogenase (AldDH) in Aspergillus nidulans. *Proc. Roy. Soc. Lond.*, *B217*:243-264, 1983.
86. Gwynne, D. I., Buxton, F. P., Gleeson, M. A., and Davis, R. W. Genetically engineered secretion of foreign proteins from Aspergillus nidulans. In *Protein Purification: Micro to Macro* (R. Burgess, ed.), Alan R. Liss, New York, 1987, pp. 355-365.
87. Upshall, A., Kumar, A. A., Bailey, M. C., Parker, M. D., Favreau, M. A., Lewison, K. P., Joseph, M. L., Maraganore, J. M., and McKnight, G. L. Secretion of active human tissue plasminogen activator from the filamentous fungus Aspergillus nidulans. *Biotechnology*, *5*:1301-1304, 1987.
88. Gray, G. L., Hayenga, K., Cullen, D., Wilson, L. J., and Norton, S. Primary structure of Mucor miehei aspartyl protease: evidence for a zymogen intermediate. *Gene*, *48*:41-53, 1986.
89. Tien, M. and Tu, C-P. D. Cloning and sequencing of a cDNA for a ligninase from Phanerochaete chryosporum. *Nature*, *326*:520-523, 1987.

90. Teeri, T. T., Lehtovaara, P., Kauppinen S., Salovuori, I., and Knowles, J. Homologous domains in Trichoderma reesei cellulolytic enzymes: gene sequence and expression of cellobiohydrolase II. *Gene, 51*:43-52, 1987.
91. Pennica, D., Holmes, W. E., Kohr, W. J., Hawkins, R. N., Vehar, G. A. Ward, C. A., Bennett, W. F., Yelverton, E., Seeburg, P. H., Heyneker, H. L., and Goeddel, D. V. Cloning and expression of human tissue type plasminogen activator cDNA in E. coli. *Nature, 301*:214-221, 1983.
92. MacKay, V. L. Secretion of heterologous proteins in Yeast. In *Biochemistry and Molecular Biology of Industrial Yeasts* (G. G. Stewart, I. Russell, R. D. Klein, and R. R. Hiebsch, eds.), CRC Press, Boca Raton, 1987, pp. 27-36.
93. Lemontt, J. F., Wei, C. M., and Dackowski, W. R. Expression of human type tissue plasminogen activator in yeast. *DNA, 4*:419-428, 1985.
94. Wang, J., Holden, D. W., and Leong, S. A. Gene transfer system for the phytopathogenic fungus Ustilago maydis. *Proc. Natl. Acad. Sci. USA, 85*:865-869, 1988.
95. Gonu, K., Limura, Y., and Hara, S. Integrative transformation of Aspergillus oryzae with a plasmid containing the Aspergillus nidulans argB gene. *Agr. Biol. Chem., 51*:2549-2555, 1987.
96. Christensen, T., Woeldike, H., Boel, E., Mortensen, S. B., Hjortshoej, K., Thim, L., and Hansen, M. T. High level expression of recombinant genes in Aspergillus oryzae. *Biotechnology, 6*:1419-1422, 1988.
97. Dickman, M. B. Whole cell transformation of the alfalfa pathogen Colleototrichum trifolii. *Curr. Genet., 14*:241-246, 1988.
98. Rodriguez, R. J. and Yoder, O. C. Selectable genes for the transformation of the fungal pathogen Glomerella cingulata f. sp. phasoeli (Colleototrichum lindemuthianum). *Gene, 54*:73-80, 1987.
99. Parsons, K. A., Chumley, F. G. and Valent, D. Genetic transformation of the fungal pathogen responsible for rice blast disease. *Proc. Natl. Acad. Sci. USA, 84*:4161-4165.
100. Dickman, M. D. and Kollatukudy, P. E. Transformation of Fusarium solani pisi using the cutinase promoter. Abstr. *Phytopathology, 77*: 1740, 1987.

5

PROTOPLAST FUSION IN FUNGI

MICHAEL E. GADAU* *Seattle Veterans Administration Medical Center, Seattle, Washington*

A. J. LINGG *University of Idaho, Moscow, Idaho*

I. INTRODUCTION

Protoplast fusion techniques have been used in both plant and fungal systems for well over a decade. Since the development of enzyme systems able to remove the cell wall of both single-celled and filamentous fungi, protoplasts have been the source of answers to questions of wall biogenesis and repair, physiology, and biochemistry. With the advent of fusion techniques, protoplasts have also been used to answer questions about genetics.

The impetus for applying fusion techniques to fungal protoplasts arose from similar work on plant protoplasts and animal cells. Lopez-Belmonte et al. [1] described the first spontaneous fungal protoplast fusion as a rare event involving *Fusarium culmorum* protoplasts. The first attempts at controlled protoplast fusion in a fungus were made with auxotrophic strains of *Geotrichum candidum* [2]. No fusion stimulant was used and the fusion frequency was characteristically low. Increases in fusion frequency were attempted by modification of plating techniques [3].

The introduction of polyethylene glycol (PEG)/Ca^{2+} as a powerful fusogenic agent in plant protoplast systems [4] spurred the interest in use of PEG/Ca^{2+} to induce fusion in other protoplast systems. The first studies using fungal protoplasts and PEG/Ca^{2+} to induce fusion [5,6] reported fusion frequencies 1000 times higher than previously achieved.

Protoplast fusion techniques have been used to bypass many natural barriers to cross-breeding in fungi. For example, protoplast fusions have been reported in numerous species of Deuteromycetes where no sexual cycle exists [7-9]. In Ascomycetes, where the sexual cycle is difficult or impossible to demonstrate, protoplast fusion has been successful in allowing genetic exchange [10,11]. Protoplast fusion allows combining of entire biosynthetic pathways which are separated by species and genus barriers, allowing introduction of novel characteristics into strains [12-14]. Additionally, protoplast fusion allows investigators to bypass mating type and incompatibility group barriers [15-17] and to study mitochondrial genetics [18,19].

The following review looks at many of the questions that can be answered utilizing protoplast fusion techniques. In the future, many of these techniques will be applied to a more diverse group of industrially important fungi.

**Present affiliation*: ICOS Corporation, Bothell, Washington.

II. WHY PROTOPLAST FUSION?

In both filamentous and nonfilamentous fungi, there are essentially four methods available to exchange genetic information. Sexual crosses can be used if the species has a sexual cycle and has the advantage of being very precise, but the strains must be of the same species and be sex-compatible. A second method takes advantage of the parasexual cycle, specifically hyphal anastomosis. A major advantage of this method is that asexual fungi can be used but this technique is again limited in that the strains must be of the same species and must be of the correct heterokaryon incompatibility group. A third method, transformation with recombinant DNA technology, is advantageous in that very precise pieces of DNA can be moved between taxonomically diverse organisms and gene dosage can be increased readily. A disadvantage in applying these techniques is that the recombinant DNA is restricted in size and the technology for protoplast formation is usually required. The fourth method and the topic of this chapter is protoplast fusion. The advantages of this technique will be discussed shortly. A disadvantage of this technique is that formation and regeneration techniques for fungal protoplasts are not uniform across fungal species and methods must be devised for each species. Additionally, markers are needed in each parent strain to distinguish heterokaryons from diploids from recombinants.

A. Genetic Exchange in Deuteromycetes and Ascomycetes

There are many fungi of economic and/or medical importance that are members of the class Deuteromycetes, e.g., *Candida* species. Members of this class of fungi do not possess or have never demonstrated a sexual stage [20], and therefore conventional crosses and selection of desirable progeny are not an option with these fungi. Protoplast fusion has been demonstrated in *C. albicans* [7,21-25], *C. tropicalis* [8,26,27], and *C. utilis* [28], and has also been successful in filamentous Deuteromycetes. For example, *Beauveria bassiana*, an entomopathogen with biological control applications, has been fused [9,29,30]. Other members of the genus *Beauveria* have also been fused. Both intraspecific and interspecific fusions have been reported in *B. amorpha*, *B. bassiana*, and *B. brongniartii* [9].

Members of the class Ascomycetes, such as *Aspergillus* and *Penicillium* species, may have a sexual cycle but it is difficult and sometimes impossible to demonstrate [20]. Protoplast fusion of either Deuteromycetes or Ascomycetes with different genetic characteristics offers the possibility of isolating fusion products with the desired characteristics of both parent strains now contained in one strain. Numerous examples of protoplast fusion between members of the class Ascomycetes are cited in this chapter.

B. Mating Type and Incompatibility Group Barriers

Protoplast fusion has been applied to organisms having mating type and incompatibility group barriers. As a consequence, genetic recombination experiments can be done between sexually or heterokaryon-incompatible strains. In *Saccharomyces cerevisiae*, mating normally occurs between heterothallic strains of the opposite mating type. van Solingen and van der Plaat [15] found that fusion of two strains of *S. cerevisiae* of mating type a resulted in fusion products that were diploids of mating type a.

Svoboda [31] fused two strains of mating type α and found the fusion products to also be of mating type α. Other investigators obtained similar results with *S. cerevisiae* [32-37] and with other organisms such as *Schizosaccharomyces pombe*. Sipiczki and Ferenczy [38] fused two strains of *S. pombe* of mating type h and found the resulting fusion products were diploid and would mate only with h^+ strains. Similar results have been reported for *Rhodosporidium toruloides* [39] and *Saccharomycopsis lipolytica* [40].

The protoplast fusion technique has also been used to successfully fuse *Aspergillus nidulans* protoplasts of different heterokaryon compatibility (h-c) groups [16,17], a condition that normally would not occur in nature. This fusion event lead to recovery of both heterokaryons and diploids. A linkage map of genes responsible for h-c groups in *A. nidulans* has been constructed by fusion of h-c-incompatible strains and recovery of recombinants by treatment of diploids with benomyl [17]. Typas [41], studying the phytopathogens *Verticillium albo-atrum* and *V. dahliae*, found that a number of incompatibility factors exist in *Verticillium* that prevent formation of heterokaryons via normal parasexual processes. However, application of the protoplast fusion technique resulted in the isolation of both heterokaryons and diploids. The same type of analysis has recently been carried out in *V. lecanii* using protoplast fusion [42]. It should be noted that there are numerous examples where protoplast fusion techniques will not allow formation of viable fungi across heterokaryon incompatibility group barriers [43,44].

C. Species and Genus Barriers

Protoplast fusion can also be used to overcome species and genus barriers producing interspecific and intergeneric hybrids. The fusion frequency, an indicator of the success of a fusion, often falls two to six orders of magnitude below that of an intraspecific fusion and is probably an indication of the amount of genetic relatedness between the parent species involved in the fusion. It should be noted that there are numerous examples of protoplast fusion techniques not allowing formation of viable fungi across species barriers [10,45].

Early studies suggested that hybrid formation with subsequent generation of nonparental types could be achieved only if closely related parental species were used [43,46]. Fusion products between *Aspergillus nidulans* and *A. rugulosus*, closely related species of the same species group, first grow as heterokaryotic colonies on minimal medium. Sectors of more vigorous growth arise from the heterokaryons and, on the basis of segregation patterns and DNA estimations, were deemed to be interspecific hybrids whose origin rested in fusion of the two parental nuclei. These fusants were shown to be diploid and/or aneuploid in nature [10,13,43,47,48]. Crosses between *Penicillium chrysogenum* and *P. cyaneo-fulvum*, two closely related species, yielded colony types that behaved similarly [46]. As a matter of terminology, interspecific or intergeneric diploids are referred to as allodiploids while aneuploids are referred to as interploids, partial alloploids, and partial heteroploids.

Fusion products from the first reported fusion experiments utilizing less closely related species behaved differently. Following the fusion of *P. chrysogenum* and *P. roqueforti* protoplasts [49], hybrid prototrophic colonies that developed on minimal medium readily segregated and produced the *P. roqueforti* parent but the *P. chrysogenum* parent could never be reisolated. However, the prototrophic nature of these colonies along with their ability to produce penicillin was suggestive of at least part of the *P. chrysogenum* genome, which was readily lost under nonselective conditions. When *Aspergillus fumigatus* and *A. nidulans* protoplasts were fused [50,51], a similar type of aneuploid situation was found, although both parental types could be reisolated from the fusion products. A problem that must be addressed when attempting such fusions is the amount of genetic and/or chromosomal homology within a genus [52].

The first documented intergeneric cross was achieved by Provost et al. [53] who fused *Candida tropicalis* and *Saccharomycopsis fibuligera* protoplasts. The fusion frequency was very low. Some of the fusion products resembled one or the other parent with respect to certain key characteristics used as taxonomic criteria, but a group of intermediate types was also found. These intermediate types were uninucleate, showed varying degrees of stability, and thus resembled aneuploids. Recently, Pina et al. [12] fused protoplasts of *Saccharomyces cerevisiae* and *Zygosaccharomyces cerevisiae* with the ultimate aim of producing a hybrid strain with characteristics of both parents such as the ability to utilize cellobiose, produce ethanol, and remain thermotolerant. As expected, the hybrids were produced at a very low frequency. The hybrids possessed characteristics of both parents (drug resistance and the ability to utilize carbon sources). The hybrids were also very stable under both selective and nonselective conditions. Although DNA determinations were not performed, these hybrids behaved as true allodiploids. These experiments provide an incentive for other intergeneric crosses.

D. Nucleus-Mitochondrial Interactions

Another possibility for using the protoplast fusion technique is the study of nucleus-mitochondria interactions and mitochondrial genetics and biogenesis. Intraspecific nonselective mitochondrial transfer was accomplished by Ferenczy and Maráz in *S. cerevisiae* [32]. One strain had the genotype α ural trp1 (E^R) and the other strain ade2 (rho^o E^R). The fusion products were characterized as respiratory competent and uninucleate; they had the expected diploid DNA level and were erythromycin-resistant. Other studies have also shown that transmission of mitochondria by protoplast fusion is possible and that recombinational processes involved in mitochondrial genetics are independent of the mating-type barriers [18]. Mitochondrial markers have been used extensively in yeast strain improvement programs in industry. These yeasts are typically diploid or of higher ploidy [54,55]. Introduction of markers in the nuclear genome is therefore difficult [55,56] and often causes deleterious effects in vigor and stability of production strains as well as production losses [55-60]. On the other hand, both mitochondrial petite and resistance mutations can readily be introduced and applied in hybrid formation [61-64].

Nonselective mitochondrial transfer in the petite-negative yeast *Kluyveromyces lactis* has proven to be the only genetic method available to study respiratory-deficient mutants of this yeast [65,66]. The technique has also proven useful in studying transmission and recombination frequencies as well as segregation patterns of mitochondrial markers in *Schizosaccharomyces pombe* [67] and *Torulopsis glabrata* [68,69].

The first attempts at controlled intraspecific selective mitochondrial transfer or the transfer of isolated mitochondria [70] utilized a mutant strain of *S. cerevisiae* carrying oligomycin-resistant mitochondria. These mitochondria were isolated and fused with another strain of *S. cerevisiae* which was rho^0 and carried different nuclear markers. The frequency of transfusion was less than 1×10^{-8} per regenerated protoplast. Several investigators have since obtained higher frequencies [71,72].

A novel approach to the problem of isolating intact mitochondria has been developed in budding yeasts. During protoplast formation small protoplasts, typically 1 μm or less in diameter, are released from the buds and often contain mitochondria but no nucleus [73]. These small, anucleate protoplasts can then easily be separated from larger nucleated protoplasts by differential centrifugation. Anucleate protoplasts were prepared from an *S. cerevisiae* strain containing mitochondria conferring erythromycin resistance. These small protoplasts were fused with a recipient rho^0 strain. Under partially selective conditions, the majority of fusion products were haploid, and contained the nucleus of the recipient strain and mitochondria conferring erythromycin resistance.

Interspecific nonselective mitochondrial transfer was carried out with strains of *Aspergillus nidulans*, *A. nidulans* var. *echinulatus*, and *A. quadrilineatus* by Croft et al. [19]. The molecular weights and restriction patterns of the mitochondrial DNA differ in the three species. Recombinant mitochondrial DNA was obtained with high frequency when the small mitochondrial genome was selected for transfer to the species containing the larger genome. After fusion of *A. nidulans* and *A. nidulans* var. *echinulatus* protoplasts, it was found that the mitochondrial DNA differed only in a small number of species-specific restriction sites and in six insertions/deletions [74]. Petite mutants of *S. cerevisiae* have been used as fusion partners with wild-type strains of *Candida pseudotropicalis*, *Hansenula anomala*, *H. capsulata*, *S. montanus*, *S. rosei*, and *Schizosaccharomyces pombe* resulting in respiratory-competent fusion products [75].

Interspecific selective mitochondrial transfer using respiratory-deficient *S. cerevisiae* protoplasts as recipients for mitochondria from *Hansenula wingei* and *Schizosaccharomyces pombe* have been reported. Restoration of the rho$^+$ character was reported following uptake of isolated foreign mitochondria [71,72] or with miniprotoplasts containing mitochondria [76].

III. STRATEGIES FOR SELECTING FUSION PRODUCTS

A. Auxotrophic Mutants

Because of the enormous problem in identifying true fusion products when wild-type parent strains are used as fusion partners, some type of selective system or medium must be employed to ensure that only fusion products are able to regenerate into colonies. The most common method to date has been the use of complementary auxotrophic mutants as fusion partners. The fusion mixture is plated on a minimal medium, one on which neither parent strain can grow but on which the fusion products can grow (Table 1).

B. Drug Resistance and Antibiotic Markers

Another common method for selection of fusion products is the use of drug resistance and antibiotic markers (Table 1). This method allows at least one fusion partner to be a wild-type organism, provided the wild-type strain is sensitive to the antibiotic or drug in question. The other member

TABLE 1. Strategies for Selecting Fusion Products

Strategy	Ref.
Antibiotic markers; drug resistance	18,61,70,73,77,78,79,80,127
Inactivated protoplast; dead donor	56,81,82,83,84
Metabolic poisons; irreversible biochemical inhibitors	81,86,87
Mitochondria-protoplast fusion; mitochondrial transfer; mitochondrial markers	18,19,32,64,66,67,69,70,71,72,73,74,76,77,78,85,108,137,138
Nutritional (auxotrophic) complementation	5,6,7,8,10,11,14,16,21,24,25,29,38,42,43,44,45,46,47,48,49,50,53,68,80,88,89,90,100,115,118,126,127,133,134,137,139,140,141,142,143,144
Protoplast-nucleus fusion	33,145
Spore color markers	91

of the fusion pair is usually auxotrophic and carries the elements responsible for conferring resistance to the particular antibiotic or drug. The medium for regeneration of fused protoplasts consists of a minimal medium to which sufficient antibiotic or drug has been added to prevent the growth of the sensitive, wild-type fusion partner. The auxotrophic fusion partner cannot grow since it is plated on a nutritionally deficient medium. It should be noted that not all alleles for drug resistance are dominant.

This method has been used extensively in yeast species, particularly *S. cerevisiae*. Spencer et al. [61] reported that mutations to antibiotic resistance are easily obtained in most industrial yeast strains by either spontaneous mutation or induction with Mn^{2+}. Chloramphenicol- and erythromycin-resistant mitochondrial mutants of *Saccharomyces diastaticus* and erythromycin mutants of *S. cerevisiae* were obtained by treatment with 8 mM $MnCl_2$. These were then fused with rho^- mutants of *S. cerevisiae* obtained by treatment with acriflavine. Respiratory-competent (rho^+), antibiotic-resistant fusion products were the end result. Rho^+ cells are easily distinguished from rho^- by their ability to ferment glycerol [76]. Rho^- mutants in *Kluyveromyces lactis* can also be obtained by treatment with ethidium bromide [66]. In *S. cerevisiae*, resistance has been obtained to chloramphenicol [18,62], cycloheximide [71,72], erythromycin [18,61,62, 71-73,77], mucidin [18], oligomycin [18,62,70], and parmomycin [18]. In *S. diastaticus*, resistance has been induced to chloramphenicol [61,64] and erythromycin [61]. Resistance has also been reported in other yeast species [72,78].

Although this technique is usually applied to yeast species, it has recently been developed for use in filamentous fungi. Bradshaw and Peberdy [79] recently fused strains of *Aspergillus nidulans* and *A. rugulosus*. Wild-type isolates were used as one of the fusion partners and auxotrophic, inhibitor-resistant strains as the other fusion partner. Resistance/sensitivity to acriflavine, carboxin, benomyl, cycloheximide, and imazilil were used as partial selective pressure to aid in detection of fusion products. In intraspecific, interspecific, and intergeneric crosses between *Cephalosporium* and *Emericellopsis* species, resistance/sensitivity to benomyl, griseofulvin, and nystatin has been used as markers for detecting and characterizing heterokaryons and hybrid formation [80].

C. Inactivated Protoplasts and Dead Donors

Another selection strategy which was originally developed for use in bacterial protoplast fusions is the use of inactivated protoplasts as one of the fusion partners (Table 1). One of the fusion partners can be inactivated or killed by use of a brief heat treatment [56,81,82] or by exposure to ultraviolet light [83,84], theoretically allowing the utilization of a wild-type organism as a fusion partner. Ferenczy and Joó [56] recently applied this technique to protoplasts of *A. nidulans*. One strain had the genotype ade$^-$, PABA$^-$ and the other fusion partner was lys$^-$, PABA$^-$.

Among the fusion products, a high proportion exhibited an altered morphology (typical of diploids or aneuploids in contrast to heterokaryons) if one of the fusion partners was heat-treated. If neither partner was heat-treated, greater than 99% of the fusion products were heterokaryons and less than 1% were diploids. On the other hand, if the ade$^-$ partner was heat-treated for 6 min at 50°C, approximately 80% of the fusion products were heterokaryons, 10% were diploids, and 10% were aneuploids. This technique not only adds selective pressure but also allows isolation of a higher proportion of desirable fusion products, namely, hybrids of diploid and aneuploid nature, vs. almost exclusively heterokaryons.

D. Isolated Nuclei

A selection strategy which has been successfully applied in *S. cerevisiae* fusions is the use of isolated nuclei as one of the fusion partners (Table 1). Ferenczy and Pesti [33] isolated nuclei of an adenine requiring respiratory-deficient mutant of *S. cerevisiae* and fused these with protoplasts of a strain auxotrophic for histidine, leucine, and uracil. Both strains were of the α mating type. The resulting fusion products were nutritionally complemented and genetically consisted of heterokaryons and stable diploids. The nuclei were isolated by lysing the protoplasts of the donor strain and then purified in a discontinuous sucrose gradient.

E. Mitochondrial Transfer and Mitochondria-Protoplast Fusion

A selection strategy which has been used extensively in both intra- and interspecific fusions in yeast species is the use of mitochondrial transfer or mitochondria-protoplast fusion (Table 1). Selection of fusion products is based on complementation of auxotrophic markers and mitochondrially inherited drug resistance markers [18,32,67,70-74,85] or restoration of the respiratory-sufficient characteristic, using a petite or mitochondrialless strain as one of the fusion partners [18,32,65,66,70-73,76,85], or both.

F. Use of Irreversible Biochemical Inhibitors

Use of irreversible biochemical inhibitors has also been used as a strategy for selection of fusion products (Table 1). Wright [86] first introduced this technique, using diethylpyrocarbonate and iodoacetamide as irreversible biochemical inhibitors. Using this strategy obviates the need for the time-consuming and tedious process of mutation and identification of auxotrophs. Both fusion partners are treated with a lethal dose of inhibitor with a different specificity, washed free of unreacted inhibitor, and then fused. Unfused parental cells and homokaryons do not have their damaged molecules replaced and die. Only heterokaryons receive the full complement of molecules needed for cell survival. The technique has been used successfully to produce viable heterokaryons in animal cells [86] and plant protoplasts [87]. Peberdy et al. [81] attempted this strategy using a variety of fungal protoplasts as fusion partners and found that the concentrations and exposure times used in the animal cell and plant protoplast experiments proved inadequate with fungi.

G. Spore Color Markers

Spore color markers have been used in protoplast fusions of both *Aspergillus* [16,43,47] and *Penicillium* [46,49,88-90] species to aid in the detection of fusion products. This detection system, however, has been used secondary to either auxotrophic and/or drug-resistant markers. Recently, spore color markers were used as the only means of selection for protoplast fusion products of *Trichoderma reesei* QM 9414, a production strain used for conversion of lignocellulosic waste to soluble sugars [91]. Wild-type *T. reesei* conidiospores are green. A white-spored mutant, isolated spontaneously, and a brown-spored mutant, obtained following treatment with UV irradiation, were used as fusion partners. Fusion products were fawn-colored and segregated spontaneously to form either white or brown conidia. If the heterokaryon was plated on a media containing d-camphor, diploids were formed which produced the wild-type green conidia.

IV. FUSOGENS

A. Polyethylene Glycol

Since the discovery of the fusogenic properties of PEG/Ca^{2+} by Kao and Michayluk [4] in 1974, PEG/Ca^{2+} has been used to induce the fusion of many types of cells and organelles, and continues to be the most widely used method of fusion. Besides the fusion of plant, fungal, and bacterial protoplasts and organelles, PEG/Ca^{2+} is the fusogen of choice in somatic cell genetic studies with mammalian cells [92,93]. The fusogenic properties have also been used to fuse recombinant *Agrobacterium tumefaciens* protoplasts with plant protoplasts [94] as well as artificially constructed liposomes containing DNA with bacterial, plant, and fungal protoplasts [95-97].

The action of PEG/Ca^{2+} as a fusogenic agent is not fully understood [93]. PEG was first used in conjunction with protoplasts in 1953 when Weibull [98] used a PEG solution to stabilize *Bacillus megaterium* protoplasts. Neither agglutination nor subsequent fusion was noted at that time. Numerous investigators have documented the necessity of small quantities (1-100 mM) of Ca^{2+} to stimulate high-frequency protoplast fusion [5,99-101]. Ferenczy et al. [6] found that PEG solutions are good osmotic stabilizers if the concentration is at least 25% (w/v) and that the effect of Ca^{2+} on fusion was significant. Despite the good aggregation of protoplasts at higher PEG concentrations, the fusion frequency was negligible without the addition of Ca^{2+}. The addition of as little as 1 mM $CaCl_2$ was effective in stimulating the fusion process. Prior to the discovery of the fusogenic properties of PEG/Ca^{2+}, high Ca^{2+} concentrations under alkaline pH conditions had been reported to cause fusion [5,90,102,103].

PEG may act as a polycation, inducing the formation of small aggregates of protoplasts [55]. Constabel and Kao [99], working with plant protoplasts, assumed that PEG acted as a molecular bridge between adjacent membranes, either directly by hydrogen bonds or indirectly via Ca^{2+}. Anné and Peberdy [5] found that a lower pH promotes formation of linkages via hydrogen bonds if Ca^{2+} is not present. They also found that the presence of other cations, namely, K^+, Na^+, or Mg^{2+}, greatly diminishes the stimulatory effect of Ca^{2+} on fusion and attributed this effect to a decrease in the amount of Ca^{2+} bound to the membranes. They also postulated that high pH may promote Ca^{2+} links with the [$^-O-CH_2-$] ends of the PEG molecules.

Optimal procedures for fungal protoplast fusion were investigated by Anné and Peberdy [90] and it was found that a solution of 30% (w/v) PEG containing 10 mM $CaCl_2$ and 50 mM glycine at pH 7.5 was optimal. Other investigators showed that the molecular weight of PEG used is critical to the fusion frequency and most protocols call for PEG 4000 or 6000 [5,6, 90,100]. Ferenczy et al. [101] studied PEG solutions with molecular weights of 400, 1540, 4000, 6000, and 20,000 for their ability to promote fusion. The best complementation frequencies were obtained with 25% (w/v) PEG 4000 or 6000 in 10-100 mM $CaCl_2$.

B. Other Polyols

Other polyols have been used successfully to induce protoplast fusions. Nagata [104] used a solution containing 15% (w/w) polyvinyl alcohol (PVA, molecular weight 500), 50 mM $CaCl_2$, and 0.3 M mannitol to induce fusion between plant protoplasts and found that PVA/Ca^{2+} induced fusion at least comparable to that of a PEG/Ca^{2+} system. PVA/Ca^{2+} has also been used to successfully fuse artificial liposomes containing tobacco mosaic virus (TMV) RNA with protoplasts of *Nicotiana tabacum* and *Vinca rosea* [97].

C. Electric Field-Mediated Protoplast Fusion

A technique for fusion which was recently developed and seems to have great potential is the application of an electric field to induce fusion of cells, liposomes, or protoplasts. Electrofusion has had its widest application in stimulating plant protoplast fusions and formation of hybridomas [105] but has also been used to stimulate *S. cerevisiae* protoplast fusions [106-108]. Two procedural steps are needed to induce electrofusion [105]. A close contact between at least two cells, liposomes, or protoplasts is established by the application of a nonuniform alternating electric field of low intensity and the fusion process is then triggered by a field pulse of high intensity and short duration. This results in a temporary structural change in the area of membrane contact. Weber et al. [109] applied a high electric field pulse to *S. cerevisiae* protoplasts which were initially induced to fuse by PEG/Ca^{2+}. They found that the fusion rate was enhanced by a factor of 200 for different genetically marked strains as compared with fusion by PEG/Ca^{2+} alone.

V. PROTOPLAST FUSION: RELATIONSHIP TO PARASEXUAL CYCLE

The Deuteromycetes or Fungi Imperfecti do not possess a sexual stage. How, then, do these fungi generate variability to adapt to an ever changing environment if all that is available to them is mitotic division? The answer was reported in 1952 when Pontecorvo and Roper [110] discovered parasexuality in *A. nidulans*. Since then, parasexual phenomena have been observed in several imperfect fungal species as well as in Basidiomycetes and in Ascomycetes other than *Aspergillus* [20,111]. The parasexual cycle, as the process has come to be called, is defined as genetic recombination in which there is no fine coordination between segregation, recombination, and chromosome reduction as at meiosis. Alternatively, it is defined as a cycle in which plasmogamy, karyogamy, and haploidization take place but not at specified times, places, or points in the life cycle of an organism.

Functionally, the parasexual cycle can be divided into a series of events [111,112]: (a) fusion of protoplasts or hyphal anastamosis to form the heterokaryotic mycelium (plasmogamy); (b) infrequent nuclear fusion (karyogamy) to form both homozygous diploid and heterozygous diploid nuclei; (c) multiplication of diploid nuclei side by side with the unfused haploid nuclei; (d) occasional mitotic crossing over during multiplication of the diploid nuclei followed by mitosis with occasional nondysjunction to form a heterokaryon with parental and recombinant genotypes; and (e) nuclear segregation leading to homokaryosis.

Protoplast fusion generates many of the same types of conditions as seen in the parasexual cycle. A number of events can occur once two protoplasts fuse. While the fusion product reverts to normal form, the nuclei can remain separate and divide mitotically forming a heterokaryotic colony [6,10,11,88,90,113]. The two genetically different nuclei in the heterokaryon may fuse, forming a stable or unstable heterozygous diploid colony or sector [10,36,88,114]. These diploid colonies can then spontaneously lose chromosomes to achieve an aneuploid or haploid state or the colonies can be induced to undergo haploidization by treatment with UV, X rays, para-fluorophenylalanine, or benomyl [25,38,39,46,68]. Tables 2 and 3 show the full range of genetic complementations possible with the protoplast fusion technique.

VI. GENETIC, CYTOLOGICAL, AND BIOCHEMICAL CLASSIFICATION OF FUSION PRODUCTS

A. Genetic and Cytological Classification

Once fusion products have developed on a osmotically stabilized regeneration medium, a genetic and cytological evaluation must ensue to determine what genetic state the fusion products are in (heterokaryon, diploid, aneuploid, haploid) and what physiological properties are expressed in the fusion products relative to the parent strains involved in the fusion.

In most fungal protoplast fusion systems, auxotrophic mutants are used as the fusion partners and detection of fusion products is based on complementation on a nutrient-deficient medium. Heterokaryons are easily detected by inoculating the fusion product back into a medium containing all the growth requirements of the parental strains. Under such nonselective conditions, heterokaryons readily segregate nuclei and only

TABLE 2. Fungal Species in Which Intraspecific Protoplast Fusion Has Been Reported

Species	Ref.
Heterokaryons	
Aspergillus amstelodami	56
A. flavus	6
A. heterothallicus	56
A. japonicus	56
A. nidulans	3,90,100
A.nidulans var. *echinulatus*	56
A. niger	6,90
A. oryzae	146
A. quadrilineatus	56
A. stellatus	56
A. unquis	56
Cochliobolus heterostrophus	45
Geotrichum candidum	2
Mucor racemosus	125,147
Oudemansiella mucida	142
Penicillium cyaneo-fulvum	11,46
P. frequentans	6
P. notatum	11
P. patulum	90
P. ramigena	6
P. roqueforti	90
Phanerochaete chrysosporium	139
Phycomyces blakesleeanus	148
Trichoderma viride	102
Heterokaryons and transient diploids or polyploids	
Aspergillus nidulans	56
Candida tropicalis	8
Cephalosporium acremonium	90,117,149
Hansenula polymorpha	150
Penicillium chrysogenum	117
Trichoderma viride	56
Heterokaryons and stable diploids	
Aspergillus awamori var. *kawachi*	126
A. fumigatus	50
A. nidulans	16,50,101,151,152
A. oryzae	85
Candida albicans	22,24,25
C. tropicalis	26,27
C. utilis	28
Penicillium album	114
P. caseicolum	114,130
P. chrysogenum	5,90
P. roqueforti	114

Table 2 (Continued)

Species	Ref.
Heterokaryons and stable diploids	
Saccharomyces cerevisiae	33
Trichoderma reesei	153
Verticillium albo-atrum	41
V. dahliae	41
Heterokaryons, stable diploids, and/or aneuploids	
Aspergillus nidulans	56
Beauveria bassiana	30
Candida albicans	21,24,25
Gibberella zeae	44
Penicillium chrysogenum	141
Phytophthora megasperma f. sp. *glycinea*	154
Saccharomycopsis lipolytica	155
Schwanniomyces alluvius	113
Trichoderma reesei	91
Verticillium lecanii	42
Heterokaryons, transient diploids and/or aneuploids, stable haploid recombinants	
Cephalosporium acremonium	143
Heterokaryons, diploids, aneuploids, and haploid recombinants	
Aspergillus nidulans	17
Stable diploids and polyploids	
Candida albicans	134
Hansenula polymorpha	156
Stable diploids	
Kluyveromyces lactis	65,66,157
Rhodosporidium toruloides	39
Saccharomyces cerevisiae	31,32,34,36,54,83,158,159
Saccharomycopsis lipolytica	40,78
Schizosaccharomyces pombe	31,36,38,67,123
Stable diploids and/or aneuploids	
Penicillium verruculosum	135,160
Pichia guillermondii	161,162
Rhodosporidium toruloides	161,163
Saccharomyces cerevisiae	35,37,68,161
S. unisporus	68
Saccharomycopsis lipolytica	129
Schizosaccharomyces pombe	164,165
Torulopsis glabrata	68
Aneuploids and haploid recombinants	
Cephalosporium acremonium	80
Emericellopsis glabra	80

Table 2 (Continued)

Species	Ref.
Unknown genetic composition	
Aspergillus flavus	100
A. niger	100
Beauveria amorpha	9
B. bassiana	9,29
B. brongniartii	9
Candida albicans	7
Cochliobolus carbonum	45
C. sativus	45
Debaryomyces formicarius	124
Geotrichum candidum	100
Hansenula anomala	150
Hansenula bimundalis	150
Penicillium frequentans	100
P. ramigena	100
Pyricularia oryzae	166,167
Rhizoctonia solani	144
Saccharomyces diastaticus	61

TABLE 3. Fungal Species in Which Interspecific or Intergeneric Protoplast Fusions Have Been Reported

Species	Ref.
Heterokaryons	
Aspergillus awamori + *A. niger*	13
A. nidulans + *A. nidulans* var. *echinulatus*	74
A. nidulans var. *echinulatus* + *A. rugulosus*	10
A. nidulans var. *echinulatus* + *A. violaceus*	10
A. quadrilineatus + *A. rugulosus*	10
A. rugulosus + *A. violaceus*	10
Penicillium chrysogenum + *P. citrinum*	89
P. chrysogenum + *P. cyclopium*	88
P. chrysogenum + *P. lanosum*	88
P. chrysogenum + *P. nigricans*	88
P. chrysogenum + *P. notatum*	90
P. roqueforti + *P. stoloniferum*	115
Saccharomyces cerevisiae + *Schwanniomyces alluvius*	113
Saccharomyces uvarum + *Schwanniomyces alluvius*	113
Heterokaryons and diploids	
Aspergillus awamori var. *kawachi* + *A. oryzae*	127
A. nidulans + *A. nidulans* var. *echinulatus*	10
A. nidulans + *A. quadrilineatus*	10
A. nidulans + *A. rugulosus*	10,13,43,48,79

Table 3 (Continued)

Species	Ref.
Heterokaryons and diploids	
A. nidulans + *A. violaceus*	10
Penicillium album + *P. caseicolum A*	114
P. chrysogenum + *P. patulum*	88
P. chrysogenum + *P. verrucosum* var. *cyclopium*	88
Verticillium albo-atrum + *V. dahliae*	41
Heterokaryons and diploids or aneuploids	
Aspergillus nidulans + *A. rugulosus*	47
A. niger + *A. rugulosus*	81
Candida utilis + *Saccharomyces cerevisiae*	118
Cephalosporium acremonium + *Penicillium chrysogenum*	117
Cephalosporium acremonium + *Paecilomyces carneus*	80
Mucor miehei + *M. pusillus*	168
Penicillium baarnense + *P. chrysogenum*	131
Penicillium chrysogenum + *P. cyaneo-fulvum*	46
P. chrysogenum + *P. roqueforti*	14,49
P. chrysogenum + *P. stoloniferum*	88,115
P. citrinum + *P. cyaneo-fulvum*	89
P. citrinum + *P. roqueforti*	131
P. cyaneo-fulvum + *P. roqueforti*	131
P. cyaneo-fulvum + *P. stoloniferum*	115,131
Saccharomyces cerevisiae + *Zygosaccharomyces fermentati*	12
Heterokaryons, allodiploids, and haploid recombinants	
Aspergillus nidulans + *A. nidulans* var. *echinulatus*	17
A. nidulans + *A. rugulosus*	17
A. nidulans var. *echinulatus* + *quadrilineatus*	17
Allodiploids and/or aneuploids	
Hansenula capsulata + *Saccharomyces diastaticus*	64
Kluyveromyces lactis + *Yarrowia lipolytica*	134
Saccharomyces cerevisiae + *Schwanniomyces castellii*	133
Aneuploids	
Aspergillus fumigatus + *A. nidulans*	50,51
A. nidulans + *A. quadrilineatus*	152
Candida albicans + *C. tropicalis*	56,140,169
Candida tropicalis + *Saccharomycopsis fibuligera*	53
Penicillium baarnense + *P. chrysogenum*	60
P. chrysogenum + *P. roqueforti*	81
Saccharomyces diastaticus + *S. uvarum*	116
Saccharomyces diastaticus + *Schizosaccharomyces pombe*	170
Schizosaccharomyces octosporous + *S. pombe*	171
Haploids	
Cephalosporium acremonium + *Cephalosporium* spp.	80
C. acremonium + *Emericellopsis glabra*	80
Emericellopsis glabra + *E. salmosynnemata*	80

Table 3 (Continued)

Species	Ref.
Unknown genetic composition	
Beauveria amorpha + *B. bassiana*	9
B. amorpha + *B. brongniartii*	9
B. bassiana + *B. brongniartii*	9
Candida guillermondii + *Saccharomycopsis lipolytica*	129
Kluyveromyces fragilis + *K. lactis*	128
Kluyveromyces lactis + *Saccharomyces* sp.	55
Lodderomyces elongisporus + *Saccharomycopsis lipolytica*	155
Saccharomyces cerevisiae + *S. uvarum*	54,137
Saccharomyces cerevisiae + *Schizosaccharomyces pombe*	31,172
Trichoderma koningii + *T. reesei*	153
T. reesei + *T. viride*	153

parental-type cells or conidia can then be isolated [6,14,49,88,90]. In some cases only one of the parent strains can be reisolated.

Several other methods are used to distinguish diploid cell types from heterokaryons. Both diploids and aneuploids typically grow faster than heterokaryons on minimal medium [89,115]. On complete medium, diploids are typically more stable than heterokaryons. Estimates on the amount of DNA per cell are often used, in conjunction with the above method or DNA-staining techniques, to classify fusion products. In yeast species, DNA estimates are most often based on the amount of DNA per 10^8 cells determined prior to DNA extraction by hemacytometer [35,68,73,116]. In filamentous fungi, DNA is usually extracted from some type of spore due to the multinucleate and nonseptate aspect of many filamentous fungi [43, 90,117]. Staining of DNA is also useful in determining if one or more nuclei are present. The most commonly used stain has been the intercholating agent acridine orange [35,43,73]. Classic fungal DNA stains such as HCl-Feulgen [25,31,41] and Giemsa [8,37] have also been used.

Fusion products are often referred to as hybrids after the heterokaryon has undergone karyogamy because of the difficulty in distinguishing between diploids and aneuploids [43,47,49,115]. The major characteristics of aneuploids are larger cell and spore size than the haploid fusion partners [25,65,89, 113,115] and a hyperhaploid amount of DNA per cell [25,65,113]. Both diploids and aneuploids can be distinguished from heterokaryons by their growth patterns in the presence of haploidizing agents. When diploid or aneuploid cells are inoculated into a complete medium containing a haploidizing agent, sectors of more vigorous growth develop. When analyzed genetically and biochemically, these sectors are found to be haploid. The most common haploidizing agents are parafluoro-DL-phenylalanine [14,38,46,80,117,118] and benomyl [methyl-1-(butylcarbamoyl)-2-benzimidazole carbamate] [14,33, 79,80,89,119]. These compounds are thought to act by inducing random chromosomal loss or reduction until the haploid chromosome level is reached [120,121]. Other haploidizing agents such as acriflavine [122], chloral hydrate [22,27,48,80], metafluoro-DL-phenylalanine [123], and UV light [118] have been used.

Hybrids and haploidizing agents can be used to determine alleles and linkage groups in fungal species. Chepurnaya et al. [124] fused three different adenine-requiring strains of *Debaryomyces formicarius* in all possible combinations. When ade17 × ade21 or ade21 × ade22 fusions were performed, complementing colonies would develop on minimal medium. If an ade17 × ade22 fusion was done, no fusion products were recovered. From this series of experiments, the authors concluded that the ade17 and ade22 mutants were allelic and at least two genes controlling the synthesis of adenine are present in *D. formicarius*. In *Mucor racemosus*, two arginine auxotrophs were found to complement, and the same conclusion as with *Debaryomyces* can be drawn [125]. Bradshaw et al. [47] produced hybrids between strains of *Aspergillus rugulosus* and mitotic master strains of *A. nidulans* with a genetic marker on each linkage group. By analyzing segregants induced by growth of the hybrids on a benomyl-containing complete medium, recombination between every pair of unlinked markers was found. The analysis also revealed that parental combinations of markers were often recovered at significantly higher frequencies than expected. The aberrant segregation suggested that the two species differ in the number of linkage groups, quite possibly in the number and size of chromosomes. The same type of analysis has been carried out in intraspecific hybrids to determine linkage relationships and complementation groups in *Candida albicans* [7,21,23,25].

Organisms that form heterokaryons upon protoplast fusion can be analyzed by linkage analysis after they have been forced to undergo karyogamy or diploidization by treatment with d-camphor. Forced diploidization has been applied to both intraspecific [91,126] and interspecific [127] fungal protoplast fusion products. If auxotrophic parental strains are used as the fusion partners, heterokaryons are typically inoculated onto a minimal media containing 0.1% (v/v) d-camphor and incubated for 7-14 days at the appropriate temperature. Diploid sectors can then be identified by faster growth rate, conidial diameter, DNA content, or other commonly used criteria.

Hybrids can be distinguished from heterokaryons, at least in yeast species, by their larger cell volumes. Hybrids are usually grown to stationary phase, photographs are taken, and the negatives are enlarged. Volumes are then calculated on the assumption that the cells are perfect ellipsoids. Comparisons of cell volume have been made in fusion hybrids of *C. albicans* [22,24,25], *C. tropicalis* [26], *Kluyveromyces lactis* [65,66], *Rhodosporidium toruloides* [39], *S. cerevisiae* [32-35,73], *Schizosaccharomyces pombe* [38], and *Schwanniomyces alluvius* [113]. The technique has also been used to analyze interspecific hybrids of *K. lactis* and *K. fragilis* [128]. Although the technique seems to have wide applicability, some hybrids have been found not to have a significantly higher cell volume than the fusion partners. Spata and Weber [129] found that hybrids of *Saccharomycopsis lipolytica* did not have significantly different cell volumes than the fusion partners but the amount of DNA per cell was at the expected diploid or aneuploid level.

Spore size or conidiospore diameter has also been used to classify fusion products as hybrids. The technique has been used most extensively in interspecific fusions of penicillia [14,46,49,88,89,115]. Anné and Peberdy [90] fused two auxotrophic and spore color mutants of *Penicillium chrysogenum*. Tentative diploid colonies were confirmed on the basis of prototrophy, color of spores, spore size, and their doubled DNA content. Haploid and diploid spores had mean diameters of 2.95 and 3.60 μm, respectively. Other penicillia hybrids have been found to have larger conidium diameters [114,117,130].

Spore color markers have also been used to characterize fusion hybrids and heterokaryons in the aspergilli [16,43,47] and penicillia [46,49,88,89]. Anné [115] fused protoplasts of *P. chrysogenum* leu met oli^R ylo and *P. stoloniferum* lys, which normally produces green spores. Spore diameters were 3.35 and 2.44 μm, respectively. Four types of hybrid colonies developed after the heterokaryons were cultivated in liquid minimal medium. Spores were characterized as prototrophic oli^R brown (4.80 μm), prototrophic oli^R green (4.75 μm), leu met oli^R green (4.63 μm), or prototrophic oli^R ylo (3.79 μm). All hybrid progeny were oligomycin-resistant. Prototrophy and green spore color, as in the wild-type *P. chrysogenum*, resulted from complementation for auxotrophy and spore color markers. The hybrid nature of the fusion products is also indicated by the larger spore diameters.

Colony morphology has been used to identify interspecific and intergeneric fusion products, especially if the species differ greatly in colony morphology [43,47,49,51,53,117,130]. Interspecific fusion products have also been classified as either heterokaryons or diploids based on colony morphology and slower growth rate of heterokaryons [46,49,81,115]. Kevei and Peberdy [43] fused strains of *A. nidulans* and *A. rugulosus* which differed greatly in colony morphology. The *A. nidulans* parents were fast-growing, had a spreading habit, produced abundant conidia, and produced cleistothecia that were generally scattered over the colony. The *A. rugulosus* parents grew more slowly, were buckled and wrinkled, produced few conidiophores, and produced abundant cleistothecia forming several layers in the mycelium. Fusion products first grew as heterokaryons which were characterized by an irregular colony shape and different size from the parental colony on complete medium. The heterokaryons gave rise to hybrids after nuclear fusion. Hybrids were characterized as having hyphae identical to those of the parental form, but they showed a pattern of branching similar to the *A. nidulans* parent, produced few white conidia, and the conidial apparatus showed a reduced structure. As the colonies aged, small clusters of conidiophores developed which were sites of spontaneous segregation and which showed the normal morphology of one or the other parent. The conidiophores were further distinguished by spores with the appropriate color markers of the parent strains. Other hybrids showed a mixture of characters of the two parents, i.e., spore color and abundant cleistothecial production typical of *A. rugulosus*, while the spreading growth habit and intense conidiation were features of *A. nidulans*.

Pigment production has been noted in diploid and/or hybrid protoplast fusion products that neither parent produced. Red-pigmented diploids have been reported in *A. nidulans* and *A. nidulans* + *A. nidulans* var. *echinulatus* fusion products [17]. Hybrid fusion products of a cross between *A. nidulans* and *A. rugulosus* have been reported to produce pigments [13]. Similarly, a hybrid of *P. chrysogenum* and *P. patulum* produced a pigment that neither parent produced [11]. A correlation was noticed when hybrids of *P. chrysogenum* and *P. roqueforti* were plated on a glucose-rich media [11]. If plated on such a media, the hybrids produced a red pigment that neither parent produced. Pigment production has been reported in hybrids where one of the parents of the fusion pair produced a pigment [88].

B. Biochemical Classification

Fusion products have been characterized biochemically by such techniques as isoenzyme banding patterns [13,60,131]. This technique has been used to characterize interspecific fusion products in both plant and fungal systems. Maliga et al. [132] showed the hybrid nature of fusion products between *Nicotiana knightiana* and *N. sylvestris* by comparing banding patterns of esterase, alcohol dehydrogenase, and glucose-6-phosphate dehydrogenase isoenzymes. Anné and Peberdy [14] fused protoplasts of *P. chrysogenum* with *P. roqueforti* and analyzed hybrids with respect to aryl esterases, catalases, succinate dehydrogenases, glutamate dehydrogenases, malate dehydrogenases, glucose-6-phosphate dehydrogenases, and alkaline and acid phosphatases. The enzyme patterns of both species were distinct in most cases. Hybrids contained parental isoenzymes in different quantities as well as novel bands, demonstrating the interaction of the two genomes in hybrids. Novel catalase and glucose-6-phosphate dehydrogenase bands were observed in several hybrid extracts.

Enzyme activities have also been used to biochemically characterize fusion products. Amylolytic, lignocellulolytic, lipolytic, and proteolytic activities have been used to determine if fusion products have different activities or levels of enzyme relative to the parent strains [53,61,91,130, 133]. Carbon assimilation and fermentative spectra have also been used to characterize fusion products and segregants and to demonstrate genome interaction [12,48,53,64,70,134-136].

Increases in penicillin yield have also been used to characterize fusion products. Kevei and Peberdy [13] fused protoplasts of *A. nidulans*, a penicillin producer, with *A. rugulosus*, a nonproducer. Among the fusion products, approximately 50% were penicillin producers. Some fusion products produced considerably more penicillin than the *A. nidulans* parent, and the authors suggested that protoplast fusion could be advantageous in stimulating secondary metabolite production.

Other metabolic products have been used to characterize fusion products. *Saccharomyces cerevisiae*, a single-cell protein source for both man and cattle, is low in methionine content. Brigidi et al. [137] recently fused DL-ethionine (a methionine analog)-resistant, petite mutants of *S. uvarum* with an *S. cerevisiae* strain containing six auxotrophic markers. The aim of the research was to obtain an *S. cerevisiae* strain that would overproduce methionine. Some of the fusion products produced exogenous L-methionine levels above that of the overproducing *S. uvarum* parent. The fact that the fusion products maintained four auxotrophic markers and had the fermentation abilities of the *S. cerevisiae* parent suggests that the fusion products were aneuploid in nature.

VII. CONCLUSIONS

Protoplasts of fungal organisms offer the possibility of performing genetic studies on organisms where previously no or few options existed. Both transformation and protoplast fusion techniques require isolation of intact protoplasts as a starting point. Protoplast fusion techniques, which have been continuously developing for over a decade, now offer many strategies in selecting for growth of fusion products over parent strains which remain unfused. This allows latitude in selecting fusion pairs.

The application of these techniques to organisms used in industrial processes offers possibilities of producing strains of higher yield and different secondary metabolites as well as answering questions about basic biology. As these techniques continue to be modified and adapted to other industrially important organisms, undoubtedly important strains will be created for human exploitation.

ACKNOWLEDGMENT

We are grateful to Dr. Jeanette Leach for helpful criticism of this manuscript. We also thank the Idaho Agricultural Experiment Station for support of our programs through project 058-H727.

REFERENCES

1. Lopez-Belmonte, F., Garcia Acha, I., and Villanueva, J. R. Observations on the protoplasts of Fusarium culmorum and on their fusion. *J. Gen. Microbiol.*, *45*:127-134, 1966.
2. Ferenczy, L., Kevei, F., and Zsolt, J. Fusion of fungal protoplasts. *Nature*, *248*:793-794, 1974.
3. Ferenczy, L., Kevei, F., and Szegedi, M. Increased fusion frequency of Aspergillus nidulans protoplasts. *Experientia*, *31*:50-52, 1975.
4. Kao, K. N., and Michayluk, M. R. A method for high-frequency intergeneric fusion of plant protoplasts. *Planta*, *115*:355-367, 1974.
5. Anné, J., and Peberdy, J. F. Conditions for induced fusion of fungal protoplasts in polyethylene glycol solutions. *Arch. Microbiol.*, *105*:201-205, 1975
6. Ferenczy, L., Kevei, F., and Szegedi, M. High-frequency fusion of fungal protoplasts. *Experientia*, *31*:1028-1030, 1975.
7. Kakar, S. N., Partridge, R. M., and Magee, P. T. A genetic analysis of Candida albicans: isolation of a wide variety of auxotrophs and demonstration of linkage and complementation. *Genetics*, *104*:241-255, 1983.
8. Fournier, P., Provost, A., Bourguignon, C., and Heslot, H. Recombination after protoplast fusion in the yeast Candida tropicalis. *Arch. Microbiol.*, *115*:143-149, 1977.
9. Shimizu, S. Protoplast fusion of insect pathogenic fungi. In: *Biotechnology in Invertebrate Pathology and Cell Culture* (K. Maramorosch, ed.), Academic Press, San Diego, 1987, pp. 401-414.
10. Kevei, F., and Peberdy, J. F. Further studies on protoplast fusion and interspecific hybridization within the Aspergillus nidulans group. *J. Gen. Microbiol.*, *130*:2229-2236, 1984.
11. Anné, J. Somatic hybridization between Penicillium species after induced fusion of their protoplasts. *Agricultura*, *25*:1-117, 1977.
12. Pina, A., Calderón, I. L., and Benitez, T. Intergeneric hybrids of Saccharomyces cerevisiae and Zygosaccharomyces fermentati obtained by protoplast fusion. *Appl. Environ. Microbiol.*, *51*:995-1003, 1986.
13. Kevei, F., and Peberdy, J. F. Interspecies hybridization after protoplast fusion in Aspergillus. In *Fungal Protoplasts: Applications in Biochemistry and Genetics* (J. F. Peberdy and L. Ferenczy, eds.), Marcel Dekker, New York, 1985, pp. 241-257.
14. Anné, J., and Peberdy, J. F. Characterization of interspecific hybrids between Penicillium chrysogenum and P. roqueforti by iso-enzyme analysis. *Trans. Br. Mycol. Soc.*, *77*:401-408, 1981.
15. van Solingen, P., and van der Plaat, J. B. Fusion of yeast spheroplasts. *J. Bacteriol.*, *130*:946-947, 1977.

16. Dales, R. B. G., and Croft, J. H. Protoplast fusion and the isolation of heterokaryons and diploids from vegetatively incompatible strains of Aspergillus nidulans. *FEMS Microbiol. Lett.*, *1*:201-204, 1977.
17. Croft, J. H. Protoplast fusion and incompatibility in Aspergillus. In *Fungal Protoplasts: Applications in Biochemistry and Genetics* (J. F. Peberdy and L. Ferenczy, eds.), Marcel Dekker, New York, 1985, pp. 225-240.
18. Maráz, A., and Subik, J. Transmission and recombination of mitochondrial genes in Saccharomyces cerevisiae after protoplast fusion. *Mol. Gen. Genet.*, *181*:131-133, 1981.
19. Croft, J. H., Dales, R. B. G., Turner, C., and Earl, A. The transfer of mitochondria between species of Aspergillus. In *Advances in Protoplast Research* (L. Ferenczy and G. L. Farkas, eds.), Akadémiai Kiadó, Budapest; Pergamon Press, Oxford, 1980, pp. 85-92.
20. Alexopoulos, C. J., and Mims, C. W. *Introductory Mycology*, 3rd ed. John Wiley and Sons, New York, 1979.
21. Kakar, S. N., and Magee, P. T. Genetic analysis of Candida albicans: Identification of different isoleucine-valine, methionine, and arginine alleles by complementation. *J. Bacteriol.*, *151*:1247-1252, 1982.
22. Pesti, M., and Ferenczy, L. Protoplast fusion hybrids of Candida albicans sterol mutants differing in nystatin resistance. *J. Gen. Microbiol.*, *128*:123-128, 1982.
23. Poulter, R., Hanrahan, V., Jeffery, K., Markie, D., Shepherd, M. G., and Sullivan, P. A. Recombination analysis of naturally diploid Candida albicans. *J. Bacteriol.*, *152*:969-975, 1982.
24. Poulter, R., Jeffery, K., Hubbard, M. J., Shepherd, M. G., and Sullivan, P. A. Parasexual genetic analysis of Candida albicans by spheroplast fusion. *J. Bacteriol.*, *146*:833-840, 1981.
25. Sarachek, A., Rhoads, D. D., and Schwarzhoff, R. H. Hybridization of Candida albicans through fusion of protoplasts. *Arch. Microbiol.*, *129*:1-8, 1981.
26. Vallin, C., and Ferenczy, L. Diploid formation of Candida tropicalis via protoplast fusion. *Acta Microbiol. Acad. Sci. Hung.*, *25*:209-212, 1978.
27. Vallin, C., and Ferenczy, L. Protoplast fusion in Candida tropicalis and the analysis of fusion products. *Acta Microbiol. Acad. Sci. Hung.*, *27*:246, 1980.
28. Delgado, J. M., and Herrera, L. S. Protoplast fusion in the yeast Candida utilis. *Acta Microbiol. Acad. Sci. Hung.*, *28*:339-345, 1981.
29. Kawamoto, H., and Aizawa, K. Fusion conditions for protoplasts of an entomogenous fungus, Beauveria bassiana. *Appl. Ent. Zool.*, *21*:624-626, 1986.
30. Gadau, M. E., and Lingg, A. J. Protoplast fusion of auxotrophic strains of Beauveria bassiana (submitted).
31. Svoboda, A. Fusion of yeast protoplasts induced by polyethylene glycol. *J. Gen. Microbiol.*, *109*:169-175, 1978.
32. Ferenczy, L., and Maráz, A. Transfer of mitochondria by protoplast fusion in Saccharomyces cerevisiae. *Nature*, *268*:524-525, 1977.
33. Ferenczy, L., and Pesti, M. Transfer of isolated nuclei into protoplasts of Saccharomyces cerevisiae. *Curr. Microbiol.*, *7*:157-160, 1982.
34. Gunge, N., and Tamaru, A. Genetic analysis of products of protoplast fusion in Saccharomyces cerevisiae. *Jap. J. Genet.*, *53*:41-49, 1978.
35. Maráz, A., Kiss, M., and Ferenczy, L. Protoplast fusion in Saccharomyces cerevisiae strains of identical and opposite mating types. *FEMS Microbiol. Lett.*, *3*:319-322, 1978.

36. Svoboda, A. Intraspecies fusion of yeast protoplasts. *Folia Microbiol.*, *22*:441-442, 1977.
37. Yamamoto, M., and Fukui, S. Fusion of yeast protoplasts. *Ag. Biol. Chem.*, *41*:1829-1830, 1977.
38. Sipiczki, M., and Ferenczy, L. Protoplast fusion of Schizosaccharomyces pombe auxotrophic mutants of identical mating-type. *Mol. Gen. Genet.*, *151*:77-81, 1977.
39. Sipiczki, M., and Ferenczy, L. Fusion of Rhodosporidium (Rhodotorula) protoplasts. *FEMS Microbiol. Lett.*, *2*:203-205, 1977.
40. Stahl, U. Zygote formation and recombination between like mating types in the yeast Saccharomycopsis lipolytica by protoplast fusion. *Mol. Gen. Genet.*, *160*:111-113, 1978.
41. Typas, M. A. Heterokaryon incompatibility and interspecific hybridization between Verticillium albo-atrum and Verticillium dahliae following protoplast fusion and microinjection. *J. Gen. Microbiol.*, *129*:3043-3056, 1983.
42. Jackson, C. W., and Heale, J. B. Parasexual crosses by hyphal anastomosis and protoplast fusion in the entomopathogen Verticillium lecanii. *J. Gen. Microbiol.*, *133*:3537-3547, 1987.
43. Kevei, F., and Peberdy, J. F. Interspecific hybridization between Aspergillus nidulans and Aspergillus rugulosus by fusion of somatic protoplasts. *J. Gen. Microbiol.*, *102*:255-262, 1977.
44. Adams, G., Johnson, N., Leslie, J. F., and Hart, L. P. Heterokaryons of Gibberella zeae formed following hyphal anastomosis or protoplast fusion. *Exp. Mycol.*, *11*:339-353, 1987.
45. Leach, J., and Yoder, O. C. Heterokaryosis in Cochliobolus heterostrophus. *Exp. Mycol.*, *6*:364-374, 1982.
46. Peberdy, J. F., Eyssen, H., and Anné, J. Interspecific hybridization between Penicillium chrysogenum and Penicillium cyaneo-fulvum following protoplast fusion. *Mol. Gen. Genet.*, *157*:281-284, 1977.
47. Bradshaw, R. E., Lee, K., and Peberdy, J. F. Aspects of genetic interaction in hybrids of Aspergillus nidulans and Aspergillus rugulosus obtained by protoplast fusion. *J. Gen. Microbiol.*, *129*:3525-3533, 1983.
48. Kevei, F., and Peberdy, J. F. Induced segregation in interspecific hybrids of Aspergillus nidulans and Aspergillus rugulosus obtained by protoplast fusion. *Mol. Gen. Genet.*, *170*:213-218, 1979.
49. Anné, J., Eyssen, H., and DeSomer, P. Somatic hybridisation of Penicillium roquefortii with P. chrysogenum after protoplast fusion. *Nature*, *262*:719-721, 1976.
50. Ferenczy, L. Some characteristics of intra- and interspecific protoplast fusion products of Aspergillus nidulans and Aspergillus fumigatus. In *Cell Genetics in Higher Plants* (D. Dudits, G. L. Farkas, and P. Maliga, eds.), Pergamon Press, Oxford, 1976, pp. 171-182.
51. Ferenczy, L., Szegedi, M., and Kevei, F. Interspecific protoplast fusion and complementation in Aspergilli. *Experientia*, *33*:184-186, 1977.
52. Peberdy, J. F. Fungal protoplast: isolation, reversion, and fusion. *Annu. Rev. Microbiol.*, *33*:21-39, 1979.
53. Provost, A., Bourguignon, C., Fournier, P., Ribet, A. M., and Heslot, H. Intergeneric hybridization in yeasts through protoplast fusion. *FEMS Microbiol. Lett.*, *3*:309-312, 1978.
54. Stewart, G. G. Application of yeast genetics within the brewing industry. A review. *J. Am. Soc. Brew. Chem.*, *36*:175-185, 1978.

55. Stewart, G. G. The genetic manipulation of industrial yeast strains. *Can. J. Microbiol.*, *27*:973-990, 1981.
56. Ferenczy, L. Fungal protoplast fusion: basic and applied aspects. In *Cell Fusion: Gene Transfer and Transformation* (R. E. Beers, Jr., and E. G. Bassett, eds.), Raven Press, New York, 1984, pp. 145-169.
57. Chang, L. T., and Terry, C. A. Intergenic complementation of glucoamylase and citric acid production in two species in Aspergillus. *Appl. Microbiol.*, *25*:890-895, 1973.
58. Fantini, A. A. Genetics and antibiotic production of Emericellopsis species. *Genetics*, *47*:161-177, 1962.
59. MacDonald, K. D., Hutchinson, J. M., and Gillet, W. A. Isolation of auxotrophs of Penicillium chrysogenum and their penicillin yields. *J. Gen. Microbiol.*, *33*:365-374, 1963.
60. Mellon, F. M. Protoplast fusion and hybridization in Penicillium. In *Molecular Genetics of Filamentous Fungi* (W. E. Timberlake, ed.), Alan R. Liss, New York, 1985, pp. 69-82.
61. Spencer, J. F. T., Laud, P., and Spencer, D. M. The use of mitochondrial mutants in the isolation of hybrids involving industrial yeast strains. II. Use in isolation of hybrids obtained by protoplast fusion. *Mol. Gen. Genet.*, *178*:651-654, 1980.
62. Spencer, J. F. T., and Spencer, D. M. The use of mitochondrial mutants in the isolation of hybrids involving industrial yeast strains. *Mol. Gen. Genet.*, *177*:355-358, 1980.
63. Spencer, J. F. T., and Spencer, D. M. Genetic improvement of industrial yeasts. *Annu. Rev. Microbiol.*, *37*:121-142, 1983.
64. Spencer, J. F. T., Spencer, D. M., Whittington-Vaughan, P., and Miller, R. Use of mitochondrial mutants in the isolation of hybrids involving industrial yeast strains. IV. Characterization of an intergeneric hybrid, Saccharomyces diastaticus × Hansenula capsulata, obtained by protoplast fusion. *Curr. Genet.*, *7*:159-164, 1983.
65. Allmark, B. M., Morgan, A. J., and Whittaker, P. A. The use of protoplast fusion in demonstrating chromosomal and mitochondrial inheritance of respiratory-deficiency in Kluyveromyces lactis, a petite-negative yeast. *Mol. Gen. Genet.*, *159*:297-299, 1978.
66. Morgan, A. J., Heritage, J., and Whittaker, P. A. Protoplast fusion between petite and auxotrophic mutants of the petite-negative yeast, Kluyveromyces lactis. *Microbios Lett.*, *4*:103-107, 1978.
67. Lückemann, G., Sipiczki, M., and Wolfe, K. Transmission, segregation, and recombination of mitochondrial genomes in zygote clones and protoplast fusion clones of yeast. *Mol. Gen. Genet.*, *177*:185-187, 1979.
68. Galeotti, C. L., Sriprakash, K. S., Batum, C. M., and Clark-Walker, G. D. An unexpected response of Torulopsis glabrata fusion products to X-irradiation. *Mut. Res.*, *81*:155-164, 1981.
69. Sriprakash, K. S., and Batum, C. Segregation and transmission of mitochondrial markers in fusion products of the asporogenous yeast Torulopsis glabrata. *Curr. Genet.*, *4*:73-80, 1981.
70. Gunge, N., and Sakaguchi, K. Fusion of mitochondria with protoplasts in Saccharomyces cerevisiae. *Mol. Gen. Genet.*, *170*:243-247, 1979.
71. Yoshida, K. Interspecific and intraspecific mitochondria-induced cytoplasmic transformation in yeasts. *Plant Cell Physiol.*, *20*:851-856, 1979.
72. Yoshida, K., and Takeuchi, I. K. Cytological studies on mitochondria-induced cytoplasmic transformation in yeasts. *Plant Cell Physiol.*, *21*: 497-509, 1980.

73. Maráz, A., and Ferenczy, L. Selective transfer of fungal cytoplasmic genetic elements by protoplast fusion. *Curr. Microbiol.*, *4*:343-345, 1980.
74. Earl, A. J., Turner, G., Croft, J. H., Dales, R. B. G., Lazarus, C. M., Lünsdorf, H., and Küntzel, H. High frequency transfer of species-specific mitochondrial DNA sequences between members of the Aspergillaceae. *Curr. Genet.*, *3*:221-228, 1981.
75. Spencer, J. F. T., and Spencer, D. M. The use of mitochondrial mutants in hybridization of industrial yeasts. III. Restoration of mitochondrial function in petites in industrial strains by fusion with respiratory-competent protoplasts of other yeast species. *Curr. Genet.*, *4*:177-180, 1981.
76. Yamashita, K., Fukuda, H., Murata, K., and Kimura, A. Transfer of mitochondria of Hansenula wingei into protoplasts of Saccharomyces cerevisiae by mini-protoplast fusion. *FEBS Lett.*, *132*:305-307, 1981.
77. Maráz, A., and Ferenczy, L. Transfer of mitochondria via anucleate protoplasts in Saccharomyces cerevisiae. *Acta Microbiol. Acad. Sci. Hung.*, *27*:247-248, 1980.
78. Matsuoka, M., Uchida, K., and Aiba, S. Cytoplasmic transfer of oligomycin resistance during protoplast fusion of Saccharomycopsis lipolytica. *J. Bacteriol.*, *152*:530-533, 1982.
79. Bradshaw, R. E., and Peberdy, J. F. Protoplast fusion in Aspergillus: selection of interspecific heterokaryons using antifungal inhibitors. *J. Microbiol. Meth.*, *3*:27-32, 1984.
80. Minuth, W., and Esser, K. Intraspecific, interspecific, and intergeneric recombination in β-lactam producing fungi via protoplast fusion. *Eur. J. Appl. Microbiol. Biotechnol.*, *18*:38-46, 1983.
81. Peberdy, J. F. Protoplast fusion- A new approach to interspecies genetic manipulation and breeding in fungi. In *Advances in Protoplast Research* (L. Ferenczy and G. L. Farkas, eds.), Akadémiai Kiadó, Budapest; Pergamon Press, Oxford, 1980, pp. 63-72.
82. Peberdy, J. F. Protoplast fusion: a tool for genetic manipulation and breeding in industrial microorganisms. *Enz. Microb. Technol.*, *2*:23-29, 1980.
83. Hockney, R. C., and Freeman, R. F. Construction of a polysaccharide-degrading brewing yeast by protoplast fusion. In *Advances in Protoplast Research* (L. Ferenczy and G. L. Farkas, eds.), Akadémiai Kiadó, Budapest; Pergamon Press, Oxford, 1980, pp. 139-143.
84. Hopwood, D. A., and Wright, H. M. Protoplast fusion in Streptomyces: fusions involving ultraviolet-irradiated protoplasts. *J. Gen. Microbiol.*, *126*:21-27, 1981.
85. Sakaguchi, K., Ochi, K., Gunge, N., and Uchida, K. Protoplast fusion. In *Molecular Breeding and Genetics of Applied Microorganisms* (K. Sakaguchi and M. Okanishi, eds.), Academic Press, New York, 1980, pp. 85-105.
86. Wright, W. E. The isolation of heterokaryons and hybrids by a selection system using irreversible biochemical inhibitors. *Exp. Cell Res.*, *112*: 395-407, 1978.
87. Nehls, R. The use of metabolic inhibitors for the selection of fusion products of higher plant protoplasts. *Mol. Gen. Genet.*, *166*:117-118, 1978.
88. Anné, J. Comparison of penicillins produced by inter-species hybrids from Penicillium chrysogenum. *Eur. J. Appl. Microbiol. Biotechnol.*, *15*:41-46, 1982.

89. Anné, J. and Eyssen, H. Isolation of inter-species hybrids of Penicillium citrinum and Penicillium cyaneo-fulvum following protoplast fusion. *FEMS Microbiol. Lett.*, *4*:87-90, 1978.
90. Anné, J., and Peberdy, J. F. Induced fusion of fungal protoplasts following treatment with polyethylene glycol. *J. Gen. Microbiol.*, *92*: 413-417, 1976.
91. Ogawa, K., Brown, J. A., and Wood, T. M. Intraspecific hybridization of Trichoderma reesei QM 9414 by protoplast fusion using colour mutants. *Enz. Microb. Technol.*, *9*:229-232, 1987.
92. O'Malley, K. A., and Davidson, R. L. A new dimension in suspension fusion techniques with polyethylene glycol. *Somat. Cell Genet.*, *3*:441-448, 1977.
93. Robinson, J. M., Roos, D. S., Davidson, R. L., and Karnovsky, M. J. Membrane alterations and other morphological features associated with polyethylene glycol-induced cell fusion. *J. Cell Sci.*, *40*:63-75, 1979.
94. Cocking, E. C. Plant cell fusion: transformations using plant and bacterial protoplasts. In *Cell Fusion: Gene Transfer and Transformation* (R. F. Beers, Jr., and E. G. Bassett, eds.), Raven Press, New York, 1984, pp. 139-144.
95. Radford, A., Pope, S., Sazci, A., Fraser, M. J., and Parish, J. H. Liposome-mediated genetic transformation of Neurospora crassa. *Mol. Gen. Genet.*, *184*:567-569, 1981.
96. Makins, J. F., and Holt, G. Liposome-mediated transformation of Streptomycetes by chromosomal DNA. *Nature*, *293*:671-673, 1981.
97. Nagata, T. Interaction of liposomes and protoplasts as a model system of protoplast fusion. In *Cell Fusion: Gene Transfer and Transformation* (R. F. Beers, Jr., and E. G. Bassett, eds.), Raven Press, New York, 1984, pp. 217-226.
98. Weibull, C. The isolation of protoplasts from Bacillus megaterium by controlled treatment with lysozyme. *J. Bacteriol.*, *66*:688-695, 1953.
99. Constabel, F., and Kao, K. N. Agglutination and fusion of plant protoplasts by polyethylene glycol. *Can. J. Bot.*, *52*:1603-1606, 1974.
100. Ferenczy, L., Kevei, F., and Szegedi, M. Fusion of fungal protoplasts induced by polyethylene glycol. In *Microbial and Plant Protoplasts* (J. F. Peberdy, A. H. Rose, H. J. Rogers, and E. C. Cocking, eds.), Academic Press, London, 1976, pp. 177-187.
101. Ferenczy, L., Kevei, F., Szegedi, M., Frankó, A., and Rojik, I. Factors affecting high-frequency fungal protoplast fusion. *Experientia*, *32*:1156-1158, 1976.
102. Binding, H., and Weber, H. J. The isolation, regeneration and fusion of Phycomyces protoplasts. *Mol. Gen. Genet.*, *135*:273-276, 1974.
103. Keller, W. A., and Melchers, G. The effect of high pH and calcium on tobacco leaf protoplast fusion. *Z. Naturforsch.*, *28c*:737-741, 1973.
104. Nagata, T. A novel cell-fusion method of protoplasts by polyvinyl alcohol. *Naturwissen.*, *65*:263-264, 1978.
105. Zimmermann, U., and Vienken, J. Electric field-mediated cell-to-cell fusion. In *Cell Fusion: Gene Transfer and Transformation* (R. F. Beers, Jr., and E. G. Bassett, eds.), Raven Press, New York, 1984, pp. 171-187.
106. Weber, H., Förster, W., Jacob, H. E., and Berg, H. Enhancement of yeast protoplast fusion by electric field effects. In *Current Developments in Yeast Research* (G. G. Stewart and I. Russell, eds.), Pergamon Press, Oxford, 1981, pp. 219-224.

107. Weber, H., Förster, W., Jacob, H. E., and Berg, H. Microbiological implications of electric field effects. III. Stimulation of yeast protoplast fusion by electric field pulses. *Z. Allg. Microbiol.*, *21*:555-562, 1981.
108. Halfmann, H. J., Röcken, W., Emeis, C. C., and Zimmermann, U. Transfer of mitochondrial function into a cytoplasmic respiratory-deficient mutant of Saccharomyces yeast by electro-fusion. *Curr. Genet.*, *6*:25-28, 1982.
109. Weber, H., Förster, W., Berg, H., and Jacob, H. E. Parasexual hybridization of yeasts by electric field stimulated fusion of protoplasts. *Curr. Genet.*, *4*:165-166, 1981.
110. Pontecorvo, G., and Roper, J. A. Genetic analysis without sexual reproduction by means of polyploidy in Aspergillus nidulans. *J. Gen. Microbiol.*, *6*:vii-viii, 1952.
111. Hastie, A. C. The genetics of conidial fungi. In *Biology of Conidial Fungi*, Vol. 2 (G. T. Cole and B. Kendrick, eds.), Academic Press, New York, 1981, pp. 511-547.
112. Pontecorvo, G. The parasexual cycle in fungi. *Annu. Rev. Microbiol.*, *10*:393-400, 1956.
113. Wilson, J. J., Khachatourians, G. G., and Ingledew, W. M. Protoplast fusion in the yeast, Schwanniomyces alluvius. *Mol. Gen. Genet.*, *186*:95-100, 1982.
114. Reymond, P., and Fevre, M. Recombination following protoplast fusion of Penicillium strains used in the dairy industry. *Enz. Microb. Technol.*, *8*:41-44, 1986.
115. Anné, J. Genetic evidence for selective chromosomal loss in interspecies hybrids from Penicillium chrysogenum + Penicillium stoloniferum. *FEMS Microbiol. Lett.*, *14*:191-196, 1982.
116. Barney, M. C., Jansen, G. P., and Helbert, J. R. Use of spheroplast fusion and genetic transformation to introduce dextrin utilization into Saccharomyces uvarum. *Am. Soc. Brew. Chem.*, *38*:1-5, 1980.
117. Chang, L. T., Terasaka, T., and Elander, R. P. Protoplast fusion in industrial fungi. *Dev. Ind. Microbiol.*, *23*:21-29, 1982.
118. Perez, C., Vallin, C., and Benitez, J. Hybridization of Saccharomyces cerevisiae with Candida utilis through protoplast fusion. *Curr. Genet.*, *8*:575-580, 1984.
119. Hamlyn, P. F., Birkett, J. A., Perez, G., and Peberdy, J. F. Protoplast fusion as a tool for genetic analysis in Cephalosporium acremonium. *J. Gen. Microbiol.*, *131*:2813-2823, 1985.
120. Hastie, A. C. Benlate-induced instability of Aspergillus diploids. *Nature*, *226*:771, 1970.
121. Lhoas, P. Growth rate and haploidization of Aspergillus niger on medium containing p-fluorophenylalanine. *Genet. Res.*, *12*:305-315, 1968.
122. Elander, R. P. Traditional versus current approaches to the genetic improvement of microbial strains. *FEMS Symp.*, *13*:353-369, 1982.
123. Thuriaux, P., Sipiczki, M., and Fantes, P. A. Genetic analysis of a sterile mutant by protoplast fusion in the fission yeast Schizosaccharomyces pombe. *J. Gen. Microbiol.*, *116*:525-528, 1980.
124. Chepurnaya, O. V., Kozhina, T. N., and Zakharov, I. A. Hybridization of yeasts of the genus Debaryomyces by protoplast fusion. *Genetika*, *17*:618-622, 1981.
125. Lasker, B. A., and Borgia, P. T. High-frequency heterokaryon formation by Mucor racemosus. *J. Bacteriol.*, *141*:565-569, 1980.

126. Ogawa, K., Ohara, H., and Toyama, N. Intraspecific hybridization of Aspergillus awamori var. kawachi by protoplast fusion. *Agr. Biol. Chem.*, *52*:237-242, 1988.
127. Ogawa, K., Ohara, H., and Toyama, N. Interspecific hybridization of Aspergillus awamori var. kawachi and Aspergillus oryzae by protoplast fusion. *Agr. Biol. Chem.*, *52*:1985-1991, 1988.
128. Whittaker, P. A., and Leach, S. M. Interspecific hybrid production between the yeast Kluyveromyces lactis and Kluyveromyces fragilis by protoplast fusion. *FEMS Microbiol. Lett.*, *4*:31-34, 1978.
129. Spata, L., and Weber, H. A study on protoplast fusion and parasexual hybridization of alkane utilizing yeasts. In *Advances in Protoplast Research* (L. Ferenczy and G. L. Farkas, eds.), Akadémiai Kiadó, Budapest; Pergamon Press, Oxford, 1980, pp. 131-137.
130. Reymond, P., Veau, P., and Fevre, M. Production by protoplast fusion of new strains of Penicillium caseicolum for use in the dairy industry. *Enz. Microb. Technol.*, *8*:45-47, 1986.
131. Anné, J., and Peberdy, J. F. Protoplast fusion and interspecies hybridization in Penicillium. In *Fungal Protoplasts: Applications in Biochemistry and Genetics* (J. F. Peberdy and L. Ferenczy, eds.), Marcel Dekker, New York, 1985, pp. 259-277.
132. Maliga, P., Lázár, G., Joó, F., Nagy, A. H., and Menczel, L. Restoration of morphogenic potential in Nicotiana by somatic hybridization. *Mol. Gen. Genet.*, *157*:291-296, 1977.
133. Tamaki, H. Genetic analysis of intergeneric hybrids obtained by protoplast fusion in yeasts. *Curr. Genet.*, *10*:491-494, 1986.
134. Groves, D. P., and Oliver, S. G. Formation of intergeneric hybrids of yeast by protoplast fusion of Yarrowia and Kluyveromyces species. *Curr. Genet.*, *8*:49-55, 1984.
135. Chung, K. C., Park, C. R., Bai, S., Chun, S. B., and Kim, K. C. Isolation and characterization of intraspecific complementing fusants of Penicillium verruculosum. *Kor. J. Appl. Microbiol. Bioeng.*, *16*:182-186, 1988.
136. Farahnak, F., Seki, T., Ryu, D. D. Y., and Ogrydziak, D. Construction of lactose assimilating and high ethanol producing yeasts by protoplast fusion. *Appl. Environ. Microbiol.*, *51*:362-367, 1986.
137. Brigidi, P., Matteuzzi, D., and Fava, F. Use of protoplast fusion to introduce methionine overproduction into Saccharomyces cerevisiae. *Appl. Microbiol. Biotechnol.*, *28*:268-271, 1988.
138. Fukuda, H., and Kimura, A. Transfer of mitochondria into protoplasts of Saccharomyces cerevisiae by mini-protoplast fusion. *FEBS Lett.*, *113*:58-60, 1980.
139. Gold, M. H., Cheng, T. M., and Alic, M. Formation, fusion, and regeneration of protoplasts from wild-type and auxotrophic strains of the white rot basidiomycete Phanerochaete chrysosporium. *Appl. Environ. Microbiol.*, *46*:260-263, 1983.
140. Kucsera, J., and Ferenczy, L. Interspecific protoplast fusion between Candida albicans and Candida tropicalis. *Acta Microbiol. Acad. Sci. Hung.*, *27*:247, 1980.
141. Pesti, M., Konszky, E., Erdei, J., Pólya, K., and Ferenczy, L. Penicillin production of Penicillium chrysogenum strains obtained by protoplast fusion. *Acta Microbiol. Acad. Sci. Hung.*, *27*:249, 1980.
142. Homolka, L. Preparation, reversion, mutagenesis, and intra-species fusion of protoplasts of Oudemansiella mucida. *Folia Microbiol.*, *33*:298-308, 1988.

143. Birkett, J. A., and Hamlyn, P. F. Protoplast fusion and genetic analysis in Cephalosporium. In *Fungal Protoplasts: Applications in Biochemistry and Genetics* (J. F. Peberdy and L. Ferenczy, eds.), Marcel Dekker, New York, 1985, pp. 207-223.
144. Hashiba, T., and Yamada, M. Intraspecific protoplast fusion between auxotrophic mutants of Rhizoctonia solani. *Phytopathology, 74*:398-401, 1984.
145. Becher, D., Conrad, B., and Böttcher, F. Genetic transfer mediated by isolated nuclei in Saccharomyces cerevisiae. *Curr. Genet., 6*:163-165, 1982.
146. Benkhemmar, O., Gaudemer, F., and Bouvier-Fourcade, I. Heterokaryosis between Aspergillus oryzae cyclopiazonic acid-defective strains: method for estimating the risk of inducing toxin production among cyclopiazonic acid-defective industrial strains. *Appl. Environ. Microbiol., 50*:1087-1093, 1985.
147. Genthner, F. J., and Borgia, P. T. Spheroplast fusion and heterokaryon formation in Mucor racemosus. *J. Bacteriol., 134*:349-359, 1978.
148. Bojnanská, A., Sipiczki, M., and Ferenczy, L. Characterization of conidiation mutants in Trichoderma viride by hyphal anastomosis and protoplast fusion. *Acta Microbiol. Acad. Sci. Hung., 27*:305-307, 1980.
149. Hamlyn, P. F., and Ball, C. Recombination studies with Cephalosporium acremonium. In *Genetics of Industrial Microorganisms* (O. K. Sebek and A. I. Laskin, eds.), American Society for Microbiology, Washington, D.C., 1979, pp. 185-191.
150. Savchenko, G. V., and Kapultsevich, Yu. G. Hybridization of yeast from genus Hansenula by protoplast fusion. In *Advances in Protoplast Research* (L. Ferenczy and G. L. Farkas, eds.), Akadémiai Kiadó, Budapest; Pergamon Press, Oxford, 1980, pp. 125-130.
151. Chase, D. S., Dales, R. B. G., and Croft, J. H. Heterokaryons and diploids in Aspergillus nidulans produced by the protoplast fusion of heterokaryon incompatible haploid strains. *Heredity, 47*:154, 1981.
152. Dales, R. B. G., and Croft, J. H. Protoplast fusion and the genetical analysis of vegetative incompatibility in Aspergillus nidulans. In *Advances in Protoplast Research* (L. Ferenczy and G. L. Farkas, eds.), Akadémiai Kiadó, Budapest; Pergamon Press, Oxford, 1980, pp. 73-84.
153. Toyama, H., Yamaguchi, K., Shinmyo, A., and Okada, H. Protoplast fusion of Trichoderma reesei using immature conidia. *Appl. Environ. Microbiol., 47*:363-368, 1984.
154. Layton, A. C., and Kuhn, D. N. Heterokaryon formation by protoplast fusion of drug-resistant mutants in Phytophthora megasperma f. sp. glycinea. *Exp. Mycol., 12*:180-194, 1988.
155. Weber, H., and Spata, L. Characterization of yeast protoplast fusion products. In *Current Developments in Yeast Research* (G. G. Stewart and I. Russell, eds.), Pergamon Press, Oxford, 1981, pp. 213-218.
156. Savchenko, G. V., Vikhansky, Y. D., and Kapultsevich, Y. G. The use of protoplast fusion method in selection of methanol-utilizing yeasts Hansenula polymorpha. *Genetika, 17*:1521-1523, 1981.
157. Morgan, A. J., Brunner, A., and Whittaker, P. A. Protoplast fusion in a petite-negative yeast, Kluyveromyces lactis. *Curr. Genet., 2*:87-93, 1980.
158. Russell, I., and Stewart, G. G. Spheroplast fusion of brewer's yeast strains. *J. Inst. Brew. London, 85*:95-98, 1979.

159. Svoboda, A. Mating reaction in yeast protoplasts. *Arch. Microbiol.*, *110*:313-318, 1976.
160. Chung, K. C., Park, C. R., Bai, S., Chun, S. B., and Kim, K. C. Intraspecific protoplast fusion of cellulolytic fungus Penicillium verruculosum. *Kor. J. Appl. Microbiol. Bioeng.*, *16*:163-167, 1988.
161. Böttcher, F., Becher, D., Klinner, U., Samsonova, I. A., and Schilowa, B. Genetic structure of yeast hybrids constructed by protoplast fusion. *Advances in Protoplast Research* (L. Ferenczy and G. L. Farkas, eds.), Akadémiai Kiadó, Budapest; Pergamon Press, Oxford, 1980, pp. 99-104.
162. Klinner, U., Böttcher, F., and Samsonova, I. A. Hybridization of Pichia guilliermondii by protoplast fusion. In *Advances in Protoplast Fusion* (L. Ferenczy and G. L. Farkas, eds.), Akadémiai Kiadó, Budapest; Pergamon Press, Oxford, 1980, pp. 113-118.
163. Becher, D., and Böttcher, F. Hybridization of Rhodosporidium toruloides by protoplast fusion. In *Advances in Protoplast Fusion* (L. Ferenczy and G. L. Farkas, eds.), Akadémiai Kiadó, Budapest; Pergamon Press, Oxford, 1980, pp. 105-111.
164. Sipiczki, M., Ducsera, J., Ulaszewski, S., and Zsolt, J. Hybridization studies by crossing and protoplast fusion with the genus Schizosaccharomyces Linder. *J. Gen. Microbiol.*, *128*:1989-2000, 1982.
165. Girgsdies, O. Sterile mutants of Schizosaccharomyces pombe: analysis by somatic hybridization. *Curr. Genet.*, *6*:223-227, 1982.
166. Shemchukhina, N. S., and Dzhavakhina, V. G. Protoplast fusion in phytopathogenic fungus Pyricularia oryzae cav. *Genetika*, *21*:716-723, 1985.
167. Zhemchuzhina, N. S., and Dzhawakhia, V. G. Protoplast fusion in phytopathogenic fungus Pyricularia oryzae cav. *Genetika*, *21*:716-723, 1985.
168. Ohnuki, T., Etch, Y., and Beppu, T. Intraspecific and interspecific hybridization of Mucor pusillus and M. miehei by protoplast fusion. *Agr. Biol. Chem.*, *46*:451-458, 1982.
169. Ferenczy, L. Protoplast fusion in yeasts. In: *Fungal Protoplasts: Applications in Biochemistry and Genetics* (J. F. Peberdy and L. Ferenczy, eds.), Marcel Dekker, New York, 1985, pp. 279-306.
170. Tamaki, H. Genetic properties of abortive products resulting from the protoplast fusion in yeasts. *Mol. Gen. Genet.*, *187*:177-179, 1982.
171. Sipiczki, M. Induced interspecific protoplast fusion in fission yeasts. *Acta Microbiol. Acad. Sci. Hung.*, *27*:247, 1980.
172. Svoboda, A. Intergeneric fusion of yeast protoplasts: Saccharomyces cerevisiae + Schizosaccharomyces pombe. In: *Advances in Protoplast Research* (L. Ferenczy and G. L. Farkas, eds.), Akadémiai Kiadó, Budapest; Pergamon Press, Ocford, 1980, pp. 119-124.

6

ISOLATION OF FUNGAL PROTOPLASTS

TERUYOSHI HASHIBA *Faculty of Agriculture, Tohoku University, Sendai, Japan*

I. INTRODUCTION

Many researchers are interested in using fungal protoplasts for physiological studies and as genetic tools. For example, protoplast generation is vital to genetic fusion, transformation, and in detection of large-plasmid DNA. In addition, protoplast fusion often plays an important role in industrial strain improvement programs. Furthermore, development of methods for genetic transformation of protoplasts by DNA uptake are essential for genetic studies to improve fungal species.

The success in the above new fields has been very dependent on protoplast technology development. However, studies on conditions of protoplast generation from fungi are still somewhat limited. Clearly, the key factors for success in protoplast isolation include temperature, pH, age, method and concentration of culture of the fungus, suitable lytic enzyme system, and an effective osmotic stabilizer system.

The aim of this chapter is to outline procedures for the isolation of fungal protoplasts in general and to describe a highly efficient procedure for the isolation of protoplasts from a strain of *Rhizoctonia solani* by using a combined enzyme system. To this end, a detailed description of the process is one of the important considerations that influence protoplast formation.

II. PROCEDURE FOR ISOLATION OF PROTOPLASTS

A. Isolation

Fungal protoplasts have been isolated from many different taxonomic groups. For detailed explanations of these isolation procedures, readers should consult other review articles [1-4]. In brief, I will describe here a protoplast isolation using *R. solani* [5,6].

The procedure for the isolation and purification of the protoplasts is shown in Fig. 1. Fresh mycelium was harvested, washed with distilled water, and suspended in 10 ml of an enzyme mixture (cellulase "Onozuka" RS, 20 mg/ml; Macerozyme R-10, 5 mg/ml; and β-glucuronidase, 0.06 ml/ml in a 100-ml Erlenmeyer flask). Flasks were placed on a reciprocal shaker at 75 strokes/min and incubated at 34°C for 3 h. Maximum release of protoplasts was attained after incubation for 3 h. The culture was filtered through a 150-μm sieve to remove mycelial fragments. This filtrate was

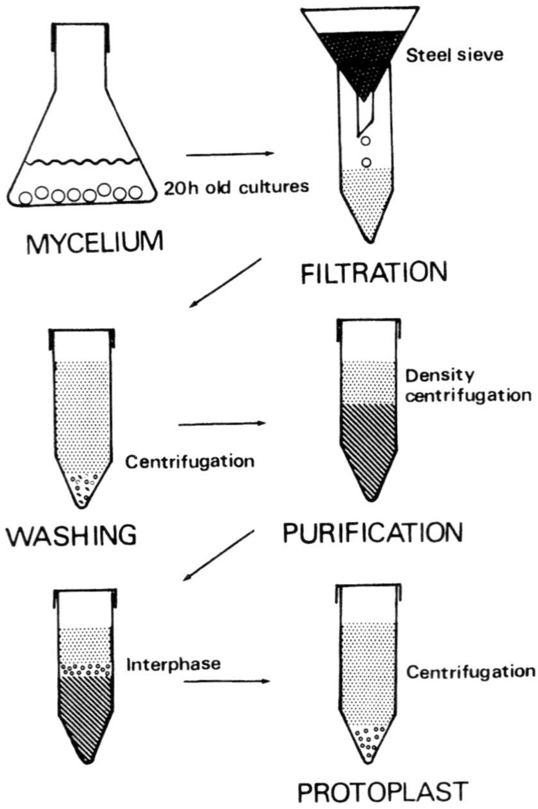

FIGURE 1. The procedure for isolation and purification of protoplasts from *R. solani*.

centrifuged twice in 0.6 M mannitol at 700 g for 5 min to remove the enzyme. The pellet consists of intact and broken protoplasts, undigested tissues, and cellular organelles (Fig. 2a), and was resuspended in 2 ml of 0.6 M mannitol.

B. Purification

The crude protoplast suspension (2 ml) was layered on 4 ml of 0.6 M sucrose in 1 mM HEPES-KOH (pH 7.0) and centrifuged in a swinging bucket rotor at 200 g for 5 min. The intact protoplasts, located at the interface of the two-phase system, were removed and resuspended in 0.6 M mannitol. An apparently homogeneous suspension of protoplasts was obtained by purifying the crude protoplast preparation with the liquid-liquid two-phase system of 0.6 M sucrose and 0.6 M mannitol (Fig. 2b).

In many fungi, most of the liberated protoplasts develop large vacuoles and float in the $MgSO_4$ stabilizer. This allows for a convenient separation of the protoplasts from the hyphal fragments by low-speed centrifugation [7-9].

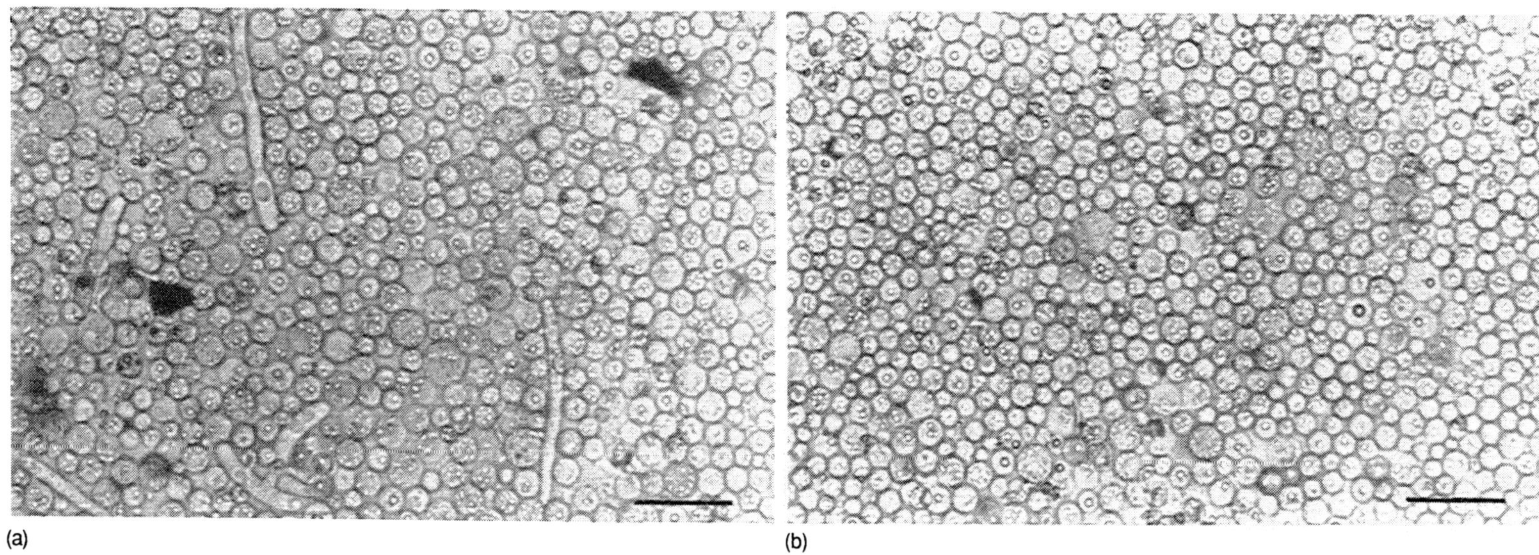

FIGURE 2. (a) Light micrograph of crude protoplasts preparation obtained from mycelium of *R. solani*. (b) Light micrograph of intact protoplasts of *R. solani* after purification with the two-phase system. Bar represents 20 μm.

C. Long-Term Preservation

The *R. solani* protoplasts remained viable in 0.6 M mannitol solution for at least 4 days at 4°C. Intact protoplasts can be also preserved in liquid nitrogen for up to 4 years or longer. Protoplasts of *R. solani* were frozen at a rate of about 1°C min followed by transfer into liquid nitrogen from a temperature of -60°C in combination with 5% polyethylene glycol in 0.6 M sucrose [10]. Long-term preservation of fungal protoplasts has been reported for *Entomophthora* sp. These protoplasts can be stored in liquid nitrogen for up to 4 months [11].

D. Nuclear Staining

A drop of protoplast suspension was placed on a glass microslide and allowed to dry quickly in a flame [5]. This procedure is most important for nuclear staining of protoplasts. The specimen was fixed in ethanol/acetic acid for 10 min, dehydrated through an ethanol series, and stained with HCl-Giemsa. Another nuclear stain for protoplasts used was 4',6-diamidino-2-phenylindole (DAPI) [12].

E. Regeneration

When a diluted suspension of purified *R. solani* protoplasts was incubated on modified potato/sucrose agar medium containing a high concentration of sucrose (0.6 M), 10-20% of the protoplasts produced visible colonies after 25°C incubation for 48-72 h. Protoplast regeneration frequency differences between strains and species are frequently found and regeneration is also affected by several physical factors including the nature of the culture medium and the osmotic stabilizer.

F. Mutagenesis and Isolation of Mutants

To develop a transformation system, the first step is to obtain a selective marker, such as an auxotrophic or drug resistance marker that can be used to screen fused protoplast populations. Several difficulties are encountered in attempts to isolate auxotrophic mutants in fungi which sporulate poorly or produce multinucleate cells.

Rochfrette et al. [13] and Keller [14] described procedures for mutant protoplast isolation in *Ophiobolus graminis* and *Claviceps purpurea*, respectively. Described here is a procedure for mutant isolation in *R. solani* using protoplast technology [15]. Protoplasts of the parental strain (5×10^6 protoplasts/ml) in 0.05 M phosphate buffer (pH 7.2) supplemented with 0.6 M mannitol were treated with *N*-methyl-*N'*-nitro-*N*-nitrosoguanidine (NTG) for 45-60 min at 25°C to give a 3-20% survival rate. The concentration of NTG used was 20 μg/ml. The treated protoplasts were centrifuged and the pellet washed with mannitol buffer solution before being suspended in fresh mannitol buffer. The protoplasts were plated on complete medium (CM) supplemented with 0.6 M mannitol. To isolate mutants, YS agar medium was used as CM [15].

Colonies that appeared after incubation for 5 days at 28°C were transferred to fresh CM plates. The resulting colonies were checked for growth on minimum medium (MM). Specific nutritional requirements of auxotrophs were determined by the methods of Holliday [16].

The reversion frequency of these mutants was 10^{-3}, an acceptable frequency. Therefore, the proposed method may be broadly applicable for genetic analysis and manipulation of other multinucleate, filamentous, or non-spore-producing fungi.

III. IMPORTANT FACTORS INFLUENCING FUNGAL PROTOPLAST ISOLATION

There are a number of important factors influencing fungal protoplast isolation, i.e., organisms, lytic enzyme systems, osmotic stabilizer, and other factors influencing protoplast isolation. The optimum conditions for protoplast isolation from a variety of species are shown in Table 1.

A. Culture Conditions of the Organism

The culture conditions used to produce mycelium for protoplast isolation is specific for each organism. Often papers have detailed isolation procedures from mycelium produced under "normal conditions." Such methods fail the test of reproducibility of protoplast yield.

A critical factor determining the number of protoplasts is the age of the culture. In some species of fungi, it is necessary to use young growing mycelia (10- to 20-h cultures) [12,17-32]. The actively growing mycelium of all fungi might have a similar cell wall construction in general. Protoplast yields from other fungi were highest when 24- to 48-h cultures were used [5,13,33-44]. However, high levels of protoplasts were also obtained from 3- to 10-day-old cultures [23,45-59]. Several workers have used mycelium derived from slowly shaken cultures for protoplast isolation.

Also, the nature of the culture medium used is an important factor for protoplast isolation. Optimum conditions of the culture medium are shown in Table 1.

Finally, several workers observed that preliminary treatment with sulfhydryl compounds leads to large-scale protoplast isolation with yeasts. However, protoplast isolation from pretreatment fungi with sulfhydryls is not widely reported [20,24,29,35,41,48,49,54,55,60,61].

B. Lytic Enzyme System

In most cases the enzymes used for fungal protoplast isolation either are commercially available or are obtained as by-products from other microorganisms. Prior to 1980, mycolytic enzymes were used for the isolation of protoplasts (see Table 1). Recently, the enzymes used for fungal protoplast production have been made of several commercial products. The commercial enzymes effective in releasing protoplasts from many filamentous fungi are snail (*H. pomatia*) digestive juices, β-glucuronidase (Sigma Chemical Co., USA), Helicase (L' Industrie Biologique Grennevilliers, France), chitinase from *Streptomyces griseus* (Sigma Chemical Co., USA), chitinase T-1 from *Trichoderma* sp. (Asahi Industries Co., Ltd., Japan), cellulase "Onozuka" RS from *Trichoderma viride* (Yakult Biochemical Co., Ltd., Japan), Novozyme 234 from *Trichoderma harzianum* (Novo Enzyme Products Ltd., UK), and Zymolyase 100T from *Arthrobacter luteus* (Kirin Brewery Co., Japan). All enzymes reported to be effective in releasing protoplasts from many fungi are shown in Table 1.

TABLE 1. Production and Isolation of Protoplasts from Different Fungal Taxonomic Groups

Species	Culture conditions of organisms[a]	Lytic enzyme systems	Osmotic stabilizers	Other factors influencing protoplast isolation[b]	Year	Ref.
Neurospora crassa	—	Hemicellulase or snail hepatic juice	2% sucrose 5-10% rhamnose or sorbose	—	1958	63
Neurospora crassa	28°C, 16-18 h	*Helix pomatia*	20% sucrose	0.03 M phosphate buffer, pH 6.0, 28°C, 5-6 h	1959	17
Penicillium italicum	Czapek medium, 28°C, 18-28 h, shaking	Snail enzyme	0.8 M mannitol	0.005 M Mcllvaine's sodium citrate-phosphate buffer pH 5.8, 28°C, 2 h	1962	18
Aspergillus nidulans	Ibid.	Ibid.	Ibid.	Ibid.	1962	18
Alternaria citri	Ibid.	Ibid.	Ibid.	Ibid.	1962	18
Fusarium culmorum	Ibid.	Ibid.	Ibid.	Ibid.	1962	18
Helminthosporium gramineum	Ibid.	Ibid.	Ibid.	Ibid.	1962	18
Trichoderma roseum	28°C, 4 days	*Streptomyces* lytic enzyme or gut juice of *Helix pomatia*	0.8 M mannitol	0.1 M phosphate buffer, pH 6.8, 25°C, 16 h, shaking	1963	45
Fusarium culmorum	Glucose-asparagine-yeast extract medium, 28°C	*Streptomyces* lytic enzyme	0.8 M mannitol	0.1 M phosphate buffer, pH 6.8, 28°C, shaking	1964	75

Geotrichum lactis	Czapek or GAE-medium	*Micromonospora* AS lytic enzyme	0.8-1.0 M $MgSO_4$	pH 5.8-8.8, 37°C, 48-72 h, shaking	1965	76
Candida utilis	25°C, shaking	Ibid.	Ibid.	Ibid.		
Fusarium culmorum	—	Strepzyme	0.8 M mannitol or NH_4Cl	0.1 M phosphate + citrate buffer pH 6.5	1966	77
Phytophytora infestans	V8 solution, 7-10 d mycelial suspension 1-2 d, 27°C	*Streptomyces* enzyme	0.4-0.6 M sucrose or mannitol	0.05 M citrate buffer, pH 5.8, 21-26°C	1966	46
Phythium spp.	YEG + glucose, 25°C 24 h	*Streptomyces* enzyme + cellulase or lypase	0.8 M $MgSO_4$	0.005 M phosphate buffer, pH 6.2, 28°C, 24 h	1967	8
Aspergillus niger	Czapek-Dox medium, 28°C, 24 h, shaking	Snail gut juice	0.7 M NaCl	31°C, 3-4 h, shaking	1968	33
Rhizopus nigricans	30°C, 6-8 h	*Helix pomatia*	0.6 M mannitol	Citrate phosphate buffer pH 6.8, 30°C, 5-6 h	1968	19
Geotrichum candidum	Sabourand's liquid medium, 48-h subculture, 30°C, 3 h	Snail gut juice	0.75 M NaCl	0.005 M Mcllvaine's sodium citrate-phosphate buffer pH 6.0, 30°C, 7-10 h	1969	64
Mortierella vinacea	Liquid medium, 25°C, 24 h, shaking	*Streptomyces* sp. no. 6 lytic enzyme	0.8 M NH_4Cl	0.2 M phosphate buffer, pH 6.0, 28°C, 4-5 h	1971	34
Aspergillus nidulans	SM medium, 29°C, 7 d, aerated	*Streptomyces* lytic enzyme	0.8 M NH_4Cl	pH 6.0, 28°C, 2 h, shaking	1971	74

TABLE 1 (Continued)

Species	Culture conditions of organisms[a]	Lytic enzyme systems	Osmotic stabilizers	Other factors influencing protoplast isolation[b]	Year	Ref.
Geotrichum lactis	30°C, shaking pretreatment DTT or β-mercaptoethanol	Snail gut juice	0.8 M mannitol	pH 7.0, 30°C	1972	60
Basidiomycetes (23 species)	Yeast-extracted liquid medium, 25°C	*Trichoderma* lytic enzyme or Helicase	0.5 M $MgSO_4$	0.05 M Na-maleate buffer, pH 5.8, 24°C, 5 h	1973	65
Ascomycetes (11 species)	Ibid.	Ibid.	Ibid.	Ibid.	1973	65
Penicillium chrysogenum	27°C, 48 h, shaking	Cytophaga (*Helix pomatia*)	0.8-0.9 M NaCl or NH_4Cl	Mcllvaine's buffer, pH 6.8, 30°C, 1-3 h, shaking	1973	35
Cephalosporium acremonium	Pretreatment DTT or 2-mercaptoethanol	*Helix pomatia*	Ibid.	Ibid.	1973	35
Pythium spp.	24 h, Pretreatment Triton x100	Helicase + *A. niger* cellulase	0.4 M NaCl	4 h	1973	20
Penicillium chrysogenum	PM medium, 28°C, 15 h, shaking	*Helix pomatia* gut juice or cellulase + culture filtrate of *Streptomyces graminofaciens*	0.55 M NaCl	0.005 M phosphate buffer, pH 5.6, 28°C, 4-5 h, shaking	1974	21

Phycomyces blakesleeanus	22°C, 8 h, shaking	Crude chitinase of *Streptomyses griseus*	0.35 M sorbitol	0.01 M phosphate buffer, pH 7.0, 25°C, 1 h	1974	22
Geotrichum candidum	Pretreatment 2-mercaptoethanol	Snail gut juice	0.8 M mannitol or glucose	pH 6.5, 2 h	1974	61
Trichoderma viride	GAE medium 28°C, 16-20 h, shaking	*Micromonospora chalcea* lytic enzyme	0.7 M $MgSO_4$ or $(NH_4)_2SO_4$	20°C, 2 h	1975	23
Coprinus cinereus	Maltose-CM medium 37°C, 24 h	Helicase or chitinase	1 M sucrose	SNC medium 37°C, 24 h	1975	36
Aspergillus nidulans	Vogel's medium N salts, 30°C, 21 h, shaking	*Streptomyces* lytic enzyme	0.4-0.6 M NH_4Cl or KCl	0.2 M phosphate buffer, pH 6.4	1976	78
Aspergillus flavus	Ibid.	Ibid.	0.4-0.6 M NH_4Cl, KCl or $(NH_4)_2SO_4$	30°C, 3 h	1976	78
Penicillium chrysogenum	Ibid.	Ibid.	0.4-0.6 M KCl or NaCl	Ibid.	1976	78
Fusarium culmorum	Ibid.	Ibid.	0.4-0.6 M NH_4Cl	Ibid.	1976	78
Aspergillus nidulans	Glucose-salts medium	*T. harzianum* lytic enzyme	0.6 M KCl, $MgSO_4$ or NH_4Cl	—	1976	79
Ustilago maydis	30°C, 18 h, shaking Pretreatment 2-mercaptoethanol	Helicase + cellulase "Onozuka" R-10	0.6 M $(NH_4)_2SO_4$	McIlvaine buffer pH 6.0, 30°C, 3 h	1976	24
Aspergillus flavus	Mineral salts medium N + glucose, 30°C, 48 h, shaking	*Streptomyces* S4 lytic enzyme	0.8 M NH_4Cl	0.2 M phosphate buffer, 30°C, 2-3 h, shaking	1976	37

TABLE 1 (Continued)

Species	Culture conditions of organisms[a]	Lytic enzyme systems	Osmotic stabilizers	Other factors influencing protoplast isolation[b]	Year	Ref.
Piricularia oryzae	Maleate-glucose medium	Helicase	21% sucrose	pH 6.0, 26°C, 24 h	1976	66
Aspergillus niger	Potato glucose agar 27°C	*Trichoderma viride* lytic enzyme + chitinase	0.4 M $MgCl_2$	TES or HEPES buffer, pH 7.0 40°C, 4-5 h	1977	25
Aspergillus nidulans	Czapek-Dox broth, 30°C, 12-14 h	*Trichoderma* lytic enzyme + snail gut lytic enzyme	0.5 M $MgSO_4$	0.05 M sodium malate buffer, pH 5.8, 1-2 h, shaking	1978	26
Helminthosporium victoriae	YG medium, 2-3 d	β-Glucuronidase type H-2	0.7 M KCl	25°C, 3 h, shaking	1979	38
Ophiobolus graminis	23°C, 2-6 d	Helicase	0.6 M KCl	Phosphate buffer pH 6.0, 23°C, 2 h	1979	13
Claviceps spp.	NL-406 culture medium 5 d, shaking	β-Glucuronidase	0.7 M KCl	30°C, 4 h	1979	47
Claviceps purpurea	—	β-Glucuronidase	0.8 M sucrose	pH 4.7	1980	67
Aspergillus nidulans	37°C, 3 d, pretreatment 2-deoxy-D-glucose	*Oerskovia* lytic enzyme	0.4 M $(NH_4)_2SO_4$	30°C, 3-4 h	1981	48
Aspergillus niger	Ibid.	Ibid.	Ibid.	Ibid.	1981	48

Aspergillus nidulans	Yeast extract and casamino acids, 28°C 19 h	Novozyme 234 + cellulase CP	0.6 M KCl	0.2 M phosphate buffer, pH 5.8 28°C, 3 h, shaking	1981	72
Aspergillus niger	Ibid.	Novozyme 234 + cellulase CP + cereflo	0.6 M KCl	Ibid.	1981	72
Aspergillus rugulosus	Ibid.	Novozyme 234 + cellulase CP	0.6 M KCl	Ibid.	1981	72
Aspergillus ochraceus	Ibid.	Novozyme 234 + cellulase CP + cereflo	0.6 M KCl	Ibid.	1981	72
Acremonium chrysogenum	Pretreatment DTT	Lytic enzyme L1	0.7 M NaCl	0.2 M citrate-phosphate buffer	1981	72
Penicillium chrysogenum	Yeast extract and casamino acids, 28°C	Novozyme 234	0.7 M NaCl	0.2 M phosphate buffer, pH 5.8	1981	72
Aspergillus parasiticus	Glucose-peptone-agar medium, 5-6 d Pretreatment β-mercaptoethanol	β-Glucuronidase	1.0 M NaCl	0.02 M potassium phosphate buffer pH 6.2	1981	49
Piricularia oryzae	Misato medium pre-culture, 28°C, 2 w, Vogel N-sucrose medium, 28°C, 24 h, shaking	*Bacillus circulans* WL12 lytic enzyme	0.6 M KCl + 2 mM $MgCl_2$	0.02 M phosphate buffer, pH 7.0, 16 h	1981	39
Mucor pusillus	YPG medium, 30°C, 13 h, shaking	Streptomyces No. 6 chitosanase + chitinase + sulfatase	0.5 M sorbitol or mannitol	0.01 M sodium phosphate buffer pH 6.0, 25°C, 2-3 h	1982	27

TABLE 1 (Continued)

Species	Culture conditions of organisms[a]	Lytic enzyme systems	Osmotic stabilizers	Other factors influencing protoplast isolation[b]	Year	Ref.
Mucor miehei	Ibid.	Ibid.	Ibid.	Ibid.	1982	27
Tricholoma matsutake	Glucose, yeast extract, and Bacto-Soytone, 23°C, 15-25 d	Cellulase R-10 + Zymolyase 5000 + β-glucuronidase	0.6 M $MgSO_4$	0.05 M malic acid -NaOH buffer, pH 5.6, 25°C, 3 h	1982	50
Rhizoctonia solani	Potato-sucrose liquid medium, 25°C, 2 d	Cellulase "Onozuka" R-10 + Macerozyme R-10 + β-glucuronidase	0.6 M mannitol	pH 5.2, 34°C, 3 h, shaking	1982	5
Aspergillus nidulans	0.1 M phosphate buffer 18 h	Chitinase + β-glucuronidase	0.6 M KCl	pH 5.8, 48 h	1982	28
Giberella fujikuroi	Czapek's broth, 27°C, 18-24 h, shaking Pretreatment EDTA	Chitinase + β-glucuronidase	0.8 M mannitol	Sodium citrate-phosphate buffer pH 6.1	1982	29
Aspergillus awamori	Glucose salts medium 24°C, 12 h	β-glucuronidase type H-2 + *Trichoderma viride* lytic enzyme	0.5 M $MgSO_4$	0.05 M malic acid pH 5.8, 30°C, 12-18 h, shaking	1982	30
Cochliobolus heterostrophus	—	β-glucuronidase	0.7 M KCl + 0.8 mannitol	24-48 h, shaking	1982	68
Gibberella zeae	NSC broth, 28°C 18-36 h	Chitinase + β-glucuronidase	0.6 M KCl	18-24 h, shaking PPM medium	1983	69

Hypomyces ochraceus	3 d	*Helix pomatia*	20% sucrose	—	1983	51
Coprinus cinereus	CY-1 medium, 28°C, 7-10 d	Cellulase "Onozuka" R-10	0.5 M $MgSO_4$	0.05 M Na-maleate buffer, pH 5.8, 28°C, 24 h	1983	52
Verticillium albo-atrum	Czapek-Dox medium 24°C, 3-4 d	Helicase + mycolase	15% sucrose	24°C	1983	70
Verticillium dahliae	Ibid.	Ibid.	Ibid.	Ibid.	1983	70
Trichoderma reesei	Potato-dextrose agar 30°C, 3-4 d	*T. viride* BA lytic enzyme	0.6 M sucrose	0.1 M acetate buffer, pH 5.0, 45°C, 6 h	1984	53
Hypoderma roseum	PL-2 medium, 28°C, 15-24 h	*Trichoderma* lytic enzyme	0.8 M $MgSO_4$	pH 4.2, 3 h	1984	40
Cunninghamella elegans	PL-2 medium, 28°C, 6-7 h	Ibid.	Ibid.	Citrate-phosphate buffer, pH 4.2, 3 h		
Fusarium oxysporum	Bacto-peptone + yeast extract + D-glucose 24°C, 7 d pretreatment 2-mercaptoethanol	Novozyme 234 + cellulase CP	0.7 M NaCl + 0.3 M sucrose	0.1 M citrate-phosphate buffer pH 5.8, 24°C, 16 h	1984	54
Lyophyllum shimeji	LM-medium 9 d	Cellulase "Onozuka" R-10 + Zymolyase 5000 + β-glucuronidase	0.6 M $MgSO_4$	0.05 M malic acid -NaOH buffer, pH 5.6, 24°C, 3 h	1984	55
Dermatophytes (6 species)	Sabouraud's medium 30°C, 48 h, pretreatment β-mercaptoethanol	Novozyme 234	0.5 M $MgSO_4$	0.01 M citrate-phosphate buffer pH 5.2, 30°C, 9-10 h	1984	41

TABLE 1 (Continued)

Species	Culture conditions of organisms[a]	Lytic enzyme systems	Osmotic stabilizers	Other factors influencing protoplast isolation[b]	Year	Ref.
Gaumannomyces graminis	Liquid medium 24°C, 7 d, shaking	Novozyme 234 + cellulase CP	0.6 M KCl	0.2 M sodium phosphate, pH 5.8, 24°C, 4 h	1984	56
Coprinus pellucidus	Preculture-YGP medium 28°C, 4 d → homogenate 18 h, shaking	Chitinase + cellulase "Onozuka" R-10 + Zymolyase 5000 + β-glucuronidase	0.5 M $MgSO_4$	0.05 M malic-acid -NaOH buffer, pH 5.8, 30°C, 3 h, shaking	1985	31
Coprinus cinereus	Ibid.	Ibid.	Ibid.	Ibid.	1985	31
Coprinus macrorhizus	MYG medium, 30°C, 24 h	Chitinase + cellulase "Onozuka" R-10 + Zymolyase 60000	0.5 M mannitol	0.05 M maleate buffer, pH 5.5, 30°C, shaking	1985	42
Verticillium albo-atrum	Czapek-Dox broth 1 d, shaking 25°C, 3d	Streptozyme lytic enzyme	0.7 M $MgSO_4$	23°C, 20 h, shaking	1985	80
Fomes arrnosus	Ibid.	Novozyme 234	0.5 M $MgSO_4$	pH 4.5, 25°C, 2 d, shaking	1985	57
Fusarium trictum	Liquid medium 25°C, 40 h, shaking	Novozyme 234	1.4 M $MgSO_4$	0.2 M phosphate buffer, pH 5.8, 30°C, 5 h, shaking	1985	12
Fusarium oxysporum	Ibid.	Ibid.	1.4 M KCl	Ibid.	1985	12

Cochliobolus heterostrophus	Liquid complete medium, 30°C, 15-18 h, shaking	β-Glucuronidase H2 + Driserase + chitinase	0.7 M NaCl	30°C, 3-4 h, shaking	1985	32
Pleurotus ostreatus	Liquid GMY medium 2-3 d	Cellulase "Onozuka" RS + Driserase + β-glucuronidase	0.7 M mannitol	30°C, 2-3 h, shaking	1986	43
Pleurotus salmoneo-stramineus	Ibid.	Ibid.	Ibid.	Ibid.	1986	43
Pseudo-cercosporella herpetrichoides	MYG liquid medium 19°C, 4-6 d	Rozyme HP150 + Driserase + cellulase CP	0.6 M KCl	0.05 M sodium maleate buffer pH 5.8, 28°C, 3 h, shaking	1987	58
Lentinus edodes	MYG liquid medium 25°C, 3-4 d	Cellulase "Onozuka" RS + chitinase	0.6 M mannitol	0.05 M succinate buffer, pH 4.6	1987	59
Alternaria alternata	CM liquid medium 25°C, 2 h	Cellulase "Onozuka" RS + Zymolyase + Mecerozyme R-10 + Driserase	0.6 M mannitol	0.1 M citrate pH 5.2, 33°C, 4 h, shaking	1987	73
Mycorrhizal fungi	Liquid complete medium, 24°C, 48 h	Cellulase "Onozuka" R-10 + Driserase	0.7 M $MgSO_4$	32°C, 4-6 h shaking	1988	44
Trichoderma harzianum	Potato dextrose agar	β-Glucuronidase type H-2 + Driserase + chitinase	0.6 M sucrose	25°C, 16 h, shaking	1988	71

[a]Culture conditions of organisms are shown in culture medium, temperature and age, pretreatment of organisms, and other conditions.

[b]Other factors influencing protoplast isolation are shown in buffer, pH and temperature, period of lytic digestion, and other conditions.

Eddy and Williamson [62] were the first to introduce the use of *Helix pomatia* digestive juice as a lytic enzyme. The snail digestive juice preparation was shown to be effective for many fungi [5,13,17,18,20,21,24,26-33, 35,36,38,43-45,49-51,55,60,61,63-71], although conditions are relatively specific for each organism. Furthermore, the addition of chitinase to β-glucuronidase (*H. pomatia* digestive juice) was also important for protoplast isolation (28,31,32,69,71). We also obtained a high yield of protoplasts from *R. solani* by using a combined enzyme system consisting of β-glucuronidase and cellulase "Onozuka" RS enzyme [5]. It is not known if chitinase enzyme is present in the cellulase "Onozuka" RS used in our study but the preparation is known to contain β-1,3-glucanase. The effect of combinations of enzymes in enhancing protoplast release was presumed to result from synergistic action of the major components. Recently, Novozyme 234 enzyme which was derived from *Trichoderma harzianum* was used to liberate protoplasts from several fungi [12,41,54,56,57,72]. While most commercial enzyme preparations consist of a mixture of several components, recent workers generally used enzymes in combinations of two or three to achieve optimum effect [5,8,20-22,24,26-32,41-44,50,54-56,58,59,69-73]. Enzymes effective in releasing protopalsts from many fungi were shown in Table 1. The optimum enzyme system must be elucidated for each fungus studied.

C. Osmotic Stabilizers

From a wide range of osmotic stabilizers, KCl, $MgSO_4$, NaCl, sucrose, and mannitol are used most frequently (Table 1). Only seven reports used other stabilizers, such as NH_4Cl, $(NH_4)_2SO_4$, $MgCl_2$, and sorbitol [22,24, 25,34,37,48,77]. The concentration of the stabilizer can also affect protoplast isolation. In most cases, 0.4-0.8 M concentrations gave the best isolation. Little evidence of mycelial digestion was obtained at molar values below 0.4 or above 0.8 [12,17,22,36,49,51,63,66,70]. Therefore, it is necessary to determine the best concentrations for each isolate (Table 1).

D. Other Factors Influencing Protoplast Isolation

Temperature and pH effects are related to the activities of the lytic enzyme systems and osmotic stabilizers. Protoplast isolation from many fungi have been achieved in the pH range 4.0-8.0. The choice of buffer in the incubation medium can also affect the yield of protoplasts. Therefore, optimum temperature, pH, buffer in incubation and period lytic digestion need to be determined for each species and strain. The optimum conditions on other fungi are seen in Table 1.

IV. CONCLUSIONS

I have described many factors that affect protoplast isolation although there is no universal technique to achieve such isolation. Efficient methods for isolation of protoplasts are vital for cell fusion and transformation. Detection of large-plasmid DNA is now reported. If broad-spectrum enzymes are developed, isolation of the protoplasts could serve a useful purpose in laboratories and industries. I hope that in the future the isolation techniques for protoplasts will allow rapid progress in fungal genetics.

Protoplast fusion now plays an important role in strain improvement of industrially important organisms. For detailed procedures for protoplast fusion and characteristics of fused protoplasts, the reader may consult other review articles [15,81-84]. These articles have described both intra- and interspecific hybridization. The methods used for transformation of protoplasts by the uptake of DNA is described in other chapters in this book.

REFERENCES

1. Villanueva, J. R., and Garcia, A. I. Production and use of fungal protoplasts. In *Methods in Microbiology*, Vol. 4 (C. Booth, ed.), Academic Press, New York, 1971, pp. 665-718.
2. Peberdy, J. F. Isolation and properties of protoplasts from filamentous fungi. In *Microbial and Plant Protoplasts* (J. F. Peberdy, A. H. Rose, H. J. Rogers, and E. C. Cocking, eds.), Academic Press, London, 1976, pp. 39-50.
3. Peberdy, J. F. Fungal protoplasts: isolation, reversion, and fusion. *Ann. Rev. Microbiol.*, *33*:21-39, 1979.
4. Davis, B. Factors influencing protoplast isolation. In *Fungal Protoplasts: Applications in Biochemistry and Genetics* (J. F. Peberdy and L. Ferenczy, eds.), Marcel Dekker, New York, 1985, pp. 45-72.
5. Hashiba, T., and Yamada, M. Formation and purification of protoplasts from Rhizoctonia solani. *Phytopathology*, *72*:849-853, 1982.
6. Hashiba, T. An improved system for biological control of damping-off by using plasmids in fungi. In *Innovative Approaches to Plant Disease Control* (I. Chet, ed.), John Wiley and Sons, New York, 1987, pp. 337-351.
7. Gascon, S., and Villanueva, J. R. Magnesium sulphate as stabilizer during liberation of yeast and mould protoplasts. *Nature*, *205*:822-823, 1965.
8. Sietsuma, J. H., Eveleigh, D. E., Haskins, R. H., and Spencer, J. F. T. "Protoplast" from Pythium sp. PRL 2142. *Can. J. Microbiol.*, *13*:1701-1704, 1967.
9. Sietsuma, J. H., and Wouters, J. T. M. Cell wall composition and protoplast formation in Geotrichum candidum. *Arch. Microbiol.*, *79*:263-273, 1971.
10. Hashiba, T., Tetsuka, Y., and Suzui, T. Technique for long-term preservation of protoplasts of fungi in liquid nitrogen. *Jap. J. Freez. Dry.*, *31*:63-67, 1985.
11. Tyrrell, D., Sohi, S. S., and Welton, M. A. Preservation of Entomophthora protoplasts in liquid nitrogen. *Can. J. Microbiol.*, *18*:1967-1968, 1972.
12. Lynch, P. T., Collin, H. A., and Issac, S. Isolation and regeneration of protoplasts from Fusarium tricinctum and F. oxysporum. *Trans. Br. Mycol. Soc.*, *85*:135-140, 1985.
13. Rochfrette, C., Tivoli, B., and Boissonet-Menes, M. Obtention de mutants resistents a des fongicides et de mutants auxotrophes obtenus a partir de protoplastes chez Ophiobolus graminis Sacc. Etude de leur agressivite et de leur aptitude a premunir le Ble. *Ann. Phytopathol.*, *11*:43-51, 1979.

14. Keller, U. Highly efficient mutagenesis of Claviceps purpurea by using protoplasts. *Appl. Environ. Microbiol.*, *46*:580-584, 1983.
15. Hashiba, T., and Yamada, M. Intraspecific protoplast fusion between auxotrophic mutants of Rhizoctonia solani. *Phytopathology*, *74*:398-401, 1984.
16. Holliday, R. A new method for the identification of biochemical mutants of micro-organisms. *Nature*, *178*:987, 1956.
17. Bachman, B. J., and Bonner, D. M. Protoplasts from Neurospora crassa. *J. Bacteriol.*, *78*:550-556, 1959.
18. Rodriguez Aguirre, M. J., and Villanueva, J. R. Production of protoplast-like structures from various species of fungi. *Nature*, *196*: 693-694, 1962.
19. Gabriel, M. Formation and regeneration of protoplasts in the mold Rhizopus nigricans. *Folia Microbiol.*, *13*:231-234, 1968.
20. Sietsuma, J. H., and DeBoer, W. R. Formation and regeneration of protoplasts of Phythium PRL 2142. In *Yeast, Mould and Plant Protoplasts* (J. R. Villanueva, I. Garcia Acha, S. Gascón, and F. Uruburu, eds.), Academic Press, London, 1973, pp. 275-283.
21. Anné, J., Eyssen, H., and De Somer, P. Formation and regeneration of Penicillium chrysogenum protoplast. *Arch. Microbiol.*, *98*:159-166, 1974.
22. Binding, H., and Weber, H. J. The isolation, regeneration and fusion of Phycomyces protoplast. *Mol. Gen. Genet.*, *135*:273-276, 1974.
23. Benitez, T., Ramos, S., and Acha, I. G. Protoplasts of Trichoderma viride. *Arch. Microbiol.*, *103*:199-203, 1975.
24. Waad, M. A. de. Formation of protoplasts from Ustilago maydis. *Antonie van Leeuwenhoek J. Microbiol. Serol.*, *42*:211-216, 1976.
25. Davis, B., D'Avillez Paixao, M. T., Deans, S. G., and Smith, J. E. Protoplast formation from giant cells of Aspergillus niger. *Trans. Br. Mycol. Soc.*, *69*:207-212, 1977.
26. Moris, N. R. Preparation of large molecular weight DNA from the fungus Aspergillus nidulans. *J. Gen. Microbiol.*, *106*:387-389, 1978.
27. Ohnuki, T., Etoh, Y., and Beppu, T. Intraspecific and interspecific hybridization of Mucor pusillus and M. Miehei by protoplast fusion. *Agr. Biol. Chem.*, *46*:451-458, 1982.
28. Issac, S., and Gokhale, A. V. Autolysis: a tool for protoplast production from Aspergillus nidulans. *Trans. Br. Mycol. Soc.*, *78*:389-394, 1982.
29. Harris, G. M. Protoplasts from Gibberella fujikuroi. *Phytopathology*, *72*:1403-1407, 1982.
30. Bobbitt, T. F., and Douglas, C. M. Microcycle conidation by regenerating protoplast of Aspergillus awamori. *Exp. Mycol.*, *6*:307-311, 1982.
31. Morinaga, T., Kikuchi, M., and Nomi, R. Formation and regeneration of protoplasts in Coprinus pellucidus and Coprinus cinereus. *Agr. Biol. Chem.*, *49*:523-524, 1985.
32. Turgeon, B. G., Garber, R. C., and Yoder, O. C. Transformation of the fungal maize pathogen Cochliobolus heterostrophus, using the Aspergillus nidulans amdS gene. *Mol. Gen. Genet.*, *201*:450-453, 1985.
33. Musilková, M., and Fenel, Z. Some factors affecting the formation of protoplasts in Aspergillus niger. *Folia Microbiol.*, *13*:235-239, 1968.
34. Peberdy, J. F. Protoplasts from Mortierella vinacea. *Trans. Br. Mycol. Soc.*, *56*:76-72, 1971.

35. Fawcett, P. A., Loder, P. B., Duncan, M. J., Beesley, T. J., and Abraham, E. P. Formation and properties of protoplasts from antibiotic-producing strains of Penicillium chrysogenum and Cephalosporium acremonium. *J. Gen. Microbiol.*, *79*:293-309, 1973.
36. Moore, D. Production of Coprinus protoplasts by use of chitinase or helicase. *Trans. Br. Mycol. Soc.*, *65*:134-136, 1975.
37. Moore, M., and Peberdy, J. F. Release and regeneration of protoplasts from the conidia of Aspergillus flavus. *Trans. Br. Mycol. Soc.*, *66*:421-425, 1976.
38. Ghabrial, S. A., Sanderlin, R. S., and Calvert, L. A. Morphology and viruslike particle content of Helminthosporium victoriae colonies regenerated from protoplasts of normal and diseased isolates. *Phytopathology*, *69*:312-315, 1979.
39. Tanaka, H., Ogasawara, N., and Kamimiya, S. Protoplast of Pyricularia oryzae p_2:preparation and regeneration into hyphal form. *Agr. Biol. Chem.*, *45*:1541-1552, 1981.
40. Dlugônski, J., Sedlaczek, L., and Jaworski, A. Protoplast release from fungi capable of steroid transformation. *Can. J. Microbiol.*, *30*: 57-62, 1984.
41. Srikantha, T., and Rao, G. R. A method for isolation of protoplasts from dermatophytes. *J. Gen. Microbiol.*, *130*:1503-1506, 1984.
42. Yanagi, S. O., Monma, M., Kawasumi, T., Hino, A., Kito, M., and Takebe, I. Conditions for isolation of and colony formation by mycelial protoplasts of Coprinus macrohizus. *Agr. Biol. Chem.*, *49*:171-179, 1985.
43. Toyomasu, T., Matsumoto, T., and Mori, K. Interspecific protoplast fusion between Pleurotus ostreatus and Pleurotus salmoneo-stramineus. *Agr. Biol. Chem.*, *50*:223-225, 1986.
44. Hêbraud, M., and Févre, M. Protoplast production and regeneration from mycorrhizal fungi and their use for isolation of mutants. *Can. J. Microbiol.*, *34*:157-161, 1987.
45. Garcia, A. I., and Villanueva, J. R. Difference in the mode of action of strepzyme and Helix pomatia enzyme preparations on Trichothecium roseum spores. *Nature*, *200*:1231, 1963.
46. Bartinicki-Garcia, S., and Lippman, E. Liberation of protoplasts from the mycelium of Phytophthora infestans. *J. Gen. Microbiol.*, *42*:411-416, 1966.
47. Robbers, J. E., Cheung, L-J., Anderson, J. A., and Floss, H. G. Synthesis of ergot alkaloids by protoplasts of *Claviceps* species. *J. Nature Products*, *42*:537-539, 1979.
48. Bos, C. J., and Slakhorst, S. M. Isolation of protoplasts from Aspergillus nidulans conidiospores. *Can. J. Microbiol.*, *27*:400-407, 1981.
49. Tyagi, J. S., Tyagi, A. K., and Venkitasubramanian, T. A. Preparation and properties of spheroplasts from Aspergillus parasitica with special reference to the de novo synthesis of aflatoxins. *J. Appl. Bacteriol.*, *50*:481-491, 1981.
50. Abe, M., Umetsu, H., Nakai, T., and Sasage, D. Regeneration and fusion of mycelial protoplasts of Tricholoma matsutake. *Agr. Biol. Chem.*, *46*:1955-1957, 1982.
51. Riemay, K. H., Ellrich, S., and Tröger, R. Protoplastenfreisetzung und regeneration des Ascomyceten Hypomyces ochraceus (Pers.) Tul. *Z. Allg. Mikrobiol.*, *23*:247-257, 1983.

52. Akamatsu, K., Kamada, T., and Takemaru, T. Release and regeneration of protoplasts from the oidia of Coprinus cinereus. *Trans. Mycol. Soc. Jpn.*, *24*:173-184, 1983.
53. Toyama, H., Yamaguchi, K., Shinmyo, A., and Okada, H. Protoplast fusion of Trichoderma reesei using immature conidia. *Appl. Environ. Microbiol.*, *47*:363-368, 1984.
54. Marriott, A. C., Archer, S. A., and Buck, K. W. Mitochondrial DNA in Fusarium oxysporum is a 46.5 kilobase pair circular molecule. *J. Gen. Microbiol.*, *130*:3001-3008, 1984.
55. Abe, M., Nakai, T., Umetsu, H., and Sasage, D. Regeneration of mycelial protoplasts from Lyophyllum shimeji. *Agr. Biol. Chem.*, *48*: 1635-1636, 1984.
56. Stanway, C. A., and Buck, K. W. Infection of protoplasts of the wheat take-all fungus Gaeumannomyces graminis var. tritici, with double-stranded RNA viruses. *J. Gen. Virol.*, *65*:2061-2065, 1984.
57. Trojanovoski, J., Huettermann, A., Haider, K., and Wessels, J. G. H. Degradation of lignin and lignin related compounds by protoplasts isolated from Fomes annosus. *Arch. Microbiol.*, *140*:326-330, 1985.
58. Hocart, M. J., Lucas, J. A., and Peberdy, J. F. Production and regeneration of protoplasts from Pseudocercosporella herpotrichoides (Fron) Deighton. *J. Phytopathol.*, *119*:193-205, 1987.
59. Kawasumi, T., Kiuchi, N., Futatsugi, Y., Ohba, K., and Yanagi, S. O. High yield preparation of Lentinus edodes ("Shiitake") protoplasts with regeneration capacity and mating type stability. *Agr. Biol. Chem.*, *51*:1649-1656, 1987.
60. Gull, K., Moore, P. M., and Trinci, A. P. J. Preparation of protoplasts from Geotrichum lactis and use of fluorescence to detect cell walls. *Trans. Br. Mycol. Soc.*, *59*:79-85, 1972.
61. Ferenczy, L., Kevei, F., and Zsolt, J. Fusion of fungal protoplasts. *Nature*, *248*:793-794, 1974.
62. Eddy, A. A., and Williamson, D. H. A method of isolating protoplasts from yeast. *Nature*, *179*:1252-1253, 1957.
63. Emerson, S., Emerson, M. R. Production, regeneration and reversion of protoplast-like structures in the osmotic strain of Neurospora crassa. *Proc. Natl. Acad. Sci. USA*, *44*:668-671, 1958.
64. Fukui, K., Sagara, Y., Yoshida, N., and Matsuoka, T. Analytical studies on regeneration of protoplasts of Geotrichum candidum by quantitative thin-layer agar plating. *J. Bacteriol.*, *98*:256-263, 1969.
65. de Vries, O. H. H., and Wessels, J. G. H. Effectiveness of a lytic enzyme preparation from Tricoderma viride in releasing spheroplasts from fungi, particularly basidiomycetes. *Antonie van Leeuwenhoek*, *39*:397-400, 1973.
66. Boissonnet-Mennes, M., and LeCof, H. Transmission du virus par fusion de protoplasts chez Pyricularia oryzae Briosi et Cav. *Physiologie Vegetale*, *14*:251-257, 1976.
67. Keller, U., Zocher, R., and Kleinkauf, H. Biosynthesis of ergotamine in protoplasts of Calviceps purpurea. *J. Gen. Microbiol.*, *118*:485-494, 1980.
68. Leach, J., and Yoder, O. C. Heterokaryosis in Cochliobolus heterostrophus. *Exp. Mycol.*, *6*:364-374, 1982.
69. Leslie, J. F. Some genetic techniques for Gibberella zeae. *Phytopathology*, *73*:1005-1008, 1983.

70. Typas, M. A. Heterokaryon incompatibility and interspecific hybridization between Verticillium albo-atrum and Verticillium dahliae following protoplast fusion. *J. Gen. Microbiol.*, *129*:3043-3056, 1983.
71. Stasz, T. E., Harman, G. E., and Weeder, N. F. Protoplast preparation and fusion in two biocontrol strains of Trichoderma harzianum. *Mycologia*, *80*:141-150, 1988.
72. Hamlyn, P. F., Bradshaw, R. E., Mellonma, F. M., Santiago, C. M., Wilson, J. M., and Peberdy, J. F. Efficient protoplast isolation from fungi using commercial enzymes. *Enz. Microb. Technol.*, *3*:321-325, 1981.
73. Tsuge, T., Hayashi, N., and Nishimura, S. Selection of auxotrophic mutants and heterokaryosis in Alternaria alternata. *Ann. Phytopathol. Soc. Jpn.*, *53*:182-190, 1987.
74. Peberdy, J. F., and Bibson, R. K. Regeneration of Aspergillus nidulans protoplasts. *J. Gen. Microbiol.*, *69*:325-330, 1971.
75. Garcia, A. I., and Villanueva, J. R. Protoplasts from conidia of Fusarium culmorum. *Can. J. Microbiol.*, *10*:99-101, 1964.
76. Gascon, S., Ochoa, A. G., and Villanueva, J. R. Production of yeast and mold protoplasts by treatment with the streptzyme of Micromonospora AS. *Can. J. Microbiol.*, *11*:573-580, 1965.
77. Garcia, A. I., Lopez-Belmonte, F., and Villanueva, J. R. Regeneration of mycelial protoplasts of Fusarium culmorum. *J. Gen. Microbiol.*, *45*:515-523, 1966.
78. Peberdy, J. F., Buckley, C. E., Daltrey, D. C., and Moore, P. M. Factors affecting protoplast release in some filamentous fungi. *Trans. Br. Mycol. Soc.*, *67*:23-26, 1976.
79. Peberdy, J. F., and Issac, S. An improved procedure for protoplast isolation in Aspergillus nidulans. *Microbios Lett.*, *3*:7-9, 1976.
80. Morehart, A. L., Blits, K. C., and Melchior, G. L. Pathogenic potential of Verticillium albo-atrum protoplasts. *Mycologia*, *77*:784-790, 1985.
81. Anné, J., and Peberdy, J. F. Protoplast fusion and interspecies hybridization in Penicillium. In *Fungal Protoplast: Applications in Biochemistry and Genetics* (J. F. Peberdy and L. Ferenczy, eds.), Marcel Dekker, New York, 1985, pp. 259-277.
82. Ferenczy, L. Microbial protoplast fusion. In *Genetics as a Tool in Microbiology* (S. W. Glover and D. A. Hopwood, eds.), Cambridge University Press, Cambridge, 1981, pp. 1-34.
83. Peberdy, J. F. Genetic recombination in fungi following protoplast fusion and transformation. In *Fungal Differentiation* (J. E. Smith, ed.), Marcel Dekker, New York, 1983, pp. 559-581.
84. Peberdy, J. F., and Hocart, M. Protoplasts as a tool in the genetics of plant pathogenic fungi. In *Genetics and Plant Pathogenesis* (P. R. Day and G. J. Jellis, eds.), Blackwell Scientific, Oxford, 1987, pp. 127-141.

7

TRANSFORMATION AND GENE MANIPULATION IN FILAMENTOUS FUNGI: AN OVERVIEW

THEO GOOSEN, CEES J. BOS, and HENK van den BROEK *Wageningen Agricultural University, Wageningen, The Netherlands*

I. INTRODUCTION

In order to study the organization of the eukaryotic genome, the structure of eukaryotic genes, and the regulation of the different molecular processes in eukaryotic cells, it is important that efficient gene manipulation methods and gene transfer systems be available. The isolation and analysis of specific eukaryotic DNA sequences is now well established. However, to determine, evaluate, and exploit their biological function, such sequences should be reintroduced into the organism from which they originate in a defined way. For this reason, the development of DNA-mediated gene transfer systems for a variety of organisms has obtained much attention. Although transformation has been well documented in bacteria (for review, see Portugal and Cohen [1]), conclusive evidence for transformation of higher [2] and lower [3] eukaryotes was obtained much later. Since then, molecular genetic manipulation techniques for many organisms have been developed, and especially those for yeast [4] have reached a level of high sophistication. The experience gained with the latter organism has contributed greatly to the development of similar techniques for filamentous fungi.

Filamentous fungi are a very heterogeneous group of lower eukaryotes [5] which exhibit highly diverse biological functions. Some, like *Aspergillus nidulans* and *Neurospora crassa*, have been extensively used as model organisms to study aspects of eukaryotic genetics, metabolism, and development. Others, like *A. niger*, *A. oryzae*, *Penicillium chrysogenum*, and *Cephalosporium acremonium*, have a long history as producers of substances of commercial interest, e.g., food products, enzymes, primary metabolites, and secondary metabolites such as organic acids and antibiotics. A third group, which comprises a wide variety of genera, is formed by the plant pathogens.

In this overview we concentrate on the different aspects of gene transfer and gene cloning in filamentous fungi. We have aimed at updating the state of the art, emphasizing the general aspects of gene manipulation and transformation rather than trying to be comprehensive, since several topics were reviewed recently [6-8]. The results discussed illustrate the great potential of filamentous fungi for both fundamental and applied studies.

II. METHODOLOGY OF TRANSFORMATION

In general, transformation of filamentous fungi is being performed according to modifications of the protocol developed for *Saccharomyces cerevisiae* [3]. In this regard, all first successful reports on the transformation of *N. crassa* [9-12], *A. nidulans* [13-16], and *Podospora anserina* [17,18] included the preparation of osmotically sensitive cells with cell wall-digesting enzymes and treatment of these cells with $CaCl_2$ and polyethylene glycol (PEG), which facilitates the uptake of DNA. Regeneration and selection for expression results in the detection of the desired transformants. More recently, other means for the introduction of foreign DNA have been reported, but until recently these other methods were not widely applied for the transformation of filamentous fungi due to the satisfying results obtained with the $CaCl_2$-PEG methods.

A. Preparation of Protoplasts and Regeneration

Filamentous fungi have thick, rigid cell walls which form a barrier to the uptake of macromolecules like exogenous DNA. Therefore, most protocols for fungal transformations are based on the use of protoplasts or spheroplasts prepared under osmotically stabilized conditions. Much attention has to be paid to the preparation and regeneration of considerable amounts of protoplasts, since the success of transformation in many cases will primarily be determined by factors involved in these processes. The preparation and regeneration of protoplasts has been studied in detail for a variety of filamentous fungal organisms and have been reviewed by Peberdy and Ferenczy [19].

A large set of commercially available enzymatic preparations, either alone or in combination, have been used to prepare fungal protoplasts for transformation experiments. These include helicase [14,20], β-glucuronidase [15,21,22,23,24], glusulase [10,11,25], cellulases [26,27, 28,29], Streptozyme [30], driselase [22,31,32], chitinase [20,22,29,30], chitosanase [30], Zymolyase [33,34], Novozym [13,15,21,35,36,37], and in some cases extracellular enzyme preparations from cell-wall grown cultures of *Trichoderma harzanium* or *Oerskovia xanthineolytica* [38]. These days, Novozym 234, a product of Novo Industries, which contains a mixture of hydrolytic enzymes like glucanases and chitinases, is preferred as the digestive enzyme in the preparation of protoplasts from almost any fungus, although sometimes the preparation is used in combination with one of the before-mentioned enzymes. It should be stated that different enzyme batches may have highly variable activities, and also the time of incubation with the digestive enzyme necessary to produce protoplasts can vary greatly from one organism to another [39].

The effectiveness of digestive enzymes is sometimes in part determined by the stage of growth of the fungus; this points to differences in cell wall composition at different stages of fungal development. For transformation of filamentous fungi, protoplasts have been prepared from preincubated and swollen conidiospores [11,32,38] or oida [29,40], germinating conidiospores [27,28,42], early logarithmic cells [39], young hyphal material grown in liquid cultures [14,15,43,44] or grown on plates with cellophane disks [13,20,45,46] or fragmented hyphal cells [34]. Hyphal cells are multinucleate and in general will be parceled to many multinucleate protoplasts, whereas spores or microconidia in general are uninucleate and each will yield one single protoplast.

Since protoplasts are osmotically sensitive cells, they should be stabilized at specific osmotic values during preparation and handling. This has been done by using 0.6-1.1 M KCl/NaCl [13,23,32,47,48], 0.5-1.3 M sorbitol [10,15,38,39,49,50], 0.5 M mannitol [31], or 1.0-1.2 M $MgSO_4$ [14,41,45, 51,52] in a buffer optimal for the digestive enzyme. For some organisms the use of sugars or sugar alcohols may reduce the protoplast yield. In some cases $MgSO_4$ is preferred for purification purposes (protoplasts tend to float upon centrifugation in high $MgSO_4$); whether protoplasts are best purified by filtration or centrifugation must be determined empirically. By washing the protoplasts, contaminating enzymes which will interfere with transformation (DNases) or regeneration (lytic enzymes) are removed. At the same time, the buffer can be adapted to the optimal conditions for the incubation of protoplasts with the transforming DNA. At this stage, protoplasts can be used directly for transformation (see below) or are stored at -80°C in convenient portions for transformation at a later stage. As indicated for *N. crassa* [50,53] and *A. nidulans* [54], addition of PEG and DMSO before freezing might be necessary to enhance stability. In many cases, the use of stored frozen protoplasts results in reduced survival and lower transformation frequencies [28,42,50], but high efficiencies have also been reported [55, T. Goosen et al., unpublished]. However, for a routine control transformation experiment a stored pool of osmotically sensitive cells can be very convenient. Viability of the protoplast suspension used can easily be monitored at different stages of the transformation procedure by plating aliquots on stabilized regeneration medium.

B. Methods for Transformation

1. Fusionogenic Method

Most published procedures for the transformation of fungal recipients are based on the incubation of osmotically stabilized and $CaCl_2$-treated protoplasts with exogenous DNA. In some reports [9,50,51] specific additives like the nuclease inhibitor aurintricarboxylic acid [56], DMSO [9,50], heparin [50,51,57], spermidine [50,51], or carrier DNA are included to increase transformation efficiency, although the basis of their effects is not always known. According to Vollmer and Yanofsky [50], excess spermidine prevents the precipitation of spheroplasts by the added DNA.

Protoplasts are incubated with the DNA on ice for a short period, before a PEG solution as fusionogenic medium is added. For some organisms the concentration, amount, and type of PEG is critical [H. W. J. van den Broek et al., unpublished]. This may in part be due to the presence of toxic compounds in some preparations. Routinely, PEG is made up freshly in buffer with or without $CaCl_2$ and osmotic stabilizer [58]. In some cases, the addition of PEG is being performed gradually [22,27,39,51].

After a second short incubation period, usually at room temperature, the mixture can be diluted with osmotically stabilized buffer; in some protocols dilution is carried out gradually. Aliquots are then spread on stabilized selective medium, either directly or in a soft agar overlay. It should be emphasized that sometimes certain stabilizers interfere with selection [14,38]. In several cases, the incubated protoplasts are washed free of PEG before being spread. Alternatively [59,60], incubated protoplasts have also been diluted in stabilized nonselective liquid medium, first incubated in a recovery broth to allow some regeneration to occur and then spread on agar selection plates. Sometimes delayed selection [61] is applied by adding the selective compound in a second overlay [56,62]. Transformants will appear as single colonies, depending on the phenotype to be selected, sometimes within a background of untransformed, slowly growing cells. In many cases so-called "abortive transformants" are present as small, ill-developed colonies among authentic transformants. Ratios between normal and abortive transformants vary from organism to organism and may in part depend on the selective markers or vectors used for transformation [32,47]. For each organism, and often for each strain, and selective marker used, the ultimate transformation conditions should be determined by trial and error.

2. Liposomes

A liposome-mediated genetic transformation protocol for fungi has been reported [63], but due to difficulties encountered in the reproducible preparation of liposomes and the positive results obtained with other methods, this procedure has not been widely accepted and applied to fungal transformation.

3. Lithium Acetate

An alternative transformation methodology for *N. crassa* which avoids the preparation of protoplasts is based on the lithium acetate procedure developed for yeast [64]. Dhawale et al. [65] developed "competence" in germinated *N. crassa* conidiospores by treating them with lithium acetate before incubating them with DNA and PEG. A short heat-shock exposure at 37°C before plating was an essential element in this protocol. Although it was announced as a reliable simplification of the *N. crassa* transformation protocol [65] with 100% viability of the cells, this procedure has not been widely adopted. Similar protocols were utilized for only a few fungi [29, 40,66] since in general no improvement with respect to the transformation frequencies could be obtained. Furthermore it was suggested that the type of transformants obtained by this procedure might differ significantly from that obtained by protoplast transformation [67,68].

4. Electroporation

A recent development, electroporation-mediated gene transfer, has been shown to be very effective for bacterial [69], lower eukaryotic [70,71], and plant [72] cells. Preliminary results using electroporation to introduce DNA into protoplasts of *A. nidulans* [56], however, were not satisfactory, although this method might be a good alternative if swollen or germinated conidiospores could be used instead of protoplasts. Recently, transformation of protoplasts of the phytopathogenic fungi *Ustilago violacea* [73] and *Fusarium solani phaseoli* [74] by electroporation was reported.

5. Physical Damage

Other very recent approaches to transformation methodologies included physically damaging cells, which was demonstrated as an ineffective method of transforming yeast [75], or bombarding cells with microprojectiles coated with DNA [76,77]. These methods have not yet been applied to filamentous fungi but might provide a future alternative.

III. SELECTABLE MARKERS FOR TRANSFORMATION

The application of recombinant DNA techniques and the development of transformation systems for filamentous fungi initially was hampered by the lack of useful selectable markers and/or the availability of the appropriate recipient. This was overcome by the isolation and application of homologous and heterologous selection markers and the development of selection systems as will be discussed below.

A. Auxotrophic Selection Markers

By using total *N. crassa* DNA from a wild-type strain, Mishra and Tatum [78] were able to complement a mutant phenotype. Although this was actually the first report on fungal transformation, it was not recognized as such. The experiments of Case et al. [9], based on the cloned *qa*-2 gene of *N. crassa*, are always cited as the first positive *N. crassa* transformation results. Soon thereafter, effective transformation of *N. crassa* was also realized with the cloned *am* [79] gene, while Kinnaird et al. [10] and Dhawale et al. [65] succeeded in obtaining a few transformants by using total genomic DNA.

The first positive report on *A. nidulans* transformation from Ballance and Turner [13] was based on the complementation of an *A. nidulans pyr*G mutant by the heterologous *N. crassa pyr*-4 gene. Soon after that, positive results were obtained with the cloned *A. nidulans amd*S [14,38], *trp*C [15], and *arg*B [16] genes. With all these systems low efficiencies (10-100 transformants per microgram vector DNA) were observed.

In the same period reports on the transformation of *Podospora anserina* [80,81] to senescence also appeared; it was shown that both purified mitochondrial plasmidlike (pl) DNA or pl-DNA cloned into pBR322 could be used to transform juvenile strains with high efficiency. In later studies [82] these results were questioned.

Since then, several homologous and heterologous auxotrophic selection markers for *N. crassa*, *A. nidulans*, *A. niger*, *Penicillium chrysogenum*, and, more recently, for many other fungi have been isolated by different procedures (see Section VII), and used for transformation studies (see Table 1). Many cloned genes have been shown to complement corresponding auxotrophic mutations of both homologous and heterologous recipients, thus facilitating the development of transformation systems for other species, provided that strains harboring such auxotrophic mutations are available.

Suppression of auxotrophic mutations, as a special type of selection of transformants in *P. anserina*, should be mentioned. Brygoo and Debuchy [44] used the cloned tRNA suppressor genes *su*4-1 and *su*8-1 to transform a leucine auxotrophic strain, which carries the non-sense mutation *leu*1-, to leucine prototrophy. By using a second suppressible non-sense mutation, *193*, which prevents spore pigmentation, genetic analysis could be performed easily.

B. Dominant Selection Markers

For many fungal organisms the isolation of auxotrophic mutants is very difficult. In some cases, e.g., industrially important filamentous fungi, plant pathogenic fungi, a change in the genetic background may not be desirable. In such cases the construction of vectors that can be selected

TABLE 1. Transformable Filamentous Fungi

Fungal species	Selection marker: Homologous	Selection marker: Heterologous
Absidia glauca		Nm^R [230]
Achlya ambisexualis		Nm^R [31]
Aspergillus awamori		Hm^R [231], *pyr*-4 [232]
A. ficuum		*amd*S [89], Hm^R [89]
A. flavus		*pyr*-4 [23]
A. nidulans	*amd*S [14,38], *arg*B [16], *ben*A [158], *oli*C [149], *pyr*G [32], *trp*C [15]	*pyr*-4 [13], Bm^R [46,90], Hm^R [91,151]
A. niger	*pyr*A/*pyr*G [37,88] *oli*C [56], *trp*C [100]	*amd*S [95], *pyr*-4 [231], *arg*B [221], Bm^R [46]
A. oryzae		*arg*B [48], *pyr*G [233]
A. terreus		*amd*S [234]
Bremia lactucae		Drug^R (A*)
Cephalosporium acremonium		$\mathrm{G418}^R$ [33], Hm^R [105]
Cercospora kikuchii	*ben*A (B*)	
Cladosporium fulvum		Hm^R [51]
Claviceps purpurea		Bm^R [46]
Cochliobolus heterostrophus	TRP-1 [97]	*amd*S [22], Hm^R [97]
Colletotrichum lindemuthianum		Hm^R [42], *amd*S [42]
C. trifolii		*tub*-2 [66], Hm^R [66]
Coprinus cinereus	TRP-1 [29,40]	
Cryphonectria parasitica		*tub*-2 (C*)
Curvularia lunata		Hm^R [222]
Fusarium oxysporum		Hm^R [41]
F. moniliforme		Hm^R (D*)
F. sambucinum		Hm^R (E*)
F. solani phaseoli		Hm^R [74]
Gaeumannomyces graminis		*tub*-2 [26]

TABLE 1 (Continued)

Fungal species	Selection marker: Homologous	Selection marker: Heterologous
Leptosphaeria maculans		*Hm*^{R} [235]
Magnaporthe grisea		*arg*B [34]
Mucor circinelloides	*leu* [135]	
Neurospora crassa	*inl* [20], *tub*-2 [59], *am* [10], *trp*-1 [12], *pyr*-4 [98], *qa*-2 [9,65)	*arg*B [208], *Hm*B (F*)
Ophiostoma ulmi		*Hm*^{R} (G*)
Penicillium chrysogenum	*oli*C [61], *trp*C [21,36]	*amd*S [90], *pyr*-4 [35], G418^R [112], *Bm*^{R}[90]
P. nalgovense		*amd*S (H*)
Phanerochaete chrysosporium		*ade*2/*ade*5 (I*)
Phycomyces blakesleeanus		G418^R [30], *Km*^{R} [85]
Podospora anserina	*ura*-5 [18,236], *su*4-1/*su*8-1 [44,118]	
Pseudocercosporella herpotrichoides	β-tubulin (J*)	*amd*S (J*), *Hm* (J*)
Schizophyllum commune	*ade*2 [27], *ade*5 [27], *ura*1 [27], *trp*1 [28]	*Hm*^{R} [151]
Septoria nodorum	*tub*A (K*)	*Hm*^{R} [223]
Trichoderma reesei		*amd*S [45], *arg*B [45], *pyr*-4 [232]
Ustilago maydis	PYR-3 [155]	G418^R [25], *Hm*^{R} [39]

Abbreviations: *Hm*^{R}: hygromycin resistance. *Bm*^{R}: bleomycin resistance. *Nm*^{R}: neomycin resistance. *Km*^{R}: kanamycin resistance. G418^R: resistance to the aminoglycosidic antibiotic G418.
*Personal communications: A, H. Judelson et al.; B, R. G. Upchurch et al.; C, A. C. L. Churchill and N. K. van Alfen; D, J. F. Leslie and M. G. Dickman; E, Y. P. Salch and M. N. Beremand; F, B. M. Tyler and B. Austin; G, J. C. Royer et al.; H, R. Geisen and L. Leistner; I, M. Alic et al.; J, E. Blakemore et al.; K, R. N. Cooley et al. See also Ref. 94.

on the basis of dominant selection is required. However, one of the main problems in obtaining a versatile dominant selection marker is the relatively low and often variable sensitivity of fungal species (and sometimes even strains within a species) to the antibiotics normally used for yeasts and higher eukaryotes. Banks [25] reported the transformation of *Ustilago maydis* to G418 resistance using a plasmid which contained the yeast LEU-2 gene and 2μ yeast DNA in addition to a bacterial G418^R gene; G418^R transformants were obtained and it was suggested that the yeast 2μ DNA promoted autonomous replication in *Ustilago*. Bull and Wootton [83] used a cosmid vector which contained the *N. crassa am* gene and the G418^R gene from Tn903, and with a two-stage selection procedure they obtained stable am$^+$ transformants of which 85% had a G418^R phenotype. The transformants showed a variable degree of resistance due to a variable copy number. Penalva et al. [33] were able to transform *C. acremonium* with low efficiency with plasmids containing the G418^R gene from Tn903. Selection of transformants was improved when expression of the G418^R gene was governed by the yeast ADCI promoter. Revuelta and Jayaram [30] as well as Suarez and Eslava [84] transformed *Phycomyces blakesleeanus* to G418 resistance with a vector containing the G418^R gene from Tn903 and a putative *Phycomyces* ARS sequence. In both cases the transformants were highly unstable. Arnau et al. [85] obtained *Phycomyces* transformants with high frequencies using the kanamycin gene from Tn5 fused to an *E. coli*-selected *Phycomyces* promoter. These latter transformants were stable even in the absence of selection. In most filamentous fungi, however, prokaryotic genes will not be expressed unless they are fused to fungal expression signals. The usefulness of gene fusions was first demonstrated with the *E. coli lacZ* gene. By fusing the *lacZ* coding region to the transcriptional and translational expression signals of the *A. nidulans trpC* [86] or *gpdA* [87] genes it could be demonstrated that expression in *A. nidulans* [86,87] occurred. Also in several other fungal species such gene fusions could be expressed [45,88-90]. This provided the basis for the construction of hybrid-dominant selective markers for hygromycin B resistance [91], neomycin resistance [85], and bleomycin resistance [46], which have been used for the development of transformation systems for a wide variety of filamentous fungi (Table 1).

Another type of selective marker is based on mutant fungal nuclear genes which confer resistance to oligomycin or benomyl. Turner et al. [92] isolated the *oliC* gene (encoding ATP synthase) from an oligomycin-resistant *A. nidulans* mutant. Transformation of a wild-type recipient with this gene resulted in weakly resistant (oli^{SR}) as well as strongly resistant (oli^R) transformants. The relative frequency of these classes depended on the method of selection. Analysis of both classes showed that in fully resistant transformants the wild-type gene was replaced by the mutant allele, whereas the weakly resistant transformants were heterozygous for wild-type and mutant gene. So, with this transformation system, gene replacements can phenotypically be distinguished. Similar systems have been developed for *A. niger* [56] and for *P. chrysogenum* [61].

The mutant β-tubulin gene which confers resistance to the drug benomyl [59] was first used as selective marker in *N. crassa*. This marker is also semidominant [59], and it can be used for selection in some heterologous fungi [26, A. C. L. Churchill and N. K. van Alfen, pers. commun., 94]. Vollmer and Yanofsky [50] used this gene in the construction of an ordered cosmid library of *N. crassa* from which genes can be isolated with the sib selection procedure. Since benomyl-resistant mutants are quite easily obtained and, due to high conservation, cloning of the mutant β-tubulin genes by heterologous hybridization is fairly easy, this gene has become a homologous transformation marker for many filamentous fungi, including *A. nidulans* [93], *A. flavus* (E. R. Seip et al., pers. commun., 94), and several plant pathogens (see Table 1).

A heterologous dominant selection marker, useful with several filamentous fungi, is the *A. nidulans amd*S gene. This gene codes for an acetamidase, which allows growth on acetamide as sole carbon and nitrogen source. Since many fungal species do not exhibit acetamidase activity, wild-type strains are potentially transformable with this gene. Kelly and Hynes [95] first demonstrated its effective use by transforming *A. niger*. In later studies also transformation of *P. chrysogenum* [96], *P. nalgiovense* [R. Geisen and L. Leistner, pers. commun., 94], *A. ficuum* [89], *T. reesei* [45], *Cochliobolus heterostrophus* [97], and *Glomerella cingulata* [42] with the *amd*S gene was achieved, albeit with low efficiencies.

C. Markers for Two-Way Selection

A very recent development is transformation with genes which have a selectable phenotype for both the wild-type and mutant allele. An example is the orotidine-5'-decarboxylase (OMP) gene (*pyr*-4 in *N. crassa* [98], *pyr*G in *A. nidulans* [32], *pyr*A in *A. niger* [37]). The pyr^+ allele allows growth in the absence of uridine, whereas the *pyr*- allele confers resistance to 5-fluoroorotic acid [99]. By using gene constructs which autoselect for gene replacements [100] we have been able to transform wild-type *A. niger* strains to $pyrA^-$ [T. Goosen et al., unpublished results]. A subsequent transformation with a similar construct containing the $pyrA^+$ allele restores the original wild-type strain. This approach allows the introduction of several genes in an otherwise unaltered genetic background by repeating these cycles and applying cotransformation (see Section VI).

The same principle is also applicable to the nitrate reductase gene, where the mutant allele confers chlorate resistance [101], and to 2-aminoadipate reductase, mutants of which grow on 2-aminoadipate as sole nitrogen source and are lysine-auxotrophic [102]. For fungi in which the *A. nidulans amd*S gene can be introduced and expressed, the *amd*S system can also be used in this way by selecting for fluoroacetamide resistance vs. acetamide utilization [239].

IV. VECTORS FOR TRANSFORMATION

A. General Cloning Vectors

Detection of transformation depends on the ability to introduce, maintain, and select vector DNA sequences in an organism. In its simplest form, such a vector consists of a selectable marker cloned in a bacterial replicon, and in this case maintenance in the host organism depends on integration into chromosomal DNA. The first plasmid-based fungal vectors, used to transform *N. crassa* or *A. nidulans*, contained either the *am* [103], *qa*-2 [9], or *pyr*-4 [13,98] gene from *N. crassa* or the *amd*S [14], *trp*C [15], or *arg*B [16] gene from *A. nidulans*, respectively. The *am* gene was initially used as a λ clone [10]. The transformation frequencies obtained were rather low (10-100 transformants/μg vector DNA) for both fungal species. This was not surprising in view of the results with *S. cerevisiae* [3] and it prompted many investigators to modify their vectors. One approach has been to increase the degree of homology between the integrating vector and the fungal genome. Toward this end, randomly selected genomic DNA fragments [34,43,104] as well as fragments originating from repeated ribosomal RNA genes [14,38,105] or other reiterated genomic sequences [47,106] have been included in certain vectors. However, except for *A. nidulans ans*1 [47,54,107] and *unt* [106], these sequences did not increase transformation frequencies significantly.

Another approach was the incorporation of sequences by which the vector should acquire the ability to replicate autonomously in the fungal host. This approach was based on results obtained with yeast [108,109]. The origin of such sequences (termed ARS sequences) has included elements from true mitochondrial plasmids [24,110), mtDNA ARS elements from aspergilli [14,111], *Penicillium* [112], *Ustilago* [113], and *Cephalosporium* [105], genomic ARS sequences isolated by virtue of functionality in yeast [30,43,47,84,106,113,114] or selected in the homologous host [43,104,115], and the 2μ DNA sequence from *S. cerevisiae* [25]. In only a few cases were such sequences found to effectively increase transformation frequencies, sometimes with alleged autonomous replication (see below).

Ballance and Turner [47] isolated a genomic sequence (*ans*1) from *A. nidulans* as an ARS element in yeast. Incorporating this sequence into *pyr*4-based plasmids increased the *A. nidulans* transformation frequency 50- to 100-fold [47], while the number of "abortive" transformants decreased drastically. Incorporation of the *ans*1 sequence in vectors carrying other selection markers gave variable results. Wernars et al. [116] reported a moderate increase in transformation frequency of *trp*C, but no significant effect at all on *arg*B, whereas Cullen et al. [54] did find elevated frequencies with the latter marker. With plasmids containing *amd*S, the location and orientation of *ans*1 within the vector was shown to determine the transformation frequencies [116]. Remarkably, in some cases *ans*1 could also exert its stimulating effect when present on a second vector [54,116], which was cotransformed with the plasmid containing the selective marker. This cotransformation phenomenon provided a convenient method for improving transformation efficiency without the need to reconstruct vectors. Gene libraries in vectors containing *ans*1 have been constructed and successfully used to isolate genes from *A. nidulans* [58,92] and *P. chrysogenum* [35]. Several possibilities for the mechanism of action of *ans*1 have been suggested [54,116].

Since the cloning capacity of cosmid and bacteriophage λ vectors is much higher than that of plasmid vectors, the number of clones necessary to represent the total fungal genome in a cosmid or phage library will be much less than in a plasmid library. This can facilitate transformation/complementation and gene isolation studies. Useful cosmid libraries have been constructed for *Aspergillus* [92,119], *Neurospora* [50], and *Podospora* [118]. Cloned DNA from such libraries has been used to transform fungal hosts. However, results obtained with such vectors might be different from those obtained with plasmid vectors with respect to integrational behavior of vector sequences and stability of the transformants [10,44,103, 117,118]. For example, it was demonstrated in experiments with *P. anserina* that the large inserts in cosmids tended to direct integration to the homologous locus [117,118], whereas plasmid-based vectors integrated at heterologous sites.

B. Promoter Probe Vectors

The rapidly increasing number of cloned fungal genes necessitated the development of techniques for the analysis of the transcriptional and translational regulatory signals of such genes at the molecular level. Van Gorcom et al. [87] developed a system for both the selection and analysis of fungal regulatory sequences. A set of three vectors was constructed in which expression signals can be fused in phase to the *E. coli lac*Z gene. The *arg*B gene was included in the vectors as a selective marker. Analysis of different signals in a constant chromosomal environment is achieved by selecting transformants with one homologously integrated copy of the constructs. Quantitative analysis of expression was performed for the weak *trp*C [86] and strong *gpd*A [87] promoters. In later studies, *lac*Z fusions were used for deletion mapping of promoter regions [120] and to study regulation of *amd*S expression [121]. A recent development is the use of the much more sensitive *E. coli* β-glucuronidase as reporter for gene activity [62, T. Goosen et al., unpublished results].

C. Expression Vectors

Many fungi, in particular members of the genus *Aspergillus*, are industrially important as producers of (extracellular) proteins. To exploit them for the production of heterologous proteins, fungal expression vectors have been developed. Such vectors, in addition to a selection marker, should contain a strong promoter, which preferably can be differentially regulated, to ensure expression of the heterologous gene. Promoters of the *A. nidulans alc*A [122] and *tpi*A [123] or the *A. niger* glucoamylase [124] gene have been used for this purpose. In all three cases expression can be modulated by growing strains on the appropriate carbon source. In order to obtain heterologous proteins extracellularly, constructs have to be provided with secretional control regions. Both signal sequences of the *A. niger* glucoamylase gene [124] and synthetic consensus signal sequences [122] have been effectively used for this purpose. At least in some cases, heterologous proteins can be secreted from aspergilli by their own signal sequences [124,125].

The details given above on the vectors for transformation of filamentous fungi clearly indicate that most attention has been given to integrating vectors. One reason, of course, is the questionable results obtained with ARS sequences and the doubts on segregation of plasmids in hyphal fungi. To create minichromosomes, which in analogy to yeast might have a more stable character, additional sequences, like telomeres and centromeres, must be isolated. Recently, the first trials on the isolation of telomeric sequences from *N. crassa* [126] and centromeric structures of *P. anserina* [118] have been reported. Integrative transformation systems for the different filamentous fungi, however, are so efficient and versatile that for many purposes the development of minichromosomes is superfluous.

V. FATE OF THE TRANSFORMING DNA

As for *S. cerevisiae* [3], the early reports on transformation of filamentous fungi with vectors consisting of a bacterial plasmid with a cloned fungal gene (*N. crassa qa*-2 [9]; *A. nidulans amd*S [14]) made mention of integration of vector sequences into the genome of the acceptor. When the *N. crassa* *pyr*-4 gene was used to transform an *A. nidulans pyr*G mutant to prototrophy, efficient integration occurred in spite of the lack of homology [13]. Since then considerable effort has been made, on the one hand, to characterize the different types of integrative transformants that could be obtained, and on the other hand, to modify vectors in such a way that they would be maintained as free replicating plasmids in the fungus, again in analogy to what has been established in yeast.

A. Replicating Vectors

Reports claiming the successful transformation of filamentous fungi with autonomously replicating plasmids have been numerous, but in many cases the evidence was poor or results from different investigators were contradictory.

Stohl and Lambowitz [110] concluded that a vector, based on the *N. crassa qa*-2 gene and the *N. intermedia* mitochondrial Labelle plasmid (pALS1), was able to replicate autonomously in *N. crassa*. The reisolated plasmids, however, frequently had lost the Labelle sequences completely [127] but had retained the autonomous replication character. With a similar Labelle-*qa*-2 construct, Kuiper and De Vries [24] also found increased transformation frequencies and frequent rearrangements of plasmids. However, they concluded that if plasmids were present, it was at less than 1 copy per 100 haploid genomes. Paietta and Marzluf [115] also claimed autonomous replication on the basis of reisolation of vectors in *E. coli*. The frequencies of reisolation were very low and again rearrangements were found. Dhawale and Marzluf [67] subsequently showed that many *qa*-2 transformants failed to pass the *qa*-2 gene through a cross which they again associated with possible autonomous replication. They did not, however, correlate instability of the transformants and reisolation of vector sequences with autonomous replication, nor were they able to unequivocally demonstrate free plasmids in Southern blots.

Although the initial transformation studies with a λ clone containing the *am* gene showed stable integration of the transforming DNA [79], Grant et al. [103] also observed extreme instability of transformants after subcloning of the *am* gene on a plasmid. Again, no monomer-sized plasmids were detected in Southern blots. Autonomous replication of multimeric plasmid forms was brought forward as the most feasible explanation, but the possibility of tandemly integrated transformed sequences which are eliminated during meiosis was not ruled out. In later reports, Selker and coworkers [129,130] indicated that rearrangements frequently occur in *Neurospora* transformants as a direct result of sequence duplications [131]; they occur just before meiosis (RIP = rearrangement induced premeiotically) and are attended by heavy methylation of cytosine residues. Recently, it was shown that RIPping involves extensive GC to AT transitions rather than deletions, reason to change the meaning of RIP to "repeat induced point mutation" [E. U. Selker et al., pers. commun., 94].

Case et al. [9] concluded that *qa*-2-based vectors exclusively show integrative transformation. In later reports [132,133] no mitotic or meiotic instability has been observed. Buxton and Radford [43] made a systematic search for *Neurospora* sequences which could promote autonomous replication, but no free plasmids were observed in the transformants. The authors performed computer modeling of plasmid loss from growing hyphal tips and discussed the conditions which an autonomously replicating plasmid in a filamentous fungus must meet to be stably maintained. They concluded that the inclusion of either a functional centromeric sequence or a natural plasmid replicon will be a prerequisite.

For several filamentous fungi autonomously replicating sequences have been sought, usually by selecting for ARS capacity [108,109] in yeast. For instance, many attempts have been made with *A. nidulans*, but without irrefutable success. Barnes and MacDonald [114] obtained unstable transformants with a vector containing such a putative *A. nidulans* ARS sequence and the *N. crassa pyr*-4 gene, which they interpreted as possible autonomous replication of multimeric plasmid forms. Similarly, Johnstone [106] described another *A. nidulans* ARS sequence (*unt*) which, upon inclusion in an *arg*B-containing vector, showed elevated transformation frequencies and resulted in many unstable transformants. On the other hand, Ballance and Turner [47] isolated an *A. nidulans* DNA sequence (*ans*1) as ARS sequence in yeast, which upon inclusion in *A. nidulans* vectors substantially increased transformation frequencies but did not promote autonomous replication. All other efforts [14,104, Van Gorcom, unpublished results] were equally unsuccessful. Goosen et al. [128] concluded that the sequences reisolated from transformants are probably excised from genomic DNA by recombination between tandemly integrated copies. Although in *P. anserina* autonomously replicating plasmids found in mitochondria have been used to construct vectors for transformation [17,81], their effectiveness for gene cloning has never been demonstrated. One report [134] mentions a replicating linear plasmid in *Podospora* composed in part of *Tetrahymena* telomeres, but this replicon is extremely unstable. That leaves the *Zygomycetes* as the only filamentous fungi in which autonomously replicating plasmids have been found. Van Heeswijck [135] clearly demonstrated the presence of freely replicating plasmids in *Mucor* not only by reisolation and Southern blotting; moreover, no integrated plasmid sequences could be detected. In addition autonomous replicating vectors have been described for the Zygomycetes *P. blakesleeanus* [30,84,85] and *Absidia glauca* [136]. Recently, the presence of a circular plasmid in specialized strains of *Schizophyllum* was reported [137].

B. Integration of Transforming DNA

Integration of the transforming DNA sequences in the genome of the recipient has proven to be the rule for almost all filamentous fungi where transformation has been demonstrated (Table 1). There is, however, a large variation in the frequency with which the different types of integration occur. Basically, one can distinguish three integration types, which are similar to those described for integrative transformation of yeast [3]. Type I arises by single crossing over between the resident and cloned copies of the gene, resulting in a linked duplication of the genes, separated by bacterial vector sequences, at the homologous locus in the acceptor. Type II occurs when the vector integrates at a different location in the chromosome. For yeast it has been demonstrated that such recombination without any apparent homology is mediated by δ sequences [138,139]. In filamentous fungi, type I and type II transformants often contain more than one integrated copy in a tandemly arranged fashion. Type III transformants are the result of double crossing over or gene conversion, which results in replacement of the resident allele by the transforming one. For fungi it has been demonstrated that in many cases integrations of different types have occurred simultaneously in a single transformant.

With *A. nidulans*, the transformation events of many different marker genes have been analyzed in detail. Tilburn et al. [14], using the *amd*S gene, predominantly found type I integrations. At least in some cases these transformants were meiotically unstable. By inclusion of a fragment of the ribosomal RNA gene repeat, type I integrations could be directed to the rDNA. Wernars et al. [38,104] used the same selection system and obtained contrasting results. Type II integrations were observed almost exclusively, with tandem arrays of many copies. This different integration behavior could be attributed to the use of a different acceptor strain. Strain-dependent integration was also demonstrated for *N. crassa* [68] (see below). By genetic analysis Wernars et al. [140] demonstrated that integration could occur at many different loci, if not at random. In this case, inclusion of ribosomal DNA repeat fragments did not lead to targeting of the integration. Unlike observations with *Neurospora*, multicopy-transformed DNA is not significantly modified as judged from digestions with restriction endonucleases which are sensitive to DNA modifications [141, Wernars et al., unpublished results]. The AmdS$^+$ phenotype of these transformants is meiotically stable, but rearrangements within multicopy insertions have been detected [38,141].

Upon transforming *A. nidulans* with *trp*C, Yelton et al. [15] obtained all three types of integrants. Moreover, they were able to diminish the fraction of type II by transforming with linearized vectors, a treatment which had no effect on the integration events with *amd*S [38]. Transformants of types I and II show some meiotic instability. Miller et al. [142], however, showed that even upon transformation with a purified DNA fragment containing the *trp*C gene, only a minority of the transformants are of type III.

Transformants obtained with the *prn* gene cluster of *A. nidulans* [143] were found to be of types I and II in equal proportions, had multiple-copy insertions, and were unstable in meiosis.

Transformation of *A. nidulans* with *arg*B [16,144,145] occurred predominantly by homologous recombination. Less than 10% of the transformants were of type II. Type I integrations were observed in 65% of the transformants, the remaining 25% were of type III. In none of the cases described were more than two copies of the transforming plasmid integrated [144]. Where Upshall [144] demonstrated that all transformants of types I and II showed a high degree of meiotic instability, Johnstone et al. [145] only mentioned stable transformants.

The first *A. nidulans* transformation system, based on the *N. crassa* *pyr*-4 gene [13], showed integration into the genome which, by virtue of lack of homology, seemed random [47]. Similar results were obtained with other heterologous markers, e.g., the *A. niger trp*C [146] and *N. crassa nit*-2 [147] genes. Introduction of a reiterated *A. nidulans* sequence which increased transformation frequency (*ans*1) in *pyr*-4 containing vectors still resulted in nonhomologous integration [47]. May et al. [148] showed that if one of the β-tubulin genes (*ben*A) is introduced into a *pyr*-4-containing plasmid, at least nine out of 10 Pyr(+) transformants have the vector integrated at the *ben*A locus. However, if (part of) the second β-tubulin gene (*tub*C) was incorporated in such a vector, only a minority (6%) of the transformants showed integration at this locus.

The majority of the transformants obtained with the semidominant *A. nidulans oli*C gene [149] were of type III (73%). The type I transformants (13%) frequently gave rise to gene conversions, whereas the type II transformants (13%) seldom did. Transformants obtained with the *A. nidulans* *pki* gene [150] (55% type III, 25% type I) and the *A. nidulans pyr*G gene [32] (30% type III, 50% type I) showed a similar high degree of targeting.

With dominant selection markers, like *hph* (hygromycin resistance) [91,151], or *Bm* (bleomycin resistance) [46], fused to *A. nidulans trp*C or *gpd*A expression signals, integration seemed exclusively of type II, and in many cases involved multiple vector DNA copies.

From the enumeration given above it is evident that there is no general rule which predicts the precise deportment of a transforming DNA sequence in *A. nidulans*. Although integration is explicit, its type may depend on the composition of the vector, the nature of the selective marker, the form in which the vector is used (linear vs. circular), and even the genetic background of the acceptor strain. Furthermore, the amount of DNA used to transform a certain number of protoplasts influences the fraction of the transformants with multiple integrated copies [32, T. Goosen and H. W. J. van den Broek, unpublished results]. The integration events which take place when unlinked cotransforming DNA sequences are involved are even less conspicuous. Cointegration of the two different sequences at homologous and nonhomologous loci, as well as integration of each sequence separately, have been observed [100,152, T. Goosen et al., unpublished results].

The picture drawn for *A. nidulans* applies equally to other filamentous fungi. With *A. niger pyr*A/*pyr*G [37,88] integration is almost exclusively at the homologous locus (types I and III); with the semidominant *oli*C3 gene [153], 80% of the transformants are of type II. When the *A. nidulans amd*S marker [95] is used to transform *A. niger*, integration of tandem arrays of vector molecules is observed. Since *A. niger* does not contain an equivalent gene, integration is random and has been employed in our laboratory to construct a set of transformed strains in which every individual chromosome has been marked by an *amd*S insertion [A. J. M. Debets et al., unpublished]. With *N. crassa qa*-2, Case et al. [9,132,133] reported stable transformants with all three types of integrations, albeit with considerable difference in the relative frequencies between experiments with different constructs; however, predominantly type II insertions were observed. Others [67,154] confirmed these results, but in addition found an unstable class of transformants (see Section V.A; autonomous replicating vectors). Linearization of the vector did not result in better targeting to the resident locus [67]. For the *N. crassa trp*-1 gene, Kim and Marzluf [68] showed that the recipient strain might also influence the integration type. In one strain they observed almost exclusively type III integration in transformants, obtained with vector DNA which was circular or linearized within the *trp*-1 sequence. When the vector was cut outside the *trp*-1 gene, up to 40% of the transformants were of type II. In another strain the same vector integrated homologously in only 20% of the transformants, irrespective of whether the vector was circular or linear. Transformation with homologous selection markers has been demonstrated to result in all three types of integration, with varying proportions in *P. chrysogenum* [21,36,61], *Coprinus cinereus* [29,40], *S. commune* [27], *U. maydis* [155], and *P. anserina* [18]. For the latter fungus it seems that cosmids [117] can be better targeted than plasmids [44] when each vector carries the same selection marker.

VI. COTRANSFORMATION

A. Cotransformation of Covalently Linked DNA Sequences

Since transformation in fungi in general occurs by integration, the incorporation of different DNA sequences in one vector enlarges the possible configurations of the integrated DNA in the resulting transformants. Nonhomologous integration can lead to inactivation of unselected sequences whereas type III integrations of the marker gene will not lead to insertion of the cotransforming DNA into the chromosome. Wernars et al. [152] found that these events may occur in as many as 40% of the transformants. Integration of covalently linked DNA sequences by homologous recombination can now take place at two loci. The frequency with which either of these two types will occur can vary greatly [87,156], depending on, among other factors, the nature of the selection marker and the nature of the cotransforming DNA. Consequently, obtaining transformants with the desired integration pattern might be very cumbersome, unless specific measures are taken which favor one or the other type of recombination.

To facilitate the selection of transformants, which have arisen by recombination at the locus of the selection marker, vectors have been constructed which carry a mutant selection marker. Hamer and Timberlake [120] showed that marker rescue by homologous integration had occurred in most *A. nidulans argB* transformants, obtained with a disrupted *argB* marker gene. This approach allows the study of cotransformed sequences in a constant chromosomal environment, albeit not necessarily its natural one. The study of promoter sequences fused to the *lacZ* reporter and integrated at the *argB* locus [87] of *A. nidulans* is an example of such a directed integration.

Of more interest, but often also more difficult to obtain, are the transformants arising from integration at the homologous chromosomal location of the cotransformed gene. Usually, selection for this event is impossible, but its occurrence can in many cases be favored by using nonhomologous genes as selection markers. With the *N. crassa pyr-4* gene in *A. nidulans*, for instance, May et al. [148] obtained 90% integration at the locus of the (nonselected) cotransformed *benA* sequences. Using a diploid *A. nidulans* acceptor strain, Fidel et al. [157] integrated *pyr-4* at one of the actin homologs. Segregation of the integrated *pyr-4* gene with respect to other genetic markers was used to map the actin gene on chromosome I. These linked transformants thus provide a means for genetic analysis of loci, which are not associated with any known phenotype.

Integration of linked sequences can also be used to replace one allele of a gene by mutant alleles constructed in vitro in a two-step procedure. In the first step, integration of the vector at the locus of the cotransformed gene creates a duplication of the wild-type and mutant alleles of that gene, interrupted by the marker gene and the plasmid sequences. Subsequent selection for loss of the marker gene will result in segregants containing either one or the other allele. Dunne and Oakley [158] used this approach to replace the *benA* gene, whereas Miller et al. [142] obtained mutants in the sporulation-specific *spoC1-C* gene in this way. Experiments with the latter gene illustrated the necessity to study expression of transformed genes in their normal chromosomal environment. Integration of *spoC1-C* at ectopic sites resulted in loss of developmentally specific expression [156].

Although the examples given before clearly indicate that cotransformation with linked sequences is very useful, it has the disadvantage that the cotransforming sequence and the marker gene inevitably integrate at the same chromosomal location. Moreover, each new DNA sequence to be cotransformed has to be covalently linked to the vector containing the selectable marker.

B. Cotransformation of Unlinked Sequences

Efficient cotransformation of physically unlinked sequences has been demonstrated in many organisms, including yeast [138], but usually at least the selection marker was located on an autonomously replicating plasmid. Wernars et al. [152] investigated cotransformation of unlinked sequences on integrating vectors in *A. nidulans* and demonstrated that in general the nonselected cotransforming vector is incorporated in a large fraction of the transformants. The frequency of cotransformation was found to be dependent on the concentrations of the two separate vectors as well as on their conformation. Later results [T. Goosen et al., unpublished results] indicated that high concentrations of cotransforming DNA increases not only the frequency of cotransformation, but also the average number of cotransformed DNA molecules which will have integrated. The general applicability of cotransformation by unlinked sequences has since been demonstrated for a variety of filamentous fungi (see Table 2), often by using *lac*Z reporter fusions as the cotransforming DNA. This provides a simple test, applicable to most filamentous fungi, since they either lack endogenous β-galactosidase activity completely, or such activity can be efficiently suppressed.

Although administered as unlinked sequences during transformation, the two plasmids are not always integrated in an unlinked fashion. Several authors [152,159, D. Agnieszka and P. Weglenski, pers. commun., 94] have reported cointegration of the sequences at one chromosomal location. Nevertheless, at least in aspergilli, homologous integration of the two transforming sequences at their respective chromosomal locations occurs frequently enough to allow the replacement of genes by an in vitro mutagenized allele [100,121,152], thus creating predetermined mutants.

Cotransformation has been used to rapidly map cloned genes on cosmids [160] by using gel-purified restriction fragments, and also to demonstrate the functionality of cloned genes when no mutants or selectable phenotypes are available. By using high amounts of the vector containing such gene, cotransformants can be obtained in which many copies have been integrated. We [P. J. M. van den Broek et al., unpublished results] used this approach with putative *A. niger g6pd* genes and obtained cotransformants with elevated levels of glucose-6-phosphate dehydrogenase, thus demonstrating the functionality of the gene. Similarly, Davis and Hynes [147] demonstrated the functional expression of the regulatory *nit*-2 gene of *N. crassa* in *are*A mutants of *A. nidulans*. Recently, the benzoate hydroxylase gene from *A. niger* was also expressed in this way [240; J. G. Boschloo et al., in press].

VII. GENE REPLACEMENTS AND GENE CONVERSIONS

The unpredictable integrational behavior of transforming and cotransforming DNA in fungi will certainly hamper gene replacement events. Nevertheless,

TABLE 2. Expression of Cloned Fungal Genes and Derivatives in Different Hosts

Gene	Fungal species	Ref.
N. crassa pyr-4	*Neurospora crassa*	98
	Aspergillus awamori	232
	A. flavus	23
	A. nidulans	13
	A. niger	37,89
	Penicillium chrysogenum	35,237
	Trichoderma reesei	232
*A. nidulans arg*B	*Aspergillus nidulans*	16
	A. niger	221
	A. oryzae	238
	Magnaporthe grisea	34
	Trichoderma reesei	45
*A. nidulans amd*S	*Aspergillus nidulans*	14,38
	A. ficuum	89
	A. niger	95
	A. terreus	234
	Cochliobolus heterostrophus	22
	Colletotrichum lindemuthianum	42
	Penicillium chrysogenum	90
	Septoria nodorum	223
	Trichoderma reesei	45
*A. nidulans gpd*A fused to *E. coli hph*[a]	*Aspergillus nidulans*	91
	A. ficuum	89
	A. niger	91
	Cladosporium fulvum	51
	Curvularia lunata	222
	Fusarium oxysporum	41
	Leptosphaeria maculans	235
	Septoria nodorum	223
*A. nidulans gpd*A fused to *E. coli lac*Z[b]	*Aspergillus nidulans*	87
	A. ficuum	89
	A. niger	88
	A. oryzae	233
	Penicillium chrysogenum	90
	Trichoderma reesei	45
*A. nidulans trp*C fused to *E. coli Bm*[c]	*Aspergillus nidulans*	46
	A. niger	46
	Claviceps purpurea	46

[a]Confers hygromycin resistance to the transformed strains [91].
[b]Gives blue-transformed colonies on X-gal-containing media [87].
[c]Confers bleomycin resistance to the transformed strains [46].

replacements are worthwhile to perform since they are valuable tools for the inactivation of genes which do not possess a selectable phenotype or for which their function is not known, or for the introduction of in vitro-modified regulatory sequences at their normal chromosomal location.

Replacements are relatively easy to obtain for only a very few genes, usually those that have a semidominant phenotype. Ward et al. showed that replacement of the ATP synthase subunit 9 gene in *A. nidulans* [149] and *A. niger* [56] by the oligomycin-resistant allele can be selected by virtue of its higher level of resistance compared to the heterozygous transformants, whereas the reverse replacement can be selected using triethyltin. Similarly, replacements of the *ben*A gene by a benomyl-resistant allele can be selected at high concentrations of benomyl. Dunne and Oakley [158] used this selection to estimate the frequencies with which gene replacements and gene conversions occur at the two copies of the *ben*A gene in a type I transformant of *A. nidulans* during mitosis. Both frequencies are of the order 2×10^{-4}, indicating that without selection these recombinants are very hard to obtain. Miller et al. [142] demonstrated the feasibility of one-step gene replacement in *A. nidulans* by transforming an *arg*B^{+}-*trp*C^{-} recipient with a functional *trp*C gene, inserted within the coding region of the *arg*B gene. Of the *trp*C transformants, 30% was simultaneously *arg*B^{-}, mainly through gene replacement. Similarly, Paietta and Marzluf [154] obtained *am* mutants of *N. crassa* with a *qa*-2 gene integrated within an *am* gene at a frequency of 8%. Wernars et al. [152] replaced the *A. nidulans* *trp*C gene in one step by a *trp*C-*lac*Z hybrid gene using cotransformation with an unlinked selection marker. The frequency was very low but the recovery of the mutants was greatly facilitated by their *lac*Z^{+} phenotype. Following a similar approach we recently constructed *trpC* mutants in *A. niger* [100], which could previously not be isolated by classical genetic methods. In those cases where replacement does not lead to a change in phenotype, a two-step procedure can be applied in *A. nidulans*. In the first step a type I integration, where the selected gene is integrated in between the two alleles of the gene to be replaced, has to be isolated. Such a transformant is then allowed to self-fertilize and from the progeny colonies are selected which have lost the selection gene. This scheme is essentially the same as described above for the *ben*A locus, but the frequencies of meiotic recombination are considerably higher. Miller et al. [142] used this approach to replace a nonessential sporulation-specific gene at a frequency of about 5%. Davis et al. [121] used a similar two-step procedure to replace the wild-type *amd*S gene by *amd*S-*lac*Z fusion constructs. Recently, similar experiments were performed with the *N. crassa* *am* [G. D. Frederick and J. A. Kinsey, pers. commun., 94] and *cpc*-1 [M. Plamann and C. Yanofsky, pers. commun., 94] genes and the *A. nidulans* *acu*D gene [P. Bowyer, pers. commun., 94].

VIII. ISOLATION OF FUNGAL GENES

A. Complementation in *E. coli* and *S. cerevisiae*

The first fungal genes isolated, e.g., *arg*B [161], *arom* [162], and *trp*C [163] of *A. nidulans* and *qa*-2 [9] and *pyr*-4 [98] of *N. crassa*, were selected by complementation of the corresponding auxotrophic mutants of *E. coli*. This approach, however, is very limited since many fungal genes contain introns, which cannot be processed by *E. coli*, and fungal expression signals might not be recognized by the prokaryotic transcription and translation machinery. In fact, for some genes it was demonstrated that transcription in *E. coli* was initiated at a site different from that in the fungus [12,164]. Surprisingly, there seems to be a "conservation" in the genes which will complement *E. coli* mutants. For instance, the gene for the trifunctional enzyme in tryptophan biosynthesis was successfully cloned from *A. nidulans* [163], *A. niger* [146], *N. crassa* [165], *S. commune* [27], *P. chrysogenum* [166], *Ph. blakesleeanus* [167], and *C. heterostrophus* [168].

To overcome the disadvantages of *E. coli*, the same approach was made in the lower eukaryote *S. cerevisiae*, which has at least the potential to splice introns [169]. In this way Berse et al. [170] succeeded in cloning the *A. nidulans argB* gene. The genetic distance between this yeast and filamentous fungi, however, is apparently so large that splicing and expression are very inefficient [171-173]. One remaining option then, is to construct cDNA libraries in expression vectors to select for the desired gene [173].

B. Selection by Hybridization or Immunodetection

For genes which are sufficiently conserved, isolation could possibly be achieved by hybridizing gene libraries with probes containing the homologous counterpart from another organism at the appropriate conditions. Indeed, this approach has been shown to be successful using gene probes which originated from other fungi. For instance, the isopenicillin N synthetase genes from *P. chrysogenum* [174] and *A. nidulans* [175] were isolated with the *C. acremonium* gene as probe. Similarly, the *N. crassa arg*-12 (with *A. nidulans argB* [176]), *A. nidulans gdhA* (with *N. crassa am*, [177]), and *A. niger pyrA* (with *N. crassa pyr*-4 [37]) genes were obtained in this way. Unfortunately, however, for many genes homology rapidly diminishes with greater evolutionary distance. Only highly conserved genes, like glycolytic enzymes, can be isolated with the corresponding *S. cerevisiae* genes [150,178,179]. In our laboratory we have succeeded in cloning a gene from *A. niger* (*g6pd*, glucose-6-phosphate dehydrogenase) with a probe from the corresponding *Drosophila* gene [P. J. M. van den Broek et al., unpublished results]. If protein sequence data were available, genes have been cloned using synthetic oligonucleotides, either as probes [10] or as primers for cDNA synthesis [180,181].

A differential hybridization technique [182] has proven to be very useful for the isolation of genes which exhibit expression in different cell types or under different culture conditions. Originally developed for the cloning of genes from *A. nidulans*, involved in conidiation [182,183], it has been successfully adapted to isolate similar genes from *N. crassa* [184], several of which have now been mapped [M. A. Nelson, pers. commun., 94], acetate-inducible genes [I. F. Connerton, pers. commun., 94] including *acu*-5 [185] from *N. crassa*, *amdS* from *A. nidulans* [186], fruiting genes from *S. commune* [187], and circadian clock genes from *N. crassa* [188].

Caddick et al. [189] isolated the *A. nidulans areA* gene by exploiting the existence of a pericentric inversion near this gene. From crosses, progeny were obtained in which the *areA* gene and the region centromere-distal to it were deleted. Differential screening of a λ library with labeled total genomic DNA from a wild-type strain and the deletion strain resulted in clones containing the *areA* gene and a considerable region centromere-distal to it.

Chromosome walking can be used in those cases where a DNA region of interest shows sufficiently tight linkage to a previously cloned gene. Schechtman [126] used this approach to isolate telomeric sequences from the right arm of chromosome V of *N. crassa*. Debuchy et al. [118] cloned a considerable region in the vicinity of the centromere of chromosome V of *P. anserina* by walking from a closely linked gene, *su*-3, and thus identified a region which is lethal at a two-copy state. With the rapid accumulation of cloned genes one might expect this technique to become increasingly important, especially when ordered cosmid libraries [50] can be used. The isolation of complete chromosomes by pulsed field electrophoresis [R. Ballario et al., pers. commun., 94], allowing the construction of chromosome-specific gene libraries, will further increase the potentials of these methods.

Finally, antibodies raised against purified proteins have been employed to isolate genes directly from cDNA expression libraries [190] or from genomic libraries by hybrid release and in vitro translation [191,192].

C. Selection on the Basis of Complementation in Filamentous Fungi

Expression of the gene of interest in the homologous organism will not suffer from problems regarding intron splicing and promoter recognition. Given the sufficiently high transformation frequencies now achieved with *A. nidulans*, *N. crassa*, *S. commune*, and a few other fungi, selection by complementation in these species has become feasible. Gene libraries based on plasmids [49,60,92,193] or cosmids [55,160], containing a fungal selectable marker, have resulted in transformants complementing numerous mutations. As was discussed before, nearly all fungal transformants contain integrated copies of the transforming DNA, sometimes in multiple copies. Although recovery of plasmid sequences by transforming *E. coli* with undigested DNA was reported for *A. nidulans* [104,194] and *N. crassa* [115], these plasmids are frequently rearranged [115,145] and sometimes even lack completely fungal sequences [20,24]. Reisolation frequencies can be substantially increased by partial digestion of the transformant DNA with a restriction enzyme and relegation [60,193]. Oakley et al., who used this approach to clone the *A. nidulans pyr*G [32] and *ribo*B [195] genes, reported that recovery of the correct plasmid might be obscured by the fact that in general *A. nidulans* transformants contain several different integrated plasmid molecules. This prompted them to modify their transformation protocol as to promote single-copy integration. When cosmid libraries were used an additional rescue method was successfully applied by Yelton et al. [119]; they packaged transformant DNA into λ phage particles and transduced *E. coli* to recover the cosmid vector. In this way they reisolated a cosmid carrying the *A. nidulans y*A gene. The efficiency of this type of rescue, however, may vary considerably for individual transformants [R. van Gorcom et al., unpublished results]. The possibility to (co)transform filamentous fungi with isolated restriction fragments [100,160] allows rapid localization of the gene within the recovered sequences.

For *N. crassa*, in which stability of transformed DNA sequences sometimes is a serious problem (see hereafter), Akins and Lambowitz [53] described a sibling selection procedure which obviates the need to recover transforming DNA from a transformant. In a few consecutive steps the *nic*-1 gene was isolated with this procedure. The efficiency of the sib selection method was considerably improved by Vollmer and Yanofsky [50]. They constructed a genomic library of *N. crassa* DNA in a cosmid vector that contained a benomyl resistance dominant selection marker and arranged a total of 3072 clones in microtiter dishes. By just three consecutive *N. crassa* transformations, numerous genes were isolated [50,196,197] using this ordered library. The large inserts in the cosmid library offer the additional opportunity to isolate tightly linked genes. This was exploited by Glass et al. [198], who cloned the mating type allele A by selecting for complementation of the linked *un*-3 mutation.

The isolation of promoter sequences from the DNA of filamentous fungi has received special attention. Van Gorcom et al. [86] showed that the *E. coli lac*Z gene can be used as reporter for gene expression in *A. nidulans*. They developed vectors for aspergilli in which fungal expression sequences can be fused to this reporter gene [87]. This system allows not only selection but also characterization of fungal promoters [120]. Recently, constructs have been made with the *E. coli uid*A gene, coding for β-glucuronidase (GUS) [62, T. Goosen et al., unpublished results]. GUS fusions have some distinct advantages over other reporter systems. Since it is a small gene, expression can be easily assayed in culture plates, in tissue sections, by spectrophotometry or fluorometry. Sensitivity of detection is very high and endogenous activities are absent in a variety of organisms. Furthermore, in contrast to β-galactosidase, β-glucuronidase can be translocated across membranes [199]. It seems, however, that the GUS enzyme is less tolerant toward amino-terminally fused proteins than β-galactosidase [T. Goosen et al., unpublished results].

Turgeon et al. [97] isolated promoter sequences from the plant pathogenic fungus *C. heterostrophus* by their ability to drive expression of a bacterial *hph* gene, thus conferring hygromycin resistance to the fungus. A slightly different approach was used for *Phycomyces* [85]. This method might be applicable to many otherwise inaccessible fungi.

IX. STRUCTURE AND EXPRESSION OF CLONED FUNGAL GENES

A. Structure and Organization of Fungal Genes

The considerable progress made in the efficiency of cloning fungal genes has also resulted in the rapid accumulation of sequence data. At the time of writing, over 230 fungal genes had been cloned from either A. nidulans (>60 genes), *N. crassa* (≈80 genes), *A. niger* (≈20 genes), and from at least 23 other fungal species. For at least one-third of them (primarily from *A. nidulans* and *N. crassa*), sequence data have been published. Due to size limitations for this overview neither a survey of cloned and sequenced genes nor a complete list of references can be given. Detailed information is available on request.

Analysis of these sequence data so far [T. Goosen et al., unpublished] has provided some insight into the general structure of fungal genes. About two-third of the sequenced genes contain introns, which in general are between 50 and 80 bp. Most introns fit the deduced consensus sequences for fungi [6] and are supposed to be involved in the splicing process.

With respect to codon usuage some bias has been found for most fungal genes, in *N. crassa* being most prominent.

Sequences responsible for the reliable expression of the gene have also been analyzed in some more detail for *A. nidulans* and *N. crassa* in particular. This concerns both the 5' nontranslated region (transcription initiation signals, translation signals) and the 3' nontranslated region (transcription termination and polyadenylation). Only a few promoter deletion studies have been performed.

Most of the fungal genes are organized as single transcription units. This implicates that coordinated expression of genes relies on the exact tuning of mRNA synthesis. In several cases, however, genes involved in the same metabolic route are clustered, probably to allow for such a coordinated expression.

B. Homologous Expression of Cloned Fungal Genes

Cloning of genes provides a means to study their expression and regulation at the molecular level. For an increasing number of fungal genes such studies are in progress. The best documented case so far is for the genes from the *N. crassa qa* gene cluster, for which both cis- [200] and trans-acting [201] elements have been identified. Baum et al. [202] used DNase foot printing and gel retardation techniques to identify 13 binding sites of the *qa*-1F activator protein in the gene cluster. In *A. nidulans*, the regulatory regions of the *trp*C [120] and *arg*B [203] genes have been analyzed by deletion mapping. In the *trp*C 5' region a putative repressor binding site was identified, whereas in the *arg*B gene a 100-bp region was shown to be involved in general amino acid control.

Apart from studies on 5'-located regulatory elements, there are increasing reports concerning expression of fungal genes after reintroduction by transformation. In many cases it has been found that in transformants in which intact copies of the gene are integrated, the level of expression and/or the response to effectors normally regulating the expression may have been altered. Transformants obtained with the *N. crassa am* gene [79] in most cases have only a fraction of the wild-type glutamate dehydrogenase activity, unless the transformed gene replaced the resident mutant allele. Miller et al. [156] demonstrated that the chromosomal region in which a gene is situated influences gene expression. When they relocated an *A. nidulans* spore-specific gene, which normally is part of the *Spo*C1 gene cluster [204], to different chromosomal positions, it was expressed at a considerable level in vegetative hyphae. Conversely, insertion of the *arg*B gene into the *Spo*C1 gene cluster markedly reduced its vegetative expression. For the *qa*-2 gene [9], *N. crassa* transformants containing one copy of the gene showed wild-type levels of enzymatic activity, irrespective of the type of integration. Da Silva et al. [205] demonstrated that the equivalent *qut*E gene in *A. nidulans* shows normal regulated expression, even when 16 copies of the gene are present.

With the *A. nidulans amd*S gene, which is normally subjected to both general and specific control mechanisms [206], we observed [K. Wernars et al., unpublished results] that the level of expression can be well below that of the wild type in some transformants, whereas in others a considerable overexpression occurred, more or less proportional to the number of integrated *amd*S genes. Kelly and Hynes [141] later showed that multiple *amd*S gene copies cause titration of the product of the positively acting regulatory gene *amd*R, thereby deregulating the expression of other genes under *amd*R control.

In the basidiomycete *Coprinus cinereus* aberrant expression of transformed genes has been observed. Mellon and Casselton [207] analyzed the expression of nonhomologously integrated *acu7* genes, introduced into *Coprinus* by cotransformation, and found that cotransformants with high copy numbers showed less than wild-type isocitrate lyase activity. In several cotransformants containing intact copies of the *acu7* gene, no expression occurred at all, indicating that the chromosomal location of a gene can be important for its expression.

C. Heterologous Expression of Cloned Fungal Genes

Genes from one filamentous fungus are in general readily expressed in a wide variety of other filamentous fungi (Table 2), albeit not always at levels normal for the homologous situation [208]. Heterologous expression is not unexpected in view of the many similarities in the structure of such genes. This finding has allowed the construction of more or less universal fungal cloning vectors [46,91] (see also Table 1). There seems, however, to be a considerable evolutionary divergence between the ascomycetes like *Neurospora* and *Aspergillus* on one hand and the filamentous basidiomycetes like *Coprinus* and *Schizophyllum* on the other [40]. Studies by De la Fuente Herce et al. [209] make mention of the absence of any detectable expression of *Aspergillus* genes in *Coprinus* cotransformants. A more striking phenomenon is that in many cases where heterologous genes are expressed, they are subject to the normal regulatory mechanisms. This was first shown by Kelly and Hynes [95] for the *A. nidulans amd*S gene, which in *A. niger* is subject to the same positive control as exerted by the *amd*R gene in *A. nidulans*. Whitehead et al. [210] showed that the *nia*D genes from both *A. nidulans* and *A. niger* are expressed in a *Penicillium* nitrate reductase-deficient mutant in a regulated fashion. Similarly, regulated expression of the acid phosphatase gene of *A. niger* has been demonstrated in *A. nidulans* [211]. Davis and Hynes [147] demonstrated that mutants in the *A. nidulans are*A gene, a regulatory gene in nitrogen metabolism, can be complemented by the *Neurospora nit-2* gene.

A special aspect of heterologous expression is its use in strain breeding. Beckman et al. [pers. commun., 94] at Eli Lilly expressed the *Streptomyces* expandase gene, which is involved in the biosynthesis of cephalosporins, in *Penicillium*. Genetic alterations in the cloned gene eventually could lead to development and production of new antibiotics.

In addition, filamentous fungi are of special interest for the production of heterologous proteins. Several species, e.g., *Aspergillus* [212], *Trichoderma* [213], have the capacity to secrete large amounts of protein. If one wants to utilize this secretion capacity with foreign products, these heterologous genes have to be expressed in such a way that the expression products are secretable. Secretion of proteins depends on the presence of a so-called signal peptide, which is constituted by the amino terminal part of a secreted protein. Although signal peptides from different origins show remarkable conservation, their effectiveness in heterologous organisms may vary considerably [171,214]. In a few cases the feasibility of expression and secretion of heterologous proteins by *A. nidulans* has been demonstrated. The *Mucor miehei* aspartyl protease [125] is secreted from *A. nidulans* by its own signal peptide. The primary translation product contains, besides a signal peptide, a prosequence. As the secreted protease is in its mature form, not only secretion but also normal processing occurs in *A. nidulans*. The bovine prochymosin [116], when expressed from the *A. niger* glucoamylase promoter, is secreted using its own signal peptide. However, replacement by the glucoamylase signal sequence improves secretion. Processing of prochymosin can occur in *A. nidulans*, since mature chymosin-like proteins have been detected in the culture medium [116]. Upshall et al. [123] reported the expression of human tissue plasminogen activator (tPA) using the promoter of the *A. nidulans tpiA* gene. Active human tPA is derived from the primary translation product by cleaving the pro sequence, after which an aminopeptidase removes four more NH_2 terminal amino acids. Both these tPA forms are active. Surprisingly, *A. nidulans* is capable of both kinds of processing, since both forms of tPA are found in the culture medium. In addition, tPA produced in *A. nidulans* is not excessively glycosylated, a problem often encountered with yeast [215]. Other examples of expressed and secreted proteins are human interferon α-2 [122], human α-1-antitrypsin, and human granulocyte-macrophage colony-stimulating factor [123]. Recently, *Trichoderma reesei* [A. Harrki, pers. commun., 94] was also used to express and secrete mammalian proteins.

The high potential of filamentous fungi as hosts for the production of proteins of industrial, diagnostic, or therapeutic importance is well documented now. Many studies on these gene expression systems have been initiated. The large commercial interest in such projects, however, causes a considerable delay in the disclosure of relevant facts and findings. This implicates that the list of examples, given in this section, is far from complete.

X. THE VALUE OF GENETIC TECHNIQUES FOR TRANSFORMATION AND GENE MANIPULATION

In several fungi gene manipulation can be performed successfully now, even if the genetics of the organism is still poorly developed. By using vectors containing heterologous dominant selection markers, transformation is possible without the requirement of any specific marker in the acceptor strain. For the transformation of industrial production organisms or phytopathogenic fungi, dominant selection markers are sometimes even to be preferred. The introduction of auxotrophic mutations by mutagenic procedures will certainly result in a change in the genetic background. Such a change might influence production or virulence properties of these organisms.

Several examples of transformation with dominant selection markers, both for industrially important fungi and for a variety of phytopathogens, have been published recently (Table 1). In all these examples recombinant plasmid DNA could be introduced and expressed in wild-type strains. In addition, it was shown that other cloned genes, from either homologous or heterologous origin, could be introduced and expressed by cotransformation. This developed methodology will certainly allow the construction of auxotrophic mutations in a defined way, e.g., by gene replacement with an in vitro-mutated or disrupted gene [100,152]. This latter procedure can still avoid the use of classical genetic techniques.

However, if one wants to study complicated processes which involve multiple genes, e.g., pathogen-host interactions or the biosynthesis of complex secondary metabolites, the need for fungal genetics will soon emerge.

For this reason we have developed materials and methods to study the genetics of the industrially exploited fungus *A. niger* [216-218]. The application of genetic techniques to industrial filamentous fungi can be of great value to analyze and characterize transformants and to construct specific strains.

A. Genetic Methods as Tools to Characterize Transformants

In order to verify integration of the transforming sequence into the genome of an acceptor strain, genetic analysis is equally important as molecular analysis. If classical methods of genetic manipulation are available, as is the case with the extensively studied *A. nidulans* and *N. crassa*, such and analysis can be performed quite easily. If, however, the species cannot reproduce sexually as is the case with *A. niger*, this analysis will be complicated and additional genetic tools will be necessary.

1. Genetic Analysis of *A. nidulans* Transformants

The very sophisticated genetics of *A. nidulans*, which employs both the sexual and parasexual cycle, allows a detailed analysis of transformation events, resulting in information on the integration specificity and the stability of the transformants. For most *A. nidulans* transformation systems analyzed, mitotic stability has been claimed to be very high [13-15,38]. The mitotic instability observed in some cases was thought to be a consequence of heterokaryosis [144] due to the multinucleate character of the protoplasts. In some cases [106,114], mitotic instability was ascribed to alleged extrachromosomal maintenance of the vector. In contrast, meiotic stability of transformants differs greatly between the various transformation systems [14,38,142]. In only a few cases has the detailed genetic analysis of *A. nidulans* transformants been reported, and such analysis was sometimes obscured by meiotic instability. Wernars et al. [140] performed such analysis on (multicopy) *amd*S$^+$ transformants and could show by parasexual analysis that this gene did not exhibit any preference for integration into one particular of the eight chromosomes of *A. nidulans*. However, in some transformants meiotic instability prevented the mapping of the integrated *amd*S gene within the chromosome by meiotic analysis and DNA rearrangements could be detected after meiosis [38].

Upshall [144] characterized a large set of *arg*B$^+$ transformants by genetic and molecular analysis and observed a high degree of mitotic stability. However, all transformants which had not arisen by gene replacement (type III) showed meiotic instability, which correlated with the type of transformation event that had occurred. As found for some of the *amd*S$^+$ transformants [140], a few *arg*B$^+$ transformants could not be analyzed due to instability; a few showed a high frequency of mitotic aneuploidy, meiotic sterility, and poor heterokaryon formation and were shown to be isolated as diploids. Apparently the location of the integrated DNA within the chromosome influences its stable inheritance. Remarkably, also when not homologously integrated, the transformed *arg*B gene was always situated on chromosome III, even in the case of the heterozygous diploids [144].

*Trp*C$^+$ transformants are mitotically stable [15]. Meiotic instability upon self-fertilization was especially observed in transformants containing tandem copies of the selection marker [142] and could result in >1% auxotrophic segregants. This type of intrachromosomal recombination or unequal crossing over has been used [142] to develop two-step gene replacement procedures for *A. nidulans*.

Dunne and Oakley [158] quantified the mitotic recombination of an interrupted duplication, created by integration of *benA*- and *pyr*-4-bearing transforming plasmid sequences at the *benA* locus of *A. nidulans*. The low frequencies of reciprocal recombination (2×10^{-4}) and gene conversion (2×10^{-4}) observed justifies the classification of the transformants as mitotically stable. Two-step replacement of chromosomal alleles with alleles carried on the transforming plasmid can nevertheless be performed with this system by selecting for loss of *pyr*-4 function on 5-fluoroorotic acid (5-FOA).

Genetic analysis of diploid transformants can provide a means for the mapping of essential genes which are not associated with any known phenotype. In diploids one of the alleles of such a gene can be deleted or interrupted, which is of course impossible in haploid strains. Segregation analysis of such diploid transformants was recently used [157] to assign the *A. nidulans* actin gene to chromosome I. Cullen et al. [54] used mitotic haploidization and sexual crossing to map the *A. nidulans ans*1 sequence, which by transformation had been linked to an oligomycin-resistant marker gene, to the centromeric region of chromosome I.

2. Genetic Analysis of *A. niger* Transformants

Genetic analysis in *A. niger* has long been hampered by the absence of a sexual cycle (*A. niger* belongs to the Fungi Imperfecti) and its ill-developed genetics. For this reason a collection of mutants with genetic markers in an isogenic background has been established [219]. These monoauxotrophic mutants have been induced by low doses of mutagen and were isolated by enrichment procedures [217,220]. Using somatic recombination, various combinations of markers have been made, finally resulting in master strains which contain markers on six linkage groups [216]. With these strains, genetic analysis of *A. niger pyrA* transformants as well as *A. niger amdS* transformants has been performed in our laboratory. Integration of the *A. niger pyrA* gene was shown to occur predominantly at the homologous *pyrA* locus on linkage group III, resulting in transformants with a high mitotic stability [37,240]. Integration of the heterologous *A. nidulans amdS* gene very often occurs in multiple copies [95, J. G. Boschloo et al., in press]. In all transformants genetically analyzed so far, the multiple-copy insert was found at one single chromosomal location, which is different in all cases [239]. The multiple-copy *amdS* transformants exhibit a high degree of mitotic stability. Similar reports have been made for other homologous or heterologous marker genes [56,151, 221]. Most transformation systems for *A. niger* have been based on (semi)dominant selection markers [56,95,151], which usually give low transformation frequencies. As many *A. niger* mutants are isolated now or can readily be obtained, more transformation systems based on homologous genes will become available. Recently we established a system based on the homologous *trpC* gene [146] and obtained very high transformation frequencies [100].

3. Genetic Analysis of Transformants in Other Fungi

Neurospora crassa qa-2 and *am* transformants have been extensively analyzed genetically in order to determine linkage to the homologous locus [67,103,133]. In doing so, Grant et al. [103] found an extreme meiotic instability for *am* transformants when they were backcrossed with an *am* deletion strain or intercrossed. Furthermore, the am^+ progeny from backcrosses gave a high proportion of *am*-auxotrophic segregants during vegetative growth. Case et al. [133] analyzed *qa*-2 transformants and reported 100% mitotic stability. Tetrad analysis of crosses of unlinked duplication-type transformants with each other and selfings of such transformants resulted in a high frequency of *qa*-2^- progeny, either with or without visible changes in Southern hybridization patterns. A "correction process," acting during meiosis and preferentially changing the number of functional *qa*-2 genes, was suggested. Also Dhawale and Marzluf [67] observed high mitotic stability of *qa*-2 transformants, but 74% of them failed to segregate through a cross. It was suggested that these nontransmissible transformants harbored autonomously replicating plasmids, which were lost during meiosis. Later reports by Selker and coworkers [129,130] demonstrated that the observed meiotic instability is caused by inactivation of duplicated sequences through RIPping (see Section V).

For *P. anserina* transformants obtained with the tRNA suppressors *su*4-1 or *su*8-1, Brygoo and Debuchy [44] used an acceptor strain which contained, besides the suppressible *leu* mutation used for selection, a suppressible mutation which prevents spore pigmentation. Genetic analysis of the transformants in ordered tetrads was then performed by scoring for restored pigmentation. In this way it was shown that transformants of *Podospora* obtained with plasmids are meiotically very stable. Transformants obtained with cosmids, on the other hand, were highly unstable when integration at the homologous locus had occurred, but stable when integration was nonhomologous. Analysis of the timing of the instability revealed that loss of the selection marker was limited to the fruiting body.

For most other filamentous fungi which have been transformed, high mitotic stability of the transformants has been reported. Only in a few cases has a certain degree of instability [35,45] or DNA rearrangement [41,42,222] been reported. For the plant pathogenic species *C. lindemuthianum* [42], *Septoria nodorum* [223], and *U. maydis* [39], the transformed DNA was shown to be stably inherited during growth in the host plant. Meiotic stability has been analyzed by genetic methods in only a limited number of fungi. For *S. commune* transformants [49], the meiotic stability was studied by fruiting the dikaryotic mycelium. In nine out of 11 transformants, the transforming DNA segregated stably as a single gene. Absolute meiotic stability was found for *C. cinereus acu*-7 transformants [40], *U. maydis* Hm^R transformants [39], and *C. heterostrophus amd*S transformants [22]. The *C. cinereus* TRP1 transformants [29] were stable through cell division, mating, fruiting body formation, and meiosis.

B. Recombination as a Genetic Tool in Strain Construction

The procedures normally used in breeding production strains (mutagenesis and selection) may result in chromosomally aberrant strains, especially if several independent mutations have to be introduced. Different genes involved in a complex process can be amplified, switched off, or deregulated in separate strains if parasexual analysis can be performed in the fungal species of interest. Improved strains can be isolated by haploidization of heterozygous diploids, and excessive mutational damage can be prevented. The high potential of recombination as a tool to construct strains which produce large amounts of a specific product has been nicely demonstrated by recombining several multicopy *ura*5 transformants of *P. anserina* to a stable high-copy strain [224].

For fungi belonging to the Ascomycetes or Basidiomycetes, genetic analysis and breeding techniques are mainly based on meiotic recombination. Many of the industrially important fungi and plant pathogenic fungi belong to the Fungi Imperfecti or to the Phycomycetes, which either lack an effective meiotic mechanism or are difficult to handle in recombination experiments. Somatic recombination can then be an alternative for genetic analysis and strain construction.

Prerequisites for somatic recombination are heterokaryosis and somatic karyogamy. Heterokaryons are formed by hyphal fusions (anastomoses) followed by the exchange of nuclei from genetically different strains. Although vegetative incompatibility can prevent the formation of heterokaryons [225] or incompatibility reactions after anastomosis can occur [226], intra- and interspecific fusions have been obtained. Since production strains often are of different origin and recurrent mutation induction can reduce the ability to form heterokaryons, other ways have been exploited to achieve somatic recombination. Protoplast fusion can be a useful tool to bypass heterokaryon incompatibility [227]. Fusion of protoplasts of two genetically different strains will unambiguously lead to heterokaryons. In some cases even fusion of protoplasts of different (related) species may lead to heterokaryons (reviewed in Ref. 228). In other studies, however, only very poorly growing fusion products were obtained (reviewed in Ref. 228). The use of protoplast fusion for somatic recombination between vegetative incompatible strains might be limited, since anastomosis might still be necessary to maintain a heterokaryon. Nevertheless, diploid conidiospores or amphihaploids (in the case of interspecific fusions) have been isolated from poorly growing heterokaryons (e.g. 229). Due to their different genetic background, unexpected and unpredictable phenomena can be encountered in subsequent recombination experiments. In conclusion, recombination between unrelated strains can be used for genetic analysis, but for breeding purposes not without due consideration.

Somatic recombination is being applied in our laboratory for the construction of *A. niger* acceptor strains for transformation [A. J. M. Debets et al., unpublished]. Such strains then contain a linked marker in addition to the selection marker. These marker combinations are also useful to construct test strains (with marked chromosomes) which can be used for genetic analysis of the transformants. Combined molecular and genetic analysis of transformants is important for a good understanding of all steps involved in transformation, integration, and stable expression of integrated sequences. For a complete description of the final strains with respect to risk assessment, breeding programs, and patent rights, such an analysis will also be a necessity.

XI. CONCLUDING REMARKS

As outlined in this chapter, DNA-mediated transformation and gene manipulation procedures for filamentous fungi have grown out to standard laboratory techniques. The development of this advanced methodology has already resulted in the coming-of-age of molecular genetics of several fungal species, most notably of course *A. nidulans* and *N. crassa* for which an enormous wealth of genetic and biochemical information already existed from classical studies. Many other fungal species will become amenable to genetic engineering soon.

The results discussed allow some generalizations with respect to transformation of filamentous fungi. Some of the transformation features summarized are highly similar to those described for the most intensively studied fungus, the unicellular yeast *S. cerevisiae*; some are, however, quite different. These similarities and dissimilarities have been used to develop the specific methodologies and strategies for the filamentous fungi.

Similarities are based on the behavior of integrative vectors in yeast. Transforming DNA can be integrated by means of single homologous crossing over, resulting in tandemly repeated sequences separated by the vector sequence or by double crossover events (gene conversion) resulting in replacement of the resident gene by the incoming gene free of vector sequences. The use of autonomously replicating vectors or artificial minichromosomes in filamentous fungi is still doubtful. These homologous recombination features have been used to develop strategies for one- and two-step gene replacements and for directed integration of DNA sequences at predetermined sites in the genome of the recipient.

Quite different from the yeast system is the limited level of homology needed for the stable integration of the transforming DNA into the genome of the fungal recipient. This limited homology allows integration at many sites in the genome. The main advantage of this feature, however, is the possibility of readily transferring cloned genes between different species. This phenomenon has been extensively used to develop transformation procedures for a great variety of fungal species. Such heterologous transformation systems will continue to be the starting point for molecular biological studies in fungi of economic importance. This gene transfer feature will also greatly facilitate studies on gene function and gene regulation in species which lack good genetic systems. Regulatory genes can be cloned directly from gene libraries by transformation and then be studied in well-characterized species such as *A. nidulans* or *N. crassa*.

Furthermore, cotransformation of unselected sequences occurs in fungi with high frequency, thus facilitating transformation studies by simplifying the vector construction work.

Finally, genes can be amplified in fungi by transformation to high copy number, frequently resulting in increased amounts of gene product. Since in general these multicopy transformants are mitotically stable, the effects of gene overexpression on the regulation of other genes can be studied (titration/antititration of regulatory gene products).

These developments have already resulted in the precise genetic engineering of several fungal species. Since fungi are of great importance in medicine, agriculture, and industry, and continue to serve as models for higher eukaryotes, the developed methodology will be used to modify the genetic composition of the organisms in a profitable way for basic studies or applicational aspects.

REFERENCES

1. Portugal, F. H., and Cohen, J. S. *A Century of DNA: A History of the Discovery of the Structure and Function of the Genetic Substance*. MIT Press, Cambridge, Mass., 1977.
2. McBride, O. W., and Ozer, H. L. Transfer of genetic information by purified metaphase chromosomes. *Proc. Natl. Acad. Sci. USA, 70*: 1258-1262, 1973.

3. Hinnen, A., Hicks, J. B., and Fink, G. R. Transformation of yeast. *Proc. Natl. Acad. Sci. USA, 75*:1929-1933, 1978.
4. Struhl, K. The new yeast genetics. *Nature, 305*:391-397, 1983.
5. Smith, J. E., and Berry, D. R. *The Filamentous Fungi*, Vol. 4. Edward Arnold, London, 1983.
6. Rambosek, J., and Leach, J. Recombinant DNA in filamentous fungi: progress and prospects. *Crit. Rev. Biotechnol., 6*:357-393, 1987.
7. Perkins, D. D. Aspects of the Neurospora genome. In *Molecular Genetics of Filamentous Fungi*. UCLA Symposia on Mol. and Cell. Biol., Vol. 34, Alan R. Liss, New York, 1985, pp. 277-294.
8. Davis, M. A., and Hynes, M. J. Regulatory genes in Aspergillus nidulans. *Trends Genet., 5*:14-19, 1989.
9. Case, M. E., Schweizer, M., Kushner, S. R., and Giles, N. H. Efficient transformation of Neurospora crassa by utilizing hybrid plasmid DNA. *Proc. Natl. Acad. Sci. USA, 76*:5259-5263, 1979.
10. Kinnaird, J. H., Keighren, M. A., Kinsey, J. A., Eaton, M., and Fincham, J. R. S. Cloning of the am (glutamate dehydrogenase) gene of Neurospora crassa through the use of a synthetic DNA probe. *Gene, 20*:387-396, 1982.
11. Schweizer, M., Case, M. E., Dykstra, C. C., Giles, N. H., and Kushner, S. R. Identification and characterization of recombinant plasmids carrying the complete qa gene cluster from Neurospora crassa including the qa-1 regulatory gene. *Proc. Natl. Acad. Sci. USA, 78*: 5086-5090, 1981.
12. Schechtman, M. G., and Yanofski, C. Structure of the trifunctional trp-1 gene from Neurospora crassa and its aberrant expression in Escherichia coli. *J. Mol. Appl. Genet., 2*:83-99, 1983.
13. Ballance, D. J., Buxton, F. P., and Turner, G. Transformation of Aspergillus nidulans by the orotidine-5'-phosphate decarboxylase gene of Neurospora crassa. *Bioch. Bioph. Res. Commun., 112*:284-289, 1983.
14. Tilburn, J., Scazzocchio, C., Taylor, G. G., Zabicky-Zissman, J. H., Lockington, R. A., and Davies, R. W. Transformation by integration in Aspergillus nidulans. *Gene, 26*:205-221, 1983.
15. Yelton, M. M., Hamer, J. E., and Timberlake, W. E. Transformation of Aspergillus nidulans by using a trpC plasmid. *Proc. Natl. Acad. Sci. USA, 81*:1470-1474, 1984.
16. John, M. A., and Peberdy, J. F. Transformation of Aspergillus nidulans using the argB gene. *Enz. Microb. Technol., 6*:386-389, 1984.
17. Tudzynski, P., Stahl, U., and Esser, K. Transformation to senescence with plasmid like DNA in the ascomycete Podospora anserina. *Curr. Genet., 2*:181-184, 1980.
18. Razanamparany, V., and Begueret, J. Positive screening and transformation of ura-5 mutants in the fungus Podospora anserina: characterization of the transformants. *Curr. Genet., 10*:811-818, 1986.
19. Peberdy, J. F., and Ferenczy, L. *Fungal Protoplasts*. Marcel Dekker, New York, 1985, pp. 1-345.
20. Feher, Z., Schablik, M., Kiss, A., Zsindely, A., and Szabo, G. Characterization of inl-positive transformants of Neurospora crassa obtained with a recombinant cosmid-pool. *Curr. Genet., 11*:131-138, 1986.
21. Picknett, T. M., Saunders, G., Ford, P., and Holt, G. Development of a gene transfer system for Penicillium chrysogenum. *Curr. Genet., 12*:449-456, 1987.

22. Turgeon, B. G., Garber, R. C., and Yoder, O. C. Transformation of the fungal maize pathogen Cochliobolus heterostrophus using the Aspergillus nidulans amdS gene. *Mol. Gen. Genet.*, *201*:450-453, 1985.
23. Woloshuk, C. P., Seip, E. R., Payne, G. A., and Adkins, C. R. Genetic transformation system for the aflatoxin producing fungus Aspergillus flavus. *Appl. Environ. Microbiol.*, *55*:86-90, 1989.
24. Kuiper, M. T. R., and De Vries, H. A recombinant plasmid carrying the mitochondrial plasmid sequence of Neurospora intermedia Labelle yields new plasmid derivatives in Neurospora. *Curr. Genet.*, *9*:471-477, 1985.
25. Banks, G. R. Transformation of Ustilago maydis by a plasmid containing yeast 2-micron DNA. *Curr. Genet.*, *7*:73-77, 1983.
26. Henson, J. M., Blake, N. K., and Pilgeram, A. L. Transformation of Gaeumannomyces graminis to benomyl resistance. *Curr. Genet.*, *14*:113-117, 1988.
27. Munoz-Rivas, A., Specht, C. A., Drummond, B. J., Froeliger, E., Novotny, C. P., and Ullrich, R. C. Transformation of the basidiomycete Schizophyllum commune. *Mol. Gen. Genet.*, *205*:103-106, 1986.
28. Specht, C. A., Munoz-Rivas, A., Novotny, C. P., and Ullrich, R. C. Transformation of Schizophyllum commune: an analysis of parameters for improving transformation frequencies. *Exp. Mycol.*, *12*:357-366, 1988.
29. Binninger, D. M., Skrzynia, C., Pukkila, P. J., and Casselton, L. A. DNA-mediated transformation of the basidiomycete Coprinus cinereus. *EMBO J.*, *6*:835-840, 1987.
30. Revuelta, J. L., and Jayaram, M. Transformation of Phycomyces blakesleeanus to G-418 resistance by an autonomously replicating plasmid. *Proc. Natl. Acad. Sci. USA*, *83*:7344-7347, 1986.
31. Manavathu, E. K., Suryanarayana, K., Hasnain, S. E., and Leung, W. C. DNA-mediated transformation in the aquatic filamentous fungus Achlya ambisexualis. *J. Gen. Microbiol.*, *134*:2019-2028, 1988.
32. Oakley, B. R., Rinehart, J. E., Mitchell, B. L., Oakley, C. E., Carmona, C., Gray, G. L., and May, G. S. Cloning, mapping and molecular analysis of the pyrG (orotidine-5'-phosphate decarboxylase) gene of Aspergillus nidulans. *Gene*, *61*:385-399, 1987.
33. Penalva, M. A., Tourino, A., Patino, C., Sanchez, F., Fernandez Sousa, J. M., and Rubio, V. Studies on transformation of Cephalosporium acremonium. In *Molecular Genetics of Filamentous Fungi*. Alan R. Liss, New York, 1985, pp. 59-68.
34. Parsons, K. A., Chumley, F. G., and Valent, B. Genetic transformation of the fungal pathogen responsible for rice blast disease. *Proc. Natl. Acad. Sci. USA*, *84*:4161-4165, 1987.
35. Cantoral, J. M., Diez, B., Barredo, J. L., Alvarez, E., and Martin, J. F. High-frequency transformation of Penicillium chrysogenum. *Biotechnology*, *5*:494-497, 1987.
36. Sanchez, F., Lozano, M., Rubio, V., and Penalva, M. A. Transformation in Penicillium chrysogenum. *Gene*, *51*:97-102, 1987.
37. Goosen, T., Bloemheuvel, G., Gysler, C., De Bie, D. A., Van Den Broek, H. W. J., and Swart, K. Transformation of Aspergillus niger using the homologous orotidine-5'-phosphate-decarboxylase gene. *Curr. Genet.*, *11*:499-504, 1987.
38. Wernars, K., Goosen, T., Wennekes, L. M. J., Visser, J., Bos, C. J., Van Den Broek, H. W. J., Van Gorcom, R. F. M., Van Den Hondel,

C. A. M. J. J., and Pouwels, P. H. Gene amplification in Aspergillus nidulans by transformation with vectors containing the amdS gene. *Curr. Genet.*, *9*:361-368, 1985.
39. Wang, J., Holden, D. W., and Leong, S. A. Gene transfer system for the phytopathogenic fungus Ustilago maydis. *Proc. Natl. Acad. Sci. USA*, *85*:856-869, 1988.
40. Mellon, F. M., Little, P. F. R., and Casselton, L. A. Gene cloning and transformation in the basidiomycete fungus Coprinus cinereus: isolation and expression of the isocitrate lyase gene acu-7. *Mol. Gen. Genet.*, *210*:352-357, 1987.
41. Kistler, H. C., and Benny, U. K. Genetic transformation of the fungal plant wilt pathogen Fusarium oxysporum. *Curr. Genet.*, *13*:145-150, 1988.
42. Rodriguez, R. J., and Yoder, O. C. Selectable genes for transformation of the fungal plant pathogen Glomerella cingulata f. sp. phaseoli (Colletotrichum lindemuthianum). *Gene*, *54*:73-82, 1987.
43. Buxton, F. P., and Radford, A. The transformation of mycelial spheroplasts of Neurospora crassa and the attempted isolation of an autonomous replicator. *Mol. Gen. Genet.*, *196*:339-344, 1984.
44. Brygoo, Y., and Debuchy, R. Transformation by integration in Podospora anserina. I. Methodology and phenomenology. *Mol. Gen. Genet.*, *200*:128-131, 1985.
45. Penttila, M., Nevalainen, H., Ratto, M., Salminen, E., and Knowles, J. A versatile transformation system for the cellulolytic filamentous fungus Trichoderma reesei. *Gene*, *61*:155-164, 1987.
46. Van Engelenburg, F., Smit, R., Goosen, T., Van Den Broek, H., and Tudzynski, P. Transformation of Claviceps purpurea using a bleomycin resistance gene. *Appl. Microbiol. Biotechnol.*, *30*:364-370, 1989.
47. Ballance, D. J., and Turner, G. Development of a high-frequency transforming vector for Aspergillus nidulans. *Gene*, *36*:321-331, 1985.
48. Gomi, K., Iimura, Y., and Hara, S. Integrative transformation of Aspergillus oryzae with a plasmid containing the Aspergillus nidulans argB gene. *Agr. Biol. Chem.*, *51*:2549-2556, 1987.
49. Froeliger, E. H., Munoz-Rivas, A. M., Specht, C. A., Ullrich, R. C., and Novotny, C. P. The isolation of specific genes from the basidiomycete Schizophyllum commune. *Curr. Genet.*, *12*:547-554, 1987.
50. Vollmer, S. J., and Yanofsky, C. Efficient cloning of genes of Neurospora crassa. *Proc. Natl. Acad. Sci. USA*, *83*:4869-4873, 1986.
51. Oliver, R. P., Roberts, I. N., Harling, R., Kenyon, L., Punt, P. J., Dingemanse, M. A., and Van Den Hondel, C. A. M. J. J. Transformation of Fulvia fulva, a fungal pathogen of tomato, to hygromycin B resistance. *Curr. Genet.*, *12*:231-233, 1987.
52. Boylan, M. T., Holland, M. J., and Timberlake, W. E. Saccharomyces cerevisiae centromere CEN11 does not induce chromosome instability when integrated into the Aspergillus nidulans genome. *Mol. Cell. Biol.*, *6*:3621-3625, 1986.
53. Akins, R. A., and Lambowitz, A. M. General method for cloning Neurospora crassa nuclear genes by complementation of mutants. *Mol. Cell. Biol.*, *5*:2272-2278, 1985.
54. Cullen, D., Wilson, L. J., Grey, G. L., Henner, D. J., Turner, G., and Ballance, D. J. Sequence and centromere proximal location of a transformation enhancing fragment ans1 from Aspergillus nidulans. *Nucl. Acids Res.*, *15*:9163-75, 1987.

55. Adrianopoulos, A., and Hynes, M. J. Cloning and analysis of the positively acting regulatory gene amdR from Aspergillus nidulans. *Mol. Cell. Biol.*, *8*:3532-3541, 1988.
56. Ward, M., Wilson, L. J., Carmona, C. L., and Turner, G. The oliC3 gene of Aspergillus niger: isolation sequence and use as a selectable marker for transformation. *Curr. Genet.*, *14*:37-42, 1988.
57. Rambosek, J. A., and Kinsey, J. A. An unstable mutant gene of the am locus of Neurospora results from a small duplication. *Gene*, *27*:101-107, 1984.
58. Ballance, D. J., and Turner, G. Gene cloning in Aspergillus nidulans: isolation of the isocitrate lyase gene acuD. *Mol. Gen. Genet.*, *202*:271-275, 1986.
59. Orbach, M. J., Porro, E. B., and Yanofsky, C. Cloning and characterization of the gene for β-tubulin from a benomyl-resistant mutant of Neurospora crassa and its use as a dominant selection marker. *Mol. Cell. Biol.*, *6*:2452-2461, 1986.
60. Nelson, M. A., Morelli, G., Carattoli, A., Romano, N., and Macino, G. Molecular cloning of a Neurospora crassa carotenoid biosynthetic gene (albino-3) regulated by blue light and the products of the white collar genes. *Mol. Cell. Biol.*, *9*:1271-1276, 1989.
61. Bull, J. H., Smith, D. J., and Turner, G. Transformation of Penicillium chrysogenum with a dominant selectable marker. *Curr. Genet.*, *13*:377-382, 1988.
62. Roberts, I. N., Oliver, R. P., Punt, P. J., and Van Den Hondel, C. A. M. J. J. Expression of the E. coli β-glucuronidase gene in filamentous fungi. *Curr. Genet.*, *15*:177-180, 1989.
63. Radford, A., Pope, S., Saczi, A., Frazer, M. J., and Parish, J. H. Liposome mediated genetic transformation of Neurospora crassa. *Mol. Gen. Genet.*, *184*:567-569, 1981.
64. Ito, H., Fukuda, Y., Murata, K., and Kimura, A. Transformation of intact yeast cells treated with alkali cations. *J. Bacteriol.*, *153*:163-168, 1983.
65. Dhawale, S. S., Paietta, J. V., and Marzluf, G. A. A new rapid and efficient transformation procedure for Neurospora. *Curr. Genet.*, *8*:77-79, 1984.
66. Dickman, M. B. Whole cell transformation of the alfalfa fungal pathogen Colletotrichum trifolii. *Curr. Genet.*, *14*:241-246, 1988.
67. Dhawale, S. S., and Marzluf, G. A. Transformation of Neurospora crassa with circular and linear DNA and analysis of the fate of the transforming DNA. *Curr. Genet.*, *10*:205-212, 1985.
68. Kim, S. Y., and Marzluf, G. A. Transformation of Neurospora crassa with the trp-1 gene and the effect of host strain upon the fate of the transforming DNA. *Curr. Genet.*, *13*:65-70, 1988.
69. Fiedler, S., and Wirth, R. Transformation of bacteria with plasmid DNA by electroporation. *Anal. Biochem.*, *170*:38-44, 1988.
70. Hashimoto, H., Morikawa, H., Yamada, Y., and Kimura, A. A novel method for transformation of intact yeast cells by electroporation. *Appl. Microbiol. Biotechnol.*, *21*:336-339, 1985.
71. McCurrach, K., Glover, L. A., and Hardman, N. Transient expression of a chloramphenicol acetyltransferase gene following transfection of Physarum polycephalum myxamoebae. *Curr. Genet.*, *13*:71-74, 1988.
72. Fromm, M., Taylor, L. P., and Walbot, V. Expression of genes transferred into monocot and dicot plant cells by electroporation. *Proc. Natl. Acad. Sci. USA*, *81*:5824-5828, 1985.

73. Bej, A. K., and Perlin, M. Apparent transformation and maintenance in Basidiomycete mitochondria of a plasmid bearing the Hygromycin (Hyg) B gene. *Genome, 30*, Suppl. 1:300, 1988.
74. Marek, E. T., Richey, M. G., Schardl, C. L., and Smith, D. A. Transformation of fungal protoplasts by electric pulse treatment. *Phytopathology, 77*:1740, 1987.
75. Constanzo, M. C., and Fox, T. D. Transformation of yeast by agitation with glass beads. *Genetics, 120*:667-670, 1988.
76. Klein, T. M., Wolf, E. D., Wu, R., and Sandford, J. C. High velocity micro projectiles for delivering nucleic acids into living cells. *Nature, 327*:70-73, 1987.
77. Johnston, S. A., Anziano, P., Shark, K., Sanford, J. C., and Butow, R. A. Transformation of yeast mitochondria by bombardment of cells with micro projectiles. *Science, 240*:1538-1541, 1988.
78. Mishra, N. C., and Tatum, E. L. Non-mendelian inheritance of DNA induced inositol independence in Neurospora. *Proc. Natl. Acad. Sci. USA, 70*:3875-3879, 1973.
79. Kinsey, J. A., and Rambosek, J. A. Transformation of Neurospora crassa with a cloned am (glutamate dehydrogenase) gene. *Mol. Cell. Biol., 4*:117-122, 1984.
80. Tudzynski, P., Stahl, U., and Esser, K. Development of a eukaryotic cloning system in Podospora anserina. I. Long lived mutants as potential recipients. *Curr. Genet., 6*:219-222, 1982.
81. Stahl, U., Tudzynski, P., Kuck, U., and Esser, K. Replication and expression of a bacterial-mitochondrial hybrid plasmid in the fungus Podospora anserina. *Proc. Natl. Acad. Sci. USA, 79*:3641-3645, 1982.
82. Sainsard-Chanet, A. and Begel, O. Transformation of yeast and Podospora: innocuity of senescence-specific DNA species. *Mol. Gen. Genet., 204*:443-451, 1986.
83. Bull, J. H., and Wootton, J. C. Heavily methylated amplified DNA in transformants of Neurospora crassa. *Nature, 310*:701-704, 1984.
84. Suarez, T., and Eslava, A. P. Transformation of Phycomyces with a bacterial gene for kanamycin resistance. *Mol. Gen. Genet., 212*:120-123, 1988.
85. Arnau, J., Murillo, F. J., and Torres-Martinez, S. Expression of Tn5-derived kanamycin resistance in the fungus Phycomyces blakesleeanus. *Mol. Gen. Genet., 212*:375-377, 1988.
86. Van Gorcom, R. F. M., Pouwels, P. H., Goosen, T., Visser, J., Van Den Broek, H. W. J., Hamer, J. E., Timberlake, W. E., and Van Den Hondel, C. A. M. J. J. Expression of an Escherichia coli β-galactosidase fusion gene in Aspergillus nidulans. *Gene, 40*:99-106, 1985.
87. Van Gorcom, R. F. M., Punt, P. J., Pouwels, P. H., and Van Den Hondel, C. A. M. J. J. A system for the analysis of expression signals in Aspergillus. *Gene, 48*:211-217, 1986.
88. Van Hartingsveldt, W., Mattern, I. E., Van Zeijl, C. M. J., Pouwels, P. H., and Van Den Hondel, C. A. M. J. J. Development of a homologous transformation system for Aspergillus niger based on the pyrG gene. *Mol. Gen. Genet., 206*:71-75, 1987.
89. Mullaney, E. J., Punt, P. J., and Van Den Hondel, C. A. M. J. J. DNA mediated transformation of Aspergillus ficuum. *Appl. Microbiol. Biotechnol., 28*:451-454, 1988.

90. Kolar, M., Punt, P. J., Van Den Hondel, C. A. M. J. J., and Schwab, H. Transformation of Penicillium chrysogenum using dominant selection markers and expression of an Escherichia coli lacZ fusion gene. *Gene, 62*:127-134, 1988.
91. Punt, P. J., Oliver, R. P., Dingemanse, M. A., Pouwels, P. H., and Van Den Hondel, C. A. M. J. J. Transformation of Aspergillus based on the hygromycin B resistance marker from Escherichia coli. *Gene, 56*:117-124, 1987.
92. Turner, G., Ballance, D. J., Ward, M., and Beri, R. K. Development of cloning vectors and a marker for gene replacement techniques in Aspergillus nidulans. In *Molecular Genetics of Filamentous Fungi*, Alan R. Liss, New York, 1985, pp. 15-28.
93. May, G. S., Tsang, M. L. S., Smith, H., Fidel, S., and Morris, N. R. Aspergillus nidulans β-tubulin genes are unusually divergent. *Gene, 55*:231-243, 1987.
94. Abstracts of the fifteenth fungal genetics conference. *Fungal Genet. Newslett., 39*:5-37, 1989.
95. Kelly, J. M., and Hynes, M. J. Transformation of Aspergillus niger by the amdS gene of Aspergillus nidulans. *EMBO J., 4*:475-479, 1985.
96. Beri, R. K., and Turner, G. Transformation of Penicillium chrysogenum using the Aspergillus nidulans amdS gene as a dominant selective marker. *Curr. Genet., 11*:639-641, 1987.
97. Turgeon, B. G., Garber, R. C., and Yoder, O. C. Development of a fungal transformation system based on selection of sequences with promoter activity. *Mol. Cell. Biol., 7*:3297-3305, 1987.
98. Buxton, F. P., and Radford, A. Cloning of the structural gene for orotidine 5'-phosphate carboxylase of Neurospora crassa by expression in Escherichia coli. *Mol. Gen. Genet., 190*:403-405, 1983.
99. Boeke, J. D., LaCroute, F., and Fink, G. R. A positive selection for mutants lacking orotidine-5'-phosphate decarboxylase activity in yeast: 5-fluoro-orotic acid resistance. *Mol. Gen. Genet., 197*:345-346, 1984.
100. Goosen, T., Van Engelenburg, F., Debets, F., Swart, K., Bos, K., and Van den Broek, H. Tryptophan auxotrophic mutants in Aspergillus niger: inactivation of the trpC gene by cotransformation mutagenesis. *Mol. Gen. Genet., 219*:282-288, 1989.
101. Cove, D. J. Chlorate toxicity in Aspergillus nidulans: the selection and characterization of chlorate resistant mutants. *Heredity, 36*:191-203, 1976.
102. Barnes, D. A., and Thorner, J. Use of the LYS2 gene for gene disruption, gene replacement and promoter analysis in Saccharomyces cerevisiae. In *Gene Manipulations in Fungi* (J. W. Bennet and L. L. Lasure, eds.), Academic Press, New York, 1985, pp. 197-226.
103. Grant, D. M., Lambowitz, A. M., Rambosek, J. A., and Kinsey, J. A. Transformation of Neurospora crassa with recombinant plasmids containing the cloned glutamate dehydrogenase (am) gene: evidence for autonomous replication. *Mol. Cell. Biol., 4*:2041-2051, 1984.
104. Goosen, T., Wernars, K., Wennekes, L. M. J., Visser, J., Bos, C. J., and Van den Broek, H. W. J. Multiple copy integration of transforming sequences in Aspergillus nidulans. *Proc. 3rd Eur. Congr. Biotechnol.*, Vol. 3, Verlag Chemie, Weinheim, 1984, pp. 311-316.
105. Skatrud, P. L., Queener, S. W., Carr, L. G., and Fisher, D. L. Efficient integrative transformation of Cephalosporium acremonium. *Curr. Genet., 12*:337-348, 1987.

106. Johnstone, I. L. Transformation of Aspergillus nidulans. *Microbiol. Sci.*, *2*:307-311, 1985.
107. Turner, G., and Ballance, D. J. Cloning vectors for Aspergillus nidulans based on ans1, a chromosomal fragment enhancing transformation frequency. *Fifth International Symposium on the Genetics of Industrial Microorganisms*, Part B (A. Alacevic, D. Hranueli, and Z. Toman, eds.), Ognjen Prica Printing Works, Karlovac, Yugoslavia, 1987, pp. 121-127.
108. Struhl, K., Stinchcomb, D. T., Scherer, S., and Davis, R. W. High frequency transformation of yeast: Autonomous replication of hybrid DNA molecules. *Proc. Natl. Acad. Sci. USA*, *76*:1035-1039, 1979.
109. Stinchcomb, D. T., Thomas, M., Kelly, J., Selker, E., and Davis, R. W. Eukaryotic DNA segments capable of autonomous replication in yeast. *Proc. Natl. Acad. Sci. USA*, *77*:4559-4563, 1980.
110. Stohl, L. L., and Lambowitz, A. M. Construction of a shuttle vector for the filamentous fungus Neurospora crassa. *Proc. Natl. Acad. Sci. USA*, *80*:1058-1062, 1983.
111. Beri, R. K., Lewis, E. L., and Turner, G. Behaviour of a replicating mitochondrial DNA sequence from Aspergillus amstelodami in Saccharomyces cerevisiae and Aspergillus nidulans. *Curr. Genet.*, *13*:479-486, 1988.
112. Stahl, U., Leitner, E., and Esser, K. Transformation of Penicillium chrysogenum by a vector containing a mitochondrial origin of replication. *Appl. Microbiol. Biotechnol.*, *26*:237-241, 1987.
113. Banks, G. R. Chromosomal DNA sequences from Ustilago promote autonomous replication of plasmids in Saccharomyces cerevisiae. *Curr. Genet.*, *7*:79-84, 1983.
114. Barnes, D. E., and MacDonald, D. W. Behaviour of recombinant plasmids in Aspergillus nidulans: structure and stability. *Curr. Genet.*, *10*:767-776, 1986.
115. Paietta, J., and Marzluf, G. A. Plasmid recovery from transformants and the isolation of chromosomal DNA segments improving plasmid replication in Neurospora crassa. *Curr. Genet.*, *9*:383-388, 1985.
116. Wernars, K. DNA mediated transformation of the filamentous fungus Aspergillus nidulans. Thesis. Agricultural University Wageningen, The Netherlands, 1986.
117. Picard, M., Debuchy, R., Julien, J., and Brygoo, Y. Transformation by integration in Podospora anserina. II. Targeting to the resident locus with cosmids and instability of the transformants. *Mol. Gen. Genet.*, *210*:129-134, 1987.
118. Debuchy, R., Coppin-Raynal, E., Le Coze, D., and Brygoo, Y. Chromosome walking toward a centromere in the filamentous fungus Podospora anserina; cloning of a sequence lethal at a two-copy state. *Curr. Genet.*, *13*:105-111, 1988.
119. Yelton, M. M., Timberlake, W. E., and Van Den Hondel, C. A. M. J. J. A cosmid for selecting genes by complementation in Aspergillus nidulans: selection of the developmentally regulated yA locus. *Proc. Natl. Acad. Sci. USA*, *82*:834-838, 1985.
120. Hamer, J. E., and Timberlake, W. E. Functional organization of the Aspergillus nidulans trpC promoter. *Mol. Cell. Biol.*, *7*:2352-2359, 1987.
121. Davis, M. A., Cobbet, C. S., and Hynes, M. J. An amdS-lacZ fusion for studying gene regulation in Aspergillus. *Gene*, *63*:199-212, 1988.

122. Gwynne, D. I., Buxton, F. P., Williams, S. A., Garven, S., and Davies, R. W. Genetically engineered secretion of active human interferon and a bacterial endoglucanase from Aspergillus nidulans. *Biotechnology, 5*:713-719, 1987.
123. Upshall, A., Kumar, A. A., Bailey, M. C., Parker, M. D., Favreau, M. A., Lewison, K. P., Joseph, M. L., Maraganore, J. M., and McKnight, G. L. Secretion of active human tissue plasminogen activator from the filamentous fungus Aspergillus nidulans. *Biotechnology, 5*:1301-1304, 1987.
124. Cullen, D., Gray, G. L., Wilson, L. J., Hayenga, K. J., Lamsa, M. H., Rey, M. W., Norton, S., and Berka, R. M. Controlled expression and secretion of bovine chymosin in Aspergillus nidulans. *Biotechnology, 5*:369-376, 1987.
125. Gray, G. L., Hayenga, K., Cullen, D., Wilson, L. J., and Norton, S. Primary structure of Mucor miehei aspartyl protease: evidence for a zymogen intermediate. *Gene, 48*:41-53, 1986.
126. Schechtman, M. G. Isolation of telomere DNA from Neurospora crassa. *Mol. Cell. Biol., 7*:3168-3177, 1987.
127. Stohl, L. L., Akins, R. A., and Lambowitz, A. M. Characterization of deletion derivatives of an autonomously replicating Neurospora plasmid. *Nucl. Acids Res., 12*:6169-6178, 1984.
128. Clutterbuck, A. J. Aspergillus nidulans genetic maps. In *Genetic Maps: A compilation of linkage and restriction maps of genetically studied organisms*, Vol. 4 (S. J. O'Brien, ed.), Cold Spring Harbor Laboratory, New York, 1984.
129. Selker, E. U., and Stevens, J. N. Signal for DNA methylation associated with tandem duplication in Neurospora crassa. *Mol. Cell. Biol., 7*:1032-1038, 1987.
130. Selker, E. U., Cambareri, E. B., Jensen, B. C., and Haack, K. R. Rearrangement of duplicated DNA in specialized cells of Neurospora. *Cell, 51*:741-752, 1987.
131. Selker, E. U., and Garrett, P. W. DNA sequence duplications trigger gene inactivation in Neurospora crassa. *Proc. Natl. Acad. Sci. USA, 85*:6870-6874, 1988.
132. Case, M. E. Gene organization and regulation in Neurospora crassa. Evidence from the cloning and transformation of the Qa gene cluster. In *Genetic Engineering in Eukaryotes* (P. F. Lurquin and A. Kleinhofs, eds.), Plenum Press, New York, 1983, pp. 7-20.
133. Case, M. E. Genetical and molecular analyses of Qa-2 transformants in Neurospora crassa. *Genetics, 113*:569-588, 1986.
134. Perrot, M., Barreau, C., and Begueret, J. Non-integrative transformation in the filamentous fungus Podospora anserina; stabilization of a linear vector by the chromosomal ends of Tetrahymena. *Mol. Cell. Biol., 7*:1725-1730, 1987.
135. Van Heeswijck, R. Autonomous replication of plasmids in Mucor transformants. *Carlsberg Res. Commun., 51*:433-444, 1986.
136. Woestemeyer, J. Transformation of the mucoraceous fungus Absidia glauca by an autonomously replicating plasmid. *Genome, 30*: Suppl. 1. 299, 1988.
137. Raper, C. A., and Horton, J. S. A circular DNA plasmid of Schizophyllum. *J. Cell. Biochem. Suppl. 13E*:31, 1989.
138. Hicks, J. B., Hinnen, A., and Fink, G. R. Properties of yeast transformation. *Cold Spring Harbor Symp. Quant. Biol., 43*:1305-1313, 1979.

139. Kingsman, A. J., Gimlich, R. L., Clarice, L., and Chinault, A. C. Sequence variations in dispersed repetitive sequences in Saccharomyces cerevisiae. *J. Mol. Biol.*, *145*:619-632, 1981.
140. Wernars, K., Goosen, T., Swart, K., and Van Den Broek, H. W. J. Genetic analysis of Aspergillus nidulans amdS-positive transformants. *Mol. Gen. Genet.*, *205*:312-317, 1986.
141. Kelly, J. M., and Hynes, M. J. Multiple copies of the amdS gene of Aspergillus nidulans cause titration of trans-acting regulatory proteins. *Curr. Genet.*, *12*:21-31, 1987.
142. Miller, B. L., Miller, K. Y., and Timberlake, W. E. Direct and indirect gene replacement in Aspergillus nidulans. *Mol. Cell. Biol.*, *5*:1714-1721, 1985.
143. Durrens, P., Green, P. M., Arst Jr., H. N., and Scazzocchio, C. Heterologous insertion of transforming DNA and generation of new deletions associated with transformation in Aspergillus nidulans. *Mol. Gen. Genet.*, *203*:544-549, 1986.
144. Upshall, A. Genetic and molecular characterization of argB-positive transformants of Aspergillus nidulans. *Curr. Genet.*, *10*:593-600, 1986.
145. Johnstone, I. L., Hughes, S. G., and Clutterbuck, A. J. Cloning an Aspergillus nidulans developmentally regulated gene by transformation. *EMBO J.*, *4*:1307-1311, 1985.
146. Kos, A. Kuijvenhoven, J., Wernars, K., Bos, C. J., Van den Broek, H. W. J., Pouwels, P. H., and Van den Hondel, C. A. M. J. J. Isolation and characterization of the Aspergillus niger trpC gene. *Gene*, *39*:231-238, 1985.
147. Davis, M. A., and Hynes, M. J. Complementation of areA negative regulatory gene mutations of Aspergillus nidulans by the heterologous regulatory gene nit-2 of Neurospora crassa. *Proc. Natl. Acad. Sci. USA*, *84*:3753-3757, 1987.
148. May, G. S., Gambino, J., Weatherbee, J. A., and Morris, N. R. Identification and functional analysis of β-tubulin genes by site specific integrative transformation in Aspergillus nidulans. *J. Cell. Biol.*, *101*:712-719, 1985.
149. Ward, M., Wilkinson, B., and Turner, G. Transformation of Aspergillus nidulans with a cloned oligomycin-resistant ATP synthase subunit 9 gene. *Mol. Gen. Genet.*, *202*:265-270, 1986.
150. De Graaff, L., Van Den Broek, H., and Visser, J. Isolation and transformation of the pyruvate kinase gene of Aspergillus nidulans. *Curr. Genet.*, *13*:315-322, 1988.
151. Cullen, D., Leong, S. A., Wilson, L. J., and Henner, D. J. Transformation of Aspergillus nidulans with the hygromycin-resistance gene Hph. *Gene*, *57*:21-26, 1987.
152. Wernars, K., Goosen, T., Wennekes, L. M. J., Swart, K., Van Den Hondel, C. A. M. J. J., and Van Den Broek, H. W. J. Cotransformation of Aspergillus nidulans, a tool for replacing fungal genes. *Mol. Gen. Genet.*, *209*:71-77, 1987.
153. Lockington, R. A., Sealy-Lewis, H. M., Scazzocchio, C., and Davies, R. W. Cloning and characterization of the ethanol utilization regulon in Aspergillus nidulans. *Gene*, *33*:137-149, 1985.
154. Paietta, J. V., and Marzluf, G. A. Gene disruption by transformation in Neurospora crassa. *Mol. Cell. Biol.*, *5*:1554-1559, 1985.

155. Banks, G. R., and Taylor, S. Y. Cloning the pyr3 gene of Ustilago maydis and its use in DNA transformation. *Mol. Cell. Biol.*, *8*:5417-5424, 1988.
156. Miller, B. E., Miller, K. Y., Roberti, K. A., and Timberlake, W. E. Position dependent and -independent mechanisms regulate cell-specific expression of the spoC1 gene cluster of Aspergillus nidulans. *Mol. Cell. Biol.*, *7*:427-434, 1987.
157. Fidel, S., Doonan, J. H., and Morris, N. R. Aspergillus nidulans contains a single actin gene which has unique intron locations and encodes a τ-actin. *Gene*, *70*:283-293, 1988.
158. Dunne, P. W., and Oakley, B. R. Mitotic gene conversion, reciprocal recombination and gene replacement at the benA β-tubulin locus of Aspergillus nidulans. *Mol. Gen. Genet.*, *213*:339-345, 1988.
159. De Graaff, L. H. The structure and expression of the pyruvate kinase gene of Aspergillus nidulans and Aspergillus niger. Thesis, Agricultural University Wageningen, The Netherlands, 1989.
160. Timberlake, W. E., Boylan, M. T., Cooley, M. B., Mirabito, P. M., O'Hara, E. B., and Willett, C. E. Rapid identification of mutation-complementing restriction fragments from Aspergillus nidulans cosmids. *Exp. Mycol.*, *9*:351-355, 1985.
161. Dmochowska, A. T., Bal, J., Bartnik, E., and Weglenski, P. Cloning of Aspergillus nidulans DNA. I. Selection and analysis of recombinant plasmids capable of complementing pyrF, argIF and proAB mutations in E. coli. *Acta Microbiol. Pol.*, *29*:213-225, 1980.
162. Kinghorn, J. R., and Hawkins, A. R. Cloning and expression in Escherichia coli K-12 of the biosynthetic dehydroquinase function of the AROM cluster gene from the eukaryote Aspergillus nidulans. *Mol. Gen. Genet.*, *186*:145-152, 1982.
163. Yelton, M. M., Hamer, J. E., De Souza, E. R., Mullaney, E. J., and Timberlake, W. E. Developmental regulation of the Aspergillus nidulans trpC gene. *Proc. Natl. Acad. Sci. USA*, *80*:7576-7580, 1983.
164. Alton, N. K., Buxton, F., Patel, V., Giles, N. H., and Vapnek, D. 5'-Untranslated sequences of two structural genes in the qa gene cluster of Neurospora crassa. *Proc. Natl. Acad. Sci. USA*, *79*:1955-1959, 1983.
165. Keesey, J. K., and DeMoss, J. A. Cloning of the trp-1 gene from Neurospora crassa by complementation of a trpC mutation in Escherichia coli. *J. Bacteriol.*, *152*:954-958, 1982.
166. Sanchez, F., Tourino, A., Traseira, S., Perez-Aranda, A., Rubio, V., and Penalva, M. A. Molecular cloning and characterization of the trpC gene from Penicillium chrysogenum. *Mol. Gen. Genet.*, *205*:248-252, 1986.
167. Revuelta, J. L., and Jayaram, M. Phycomyces blakesleeanus trp1 gene: organization and functional complementation in Escherichia coli and Saccharomyces cerevisiae. *Mol. Cell. Biol.*, *7*:2664-2670, 1987.
168. Turgeon, B. G., Macrae, W. D., Garber, R. C., Fink, G. R., and Yoder, O. C. A cloned tryptophan synthesis gene from the ascomycete Cochliobolus heterostrophus functions in Escherichia coli, yeast and Aspergillus nidulans. *Gene*, *42*:79-88, 1986.
169. Ng, R., and Abelson, J. Isolation and sequence of the gene for actin in Saccharomyces cerevisiae. *Proc. Natl. Acad. Sci. USA*, *77*:3912-3916, 1980.

170. Berse, B., Dmochowska, A., Skrzypek, M., Weglenski, P., Bates, M. A., and Weiss, R. L. Cloning and characterization of the ornithine carbamoyltransferase gene from Aspergillus nidulans. *Gene, 25*:109-117, 1983.
171. Innis, M. A., Holland, M. J., McCabe, P. C., Cole, G. E., Wittman, V. P., Tal, R., Watt, K. W. K., Gelfand, D. H., Holland, J. P., and Meade, J. H. Expression, glycosilation and secretion of an Aspergillus glucoamylase by Saccharomyces cerevisiae. *Science, 228*:21-26, 1985.
172. Penttila, M. E., Nevalainen, K. M. H., Raynal, A., and Knowles, J. K. C. Cloning of Aspergillus niger genes in yeast. Expression of the gene coding Aspergillus β-glucosidase. *Mol. Gen. Genet., 194*:494-499, 1984.
173. McKnight, G. L., Kato, H., Upshall, A., Parker, M. D., Saari, G., and O'Hara, P. J. Identification and molecular analysis of a third Aspergillus nidulans alcohol dehydrogenase gene. *EMBO J., 4*:2093-2099, 1985.
174. Carr, L. G., Skatrud, P. L., Scheetz, M. E., Queener, S. W., and Ingolia, T. D. Cloning and expression of the isopenicillin N synthetase gene from Penicillium chrysogenum. *Gene, 48*:257-266, 1986.
175. Ramon, D., Carramolino, L., Patino, C., Sanchez, F., and Penalva, M. A. Cloning and characterization of the isopenicillin N synthetase gene mediating the formation of the β-lactam ring in Aspergillus nidulans. *Gene, 57*:171-181, 1987.
176. Flint, H. J., and Wilkening, J. Cloning of the arg-12 gene of Neurospora crassa and regulation of its transcript via cross-pathway amino-acid control. *Mol. Gen. Genet., 203*:110-116, 1986.
177. Gurr, S. J., Hawkins, A. R., Drainas, C., and Kinghorn, J. R. Isolation and identification of the Aspergillus nidulans gdhA gene encoding NADP-linked glutamate dehydrogenase. *Curr. Genet., 10*:761-766, 1986.
178. Punt, P. J., Dingemanse, M. A., Jacobs-Meijsing, B. J. M., Pouwels, P. H., and Van Den Hondel, C. A. M. J. J. Isolation and characterization of the glyceraldehyde-3-phosphate dehydrogenase gene of Aspergillus nidulans. *Gene, 69*:49-57, 1987.
179. Clements, J. M., and Roberts, C. F. Molecular cloning of the 3-phosphoglycerate kinase (PGK) gene from Aspergillus nidulans. *Curr. Genet., 9*:293-298, 1985.
180. Germann, U. A., and Lerch, K. Isolation and partial nucleotide sequence of the laccase gene from Neurospora crassa: amino acid sequence homology of the protein to human ceruloplasmin. *Proc. Natl. Acad. Sci. USA, 83*:8854-8858, 1986.
181. Muenger, K., Germann, U. A., and Lerch, K. Isolation and structural organization of the Neurospora crassa copper metallothionein gene. *EMBO J., 4*:2665-2668, 1985.
182. Zimmermann, C. R., Orr, W. C., Leclerc, R. F., Barnard, E. C., and Timberlake, W. E. Molecular cloning and selection of genes regulated in Aspergillus development. *Cell, 21*:709-715, 1980.
183. Orr, W. C., and Timberlake, W. E. Clustering of spore-specific genes in Aspergillus nidulans. *Proc. Natl. Acad. Sci. USA, 79*:5976-5980, 1982.
184. Berlin, V., and Yanofski, C. Isolation and characterization of genes differentially expressed during conidiation of Neurospora crassa. *Mol. Cell. Biol., 5*:849-856, 1985.

185. Thomas, G. H., Connerton, I. F., and Fincham, J. R. S. Molecular cloning, identification and transcriptional analysis of genes involved in acetate utilization in Neurospora crassa. *Mol. Microbiol.*, *2*:599-606, 1988.
186. Hynes, M. J., Corrick, C. M., and King, J. A. Isolation of genomic clones containing the amdS gene of Aspergillus nidulans and their use in the analysis of structural and regulatory mutations. *Mol. Cell. Biol.*, *3*:1430-1439, 1983.
187. Dons, J. J. M., Mulder, G. H., Rouwendal, G. J. A., Springer, J., Bremer, W., and Wessels, J. G. H. Sequence analysis of a split gene involved in fruiting from the fungus Schizophyllum commune. *EMBO J.*, *3*:2101-2106, 1984.
188. Loros, J. J., Denome, S. A., and Dunlap, J. C. Molecular cloning of genes under control of the circadian clock in Neurospora, *Science*, *243*:385-390, 1989.
189. Caddick, M. X., Arst Jr., H. N., Taylor, L. H., Johnson, R. I., and Brownlee, A. G. Cloning of the regulatory gene areA mediating nitrogen metabolite repression in Aspergillus nidulans. *EMBO J.*, *5*:1087-1090, 1986.
190. Hager, K. M., Mandala, S. M., Davenport, J. W., Speicher, D. W., Benz Jr., E. J., and Slayman, C. W. Amino acid sequence of the plasma membrane ATPase of Neurospora crassa: deduction from genomic and cDNA sequences. *Proc. Natl. Acad. Sci. USA*, *83*:7693-7697, 1986.
191. Arends, H., and Sebald, W. Nucleotide sequence of the cloned mRNA and gene of the ADP/ATP carrier from Neurospora crassa. *EMBO J.*, *3*:377-382, 1984.
192. Teeri, T., Salovuori, I., and Knowles, J. The molecular cloning of the major cellulase gene from Trichoderma reesei. *Biotechnology*, *1*:696-699, 1983.
193. Buxton, F. P., Gwynne, D. I., Garven, S., Sibley, S., and Davies, R. W. Cloning and molecular analysis of the ornithine carbamoyltransferase gene of Aspergillus niger. *Gene*, *60*:255-266, 1987.
194. Goosen, T., Wernars, K., Vos, J. C., Wennekes, L. M. J., Van den Hondel, C. A. M. J. J., and Van den Broek, H. W. J. Isolation of transforming DNA sequences from Aspergillus nidulans AmdS+ transformants in E. coli. In *DNA mediated transformation of the filamentous fungus Aspergillus nidulans*. Wernars, K. thesis, Agricultural University Wageningen, The Netherlands, 1986.
195. Oakley, C. E., Weil, C. F., Kretz, P. L., and Oakley, B. R. Cloning of the riboB locus of Aspergillus nidulans. *Gene*, *53*:293-298, 1987.
196. Stewart, V., and Vollmer, S. J. Molecular cloning of nit-2, a regulatory gene required for nitrogen metabolite repression in Neurospora crassa. *Gene*, *46*:291-295, 1986.
197. Fu, Y., and Marzluf, G. A. Molecular cloning and analysis of the regulation of nit-3, the structural gene for nitrate reductase in Neurospora crassa. *Proc. Natl. Acad. Sci. USA*, *84*:8243-8247, 1987.
198. Glass, N. L., Vollmer, S. J., Staben, C., Grotelueschen, J., Metzenberg, R. L., and Yanofsky, C. DNAs of the two mating-type alleles of Neurospora crassa are highly dissimilar. *Science*, *241*:570-573, 1988.
199. Jefferson, R. A., Kavanagh, A. A., and Bevan, M. W. GUS fusion: β-glucuronidase as a sensitive and versatile gene fusion in higher plants. *EMBO J.*, *6*:3901-3907, 1987.

200. Geever, F. R., Murayma, T., Case, M. E., and Giles, N. H. Rearrangement mutations on the 5' side of the Qa-2 gene of Neurospora implicate two regions of Qa-1F activator-protein interaction. *Proc. Natl. Acad. Sci. USA*, *83*:3944-3948, 1986.
201. Huiet, L. Molecular analysis of the Neurospora Qa-1 regulatory region indicates that two interacting genes control Qa gene expression. *Proc. Natl. Acad. Sci. USA*, *81*:1174-1178, 1984.
202. Baum, J. A., Geever, R., and Giles, N. H. Expression of Qa-1F activator protein: Identification of upstream binding sites in the Qa gene cluster and localization of the DNA binding domain. *Mol. Cell. Biol.*, *7*:1256-1266, 1987.
203. Goc, A., and Weglenski, P. Regulatory region of the Aspergillus nidulans argB gene. *Curr. Genet.*, *14*:425-429, 1988.
204. Gwynne, D. I., Miller, B. L., Miller, K. Y., and Timberlake, W. E. Structure and regulated expression of the SpoC1 gene cluster from Aspergillus nidulans. *J. Mol. Biol.*, *180*:91-109, 1984.
205. Da Silva, A. J. F., Whittington, H., Clements, J., Roberts, C., and Hawkins, A. R. Sequence analysis and transformation by the catabolic 3-dehydroquinase qutE gene from Aspergillus nidulans. *Biochem. J.*, *240*:481-488, 1986.
206. Hynes, M. J. Multiple independent control mechanisms affecting the acetamidase of Aspergillus nidulans. *Mol. Gen. Genet.*, *161*:59-65, 1978.
207. Mellon, F. M., and Casselton, L. A. Transformation as a method of increasing gene copy number and gene expression in the basidiomycete fungus Coprinus cinereus. *Curr. Genet.*, *14*:451-456, 1988.
208. Weiss, R. L., Puetz, D., and Cybis, J. Expression of Aspergillus genes in Neurospora. In *Gene Manipulations in Fungi* (J. W. Bennett and L. L. Lasure, eds.), Academic Press, New York, 1985, pp. 280-292.
209. De La Fuente Herce, A., Mellon, F. M. and Casselton, L. A. Gene transfer between basidiomycete and ascomycete fungi by cotransformation. *Heredity*, *60*:312-313, 1988.
210. Whitehead, M., Unkles, S. E., Ramsden, M., Campbell, E., Gurr, S. J., Spence, D., and Kinghorn, J. R. Transformation of a nitrate reductase deficient mutant of Penicillium chrysogenum with the corresponding Aspergillus niger and Aspergillus nidulans genes. *Mol. Gen. Genet.*, *216*:408-411, 1989.
211. Macrea, W. D., Buxton, F. P., Sibley, S., Garven, S., Gwynne, D. I., Davies, R. W., and Arst Jr., H. N. A phosphate-repressible acid phosphatase gene from Aspergillus niger: its cloning, sequencing and transcriptional analysis. *Gene*, *71*:339-348, 1988.
212. Barbesgaard, P. Industrial enzymes produced by members of the genus Aspergillus. In *Genetics and Physiology of Aspergillus* (J. E. Smith and J. A. Pateman, eds.), Academic Press, London, 1977, pp. 391-404.
213. Montencourt, B. S. Trichoderma reesei cellulases. *Trends Biotechnol.*, *1*:156-161, 1983.
214. Smith, R. A., Duncan, M. J., and Moir, D. T. Heterologous protein secretion in yeast. *Science*, *229*:1219-1224, 1985.
215. Lemontt, J. F., Wei, C. M., and Dackowski, W. R. Expression of active human uterine tissue plasminogen activator in yeast. *DNA*, *4*:419-428, 1985.

216. Bos, C. J., Debets, A. J. M., Swart, K., Huybers, A., Kobus, G., and Slakhorst, S. M. Genetic analysis and the construction of master strains for assignment of genes to six linkage groups in Aspergillus niger. *Curr. Genet.*, *14*:437-443, 1988.
217. Bos, C. J. Induction and isolation of mutants in fungi at low mutagen doses. *Curr. Genet.*, *12*:471-474, 1987.
218. Debets, A. J. M., Swart, K., and Bos, C. J. Mitotic mapping in linkage group V of Aspergillus niger based on the selection of auxotrophic recombinants by Novozym enrichment. *Can. J. Microbiol.*, *35*:982-988. 1989.
219. Bos, C. J. Mutants collection and master strains of Aspergillus niger. *Fungal Genet. Newslett.*, *34*:27, 1985.
220. Bos, C. J. In *Induced mutation and somatic recombination as tools for genetic analysis and breeding of imperfect fungi*. Thesis, Agricultural University Wageningen, The Netherlands, 1985.
221. Buxton, F. P., Gwynne, D. I., and Davies, R. W. Transformation of Aspergillus niger using the argB gene of Aspergillus nidulans. *Gene*, *37*:207-214, 1985.
222. Osiewacz, H. D., and Weber, A. DNA mediated transformation of the filamentous fungus Curvularia lunata using a dominant selectable marker. *Appl. Microbiol. Biotechnol.*, *30*:375-380, 1989.
223. Cooley, R. N., Shaw, R. K., Franklin, F. C. H., and Caten, C. E. Transformation of the phytopathogenic fungus Septoria nodorum to hygromycin B resistance. *Curr. Genet.*, *13*:383-390, 1988.
224. Turcq, B., and Begueret, J. The ura5 gene of the filamentous fungus Podospora anserina: nucleotide sequence and expression in transformed strains. *Gene*, *53*:201-209, 1987.
225. Caten, C. E., and Jinks, J. L. Heterokaryosis significance in wild homothallic Ascomycetes and Fungi Imperfecti. *Trans. Br. Mycol. Soc.*, *49*:81-93, 1966.
226. Garnjobst, L. Genetic control of heterokaryosis in Neurospora crassa. *Am. J. Bot.*, *40*:607-613, 1953.
227. Dales, R. B. G., and Croft, J. H. Protoplast fusion and the genetical analysis of vegetative incompatibility in Aspergillus nidulans. In *Advances in Protoplast Research* (L. Ferenczy and J. L. Farkas, eds.), Pergamon Press, Oxford, 1979, pp. 73-84.
228. Anne, J. Protoplasts of filamentous fungi in genetics and metabolite production. *Proc. 6th Int. Symp. Protoplasts*, Basel, 1983, pp. 167-178.
229. Anne, J., and Peberdy, J. F. Induced fusion of fungal protoplasts following treatment with polyethylene glycol. *J. Gen. Microbiol.*, *92*:413-417, 1976.
230. Woestemeyer, J. Burmester, A., and Weigel, C. Neomycin resistance as a dominant selection marker for transformation of the zygomycete Absidia glauca. *Curr. Genet.*, *12*:625-627, 1987.
231. Punt, P. J., Van Hartingsveldt, W., Mattern, I. E., Pouwels, P. H., and Van Den Hondel, C. A. M. J. J. Gene transfer and gene expression in Aspergillus species. *Proc. 4th European Congress on Biotechnology*, Vol. 1 (O. M. Neijssel, R. R. van der Meer, and K. Ch. A. M. Luyben, eds.), Elsevier, Amsterdam, 1987, pp. 445-448.
232. Cullen, D., and Leong, S. Recent advances in the molecular genetics of industrial filamentous fungi. *Trends Biotechnol.*, *4*:285-288, 1986.
233. Mattern, I. E., Unkles, S., Kinghorn, J. R., Pouwels, P. H., and Van Den Hondel, C. A. M. J. J. Transformation of Aspergillus oryzae

234. Upshall, A. Filamentous fungi in biotechnology. *Biotechniques, 4*:158-166, 1986.
235. Farman, M. L., and Oliver, R. P. The transformation of protoplasts of Leptosphaeria maculans to hygromycin B resistance. *Curr. Genet., 13*:327-330, 1988.
236. Begueret, J., Razanamparany, V., Perrot, M., and Barreau, C. Cloning gene ura5 for the orotidylic acid pyrophosphorylase of the filamentous fungus Podospora anserina: transformation of protoplasts. *Gene, 32*:487-492, 1984.
237. Diez, B., Cantoral, J. M., Barredo, J. L., and Martin, J. F. Selection and characterization of pyrG mutants of Penicillium chrysogenum lacking orotidine-5'-phosphate decarboxylase and complementation by the pyr-4 gene of Neurospora crassa. *Curr. Genet., 12*:277-282, 1987.
238. Hahm, Y. T. and Batt, C. A. Genetic transformation of an argB mutant of Aspergillus oryzae. *Appl. Environ. Microbiol., 54*:1610-1611, 1988.
239. Debets, A. J. M., Swart, K., Holub, E. F., Goosen, T., and Bos, C. J. Genetic analysis of amdS transformants of Aspergillus niger and their use in chromosome mapping. *Mol. Gen. Genet., 222*:284-290, 1990.
240. Van Gorcom, R. F. M., Boschloo, J. G., Kuÿvenhoven, A., Lange, J., Van Vark, A. J., Bos, C. J., Van Balken, J. A. M., Pouwels, P. H., and Van den Hondel, C. A. M. J. J. Isolation and molecular characterization of the benzoate-*para*-hydroxylase gene (bphA) of Aspergillus niger: a member of a new gene family of the cytochrome P_{450} superfamily. *Mol. Gen. Genet., 223*:192-197, 1990.

8

THE APPLICATION OF GENETIC ENGINEERING TO STRAIN IMPROVEMENT OF β-LACTAM-PRODUCING FILAMENTOUS FUNGI

SHU-JEN D. CHIANG and RICHARD P. ELANDER *Industrial Division, Bristol-Myers Squibb Company, Syracuse, New York*

I. INTRODUCTION

Industrial production of cephalosporins and penicillins by fermentation is an excellent example of development of commercial biotechnology over the past four decades. During this period, hundreds-fold improvement in the antibiotic productivity of the β-lactam-producing filamentous fungi has been achieved by random mutation and selection strategies [1]. Historically, the strategies used for screening for better producers following random mutagenesis and selection were based on empirical experience. In recent years, however, more rational selection techniques have been developed, such as selection for mutants resistant to analogs of amino acids which are precursors or stimulants of β-lactam biosynthesis, and for mutants resistant to metallic ions (such as mercuric or cupric ions) that complex with antibiotics generated by β-lactam-producing microorganisms [1]. In addition, protoplast fusion methods have also been used successfully to obtain improved strains [2].

Recent developments in the molecular cloning of the β-lactam biosynthetic genes can facilitate investigation of the mechanisms of the biochemical and genetic regulation of the β-lactam biosynthetic pathways in *Cephalosporium acremonium* (*Acremonium chrysogenum*) and *Penicillium chrysogenum* [3-6]. Advances in the understanding of the above mechanisms would provide new strategies for both strain improvement and novel antibiotic development. It has been demonstrated that high-cephalosporin C-producing strains of *C. acremonium* produce higher levels of three enzymes of the cephalosporin C biosynthetic pathway compared to a low-producing strain [7,8]. This result indicates that some of the genes coding for cephalosporin C biosynthetic enzymes in *C. acremonium* may be expressed under the regulation of a common effector [7]. Therefore, it will be important to identify and clone the structural gene of this regulatory protein. In addition, cloning of genes encoding enzymes involved in rate-limiting steps ("bottlenecks") in the biosynthesis of β-lactam antibiotics or their precursors is also important. Construction of strains with multiple copies of these cloned genes would lead to amplification of enzymes important in β-lactam biosynthesis, thereby resulting in improved strains for high-level β-lactam production.

In this chapter, recent progress in the cloning of β-lactam biosynthetic genes from filamentous fungi and their applications will be reviewed and discussed.

II. β-LACTAM BIOSYNTHETIC PATHWAYS

Recently, several excellent reviews covered in detail the elucidation of β-lactam biosynthetic pathways [9,10]. The current view of the overall biosynthetic scheme for cephalosporins and penicillins is depicted in Fig. 1. The nomenclature of genes encoding enzymes involved in each step of β-lactam biosynthesis follows the recommendation by Ingolia and Queener [11].

The biosynthesis of penicillins by *Aspergillus nidulans* and *P. chrysogenum*, and the biosynthesis of cephalosporin C by *C. acremonium* all begin with the condensation of three amino acids, L-α-aminoadipic acid, L-cysteine, and L-valine, to form the tripeptide δ-(L-α-aminoadipyl)-L-cysteinyl-D-valine (LLD-ACV) by the enzyme complex ACV synthetase. The LLD-ACV tripeptide is converted to form the penam nucleus, isopenicillin N, by isopenicillin N synthetase (IPNS). After the common intermediate isopenicillin N (IPN), the biosynthesis of penicillins is separate and distinct from the biosynthesis of cephalosporins. The genes coding for enzyme involved in the biosynthesis of the common intermediate LLD-ACV and isopenicillin N are designated *pcb* (penicillin/cephalosporin biosynthesis). The genes coding for enzymes involved in the biosynthesis of cephalosporins after branching are designated *cef* and those involved in the biosynthesis of penicillins are designated *pen* [11].

Demain proposed that in *P. chrysogenum* the L-α-aminoadipyl side chain of isopenicillin N can be exchanged for the phenylacetyl group from phenylacetyl coenzyme A (CoA) to form the final product, penicillin G [12]. However, there is also evidence that the L-α-aminoadipyl side chain can be removed to yield 6-aminopenicillanic acid (6-APA) and penicillin G can be formed by the addition of phenylacetyl side chain to 6-APA [13,14]. At the present time, it is not clear which of the two reactions accounts for the biosynthesis of penicillin G. Recently, the acyl CoA:6-APA acyltransferase was purified and reported to form penicillin G from 6-APA more efficiently than from isopenicillin N [15]. This result still cannot rule out either of the two reactions.

In *C. acremonium*, the IPN epimerase catalyzes the conversion of the L-α-aminoadipyl side chain of isopenicillin N into the D-α-aminoadipyl side chain of penicillin N. The five-membered thiazolidine ring of penicillin N is then expanded to give the six-membered dihydrothiazine ring of deacetoxycephalosporin C (DAOC) by deacetoxycephalosporin C synthetase (DAOCS or expandase). The next enzyme in the pathway, deacetylcephalosporin C synthetase (DACS or hydroxylase), catalyzes hydroxylation of deacetoxycephalosporin C at the C-3' methyl group to yield deacetylcephalosporin C (DAC). Scheidegger et al. provided evidence that ring expansion and hydroxylation activity both reside on the same protein in *C. acremonium* [16]. The last step in cephalosporin C synthesis, catalyzed by acetyltransferase, involves the transfer of an acetyl moiety from acetyl CoA to the C-3' hydroxyl moiety of deacetylcephalosporin C.

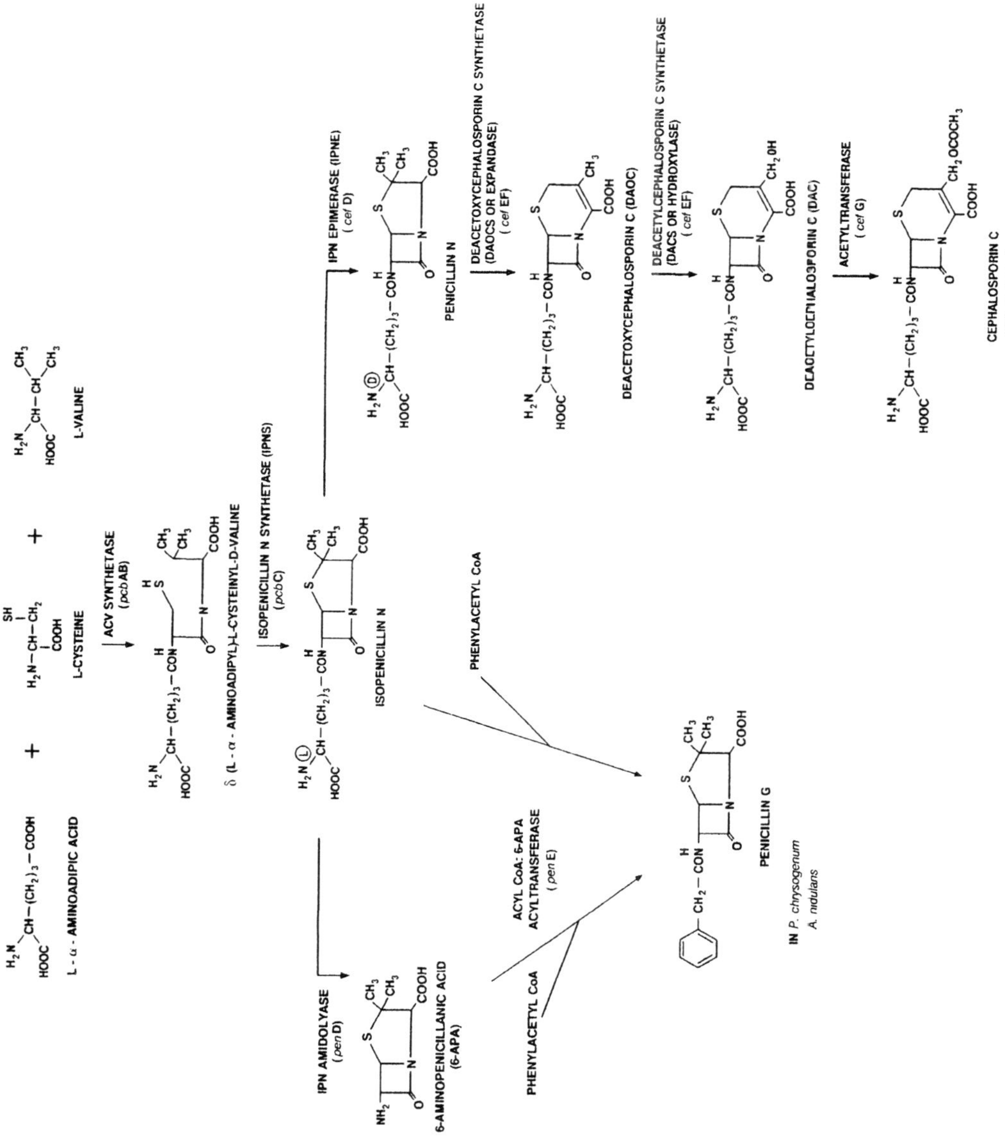

FIGURE 1. Biosynthetic pathways for penicillin G and cephalosporin C.

III. CLONING OF β-LACTAM BIOSYNTHETIC GENES

Since DNA transformation systems developed in *C. acremonium* and *P. chrysogenum* are relatively inefficient when compared with bacterial systems [17-20], the self-cloning of genes involved in cephalosporin and penicillin biosynthesis into a mutant strain which is defective in one of the β-lactam biosynthetic steps and screening for complementation to restore the β-lactam producing phenotype is not feasible at this time. However, the "reverse genetics" approach has been used successfully in the cloning of a number of β-lactam biosynthetic genes [6]. This approach starts with purification of a desired enzyme, determination of a partial amino acid sequence, and synthesis of oligonucleotide probes based on the experimentally derived amino acid sequence. A genomic DNA library of *C. acremonium* or *P. chrysogenum* is then constructed in *Escherichia coli* and screened for clones that hybridize with the DNA probes. Further characterization of the plasmid DNA purified from this clone will identify the gene encoding this amino acid sequence.

A. Cloning and Characterization of the Isopenicillin N Synthetase Genes

A number of IPNS genes (*pcb*C) have been cloned from different fungal cells [21-25]. The first IPNS gene was cloned by the reverse genetics approach from *C. acremonium* as shown in Fig. 2 [21]. The *Cephalosporium* IPNS (cIPNS) gene encodes a polypeptide of 338 amino acids with a relative molecular mass of 38,416. When the open reading frame of the cIPNS gene was cloned into an *E. coli* expression vector and introduced into *E. coli* the transformant produced the recombinant cIPNS protein at an amount equivalent to 20% of the total cellular protein [21]. The purified recombinant cIPNS protein converts LLD-ACV to isopenicillin N with identical kinetics to the authentic protein, which can also convert ACV tripeptide analog substrates to unusual β-lactam antibiotics [26]. One most striking result is the conversion of the allylglycine peptide to the seven-membered ring-containing homocephem illustrated in Fig. 3 [26].

Subsequently, a number of different IPNS genes were cloned from *A. nidulans* and *P. chrysogenum* on the basis of the reverse genetics approach or the cloned cIPNS gene as a DNA probe [22-24]. On the other hand, three IPNS genes were cloned from three different *Streptomyces* species [24,27,28]. The IPNS genes of *P. chrysogenum* (pIPNS) and *A. nidulans* (aIPNS) were located in an open reading frame of 996 base pairs (bp) which encodes a 331-amino-acid polypeptide that is seven amino acids smaller than the cIPNS protein. None of the fungal IPNS genes contain introns (regions containing nonprotein coding sequences) within the open reading frame. The three fungal IPNS genes are similar to one another in the nucleotide and predicted amino acid sequences, and the degree of similarity ranges from 73 to 79% (Figs. 4 and 5).

Recently, Barredo et al. [25] reported a sequence comparison between two pIPNS genes and revealed that there is a two-nucleotide difference in the pIPNS gene of a low-penicillin-producing strain AS-P-78 compared to a pIPNS gene of a high-producing strain 23X-80-269-37-2. This results in a difference of one amino acid in the predicted amino acid sequence of the pIPNS protein at amino acid position 195 (Fig. 4, pIPNS numbering, tyrosine vs. isoleucine). Ramon et al. [23] suggested that all the IPNS genes may have arisen from a common ancestral gene based on the high

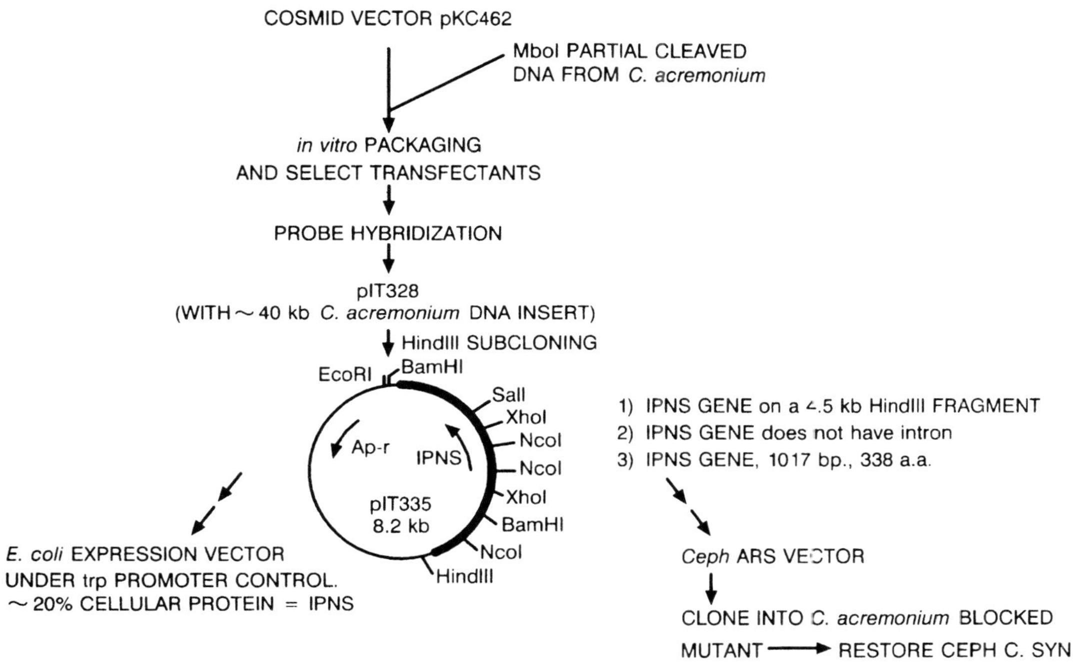

FIGURE 2. Cloning of the IPNS gene from *Cephalosporium acremonium*. (From Ref. 21.)

degree of similarity between the three fungal IPNS genes. If this hypothesis is correct, the nucleotide sequences from the wild-type or progenitor strain should have a higher degree of similarity to the ancestral IPNS gene. Since the IPNS genes cloned from the wild-type *A. nidulans*, *C. acremonium*, and a low producer of *P. chrysogenum* all have the same tyrosine amino acid at position 195, the different amino acid (isoleucine) in the high-penicillin-producing strain could have resulted from consecutive mutagenic treatments during the long-term strain improvement process.

Using transverse alternating field electrophoresis (TAFE), a molecular karyotype was established for *C. acremonium* [29]. Eight chromosomal bands were detected. Southern hybridization with the cloned cIPNS gene as a probe revealed that the cIPNS gene is located on chromosome VI and is present at one gene copy per chromosome [29]. The karyotype of *P. chrysogenum* has not been established by TAFE method. Southern hybridization revealed that the wild-type and a number of high-penicillin-producing strains of *P. chrysogenum* have the same intensity of hybridization signal when using the pIPNS gene as a probe (Chiang, unpublished data). However, Smith et al. recently demonstrated that the number of copies of the pIPNS gene ranges from eight to 16 in a high-penicillin-producing strain BW1890 [30]. This phenomenon has not been observed in other high-penicillin-producing strains.

With the availability of the cloned IPNS gene and the ability to overproduce its protein in *E. coli*, information regarding the biochemical properties of this enzyme can be readily obtained through site-directed mutagenesis.

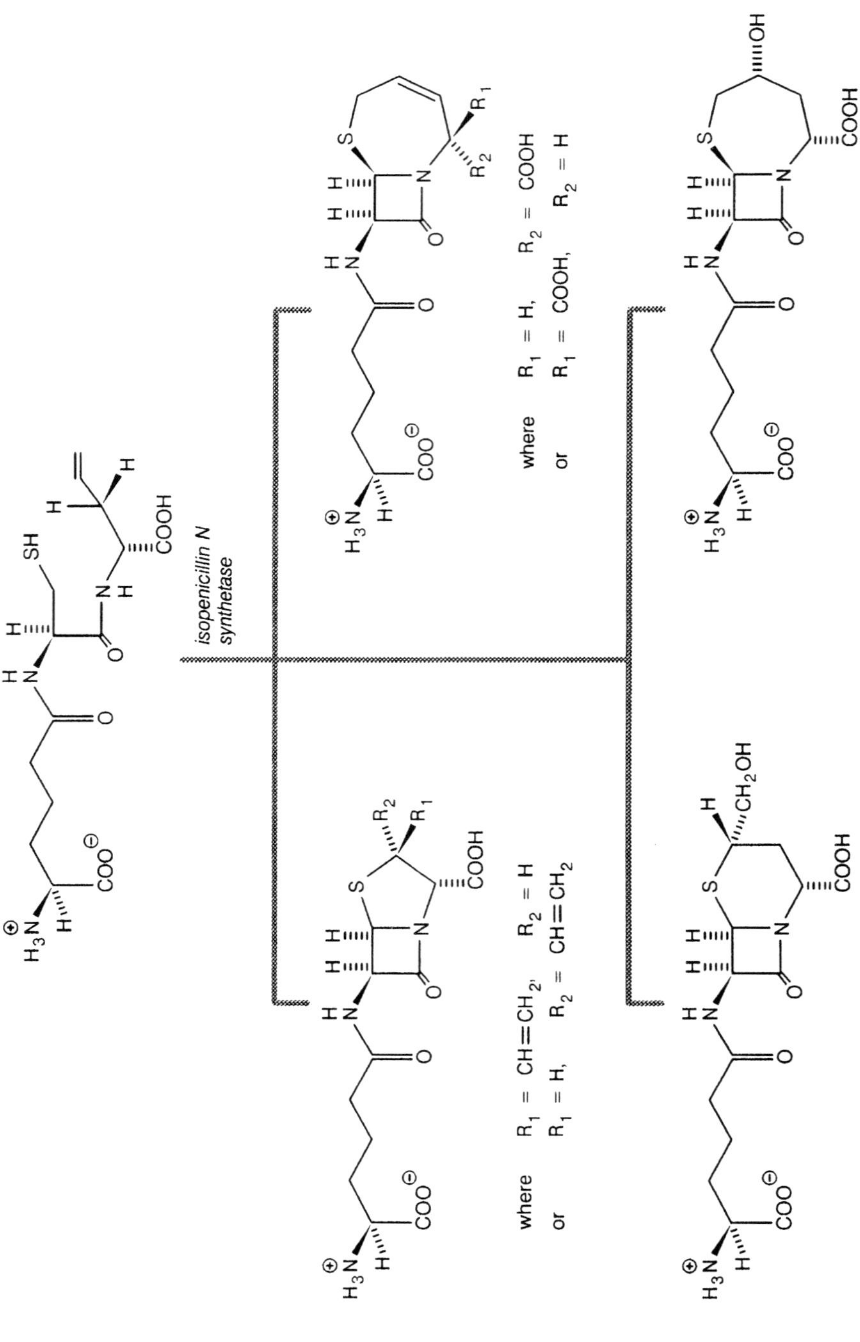

FIGURE 3. Structures of new β-lactam antibiotics produced by the action of isopenicillin N synthetase on an allyl analog of ACV tripeptide. (From Ref. 26.)

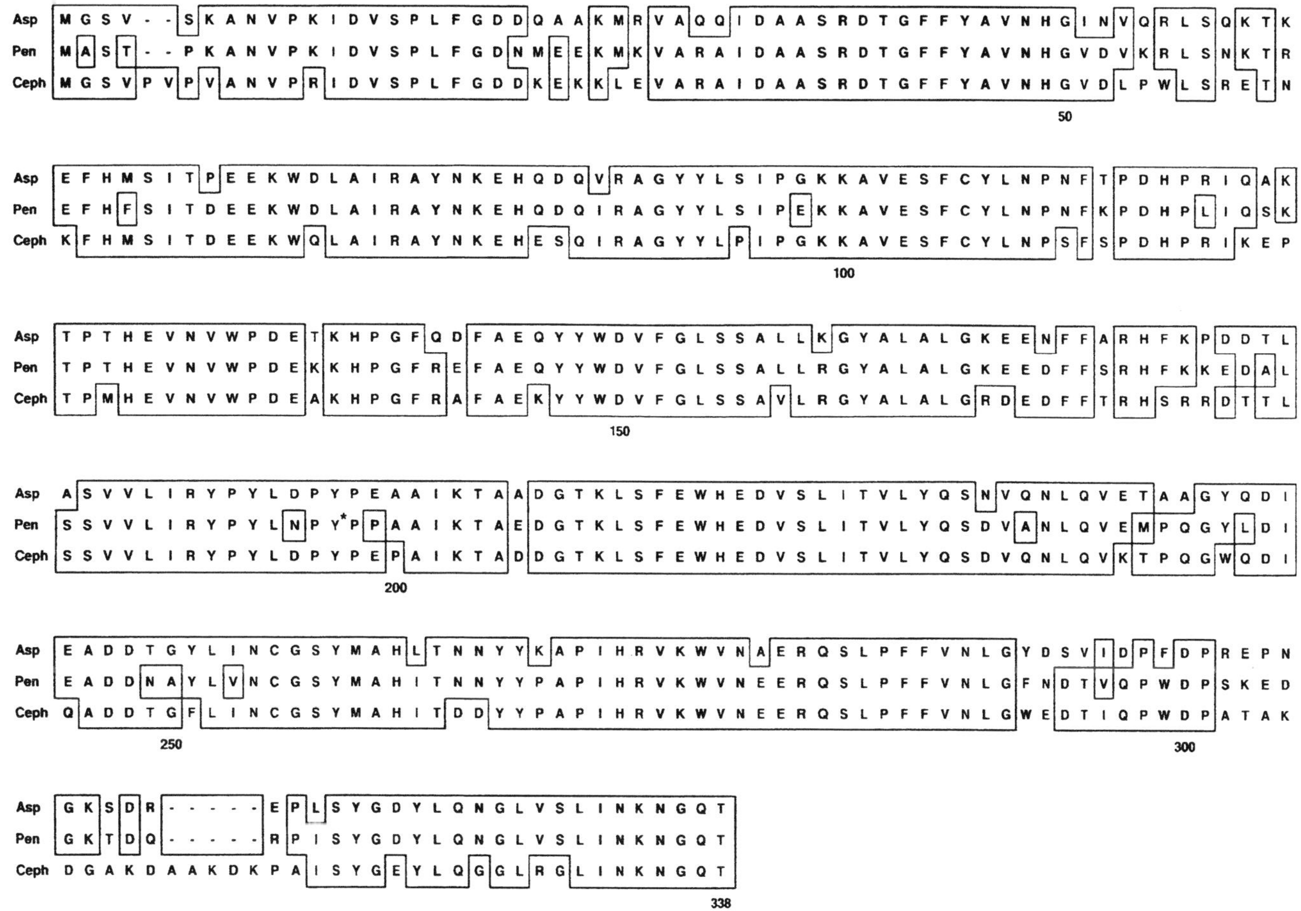

FIGURE 4. Amino acid sequence comparison of the fungal IPNS genes. Asp, *Aspergillus nidulans* [23,24]; Pen, *Penicillium chrysogenum* [22,25]; and Ceph, *Cephalosporium acremonium* [21]. All of the matching residues are enclosed in boxes. Hyphens indicate deletions. *At this position, it is Y (tyrosine) in the pIPNS gene of a low-penicillin-producing strain AS-P-78 [25]; and it is I (isoleucine) in the pIPNS gene of a high-producing strain 23X-80-269-37-2 [21].

	Asp	Pen	Ceph	
Asp	\	78.8	72.6	Amino
Pen	75.7	\	74.7	Acid
Ceph	68.8	72.7	\	Similarities

Nucleotide Similarities

FIGURE 5. Pairwise comparison of nucleotide (lower left) and amino acid (upper right) similarities between the fungal IPNS genes. Asp, *Aspergillus nidulans*; Pen, *Penicillium chrysogenum*; and Ceph, *Cephalosporium acremonium*. (From Ref. 11.)

Previous biochemical studies have shown that sulfhydryl-reactive alkylating agents inactivated the IPNS enzyme, suggesting the involvement of at least one cysteine residue in enzyme activity [31]. The fungal IPNS proteins contain two cysteine residues and the location of these two cysteines is conserved in all three fungal IPNS proteins (Fig. 4, Cys-106 and Cys-255 in the cIPNS numbering). The cysteine residues in the cIPNS were changed to serine residues, one at a time or in combination, by site-directed mutagenesis of the *E. coli*/cIPNS expression vector. Serine was chosen because the shape of the molecule is very similar to that of cysteine. Mutation of Cys-255 reduces the specific activity by nearly 50%. However, mutation of the other cysteine residue, Cys-106, reduces the specific activity to 5% of that of the wild-type enzyme and the K_m of the protein increases about fivefold either alone or in combination with the Ser-255 mutation [31]. It appears likely that the Cys-106 residue is involved in substrate binding. The fact that the doubly mutated cIPNS protein still has residual activity in the absence of any cysteine residues indicates that a free sulfhydryl moiety is important but not essential for IPNS activity [31].

B. Cloning of the Deacetoxycephalosporin C Synthetase/Deacetylcephalosporin C Synthetase Gene

The deacetoxycephalosporin C synthetase (DAOCS or expandase) and the deacetylcephalosporin C synthetase (DACS or hydroxylase) activities were copurified as a monomeric protein from *C. acremonium* [16,32]. A partial amino acid sequence of this protein was determined and the reverse genetics approach was used to clone the DAOCS/DACS gene (*cef*EF) [33]. The nucleotide sequence of the DAOCS/DACS gene revealed one open reading frame of 999 bp that encodes a polypeptide of 332 amino acids in length. When this gene was expressed in *E. coli*, both DAOCS and DACS activities were identified in crude cell extracts, providing unambiguous proof of the bifunctional nature of the DAOCS/DACS polypeptide from *C. acremonium*. However, in two *Streptomyces* species, DAOCS and DACS were purified as two separate enzymes [34,35]. The *cef*E gene cloned from *S. clavuligerus* encodes a polypeptide with DAOCS activity only [36].

Although the predicted amino acid sequence of the DAOCS/DACS gene has only slightly greater than 11% similarity to the cIPNS gene, certain areas showed higher similarity. In particular a 10-amino-acid region containing a cysteine residue at position 100 of the DAOCS/DACS is 50% identical to the corresponding region containing the Cys-106 of the cIPNS [33]. As discussed above, the Cys-106 residue of the cIPNS protein is involved in substrate binding. Perhaps the Cys-100 of the DAOCS/DACS protein is also involved in substrate binding. This hypothesis could be tested by site-directed mutagenesis of the Cys-100 codon of the DAOCS/DACS gene.

On analysis of *C. acremonium* chromosomes by TAFE, the DAOCS/DACS gene probe hybridized with chromosome II. Thus, unlike *Streptomyces* in which antibiotic biosynthetic genes are clustered, the fungal genes involved in cephalosporin C biosynthesis are not located on a single chromosome, but are located on at least two or more chromosomes [5,6].

C. Cloning of the Acyl CoA:6-APA Acyltransferase Gene

The reverse genetics approach was also used to clone the acyl CoA:6-APA acyltransferase gene (*pen*E) from *P. chrysogenum* [37]. The acyltransferase gene was identified on a cosmid clone in which the acyl CoA:6-APA acyltransferase gene is about 1.5 kb downstream of the 3' end of the pIPNS gene, indicating that these two genes are clustered and located on the same chromosome. In addition, the acyltransferase and the pIPNS genes are also transcribed in the same direction. The nucleotide sequence revealed that the acyltransferase gene contains at least two introns [37].

IV. STRAIN IMPROVEMENT BY AMPLIFICATION OF THE CLONED β-LACTAM BIOSYNTHETIC GENES

A. Development of Fungal Cloning Vectors

The cloning vector serves as the vehicle to replicate and amplify the inserted DNA fragment. The vector must contain its own replication apparatus to be autonomous from the host chromosomal DNA, a selectable marker for identification, and at least one unique restriction enzyme recognition site so that the foreign DNA fragment can be inserted and removed intact. Since transformation systems have been best developed for *E. coli*, most fungal cloning vectors contain the replication origin from *E. coli* and a fungal autonomous replicating sequence (ARS). Therefore, the initial cloning and identification of the gene can be accomplished in *E. coli*. The recombinant DNA can then be transformed without further DNA manipulation into the fungal cells.

Intensive efforts to identify fungal ARS sequences have not yet led to the construction of a truly autonomous vector for gene cloning in fungi [38,39]. For example, fungal vectors containing a chromosomal sequence from *C. acremonium* originally isolated as an ARS function in yeast *Saccharomyces cerevisiae* do not autonomously replicate in *C. acremonium* [17].

When recombinant DNA is introduced into fungal cells, it integrates as either single or multiple copies into the host chromosomal DNA by homologous or heterologous recombination [38-43]. The stable transformants can be maintained in the absence of selective pressure [40-42]. Such stability is of key importance for strains intended for use in industrial fermentations.

Multiple integration of the same gene may allow for a higher expression of the gene product due to a gene dosage effect.

Selection of transformants from a background of nontransformed cells depends on expression of genes conferring readily selectable dominant phenotypes. High-frequency transformation has been achieved using vectors carrying the *pyr*4 gene of *Neurospora crassa* to complement a *pyr*G mutant of *P. chrysogenum* (the *N. crassa pyr*4 and the *P. chrysogenum pyr*G genes both code for orotidine-5'-phosphate decarboxylase [20]). However, appropriate auxotrophic mutant strains do not exist for most industrial high-β-lactam-producing strains. Thus, transformation with plasmids containing metabolic genes is usually not practical and dominant selectable markers that do not require corresponding mutant strains are therefore of particular value. The benomyl resistance gene (*bml*) has been useful [44]. Plasmids containing antibiotic resistance genes, e.g., resistant to G418, hygromycin, or phleomycin, fused to fungal promoters have been incorporated into a number of fungal cloning vectors [18,19,45].

B. Development of DNA-Mediated Transformation Systems for *Cephalosporium* and *Penicillium* Strains

DNA-mediated transformation systems for *A. nidulans*, *C. acremonium*, and *P. chrysogenum* have been developed along the same lines as the yeast transformation systems [17-20,39-42]. The protoplasts formed from germinated conidia or mycelia in the presence of Novozyme 234, a commercially available hydrolytic enzyme mixture secreted by *Trichoderma harzianum* [46], can be transformed with DNA in the presence of polyethylene glycol (PEG) and $CaCl_2$. DNA can also be introduced into fungal protoplasts by electroporation [47]. It is necessary to optimize the protoplast condition for each strain of interest. The fungal transformation frequency is very low and the transformed DNA is often integrated into the host chromosome. Transformation of *N. crassa* using lithium acetate-treated cells has been developed [48] and may be applied to other fungal species.

C. Strain Improvement

The cloned β-lactam biosynthetic genes can be transformed into *C. acremonium* and *P. chrysogenum* to provide additional copies of these genes. If the cloned gene corresponds to a rate-limiting step, increased production of cephalosporin C and penicillins can be obtained.

Transformation of wild-type and high-cephalosporin C-producing strains of *C. acremonium* with a plasmid containing the cIPNS gene has been performed. Most of the transformants showed equivalent cephalosporin C titers to the parental strains, even though cell extracts of these transformants had three- to fourfold more IPNS enzymatic activity than the parental strains. This result suggested that IPNS may not be the rate-limiting step in cephalosporin C biosynthesis in these two strains [49].

This high-cephalosporin C-producing strain of *C. acremonium* was transformed with a plasmid containing the DAOCS/DACS gene. By screening a small number of transformants, one of the transformants produced approximately 40% more cephalosporin C than its parental strain in shake-flask fermentations [50]. Extracts of this transformant contained nearly twofold more DAOCS/DACS activity than corresponding extracts of its parent. The parental strain accumulated a significant amount of penicillin N in fermentation medium, whereas the amount of penicillin N was markedly reduced in the recombinant strain. Reduction of penicillin N and increase of cephalosporin C are correlated with an increase of DAOCS/DACS activity in the recombinant strain. Apparently, accumulation of intermediate penicillin N indicates that DAOCS may be rate limiting for cephalosporin C biosynthesis in this strain [16,50].

V. OTHER APPLICATIONS OF THE CLONED β-LACTAM BIOSYNTHETIC GENES

As described above, the DAOCS activity may be rate limiting for cephalosporin C biosynthesis in at least some of the *C. acremonium* strains. The low DAOCS activity can be caused by inefficient expression of the gene due to promoter strength or negative regulation by a repressor. This can be solved by replacing the authentic promoter with a strong promoter which is not negatively regulated by the same repressor and can be recognized efficiently by the RNA polymerase of *C. acremonium* during the antibiotic production phase. Ramos et al. showed that the cIPNS enzyme was formed early in the fermentation (the highest specific activities were obtained between 24 and 48 h of fermentation) and declined in the latter part of the fermentation (after 96 h) at which time cephalosporin C production was still at a maximal rate [7]. These data suggest that cIPNS was overproduced and, therefore, the promoter may be expressed very efficiently. Replacing the promoter with the cIPNS promoter could be easily accomplished.

Alternatively, the low DAOCS activity can be caused by low substrate affinity of the enzyme. The problem of low substrate affinity can be overcome by protein engineering via site-directed mutagenesis. Large quantities of recombinant proteins can be obtained from *E. coli* and used for determination of the substrate binding site of the DAOCS/DACS enzyme. Using this information, it is feasible to mutagenize specific regions of enzyme and alter enzyme properties to give higher substrate affinity and/or broader substrate specificity.

The purified recombinant cIPNS protein obtained from *E. coli* has identical properties to the authentic protein, which can convert ACV tripeptide analog substrates to unusual β-lactam antibiotics [26]. Bioconversion of analogs of penicillin N to novel cephalosporins using recombinant DAOCS/DACS proteins has not been reported.

Site-directed mutagenesis of the cloned β-lactam biosynthetic genes could result in enzymes with broader substrate specificity which would generate novel β-lactams more efficiently and could lead to the eventual commercial manufacture of new β-lactams via immobilized biosynthetic enzymes. Moreover, in vivo production of novel β-lactams would be feasible by returning these modified genes to mutant strains of *C. acremonium*.

VI. CONCLUSION

Several genes encoding key enzymes involved in the biosynthesis of cephalosporins and penicillins have been cloned, characterized, and overproduced in *E. coli*. The remainder of the genes involved in β-lactam biosynthesis will be cloned in the near future. With an understanding of the genetics and biochemistry of β-lactam biosynthesis, it may be possible to generate high-β-lactam-producing strains through rational genetic engineering approaches. Moreover, novel β-lactam antibiotics could be synthesized in vitro or in vivo using engineered enzymes which have broader substrate specificity and better enzyme activity.

REFERENCES

1. Elander, R. P., Microbial screening, selection and strain improvement. In *Basic Biotechnology* (J. D. Bu'Lock and B. Kristiansen, eds.). Academic Press, London, 1987, pp. 217-251.
2. Chang, L. T., Terasaka, D. T., and Elander, R. P. Protoplast fusion in industrial fungi. *Dev. Ind. Microbiol.*, *23*:21-30, 1982.
3. Elander, R. P. Applied genetics and molecular biology of industrial microorganisms. In *Advanced Biochemical Engineering* (H. R. Bungay and G. Belfort, eds.), John Wiley and Sons, New York, 1987, pp. 167-186.
4. Martin, J. F. Cloning of genes involved in penicillin and cephalosporin biosynthesis. *Trends Biotechnol.*, *5*:306-308, 1987.
5. Fayerman, J. T. New developments in gene cloning in antibiotic producing microorganisms. *Biotechnology, 4*:786-789, 1986.
6. Miller, J. R., and Ingolia, T. D. Cloning and characterization of beta-lactam biosynthetic genes. *Mol. Microbiol.*, *3*:689-695, 1989.
7. Ramos, F. R., Lopez-Nieto, M. J., and Martin, J. F. Coordinate increase of isopenicillin N synthetase, isopenicillin N epimerase and deacetoxycephalosporin C synthetase in a high cephalosporin-producing mutant of Acremonium chrysogenum and simultaneous loss of the three enzymes in a non-producing mutant. *FEMS Microbiol. Lett.*, *35*:123-127, 1986.
8. Shen, Y. Q., Wolfe, S., and Demain, A. L. Levels of isopenicillin N synthetase and deacetoxycephalosporin C synthetase in Cephalosporium acremonium producing high and low levels of cephalosporin C. *Biotechnology, 4*:61-64, 1986.
9. Jensen, S. E. Biosynthesis of cephalosporins. *CRC Crit. Rev. Biotechnol.*, *3*:277-301, 1986.
10. Nüesch, J., Heim, J., and Treichler, H. J. The biosynthesis of sulfur-containing β-lactam antibiotics. *Ann. Rev. Microbiol.*, *41*:51-75, 1987.
11. Ingolia, T. D., and Queener, S. W. Beta-lactam biosynthetic genes. *Med. Res. Rev.*, *9*:245-264, 1989.
12. Demain, A. L. Biosynthesis of β-lactam antibiotics. In *Antibiotics Containing the β-Lactam Structure I* (A. L. Demain and N. A. Solomon, eds.), Springer-Verlag, Berlin, 1983, pp. 189-228.
13. Batchelor, F. R., Doyle, F. P., Nayler, J. H. C., and Rounson, G. N. Synthesis of penicillin:6-aminopenicillanic in penicillin fermentations. *Nature, 183*:257-258, 1959.
14. Erickson, R. C., and Dean, L. D. Acylation of 6-APA by Penicillium chrysogenum. *Appl. Microbiol.*, *14*:1047-1048, 1966.
15. Alvarez, E., Cantoral, J. M., Barredo, J. L., Diez, B., and Martin, J. F. Purification to homogeneity and characterization of acyl coenzyme A:6-aminopenicillanic acid acyltransferase of Penicillium chrysogenum. *Antimicrob. Agents Chemother.*, *31*:1675-1682, 1987.
16. Scheidegger, A., Küenzi, M. T., and Nüesch, J. Partial purification and catalytic properties of a bifunctional enzyme in the biosynthetic

pathway of β-lactams in Cephalosporium acremonium. *J. Antibiot.*, *37*:522-531, 1984.
17. Queener, S. W., Ingolia, T. D., Skatrud, P. L., Chapman, J. L., and Kaster, K. R. A system for genetic transformation of Cephalosporium acremonium. In *Microbiology 1985* (L. Leive, P. F. Bonventre, J. A. Morello, S. Schlesinger, S. D. Silver, and H. C. Wu, eds.), American Society for Microbiology, Washington, D. C., 1985, pp. 468-472.
18. Skatrud, P. L., Queener, S. W., Carr, L. G., and Fisher, D. L. Efficient integrative transformation of Cephalosporium acremonium. *Curr. Genet.*, *12*:337-348, 1987.
19. Isogai, T., Yoshida, M., Tanaka, M., and Aoki, H. Transformation of Acremonium chrysogenum and Saccharomyces cerevisiae using an antibiotic resistance marker. *Agr. Biol. Chem.*, *51*:2321-2329, 1987.
20. Cantoral, J. M., Diez, B., Barredo, J. L., Alvarez, E., and Martin, J. F. High-frequency transformation of Penicillium chrysogenum. *Biotechnology*, *5*:494-497, 1987.
21. Samson, S. M., Belagaje, R., Blankenship, D. T., Chapman, J. L., Perry, D., Skatrud, P. L., VanFrank, R. M., Abraham, E. P., Baldwin, J. E., Queener, S. W., and Ingolia, T. D. Isolation, sequence determination and expression in Escherichia coli of the isopenicillin N synthetase gene from Cephalosporium acremonium. *Nature*, *318*:191-194, 1985.
22. Carr, L. G., Skatrud, P. L., Scheetz, M. E. II, Queener, S. W., and Ingolia, T. D. Cloning and expression of the isopenicillin N synthetase gene from Penicillium chrysogenum. *Gene*, *48*:257-266, 1986.
23. Ramon, D., Carramolino, L., Patino, C., Sanchez, F., and Penalva, M. A. Cloning and characterization of the isopenicillin N synthetase gene mediating the formation of the β-lactam ring in Aspergillus nidulans. *Gene*, *57*:171-181, 1987.
24. Weigel, B. J., Burgett, S. G., Chen, V. J., Skatrud, P. L., Frolik, C. A., Queener, S. W., and Ingolia, T. D. Cloning and expression in Escherichia coli of isopenicillin N synthetase genes from Streptomyces lipmanii and Aspergillus nidulans. *J. Bacteriol.*, *170*:3817-3826, 1988.
25. Barredo, J. L., Cantoral, J. M., Alvarez, E., Diez, B., and Martin, J. F. Cloning, sequence analysis and transcriptional study of the isopenicillin N synthetase of Penicillium chrysogenum As-P-78. *Mol. Gen. Genet.*, *216*:91-98, 1989.
26. Baldwin, J. K., Killin, S. J., Pratt, A. J., Sutherland, J. D. Turner, N. J., Crabbe, M. J. C., Abraham, E. P., and Willis, A. Purification and characterization of cloned isopenicillin N synthetase. *J. Antibiot.*, *40*:652-659, 1987.
27. Leskiw, B. K., Aharonowitz, Y., Mevarech, M., Wolfe, S., Vining, L. C., Westlake, D. W. S., and Jensen, S. E. Cloning and nucleotide sequence determination of the isopenicillin N synthetase gene from Streptomyces clavuligerus. *Gene*, *62*:187-196, 1988.
28. Shiffman, D., Mevarech, M., Jensen, S. E., Cohen, G., and Aharonowitz, Y. Cloning and comparative sequence analysis of the gene coding for isopenicillin N synthetase in Streptomyces. *Mol. Gen. Genet.*, *214*:562-569, 1988.
29. Skatrud, P. L., and Queener, S. W. An electrophoretic molecular karyotype for Cephalosporium acremonium. *Gene*, *78*:331-338, 1989.

30. Smith, D. J., Bull, J. H., Edwards, J., and Turner, G. Amplification of the isopenicillin N synthetase gene in a strain of Penicillium chrysogenum producing high levels of penicillin. *Mol. Gen. Genet.*, *216*:492-497, 1989.
31. Samson, S. M., Chapman, J. L., Belagaje, R., Queener, S. W., and Ingolia, T. D. Analysis of the role of cysteine residues in isopenicillin N synthetase activity by site-directed mutagenesis. *Proc. Natl. Acad. Sci. USA*, *84*:5705-5709, 1987.
32. Dotzlaf, J. E., and Yeh, W. K. Copurification and characterization of deacetoxycephalosporin C synthetase/hydroxylase from Cephalosporium acremonium. *J. Bacteriol.*, *169*:1611-1618, 1987.
33. Samson, S. M., Dotzlaf, J. E., Slisz, M. L., Becker, G. W., VanFrank, R. M., Veal, L. E., Yeh, W. K., Miller, J. R., Queener, S. W., and Ingolia, T. D. Cloning and expression of the fungal expandase/hydroxylase gene involved in cephalosporin biosynthesis. *Biotechnology*, *5*:1207-1214, 1987.
34. Jensen, S. E., Westlake, D. W. S., and Wolfe, S. Deacetoxycephalosporin C synthetase and deacetoxycephalosporin C hydroxylase are two separate enzymes in Streptomyces clavuligerus. *J. Antibiot.*, *38*:263-265, 1985.
35. Cortes, J., Martin, J. F., Castro, J. M., Laiz, L., and Liras, P. Purification and characterization of a 2-oxoglutarate-linked ATP-independent deacetoxycephalosporin C synthetase of Streptomyces lactamdurans. *J. Gen. Microbiol.*, *133*:3165-3174, 1987.
36. Kovacevic, S., Weigel, B. J., Tobin, M. B., Ingolia, T. D., and Miller, J. R. Cloning, characterization and expression in Escherichia coli of the Streptomyces clavuligerus gene encoding deacetoxycephalosporin C synthetase. *J. Bacteriol.*, *171*:754-760, 1989.
37. Veenstra, A. E., Van Solingen, P., Huininga-Muurling, H., Koekman, B. P., Groenen, M. A. M., Smaal, E. B., Kattevilder, A., Alvarez, E., Barredo, J. L., and Martin, J. F. Cloning of penicillin biosynthetic genes. In *Genetics and Molecular Biology of Industrial Microorganisms* (C. L. Hershberger, S. W. Queener, and G. Hegeman, eds.), American Society for Microbiology, Washington, D.C., 1989, pp. 262-269.
38. Saunders, G., Tuite, M. F., and Holt, G. Fungal cloning vectors. *Trends Biotechnol.*, *4*:93-98, 1986.
39. Tilburn, J., Scazzocchio, C., Taylor, G. G., Zabicky-Zissman, J. H., Lockington, R. A., and Davis, R. W. Transformation by integration in Aspergillus nidulans. *Gene*, *26*:205-221, 1983.
40. Haynes, M. J. Transformation of filamentous fungi. *Exp. Mycol.*, *10*:1-8, 1986.
41. Cullen, D., and Leong, S. Recent advances in the molecular genetics of industrial filamentous fungi. *Trends Biotechnol.*, *4*:285-288, 1986.
42. Fincham, J. R. S. Transformation in fungi. *Microbiol. Rev.*, *53*:148-170, 1989.
43. Timberlake, W., and Marshall, M. A. Genetic engineering of filamentous fungi. *Science*, *244*:1313-1317, 1989.
44. Orbach, M. J., Porro, E. B., and Yanofsky, C. Cloning and characterization of the gene for β-tubulin from a benomyl-resistant mutant of Neurospora crassa and its use as a dominant selectable marker. *Mol. Cell. Biol.*, *6*:2452-2461, 1986.
45. Gatignol, A., Baron, M., and Tiraby, G. Phleomycin resistance encoded by the ble gene from transposon Tn5 as a dominant selectable marker in Saccharomyces cerevisiae. *Mol. Gen. Genet.*, *207*:342-348, 1987.

46. Kitamato, Y., Kono, R., Tokimato, K., Mori, N., and Ichikawa, Y. Production of lytic enzymes against cell walls of basidiomycetes from Trichoderma harzianum. *Trans. Mycol. Soc. Jpn.*, *26*:69-79, 1984.
47. Lynch, P. T., Isaac, S., and Collin, H. A. Electrofusion of protoplasts from Aspergillus nidulans. *FEMS Microbiol. Lett.*, *59*:225-228, 1989.
48. Dhawale, S. S., Paietta, J. V., and Marzluf, G. A. A new rapid and efficient transformation procedure for Neurospora, *Curr. Genet.*, *8*:77-79, 1984.
49. Skatrud, P. L., Fisher, D. L., Ingolia, T. D., and Queener, S. W. Improved transformation of Cephalosporium acremonium. In *The Proceedings of the Fifth International Symposium on the Genetics of Industrial Microorganisms* (M. Alacevic, D. Hranueli, and Z. Toman, eds.), Pliva, Zagreb, 1987, pp. 111-119.
50. Skatrud, P. L., Tietz, A. J., Ingolia, T. D., Cantwell, C. A., Fisher, D. L., Chapman, J. L., and Queener, S. W. Use of recombinant DNA to improve production of cephalosporin C by Cephalosporium acremonium. *Biotechnology*, *7*:477-485, 1989.

9

THE FUNGAL MITOCHONDRIAL GENOME—A BROADER PERSPECTIVE

MICHAEL E. S. HUDSPETH *Northern Illinois University, DeKalb, Illinois*

I. INTRODUCTION

Fungi have played a central role in unraveling the intricacies of mitochondrial biogenesis. The ease with which *Saccharomyces cerevisiae* and, to a lesser extent, its filamentous counterparts *Neurospora crassa* and *Aspergillus nidulans* can be genetically manipulated has been largely responsible for the transition of this field from an emphasis on mitochondrial genome organization to one addressing the complex problems of nucleocytoplasmic interactions. Indeed, the more recent endeavors of relocating ubiquitously mitochondrially encoded genes in the nucleus have begun to directly question the evolutionary necessity of maintaining an organellar genome [1].

In recent years several insightful reviews and commentaries encompassing various aspects of the fungal mitochondrial genome have appeared. These range from the relatively broad-based [2-7] to the species-specific [8-11], and to the more topical aspects of nucleocytoplasmic interactions, protein targeting, RNA splicing, and intronic sequences [12-17].

From a truly mycological perspective, however, we remain largely ignorant of the details of mitochondrial genetic structure and function. Apart from the extensive sequence data of five ascomycetous genomes (*S. cerevisiae*, *A. nidulans*, *N. crassa*, *Podospora anserina*, and *Schizosaccharomyces pombe*), only limited organizational detail is known. Although the variety of encoded information, expression, organization, and genetic code variation inherent in just these five systems poignantly suggests that the expansion of research initiatives into the fungal kingdom in its broadest sense will continue to produce interesting results, such efforts have only been marginally undertaken. These endeavors, hindered by a paucity of manageable genetic systems, are initially expected to continue their focus on the structural aspects of gene and genome structural organization. Thus, rather than attempting to summarize in detail information already available in review form, an integrated review of what is known of the structural organization of the "other" largely ignored fungal mitochondrial genomes, in the context of a modest updating of the genomes of the major ascomycete systems, would be more useful.

Finally, in line with the observations of Kohli [18] and Wolf and del Giudice [7] that a consistent nomenclature for mitochondrial genes is lacking, the genetic loci in this chapter use a common three-letter code derived

TABLE 1. Nomenclature for Mitochondrial Genes

Gene	Equivalent		Function
Translational apparatus			
rnl	L-RNA		Large ribosomal subunit RNA
rns	S-RNA		Small ribosomal subunit RNA
tsl			tRNA synthesis locus
*var*1			Small ribosomal subunit protein
S-5			Small ribosomal subunit protein
Respiratory chain			
*cox*1	COI, *oxi*3		Cytochrome c oxidase subunit 1
*cox*2	COII, *oxi*1		Cytochrome c oxidase subunit 2
*cox*3	COIII, *oxi*2		Cytochrome c oxidase subunit 3
cob	*cytb*		Apocytochrome b
*ndh*1	*urf*1	*ND*1	NADH dehydrogenase subunit 1
*ndh*2	*urf*2	*ND*2	NADH dehydrogenase subunit 2
*ndh*3	*urf*3	*ND*3	NADH dehydrogenase subunit 3
*ndh*4	*urf*4	*ND*4	NADH dehydrogenase subunit 4
*ndh*4l	*urf*4L	*ND*4L	NADH dehydrogenase subunit 4L
*ndh*5	*urf*5	*ND*5	NADH dehydrogenase subunit 5
*ndh*6	*urf*6	*ND*6	NADH dehydrogenase subunit 6
ATP synthase			
*atp*6	*oli*2,4		ATP synthase subunit 6
*atp*8	*urfA*6L, *aap*1		ATP synthase subunit 8
*atp*9	*mal*, *oli*1,3		ATP synthase subunit 9
General			
ORF			Open reading frame
URF			Unidentified reading frame

Source: Modified from Refs. 7 and 18.

from their gene product. A list of abbreviations used and their common equivalents is found in Table 1.

II. AN OVERVIEW OF NONFUNGAL SYSTEMS

Despite major organizational differences prevalent in the mitochondrial genomes of the four eukaryotic kingdoms, several generalizations can be made. As a rule the genomes are circular and encode a limited set of ubiquitous gene products which are constituents either of the mitochondrial translational apparatus, or of the inner membrane respiratory chain and oxidative phosphorylation complexes. A minimum set of 22 tRNAs, sufficient to translate all 20 amino acids, and the small and large ribosomal RNAs (*rns* and *rnl*) comprise the mtDNA-encoded translational apparatus. Encoded membrane proteins usually include seven subunits of the NADH dehydrogenase complex (*ndh*), apocytochrome b of the cytochrome b,c_1 complex (*cob*), subunits 1-3 of cytochrome c oxidase (*cox*), and subunits 6 and 8 of ATP synthase (*atp*). All other genes related to mitochondrial function are usually nuclear-encoded. The following brief descriptions of different systems highlight unique organizational and genetic features at variance with these generalizations, and are intended to provide a frame of reference for the fungal systems.

Animal genomes usually range in size from 16 to 18 kb with a lower limit of 14.1 kb for the amphibian *Siredon* [19] and an upper limit of 34 kb for the deep-sea scallop *Placopecten* [20]. The complete DNA sequence of human [21], bovine [22], mouse [23], rat [24], *Xenopus* [25], chicken [26], sea urchin [27-29], *Drosophila* [30], and *Ascaris* [31] confirm previous contour length size estimates. As animal mitochondrial genomes are intronless, the primary source of size variation occurs in the region of the replication origin. The unusual larger genomes, as in the scallop, arise from reiteration of sequences. As suggested by their usually small sizes, these genomes are extremely compact with overlapping or virtually contiguous genes. The genic complements are identical (exclusive of the absence of *atp8* in *Ascaris*), but with variations in gene orders resulting from a number of inversions and/or transpositions.

Plant mtDNAs exhibit the largest size variation and typically exist as subgenomic molecules. Restriction analyses set a lower limit of 218 kb in a *Brassica* species [32], and reassociation kinetics an upper limit of 2400 kb in members of the Cucurbitae [33]. Additional genes encoded include the α subunit of ATP synthase, *atp*9, 5S RNA, and at least one ribosomal protein. The DNA sequence analysis of cDNA clones demonstrated that RNA editing plays a role in the expression of plant mitochondrial genes [34-36]. While the complete sequence of a genome is expected to provide additional genes, and the inclusion of at least some introns in specific genes will account for some of the extreme size variations of these genomes, it is unlikely that a significant number of novel genes will be encoded to explain the variation. The population of subgenomic molecules is a consequence of intramolecular recombination between direct repeats [32], with the number of molecules a function of the number and types of repeats.

The diversity of the Protista is reflected in the organizational patterns of their mtDNAs. Trypanosome mtDNAs (kinetoplast or kDNAs) are organized as a catenated network of mini- and maxicircles with the maxicircles encoding the mitochondrial genes [37]. The ciliates *Paramecium* and *Tetrahymena* [38,39] as well as the chlorophyte *Chlamydomonas* [40] have linear genomes, while *Acanthamoeba* [41] and *Isospora* [43] have the typical circular conformation. Extensive sequence data from trypanosome mtDNAs [37] failed to identify a single tRNA, and in *Chlamydomonas* [43] only three appear to be encoded in the genome. Similarly, only eight tRNAs appear to be present in ciliates [44]. Furthermore, *Chlamydomonas* lacks *cox*2 and *cox*3, and the analysis of *Paramecium cox*2 shows a complete lack of homology of its amino half with all documented *cox*2 sequences—including universally conserved residues [45,46]. Finally, in the trypanosomes, RNA editing of transcripts is so extensive that for some loci the DNA sequences are unrecognizable as the appropriate genetic loci [37,47].

III. PHYSICAL CHARACTERISTICS OF FUNGAL mtDNA

Mitochondrial genomes of the fungi and their allies are best described as circular AT-rich molecules ranging in size from a compact 17.6 kb in

Schizosaccharomyces pombe EF [48] to 175 kb in *Agaricus bitorquis* [49]. Linear conformations have been identified only for *Hansenula mrakii* [50], *Candida rhagii* [51], and *Physarum polycephalum* [52]. The AT-rich genomes, ostensibly the result of an inability to efficiently excise uracil residues derived from cytosine demethylations, have proven useful in the isolation of large quantities of purified mtDNAs by the use of the fluorescent AT-specific DNA-binding dyes DAPI or bisbenzimide (Hoechst 33258) [53-55]. The origin and nature of the ninefold kingdom-wide genomic size variation can only be partially explained by sequence reiterations, differential coding sequences, and the presence of intronic sequences.

A compendium of physical parameters is found in Table 2.

A. The Ascomycetes

The mitochondrial genome organization of *S. cerevisiae* (and *S. carlsbergensis*) was recently reviewed [10,56]. Structurally these 74- to 85-kb exceptionally AT-rich DNAs consist of long stretches of AT spacers punctuated by coding sequences and about 200 identified short GC-rich clusters. The AT distribution ranges from 75% in coding regions, with the exception of *var*1 [57], to 85% for intergenic regions. GC clusters are primarily confined to intergenic sequences (inclusive of open reading frames, or ORFs, and *ori*/*rep* sequences), some intronic regions, and *var*1. Interstrain size variation is readily accounted for by the presence of optional introns, intergenic ORFs, GC clusters, and mini-inserts such as those identified in *rns* [58,59] and *var*1 [60]. Remarkably, even though a maximum of only one-third of the genomes are coding sequences, the remaining sequences are also highly conserved between strains. The 80% genic homology between *S. cerevisiae* and the closely related 20-kb genome of *Torulopsis glabrata* [61] suggests that the primary size variation in these and probably other less characterized related yeasts is simply the result of the absence of introns and extensive spacer regions. Finally, clusters of GC-rich *Sac*I-like palindromic sequences have been identified in *Kluyveromyces lactis* [62].

Extensive sequence analyses of the *Aspergillus*, *Neurospora*, and *Podospora* genomes indicate that, like *S. cerevisiae*, spacer regions of the filamentous ascomycetes are AT-rich relative to coding regions. GC-rich clusters, however, have been identified in *Neurospora*. Fifty to one hundred of these short palindromic sequences, containing *Pst*I-like recognition sequences [63], are interspersed in the genome [64] and flank most genic regions [63-67]. While their structure is suggestive of a role in RNA processing [63], transcriptional analyses [68-70] and their absence in other filamentous genomes indicate that this is unlikely, and that their function is not basic to the structural organization of the filamentous ascomycete mtDNA.

B. The Basidiomycetes

Although no detailed structural analysis of a basidiomycete genome is yet available, characterization of different species in three genera suggest that the nearly fourfold genome size differential is the result of a series of insertion/deletion events. Restriction endonuclease and DNA hybridization data in *Suillus* (Family Boletaceae) suggest that the virtual continuum of sizes (36-121 kb) which encompasses 15 species is a direct result of the accumulation of multiple insertion events distributed throughout the

TABLE 2. Size Estimates of Fungal Mitochondrial Genomes

Taxon	Mitochondrial genome size	Ref.
Myxomycetes		
Physarum polycephalum	69	52
Oomycetes		
Achlya ambisexualis	49.8-50.7	79
A. heterosexualis	50.7-51.2	79
A. klebsiana	50.7	181
Aplanopsis terrestris	45	77
Apodachlya pyrifera	40	77
A. brachynema	36.4	77
Leptolegnia caudata	52.4	77
Leptomitus lacteus	37	78
Phytophthora cryptogea	40.7	77
P. infestans	36.2	76
P. megasperma	41-45.3	82,182
P. parasitica	39.5	182
Pythium diclinum	70	80
P. irregulare	60	77
P. paddicum	61.3	77
P. torulosum	73	80
P. ultimum	57	78
Saprolegnia sp.	44.5	183
S. ferax	46.5	78
Sapromyces elongatus	53	77
Hyphochytridiomycetes		
Hypochytrium catenoides	54	184
Rhizidiomyces apophysatus	50	184
Chytridiomycetes		
Allomyces arbuscula	49.2	185
A. macrogynus	56	87
Blastocladiella emersonii	35.5	186
Zygomycetes		
Phycomyces blakesleeanus	25.6	78
Ascomycetes (Yeasts)		
Brettanomyces anomalus	57.7	187
B. custersianus	28.5	187
B. custersii	101.1	187
B. naardenensis	41.7	187
Candida albicans	41	188
C. maltosa	52	189
C. parapsilosis	34.5	190
C. rhagii	30	51
Dekkera bruxellensis	85	187
D. intermedia	73.2	187
Eeniella nana	34.5	187

Table 2 (Continued)

Taxon	Mitochondrial genome size	Ref.
[Ascomycetes (Yeasts)]		
Hanseniaspora vinea	27.6	191
Hansenula mrakii	55	50
H. petersonii	42	192
H. wingei	25.5	190
Kloeckera africana	27.1	193
Kluyveromyces lactis	37	194
Pachytichospora transvaalensis	41.4	191
Saccharomyces cerevisiae	68-76	195
S. exiguus	23.4	191
S. telluris	34.8	191
S. unisporus	27.3	191
Saccharomycopsis lipolytica	44-45.5	194,196
Schizosaccharomyces pombe	17.3-22.3	197,198
Torulopsis glabrata	18.9-20.3	193
Ascomycetes (Filamentous)		
Aspergillus nidulans	31.5	199
A. nidulans var *echinulatus*	38	200
A. nidulans var *guadrilineatus*	31	200
Cephalosporium acremonium	26.7	201
Claviceps purpurea	45	202
Cochliobolus heterostrophus	115	203
Neurospora crassa	60	204
Podospora anserina	94	205
Penicillium chrysogenum	48.3-49.2	206,207
Basidiomycetes		
Agaricus bitorquis	148-172	49
A. brunnescens	136	74
Coprinus cinereus	43.3	73
C. stercorarius	91.1	73
Gyrodon lividus	34	72
G. merulioides	52	72
Paragyrodon sphaerosporus	39	72
Schizophyllum commune	49.9	208
Suillus americanus	66	71
S. brevipes	68	71
S. cavipes	36	71
S. grevillei	85	71
S. grisellus	121	71
S. luteus	104-106	71
S. neoalbidipes	84	71
S. ochraceoroseus	41	71
S. placidus	62	71
S. sinuspaulianus	50	71
S. subalutaceous	58	71

Table 2 (Continued)

Taxon	Mitochondrial genome size	Ref.
S. spectabilis	50	71
S. spraguei	104	71
S. tomentosus	43	71
S. tridentinus	68	71
S. variegatus	41	71
S. viscidus	104-108	71
Ustilago cynodontis	76.5	41
Xerocomus chrysenteron	42	72

genomes [71]. Furthermore, phylogenetic analyses support an evolutionary trend within the genus toward mtDNA expansion [72]. In the Agaricaceae Weber et al. [73] noted minimal correspondence of restriction sites between *Coprinus cinereus* (43 kb) and *C. stercorarious* (91 kb). Cross-hybridization of these genomes generated only 17-20 kb of homologous sequences with the T_m values implying a 15% sequence divergence between the hybridizing complexes. Lesser extremes are shown by the hybridization of 60% of the restriction fragments of *Agaricus bitorquis* (150-175 kb) with *A. brunnescens* (136 kb) [49,74]. While basidiomycete size variations are undoubtedly due in part to optional intronic sequences as in *Suillus* [72], repeated sequences as in *Agaricus* [74], or potential differences in encoded genes as in *Coprinus* [73], the origin and nature of the majority of additional, presumably intergenic spacer, sequences is unknown.

C. The Oomycetes

The mtDNAs of the Oomycetes fall into two organizational patterns. *Achlya*-like genomes [75] contain a large inverted repeat (IR) which is conspicuously absent in *Phytophthora*-like genomes [76]. The IR minimally encodes both ribosomal cistrons and serves as the focal point for intramolecular recombination events that maintain an equimolar concentration of two genomic orientational isomers. Of 13 oomycete genera (Table 2), encompassing five orders, only three lack the IR: *Phytophthora* in the Peronosporales [76], and *Apodachlya* [77] and *Leptomitus* [78] in the Leptomitales. The loss of the IR in *Phytophthora*, but its persistence in other members of the Pythiales, suggests that IR losses are a secondarily derived character [78].

Size variation in the Oomycetes appears to depend in part on the type of organizational pattern present. In *Achlya*-like genomes a direct correlation of the genome size with the size of the IR [77] suggests that genomic complexity is relatively constant (37-40 kb) and that genome size variations are primarily a function of IR size. Comparative restriction analyses and DNA hybridization data [78] experimentally confirm that sequences present in single copy in some systems are indeed integrated into the IR of others. Furthermore, the number of insertion/deletion events in these genomes is limited [79,80]. Conversely, in the absence of the IR, the 36- to 45-kb range in *Phytophthora* results in part from the accretion of multiple insertion events [81] and, in at least one instance, a short duplication [82].

IV. FUNGAL MITOCHONDRIALLY ENCODED GENES

Fungal mtDNAs commonly encode the same small repertoire of conserved loci found in most eukaryotes. An unusual feature of many of these genes, however, is their extremely mosaic nature—a result of intronic sequences. As a prelude to a condensed summary of the structural features of both the ubiquitous and infrequently mitochondrially encoded genes, an extensive survey of fungi (in which four or more mt genes have been presumptively identified) is given in Table 3. It should be noted that the majority of these genes were detected by DNA hybridization using heterologous probes derived primarily from *S. cerevisiae*, *N. crassa*, and *A. nidulans*. For the most part, especially when purely exonic probes of known DNA sequence are used, subsequent DNA sequencing has confirmed the validity of this approach. However, inappropriate identifications of gene loci have occurred using probes from evolutionarily diverse sources under conditions of lowered stringency [e.g., 83 vs. 84, 85 vs. 86].

Variation in mitochondrially encoded proteins primarily involve the presence or absence of *ndh* and ribosomal protein genes, the presence or absence of an active *atp*9 locus, and the presence of intronic sequences. Based on the detailed analysis of the extensively sequenced Ascomycetes a dichotomy of coding may exist in this class. Current data suggest that *ndh* loci are absent in yeasts (but see *Brettanomyces*; Table 3) but present in filamentous forms. Similarly, a single, but different, ribosomal protein appears in both forms (except in *S. pombe*), and, while *atp*9 sequences may appear in both, the functional subunit is housed in the nucleus of the filamentous forms. The limited sequence data for other fungi coupled with the potential vagaries of DNA transfer hybridization data prohibit similar meaningful statements for other classes. However, it should be noted (Table 3) that *atp*9 probes generally hybridize, tentative *var*1 homologies have been identified in the Basidiomycetes and the Oomycetes [73,79], and *ndh*1 homology has been detected in the chytridiomycete *Allomyces* [87]. None of these hybridizations, however, have been confirmed by either DNA sequence analysis or protein isolation.

A major feature of many mitochondrial genes, and a current area of intensive research, is the presence of introns. However, while the pervasiveness of intronic sequences makes them a critical feature of any discussion of fungal mitochondrial genomes, they have been recently reviewed in varying detail [5,14,16,17,88-90] and are dealt with only briefly here.

Two types of introns, group I and group II, along with several subtypes, are distinguishable by their secondary structural characteristics. The majority class, group I, includes the well-studied nuclear rDNA introns of *Tetrahymena* and *Physarum*, which are also found in chloroplast, nonmetazoan mitochondrial, and T-even bacteriophage genomes. Structurally they are characterized by four highly conserved sequence regions which, along with a position conserved pair of complementary regions, facilitate the alignment of flanking exons for splicing. In group II introns a conserved secondary structure is also maintained, but the conserved sequences are limited to the intronic 3' end.

Both types of introns frequently contain long ORFs. In group I the ORF diagnostically encodes a protein with a repeated decapeptide separated by approximately 100 amino acids. These proteins when present may be required for the precise excision of the intron itself and, in at least one instance, the excision of intronic sequences from a second gene's transcript.

TABLE 3. Mitochondrial Genes of Various Fungi

	rns/l		*cox*			*cyt*	*atp*			*ndh*						
Taxon	s	l	1	2	3	b	6	8	9	1	2	3	4	4L	5	6
Ascomycetes (Filamentous)																
Aspergillus nidulans	+	+	+	+	+	+	+	+	+	+	+	+	+	+	+	+
Cephalosporium acremonium		+	+	+	+	+				+						
Neurospora crassa	+	+	+	+	+	+	+	+	+	+	+	+		+	+	
Saccharomycopsis lipolytica	+	+					+		+							
Podospora anserina	+	+	+	+	+	+	+	+		+	+	+	+	+	+	+
Ascomycetes (Yeasts)																
Brettanomyces (3 sp)	+	+	+	+	+	+	+		+	+						
B. custersii	+	+	+	+	+	+	+		+	+						
Candida rhagii	+	+	+	+		+										
C. maltosa		+	+	+		+	+		+							
Dekkera (2 sp)	+	+	+	+	+	+	+		+							
Eeniella nana	+	+	+	+	+	+	+		+							
Hansenula mrakii	+	+	+	+		+										
H. petersonii	+	+	+	+	+	+	+		+							
Kloeckera africana	+	+	+	+	+	+	+		+							
Kluyveromyces lactis	+	+	+			+	+		+							
Saccharomyces cerevisiae	+	+	+	+	+	+	+	+	+							
S. exiguus	+	+	+	+	+	+	+		+							
Schizosaccharomyces pombe	+	+	+	+	+	+	+	+	+							
Torulopsis glabrata	+	+	+	+	+	+	+	+	+							
Basidiomycetes																
Agaricus brunnescens	+	+	+			+	+	+								
Coprinus (sp)	+	+	+	+	+	+	+		+							
Gyrodon (sp)	+	+	+	+	+	+	+		+							
Paragyrodon sphaerosporus	+	+	+	+	+	+	+		+							
Suillus (15 sp)	+	+	+	+	+	+	+		+							
Xerocomus chrysenteron	+	+	+	+	+	+	+		+							
Chytridiomycetes																
Allomyces arbuscula	+	+	+	+	+	+	+		+							
A. macrogynus	+	+	+	+	+	+	+		+	+						
Blastocladiella emersonii	+	+	+	+	+	+	+		+							
Oomycetes																
Achlya (2sp)	+	+	+	+	+	+	+		+							
Aphanomyces euteiches	+	+	+	+	+	+			+							
Lagenidium giganteum	+	+	+	+	+	+			+							
Leptomitus lacteus	+	+	+	+	+	+										
Phytophthora megasperma	+	+	+	+	+	+			+							
P. parasitica	+	+			+	+										
Saprolegnia ferax	+	+	+	+	+	+			+							
Pythium ultimum	+	+	+	+	+	+			+							

The *S. cerevisiae rnl* intron encodes for a double-stranded endonuclease involved in its transfer to intronless strains, while the *rnl* ORFs of the filamentous ascomycetes likely encode a ribosomal protein. Additional double-stranded endonucleases have been identified for *cob* and *cox*1 introns [91, 92]. ORFs of the lesser studied group II introns encode proteins bearing a striking homology to viral reverse transcriptases.

Finally, while the in vivo splicing of mitochondrial primary transcripts is dependent in great part on nuclear-encoded proteins, in vitro self-splicing, albeit at a much reduced rate, has been demonstrated for representatives of group I. As anticipated the reverse self-catalysis was recently demonstrated [93] and consequently suggests a means of intron dissemination.

A. Genes of the Translational Apparatus

1. Transfer RNA Genes

Our understanding of the genic organization and functioning of the fungal tRNA loci are totally derived from the Ascomycetes. The majority of these loci are found in clusters with one or more clusters in the vicinity of the *rnl* gene. The characteristic 3'CCA terminus is absent from all genic sequences but present in mature tRNAs. A minimum number of 24 tRNAs includes both initiator Met (fMet, Met-1) and elongator Met (Met-2) loci, plus at least one tRNA species for the remaining amino acids. No tRNA intronic sequences have been located.

Extensive tRNA data are available from four yeasts and three filamentous ascomycetes. The 24 loci in *S. cerevisiae* [10] include isoaccepting species for Arg, Ser, and Thr, with Thr-1 the only locus encoded on the opposite strand. An additional Lys tRNA has been identified [94], but this species does not appear to be mitochondrially encoded, cannot be charged, and is presumably not involved in protein synthesis. The closely related *T. glabrata* genome [95] contains 23 identified tRNA genes which, like *S. cerevisiae*, include isoaccepting species for Ser and Thr (Arg-1 has not yet been located). In this instance, however, both Thr loci are encoded on the same strand. Based on the close phylogenetic relationship with *S. cerevisiae*, *K. lactis* is presumed to contain an identical 24 loci. Twenty-one of these have been sequenced (Val and isoaccepting species for Arg and Thr remain unidentified) and all are encoded on the same strand [62]. A direct repeat of the 3' terminal 21 bases of Pro is located 65 bp downstream of the prolyl tRNA locus. Twenty-five tRNA loci were reportedly identified in the more distantly related genome of *S. pombe* [96]. Unlike the aforementioned yeasts, *S. pombe* (see Section V) utilizes an unmodified universal genetic code.

Among the filamentous ascomycetes *A. nidulans* [9] encodes either 27 or 28 loci. The ambiguity resides in an as yet unlocated Arg for the frequently used CGN codons. This locus is presumably housed in the remaining few hundred unsequenced nucleotides. Isoaccepting species are present for Gly, Leu, Met-2, and Ser. Additional Cys and Asn genes are found incorporated into 153- and 300-bp duplications, respectively, with the Asn repeat also including a 36-codon N-terminal duplication of *urfa*3. The identified *N. crassa* genes [8] are similar to those of *A. nidulans* except that the duplications are for both Met species and Cys [97]. Reminiscent of the *A. nidulans* Asn-*urfa*3 duplication, the Met-2 repeat also includes the N-terminal region of *ndh*2 [98]. Recombination between the Met-2 loci has been shown to generate subgenomic molecules in a class of stopper mutants [99]. In *P. anserina* [100], 24 loci have been identified with Cys, Glu, Gly, and fMet sequences yet to be located. Isoacceptors are known for Arg, Leu, Lys, Ser, and Met [101-103], with duplications noted for Met-2 and Val [100]. As in *N. crassa* the Met-2 loci are involved in a recombinational event which in *P. anserina* leads to the generation of one of the senescent plasmids.

A tRNA synthesis locus (*tsl*) has been identified in *S. cerevisiae*. Initially identified by the accumulation of precursor tRNAs in a *rho*$^-$ petite, *tsl* was subsequently localized to a 780-bp region of the genome [104,105]. This AT-rich segment of mtDNA contains a short GC-rich element and encodes a 450-bp transcript [106,107]. In association with a nuclear-encoded protein the *tsl* RNA confers an *E. coli* RNAse P-like activity for the 5' processing of mt tRNAs [108]. DNA sequence homologies suggest the presence of similar loci in *T. glabrata* [109] and *K. lactis* [62]. There is no current evidence for a *tsl*-like locus in any of the filamentous fungi.

2. Ribosomal RNA Genes

The rRNA loci of fungal mitochondrial genomes are typically encoded as single-copy genes. While exceptions are rare in all but the Oomycetes, duplicate copies, when present, are invariably incorporated as part of a large IR. A partial duplication of *rnl* is found in *Kloeckera africanus* [110] and apparently complete duplications of both loci in all the IR-containing Oomycetes [75,78]. Secondary structural models of all sequenced ribosomal genes resemble those of the equivalent prokaryotic loci [111,112].

The deduced *rnl* gene products contain significantly longer 5' and 3' termini than their mammalian counterparts. The 5' extensions in *S. cerevisiae* [113], *S. pombe* [96], *A. nidulans* [114-116], *N. crassa* [117], and *P. anserina* [118] can be shown to form secondary structural similarities with the 5.8S-like bacterial 23S rRNA regions. Secondary structures derived from the 3' ends, however, have a less convincing homology to the prokaryotic 4.5S-like region.

Intronic sequences are frequently associated with *rnl*. When present in *S. cerevisiae* strains (ω^+) an intron-encoded protein, which displays endonucleolytic activity when cloned into *E. coli*, is postulated to be involved in the highly polarized transmission of the intronic sequences to progeny of ω^+/ω^- matings [119].

Introns are usually found in the equivalent position in the filamentous Ascomycetes. These also contain a long ORF, which in *N. crassa* [120], *A. nidulans* [9], and *P. anserina* [118] probably encodes the ribosomal protein S-5. An additional upstream intron in *P. anserina* contains an ORF similar to that of *S. cerevisiae* ω^+ strains. Indeed, like *S. cerevisiae* ω^- strains, *P. anserina* race M contains an *rnl* deletion approximating the size of the *S. cerevisiae* intron [121].

Complete sequences of *rns* genes have been obtained for *S. cerevisiae* [58,59], *A. nidulans* [111], *P. anserina* [103], and *S. pombe* (Trinkl et al., in Ref. 7). Like the *rnl* locus the *S. cerevisiae rns* gene also contains strain-dependent, GC-rich mini-inserts [59].

3. Ribosomal Protein Genes

Only two mitochondrial ribosomal proteins, both constituents of the small subunit, are known to be encoded in fungal mtDNAs. *var1* is found in the closely related yeasts *S. cerevisiae* [122,123], *Saccharomyces exiguus* [124], and *T. glabrata* [125], but is apparently absent in the distantly related genome of *S. pombe* [96]. S-5 has been identified in *N. crassa* [120,126] and, by DNA sequence-deduced amino acid homology, is probably present in the other filamentous ascomycetes *A. nidulans* [9] and *P. anserina* [118]. Hybridization data suggest that a *var*1-like sequence may also be present in the mt genome of one species of the basidiomycete *Coprinus*, but in the nuclear genome of another [73]. Similar data suggest the presence of *var*1 sequences in the mtDNA of the oomycete *Achlya* [79].

The *var*1 locus of *S. cerevisiae* (reviewed in Ref. 127) is unusual in both its extremely high AT content and the strain-dependent electrophoretic polymorphism exhibited by its product. The shortest allele (40.0) is 89.6% AT with nearly 40% of the GC content confined to a short, 46-bp, GC-rich cluster [57]. The codon usage for the locus resembles more closely that of intronic ORFs than ubiquitous genes and has led to speculation that the origin of *var*1 is linked to the evolution of intergenic sequences. The electrophoretic variants are readily explainable by the expansion of the locus via three discrete site-specific in-frame insertion elements [60]. The first of these, *a*, is a 46-bp inverted duplication of the common GC cluster whose insertion downstream from the common cluster is accompanied by the addition of two flanking adenosines. The other insertions, b_1 and b_2, consist of the repeated trinucleotide AAT, and are inserted concomitantly with each other but independently of $\alpha . \beta_1$, $(AAT)_4$, is inserted downstream from the optional inverted GC cluster, and b_2, $(AAT)_6$, further downstream from b_1. The *b* inserts expand the protein by inserting additional asparagine residues into two asparagine-rich regions. Rarer partial b_1 elements (b_p) with shorter AAT repeats provide additional variation. The insertion of any or all elements has no apparent effect on growth despite lengthening the polypeptide by up to 26 residues.

Except for the absence of the common GC cluster, the *var*1 locus of *T. glabrata* is highly homologous to that of *S. cerevisiae* [125]. As such this locus is a remarkable 93% AT.

B. Genes of the Respiratory Chain

1. Cytochrome c Oxidase Subunits

The *cox* loci have been analyzed in several systems. With the exception of *N. crassa* [66], *cox*1 is usually a mosaic gene with multiple optional introns allowing it to extend for over 10 kb in *S. cerevisiae* [128] and 20 kb in *P. anserina* [86]. Up to seven introns can be present in *S. cerevisiae* with several of these held in common with the filamentous ascomycetes. Unlike the yeasts, no initiation codon has been clearly determined for the filamentous ascomycetes. Comparison of *A. nidulans* [129] and *N. crassa* [66] predicted amino terminal residues with those of *S. cerevisiae*, originally suggested the presence of a short upstream exon, the use of AUU or AUA as an initiation codon, or initiation via an AUGA quadruplet found in close proximity to residues homologous to the amino terminal residues in *S. cerevisiae* [9]. The use of the quadruplet initiation codon is supported by the probable use of AUAA in *cox*1 of *Drosophila yakuba* [30] and further strengthened by the conservation of the AUGA quadruplet in *P. anserina* [130]. RNA processing data have eliminated the likelihood of the additional upstream exon in *A. nidulans* [131], and the *N. crassa* transcript, likewise unprocessed, includes complete sequences from the upstream *urfu* as well as the overlapping downstream *urfn* sequences [66, 132]. In the oomycete *Phytophthora*, the only nonascomycetous sequence available, *cox*1 is AUG initiated and intronless [133]. Finally, indirect evidence suggests the presence of a *cox*1 intron in the basidiomycete *Suillus* [72].

In contrast to metazoan systems fungal (ascomycetous) *cox*2 proteins are processed. Comparisons of *S. cerevisiae* [134-136] and *N. crassa* [67, 137] deduced genic sequences with the actual N-terminal regions of their proteins provide evidence for the posttranslational cleavage of the amino terminal sequence of the *cox*2 product. Similar data exist by deduced amino acid sequence analogy for the yeast *Hansenula saturnus* [138], *A. nidulans* [9,131], *P. anserina* [139], and the oomycete *Phytophthora* [140]. Introns are absent in all fungal *cox*2 sequences reported with the exception of the single intron in *Podospora*.

Sequences of *cox*3 are available for the ascomycetes *S. cerevisiae* [141], *S. pombe* (in Ref. 96), *A. nidulans* [115], *N. crassa* [65], *P. anserina* [103], and the basidiomycete *Schizophyllum commune* [142]. No unusual or intronic sequences have been noted.

2. Apocytochrome b

The DNA sequence of *cob* was established from the mtDNAs of five ascomycetes and one oomycete. Two to five introns are found in various *S. cerevisiae* strains [143,144], none in either of two *K. lactis* strains [145], one each in *S. pombe* [146] and *A. nidulans* [147], and two in *N. crassa* [148]. As postulated by Church and Gilbert [149], convincing evidence now exists for the encoding of functional *cob* maturases in the ORFs of *S. cerevisiae* introns 2, 3, and 4 (see, e.g., Refs. 150-152). The maturase of intron 4 was also convincingly shown to be required for the excision of intron 4 of *cox*1 [153]. The sequenced *cob* locus from the oomycete *Phytophthora* is intronless [154].

3. NADH Dehydrogenase Subunits

The *ndh* genes are present in mtDNA of the filamentous ascomycetes but appear to have been excluded from yeast mtDNAs (10, in Ref. 96). In *N. crassa* at least six subunits of the NADH dehydrogenase complex have been identified as mitochondrial translation products [155] with five of these loci located by sequence. Single introns have been found in *ndh*1 [68] and *ndh*4L [156], two in *ndh*5 [156], none in *ndh*2, and at least one in the 5' region of the partially sequenced *ndh*3 [98,100]. Interestingly, the stop codon of *ndh*4L overlaps by a single nucleotide with the start codon of *ndh*5 as does that of *ndh*2 with *ndh*3. The *ndh*4L-*ndh*5 loci are cotranscribed and probably cotranslated [156].

In *P. anserina* all seven mitochondrially encoded *ndh* loci have been identified and sequenced. Five of the genes are mosaic. Four introns are located in the 10-kb *ndh*1 locus [157], four in *ndh*5 [158], and one each in *ndh*3 and *ndh*4 [100]. In *ndh*4L a single intron is present in race s, while in race A *ndh*4L is duplicated within the intron, and the duplication in turn contains a second intron [102,158]. Both *ndh*2 and *ndh*6 (URFC) are intronless [100,158]. As in *N. crassa ndh*2 and *ndh*3 overlap stop-and-start codons [100], as do *ndh*4L and *ndh*5 [158].

All seven of the *ndh* loci appear to be present in *A. nidulans* [131, 159,160] and all are devoid of introns [9]. The *ndh*4L and *ndh*5 loci are cotranscribed and contain the same single-base overlap found in the other filamentous ascomycetes [160].

The *ndh*1 locus of *C. acremonium* is intronless [161].

C. ATP Synthase Subunits

Genic regions for *atp*6, *atp*8, and *atp*9 are invariably found in fungal mtDNAs with intronic sequences identified in some *atp*6 genes. This locus is punctuated by two introns in *N. crassa* [162], one in *P. anserina* [100], and none in *A. nidulans* [115,163], *S. cerevisiae* [164], and *S. pombe* (in Ref. 96). Analysis of the protein products of these three loci in *S. cerevisiae* confirms the presence of a processed leader sequence for subunit 6—an amino extension absent in vertebrate systems [165].

The genetic and sequence identification of *atp*8 in *S. cerevisiae* [166] was used to identify *urfb*/*urfx* [9,115] as the equivalent gene in *A. nidulans*. *N. crassa* [162], *P. anserina* [167], and *S. pombe* (in Ref. 96) sequences were subsequently located.

The *atp*9 locus is of special evolutionary interest. Functional sequences are present in *S. cerevisiae* [10], and presumably other yeasts, while the undisturbed mtDNA sequences remain apparently translationally silent in the filamentous *N. crassa* [168] and *A. nidulans* [169]. The presence of functional nuclear genes in these systems has been established [170,171], yet Dyson et al. [131] recently identified a processed mitochondrial *atp*9 mRNA containing *atp*9 sequences in *A. nidulans*. In the Oomycetes [154] a mitochondrial *atp*9 locus was sequenced in *Phytophthora*, but whether it represents a functional or silent locus has not been established. Despite the virtually universal hybridization of ascomycete *atp*9 sequences with other fungal mtDNAs (Table 3), the question of mitochondrial genome *atp*9 expression in the other fungal systems remains largely unanswered.

V. GENETIC CODE VARIATION

Nonstandard uses of the universal genetic code are a common, but not universal, feature of mitochondrial translational systems. Recent reviews on this topic (see, e.g., Refs. 172 and 173) address the nature and extent of code variation as well as evolutionary mechanisms for codon reassignments. The summary of nonstandard uses listed in Table 4 for fungal mitochondrial genomes is subject to the limitations noted in the text.

Following the initial identification of codon reassignments in the human mitochondrial genome [174] it became apparent that the universal termination codon UGA had also been reassigned to Trp in *S. cerevisiae*. Comparison of the predicted *cox*2 sequence with the bovine amino acid sequence indicated that four of five *S. cerevisiae* UGA codons corresponded to equivalent bovine Trp residues [135]. Subsequently, Sibler et al. [175] isolated a tryptophanyl t-RNA with an appropriate U*CA anticodon, and N. C. Martin et al. [176] confirmed the mitochondrial origin of this species with the DNA sequence.

Additional *S. cerevisiae* codon variations were identified for the CUN family and AUA. Reassignment of the rarely used CUN from Leu to Thr is based on the comparison of DNA sequence [177,178] and protein data

TABLE 4. Codon Variation in Fungal Mitochondrial Genomes

Taxon	UGA	AUA	CUN
Universal	Ter	Ile	Leu
Ascomycetes (Yeasts)			
Hansenula saturnus	Trp	?[a]	?
Kluyveromyces lactis	Trp	0[b]	0
Saccharomyces cerevisiae	Trp	Met	Thr
Schizosaccharomyces pombe	0	Ile	Leu
Torulopsis glabrata	Trp	Met	Thr
Ascomycetes (Filamentous)			
Aspergillus nidulans	Trp	Ile	Leu
Cephalosporium acremonium	Trp	?	?
Neurospora crassa	Trp	Ile	Leu
Podospora anserina	Trp	Ile	Leu
Basidiomycetes			
Schizophyllum commune	0	?	?
Oomycetes			
Phytophthora megasperma	0	?	?

[a]Insufficient data to confirm codon reassignment.
[b]Codons absent from sequenced genes.

[170] for the *atp*9 gene. Confirmation resulted from the DNA sequence of an unusual tRNA which contained an extra residue in the anticodon loop [179]. As such, this tRNA could potentially translate either CUN or UAA. However, UAA is the *S. cerevisiae* mitochondrial termination codon. The infrequent use of AUA as a ubiquitous gene codon allowed its codon reassignment (from Ile to Met) to escape detection until the DNA sequencing of *var*1 where it is a frequently used codon [57]. Amino acid composition data and a short protein sequence of the amino terminus of *var*1 protein compared with the deduced protein sequence revealed the codon reassignment. Further confirmation resulted from the subsequent sequencing of a methionyl tRNA which contained an additional unpaired nucleotide in the pseudo-U stem theoretically allowing the translation of both AUG and AUA as Met [175].

Evidence for code variations in other yeasts is not as complete. In *T. glabrata* the *var*1 sequence [125] dictates the use of UGA as Trp, but only the likely use of AUA as Met and CUN as Thr. The latter reassignment is supported by the DNA sequence of a tRNA locus highly homologous to *S. cerevisiae* threonyl tRNA, but with two additional residues in the anticodon loop [109]. Sequence of the *H. saturnus cox*2 gene also dictates the use of UGA as Trp, and suggests the universal assignments of CUU to Leu and AUA to Ile [138]. However, as noted by Fox [172], Thr and Leu are poorly conserved residues and the use of AUA occurs only once. In *K. lactis* the sequence of *cob* uses UGA as Trp, but neither CUN nor AUA codons are encountered [145]. Two CUN codons are present in the *rnl* intron of *K. thermotolerans* [180], but in the absence of sufficient comparative DNA or protein data no firm conclusions can be drawn.

In contrast to the other yeasts the phylogenetically distant *S. pombe* uses the universal genetic code for all its ubiquitous genes [146]. UGA is found only in intronic ORFs, and, while these are homologous to some *A. nidulans* ORFs, no translational products have been identified. Furthermore, the DNA sequence of the tryptophanyl tRNA suggests that it is unlikely to translate UGA [5].

With the exception of the UGA Trp reassignment all filamentous ascomycetes examined to date use the standard codon assignments. The Trp reassignment was first noted in *cox*3 for *N. crassa* [65], in *cob* for *A. nidulans* [147], and in *cox*1 for *P. anserina* [167]. In *N. crassa* further comparison of genic sequences with amino acid composition data of *cox*3, combined with partial amino acid sequence data of the *cox*1 protein [66], revealed no additional reassignments. Similarly, comparisons of deduced protein sequences from extensive *A. nidulans* and *P. anserina* data with those of *N. crassa* indicate that UGA is the only nonstandard codon usage in these systems. A more tentative but similar conclusion can be drawn from the sequence of a single *C. acremonium* gene (*ndh*1) [161].

In the nonascomycetous fungi data exist for only two organisms. DNA sequence data are available for the *cox*3 locus of the basidiomycete *Schizophyllum* [142], and for the *cox*1, *cox*2, *cob*, and *atp*9 loci of the oomycete *Phytophthora* [133,140,154]. Despite the high AT content of all these genes, Trp is exclusively encoded by UGG with UGA absent from all identified reading frames. Thus, in the absence of any other apparent nonstandard reassignments these organisms appear to use the universal genetic code.

VI. CONCLUDING REMARKS

The foregoing serves to emphasize that the current concept of the organization and expression of the typical fungal mitochondrial genome is extremely narrow and is derived almost solely from the Ascomycetes. Even so, the variability in organizational patterns, encoded genes, and genetic code variations is more pronounced than that of the metazoa. Indeed, a true measure of the extent of fungal mtDNA diversity can only be obtained by in-depth analyses of representative organisms from each of the major fungal classes. Generation of a sufficient DNA sequence database for each of the classes of fungi and allies would provide considerable insight not only into the evolution of the "generic" fungus, but also into the parallel evolution of mitochondrial genetic systems and respiratory chain complexes in the most biochemically diverse group of eukaryotes. There is little doubt that *S. cerevisiae* and probably *S. pombe*, *A. nidulans*, and *N. crassa* will continue to contribute in the broadest sense to a more complete understanding of the functioning of the mitochondrion in general. There is, however, a legitimate concern that meaningful knowledge of mitochondrial function in other fungal groups may be drastically limited by the "economical" application of polynucleotide chain reaction technology.

ACKNOWLEDGMENTS

The author is indebted to the many colleagues who provided reprints of recent work, and to the editor for his patience.

REFERENCES

1. Nagley, P., and Devenish, R. J. Leading organellar proteins along new pathways: the relocation of mitochondrial and chloroplast genes to the nucleus. *TIBS*, *14*:31-35, 1989.
2. Clark-Walker, G. D. Basis of diversity in mitochondrial DNAs. In *The Evolution of Genome Size* (T. Cavalier-Smith, ed.), John Wiley and Sons, Sussex, UK, 1985, pp. 277-297.
3. Grossman, L. L. and Hudspeth, M. E. S. Fungal mitochondrial genomes. In *Gene Manipulations in Fungi* (J. W. Bennet and L. Lasure, eds.), Academic Press, New York, 1985, pp. 65-103.
4. Tzagoloff, A., and Meyers, A. M. Genetics of mitochondrial biosynthesis. *Annu. Rev. Biochem.*, *55*:249-286, 1986.
5. Scazzocchio, C. The natural history of mitochondrial genomes. In *Evolutionary Biology of the Fungi* (A. D. M. Rayner, C. M. Brasier, and D. Moore, ed.), Cambridge University Press, New York, 1987, pp.53-73.
6. Attardi, G., and Schatz, G. Biogenesis of mitochondria. *Annu. Rev. Cell Biol.*, *4*:289-33, 1988.
7. Wolf, K., and del Giudice, L. The variable mitochondrial genome of ascomycetes: organization, mutational alterations, and expression. *Adv. Genet.*, *25*:185-308, 1988.
8. Breitenberger, C. A., and RajBhandary, U. L. Some highlights of mitochondrial research based on analyses of Neurospora crassa mitochondrial DNA. *TIBS*, *10*:478-483, 1985.
9. Brown, T. A., Waring, R. B., Scazzocchio, C., and Davies, R. W. The Aspergillus nidulans mitochondrial genome. *Curr. Genet.*, *9*:113-117, 1985.
10. de Zamaroczy, M., and Bernardi, G. Sequence organization of the mitochondrial genome of yeast: a review. *Gene*, *37*:1-17, 1985.
11. Kuck, U. Mitochondrial DNA rearrangements in Podospora anserina. *Exp. Mycol.*, *13*:111-120, 1989.
12. Hurt, E. C., van Loon, A. P. G. M. How proteins find mitochondria and intramitochondrial compartments. *TIBS*, *11*:204-207, 1986.
13. Pfanner, N. Hartl, F.-U., and Neupert, W. Import of proteins into mitochondria: a multi-step process. *Eur. J. Biochem.*, *175*:205-212, 1988.
14. Grivell, L. A. Nucleo-mitochondrial interactions in yeast mitochondrial biogenesis. *Eur. J. Biochem.*, *182*:477-493, 1989.
15. Grivell, L. A., and Schweyen, R. J. RNA splicing in yeast mitochondria: taking out the twists. *TIG*, *5*:39-41, 1989.
16. Scazzocchio, C. Group I introns: do they only go home? *TIG*, *5*:168-172, 1989.
17. Lambowitz, A. M. Infectious introns. *Cell*, *56*:323-326, 1989.
18. Kohli, J. Genetic nomenclature and gene list of the fission yeast Schizosaccharomyces pombe L. *Curr. Genet.*, *11*:575-590, 1987.
19. Wolstenholme, D. R., and Dawid, I. R. A size difference between mitochondrial DNA molecules of urodele and anuran amphibia. *J. Cell Biol.*, *39*:222-228, 1968.
20. Snyder, M., Fraser, A. R., LaRoche, J., Gartner-Kepkay, K. E., Zouros, E. Atypical mitochondrial DNA from the deep-sea scallop Placopecten magellanicus. *Proc. Natl. Acad. Sci. USA*, *84*:7595-7599, 1987.

21. Anderson, S., Bankier, A. T., Barrell, B. G., de Bruijn, M. H. L., Coulson, A. R., Drouin, J., Eperon, I. C., Nierlich, D. P., Roe, B. A., Sanger, F., Schreier, P. H., Smith, A. J. H., Staden, R., and Young, I. G. Sequence and organization of the human mitochondrial genome. *Nature, 290*:457-465, 1981.
22. Anderson, S., de Bruijn, M. H. L., Coulson, A. R., Eperon, I. C., Sanger, F., and Young, I. G. Complete sequence of bovine mitochondrial DNA: conserved features of the mammalian mitochondrial genome. *J. Mol. Biol., 156*:683-717, 1982.
23. Bibb, M. J., Van Etten, R. A., Wright, C. T., Walberg, M. W., and Clayton, D. A. Sequence and gene organization of mouse mitochondrial DNA. *Cell, 26*:167-180, 1981.
24. Gadaleta, G., Pepe, G., De Candia, G. Quagliariello, C., Sbisa, E., and Saccone, C. The complete nucleotide sequence of the Rattus norvegicus genome: Cryptic signals revealed by comparative analysis between vertebrates. *J. Mol. Evol., 28*:497-516, 1989.
25. Roe, B. A., Ma, D. P., Wilson, R. K., and Wong, J. F. H. The complete nucleotide sequence of the Xenopus laevis mitochondrial genome. *J. Biol. Chem., 260*:9759-9774, 1985.
26. Desjardins, P., and Morais, R. Sequence and gene organization of the chicken mitochondrial genome: a novel gene order in higher vertebrates. *J. Mol. Biol., 212*:599-634, 1990.
27. Jacobs, H. T., Elliott, D. J., Math, V. B., and Farquharson, A. Nucleotide sequence and gene organization of sea urchin mitochondrial DNA. *J. Mol. Biol., 202*:185-217.
28. Jacobs, H. T., Elliott, D. J., Math, V. B., Farquharson, A. Correction: nucleotide sequence and gene organization of sea urchin mitochondrial DNA. *J. Mol. Biol., 211*:663, 1990.
29. Cantatore, P., Roberti, M., Rainaldi, G., Gadaleta, M. N., and Saccone, C. The complete nucleotide sequence, gene organization, and genetic code of the mitochondrial genome of Paracentrotus lividus. *J. Biol. Chem., 264*:10965-10975, 1989.
30. Clary, D. O., and Wolstenholme, D. R. The mitochondrial DNA molecule of Drosophila yakuba: nucleotide sequence, gene organization, and genetic code. *J. Mol. Evol., 22*:252-271, 1985.
31. Wolstenholme, D. R., MacFarlane, J. L., Okimoto, R., Clary, D. O., and Wahleithner, J. A. Bizarre tRNAs inferred from DNA sequences of mitochondrial genomes of nematode worms. *Proc. Natl. Acad. Sci. USA, 84*:1324-1328, 1987.
32. Palmer, J. D., and Shields, C. R. Tripartite structure of the Brassica campestris mitochondrial genome. *Nature, 307*:437-440, 1984.
33. Ward, B. L., Anderson, R. S., Bendich, A. J. The mitochondrial genome is large and variable in a family of plants (Cucurbitae). *Cell, 25*:793-803, 1981.
34. Covello, P. S., and Gray, M. W. RNA editing in plant mitochondria. *Nature, 341*:662-666, 1989.
35. Gualberto, J. M., Lamattina, L., Bonnard, G., Weil, J.-H., and Grienenberger, J.-M. RNA editing in wheat mitochondria results in the conservation of protein sequences. *Nature, 341*:660-662, 1989.
36. Hiesel, R., Wissinger, B., Schuster, W., and Brennicke, A. RNA editing in plant mitochondria. *Science, 246*:1632-1634, 1989.
37. Simpson, L., and Shaw, J. RNA editing and the mitochondrial cryptogenes of kinetoplastid protozoa. *Cell, 57*:355-366, 1989.

38. Goddard, J. M., and Cummings, D. J. Structure and replication of mitochondrial DNA from Paramecium aurelia. *J. Mol. Biol.*, *97*:593-609, 1975.
39. Goldbach, R. W., Bollen-DeBoer, J. E., Van Bruggen, E. F. J., and Borst, P. Replication of the linear mitochondrial DNA of Tetrahymena pyriformis. *Biochim. Biophys. Acta*, *562*:400-417, 1979.
40. Grant, D., and Chiang, K.-S. Physical mapping and characterization of Chlamydomonas mitochondrial DNA molecules: their unique ends, sequence homogeneity, and conservation. *Plasmid*, *4*:82-96, 1980.
41. Mery-Drugeon, E., Crouse, E. J., Schmitt, J. M., Bohnert, H. J., and Bernardi, G. The mitochondrial genomes of Ustilago cynodontis and Acanthameba castellanii. *Eur. J. Biochem.*, *114*:577-583, 1981.
42. Borst, P., Overdulve, J. P., Weyers, P. J., Fase-Fowler, F., Van Den Berg, M. DNA circles with cruciforms from Isospora (Toxoplasma) gondii. *Biochim. Biophys. Acta*, *781*:100-111, 1984.
43. Boer, P. H., and Gray, M. G. Transfer RNA genes and the genetic code in Chlamydomonas reinhardtii mitochondria. *Curr. Genet.*, *14*:583-590, 1988.
44. Suyama, Y. Two dimensional polyacrylamide gel electrophoresis analysis of Tetrahymena mitochondrial tRNA. *Curr. Genet.*, *10*:411-420, 1986.
45. Mahalingam, R. Seilhamer, J. J., Pritchard, A. E., and Cummings, D. J. Identification of Paramecium mitochondrial proteins using antibodies raised against fused mitochondrial gene products. *Gene*, *49*:129-138, 1986.
46. Rao, J. K. M., and Argos, P. A conformational preference parameter to predict helices in integral membrane proteins. *Biochim. Biophys. Acta*, *869*:197-214, 1986.
47. Benne, R. RNA editing in trypanosome mitochondria. *Biochim. Biophys. Acta*, *1007*:131-139, 1989.
48. Zimmer, M., Luckemann, G., Lang, B. F., and Wolf, K. The mitochondrial genome of the fission yeast Schizosaccharomyces pombe. 3. Gene mapping in strain EF1 (CBS 356) and analysis of hybrids between the strains EF1 and ade 7-50h$^-$. *Mol. Gen. Genet.*, *196*:473-481, 1984.
49. Hintz, W. E., Mohan, M., Anderson, J. B., and Horgen, P. A. The mitochondrial DNAs of Agaricus: heterogeneity in A. bitorquis and homogeneity in A. brunnescens. *Curr. Genet.*, *9*:127-132, 1985.
50. Wesolowski, M., and Fukuhara, H. Linear mitochondrial deoxyribonucleic acid from the yeast Hansenula mrakii. *Mol. Cell. Biol.* *1*:387-393, 1981.
51. Kovac, L., Lazowska, J., and Slonimski, P. P. A yeast with linear molecules of mitochondrial DNA. *Mol. Gen. Genet.*, *197*:420-424, 1984.
52. Kawano, S., Suzuki, T., and Kuroiwa, T. Structural homogeneity of mitochondrial DNA in the mitochondrial nucleoid of Physarum polycephalum. *Biochim. Biophys. Acta*, *696*:290-298, 1982.
53. Williamson, D. H., and Fennell, D. J. The use of a fluorescent DNA binding agent for detecting and separating yeast mitochondrial DNA. In *Methods in Cell Biology* (D. M. Prescott, ed.), Academic Press, New York, 1975, pp. 335-351.
54. Hudspeth, M. E. S., Shumard, D. S., Tatti, K. M., and Grossman, L. I. Rapid purification of yeast mitochondrial DNA in high yield. *Biochim. Biophys. Acta*, *610*:221-228, 1980.

55. Garber, R. C., and Yoder, O. C. Isolation of DNA from filamentous fungi and separation into nuclear, mitochondrial, ribosomal, and plasmid components. *Anal. Biochem.*, *135*:416-422, 1983.
56. de Zamaroczy, M., and Bernardi, G. The GC clusters of the mitochondrial genome of yeast and their evolutionary origin. *Gene*, *41*:1-22, 1986.
57. Hudspeth, M. E. S., Ainley, W. M., Shumard, D. S., Butow, R. A., and Grossman, L. I. Location and structure of the var1 gene on yeast mitochondrial DNA: nucleotide sequence of the 40.0 allele. *Cell*, *30*: 617-626, 1982.
58. Sor, F., and Fukuhara, H. Sequence nucleotidique du gene de l'ARN ribosomique 15S mitochondrial de la levure. *C. R. Acad. Sci. Paris Ser. D*, *291*:933-936, 1980.
59. Sor, F., and Fukuhara, H. Nature of an inserted sequence in the mitochondrial gene coding for the 15S ribosomal RNA of yeast. *Nuc. Acids Res.*, *10*:1625-1633, 1982.
60. Hudspeth, M. E. S., Vincent, R. D., Perlman, P. S., Shumard, D. S., Treisman, L. O., and Grossman, L. I. The expandable var1 gene of yeast mitochondrial DNA: In-frame insertions can explain the strain-specific protein size polymorphisms. *Proc. Natl. Acad. Sci. USA*, *81*: 3148-3152, 1984.
61. de Zamaroczy, M., and Bernardi, G. The AT spacers and the var1 genes from the mitochondrial genomes of Saccharomyces cerevisiae and Torulopsis glabrata: evolutionary origin and mechanism of formation. *Gene*, *54*:1-22, 1987.
62. Wilson, C., Ragnini, A., and Fukuhara, H. Analysis of the regions coding for transfer RNAs in Kluyveromyces lactis mitochondrial DNA. *Nucl. Acids Res.*, *17*:4485-4491, 9189.
63. Yin, S., Heckman, J., and RajBhandary, U. L. Highly conserved GC rich palindromic DNA sequences which flank tRNA genes in Neurospora crassa mitochondria. *Cell*, *26*:326-332, 1981.
64. Taylor, J. W., and Smolich, B. D. Molecular cloning and physical mapping of the Neurospora crassa 74-OR23-1A mitochondrial genome. *Curr. Genet.*, *9*:597-604, 1985.
65. Browning, K. S., and RajBhandary, U. L. Cytochrome oxidase subunit III gene in Neurospora crassa mitochondria. Location and sequence. *J. Biol. Chem.*, *257*:5253-5256, 1982.
66. Burger, G., Scriven, C., Machleidt, W., and Werner, S. Subunit 1 of cytochrome oxidase from Neurospora crassa: nucleotide sequence of the coding gene and partial amino acid sequence of the protein. *EMBO J.*, *1*:1385-1391, 1982.
67. Macino, G., and Morelli, G. Cytochrome oxidase subunit 2 gene in Neurospora crassa mitochondria. *J. Biol. Chem.*, *258*:13230-13235, 1983.
68. Burger, G., and Werner, S. The mitochondrial URF1 gene in Neurospora crassa has an intron that contains a novel type of URF. *J. Mol. Biol.*, *186*:231-242, 1985.
69. Breitenberger, C. A., Browning, K. S., Alzner-DeWeerd, B., and RajBhandary, U. L. RNA processing in Neurospora crassa mitochondria: use of transfer RNA sequences as signals. *EMBO J.*, *4*:185-189, 1985.
70. De Vries, H., Haima, P., Brinker, M., and De Jonge, J. C. The Neurospora mitochondrial genome. The region coding for the polycistronic cytochrome oxidase subunit I transcript is preceded by a transfer RNA gene. *FEBS. Lett.*, *179*:337-342, 1985.

71. Bruns, T. D., Palmer, J. D., Shumard, D. S., Grossman, L. I., and Hudspeth, M. E. S. Mitochondrial DNAs of Suillus: three fold size change in molecules that share a common gene order. *Curr. Genet.*, *13*:49-56, 1988.
72. Bruns, T. D., and Palmer, J. D. Evolution of mushroom mitochondrial DNA: Suillus and related genera. *J. Mol. Evol.*, *28*:349-362, 1989.
73. Weber, C. A., Hudspeth, M. E. S., Moore, G. P., and Grossman, L. I. Analysis of the mitochondrial genomes of two basidiomycetes, Coprinus cinereus and Coprinus stercorarius. *Curr. Genet.*, *10*:515-525, 1986.
74. Hintz, W. E., Anderson, J. B., and Horgen, P. A. Physical mapping of the cultivated mushroom Agaricus brunnescens (= A. bisporus). *Curr. Genet.*, *14*:43-49, 1988.
75. Hudspeth, M. E. S., Shumard, D. S., Bradford, C. J. R., and Grossman, L. I. Organization of Achlya mtDNA: a population with two orientations and a large inverted repeat containing the rRNA genes. *Proc. Natl. Acad. Sci. USA*, *80*:142-146, 1983.
76. Klimczak, L. J., and Prell, H. H. Isolation and characterization of mitochondrial DNA of the oomycetous fungus Phytophthora infestans. *Curr. Genet.*, *8*:323-326, 1984.
77. McNabb, S. A., and Klassen, G. R. Uniformity of mitochondrial complexity in Oomycetes and the evolution of the inverted repeat. *Exp. Mycol.*, *12*:233-242, 1988.
78. Hudspeth, M. E. S. Unpublished results.
79. Shumard, D. S., Grossman, L. I., and Hudspeth, M. E. S. Achlya mitochondrial DNA: gene localization and analysis of inverted repeats. *Mol. Gen. Genet.*, *202*:16-23, 1986.
80. McNabb, S. A., Boyd, D. A., Belkhiri, A., Dick, M. W., and Klassen, G. R. An inverted repeat comprises more than three-quarters of the mitochondrial genome in two species of Pythium. *Curr. Genet.*, *12*:205-208, 1987.
81. Forster, H., Kinscherf, T. G., Leong, S. A., and Maxwell, D. P. Estimation of relatedness between Phytophthora species by analysis of mitochondrial DNA. *Mycologia*, *80*:466-478, 1988.
82. Shumard-Hudspeth, D. S., and Hudspeth, M. E. S. Genic rearrangements in Phytophthora mitochondrial DNA. *Curr. Genet.*, *17*:413-415, 1990.
83. Johnson, B. J. B., Hill, G. C., Fox, T. D., and Stuart, K. The maxicircle of Trypanosoma brucei kinetoplast DNA hybridizes with a mitochondrial gene encoding cytochrome oxidase subunit II. *Mol. Biochem. Parasitol.*, *5*:381-390, 1982.
84. Benne, R., DeVries, B. F., Van den Burg, J., and Klaver, B. The nucleotide sequence of a segment of Trypanosoma brucei mitochondrial maxi-circle DNA that contains the gene for apocytochrome b and some unusual unassigned reading frames. *Nucl. Acids Res.*, *11*:6925-6941, 1983.
85. Wright, R. M., Horrum, M. A., and Cummings, D. F. Are mitochondrial structural genes amplified during senescence in Podospora anserina? *Cell*, *29*:505-515, 1982.
86. Jamet-Vierny, C., Begel, O., and Belcour, L. A $20X10^3$-base mosaic gene identified on the mitochondrial chromosome of Podospora anserina. *Eur. J. Biochem.*, *143*:389-394, 1984.

87. Borkhardt, B., Brown, T. A., Thim, P., and Olson, L. W. The mitochondrial genome of the aquatic phycomycete Allomyces macrogynus. Physical mapping and mitochondrial DNA instability. *Curr. Genet.*, *13*:41-47, 1988.
88. Burke, J. M. Molecular genetics of Group I introns: RNA structure and protein factors required for splicing—a review. *Gene*, *73*:273-294, 1988.
89. Cech, T. R. Conserved sequences and structures of group I introns: building an active site for RNA catalysis. *Gene*, *73*:259-271, 1988.
90. Cech, T. R., and Bass, B. L. Biological catalysis by RNA. *Annu. Rev. Biochem.*, *55*:599-629, 1986.
91. Delahodde, A., Goguel, V., Becam, A. M., Creusot, F., Perea, J., Banroques, J., and Jacq, C. Site-specific DNA endonuclease and RNA maturase activities of two homologous intron-encoded proteins from yeast mitochondria. *Cell*, *56*:431-441, 1989.
92. Wenzlau, J. M., Saldanha, R. J., Butow, R. A., and Perlman, P. S. A latent intron-encoded maturase is also an endonuclease needed for intron mobility. *Cell*, *56*:421-430, 1989.
93. Woodson, S. A., and Cech, T. R. Reverse self-splicing of the Tetrahymena group I intron: implication for the directionality of splicing and for intron transposition. *Cell*, *57*:335-345, 1989.
94. Martin, R. P., Schneller, J. M., Stahl, A. J. C., and Dirheimer, G. Import of nuclear deoxyribonucleic acid coded lysine-accepting transfer ribonucleic acid (anticodon CUU) into yeast mitochondria. *Biochemistry*, *18*:4600-4605, 1979.
95. Clark-Walker, G. D., and Sriprakash, K. S. Map location of transcripts from Torulopsis glabrata mitochondrial DNA. *EMBO J.*, *2*:1465-1472, 1983.
96. Lang, B. F. Cedergren, R., and Gray, M. W. The mitochondrial genome of the fission yeast, Schizosaccharomyces pombe: sequence of the large-subunit ribosomal RNA gene, comparison of potential secondary structure in fungal mitochondrial large-subunit rRNAs and evolutionary considerations. *Eur. J. Biochem.*, *169*:527-537, 1987.
97. Agsteribbe, E., Hartog, M., and de Vries, H. Duplication of the $tRNA^{Met}$ and $tRNA^{Cys}$ genes and of fragments of a gene encoding a subunit of the NADH dehydrogenase complex in Neurospora crassa mitochondrial DNA. *Curr. Genet.*, *15*:57-62, 1989.
98. De Vries, H., Alzner-DeWeerd, B., Breitenberger, C. A., Chang, D. D., de Jonge, J. C., and RajBhandary, U. L. The E35 stopper mutant of Neurospora crassa: precise location of deletion endpoints in mitochondrial DNA and evidence that the deleted DNA codes for a subunit of NADH dehydrogenase. *EMBO J.*, *5*:779-785, 1986.
99. Gross, S. R., Hsiah, J.-S., and Levine, P. H. Intramolecular recombination as a source of mitochondrial chromosome heteromorphism in Neurospora. *Cell*, *38*:233-239, 1984.
100. Cummings, D. J., Domenico, J. M. Sequence analysis of mitochondrial DNA from Podospora anwerina: pervasiveness of a class I intron in three separate genes. *J. Mol. Biol.*, *204*:815-839, 1988.
101. Turker, M. S., Domenico, J. M., and Cummings, D. J. Excision-amplification of mitochondrial DNA during senescence in Podospora anserina: a potential role for an 11 base-pair consensus sequence in the excision process. *J. Mol. Biol.*, *198*:171-185, 1987.

102. Vierny-Jamet, C. Senescence in Podospora anserina: a possible role for nucleic acid interacting proteins suggested by the sequence analysis of a mitochondrial DNA region specifically amplified in senescent cultures. *Gene, 74*:387-398, 1988.
103. Cummings, D. J., Domenico, J. M., Nelson, J., and Sogin, M. L. DNA sequence, structure, and phylogenetic relationship of the small subunit rRNA coding region of mitochondrial DNA from Podospora anserina. *J. Mol. Evol., 28*:232-241, 1989.
104. Martin, N. C., Underbrink-Lyon, K., and Miller, D. L. Identification and characterization of a yeast mitochondrial locus necessary for tRNA biosynthesis. In *Mitochondrial Genes* (P. Slonimski, P. Borst, and G. Attardi, eds.), Cold Spring Harbor Laboratory, New York, 1982, pp. 264-267.
105. Underbrink-Lyon, K., Miller, D. L., Ross, N. A., Fukuhara, H., and Martin, N. C. Characterization of a yeast mitochondrial locus necessary for tRNA biosynthesis. Deletion mapping and restriction mapping studies. *Mol. Gen. Genet., 191*:512-518, 1983.
106. Miller, D. L., and Martin, N. C. Characterization of the yeast mitocondrial locus necessary for tRNA biosynthesis: DNA sequence analysis and identification of a new transcript. *Cell, 34*:911-917, 1983.
107. Martin, N. C., Miller, D. L., Underbrink, K., and Ming, X. Structure of the primary transcript of the yeast mitochondrial tRNA fMet gene: implications for the function of the tRNA synthesis locus. *J. Biol. Chem., 260*:1479-1483, 1985.
108. Hollingsworth, M. J., and Martin, N. C. RNase P activity in the mitochondria of Saccharomyces cerevisiae depends on both mitochondrion- and nucleus-encoded components. *Mol. Cell. Biol., 6*:1058-1064, 1986.
109. Clark-Walker, G. D., McArthur, C. R., and Sriprakash, K. S. Location of transcriptional control signals and transfer RNA sequences in Torulopsis glabrata mitochondrial DNA. *EMBO J., 4*:465-473, 1985.
110. Clark-Walker, G. D. McArthur, C. R., and Sriprakash, K. S. Partial duplication of the large ribosomal RNA sequence in an inverted repeat in circular mitochondrial DNA from Kloeckera africana: Implications for mechanisms of the petite mutation. *J. Mol. Biol., 147*:399-415.
111. Kochel, H. G., and Kuntzel, H. Nucleotide sequence of the Aspergillus nidulans mitochondrial gene coding for the small ribosomal subunit RNA: homology to E. coli 16S rRNA. *Nucl. Acids Res., 9*:5689-5696, 1981.
112. Noller, H. F. Structure of ribosomal DNA. *Annu. Rev. Biochem., 53*:119-162, 1984.
113. Sor, F., Fukuhara, H. Complete DNA sequence coding for the large ribosomal RNA of yeast mitochondria. *Nucl. Acids Res., 11*:339-348, 1983.
114. Kochel, H. G., and Kuntzel, H. Mitochondrial L-rRNA from Aspergillus nidulans: potential secondary structure and evolution. *Nucl. Acids Res., 10*:4795-4801, 1982.
115. Netzker, R., Kochel, H. G., Basak, N., and Kuntzel, H. Nucleotide sequence of Aspergillus nidulans mitochondrial genes coding for ATPase subunit 6, cytochrome oxidase subunit 3, seven unidentified proteins, four tRNAs and L-rRNA. *Nucl. Acids Res., 10*:4783-4794, 1982.
116. Dyson, N. J., Brown, T. A., Waring, R. B., Davies, R. W. The mitochondrial ribosomal RNA molecules of Aspergillus nidulans. *Gene, 75*:109-118, 1989.

117. Heckman, J. E., and RajBhandary, U. L. Organization of tRNA and rRNA genes in N. crassa mitochondria: Intervening sequence in the large rRNA gene and strand distribution of the RNA genes. *Cell, 17*:583-595, 1979.
118. Cummings, D. J., Domenico, J. M., and Nelson, J. DNA sequence and secondary structures of the large subunit rRNA coding regions and its two class I introns of mitochondrial DNA from Podospora anserina. *J. Mol. Evol., 28*:242-255.
119. Colleaux, L., d'Auriol, L., Betermier, M., Cottarel, G., Jacquier, A., Galibert, F., and Dujon, B. Universal code equivalent of a yeast mitochondrial intron reading frame is expressed in E. coli as a specific double strand endonuclease. *Cell, 44*:521-533, 1986.
120. Burke, J. M., and RajBhandary, U. L. Intron within the large rRNA gene of N. crassa mitochondria: A long open reading frame and a consensus sequence possibly important in splicing. *Cell, 31*:509-520, 1982.
121. Kuck, U., Kappelhoff, B., and Esser, K. Despite mtDNA polymorphism the mobile intron (p1DNA) of the COI gene is present in ten different races of Podospora anserina. *Curr. Genet., 10*:59-67, 1985.
122. Groot, G. S. P., Mason, T. L., and Van Harten-Loosbroek, N. Var1 is associated with the small ribosomal subunit of mitochondrial ribosomes in yeast. *Mol. Gen. Genet., 174*:339-342, 1979.
123. Terpstra, P., Zanders, E., and Butow, R. A. The association of var1 with the 38S mitochondrial ribosomal subunit of yeast. *J. Biol. Chem., 254*:12653-12661, 1979.
124. Clark-Walker, G. C., McArthur, C. R., and Sriprakash, K. S. Order and orientation of genic sequences in circular mitochondrial DNA from Saccharomyces exiguus: implications for evolution of yeast mtDNAs. *J. Mol. Evol., 19*:333-341, 1983.
125. Ainley, W. M., MacReadie, I. G., and Butow, R. A. Var1 gene on the mitochondrial genome of Torulopsis glabrata. *J. Mol. Biol., 184*: 565-576, 1985.
126. La Polla, R. J., and Lambowitz, A. M. Mitochondrial ribosome assembly in Neurospora crassa. Purification of the mitochondrially synthesized ribosomal protein, S-5. *J. Biol. Chem., 256*:7064-7067, 1981.
127. Butow, R. A., Perlman, P. S., and Grossman, L. I. Organization and evolution of the unusual var1-gene of yeast mitochondrial DNA. *Science, 228*:1496-1501, 1985.
128. Bonitz, S. D., Coruzzi, G., Thalenfeld, B. E., and Tzagoloff, A. Assembly of the mitochondrial membrane system. Structure and nucleotide sequence of the gene coding for subunit 1 of yeast cytochrome oxidase. *J. Biol. Chem., 255*:11927-11941, 1980.
129. Waring, R. B., Brown, T. A., Ray, J. A. Scazzocchio, C., and Davies, R. W. Three variant introns of the same general class in the mitochondrial gene for cytochrome oxidase subunit 1 in Aspergillus nidulans. *EMBO J., 3*:2121-2128, 1984.
130. Kuck, U., Osiewacz, H. D., Schmidt, U., Kappelhoff, B., Schulte, E., Stahl, U., and Esser, K. The onset of senescence is affected by DNA rearrangement of a discontinuous mitochondrial gene in Podospora anserina. *Curr. Genet., 9*:373-382, 1985.
131. Dyson, N. J., Brown, T. A., Ray, J. A., Waring, R. B., Scazzocchio, C., and Davies, R. W. Processing of mitochondrial RNA in Aspergillus nidulans. *J. Mol. Biol., 208*:587-599, 1989.

132. Burger, G., and Werner, S. Mitochondrial gene URFN of Neurospora crassa codes for a long polypeptide with highly repetitive structure. *J. Mol. Biol.*, *191*:589-599, 1986.
133. Sachay, D. J., and Hudspeth, M. E. S. Unpublished data.
134. Coruzzi, G., and Tzagoloff, A. Assembly of the mitochondrial membrane system. DNA sequence of subunit 2 of yeast cytochrome oxidase. *J. Biol. Chem.*, *254*:9324-9330, 1979.
135. Fox, T. D. 5 TGA stop codons occur within the translated sequence of the yeast mitochondrial gene for cytochrome c oxidase subunit II. *Proc. Natl. Acad. Sci. USA*, *76*:6534-6538, 1979.
136. Sevarino, K. A., and Poyton, R. O. Mitochondrial membrane biogenesis: identification of a precursor yeast cytochrome c oxidase subunit II, an integral polypeptide. *Proc. Natl. Acad. Sci. USA*, *77*:142-146, 1980.
137. van den Boogaart, P., van Dijk, S., and Agsteribbe, E. The mitochondrially made subunit 2 of Neurospora crassa cytochrome aa_3 is synthesized as a precursor protein. *FEBS Lett.*, *147*:97-100, 1982.
138. Lawson, J. E., and Deters, D. W. Identification and isolation of the cytochrome oxidase subunit II gene in mitochondria of the yeast Hansenula saturnus. *Curr. Genet.*, *9*:345-350, 1985.
139. Cummings, D. J., Michel, F., Domenico, J. M., and McNally, K. L. Mitochondrial DNA sequence analysis of the cytochrome oxidase subunit II gene from Podospora anserina: a group IA intron with a putative alternative splice site. *J. Mol. Biol.*, *212*:287-294, 1990.
140. Sachay, D. J., Shumard-Hudspeth, D. S., and Hudspeth, M. E. S., unpublished data.
141. Thalenfeld, B. E., and Tzagoloff, A. Assembly of the mitochondrial membrane system. Sequence of the oxi2 gene of yeast mitochondrial DNA. *J. Biol. Chem.*, *255*:6173-6180, 1980.
142. Phelps, L. G., Burke, J. M., Ullrich, R. C., and Novotny, C. P. Nucleotide base sequence of the mitochondrial COIII gene of Schizophyllum commune. *Curr. Genet.*, *14*:401-403, 1988.
143. Nobrega, F. G., and Tzagoloff, A. Assembly of the mitochondrial membrane system. DNA sequence and organization of the cytochrome b gene in Saccharomyces cerevisiae D273-10B. *J. Biol. Chem.*, *255*:9828-9837, 1980.
144. Grosch, G., Schmelzer, C., and Mathews, S. Physical map of the COB region in mitochondrial DNA of Saccharomyces cerevisiae. *Curr. Genet.*, *3*:65-71, 1981.
145. Brunner, A., and Coria, R. Cloning and sequencing of the gene for apocytochrome b of the yeast Kluyveromyces lactis strains WM27 (NRRL Y-17066) and WM37 (NRRL Y-1140) *Yeast*, *5*:209-218, 1989.
146. Lang, B. F., Ahne, F., Bonen, L. The mitochondrial genome of the fission yeast Schizosaccharomyces pombe: the cytochrome b gene has an intron closely related to the first two introns in the Saccharomyces cerevisiae cox1 gene. *J. Mol. Biol.*, *184*:353-366, 1985.
147. Waring, R. B., Davies, R. W., Lee, S. Grisi, E., Berks, M. M., and Scazzocchio, C. The mosaic organization of the apocytochrome b gene of Aspergillus nidulans revealed by DNA sequencing. *Cell*, *27*:4-11, 1981.
148. Citterich, M. H., Morelli, G., and Macino, G. Nucleotide sequence and intron structure of the apocytochrome b gene of Neurospora crassa mitochondria. *EMBO J.*, *2*:1235-1242, 1983.

149. Church, G. M., and Gilbert, W., Yeast mitochondrial intron products required in trans for RNA splicing. In *Mobilization and Reassembly of Genetic Information* (D. R. Joseph et al., eds.), Academic Press, New York, pp. 379-394, 1980.
150. Lazowska, J., Jacq, C., Slonimski, P. Sequence of introns and flanking exons in the wild type and box3 mutants of the mitochondrial cytochrome b gene reveals an interlaced splicing protein coded by an intron. *Cell, 22*:333-348, 1980.
151. Lazowska, J. Claisse, M., Gargouri, A., Kotylak, Z., Spyridakis, Z., and Slonimski, P. Protein encoded by the third intron of cytochrome b gene in Saccharomyces cerevisiae is an mRNA maturase: Analysis of mitochondrial mutants, RNA transcripts, proteins and evolutionary relationships. *J. Mol. Biol., 205*:275-289, 1989.
152. Anziano, P. Q., Hanson, D. K., Mahler, H. R., and Perlman, P. S. Functional domains in introns: Trans-acting and cis-acting regions of intron 4 of the cob gene. *Cell, 30*:925-932, 1982.
153. Labouesse, M., Netter, P., and Schroeder, R. Molecular basis of the "box-effect." A maturase deficiency leading to the absence of splicing of two introns located in two split genes of yeast mitochondrial DNA. *Eur. J. Biochem., 144*:85-93, 1984.
154. Clary, S. A., and Hudspeth, M. E. S. Unpublished data.
155. Ise, W., Haiker, H., and Weiss, H. Mitochondrial translation of subunits of the rotenone-sensitive NADH: ubiquinone reductase in Neurospora crassa. *EMBO J., 4*:2075-2080, 1985.
156. Nelson, M. A., and Macino, G. Structure and expression of the overlapping ND4L and ND5 genes of Neurospora crassa mitochondria. *Mol. Gen. Genet., 206*:307-317, 1987.
157. Cummings, D. J., Domenico, J. M., and Michel, F. DNA sequence and organization of the mitochondrial ND1 gene from Podospora anserina: analysis of alternate splice sites. *Curr. Genet., 14*:253-264, 1988.
158. Cummings, D. J., Michel, F., Domenico, J. M., and McNally, K. L. DNA sequence analysis of the mitochondrial ND4L-ND5 gene complex from Podospora anserina: duplication of the ND4L gene within its intron. *J. Mol. Biol., 212*:269-286, 1990.
159. Brown, T. A., Davies, R. W., Ray, J. A., Waring, R. B., and Scazzocchio, C. The mitochondrial genome of Aspergillus nidulans contains reading frames homologous to the human URFs 1 and 4. *EMBO J., 2*:427-435, 1983.
160. Brown, T. A., Constable, A., Ray, J. A., Waring, R. B., Scazzocchio, C., and Davies, R. W. Nucleotide sequence of the Aspergillus nidulans mitochondrial gene for subunit 5 of NADH dehydrogenase. *Nucl. Acids Res., 17*:4371, 1989.
161. Penalva, M. A., and Garcia, J. L. The subunit I of the respiratory-chain in NADH dehydrogenase from Cephalosporium acremonium: the evolution of a mitochondrial gene. *Curr. Genet., 10*:797-801, 1986.
162. Morelli, G., and Macino, G. Two intervening sequences in the ATPase subunit 6 gene of Neurospora crassa: a short intron (93 base-pairs) and a long intron that is stable after excision. *J. Mol. Biol., 178*:491-507, 1984.
163. Grisi, E., Brown, T. A., Waring, R. B., Scazzocchio, C., and Davies, R. W. Nucleotide sequence of a region of the mitochondrial genome of Aspergillus nidulans including the gene for ATPase subunit 6. *Nucl. Acids Res., 11*:3531-3539, 1982.

164. Macino, G., and Tzagoloff, A. Assembly of the mitochondrial membrane system. Sequence analysis of a yeast mitochondrial ATPase gene containing the oli-2 and oli-4 loci. *Cell, 20*:507-517, 1980.
165. Michon, T., Galante, M., and Velours, J. NH_2-terminal sequence of the isolated yeast ATP synthase subunit 6 reveals posttranslational cleavage. *Eur. J. Biochem., 172*:621-625, 1988.
166. Macreadie, I. G., Novitski, C. E., Maxwell, R. J., John, U., Ooi, B.-G., McMullen, G. L. L., Lukins, H. B., Linnane, A. W., and Nagley, P. Biogenesis of mitochondria: the mitochondrial gene (aapl) coding for mitochondrial ATPase subunit 8 in Saccharomyces cerevisiae. *Nucl. Acids Res., 11*:4435-4451, 1983.
167. Cummings, D. J., MacNeil, A. I., Domenico, J., and Matsura, E. T. Excision-amplification of mitochondrial DNA during senescence in Podospora anserina. DNA sequence analysis of three unique plasmids. *J. Mol. Biol., 185*:659-680, 1985.
168. van den Boogaart, P., Samallo, J., and Agsteribbe, E. Similar genes for a mitochondrial ATPase subunit in the nuclear and mitochondrial genomes of Neurospora crassa. *Nature, 298*:187-189, 1982.
169. Brown, T. A., Ray, J. A., Waring, R. B., Scazzocchio, C., and Davies, R. W. A mitochondrial reading frame which may code for a second form of ATPase subunit 9 in Aspergillus nidulans. *Curr. Genet., 8*:489-492, 1984.
170. Sebald, W., Hoppe, J., and Wachter, E. Amino acid sequence of the ATPase proteolipid from mitochondria, chloroplasts and bacteria (wild type and mutants). In *Function and Molecular Aspects of Biomembrane Transport* (E. Quagliariello, F. Palmieri, S. Papa, and M. Klingenberg, eds.), Elsevier, Amsterdam, 1979, pp. 63-74.
171. Turner, G., Imam, G., and Kuntzel, H. Mitochondrial ATPase complex of Aspergillus nidulans and the dicyclohexylcarbodiimide-binding protein. *Eur. J. Biochem., 97*:565-571, 1979.
172. Fox, T. D. Natural variation in the genetic code. *Ann. Rev. Genet., 21*:67-91, 1987.
173. Osawa, S., and Jukes, T. H. Codon reassignment (codon capture) in evolution. *J. Mol. Evol., 28*:271-278, 1989.
174. Barrell, B. G., Bankier, A. T., and Drouin, J. A different genetic code in human mitochondria. *Nature, 282*:189-194, 1979.
175. Sibler, A.-P., Dirheimer, G., and Martin, R. P. Yeast mitochondrial tRNAIle and tRNAMet: Nucleotide sequence and codon recognition patterns. *Nucl. Acids Res., 13*:1341-1346, 1985.
176. Martin, N. C., Pham, H. D., Underbrink-Lyon, K., Miller, D. L., and Donelson, J. E. Yeast mitochondrial tRNA Trp can recognize the nonsense codon UGA. *Nature, 285*:579-581, 1980.
177. Hensgens, L. A. M., Grivell, L. A., Borst, P., and Bos, J. L. Nucleotide sequence of the mitochondrial structural gene for subunit 9 of the yeast ATPase complex. *Proc. Natl. Acad. Sci. USA, 76*:1663-1667, 1979.
178. Macino, G., and Tzagoloff, A. Assembly of the mitochondrial membrane system. The DNA sequence of a mitochondrial ATPase gene in Saccharomyces cerevisiae. *J. Biol. Chem., 254*:4617-1623, 1979.
179. Li, M., and Tzagoloff, A. Assembly of the mitochondrial membrane system. Sequences of yeast mitochondrial valine and an unusual threonine tRNA gene. *Cell, 18*:47-54, 1979.

180. Jacquier, A., and Dujon, B. The intron of the mitochondrial 21S rRNA gene: Distribution in different yeast species and sequence comparison between Kluyveromyces thermotolerans and Saccharomyces cerevisiae. *Mol. Gen. Genet.*, *192*:487-499, 1983.
181. Boyd, D. A., Hobman, T. C., Gruenke, S. A., and Klassen, G. R. Evolutionary stability of mitochondrial DNA organization in Achlya. *Can. J. Biochem. Cell Biol.*, *62*:571-576, 1984.
182. Forster, H., Kinscherf, T. G., Leong, S. A., and Maxwell, D. P. Molecular analysis of the mitochondrial genome of Phytophthora. *Curr. Genet.*, *12*:215-218, 1987.
183. Clark-Walker, G. D., and Gleason, F. H. Circular DNA from the water mold Saprolegnia. *Arch. Mikrobiol.*, *92*:209-216, 1973.
184. McNabb, S. A., Eros, R. W., and Klassen, G. R. Presence and absence of large inverted repeats in the mitochondrial DNA of Hyphochytridiomycetes. *Can. J. Bot.*, *66*:2377-2379, 1988.
185. Borkhardt, B., Pedersen, M. B., and Olson, L. W. Mitochondrial DNA in the aquatic fungus Allomyces arbuscula. *Curr. Genet.*, *12*: 149-156, 1987.
186. Borkhardt, B., and Olson, L. W. The mitochondrial genome of the aquatic phycomycete Blastocladiella emersonii. *Curr. Genet.*, *11*:139-143, 1986.
187. Hoeben, P., and Clark-Walker, G. D. An approach to yeast classification by mapping mitochondrial DNA from Dekkera/Brettanomyces and Eniella genera. *Curr. Genet.*, *10*:371-379, 1986.
188. Wills, J. W., Troutman, W. B., and Riggsby, W. S. Circular mitochondrial genome of Candida albicans contains a large inverted duplication. *J. Bacteriol.*, *164*:7-13, 1985.
189. Kunze, G., Bode, R., and Birnbaum, D. Physical mapping and genome organization of mitochondrial DNA from Candida maltosa. *Curr. Genet.*, *10*:527-530, 1986.
190. O'Connor, R. M., McArthur, C. R., and Clark-Walker, G. D. Closed circular DNA from mitochondria-enriched fractions of four petite-negative yeasts. *Eur. J. Biochem.*, *53*:137-144, 1975.
191. Clark-Walker, G. D., McArthur, C. R., and Daley, D. J. Does mitochondrial DNA length influence the frequency of spontaneous petite mutants in yeasts? *Curr. Genet.*, *4*:7-12, 1981.
192. Falcone, C. The mitochondrial DNA of the yeast Hansenula petersonii: genome organization and mosaic genes. *Curr. Genet.*, *8*:449-455, 1984.
193. Clark-Walker, G. D., and Sriprakash, K. S. Sequence rearrangements between mitochondrial DNAs of Torulopsis glabrata and Kloeckera africana identified by hybridization with six polypeptide encoding regions from Saccharomyces cerevisiae mitochondrial DNA. *J. Mol. Biol.*, *151*:367-387, 1981.
194. Wesolowski, M., Algeri, A., and Fukuhara, H. Gene organization of the mitochondrial DNA of yeasts: Kluyveromyces lactis and Saccharomycopsis lipolytica. *Curr. Genet.*, *3*:157-162, 1981.
195. Kuck, U., Stahl, U., Lhermitte, A., and Esser, K. Isolation and characterization of mitochondrial DNA from the alkane yeast Saccharomycopsis lipolytica. *Curr. Genet.*, *2*:97-101, 1980.
196. Sanders, J. P. M., Verbeet, M. P., Meijlink, F. C. P. W., Heyting, C., and Borst, P. The construction of the physical maps of three different Saccharomyces mitochondrial DNAs. *Mol. Gen. Genet.*, *157*: 271-280, 1977.

197. Wolf, K., Lang, B., Del Giudice, L., Anziano, P. Q., and Perlman, P. S. Schizosaccharomyces pombe—a short review of a short mitochondrial genome. In *Mitochondrial Genes* (P. Slonimski, P. Borst, and G. Attardi, eds.), Cold Spring Harbor Laboratory, New York, pp. 355-360.
198. Wolf, K. Mitochondrial genome divergence in the petite negative yeast Schizosaccharomyces pombe. *Endocytobiology, 2*:229-240, 1983.
199. Stepien, P. P., Bernard, U., Cooke, H. J., and Kuntzel, H. Restriction endonuclease cleavage map of mitochondrial DNA from Aspergillus nidulans. *Nucl. Acids Res., 5*:317-330, 1978.
200. Turner, G., Earl, A. J., and Greaves, D. R. Interspecies variation and recombination of mitochondrial DNA in the Aspergillus nidulans species group and the selection of species-specific sequences by nuclear background. In *Mitochondrial Genes* (P. Slonimski, P. Borst, and G. Attardi, eds.), Cold Spring Harbor Laboratory, New York, pp. 411-414, 1982.
201. Minuth, W., Tudzynski, P., and Esser, K. Extrachromosomal genetics of Cephalosporium acremonium: I. Characterization and mapping of mitochondrial DNA. *Curr. Genet., 5*:227-231, 1982.
202. Tudzynski, P., Duvell, A., and Esser, K. Extrachromosomal genetics of Claviceps purpurea. I. Mitochondrial DNA and mitochondrial plasmids. *Curr. Genet., 1*:145-150, 1983.
203. Garber, R. C., and Yoder, O. C. Mitochondrial DNA of the filamentous ascomycete Cochliobolus heterostrophus. Characterization of the mitochondrial chromosome and population genetics of a restriction enzyme polymorphism. *Curr. Genet., 8*:621-628, 1985.
204. Bernard, U., Goldthwaite, C., and Kuntzel, H. Physical map of Neurospora crassa mitochondrial DNA and its transcription unit for ribosomal RNA. *Nucl. Acids Res., 3*:3101-3108, 1976.
205. Cummings, D. J., Belcour, L., Grandchamp, C. Mitochondrial DNA from Podospora anserina. Isolation and characterization. *Mol. Gen. Genet., 171*:229-238, 1979.
206. Saunders, G., Rogers, M. E., Adlard, M. W., and Holt, G. Chromatographic resolution of nucleic acids extracted from Penicillium chrysogenum. *Mol. Gen. Genet., 194*:343-345, 1984.
207. Smith, T. M. Saunders, G., Stacey, L. M., and Holt, G. Restriction endonuclease map of mitochondrial DNA from Penicillium chrysogenum. *J. Biotechnol., 1*:37-46, 1984.
208. Specht, C. A., Novotny, C. P., and Ullrich, R. C. Isolation and characterization of mitochondrial DNA from the basidiomycete Schizophyllum commune. *Exp. Mycol., 1*:336-343, 1983.
209. Clark-Walker, G. D., and Sriprakash, K. S. Analysis of a five gene cluster and unique orientation of large genic sequences in Torulopsis glabrata mitochondrial DNA. *J. Mol. Evol., 19*:342-345, 1983.

10

THE STRUCTURE AND FUNCTION OF PLASMIDS IN FILAMENTOUS FUNGI

TIM RUBIDGE *Chemical Defence Establishment, Salisbury, Wiltshire, England*

I. INTRODUCTION

Filamentous fungi and other more structurally complex microbes have been studied for a longer time than have bacteria but their size and complexity are drawbacks to their selection as experimental systems. Nowhere is this more true than in the development of molecular genetics and biotechnology. Prokaryotic systems, especially *Escherichia coli* and *Bacillus subtilis*, are not only easier to handle but also there is a relative wealth of experimental data on which to build. If the recent advances commonly termed genetic engineering can be resolved into the directed manipulation of isolated genes and purified genetic material (both DNA and RNA), then it will have been the discovery of the ability of bacterial plasmids to transform cells and bring about the expression of recombinant genes that made the most prominent contribution.

Plasmids have been defined as "circular, double-stranded extrachromosomal DNA molecules that contain specialized genes and have the ability to be replicated in the bacterial cell. . . . Often associated with plasmids are transposons and insertion sequences" [1]. At one time described exclusively in bacteria, it is now clear that plasmidlike DNA species (plDNA) can be detected in a variety of eukaryotic organisms including fungi and higher plants. Not unexpectedly, the pattern of plDNA in the higher organisms shows greater complexity than that described generally for bacteria and reflects the greater complexity of the eukaryotic genome; the DNA is organized into chromosomes within the nucleus and also present in mitochondria and chloroplasts. A broader definition is required to encompass such plDNAs.

This chapter reviews the distribution and characterization of plasmids in the filamentous fungi. For this purpose plasmids are considered to be circular or linear, double-stranded DNA species existing and replicating, for at least part of the fungus life cycle, separate from nuclear and organelle chromosomes.

II. PLASMIDS IN *PODOSPORA* SPECIES

In 1978, Stahl and his colleagues claimed the initial description of plDNA in a filamentous fungus [2]. DNA was extracted from *Podospora anserina* hyphae preparations and purified by centrifugation through a cesium chloride-ethidium bromide gradient. Apart from mitochondrial DNA (mtDNA), and that derived from the nucleus, a faint band was visible at a point in the gradient equivalent to a greater guoyant density. This band, presumed to consist of plDNA, was shown by electron microscopy to contain molecules predominantly in a covalently closed circular (ccc) form. From a determination of the contour length of open circular form (oc) plDNA the size of the molecule was calculated to be in the region of 2.5 kilobase pairs (kbp). The presence of plDNA in senescent cultures and its absence in juvenile cultures suggested an association of plDNA with senescence.

Non-Mendelian genetic control of senescence in *P. anserina* was by this time well known and the involvement of an unstable mitochondrial genome had been clearly established [3]. It was perhaps no surprise therefore that a year later the mitochondrial origin of a dense species of extrachromosomal DNA isolated from senescent hyphae was reported [4]. Although plDNA existed in a broad distribution of sizes (determined by contour length), restriction with *Hae*III endonuclease gave rise only to a fragment of 2.5 kbp. Restriction of plDNA with *Alu*I gave rise to 10 distinct fragments totaling 2.57 kbp. From other data it was concluded that in senescent hyphae there was a deletion of DNA from the mitochondrial chromosome. Included among the deleted material was a fragment of some 2.6 kbp which became amplified and tandemly repeated forming circular, self-replicating plasmids.

Further investigation revealed that plDNA existed as molecules other than the 2.6-kbp circle originally observed [5,6]. At least six species of plDNA in senescent cultures have now been defined [7]. The technique of heterologous hybridization between known protein-coding regions of mitochondrial DNA isolated from yeast petite mutants and the mitochondrial chromosome of *P. anserina* has been used to identify genes within the latter's mitochondrial genome. Further, by studying the homology between species of plDNA and the coding regions of the mitochondrial chromosome, it has become possible to determine some of the genetic functions of the plDNA [6,8].

A model linking senescence in *P. anserina* with plDNA was described by Esser and his colleagues [9] who proposed three phases in the aging of cultures. The first, or juvenile, phase was characterized by mitochondria containing only the intact mitochondrial chromosome. No free plDNA was detectable. This phase was followed by the presenescent phase when free circles of plDNA were observed. it was suggested that the emergence of ccc plDNA resulted from one of two possible processes. Either mitochondrial chromosome sequences homologous to those of plDNA were deleted and subsequently amplified as autonomously replicating tandem repeats, or integrated plDNA sequences functioned as polymerase templates from which circularized, tandem repeats were formed. The final, senescent stage represented the increased amplification of plDNA (also termed senDNA). Disintegration of the mitochondrial chromosome occurred and death of the culture ensued.

The transmissibility of senescence was demonstrated in the ability of ccc plDNA isolated from senescent hyphae to transform protoplasts derived from juvenile hyphae [9,10]. Some drugs with effects directed toward mitochondria have activity in reversing senescence or preventing its onset [11]. It is presumed that ethidium bromide, in selecting for the small number of intact mitochondrial chromosomes present in a population of senescent hyphae, thus suspends the onset of senescence. The same workers had previously showed the similar effect of chloramphenicol [12]. Cycloheximide, with a cytoplasmically directed mode of action, was without effect. No drug has been shown to be able to prevent irreversibly the onset of senescence in rejuvenated hyphae unless it is maintained within the culture medium.

Recent advances in the study of senescence in *P. anserina* have focused on the mapping and sequence analysis of the plDNA species and the juvenile mitochondrial chromosome [13-16].

A detailed discussion of these developments is more appropriate to a review of senescence per se. However, it is becoming clear that the infectivity of the senescence phenomenon and the presence within plDNA species of apparently mobile introns are closely linked.

The excision/amplification of DNA sequences from the mitochondrial chromosome—the earliest detectable event in senescence—has been linked to the introns discovered to be present within the senescence-associated plDNAs. Some mitochondrial genes are not represented on plDNA species and are thus deleted from the genome as the fungus enters the later senescent phases [17]. Senescence might then be a reflection of the loss of one or more genes essential in oxidative phosphorylation and energy metabolism. Another possibility has been suggested, i.e., excision and splicing might have an adverse effect on translation and transcription [13]. Although the similarity between senescent plDNA species and transposons is an attractive analysis, no data have emerged to indicate what stimulates the mitochondrial sequences to become mobile and thus be deleted from the mitochondrial chromosome. Experiments showing that the 9.8-kbp monomer plDNA is transposed to the nucleus [18] cannot be repeated [19].

Bockelman and Esser [20] pointed out the similarity of senescence in *P. curvicolla*, stopper mutants of *Neurospora crassa* [21], and the ragged mutation in *Aspergillus amstelodami* [22]. As opposed to the effect in *P. anserina* where senescence leads to the death of the entire colony, that in *P. curvicolla* is characterized by the survival of single hyphae at a distance behind the margin of the senescent colony. These give rise to a series of short bursts of renewed vigorous growth. Cultures of *P. curvicolla* were therefore screened for plDNA species analogous to that described in *P. anserina* [20].

It was established that there was less mtDNA in the senescent mycelium compared to the juvenile. Additionally, there was a new species of DNA of increased buoyant density detectable using the technique of density gradient centrifugation. The patterns of fragments yielded by restriction endonuclease digestion of the plDNA isolated from three independently aged mycelia were consistent with them, comprising tandemly repeated sequences of similar but discretely differing length (10.85, 9.01, and 10.50 kbp). Electron microscopy of one species of plDNA confirmed the presence of covalently closed circular and open circular molecules which could be interpreted as monomers, dimers, and trimers. Careful comparison of the restriction enzyme digestion profiles of the three plDNAs with the circular restriction map of *P. curvicolla* mtDNA showed that each plDNA monomer subunit overlapped and all originated in the same region of the mitochondrial chromosome. There was thus a high degree of homology between the plDNAs, which nevertheless differed in their excision and junction points.

All the plDNA species were found to hybridize to mtDNA sequences containing the *lrRNA* gene. The sequence associated with the *COI* gene, important with the most abundant senescent *P. anserina* plDNA, did not hybridize to any *P. curvicolla* plDNA and was located at a point remote from the *lrRNA* gene on the mitochondrial chromosome. No homology could be detected between plasmids of *P. curvicolla* and the most abundant 2.5-kbp monomer plasmid of *P. anserina*. Whether the phenotypic differences between patterns of senescence in the two fungi reflect the different genetic organization of their plasmids is not yet clear. However, it is tempting to suggest that the different genetic organizations indicate that it is the excision/amplification of mitochondrial genes with the consequently disrupted mitochondrial genome which promotes senescence, not the amplification of specific genes.

III. PLASMIDS IN *NEUROSPORA* SPECIES

A variety of mutant strains of *N. crassa* show abnormal growth phenotypes. Among them are strains with altered mitochondrial genomes, some showing amplifications of specific regions of the mitochondrial chromosome strongly reminiscent of the senescent phenomenon in *Podospora*. The structure of mitochondrial DNA in "stopper" mutant strains has been studied [23]. It was shown that a spectrum of defective mitochondrial chromosomes existed within the mutant group. The characteristic feature of the altered mitochondrial genome was the loss of large segments of the DNA. Those portions that were retained, representing possibly half of normal mitochondrial DNA, possessed a variety of amplified regions. *N. crassa* strain E35 also shows the stop-start growth phenotype. Again the mitochondrial genome was shown to be unstable [24]. In this case the primary event was the deletion of a sequence of some 4 kbp from the normal mitochondrial chromosome. When growth of the strain was prolonged a third species of circular DNA could be identified within the mitochondria of contour length 8 μm (~25 kbp). It was suggested that the 4-kbp deletion brought about a premature termination of replication so that the amplified regions became retained as circular, autonomous, plasmidlike DNAs.

The *poky* mutation of *N. crassa*, characterized by defective mitochondrial ribosome assembly, has been shown to have an origin in organization of the mitochondrial genome [25]. In a subcultured strain (strain IIa), rather than showing additional and discrete, aberrant DNA structures, there was an amplified sequence retained integrated within the chromosome. The monomeric unit of the reiterated sequence was 2.2 kbp. Strain IIb, culture bearing the related *exn*-1 mutation, showed a similar effect but the amplified sequence was of smaller monomeric length—1.9 kbp [25]. With strain HI-10 an amplified sequence of 28 kbp was released from the mitochondrial chromosome. In each case the relevant sequence included regions important in coding for rRNA.

Species of *Neurospora* may also show plasmidlike DNA molecules distinct from those observed in the mitochondria of mutants with an unstable mitochondrial chromosome. Studies have been reported of a plDNA found in a wild-type strain of *N. crassa* (Mauriceville-1c, FGSC 2225) [27]. Using agarose and polyacrylamide gel electrophoresis and electron microscopy they were able to show that it existed in a ccc form as head-to-tail polymers of a monomer unit 3.6 kbp in length. The largest molecule measured comprised six monomer units. The restriction map of the monomer unit showed two unusual arrays of restriction endonuclease recognition sites; within one 1-kbp region there were six sites for *Eco*RI. Within a quite distinct region extending to 0.4 kbp at least five *Pst*I recognition sites were located. No homology between the plasmid DNA and mtDNA was found. Careful digestion of contaminating extramitochondrial DNA present in preparations of mitochondria allowed the conclusion that the plasmid was located within the intact organelles. No function could be ascribed to the plDNA although a protein (equivalent to the monomer length of the plasmid) was detected which suggested that the plasmid was transcribed in vivo.

The structure of the plasmid containing the multiple *Pst*I sites has been established [28] and eight-clustered recognition sites for the endonuclease clustered have been identified. It has been suggested that the detection of a transcript only in the plasmid bearing such sequences might reflect a function for the *Pst*I elements transcription, perhaps as an enzyme recognition site.

Species of plDNA have also been isolated from *N. intermedia* strains Fiji N6-6 and P405 Labelle [29]. The plasmids were shown to exist as monomeric units of ccc DNA of some 5.2 and 4.1 kbp, respectively. Oligomeric plasmids containing up to six monomer units were observed. Using the Southern blotting technique no homology between the plasmids themselves or between the individual plasmids and mtDNA could be detected [30]. A plasmid closely related to that detected in the Mauriceville strain 2225 was also isolated from *N. intermedia* strain Varkud-1c and was reported to be similarly transcribed in vivo. No other products of transcription have been reported for *Neurospora* plasmids. All of the plasmids, however, exist within the mitochondria as arrays of oligomeric, head-to-tail tandem repeats [31].

It has been suggested that a sequence organization similar to that comprising the *Pst*I elements could have had value in previous transposition events leading eventually to a fixing of the elements within the plasmid when transposing ability was lost [28,32,33].

The complete 3581-nucleotide sequence of the Mauriceville-1c plasmid has been established [31]. The same group reported the sequencing of that region of the Varkud-1c plasmid isolated from *N. intermedia* strain Varkud-1c that corresponded to the open reading frame identified for the Mauriceville-1c strain plasmid transcript [34]. The relatedness of the two plasmids, previously revealed by analysis of the restriction patterns and Southern hybridization experiments, was confirmed. A similar reading frame was identified and from over 2133 bases only 34 base substitutions were detected. It was argued that the close similarity between the sequences of the two reading frames was unlikely to be accounted for by chance alone. Rather it was likely that the protein encoded by the plasmid-borne gene was important either in maintaining the plasmids themselves or in conferring some selective advantage on the host strains. If the latter were to be the case; then the selective advantage must be one which ceases to be significant under the laboratory conditions for these plasmids have not been isolated from laboratory strains but are often found in strains drawn from nature [31].

Additional plasmids have been reported in species of *Neurospora* [35,36]. Plasmids, apparently organized as head-to-tail, oligomeric circles, have been isolated from four of seven isolates of *N. tetrasperma* Hanalei. Southern hybridization experiments using cloned plasmid showed some homology between the *N. tetrasperma* plasmid and that obtained from the Fiji N6-6 strain of *N. intermedia*. No clear homology was shown between the probe and other plasmids, but a plasmid homologous both to the *N. tetrasperma* Hanalei plasmid and to an *N. intermedia* Fiji plasmid was observed in half of the isolates of *N. crassa* [36].

IV. PLASMIDS IN *FUSARIUM* SPECIES

The first report that plDNA might exist in species of *Fusarium* was made in 1978 [37]. The initial experiments suggested that the possession of plDNA was also a property of *Rhizoctonia solani*. The same group further observed the apparent loss of resistance to nifuroxime in *F. oxysporum* f. sp. *lycopersici* [38]. Although they remarked that their findings were in themselves insufficiently complete to conclude that the existence of a plDNA molecule was proved; it was judged likely that the loss of resistance was associated with the loss of a cytoplasmically located, circular DNA molecule of 46.7 kbp. Through the exhaustive use of controls during the purification and characterization of the extranuclear DNA, the circular molecule was subsequently shown to be the mitochondrial chromosome [39].

The importance of *Fusarium* species in plant disease encouraged researchers to look further in this genus for plDNA. The examination of 18 strains of *F. oxysporum* f. sp. *conglutinans* for the presence of plDNA has been described [40,41]. In contrast to the above findings [39], the mitochondrial chromosome appeared to exist predominantly as a linear strand of DNA. In addition they detected linear, double-stranded DNA molecules located in the mitochondria of each strain studied. In all cases the size of this material was about 1.9 kbp. Restriction enzyme mapping revealed that the plasmid pFOXC2, isolated from a race 2 culture, had inverted, terminally repeated sequences and a 5' terminus resistant to the action of λ-exonuclease. When all but 0.1 kbp of this plasmid was cloned into pUC12 and prepared as a probe, no homology could be detected between the recombinant plasmid and either the mitochondrial chromosome or the nuclear DNA.

A role for the plasmid in the pathogenicity of the races of *F. oxysporum* f. sp. *conglutinans* was tentatively identified [41]. Strains of race 2, pathogenic on radish, have plasmids homologous to cloned pFOXC2. Plasmids isolated from strains of races 1 and 5, pathogenic on cabbage, show no homology to cloned pFOXC2 but a marked homology to each other. The investigators suggested that even if the genes borne on the plasmids were not implicated in host-specific pathogenicity, then at least they might have a value in taxonomy.

Two species of plDNA were isolated from *F. solani* f. sp. *cucurbitae* [42]. Although larger than pFOXC2 (about 8 and 10 kbp), their structure was similar in being linear, double-stranded molecules with blocked 5' termini. However, one plasmid showed a high degree of homology with the mitochondrial DNA. The presence of the plasmids could not be correlated with either the fungus's pathogenicity or with its growth characteristics. In this latter regard it may be significant that *Nectria haematococca*, the teleomorph of *F. solani*, shows sudden changes in morphology which are controlled by extranuclear genetic factors. These phenomena were discussed before the implication of plasmids in fungal senescence was fully appreciated [43]. Nevertheless attention was drawn to the diverse range of fungi showing sudden morphological changes which might be under the control of extranuclear genetic factors. In the light of recent findings perhaps the time is ripe for these species (including *Pestalozzia annulata*, *Curvularia pallescens*, and *Hypomyces ipomoeae*) to be examined for the presence of plDNA.

A survey of 23 strains from 13 species of *Fusarium* has been described [44]. Only one strain (of *F. merismoides*) showed evidence for the occurrence of plDNA and although no evidence was presented for its structure it was located within the mitochondria.

V. PLASMIDS IN OTHER FILAMENTOUS FUNGI

The observation that *Rhizoctonia solani* might contain a DNA plasmid [37] was confirmed in 1984 [45]. Working with three weakly pathogenic strains the investigators were able to isolate a double-stranded, linear DNA plasmid of size 2.6 kbp. The DNA did not sediment with mitochondria in the preparations from protoplasts and may have been a cytoplasmic plasmid.

Further linear plasmids were isolated from *Claviceps purpurea* [46]. Four bands of DNA were detected with gel electrophoresis with apparent sizes of 6.6, 5.3, 10, and 1.1 kbp and were named pl1, pl2, pl3, and pl4, respectively. Because they were consistently only minor components of the population, the latter two were studied no further. It was concluded that the plasmids were mitochondrial in origin since they were efficiently prepared from DNAase-treated intact mitochondria and showed buoyant densities similar to that of mtDNA. However, pl1 showed no homology with mtDNA in Southern blot experiments. It was suggested, but not determined experimentally, that pl2 would be equally dissimilar since the two plasmids were shown in a different experiment to have a high degree of homology with each other. Linear plasmids have been shown to be possessed by at least five other wild-type strains of *C. purpurea* [47]. Plasmids from three strains showed no homology with the mitochondrial chromosome but the two others did. The authors suggested that the plasmids structurally resembled transposons and the homology with the mitochondrial DNA indicated an inserted sequence.

A linear plasmid DNA has been isolated from an unstable mutant strain of *Ascobolus immersus* [48]. The plasmid was maternally inherited and approximately 6 kbp in size. The pattern of renaturing after alkaline denaturation suggested that the ends of the molecules were inverted repeats. Electron microscopy of the renatured molecules allowed the size of the terminal repeated sequences to be measured as 1.2 kbp. The complexity of the organization of DNA within the organelles of eukaryotes and the small differences in buoyant density between the different DNA components has often forced workers to select bisbenzimide rather than ethidium bromide as the intercalating fluorescent dye when purifying plasmids and mtDNA. In the case of the mushroom *Agaricus bitoquis* [49,50], the two plasmid species, pEM and pMPJ, could not be separated from the mtDNA using bisbenzimide centrifugation. Electron microscopy and agarose gel electrophoresis revealed the plasmids to be linear and of sizes 7.35 (pEM) and

3.65 (pMPJ) kbp, respectively. They were characterized as mitochondrial plasmids. The manner in which purified pEM renatured suggested the presence of terminal inverted repeat sequences. No homology could be detected between the plasmids themselves or between either of the plasmids and the nuclear or mtDNA. On the basis of the pattern of digestion of pEM with restriction endonucleases it was concluded that pEM represented two distinct species of plasmid DNA, both about 7.5 kbp. These species were termed pEM1 and pEM2. The latter could be distinguished by its possession of a unique nucleotide sequence recognized by the endonuclease *Pvu*II which was absent from both pEM1 and pMPJ. An additional complexity in the extrachromosomal DNA pattern of *A. bitorquis* was the appearance in low numbers of small circles of DNA (each less than 1 kbp) within preparations of mtDNA.

Linear plasmids were observed in the morel *Morchella conica* but not in the related *M. esculenta* [51]. Two species of DNA were detected of size 6.09 and about 8 kbp. The intracellular localization of these species was not determined. The former plasmid was concluded to possess terminal inverted repeats of about 0.75 kbp from the appearance of stem loop structures during renaturation after the DNA had been subjected to alkaline denaturation.

The mitochondrial DNA of "ragged" mutants of *Aspergillus amstelodami* has been examined and amplified DNA sequences could be demonstrated similar in many ways to the plDNA of senescent *P. anserina* [52]. In DNA preparations from the mitochondria of ragged mutants a very high molecular weight species of DNA could be detected quite separate from the mitochondrial chromosome. Digestion with restriction endonucleases revealed the structure of this DNA to be that of an array of tandem, head-to-tail oligomers. The monomer was a 0.82- to 0.84-kbp sequence homologous to a discrete section of the mitochondrial chromosome. It was shown that within these amplified regions of the mitochondrial chromosome there are sequences which could form hairpin loops and perhaps function of origins for replication [53]. It has been suggested that the phenomenon of amplification is a more widespread one than might be predicted from the reports of senescencelike effects. Only when the amplified region contained a sequence capable of acting as an origin of replication would the amplification be maintained. Additionally, they argued, in a facultative anaerobe such as the yeast petite mutant the possible preferential replication of the amplified sequences could proceed until little or none of the unamplified mitochondrial genome remained. At this point no oxidative phosphorylation would be possible. The consequences to an obligate aerobe (e.g., *Podospora anserina*) would be fatal.

In the cephalosporin producer *Cephalosporium acremonium*, there is a species of extrachromosomal DNA which cannot be prepared from intact mitochondria and is probably located free within the cytoplasm [54]. Although data are sparse the plasmid preparation appeared as a range of sizes (as shown by restriction digests) and buoyant densities (as shown by density gradient centrifugation).

Garber and Yoder [55] described in 1983 their methods for the isolation of DNA from filamentous fungi and its separation into nuclear, ribosomal, and mitochondrial components. A year later they described the discovery of plDNA within a strain of *Cochliobolus heterostrophus* [56,57]. Undigested mtDNA showed, apart from high molecular weight DNA, a series of bands of lower molecular weight. When the mtDNA was digested with a restriction endonuclease (*Bam*HI, *Eco*RI, *Pvu*II, or *Hinc*II), the series of smaller bands was replaced with a single, intense band of DNA equivalent to 1.9 kbp. When undigested mtDNA was examined by electron microscopy it became clear that the plasmids were composed of oc and ccc DNA formed from up to 17 monomer units of about 1.9 kbp each. Hybridization experiments established that the 1.9-kbp monomer was homologous to a single site on the mitochondrial chromosome of all strains tested but to no portion of the nuclear chromosome. No homology was observed between the *C. heterostrophus* plasmid monomer and plasmids isolated from *P. anserina* or *N. crassa*. Although no function could be ascribed to the plasmid, a phenomenon with some similarity to senescence was observed in cultures of the strain from which the plDNA was isolated.

VI. CLASSIFYING PLASMIDS IN FILAMENTOUS FUNGI

Plasmids have been described in 15 species of filamentous fungi belonging to 10 genera. Fungal plasmids have been grouped in two classes depending on whether they were independent of or derived from high molecular weight DNA [58]. However, many of the oligomeric, head-to-tail plasmids have a high molecular weight. Those which do not show close homology with the mitochondrial chromosome fall into two distinct classes; some are double-stranded ccc DNA whereas the remainder are linear. It seems appropriate to consider there to be three groups of fungal plasmids distinguishable according to their structure (linear or circular) and their homology with the mitochondrial chromosome:

Group 1. Covalently closed circular, oligomeric plasmids homologous to regions of the mitochondrial chromosome and implicated in senescence-like phenomena (Table 1).

Group 2. Linear, monomeric plasmids showing no homology with the mitochondrial chromosome (Table 2).

Group 3. Covalently closed circular, oligomeric plasmids showing no homology with the mitochondrial chromosome (Table 3).

A. Group 1 Plasmids

Group 1 plasmids (Table 1) are implicated in senescence or similar growth-related phenomena. It can be assumed that the possession of a plasmid from one of the other groups confers at least a small selective advantage on the host strain. However, since they are seldom isolated from strains with a long history of laboratory maintenance, they cannot bear genes essential for growth; the selective advantage must only apply in nature. A culture in which a group 1 plasmid can be observed represents a culture with a perhaps terminally impaired ability to maintain growth. It is difficult to establish an evolutionary role for such a process. The result of a terminal effect, such as senescence in *P. anserina*, is the amplification of a portion of the mitochondrial genome; in the absence of a demonstration of infectivity in vivo the selective value of this effect is unclear. Until it is possible to establish what stimulates the amplification/excision of plDNAs in senescence or, more importantly, what mechanism prevents the onset of senescence in juvenile hyphae, it is unlikely that the role of these phenomena will be illuminated.

TABLE 1. Group 1 Plasmids (Covalently Closed Circular, Oligomeric Plasmids Implicated in Senescencelike Phenomena)

Organism	Size[a]	Loc[b]
Podospora anserina	< 9.8	mt
Podospora curvicolla	~ 10	mt
Cochliobolus heterostrophus	1.9	mt
Aspergillus amstelodami	~ 0.8	mt
Neurospora crassa	< 25	mt

[a]Size in kbp.
[b]Location of the plasmid: mt, mitochondrial; cyt, cytoplasmic
Note: All data are referenced in the text.

It has been suggested that the identification of introns within the nucleotide sequence of plasmids of senescent *P. anserina* and the striking similarity of one with the ribosomal intron in *Tetrahymena* recalls an insertion event common to the evolution of both organisms [13]. If this is so then it might be reasonable to retain the senescent plasmids as a separate group distinct from plasmids capable of a continued autonomous existence; their appearance represents a breakdown in the unstable relationship between the host mitochondrial chromosome and the inserted nucleotide sequence. It is equally possible that the insertion event has no relationship to the senescent phenomenon.

The presumed evolutionary event suggests a link between fungal plasmids and insertion sequences (transposons). Other structural coincidences of transposons and fungal plasmids can be seen with the group 2 plasmids.

B. Group 2 Plasmids

Group 2 fungal plasmids (Table 2) can be characterized as linear, double-stranded DNA molecules with inverted terminal repeats and blocked 5' termini showing no homology with the mitochondrial chromosome. They have structural similarities to the linear plasmids of higher plants [59], bacterial transposons, and viruses such as the adenoviruses [60] and the bacteriophage Φ29 [61] and Cp-1 [62].

The role of the S1 and S2 linear plasmids of *Zea mays* in sterility was reviewed by Tudzynski [17]. He pointed to their homology with sequences within the mitochondrial DNA and their ability to integrate within and be lost from the mitochondrial chromosome. The plasmids bear terminal inverted repeats and are structurally similar to prokaryotic transposons [63]. Except for two of the plasmids isolated from *Claviceps purpurea* [47], none of the group 2 fungal plasmids show any homology with the mitochondrial chromosome.

The group 2 plasmids possess the inverted terminal repeat sequences which are characteristic of transposons. The possible role of transpositional event in the evolution of fungal senescence and the group 1 plasmids has already been discussed. However, were there to be a transposonlike function for the group 2 plasmids it would be reasonable to suppose that a

TABLE 2. Group 2 Plasmids (Linear, Monomeric Plasmids Showing No Homology with mtDNA)

Organism	Size	Loc
Rhizoctonia solani	2.617	ND
Agaricus bitorquis	7.35	mt
	3.65	mt
Morchella conica	~8	ND
	6.09	ND
Claviceps purpurea[a]	6.6	mt
	5.3	mt
Fusarium merismoides[b]	>1.4	mt
Fusarium oxysporum	1.9	mt
Fusarium solani[c]	8	mt
	10	mt
Ascobolus immersus[d]	6	

[a]It has been suggested that two plasmid-bearing strains carry integrated plasmid copies within the mitochondrial genome.
[b]The linear plasmid structure in this organism has not been determined.
[c]This organism might show plasmid involvement with a senescencelike phenomenon.
[d]Homology with mtDNA was not examined.

degree of homology would be detected between the plasmids and the chromosome of the mitochondria within which they are found. This has not been shown to be the case. If the resemblance between this group of plasmids and transposons is other than a structural coincidence, then the transposable elements have lost their ability to insert within the host chromosome while retaining the capability for independent replication. There is some evidence to support this hypothesis in *Ascobolus immersus*. The strains showing genetic instability and from which group 2 plasmids may be isolated [48] also possess unstable insertion elements [64].

At the 5' terminus of the linear adenovirus DNA is a covalently bound protein implicated in replication [60]. The same is true for the bacteriophages Φ29 and Cp-1 [61,62]. Perhaps significantly, terminal inverted repeats have been identified in the nucleotide sequences of the bacteriophages. No data are available concerning the infectivity of group 2 plasmids and it would be premature to suggest that the similar presence of a protein bound to the 5' termini demonstrates a virus origin for the plasmids. However, the lack of homology of the plasmids with the mitochondrial chromosome is not inconsistent with this possibility. The linear plasmids are capable of autonomous replication and it is likely that, whatever their origin, the terminal protein is involved in priming replication.

In summary, the group 2 plasmids show structural similarities with diverse other linear, extrachromosomal DNA molecules but show no functional

TABLE 3. Group 3 Plasmids (Covalently Closed Circular Oligomeric Plasmids Showing No Homology to mtDNA)

Organism		Size	Loc
Neurospora crassa Mauriceville		3.6	mt
Neurospora intermedia	Labelle	4.1	mt
	Fiji	5.2	mt
	Varkud	ND	mt
Neurospora tetrasperma Hanalei		5.0	mt
Cephalosporium acremonium[a]		ND	cyt

[a]Homology with mtDNA was not examined.

relationship with any one. No data have been published clearly identifying a role for one of the plasmids as a whole or even for a gene borne thereon. Some evidence pointing to a function was established when linear plasmids were isolated from weakly pathogenic, slow-growing strains of *Rhizoctonia solani* but not from pathogenic isolates which grew normally [45]. Unfortunately, the localization of the plasmids was not determined. Hybridization experiments would doubtless be valuable in establishing whether the pathogenic strains harbor integrated copies of the linear plasmids within their chromosomal genomes. Correlation of host specificity of strains of *Fusarium oxysporum* f. sp. *conglutinans* with the group 2 plasmids has been reported [41].

C. Group 3 Plasmids

Group 3 plasmids (Table 3) are covalently closed circular DNA molecules showing no homology with the mitochondrial chromosome. So far they have been detected only in species of *Neurospora* and in *Cephalosporium acremonium*. They represent the closest approach in filamentous fungi to the plasmids of bacteria and it is appropriate that those isolated from *Neurospora* have received much attention from workers developing fungal-bacterial shuttle vectors for biotechnological investigations. (Since it is beyond the scope of this chapter, the attention of the interested reader is drawn to the work of Mary E. Case, University of Georgia, and Alan M. Lambowitz, Saint Louis University School of Medicine, for further information on this line of research.)

Insufficient data are so far available to judge the future significance, application, and value of the plasmid species isolated from *C. acremonium* [54]. It appears to be the plasmid species uniquely isolated from the fungal cytoplasm rather from within the intact mitochondria. This fact, which separates it from all other plasmid species described in filamentous fungi, should facilitate its manipulation in vivo, and its structural resemblance to the double-stranded ccc plasmids familiar to genetic engineering in bacteria makes it an ideal candidate for further study.

VII. CONCLUSIONS

Plasmids isolated from filamentous fungi fall into three distinct groups defined by their genetic and structural properties. It is probable that the three groups represent the diversity of their evolutionary origins. The first group, often implicated in senescence, arise from the host fungus itself through the instability of its mitochondrial genome. The resulting amplified sequences detected as autonomous plasmids indicate the presence within the plasmids of a nucleotide sequence able to function as an origin of replication. The second group of linear plasmids resemble insertion sequences but have possibly developed from an infection of the evolutionary antecedents of mitochondria with a viruslike particle. The third group most closely resemble circular bacterial plasmids and may also be descended from DNA introduced into the host strains. With the exception of that isolated from the cytoplasm of *Cephalosporium acremonium*, they have been observed only in the mitochondria of species of *Neurospora*.

REFERENCES

1. Lin, E. C. C., Goldstein, R., and Syvanen, M. *Bacteria, Plasmids, and Phages: An Introduction to Molecular Biology*. Harvard University Press, Cambridge, Mass., 1984, p. 164.
2. Stahl, U., Lemke, P. A., Tudzynski, P., Kuck, U., and Esser, K. Evidence for plasmid like DNA in a filamentous fungus, the ascomycete Podospora anserina. *Mol. Gen. Genet.*, *162*:341-342, 1978.
3. Munkres, K. D. Aging of fungi. *Rev. Biol. Res. Aging*, *2*:49-43, 1985.
4. Cummings, D. J., Belcour, L., and Grandchamp, C. Mitochondrial DNA from Podospora anserina and the occurrence of multimeric circular DNA in senescent cultures. *Proc. Alfred Benzon Symp.*, *13*:268-277, 1978.
5. Belcour, L., Begel, O., Mosse, M.-O., and Vierny, C. Mitochondrial DNA amplification in senescent cultures of Podospora anserina: variability between the retained, amplified sequences. *Curr. Genet.*, *3*:13-21, 1981.
6. Wright, R. M., Horrum, M. A., and Cummings, D. J. Are mitochondrial structural genes selectively amplified during senescence in Podospora anserina? *Cell*, *29*:309, 315, 1982.
7. Kuck, U., and Esser, K. Genetic map of mitochondrial DNA in Podospora anserina. *Curr. Genet.*, *5*:143-147, 1982.
8. Cummings, D. J., MacNeil, I. A., Domenico, J., and Matsuura, E. T. Excision-amplification of mitochondrial DNA during senescence in Podospora anserina. *J. Mol. Biol.*, *185*:659-680, 1985.
9. Esser, K., Tudzynski, P., Stahl, U., and Kuck, U. A model to explain senescence in the filamentous fungus Podospora anserina by the action of plasmid like DNA. *Mol. Gen. Genet.*, *178*:213-216, 1980.
10. Tudzynski, P., and Esser, K. Chromosomal and extrachromosomal control of senescence in the ascomycete Podospora anserina. *Mol. Gen. Genet.*, *173*:71-84, 1979.
11. Koll, F., Begel, O., Keller, A.-M., Vierny, C., and Belcour, L. Ethidium bromide rejuvenation of senescent cultures of Podospora anserina: Loss of senescence-specific DNA and recovery of normal mitochondrial DNA. *Curr. Genet.*, *8*:127-134, 1984.

12. Belcour, L., and Begel, O. Life-span and senescence in Podospora anserina: effect of mitochondrial genes and functions. *J. Gen. Microbiol.*, *119*:505-515, 1980.
13. Michel, F., and Cummings, D. J. Analysis of class I introns in a mitochondrial plasmid associated with senescence of Podospora anserina reveals extraordinary resemblance to the Tetrahymena ribosomal intron. *Curr. Genet.*, *10*:69-79, 1985.
14. Osiewacz, H. D., and Esser, K. DNA sequence analysis of the plasmid of Podospora anserina. *Curr. Genet.*, *7*:219-223, 1983.
15. Osiewacz, H. D., and Esser, K. The mitochondrial plasmid of Podospora anserina: a mobile intron of a mitochondrial gene. *Curr. Genet.*, *8*:299-305, 1984.
16. Wright, R. M., and Cummings, D. J. Transcription of a mitochondrial plasmid during senescence in Podospora anserina. *Curr. Genet.*, *7*: 457-464, 1983.
17. Tudzynski, P. DNA plasmids in eukaryotes with emphasis on mitochondria. *Progr. Bot.*, *44*:297-307, 1982.
18. Wright, R. M., and Cummings, D. J. Integration of mitochondrial gene sequences within the nuclear genome during senescence in a fungus. *Nature*, *302*:86-88, 1983.
19. Koll, F. Does nuclear integration of mitochondrial sequences occur during senescence in Podospora? *Nature*, *324*:297-599, 1986.
20. Bockelman, B., and Esser, K. Plasmids of mitochondrial origin in senescent mycelia of Podospora curvicolla. *Curr. Genet.*, *10*:803-810, 1986.
21. Bertrand, H., and Pittenger, T. H. Isolation and classification of extranuclear mutants in Neurospora crassa. *Genetics*, *71*:521-533, 1972.
22. Caten, C. E. Vegetative incompatibility and cytoplasmic infection in fungi. *J. Gen. Microbiol.*, *72*:221-229, 1972.
23. Bertrand, H., Collins, R. A., Stohl, L. L., Goewert, R. R., and Lambowitz, A. M. Deletion mutants of Neurospora crassa mitochondrial DNA and their relationship to the "stop-start" growth phenotype. *Proc. Natl. Acad. Sci. USA*, *77*:6032-6036, 1980.
24. de Vries, H., de Jonge, J. C., van't Sant, P., Agsteribbe, E., and Arnberg, A. A "stopper" mutant of Neurospora crassa containing two populations of aberrant mitochondrial DNA. *Curr. Genet.*, *3*:205-211, 1981.
25. Manella, C. A., Goewert, R. R., and Lambowitz, A. M. Characterization of variant Neurospora crassa mitochondrial DNAs which contain tandem reiterations. *Cell*, *18*:1197-1207, 1979.
26. Collins, R. A., and Lambowitz, A. M. Characterisation of variant Neurospora crassa mitochondrial DNA which contains tandem reiterations of a 1.9 kb sequence. *Curr. Genet.*, *4*:131-133, 1981.
27. Collins, R. A., Stohl, L. L., Cole, M. D., and Lambowitz, A. M. Characterization of a novel plasmid DNA found in mitochondria of N. crassa. *Cell*, *24*:443-452, 1981.
28. Nargang, F. E., Bell, J. B., Stohl, L. L., and Lambowitz, A. M. A family of repetitive palindromic sequences found in Neurospora mitochondrial DNA is also found in a mitochondrial plasmid DNA. *J. Biol. Chem.* *258*:4257-4260, 1983.
29. Collins, R. A., and Lambowitz, A. M. Structural variations and optional introns in the mitochondrial DNAs of Neurospora strains isolated from nature. *Plasmid*, *9*:53-70, 1983.

30. Stohl, L. L., Collins, R. A., Cole, M. D., and Lambowitz, A. M. Characterization of two new plasmid DNAs found in wild-type Neurospora intermedia strains. *Nuc. Acids Res.*, *10*:1439-1458, 1982.
31. Nargang, F. E., Bell, J. B., Stohl, L. L., and Lambowitz, A. M. The DNA sequence and genetic organization of a Neurospora mitochondrial plasmid suggest a relationship to introns and mobile elements. *Cell*, *38*:441-453, 1984.
32. Sor, F., and Fukuhara, H. Nature of an inserted sequence in the mitochondrial gene coding for the 15S ribosomal RNA of yeast. *Nuc. Acids Res.*, *10*:1625-1633, 1982.
33. Yin, S., Heckman, J., and RajBhandary, U. L. Highly conserved GC-rich palindromic DNA sequences flank tRNA genes in Neurospora crassa mitochondria. *Cell*, *26*:325-332, 1981.
34. Nargang, F. E. Conservation of a long open reading frame in two Neurospora mitochondrial plasmids. *Mol. Biol. Evol.*, *3*:19-28, 1986.
35. Natvig, D. O., May, G., and Taylor, J. W. Distribution and evolutionary significance of mitochondrial plasmids in Neurospora spp. *J. Bacteriol.*, *159*:288-293, 1984.
36. Taylor, J. W., Smolich, B. D., and May, G. In evolutionary comparison of homologous mitochondrial plasmid DNAs from three Neurospora species. *Mol. Gen. Genet.*, *201*:161-167, 1985.
37. Martini, G., Grimaldi, G., and Guardiola, J. Extrachromosomal DNA in phytopathogenic fungi. In *Genetic Engineering* (H. W. Boyer and S. Nicosia, eds.), Elsevier, Amsterdam, 1978, pp. 187-200.
38. Guardiola, J., Grimaldi, G., Constantino, P., Micheli, G., and Cervone, F. Loss of nitrofuran resistance in Fusarium oxysporum is correlated with loss of a 46.7 kb circular DNA molecule. *J. Gen. Microbiol.*, *128*:2235-2242, 1982.
39. Marriott, A. C., Archer, S. A., and Buck, K. W. Mitochondrial DNA in Fusarium oxysporum is a 46.5 kilobase pair circular molecule. *J. Gen. Microbiol.*, *130*:3001-3008, 1984.
40. Kistler, H. C., and Leong, S. A. Linear plasmid-like DNA in Fusarium oxysporum. *J. Cell Biochem.*, *Supp. 9C*:170, 1985.
41. Kistler, H. C., and Leong, S. A. Linear plasmid-like DNA in the plant pathogenic fungus Fusarium oxysporum f. sp. conglutinans. *J. Bacteriol.*, *167*:587-593, 1986.
42. Samac, D. A., and Leong, S. A. Plasmid-like DNAs from Fusarium solani. *J. Cell. Biochem.*, *Supp. 10C*:27, 1986.
43. Daboussi-Bareyre, M. J., and Parisot, D. Nucleocytoplasmic interactions implicated in differentiation in Nectria haematococca. In *Fusarium: Diseases, Biology, and Toxonomy* (P. E. Nelson, T. A. Tousson, and R. J. Cook, eds.), Pennsylvania State University Press, University Park, 1981, pp. 306-317.
44. Rubidge, T. Survey of Fusarium species for plasmid-like DNA and some evidence for its occurrence in a strain of F. merismoides. *Trans. Br. Mycol. Soc.*, *87*:463-466, 1986.
45. Hashiba, T., Homma, Y., Hyakumachi, M., and Matsuda, I. Isolation of a DNA plasmid in the fungus Rhizoctonia solani. *J. Gen. Microbiol.*, *130*:2067-2070, 1984.
46. Tudzynski, P., Duvell, A., and Esser, K. Extrachromosomal genetics of Claviceps purpurea. I. Mitochondrial DNA and mitochondrial plasmids. *Curr. Genet.*, *7*:145-150, 1983.

47. Tudzynski, P., and Esser, K. Extrachromosomal genetics of Claviceps purpurea. II. Plasmids in various wild strains and integrated sequences in mitochondrial genomic DNA. *Curr. Genet.*, *10*:463-467, 1986.
48. Francou, F. Isolation and characterization of a linear DNA molecule in the fungus Ascobolus immersus. *Mol. Gen. Genet.*, *184*:440-444, 1981.
49. Horgen, P. A., Hohan, M., Meyer, R. J., and Anderson, J. B. Plasmid-like DNAs in mushrooms. *J. Cell Biochem.*, *Suppl. 9C*:170, 1985.
50. Mohan, M., Meyer, R. J., Anderson, J. B., and Horgen, R. A. Plasmid-like DNAs in the commercially important mushroom genus Agaricus. *Curr. Genet.*, *8*:615-619, 1984.
51. Meinhardt, F., and Esser, K. Linear extrachromosomal DNA in the morel Morchella conica. *Curr. Genet.*, *8*:15-18, 1984.
52. Lazarus, C. M., Earl, A. J., Turner, G., and Kuntzel, H. Amplification of a mitochondrial DNA sequence in the cytoplasmically inherited "ragged" mutant of Aspergillus amstelodami. *Eur. J. Biochem.*, *106*:633-641, 1982.
53. Lazarus, C. M., and Kuntzel, H. Anatomy of amplified mitochondrial DNA in "ragged" mutants of Aspergillus amstelodami: excision points within protein genes and a common 215 bp segment containing a possible origin of replication. *Curr. Genet.*, *4*:99-107, 1981.
54. Minuth, W., Tudzynski, P., and Esser, K. Extrachromosomal genetics of Cephalosporium acremonium. *Curr. Genet.*, *5*:227-231, 1982.
55. Garber, R. C., and Yoder, O. C. Isolation of DNA from filamentous fungi and separation into nuclear, mitochondrial, ribosomal, and plasmid components. *Anal. Biochem.*, *135*:416-422, 1983.
56. Garber, R. C., Turgeon, B. G., and Yoder, O. C. A mitochondrial plasmid from the plant pathogenic fungus Cochliobolus heterostrophus. *Mol. Gen. Genet.*, *196*:301-310, 1984.
57. Garber, R. C., and Yoder, O. C. Mitochondrial DNA of the filamentous ascomycete Cochliobolus heterostrophus. *Curr. Genet.*, *8*:621-628, 1984.
58. Tudzynski, P., and Esser, K. Mitochondrial DNA for gene cloning in eukaryotes. In *Gene Manipulations in Fungi* (J. W. Bennett and L. L. Lasure, eds.), Academic Press, New York, 1985, pp. 403-416.
59. Pring, D. R., and Lonsdale, D. M. Molecular biology of higher plant mitochondrial DNA. *Int. Rev. Cytol.*, *97*:1-46, 1985.
60. Smart, J. E., and Stillman, B. W. Adenovirus terminal protein precursor: partial amino acid sequence and the site of covalent linkage to virus DNA. *J. Biol. Chem.*, *257*:13499-13506, 1982.
61. Salas, M. A new mechanism for the initiation of replication of Φ29 and adenovirus DNA: priming by the terminal protein. *Curr. Top. Microbiol. Immunol.*, *109*:89-96, 1983.
62. Garcia, E., Gomez, A., Ronda, C., Escarmis, C., and Lopez, R. Pneumococcal bacteriophage Cp-1 contains a protein tightly bound to the 5' termini of its DNA. *Virology*, *128*:92-104, 1983.
63. Kim, B. D., Mans, R. J., Conde, M. F., Pring, D. R., and Levings, C. S. Physical mapping of homologous segments of mitochondrial episomes from S male-sterile maize. *Plasmid*, *7*:1-4, 1982.
64. Decaris, B. Intragenic location of an unstable insertion element with gene b5 of the fungus Ascobolus immersus. *Mol. Gen. Genet.*, *184*:434-439, 1981.

Part II

COMMERCIAL APPLICATIONS

11

THE BIOCHEMISTRY OF SOME FUNGAL POLYSACCHARIDES WITH INDUSTRIAL POTENTIAL

BRIAN J. CATLEY *Heriot-Watt University, Edinburgh, Scotland*

I. INTRODUCTION

The three fungal polysaccharides reviewed are pullulan, scleroglucan (also known as schizophyllan and by a number of trade names), and a complex β-D-glucan. The first two are soluble and secreted by the cell. They have been selected because they are examples of a small but expanding group of polysaccharides that are useful in industry, the commercial sector, and medicine [1-4]. The majority of polysaccharides that find application outside of the laboratory are of plant or bacterial origin. An advantage of using fungi for the production of polymers is their ability to grow at low pH thus severely limiting most bacterial contaminations. On a large scale this is useful as growth conditions do not have to be rigorously aseptic.

What are the properties of polysaccharides that can be exploited? To answer that, it is necessary to know something of their molecular structure and the way in which they interact with water. They have been produced by cells, not for the benefit of industry but for the protection of the cytoplasmic membrane or to enable cells to adhere to one another or selected surfaces. The immobilization of water creates gels or viscous slimes that prevent the dehydration of cells and participate in the selection of molecules passing from the environment to transport proteins on the external face of the cytoplasmic membrane. To create gels requires the formation of a molecular network within which there reside domains of temporarily immobilized clusters of water molecules. That suggests that some sort of crosslinking takes place. The subtlety of natural polysaccharide networks is that components are not permanently associated with one another. Unlike the rigid covalent crosslinking produced in polyacrylamide or Sephadex gels which, once formed, are relatively chemically stable and of fixed dimensions, the natural gels can be manipulated by the cell or can be disrupted by physical forces such as shearing. Thus, given some control of the enzyme content of a cell, there is the prospect of being able to tailor these molecules to produce interesting and useful properties.

An understanding of the formation of these networks lies in appreciating why individual polysaccharide chains adopt particular shapes or conformations and the factors that determine whether they associate with water, thereby becoming soluble, or with themselves or other polymers, thus becoming insoluble. It is a curious property of some polysaccharides that they are capable of both types of association within the same molecule. Like proteins they have domains where a particular folding furnishes a particular shape and so a particular property. Folding is orchestrated by the shape and chemistry of neighboring sugar residues in the chain and the glycosidic bond between them. Like proteins, the primary structure dictates secondary, tertiary, and quaternary folding and association patterns.

Unlike proteins, however, polysaccharides have simpler primary structures. They may comprise but one type of residue and linkage. Many are composed of repeating blocks of disaccharide; occasionally they are larger. Regular sequences produce regular shapes or conformations. These may be helical. An extreme case of this category is an extended ziz-zag ribbon, a sort of two-dimensional molecule which could be thought of as a fully extended coil. Examples of helical and ribbonlike molecules are, respectively, amylose and cellulose. Both are glucans, both are linked 1,4-. The difference in the configuration of the bond at the 1 position, α- or β-, respectively, produces different sets of possible folding patterns leading to the two distinct conformations. Occasionally, the ribbon is regularly buckled. They are often charged anionic polysaccharides associated with divalent cations. Examples of these are the pectins and alginates of plant origin. Conformation may be neither helical nor ribbon but a so-called random coil. This shape is adopted largely through an inability to form physical bonds with neighboring sugar residues or because of a more than usual flexible glycosidic bond between them. An example of the latter is the 1,6 bond between pyranose rings.

In many circumstances helical and ribbon structures can line up with one another and form interchain physical bonds that maintain the association. And with that comes the possibility of network formation and trapping of water. Two neighboring molecules may align ordered portions of their structure to produce an ordered junction zone. In areas of polymer where the conformation is irregular association cannot occur. Alternation of ordered and disordered sections of polymers allow a series of junction zones to assemble, thus creating a network of cavities within which the water molecules are trapped. Depending on the nature of the junction zones they are more or less easily parted by shear forces in their neighborhood. Destruction of the network releases immobilized water and reduces the viscosity of the solution. It is the extent and strength of crosslinks that confer rheological properties on the solution of polymer.

More detailed and rigorous explanations of carbohydrate conformations, physical bonds, and water structure can be found in reviews by Morris [5] or Rees [6]. The two soluble polysaccharides described in this chapter, pullulan and scleroglucan, do not, as far as can be seen, have structures that could obviously generate junction zones. Nevertheless they produce viscous solutions, exceptionally so in the case of scleroglucan, and exhibit shear-thinning characteristics. They are homopolymers of apparently uniform structure with a repeating trisaccharide (pullulan) or tetrasaccharide unit (scleroglucan). As will be seen, they have quite different abilities of association and must have, though somewhat obscurely, the ability to create intermolecular networks and immobilize water.

II. PULLULAN

A. Introduction

Pullulan (Fig. 1) was first isolated and characterized by Bender et al. [7] in 1958. Very soon it became a substrate for the assay of α-1,6-glucanases and, as most α-amylases could not attack the α-1,4-glucosidic bonds, it was useful when these enzymes were present as contaminants. It has remained of great value in research. However, more recently, pullulan has been examined with regard to its industrial potential and the Hayashibara Biochemical Laboratories Inc. (Japan) have done much to stimulate this exploration.

Most of the research conducted on pullulan between 1958 and 1978 has been reviewed [8]. This section of the chapter summarizes reports appearing in the last decade.

B. Structure

There is reasonable evidence in the literature to support the periodic structure shown in Figure 1. The determination of the structure leading to the polymaltotriosyl model has been reviewed [8,9]. The discovery of amylases that specifically cleave the two α-1,4-glucosidic bonds [10] has confirmed this and also provided elegant means of producing panose (6^2-α-glucosyl maltose) and isopanose (6^1-α-maltosyl glucose). The occasional maltotetraosyl residue that replaces the maltotriosyl appears to be randomly distributed throughout the polymer [11]. The susceptibility of pullulan at these replacement sites to attack by amylases, normally unable to cleave the α-1,4 bonds, makes it important to know their frequency of occurrence as viscosity or molecular weight will be affected. The observation [9] that an exoglucosidase capable of completely hydrolyzing glycogen, and therefore capable of cleaving α-1,4 and α-1,6 bonds, could not completely hydrolyze pullulan raised a question as to whether the linear regular structure was sufficient to describe all of the molecule. Absence of complex formation with cetyltrimethylammonium hydroxide and lack of binding to DEAE-Sephadex, both indications of the presence of anionic polysaccharides, ruled out blocking groups such as phospho- or other polybasic acid esters. The possibility remains that there are infrequent branched domains within the molecule that are rare enough to escape present analyses and yet are of sufficiently different configuration to resist the action of glucosidases.

There is a report [12] that pullulan may be terminally attached to a nonpullulan like glucan of degree of polymerization between 52 and 125. It has also been suggested [13] that the polymaltotriosyl structure is covalently bound to a β-1,3-glucan moiety raising the possibility that in some yeasts there may be covalent bonding with the cell wall architecture.

The interest in polymers lies in their higher orders of structure as well as in the primary sequence which dictates folding patterns. The prediction of pullulan configuration in aqueous solutions has been attempted, but the approximations and assumptions required for the model-building procedures [14,15] means that results are still uncertain. Using alternative approaches the polymer's molecular characteristics were measured by light scattering, viscometry, and gel permeation chromatography [16]. From these data the Mark-Houwink Sakurada viscosity equation was determined

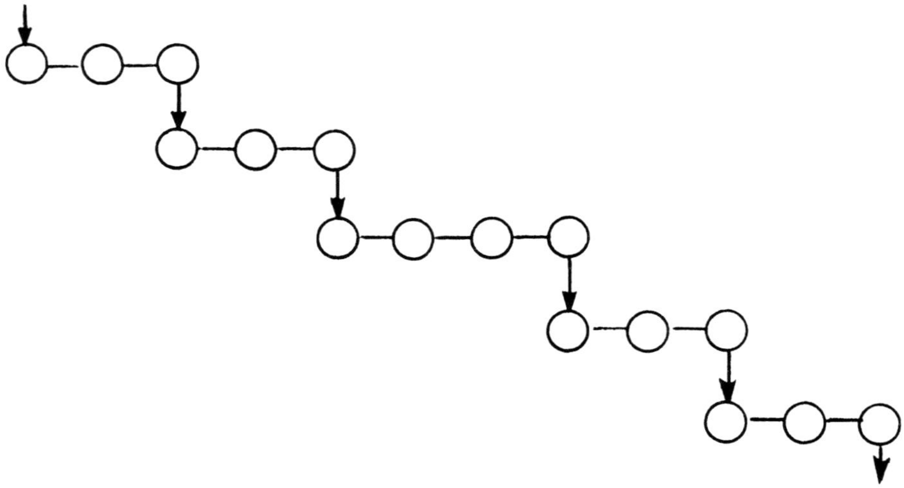

FIGURE 1. Pullulan. A section of pullulan structure including the occasional maltotetraosyl moiety. D-Glucosyl residues are represented by O, 1,4-α-D-glucosidic bonds by —, and 1,6-α-D-glucosidic bonds by ↓.

for polysaccharides in the molecular weight range 48×10^3 to 2.2×10^6 and the general conclusion reached that in aqueous solution pullulan behaved like a flexible chain.

The molecular size of pullulan had been shown to vary [17], the probable reason being attributed to postpolymerization cleavage of the polymer by amylases during the course of fermentation. In further support of change in polymer size it was found [18] that shear-dependent viscosity dropped at the end of fermentation runs when pullulan concentrations were at their highest. Toward the beginning of fermentation the molecular weight was in the region of 2×10^6 but fell to about a twentieth of that at the end. Molecular weight was also effected by the speed of the fermenter impeller which was thought to be one of the factors influencing the morphological form of the organism. Clearly it is important, as the authors observe, to separate the influences of morphology and product concentration on broth rheology.

C. Assay

A survey of the literature reveals that the assays used for the determination of pullulan vary considerably in their specificity and therefore in their accuracy. Polymaltotriose is not the only polysaccharide elaborated by

A. pullulans. The nature of these polymers is described elsewhere in this chapter but their composition is such that they may be included in some estimations. Precipitation with alcohol is usually the first step and is necessary to remove low molecular weight sugars present in the culture. Weighing the dried precipitate or using a general assay for carbohydrate such as the phenol-sulfuric acid procedure [19] will produce overestimates as non-pullulan polysaccharides cannot be distinguished.

Another procedure [20] especially useful in measuring the cellular capacity to produce pullulan is to include [^{14}C]glucose in the culture medium whence labeled polymer is secreted. Distinction between unassimilated low molecular weight and elaborated high molecular weight label is made by drying samples on small squares of filter paper and extracting them with 66% aqueous ethanol. Extracted papers, the low molecular weight label having been removed, are now counted by liquid scintillation procedures. However, it is an assay for labeled secreted polymer which may not be pullulan.

The only specific assay for pullulan is to first hydrolyze the polymer with pullulanase [20] and then both measure released oligosaccharides and show that the product is maltotriose. A neat, quick, and simplified microassay extending the above labeling protocol but combining the necessary treatment with pullulanase was reported by Finkelman and Vardanis [21]. Only 10 μl of cell suspension was needed.

A rapid but nonspecific method that has been used to monitor shake cultures for pullulan elaboration employed turbidity [22]. Suspensions of culture were shaken with equal volumes of ethanol and the density of suspended precipitate used as a measure of polymer concentration.

Pullulan can now be measured rapidly and conveniently in situ by a probe. The combination of coimmobilized glucoamylase and glucose oxidase attached to the tip of a modified oxygen electrode together with prior hydrolysis with pullulanase allowed concentrations of pullulan in the range of 0.001-0.01% to be measured within 90 s [23].

D. Industrial Potential

The practicality of pullulan as a biodegradable material that can be manufactured in a variety of forms was reviewed by Yuen [24]. Film requiring no stabilizing plasticizer can be cast. Possessing excellent physical properties, e.g., impermeability to oxygen, without smell or taste, nontoxic and heat-sealable, it has potential for the packaging of food and drugs. Water solubility is both an advantage and disadvantage but its susceptibility to biodegradation commends it to the environmental lobby and as a plastic produced from renewable resources it has an advantage over petroleum-derived products. Cast as a fiber it has a shining gloss similar to that of rayon and a tensile strength on the same order as that of nylon. Pullulan fibers can be mixed with natural fibers to produce specialized papers. It can be used in the manufacture of adhesive.

In solution it has been used as a flocculating agent in the separation of clay slimes [25], and both chemical derivatives and oligosaccharides produced by controlled hydrolysis are finding increasing commercial use [26]. It is, however, not yet in the same league as other mass-produced microbial polysaccharides, e.g., xanthan. Nevertheless its potential has influenced research in recent years and much is now known about the nutritional regimes and fermentation conditions that promote a satisfactory yield of polymer from readily available carbon sources.

E. Factors Affecting Production

Initial investigations [27] into the rate of elaboration of pullulan at different points in the growth cycle suggested a relationship between morphological form, yeast or mycelial, and polysaccharide production. A subsequent investigation [28] in which the two forms, having been separated by filtration through nylon mesh, were individually assayed showed that there were considerable but not absolute differences. At all stages of the growth cycle the yeast form was up to five times the better producer. It was also clear that the rates for each form varied from point to point in the growth cycle. Clearly, it is important to know what environmental factors determine the morphological form adopted by the organism. These can be separated into nutritional and nonnutritional factors.

1. Nutritional Factors

The effect of a variety of nitrogen sources, e.g., urea, amino acids, ammonium, and nitrate ion, was studied by Park [29-31]. No simple correlation could be found and the Nickerson model, in which the presence of thiol-containing compounds prevents the formation of mycelia and promotes yeast growth, may not be tenable. However, what has emerged from these studies, and was apparent in other conversions, is that cell density may play an important role in favoring one form or another. Low cellular concentrations promote mycelial development [28] but above a threshold there is a simple relationship between cell density and yeast development. The threshold can be influenced by nutritional factors, particularly nitrogen sources. Moreover, the influence of inorganic nitrogen is not on the cell proximity effect itself but on the tolerance of cells to it.

That microbial population in some way controls its own morphology suggests the production of a morphogen. No substance has been found and indeed the effect may be due not to production but to removal of some substance. It is interesting to note that a similar transition in *Candida albicans* is effected by ethanol [32] and that ethanol is both produced and consumed by mycelial and yeast cells of *A. pullulans* [33], respectively.

The idea that cells are responsive to a regulator was extended by Gadd and Cooper to include commitment [34,35]. A yeast-to-mycelial transition could be induced by yeast extract. Removal of extract, however, did not halt the conversion. Moreover, progeny of the mycelial cells did not germinate even when innoculated into fresh medium. The same authors provide a word of caution. Considerable variation occurs on solid medium between different strains and observations should be regarded as strain-specific.

2. Nonnutritional Factors

One of the major influences on the balance between yeast and hyphal growth, and therefore on the production of pullulan, is environmental pH. In the range 3-6 the effect of pH is not to change the total amount of biomass but to alter the proportions of the two forms produced during the initial growth period [36]. This balance is determined early in the fermentation and certainly before commencement of exponential growth.

Subsequent alterations of pH do not seem to alter the proportions. The importance of initial pH had already been noted by Ono et al. [37] and described previously [38]. The lower the pH, the smaller was the proportion of yeast cells. Thus, at pH 3.5 about 15% of the organism was single cells whereas at pH 6.3 the proportion had risen to 63%. The rates of pullulan accumulation were in line with this, the more productive being in the more alkaline medium [37].

The results of Heald and Kristiansen [36] seemed at odds with earlier studies on the effect of pH on pullulan elaboration [38]. In studying the influence of environmental acidity a distinction must be made between the effect of pH on the morphological balance and the ability of the yeast to produce pullulan. Early studies had shown that an acid pH favored production and that an adjustment from pH 5 to 7.5 was sufficient to inhibit elaboration. This would seem to be counteracted by the more acidic environment decreasing the proportion of yeast. Reexamination of the effect of pH on the ability of the yeast cell to produce polysaccharide, but over a longer period, resolved this paradox [39]. Cells were now examined over a much longer period of time (4 h). As before there was an initial effect. Cells maintained at pH 7.5 did not elaborate whereas those at 5.0 produced. However, after 90 min pullulan began to appear in the alkaline medium and the rate of accumulation rose to match that of cells in the more acidic milieu. Cells rapidly adjusted after the first hour from pH 7.5 to 5.0 did not show an immediate ability to elaborate polysaccharide but again exhibited a lag period. In contrast, cells elaborating pullulan at pH 4 that were shifted to pH 8 showed immediate cessation of polysaccharide synthesis followed once more by a period of about 1 h before accumulation was restored to the previous rate. The assimilation rate of glucose by cells was not affected by these pH values from which it was presumed that proton concentration in the pH range 7.5-5.0 had no effect on transport processes. Thus, although there may be short-term (1-2 h) effects of pH on individual cells, there seems to be little or no effect in the long term in fermentation runs of a few days and the proportion of morphological forms is the important factor.

Bermejo et al. [40] also studied the control on morphology exerted by proton concentration. Providing that the initial pH was about 6, growth in a glucose medium was limited by nitrogen, and buffer capacity was low, there was a proliferation of chlamydospores. Increasing buffer capacity or maintaining the pH at 6 inhibited this transition. Considerable changes in the chemical composition of the cell walls were observed during the change and claims were made that an outer layer contained pullulan but the analytical techniques used were not sufficient to justify this [41].

At very low values of pH, about 2, cells produce little or no pullulan. Nevertheless growth at this acidity is reasonable. This observation led [42] to investigations of polysaccharide production in bistaged fermentations. Initially growth was conducted at pH 2 and biomass allowed to accumulate in the absence of pullulan elaboration. The pH was subsequently raised to the 4-5 region when polysaccharide accumulation commenced. With sucrose as carbon source the maximum polysaccharide concentration was increased from 17 g/liter obtained in the control unstaged to 27 g/liter in the staged fermentation. Comparisons of controlled, low-pH, first-stage cultures and those naturally obtained with a marginally lower acidity showed little difference in final yields. An advantage of this procedure is that the initial acidic stage could be operated under nonaseptic conditions, since most bacteria cannot grow under these circumstances. In addition, the high concentration of cells in the second stage reduced the effect of bacterial contamination.

Another consequence of the bistaged program is that, judging by the rheological properties of the solution, the molecular weight of pullulan is low.

It is apparent that the acidity of the culture medium profoundly influences the morphology of *A. pullulans*. The mechanism by which protons effect changes is poorly understood. It would appear that changes in external pH are in some way transmitted to the cytoplasm. There is now [43] experimental evidence for this arising from investigations of *C. albicans*. Variation in the external pH of the cell from 4.5 to 6.5 can produce up to a 40-fold change in the cytoplasmic proton concentration of some strains. This, in turn, is thought to control the concentration of key metabolic messengers such as cyclic AMP and calcium ion.

Whereas most investigations of *A. pullulans* morphology and pullulan production have been made in shake culture, the stirred fermentor is perhaps more appropriate. McNeil and Kristiansen [18] examined the effect of impeller speed on biomass, polysaccharide elaboration, morphology, and shear-dependent viscosity. Sitrring speed influenced morphology and pullulan concentration rather than biomass accumulation. This was most likely due to the level of dissolved oxygen which is known to have a considerable effect on morphology. The greater the concentration of oxygen, it was observed, the better the yield of pullulan. The shear-dependent viscosity dropped toward the end of fermentation, a decrease due to a reduction in the molecular weight of pullulan from 2×10^6 to 10^5.

Very few investigations of attempts to inhibit specific metabolic pathways that might be expected to alter the distribution of assimilated carbohdyrate have been reported. A possible effect of ammonium ion on glycolysis was already described. Attenuation of the respiratory pathway through mutation with ethidium bromide produced "yeasty" mutants with low levels of cytochrome proteins [44]. A more recent attempt is that by Finkelman and Vardanis [45] who used fluoroacetate to inhibit metabolic activity of the tricarboxylic acid cycle. At a concentration of 1 mM the yield of pullulan increased by a factor of 1.10 and rose to 1.215 at 10 mM. Glucose assimilation was reduced by 75%. Moreover, the proportion of pullulan to other polysaccharide was increased from about 75% to almost 90%, suggesting that pullulan and the other polymer(s) are controlled by different cellular parameters.

F. Nutrition of *A. pullulans*

1. Nitrogen Sources

The effect of initial ammonium ion concentration in batch culture, first reported by Catley [38], has been reexamined. Exopolysaccharide production did not commence before removal of ammonium ions from the medium [46]. This was confirmed by continuous culture studies [39]. Increased ammonium ion concentration enhanced carbon flow into biomass with a corresponding decrease in polysaccharide levels. At 2.5 g/liter polymer synthesis was almost negligible. The precise effect of ammonium ion was not clear. The data of Bulmer et al. [39] support the notion that pullulan is a secondary metabolite. Reports on the effect of ammonium ion on catabolic pathways in yeast agree on the influence but differ in the proposed mechanism. Rothman and Cabib [47] suggested that in *Saccharomyces cerevisiae* there was a stimulation of glycolysis through activation of phosphofructokinase and that loss of this enhancement through removal of ammonium ion allowed glucose to be directed toward polysaccharide production, in this case glycogen. Saita and Slaughter [48] proposed an alternative, namely, that stimulation was of protein synthesis. A combination of both these events was proposed by Lloyd et al. [49].

It is clear that the effect of ammonium ion on pullulan elaboration is complex. Examination of the incorporation of assimilated glucose into cellular material and secretion of pullulan in its absence and presence [39] showed that for both yeast and mycelial cells over a period of 6 h there was no change in extracellular polysaccharide production, but in the presence of nitrogen source there was a large increase in the diversion of glucose into cellular mass. One explanation of this is that in the presence of the ion old cells continue to secrete pullulan but new cells formed in the period of labeling do not. Cycloheximide, an inhibitor of protein synthesis, had no effect on elaboration over the first 2 h but synthesis ceased after 3 h as did the ability of the cells to utilize glucose.

2. Carbon Sources

Yields of pullulan from mono- or disaccharides can reach 30%, but these figures are modest by commercial standards. Conversions of 50% were attainable using starch syrups [24], i.e., mixtures of α-1,4/1,6-linked oligosaccharides. Moreover, the cheap and available sources of these substrates, hydrolysates of starches from corn, maize, or potato, made them all the more attractive. By selection of strains and optimization of substrate concentration yields of up to 75% were obtained. The extent of starch hydrolysis had a strong influence on the overall yield. Too high a proportion of mono- or disaccharides or large oligosaccharides arising from insufficient hydrolysis led to a reduction in pullulan production.

Although carbon substrates used in industrial fermentations are usually starch hydrolysates, alternatives have been sought. Peat hydrolysate [50] is the liquid phase resulting from the autoclaving of untreated peat in dilute acid, e.g., 1.5% H_2SO_4. It contains up to 5% total carbohydrate comprising mainly glucose, xylose, galactose, rhamnose, mannose, and arabinose. Comparative studies using an equivalent amount of glucose showed very similar conversion rates. After 7 days the maximum polysaccharide concentration was about 14 g/liter. Different strains of *A. pullulans* produced different fungal physiologies when grown using peat substrate, but there seemed to be little change in biomass or polysaccharide production. Later studies [51] showed that additions of $(NH_4)_2SO_4$ and K_2HPO_4 as sources of nitrogen and phosphorus were not necessary, there being a sufficient amount of these in the peat extract. 0.05% NaCl and 0.02% $MgSO_4$ together with antifoam were the only additions necessary. The initial pH of the peat medium was adjusted to 6 with $Ca(OH)_2$ and the yield was again in the region of 12-14 g/liter. Antifoam, required in large-scale fermentation, was found to reduce pullulan accumulation. Thus at 0.1% concentration the product fell by as much as one-third. Moreover, the scale of inhibition varied between antifoams. Although the yields of polysaccharide from peat do not compare with those from starches, the very low cost of peat hydrolysate, a waste product from the manufacture of peat briquettes, makes it an attractive proposition as a carbon source. Examination of fermentation parameters [52] demonstrated that if commencement was started as a simple batch growth and was then followed by a feed-batch regime, then overall productivity was significantly better that simple batch or continuous fermentations.

The effect of raising glucose concentrations in shake cultures to 0.5 M (9.1%) had been shown to reduce the yield of pullulan [20]. This observation was confirmed and extended using feed-batch fermentations where the inhibitory effect of high carbohydrate concentration was overcome and yields of 58 g/liter were obtained from 10% sucrose. Most of the sucrose was hydrolyzed by secreted invertase before entering the cell, all of the glucose being assimilated before the fructose was taken up [20,53].

Another waste product that might provide a suitable source of carbohydrate for pullulan production is lactose. It had been thought that *A. pullulans* did not possess the necessary constitutive enzymes to utilize the disaccharide. However, adaption by multiple transfer produced an organism that was able to grow [54]. After eight transfers, three strains were isolated that showed considerable enhancement of biomass. Two of these exhibited modest pullulan production but the third showed an improvement of more than 200-fold producing final concentrations of exopolysaccharide of 10.5 g/liter. Mixed cultures of *Ceratocystis ulmi* and the adapted *A. pullulans* increased this level by 50%, an improvement presumably due to a β-glucosidase secreted by *C. ulmi* that hydrolyzed lactose.

G. Genetic Content

Other lines of investigation into factors diverting assimilated sugar to pullulan have been into the relationship between genetic content, the ploidy, and polysaccharide yield. In general, polyploid strains, produced by the exposure of haploid strains to colchicine, are more efficient. The greatest synthetic activity [55] per unit of carbon source assimilated and biomass produced was a diploid strain providing 75% more than the haploid. Pullulan secreted per unit area of cell surface also increased. Thus, compared with the parent haploid, elaboration from the diploid was again some 1.5- to 1.8-fold greater. A tetraploid produced in the same investigation increased elaboration by 3.4-fold. Polysaccharide produced from the maximum yielding diploid strain was 12.6 g/liter compared to 7.2 g/liter. By contrast the biomass was little changed. Respiration studies [56] suggested that diploid strains not only diverted more assimilated carbohydrate to pullulan but less to oxidative metabolism.

Growth on a variety of carbon sources revealed [57] little difference in the ability of the haploid, diploid, or tetraploid strains to utilize glucose, sucrose, or maltose. In all cases maltose was the best substrate. This finding is at variance with the data of Bender et al. [7], Ueda et al. [58], and Catley [20] and indicates that strain differences may respond differently to carbon uptake. Comparisons of nitrogen sources such as urea, nitrate, and ammonium ion also revealed considerable differences. As in previous investigations, the superiority of oligosaccharides as efficient carbon sources for pullulan elaboration was demonstrated, molasses and cornflour hydrolysate being particularly effective with the strains used. However, the yields of 23 g/liter were not as high as those previously reported by Yuen [24].

One feature of the polyploid strains was their apparent unresponsiveness to environmental changes when compared to the parent haploid [59]. Changing pH was less influential. A decrease in aeration rate lowered the synthetic ability of the tetraploid but increased that of the haploid.

This may have been related to the change in mass/surface ratios. Unfortunately, there was no report of the yeast/mycelial ratio, which might have helped correlate the results.

Polysaccharide produced by polyploid strains was shown to have the basic polymaltotriosyl structure but the molecular weight varied [60], appearing to be related to the chromosome content. That derived from the haploid strain was highest and there may be a relationship between amylase content and the strain used.

Other mutagens besides colchicine have been investigated. A diploid strain was isolated after UV irradiation that secreted a little over twofold more polysaccharide than the parent haploid [61]. Methods for selecting mutant overproducers have been developed that depend on screening for heavy precipitates upon addition of ethanol to the culture broth [22].

H. Other Polysaccharides Produced by *A. pullulans*

The growth of *A. pullulans* on most carbohydrate substrates produces not only pullulan but also β-glucan(s) and anionic polysaccharides. The proportions of these depend on both the carbon source and the strain of the organism. Early work on the composition of these polysaccharides was reviewed [1]. The acidic component is usually separated from the neutral polymer by complexing with a cetyl quaternary ammonium ion, e.g., cetyltrimethylammonium bromide. Using this method Leal-Serrano et al. [62] isolated a glucan comprising 32% β-1,6 and 68% β-1,3 bonds. Acidity was conferred on the molecule by esterification of sugar residues with malic acid which accounted for about 10% by weight of the polysaccharide. Traces of phosphate were also found. The maximum yield, obtained with growth on 3% glucose, was 2.1 g/liter, considerably less than that of pullulan.

A similar acidic polysaccharide was isolated by Hamada and Tsujisaka [63] who described a structure comprising a β-1,3-D-linked backbone with single glucosyl residues attached by β-1,6 bond on every three out of four residues of the main chain. The β-1,6-glucosidic bond content was thus somewhat higher than the previous report and the acidic moiety was identified as sulfoacetic acid located on some, but not all, of the side chains.

I. Melanin

A characteristic of the growth of *A. pullulans* is the diversity of morphological forms. Of these both hyphae and chlamydospores are associated with the production of dark green, brown, or black pigments, the melanins. That this pigmentation can detach from the surface of the cell and enter the extracellular environment is of concern to those engaged in commercial production.

Melanin is a term used to describe a range of polymers possessing related but different structures and often comprising a protein moiety with a chromophore. Spectral and amino acid analyses allow rough distinctions to be made between the colored forms. These characteristics, together with the molecular weight distribution, have been shown to change during the growth and development of the organism [64].

The role of melanin in cellular function is thought to be as a protection against UV radiation, heavy-metal toxicity, and enzymic attack. It may also participate in the adherence of fungal cells to the root cells of plants.

Cellular melanin production appears at the cessation of exponential growth [65] and is thus like a secondary metabolite. Production was concomitant with the transition of swollen cells into true chlamydospores and a rapid darkening of the culture occurred. The release of melanin from cells confers a cloudiness on the culture supernatant, and observing the appearance of protein in the culture broth before pigment appeared, suggested that it may be a melanin precursor. The granular nature of some of the extracellular melanin was large enough for retention by a 0.45-μm-diameter pore membrane.

III. SCLEROGLUCAN

A. Introduction

Scleroglucan, a neutral polysaccharide comprising a linear chain of β-1,3-linked D-glucosyl residues with a single β-1,6-linked D-glucosyl residue attached to every third sugar of the backbone (Fig. 2), belongs to a class of polysaccharides secreted by a number of species of plant pathogenic fungi. Of particular interest, it is the polysaccharides from *Sclerotium glucanicum* and *Sclerotium rolfsii* that have been widely studied for their commercial potential, particularly the latter, which was extensively examined in the context of oil recovery. The same polysaccharide was also identified as a component of the fungal wall of *Schizophyllum commune* and for that reason is often referred to as schizophyllan.

B. Structure

Unlike pullulan, the molecular orientation in aqueous solution assumes a more regular conformation. Like the related bacterial polysaccharide curdlan, the structure of which is the simpler β-1,3-D-glucan, it self-associates through physical bonds between individual chains to produce a triple helix [66]. For curdlan, the association does not stop there and microfibrils comprising many chains of oriented glucan molecules are formed and gradually precipitate from solution. The protruding β-1,6-linked residues of the scleroglucan prevent intermolecular association beyond the stage of the triple helix and thus prevent the formation of microfibrils and precipitation. Nevertheless the stiffened linear structure confers viscosity on its aqueous solution and it is the particular nature of this viscosity, its pseudoplasticity or shear thinning, that attracts commercial attention. Dilute solutions are exceptionally viscous. Concentrations of 0.1-0.2% in water can be used to stabilize 5-10% aqueous suspensions of powders such as zinc oxide [67] and there are numerous other applications in the food, industrial, and pharmaceutical industries. One of the more recent uses is in the thickening of drilling muds and its role in enhanced oil recovery. These will be discussed in the next section.

Dissociation of the triple helix is thought to occur at pH >12 where a sharp reduction in viscosity is observed. Concurrent measurements of optical rotation demonstrated a conformational change [67]. The viscosity change was irreversible and suggested that below pH 12 the triple helix remained intact and the molecule adopted an extended conformation. At higher concentrations of alkali the association broke down and individual

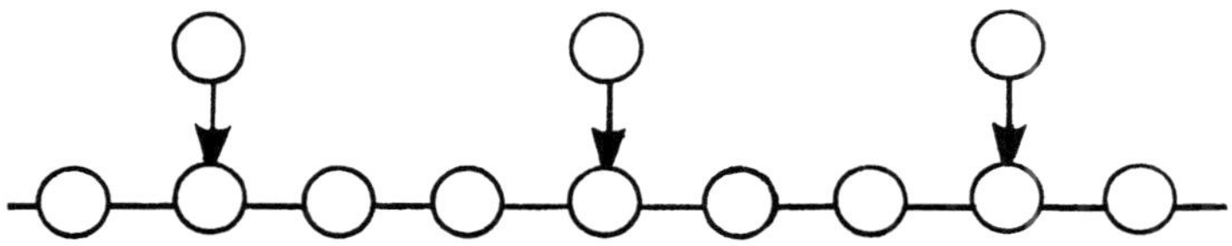

FIGURE 2. Scleroglucan. A section of scleroglucan structure. D-Glucosyl residues are represented by O, 1,3-β-D-glucosidic bonds by —, and 1,6-β-D-glucosidic bonds by ↓.

chains appeared in solution. Upon lowering the pH the chains either did not reassociate or were not matched the length of the molecule but associated in randomly selected areas. This permanent change was explored by measuring the susceptibility of scleroglucan to enzymic hydrolysis [68]. At neutrality the polysaccharide was not hydrolyzed by Zymolyase, a β-1,3-D-glucanase from *Arthrobacter luteus*. Exposure to 0.2M NaOH for 20 min at room temperature followed by a fast return (30 s) to neutrality was sufficient to render 38% of glucosidic bonds accessible to enzymic hydrolysis. Analysis of the product revealed why degradation was not more extensive. A considerable proportion (42%) of the polysaccharide remained intact and it was concluded that intractable parts of the polymer were still in triple helical formation whereas the susceptible domains comprised dissociated strands.

At neutrality and below 10°C the viscosity of aqueous solutions increased dramatically with, once again, an increase in optical rotary power [69]. Gel formation was found at lower temperatures than that of curdlan, which solidifies when aqueous suspensions are heated to a little over 50°C. Again, the introduction of the 1,6-linked glucosyl stub has profound effects.

In fact there are two transitions of conformation in alkaline media. Below 0.04 M NaOH a reversible change was detected [70] between two triple helical conformations. The change is probably an alteration in the disposition of protruding glucosyl stubs together with the attendant associated water molecules producing an adjustment to the hydration characteristics of the triple-helix conformation. Much the same order-disorder transitions have also been observed at neutrality [69] in what is presumably the same structure, namely, schizophyllan. These occured at 7°C in H_2O and at 18°C in D_2O. As with the reversible alterations in alkali the helix remained intact throughout the change, which is highly cooperative. Again the reason for change is a rearrangement of ordered water layers around the polymer. At temperatures below the change neighboring glucosyl stubs are in effect in contact with one another through a short chain of water molecules. It is proposed that spirals of water molecules encompass the triple-helix cylinder connecting the protruding stubs and limiting their rotation about the β-1,6-glucosidic linkage. Transition to the less ordered state at the higher temperature is perceived as an increase in freedom of the side chains and their loss of contact with one another through the water molecules. At neutrality dissociation of the triple helix itself does not occur below 135°C.

Partial fragmentation of schizophyllan occurs when a succession of shear forces are applied to the polymer by multiple passage under high pressure through a capillary [71]. The molecular weight first decreased rapidly and then approached a constant value. The rate of decrease was related to shear rate and enhanced at higher concentrations of polysaccharide. Depolymerization of the individual β-1,3-glucan chains was proposed rather than dissociation of the triple helix.

The viscosity of scleroglucan in dimethyl sulfoxide [66] is similar to that in aqueous solution of pH >12, suggesting that the recognized facility of the solvent to dissolve glucans lies in its ability to dissociate the interchain hydrogen bonding.

C. Industrial and Medical Potential

Scleroglucan is attracting interest for two quite separate reasons: as a component of the enhanced oil recovery process and as an antitumor treatment.

1. Enhanced Oil Recovery

The process of enhanced recovery is the flushing out of residual oil from porous rock with water. The rheological characteristics of the purging aqueous phase must provide a broad advancing bank of water that will not channel through the more permeable pores of the reservoirs and which will assume viscosity at a slow flow rate but be capable of pumping and filtration, i.e., adopt a reduced viscosity at a high shear rate [72]. The rock, usually sandstone or limestone, is very porous. Thus, a cubic foot may contain up to 3.5 liters of oil and 1 liter of water distributed throughout micropores presenting more than an 0.5 hectare of surface. The conditions under which these polymers must function are severe. Performance is required at 100°C or more in high salinity, various chemical environments, and in the presence of potentially damaging microbial populations. In this environment it must be structurally stable for a period of years. Currently two classes of compound are in use: the xanthans and the polyacrylamides. Xanthan polysaccharides are secreted by the bacterium *Xanthomonas campestris*. That solutions of scleroglucan possess much the same pseudoplastic properties as xanthan, can withstand the same demanding conditions, and, because of their neutrality, are not as susceptible to ionic environments as the anionic xanthans make it an attractive candidate for enhanced oil recovery.

2. Antitumor Activity

Administered intramuscularly at a rate of 40 mg per weekly dose, and in conjunction with radiotherapy, Okamura et al. [73] demonstrated a significant reduction in cervical cancer. Schizophyllan is thought to have various immunological properties that modify biological responses, e.g., induction of natural killer cells and the activation of macrophages without the aid of T cells. Similarly potent antitumor activity against mouse sarcomas was found to occur with both intraperitoneal and intralesional injections at doses of 100 and 20 μg per mouse per day, respectively [74].

The structure of the glucans used in these studies is not always defined. Products of *Schizophyllum commune* are often referred to as β-1,3-D-glucans and may not possess the structural features of the glucosyl stub on every third residue. Nevertheless the integrity of the triple helix appears to be important. Heated samples, where dissociation of the chains had occurred, were found to be much less effective; about a 10-fold increase in dose is required to elicit the response produced by native polysaccharide [75].

D. Production

Scleroglucan is produced on a large scale in submerged aerobic fermentation with glucose or corn-steep liquor as carbon source and nitrate as nitrogen supply [76]. The temperature is 30°C and the pH of the culture, initially 4.5, falls to around 2.5 in the first stages of growth with the accumulation of oxalic acid. The low pH means that the fermentation conditions do not have to be fully aseptic. Elaboration of scleroglucan proceeds with mycelial growth and the broth becomes very viscous. After about 60 h the polysaccharide concentration is 20-30 g/liter. The high viscosity reduces oxygen transfer and inhibits the growth of contaminating aerobes. Homogenizing and filtration separate cellular mass from the solution of scleroglucan. Depending on the required properties of the polysaccharide solution different strains of *Sclerotinia* are selected to produce polymers of different structures and size [77]. The polysaccharide from *S. gluccnicum* has a structure depicted in Fig. 2 and a degree of polymerization of about 110 whereas that from the commercially favored *S. rolfsii* may have β-1,6-linked glucan side chains that are larger than a single residue and a degree of polymerization on the order of 800 [78].

IV. YEAST GLUCAN

An interesting development in the exploitation of fungal polysaccharides is the use of not elaborated soluble polysaccharide but rather the insoluble skeletal portion of the yeast cell wall [79].

Yeast walls comprise a fibrillar network that determines the shape and size of the cell and an infilling gel that provides restricted access to the cytoplasmic membrane. The fibrillar insoluble exoskeleton consists of families of β-1,3/1,6-D-glucans some of which are branched and which between them create a three-dimensional network around the cell. The infilling gel is mannoprotein and some β-D-glucan. Vigorous alkaline extraction of whole cells destroys the cytoplasmic membrane, releases cytoplasmic contents, and solubilizes the mannoprotein. The insoluble residue is the β-D-glucan skeleton, a potentially useful molecular bag of uniform size and low toxicity that allows its use in foods and medical therapy for the encapsulation and controlled release of flavor and drugs, and for the enhancement of oxygen transfer in fermentations [80]. Its ability to influence water structure contained in and around it together with a resistance to intestinal hydrolases suggests a role as a noncalorific food thickener. Another possible use is as a solid phase in the resolution of optical isomers.

The glucan spheres can be manipulated in both size and porosity [81]. The proportion of β-1,6-D-glucosidic linkages in the cell wall influence these properties and can be controlled by extraction procedures. Controlled hydrolysis by β-1,3-D-glucanases removed portions of the wall and considerably altered the hydrodynamic properties of the spheres. Genetic manipulation can also alter the proportions of β-1,6 linkages, all of which means that there is a potential for adjusting the properties of the glucan to the required function.

Other commercial advantages appear when the procedures of fermentation and product recovery are considered. Large-scale fermentations produce large amounts of biomass which are normally discarded but which represent a loss of materials and energy. Thus the conversion of spent cells to a useful product is an attractive proposition. The other advantage is in product isolation. Extracellular polysaccharides in dilute solution require large volumes of costly solvent for precipitation and recovery steps. The greatly reduced volume of cell mass with the consequent much smaller volumes of extractant and no subsequent precipitation reduces the cost of yeast glucan preparation.

REFERENCES

1. Berkeley, R. C. W., Gooday, G. W., and Ellwood, D. C. (eds.). *Microbial Polysaccharides and Polysaccharidases*, Academic Press, London, 1979.
2. Jeanes, A. Dextrans and pullulans: industrially significant α-D-glucans. In *Advances in Chemistry Symposia*, American Chemical Society Symposium No. 45 (P. A. Sandford and A. Laskin, eds.), American Chemical Society, Washington, D. C., 1977, pp. 284-297.
3. Yalpani, M. (ed.). *Progress in Biotechnology, Vol. 3, Industrial Polysaccharides, Genetic Engineering, Structure/Property Relations and Applications*. Elsevier, Amsterdam, 1987.
4. Phillips, G. O., Williams, P. A., and Wedlock, D. J. (eds.). *Gums and Stabilisers for the Food Industry, Vol. 4*. IRL Press, Oxford, 1987.
5. Morris, E. R. Molecular interactions in polysaccharide gelation. *Br. Polym. J.*, *18*:14-21, 1986.
6. Rees, D. A. *Outline Studies in Biology, Polysaccharide Shapes*. Chapman and Hall, London, 1977.
7. Bender, H., Lehmann, J., and Wallenfels, K. Pullulan, ein extracellulares Glucan von Pullularia pullulans. *Biochem. Biophys. Acta*, *36*:309-316, 1959.
8. Catley, B. J. Pullulan synthesis by Aureobasidium pullulans. In *Microbial Polysaccharides and Polysaccharidases* (R. C. W. Berkeley, G. W. Gooday, and D. C. Ellwood, eds.), Academic Press, London, 1979, pp. 69-84.
9. Catley, B. J., Ramsay, A., and Servis, C. Observations on the structure of the fungal extracellular polysaccharide, pullulan. *Carbohydr. Res.*, *153*:79-86, 1986.

10. Sakano, Y., Kogure, M., Kobayashi, T., Tamura, M., and Suekane, M. Enzymic preparation of panose and isopanose from pullulan. *Carbohydr. Res., 61*:175-179, 1978.
11. Carolan, G., Catley, B. J., and McDougal, F. J. The location of tetrasaccharide units in pullulan. *Carbohydr. Res., 114*, 237-243, 1983.
12. Zemek, J., Augustin, J., Kuniak, L., and Kučár, S. Metabolism of α-glucans in Aureobasidium pullulans. *Biologia (Bratislava), 35*:173-179, 1980.
13. Elinov, N. P., Marikhin, V. A., Dranishnikov, A. N., Myasnikova, L. P., and Maryukhta, Y. B. Peculiarities of the glucan produced by a culture of Aureobasidium (Pullularia) pullulans. *Doklady Akad. Nauk. SSSR, 221*:213-216, 1975.
14. Buliga, G. S., and Brant, D. A. Temperature and molecular weight dependence of the unperturbed dimensions of aqueous pullulan. *Int. J. Biol. Macromol., 9*:71-76, 1987.
15. Buliga, G. S., and Brant, D. A. Theoretical interpretation of the unperturbed aqueous solution configuration of pullulan. *Int. J. Biol. Macromol., 9*:77-86, 1987.
16. Kato, T., Okamoto, T., Tokuya, T., and Takahashi, A. Solution properties and chain flexibility of pullulan in aqueous solution. *Biopolymers, 21*:1623-1633, 1982.
17. Catley, B. J. Pullulan, a relationship between molecular weight and fine structure. *FEBS Lett., 10*:190-193, 1970.
18. McNeil, B., and Kristiansen, B. Influence of impeller speed upon the pullulan fermentation. *Biotechnol. Lett., 9*:101-104, 1987.
19. Dubois, M., Gilles, K. A., Hamilton, J. K., Rebers, P. A., and Smith, F. Colorimetric method for the determination of sugars and related substances. *Anal. Chem., 28*:350-357, 1956.
20. Catley, B. J. Utilization of carbon sources by Pullularia pullulans for the elaboration of extracellular polysaccharides. *Appl. Microbiol., 22*:641-649, 1971.
21. Finkelman, M. A. J., and Vardanis, A. Simplified microassay for pullulan synthesis. *Appl. Environ. Microbiol., 43*:483-485, 1982.
22. Imshenetskii, A. A., and Kondrateva, T. F. Rapid method of quantitative determination of pullulan in culture fluid of Pullularia pullulans. *Mikrobiologiya, 47*:566-568, 1978.
23. Rennenberg, R., Kaiser, G., Scheller, F., and Tsujisaka, Y. Enzyme sensor for pullulan and pullulanase activity. *Biotechnol. Lett., 7*:809-812, 1985.
24. Yuen, S. Pullulan and its applications. *Process Biochem.*, November 7-22, 1974.
25. Zajic, J. E., and LeDuy, A. Flocculant and chemical properties of a polysaccharide from Pullularia pullulans. *Appl. Microbiol., 25*:628-635, 1973.
26. Preparation of α-glycosyl pullulans. *Chem. Abstr., 96*:179449, 1982.
27. Catley, B. J. The rate of elaboration of the extracellular polysaccharide, pullulan, during growth of Pullaria pullulans. *J. Gen. Microbiol., 78*:33-38, 1973.
28. Catley, B. J. The extracellular polysaccharide, pullulan, produced by Aureobasidium pullulans: a relationship between elaboration rate and morphology. *J. Gen. Microbiol., 120*:265-268, 1980.

29. Park, D. Inorganic nitrogen nutrition and yeast-mycelial dimorphism in Aureobasidum pullulans. *Trans. Br. Mycol. Soc.*, *78*:385-388, 1982.
30. Park, D. Amino acid nutrition and yeast-mycelial dimorphism in Aureobasidium pullulans. *Trans. Br. Mycol. Soc.*, *79*:170-172, 1982.
31. Park, D. Population density and yeast-mycelial dimorphism in Aureobasidium pullulans. *Trans. Br. Mycol. Soc.*, *82*:39-44, 1984.
32. Pollack, J. H., and Hashimoto, T. Ethanol-induced germ tube formation in Candida albicans. *J. Gen. Microbiol.*, *131*:3303-3310, 1985.
33. Carolan, G., Catley, B. J., and Kelly, P. J., Ethanol production during the growth cycle of Aureobasidium pullulans. *Biochem. Soc. Trans.*, *4*:1021-1022, 1976.
34. Cooper, L. A., and Gadd, G. M. The induction of mycelial development in Aureobasidium pullulans (IMI 45533) by yeast extract. *Antonie van Leeuwenhoek*, *50*:249-260, 1984.
35. Cooper, L. A., and Gadd, G. M. Strain and medium related variability in the yeast-mycelial transition of Aureobasidium pullulans. *FEMS Lett.*, *23*:47-50, 1984.
36. Heald, P. J., and Kristiansen, B. Synthesis of polysaccharides by yeast-like forms of Aureobasidium pullulans. *Biotechnol. Bioeng.*, *27*: 1516-1519, 1985.
37. Ono, K., Yasuda, N., and Ueda, S. Effect of pH on pullulan elaboration by Aureobasium pullulans S-1. *Agr. Biol. Chem.*, *41*:2113-2118, 1977.
38. Catley, B. J. Role of pH and nitrogen limitation in the elaboration of the extracellular polysaccharide pullulan by Pullularia pullulans. *Appl. Microbiol.*, *22*:650-654, 1971.
39. Bulmer, M. A., Catley, B. J., and Kelly, P. J. The effect of ammonium ions and pH on the elaboration of the fungal extracellular polysaccharide, pullulan, by Aureobasidium pullulans. *Appl. Microbiol. Biotechnol.*, *25*:362-365, 1987.
40. Bermejo, J. M., Dominquez, J. B., Goñi, F. M., and Uruburu, F. Influence of pH on the transition from yeast-like cells to chlamydospores in Aureobasidium pullulans. *Antonie van Leeuwenhoek*, *47*:385-392, 1981.
41. Dominquez, J. B., Goñi, F. M., and Uruburu, F. The transition from yeast-like to chlamydospore cells in Pullularia pullulans. *J. Gen. Microbiol.*, *108*:111-117, 1978.
42. Lacroix, C., Le Duy, A., Noel, G., and Choplin, L. The effect of pH on the batch fermentation of pullulan from sucrose medium. *Biotechnol. Bioeng.*, *27*:202-207, 1985.
43. Stewart, E., Gow, N. A. R., and Bowen, D. V. Cytoplasmic alkanization during germ tube formation in Candida albicans. *J. Gen. Microbiol.*, *134*:1079-1087, 1988.
44. Kelly, P. J., and Catley, B. J. The effect of ethidium bromide mutagenesis on dimorphism, extracellular metabolism and cytochrome levels in Aureobasidium pullulans, *J. Gen. Microbiol.*, *102*:249-254, 1977.
45. Finkelman, M. A. J., and Vardanis, A. Enhancement of pullulan elaboration by fluoracetate. *Biotechnol. Lett.*, *4*:393-396, 1982.
46. Seviour, R. J., and Kristiansen, B. Effect of ammonium ion concentration on polysaccharide production by Aureobasidium pullulans in batch culture. *Eur. J. Appl. Microbiol. Biotechnol.*, *17*:178-181, 1983.
47. Rothman, L. B., and Cabib, E. Regulation of glycogen synthesis in the intact yeast cell. *Biochemistry*, *8*:3332-3341, 1969.

48. Saita, M., and Slaughter, J. C. Acceleration of the rate of fermentation by Saccharomyces cerevisiae in the presence of ammonium ion. *Enz. Microbiol. Technol.*, *6*:375-378, 1984.
49. Lloyd, D., Kristensen, B., and Degn, H. Glycolysis and respiration in yeasts. *Biochem. J.*, *212*:749-754, 1983.
50. Le Duy, A., and Boa, J. M. Pullulan production from peat hydrolysate. *Can. J. Microbiol.*, *29*:143-146, 1983.
51. Boa, J. M., and Le Duy, A. Peat hydrolysate medium optimization for pullulan production. *Appl. Environ. Microbiol.*, *48*:26-30, 1984.
52. Boa, J. M., and Le Duy, A. Pullulan from peat hydrolysate, fermentation kinetics. *Biotechnol. Bioeng.*, *30*:463-470, 1987.
53. Shin, Y. C., Kim, Y. H., Lee, H. S., Kim, Y. N., and Byun, S. M. Production of pullulan by a fed-batch fermentation. *Biotechnol. Lett.*, *9*:621-624, 1987.
54. Le Duy, A., Yarmoff, J., and Chagraoui, A. Enhanced production of pullulan from lactose by adaptation and by mixed culture techniques. *Biotechnol. Lett.*, *5*:49-54, 1983.
55. Imshenetskii, A. A., and Kondrateva, T. F. More active synthesis of the polysaccharide pullulan by polyploid cultures of Pullularia pullulans. *Microbiology* (translation of *Mikrobiologiya*), *48*:257-260, 1979.
56. Imshenetskii, A. A., and Kondrateva, T. F. Respiration activity in Pullularia pullulans diploid strains. *Microbiology* (translation of *Mikrobiologiya*), *49*:227-230, 1980.
57. Imshenetskii, A. A., Kondrateva, T. F., and Smutko, A. N. Influence of carbon and nitrogen sources on pullulan biosynthesis by polyploid strains of Pullularia pullulans. *Microbiology* (translation of *Mikrobiologiya)*, *50*:73-76, 1981.
58. Ueda, S., Fujita, K., Komatsu, K., and Nakashima, Z. Polysaccharide produced by the genus Pullularia. 1. Production of polysaccharide by growing cells. *Appl. Microbiol.*, *11*, 211-215, 1963.
59. Imshenetskii, A. A., Kondrateva, T. F., and Smutko, A. N. Influence of the acidity of the medium, conditions of aeration, and temperature on pullulan biosynthesis by polyploid strains. *Microbiology* (translation of *Mikrobiologiya*), *50*:330-333, 1981.
60. Imshenetskii, A. A., Kondrateva, T. F., Kudryashev, L. I., Yarovaya, S. M., Smutko, A. N., and Alekseeva, G. S. A comparative study of pullulans synthesised by strains of Pullularia (Aureobasidium) pullulans of differing levels of ploidy. *Microbiology* (translation of *Mikrobiologiya*), *52*:640-644, 1983.
61. Imshenetskii, A. A., Kondrateva, T. F., and Smutko, A. N. Spontaneous and ultra-violet induced variability of the pullulan-synthesizing activity of Pullularia (Aureobasidium) pullulans strains with different levels of ploidy. *Microbiology* (translation of *Mikrobiologiya*), *51*:765-767, 1982.
62. Leal-Serrano, G., Ruperez, P., and Leal, J. A. Acidic polysaccharide from Aureobasidium pullulans. *Trans. Br. Mycol. Soc.*, *75*:57-62, 1980.
63. Hamada, N., and Tsujisaka, Y. The structure of the carbohydrate moiety of an acidic polysaccharide produced by Aureobasidium sp.K-1. *Agr. Biol. Chem.*, *47*:1167-1172, 1983.
64. Tarasov, B. P., Yurlova, N. A., and Elinov, N. P. Melanins formed by a culture of Aureobasidium (Pullularia) pullulans. Arnaud (de Bary), 1910. *Khimiya Prirodnykh Soedinenii*, *2*:254-261, 1977.

65. Gadd, G. M. Melanin production and differentiation in batch cultures of the polymorphic fungus Aureobasidium pullulans. *FEMS Microbiol. Lett.*, *9*:237-240, 1980.
66. Norisuye, T., Yanaki, T., and Fujita, H. Triple helix of a Schizophyllum commune polysaccharide in aqueous solution. *J. Polym. Sci.*, *18*:547-558, 1980.
67. Bluhm, T. L., Deslandes, Y., Marchessault, R. H., Perez, S., and Rinaudo, M. Solid-state and solution conformation of scleroglucan. *Carbohydr. Res.*, *100*:117-130, 1982.
68. Catley, B. J., and Fraser, M. E. Susceptibility of scleroglucan to the (1,3)-β-D-glucanase zymolyase. *Carbohydr. Res.*, *183*:83-88, 1988.
69. Itou, T., Teramoto, A., Matsuo, T., and Suga, H. Isotope effect on the order-disorder transition in aqueous schizophyllan. *Carbohydr. Res.*, *160*:243-257, 1987.
70. Bo, S., Milas, M., and Rinaudo, M. Behaviour of scleroglucan in aqueous solution containing sodium hydroxide. *Int. J. Biol. Macromol.*, *9*:153-157, 1987.
71. Kojima, T., Tabata, K., Ikumoto, T., and Yanaki, T. Depolymerization of schizophyllan by controlled hydrodynamic shear. *Agr. Biol. Chem.*, *48*:915-921, 1984.
72. Gabriel, A. Economic value of biopolymers and their use in enhanced oil recovery. In *Microbial Polysaccharides and Polysaccharidases* (R. C. W. Berkeley, G. W. Gooday, and D. C. Ellwood, eds.). Academic Press, London, 1979, pp. 191-204.
73. Okamura, K., Suzuki, M., Chihara, T., Fujiwara, A., Fukuda, T., Goto, S., Ichinohe, K., Jimi, S., Kasamatsu, T., Kawai, K., Mizuguchi, K., Mori, S., Nakano, H., Noda, K., Sekiba, K., Suzuki, K., Suzuki, T., Takahashi, I., Takenchi, K., Takenchi, S., Yajima, A., and Ogawa, N. Clinical evaluation of schizophyllan combined with irradiation in patients with cervical cancer. *Cancer*, *58*:865-872, 1986.
74. Shinohara, H., Ohno, N., and Yadomae, T. Antitumour activity and structural characteristics of a (1,3)-β-D-glucan extracted with cold alkali from sclerotia of Sclerotinia. *Chem. Pharm. Bull.*, *36*:819-823, 1988.
75. Yanaki, T., Itoh, W., and Tabata, K. Connection between antitumour activity of schizophyllan and its triple helix. *Agr. Biol. Chem.*, *50*:2415-2416, 1986.
76. Donche, A. Scleroglucan for enhanced oil recovery. In *Biotech '85 Conference*, Geneva, Switzerland. Online Publications, Pinner, U.K., 1985, pp. 429-436.
77. Lecacheux, D., Mustiere, Y., Parnas, R., and Brigand, G. Molecular weight of scleroglucan and other extracellular microbial polysaccharides by size-exclusion chromatography and low angle laser light scattering. *Carbohydr. Polym.*, *6*:477-492, 1986.
78. Sandford, P. A. Exocellular, microbial polysaccharides. *Adv. Carbohydr. Chem. Biochem.*, *36*:265-313, 1979.
79. Jamas, S., Sinskey, A. J., and Rha, C. K. Controlled biosynthesis of yeast glucan particles. In *Progress in Biotechnology 3, Industrial Polysaccharides, Genetic Engineering, Structure/Property Relationships and Applications* (M. Yalpani, ed.), Elsevier, Amsterdam, 1987, pp. 65-70.
80. Sinskey, A., Jamas, S., Easson, D., and Rha, C. K. Biopolymers and modified polysaccharides. In *Biotechnology in Food Processing*

(S. K. Harlander and T. P. Labuza, eds.), Noyes, Park Ridge, N.J., 1986, pp. 73-113.

81. Jamas, S., Rha, C. K., and Sinskey, A. J. Morphology of yeast cell wall as affected by genetic manipulation of β(1,6)-glucosidic linkages. *Biotech. Bioeng.*, *28*:769-784, 1986.

12

CHITIN BIOSYNTHESIS IN FUNGI

JOSE RUIZ-HERRERA* *Centro de Investigación y de Estudios Avanzados, Mexico City, Mexico*

RAFAEL SENTANDREU RAMON and J. PEDRO MARTINEZ *Facultat de Farmacia, Universitat de Valencia, Valencia, Spain*

I. INTRODUCTION

The polysaccharide we now know as chitin was first isolated in 1811 by Henry Braconot from mushrooms. It received his attention because as cellulose it was alkali-resistant. In 1823, Odier isolated the same compound from the elytra of May beetles, and named it "chitine" or chitin (from the greek χιτων, frock, tunic), and recalled its protective role in animals similar to that of cellulose in plants.

Much time has passed since those pioneer studies. In all these years great efforts have been made to study the distribution of this polysaccharide, to analyze its chemical composition and macromolecular structure, and to determine its mechanism of biosynthesis. Chitin is said to be the second most abundant organic compound in nature, yielding only to cellulose. It is an unbranched polysaccharide made of ~2000 β-1-4-linked 2-acetamido-2-deoxy-D-glucose (*N*-acetylglucosamine, GlcNAc) residues. Chains of the polysaccharide associate through hydrogen bonding in three different crystalline structures (α, β, and γ) according to the polarity acquired by the sugar chains; of these the most abundant in nature is the α form. Association of around 20-400 chains constitutes the microfibrils characteristic of chitin. Chitin is a compound restricted to the eukaryotic world, being present in protozoa, algae, fungi, and higher animals, with the exception of echinoderms and chordates, but absent in bacteria. In fungi, chitin is the most important structural component of the cell wall, being responsible for maintaining the shape, strength, and integrity of this structure which endows them with their astonishing capacity to colonize all kinds of ecological niches.

From a chemical point of view, the final step in chitin biosynthesis can be described as a simple transglycosylation reaction through which a sugar donor transfers *N*-acetylglucosaminyl residues to a sugar acceptor. But the insolubility of the crystalline chitin, the fact that it accumulates outside the permeability barrier of the cell, and its precise deposition in

---

**Present affiliation*: Instituto de Investigación en Biología Experimental, Universidad de Guanajuato, Guanajuato, Mexico.

time and space at the growing sites of the cell all speak about the complexity of the finely regulated machinery developed by fungi in the course of evolution to synthesize this compound which is indispensable for their survival.

In recent years great advances have been made in the comprehension of the mechanisms involved in chitin synthesis, and in the following pages we will try to summarize the present status of our knowledge on the mechanisms operating in vitro and in vivo for the biosynthesis of chitin.

II. MECHANISMS OF CHITIN BIOSYNTHESIS

A. Properties of Chitin Synthetase

Properties of chitin synthetases have been studied in crude or partially purified preparations obtained from several fungi; there are no reports of studies performed with pure samples of the enzyme. Despite this, a substantial amount of knowledge on the enzymatic reaction has been obtained with these kinds of preparations. It has been observed that several properties are held in common by enzymes obtained from different sources.

1. Glycosyl Donor

From the initial studies by Glaser and Brown [1] on *Neurospora crassa*, it was established that uridine diphosphate-*N*-acetylglucosamine (UDP-GlcNAc) was the substrate for chitin synthetase. Further studies with other fungi have confirmed the role of this nucleotide sugar in chitin biosynthesis, i.e., *Allomyces macrogynus* [2], *Blastocladiella emersonii* [3], *Mucor rouxii* [4], and *Cochliobolus miyabeanus* [5]. No report exists in which chitin biosynthesis in vitro was obtained using a different nucleotide sugar (see Ref. 6 for a review).

2. Divalent Metal Requirement

Glaser and Brown [1] demonstrated that chitin synthetase from *N. crassa* was stimulated by Mg^{2+} with optimal concentrations around 5 mM. Similar results were obtained with the enzyme from *M. rouxii* [7], *Saccharomyces carlsbergensis* [8], *Aspergillus flavus* [9], and *Aspergillus nidulans* [10]. Other authors [11,12] reported no stimulation by Mg^{2+} of the enzyme from *Cunninghamella elegans* and *Mortierella vinacea*. These discrepancies were explained when preparations from *M. rouxii*, partially stimulated by Mg^{2+}, were dialyzed first in EDTA and later in water to remove all traces of divalent metal ions. Under these conditions, dialyzed samples displayed an absolute requirement for Mg^{2+} to be active [6]. These results suggested that partial or no stimulation of chitin synthetase by Mg^{2+} in some fungal preparations was due to partial contamination with the metallic ion. Mn^{2+} can substitute for Mg^{2+}, although less efficiently, in several systems [3,8,9,13]. In *M. rouxii*, Mn^{2+} was more efficient than Mg^{2+}, but in a narrower range of concentrations, apparently because it sensitized the enzyme to proteolytic destruction [14]. It was reported that for chitin synthetase 2 (see below) from *Saccharomyces cerevisiae*, Co^{2+} is more efficient than Mg^{2+} [15], but this result was not confirmed by Orlean [16]. Co^{2+} was found to be as efficient as Mg^{2+} in *Chaenophora cucurbitarum* and *Phascolomyces articulosus* [13]. Other divalent metals (Zn^{2+}, Ca^{2+}) are either much less efficient [13] or inhibitory [14].

3. Kinetic Characteristics

Kinetic analysis of all carefully studied chitin synthetases [7,9,10,14,17] revealed nonconventional kinetics; reciprocal plots of 1/v vs. 1/s are sigmoidal, apparently due to a cooperative effect. Accordingly, only approximate values of apparent K_m can be calculated. These values, which oscillate from 0.5 to 3.0 mM, may appear to be very high but are of the same order of magnitude as UDP-GlcNAc concentrations found in the cytoplasm of fungi [18-20]. Hill's coefficients, which may be indicative of the number of active sites on the enzyme surface, oscillate between 1.0 for chitin synthetase from *A. nidulans* [10] and *M. rouxii* [6], 1.9 for the enzyme from *Agaricus bisporus* [21], and 4.0 for the enzyme from *Coprinus cinereus* [22].

Optimal pH values for chitin synthetases are around neutrality; extreme values reported are 5.8 for the enzyme from *Mortierella vinacea* [12] and 8.0 for *C. cinereus* chitin synthetase [17]. Optimum temperatures vary from 15 to 40°C, the minimum (15°C) corresponding to chitin synthetase from *A. flavus* [9], and the maximum (40°C) to the enzyme from *S. cerevisiae* [16]. Analysis of the effect of temperature on chitin synthetase activity in chitosomes from *M. rouxii* by the procedure of Arrhenius revealed a break point in the plot at 6.5°C. Energy of activation was 27,752 cal/mol below this temperature and 14,958 cal/mol above it [23]. Similarly, Arrhenius plots of chitin synthetase from *Schizophyllum commune* revealed break points at 10-12°C and 18-20°C depending on the samples used [24]; and 16 and 25°C for chitin synthetase from *C. cinereus* [25]. These data probably represent thermal phase transitions of the lipids which constitute the hydrophobic environment of the enzyme, from the liquid to the solid crystalline state, at specific temperatures.

4. Activators of Chitin Synthetase

As initially described by Glaser and Brown [1] for *N. crassa*, free *N*-acetylglucosamine (GlcNAc) stimulates chitin synthetase from most fungal preparations studied thus far (see Ref. 6). GlcNAc does not act as a sugar acceptor for chain elongation, nor is it incorporated randomly into chitin [14]. Kinetic analysis of its mode of action suggested that it behaved as an allosteric activator reducing the cooperative behavior of the enzyme [8,14,26], but its role in vivo is uncertain as high concentrations are required for activation (K_a = 4-13 mM). Due to the peculiar kinetics of chitin synthetase (see above), it has been suggested that UDP-GlcNAc rather than GlcNAc is the real allosteric activator of chitin synthetase [7,14]. Chitodextrins have been described as stimulators of chitin synthetase in several systems [1-5,11,17,27], though contradictory results have been obtained in other fungal preparations [12,28]. Since there is no incorporation of these compounds into chitin, we proposed [6] that their apparently stimulatory effect is due to their use as an alternative substrate for contaminating chitinases present in crude preparations.

ATP at low concentrations stimulates chitin synthetase from *M. rouxii* [7] and *Phycomyces* [29]. For *S. carlsbergensis* chitin synthetase, Vagabov et al. [30] demonstrated that concentrations of ATP smaller than 1 mM activated the enzyme with a K_a of 1.26 mM. Higher concentrations were inhibitory, whereas 1 mM CTP, UTP, PPP, and PPi were all ineffective. Nonhydrolyzable ATP analogs were also stimulatory, but with higher K_a (~10 mM). These results led the investigators to conclude that the effect of ATP was not brought about by a covalent modification of chitin synthetase but rather by a conformational change induced by binding of ATP to the surface of the enzyme. This stimulatory effect was lost when the enzyme was solubilized with digitonin, suggesting that ATP probably exerts its binding effect not to the catalytic peptide, but to a regulatory protein which remains bound to the membrane, and is not solubilized by the digitonin treatment [31].

Slight stimulation of chitin synthetase by several lipid fractions was originally demonstrated in *M. rouxii* cell-free extracts [7]. Phospholipid stimulation of digitonin-solubilized chitin synthetase has been described for several systems [24,25,32], but the specificity of this effect is open to question. For the enzyme from *S. commune* all phospholipids tested were effective, but phosphatidylinositol was the most efficient one [24]. Similarly, for chitosomal subunits from *M. rouxii* (see below) phosphatidylinositol was the most efficient phospholipid tested (J. Ruiz-Herrera and S. Bartnicki-Garcia, unpublished observations), whereas for chitin synthetase from *C. cinereus*, it was observed that efficiency of phosphatidylcholine was affected by the fatty acids substituting in the phospholipid molecule [25]. These results suggest that the hydrophobic environment of chitin synthetase plays an important role in the catalytic reaction.

In this sense it is interesting that aliphatic alcohols stimulate chitin synthetase activity (J. Ruiz-Herrera and S. Bartnicki-Garcia, unpublished observations).

5. Inhibitors of Chitin Synthetase

The most efficient and specific inhibitors of chitin synthetase are two families of nucleoside-peptide antibiotics named polyoxins, produced by *Streptomyces cacaoi* var. *asoensis* [33], and nikkomycins, produced by *Streptomyces tendae* [34]. These compounds are analogs of UDP-GlcNAc (Fig. 1), differing in the nature of some radicals, which competitively inhibit the chitin synthetase from all systems studied so far [e.g., 8,9,13, 14,35,36], with K_i values two to three orders of magnitude smaller than the corresponding K_m for UDP-GlcNAc. Kinetic analysis performed to study their mode of action revealed that polyoxins occupy the uridine-binding site of the enzyme, and that an amino group (placed in a box in Fig. 1) and a free carboxyl group (R_2 in Fig. 1) are required for their inhibitory activity [37,38]. Polyoxins and nikkomycins are effective in vivo but MIC (minimal inhibitory concentration) values are variable for different fungal strains, and generally higher than their K_i values. These differences apparently are due to their rate of uptake, which appears to involve a dipeptide permease. Analysis of the effect of several dipeptides and tripeptides on the uptake of nikkomycins and polyoxins by *M. rouxii* led to the suggestion that entrance of these nucleoside-peptide inhibitors involves the operation of the γ-glutamyl cycle [39].

Several nucleotides when present at millimolar concentrations inhibit chitin synthetase [8,9,14,30,37], but from all these, UDP, the byproduct of the chitin synthetase reaction, had the lowest K_i: 1.4 mM for the enzyme from *Piricularia oryzae* [37] and 0.4 mM for that from *M. rouxii* [14]. From their kinetic studies Hori et al. [37] concluded that UDP binds to the pyrimidine-specific site of the enzyme. The fact that complex kinetics of inhibition occurs [9,14] suggests that allosteric interactions take place in the process of inhibition.

Calcofluor white and Congo red, which affect the correct crystallization of chitin in vivo [40-42] inhibit chitin synthesis in vitro [43,44] when added

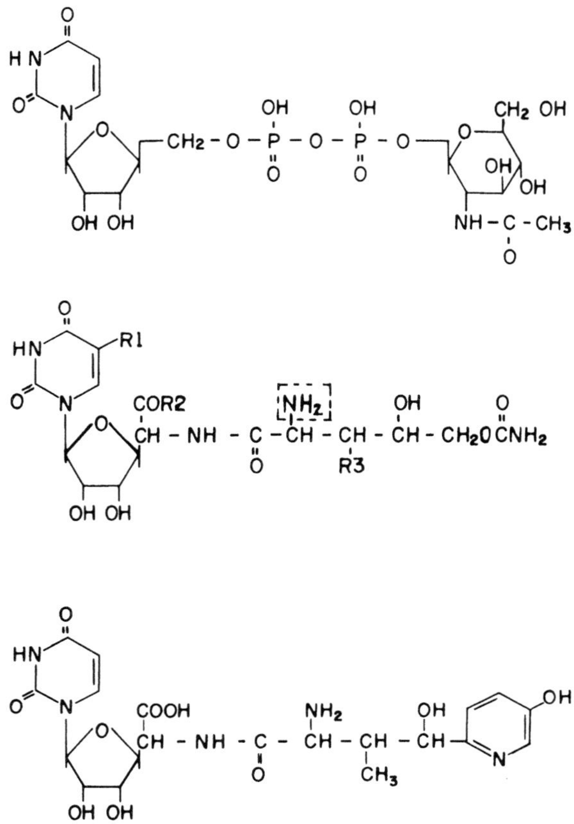

FIGURE 1. Comparative structures of the chitin synthetase substrate (UDP-GlcNAc) (top), and the competitive inhibitors polyoxin (middle) and nikkomycin Z (bottom).

at high concentrations (K_i = 0.3-0.6 g/liter). Its mode of action, however, is not known since it does not inhibit chitin synthesis in vivo [44].

Other inhibitors of chitin synthetase probably affect the hydrophobic environment of the enzyme. Of these, the most interesting ones are polyene antibiotics which interfere with enzyme activity [26,45], most likely binding to sterols which form part of the hydrophobic environment of the enzyme (see below). Sterols themselves inhibit chitin synthetase activity. We demonstrated that they become incorporated into chitosomes in vitro (J. Ruiz-Herrera and S. Bartnicki-Garcia, unpublished observations) and thus probably alter the fluidity of the membrane of the organelles. In this sense we can mention that chitin synthetase is extremely sensitive to the action of all kind of detergents and other tensioactive agents.

Diflubenzuron is a member of the insecticidal group of agents known as benzylphenyl ureas which inhibit chitin synthesis in vivo. It was considered at some time an inhibitor of chitin synthetase, but now it has been demonstrated to have no in vitro effect [46]. It was suggested that its mode of action is to inhibit UDP-GlcNAc transport through cell membranes [47].

B. Activation and Inactivation of Chitin Synthetase

Early reports for the presence of an inhibitor of chitin synthetase in crude preparations of *S. carlsbergensis* [8] were later interpreted as the interplay between a proteolytic activator of chitin synthetase and its inhibitor [48]. The endogenous protease involved in this process was identified as proteinase B [49], a vacuolar protease previously described in *S. cerevisiae* [50,51]. Although the physiological role of this vacuolar protease seems improbable since proteinase B-less mutants of yeasts are not affected in chitin synthesis in vivo [52,53], these data revealed the important property of chitin synthetase to be activated by limited proteolysis [6,54]. This property seems to be shared to a variable degree by chitin synthetases from nearly all fungal sources [6]. It is interesting to note that although the proteolytic activation process is a rather nonspecific phenomenon, there are differences in the efficiency of several proteases [54-57].

The mechanism of activation of chitin synthetase by limited proteolysis remains unknown. The most favored hypothesis is that the enzyme exists as a zymogen [54], although other possibilities may be considered, mainly that the inactive enzyme represents the binding between an active polypeptide and an inhibitor. In the latter case, the inhibitory protein would be the target for the activating protease. Interestingly, purification of trypsin-activated chitosomes from *N. crassa* revealed the same polypeptide composition as inactive ones with the exception of a 38-kDa polypeptide which was transformed into two new ones of molecular masses 18 and 21 kDa [58]. It is possible that the 38-kDa peptide may correspond to a regulatory protein which is broken down by the activating protease. The possibility that in fact the actual target for activating proteases is not chitin synthetase itself but rather a regulatory polypeptide is suggested by recent results obtained with chitin synthetase from the sporangiophores of *P. blakesleeanus*. This enzyme is activated by trypsin, and the process is inhibited by iodoacetamide which does not affect trypsin or preactivated chitin synthetase activities [59].

It is a general observation that high concentrations of exogenous proteases or a prolonged proteolytic treatment bring about chitin synthetase destruction [6,56,60]. On these bases, Hasilik [60] proposed a model to explain transient chitin synthetase activity in vivo by a mechanism which involved successive activation and destruction of the enzyme by a proteolytic event. From *M. rouxii*, a protease has been purified which only destroys but does not activate chitin synthetase [57].

Chitin synthetase obtained from *Phycomyces* sporangiophores is activated by light in vitro [29,59,61]. This process was not blocked by any of the normally used protease inhibitors, but only by iodoacetamide [59].

The role of proteases as the physiological activators of chitin synthetase is uncertain. We have demonstrated that chitin synthetase present in semipurified preparations (but not pure chitosomes) obtained from the mycelium of *Phycomyces blakesleeanus* is activated by calcium and calmodulin. Activation of the enzyme by these substances was not prevented by protease inhibitors, but activation by calmodulin was blocked by the calmodulin-antagonistic agent trifluoperazine [62]. In this sense it is interesting to note that different calmodulin-regulated enzymes may be activated in vitro by trypsin [63]. The similarity with chitin synthetase is so striking that the possibility that protease activation of this enzyme merely mimics the natural process appears a likely possibility.

C. Chitin Synthetase Structure, Enzyme Purification, and Genetic Control

When fungal cells are broken and subjected to fractionation by different techniques, the minimal structures which conserve chitin synthetase activity are the vesicles named chitosomes [14,55,64]. Using crude fungal preparations, several detergents were employed to solubilize chitin synthetase. Digitonin was found to be the most effective one to extract active enzyme from *C. cinereus* [17], *S. cerevisiae* [32], *Candida albicans* [65], and *S. commune* [24]. Digitonin was also employed to dissociate chitosomes from *M. rouxii* [66] and *A. bisporus* [26] into chitin synthetase subunits. The dissociated forms of chitin synthetase which have been obtained have a similar mass, ~500 kDa, and sedimentation coefficients of 16–18S. Using these preparations as starting material, chitin synthetase was purified from *C. cinereus* by a method which employed in the last step chromatography in a Cu^+-chelate column [67], and from *S. cerevisiae* by trapping the enzyme in its insoluble product [68]. Polyacrylamide gel electrophoresis from both preparations revealed a major polypeptide with similar molecular mass: 67 kDa for *C. cinereus* and 63 kDa for *S. cerevisiae* enzymes, plus other minor polypeptides. In neither case was it demonstrated whether the single major polypeptide was active by itself or whether the whole aggregate was responsible for the observed activity. Thus the possibility that the concerted operation of several polypeptides is required for catalysis remains open.

Recently, a form of chitin synthetase of smaller molecular mass was obtained from chitosomes of *M. rouxii* by treatment with low concentrations of Brij 36T (E. Lopez-Romero and T. Arredondo, unpublished observations). The solubilized aggregate has a molecular mass of 250 kDa and shows an absolute requirement of phosphatidylinositol for activity.

Analysis of purified chitosomal preparations from *N. crassa* revealed the presence of four major polypeptides with molecular masses of 22, 30, 38, and 53 kDa plus six minor polypeptides of molecular masses 20, 24, 25, 26, 27, and 28 kDa. As described above, trypsin-activated chitosomes showed the same polypeptide pattern with the exception of the 38-kDa polypeptide, and the appearance of two new polypeptides of molecular masses 18 and 21 kDa [58]. Preparations of chitosomes from *M. rouxii* have been shown to reproducibly contain polypeptides of molecular masses 60, 55, and 18 kDa [68a], the 55-kDa polypeptide being the most abundant. This polypeptide also appears commonly in 16S chitosomal subunits separated in sucrose gradients or purified by affinity chromatography in columns prepared with monoclonal antibodies or polyoxin D as binding agents (J. Ruiz-Herrera and S. Bartnicki-Garcia, unpublished).

Two chitin synthetases have been described in *S. cerevisiae*: chs1 and chs2, which differ in kinetic properties, optimal temperature and pH, and divalent metal requirements [16]. Genes coding for both enzymes have been cloned. A gene coding for chs1, which is the most active in in vitro assays, was cloned by complementation of chitin synthetase-less mutants [69]. It has an open reading frame of 3.4 kbp, which encode a protein with a molecular mass of 130 kDa, and no consensus splice sites. As described above, the major protein band observed in purified preparations of chitin synthetase from the yeast has a molecular weight of 63 kDa;

accordingly, if this is the product from the cloned gene, it must suffer an extensive proteolytic degradation during the activation process. Protein and DNA sequencing will throw some light on this item. Disruption of the gene with a ura3 insert did not affect chitin synthesis in vivo or cell viability [69]. This result is surprising since the recipient strains had been selected originally as temperature-sensitive chitin synthetase-less mutants; nevertheless it demonstrated that the *CHS*1 gene was dispensable. By use of a strain containing the disrupted *CHS*1 gene as a recipient, the gene corresponding to the second chitin synthetase (chs2) was cloned by selection of chitin synthetase overproducing strains transformed with a multicopy plasmid. Disruption of the *CHS*2 gene with a ura3 insert was lethal for the cells, evidencing that chs2 is probably the enzyme responsible for chitin synthesis in vivo in *Saccharomyces* [70]. What is the role of the more studied and more active chs1 enzyme remains uncertain. Interestingly, both *CHS* genes have no similarity in their restriction maps and do not hybridize. More recently, Bulawa and Osmond (*Proc. Natl. Acad. Sci. USA*, *87*:7424-7428, 1990) demonstrated that chitin synthetase II is also dispensable and null mutants of the *CHS*2 gene contain normal amounts of chitin and display nonzymogenic chitin synthetase activity.

No information exists thus far on the genetic control of chitin synthetases from filamentous fungi. Such information is necessary since chitin is only a small component of *Saccharomyces* cell walls which accumulates in the budding scars; whereas on the other hand, it is the most important component of the cell wall of most mycelial fungi. Also, it is known that yeast chitin is made of very short chains of about 116 residues [68], whereas microfibrils synthesized in vitro by chitin synthetase from *Mucor* contain at least 1800-2300 units [71]. Large differences in the synthesis and regulation of the enzymes are thus expected between both types of organisms. In this sense it is interesting that a group of mutants of *P. blakesleeanus* resistant to nikkomycin which are affected in the catalytic center of chitin synthetase have been isolated and characterized [72]. Their possible use for cloning of the chitin synthetase gene from this organism seems feasible especially since a transformation system in this fungus was recently developed [73].

D. Mechanism of Chitin Synthesis

The final step in chitin biosynthesis may be considered a transglycosylation reaction by which *N*-acetylglucosaminyl radicals are sequentially added to a growing acceptor chain (Fig. 2). Although true in general, this concept is an oversimplification, and the process may be more sophisticated. Unfortunately, because of the insolubility of the product and the general interest raised by the enzyme itself, scant work has been made on the mechanism of chitin synthesis.

As mentioned above, UDP-GlcNAc appears to be the universal donor of GlcNAc residues for the reaction; however, since the polysaccharide is made of β-1-4-joined sugars, it is likely that the chain grows by the addition of two GlcNAc units at a time. Whether the enzyme has two active sites of different configuration or there exists a transient diacetylchitobiosyl derivative at the surface of the enzyme remains unknown.

The existence of a primer necessary to initiate the growth of a poly-GlcNAc chain has been a matter of debate. Camargo et al. [3] suggested that the stimulatory effect of GlcNAc on chitin biosynthesis was due to its role as a primer. This claim was not confirmed by later work [2,7,14], which showed that free GlcNAc was not incorporated into chitin. Chito-

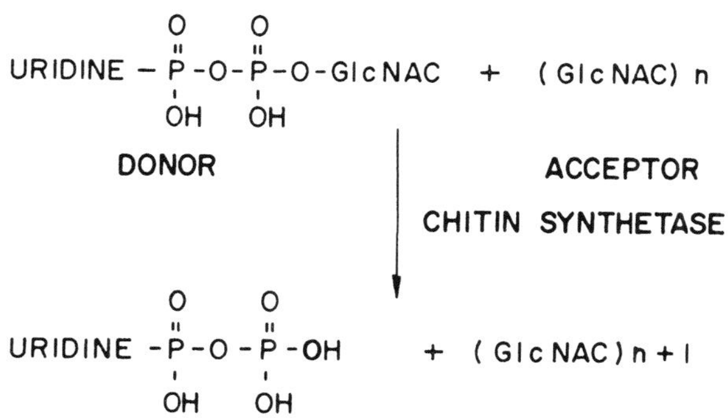

FIGURE 2. Schematic representation of the transglycosylation reaction involved in chitin biosynthesis.

dextrins were also considered as primers since they may be stimulatory in some systems [1,27]; however, they have not been demonstrated to be incorporated into the product. Mills and Cantino [74] reported that during chitin synthesis by γ particles from *B. emersonii*, GlcNAc was transferred to a diglycosyldiglyceride molecule, and that as the reaction proceeded, labeled lipid was incorporated into insoluble chitin, thus behaving as the acceptor required for chain initiation. Different results were obtained by Bartnicki-Garcia et al. [75] who found that chitin microfibrils synthesized in vitro were covalently bound to a peptide fraction containing mostly basic amino acids, suggesting that a protein was the normal acceptor for chitin biosynthesis.

In this regard it is interesting to note that chitin microfibrils synthesized in vitro by preparations from *M. rouxii* were heavily labeled by a gold-tagged antibody prepared against a major chitosomal protein (C. Reynaga, E. Lopez-Romero, and A. Carabez-Trejo, unpublished).

Biosynthesis of a great number of polysaccharides involves lipid carriers [76,77]. The possibility for the involvement of lipid carriers in chitin biosynthesis has been entertained, but most evidence is negative. Thus all attempts to extract labeled lipids from incubation mixtures containing purified chitosomes or chitosomal subunits and radioactive UDP-GlcNAc have been unsuccessful [6]. Moreover, exchange reactions between uridine-labeled UMP or UDP and UDP-GlNAc during chitin-synthesizing reactions, indicative of the formation of a high-energy intermediate, gave negative results. Also, addition of labeled dolichyl-P-O-P-O-GlcNAc or dolichyl-P-O-P-O-$(GlcNAc)_2$ to chitin synthetase reactions proceeding in the presence of unlabeled UDP-GlcNAc failed to produce any radioactively labeled chitin [6].

It is known that there are three crystalline forms of chitin; the one present in fungi is α-chitin, characterized by the fact that its microfibrils are made of antiparallel chains. It must be recalled that the mechanism of a transglycosylation reaction involves the transfer of a glycosyl unit to the nonreducing end of the acceptor molecule; thus chitin chains must grow in the $1 \rightarrow 4$ direction. Growth in the opposite direction has never been reported. Accordingly, the process of synthesis of complete chitin chains must precede the process of crystallization by which the chains associate in antiparallel fashion and establish H bonding. This hypothesis appears to be supported by several observations. Dipolar planar molecules such as Calcofluor white interfere with the correct crystallization of chitin in vivo [40-42]. Nascent chitin is very sensitive to chitinase digestion, whereas the preformed polysaccharide is very resistant to enzyme attack [78,79]. Finally, chitin synthesized in vitro in the presence of Calcofluor white is very sensitive to chitinase degradation and is not crystalline as revealed by X-ray diffraction. Upon drying, the polysaccharide becomes both resistant to chitinase and crystalline [80].

In regard to the synthesized products, early investigations did not establish their physical nature because incorporation of only minute amounts of radioactive precursors were obtained and crude membranous preparations were used. The identity of the product was established only by its insolubility in acid and alkali, and its susceptibility to chitinase. The possibility that the reaction represented merely growth of preexisting molecules could not be eliminated [1-3,8]. In 1974 we solubilized chitin synthetase by exposure of membranous fractions from *M. rouxii* to high concentrations of UDP-GlcNAc. With the resulting sample, chitin microfibrils were synthesized in vitro for the first time and identified by X-ray diffraction [81]. On the basis of this finding some other workers were able to synthesize chitin microfibrils in vitro using preparation from *N. crassa* [55], *A. bisporus* [21,55], *S. cerevisiae* [32,55], *B. emersonii* [74], and *S. commune* [82]. In all these systems the product synthesized by chitin synthetase was made of chains of GlcNAc which crystallized into microfibrils, although the size of the product was variable; thus as described above, yeast chitin chains are very short and made of only ~116 sugar moieties [68], whereas the chains synthesized by filamentous fungi contain ~2000 sugar units [71].

III. SUBCELLULAR LOCALIZATION OF CHITIN SYNTHETASE

A. Fractionation of Cell-Free Extracts

When fungal cells are broken and several fractions are separated by differential centrifugation, it is a common experience to find chitin synthetase activity in almost all subcellular fractions including cell walls, cytoplasmic membrane, mitochondrial and microsomal fractions, and the high-speed supernatant. This peculiar behavior of chitin synthetase is probably due to the artifactual adsorption of the enzyme to all kinds of cellular fragments. More recent and careful studies reduced the possible locations of the enzyme to two subcellular compartments: plasmalemma and specialized vesicles called chitosomes. These two locations in fact are not exclusive if one considers that, according to the flow membrane hypothesis, the precursors for the plasma membrane are secretory vesicles.

Artifactual aggregation of chitin synthetase to different membrane fractions can be illustrated by results obtained on the separation of chitin synthetase in sucrose gradients [83]. More than 80% of the chitin synthetase present in crude extracts from *M. rouxii* sedimented as a sharp peak with a buoyant density of 1.146. Small amounts of the enzyme appeared in two turbid bands of higher buoyant density (d = 1.20 and 1.22, respectively). The relative amounts of chitin synthetase in these two turbid bands increased proportionally to the amount of extract loaded into the gradients, suggesting that they represent nonspecific aggregates. Moreover, change in buffer composition produced massive aggregation of membranes and change in the sedimentation properties of chitin synthetase.

Chitin synthetase was located bound to plasmalemma in two yeasts, *S. cerevisiae* [84] and *C. albicans* [85]. In these studies, plasmalemma was purified from protoplasts after reinforcing with concanavalin A, thus producing open sheets. When these samples were incubated with radioactive UDP-GlcNAc they synthesized chitin, revealed by fluorescence with Calcofluor white or by autoradiography [86]. On the other hand, Hanson and Marple [87] described that 95% of the chitin synthetase from *S. cerevisiae* was located in membranes lighter than plasmalemma. In these studies concanavalin A and anti-concanavalin A antibodies were bound sequentially to the protoplasts, and after rupture, plasmalemma was immunoprecipitated with *Staphylococcus* protein A cells. More recently it was described that if protoplasts of *S. cerevisiae* are broken by a method which minimizes rupture of vacuoles, only 26% of the chitin synthetase appears associated with the plasma membrane fraction, the rest appearing in the form of chitosomes [88]. Similar results were obtained when extracts obtained by ballistic disruption of *S. cerevisiae* were subjected to isopycnic centrifugation in sucrose gradients [89]. In *C. albicans* protoplasts broken isoosmotically to prevent vacuole disruption, a major amount of chitin synthetase was associated with chitosomes [90].

Intracellular location of chitin synthetase was also assessed by a different procedure. Germinated spores of *M. rouxii* were permeabilized with a mixture of toluene-ethanol and incubated in the presence of [^{3}H]UDP-GlcNAc. Under these conditions cells are depleted of energy and the product is accumulated at the same site where it is synthesized. When these cells were subjected to autoradiographic analysis, it was demonstrated that chitin was accumulated in the cytoplasm of the spores [91].

B. Chitosomes

Chitosomes are specialized microvesicles containing most of the chitin synthetase from the fungal cells. They are the smallest entities with chitin synthetase activity which can be isolated from fungal cell-free extracts. Chitosomes were originally isolated from cell-free extracts of the yeast form of *M. rouxii* [14,64], by a procedure involving column chromatography, ribonuclease digestion, and sedimentation in sucrose gradients. By this lengthy procedure ~30-50% of the chitin synthetase originally present in the extracts was found in chitosomes. However, as described above, isopycnic sedimentation of crude extracts revealed that chitosomes contained more than 80% of the chitin synthetase of the cell [83]. Later on chitosomes were isolated from a large number of fungi belonging to different groups containing chitin in their cell walls (Table 1). When negatively stained preparations are observed under the electron microscope, chitosomes appear in two different morphologies depending on whether or not the stain penetrates the vesicles (Fig. 3). Impermeable chitosomes ("proctoids") are spheroidal structures with a central depression and characteristic furrows. "Cycloids" are forms in which the stain has penetrated and reveals some internal structure and a delimiting membrane 6-12 nm thick. In thin sections chitosomes appear as hollow sacs with a unit membrane 6.5-7.0 nm thick, thinner and more difficult to stain than the rest of the fungal membranes (Fig. 3B). Chitosomes are uniform in both size (40-100 nm in diameter) and buoyant density (1.125-1.145), according to the species. In *N. crassa* a distinct difference in buoyant density was observed between chitosomes from the wild-type and the wall-less slime strain [92]. Whether this difference is related with the absence of chitin synthetase activity in vivo in the slime variant remains unknown. Sedimentation coefficient of chitosomes from

TABLE 1. Fungal Species from Which Chitosomes Have Been Isolated

Taxonomic group	Species	Ref.
Chytridiomycetes	*Allomyces macrogynus*	55
	Blastocladiella emersonii	109
Zygomycetes	*Mucor rouxii* (yeasts)	64
	M. rouxii (mycelium)	55,101
	Gilbertella persicaria	
	Phycomyces blakesleeanus (sporangiospores)	101
	P. blakesleeanus (mycelium)	101
Hemiascomycetes	*Saccharomyces cerevisiae*	55,88
Euascomycetes	*Neurospora crassa*	55
	N. crassa (slime variant)	58,96
Homobasidiomycetes	*Agaricus bisporus* (mycelium)	21
	A. bisporus (fruiting body)	55
Deuteromycetes	*Candida albicans*	90

M. rouxii was found to be 105S [14], whereas chitosomes from *Agaricus bisporus* gave a value of 95S [26].

When chitosomes are treated with a high concentration of trypsin they suffer an interesting alteration in morphology [55]. The chitosomal membrane thickens and reveals two layers, a thin fuzzy outer one and an internal thicker layer. In most cases an internal pouch is revealed by contraction of the internal layer.

The cell breakage method originally used for the isolation of chitosomes was ballistic disruption with glass beads [6,55,64]. The use of this method was criticized, and suggestions were made that chitosomes could be vesiculized pieces of different cell membranes [93]. However, chitosomes have also been isolated by milder breakage techniques, e.g., low shear disruption of *N. crassa* and *M. rouxii* mycelium with a razor blade chopper in a hypertonic medium [94]. Other methods for chitosome recovery have included pressure extrusion and use of a microsyringe in the case of the giant sporangiophore of *Phycomyces* [95]. When the slime variant of *N. crassa* was broken by an osmotic shock or by the use of glass beads in the MSK cell homogenizer, equal yields of identical looking chitosomes were obtained [96]. Finally, it must be indicated that chitosomes have also been isolated after isoosmotic breakage with polycations of the slime variant of *N. crassa* [58], and protoplasts from *C. albicans* [90] and *S. cerevisiae* [91]. Chitosomes from the slime variant of *N. crassa* could be separated from secretory vesicles by sucrose density centrifugation, both rate zonal and isopycnic [58].

All these observations, and the fact that chitosomes do not exhibit enzymatic activities normally associated with plasma membrane, mitochondria, endoplasmic reticulum, or vacuoles [14,58,83], demonstrate that chitosomes comprise a unique population of vesicles.

The average protein/lipid ratio of chitosomes from *M. rouxii* was found to be 2 [97], whereas this ratio was 1.4 for the chitosomes from *A. bisporus*

[98]. Analysis of the lipids from the chitosomes of *M. rouxii* showed about equal amounts of polar and neutral lipids—55 and 45%, respectively [97]—whereas 68% polar lipids were found in chitosomes from *A. bisporus* [98]. Sterols and sterol esters were the most abundant neutral lipids and phosphatidylcholine and phosphatidylserine the most abundant polar lipids in *M. rouxii* chitosomes [97]. Significant qualitative differences were found in the sterol components present in chitosomes and whole cells of the fungus [99]. Other noticeable differences between chitosomal and total cellular lipids from *M. rouxii* were the absence of phosphatidylserine and the abundance of glycolipids in chitosomes. The most abundant sugar in chitosomal lipids was galactose [97]. Further studies of these glycolipids showed the presence of a series of galactolipids with increasing number of galactosyl units (Lopez-Romero and J. Ruiz-Herrera, unpublished). The role of these glaactolipids in chitin biosynthesis remains unknown, although it is important to recall that a diglucosyldiglyceride was reported to be involved as a lipid acceptor in chitin biosynthesis by γ particles from *B. emersonii* [74] as described above. Chitosomes from *A. bisporus* did contain phosphatidylserine; and phosphatidic acid and phosphatidylethanolamine were the most abundant polar lipids in these organelles [98].

Polypeptide analysis of chitosomes has been a difficult task because of the exceedingly small amounts of chitin synthetase in fungal cells (less than 0.01% of the total cellular protein) and the difficulty in eliminating contaminating structures. As described above, chitosomes from the slime variant of *N. crassa* are enriched in polypeptides with molecular masses of 53, 38, 30, and 22 kDa [58], and chitosomes from *M. rouxii* contain polypeptides of molecular masses 60, 55, and 18 kDa [68a].

Chitosomes have a negative net charge and are precipitated by polycations [66]. Table 2 summarizes the general properties of chitosomes.

Chitin synthetase activity of chitosomes obtained from the yeast form of *M. rouxii* is very stable, even when maintained at room temperature. On the other hand, enzymatic activity of mycelial cells is extremely unstable, apparently due to the high concentration of inactivating proteases present in the mycelial form [57,100]. When the standard method of purification of chitosomes was modified breaking the cells in the presence of sucrose, precipitating chitosomes with ammonium sulfate, and eliminating the step of ribosome digestion by ribonuclease, fair yields of active chitosomes were obtained from mycelia of *M. rouxii*, *P. blakesleeanus*, and *G. persicaria* [101].

Chitosomes have provided a unique opportunity to study the mechanism of microfibril assembly since it is one of the few examples where both the enzyme complex and the product can be simultaneously analyzed by biochemical and ultrastructural methods [64]. Most chitosomal preparations are inactive unless treated with a protease. As described above, proteolytic activation is a rather unspecific process, but the sensitivity to different proteases varies with the fungal species. For example, acid proteases are more effective on chitosomes from *Mucor* [14,55]; whereas trypsin and chymotrypsin are more efficient activators of chitosomes from *Saccharomyces* [55,88], and pronase is more effective on chitosomes from *A. bisporus* [21].

When chitosomes are incubated with UDP-GlcNAc and activators (GlcNAc and a protease), they suffer a series of irreversible transformations (Figs. 3C and D). Inside some chitosomes delicate microfibrils can be

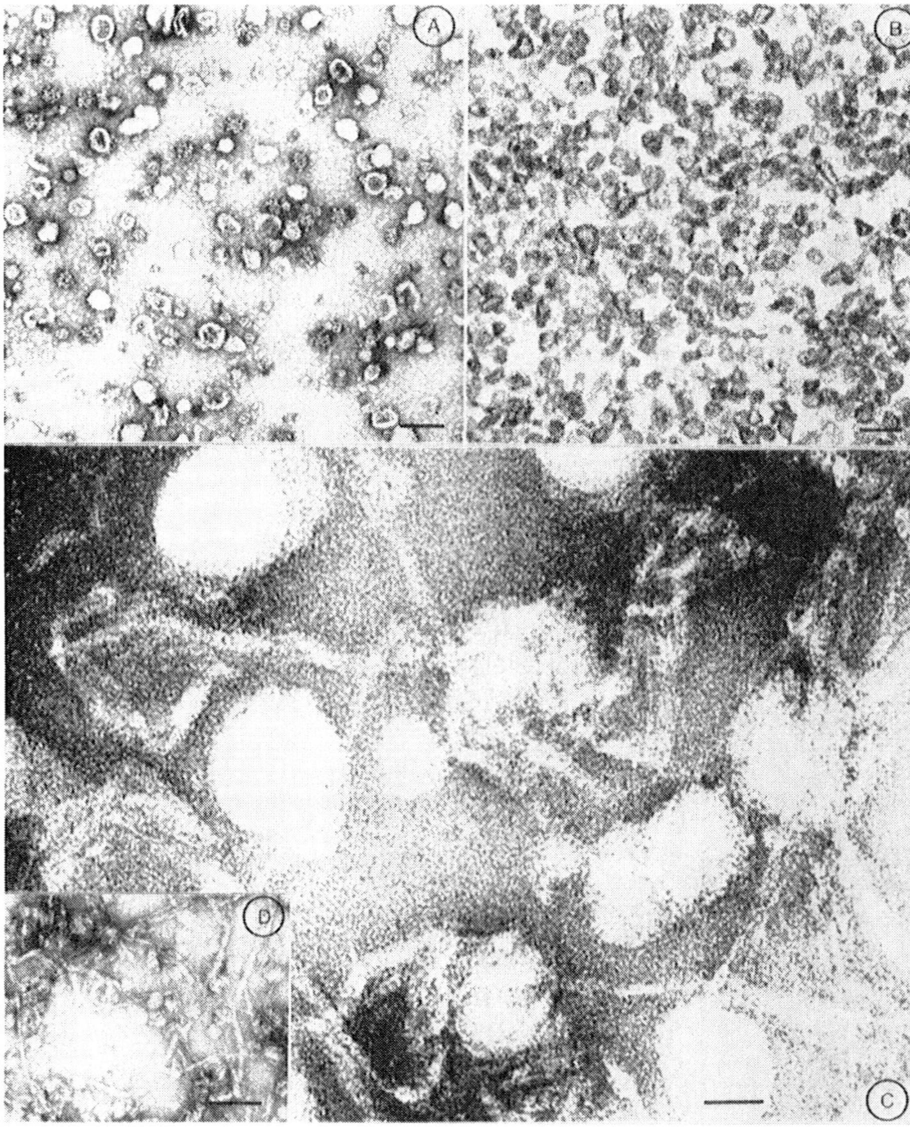

FIGURE 3. Chitosomes and chitin microfibrils. (A) Field of chitosomes obtained from the yeast form of *Mucor rouxii* observed in negative staining. (B) Thin section of chitosomes from *M. rouxii* to illustrate the appearance of the chitosomal membrane. (C) Appearance in negative staining of chitosomes, fibroids, and straight chitin microfibrils after incubation of chitosomes

TABLE 2. Properties of Chitosomes

Parameter	Value
Mean diameter	40-100 nm
Sedimentation coefficient	95-105S
Specific gravity	1.125-1.145
Membrane thickness	6.5-7.0 nm
Protein content	58-66%
Lipid content	33-42%

observed, whereas others accumulate thicker microfibrils in the form of coils or baskets (fibroids). Incubations carried out at low temperatures give rise mostly to these fibroid structures which are made of chitin as demonstrated by X-ray diffraction [64].

In most cases the chitosomal membrane is either broken or lost, but in some cases remnants of the membrane remain adhered to the chitin coils. It is a general observation that straight microfibrils emerge from the fibroids. After prolonged incubation, microfibrils associate giving rise to thicker fibers. Since fibroids are not found in cell walls, it has been hypothesized that they are produced by chitin crystallization within the confinement of the chitosome, probably because of the premature activation of the enzyme. The fact that these structures are made of twisted microfibrils is a strong evidence that the microfibril is the normal product of chitin crystallization.

The observation that chitosomes synthesize thick microfibrils 12-18 nm in diameter [102] suggests that they are equipped with several chitin synthetase units, each one responsible for the synthesis of one sugar chain, which upon hydrogen bonding give rise to a microfibril.

Chitosomes can be dissociated by several detergents, but chitin synthetase activity is preserved only when treated with digitonin [26,66] or with very low concentrations of Brij 36T (E. Lopez-Romero and T. Arredondo, unpublished observations). Digitonin treatment causes the disintegration of chitin synthetase units which sediment as a sharp band in sucrose gradients containing digitonin, with a sedimentation coefficient of 16S. This value corresponds to a molecular mass ~500 kDa. The 16S subunits from *M. rouxii* measure 7-12 nm [66]; whereas those from *A. bisporus* measure 10-20 nm [26]. Brij 36T dissociates chitosomes into smaller subunits with a molecular mass of 250 kDa.

Centrifugation of chitosomal subunits in sucrose gradients devoid of digitonin gives rise to their reaggregation into structures with higher sedimentation coefficients. *M. rouxii* subunits reconstitute into open planar, or vesiculoid structures. In thin sections, these structures revealed tripartite

---

from *Phycomyces blakesleeanus* with UDP-GlcNAc, GlcNAc, and trypsin. (D) Field of chitosomes, fibroids, and microfibrils obtained after incubation of chitosomes from *Saccharomyces cerevisiae* as in Part D. Negative staining. Magnification bars: A, B, and D, 100 nm; C, 25 nm. (Photographs in Parts A, B, and D by C. E. Bracker, S. Bartnicki-Garcia, and J. Ruiz-Herrera; in Part C, by B. Chavez, L. Herrera-Estrella, and J. Ruiz-Herrera.)

membrane profiles, some of them straight and some nearly vesicular [94]. In *A. bisporus* similar experiments produced populations of chitin synthetase-containing structures with the same sedimentation coefficient as the original chitosomes. Microscopic observations of these reaggregated structures revealed the presence of typical chitosomes [26].

Properties of chitin synthetase present in the 16S subunits are about the same as those displayed by intact chitosomes [26,103], including the necessity for proteolytic treatment to become active.

When chitosomal subunits are incubated with substrate and activators, they synthesize chitin microfibrils in the form of short needles, different in morphology from the long, slender ribbons synthesized by intact chitosomes [66]. Apparently this morphology is due to the aggregation of thinner microfibrils that occurs after prolonged incubations, since after short times of incubation, very thin microfibrils were found associated with chitosomal subunits from *A. bisporus* [26]. The tiniest microfibrils observed in these preparations were 1 nm thick, which suggests that they are formed by two GlcNAc chains. According to these results, the minimal chitin synthetase unit would be the 250-kDa aggregates obtained by dissociation with Brij 36T.

Chitosomes have been considered the conveyors of chitin synthetase en route to the sites of the cell where chitin synthesis and deposition takes place. Vesicles with the same characteristics of chitosomes, i.e., size and membrane width, have been observed in cell sections of almost all fungal cells [102-104]. These vesicles are accumulated at those sites where cell wall synthesis takes place: hyphal apices and developing septa. Microvesicles with similar characteristics to chitosomes are also observed inside multivesicular bodies [64].

The origin of chitosomes remains unknown, but several possibilities have been entertained [94]. Considering the results obtained in the reaggregation studies described above, it was suggested that chitosomes could originate by the aggregation of chitosomal subunits synthesized in the cytoplasm of the fungal cells. However, this possibility seems unlikely since chitosomes are not made solely of protein; in their structure intervene substantial amounts of lipids which play an important role providing the hydrophobic environment necessary for the catalytic activity of chitin synthetase (see above). A second possibility would be that they were vesicles originated from the endoplasmic reticulum (ER). This is also an unlikely possibility since chitosomes do not contain the normal enzymatic markers present in the ER. A third hypothesis would be that chitosomes originated from Golgi apparatus or its equivalent, accumulated close to the cell sites where cell wall synthesis takes place; it is pertinent to recall that in the giant sporangiohore of *P. blakesleeanus* chitin synthetase is accumulated in the growing apical region [61].

In this regard the results obtained with *B. emersonii* are important. Zoospores of the fungus encyst producing a chitinous cell wall, and at this time the specialized cytoplasmic organelles named γ particles start releasing numerous small vesicles (~80 nm in diameter) which migrate to the surface of the zoospore and fuse with the plasmalemma, coinciding with the formation of the primary cyst wall [107]. Since γ particles contain the chitin synthetase of the zoospore in an inactive state [108], it was hypothesized that those vesicles were equivalent to chitosomes and responsible for the transport of the enzyme to the cell surface. It was later demonstrated [109] that incubation of γ particles with the high-speed supernatant of a zoospore lysate induced the liberation of chitosomelike vesicles 70-120 nm in diameter with a buoyant density of 1.18. When these vesicles were incubated with substrate and activators, they synthesized chitin microfibrils. Images of microfibrils coming from single vesicles were observed in shadow cast preparations. These results suggest that some of the high-density populations of chitin synthetase observed when cell-free extracts of other fungi are centrifuged in sucrose gradients [83] may represent the membranous precursors of chitosomes.

IV. CHITIN BIOSYNTHESIS IN VIVO

A. Chitin Synthesis and Chitin Fibrillogenesis

The mechanism of chitin microfibril formation in vivo remains unknown. As discussed above, recent evidence has accumulated which suggests that the process of GlcNAc chain biosynthesis and the assembly of microfibrils are processes separated in time. Herth [40] demonstrated that two dyes which have great affinity for structural polysaccharides—Congo red and Calcofluor white—interfered with the correct crystallization of chitin microfibrils of the lorica from the chrysoflagellate alga *Poterioochromonas stipitata*. In the presence of either dye, instead of the normal long microfibrils, irregular meshworks of extremely thin and short subelementary fibrils were formed, mixed with patches of rodlike elements. These observations were later extended by the author to the ciliate protozoan *Eufolliculina uhligi* with similar results [110]. These observations were taken as evidence that molecules of the dyes introduced within the nascent chains of chitin, interfering with their correct assembly into microfibrils. These results, which were obtained with organisms synthesizing β-chitin, made of parallel GlcNAc chains, were extended to fungi which instead synthesize α-chitin, made of antiparallel chains. When protoplasts of *Candida albicans* are regenerated, microfibrillar networks appear homogeneously over the cell surface, whereas in the presence of Calcofluor white scattered ribbons with a low degree of organization were observed [41].

Chitin synthesized in vitro by cell-free extracts from *S. commune* in the presence of Calcofluor white lacked crystallinity, as revealed by X-ray diffraction, as long as the samples were kept wet. Only after drying did the material acquire a crystalline configuration [80]. This abnormal chitin was also highly susceptible to chitinase degradation, becoming more resistant to the enzymatic attack after drying. Nascent chitin synthesized by other enzymatic systems was found also to be very susceptible to degradation by chitinases and to become more resistant to hydrolisis as it ages [78,79]. Susceptibility of in vitro-synthesized nascent chitin to a specific chitin deacetylase was demonstrated in cell-free extracts from *M. rouxii* [71,111].

That the existence of this gap between chitin synthesis and crystallization is not an artifact from the in vitro systems or induced by the addition of Congo red and Calcofluor white was demonstrated by Vermeulen and Wessels [112], who found that newly synthesized chitin present in the growing apices of *S. commune* had different properties than the polysaccharide present in the lateral walls of the hyphae; this apical chitin was highly susceptible to chitinase and partially solubilized by dilute acids, whereas older chitin was resistant to both treatments.

Two mechanisms have been suggested for the in vivo synthesis of structural polysaccharides. The first mechanism invokes the operation of membrane-bound glycosyltransferases by a vectorial mechanism, in such a way that their active sites react with the nucleotide precursors at the cytoplasmic face of the membrane and extrude the polysaccharide to the outer medium. Using isolated plasma membrane preparations from *Saccharomyces*, it was demonstrated that synthesized chitin accumulated at the outer face of the membranes labeled with concanavalin A [113]. However, the possible entrapment of other chitin-synthesizing structures in these preparations cannot be completely ruled out. In the centric diatoms *Cyclotella cryptica* and *Thalassiosira fluviatilis*, Herth and his coworkers [110, 114] described the presence of specialized regions underlying the pores present in their silica valves, which give rise to long fibrillar appendages. These fibrils are characteristic of this group of microorganisms and are made of β-chitin. In thin sections those specialized regions showed conical invaginations in the plasma membrane characterized by dense layer coats. It was suggested that this dense membrane region contained the chitin synthetase molecules responsible for the formation of the chitin fibrils. Cytoplasmic vesicles accumulated in the underlying cytoplasm seemed to contribute to the formation of the specialized structures. In the case of cellulose, operation of transglycosylases embedded in the plasmalemma has been suggested by the observation of aggregates of particles, some in the form of rosettes, associated with microfibril impressions on the plasmalemma of algae and plant cells subjected to freeze-etching [115,116]. It is unfortunate that no biochemical evidence supports the hypothesis that the rosettes indeed represent cellulose-synthesizing enzymes, or that the impressions come from cellulose microfibrils [117].

Intracellular synthesis of structural polysaccharides followed by their secretion to the extracellular medium has been demonstrated in several organisms. In the alga *Pleurochrysis scherffelii* cellulose microfibril biosynthesis occurs in Golgi cisternae, and after arrangement in the form of scales, the microfibrils are released by an exocytic process [118]. During the encystment of *Polysphondylium pallidum* cellulose accumulation occurs in cytoplasmic vesicles which move to the cell surface to discharge their contents [119]. In the case of chitin, Mulisch et al. [120] and Herth [110] described that in the ciliate *Eufolliculina uhligi* the polysaccharide accumulates in an amorphous stage inside cytoplasmic vesicles and crystallizes in the form of β-chitin microfibrils after being discharged into the medium. The authors suggested that probably the presence of other compounds prevented crystallization of chitin inside the vesicles. In sections of *M. rouxii* we observed labeling by wheat germ lectin tagged with colloidal gold; and insoluble intracellular radioactive accumulation in sporangiophores of *Phycomyces* injected with ^{3}H-labeled UDP-GlcNAc [77], suggesting that chitin biosynthesis may be initiated in the chitosomes while in transit to the cell surface.

B. Chitin Deposition and Cell Wall Organization and Expansion

The concept that nascent chitin has different physicochemical properties than crystallized chitin agrees with the observation that the growing region of fungal cell wall has different mechanoelastic properties than nongrowing cell wall. Thus, it is a normal observation that fungal hyphae lyse at the apex when they are subjected to an osmotic shock, and that protoplasts are released at the hyphal tips when fungal cells are incubated with low concentrations of lytic enzymes.

Ahlquist and Gamow [121] measured the mechanical properties of the cell wall along the sporangiophores of *Phycomyces*, and demonstrated that extensibility was maximal at the tip and minimal at the base of the growing region. Dead sporangiophores had a homogeneously stiff cell wall. Ahlquist et al. [122] obtained a correlation between the extensibility of the cell wall from *Phycomyces* and a model which incorporated the concept of chitin microfibrils embedded in a Newtonian viscous fluid. Saunders and Trinci [123] also concluded that the shape of hyphal tips is established by the relative elasticity of the cell wall, establishing that the elasticity of the nascent wall decreases rapidly with time until it becomes rigidified.

What are the mechanisms which are involved in cell wall rigidification? In general, two concepts have been developed to explain this phenomenon. According to earlier ideas, Bartnicki-Garcia [124] proposed that the newly deposited cell wall was inherently rigid but that it was maintained in a plastic state by the controlled action of lytic enzymes; whereas Wessels [125,126] suggested that rigidity of the cell wall develops as the result of secondary processes, mainly chitin (or cellulose, or glucan) crystallization and establishment of covalent bonding between different components, in the wall itself. Both proposals are sustained by a body of evidence. Lytic enzymes have been demonstrated in a variety of fungi, and their operation in hyphal fusion in mating, heterokarion formation, and clamp connecting is evidence that they are present in hyphal growing zones. There is also evidence that once rigidified, the cell wall may become plastic again in response to different stimuli, i.e., in branching. Weakening of the growing zone of the sporangiophores of *P. blakesleeanus* induced by illumination was observed [127] and demonstrated to be due to chitin breakage [61]. On the other hand, the existence of a nascent stage of chitin with physicochemical properties different from those of the crystalline polysaccharide, and which is more susceptible to chitinolytic attack, is a proven fact. During this stage chitin may be modified enzymatically by a deacetylase to form chitosan [71,111,128]. It was observed that stage I sporangiophores from *Phycomyces* can be stained by fluorescent wheat germ lectin only at the tip, but when they are treated with HNO_2 (which degrades chitosan), the lectin binds to the whole sporangiophore, suggesting that part of the chitin deposited at the apex of the cell is deacetylated into chitosan as it moves down to the lateral wall [61].

Experimental evidence suggested that chitin and glucan may form complexes [129,130] through covalent linkages. Thus, in *S. commune* and *S. cerevisiae* the existence of covalent interactions between chitin and β-D-glucan via amino acid residues has been described [131,132]. These alkali-insoluble complexes derive from water-soluble β-1-3 glucans and nascent chitin deposited at the hyphal apices [133,134].

Gopal et al. [135] suggested the existence of a 1,3-β/1,6-β-glucan/chitin complex adjacent to the plasma membrane in *C. albicans*. Evidence for a covalent linkage between chitin and 1,6-β-glucan through a glycosidic linkage at positions 6 of the GlcNAc and 1 of the glucose moieties was recently reported [136]. Unfortunately, the catalytic systems which seemingly are responsible for the formation of these bonds are still unknown.

Formation of linkages between chitin and β-glucans may account for the insolubilization (or changes in the solubilization characteristics) of these latter polymers. Glucans, in regenerating protoplasts and growing hyphae, become insoluble in water or alkali only after polymerization of chitin [131,133,137]. Moreover, inhibition of chitin synthesis by polyoxin D [137, 138] abolishes insolubilization of glucan chains in the growing walls of regenerating protoplasts. As the wall structure matures, the stiffening effect of an increasing degree of crosslinking between wall polymers counterbalances the increasing flexibility due to crosslinking, which reduces crystallinity.

The situation in the Zygomycetes which do not contain glucan in their hyphal walls appears to be different. In these forms, chitin microfibrils coated with partly deacetylated chitin chains (chitosan) would form a rigid structure by establishing ionic interactions with several anionic wall polymers [126,139].

Other data suggest the association of structural components with glycoproteins. Thus, in the presence of Calcofluor white, cells of *Candida* accumulate covalently linked wall mannoproteins in high amounts in the periplasmic space, being finally released to the surrounding medium [140]. Inhibition of protein association with the structural polysaccharides and blockage of the wall normal morphogenetic process by the dye suggest that in vivo the mannoproteins might regulate formation of crosslinks between polymers. This interpretation is in agreement with the observation that the degree of crystallinity of the glucan in vivo is very low. Experiments carried out with regenerating protoplasts have shown that noncovalently linked wall mannoproteins are present in the newly formed abnormal walls from the very beginning of the regeneration process and before other wall proteins are incorporated [140-142]. In vitro experiments have shown that partially regenerated walls isolated from protoplasts [138] strongly retain noncovalently bound wall mannoproteins.

Another interesting observation is that some of these mannoproteins exhibit affinity not only for the structural polysaccharides but also for other high molecular weight mannoproteins, which are thought to be covalently linked to glucan [143] and chitin [143a]. Nothing is known yet about the physiological significance of these "bifunctional" mannoproteins. One hypothesis is that they could be related to the process of wall assembly, by modulating the spontaneous interaction of different wall constituents. The characteristics of the domains involved in the interaction between the two types of mannoproteins are unknown, but they may be localized in their polypeptidic moieties, as the molecules lacking N-linked carbohydrate chains still retain this interplay capability [143].

Based on all these data, we suggested [77] that both Bartnicki-Garcia's and Wessels's proposals may be joined in a general hypothesis which considers the normal operation of the so-called steady-state model [126] for cell wall growth, under the regulatory control of lytic enzymes [124].

C. Polarization of Chitin Deposition and Wall Growth

Chitin deposition in spherical cells occurs isodiametrically, whereas in hyphae it takes place at the apex [144]. This differential deposition of chitin apparently is in turn due to the selective movement of chitosomes to those sites of the cell where wall growth takes place. The operation of this highly regulated mechanism is responsible for cell shape. The specialized mechanism of hyphal growth depends on the different mechanoelastic properties of the hyphal tip described above and to the characteristic rate of area expansion in the apex. It has been demonstrated that the specific rate of area expansion of the apical wall at a specific point depends on a trigonometric function relating the angle between the longitudinal axis of the hypha and the particular point on the tip wall [145,146]. Since the shape of hyphal apices approximates that of half-ellipsoids of revolution, it appears that wall expansion depends on the cotangent of the indicated angle [146]. The mechanism responsible for maintaining this wall growth gradient in apically growing cells remains unknown. It has been observed that fungal cells treated with cycloheximide continue synthesizing chitin for several hours [147,148]. Under these conditions the pattern of chitin deposition changes and occurs subapically in *Aspergillus nidulans*, giving rise to subapical cell thickening [147]. These results suggested that chitin synthetase present in the older parts of the cell was not irreversibly inactivated and that cycloheximide somehow permitted its reactivation. With this background Sietsma et al. [149] suggested that chitin synthetase was organized as a patch at the tip of the hyphae and moved along with the expanding cell, probably stabilized and driven by a bundle of microtubules. On the other hand, we suggested that each chitosome is responsible for the formation of a microfibril, which, having reached the apical pole of the cell and performed its task rapidly decays [150]. Since most chitosomes would fuse at the apical pole, their own operation would provide the mechanistic explanation for the observed apical gradient of chitin biosynthesis. In support for this hypothesis we demonstrated that chitin biosynthesis is immediately stopped in germlings of *P. blakesleeanus* by the addition of the calcium ionophore A23187. This effect of the ionophore was not due to metabolic alterations or inhibition in chitin synthetase formation or activity, but was traced back to the inhibition in the formation and migration of apical vesicles [151].

In regard to the delayed effect of cycloheximide on chitin synthesis, it may be attributed to the presence of a chitin synthetase pool in the cell cytoplasm as was demonstrated for *M. rouxii* [77]. We observed that exhaustion of chitin synthesis in cells incubated with cycloheximide correlates with a drop in the number of apical vesicles (A. Obregon and J. Ruiz-Herrera, unpublished data). Bartnicki-Garcia et al. [152] developed an ingenious model by computer simulation which illustrates that the shape of the cell is determined by the relative position of a "vesicle supply center" (VSC). According to this model, cells grow isodiametrically if the VSC is present in the center of the cells. Movement of the VSC to the surface of the cell distorts the spherical shape and eventually generates typical cylindrical cells with round tips. The model yielded the equation $y = x \cot(xVo/N)$, which defines cell shape depending on two parameters, the number of vesicles released in the unit of time (N) and the rate of displacement of the VSC (V_0). By fitting their theoretical data with published electron micrographs, the authors obtained good correlations and established the possible identity of the vesicle supply center with the Spitzenkörper. This model may solve the problem relating to the mechanisms involved in vesicle migration, considering that it would be simpler to displace a single organelle than the extremely high number of vesicles required for cell wall growth. For example, Collinge and Trinci [153] calculated that 40,000 vesicles are necessary to sustain 1 min of growth of *N. crassa* hyphae.

In regard to the mechanisms involved in vesicle (or VSC) migration to the cell surface, two main candidates have been considered: the cytoskeleton, and electrical or ionic currents. In support for a role of microtubules in hyphal growth we may cite the observation that the anti-microtubular agent benzimidazole-2-ylcarbamate affected microtubular arrangement and vesicle distribution in *Fusarium acuminatum* [154,155]; whereas the alterations in the pattern of wall synthesis in several fungi by cytochalasins point to a role of microfilaments in vesicle migration [156-158].

In support of the suggestion that electrical currents and ion gradients are the driving force involved in vesicle migration, we may cite that trans-hyphal currents, positive entering at the apex and leaving at the subapical region, have been registered in a large number of fungal species [159-163]. It has been suggested that the electric field may charge the cytoplasm or the membrane, displacing the vesicles by electrophoresis [164] or by an electroosmotic flow [165]. It was also suggested that perhaps more important than the electrical current are the ionic gradients established, which may be the leading force for the displacement of vesicles [162]. The suggested ions to play this role are H^+ [162] or potassium or calcium [166-167]. Calcium gradients in different apical-growing cells, including fungi, were demonstrated by different techniques [168,169], and the collapse in the calcium gradient induced by the calcium ionophore A23187 [170] was shown to stop apical growth and originate a disorganization and reduction in the number of apical vesicles [170,171]. As described above, we demonstrated that the calcium ionophore stops chitin biosynthesis in *Phycomyces* germlings, and inhibits the formation and migration of apical vesicles [151].

A compromise hypothesis to include all these observations may be that the cytoskeleton of the cell would be responsible for the positioning of the "vesicle formation center" and that vesicle pinpointing to specific sites of the cell surface involved the operation of electrical or ionic currents.

V. CONCLUSIONS

Chitin and cellulose are the skeletal components used by a great number of eukaryotic organisms to form their protective structures. In fungi, as described in the preceding pages, chitin has filled the need to build a resistant cell wall. It must be accepted that our knowledge on the mechanisms of synthesis of the polysaccharide both in vivo and in vitro is still fragmentary, but much has been gained in recent years. Although we have been unable to isolate the polypeptide(s) responsible for catalytic activity, we have acquired wide knowledge on the kinetics, stimulation, and inhibition of chitin synthetase; and it may be predicted that with the cloning of the structural genes, knowledge of the enzyme at the molecular level will be achieved in a few more years. Isolation and study of chitosomes have provided valuable information on the mechanism of chitin fibrillogenesis. We know that the enzyme is maintained intracellularly in an inactive state to avoid its premature activity and that it travels to its final destination following the normal exocytic route inside specialized vesicles, the chitosomes. During its transit or until it reaches the cell surface, the transglycosylase becomes activated and synthesizes GlcNAc chains in a nascent stage thus allowing that they be covalently modified, and become associated with other polysaccharides and glycoproteins in the wall matrix itself, before giving rise to the typical microfibrils. This process of maturation—and we are still ignorant of the intimate mechanisms—occurs at the hyphal apices, transforming a plastic structure into a rigid composite by an exquisite process of self-assembly which has been conserved along the evolution of fungi, illustrating how it has fit the necessities for the specialized pattern of growth of these organisms.

ACKNOWLEDGMENTS

The original work from J. Ruiz-Herrera was supported in part by CONACyT and DIGICySA from the Secretaria de Educacion Publica, Mexico. JRH is National Investigator, Step III, Mexico. Thanks are given to Mrs. Carmen Medrano for careful typing of the manuscript.

REFERENCES

1. Glaser, L., and Brown, D. H. The synthesis of chitin in cell-free extracts of Neurospora crassa. *J. Biol. Chem.*, *228*:729-742, 1957.
2. Porter, C. A. and Jaworski, E. G. The synthesis of chitin by particulate preparations of Allomyces macrogynus. *Biochemistry*, *5*:1149-1154, 1966.
3. Camargo, P. E., Dietrich, C. P., Sonneborn, D., and Strominger, J. L. Biosynthesis of chitin in spores and growing cells of Blastocladiella emersonii. *J. Biol. Chem.*, *242*:3121-3128, 1967.
4. McMurrough, I., Flores-Carreon, A., and Bartnicki-Garcia, S. Pathway of chitin synthesis and cellular localization of chitin synthetase in Mucor rouxii. *J. Biol. Chem.*, *246*:3999-4007, 1971.
5. Nanba, H., and Kuroda, H. Studies on fungicides X. Biosynthesis of β-glucan and chitin-like substance of cell wall from Cochliobolus miyabeanus. *Chem. Pharm. Bull.*, *22*:610-616, 1974.
6. Ruiz-Herrera, J. Synthesis of chitin microfibrils in vitro. In *Cellulose and Other Natural Polymer Systems* (R. M. Brown, ed.), Plenum Press, New York, 1982, pp. 207-223.
7. McMurrough, I., and Bartnicki-Garcia, S. Properties of a particulate chitin synthetase from Mucor rouxii. *J. Biol. Chem.*, *246*:4008-4016, 1971.
8. Keller, F. A., and Cabib, E. Chitin and yeast budding. Properties of chitin synthetase from Saccharomyces carlsbergensis. *J. Biol. Chem.*, *246*:160-166, 1971.
9. Lopez-Romero, E., and Ruiz-Herrera, J. Synthesis of chitin by particulate preparations from Aspergillus flavus. *Antonie van Leeuwenhoek*, *42*:261-276, 1976.
10. Ryder, N. S., and Peberdy, J. F. Chitin synthase in Aspergillus nidulans: properties and proteolytic activation. *J. Gen. Microbiol.*, *99*:69-76, 1977.
11. Moore, P. M., and Peberdy, J. F. A particulate chitin synthase from Aspergillus flavus Link: the properties, location and levels of activity in mycelium and regenerating protoplast preparations. *Can. J. Microbiol.*, *22*:915-921, 1976.
12. Peberdy, J. F., and Moore, P. M. Chitin synthase in Mortierella vinacea, properties, cellular location and synthesis in growing cultures. *J. Gen. Microbiol.*, *90*:228-236, 1975.

13. Manocha, M. S., and Begum, A. Properties of chitin synthase from mucoraceous hosts of a mycoparasite. *Can. J. Microbiol.*, *31*:6-12, 1985.
14. Ruiz-Herrera, J., Lopez-Romero, E., and Bartnicki-Garcia, S. Properties of chitin synthetase in isolated chitosomes from yeast cells of Mucor rouxii. *J. Biol. Chem.*, *252*:3338-3343, 1977.
15. Sburlati, A., and Cabib, E. Chitin synthetase 2, a presumptive participant in septum formation in Saccharomyces cerevisiae. *J. Biol. Chem.*, *261*:15147-15152, 1986.
16. Orlean, P. Two chitin synthases in Saccharomyces cerevisiae. *J. Biol. Chem.*, *262*:5732-5739, 1987.
17. Gooday, G. W., and Rousset-Hall, A. Properties of chitin synthetase from Corpinus cinereus. *J. Gen. Microbiol.*, *89*:137-145, 1975.
18. Edson, C. M., and Brody, S. Biochemical and genetic studies on galactosamine metabolism in Neurospora crassa. *J. Bacteriol.*, *126*:799-805, 1976.
19. Schmit, J. C., Edson, C. M., and Brody, S. Changes in glucosamine and galactosamine levels during conidial germination in Neurospora crassa. *J. Bacteriol.*, *122*:1062-1070, 1975.
20. Leal-Morales, C., and Ruiz-Herrera, J. Alterations in the biosynthesis of chitin and glucan in the slime mutant of Neurospora crassa. *Exp. Mycol.*, *9*:28-38, 1985.
21. Hanseler, E., Nyhlen, L. E., and Rast, D. M. Isolation and properties of chitin synthetase from Agaricus bisporus mycelium. *Exp. Mycol.*, *7*:17-30, 1983.
22. Rousset-Hall, A., and Gooday, G. W. A kinetic study of solubilized chitin synthetase preparations from Coprinus cinereus. *J. Gen. Microbiol.*, *89*:146-154, 1975.
23. Lopez-Romero, E. Estudio sobre la biosintesis de quitina en Aspergillus flavus y Mucor rouxii. Ph.D. thesis, Instituto Politecnico Nacional, Mexico, 1976, pp. 160-165.
24. Vermeulen, C. A., and Wessels, J. G. H. Evidence for a phospholipid requirement of chitin synthase in Schizophyllum commune. *Curr. Microbiol.*, *8*:67-71, 1983.
25. Montgomery, G. W. G., and Gooday, G. W. Phospholipid-enzyme interactions of chitin synthase of Coprinus cinereus. *FEMS Microbiol. Lett.*, *27*:29-33, 1985.
26. Hanseler, E., Nyhlen, L. E., and Rast, D. M. Dissociation and reconstitution of chitosomes. *Biochim. Biophys. Acta*, *745*:121-133, 1983.
27. Jaworski, E. G., Wang, L. C., and Carpenter, W. D. Biosynthesis of chitin in cell-free extracts of Venturia inaequalis. *Phytopathology*, *55*:1309-1312, 1965.
28. Moore, P. M., and Peberdy, J. F. Biosynthesis of chitin by particulate fractions from Cunninghamella elegans. *Microbios*, *12*:29-39, 1975.
29. Jan, Y. N. Properties and cellular localization of chitin synthetase in Phycomyces blakesleeanus. *J. Biol. Chem.*, *249*:1973-1979, 1974.
30. Vagabov, V. M., Lupashin, V. V., Knyazeva, I. N., and Kulaev, I. S. Study of the influence of ATP on the activity of membrane-bound chitin synthase of the yeast Saccharomyces carlsbergensis. Translated from *Biokhimiya*, *51*:1802-1810, 1986.
31. Vagabov, V. M., Knyazeva, I. N., Lupashin, V. V., Krupyanko, V. I., and Kulaev, I. S. Study of chitin synthase from the yeast

Saccharomyces carlsbergensis. Effect of ATP on solubilized enzyme. Translated from *Biokhimiya, 52*:534-542, 1987.
32. Duran, A., and Cabib, E. Solubilization and partial purification of yeast chitin synthase. Confirmation of the zymogenic nature of the enzyme. *J. Biol. Chem., 253*:4419-4425, 1978.
33. Endo, A., Kakiki, K., and Misato, T. Mechanism of action of the antifungal agent polyoxin D. *J. Bacteriol., 104*:189-196, 1970.
34. Dahn, U., Hagenmaier, H., Hohne, H., Konig, W. A., Wolf, G., and Zahner, H. Stffwechselfprodukte von Mikroorganismen, 154 Mitteilung. Nikkomycin, ein neuer Hemmstoff der Chitinsynthese bei Pilzen. *Arch. Microbiol., 107*:143-160, 1976.
35. Endo, A., and Misato, T. Polyoxin D, a competitive inhibitor of UDP-N-acetyl-glucosamine: chitin N-acetyl glucosaminyl transferase in Neurospora crassa. *Biochem. Biophys. Res. Commun., 37*:718-722, 1969.
36. Ohta, N., Kakiki, K., and Misato, T. 1970. Studies on the mode of action of polyoxin D. Part II. Effect of polyoxin D on the synthesis of cell wall chitin. *Agr. Biol. Chem., 34*:1224-1234, 1970.
37. Hori, M., Kakiki, K., and Misato, T. Further study on the relation of polyoxin structure to chitin synthetase inhibition. *Agr. Biol. Chem., 38*:691-698, 1974.
38. Hori, M., Kakiki, K., and Misato, T. Interaction between polyoxin and active center of chitin synthetase. *Agr. Biol. Chem., 38*:699-705, 1974.
39. Furter, R., and Rast, D. M. A comparison of the chitin synthase-inhibitory and antifungal efficacy of nucleoside-peptide antibiotics: structure-activity relationship. *FEMS Microbiol. Lett., 28*:205-211, 1985.
40. Herth, W. Calcofluor white and Congo red inhibit chitin microfibril assembly of Poteriochromonas: evidence for a gap between polymerization and microfibril formation. *J. Cell Biol., 87*:442-450, 1980.
41. Elorza, M. V., Rico, H., and Sentandreu, R. Calcofluor white alters the assembly of chitin fibrils in Saccharomyces cerevisiae and Candida albicans cells. *J. Gen. Microbiol., 129*:1577-1582, 1983.
42. Vannini, G. L., Poli, F., Donini, A., and Pancaldi, S. Effects of Congo red on cell wall synthesis and morphogenesis in Saccharomyces cerevisiae. *Plant Sci. Lett., 31*:9-17, 1983.
43. Selitrennikoff, C. P. Calcofluor white inhibits Neurospora chitin synthetase activity. *Exp. Mycol., 8*:269-272, 1984.
44. Roncero, C., and Duran, A. Effect of Calcofluor white and Congo red on fungal cell wall morphogenesis: in vivo activation of chitin polymerization. *J. Bacteriol., 163*:1180-1185, 1985.
45. Rast, D. M., and Bartnicki-Garcia, S. Effects of amphotericin B, nystatin, and other polyene antibiotics on chitin synthase. *Proc. Natl. Acad. Sci. USA, 78*:1233-1236, 1981.
46. Cohen, E. Inhibition of chitin synthesis in insect systems. In *Chitin in Nature and Technology* (R. Muzzarelli, C. Jeuniaux, and G. W. Gooday, eds.), Plenum Press, New York, 1986, pp. 139-145.
47. Mitsui, T. Mode of inhibition of chitin synthesis by diflubenzuron. In *Chitin in Nature and Technology* (R. Muzzarelli, C. Jeuniaux, and G. W. Gooday, eds.), Plenum Press, New York, 1986, pp. 193-196.
48. Cabib, E., and Farkas, V. The control of morphogenesis: an enzymatic mechanism for the initiation of septum formation in yeast. *Proc. Natl. Acad. Sci. USA, 68*:2052-2056, 1971.

49. Ulane, R. E., and Cabib, E. The activating system of chitin synthetase from Saccharomyces cerevisiae. Purification and properties of the activating factor. *J. Biol. Chem.*, *251*:3367-3374, 1976.
50. Lenney, J. F., and Dalbec, J. M. Purification and properties of two proteinases from Saccharomyces cerevisiae. *Arch. Biochem. Biophys.*, *129*:407-409, 1967.
51. Schott, E. H., and Holzer, H. Purification and some properties of tryptophan synthase inactivase II from yeast. *Eur. J. Biochem.*, *42*:61-66, 1974.
52. Wolf, D. H., and Ehman, C. Isolation of yeast mutants lacking proteinase B activity. *FEBS Lett.*, *92*:121-124, 1978.
53. Zubenko, G. S., Mitchell, A. P., and Jones, E. W. Septum formation, cell division, and sporulation in mutants of yeast deficient in proteinase B. *Proc. Natl. Acad. Sci. USA*, *76*:2395-2399, 1979.
54. Cabib, E. The synthesis and degradation of chitin. *Adv. Enzymol.*, *59*:59-101, 1987.
55. Bartnicki-Garcia, S., Bracker, C. E., Reyes, E., and Ruiz-Herrera, J. Isolation of chitosomes from taxonomically diverse fungi and synthesis of chitin microfibrils in vitro. *Exp. Mycol.*, *2*:173-192, 1978.
56. Arroyo-Begovich, A., and Ruiz-Herrera, J. Proteolytic activation and inactivation of chitin synthase from Neurospora crassa. *J. Gen. Microbiol.*, *113*:339-345, 1979.
57. Diaz, S., and Ruiz-Herrera, J. Purification of an acid protease from Mucor rouxii that inactivates chitin synthetase. *Antonie van Leeuwenhoek*, *53*:279-291, 1987.
58. Ruiz-Herrera, J., Martinez, J. P., Casanova, M., Gil, M. L., and Sentandreu, R. Separation of chitosomes and secretory vesicles from the "slime" variant of Neurospora crassa. *Arch. Microbiol.*, *149*:156-162, 1987.
59. Reyna, G., and Ruiz-Herrera, J. Fotoactivacion in vitro de la quitina sintetasa de los esporangioforos de Phycomyces blakesleeanus. *Rev. Mex. Micol.*, *3*:249-263, 1987.
60. Hasilik, A. Inactivation of chitin synthase in Saccharomyces cerevisiae. *Arch. Microbiol.*, *101*:295-301, 1974.
61. Herrera-Estrella, L., and Ruiz-Herrera, J. Light response in Phycomyces blakesleeanus: evidence for roles of chitin biosynthesis and breakdown. *Exp. Mycol.*, *7*:362-369, 1983.
62. Martinez-Cadena, G., and Ruiz-Herrera, J. Activation of chitin synthetase from Phycomyces blakesleeanus by calcium and calmodulin. *Arch. Microbiol.*, *148*:280-285, 1987.
63. Meijer, L., and Guerrier, P. Activation of calmodulin-dependent NAD kinase by trypsin. *Biochim. Biophys. Acta*, *702*:143-146, 1982.
64. Bracker, C. E., Ruiz-Herrera, J., and Bartnicki-Garcia, S. Structure and transformation of chitin synthetase particles (chitosomes) during microfibril synthesis in vitro. *Proc. Natl. Acad. Sci. USA*, *73*:4570-4574, 1976.
65. Braun, P. C., and Calderone, R. A. Regulation and solubilization of Candida albicans chitin synthetase. *J. Bacteriol.*, *140*:666-670, 1979.
66. Ruiz-Herrera, J., Bartnicki-Garcia, S., and Bracker, C. E. Dissociation of chitosomes by digitonin into 16S subunits with chitin synthetase activity. *Biochim. Biophys. Acta*, *629*:201-216, 1980.
67. Montgomery, G. W. G., Adams, D. J., and Gooday, G. W. Studies on the purification of chitin synthase from Coprinus cinereus. *J. Gen. Microbiol.*, *130*:291-297, 1984.

68. Kang, M. S., Elango, N., Mattio, E., Au-Young, J., Robbins, P. W., and Cabib, E. Isolation of chitin synthase from Saccharomyces cerevisiae: purification of an enzyme by entrapment in the reaction product. *J. Biol. Chem.*, *259*:14966-14972, 1984.
68a. Flores-Martínez, A., López-Romero, E., Martinez, J. P., Bracker, C. E., Ruiz-Herrera, J., and Bartnicki-Garcia, S. Protein composition of purified chitosomes of Mucor rouxii. *Exp. Mycol.*, *14*:160-168, 1990.
69. Bulawa, C. E., Slater, M., Cabib, E., Au-Young, J., Sburlati, A., Adair, W. L., and Robbins, P. W. The Saccharomyces cerevisiae structural gene for chitin synthase is not required for chitin synthesis in vivo. *Cell*, *46*:213-225, 1986.
70. Silverman, S. J., Sburlati, A., Slater, M. L., and Cabib, E. Chitin synthase 2 is essential for septum formation and cell division in Saccharomyces cerevisiae. *Proc. Natl. Acad. Sci. USA*, *85*:4735-4739, 1988.
71. Calvo-Mendez, C., and Ruiz-Herrera, J. Biosynthesis of chitosan in membrane fractions from Mucor rouxii by the concerted action of chitin synthetase and a particulate deacetylase. *Exp. Mycol.*, *11*:128-140, 1987.
72. Cubero, B., and Cerda-Olmedo, E. Mutantes de la sintetasa de la quitina de Phycomyces blakesleeanus. *IV Reunion conjunta de Micologia. Resumenes*. 1988, p. 132.
73. Suarez, T., and Eslava, A. P. Transformation of Phycomyces with a bacterial gene for kanamycin resistance. *Mol. Gen. Genet.*, *212*:120-123, 1988.
74. Mills, G. L., and Cantino, E. The lipid composition of the Blastocladiella emersonii γ-particle and the function of γ-particle lipid in chitin formation. *Exp. Mycol.*, *2*:99-109, 1978.
75. Bartnicki-Garcia, S., Lippman, E., and Heick, J. Evidence for a polypeptide acceptor in fungal chitin biosynthesis. *Abstracts of the Annual Meeting of the American Society for Microbiology*, 1979, p. 106.
76. Kukuruzinska, M. A., Bergh, M. L. E., and Jackson, B. J. Protein glycosylation in yeast. *Annu. Rev. Biochem.*, *56*:915-944, 1987.
77. Ruiz-Herrera, J., and Sentandreu, R. Fungal cell wall synthesis and assembly. *Advances in Medical Mycology*, Vol. 13 (M. R. McGinnis and M. Borgers, eds.), Springer-Verlag, New YOrk, 1989, pp. 168-217.
78. Molano, J., Polacheck, I., Duran, A., and Cabib, E. An endochitinase from wheat germ: activity on nascent and preformed chitin. *J. Biol. Chem.*, *254*:4901-4907, 1979.
79. Lopez-Romero, E., Ruiz-Herrera, J., and Bartnicki-Garcia, S. The inhibitory protein of chitin synthetase from Mucor rouxii is a chitinase. *Biochim. Biophys. Acta*, *702*:233-236, 1982.
80. Vermeulen, C. A., and Wessels, J. G. H. Chitin synthesis by a fungal membrane preparation. Evidence for a transient non-crystalline state of chitin. *Eur. J. Biochem.*, *158*:411-415, 1986.
81. Ruiz-Herrera, J., and Bartnicki-Garcia, S. Synthesis of cell wall microfibrils in vitro by a "soluble" chitin synthetase from Mucor rouxii. *Science*, *186*:357-359, 1974.
82. Vermeulen, C. A., Raeven, M. B. J. M., and Wessels, J. G. H. Localization of chitin synthase activity in subcellular fractions of Schizophyllum commune protoplasts. *J. Gen. Microbiol.*, *114*:87-97, 1979.
83. Ruiz-Herrera, J., Bracker, C. E., and Bartnicki-Garcia, S. Sedimentation properties of chitosomes from Mucor rouxii. *Protoplasma*, *122*:178-190, 1984.
84. Duran, A., Bowers, B., and Cabib, E. Chitin synthetase zymogen is attached to the yeast plasma membrane. *Proc. Natl. Acad. Sci. USA*, *72*:3952-3955, 1975.

85. Braun, P. C., and Calderone, R. A. Chitin synthesis in Candida albicans: comparison of yeast and hyphal forms. *J. Bacteriol.*, *135*: 1472-1477, 1978.
86. Duran, A., Cabib, E., and Bowers, B. Chitin synthetase distribution on the yeast plasma membrane. *Science*, *203*:363-365, 1979.
87. Hanson, B., and Marple, S. The localization of chitin synthetase in yeast. *Abstracts of the Annual Meeting of the Society for Complex Carbohydrates*, 1983, p. 62.
88. Flores-Martinez, A., and Schwencke, J. Chitin synthetase activity is bound to chitosomes and to the plasma membrane in protoplasts of Saccharomyces cerevisiae. *Biochim. Biophys. Acta*, *946*:328-336, 1988.
89. Leal-Morales, C. A., Bracker, C. E., and Bartnicki-Garcia, S. Localization of chitin synthetase in cell-free homogenates of Saccharomyces cerevisiae: chitosomes and plasma membrane. *Proc. Natl. Acad. Sci. USA*, *85*:8516-8520, 1988.
90. Gozalbo, D., Dubon, F., Schwencke, J., and Sentandreu, R. Characterization of chitosomes in Candida albicans protoplasts. *Exp. Mycol.*, *11*:331-338, 1987.
91. Sentandreu, R., Martinez-Ramon, A., and Ruiz-Herrera, J. Localization of chitin synthase in Mucor rouxii by an autoradiographic method. *J. Gen. Microbiol.*, *130*:1193-1199, 1984.
92. Martinez, J. P., Gimenez, G., and Bartnicki-Garcia, S. Sedimentation properties of chitosomal chitin synthetase from the wild-type strain and the "slime" variant of Neurospora crassa. *Biochim. Biophys. Acta*, *990*:45-52, 1989.
93. Farkas, V. Biosynthesis of cell walls of fungi. *Microbiol. Rev.*, *43*: 117-144, 1979.
94. Bartnicki-Garcia, S., Ruiz-Herrera, J., and Bracker, C. E. Chitosomes and chitin synthesis. In *Fungal Cell Walls and Hyphal Growth* (G. W. Gooday and A. P. J. Trinci, eds.), Cambridge University Press, Cambridge, 1979, pp. 149-168.
95. Herrera-Estrella, L., Chavez, B., and Ruiz-Herrera, J. Presence of chitosomes in the cytoplasm of Phycomyces blakesleeanus and the synthesis of chitin microfibrils. *Exp. Mycol.*, *6*:385-388, 1982.
96. Bartnicki-Garcia, S., Bracker, C. E., Lippman, E., and Ruiz-Herrera, J. Chitosomes from the wall-less "slime" mutant of Neurospora crassa. *Arch. Microbiol.*, *139*:105-112, 1984.
97. Hernandez, J., Lopez-Romero, E., Cerbon, J., and Ruiz-Herrera, J. Lipid analysis of chitosomes, chitin synthesizing microvesicles from Mucor rouxii. *Exp. Mycol.*, *5*:349-356, 1981.
98. Weete, J. D., Furter, R., Hanseler, E., and Rast, D. M. Cellular and chitosomal lipids of Agaricus bisporus and Mucor rouxii. *Can. J. Microbiol.*, *31*:1120-1126, 1985.
99. Lopez-Romero, E., Monzon, E., and Ruiz-Herrera, J. Sterol composition of chitosomes from yeast cells of Mucor rouxii: comparison with whole cells. *FEMS Microbiol. Lett.*, *30*:369-372, 1985.
100. Ruiz-Herrera, J., and Bartnicki-Garcia, S. Proteolytic activation and inactivation of chitin synthetase from Mucor rouxii. *J. Gen. Microbiol.*, *97*:241-249, 1976.
101. Martinez-Cadena, G., Lopez-Romero, E., Acosta, I., Gonzales, C., and Ruiz-Herrera, J. Stabilization of chitin synthetase and purification of chitosomes from several mycelial Mucorales. *Antonie van Leeuwenhoek*, *53*:171-181, 1987.

102. Ruiz-Herrera, J., Sing, V. O., Van Der Woude, W. J., and Bartnicki-Garcia, S. Microfibril assembly by granules of chitin synthetase. *Proc. Natl. Acad. Sci. USA, 72*:2706-2710, 1975.
103. Hernández-Delgadillo, R., and Ruiz-Herrera, J. Propiedades de la quitina sintetasa presente en las subunidades 165 de Mucor rouxii. *Rev. Mex. Micol.*, 5:33-41, 1989.
104. Girbardt, M. Die Ultrastruktur der Apicalregion von Pilzhyphen. *Protoplasma, 67*:413-441, 1969.
105. Grove, S. N., and Bracker, C. E. Protoplasmic organization of hyphal tips among fungi: vesicles and Sptizenkorper. *J. Bacteriol., 104*:989-1009, 1970.
106. Grove, S. N., Bracker, C. E., and Morre, D. J. An ultrastructural basis for hyphal tip growth in Pythium ultimum. *Am. J. Bot., 57*:245-266, 1970.
107. Truesdell, L. C., and Cantino, E. C. Decay of gamma particles in germinating zoospores of Blastocladiella emersonii. *Arch. Mikrobiol., 70*:378-392, 1970.
108. Myers, R. B., and Cantino, E. C. X. Enzymology of the gamma particle. In *The Gamma Particle*. Monographs in Developmental Biology, Vol. 8 (A. Wolsky, ed.), S. Karger, Basel, 1974, pp. 77-90.
109. Mills, G. L., and Cantino, E. C. Chitosome-like vesicles from gamma particles of Blastocladiella emersonii synthesize chitin. *Arch. Microbiol., 130*:72-77, 1981.
110. Herth, W., Mulisch, M., and Zugenmaier, P. Comparison of chitin fibril structure and assembly in three unicellular organisms. In *Chitin in Nature and Technology* (R. Muzzarelli, C. Jeuniaux, and G. W. Gooday, eds.), Plenum Press, New York, 1986, pp. 107-120.
111. Davis, L. L., and Bartnicki-Garcia, S. Co-ordination of chitosan and chitin synthesis in Mucor rouxii. *J. Gen. Microbiol., 130*:2095-2102, 1984.
112. Vermeulen, C. A., and Wessels, J. G. H. Ultrastructural differences between wall apices of growing and non-growing hyphae of Schizophyllum commune. *Protoplasma, 120*:123-131, 1984.
113. Cabib, E., Bowers, B., and Roberts, R. L. Vectorial synthesis of a polysaccharide by isolated plasma membranes. *Proc. Natl. Acad. Sci. USA, 80*:3318-3321, 1983.
114. Hearth, W. A specialized chitin-fibril-synthesizing apparatus in the centric diatom Cyclotella. *Naturwissenschaften, 65*:260, 1978.
115. Giddings, T. H., Brower, D. C., and Stahelin, A. Visualization of particle complexes in the plasma membrane of Micrasterias denticulata associated with the formation of cellulose fibrils in primary and secondary cell walls. *J. Cell Biol., 84*:327-338, 1980.
116. Mueller, S. C., and Brown, R. M. Evidence for an intramembrane component associated with a cellulose microfibril-synthesizing complex in higher plants. *J. Cell Biol., 84*:315-326, 1980.
117. Maclachlan, G., and Fevre, M. An overview of cell wall biosynthesis. In *The Cytoskeleton in Plant Growth and Development* (C. W. Lloyd, ed.), Academic Press, London, 1982, pp. 127-148.
118. Brown, R. M., Franke, W. W., Kleining, H., Falk, H., and Sitte, P. Cellulosic wall component produced by the Golgi apparatus of Pleurochrysis scherffelii. *Science, 166*:894-896, 1969.
119. Choi, A. H. C., and O'Day, D. H. Calcofluor staining of cellulose during microcyst differentiation in wild-type and mutant strains of Polysphondylium pallidum. *J. Bacteriol., 157*:291-296, 1984.

120. Mulisch, M., Herth, W., Zugenmaier, P., and Hausmann, K. Chitin fibrils in the lorica of the ciliate Eufolliculina uhligi: ultrastructure, extracellular assembly and experimental inhibition. *Biol. Cell.*, *49*:169-178, 1983.
121. Ahlquist, C. N., and Gamow, R. I. Phycomyces: mechanical behavior of stage-II and stage-IV sporangiophores. *Plant Physiol.*, *51*:586-587, 1973.
122. Ahlquist, C. N., Iverson, S. L., and Jhasman, N. E. Cell wall structure and mechanical properties of Phycomyces. *J. Biomech.*, *8*:357-362, 1975.
123. Saunders, P. T., and Trinci, A. P. J. Determination of tip shape in fungal hyphae. *J. Gen. Microbiol.*, *110*:469-473, 1979.
124. Bartnicki-Garcia, S. Fundamental aspects of hyphal morphogenesis. In *Microbial Differentiation* (J. O. Ashworth and J. E. Smith, eds.), Cambridge University Press, Cambridge, 1973, 245-267.
125. Wessels, J. G. H. Apical hyphal wall extension. Do lytic enzymes play a role? In *Microbial Cell Wall Synthesis and Autolysis* (C. Nombela, ed.), Elsevier, Amsterdam, 1984, pp. 31-42.
126. Wessels, J. G. H. Cell wall synthesis in apical hyphal growth. *Int. Rev. Cytol.*, *104*:37-79, 1986.
127. Ortega, J. K. E., Gamow, I., and Ahlquist, C. N. Phycomyces: a change in mechanical properties after a light stimulus. *J. Plant Physiol.*, *55*:333-337, 1975.
128. Davis, L. L., and Bartnicki-Garcia, S. Chitosan synthesis by the tandem action of chitin synthetase and chitin deacetylase from Mucor rouxii. *Biochemistry*, *23*:1065-1073, 1984.
129. Sietsma, J. H., and Wessels, J. G. H. Chemical analysis of the hyphal wall of Schizophyllum commune. *Biochim. Biophys. Acta*, *496*:225-239, 1977.
130. Wang, M. C., and Bartnicki-Garcia, S. Structure and composition of walls of the yeast form of Verticillium alboatrum. *J. Gen. Microbiol.*, *64*:41-54, 1970.
131. Sietsma, J. H., and Wessels, J. G. H. Evidence for covalent linkages between chitin and β-glucan in a fungal wall. *J. Gen. Microbiol.*, *114*:99-108, 1979.
132. Mol, P. C., and Wessels, J. G. H. Linkages between glucosa-minoglycan and glucan determine alkali-insolubility of the glucan in walls of Saccharomyces cerevisiae. *FEMS Microbiol. Lett.*, *41*:95-99, 1987.
133. Wessels, J. G. H., Sietsma, J. H., and Sonnenberg, A. S. M. Wall synthesis and assembly during hyphal morphogenesis in Schizophyllum commune. *J. Gen. Microbiol.*, *129*:1607-1616, 1983.
134. Sonnenberg, A. S. M., Sietsma, J. H., and Wessels, J. G. H. Spatial and temporal differences in the synthesis of (1-3)-β and (1-6)-β linkages in a wall glucan of Schizophyllum commune. *Exp. Mycol.*, *9*:141-148, 1985.
135. Gopal, P., Sullivan, P. A., and Shepherd, M. G. Isolation and structure of glucan from regenerating spheroplasts of Candida albicans. *J. Gen. Microbiol.*, *130*:1217-1225, 1984.
136. Surarit, R., Gopal, P. K., and Shepherd, M. G. Evidence for a glycosidic linkage between chitin and glucan in the cell wall of Candida albicans. *J. Gen. Microbiol.*, *134*:1723-1730, 1988.
137. Sonnenberg, A. S. M., Sietsma, J. H., and Wessels, J. G. H. Biosynthesis of alkali-insoluble cell wall glucan in Schyzophilum commune protoplasts. *J. Gen. Microbiol.*, *1128*:2667-2674, 1982.

138. Murgui, A., Elorza, M. V., and Sentandreu, R. Tunicamycin and papulacandin B inhibit incorporation of specific mannoproteins into the walls of Candida albicans regenerating protoplasts. *Biochim. Biophys. Acta, 884*:550-558, 1986.
139. Datema, R., Ende, H., and Wessels, J. G. H. The hyphal wall of Mucor mucedo. 1. Polyanionic polymers. *J. Biochem., 80*:611-619, 1977.
140. Murgui, A., Elorza, M. V., and Sentandreu, R. Effect of papulacandin B and calcofluor white on the incorporation of mannoproteins in the cell wall of Candida albicans blastopores. *Biochim. Biophys. Acta, 84*:215-222, 1985.
141. Elorza, M. V., Murgui, A., and Sentandreu, R. Dimorphism in Candida albicans: contribution of mannoproteins to the architecture of the yeast and mycelial cell walls. *J. Gen. Microbiol., 131*:2209-2216, 1985.
142. Elorza, M. V., Murgui, A., Rico, H., Miragall, F., and Sentandreu, R. Formation of new cell wall by protoplasts of Candida albicans: effect of papulacandin B, tunicamycin and nikkomycin. *J. Gen. Microbiol., 133*:2315-2325, 1987.
143. Elorza, M. V., Marcilla, A., and Sentandreu, R. Wall mannoproteins of the yeast and mycelial cells of Candida albicans: nature of the glycosidic bonds and polydispersity of their mannan moieties. *J. Gen. Microbiol., 134*:2393-2403, 1988.
143a. Marcilla, A., Elorza, M. V., Mormeneo, S., Rico, H., and Sentandreu, R. Candida albicans mycelial wall structure: supramolecular complexes released by zymolyase, chitinase and β mercaptoethanol. *Arch. Microbiol.* (in press).
144. Bartnicki-Garcia, S., and Lippman, E. Fungal morphogenesis: cell wall construction in Mucor rouxii. *Science, 165*:302-304, 1969.
145. Green, P. B. Morphogenesis of the cell and organ axis-biophysical models. *Brookhaven Symp. Biology, 25*:166-190, 1974.
146. Trinci, A. P. J., and Saunders, P. T. Tip growth of fungal hyphae. *J. Gen. Microbiol., 103*:243-248, 1977.
147. Katz, D., and Rosenberger, R. F. Hyphal wall synthesis in Aspergillus nidulans: effect of protein synthesis inhibition and osmotic shock on chitin insertion and morphogenesis. *J. Bacteriol., 108*:184-190, 1971.
148. Sentandreu, R., and Ruiz-Herrera, J. In situ study of the localization and regulation of chitin synthetase in Mucor rouxii. *Curr. Microbiol., 1*:77-80, 1978.
149. Sietsma, J. H., Vermeulen, C. A., and Wessels, J. G. H. The role of chitin in hyphal morphogenesis. In *Chitin in Nature and Technology* (R. Muzzarelli, C. Jeuniaux, and G. W. Gooday, eds.), Plenum Press, New York, 1986, pp. 63-69.
150. Ruiz-Herrera, J. The role of chitosomes in the apical growth of fungi. In *Microbial Cell Wall Synthesis and Autolysis* (C. Nombela, ed.), Elsevier, Amsterdam, 1984, pp. 113-120.
151. Ruiz-Herrera, J., Valenzuela, C., Martinez-Cadena, G., and Obregon, A. Alterations in the vesicular pattern and wall growth of Phycomyces induced by the calcium ionophore A23187. *Protoplasma, 148*:15-25, 1989.
152. Bartnicki-Garcia, S., Hergert, F., and Gierz, G. A novel computer model for generating cell shape: applications to fungal morphogenesis. In *Biochemistry of Cell Walls and Membranes of Fungi* (P. Kohn, ed.), Springer-Verlag, Berlin, 1989, pp. 43-60.

153. Collinge, A. J., and Trinci, A. P. J. Hyphal tips of wild-type and spreading colonial mutants of Neurospora crassa. *Arch. Mikrobiol.*, *99*:353-368, 1974.
154. Howard, R. J., and Aist, J. R. Hyphal tip cell ultrastructure of the fungus Fusarium: improved preservation by freeze-substitution. *J. Ultrastruct. Res.*, *66*:224-234, 1979.
155. Howard, R. J., and Aist, J. R. Cytoplasmic microtubules and fungal morphogenesis: ultrastructural effects of methyl benzimidazole-2-ylcarbamate determined by freeze-substitution of hyphal tip cells. *J. Cell. Biol.*, *87*:55-64, 1980.
156. Allen, E. D., Aiuto, R., and Sussman, A. S. Effects of cytochalasins on Neurospora crassa. I. Growth and ultrastructure. *Protoplasma*, *102*:63-75, 1980.
157. Grove, S. N., and Sweigard, J. A. Cytochalasin A inhibits spore germination and hyphal tip growth in Gilbertella persicaria. *Exp. Mycol.*, *4*:239-250, 1980.
158. Herr, F. B., and Heath, M. C. The effects of antimicrotubule agents on organelle positioning in the cowpea rust fungus Uromyces phaseoli var. vignae. *Exp. Mycol.*, *6*:15-24, 1982.
159. Slayman, C. L., and Slayman, C. W. Measurement of membrane potentials in Neurospora. *Science*, *136*:876-877, 1962.
160. Armbruster, B. L., and Weisenseel, M. H. Ionic currents traverse growing hyphae and sporangia of the mycelial water mould Achlya debaryana. *Protoplasma*, *115*:65-69, 1983.
161. Kropf, D. L., Lupa, M. D., Caldwell, J. C., and Harold, F. M. Cell polarity: endogenous ion currents precede and predict branching in the water mould Achlya. *Science*, *220*:1385-1387, 1983.
162. Gow, N. A. R. Transhyphal electrical currents in fungi. *J. Gen. Microbiol.*, *130*:3313-3318, 1984.
163. Horwitz, B. A., Weisenseel, M. H., Dorn, A., and Gressel, J. Electric currents around growing Trichoderma hyphae, before and after autoinduction of conidiation. *Plant Physiol.*, *74*:912-916, 1984.
164. Jaffe, L. F. Electrophoresis along cell membranes. *Nature*, *265*:600-602, 1977.
165. Jennings, D. H. Membrane transport and hyphal growth. In *Fungal Walls and Hyphal Growth* (J. H. Burnett and A. P. J. Trinci, eds.), Cambridge University Press, Cambridge, 1979, pp. 279-294.
166. Weisenseel, M. H., and Jaffe, L. F. The major growth current through lily pollen tubes enters as K^+ and leaves as H^+. *Planta*, *133*:1-7, 1976.
167. Nuccitelli, R. Ooplasmic segregation in the Pelvetia egg is accompanied by a membrane-generated electrical current. *Dev. Biol.*, *62*:13-33, 1978.
168. Galpin, M. F. J., Jennings, D. H., Oates, K., and Hobot, J. Localization by X-ray microanalysis of soluble ions, particularly potassium and sodium in fungal hyphae. *Exp. Mycol.*, *2*:258-269, 1978.
169. Reiss, H. D., and Herth, W. Calcium gradients in tip growing plant cells visualized by chlorotetracycline fluorescence. *Planta*, *146*:615-621, 1979.
170. Reiss, H. D., and Herth, W. Calcium ionophore A-23187 affects localized wall secretion in the tip of pollen tubes of Lilium longiflorum. *Planta*, *145*:225-232, 1979.
171. Herth, W. Ionophore A-23187 stops tip growth, but not cytoplasmic streaming in pollen tubes of Lilium longiflorum. *Protoplasma*, *96*:275-282, 1978.

13

PRODUCTION AND BIOSYNTHESIS OF AMINO ACIDS BY FUNGI

EVA M. KUBICEK-PRANZ and CHRISTIAN P. KUBICEK *Institute of Biochemical Technology and Microbiology, University of Technology, Vienna, Austria*

I. INTRODUCTION

Fungi, uni- or multicellular, can form all the amino acids commonly found in proteins from carbon intermediates and an inorganic (or organic) nitrogen source. In this sense they represent the simplest eukaryotic model system which still possesses the full capacity of amino acid biosynthesis, which in part has been lost in higher eukaryotic tissues. A comprehensive treatise of amino acid biosynthesis (and eventual overproduction) by fungi therefore should yield valuable information on the evolution of eukaryotic metabolism.

However, although this seems to justify a review on this particular topic, we have to be aware that fungi (at least the filamentous ones) are only rarely studied for their own sake: among the most commonly occurring reasons for investigating the biochemical properties of a fungus are (a) exploration of metabolic regulation of the overproduction of chemicals useful in the food, pharmaceutical, or fermentation industry; (b) investigations on the metabolic or genetic reasons for fungal infections; (c) basic investigations on model organisms with useful genetic systems (i.e., *Saccharomyces cerevisiae*, *Neurospora crassa*, *Aspergillus nidulans*). Consequently, one has to be aware that data on amino acid biosynthesis by filamentous fungi will undoubtedly be limited, and sufficient information may have accumulated only for genetically accessible fungi.

This chapter deals with amino acid biosynthesis, its regulation, and what is known about the possibility of producing amino acids with fungi. It was our attempt to focus on the current highlights and to underline the open problems rather than to comprehensively report on the existing literature. Furthermore, wherever possible, we tried to report specifically on what is known for the *filamentous fungi*, and use the data obtained for *yeasts* only when respective data for other fungi were completely lacking or inadequate. Also, whenever properties of individual enzymes were to be reported, our credit was primarily given to those references dealing with a purified protein. We excluded the numerous data on demonstration of and kinetic measurements with enzymes from fungal cell-free extracts, and only included a few of them when other equally important information was lacking.

Finally, we limited our review exclusively to the biosynthesis of amino acids occurring in proteins (including glutamine and asparagine) and excluded those parts of biosynthesis which deal with the reduction of inorganic compounds (NO_3^-, SO_4^{2-}) and well-reviewed compounds (1,2).

II. BIOSYNTHESIS OF AMINO ACIDS BY FUNGI

A. Primary Amino Acids and NH_4^+ Assimilation

Glutamate, aspartate, and alanine are generally termed primary amino acids, since they can be formed directly from intermediates of primary metabolism (α-ketoglutarate, oxaloacetate, and pyruvate), and because they themselves serve as initial intermediates in the biosynthesis of the other intermediates. Figure 1 shows the overall relationship of amino acid biosynthetic pathways in fungi, which demonstrates the key role of the so-called primary amino acids.

Knowledge of glutamine and asparagine biosynthesis, i.e., biosynthesis of the aminated forms of glutamate and aspartate, respectively, is essential since both components are tightly linked to nitrogen assimilation in fungi, and especially glutamine participates in numerous biosynthetic reactions as a donor of ammonium (carbamoyl phosphate, histidine, tryptophan, see later; but also purines and pyrimidines).

The assimilation of ammonium to synthesize glutamate, *the* initial "ammonium fixation product," by fungal cells is now known to involve two alternative pathways (Fig. 2): one is a reaction catalyzed by the NADP-specific glutamate dehydrogenase (EC 1.4.1.4), whereas the other proceeds via the glutamine synthase/glutamate synthase coupled reaction.

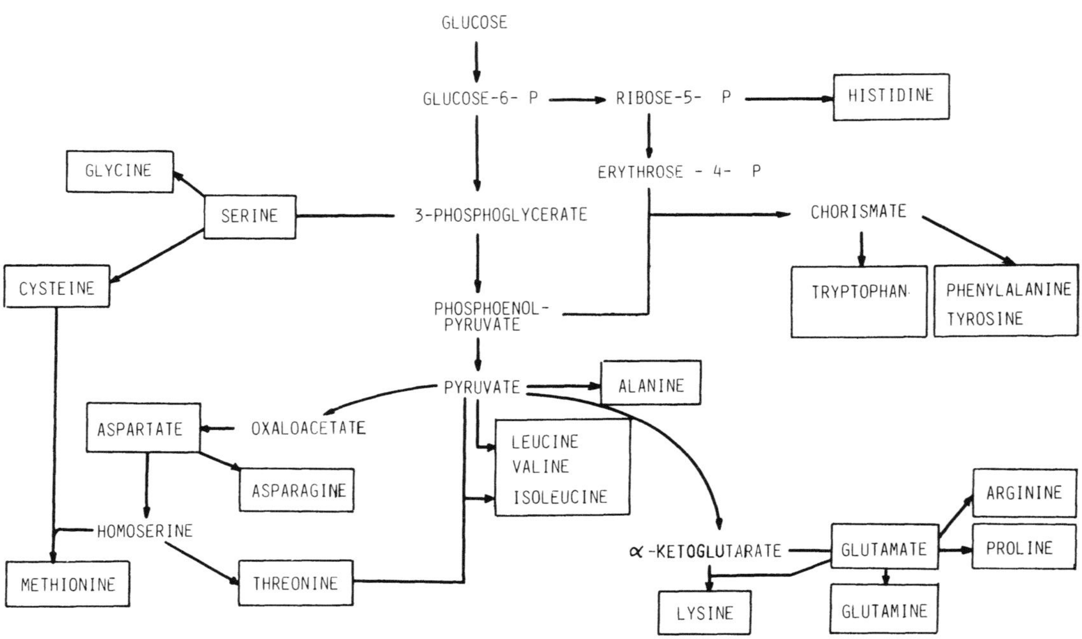

FIGURE 1. Metabolic relationship of biosynthesis of various amino acids in fungi.

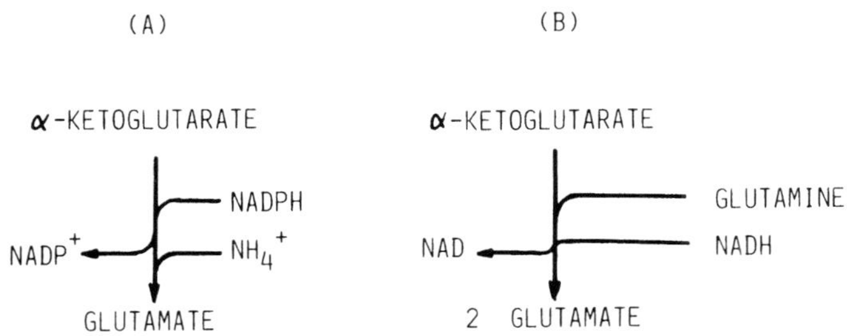

FIGURE 2. Alternative pathways of glutamate biosynthesis in fungi: (A) NADPH-glutamate dehydrogenase reaction; (B) GOGAT reaction.

In some fungi evidence indicates that in the presence of excess NH_4^+, the glutamate dehydrogenase pathway is mainly responsible for ammonium assimilation, whereas the glutamine synthase/glutamate synthase pathway becomes predominant in the presence of limiting amounts of NH_4^+ [3-7]. The molecular basis of this effect has been ascribed to nitrogen regulation of NADP-specific glutamate dehydrogenase synthesis by the type and concentration of available nitrogen [8].

The NADP-specific glutamate dehydrogenase has been characterized from several fungi [9-13]. The *N. crassa* protein is remarkably well understood: while native, it is a hexamer of 288 kDa, composed of similarly sized 48.8-kDa subunits consisting of 452-amino-acid residues [14,15]. Both amino acid and gene sequence have been reported [14-16]. The enzyme shows strong sequence homology with the enzyme from vertebrates [17]. The effect of amino acid replacement on its structure and activity has been examined in detail [18], and its allosteric kinetics are described in detail in [19]. There is little evidence for kinetic regulation of NADP-specific glutamate dehydrogenase, except for demonstration that NH_4^+ ion acts as both substrate and activator, thereby involving two different binding sites [20]. As already mentioned, regulation of the enzymes activity occurs mainly at the level of enzyme formation [8]; under certain conditions the enzyme can account for as much as 0.5% of total cellular protein. In *Coprinus cinereus*, genetic evidence was reported that both NH_4^+ and acetyl-CoA act as the intracellular inducers of glutamate dehydrogenase synthesis [21].

The NADP-specific glutamate dehydrogenase has been implicated in the mechanism of nitrogen control in fungi, whereby when complexed with NH_4^+ it may act as a repressor of synthesis of various enzymes concerned with degradation of organic nitrogen sources. The reader is referred to the work of Pateman and colleagues for a more detailed description of this hypothesis [1,22].

Glutamine synthase (EC 6.3.1.2) was also investigated from several fungi [23-25]. In *N. crassa* there appear to be two genetically distinct glutamine synthase proteins [26,27]. Upon growth on media containing an excess of nitrogen, the enzyme is structured in an octameric form composed of monomers with a molecular weight of 48 kDa, whereas upon NH_4^+ limitation a tetramer of slightly smaller subunits was observed [26,28]. These subunits have been termed β and α, respectively [26-28]. Hybrid forms of glutamine synthase have been found under conditions of nonlimiting but nonexcess nitrogen supply [26].

The bacterial glutamine synthase is a classical example of metabolic control by enzyme modification, i.e., here adenylation [29]. For fungi, however, this point has only been studied with the *Sclerotia sclerotiorum* enzyme, where a similar type of control was established [25]. Glutamine synthase is furthermore subject to regulation by various effectors, which may be considered as potential end products of further glutamine metabolism, i.e., AMP, CTP, GTP, NAD, anthranilate, histidine, glycine, and serine [30]. The latter two amino acids, which by first sight do not seem to fit into the picture of "glutamine end products," seem to be related to glutamine metabolism in a more complex manner, as has been demonstrated by [31]. The action of these effectors may also depend on their complexation state with either Mg^{2+} or Mn^{2+}, as was shown for the enzyme from *A. niger* [32], suggesting some regulatory influence of the fungal divalent metal ion content.

Glutamine synthase was shown to be subject to nitrogen regulation, which occurs here at the level of mRNA formation [33]. Oxidative degradation by intracellular proteases as a means for turning over the protein was demonstrated in *N. crassa* [34].

Glutamate synthase (GOGAT, EC 1.4.7.1) was purified from *N. crassa* [35] and *S. sclerotiorum* [25], but so far only kinetic data have been reported. The enzyme uses NADH (instead of NADPH as the bacterial enzyme [36]) and is markedly inhibited by tricarboxylic acids [25]. Regulation of enzyme synthesis takes place in response to glutamate requirements under conditions of low NH_4^+ supply [4,7].

Biosynthesis of aspartate occurs by transamination of oxaloacetate from glutamate. This reaction is catalyzed by glutamate oxaloacetate aminotransferase (EC 2.6.1.1). The enzyme was purified from *N. crassa* [37] and *A. nidulans* [1]. There appears to be both a cytoplasmic as well as a mitochondrial isoenzyme, but only the first appears to be important for aspartate biosynthesis. The enzyme contains tightly bound pyridoxal-phosphate [38]. The previous view of Munkres [39] that glutamate oxaloacetate aminotransferase and malate dehydrogenase activities are due to the same enzyme was disproven [1,37].

Asparagine synthesis from aspartate involves asparagine synthase (EC 6.3.1.1). The enzyme has not yet been studied from filamentous fungi, but the presence of two asparagine synthases in *S. cerevisiae* was reported [40]. Complementation and dominance data from an *N. crassa* asparagine synthase mutant indicate that the enzyme is probably multimeric like the mammalian enzyme [41].

Little is currently known about alanine biosynthesis in fungi, but a glutamate pyruvate aminotransferase (EC 2.6.1.2) was described from *N. crassa* and *A. nidulans* [1,38]. The enzyme was also demonstrated in cell-free extracts from several other fungi [40-44].

B. Secondary Amino Acids

1. Amino Acids from Glycolytic Intermediates

Three amino acids are derived from three carbon glycolytic intermediates, i.e., phosphoglycerate or its dephosphorylated form, namely, serine, cysteine, and glycine. Their biosynthetic pathway and metabolic interrelationship is given in Fig. 3. As can be seen, serine biosynthesis as a prerequisite for both cysteine and glycine biosynthesis may proceed via two pathways, involving either phosphoglycerate or glycerate. Most

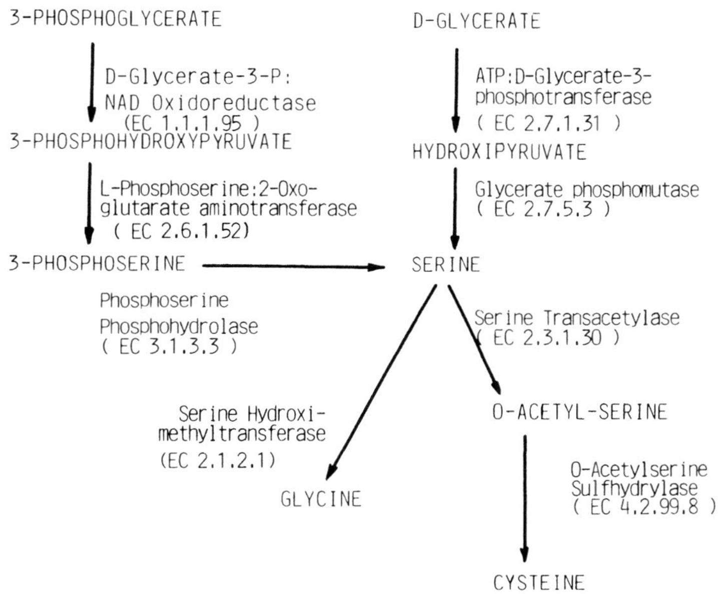

FIGURE 3. Alternative pathways of serine, glycine, and cysteine biosynthesis in fungi. Metabolic intermediates are indicated in block letters; normal lettering indicates respective enzymes (and corresponding EC numbers so far known).

of the present evidence for this comes from the use of tracer analyses and the assay of enzymes in cell-free extracts [45]. Genetic evidence is meager: six serine-requiring mutants from *N. crassa* have been isolated (*ser* 1-6), but only for one of them (*ser*-3) has the gene-enzyme relationship been established and the enzyme (phosphoserine phosphatase, EC 3.1.3.3) purified [46]. The relative participation of the two pathways of serine biosynthesis has been shown to depend on the nutritional state, i.e., the phosphorylated pathway appears to be more active during growth on media containing glucose as the carbon source [45].

Serine itself can serve as the substrate for two further amino acids, i.e., glycine and cysteine. Filamentous fungi appear to contain two alternatives for glycine biosynthesis, only one, however, being related to serine metabolism (serine hydroxymethyltransferase, EC 2.1.2.1), whose enzymology or gene has so far not been investigated. Alternatively, glycine can be formed by direct transamination via glyoxalate aminotransferase (EC 2.6.1.4), an enzyme which has also not yet been purified, but demonstrated in cell-free extracts from several fungi [42,43,49].

Biosynthesis of cysteine via serine involves a two-step enzymic reaction which involves *O*-acetylation by serine transacetylase (EC 2.3.1.30) and deacylation by *O*-acetylserine sulfhydrylase (EC 4.2.99.8). Gene-enzyme relationships are so far available for *A. nidulans* [50-56], and also some data on the regulation and enzymology of cysteine synthase from yeast [57] and the *O*-acetylserine sulfhydrylase from *Cephalosporium acremonium* [58] and *S. cerevisiae* [59] have been reported. It is noteworthy that in

S. cerevisiae O-acetylserine and *O*-acetylhomoserine sulfhydrylase appear to be one and the same enzyme [60].

The cysteine synthase reaction appears to be due to two different enzymes, one of which is *O*-acetylhomoserine transacetylase (see methionine biosynthesis), whereas the other appears to be specific for serine [61,62]. Mutants of *A. nidulans* lacking the *O*-acetylserine transacetylase have been described [53]. Studies with double mutants of *A. nidulans* suggested some yet to be defined regulatory relationship between both transacetylases which appears logical in terms of the possibility of cysteine biosynthesis also via cystathionine (see methionine biosynthesis) [61].

Regulation of cysteine biosynthesis is intimately coupled to methionine. Hence, cysteine synthase is both feedback-inhibited and repressed by methionine and cysteine [61].

2. Branched Chain Amino Acids

The biosynthesis of valine, isoeleucine, and leucine occurs by a common biosynthetic pathway, from which leucine branches off at the level of α-ketoisovalerate (Fig. 4). The steps common to all three amino acids are catalyzed by a set of four enzymes with dual substrate specificities, which are organized as a multienzyme complex within the mitochondrial matrix of about 400 kDa [63-65]. Genetic evidence for such a structural organization was also obtained by the isolation of "structural mutants" of *N. crassa* which contain activities of all four enzymes within the complex, but are unable to organize them to the complex and therefore require valine and isoleucine for growth [66]. Only limited work has been done on the characterization of the enzymes involved in the valine-isoeleucine-biosynthetic complex, but some effort has been undertaken with α-hydroxyacid synthetase, owing to its potential regulatory role (see below). The enzyme appears to consist of a basic subunit of 68 kDa, which assembles to a dimer and probably also octamer of 120 and 500 kDa, respectively [67,68]. The monomeric enzyme, but not the dimeric, is susceptible to inhibition by valine [67]. Valine sensitivity was also reported for the enzyme from *P. chrysogenum* and *Cephalosporium acremonium*, however, using permeabilized cells and crude cell-free extracts as enzyme source [69,70]. Gene-enzyme relationships have been established for the enzymes comprising the complex in *N. crassa* [66]. The provision of α-ketobutyrate as a substrate for isoleucine biosynthesis occurs by the deamination of threonine by threonine deaminase. A structural gene mutation for this enzyme was identified in *A. nidulans* [71]. Properties of the enzyme have only been examined in cell-free extracts from *N. crassa*, which indicated that it is inhibited by isoelucine [72]. In contrast to the other enzymes of the valine-isoeleucine-biosynthetic pathway, threonine deaminase is located in the cytosol [72].

Leucine biosynthesis initiates by the condensation of α-ketoisovalerate and acetyl-CoA, catalyzed by α-isopropylmalate synthase. An essentially homogeneous preparation was obtained from *N. crassa* [73] and its amino acid composition determined [74]. The synthetase is apparently a trimer of 143 kDa, composed of identical subunits [74]. It is inhibited by leucine in a complex manner, involving allosteric conformational transitions and depending on pH. Interestingly, quite contrary data have been published on the *S. cerevisiae* enzyme, which is apparently a dimer of 130 kDa, and contains 2 gram-atoms of Zn per subunit [75]. This discrepancy between yeast and *Neurospora* has not yet been explained.

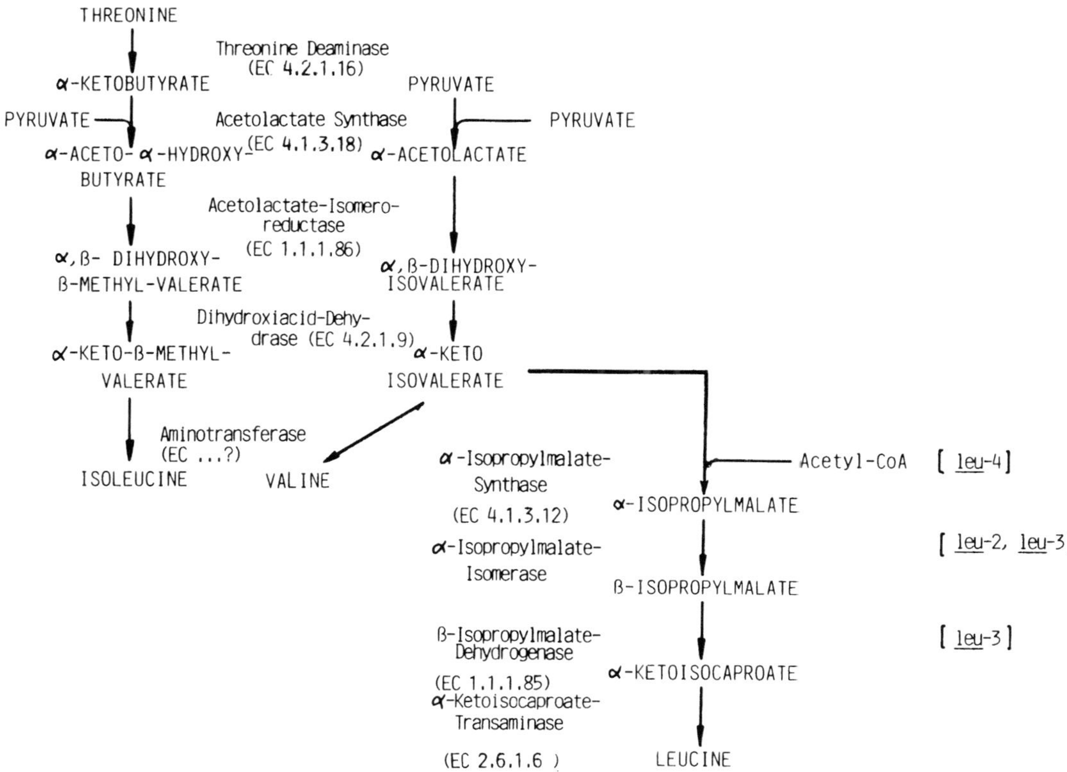

FIGURE 4. Biosynthetic pathway leading to valine, leucine, and isoleucine in fungi. Lettering was carried out as explained in the legend to Fig. 3. The genetic loci from *Neurospora crassa* so far known are specified in brackets.

Isomerization of α-propylmalate to β-propylmalate occurs in the mitochondrium, and the respective isomerase was partially purified from *N. crassa* [76]. β-Isopropylmalate, however, is transported into the cytosol, where further leucine biosynthesis commences: α-hydroxy-β-carboxyisocaproate dehydrogenase and leucine aminotransferase have not been investigated in *N. crassa* or other fungi, but there is some evidence that the transaminase is not specifically confined to leucine biosynthesis [38]. Four cistrons were demonstrated to be involved in leucine biosynthesis in *N. crassa-leu*-1, *leu*-2, *leu*-3, *and leu*-4 [77]; three of them are identical to the structural genes of β-isopropylmalate dehydrogenase, β-isopropylmalate isomerase, and α-isopropylmalate synthase (*leu*-1, 2, 4), whereas *leu*-3 appears to be a regulatory gene of β-isopropylmalate dehydrogenase and β-isopropylmalate isomerase synthesis (see below) [77-79].

Regulation of branched chain amino acid biosynthesis in filamentous fungi is exerted at several levels: valine and isoleucine exert multivalent repression on the enzymes of their biosynthesis [80], whereas leucine represses isopropylmalate synthase [77]. Leucine biosynthesis further is under general amino acid control [81], whereas corresponding data concerning valine and isoleucine biosynthesis are still lacking. Fine control of metabolic flow occurs by inhibition of threonine deaminase by isoleucine [72] and α-hydroxyacid synthetase by valine [67-70] as well as isoleucine [70]. Leucine biosynthesis is subject to control by leucine of α-isopropylmalate synthase activity [73]. It should be stressed that the latter clearly requires the mitochondrial uptake of cytosolically arising leucine, which has, however, not been dealt with thus far.

Gross and coworkers [77,79,82] originally found that the synthesis of the isomerase and the dehydrogenase, and probably also of the branched chain amino acid permease, is induced by α-isopropylmalate, which involves a *leu*-3 gene product [72,82,83]. Further evidence was obtained that the α-isopropylmalate *leu*-3 gene product is necessary for full induction of at least four genes specifying the structure of the enzymes of the isoleucine-valine pathway [72], suggesting a central role of *leu*-3 in the regulation of branched chain amino acid synthesis in *N. crassa*.

3. Amino Acids Derived from Glutamate

Proline: Proline is the amino acid most directly derived from glutamate, which involves three enzymatic steps (glutamate kinase, EC 1.5.1.12; glutamyl phosphate reductase, EC 1.2.1.41; and pyrolline-5-carboxylate reductase, EC 1.5.1.2). In *N. crassa* and *S. cerevisiae* these enzymes are products of three genes (*prol*3, *prol*4, *prol*1; *PRO*1, *PRO*2, *PRO*3; see Fig. 5 [84,85]). A cytosolic location of all three enzymes was established [84]. Furthermore, proline has been implicated as an intermediate in arginine catabolism via ornithine aminotransferase (EC 2.6.1.13), which is reviewed in [84] and shall not be dealt with here.

The key intermediate in proline biosynthesis is glutamate-γ-semialdehyde, which is formed from glutamate by phosphorylation and the subsequent reduction catalyzed by two enzymes, which have, however, not yet been purified from filamentous fungi. Glutamate-γ-semialdehyde is in rapid equilibrium with Δ^1-pyrolline-5-carboxylate [86], which is the in vivo substrate for further reduction to proline. Pyrolline-5-carboxylate reductase from *N. crassa* was partially purified [87].

Genetic evidence obtained with bacteria suggests control of proline biosynthesis by one of the enzymes prior to glutamate-γ-semialdehyde [88]; in *N. crassa*, however, the only control mechanism so far described is repression of pyrolline-5-carboxylate reductase by proline [87].

Arginine: Arginine biosynthesis in fungi can be divided into three functional processes (Fig. 6): the biosynthesis of ornithine, the biosynthesis of carbamoyl phosphate, and the final conversion of both to arginine. A thorough and excellent review on this subject with special reference to *S. cerevisiae* and *N. crassa* was recently presented by Davis [84].

Biosynthesis of ornithine in filamentous fungi, unlike that in yeast, takes place in the mitochondrion [84,89,90]. It is initiated by the formation of acetylglutamate from glutamate and acetyl-CoA by an enzyme bound to the inner mitochondrial membrane, *N*-acetylglutamate synthase. Its properties are given in Fig. 6: *N*-acetylglutamate is further converted to its semialdehyde via an unstable, phosphorylate intermediate by two enzymes—*N*-acetylglutamate kinase and *N*-acetylglutamyl phosphate reductase—localized in the mitochondrial matrix [92-94]. Both enzymes are the product of a single gene in *N. crassa* [*arg*-6], whose mRNA enables the translation of a 90-kDa polypeptide, which becomes intramitochondrially cleaved into

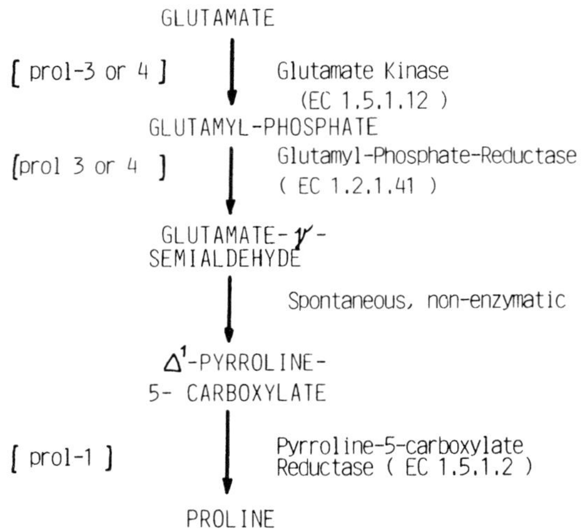

FIGURE 5. Biosynthesis of proline in fungi. Enzymes and genetic loci are given as explained in the legend to Fig. 4.

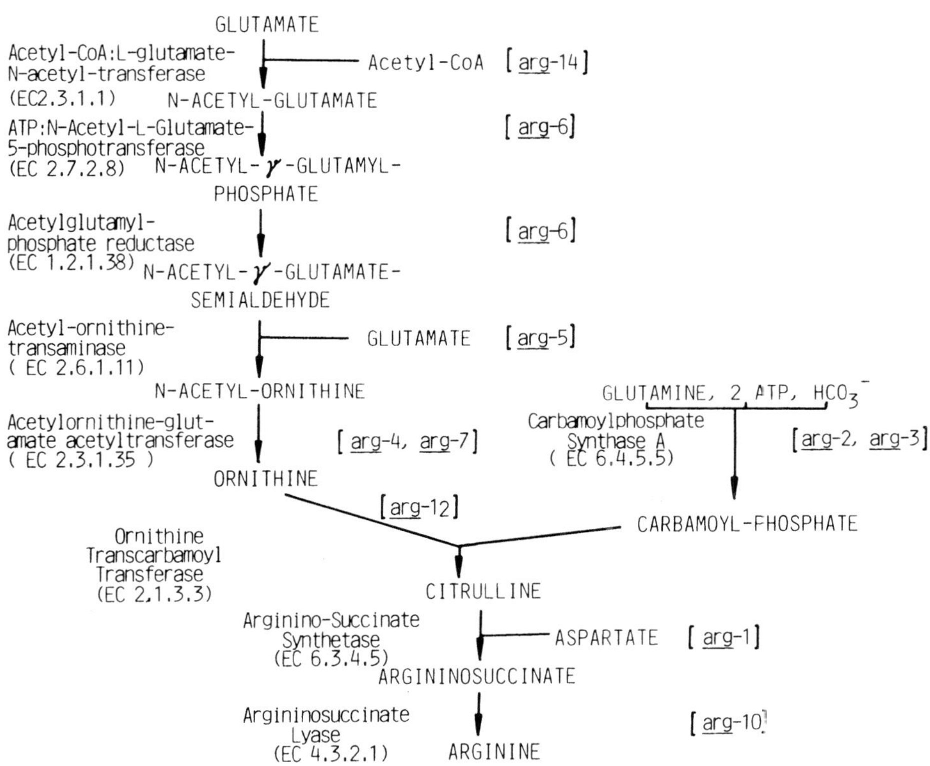

FIGURE 6. Biosynthesis of arginine in fungi. Enzymes and genetic loci are given as explained in the legend to Fig. 4.

two polypeptides of 51 and 40 kDa, which form the subunits of the kinase octamer and reductase dimer, respectively [84,92]. Blockage of mitochondrial protein import gave rise to the accumulation of the 90-kDa precurser in the cytosol [95]. The aggregation of four reductase dimers with one kinase octamer in the mitochondrial matrix was proposed—although not yet proven—to explain intermediate channeling and protection of the labile acetylglutamyl-γ-phosphate [84]. The further conversion of *N*-acetylglutamate semialdehyde to ornithine involves transamination using glutamate as amino donor and transfer of the acetyl group to a new glutamate molecule (cf. Fig. 6). Both enzymes—*N*-acetylornithine transaminase and *N*-acetylornithine: glutamylacetyltransferase—have not yet been characterized. Their genetic loci have been determined in *N. crassa*: *arg*-5 and *arg*-4 (*arg*-7) [85]. The involvement of a third enzyme—acetylornithase—is still speculative [96].

The second substrate required for arginine biosynthesis is carbamoyl-phosphate (Fig. 6). Its synthesis occurs mitochondrially from glutamine, HCO_3^-, and Mg-ATP by carbamoyl phosphate synthase. It should be noted at this point that fungi contain two different carbamoyl phosphate synthases, one of which is located in the nucleus and participates in the biosynthesis of pyrimidine [92]. This latter enzyme will not be dealt with here. The carbamoyl phosphate synthase associated with arginine biosynthesis consists of two different subunits specified by two unlinked genes (*arg*-2 and *arg*-3, [98]), which is a frequently occurring property of glutamine and amidotransferases [99]. The small, 45-kDa subunit functions in splitting glutamine in the course of the reaction, whereas the larger, 125-kDa subunit catalyzes the other reactions [100,101]. Both polypeptides enter the mitochondrion separately and aggregate only after insertion into the mitochondrial matrix [101]. The larger subunit is apparently synthetized as a larger precurser, whose C-terminal extension is involved in the localization or maturation of this protein [102]. The holoenzyme is a dimer of 176 kDa. It is a highly unstable enzyme, and its purification in an active form has not yet been achieved. The purification of the larger subunit from *N. crassa* was recently reported [94].

The last sequence of arginine biosynthesis requires the conversion of ornithine to arginine, which involves three enzymes: ornithine carbamoyl transferase, argininosuccinate synthetase, and argininosuccinate lyase (cf. Fig. 6). Ornithine carbamoyltransferase, which catalyzes the formation of citrulline from ornithine and carbamoyl phosphate, is still localized in the mitochondrion. The enzyme purified from *N. crassa* is a trimer of 110 kDa composed by three identical subunits which are products of a single gene (*arg*-12 [103]). Citrulline leaves the mitochondrion, but it is not known whether by active transport or facilitated diffusion. Argininosuccinate synthetase is a tetramer with identical subunits of 49 kDa, which is also a product of a single gene (*arg*-1 [104]). A comparison of the enzyme from yeast with that from mammalian tissues indicates several similarities, indicating considerable conservation throughout evolution [105]. Argininosuccinate lyase exhibits a native M_r of 176 kDa, but the number of subunits is known [106]. Genetic evidence, based on mutations within the *arg*-10 gene from *N. crassa*, would favor the existence of multiple subunit structures [107].

Regulation of arginine biosynthesis occurs both on the level of gene expression and on the level of enzyme inhibition or activation. Regarding gene expression, it has been shown that all enzymes involved in the whole pathway respond to the general amino acid control [108,109], whereas in no case could enzyme repression by arginine be demonstrated. Kinetic modulation of metabolic flux occurs at several points: *N*-acetylglutamate synthase, *N*-acetylglutamate kinase, and argininosuccinate synthase are feedback-inhibited by arginine [84,110]. It is obvious that for the first two enzymes the physiological importance of this observation requires arginine to enter the mitochondrion; however, studies with mutants of *N. crassa* which display reduced feedback sensitivity to arginine established that their ornithine synthesis is uncoupled from arginine biosynthesis [111]. Flux analysis in *S. cerevisiae* and *N. crassa* showed that ornithine biosynthesis is limited at the level of *N*-acetylglutamate kinase [112,113]. Also, flux analysis established the control of metabolic flow from ornithine to arginine by argininosuccinate synthase [112]. It is interesting to note that, unlike the textbook example of allosteric proteins, bacterial carbamoyl phosphate synthase, the fungal carbamoyl phosphate synthase is only slightly involved in the control of arginine biosynthesis.

Lysine: The biosynthesis of lysine in fungi is rather unique as there are two independent pathways: in bacteria, green plants, and lower fungi, lysine is synthesized via diaminopimelic acid, whereas in higher fungi, the α-aminoadipate pathway prevails. The distribution pattern of these pathways was analyzed by Vogel [113] and LeJohn [114]. So far, no organism has been described in which both pathways exist together. The diaminopimelic pathway has so far not been investigated at all in fungi, and the interested reader is refered to reviews on its enzymology in bacteria [115].

The α-aminoadipate pathway of lysine biosynthesis is depicted in Fig. 7. It is initiated by the condensation of acetyl-CoA and α-ketoglutarate by the enzyme homocitrate synthase. This enzyme has so far not been purified from any fungus, except for a cytosolic isoenzyme in *P. chrysogenum* [116]. It appears to be the product from a single gene in *Neurospora* [117], but not in *S. lipolytica* [118]. It exhibits allosteric kinetics with respect to acetyl-CoA and is inhibited by lysine [119,120]; this inhibition is reversed at pH 7 and below with the *Penicillium chrysogenum* enzyme [116,121,122]. The yeast enzyme also exhibits a complex susceptibility to inhibition by free CoA and Zn^{2+} [123], which is now shown by the *Penicillium* enzyme [116].

The steps leading from homocitrate to α-aminoadipate have not yet been characterized from a filamentous fungus. The enzymes and their mitochondrial compartmentation were demonstrated in yeast [124,125].

Once in the cytosol, α-aminoadipate is converted to α-aminoadipate semialdehyde by a series of three reactions, catalyzed by a single enzyme which in yeast is coded for by two genes [125-127]. The enzyme aminoadipate reductase, has not yet been purified from any fungus. The partial purified preparation from *P. chrysogenum* resembles the yeast enzyme in its kinetic properties, and additionally is inhibited by low concentrations of lysine [128]. There are some indications that the yeast enzyme might be a heterooligomer [127], but the genetics of these findings have not been made clear.

Further conversion to lysine requires the condensation of α-aminoadipate semialdehyde and glutamate to saccharopine and its consequent cleavage to lysine and α-ketoglutarate. The enzymes involved—saccharopine reductase and saccharopine dehydrogenase—have not yet been studied from filamentous fungi, although the loci of their genes in *Neurospora* have been mapped

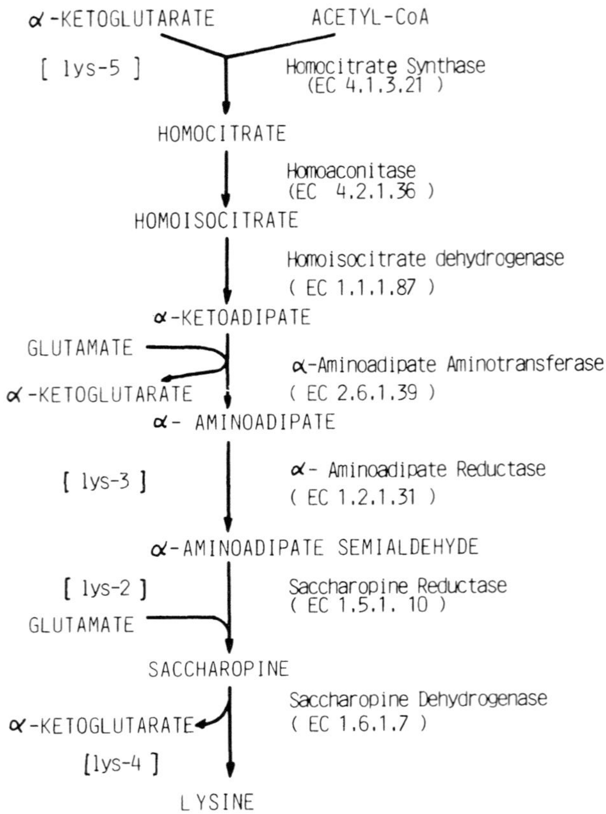

FIGURE 7. Biosynthesis of lysine by the α-aminoadipate pathway in fungi. Enzymes and genetic loci are given as explained in the legend to Fig. 4.

[129,130], but the purified enzymes have been studied from *S. cerevisiae* [131,132].

Regarding compartmentation, it was demonstrated in yeast that the first half of the pathway takes place in the mitochondrion, whereas the latter part—starting with α-aminoadipate—occurs in the cytoplasm [124]. In *P. chrysogenum*, a mitochondrial and a cytosolic isoenzyme of homocitrate synthase have been demonstrated [116].

Regulation of lysine biosynthesis via the α-aminoadipate pathway occurs on the level of both gene expression and fine control: general amino acid control appears to be the main mechanism operating, affecting mostly the enzymes of the second half of the pathway, α-aminoadipate reductase, saccharopine reductase, and saccharopine dehydrogenase [133]. Lysine repression was only once observed with *Neurospora* homocitrate synthase [134], whereas the *P. chrysogenum* enzyme's biosynthesis evidently is unaffected by lysine [121,133]. None of the other enzymes is controlled by lysine.

On the other hand, fine control by lysine has been demonstrated for homocitrate synthase from *Neurospora* in vitro [134] and from *P. chrysogenum* in vivo [116,121]. Interestingly, lysine inhibition of homocitrate synthase from *P. chrysogenum* in vitro is strongly pH dependent in crude cell-free extracts or in purified preparations [116]. A further control point may operate at the level of α-aminoadipate reductase, which is inhibited by low concentrations of lysine in vivo and in vitro in *P. chrysogenum* [128], findings which had already been suggested for yeast [135]. There is also some indication that homocitrate synthase may be regulated by the ratio of acetyl-CoA to CoA [123], but owing to the uncertainty of its location in fungi—the relevance of this is yet unclear.

4. Amino Acids Derived from Aspartate

Threonine: The biosynthesis of threonine in *N. crassa* resembles the pathway established in other organisms, which is depicted in Fig. 8, and which involves five enzymatic steps [136]. Gene-enzyme relationships have, however, only been established for *N. crassa* mutant *thr*-3, which specifies threonine synthase [137]. Among the enzymes involved, purified proteins have only been studied of homoserine dehydrogenase (EC 1.1.1.3) and threonine synthase (EC 4.2.99.2) [137,138]. These studies allow the conclusion that the enzymes exhibit properties largely resembling those reported for other organisms [139]. Homoserine dehydrogenase does not occur in isozymic forms as has been reported for *E. coli* [140,141].

Regulation of threonine biosynthesis is known to involve both inhibition and repression of enzymes by pathway end products; in yeast, the existence of multiple aspartokinases, which are differentially sensitive to inhibition by threonine, has been deduced [142]. Respective data from filamentous fungi are so far lacking. Both threonine synthase and homoserine dehydrogenase are neither inhibited nor repressed by homoserine, threonine, methionine, or isoleucine [137,138]. The regulation of threonine biosynthesis in filamentous fungi hence strongly requires further investigation.

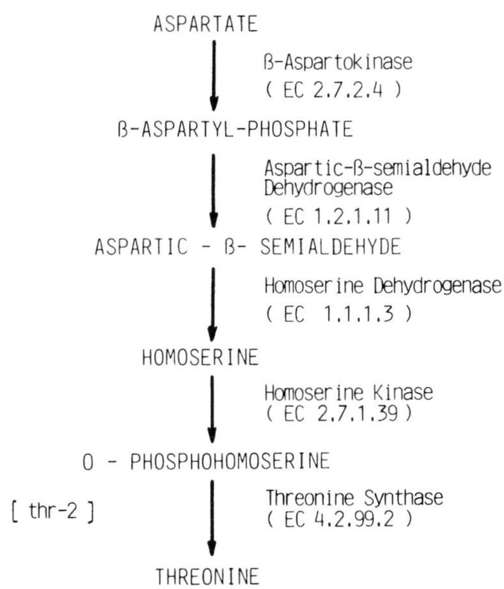

FIGURE 8. Biosynthesis of threonine in fungi. Enzymes and genetic loci are given as explained in the legend to Fig. 4.

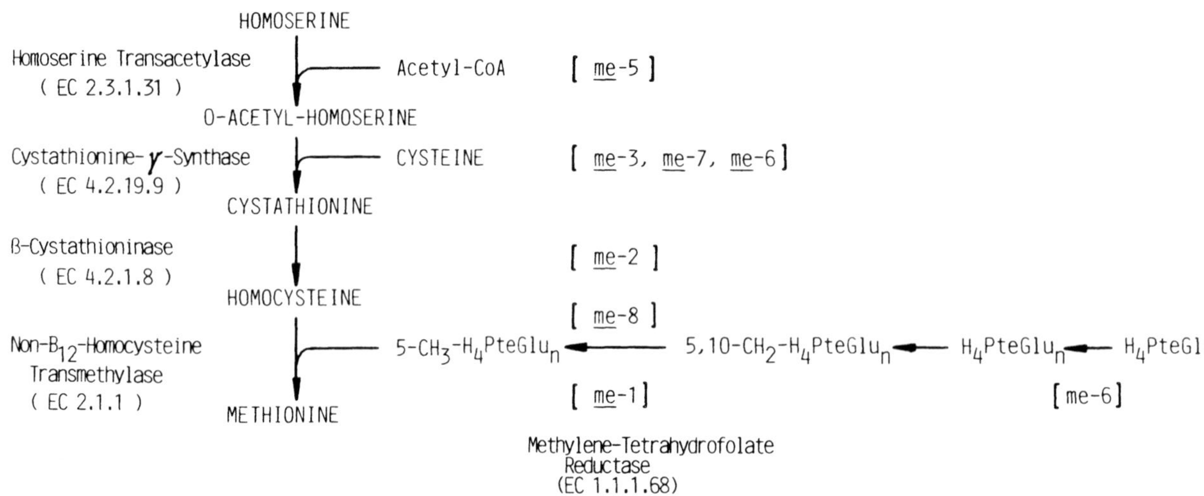

FIGURE 9. Biosynthesis of methionine in fungi. Enzymes and genetic loci are given as explained in the legend to Fig. 4. $H_4PteGlu_n$, tetrahydrofolatepolyglutamate (n = number of glutamate residues).

Methionine: There may be two biosynthetic routes leading to homocysteine, one of the intermediates in methionine synthesis in *N. crassa* or *A. nidulans*. One route involves the direct reaction of homoserine with H_2S [62], whereas the other proceeds via cystathionine [143] (Fig. 9).

The "cystathionine pathway" initiates from homoserine by its acetylation by homoserine transacetylase. The *Neurospora* enzyme has only been moderately purified [140], but recently the *S. cerevisiae* enzyme was purified 3500-fold and yet was not homogeneous [59]. The *Neurospora* enzyme is the product of one gene (*met*-5) [144], and not inhibited by methionine [145]. The *Saccharomyces* enzyme was also insensitive to methionine, but slightly inhibited by adenosylmethionine and *S*-adenosylhomoserine, although the regulatory significance of this is unclear [59].

Cystathionine-γ-synthase catalyzes the condensation of acetylhomoserine and cysteine. It is the product of two genes in *Neurospora* (*met*-3, *met*-7) [146]; incomplete studies suggest that it is a heterooligomer of around 350 kDa [145]. It is inhibited by adenosylmethionine both in vitro and in vivo [147,148], and is activated by methyltetrahydrofolatepolyglutamate ($5\text{-}CH_3H_4PteGlu_n$) [149]. Two mutants—*met*-1 and *met*-6—are unable to synthetize methionine because of their lack of methylenetetrahydrofolate reductase [150] and polyglutamatefolate conjugase [151]. Cystathionine-γ-synthase is inactive in the complete absence of $5\text{-}CH_3H_4PteGlu_n$, whose biosynthesis must therefore be considered to be obligatorily linked to methionine biosynthesis. However, the enzymes catalyzing its biosynthesis have scarcely been studied; methylenetetrahydrofolate reductase from *Neurospora* is a flavoprotein and inhibited by adenosylmethionine [149]; preliminary data are available indicating that the polyglutamate residue is attached by more than a single enzyme [154].

Cleavage of cystathionine to homocysteine and serine is catalyzed by β-cystathioninase, a pyridoxal phosphate-requiring enzyme, specified by the *met*-2 gene of *N. crassa* [155]. The situation is, however, complicated by the findings of a second enzyme, γ-cystathioninase, which forms cysteine and homoserine [155], the probability that these enzymes occur in isozymic forms [156], and the fact that like most pyridoxal phosphate-catalyzed β-eliminations, the β-cystathioninase is not specific for cystathionine [143]. Pieniazek et al. [157] suggested that the β-cystathioninase may instead be important for the biosynthesis of cystathionine and cysteine. It is possible that different fungi developed different routes of methionine biosynthesis.

O-Acetylhomoserine sulfhydrylase has been identified in *Neurospora* and *Aspergillus* [158], and partially purified from *N. crassa* [159]. Its operation provides a means for bypassing the cystathionine cleavage reaction. However, in *Neurospora*, the enzyme is formed constitutively, even in *met*-3 and *met*-7 mutants, yet the latter fail to grow on minimal medium [159]. It is therefore unlikely that this enzyme is involved in methionine biosynthesis in *Neurospora*, and Flavin [143] suggested its function in a sulfur storage and retrieval system.

No formation of cystathionine from cysteine—as has been demonstrated in bacteria—has been identified in fungi [160], which is supported by the lack of growth of methionine auxotrophs of *A. nidulans* upon supply of cysteine [1].

The final step in methionine biosynthesis involves the transfer of the methyl group which is supplied by $CH_3\text{-}H_4PteGlu_n$ [143]. As in plants and some bacteria, fungi catalyze this reaction by an enzyme not requiring vitamin B_{12} [149]. The enzyme is coded for by *met-8* in *Neurospora* [149], but its purification has so far only been reported from *S. cerevisiae* [162]. It has an M_r of 75 kDa and utilizes the di- and triglutamate conjugate, but not the monoglutamate.

Regulation of methionine biosynthesis has so far only been elucidated at the level of fine control. No indication for methionine repression of enzyme synthesis has so far been obtained as far as its own biosynthesis is concerned; however, data are available suggesting that methionine represses enzymes of sulfur mobilization and uptake [163]. No data are yet available indicating whether methionine is subject to regulation by the general amino acid control.

Fine control is mainly exerted by allosteric regulation of cystathionine-γ-synthase and methylenetetrahydrofolate reductase by adenosylmethionine [143,145]; in the first case, this is counteracted by CH_3-H_4PteGlu$_n$ activation [149,153]. Free methionine is without regulatory importance for any enzyme of the pathway.

5. Aromatic Amino Acids

The biosynthesis of the aromatic amino acids phenylalanine, tyrosine, and tryptophan occurs by the same metabolic reactions in all microorganisms so far investigated. This involves the condensation of phosphoenol pyruvate with erythrose-4-phosphate to 3-deoxy-D arabinoheptulosonic-7-phosphate (DAPH) and its subsequent conversion to chorismate via the so-called shikimic acid pathway. From chorismate, phenylalanine and tyrosine are synthesized via prephenate, whereas the biosynthesis of tryptophan occurs by reaction of chorismate with glutamine to anthranilate, and its conversion to indolglycerol phosphate and finally to tryptophan (Fig. 10). Whereas the metabolic reactions involved are largely similar to those occurring in bacteria, fungi generally display large portions of these biosynthetic pathways organized as multifunctional enzymes, i.e., the *arom* multienzyme complex [164] and the tryptophan synthase multienzyme complex [165].

The enzymes catalyzing the reactions of the shikimate pathway are shown in Fig. 10. The condensation reaction is catalyzed by DHAP synthase, which occurs in three isozymic forms in *N. crassa*, which are encoded by three different genes [166]. Each of these enzymes exhibits allosteric kinetics with its substrates, phosphoenol pyrubate and erythrose-4-phosphate, and is inhibited specifically by either phenylalanine, tyrosine, or tryptophan [166]. This contrasts with the situation in yeast where only two isoenzymes, sensitive against tyrosine and phenylalanine, are found [167].

The five successive steps of the shikimate pathway are assembled on a multifunctional protein, the *arom* multifunctional enzyme [164]. It was described in *N. crassa* [168], *A. nidulans* [169,170], and *S. cerevisiae* [170]. It consists of two identical pentafunctional polypeptides encoded by a single gene [171]. Five subregions of the polypeptide chain fold into five essentially autonomous structural domains, each of which carries one of the five component enzyme activities [164]. The isolation of these functional domains by limited proteolysis has been achieved for a bifunctional 63-kDa fragment carrying both the 3-dehydroquinase and shikimate dehydrogenase activities [172], and a 74-kDa subfragment of a larger 120-kDa polypeptide, containing EPSP synthase activity [173]. A comparison of the amino acid sequences of individual domains from the *S. cerevisiae arom* enzyme complex with those of the monofunctional enzymes from *E. coli*,

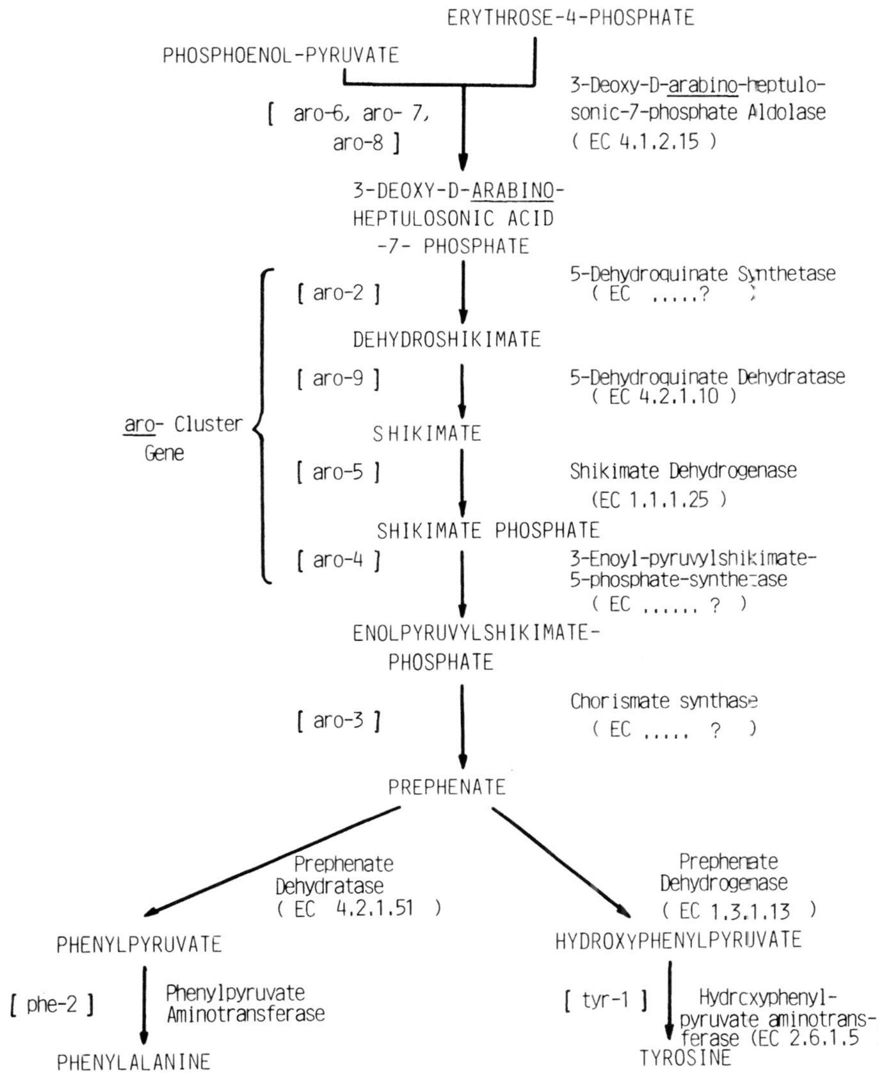

FIGURE 10. Shikimate pathway of aromatic amino acid biosynthesis in fungi. Enzymes and genetic loci are indicated as described in the legend to Fig. 4.

which are not organized as a multienzyme complex, emphasize that the fungal enzyme is built up of domains of very similar tertiary structure to those of the bacterial monomer enzymes [174-177]. These data suggest that the *arom* enzyme has a very well-defined quaternary structure. The whole topic was recently reviewed [164].

The last step of the shikimate pathway is catalyzed by chorismate synthase. The enzyme from *N. crassa* was investigated in some detail: it is a homooligomer consisting of 55-kDa subunits, which carry the diaphorase as well as the synthase activity [178].

The biosynthetic pathway leading from chorismate to phenylalanine and tyrosine involves further conversion of chorismate to prephenate by the enzyme chorismate mutase, and then branches to yield phenylpyruvate or 4-hydroxyphenylpyruvate, which serve as ultimate acceptors for the transamination to the corresponding amino acids (Fig. 10).

Chorismate mutase appears to occur in several isoenzymes in some fungi [179], whereas only a single enzyme has been demonstrated in *N. crassa* or *Claviceps paspali* [179,180]. The *N. crassa* enzyme and one of the isoenzymes from *Penicillium dupontii* and *P. chrysogenum* are inhibited by phenylalanine and tyrosine, but activated by tryptophan [179]. The enzymes catalyzing the further steps have so far not been characterized. Prephenate dehydratase and prephenate dehydrogenase are inhibited by the end products of the pathway they are involved in, i.e., phenylalanine and tyrosine, respectively [180]. It is not known whether the transaminating reaction is carried out by a single or two separate enzymes. Some microorganisms have been reported to catalyze the direct hydroxylation of phenylalanine to tyrosine, and this enzyme—phenylalanine hydroxylase—was detected in *A. nidulans* [181]; it is, however, uncertain whether this enzyme is involved in the biosynthesis of tyrosine.

Tryptophan biosynthesis branches off from chorismate by its condensation with glutamine to anthranilate (Fig. 11). Similar to the situation with the shikimate pathway, this pathway—although similar in its reactions to the one found in prokaryotes—is organized in a distinctly different pattern in fungi, and in general appears to be more highly organized in eukaryotic organisms. Gene-enzyme relationships have been defined for *N. crassa*, *A. nidulans*, *Coprinus radiatus*, and for the yeasts *S. cerevisiae* and *Schizosaccharomyces pombe* [182-188], which are shown in Fig. 11. In the filamentous fungi and *S. pombe*, the five enzymes involved in the pathway are encoded by four genes. Three of the enzymes [anthranilate synthase, phosphoribosyl (PR)-anthranilate isomerase, and indoleglycerol phosphate synthase] are organized in a multienzyme complex, which is composed by two different subunits α and β, coded for by two different genes (*TRP*1 and *TRP*2) [165]. The β unit is a trifunctional peptide that contains the active sites for the two reactions leading from PR-anthranilate to indole glycerol phosphate (PR-anthranilate isomerase and indoleglycerol phosphate synthase), as well as a further domain—located at the N terminus—that, when associated with the α unit to a $\alpha_2\beta_2$ complex, carries out the anthranilate synthase reaction ("G" domain) [189].

The two other enzymes of the pathway (PR-transferase and tryptophan synthase) are independent enzymes encoded by single, independent genes in *A. nidulans*, *C. radiatus*, and *S. pombe* [183,187,190], suggesting identical functional organization of the enzymes in these fungi to that in *N. crassa*. Using ultracentrifugation through sucrose gradients, Hütter and DeMoss [184] and Henke [182] examined the sedimentation pattern of the first four reactions of the tryptophan biosynthetic pathway in a number of different fungi from different taxonomic classes. They found a pattern distribution of possible phylogenetic significance, which was reviewed in detail [165,185].

Only *TRP*1 (and the corresponding *TRPC*) have so far been cloned and sequenced in *N. crassa* and *A. nidulans*, respectively [186,191].

Concerning regulation, the pathway of biosynthesis of aromatic amino acids in filamentous fungi is subject to various control mechanisms. On the level of gene expression, general amino acid control is dominating

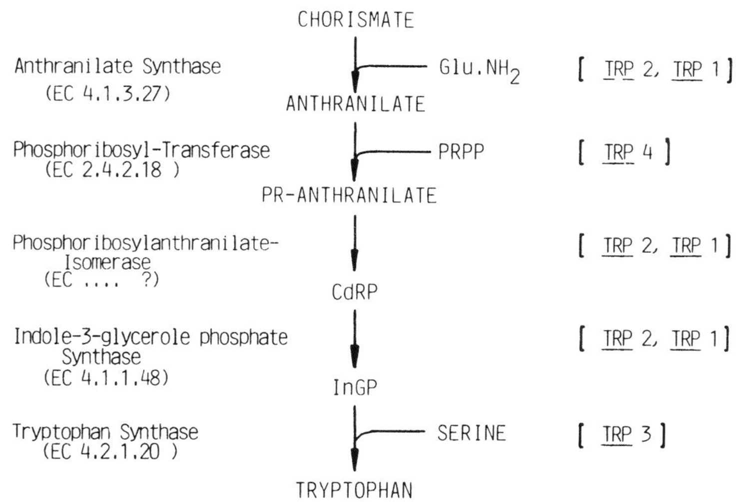

FIGURE 11. Biosynthesis of tryptophan from chorismate. Enzymes and genetic loci are indicated as described in the legend to Fig. 4. Glu.NH_2, glutamine; PRPP, phosphoribosylpyrophosphate; PR, phosphoribosyl-; CdRP, 1-(*o*-carboxyphenylamino)-1-deoxyribulose-5-phosphate; InGP, indole-3-glycerol phosphate.

[81,165], and there is little evidence for end product repression of any of the enzymes of phenylalanine, tyrosine, or tryptophan biosynthesis [167,178,180]. Several mechanisms, on the other hand, cooperate in ensuring fine control of metabolic flux. As was discussed above, the initial enzyme, DHAP synthase, occurs in isoenzymes, of which everyone is subject to feedback inhibition by one of the end product amino acids. Also these isoenzymes respond cooperatively to their substrate supply [167]. A further control point may operate at the *arom* complex, which has been reported to be activated by heptulosonate-7-phosphate [192]. These findings, however, deserve reinvestigation since they were obtained with a partially proteolyzed enzyme [164]. Finally, chorismate synthase was reported to be inhibited by ADP [178]. Within the phenylalanine-tyrosine branch, further regulation is exerted by end product inhibition of chorismate mutase, which on the other hand is activated by tryptophan [179], and by end product inhibition of prephenate dehydrogenase as well as prephenate dehydratase [180]. Tryptophan biosynthesis is also controlled by feedback inhibition at the level of anthranilate synthase [193,194].

6. Histidine

The biosynthesis of histidine involves a complex pathway, starting with phosphoribosylpyrophosphate to which ATP is attached (Fig. 12). The adenine nucleus of ATP is then modified and cleaved to yield imidazole glycerol phosphate, which subsequently becomes oxidized, transaminated, and dephosphorylated to yield histidine. Filamentous fungi were among the first organisms used to investigate histidine biosynthesis by studying the accumulation of imidazolelike compounds in various histidine auxotrophs

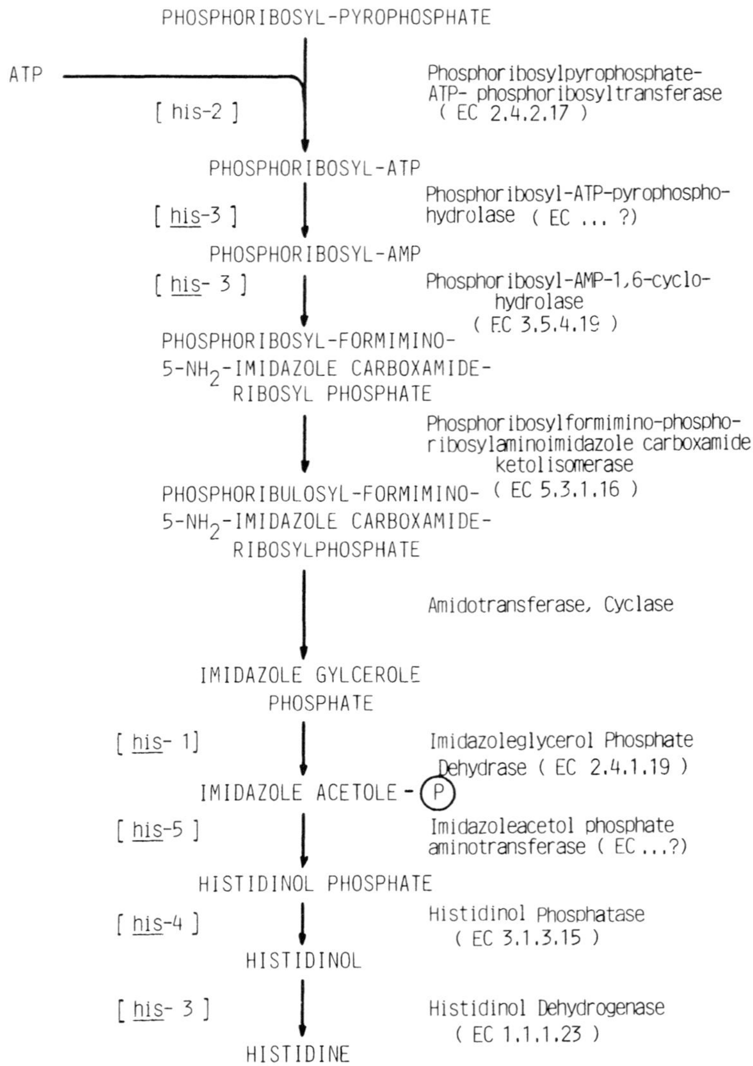

FIGURE 12. Biosynthesis of histidine by fungi. Enzymes and genetic are indicated as described in the legend to Fig. 4.

of *N. crassa* [145,196]. As a consequence, the gene-enzyme relationships of the 10 enzymes catalyzing the pathway are relatively well understood in *N. crassa* and *S. cerevisiae* [197,198], although the enzymology still contains many white spots.

The first step—the formation of phosphoribosyl-ATP—is catalyzed by phosphoribosyltransferase, which is the product of *his*-2 in *N. crassa* [199]. The enzyme is a major control point of histidine biosynthesis in several microorganisms [200,201], owing to its sensitivity to feedback inhibition by histidine. Although this was established for *S. cerevisiae*

[202], the enzyme from filamentous fungi has not yet been purified and investigated in this respect.

The second reaction in histidine biosynthesis—the hydrolysis of phosphoribosyl-ATP to phosphoribosyl-AMP—is one of three activities contained in the multifunctional protein encoded by the *his*-3 locus in *N. crassa* [204], the equivalent gene in *A. nidulans* [203], and the *HIS*4 locus from *S. cerevisiae* [198], which also catalyzes the third and tenth reactions in histidine biosynthesis (1-*N*-5'-phosphoribosyl-AMP-cyclohydrolase and histidinol dehydrogenase). A purified preparation from *N. crassa* exhibits an M_r of 130 kDa and appears to be composed of two identical subunits [205].

The enzymes catalyzing the conversion of imidazole glycerol phosphate to histidinol are products of *his*-1, *his*-5, and *his*-4 genes in *N. crassa*, from which they have also partially been purified [206-208].

As is evident from this paragraph, fine control and compartmentation of histidine biosynthesis in filamentous fungi has so far not been documented; regulation of synthesis of the enzymes involved in histidine biosynthesis, however, has been shown to be subject to general amino acid control [209-211]. It should be noted at this point that the general amino acid control of histidine biosynthesis has been a model system to study the molecular biology of general amino acid control in *S. cerevisiae*. The interested reader is refered to the reviews of Hinnebusch [212-214].

III. AMINO ACID POOLS AND METABOLIC COMPARTMENTATION

Fungi, in common with other microorganisms, contain all of their amino acids—besides their occurrence in proteins and peptides—in free, soluble pools (for a thorough review, see Ref. 215). The average soluble amino acid composition of some fungal cells is given in Table 1. Although environmental conditions sometimes evoke profound changes in the relative proportions of individual amino acids [122,216-219], it may be generalized that glutamate pools are always higher than those of other amino acids [13,216, 219]. Also, in most cases, glutamine, aspartate, and alanine are also found in relatively high amounts [216-219]. A notable finding is the appearance of strikingly high concentrations of γ-aminobutyric acid in some fungi (Table 1). Brody suggested a role of γ-aminobutyrate in glutamate metabolism, which is reviewed in Ref. 220.

It is now well established that internal amino acid pools are sequestered in at least two compartments, one with high and the other with low turnover [221-223]. Especially the basic amino acids arginine, lysine, and ornithine—which can also display high cellular concentrations (cf. Table 1)—were found to mainly occur in a pool of very low turnover [221,223]. This pool was later discovered to be the fungal vacuole [223]. Isolation of vacuoles from yeast [222,224] and filamentous fungi [225,226] largely confirmed this compartmentation of amino acids. A typical distribution, found for *S. cerevisiae* but presumably also valid for filamentous fungi, is depicted in Fig. 13. It must be noted, however, that the study of amino acid composition of isolated vacuoles is subject to some criticism because of eventual leakage from the organelles and because of metabolic changes which occur during preparation of the vacuoles, which may lead to erroneous values for some amino acids. Therefore, the most reliable data are obtained with

TABLE 1. Pools of Free Amino Acids Found in Some Fungi[a]

Amino acid	*N. crassa*	*A. niger*	*T. aureoviride*	*P. chrysogenum*
Glutamate	207	12.2	36.9	32.6
Glutamine	43.8	7.3	ND	9.5
Aspartate	18.2	1.3	5.2	6.3
Asparagine	b	0.5	ND	ND
Alanine	86.7	12.1	34.3	26.8
Glycine	3.1	2.5	8.8	3.6
Valine	3.1	3.3	ND	2.3
Leucine	2.3	0.7	2.9	0.6
Isoleucine	2.4	3.1	1.6	0.1
Arginine	10.6	46.7	7.2	0.2
Lysine	5.2	4.1	4.3	0.3
Ornithine	0.9	26.3	9.6	0.2
γ-Aminobutyrate	ND	10.4	ND	1.8
Threonine	23.5	3.5	3.2	4.7
Serine	21.2	1.7	4.6	4.2
Histidine	10.9	6.5	2.1	ND
Tryptophan	2.2	0.1	ND	ND
Tyrosine	4.2	0.9	1.5	0.2
Phenylalanine	1.4	0.9	1.9	ND
Proline	ND	ND	ND	2.6
Methionine	ND	0.3	0.5	0.1
Cysteine	ND	ND	1.3	ND

[a]Values quoted were taken from *Neurospora crassa* [216] (conidia), *Aspergillus niger* [217] (rapidly growing mycelia), *Trichoderma aureoviride* [218] (mycelia growing in continuous culture at 0.1 h^{-1}), and *Penicillium chrysogenum* [122] (fast-growing mycelia). All values are expressed as μmol amino acid/g fungal dry weight, except for *P. chrysogenum* values, which are given as % total ninhydrin-positive material.
[b]Glutamine value also includes the value for asparagine.
Note: ND, not detected.

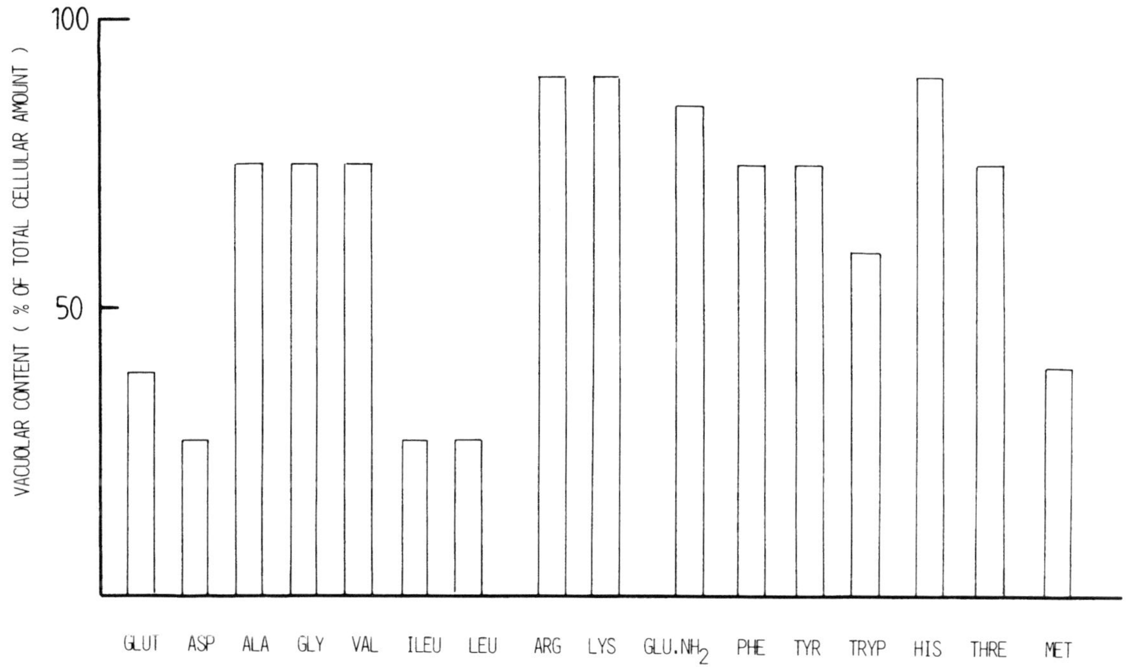

FIGURE 13. Percentage of total cellular pool of individual amino acids present in the vacuole. Data given are means for different fungi, taken from [222,224-226].

radioactive tracer methods, as was described in detail by Davis [223]. Briefly, it measures the dilution of specific radioactivity of an exogenously added amino acid in the whole cell and in the cell protein. Since only the cytosolic pool is used for protein synthesis, the ratios of specific radioactivity of protein-bound amino acid to total soluble amino acid can be used to assess the proportion of this amino acid sequestered in the cytosol and in the vacuole [223]. Clearly, this method neglects uptake of the amino acid into further compartments, such as mitochondria, but has nevertheless proven very valuable to estimate compartmented pool sizes of certain amino acids [223]. An additional tool was introduced by Kanamori et al. [221], who used nuclear magnetic resonance spectroscopy, and who established the predominant cytosolic location of glutamine, alanine, citrulline, and proline in *N. crassa*.

Uptake of amino acids into the vacuole occurs by active transport, driven by an ATP-generated proton gradient [227]. The *N. crassa* vacuolar membrane ATPase was investigated [228]. Both the amino acid specificity and the susceptibility to inhibitors distinguish this ATPase from the enzymes present in the plasma membrane or the mitochondria [228]. Efflux of vacuolar amino acids under conditions of nitrogen starvation and reduction of intracellular glutamine has also been shown to be energy (i.e., higher energy charge)-dependent [229,230].

The problem of retention and/or charge neutralization of the high concentrations of basic amino acids in the vacuole is still not clear: whereas earlier work favored a participation of vacuolar polyphosphates [225,231], more recent data argue against a stoichiometric polyphosphate-basic amino acid balance [232,233]. More detailed chemical analyses of vacuoles increase the probability that only some part of the amino acids is associated with the polyphosphates and that small anions must be transported into the vacuole to retain electroneutrality [233].

IV. GENERAL CONTROL OF AMINO ACID BIOSYNTHESIS

A. General Amino Acid Control

When fungal cells are starved for an individual amino acid or are subjected to concentrated supplements of certain amino acids, derepression of certain amino acid biosynthetic enzymes occurs [212-214]. This response is believed to help redressing amino acid imbalances in the cytosol [234]. The mechanism underlying this phenomenon is termed "general amino acid control" in *S. cerevisiae* or "cross-pathway control" in *Neurospora* and other filamentous fungi [209,234], and a full description of its molecular and biochemical properties was recently presented for yeast [212-214]. In this case, a complex cascade of regulatory events was revealed, which schematically is depicted in Fig. 14: the regulatory protein, decoded by gene *GCN*4 [135], interacts with the enzyme's structural genes. At least one copy of the hexanucleotide TGACTC (or a closely related sequence) in the 5'-nontranscribed region of the affected genes is required for general amino acid control to occur [236]. The mRNA of *GCN*4 has a long leader which in wild-type cells leads to poor translation of the coding sequence [235-237]. The negative effect on leader requires an intact second gene, *GCD*1 [238], which in turn is under the control of three further genes, *GCN*1, *GCN*2, and *GCN*3 [239,240]. Hence, summarizing the mechanism of general amino acid control, *GCN*1, *GCN*2, and *GCN*3 are activated by

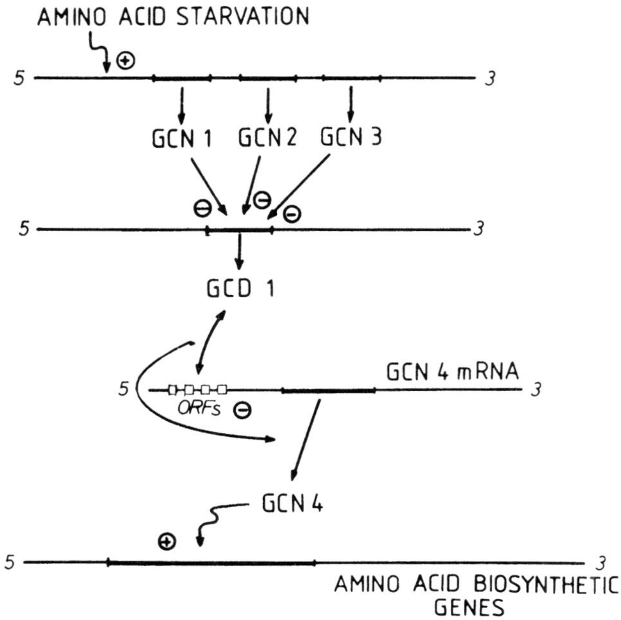

FIGURE 14. Model summarizing regulatory interactions among the upstream open reading frames (ORFs) of the *GCN*4 mRNA and the negative regulator *GCD*1 and the positive regulators *GCN*1, *GCN*2, and *GCN*3. ⊕ indicates activation, ⊖ indicates inhibition of transcription (in the case of *GCN*1, 2, or 3, *GCD*1 or amino acid biosynthetic genes) or translation (in the case of *GCN*4 mRNA). Please note that the neighborhood of *GCN*1-3 genes is artificially drawn and does not mean that these genes are clustered in fungi.

a particular amino acid limitation to block action of *GCD*1; this permits *GCN*4 mRNA to be translated and the *GCN*4 protein to stimulate transcription of amino acid biosynthetic genes.

The metabolic signal for general amino acid control to take place may be the transfer RNA charging ratio or the amount of uncharged transfer RNA [241]. Such an assumption has been proven by mutations in the amino acyl-tRNA synthetases, which can evoke general amino acid control despite of normal or high pools of amino acids. Derepression upon starvation for an amino acid occurs before the cellular amino acid pool charges significantly [240]. This is, however, easily understood, considering that only changes in metabolically active amino acid pools, i.e., cytosolic amino acid pools, should be of relevance to control (cf. Ref. 240).

In *N. crassa*, mutations of only one general amino acid control gene, *cpc*-1, are known so far [242]. A range of functional impairment, similar to the situation in yeast, is suggested by variation in severity of effect of different *cpc*-1 alleles. Thus, further investigations are needed to establish the similarity of this general control phenomenon in yeast and multicellular fungi. Recently, *cpc*-1 was cloned from *Neurospora crassa*, and some aspects of its structure and regulation were elucidated [242]:

the transcript contains three open reading frames, which would produce two peptides 20 and 41 amino acids in length. The deduced amino acid sequence of the *cpc* polypeptide contains segments similar to the DNA-binding and transcriptional activation domains of *GCN*4 from yeast. Hence, the positive regulator proteins involved in general amino acid control in yeast and *N. crassa* share basic similarities.

V. INDUSTRIAL PRODUCTION OF AMINO ACIDS BY FUNGI

The industrial production of amino acids was—and still is—a domain of applied bacterial genetics, initiated and forwarded in Japan [243-245]. Today, almost all of the 20 common amino acids are manufactured commercially, most of them via bacterial fermentations. Current trends are focused toward (a) utilizing new and cheap carbohydrate sources; (b) using biosynthetic pathway intermediates as substrates; (c) employing immobilized cells or immobilized enzymes [245]. The potential application of the amino acids so produced is wide, including food supplementation, feedstuffs, flavoring agents, cosmetics, and starting materials for pharmaceutical purposes. On a global basis, two-thirds of world production is used for food purposes [245].

Filamentous fungi have so far only rarely been considered as amino acid producer organisms. This is less likely due to the biochemical differences in amino acid biosynthesis between fungi and bacteria than to the fact that their growth rate is lower. Even if a fungus were able to produce the same amount of a given amino acid per liter of fermentation broth and per gram of biomass, it certainly would do so in a longer fermentation time. Therefore, the ability of fungi to overproduce certain amino acids has found limited interest so far, and then yeasts were preferred in most cases (see below). This situation may change when immobilized cells—or even enzymes—come into use.

Virtually, only tryptophan, lysine, and methionine have so far been produced by fungi (cf. Table 2).

A. Tryptophan

Tryptophan overproduction is not achieved in significant amounts with bacteria and therefore is still synthesized chemically. A *Hansenula polymorpha* strain, resistant to 5-fluorotryptophan, was reported to excrete tryptophan [244]. Preferably, tryptophan is produced from added precursers such as anthranilic acid or indole, as was reported for *H. anomala* [247-249] and *Claviceps purpurea* [250].

B. Lysine

Because of its nature as an essential amino acid for animal diet and its deficiency in grain proteins, lysine is one of the major amino acids produced commercially [245]. Although some fungi can produce up to 2-3 g/liter lysine, these are not used commercially since this is far below the amount which can be obtained by bacterial fermentations (cf. Table 2). Particularly thialysine resistance has been used to select for deregulated, hyperproducing yeast strains [251-253]. An attractive alternative has been introduced by using whole cells of *Cryptococcus laurentii*, *Candida humicola*, and *Tricho-*

TABLE 2 Production of Amino Acids by Fungi[a]

Amino acid			Yield (g/liter)
Tryptophan	*Hansenula anomala*	Anthranilate	6
		Indole	14
	Claviceps purpurea		
Lysine	*Candida pelliculosa*		3.2
	Saccharomycopsis lipolytica		
	Ustilago mayidis		2.5
	Sphacelotheca sorgii		
Methionine	*Saccharomycosis lipolytica*		
	Hanseniospora valbyensis		
Glutamate	*Aspergillus orycae*		
	Penicillium chrysogenum		
	Rhodotorula glutinis		
Alanine	*Ustilago mayidis*		20
Threonine	*Candida guilliermondii*		4
	Ustilago mayidis		0.3
Phenylalanine	*Torula* sp.		

[a]For original references see [243-245].

sporum cutaneum, which are able to hydrolyze exogenous α-aminocaprolactam [254].

C. Methionine

The fermentative production of methionine has found to be uneconomic, since this amino acid can also be produced chemically [245]. Strains of *Hanseniospora valbyensis*, which were selected for resistance to ethionine, *N*-acetylnorleucine, polymyxin, or bacitracin, were shown to excrete methionine into the medium [255]. Alternative research has been devoted to the use of methionine-rich microorganisms as food supplements: success in this way has been reported for *C. tropicalis* strains with higher sulfate requirement [256], methanol-assimilating yeasts [255], and ethionine-resistant strains of *S. cerevisiae* and *S. lipolytica* [257].

D. Other Amino Acids

A small number of other amino acids have been reported to be produced by fungi, i.e., glutamate [258], alanine [259], and threonine [260] (cf. Table 2). Phenylalanine is produced commercially with certain *Torula* strains for special use in the synthesis of the artificial sweetener aspartame [261].

VI. CONCLUSIONS

This chapter was written in an attempt to assess and compare the present state of knowledge on the biosynthesis and production of amino acids by filamentous fungi. At this point it appears appropriate to draw some general conclusions. As we have seen, knowledge on amino acid biosynthesis in fungi is strongly heterogeneous; in some cases (i.e., arginine, tryptophan, glutamate), the status quo is at least satisfying, and the plethora of new biochemical information, which far exceeds the interest of molecular mycologists alone, justifies the detailed dedication to this topic. In other cases, however, information is so meager that even data obtained with bacteria had to be cited as the only source of information (i.e., threonine, serine, asparagine). Although the biochemical pathways may be identical, it should have become clear in this chapter that the regulation of amino acid biosynthesis in filamentous fungi to some extent deviates from that observed in yeast or bacteria; in particular, the occurrence of feedback repression and enzyme multiplicity appears to be less common in filamentous fungi. This should justify investigating the regulation also of those biosynthetic pathways whose structure and metabolic flux has been safely elucidated in bacteria or yeast.

Compartmentation of amino acid biosynthesis is another problem solved only for a minority of amino acids. Compartmentation of amino acid biosynthesis may exert a strong influence on metabolic regulation (as seen, for example, from the differences between yeast and *Neurospora*). Since the respective methods are now well documented, it is surprising that our present knowledge on compartmentation of most amino acid biosynthetic pathways is very poor. Work in this area will undoubtedly be of primary importance for our understanding of the pathways involved.

Finally, as we have seen, fungi are generally not considered as potential amino acid producer organisms, since bacteria can do the same job in a shorter time. However, the filamentous fungi are superior in their ability to grow on a variety of polysaccharides, wood wastes, or other cheap materials; their potential to produce amino acids may therefore eventually become important, if such carbon sources are to be used for amino acid production. Furthermore, the general amino acid control may be useful in turning on/turning off of amino acid biosynthesis in fungi, which may be attractive in an industrial process. From the data presented in this chapter, no argument is apparent as to why fungi should not be able to accumulate particular amino acids in amounts comparable to bacteria. Yet, where it has been tried, the final yields were significantly below all obtained with bacteria. Our present knowledge does not allow a conclusive explanation for this. However, we must be aware that in fungi, amino acids are predominantly secreted into the vacuole and not into the surrounding medium; the capacity of the vacuole and the rate of release from it (or the existence of a bypassing mechanism) may therefore be important parameters for amino acid production by fungi.

ACKNOWLEDGMENTS

The authors gratefully acknowledge the peaceful and undisturbed atmosphere of Ca'Ballarin, Venezia, and Spitzhof at Traunstein, Lower Austria, where most sections of this chapter were written. The authors' own work on lysine and glutamate biosynthesis was supported by Fond zur Förderung Wissenschaftlicher Forschung, and Biochemie GesmbH, Kundl, Tyrol.

REFERENCES

1. Pateman, J. A., and Kinghorn, J. R. Nitrogen metabolism. In *The Filamentous Fungi*, Vol. 3 (J. E. Smith and D. E. Berry, eds.), Edward Arnold, London, 1975, pp. 159-237.
2. Hermann, K. M., and Somerville, R. L. *Amino Acid Biosynthesis and Regulation*. Addison-Wesley, Reading, Mass., 1983.
3. Limon-Lason, J., Lara, M., Resendiz, B., and Mora, J. Regulation of glutamine synthetase in fed batch cultures of Neurcspora crassa. *Biochem. Biophys. Res. Commun.*, *78*:1234-1240, 1977.
4. Dunn-Coleman, N. S., Robey, E. A., Tomsett, A. B., and Garrett, R. H. Glutamate synthase levels in Neurospora crassa mutants altered with respect to nitrogen metabolism. *Mol. Cell. Biol.*, *1*:158-164, 1981.
5. Martin, F., Stewart, G. R., Genetet, I., and Lecaton, F. Assimilation of $^{15}NH_4$ by beech (Fagus sylvatica L.) ectomycorrhizas. *New Phtyologist*, *102*:85-94, 1986.
6. Casper, P., Bode, R., and Birnbaum, D. Untersuchungen zur Regulation der Ammoniumassimilation von Candida maltosa. *J. Basic Microbiol.*, *25*:95-101, 1985.
7. Kusnan, M. B., Berger, M. G., and Fock, H. P. The involvement of glutamine synthetase/glutamate synthase in ammonia assimilation by Aspergillus nidulans. *J. Gen. Microbiol.*, *133*:1235-1242, 1987.
8. Hernandez, G., Sanchez-Pescador, R., Palacios, R., and Mora, J. Nitrogen source regulates glutamate dehydrogenase NADP synthesis in Neurospora crassa. *J. Bacteriol.*, *154*:524-528, 1983.
9. Blumenthal, K. M., and Smith, E. L. NADP-specific glutamate dehydrogenase from Neurospora crassa. I. Isolation, subunits, amino acid composition, sulfhydryl groups and identification of a lysine residue reactive with pyridoxal-phosphate and N-ethylmaleimide. *J. Biol. Chem.*, *248*:6002-6008, 1973.
10. Al-Gharawi, A., and Moore, D. Factors affecting the amount and the activity of the glutamate dehydrogenases of Coprinus cinereus. *Biochim. Biophys. Acta*, *496*:95-102, 1977.
11. Agrawal, A. K., and Mohan Rao, V. K. Kinetic studies on glutamate dehydrogenase from Aspergillus ochraceus. *Ind. J. Biochem. Biophys.*, *21*:386-388, 1984.
12. Queener, S. W., McDermott, J., and Radue, A. B. Glutamate dehydrogenase specific activity and Cephalosporin C synthesis in the M 8650 series of Cephalosporium acremonium mutants. *Antimicrob. Ag. Chemother.*, *7*:646-651, 1975.
13. Jaklitsch, W. M., Rohr, M., and Kubicek, C. P. Glutamate pools and glutamate dehydrogenase regulation in relation to penicillin biosynthesis in strains of Penicillium chrysogenum. *Exp. Mycol.*, *9*:310-317, 1985.
14. Wooton, J. C., Chambers, G. K., Holder, A. A., Baron, A. J., Taylor, J. G., Fincham, J. R., Blumenthal, K. M., Moon, K., and Smith, E. L. Amino acid sequence of NADP-specific glutamate dehydrogenase from Neurospora crassa. Homology with vertebrate enzymes. *Proc. Natl. Acad. Sci. USA*, *71*:4361-4365, 1974.
15. Holder, A. A., Wooton, J. C., Baron, A. J., Chambers, G. K., and Fincham, J. R. S. The amino acid sequence of Neurospora NADP-specific glutamate dehydrogenase. Peptic and chymotryptic peptides and the complete sequence. *Biochem. J.*, *149*:757-773, 1975.

16. Kinnaird, J. H., and Fincham, J. R. S. The complete nucleotide sequence of the Neurospora crassa am (glutamate dehydrogenase) gene. *Gene, 26*:253-260, 1983.
17. Blumenthal, K. M., Moon, K., and Smith, E. L. NADP-specific glutamate dehydrogenase of Neurospora. II. Sequences of the cyanogen bromide peptides; the complete sequence of the enzyme. *J. Biol. Chem., 250*: 3644-3654, 1975.
18. Brett, M., Chambers, G. K., Holder, A. A., Fincham, J. R. S., and Wootton, J. C. Mutational amino acid replacements in Neurospora crassa NADP-specific glutamate dehydrogenase. *J. Mol. Biol., 106*:1-22, 1976.
19. West, D. J., Tuveson, R. W., Barratt, R. W., and Fincham, J. R. S. Allosteric effects in NADP-specific glutamate dehydrogenase from Neurospora. *J. Biol. Chem., 242*:2134-2138, 1967.
20. Wootton, J. C. Re-assessment of ammonium-ion affinities of NADP-specific glutamate dehydrogenases. Activation of the Neurospora crassa enzyme by ammonium and rubidium ions. *Biochem. J., 209*:527-531, 1983.
21. Moore, D. Evidence that the NADP-linked glutamate dehydrogenase of Coprinus cinereus is regulated by acetyl CoA and ammonium levels. *Biochim. Biophys. Acta, 661*:247-254, 1981.
22. Pateman, J. A., Kinghorn, J. R., Dunn, E., Forbes, E. Ammonium regulation in Aspergillus nidulans. *J. Bacteriol., 114*:943-950, 1973.
23. Palacios, R. Neurospora crassa glutamine synthetase. Purification by affinity chromatography and characterization of subunit structure. *J. Biol. Chem., 251*:4787-4791, 1976.
24. Punekar, N. S., Vaidyanathan, C. S., and Appaji Rao, N. Studies on Aspergillus niger glutamine synthetase: regulation of enzyme levels by nitrogen sources and identification of active site residues. *J. Biosci., 6*:17-35, 1984.
25. Rachim, M. A., and Nicholas, D. J. D. Glutamine synthetase and glutamate synthase from Sclerotinia sclerotiorum. *Phytochemistry, 24*: 2541-2548, 1985.
26. Davila, G., Lara, M., Guzman, J., and Mora, J. Relation between structure and function of Neurospora crassa glutamine synthetase. *Biochim. Biophys. Res. Commun., 92*:134-140, 1980.
27. Sanchez, F., Calva, E., Campomanes, M., Blanco, L., Guzman, J., Saborio, J. L., and Palacios, R. Heterogeneity of glutamine synthase polypeptides in Neurospora crassa. *J. Biol. Chem., 255*:2231-2234, 1980.
28. Davila, G., Brom, S., Mora, Y., Palacios, R., and Mora, J. Genetic and biochemical characterization of glutamine synthase from Neurospora crassa glutamine auxotrophs and their revertants. *J. Bacteriol., 153*: 993-1000, 1983.
29. Ginsburg, A., and Stadtman, E. R. Glutamine synthetase from Escherichia coli: structure and regulation. In *Subunit Enzymes. Biochemistry and Function* (K. E. Ebner, ed.), Marcel Dekker, New York, 1975, pp. 43-84.
30. Kapoor, M., and Bray, D. Feedback inhibition of glutamine synthetase of Neurospora crassa by NAD. *Biochemistry, 7*:3583-3589, 1968.
31. Hernandez, G., Mora, Y., and Mora, J. Regulation of glutamine synthetase by glycine and serine in Neurospora crassa. *J. Bacteriol., 165*:133-138, 1986.

32. Sanchez, F., Campomanes, M., Quinto, C., Hansberg, W., Mora, J., and Palacios, R. Nitrogen source regulates glutamine synthetase mRNA levels in Neurospora crassa. *J. Bacteriol.*, *136*:880-885, 1978.
33. Punekar, N. S., Vaidyanathan, C. S., and Appaji Rao, N. Role of Mn^{2+} and Mg^{2+} in catalysis and regulation of Aspergillus niger glutamine synthetase. *Ind. J. Biochem. Biophys.*, *22*:142-151, 1985.
34. Aguirre, J., and Hansberg, W. Oxidation of Neurospora crassa glutamine synthetase. *J. Bacteriol.*, *166*:1040-1045, 1986.
35. Hummelt, G., and Mora, J. NADH-dependent glutamate synthase and nitrogen metabolism in Neurospora crassa. *Biochim. Biophys. Res. Commun.*, *92*:127-133, 1980.
36. Tempest, D. W., Meers, J. L., and Brown, C. M. Synthesis of glutamate in Aerobacter aerogenes by a hitherto unknown route. *Biochem. J.*, *117*:405-407, 1970.
37. Kitto, G. B., Kootke, M. E., Bertland, L. H., Murphy, W. H., and Kaplan, N. O. Studies on malate dehydrogenase and aspartate aminotransferases from Neurospora crassa. *Arch. Biochem. Biophys.*, *121*: 224-231, 1967.
38. Fincham, J. R., and Boulter, A. B. Effect of amino acid on transaminase production in Neurospora crassa: evidence for four different enzymes. *Biochem. J.*, *62*:72-77, 1956.
39. Munkres, K. D. Physicochemical identity of Neurospora malate dehydrogenase and aspartate aminotransferase. *Arch. Biochem. Biophys.*, *112*:347-354, 1965.
40. Ramos, F., and Wiame, J. Two asparagine synthetases in Saccharomyces cerevisiae. *Eur. J. Biochem.*, *108*:373-377, 1980.
41. McPhee, K. G., Nelson, R. E., and Schuster, S. M. Neurospora crassa mutants deficient in asparagine synthetase. *J. Bacteriol.*, *156*:475-478, 1983.
42. Galbraith, J. C., and Smith, J. E. Sporulation of Aspergillus niger in submerged liquid culture. *J. Gen. Microbiol.*, *59*:31-45, 1969.
43. Abelson, P. H., and Vogel, H. J. Amino acid biosynthesis in Torulopsis utilis and Neurospora crassa. *J. Biol. Chem.*, *213*:355-364, 1955.
44. Kubicek, C. P., and Röhr, M. Influence of manganese deficiency on enzyme synthesis and citric acid accumulation by Aspergillus niger. *Eur. J. Appl. Microbiol.*, *4*:167-173, 1977.
45. Sojka, G. A., and Garner, H. R. The serine biosynthetic pathway in Neurospora crassa. *Biochim. Biophys. Acta*, *148*:42-47, 1967.
46. Beck, F. W. J. Neurospora crassa phosphoserine phosphatase: the effect of pH on the hydrolysis of phosphoserine by enzymes present in wild type and ser (JBM5) extracts and partial purification and characterization of the wild type phosphoserine phosphatase. M.S. thesis, California State University, Northridge.
47. Cessins, E. A., P.-Y. Chan, and Combepine, G. One-carbon metabolism in Neurospora crassa wild-type and in mutants partially deficient in serine hydroxylmethyltransferase. *Biochem. J.*, *160*:305-314, 1976.
48. Cossins, E. A., and Pang, S. H. Y. Loss of cytosolic serine hydroxymethyltransferase in a formate mutant of N. crassa. *Experientia*, *36*: 289-290, 1980.
49. Franke, W., and Jilge, G. Enzymatic transformations of C2 acids by microorganisms. II. Glyoxalate-α-ketoglutarate condensing enzyme from Aspergillus niger. *Arch. Microbiol.*, *39*:88-94, 1961.

50. Guranowski, A., and Paszewski, A. Metabolism of 5'-methylthioadenosine in Aspergillus nidulans. An alternative pathway for methionine synthesis via utilization of the nucleoside thiomethylgroup. *Biochim. Biophys. Acta, 717*:289-294, 1982.
51. Paszewski, A., and Grabski, J. Studies on the β-cystathionase and O-acetylhomoserine sulfhydrylase as the enzymes of alternative methionine biosynthetic pathways in Aspergillus nidulans. *Acta Biochim. Pol., 20*:159-168, 1973.
52. Paszewski, A., and Grabski, J. Regulation of S-amino acid biosynthesis in Aspergillus nidulans: role of cysteine and homocysteine as regulatory effectors. *Mol. Gen. Genet., 132*:307-320, 1974.
53. Pieniazek, N. J., Bal, J., Balbin, E., and Stepien, P. P. An Aspergillus nidulans mutant lacking serine transacetylase: evidence for two pathways of cysteine biosynthesis. *Mol. Gen. Genet., 132*:363-366, 1974.
54. Pieniazek, N. J., Stepien, P. P., and Paszewski, A. An Aspergillus nidulans mutant lacking cystathionine β-synthase: identity of L-serine sulfhydrylase with cystathionine β-synthase and its distinctness from acetyl-L-serine sulfhydrylase. *Biochim. Biophys. Acta, 297*:37-47, 1973.
55. Pieniazek, N. J., Kowalska, I. M., and Stepien, P. P. Deficiency in methionine adenosyltransferase resulting in limited repressibility of methionine biosynthetic enzymes in Aspergillus nidulans. *Mol. Gen. Genet., 126*:367-374, 1973.
56. Paszewski, A., Prazmo, W., Nadolska, J., and Regulski, M. Mutations affecting the sulfur assimilation pathways in Aspergillus nidulans: their effect on sulphur amino acid metabolism. *J. Gen. Microbiol., 130*:1113-1121, 1984.
57. Morzycka, E., and Paszewski, A. Cysteine and homocysteine synthesis in Saccharomycopsis lipolytica. Identification and characterization of two cysteine synthases. *Acta Biochim. Pol., 29*:81-93, 1982.
58. Döbeli, H., and Nüesch, J. Regulatory properties of O-acetylserine sulfhydrylase of Cephalosporium acremonium: evidence of an isoenzyme and its importance in cephalosporin C synthesis. *Antimicrob. Ag. Chemother., 18*:111-117, 1980.
59. Yamagata, S. Partial purification and some properties of homoserine O-acetyltransferase of a methionine auxotroph of Saccharomyces cerevisiae. *J. Bacteriol., 169*:3458-3463, 1987.
60. Yamagata, S., Takeshima, K., and Naiki, N. Evidence for the identity of O-acetylserine sulfhydrylase with O-acetylhomoserine sulfhydrylase in yeast. *J. Biochem., 75*:1221-1229, 1974.
61. Wiebers, J. L., and Garner, H. R. Homocysteine and cysteine synthases of Neurospora crassa. Purification, properties, and feedback control activity. *J. Biol. Chem., 242*:12-23, 1967.
62. Wiebers, J. L., and Garner, H. R. Acyl derivatives of homoserine as substrates for homocysteine synthesis in Neurospora crassa, yeast and Escherichia coli. *J. Biol. Chem., 242*:5644-5649, 1967.
63. Kiritani, K., Narise, S., Berquist, A., and Wagner, R. P. The overall synthesis in vitro of valine from pyruvate by Neurospora homogenates. *Biochim. Biophys. Acta, 100*:432-443, 1965.
64. Bergquist, A., Eakin, E. A., Murali, D. K., and Wagner, R. P. A pyruvate-valine enzyme complex that is dependent upon the metabolite state of the mitochondria. *Proc. Natl. Acad. USA, 71*:4352-4355, 1974.

65. Cassady, W. E., Leiter, E. H., Bergquist, A., and Wagner, R. P. Separation of mitochondrial membranes of Neurospora crassa. II. Submitochondrial localization of the isoleucine-valine biosynthetic pathway. *J. Cell. Biol.*, *53*:66-72, 1972.
66. Wagner, R. P., Bergquist, A., Barbee, T., and Kiritani, K. Genetic blocks in the isoleucine-valine pathway of Neurospora crassa. *Genetics*, *49*:865-882, 1964.
67. Tanaka, H., and Kuwana, H. A basal unit of valine-sensitive acetolacetate synthase of Neurospora crassa. *Biochem. Biophys. Res. Commun.*, *123*:418-423, 1984.
68. Glatzer, L., Eakin, E., and Wagner, R. P. Acetohydroxy acid synthetase with a pH optimum of 7.5 from Neurospora crassa mitochondria: characterization and partial purification. *J. Bacteriol.*, *112*:453-464, 1972.
69. Goulden, S. A., and Chattaway, F. W. End product control of acetohydroxyacid synthetase by valine in Penicillium chrysogenum Q176 and a high penicillin yielding mutant. *J. Gen. Microbiol.*, *59*:111-118, 1968.
70. Matsumura, S., and Suzuki, M. Feedback regulation of acetohydroxy acid synthetase in Cephalosporium acremonium M8650 and a high cephalosporin-producing mutant. *Ag. Biol. Chem.*, *50*:505-507, 1986.
71. MacDonald, D. W., Arst, H. N., and Cove, D. J. The threonine dehydratase structural gene in Aspergillus nidulans. *Biochim. Biophys. Acta*, *362*:60-66, 1974.
72. Olshan, A. R., and Gross, S. R. Role of the leu-3 cistron in the regulation of the synthesis of isoeleucine and valine biosynthetic enzymes of Neurospora. *J. Bacteriol.*, *118*:374-384, 1974.
73. Webster, E. R., and Gross, S. R. The α-isopropylmalate synthetase of Neurospora. I. The kinetics and end product control of α-isopropylmalate synthetase function. *Biochemistry*, *4*:2309-2318, 1965.
74. Webster, R. E., Nelson, C. A., and Gross, S. R. The α-isopropylmalate synthetase of Neurospora. II. The relation between structure and complementation interactions. *Biochemistry*, *4*:2319-1227, 1965.
75. Roeder, P. R., and Kohlhaw, G. B. α-Isopropylmalate synthase from yeast. A zinc metallo enzyme. *Biochim. Biophys. Acta*, *613*:482-487, 1980.
76. Reichenberger, V. E. Jr., and Gross, S. R. Structural features of normal and complemented forms of the Neurospora isopropylmalate isomerase. *J. Bacteriol.*, *133*:802-810, 1978.
77. Gross, S. R. Genetic regulatory mechanisms in the fungi. *Annu. Rev. Genet.*, *3*:395-424.
78. Kashmiri, S. V. S., and Gross, S. R. Mutations affecting the regulation of the enzymes of leucine synthesis in Neurospora. *Genetics*, *64*:423-440, 1970.
79. Gross, S. R. The regulation of synthesis of the leucine biosynthetic enzymes of Neurospora. *Proc. Natl. Acad. Sci. USA*, *54*:1538-1546, 1965.
80. Jobbagy, A. J., and Wagner, R. P. Repression of the branched-chain amino acid biosynthetic enzymes in developing conidia and mycelia of Neurospora. *Can. J. Biochem.*, *52*:822-829, 1974.
81. Barthelmess, I. Regulation of amino acid synthetic enzymes in Neurospora crassa in the presence of high concentrations of amino acids. *Mol. Gen. Genet.*, *203*:533-537, 1986.

82. Polacco, J., and Gross, S. R. The product of the leu-3 cistron as a regulatory element for the production of the leucine biosynthetic enzymes of Neurospora. *Genetics*, *74*:443-459, 1973.
83. Reichenberger, V. E., Fischer, M., and Gross, S. R. Regulation of isopropylmalate isomerase synthesis in Neurospora crassa. *J. Bacteriol.*, *133*:794-801, 1978.
84. Davis, R. H. Compartmental and regulatory mechanisms in the arginine pathways of Neurospora crassa and Saccharomyces cerevisiae. *Microbiol. Rev.*, *50*:280-313, 1986.
85. Perkins, D. D., Radford, A., Newmeyer, D., and Björkman, M. Chromosomal loci of Neurospora crassa. *Microbiol. Rev.*, *46*:426-570, 1982.
86. Vogel, H. J., and Bonner, D. M. On the glutamate-proline-ornithine interrelation in Neurospora crassa. *Proc. Natl. Acad. Sci USA*, *40*: 688-694, 1954.
87. Yura, T., and Vogel, H. J. Pyrroline-5-carboxylate reductase of Neurospora crassa. Partial purification and some properties. *J. Biol. Chem.*, *234*:335-338, 1959.
88. Baich, A., and Pierson, D. J. Control of proline biosynthesis in Escherichia coli. *Biochim. Biophys. Acta*, *104*:397-404, 1965.
89. Weiss, R. L., and Davis, R. H. Intracellular localization of enzymes of arginine metabolism in Neurospora crassa. *J. Biol. Chem.*, *248*:5403-5408, 1973.
90. Jauniaux, J.-C., Urrestarazu, L. A., and Wiame, J. M. Arginine metabolism in Saccharomyces cerevisiae: subcellular localization of the enzymes. *J. Bacteriol.*, *133*:1096-1107, 1978.
91. Hinde, R. W., Jacobson, J. A., Weiss, R. L., and Davis, R. H. N-Acetyl-L-glutamate synthase of Neurospora crassa. Characteristics, localization, regulation and genetic control. *J. Biol. Chem.*, *261*:5848-5852, 1986.
92. Wandinger, A. U., Wolf, E. C., Weiss, R. L., and Davis, R. H. Acetylglutamate kinase-acetylglutamylphosphate reductase complex of Neurospora crassa: evidence for two polypeptides. *J. Biol. Chem.*, *260*:5974-5978, 1985.
93. Wolf, E. C., and Weiss, R. L. Acetylglutamate kinase. A mitochondrial feedback-sensitive enzyme of arginine biosynthesis in Neurospora crassa. *J. Biol. Chem.*, *255*:9189-9195, 1980.
94. Wandinger-Ness, A. U., Ness, S. A., and Weiss, R. L. Simultaneous purification of three mitochondrial enzymes. Acetylglutamate kinase, acetylglutamyl-phosphate reductase and carbamoyl phosphate synthetase of Neurospora crassa. *J. Biol. Chem.*, *261*:4820-4827, 1986.
95. Wandinger-Ness, A. U., and Weiss, R. L. A single precurser protein for two separable mitochondrial enzymes in Neurospora crassa. *J. Biol. Chem.*, *262*:5823-5830, 1987.
96. De Deken, R. H. Biosynthese de l'arginine chez la levure. I. Le sort de la N-acetylornithine. *Biochim. Biophys. Acta*, *78*:606-616, 1963.
97. Makoff, A. J., and Radford, A. Genetics and biochemistry of carbamoyl phosphate biosynthesis and its utilization in the pyrimidine biosynthetic pathway. *Microbiol. Rev.*, *42*:307-328, 1978.
98. Davis, R. H., Ristow, J. L., and Ginsburg, C. L. Independent localization and regulation of carbamyl phosphate synthase A polypeptides in Neurospora crassa. *Mol. Gen. Genet.*, *181*:215-221, 1981.

99. Ratner, S. Enzymes of arginine and urea synthesis. *Adv. Enzymol.*, *39*:1-90, 1973.
100. Ness, S. A., and Weiss, R. L. Carbamoyl phosphate synthetases from Neurospora crassa. Immunological relatedness of the enzymes from Neurospora, bacteria, yeast and mammals. *J. Biol. Chem.*, *260*: 14355-14362, 1985.
101. Davis, R. H., Ristow, J. L., Hanson, B. A. Carbamyl phosphate synthetase A of Neurospora crassa. *J. Bacteriol.*, *141*:144-154, 1980.
102. Ness, S. A., and Weiss, R. L. Carboxyterminal sequences influence the import of mitochondrial protein precursors in vivo. *Proc. Natl. Acad. Sci. USA*, *84*:6692-6696, 1987.
103. Bates, M., Weiss, R. L., and Clarke, S. Ornithine transcarbamylase from Neurospora crassa: purification and properties. *Arch. Biochem. Biophys.*, *239*:172-182, 1985.
104. Wampler, D. E., and Fairley, J. L. Argininosuccinate synthetase of Neurospora crassa. *Arch. Biochem. Biophys.*, *121*:580-586, 1967.
105. Hilger, F., Simon, J. P., and Stalon, V. Yeast arginosuccinate synthetase. Purification; structural and kinetic properties. *Eur. J. Biochem.*, *94*:153-163, 1979.
106. Cohen, G. G., and Bishop, J. O. Purification of argininosuccinase from Neurospora crassa and comparison of some properties of the wild type enzyme and an enzyme formed by inter-allelic complementation. *Genet. Res.*, *8*:243-252, 1966.
107. Newmeyer, D. Genes influencing the conversion of citrulline to argininosuccinate in Neurospora crassa. *J. Gen. Microbiol.*, *28*:215-230, 1962.
108. Flint, H. J., and Kemp, B. F. General control of arginine biosynthetic enzymes in Neurospora crassa. *J. Gen. Microbiol.*, *124*:129-140, 1981.
109. Flint, H. J., Dible, S., and Kacser, H. Derepression of enzyme synthesis in response to arginine limitation in Neurospora crassa. *J. Gen. Microbiol.*, *131*:2891-2900.
110. Cybis, J., and Davis, R. H. Organization and control in the arginine biosynthetic pathway of Neurospora. *J. Bacteriol.*, *123*:196-202, 1975.
111. Weiss, R. L., and Lee, C. A. Isolation and characterization of Neurospora crassa mutants impaired in feedback control of ornithine synthesis. *J. Bacteriol.*, *141*:1305-1311, 1980.
112. Flint, H. J., Tateson, R. W., Barthelmess, I. B., Porteous, D. J., Donachie, W. D., and Kacser, H. J. Control of flux in the arginine pathway of Neurospora crassa. Modulations of enzyme activity and concentration. *Biochem. J.*, *200*:231-246, 1981.
113. Vogel, H. J. Two modes of lysine synthesis among lower fungi: evolutionary significance. *Biochim. Biophys. Acta*, *41*:172-173, 1960.
114. LeJohn, H. B. Enzyme regulation, lysine pathways and cell wall structures as indicators of major lines of evolution in fungi. *Nature*, *231*: 164-168, 1971.
115. Patte, J. C. Diaminopimelate and lysine. In *Amino Acids: Biosynthesis and Genetic Regulation* (K. M. Hermann and E. R. Somerville, ed.), Addison-Wesley, Reading, Mass., 1983, pp. 213-233.
116. Jaklitsch, W. M., and Kubicek, C. P. Homocitrate synthase from Penicillium chrysogenum. Localization, purification of the cytosolic isoenzyme and sensitivity to lysine. *Biochem. J.* *269*:247 (1990).
117. Nichoalds, G. E., Bromley, M. A., and Broquist, H. P. Glutarate and the lysine requirement of Neurospora crassa STL-7. *Fed. Proc.*, *30*:463, 1971.

118. Gaillardin, C., Fournier, P., Sylvestre, G., and Heslot, H. Mutants of Saccharomycopsis lipolytica defective in lysine catabolism. *J. Bacteriol.*, *125*:48-55, 1976.
119. Gaillardin, C., Poirier, L., and Heslot, H. A kinetic study of homocitrate synthetase activity in the yeast Saccharomycopsis lipolytica. *Biochim. Biophys. Acta*, *422*:390-406, 1976.
120. Luengo, J. M., Revilla, G., Lopez-Nieto, M. J., Villanueva, J. R., and Martin, J. F. Regulation of homocitrate synthase by lysine in relation to penicillin G synthesis. *Adv. Biotechnol.*, *3*:161-165, 1981.
121. Masurekar, P. S., and Demain, A. L. Insensitivity of homocitrate synthase in extracts of Penicillium chrysogenum to feedback inhibition by lysine. *Appl. Microbiol.*, *28*:265-270, 1974.
122. Jaklitsch, W. M., Hampel, W., Rohr, M., Kubicek, C. P., and Gamerith, G. α-Aminoadipate pool concentration and penicillin biosynthesis in strains of Penicillium chrysogenum. *Can. J. Microbiol.*, *32*:473-480, 1986.
123. Tracy, J., and Kohlhaw, G. A reversible, coenzyme A mediated inactivation of biosynthetic condensing enzymes in yeast: a possible regulatory mechanism. *Proc. Natl. Acad. Sci. USA*, *72*:1664-1667, 1975.
124. Betterton, H., Fjellstedt, T., Matsuda, M., Ogur, M., Tate, R. Localization of the homocitrate pathway. *Biochim. Biophys. Acta*, *170*:459-461, 1968.
125. Bhattacharjee, J. K. α-Aminoadipate pathway for the biosynthesis of lysine in lower eukaryotes. *CRC Crit. Rev. Microbiol.*, *12*:131-151, 1985.
126. Sagisaka, S., and Shimura, K. Mechanism of activation and reduction of α-aminoadipate by yeast enzyme. *Nature*, *188*:1189-1190, 1960.
127. Barnes, D. A., and Thorner, J. Use of the LYS 2 gene for gene disruption, gene replacement, and promoter analysis in Saccharomyces cerevisiae. In *Gene Manipulations in Fungi* (J. W. Bennett and L. L. Lasure, eds.), Academic Press, Orlando, 1985, pp. 197-227.
128. Affenzeller, K., Jaklitsch, W. M., Hönlinger, C., and Kubicek, C. P. Lysine biosynthesis in Penicillium chrysogenum is regulated by feedback inhibition of α-aminoadipate reductase. *FEMS Microbiol. Lett.* (in press).
129. Ahmad, M., Ahmad, M. U., and Zaman, A. Mapping of locus lysine-2 in Neurospora crassa. *Proc. Pak. Acad. Sci.*, *3*:1-12, 1966.
130. Ahmad, M., Mozmadar, A., Baset, A., Fayaz, M., Rahman, M. A., and Saha, B. C. Studies on the organization of genes controlling lysine biosynthesis in Neurospora crassa. III. Studies on the organization of loci lys-3 and lys-4. *Pak. J. Bot.*, *11*:179-184, 1979.
131. Storts, D. R., and Bhattacharjee, J. K. Purification and properties of saccharopine dehydrogenase (glutamate forming) in the Saccharomyces cerevisiae lysine biosynthetic pathway. *J. Bacteriol.*, *169*:416-418, 1987.
132. Ogawa, H., and Fujioka, M. Purification and characterization of saccharopine dehydrogenase from bakers yeast. *J. Biol. Chem.*, *253*:3666-3673, 1978.
133. Jaklitsch, W. M., Röhr, M., and Kubicek, C. P. Lysine biosynthesis in Penicillium chrysogenum: regulation by general amino acid control and absence of lysine repression. *Exp. Mycol.*, *11*:141-149, 1987.

134. Hogg, R., and Broquist, H. P. Homocitrate formation in Neurospora crassa: relation to lysine biosynthesis. *J. Biol. Chem.*, *243*:695-702, 1968.
135. Tucci, A. F., and Ceci, L. N. Homocitrate synthase from yeast. *Arch. Biochem. Biophys.*, *153*:742-751, 1972.
136. Teas, H. J., Horowitz, N. H., and Fling, M. Homoserine as a precurser of threonine and methionine in Neurospora crassa. *J. Biol. Chem.*, *172*:651-658, 1948.
137. Flavin, M., and Slaughter, C. Purification and properties of threonine synthetase from Neurospora crassa. *J. Biol. Chem.*, *235*:1103-1108, 1960.
138. Jenkins, M. B., and Woodward, V. W. Purification and properties of homoserine dehydrogenase from Neurospora crassa. *Biochim. Biophys. Acta*, *212*:21-32, 1970.
139. Greenberg, D. M. Biosynthesis of amino acids and related compounds. In *Metabolic Pathways* (3rd ed.) (D. M. Greenberg, ed.), Academic Press, New York, 1969, pp. 238-315.
140. Patte, J. C., Le Bras, G., and Cohen, G. N. Regulation by methionine of the synthesis of a third aspartokinase and a second homoserine dehydrogenase in E. coli. *Biochim. Biophys. Acta*, *136*:245-257, 1967.
141. Patte, J. C., Truffa-Bachi, P., and Cohen, G. N. The threonine-sensitive homoserine dehydrogenase and aspartokinase of Escherichia coli. *Biochim. Biophys. Acta*, *128*:426-439, 1966.
142. Robichon-Szulmajster, H. D., and Corrivaux, D. Regulations matebo-liques de la biosynthese de la methionine et de la threonine chez Saccharomyces cerevisiae. I. Repression et retro-inhibition de l'aspartokinase. *Biochim. Biophys. Acta*, *73*:248-256, 1963.
143. Flavin, M. Methionine biosynthesis. In *Metabolism of Sulfur Compounds* (D. M. Greenberg, ed.), Academic Press, New York, 1975, pp. 457-503.
144. Nagai, S., and Flavin, M. Acetylhomoserine. An intermediate in the fungal biosynthesis of methionine. *J. Biol. Chem.*, *242*:3884-3895, 1967.
145. Kerr, D. S., and Flavin, M. The regulation of methionine synthesis and the nature of cystathionine-γ-synthase in Neurospora. *J. Biol. Chem.*, *245*:1842-1855, 1970.
146. Dulavier-Klutchko, C., and Flavin, M. Enzymatic synthesis and cleavage of cystathionine in fungi and bacteria. *J. Biol. Chem.*, *240*:2537-2549, 1965.
147. Kerr, D., and Flavin, M. Inhibition of cystathionine-γ-synthase by S-adenosylmethionine: a control mechanism for methionine synthesis in Neurospora. *Biochim. Biophys. Acta*, *177*:177-179, 1969.
148. Metzenberg, R. L. Repair of multiple defects of a regulatory mutant of Neurospora by high osmotic pressure and by reversion. *Arch. Biochem. Biophys.*, *125*:532-541, 1968.
149. Selhub, J., Savin, M. A., Sakami, W., and Flavin, M. Synchronization of converging metabolic pathways: activation of the cystathionine-γ-synthase of Neurospora crassa by methyltetrahydrofolate. *Proc. Natl. Acad. Sci. USA*, *68*:312-314, 1971.
150. Burton, E. G., and Metzenberg, R. L. Regulation of methionine biosynthesis in Neurospora crassa. *Arch. Biochem. Biophys.*, *168*:219-229, 1975.

151. Selbhub, J. Defects of four types of Neurospora crassa mutants that require methionine and are unable to utilize homocysteine. Ph.D. thesis, Case Western Reserve University, Cleveland, *Diss. Abstr. Int.*, *31*:3167 B-3168 B, 1970.
152. Selbhub, J. Burton, E. G., and Sakami, W. Identification of three enzymes specifically involved in the de novo methionine methyl biosynthesis in Neurospora crassa. *Fed. Proc.*, *28*:352, 1969.
153. Salem, A. R., and Foster, M. A. A role for folic acid conjugase in the regulation of methionine synthesis by Coprinus lagopus. *Biochim. Biophys. Acta*, *252*:597-600, 1971.
154. Cossins, E. A., and Chan, P. Y. Polypolyglutamate synthetase activities of Neurospora crassa: nature of products formed by soluble and particulate enzymes in the wild type and polyglutamate deficient mutants. *Adv. Exp. Med. Biol.*, *163*:183-197, 1983.
155. Flavin, M., and Slaughter, C. Cystathionine cleavage enzymes of Neurospora. *J. Biol. Chem.*, *239*:2212-2222, 1964.
156. Burton, E. G., and Metzenberg, R. L. Novel mutation causing derepression of several enzymes of sulfur metabolism in Neurospora crassa. *J. Bacteriol.*, *109*:140-151, 1972.
157. Pieniazek, N., Bol, J., Balbin, E., and Stepien, P. P. An Aspergillus nidulans mutant lacking serine transacetylase: evidence for two pathways of cysteine biosynthesis. *Mol. Gen. Genet.*, *132*:363-366, 1974.
158. Pieniazek, N. J., Stepien, P. P., and Paszewski, A. An Aspergillus nidulans mutant lacking cystathionine-β-synthase: identity of serine sulfhydrylase and its distinctness from O-acetyl-L-serine-sulfhydrylase. *Biochim. Biophys. Acta*, *297*:37-47, 1973.
159. Kerr, D. S. O-Acetylhomoserine sulfhydrylase from Neurospora. Purification and consideration of its function in homocysteine and methionine synthesis. *J. Biol. Chem.*, *246*:95-102, 1971.
160. Dulavier-Klutchko, C., and Flavin, M. Enzymatic synthesis and cleavage of cystathionine in fungi and bacteria. *J. Biol. Chem.*, *240*:2537-2549, 1965.
161. Jones, K. M., Guest, H., and Wood, J. A. Folic acid and the synthesis of methionine by extracts of Escherichia coli. *Biochem. J.*, *79*:566-576, 1961.
162. Burton, E., and Sakami, W. Methionine biosynthesis (yeast). *Meth. Enzymol.*, *17B*:388-392, 1971.
163. Metzenberg, R. L. Genetic regulatory mechanisms in Neurospora. *Annu. Rev. Genet.*, *6*:111-131, 1972.
164. Coggins, J. R., Duncan, K., Anton, I. A., Boocock, M. R., Chadhuri, S., Lambert, J. M., Lewendon, A., Millar, G., Mousdale, D. M., and Smith, D. D. S. The anatomy of a multifunctional enzyme. *Biochem. Soc. Trans.*, *15*:754-759, 1987.
165. Hütter, R., Niederberger, P., and DeMoss, J. A. Tryptophan biosynthetic genes in eucaryotic microorganisms. *Ann. Rev. Microbiol.*, *40*:55-77, 1986.
166. Doy, C. H. The nature of 3-deoxy-D-arbanino-heptulosonate-7-phosphate synthase in extracts of wild type Neurospora crassa: a reaction controlled by two activating substrates and three allosteric negative modifiers. *Biochim. Biophys. Acta*, *159*:352-366, 1968.
167. Bode, R., Melo, C., and Birnbaum, D. Regulatory properties of 3-deoxy-D-arabino-he;tulosonate-7-phosphate synthase isozymes from Candida maltosa. *J. Basic Microbiol.*, *25*:3-11, 1985.

168. Lumsden, J., and Coggins, J. R. The subunit structure of the arom multienzyme complex of Neurospora crassa. A possible pentafunctional polypeptide chain. *Biochem. J., 161*:599-607, 1977.
169. Kinghorn, J. R., and Hawkins, A. R. Cloning and expression in Escherichia coli K-12 of the biosynthetic dehydroquinase function of the arom cluster gene from the eukaryote Aspergillus nidulans. *Mol. Gen. Genet., 186*:145-152, 1982.
170. Duncan, K., Lewendon, A., and Coggins, J. R. The complete amino acid sequence of Escherichia coli 5-enolpyruvylshikimate 3-phosphate synthase. *FEBS Lett., 170*:59-63, 1984.
171. Lumsden, J., and Coggins, J. R. The subunit structure of the arom multienzyme complex of Neurospora crassa. Evidence from peptide maps for the identity of the subunits. *Biochem. J., 169*:441-444, 1978.
172. Smith, D. D. S., and Coggins, J. R. Isolation of a bifunctional domain from the pentafunctional arom enzyme complex of Neurospora crassa. *Biochem. J., 213*:405-415, 1983.
173. Coggins, J. R., Boocock, M. R., Campbell, M. S., Chaudhuri, S., Lambert, J. M., Lewendon, A., Mousdale, D. M., and Smith, D. D. S. Functional domains involved in aromatic amino acid biosynthesis. *Biochem. Soc. Trans., 13*:299-303, 1985.
174. Duncan, K., Edwards, M. R., and Coggins, J. R. The pentafunctional arom enzyme of Saccharomyces cerevisiae is a mosaic of monofunctional domains. *Biochem. J., 246*:375-386, 1987.
175. Millar, G., and Coggins, J. R. The complete amino acid sequence of 3-dehydroquinate synthase of Escherichia coli K-12. *FEBS Lett., 200*:11-17, 1986.
176. Millar, G., Lewendon, A., Hunter, M. G., and Coggins, J. R. The cloning and expression of the aroL gene from Escherichia coli K-12. Purification and complete amino acid sequence of shikimate kinase II, the aro L-gene product. *Biochem. J., 237*:427-436, 1986.
177. Duncan, K., Chaudhuri, S., Campbell, M. S., and Coggins, J. R. The overexpression and complete amino acid sequence of Escherichia coli 3-dehydroquinase. *Biochem. J., 238*:475-483, 1986.
178. Gaertner, F. H., and Cole, K. W. Properties of chorismate synthase in Neurospora crassa. *J. Biol. Chem., 248*:4602-4609, 1973.
179. Woodin, T. S., and Nishioka, L. Chorismate mutase isozyme patterns in three fungi. *Biochim. Biophys. Acta, 309*:224-231, 1973.
180. Lingens, F., Goebel, W., and Uessler, H. Regulation der Biosynthese der aromatischen Aminosäuren in Claviceps paspali. *Eur. J. Biochem., 2*:442-447, 1967.
181. Sinha, U. Aromatic amino acid biosynthesis and p-fluorophenylalanine resistance in Aspergillus nidulans. *Genet. Res., 16*:261-272, 1967.
182. Henke, H. Untersuchungen an Tryptophan-Biosynthese Enzymen aus Coprinus und anderen Basidiomyceten. Thesis 4825. ETH-Zürich, Zürich, Juris.
183. Henke, H., Guerdoux, J. L., and Hütter, R. Relations entre les genes et les enzymes de la biosynthese du tryptophan chez le Coprinus radiatus Fr. ex Bolt. *C.R. Acad. Sci. Ser., D276*:2001-2004, 1973.
184. Hütter, R., and DeMoss, J. A. Enzyme analysis of the tryptophan pathway in Aspergillus nidulans. *Genetics, 55*:241-247, 1967.
185. Hütter, R., and DeMoss, J. A. Organization of the tryptophan pathway: a phylogenetic study of the fungi. *J. Bacteriol., 94*:1896-1907, 1967.

186. Mullaney, E. J., Hamer, J. E., Robert, K. A., Yelton, M. M., and Timberlake, W. E. Primary structure of the trpC gene from Aspergillus nidulans. *Mol. Gen. Genet.*, *199*:37-45, 1985.
187. Schweingruber, M. E., and Dietrich, R. Gene-enzyme relationships in the tryptophan pathway of Schizosaccharomyces pombe. *Experientia*, *29*:1151-1153, 1973.
188. Thuriaux, P., Heyer, W.-D., and Strauss, A. Organisation of the complex locus trp 1 in the fission yeast Schizosaccharomyces pombe. *Curr. Genet.*, *6*:13-18, 1982.
189. Walker, M., and DeMoss, J. A. Purification and characterization of the trifunctional β-subunit of anthranilate synthase from Neurospora crassa. *J. Biol. Chem.*, *258*:3571-3575, 1983.
190. Roberts, C. E. Complementation analysis of the tryptophan pathway in Aspergillus nidulans. *Genetics*, *55*:233-239, 1967.
191. Schechtman, M. G., and Yanofsky, C. Structure of the trifunctional trp-1 gene from Neurospora crassa and its aberrant expression in E. coli. *J. Mol. Appl. Genet.*, *2*:83-99, 1983.
192. Welch, G. R., and Gaertner, F. H. Coordinate activation of a multienzyme complex by the first substrate. Evidence for a novel regulatory mechanism in the polyaromatic pathway of Neurospora crassa. *Arch. Biochem. Biophys.*, *172*:476-489, 1976.
193. De Moss, J. A. Biochemical diversity in the tryptophan pathway. *Biochim. Biophys. Res. Commun.*, *18*:850-857, 1965.
194. Lester, G. In vivo regulation of intermediate reactions in the pathway of tryptophan biosynthesis in Neurospora crassa. *J. Bacteriol.*, *96*:1768-1773, 1968.
195. Ames, B. N., and Mitchell, H. K. The paper chromatography of imidazoles. *J. Am. Chem. Soc.*, *74*:252-253, 1952.
196. Ames, B. N., and Mitchell, H. K. The biosynthesis of histidine: imidazole glycerol phosphate, imidazole-acetol phosphate and histidinol phosphate. *J. Biol. Chem.*, *212*:687-696, 1955.
197. Ahmed, M., and Catcheside, D. G. Physiological diversity among tryptophan mutants in Neurospora crassa. *Heredity*, *15*:55-64, 1960.
198. Henry, S., Klig, L. S., and Loewy, B. S. The genetic regulation and coordination of biosynthetic pathways in yeast: amino acid and phospholipid synthesis. *Annu. Rev. Genet.*, *18*:207-231, 1984.
199. Ahmed, A., Case, E. A., and Giles, H. G. The nature of complementation among mutants in the histidine-3 region of Neurospora crassa. *Brookhaven Symp. Biol.*, *17*:53-65, 1964.
200. Ames, B. N., Martin, R. G., and Garry, B. J. The first step in histidine biosynthesis. *J. Biol. Chem.*, *236*:2019-2026, 1961.
201. Martin, R. G. The first enzyme in histidine biosynthesis: the nature of feedback inhibition by histidine. *J. Biol. Chem.*, *238*:257-269, 1963.
202. Rasse-Messenguy, F., and Fink, G. Feedback resistant mutants of histidine biosynthesis in yeast. In *Genes, Enzymes and Populations* (A. M. Srb, ed.), Plenum Press, New York, 1973, pp. 85-95.
203. Catcheside, D. G. Complementation among histidine mutants of Neurospora crassa. *Proc. Roy. Soc.*, *153*:179-194, 1960.
204. Berlyn, M. Gene-enzyme relationships in histidine biosynthesis in Aspergillus nidulans. *Genetics*, *57*:561-570, 1967.
205. Minson, A. C., and Creaser, E. H. Purification of a trifunctional enzyme, catalyzing three steps of the histidine pathway from Neurospora crassa. *Biochem. J.*, *114*:49-56, 1969.

206. Ames, B. N. The biosynthesis of histidine; L-histidinol phosphate phosphatase. *J. Biol. Chem.*, *226*:583-593, 1957.
207. Ames, B. N., and Horecker, B. L. The biosynthesis of histidine. Imidazole acetol phosphate transaminase. *J. Biol. Chem.*, *220*:113-128, 1956.
208. Ames, B. N. The biosynthesis of histidine. D-Erythrose-imidazole glycerol phosphate dehydrase. *J. Biol. Chem.*, *228*:131-143, 1957.
209. Carsiotis, M., and Jones, R. F. Cross-pathway regulation: tryptophan mediated control of histidine and arginine biosynthetic enzymes in Neurospora crassa. *J. Bacteriol.*, *119*:889-892, 1974.
210. Carsiotis, M., Jones, R. F., and Wesseling, A. C. Cross-pathway regulation: histidine mediated control of histidine, tryptophan, and arginine biosynthetic enzymes in Neurospora crassa. *J. Bacteriol.*, *119*:893-898, 1974.
211. Carsiotis, M., and Lacy, A. M. Increased activity of tryptophan biosynthetic enzymes in histidine mutants of Neurospora crassa. *J. Bacteriol.*, *89*:1472-1477, 1965.
212. Hinnebusch, A. G. The general control of amino acid biosynthetic genes in the yeast Saccharomyces cerevisiae. *CRC Crit. Rev. Biochem.*, *21*:277-317, 1986.
213. Hinnebusch, A. G. Mechanisms of gene regulation in the general control of amino acid biosynthesis in Saccharomyces cerevisiae. *Microbiol. Rev.*, *52*:248-273, 1988.
214. Hinnebusch, A. G. Novel mechanisms of translational control in Saccharomyces cerevisiae. *Trends in Genetics*, *4*:169-174, 1900.
215. Holden, J. T. The composition of microbial amino acid pools. In *Amino Acid Pools* (J. T. Holden, ed.), Amsterdam, Elsevier, 1962, pp. 73-108.
216. Schmit, J. C., and Brody, S. Neurospora crassa conidial germination. Role of endogenous amino acid pools. *J. Bacteriol.*, *124*:232-242, 1975.
217. Kubicek, C. P., Hampel, W., and Röhr, M. Manganese deficiency leads to elevated amino acid pools in citric acid accumulating Aspergillus niger. *Arch. Microbiol.*, *123*:73-79, 1979.
218. Pitt, D. E., and Bull, A. T. Free amino acid pools of Trichoderma aureoviride during conditions of glucose-limited growth and glucose starvation. *Arch. Microbiol.*, *130*:180-184, 1981.
219. Jaworski, A., Diugonski, J., Wilmaska, D., and Sedlacek, L. Changes in the cellular content of the pool constituents of Monosporium olivaceum: a steroid hydroxylating mould. *Acta Microbiol. Pol.*, *25*:329-335, 1976.
220. Schmit, J. C., and Brody, S. Biochemical genetics of Neurospora crassa conidial germination. *Microbiol. Rev.*, *40*:1-41, 1976.
221. Kanamori, K., Legerton, T. L., Weiss, R. L., and Roberts, J. D. Nitrogen-15 spin lattice relaxation times of amino acids in Neurospora crassa as a probe of intracellular environment. *Biochemistry*, *21*:4916-4920, 1982.
222. Messenguy, F., Colin D., and Ten Have, J.-P. Regulation of compartmentation of amino acid pools in Saccharomyces cerevisiae and its effect on metabolic control. *Eur. J. Biochem.*, *108*:439-447, 1980.
223. Subramanian, K. N., Weiss, R. L., and Davis, R. H. Use of external, biosynthetic, and organellar arginine by Neurospora. *J. Bacteriol.*, *115*:284-290, 1973.
224. Keller, U., Madry, N., Kleinkauf, H., and Glund, K. Isolation and characterization of vacuoles from the ergot fungus Claviceps purpurea. *Appl. Environ. Microbiol.*, *47*:710-714, 1984.

225. Vaughn, L. E., and Davis, R. H. Purification of vacuoles from Neurospora crassa. *Mol. Cell. Biol.*, *1*:797-806, 1981.
226. Martinoia, E., Heck, U., Boller, T., Wiemken, A., and Matile, P. Some properties of vacuoles isolated from Neurospora crassa slime variant. *Arch. Microbiol.*, *120*:31-34, 1979.
227. Bowman, E. J., and Bowman, B. J. Identification and properties of an ATPase in vacuolar membranes of Neurospora crassa. *J. Bacteriol.*, *151*:1326-1337, 1982.
228. Bowman, E. J. Comparison of the vacuolar membrane ATP-ase of Neurospora crassa with the mitochondrial and plasma membrane ATP-ases. *J. Biol. Chem.*, *258*:15238-15244, 1983.
229. Legerton, T. L., and Weiss, R. L. Mobilization of sequestered metabolites into degradative reactions by nutritional stress in Neurospora. *J. Bacteriol.*, *138*:909-914, 1979.
230. Legerton, T. L., and Weiss, R. L. Mobilization of vacuolar arginine in Neurospora crassa. Mechanism and role of glutamine. *J. Biol. Chem.*, *259*:8875-8879, 1984.
231. Cramer, C. L., Vaughn, L. E., and Davis, R. H. Basic amino acids and inorganic polyphosphates in Neurospora crassa: independent regulation of vacuolar pools. *J. Bacteriol.*, *142*:945-952, 1980.
232. Cramer, C. L., and Davis, R. H. Polyphosphate-cation interaction in the amino acid containing vacuole of Neurospora crassa. *J. Biol. Chem.*, *259*:5125-5157, 1984.
233. Horak, J. Amino acid transport in eucaryotic microorganisms. *Biochim. Biophys. Acta*, *864*:223-256, 1986.
234. Niederberger, P., Miozarri, G., and Hütter, R. Biological role of the general control of amino acid biosynthesis in Saccharomyces cerevisiae. *Mol. Cell. Biol.*, *1*:584-593, 1981.
235. Hope, I. A., and Struhl, K. GCN 4 protein; synthesized in vitro, binds HIS 3 regulatory sequences: implications for general amino acid control of amino acid biosynthetic genes in yeast. *Cell*, *43*:177-188, 1985.
236. Donahue, T. F., Daves, R. S., Lucchini, G., and Fink, G. R. A short nucleotide sequence required for regulation of HIS 4 by the general control system in yeast. *Cell*, *32*:89-98, 1983.
237. Hinnebusch, A. G. Evidence for translational regulation of the activator of general amino acid control in yeast. *Proc. Natl. Acad. Sci. USA*, *81*:6442-6446, 1984.
238. Penn, M. D., Galgoci, B., and Greer, H. Identification of AAS genes and their regulatory role in general amino acid control of amino acid biosynthesis in yeast. *Proc. Natl. Acad. Sci. USA*, *80*:2704-2708, 1983.
239. Hinnebusch, A. G., and Fink, G. R. Positive regulation in the general amino acid control of Saccharomyces cerevisiae. *Proc. Natl. Acad. Sci. USA*, *80*:5374-5378, 1983.
240. Penn, M. D., Thireos, G., and Greer, H. Temporal analysis of general control of amino acid biosynthesis in Saccharomyces cerevisiae: role of positive regulatory genes in initiation and maintenance of mRNA derepression. *Mol. Cell. Biol.*, *4*:520-528, 1984.
241. Messenguy, F., and Delforge, J. Role of transfer ribonucleic acids in the regulation of several biosyntheses in Saccharomyces cerevisiae. *Eur. J. Biochem.*, *67*:335-339, 1976.

242. Barthelmess, I. B. Mutants affecting amino acid cross-pathway control in Neurospora crassa. *Genet. Res.*, *39*:169-185, 1982.
243. Soda, K., Tanaka, H., and Esaki, N. In *Biotechnology: Microbial Products, Biomass, Primary Metabolites* (H. Dellweg, ed.), Verlag Chemie, Weinheim, 1983, pp. 467-478.
244. Hirose, Y., and Okada, H. In *Microbial Technology*, Vol. 1 (H. J. Peppler and D. Perlman, eds.), Academic Press, New York, 1979, pp. 211-240.
245. Yoshinaga, F., and Nakamori, S. Production of amino acids. In *Amino Acids: Biosynthesis and Regulation* (K. M. Hermann and R. L. Sommerville, eds.), Addison-Wesley, Reading, Mass., 1983, pp. 405-429.
246. Denenu, E., and Demain, A. L. Regulation of tryptophan by DL-fluorotryptophan resistant mutant strains of a methanol utilizing yeast, Hansenula polymorpha, DL-1. *Yeast*, *24*:37-38, 1975.
247. Enatsu, T., Kiyoi, M., Matsushima, H., and Terui, G. Studies on the fermentative production of tryptophan from anthranilic acid by yeast. V. Investigation of auxotrophic haploid mutants. *J. Ferment. Technol.*, *41*:500-508, 1963.
248. Ebihara, Y., Niitsu, H., and Terui, G. Fermentative production of tryptophan from indole by Jansenula anomala. *J. Ferment. Technol.*, *47*:733-738, 1969.
249. Denenu, E., and Demain, A. L. Derivation of aromatic amino acid mutants from a methanol-utilizing yeast, Hansenula polymorpha. *Appl. Environ. Microbiol.*, *41*:1088-1096, 1981.
250. Malin, B., and Westhead, J. Production of L-tryptophan by submerged culture. *J. Biochem. Microbiol. Technol. Eng.*, *1*:49-57, 1959.
251. Haidaris, C. G., and Ghattacharjee, J. K. Lysine production by thialysine-resistant mutants of Saccharomyces cerevisiae. *J. Ferment. Technol.*, *56*:189-192, 1978.
252. Gaillardin, C. M., Sylvestre, G., and Heslot, H. Studies on an unstable phenotype induced by UV irradiation. The lysine excreting (lex^-) phenotype of the yeast Saccharomycopsis lipolytica. *Arch. Microbiol.*, *104*:89-94, 1975.
253. Takenouchi, E., Yamamoto, T., Nikolova, D. T., Tanaka, H., and Soda, K. Lysine production by S-(β-aminoethyl)-L-cysteine resistant mutants of Candida pelliculosa. *Ag. Biol. Chem.*, *43*:727-734, 1979.
254. Fukumura, T. Conversion of D- and DL-α-amino-∂-caprolactam into L-lysine using both yeast cells and bacterial cells. *Ag. Biol. Chem.*, *41*:1327-1330, 1977.
255. Shay, L. K., Wegner, E. H., and Reiter, S. E. Use of mutagenesis to enhance the methionine content of methanol-assimilating yeast. *Dev. Ind. Microbiol.*, *14*:305-311, 1983.
256. Okanishi, M., and Gregory, K. F. Isolation of mutants of Candida tropicalis with increased methionine content. *Can. J. Microbiol.*, *16*:1139-1143, 1970.
257. Morzycka, E., Sawnor-Korszynska, D., Paszewski, A., Grabski, J., and Raczynska-Bojanowska, K. Methionine overproduction in Saccharomycopsis lipolytica. *Appl. Environ. Microbiol.*, *32*:125-130, 1976.
258. Konoshita, S., Udaka, S., and Shimono, M. Studies on amino acid fermentation. Part I. Production of L-glutamic acid by various microorganisms. *J. Gen. Appl. Microbiol.*, *3*:193-205, 1957.
259. Dulaney, E. L., Jones, C. A., and Dulaney, D. L. Amino acid accumulation, principally alanine, by auxotrophs of Ustilage mayidis. *Dev. Ind. Microbiol.*, *5*:242-249, 1964.

260. Tsukada, Y., and Sugimori, T. Induction of auxotrophic mutants from Candida species and their application to L-threonine fermentation. *Agr. Biol. Chem.*, *35*:1-7, 1971.
261. Anonymous. Japanese make aspartame sweetener microbially via two bacteria, one fungus. *McGraw-Hill Biotechnol. Newswatch*, *3*(18):3, 1983.

14

ORGANIC ACIDS OF FUNGI

RAMUNAS BIGELIS *Amoco Research Center, Amoco Technology Company, Naperville, Illinois*

DILIP K. ARORA *Banaras Hindu University, Varanasi, India*

I. INTRODUCTION

Fungi produce a wide variety of organic acids. However, only citric acid, gluconic acid, and itaconic acid are produced in significant quantities by large-scale, commercial processes. These organic acids are emphasized in the following discussion, although organic acids that have been produced by industrial fungal fermentation in the past or may be of commercial importance in the future are also considered.

II. ORGANIC ACIDS

A. Citric Acid

Citric acid is a major commodity chemical manufactured by industrial fungal fermentation. More than 350,000 tons are produced worldwide [1]. Since early in this century, when the first unsuccessful attempts were made to produce citric acid with *Penicillium* cultures, citric acid manufacture has evolved into an industry that uses advanced large-scale fermentation and recovery technologies. Review articles provide substantial detail on the citric acid fermentation, although many of the production organisms and much of the processing technology are proprietary [1-19].

1. *Aspergillus niger*

Most industrial citric acid fermentations use genetically improved strains of *Aspergillus niger*. Since this filamentous fungus lacks a sexual cycle, parasexual recombination, diploidization, and heterokaryon formation have been explored as possible means for strain improvement; but most industrial strain development programs have probably relied on mutagenesis and screening for the derivation of efficient producers of citric acid [14].

Industrial strains of *A. niger* originally were grown by surface culture in shallow pans, but today about 80% of world production is by submerged culture, often in large, computer-controlled fermentors with aeration and mixing. Though the former method requires two-thirds less energy and is simpler, submerged fermentation is more easily automated, requires less space, takes less time, and is more easily controlled at various stages. In addition, it uses raw materials more efficiently and yields more product. Another process, the koji fermentation, which is usually performed on solid substrates such as wheat bran residues, is also used with *A. niger*, but to a limited extent mainly in Japan [1,12,20].

After inoculation of deep-tank fermentors with conidia prepared by surface culture, *A. niger* grows in the form of spherical mycelial aggregates (pellets less than 1.0 mm in diameter), provided levels of manganese (in ppb) and acidity are meticulously controlled. A growth phase during which sugar is consumed is then followed by a production phase during which sugar is converted to citric acid and very little growth occurs [1,12]. Table 1 outlines the general features of a typical submerged fermentation with *A. niger*, and Fig. 1 shows key variables as a function of the age of the fermentation. Adherence to these conditions results in substantial citric acid yields with few byproducts such as oxalic or gluconic acid. However, the yield of citric acid and the rate of its production are strain-dependent and the fermentation conditions for each strain must be optimized.

TABLE 1. General Features of a Typical Submerged Citric Acid Fermentation with *A. niger*

Component	Features
Fermentor	Corrosion-resistant, coated deep vessels of stirred or airlift design
Carbon source	Decationized sugar solutions (15-25% w/v), usually beet or cane molasses, sucrose, glucose, or fructose
Nitrogen source	Optimized levels of $(NH_4)_2SO_4$ or NH_4NO_3
Phosphate	Limiting levels when trace metals unrestricted
Trace metals	Concentrations strictly controlled, especially those of manganese, iron, copper, and zinc
pH	Between 1.5 and 2.8, but higher for molasses medium
Temperature	28-33°C
Aeration	Critical requirement for high dissolved oxygen levels
Fermentation time	Usually 5-8 days, but may be up to 10-15 days depending on raw material
Conversion of glucose to citric acid	70-90% (kg citric acid monohydrate per 100 kg of supplied carbohydrate)
Other agents sometimes added	Antifoam; chelating agents, e.g., sodium or potassium ferrocyanide; lower alcohols

Source: Ref. 19.

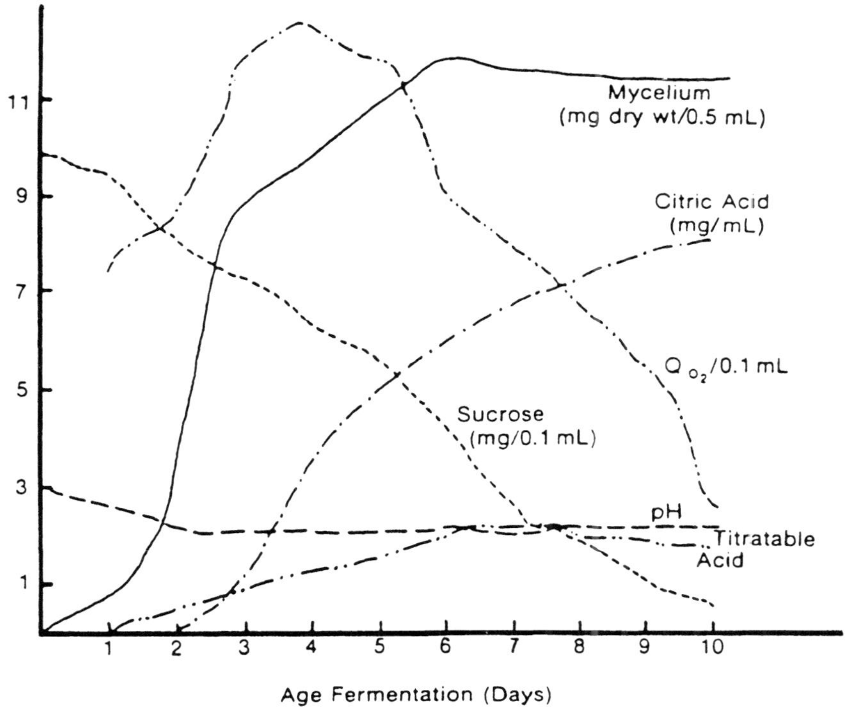

FIGURE 1. Changes associated with the production of citric acid by submerged fermentation with *Aspergillus niger*. (From Ref. 21.)

Several methods for citric acid recovery have been reported, including calcium precipitation and solvent extraction processes, as well as direct crystallization methods [10,12,16,22]. In the classical method (Fig. 2), citric acid is usually precipitated from heated fermentation mash with calcium hydroxide at about pH 5.8 after removal of mycelia by filtration. This procedure is followed by filtration in a rotary vacuum filter and then treatment with concentrated sulfuric acid at 60°C or less. The solution is concentrated, washed free of calcium sulfate, and further purified with activated charcoal or by ion exchange chromatography. The solution can then be concentrated by evaporation under vacuum to yield a crystalline product. The monohydrate is formed at temperatures above 36°C. A series of recrystallization steps often follow to remove impurities. Finally, centrifugation, washing, and drying yield citric acid that can be sieved and packaged. In solvent extraction methods, an organic solvent or mixtures of solvents are used to selectively transfer citric acid from one aqueous system to another, thus leaving behind unwanted byproducts.

The metabolic pathway for citric acid accumulation by *A. niger* strains involves breakdown of carbon compounds via glycolysis, formation of oxaloacetate from pyruvate by the pyruvate carboxylase reaction, and then formation of citric acid via the reactions of the tricarboxylic acid cycle. This pathway is based on investigations using radioactively labeled precursors [23,24] and numerous physiological studies [1,12]. An abbreviated

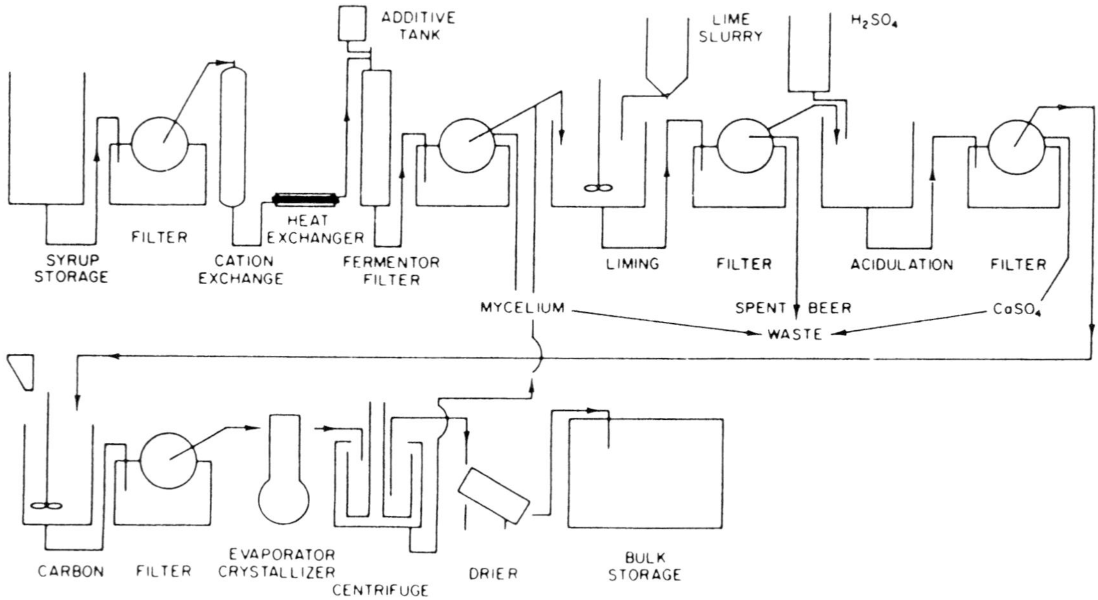

FIGURE 2. Flow diagram for the manufacture of citric acid. (From Ref. 4.)

metabolic scheme of Kubicek and Röhr [1] for citric acid accumulation is shown in Fig. 3.

The key regulatory enzymes that play a role in citric acid production by *A. niger* have been identified as phosphofructokinase, pyruvate carboxylase, citrate synthase, isocitrate dehydrogenase, and α-ketoglutarate dehydrogenase. However, no overall mechanism that explains both citric acid accumulation and the pattern of carbon flow as shown in Fig. 3 has been developed. Such a model would also have to consider the effect of trace metals, primarily manganese, which play a central role in citric acid overproduction [1,12,15]. Manganese deficiency is essential for citric acid accumulation. Phenomena associated with manganese deprivation of *A. niger* (Table 2) may be directly related to citric acid accumulation or some may be only distant links in a chain of events that lead to such significant perturbation of citric acid metabolism. Models for citric acid accumulation will have to take into account the metabolic patterns shown in Fig. 3, some or all of the physiological alterations shown in Table 2, and the observations made by those who perform the large-scale industrial fermentations with mutants of *A. niger*. Certainly, new research approaches will contribute to progress in this area. Modern genetic approaches and recombinant DNA technology can be expected to play an important role in both elucidating industrial citric acid processes and improving them [14,19].

Citric acid has many uses and the major ones are listed in Table 3 [1,17-18,31-33]. About 70% of the total output is used in the food industry as an acidulant, flavor enhancer, preservative, chelating agent, antioxidant, stabilizer, buffer, emulsifier, and processing aid. About 10% of all citric acid is used in cosmetics and pharmaceuticals, and the remainder for other purposes. New uses can be expected. For example, the application of citric acid in liquid wash products has increased significantly in recent years, especially in the United States.

2. Yeasts

Citric acid has also been produced with yeasts, mainly species of *Candida*, primarily *Candida lipolytica* or its sexual form *Yarrowia* (*Saccharomycopsis*) *lipolytica* [1,7,11-12,14-16,34-40]. Hydrocarbons or carbohydrates can serve as the carbon sources. Nitrogen limitation promotes a transition from a growth phase to a citric acid accumulation phase. Strains and culture conditions are chosen to avoid production of isocitric acid, an undesirable byproduct. *C. lipolytica* ATCC 8661 can produce 56.8 g of citric acid per liter from 12- and 18-carbon *n*-alkanes and up to 110.4 g per liter from hexadecane [36]. Strain ATCC 20228 has been used to synthesize citric acid from hydrocarbons in a continuous process [38]. The best yields of citric acid in such industrial fermentations are claimed to exceed 130% by weight. Recent yeast research has focused instead on the conversion

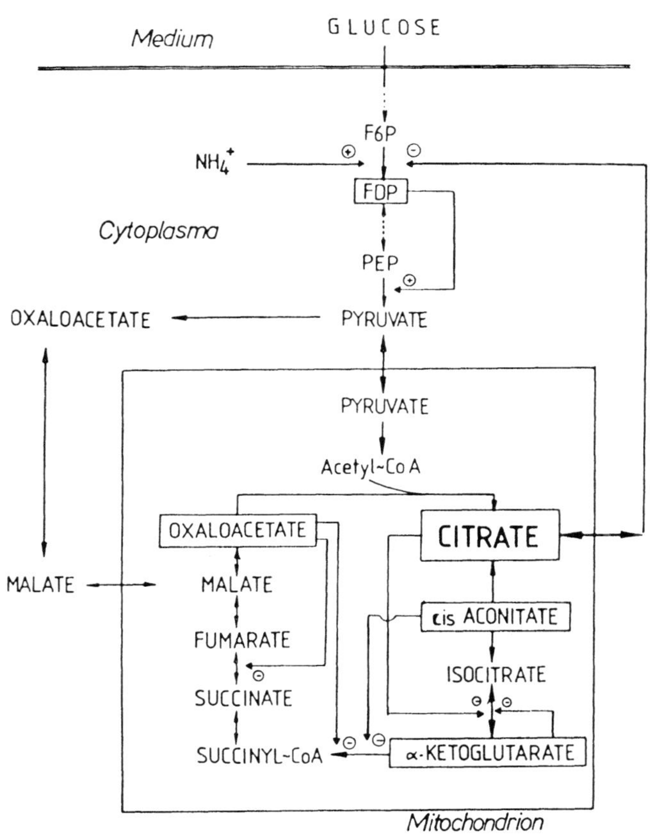

FIGURE 3. Metabolic scheme of interrelationships in citric acid accumulation: (+) indicates activation by the respective metabolite and (-) indicates inhibition. (From Ref. 1.)

TABLE 2. Metabolic Alterations Induced by Manganese Deficiency in *Aspergillus niger* During Citric Acid Fermentation

Alteration	Ref.
Accumulation of amino acids and NH_4^+	25
Elevated proteinase activity	26
Elevated rate of protein degradation	26
Increased monosome portion of ribosomes	26
Decreased cellular phospholipid content	27
Plasma membrane phospholipid and fatty acid alterations	28
Cell wall chitin elevation and decrease in β-1,3-glucan	29
Crippled, branched mycelial development, single giant cells	29
Increased glycolytic flux on the expense of the pentose phosphate pathway	30

Source: Ref. 1.

TABLE 3. Some Major Applications of Citric Acid

Foods and Beverages
Soft drinks, syrups, fruit and vegetable juices, wine and wine-based drinks, ciders, jams, jellies, gelatin desserts, candies, processed fruits and vegetables, processed seafoods and meats, dairy products, antioxidant of fats and oils
Pharmaceuticals
Effervescent powders and tablets, acidulant and buffering agent in medicine, blood anticoagulant, contraceptive jellies and ointments
Cosmetics
Antioxidant, buffer, acidifier, sequestrant in rinses, shampoos, lotions, and various cosmetics and toiletries
Detergents
Builder, biodegradable phosphate alternative in home and industrial wash products and cleaners
Industrial Applications
Chelating agent, metal cleaning, rust removal, metal electroplating, water conditioning, sulfur dioxide removal; also used in industries involved in textiles, photography, cement manufacture, tanning, mining, adhesive production, and secondary oil recovery; esters serve as plasticizers in plastic film, coating, and adhesive manufacture

of carbohydrates and waste organic matter to citric acid owing to unstable oil prices.

B. Gluconic Acid

Gluconic acid is manufactured by fermentation or electrochemical oxidation. More than 50,000 tons of the acid and its salts are produced annually worldwide. Whereas early fungal fermentation technologies employed species of *Penicillium*, improved strains of *A. niger* are now used in submerged fermentation with recycled mycelia. An alternate, less common industrial process exploits the ability of *Acetobacter* (*Gluconobacter*) *suboxydans* to make this organic acid from hydrolyzed starch [5,9,41-47].

During the gluconic acid fermentation glucose is converted to δ-gluconolactone by *A. niger* glucose oxidase, a flavoprotein. The lactone spontaneously hydrolyzes to gluconate. In essence, the fermentation conditions are designed for maximum glucose oxidase expression and activity in the mycelia. A number of variations of this fermentation are performed. *A. niger* is cultivated under high aeration for a time period as short as 19 h, but usually 60-70 h at 30-33°C and pH 4.5-7 is employed. A conidial suspension or actively growing mycelia are used as the inoculum. The medium contains D-glucose and other nutrients. Nitrogen is restricted to about 20 mmol/liter and the phosphorus concentration is also limited. In the calcium gluconate fermentation a typical medium contains 15% glucose and salts. In some such fermentations, the incremental addition of a sterile slurry of $CaCO_3$ during gluconic acid formation maintains the pH above 3.5. $CaCO_3$ levels are adjusted to avoid precipitation of calcium gluconate during the fermentation. In the sodium gluconate fermentation, the medium contains 24% glucose and other nutrients. NaOH can be added during the course of the fermentation and the pH maintained at 6.0-6.5. In most gluconate fermentations, the mycelia may be reused twice, and recycled cultures show no lag in production of gluconic acid after fresh medium is added. Sterility of the medium becomes especially important if the medium is reused several times. When levels of trace elements are controlled and sufficient manganese is present, byproduct formation is minimal, and final yields of gluconic acid reach 97-99% of the theoretical maximum [41-47].

Product recovery methods are adapted to the type of fermentation and the desired gluconic acid derivative. $CaCO_3$ or NaOH is used to recover the calcium or sodium salt of gluconic acid, which then can be treated with sulfuric acid to obtain the free acid. The preparation of free acid is vacuum evaporated to a viscous liquor, from which δ-gluconolactone crystallizes. Alternatively, the acid and the lactone forms can be prepared by passage of sodium or calcium gluconate over an ion exchange column. Since sodium gluconate is now the main product of gluconic acid manufacture, free gluconic acid and δ-gluconolactone are produced from the sodium salt by this chromatographic procedure. If crude sodium gluconate salt is desired, fermentation fluid can be spray-dried directly after treatment with NaOH (Fig. 4) [8,41-42,44-47].

Disrupted spent mycelia have been used for the recovery and processing of glucose oxidase, which is formed during gluconic acid fermentation. In the future, commercial methods may employ an immobilized glucose oxidase. Expression of the cloned gene for glucose oxidase in a suitable fungal host could benefit the production of both the enzyme and its product, gluconic acid [45,47].

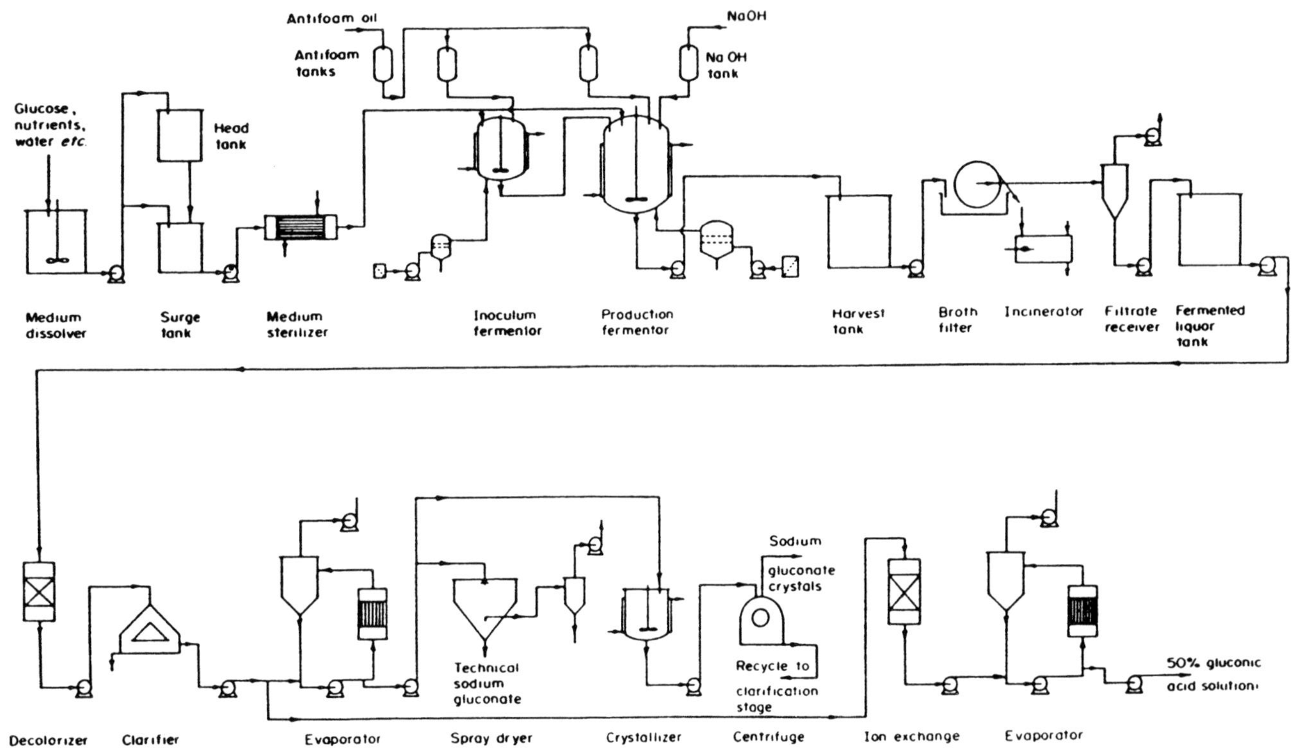

FIGURE 4. Possible layout for the production of sodium gluconate or gluconic acid by submerged fermentation using *Aspergillus niger*. (From Ref. 47.)

Gluconic acid is employed as a cleaning agent while δ-gluconolactone is used as a slow-acting acidulant in foods such as baking mixes, sausages, and tofu. The sodium salt, which is the major commercial product, finds wide application as a sequestering agent for calcium or iron. Iron(II) gluconate serves as an iron source in treatment of anemia while calcium gluconate is used to treat calcium deficiencies via oral or intravenous therapy [18,43,45,47].

C. Itaconic Acid

Itaconic acid is an α-substituted methylacrylic acid formed via dehydration and decarboxylation of *cis*-aconitic acid, an intermediate of the tricarboxylic acid cycle [48] (Fig. 5). In commercial itaconic acid production, industrial strains of *Aspergillus terreus* are grown in aerated stainless steel fermentors at pH 2-3 and at 35-40°C for 3-7 days. The yield of itaconic acid from carbohydrate, whose initial concentration is 10-20%, generally approaches 55-65% of the theoretical maximum. Itaconic acid is produced during the stationary phase of the fermentation (Fig. 6). Glucose, sucrose, and beet or cane molasses have been employed as a carbon source and ammonium sulfate or nitrate can serve as a nitrogen source. Magnesium and calcium are essential supplements and copper and zinc levels must be optimized. Some substances are detrimental to itaconic acid production. These include certain volatile aliphatic carboxylic acids found in beet molasses and excessive iron and potassium [2,47-54].

Direct crystallization of itaconic acid from clarified *A. terreus* culture medium yields commercial grade material, though recrystallization from water is necessary for purer grades (Fig. 7). Precipitation with lead carbonate under basic conditions or solvent extraction with tributyl phosphate have also been employed [47,53-54].

The main commercial products obtained via the *A. terreus* process are itaconic acid and its dimethyl and di-*n*-butyl esters. Quantities measured in thousands of tons are produced annually, mainly in the United States, Western Europe, the USSR, and Japan. Itaconic acid is used to improve the adhesive properties of acrylic emulsion paints and printer's inks. Pyrrolidone derivatives made from itaconic acid have potential applications as grease thickeners, detergents, and herbicides. The key features that make itaconic acid valuable in polymer systems are its two carboxyl groups

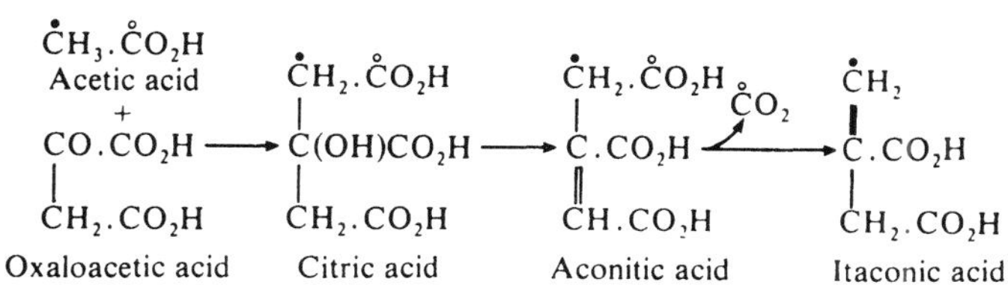

FIGURE 5. Theoretical labeling of itaconic acid metabolized from acetate via the tricarboxylic acid cycle in *A. terreus*. The open and filled circles show the expected labeling from [1-^{14}C]acetate and [2-^{14}C]acetate, respectively. (From Ref. 48.)

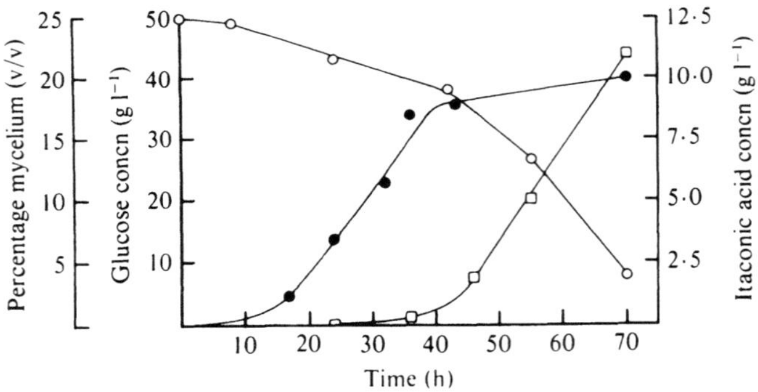

FIGURE 6. Time course of glucose utilization, growth, and itaconic acid production by *Aspergillus terreus*. Flasks containing a medium with glucose as a carbon source and NH_4^+ as a nitrogen were inoculated with a spore suspension and incubated on a shaker-incubator at 30°C. Glucose utilization (○), mycelium concentration (●), and itaconic acid concentration (□) were measured. (From Ref. 48.)

and methylene group. Polymerization occurs via the methylene groups, thus permitting the free carboxyl groups to confer desirable properties on the polymer. Itaconic acid and its ester derivatives are used to make styrene-butadiene/itaconic acid and acrylonitrile/itaconic acid copolymers. Both types of copolymers have superior adhesive and dyeing properties and are used in paper coatings and carpet backing for this reason [47, 53-54].

D. Malic Acid

DL-Malic acid is made by chemical synthesis in Western countries, though fungal fermentation systems for L-malic acid have been explored [14,55,56]. In China, some L-malic acid is manufactured from fumaric acid with immobilized *Candida rugosa* cells at an overall yield of 82-85% [57] and in Japan some methods employ fumarase from immobilized or intact cells of *Brevibacterium* spp. [58-60].

Many fungi are prolific producers of L-malic acid. Significant yields of malic acid have been obtained with strains of *Schizophyllum commune* [61], *Paecilomyces varioti* [62,63], a mixed culture of *Rhizopus arrhizus* containing a *Proteus vulgaris* isolate with high levels of fumarase activity [64], and a *Candida brumptii* strain able to convert *n*-paraffins to L-malic acid [65]. In addition, species of *Pichia* [66], *Candida*, and *Pullularia* [7], as well as species of *Aspergillus* and *Penicillium* [67], can efficiently convert carbohydrate to this organic acid.

Recent detailed studies with *Aspergillus flavus* have used biochemical, electron microscopic, and ^{13}C nuclear magnetic resonance (NMR) analyses to examine malic acid as well as succinic and fumaric acid production [68-70]. The results indicate that L-malic acid production is enhanced by aerobic conditions in a medium containing high glucose levels and limiting nitrogen. Moreover, it is apparent from this work that in *A. flavus*,

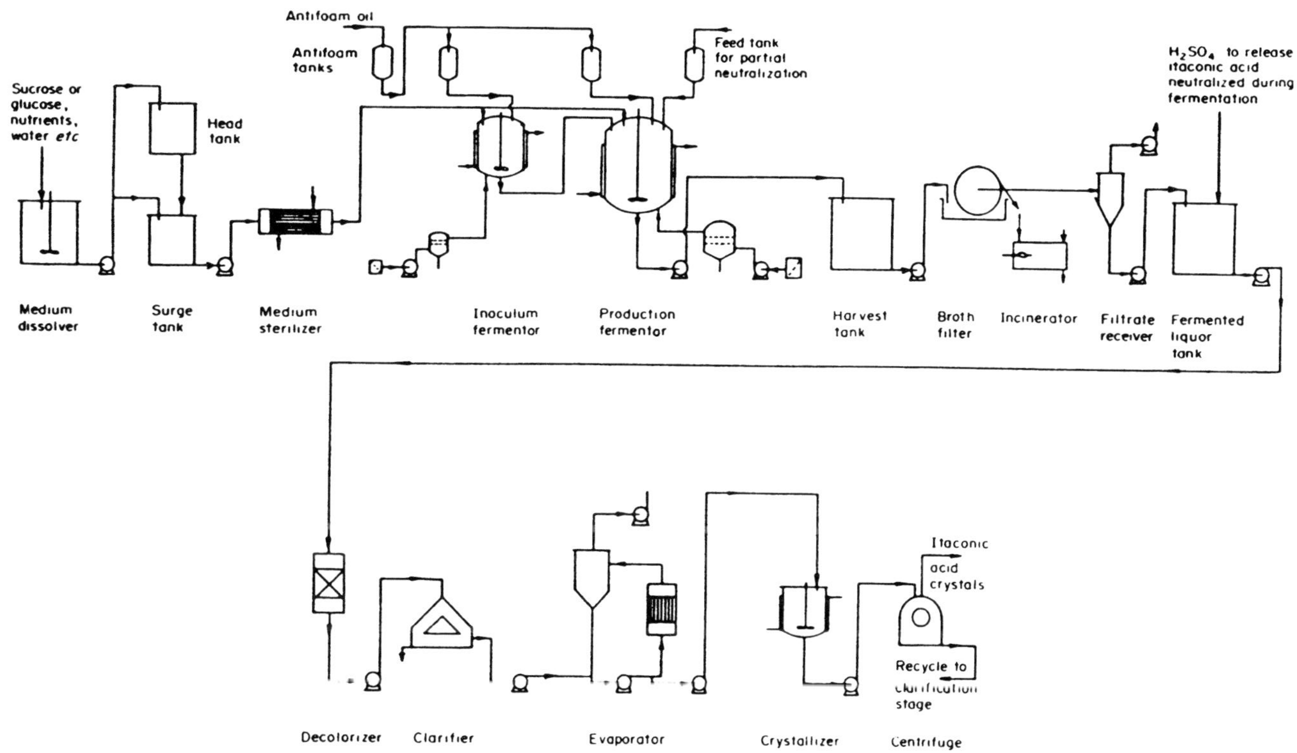

FIGURE 7. Possible layout of a plant designed to produce itaconic acid from sucrose or glucose. (From Ref. 47.)

L-malic acid is synthesized from pyruvate primarily via oxaloacetate in the cytosol and does not directly involve mitochondrial metabolism or the glyoxalate cycle. As Fig. 8 shows, only [3-^{13}C]malic acid is formed from [1-^{13}C]glucose after incubation with *A. flavus* cultures, confirming the operation of the reductive pathway as the main route for malic acid accumulation in this organism [70].

Malic acid is added to foods and beverages, serving as an acidulant and having a preservative function. It provides a pleasant tartness and smooth flavor. Recent findings show that malic acid enhances the sweetness of aspartame, allowing a reduction in the levels of this dipeptide sweetener in malic acid-supplemented soft drinks [18].

E. Fumaric Acid

Fumaric acid was once manufactured by large-scale fermentation with a species of *Rhizopus* [2,14,71] but is now made by catalytic oxidation of benzene or butane [72]. *Rhizopus nigricans*, *Rhizopus japonicus*, and *Rhizopus arrhizus* have been studied extensively [73-75] and recent results with *R. arrhizus* indicate that a fermentation using glucose can produce fumaric acid at a yield approaching 60% [5,46]. Surfactants and vegetable oils can increase the rate of fumaric acid accumulation by *R. arrhizus* [76].

Physiological [77,78] and ^{13}C NMR studies [79] indicate that in some species of *Rhizopus* fumarate is synthesized from pyruvate by carboxylation yielding oxaloacetate, which is then converted to malate and further on to fumarate by the reductive reactions of the TCA cycle. The mechanism points to the central role of pyruvate carboxylase in fumaric acid production and shares features with the main route for L-malic acid production in *A. flavus* shown in Fig. 8 [70].

Other studies have demonstrated that the yeasts *Candida hydrocarbofumarica* [80] and *C. blankii* [81] can produce fumaric acid from hydrocarbons at yields of up to 84 and 65%, respectively.

Fumaric acid serves as a food acidulant and industrial chemical. It is used as a food ingredient in fruit juices and aseptic package drinks, gelatin desserts, pie fillings, baked goods, jellies and jams, wine, egg white foams, and salt substitutes [71,82].

F. Lactic Acid

Lactic acid is manufactured chemically or by deep-tank fermentation with homofermentative lactic acid bacteria [83,84]. The bacterial fermentation can achieve yields of 93-95% when glucose serves as the carbon source. The *Rhizopus oryzae* fermentation can attain comparable yields and fermentation rates on glucose and even starch [85,86]. Early industrial processes for lactic acid production employed this organism in rotary fermentors [87-91]. The *Rhizopus* fermentation has less complex nutritional requirements and offers other advantages [9,83] but has not achieved commercial success. Nevertheless, the interest in lactic acid production with *R. oryzae* has been renewed [92]. For example, recent studies have used *Rhizopus* spp. or *R. oryzae* immobilized in porous materials or gels to produce lactic acid [93-96].

Mixed cultures of bacteria and fungi have been used to produce lactic acid. A lactic acid-producing strain of *Streptococcus lactis* has been coimmobilized in calcium alginate with an amylase-producing strain of *Aspergillus*

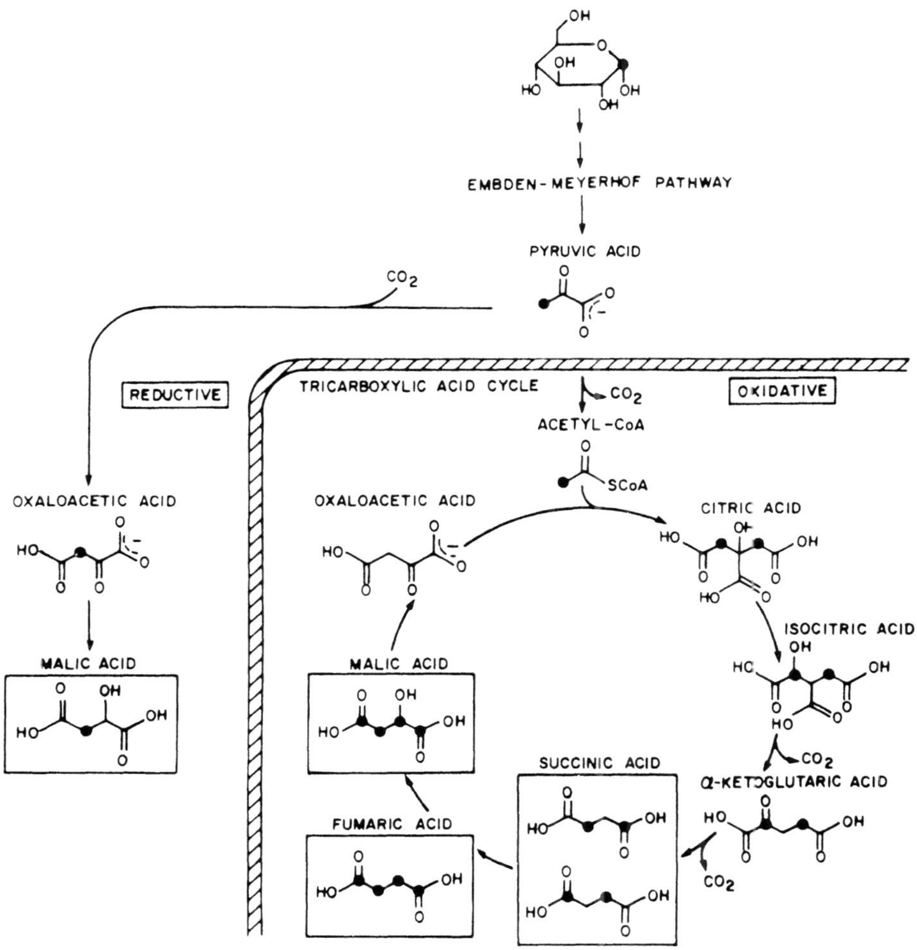

FIGURE 8. Patterns of [^{13}C]malic acid produced from [1-^{13}C]glucose by glycolytic and tricarboxylic acid (TCA) cycle reactions in *Aspergillus flavus*. Glucose is assumed to be converted to pyruvate via the common glycolytic sequence. The stereochemistry of the citrate synthase reaction is presumed to be similar to that of other eukaryotic organisms. The oxidative reactions of the TCA cycle are located in the mitochondria while the reductive reactions of the TCA cycle are present in the cytosol. Only [3-^{13}C]malic acid is formed from [1-^{13}C]glucose, indicating that the reductive pathway is important in L-malic acid production in *A. flavus*. (From Ref. 70.)

awamori. Such a coimmobilized culture system, in this case a mixture of an aerotolerant anaerobe and an aerobe, generates 25 g/liter of lactic acid from 50 g/liter of starch [97].

The multitude of food and industrial uses of lactic acid have been reviewed elsewhere [83,98]. The main application of lactic acid is as a food acidulant or preservative, although a significant amount is used in cosmetics and pharmaceuticals. Some lactic acid is converted to steroyl-2-

lactylate for use as an emulsifier. Other industrial applications involve not only lactic acid, but also salts, esters, and polymers that can be derived from it.

III. OTHER ORGANIC ACIDS PRODUCED BY FUNGI

A. Mycophenolic Acid

This compound is produced by *Penicillium stoloniferum*. Polyene antibiotic-resistant mutants of this fungus with altered yields of mycophenolic acid and ergosterol have been described [99]. The precursor of both compounds is farnesyl pyrophosphate, which also serves as the precursor for ergosterol, the major sterol in yeasts and other fungi.

Mycophenolic acid has antifungal, antibacterial, antiviral, and antitumor activity. It has also been reported to have potential for the treatment of psoriasis [99].

B. Succinic and Epoxysuccinic Acid

Succinic acid is produced by *Candida brumptii* by a fermentation process or by *Rhizopus* sp. in mixed fermentation with bacteria [100]. Since synthetic succinic acid is easy to produce, the fermentative process is not industrially exploited. Succinic acid is used as a flavoring agent and general industrial chemical. Epoxysuccinic acid is produced by oxidation of fumaric acid. This acid has been found among the metabolites of *Paecilomyces* sp. and *Aspergillus fumigatus*.

C. Kojic Acid

Kojic acid is produced by surface culture of members of the *Aspergillus flavus-oryzae* and *A. Tamorii* group. It is an unusual organic acid being a pyrone compound lacking a carboxyl group. The production procedures are generally similar to those mentioned for citric and itaconic acid. Kojic acid has been shown to have antibiotic, insecticidal, and growth-promoting activities.

D. Other Minor Organic Acids

Many other organic acids of minor importance can be produced by a fermentation process but are commonly obtained by chemical synthesis. For example, tartaric acid is produced by *Aspergillus niger*, *A. griseus*, and *Penicillium notatum*. It is commonly used as a food additive. Ascorbic acid-like substances such as isoascorbic acid (erythrobic acid) are produced by a number of fungi, including species of *Torula*, *Aspergillus*, *Penicillium*, and *Fusarium*. These acids are used as food preservatives. Pyruvic acid is produced by *Debaryomyces condertii* and *S. commune* in significant amounts, as is α-ketoglutaric acid by *C. lipolytica*. Other organic acids synthesized by fungi have been discussed elsewhere [14,101-105].

REFERENCES

1. Kubicek, C. P., and Röhr, M. Citric acid fermentation. *CRC Crit. Rev. Biotechnol.*, *3*:331-373, 1986.

2. Perlman, D., and Sih, C. J. Fungal synthesis of citric, fumaric, and itaconic acids. *Prog. Ind. Microbiol., 2*:169-194, 1960.
3. Martin, S. M. Production of organic acids by moulds. In *Biochemistry of Industrial Microorganisms* (C. Rainbow and A. H. Rose, eds.), Academic Press, London, 1963, pp. 415-451.
4. Lockwood, L. B., and Schweiger, L. B. Citric and itaconic acid fermentations. In *Microbial Technology* (H. J. Peppler, ed.), Reinhold, New York, 1967, pp. 183-199.
5. Smith, J. E., Nowakowska-Waszczuk, A., and Anderson, J. G. Organic acid production by mycelial fungi. In *Industrial Aspects of Biochemistry* (B. Spencer, ed.), Elsevier, Amsterdam, 1974, pp. 297-317.
6. Berry, D. R., Chmiel, A., and Obaidi, Z. A. Citric acid production by Aspergillus niger. In *Genetics and Physiology of Aspergillus* (J. E. Smith and J. A. Pateman, eds.), Academic Press, London, 1977, pp. 405-426.
7. Miall, L. M. Organic acids. In *Primary Products of Metabolism. Economic Microbiology*, Vol. 2 (A. H. Rose, ed.), Academic Press, New York, 1978, pp. 47-119.
8. Lockwood, L. B. Organic acid production. In *The Filamentous Fungi. Industrial Mycology*, Vol. 1 (J. E. Smith and D. R. Berry, eds.), Edward Arnold, London, 1975, pp. 140-157.
9. Lockwood, L. B. Production of organic acids by fermentation. In *Microbial Technology. Microbial Processes*, 2nd ed., Vol. 1 (H. J. Peppler and D. Perlman, eds.), Academic Press, New York, 1979, pp. 353-387.
10. Bouchard, E. F., and Merritt, E. G. Citric Acid. In *Kirk-Othmer Encyclopedia of Chemical Technology*, 3rd ed., Vol. 6 (M. Grayson and D. Eckroth, eds.), John Wiley and Sons, New York, 1979, pp. 150-179.
11. Kapoor, K. K., Chaudhary, K., and Tauro, P. Citric acid. In *Prescott and Dunn's Industrial Microbiology*, 4th ed. (G. Reed, ed.), Avi, Westport, CT, 1982, pp. 709-747.
12. Röhr, M., Kubicek, C. P., and Kominek, J. Citric acid. In *Biotechnology. Microbial Products, Biomass and Primary Products*, Vol. 3 (H. Delweg, ed.), Verlag-Chemie, Weinhem, 1983, pp. 419-454.
13. Abou-Zeid, A.-A. A., and Ashy, M. A. Production of citric acid: A review, *Agr. Wastes, 9*:51-76, 1984.
14. Bigelis, R. Primary metabolism and industrial fermentations. In *Gene Manipulations in Fungi* (J. W. Bennett and L. L. Lasure, eds.), Academic Press, New York, 1985, pp. 357-401.
15. Milsom, P. E., and Meers, J. L. Citric acid. In *Comprehensive Biotechnology. The Principles, Applications and Regulations of Biotechnology in Industry, Agriculture, and Medicine*, Vol. 3 (H. W. Blanch, S. Drew, and D. I. C. Wang, eds.), Pergamon Press, Oxford, 1985, pp. 665-680.
16. Milsom, P. E. Organic acids by fermentation, especially citric acid. In *Food Biotechnology*, Vol. 1 (R. D. King and P. S. J. Cheetham, eds.), Elsevier, London, 1987, pp. 273-307.
17. Bigelis, R., and Lasure, L. L. Fungal enzymes and primary metabolites used in food processing. In *Food and Beverage Mycology*, 2nd ed. (L. R. Beuchat, ed.), Avi, Westport, CT, 1987, pp. 473-516.
18. Bigelis, R. Role of fungal metabolites in food processing. In *Handbook of Applied Mycology, Vol. 3: Food and Feeds* (Dilip K. Arora, K. G. Mukjeri, and E. H. Marth, eds.), Marcel Dekker, New York, (1991).

19. Bigelis, R. Industrial products of biotechnology: Application of gene technology. In *Biotechnology. A Comprehensive Treatise, Vol. 7b, Gene Technology* (S. Jolly and G. Jacobson, eds.), Verlag-Chemie, Weinheim, 1989, pp. 229-259.
20. Sodeck, G., Modl, J., Kominek, J., and Salzbrunn, W. Production of citric acid according to the submerged fermentation process. *Proc. Biochem.*, *16*(6):9-11, 1981.
21. Villet, R. Citric acid. In *Biotechnology for Producing Fuels and Chemicals from Biomass, Vol. 2, Fermentation Chemicals from Biomass*, Solar Energy Research Institute, Golden, CO, 1981, pp. 25-31.
22. Kertes, A. S., and King, C. J. Extraction chemistry of fermentation product carboxylic acids. *Biotechnol. Bioeng.*, *28*:269-282, 1986.
23. Cleland, W. W., and Johnson, M. J. Tracer experiments on the mechanism of citric acid formation by Aspergillus niger. *J. Biol. Chem.*, *208*:679-689, 1954.
24. Shu, P., Funk, A., and Neish, A. C. Mechanism of citric acid formation from glucose by Aspergillus niger. *Can. J. Biochem.*, *32*:68-80, 1954.
25. Kubicek, C. P., Hampel, W., and Röhr, M. Manganese deficiency leads to elevated amino acid pool in citric acid accumulating Aspergillus niger. *Arch. Microbiol.*, *123*:73-79, 1979.
26. Ma, H., Kubicek, C. P., and Röhr, M. Metabolic effects of manganese deficiency in Aspergillus niger: Evidence for increased protein degradation. *Arch. Microbiol.*, *141*:266-268, 1985.
27. Orthofer, R., Kubicek, C. P., and Röhr, M. Lipid levels and manganese deficiency in citric acid producing strains of Aspergillus niger. *FEMS Microbiol. Lett.*, *5*:403-406, 1979.
28. Meixner, O., Mischak, H., Kubicek, C. P., and Röhr, M. Effect of manganese deficiency on plasma membrane lipid composition and glucose uptake in Aspergillus niger. *FEMS Microbiol. Lett.*, *26*:271-274, 1985.
29. Kisser, M., Kubicek, C. P., and Röhr, M. Influence of manganese on morphology and cell wall composition of Aspergillus niger during citric acid fermentation. *Arch. Microbiol.*, *128*:26-33, 1980.
30. Kubicek, C. P. and Röhr, M. Influence of enzyme synthesis and citric acid accumulation in Aspergillus niger. *Eur. J. Appl. Microbiol.*, *4*: 167-175, 1977.
31. Atticus. Citric acid forges ahead as an industrial chemical. *Chem. Age India*, *26*:49-54, 1975.
32. Fricke, H., and Jensen, S. B. Citric acid: A versatile food additive. *Processing Ind.*, *44*(528):37, 39-40, 44, 1975.
33. Andres, C. Acidulants. Flavor, preserve, texturize, and leaven foods. *Food Processing*, *46*(5):52-54, 1985.
34. Stottmeister, U., Behrens, U., Weissbrodt, E., Barth, G., Franke-Rinker, D., and Schulze, E. Nutzung von Paraffinen und anderen Nichtkohlenhydrat-Kohlenstoffquellen zur mikrobiellen Citronensauresynthese. *Z. Allg. Mikrobiol.*, *22*:399-424, 1982.
35. Takeda Chemical Industries, Ltd. Belg. Patent 744,416 (1970).
36. Pfizer, Inc. Br. Patent 1,203,006 (1970).
37. Pfizer, Inc. U.S. Patent 3,717,549 (1973).
38. Pfizer, Inc. Br. Patent 1,369,295 (1974).
39. Kyowa Hakko Kogyo Co., Ltd. U.S. Patent 4,322,498 (1982).
40. Takeda Chemical Industries, Ltd. U.S. Patent 4,389,484 (1983).

41. Blom, R. H., Pfeiffer, V. F., Moyer, A. J., Traufler, D. H., Conway, H. F., Crocker, C. K., Farison, R. E., and Hannibal, D. V. Sodium gluconate production. *Ind. Eng. Chem.*, *44*:435-440, 1952.
42. Underkofler, L. A. Gluconic acid. In *Industrial Fermentations* (L. A. Underkofler and R. J. Hickey, eds.), Chemical Publishing, New York, 1954, pp. 446-469.
43. Feldberg, C. Glucono-delta-lactone. *Cereal Sci. Today*, *4*:96-99, 1959.
44. Ward, G. E. Production of gluconic acid, glucose oxidase, fructose, and sorbose. In *Microbial Technology*, 1st ed. (H. J. Peppler, ed.), Reinhold, New York, 1967, pp. 200-221.
45. Röhr, M., Kubicek, C. P., and Kominek, J. Gluconic acid. In *Biotechnology. Microbial Products, Biomass, and Primary Products*, Vol. 3 (H. Dellweg, ed.), Verlag-Chemie, Weinheim, 1983, pp. 456-465.
46. Buchta, K. Organic acids of minor importance. In *Biotechnology. Microbial Products, Biomass, and Primary Products*, Vol. 3 (H. Dellweg, ed.), Verlag-Chemie, Weinheim, 1983, pp. 467-478.
47. Milsom, P. E., and Meers, J. L. Gluconic and itaconic acids. In *Comprehensive Biotechnology. The Principles, Applications, and Regulations of Biotechnology in Industry, Agriculture and Medicine*, Vol. 3 (H. W. Blanch, S. Drew, and D. I. C. Wang, eds.), Pergamon Press, Oxford, 1985, pp. 665-680.
48. Winskill, N. Tricarboxylic acid cycle activity in relation to itaconic acid biosynthesis by Aspergillus terreus. *J. Gen. Microbiol.*, *129*:2877-2883, 1983.
49. Pfeiffer, V. F., Vojnovich, C., and Heger, E. N. Itaconic acid by fermentation with Aspergillus terreus. *Ind. Eng. Chem.*, *44*:2975-2980, 1952.
50. Lockwood, L. B. Itaconic acid. In *Industrial Fermentations*, Vol. 1 (L. A. Underkofler, ed.), Chemical Publ., New York, 1954, pp. 488-497.
51. Kobayashi, T. Itaconic acid fermentation. *Proc. Biochem.*, *2*(g):61-65, 1967.
52. Jacubowska, J. Itaconic and itatartaric acid biosynthesis. In *Genetics and Physiology of Aspergillus* (J. E. Smith and J. A. Pateman, eds.), Academic Press, 1977, pp. 425-451.
53. Tate, B. E. Itaconic acid and derivatives. In *Kirk-Othmer Encyclopedia of Chemical Technology*, Vol. 13, 3rd ed. (M. Grayson and D. Ekroth, eds.), John Wiley and Sons, New York, 1981, pp. 865-873.
54. Luskin, L. S. Acid monomers. In *Functional Monomers, Their Preparation, Polymerization, and Application*, Vol. 2, 1st ed. (R. H. Yocum and E. B. Nyquist, eds.), Marcel Dekker, New York, 1974, pp. 357-554.
55. Irwin, W. E., Lockwood, L. B., and Zienty, M. F. Malic acid. In *Kirk-Othmer Encyclopedia of Chemical Technology*, Vol. 12, 2nd ed. (A. Standen, H. F. Mark, J. J. McKetta, Jr., and D. F. Othmer, eds.), John Wiley and Sons, New York, 1967, pp. 837-849.
56. Berger, S. E. Hydroxy carboxylic acids. In *Kirk-Othmer Encyclopedia of Chemical Technology*, Vol. 13, 3rd ed. (H. F. Mark, D. F. Othmer, C. G. Overberger, and G. T. Seaborg, eds.), John Wiley and Sons, New York, 1981, pp. 103-121.
57. Zhang, S. Z. Industrial applications of immobilized biomaterials in China. In *Enzyme Engineering* (I. Chibata, S. Fukui, and L. B. Wingard, Jr., eds.), Plenum Press, New York, 1982, pp. 265-279.

58. Yamamoto, K., Tosa, T., Yamashita, K., and Chibata, I. Continuous production of L-malic acid by immobilized Brevibacterium ammoniagenes cells. *Eur. J. Appl. Microbiol.*, *3*:169-183, 1976.
59. Takata, I., Yamashita, K., Tosa, T., and Chibata, I. Immobilization of Brevibacterium flavum with carrageenan and its application for continuous production of L-malic acid. *Enz. Microb. Technol.*, *2*:30-36, 1980.
60. Yukawa, H., Yamagata, H., and Terasawa, W. Production of L-malic acid by the cell reusing process. *Proc. Biochem.*, *21*(5):164-166, 1986.
61. Tachibana, S., and Murakami, T. L-malate production from ethanol and calcium carbonate by Schizophyllum commune. *J. Ferment. Technol.*, *51*:858-864, 1973.
62. Takao, S., Tanida, M., and Kuwabara, H. L-malic acid production from non-sugar sources by Paecilomyces varioti. *J. Ferment. Technol.*, *55*:196-199, 1977.
63. Takao, S., Takahashi, T., Tanida, M., and Takahashi, M. Purification and some properties of isocitrate lyase from Paecilomyces varioti with high malic acid production. *J. Ferment. Technol.*, *62*:577-582, 1984.
64. Takao, S., and Hotta, K. L-malic acid fermentation by mixed culture of Rhizopus arrhizus and Proteus vulgaris. *Agr. Biol. Chem.*, *41*:945-950, 1977.
65. Sato, S., Nakahara, T., and Minoda, Y. Enzymatic studies on L-malic acid production from n-paraffins by Candida brumptii IFO-0731. *Agr. Biol. Chem.*, *41*:1903-1907, 1977.
66. Rossi, J., and Clementi, F. L-malic acid production by polyacrylamide gel entrapped Pichia membranaefaciens. *Biotechnol. Lett.*, *7*:329-334, 1985.
67. Beuchat, L. R. *Food and Beverage Mycology*, Avi, Westport, CT, 1978, pp. 368-396.
68. Peleg, Y., Stieglitz, B., and Goldberg, I. Malic acid accumulation by Aspergillus flavus. I. Biochemical aspects of acid biosynthesis. *Appl. Microbiol. Biotechnol.*, *28*:69-75, 1988.
69. Peleg, Y., Rahamim, E., Kessel, M., and Goldberg, I. Malic acid accumulation by Aspergillus flavus. II. Crystals and hair-like processes formed by A. flavus in a L-malic acid production medium. *Appl. Microbiol. Biotechnol.*, *28*:76-79, 1988.
70. Peleg, Y., Barak, A., Scrutton, M. C., and Goldberg, I. Malic acid accumulation by Aspergillus flavus. III. ^{13}C NMR and isoenzyme analysis. *Appl. Microbiol. Biotechnol.*, *30*:176-183, 1989.
71. Foster, J. W. Fumaric acid. In *Industrial Fermentations*, Vol. 1 (L. A. Underkofler and R. J. Hickey, eds.), Chemical Publ., New York, 1954, pp. 470-487.
72. Robinson, W. D., and Mount, R. A. Malic anhydride, maleic acid, and fumaric acid. In *Kirk-Othmer Encyclopedia of Chemical Technology*, Vol. 13, 3rd ed. (H. F. Mark, D. F. Othmer, C. G. Overberger, and G. T. Seaborg, eds.), John Wiley and Sons, New York, 1981, pp. 770-793.
73. Foster, J. W., and Waksman, S. A. Fumaric acid formation associated with sexuality in a strain of Rhizopus nigricans. *Science*, *89*:37, 1939.
74. Waksman, S. U.S. Patent 2,326,986 (1944).
75. Rhodes, R. A., Moyer, A. J., Smith, M. L., and Kelley, S. E. Production of fumaric acid by Rhizopus arrhizus. *Appl. Microbiol.*, *7*:74-80, 1959.

76. Goldberg, I., and Steiglitz, B. Improved rate of fumaric acid production by Tweens® and vegetable oils in Rhizopus arrhizus. *Biotechnol. Bioeng.*, *27*:1067-1069, 1985.
77. Romano, A. H., Bright, M. M., and Scott, W. E. Mechanism of fumaric accumulation in Rhizopus nigricans. *J. Bacteriol.*, *93*:600-604, 1967.
78. Overman, S. A., and Romano, A. H. Pyruvate carboxylase of Rhizopus nigricans and its role in fumaric acid production. *Biochem. Biophys. Res. Commun.*, *37*:457-463, 1969.
79. Kenealy, W., Zaady, E., du Preez, J. C., Steiglitz, B., and Goldberg, I. Biochemical aspects of fumaric acid accumulation by Rhizopus arrhizus. *Appl. Environ. Microbiol.*, *52*:128-133, 1986.
80. Furakawa, T., Nakahara, T., and Yamada, K. Studies on utilization of hydrocarbons by microorganisms. Part 19. Influence of several factors in fumaric acid production from n-paraffins by Candida hydrocarbofumarica. *Agr. Biol. Chem.*, *34*:1402-1406, 1970.
81. Furukawa, T., Rodrigues de Miranda, L., and Matsuyoshi, T. Fermentative production of fumaric acid from n-paraffins by Candida blankii. *J. Ferment. Technol.*, *56*:546-549, 1978.
82. Anonymous. *Food Grade Fumaric Acid: Versatile, High Efficiency Food Acid*, Monsanto, St. Louis, 1985.
83. Vickroy, T. B. Lactic acid. In *Comprehensive Biotechnology. The Principles, Applications and Regulations of Biotechnology in Industry, Agriculture and Medicine*, Vol. 3 (H. W. Blanch, S. Drew, and D. I. C. Wang, eds.), Pergamon Press, Oxford, 1985, pp. 761-776.
84. Schopmeyer, H. H. Lactic acid. In *Industrial Fermentations* (L. A. Underkofler and R. J. Hickey, eds.), Chemical Publ., New York, 1954, pp. 391-419.
85. Snell, R. L., and Lowery, C. E. Calcium (L+) lactic acid production. U.S. Patent 3,125,944 (1964).
86. Bernhauser, K., Rauch, J., and Miksch, J. H. Acid formation of species of Rhizopus II. Lactic acid formation in submerged cultures. *Biochem. Z.*, *320*:178-188, 1950.
87. Ward, G. E., Lockwood, L. B., May, O. E., and Herrick, H. T. Biochemical studies in the genus Rhizopus. I. The production of dextrolactic acid. *J. Am. Chem. Soc.*, *58*:1286-1288, 1936.
88. Lockwood, L. B., Ward, G. E., and May, O. E. The physiology of Rhizopus oryzae. *J. Agr. Res.*, *53*:849-857, 1936.
89. Ward, G. E., Lockwood, L. B., Tabenkin, B., and Wells, P. A. Rapid fermentation process for dextrolactic acid. *Ind. Eng. Chem.*, *30*:1233-1235, 1938,
90. Ward, G. E., Lockwood, L. B., and May, O. E. Fermentation process for the production of dextrolactic acid. U.S. Patent 2,132,712 (1938).
91. Prescott, S. C., and Dunn, C. G. *Industrial Microbiology*, 3rd ed., McGraw-Hill, New York, 1959.
92. Hang, Y. D. Direct fermentation of corn to L(+)-lactic acid by Rhizopus oryzae. *Biotechnol. Lett.*, *11*:299-300, 1989.
93. Iwata Kagaku Kogyo. Lactic acid production by contacting microorganisms, immobilized in gel carrier, with sugar solution. Jap. Patent 60006196 (1985).
94. Kanegafuchi Chem. KK. Organic acid preparation using Aspergillus or Rhizopus on foamed polymer-type or porous material support. Jap. Patent 61209596 (1986).

95. Bailey, R. B., Joshi, D. K., Michaels, S. L., and Wisdom, R. A. Production of lactic acid from whey by continuous fermentation and lactic acid purification by solvent extraction. U.S. Patent 4,698,303 (1987).
96. Hang, Y. D., Hammaci, H., and Woodams, E. E. Production of L(+)-lactic acid by Rhizopus oryzae immobilized in calcium alginate gels. *Biotechnol. Lett.*, *11*:119-120, 1989.
97. Kurosawa, H., Ishikawa, H., and Tanaka, H. L-lactic acid production from starch by coimmobilized mixed culture system of Aspergillus awamori and Streptococcus lactis. *Biotechnol. Bioeng.*, *31*:183-187, 1988.
98. Anonymous. Lactic acid and lactates, Purac, Inc., Arlington Heights, Il, 1986.
99. Vegina, C., and Singh, K. Transformation of organic compounds by fungal spores. In *The Filamentous Fungi*, Vol. 3 (J. E. Smith and D. R. Berry, eds.), John Wiley and Sons, New York, 1975, pp. 158-192.
100. Sasaki, Y., Takao, S., and Khotta. Conservation of fumaric acid fermentation by the association of Rhizopus and bacteria. *Agr. Biol. Chem.*, *48*:778-781, 1970.
101. Miller, M. W. *The Pfizer Handbook of Microbial Metabolites*. McGraw-Hill, New York, 1961.
102. Shibata, S., Natori, S., and Udagawa, S. *List of Fungal Products*, University of Tokyo Press, Tokyo, 1964.
103. Turner, W. B. *Fungal Metabolites*, Academic Press, London, 1971.
104. Laskin, A. I., and Lechavalier, H. A. *Handbook of Microbiology. Microbial Products*, Vol. 3, CRC Press, Boca Raton, 1974.
105. Yamada, K. Recent advances in industrial fermentation in Japan. *Biotechnol. Bioeng.*, *19*:1563-1621, 1977.

15

POTENTIAL FOR FUNGAL LIPIDS IN BIOTECHNOLOGY

JOHN D. WEETE and SHAILESH GANDHI *Auburn University, Auburn, Alabama*

I. INTRODUCTION

Lipids include a wide variety of substances that (a) are soluble in organic solvents (e.g., chloroform, benzene, etc.), (b) are insoluble in water, (c) contain long hydrocarbon chains, and (d) are derived from living organisms [1]. Studies of fungal lipids began in the late nineteenth century with the recognition that certain strains and species of both yeasts and filamentous fungi are capable of producing relatively high amounts of fat or oil, i.e., diethylether-extractable substances consisting mostly of triacylglycerols. Research in this area spanned the first half of the twentieth century, focusing on the identification of potentially high-fat-producing (oleaginous) fungi, defining the nutritional and environmental conditions favoring fat production, and assessing the efficiency of substrate utilization [2]. The early objectives of this research, at least in part, were to determine if fungal fat could be substituted for that from plant sources as a dietary supplement and if commercial production of fungal fat was feasible and economical.

In later years, the emphasis shifted from fat production to identifying lipophilic products of various biological systems and elucidating pathways for their biosynthesis. This was made possible by the use of more efficient extraction procedures, the development of sophisticated separation techniques (thin-layer and gas-liquid chromatography) and mass spectrometry for identification, and the availability of radioactive tracers. Not surprisingly, it was found that fungi produce many of the same types of lipids as plants and animals, but some unique ones as well. More recently, techniques of molecular biology were used to identify genes coding for enzymes of fatty acid synthesis in several fungi [3]. Although the fat from some oleaginous fungi is nutritionally similar to vegetable oil, high production costs have limited its availability. It now seems that fungal fat production has come full circle. Not only is there renewed interest in fungal fat per se, but there is particular interest in specific fungal lipid products, such as those having biological activity or surfactant properties. Numerous review articles have been published on the biotechnology of natural products, some of which have included fungal lipids [4-7]. This chapter will give a brief overview of the potential of fungal lipids in biotechnology, focusing specifically on recently published information on the production of total fat (triacylglycerols), fatty acids, sterols, and extracellular lipids with surfactant properties. Lipases of fungal origin are also briefly reviewed.

II. FUNGAL LIPIDS

A. Fatty Acids

Fatty acids in fungi occur as a homologous series ranging in carbon chain length from C_{10} to C_{26}, with the even-numbered carbon chains predominating (Table 1). Palmitic acid ($C_{16:0}$) is the most abundant saturated homolog at 10-25% of the total fatty acids [8]. In some fungi, such as certain Mucorales, myristic acid ($C_{14:0}$) is also one of the major fatty acids.

The distribution of unsaturated fatty acids varies more with the fungal species and culture conditions than that of the saturated fatty acids. Oleic acid ($C_{18:1}$ $\Delta 9$) is one of the more abundant fatty acids in fungi, but it may be accompanied by small amounts of the corresponding trans isomer (elaidic), positional isomers, as well as the corresponding monoenes of other homologs (Table 2). The cis Δ^7 isomer of $C_{18:1}$ is found in approximately equal relative proportions as the Δ^9 isomer in the Hyphochrytriomycetes and Chytridiomycetes [9]. The true yeasts have palmitoleic acid ($C_{16:1}$) as the principal or at least a major fatty acid, along with $C_{18:1}$, but no polyenes [10]. Examples of species with high $C_{16:1}$ content are *Saccharomyces cerevisiae*, *Hanseniaspora valbyensis*, *Brettanomyces lambricus*,

TABLE 1. Some Naturally Occurring Straight-Chain Saturated Acids

Number of carbon atoms	Systematic name	Common name
2	*n*-Ethanoic	Acetic
3	*n*-Propanoic	Propionic
4	*n*-Butanoic	Butyric
8	*n*-Octanoic	Caprylic
10	*n*-Decanoic	Capric
12	*n*-Dodecanoic	Lauric
14	*n*-Tetradecanoic	Myristic
16	*n*-Hexadecanoic	Palmitic
18	*n*-Octadecanoic	Stearic
20	*n*-Eicosanoic	Arachidic
22	*n*-Docosanoic	Behenic
24	*n*-Tetracosanoic	Lignoceric
26	*n*-Hexacosanoic	Cerotic
28	*n*-Octacosanoic	Montanic

TABLE 2. Some Naturally Occurring Mono- and Diunsaturated Fatty Acids

Number of carbon atoms	Systematic name	Common name
Monoenoic acids		
16	*trans*-3-Hexadecenoic	—
	cis-5-Hexadecenoic	—
	cis-7-Hexadecenoic	—
	cis-9-Hexadecenoic	Palmitoleic
	cis-11-Hexadecenoic	Palmitvaccenic
18	*cis*-9-Octadecenoic	Oleic
	trans-9-Octadecenoic	Elaidic
	cis-11-Octadecenoic	Vaccenic
22	*cis*-13-Docosenoic	Erucic
Dienoic acids		
16	*cis,cis*-7,10-Hexadecadienoic	—
	cis,cis-9,12-Hexadecadienoic	—
18	*cis,cis*-6,9-Octadecadienoic	—
	cis,cis-9,12-Octadecadienoic	Linoleic

Debaryomyces nilssonii, *Kloeckera apriculata*, *Kluyveromyces polysporus*, *Nadsonia fulvescens*, *Torulopsis colliculosa*, *T. dattila*, and *Trigonopsis variabilis*.

Linoleic acid ($C_{18:2}\ \Delta^{9,12}$) is generally the most abundant fatty acid, except in "lower" fungi (Mastigomycotina) where it is a relatively minor fatty acid, and in many of the true yeasts, such as *S. cerevisiae*, where it is not found (Table 2). α-Linolenic acid ($C_{18:3}\ \Delta^{9,12,15}$) is produced by most but not all fungi. When present, it may be a major fatty acid but is rarely the most abundant. Zygomycetes and the Mastigomycotina produce mixtures of the α and the corresponding γ isomer, γ-linolenic acid ($C_{18:3}\ \Delta^{6,9,12}$) [9,11] (Table 3). A C_{18} tetraene was detected in *Thamnidium elegans* cultured at low temeprature [12].

The lower fungi, including some Zygomycetes, also produce C_{20} fatty acids containing one to five methylene-interrupted double bonds, the most widespread of which is the $\omega 6$ fatty acid, arachidonic acid ($C_{20:4}\ \Delta^{5,8,11,14}$) (Table 3). *cis*-5,8,11,14,17-Eicosapentaenoic (EPA, $C_{20:5}$) (Table 3) is an $\omega 3$ fatty acid that has been detected in some Oomycetes, particularly species of the genera *Pythium* and *Phytophthora* (pythiaceous fungi). EPA is also found in certain members of the family Entomophthoraceae [13]. There are a few reports of polyunsaturated C_{22} fatty acids in some of the pythiaceous fungi [11], and certain marine fungi such as *Thraustochytrium roseum*, *R. aureum*, *Schizochytrium aggregatum*, and *Dermocystidum* sp. [14], but they should be confirmed using modern analytical techniques.

A variety of hydroxy fatty acids have been detected in fungi, but they are not generally considered to be major fungal products (see below). For example, 2% of *S. cerevisiae* cell wall lipid consists of saturated, unsaturated, and 2-hydroxy homologs ranging from 10 to 26 carbon atoms

TABLE 3. Some Naturally Occurring Tri-, Tetra-, Penta-, and Hexaenoic Fatty acids

Number of carbon atoms	Systematic name	Common name
Trienoic acids (methylene-interrupted)		
16	All-*cis*-7,10,13-hexadecatrienoic	—
18	All-*cis*-6,9,12-octadecatrienoic	γ-Linolenic
	All-*cis*-9,12,15-octadecatrienoic	α-Linolenic
20	All-*cis*-8,11,14-eicosatrienoic	Dihomo-γ-linolenic
Tetraenoic acids		
16	All-*cis*-4,7,10,13-hexadecatetraenoic	—
20	All-*cis*-5,8,11,14-eicosatetraenoic	Arachidonic
Pentaenoic acids		
20	All-*cis*-5,8,11,14,17-eicosapentaenoic	EPA
22	All-*cis*-7,10,13,16,19-docosapentaenoic	Clupanodonic
Hexaenoic acids		
22	All-*cis*-4,7,10,13,16,19-docosahexaenoic	DHA

in chain length. They are associated with the glycosphingolipid fraction, and 2-hydroxy C_{26} is one of the principal components [15]. An example of a hydroxy fatty acid being a major component of oil from a fungus is in the sclerotia of the ergot fungus *Claviceps purpurea*, where 12-hydroxy-cis-9-octadecenoic acid (ricinoleic acid) may comprise up to 44% of total fatty acids. 9,10-Dihydroxyoctadecanoic acid composed 64% of the fatty acids in *C. sulcata* [16]. These hydroxy acids occur as components of triacylglycerols and estolides, the latter of which may contain up to six fatty acid residues. Laetisaric acid[8-hydroxy-9(*Z*), 12(*Z*)-octadecadienoic acid] was identified as a product of the basidiomycete *Laetisaria arvalis* which, after degradation to α-hydroxydodecanoic acid, exhibits antifungal activity [17].

Certain yeasts and yeastlike fungi produce several hydroxy fatty acids that occur as components of extracellular glycolipids, some of which may be acetylated (see Section V.C) [18,19]. Some of these include 3-D-hydroxypalmitic acid in *Saccharomyces malanga* [20], 8,9,13-triacetoxy-docosanoic and 13-oxo-8,9-diacetoxydocosanoic acids in a yeastlike fungus closely related to *Torulopsis fujisanesis* [21] and a *Rhodotorula* species [19,22]. Also, 17-L-hydroxyoctadecanoic and 17-L-hydroxyoctadecenoic

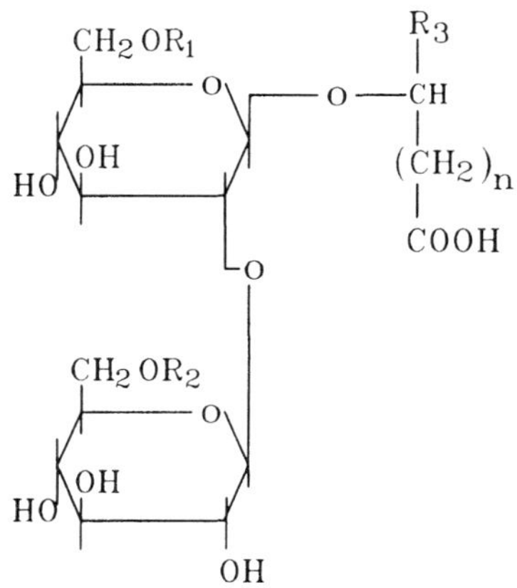

Sophorose glycolipid

R_1, R_2 = H or acetyl; R_3 = H or CH_3, n = 14–16

FIGURE 1. Sophorose glycolipid.

acids have been identified as components of sophorosides (Fig. 1) produced by *T. magnaliae* [23], and 13-hydroxydocosanoic acid in sophorosides from *Candida bogoriensis* [18].

cis-9,10-Epoxyoctadecanoic acid is known as a fungal product and, except for one known exception, seems to be restricted to the rust fungi where it may compose up to 78% of the total fatty acids in spore oil [24,25 and references cited therein]. This fatty acid was also detected in *Claviceps sulcata* and an unidentified *Claviceps* species.

B. Sterols

Ergosterol is often regarded as "the fungal sterol," and indeed it is the major sterol of most fungal species [8,26,27] (Fig. 2). With the exceptions noted below, this is true for the Zygomycetes, Ascomycetes, and most Basidiomycetes. The most prolific producer of ergosterol known is *S. cerevisiae*, some strains of which may produce up to 2% of their dry weight as this sterol and some as much as 10% [28]. Other fungi generally produce up to 5.0 μg sterol/mg dry weight, 80-100% of which may be ergosterol.

In recent years it has become apparent that ergosterol is not produced by certain fungal taxa. For example, the lower fungi (Mastigomycotina) produce mainly cholesterol (Fig. 2) and its C-24 alkyl and/or alkylidene derivatives. Fucosterol (Fig. 2) is the principal sterol of many Oomycetes and may be accompanied by cholesterol and 24-methylene cholesterol. Cholesterol is produced by some but not all Chytridiomycetes which, along with Hypochytriomycetes, may also have 24-alkyl rather than 24-alkylidene derivatives as major sterols [9]. Cholesterol is the principal sterol of some Chytridiomycetes at 73-98% of the total. Pythiaceous fungi do not produce sterols.

Some fungi, such as *Taphrina*, *Protomyces*, and *Terfezia* species, produce brassicasterol (Fig. 2) as the predominant sterol and little or

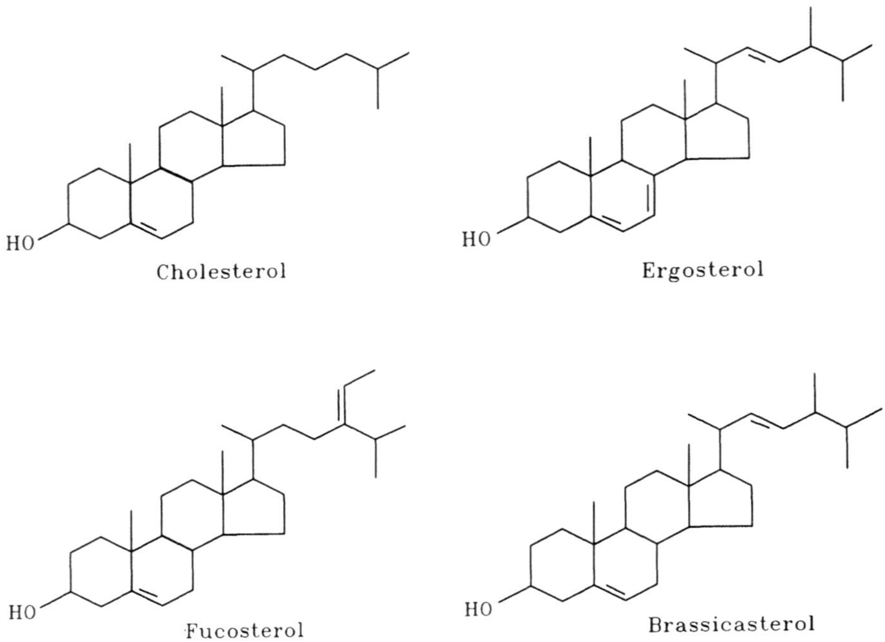

FIGURE 2. Some sterols found in fungi.

no ergosterol [29]. Interestingly, the two former genera belong to the same subclass as the yeast *S. cerevisiae*, i.e., Hemiascomycetes. Rust fungi (Uredinales) are also unique with respect to sterols in that they do not produce ergosterol but instead they have $C_{29}\,\Delta^{7}$ and/or $C_{29}\,\Delta^{7,24(28)}$ as major sterols [8,27].

C. Extracellular Glycolipids

There are a variety of extracellular glycolipids produced by fungi. Species of the yeast genus *Torulopsis* and species of the order Ustilaginales are the most thoroughly studied in this regard (Table 4) [8,30]. The most well-known group of such lipids are the sophorosides that consist of the disaccharide sophorose and long-chain fatty acids with a hydroxyl function at the terminal or penultimate carbon (Fig. 1). The carbonyl function of the fatty acid may form a lactone with the C-4 hydroxyl group of the disaccharide. Sophorosides from *Torulopsis* species have a density greater than that of water, and are composed of six to nine ionic and nonionic components. More recently, Asmer et al. [31] identified the 1',4''-lactone diacetate as the major sophorolipid (62%) of *T. bombicola* 22214, cultured in medium containing glucose and oleic acid, with 17-hydroxyoctadecanoic acid as the fatty acid component. With oleic acid as the only carbon source, the fatty acid moiety of the major lactone sophorolipid had a double bond.

Another class of extracellular glycolipids is the so-called ustilagic acids produced by *Ustilago zeae* PRL-119. These lipids contain 15,16-dihydroxy- or 2,15,16-trihydroxyhexadecanoic acid, and short-chain fatty

TABLE 4. Extracellular Glycolipids of Fungi

Glycolipid	Fungi	Ref.
Polyesters (mannitol, arabinol, xylitol linked to 3-D-hydroxypalmitic and 3-D-hydroxystearic acids)	*Rhodotorula graminis*, *R. glutinis*	22
Other polyol esters	*Saccharomyces cerevisiae*	86
	Agaricus bisporus	87
Acylated sugars		
Acyl glucose	*A. bisporus*	87
Acyl trehalose	*Pullularia pullulans*	88
	Claviceps purpurea	89
4-*o*-(2,3,4,6-tetracyl-*o*-manopyranosyl)-D-erythritol (C_{12} to C_{18} fatty acids)	*Ustilago maydis*	90
Ustilagic acids (see text, Section II.C, (Fig. 3)	*Ustilago zeae*, *U. nuda*	33
Sophorosides (see Section II.C, Fig. 1)	*Torulopsis* sp.	30
	Candida bogoriensis	18

acids glycosidically linked to cellobiose at the 2' and 6 positions (Fig. 3) [32,33]. Other acylated sugars and polyols such as 4-*o*-(2,3,4,6-tetracyl-*o*-mannopyranosyl)-D-erythritol have been identified as fungal products, but they do not appear to have attracted attention from the biotechnological point of view (see Table 4, Section IV.E, and Ref. 8 for review).

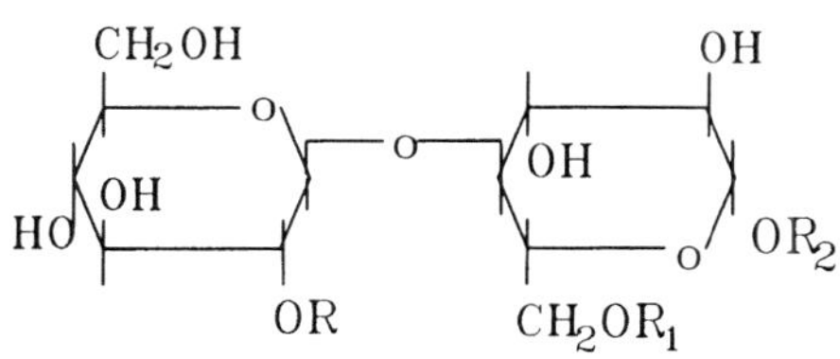

USTILAGIC ACIDS

R = Acetyl, L-3-hydroxy hexanoic, L-3-octanoic or hexanoic acids

R = 15,16-dihydroxy hexadecanoic, 2,15,16-trihydroxy, hexadecanoic, or decanoic acids

FIGURE 3. Ustilagic acids.

III. FUNGAL GROWTH AND LIPID PRODUCTION

As with bacteria, batch fungal cultures progress through identifiable stages with time. Cells may experience a lag phase immediately after inoculation of the medium where essentially no growth occurs. This is followed by a logarithmic or balanced growth phase limited only by the genetic potential of the species under the existing culture conditions, and then a linear phase where the growth rate is limited by depletion of a nutrient(s) or perhaps the accumulation of a growth-limiting substance(s). This is followed by the stationary phase where there is no net cell proliferation due to exhaustion of an essential nutrient and/or buildup of inhibitory substances (see Fig. 4).

De novo synthesis of fatty acids constitutes the bulk of lipid production, but fatty acids per se generally do not accumulate in cells. Instead, products of the fatty acid synthetase, which is mainly the CoA derivative of $C_{16:0}$, may be elongated, desaturated, and/or functionalized (e.g., hydroxylation) and derivatized to form the various acyl lipids. Some of the acyl lipids found in fungi include triacylglycerols, phospholipids, sphingolipids, and steryl esters, as well as the extracellular glycolipids noted above. The relative abundance of these lipids depends on several interacting factors, such as (a) the species, (b) cultivation conditions including temperature, pH, and aeration, (c) stage of culture development, (d) nutrition, and (e) function of the specific lipid. Phospholipids and triacylglycerols are typically the major lipids in fungi, the former being found mainly in membranes and the latter regarded as storage lipid. The amount of phospholipid formed is closely linked to membrane formation and therefore would be expected to be under rigid control and associated with growth. On the other hand, triacylglycerol formation might be generally viewed as a reflection of fatty acids produced beyond those required for phospholipid synthesis, and thus membrane formation and growth. During the stationary phase, the level of triacylglycerols may either remain constant, be reduced, or in some species accumulate. Oleaginous fungi may produce up to 70% of their dry weight as lipid. Lipid production by oleaginous yeasts is a two-stage process beginning with biomass buildup followed by lipid accumulation that is initiated by the depletion of nitrogen from the medium [4]. A representative example of this is *Rhodotorula glutinis*, as shown in Fig. 4. It is well known that a high carbon/nitrogen (C/N) ratio in the culture medium favors fat (triacylglycerol) production, although limitation of nutrients other than nitrogen may have the same result. The example in Fig. 4 illustrates that dry matter production occurring during the stationary phase in the absence of nitrogen (or low nitrogen) in the medium is due to the assimilation of carbon substrate from the medium into lipid.

The biochemical basis for lipid accumulation by oleaginous yeasts has been determined by the elegant work of Ratledge and his associates (see Ref. 5 and references cited therein). These yeasts possess cytosolic ATP:citrate lyase and an AMP-dependent isocitrate dehydrogenase which catalyzes the oxidative decarboxylation of isocitrate to α-ketoglutarate and carbon dioxide in the TCA cycle. Depletion of nitrogen from the medium results in decreased intracellular AMP, which in turn leads to a decrease in the activity of AMP-dependent isocitrate dehydrogenase. Isocitrate and citrate thus accumulate, and the latter is transported from the mitochondria into the cytoplasm where ATP:citrate lyase catalyzes its cleavage to acetyl-CoA and oxaloacetate. Acetyl-CoA is then available for

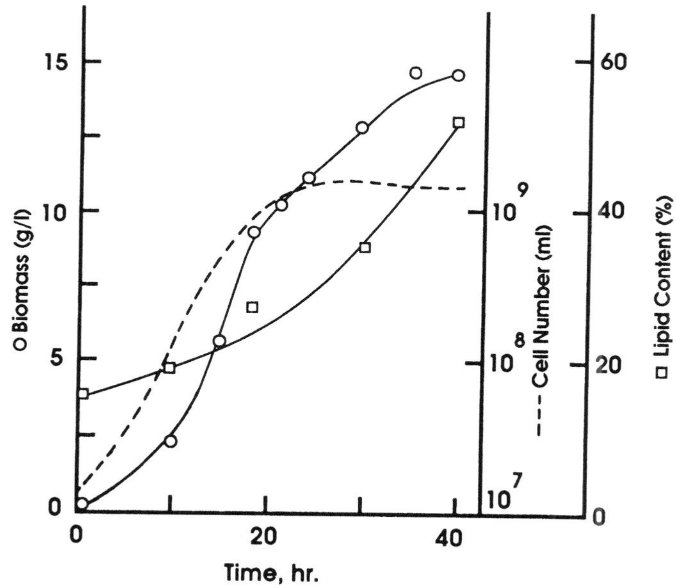

FIGURE 4. Lipid production by *Rhodotorula glutinis* in nitrogen-limited batch culture. (Redrawn from Ref. 34.)

de novo fatty acid biosynthesis in the cytosol. Thus, oleaginous fungi continue to assimilate the carbon substrate into fatty acids that are, in turn, used to form triacylglycerols, which is the basis for "fat" production.

Pan and Rhee [34] reported that lipid accumulation in *R. glutinis* is controlled by kinetic factors. For example, the maximum specific growth rate during the growing phase of nitrogen-limited batch cultures was 0.26/h when calculated on both a dry weight and cell number basis, and 0.026/h and 0.0/h during the lipid-accumulating phase, respectively. *Cryptococcus albidus* (formally *C. terricolus*) is an unusual oleaginous yeast in that it can accumulate lipid under both nitrogen-deficient and nitrogen-sufficient conditions [35] which Pan and Rhee [34] attributed to the low specific growth rate. In this yeast, for example, 34% lipid was produced under nitrogen-limited batch culture conditions with yield coefficients (see below) of 0.15 and 0.11 g lipid/g glucose for nitrogen-limited and nitrogen excess conditions, respectively. Other factors favoring lipid accumulation in this yeast were (a) the use of inorganic nitrogen rather than organic sources, (b) the use of glucose, maltose, or mannitol as carbon sources, (c) cultivation temperature of 20°C instead of 25 or 30°C, and (d) citrate stimulation of lipid production.

Lipid accumulation in oleaginous fungi can also be achieved using continuous culture techniques with medium having a high C/N ratio (e.g., 50:1) and growth rate of about 30% of maximum [5]. Lipid production may be increased by *n*-alkanes such as hexadecane as the carbon source, which is attributed to a slower growth rate of the fungus than with glucose as the source and that lipid (fatty acid) synthesis is not controlled by growth [36]. Ykema et al. [37] developed a mathematical model for lipid production

TABLE 5. Examples of Oleaginous Yeasts and Filamentous Fungi Capable of Producing Very High Lipid Contents

Fungal species	Lipid content (% dry weight)	Ref.
Yeasts		
Candida curvata	58	91
Cryptococcus albidus	63	92
Geotrichum candidum	50	93
Rhodotorula gracilis	60	94
Trichosporon pullulans	65	92
Endomycopsis vernalis	65	95
Filamentous fungi		
Mortierella vinacea	66	96
Pythium ultimum	48	49
Gibberella fujikuroi	48	97
Penicillium spinulosum	64	98
P. lilacinum	54	99
Epicoccum nigrum KSC	53	100

by oleaginous yeasts in continuous cultures based on the C/N ratio of the growth medium such that optimum fermentor operation conditions can be predicted and costs of microbial lipid minimized.

Oleaginous fungi have been listed by several authors [4,8,38]. Overall, oleaginous fungi can accumulate from about 40 to 70% lipid with some of the most prolific lipid-producing yeasts being *Cryptococcus albidus*, *Endomycopsis vernalis*, *Rhodotorula graminis*, and *Trichosporon pullulans*, all of which are capable of producing 65% or more lipid. Some of the high-fat-producing filamentous fungi are *Mortierella isabellina*, *M. vinacea*, *Mucor cincinelloides*, and *Penicillium spinulosum* at 65% or more lipid (Table 5).

The efficiency of production (yield coefficient) is based on grams lipid produced per gram carbon substrate utilized. Generally, oleaginous yeasts seem to be more efficient fat producers than the high-fat-producing filamentous species in that up to 22% conversion of glucose has been achieved with yeasts [5].

Phospholipids typically compose 1 to 2% of fungal biomass on a dry weight basis, but higher and lower values have been reported. The principal phospholipids of essentially all fungi are phosphatidylcholine (PC) and phosphatidylethanolamine (PE), which together may represent as much as 85% of the total phosphatides (Fig. 5). Other phospholipids typically found in fungi are phosphatidylserine (PS) and phosphatidylinositol (PI), which generally occur in approximately equal proportions in filamentous fungi, but PI is particularly high in *Saccharomyces* species. Phosphatidylglycerol (PG), diphosphatidylglycerol (cardiolipin, CL), and phosphatidic acid (PA) are also found in fungi (Fig. 5) [8].

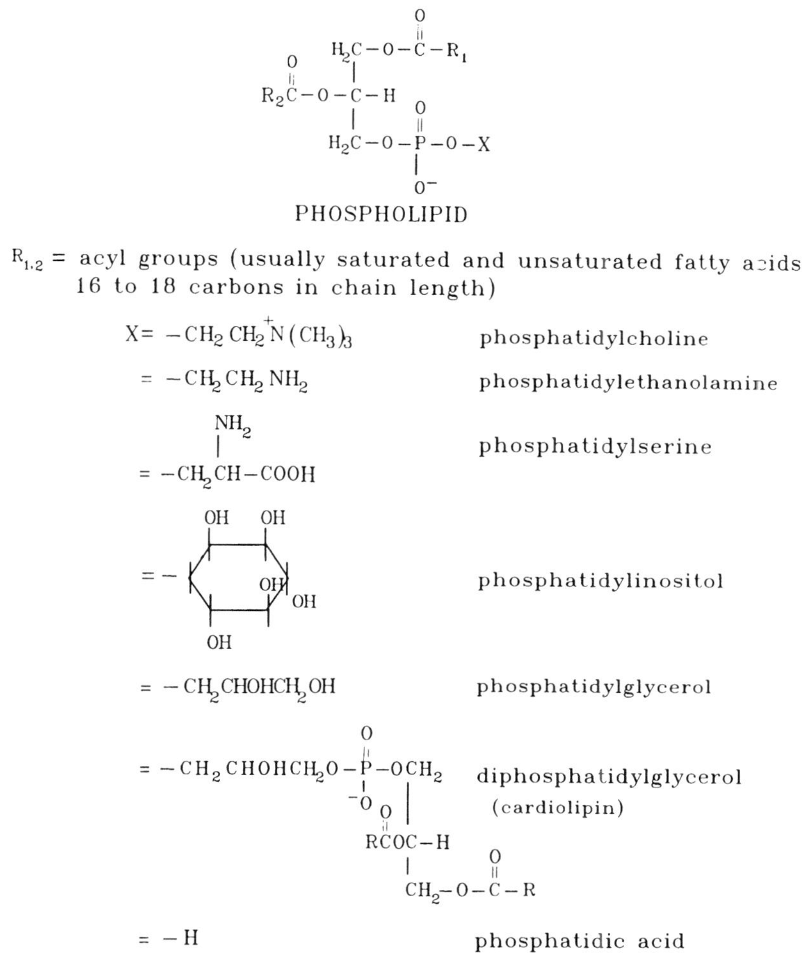

FIGURE 5. Phospholipids of fungi.

IV. PRODUCTION OF SPECIFIC LIPIDS

Studies on fungal lipids have led to the identification of species and strains that can produce specific lipids, having biological activity or surfactant properties, at high yield and cost-efficient production potential. Particular attention has been directed to polyenoic fatty acids produced by some of the "lower" fungi. For review, the following is an outline of polyunsaturated fatty acid biosynthesis: The product of de novo fatty acid biosynthesis is typically $C_{16:0}$. This saturated fatty acid is elongated to $C_{18:0}$ which in turn is desaturated to oleic acid, ($C_{18:1\Delta^9}$) which is desaturated, and the subsequent products go through a series of sequential two-carbon-unit elongations and desaturations to give the array of polyunsaturated fatty acids (PUFA) having 20 and 22 carbon chain lengths and three to six double bonds found in nature (Fig. 6). These fatty acids

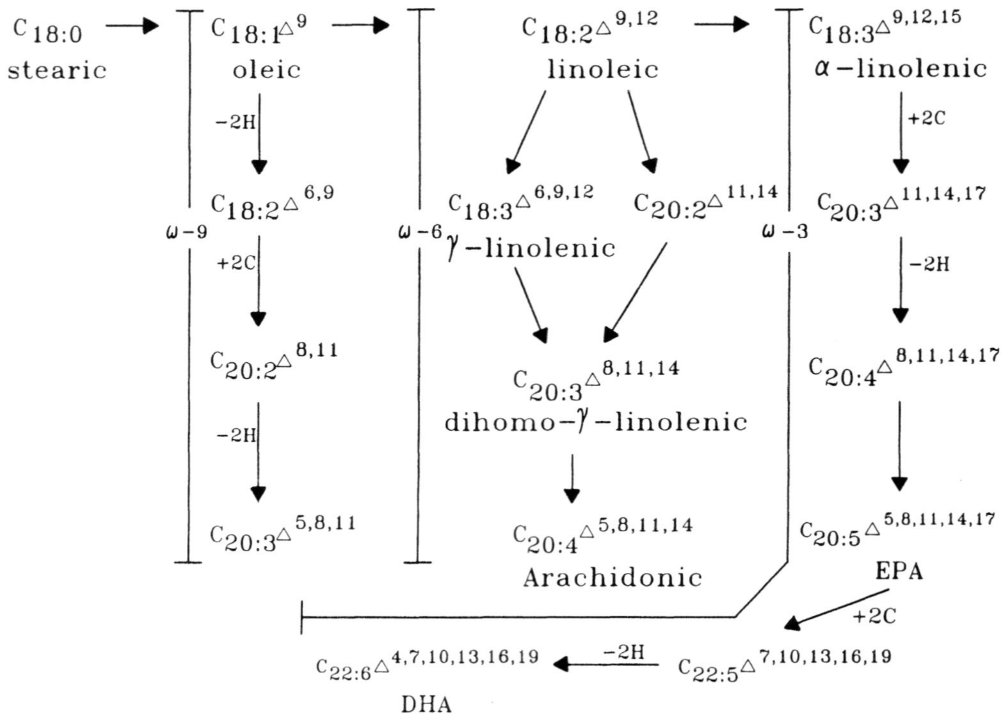

FIGURE 6. Pathways of polyunsaturated fatty acid biosynthesis.

are grouped into three series according to the location of the double bond nearest the terminal (ω or n) carbon atom, the ω9 (n9), ω6 (n6), and ω3 (n3) series as shown in Fig. 6. Generally, animals desaturate fatty acids in the carboxyl direction of the molecule and higher plants desaturate in the direction of the terminal carbon atom. Plants carry out the desaturation sequence $\Delta^{9} \rightarrow \Delta^{9,12} \rightarrow \Delta^{9,12,15}$, and from endogenous or dietary fatty acids of vegetable origin, animals carry out the transformations outlined in the ω9 and ω6 series (Fig. 6). Generally, the "higher" fungi carry out the desaturation sequence described for plants, but the "lower" fungi possess the pathway for arachidonic acid biosynthesis (ω6) and many can produce eicosapentaenoic acid via the ω3 pathway. Thus, many fungi possess PUFA biosynthetic pathways characteristic of both plants and animals. Many algae and protozoans also possess both the ω3 and ω6 pathways for PUFA synthesis through docosahexaenoic acid.

A. γ-Linolenic Acid (GLA)

This ω6 fatty acid is of pharmacological interest because it is a precursor to prostaglandins of the PGE_1 series. GLA is currently obtained commercially from the seed oils of *Oenothera* (evening primrose) and *Ribes* species. It has been known for many years that members of the Mucorales produce GLA [11], but only with the more widespread use of high-resolution glass capillary columns in gas chromatography has it become apparent that most

Zygomycetes and phylogenetically less advanced fungi produce mixtures of the α and γ isomers of $C_{18:3}$, with the latter predominating in most such species [9]. There are currently two commercial fermentation processes for the production of GLA, one using *Mortierella* in Japan and the other *Mucor javanicus* in the United Kingdom. In liquid medium, 80-156 g dry biomass/liter containing 37-42% lipid, for a yield of 30-83 g/liter, containing 4-11% GLA, can be achieved with several strains of *Mortierella* [39]. Total oil production and the GLA content of *M. javanicus* is roughly comparable to that of the seed oils with relatively little of the corresponding α isomer.

As a component of the acyl lipids, GLA remains in the cells. Using *M. ambiguus* (IFO 6742), Fukuda and Morikawa [40] showed that (a) this fungus may be suitable for immobilization, (b) certain nonionic surfactants promote excretion of GLA from the cells, and (c) improved yields and productivity can be achieved by repeated batch cultivation (also see Section IV.C).

B. Arachidonic Acid (AA)

This ω6 fatty acid is also of pharmacological interest because it is a precursor to prostaglandins of the PGE_2 series. AA is produced by some of the Zygomycetes and most of the phylogenetically less advanced fungi. Yamada et al. [41] screened several *Mortierella* species for AA production and found *M. elongata* IS-5 to be the most proficient of those surveyed, first showing 22 mg AA/g dry weight and then 80 mg/g dry weight [42]. *Penicillium cyaneum* IAM 7302 is another fungus studied in this regard, but it produces only 0.2 mg AA/g dry weight when grown on alkanes [43]. Radwan and Soliman [44] recently reported that several fungi, including species of the genera *Aspergillus*, *Pencillium*, *Fusarium*, *Paecilomyces*, *Trichoderma*, *Cladosporium*, and *Stachybotrys*, which are not typically considered to be AA producers, produce lipids enriched in AA when cultured in medium containing fatty acids as the sole carbon source. Otherwise, there appears to have been relatively little effort to find high AA-producing species and strains, or research on methods of enhancing AA production. See below for AA production by *Pythium ultimum* and *Mortierella* (Section IV.D).

C. Dihomo-γ-Linolenic Acid (all *cis*-8,11,14-eicosatrienoic acid, DHGLA)

This fatty acid also possesses unique biological activities and is a precursor of certain prostaglandins and thromboxanes. It was detected as a minor component of several fungi [45], but more recently a high-DHGLA-producing isolate of *Mortierella alpina* (IS-4) was identified [46]. When cultivated in a 50-liter fermentor containing sucrose, yeast extract, and a nonoil fraction from the fungus, 107 mg DHGLA/g dry weight (2.17 g/liter of culture broth), which was 23.1% of the total fatty acids, was produced. Under these conditions, 53.5 mg AA and 4.1 mg GLA/g dry mycelia were also produced.

D. Polyenoic ω3 Fatty Acids

In recent years, there has been considerable interest in the possible health benefits of eating certain fish, or oils therefrom, because of the presence of long-chain polyenoic ω3 fatty acids, mainly eicosapentaenoic acid (EPA) and docosahexaenoic acid (DHA). As noted above, these fatty acids have been reported as fungal products, but they appear to be restricted to certain less advanced fungi (see above), and there appears to have been no serious attempt to scale up production of these fatty acids using fungi.

Mortierella alpina produces 26.6 mg EPA/g dry weight of cells, and its production is favored by low temperatures [41,42]. Recently, Shimizu et al. [47] described how AA-producing fungi of the genus *Mortierella* can convert oil containing α-linolenic acid to one containing EPA. A medium containing the oil, glucose, and yeast extract, with cultivation temperature of 28°C, favored conversion, with linseed oil being most suitable for EPA production. Of 20 strains and *Mortierella* species tested, the best conversion of α-$C_{18:3}$ to EPA was by *M. alpina* 20-17, which converted 5.1% of the triene to EPA, with EPA production reaching 1.35 g/liter of culture broth (41.5 mg/g dry mycelia). The lipid also contained 12.3% AA. Gellerman and Schlenk [48] showed that EPA is formed by desaturation of AA in the Oomycete *Saprolegnia parasitica*.

Pythium ultimum isolates may produce up to 48% lipid [49], and it was found in this laboratory that an isolate of this species produces relatively high amounts of AA (16%) and EPA (11%) in laboratory batch culture, but low amounts of GLA (1%). Cultivation at 25°C favored AA production.

E. Extracellular Glycolipids—Biosurfactants

There is an increasing interest in the development and use of surfactants, including those of biological origin. Recently, the surfactant properties of the sophorosides produced by *T. bombicola* were described [50,51]. *Torulopsis bombicola* ATCC 22214 produces 70 g/liter of the extracellular glycolipid, or 35% of the weight of the substrate, when cultured on a medium containing a carbohydrate and vegetable oil as carbon substrates [52]. More recently, Asmer et al. [31] achieved 90% yield of sophorolipid by this yeast. The production of these extracellular lipids can be improved by the stepwise addition of fatty acid esters or alkanes to the medium. Sophoroside production as described was more economical than any other microbial surfactant [52]. Most of the glycolipid was produced in the late exponential (linear) phase of growth (Fig. 7). More recently, it was shown that *T. petrophilum* ATCC 20225, originally isolated as an organism capable of degrading crude oil fractions, also produces sophoroside glycolipids but at only 10 g glycolipid per liter, which had no ability to emulsify, with 10% glucose, 0.5% sunflower oil, and 0.5% yeast extract as the medium. This glycolipid was the same as that produced by *T. bombicola* [53]. This species also produces a protein emulsifier. When grown on water-insoluble vegetable oil in the medium (two-phase system), glycolipid production was favored, and when cultured in a single-phase medium, with no requirement for an emulsifier, a potent extracellular protein emulsifier was formed. *T. apicola* IMET 43747 produces both water-soluble and crystalline glycolipids depending on the culture conditions, i.e., medium pH less than 2 favored the water-soluble form and pH 3 favored production of large amounts of crystalline product [54]. Maximum yields were 0.46 g/g substrate when the yeast was cultured with glucose and vegetable oils as the carbon sources. Eighty percent of the glycolipid consisted of an unidentified nonionic component. (See Ref. 30 for a description of various culture conditions for extracellular glycolipid production.)

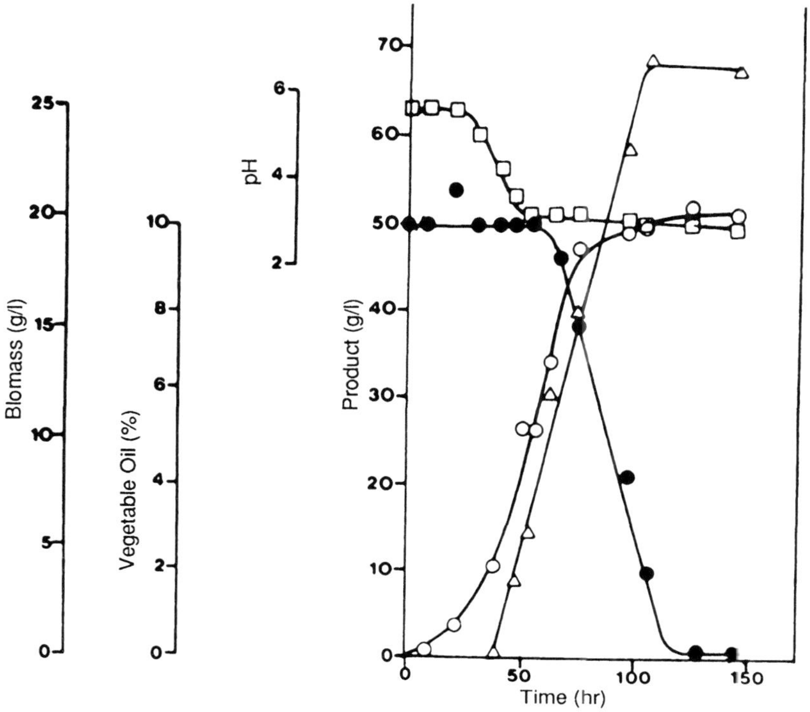

FIGURE 7. Biosurfactant production by *Torulopsis bombicola* cultured in medium containing 10% glucose and 9.5% sunflower oil. (Redrawn from Ref. 52.)

Little research has been conducted on the production of extracellular glycolipids. However, Frautz et al. [55] showed that *U. maydis* ATCC 14826 cells produce cellobiose lipids under nitrogen limitation and as resting cells using coconut oil to initiate glycolipid formation. These lipids contained hexanoic, dodecanoic, myristic, and 15,16-dihydroxyhexadecanoic acids.

V. LIPASES

Extracellular lipases are glycoproteins produced by a variety of fungi including certain species of *Aspergillus*, *Mucor*, *Rhizopus*, *Penicillium*, *Geotrichum* and species of the yeast genera *Torulopsis* and *Candida*. Generally, these lipases catalyze the degradation of triacylglycerols to monoacylglycerols as shown in Fig. 8. A variety of commercial applications has been found for microbial lipases, e.g., degradation and modification of fats, synthesis of glycerides, and formation of flavor esters [60,61]. Lipases from *Mucor*, *Rhizopus*, and *Candida* are currently used commercially [56] (see Section V.D). Fungal lipases have molecular weights from 27,000 to 44,000 [57-59], pH optima in the neutral or slightly acidic range, and optimum temperature for activity of 30-45°C. The characteristics of fungal lipases have been reviewed by Antonian [60].

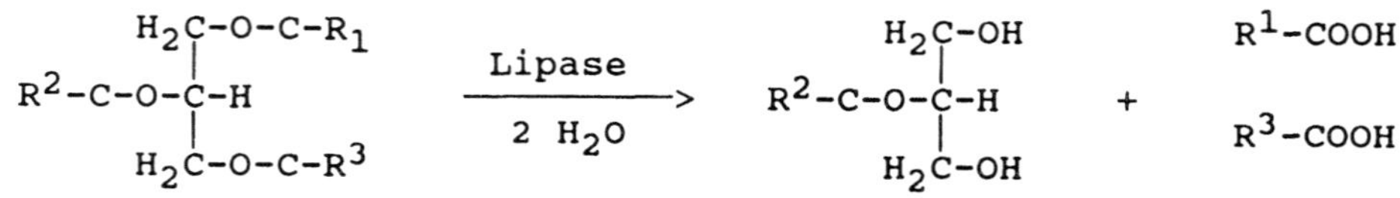

TRIACYLGLYCEROL MONOACYLGLYCEROL FATTY ACIDS

FIGURE 8. Reactions catalyzed by extracellular fungal lipases.

A. Factors Affecting Production

Recent studies showed that lipase production varies widely according to species, growth conditions, and stage of culture development. For example, maximum lipolytic activity in *Rhizopus microsporus* cultures was observed at the beginning of the stationary phase [61]. The optimal temperature for biomass production and lipase accumulation was 38°C. In *R. oligosporum*, greater lipase production occurred at 25°C [62], where lipase production was greater under stationary rather than shake culture conditions. Lipase production was not stimulated in this species by carbyhydrates, vegetable oils, protein or amino acids, but it was stimulated by Tween surfactants. On the other hand, olive oil increased production of lipases in *Beauveria bassiana* [63]. Lipase synthesis in *Candida rugosa* was also increased in the presence of fats, especially if fats and steroids were present simultaneously [64]. Extracellular lipase production in *Saccharomycopsis lipolytica* was increased by a protein fraction from soybean [65]. Studies by Akhtar et al. [66] indicated that olive oil as well as glucose induced maximum lipase production in *Mucor hiemalis*. Yet in other studies with *Aspergillus* and *Syncephalastrum* species, increased lipase production was observed when calcium and sodium citrate were added to growth medium [67,68].

B. Location of Lipases

Relatively little is known about the intracellular location of lipases in fungi. In *S. cerevisiae*, lipase activity is associated with the plasma membrane [69], whereas it is associated with the cell wall in *Candida lipolytica*, *Saccharomycopsis lipolytica*, and *Geotrichum candidum* [57,70]. In *C. paralipolytica*, it was shown by electron microscopy that lipase is formed on the outer surface of the cytoplasmic membrane, and then transported across the cytoplasmic space and cell wall into the culture medium [71]. Lipase is located in the periplasmic space of *Oospora laetis* [72]. In *G. candidum*, lipases are found in the periplasmic space, but also around the cell wall and membranes [73], and are released from the cell after a period of induced synthesis.

C. Reaction Specificity

Lipases show both substrate and positional specificity, with most microbial lipases being 1,3-specific [58,74]. The rate of hydrolysis depends on the fatty acid. For example, triacylglycerols with short-chain saturated or unsaturated fatty acids are hydrolyzed more rapidly than those composed of long-chain fatty acids having two or three double bonds. Table 6 gives examples of the positional and fatty acid selectivity of lipases from several fungi.

TABLE 6. Positional and Fatty Acid Selectivities of Some Fungal Lipases

Organism	Positional selectivity	Fatty acid selectivity	Ref.
Neurospora crassa	1,3	16:0; 18:0	59
Humicola languinosa	1,3	12:0	84
Candida rugosa	Nonspecific	18 (cis 9)	101
Aspergillus niger	Nonspecific	10:0;12:0	101
Geotrichum candida	Nonspecific	18 (cis 9)	101
Mucor miehei	1,3	12:0	101
Rhizopus arrhizus	1,3	16:0; 18:0	102

D. Industrial Applications

A variety of industrial applications have been found for fungal lipases, including the following:

1. Lipases obtained from fungi such as *Aspergillus*, *Rhizopus*, *Mucor*, *Geotrichum*, *Penicillium*, and other species have been used to catalyze the synthesis of naturally occurring flavor esters from various substrates [75,76]. For example, immobilized lipases can be used in the synthesis of isoamylacetate and ethyl butyrate [77,78]. Lipases have also been used to catalyze the synthesis of terpene alcohol esters [79].
2. The utility of fat from various sources can be increased by lipase-catalyzed modification of fat composition and properties [80].
3. Lipolytic systems are used in the dairy industry to free fatty acids that alter flavor intensity of different types of cheese [81].
4. Modification of butterfat emulsions with lipase can impart flavor character to food products where such emulsions are used [82,83].
5. Lipase is one of the enzymes used to supplement animal feeds for stimulating fat metabolism in young and stressed animals [81].
6. Lipases from *Mumicola languinosa* are used to hydrolyze fat and beef tallow [84].

VI. CONCLUSIONS

There are numerous fungal species of both the filamentous and yeast types with high-fat-producing potential, but the number of oleaginous fungi seems relatively small considering the total number of known fungal species. Despite efforts by investigators over the past century, fungal fat or oil has not appeared on the market as an alternative to vegetable oils mainly because of production economics, i.e., oils from plant sources such as soybean can be produced (plant cultivation, seed harvest, oil processing) at lower costs and on a much larger economic scale than is possible with fermentation processes involving fungi. The major economic barrier to producing fungal oil on a commercial scale is finding a suitable and inexpensive carbon source, such as waste products generated by industry, and developing the fermentation technology for cost-effective conversion of such waste products to fungal oil. Some of the oleaginous fungi are capable of utilizing hydrocarbons and a variety of carbohydrates for lipid production.

Maximizing fat production in fungi is a two-phase process, which first involves biomass production that requires balanced nutrition, and then involves assimilation of the carbon source into oil which is favored by a high C/N ratio. In the oleaginous species studied in this regard, the overall partitioning of carbon into protein and lipid is controlled by energy charge, and specifically whether isocitrate dehydrogenase requires AMP as a positive effector. That is, oleaginous yeasts have AMP-dependent isocitrate dehydrogenase and, if reduced in activity due to nitrogen depletion, carbon is preferentially partitioned into triacylglycerols via citrate.

Despite the fact that fungal oil has not developed into a commercially successful product, fungi offer considerable promise for the production of high-value specialty lipids such as those with specific biological activity, or those that are precursors to such compounds. Some of these lipids include γ-linolenic acid, arachidonic acid, dihomo-γ-linolenic acid, and polyenoic $\omega 3$ fatty acids such as EPA. With the production of only γ-linolenic acid beginning to reach the commercial stage, increasing attention is being given to identifying species and strains and developing fermentation procedures for the production of these specialty lipids of fungal origin. Considering other possible sources of these lipids, specialty lipid production through fermentation may be more economically feasible with fungi. The production of such lipids can be enhanced by adding specific lipid substrates to the growth medium. Another area with potential in biotechnology is the production of fungal extracellular glycolipids with surfactant properties, such as the sophorosides and bioemulsifiers. The amounts and properties of these substances can be partially controlled by the presence and nature of lipids added to the culture media as second carbon substrates. Furthermore, *S. cerevisiae* is a well-known source of ergosterol which is converted to vitamin D_2 upon UV irradiation. Fungi are also sources of a variety of other sterols that can be transformed to steroids with biological activities. Some of these transformations may be carried out by fungal fermentations.

Lipases are a group of the most commercially successful products of fungal origin from a biotechnology point of view. Certain fungi are sources of lipases that are most amenable to commercial production, and numerous applications are being found for these enzymes. More recently, the lipase gene from *Rhizomucor miehei* was cloned and sequenced [85]. It appears that the techniques of molecular biology are being applied to better understanding the lipase mode of action which could ultimately lead to an ability to synthesize semisynthetic lipases with improved stability and desired substrate specificity [60].

ACKNOWLEDGMENTS

This chapter was written with the support of the Alabama Agricultural Experiment Station (Hatch Project ALA-00773). We are grateful to Ms. Kathryn McCullough for typing the manuscript. Alabama Agricultural Experiment Station publication No. 6-892111P.

REFERENCES

1. Kates, M. *Laboratory Techniques in Biochemistry and Molecular Biology*, Elsevier, Amsterdam, 1986, pp. 1-464.
2. Woodbine, M. Microbial fat: microorganisms as potential fat producers. In *Progress in Industrial Microbiology*, Vol. 1 (D. J. D. Hockenhull, ed.), Heywood and Co., London, 1959, pp. 179-245.
3. Schweizer, E. Genetic control of fatty acid biosynthesis in yeasts and other lower fungi. *JAOCS, 64*:1250, 1987.
4. Ratledge, C. Lipids. In *Biotechnology: A Comprehensive Treatise*, Vol. 4 (H. Pape and H. T. Rehm, ed.), 1986, pp. 185-213.
5. Ratledge, C. Microbial oils and fats: an assessment of their commercial potential. *Prog. Ind. Microbiol., 16*:119-206, 1982.
6. Ratledge, C. Lipids in biotechnology: a wonderland for the microbial physiologist. *JAOCS, 64*:1647-1656, 1987.
7. Rattray, J. B. M. Biotechnology and the fats and oils industry: an overview. *JAOCS, 61*:1701-1712, 1984.
8. Weete, J. D. *Lipid Biochemistry of Fungi and Other Organisms*, Plenum Press, New York, 1980, pp. 1-388.
9. Weete, J. D., Fuller, M., Huang, M. Q., and Gandhi, S. Fatty acids and sterols of selected Hypochytriomycetes and Chytridiomycetes. *Exp. Mycol., 13*:183-195, 1989.
10. Hunter, K., and Rose, A. H. Yeast lipids and membranes. In *The Yeasts*, Vol. 2 (A. H. Rose, and J. S. Harrison, eds.), Academic Press, New York, 1971, pp. 8-270.
11. Shaw, R. The occurrence of γ-linolenic acid in fungi. *Biochim. Biophys. Acta, 98*:230-237, 1965.
12. Manchoa, M. S., and Campbell, C. D. Effect of growth temperature on the fatty acid composition of Thamnidium elegans link. *Can. J. Microbiol., 24*:670-674, 1978.
13. Tyrrell, D. The fatty acid compositions of 17 Entomopthora isolates. *Can. J. Microbiol., 13*:755-760, 1967.
14. Ellenbogen, B. B., Aaronson, S., Goldstein, S., and Belsky, M. Polyunsaturated fatty acids of aquatic fungi: possible phylogenetic significance. *Comp. Biochem. Physiol., 29*:805-811, 1969.
15. Nurminen, R., and Soumalaimen, H. Occurrence of long-chain fatty acids and glycolipids in the cell envelope fractions of baker's yeast. *Biochem. J., 125*:963-969, 1971.
16. Morris, L. T. Fatty acid composition of Claviceps species. Occurrence of (+)-threo-9,10-dihydroxystearic acid. *Lipids, 3*:260-261, 1968.
17. Bowers, W. S., Evans, P. H., and Katayama, M. Synthetic optimization of laetisaric acid for fungicidal activity. *J. Agr. Food Chem., 35*:1043-1046, 1987.
18. Tulloch, A. P., Spencer, J. F. T., and Drenema, M. H. A new hydroxy fatty acid sophoroside from Candida bogoriensis. *Can. J. Chem., 46*:345-348, 1968.
19. Stodola, F. H., Deinema, M. H., and Spencer, J. F. T. Extracellular lipids of yeasts. *Bacteriol. Rev., 31*:194-213, 1976.
20. Kurtzman, C. P., Vesonder, R. F., and Smiley, M. H. Formation of extracellular 3-D-hydroxy palmitic acid by Saccharomycopsis malanga. *Mycologia, 66*:580-587, 1974.
21. Stodola, F. H., Vesonder, R. F., and Wickerham, L. H. 8,9,13-Triacetoxydocosanoic acid, and extracellular lipid produced by a yeast. *Biochemistry, 4*:1390-1394, 1965.

22. Tulloch, A. P., and Spencer, J. E. T. Extracellular glycolipids of Rhodotorula species. The isolation and synthesis of 3-D-hydroxypalmitic and 3-D-hydroxystearic acids. *Can. J. Chem.*, *42*:830-835. 1964.
23. Gorin, P. A. J., Spencer, J. F. T., and Tulloch, A. P. Hydroxy fatty acid glycosides of sophorose from Torulopsis magnoliae. *Can. J. Chem.*, *39*:846-855, 1961.
24. Carmack, C. L., Weete, J. D., and Kelley, W. D. Lipids of Cronartium fusiforme aeciospores. *Physiol. Pl. Pathol.*, *8*:43-49, 1976.
25. Weete, J. D., and Kelley, W. D. Mycelial lipids as an aid in identifying fungi in culture—Cronartium fusiforme. *Lipids*, *12*:398-401, 1977.
26. Weete, J. D. Sterols in the fungi. Their distribution and biosynthesis. *Phytochemistry*, *12*:1843-1864, 1973.
27. Weete, J. D. Structure and function of sterols in fungi. *Adv. Lipid Res.*, *23*:115-167, 1989.
28. Parks, L. W. Metabolism of sterols in yeast. *CRC Crit. Rev.*, *6*:301-341, 1978.
29. Weete, J. D., Kulifaj, M., Montant, C. Nes. W. R., and Sancholle, M. Distribution of sterols in fungi. II. Brassicasterol in Tuber and Terfezia species. *Can. J. Microbiol.*, *31*:1127-1130, 1985.
30. Spencer, J. F. T., Spencer, D. M., and Tulloch, A. P. Extracellular glycolipids of yeasts. In *Economic Microbiology*, Vol. 3 (A. H. Rose, ed.), Academic Press, London, 1979, pp. 523-540.
31. Asmer, H-J., Lang, S., Wagner, F., and Wray, V. Microbial production, structure elucidation and bioconversion of sophorose lipids. *JAOCS*, *65*:1460-1466, 1988.
32. Lemieux, R. U., Haskins, R. H., Thorn, I. A., and Brice, C. Biochemistry of Ustilaginales II. Isolation and partial characterization of ustilagic acid. *Can. J. Chem.*, *29*:409-414, 1951.
33. Bhattacharjee, S. S., Haskins, R. H., and Gorin, P. A. J. Location of acyl groups on two partly acylated glycolipids from strains of Ustilago (smut fungi). *Carbohydr. Res.*, *13*:235-246, 1970.
34. Pan, J. G., and Rhee, J. S. Kinetic and energetic analysis of lipid accumulation in batch culture of Rhodotorula glutinis. *J. Ferment. Technol.*, *64*:557-560, 1986.
35. Hansson, L., and Dostalek, M. Influence of cultivation conditions on lipid production by Cryptococcus albidus. *Appl. Microbiol. Biotechnol.*, *24*:12-18, 1986.
36. Hall, M. J., and Ratledge, C. Lipid accumulation in an oleaginous yeast (Candida 107) growing on glucose under various conditions in a one- and two-stage continuous culture. *Appl. Environ. Microbiol.*, *33*:577-584, 1977.
37. Ykema, A., Verbree, E. C., van Verseveld, H. W., and Smit, H. Mathematical modelling of lipid production by oleaginous yeasts in continuous culture. *Antonie van Leeuwenhoek*, *52*:491-506, 1986.
38. Shifrin, M. In *Biotechnology for the Fats and Oil Industry* (C. Ratledge, P. S. S. Dawson and J. M. B. Rattray, eds.), JAOCS, Champaign, IL, 1984, pp. 145-161.
39. Suzuki, O. Production of γ-linolenic acid by fungi. *JAOCS*, *64*:1251, 1987.
40. Fukuda, H., and Morikawa, H. Enhancement of γ-linolenic acid production by Mucor ambiguus with non-ionic surfactants. *Appl. Microbiol. Biotechnol.*, *27*:15-20, 1987.
41. Yamada, H., Shimiza, S., and Shinmem, Y. Production of arachidonic acid by Mortierella elongata IS-5. *Agr. Biol. Chem.*, *51*:785-790, 1987.

42. Yamada, H., Shimizu, S., Shinmen, Y., Kawashima, ., and Akimoto, K. Production of arachidonic acid and eicosapentaenoic acid by microorganisms. *JAOCS, 64*:1254, 1987.
43. Iizuka, J., Ohtomo, R., and Yoshida, K. Production of arachidonic acid by a hydrocarbon-utilizing strain of Penicillium cyaneum. *Eur. J. Microbiol. Biotechnol., 7*:173-180, 1979.
44. Radwan, S. S., and Soliman, A. H. Arachidonic acid from fungi utilizing fatty acids with shorter chains as sole sources of carbon and energy. *J. Gen. Microbiol., 134*:387-393, 1988.
45. Shimizu, S., Kauashima, H., Shinmen, Y., Abimoto, K., and Yamada, H. Production of eicosapentaenoic acid by Mortierella fungi. *JAOCS, 65*: 1455-1459, 1988.
46. Shimizu, S., Akimoto, K., Kawashima, J. Shinmen, Y., and Yamada, H. Production of dihomo-γ-linoleic acid by Mortierella alpina IS-4. *JAOCS, 66*:237-241, 1989.
47. Shimizu, S. Kawashima, H., Akimota, K. Shinmen, Y., and Yamada, H. Microbial conversion of an oil containing α-linolenic acid to an oil containing eicosapentanoic acid. *JAOCS, 66*:342-347, 1989.
48. Gellerman, J. L., and Schlenk, H. Methyl-directed desaturation of arachidonic to eicosapentaenoic acid in the fungus, Saprolegnia parasitica. *Biochim. Biophys. Acta, 573*:23-30, 1979.
49. Bowman, R. D., and Mumma, R. O. The lipids of Pythium ultimum. *Biochim. Biophys. Acta, 144*:501-510, 1967.
50. Inoue, S., and Ito, S. Sophorolipids from Torulopsis bombicola as microbial surfactants in alkane fermentation. *Biotechnol. Lett., 4*:3-8, 1982.
51. Ito, S., and Inoue, S. Sophorolipids from Torulopsis bombicola. Possible relation to alkane uptake. *Appl. Environ. Microbiol., 43*:1278-1283, 1982.
52. Cooper, D. G., and Paddock, D. A. Production of biosurfactant from Torulopsis bombicola. *Appl. Environ. Microbiol., 47*:173-176, 1984.
53. Cooper, D. G., and D. A. Paddock. Torulopsis petrophilum and surface activity. *Appl. Environ. Microbiol., 46*:1426-1429, 1983.
54. Stuwer, O., Hommel, R., Haferburg, D., and Kleber, H. P. Production of crystalline surface-active glycolipids by a strain of Torulopsis apicola. *J. Biotechnol., 6*:259-269, 1987.
55. Frautz, B., Lang, S., and Wagner, F. Formation of cellobiose lipids by growing and resting cells of Ustilago maydis. *Biotechnol. Lett., 8*:757-762, 1986.
56. Crueger, W., and Crueger, A. Enzymes. In *Industrial Microbiology* (T. D. Brock, ed.), Sunderland, Mass., 1984, pp. 161-186.
57. Ota, Y., Gome, K., Kato, S., Sugiura, T., and Minoda, Y. Purification and some properties of cell-bound lipases from Saccharomycopsis lipolytica. *Agr. Biol. Chem., 46*:2885-2893, 1982.
58. Omar, I. C., Hayashi, M., and Nagai, S. Purification and some properties of a thermostable lipase from Mumicola lanuginosa No. 3. *Agr. Biol. Chem., 51*:3745, 1987.
59. Kundu, M., Basu, J., Guchhait, M., and Chakrabarti, P. Isolation and characterization of an extracellular lipase from the conidia of Neurospora crassa. *J. Gen. Microbiol., 133*:149-153, 1987.
60. Antonian, E. Recent advances in the purification, characterization, and structure determination of lipases. *Lipids, 23*:1101-1106, 1988.

61. Zubenko, T. F., Sultanova, I. G., and Lakirov, M. Z. Growth and lipolytic activity of Rhizopus microsporus cultured at different temperatures. *Mikrobiologiya, 53*:50-53, 1984.
62. Nahas, E. Control of lipase production by Rhizopus oligosporus under various growth conditions. *J. Gen. Microbiol., 134*:227-233, 1988.
63. Hegedus, D. D., and Khachatourians, G. G., Production of extracellular lipase by Beauveria bassiana. *Biotechnol. Lett., 10*:637-642, 1988.
64. Valero, F., Ayats, F., Santin, J. L., and Poch, M. Lipase production by Candida rugosa: fermentation behaviour. *Biotechnol. Lett., 10*:741-744, 1988.
65. Ota, Y., Morimoto, Y., Sugiura, T., and Minoda, Y. Soybean fraction increasing the extracellular lipase production by Saccharomycopsis lipolytica (Candida paralipolytica). *Agr. Biol. Chem., 42*:1937-1938, 1978.
66. Akhtar, M. W., Mirza, A. Q., and Chughati, M. I. D. Lipase induction in Mucor hiemalis. *Appl. Environ. Microbiol., 40*:257-263, 1980.
67. Chander, M., Batish, V. K., Sannabhadti, S. S., and Srinibasan, R. A. Factors affecting lipase production in Aspergillus wentii. *J. Food Sci., 45*:598-600 (1980).
68. Chopra, A. K., and Chander, H. Factors affecting lipase production in Synephalastrum racemosum. *J. Appl. Bacteriol., 54*:163-169, 1983.
69. Nurminen, T., and Suomalaimen, H. The lipolytic activities of the isolated cell envelope fractions of Baker's yeast. *Biochem. J., 118*:759-763, 1970.
70. Zvyagintseva, I. S., and Ruban, E. L. Lipase activity of protoplasts of yeast Candida lipolytica. *Mikrobiologiya, 41*:973-975, 1972.
71. Zvyagentseva, I. S., Dmitriev, V. V., Ruban, E. L., and Fikhte, B. A. Localization of extracellular lipase in Candida paralipolytica. *Mikrobiologiya, 49*:417-420, 1980.
72. Davranov, K. D. Alimdzhanova, M. I., and Bezborodov, A. M. Lipase activity of Oospora lactis protoplasts. *Mikrobiologiya, 49*:421-426, 1980.
73. Tsujisaka, Y., Iwai, M., Fukumoto, J., and Okamato, Y. Induced formation of lipase by Geotrichum candidum Link. *Agr. Biol. Chem., 37*:837-842, 1973.
74. Muderhwa, J. M., Ratomahenina, R., Pina, M., Graille, J., and Galzy, P. Purification and properties of the lipase from Candida deformans (Zach) Langeron and Guerra. *JAOCS, 62*:1031-1036, 1985.
75. Okumura, S., Iwai, M., and Tsujisaka, Y. Synthesis of various kinds of esters by four microbiol lipases. *Biochim. Biophys. Acta, 575*:156-165, 1979.
76. Langrand, G. Triantaphylides, C., and Barrati, J. Lipase catalyzed formation of flavor esters. *Biotechnol. Lett., 10*:549-554, 1988.
77. Gillies, B., Yamazaki, H., and Armstrong, D. W. Production of flavor esters by immobilized lipase. *Biotechnol. Lett., 9*:709-714, 1987.
78. Jensen, B. H., Galluzzo, D. R., and Jensen, R. G. Studies on free and immobilized lipases from Mucor miehei. *JAOCS, 65*:905-910, 1988.
79. Iwai, M., Okumura, S., and Tsujisaka, Y. Synthesis of terpene alcohol esters by lipases. *Agr. Biol. Chem., 44*:2731-2732.
80. Macrae, A. R. Lipase catalyzed interesterification of oils and fats. *JAOCS, 60*:291, 1983.
81. Peppler, H. J., and Reed, G. Enzymes in food and feed processing. In *Biotechnology, Vol. 7a. Enzyme Technology* (H. J. Rehm and G. Reed, eds.), VCH Publ., New York, 1987, pp. 547-603.

82. Nelson, J. H., Hensen, R. G., and Pitas, R. E. Pregastric esterase and other oral lipases: a review. *J. Dairy Sci.*, *60*:327-362, 1977.
83. Arnold, R. G., Shahani, K. M., and Pitas, R. E. Application of lipolytic enzymes to flavor development in dairy products. *J. Dairy Sci.*, *58*:1127-1143, 1975.
84. Omar, C. I., Nishio, N., and Nagai, S. Fat hydrolysis and esterification by a lipase from Mumicola lanuginosa. *Agr. Biol. Chem.*, *51*:2153-2159, 1987.
85. Boel, E., Huge-Jensen, B., Christensen, M., Thim, L., and Fiil, N. P. Rhizomucor miehei triglyceride lipase is synthesized as a precursor. *Lipids*, *23*:701-706, 1988.
86. Brennan, P. J., Flynn, M. P., and Griffin, P. F. S. Acyl glucoses in Escherichia coli, Saccharomyces cerevisiae and Agaricus bisporus. *FEBS Lett.*, *8*:322-324, 1970.
87. Brennan, P. J., Griffin, P. F. S., Losel, D. M., and Tyrrell. The lipids of fungi. In *Progress in the Chemistry of Fats and Other Lipids*, Vol. 14, Part 2 (R. T. Holman, ed.), Pergamon Press, New York, 1974, pp. 1-89.
88. Merdinger, E., Kohn, P., and McClain, R. C. Composition of lipids in extracts of Pullularia pullulans. *Can. J. Microbiol.*, *14*:1021-1027, 1968.
89. Cooke, R. C., and Mitchell, D. T. Carbohydrate physiology of sclerotia of Claviceps purpurea during dormancy and germination. *Trans. Br. Mycol. Soc.*, *54*:93-99, 1970.
90. Fluharity, A. L., and O'Brien, J. S. A mannose- and erythritol-containing glycolipid from Ustilago maydis. *Biochemistry*, *8*:2627-2632, 1969.
91. Monad, N. J., and Hammand, E. J. Oil production by fermentation of lactose and the effect of temperature on the fatty acid composition. *JAOCS*, *55*:683-688, 1978.
92. Eroshin, V. K., and Krylova, N. I. Efficiency of lipid synthesis by yeasts. *Biotechnol. Bioeng.*, *25*:1693-1700, 1983.
93. Hesse, A. Industrial biosynthesis. Part I. Fats. *Adv. Enzymol.*, *9*:653-704, 1949.
94. Enebo, L., Anderson, L. G., and Lundin, H. Microbiological fat synthesis by means of Rhodotorula yeast. *Arch. Biochem. Biophys.*, *11*:383-395, 1946.
95. Witter, B., Debuch, H., and Steiner, M. Die Lipide von Endomycopsis vernalis bei verschiedener Stidkstoff-Ernabrung. The lipids of Endomycopsis vernalis when grown with different nitrogen supply. *Arch. Microbiol.*, *101*:321-335, 1974.
96. Chesters, C. G. C., and Peberdy, J. F. Nutritional factors in relation to growth and fat synthesis in Mortierella vinacea. *J. Gen. Microbiol.*, *41*:127-134, 1965.
97. Borrow, A., Jefferys, E. G., Kessell, R. H. J., Lloyd, E. C., Lloyd, P. B., and Nixon, I. S. The metabolism of Gibberella fujikuroi in stirred culture. *Can. J. Microbiol.*, *7*:227-276, 1961.
98. Khan, A. N., and Walker, T. K. Fat production from cane molasses by Penicillium spinulosum: a study on substances in cane molasses interfering with fat production. *Can. J. Microbiol.*, *7*:895-899, 1961.
99. Phillip, S. E., and Walker, T. K. Mycological formation of fat V. Factors which influence the formation of fat in surface cultures of Penicillium lilacinum Thom. *J. Sci. Food Agr.*, *9*:223-227, 1958.

100. Duncan, B., Shah, D. N., and Herald, A. C. Fat production in Epicoccum nigrum. *Mycologia, 68*:412-418, 1976.
101. Sonnet, P. E. Lipase selectivities. *JAOCS, 65*:900-904, 1988.
102. Benzonana, G. Some properties of an exocellular lipases from Rhizopus arrhizus. *Lipids, 9*:166-172, 1974.

16

CAROTENOID BIOSYNTHESIS AND ITS REGULATION IN FUNGI

PETER M. BRAMLEY and ALAN MACKENZIE* *Royal Holloway and Bedford New College, Egham, Surrey, England*

I. INTRODUCTION

The carotenoids represent one of the most important and widespread groups of naturally occurring pigments. They are responsible for most of the yellow to red colors in flowers and fruits [1] and are present in many nonphotosynthetic organisms, including the fungi. Carotenoids are also found in animals, but are of dietary origin, although in this case they can undergo subsequent structural modifications [2,3].

The reason for their occurrence in fungi is not fully understood. They have been widely implicated in photoprotection [4-6], and are metabolized to the sex hormone trisporic acid in the Mucorales [7-9]. In addition, their presence in zygospores is probably because they are precursors of sporopollenin [10]. It is also possible that carotenoids may be formed as nonosmotic, carbon sinks in certain fungi, especially in the stationary phase of the growth cycle.

Synthetic carotenoids have been used for many years as colorants in industry [11]. They are also used, with varying levels of success, as protectors from photodermatoses in humans [12-14] and in other photosensitive skin diseases [15-17]. Carotenoid metabolites, the retinoids, e.g., vitamin A, are also of considerable commercial value as anticancer agents [18]. In more recent years there has been an increasing demand for carotenoids from natural sources such as marine algae [19] and fungi [20,21]. In order to make these naturally occurring sources compete commercially with the synthetic products, efforts are under way to improve yields, by strain improvement [22] and/or alteration of growth conditions [21]. In order to proceed on a rational basis, it is necessary to understand the regulatory mechanisms which control the formation of these pigments in response to environmental and developmental changes. The diversity of structures among the 500-plus carotenoids identified to date highlights the complexity of the biosynthetic pathways and related regulatory processes. The aim of this chapter is to describe our current understanding of carotenogenesis and its control in fungi and to discuss current and future experimental approaches which are aimed at increasing our knowledge of this area of fungal metabolism.

**Present affiliation*: Eton College, Eton, Berkshire, England.

II. STRUCTURE AND NOMENCLATURE

Carotenoids are isoprenoids that generally consist of eight isoprene units joined together (i.e., C_{40}), so that the linking of the units is reversed at the center of the symmetrical molecule (Fig. 1). The main feature of carotenoids is a polyene chain, which may extend from three to 15 conjugated double bonds. Cyclization of the carbon skeleton at one or both ends also occurs, while xanthophylls are formed from the hydrocarbon carotenes by the introduction of oxygen functions, especially hydroxy, oxo, and epoxy groups. Common carotenoids have well-established trivial names, but there is also an IUPAC-recommended semisystematic nomenclature [23,24], which describes all structures by using seven different end groups (β, γ, ε, κ, φ, χ, ψ; Fig. 1). The end groups of the fully unsaturated all-trans polyene chain are used to form the C_{40} carotene stem names, e.g., β,β-carotene and ψ,ψ-carotene. The systematic and trivial names of the carotenoids mentioned in this chapter are listed in Table 1, while the structures of some carotenoids which either commonly occur or are unique to fungi are shown in Fig. 2. The structures of all known carotenoids are given by Straub [25], while regular reviews of advances in carotenoid chemistry can be found in publications of The Chemical Society [e.g., 26,27]. An international symposium on carotenoids is organized triennially, and the proceedings are subsequently published [e.g., 28-30].

III. ISOLATION, PURIFICATION, AND IDENTIFICATION

Only an outline of the general methods for the extraction and identification of carotenoids will be presented, since detailed procedures can be found in several reviews [31-33].

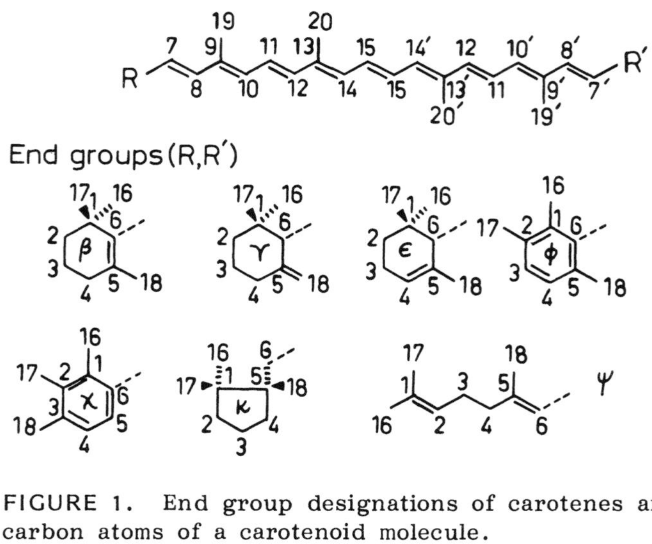

FIGURE 1. End group designations of carotenes and the numbering of carbon atoms of a carotenoid molecule.

TABLE 1. Trivial and Semisystematic Names of Carotenoids Mentioned in the Text[a]

Trivial name	Semisystematic name
Aleuriaxanthin	1',16'-Didehydro-1',2'-dihydro-β,ψ-caroten-21-ol
Astaxanthin	3,3'-Dihydroxy-4,4'-diketo-β-carotene
Canthaxanthin	4,4'-Diketo-β-carotene
α-Carotene	β,ε-Carotene
β-Carotene	β,β-Carotene
γ-Carotene	β,ψ-Carotene
ε-Carotene	ε,ψ-Carotene
δ-Carotene	7,8,7',8'-Tetrahydro-ψ,ψ-carotene
β-Cryptoxanthin	3-Hydroxy-β,β-carotene
3,4-Didehydrolycopene	3,4-Didehydro-ψ,ψ-carotene
Echinenone	β,β-Caroten-4-one
Lycopene	ψ,ψ-Carotene
Lycopersene	7,8,11,12,15,7',8',11',12',15'-Decahydro-ψ-ψ-carotene
Neurosporaxanthin	4'-Apo-β-caroten-4'-oic acid
Neurosporene	7,8-Dihydro-ψ-ψ-carotene
Phillipsiaxanthin	1,1'-Dihydroxy-3,4,3',4'-tetradehydro-1,2,1',2'-tetrahydro-ψ,ψ-carotene-2,2'-dione
Phoenicoxanthin	3-Hydroxy-β,β-caroten-4,4'-dione
Phytoene	7,8,11,12,7',8',11',12'-Octahydro-ψ,ψ-carotene
Phytofluene	7,8,11,12,7',8'-Hexahydro-ψ,ψ-carotene
Plectianaxanthin	3',4'-Didehydro-1',2'-dihydro-β,ψ-carotene-1'2'-diol
Rhodoxanthin	4',5'-Didehydro-4,5'-retro-β,β-carotene-3,3'-dione
Rubixanthin	β,ψ-Caroten-3-ol
7,8,11,12-Tetrahydrolycopene	7,8,11,12-Tetrahydro-ψ,ψ-carotene
Torularhodin	3',4'-Didehydro-β,ψ-caroten-16'-oic acid
Torulene	3',4'-Didehydro-β,ψ-carotene
α-Zeacarotene	7',8'-Dihydro-ε,ψ-carotene
β-Zeacarotene	7',8'-Dihydro-β-ψ-carotene
Zeaxanthin	β,β-Carotene-3,3'-diol

[a]End groups and numbering of the carbons of a carotenoid molecule are shown in Fig. 1.

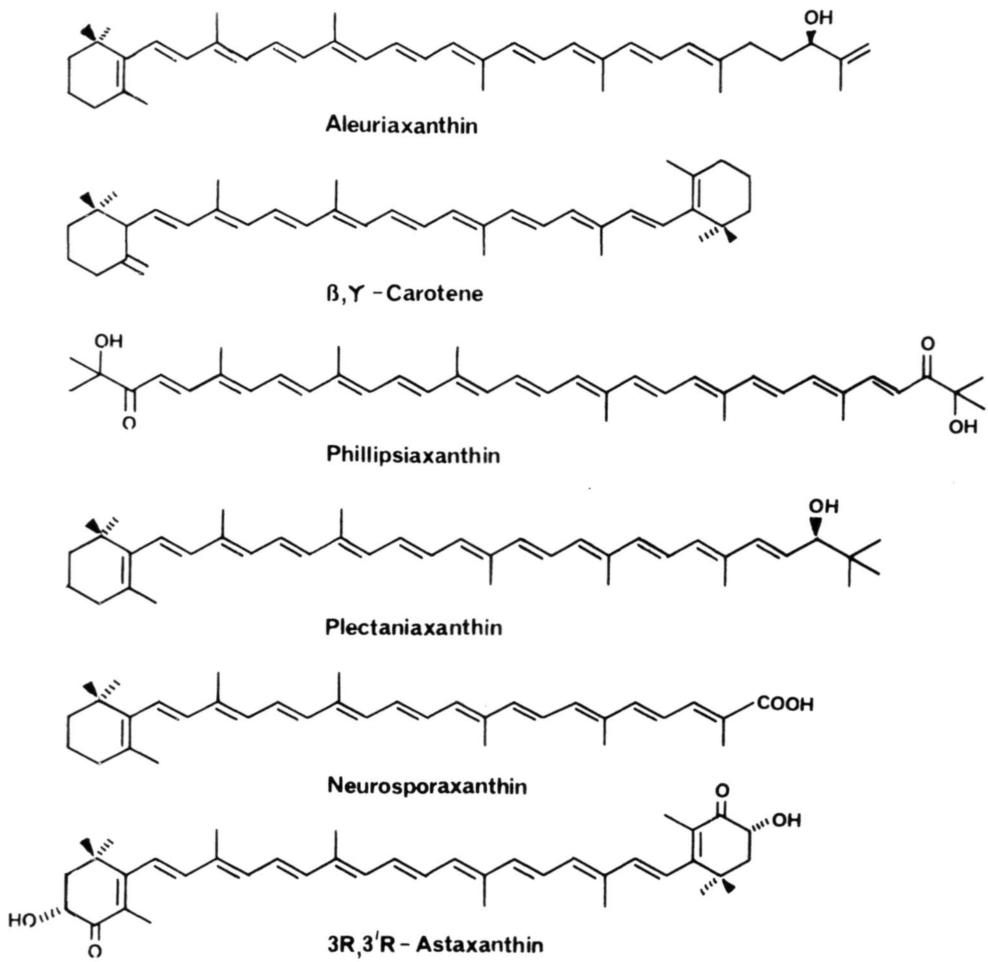

FIGURE 2. Structures of unique fungal carotenoids.

A. Isolation

Carotenoids and other polyenes are unstable toward heat, light, oxygen, acids, and, in some cases, bases. Therefore, a range of precautions must be observed during isolation and purification in order to minimize losses. The production of artifacts, e.g., cis isomers and oxidation products, must also be avoided in order to be certain that the carotenoid is structurally the same as that originally present in the fungus. All procedures should be carried out as rapidly as possible, preferably in subdued light or darkness. The addition of antioxidants at the time of extraction can be beneficial and storage of samples in an inert atmosphere (nitrogen or argon) at -20°C or below is strongly recommended.

Carotenoids can be easily extracted from mycelia using lipid solvents with a form of mechanical disruption (e.g., Waring blender, Ultra-Turrax homogenizer, pestle and mortar). Extraction from yeast cells, however, is more difficult and may require strong alkali to destroy cell walls [34]. Water-soluble solvents such as acetone and methanol are highly effective with fresh samples. After dehydration, the tissue can be extracted with petroleum ether, diethyl ether, or chloroform. The purest solvents available should always be used. The removal of saponifiable lipids from the extract is routinely carried out [31,33], but care must be taken to ensure that the carotenoids are stable to alkali. For example, tissues containing astaxanthin and related compounds must never be saponified in the normal way, and alternative procedures should be adopted [35,36].

B. Purification

Separation of carotenoids in an extract is achieved by chromatography. Typically, the mixture is separated into fractions by column chromatography with adsorbents such as alumina, magnesium oxide, silica, and cellulose. The column is developed with solvent mixtures of increasing polarity.

Thin-layer chromatography (TLC) is the most versatile method for purifying carotenoids. The most common adsorbents are silica gel (Kieselgel), aluminium oxide, magnesium oxide, and Kieselguhr. Commercially available thin layers provide better resolution and reproducibility than laboratory-made varieties. Extensive reviews cover most of the TLC separations [31,33,37,38]. It is especially important to minimize the effects of light and oxygen during the TLC of carotenoids, as adsorption to thin layers increases the chance of degradation due to photooxidation.

High-performance liquid chromatography (HPLC) is rapidly becoming the most popular method for the analysis of carotenoids and several reviews are available on this topic [39-41]. It permits rapid, routine, and quantitative analyses at the nanogram level and allows recovery of the sample. The recent advent of diode array detectors has been especially beneficial, as these enable spectra of each carotenoid to be obtained during elution. Both normal and reverse phase HPLC columns are used (Table 2).

C. Identification and Quantitation

The unequivocal identification of carotenoids requires at least the absorption and mass spectra to be obtained, as well as cochromatography with an authentic sample. All carotenoids absorb light strongly in the visible, or UV, regions of the spectrum and the positions of the maxima (usually three; Table 3) and the fine structures of the spectra (Fig. 3) are convenient parameters for tentative identification. Detailed discussions on the effects of structural features and solvents on the absorption spectra of carotenoids can be found in reviews [31,33]. Mass spectra of carotenoids (from as little as 10-20 μg of pure sample) exhibit characteristic fragmentation patterns and, with a few exceptions, reveal the molecular ions [52,53]. Nuclear magnetic resonance (NMR) spectroscopy is also applied to carotenoids [54,56,57], particularly for determining the location of cis double bonds.

A spectrophotometric method for the quantitation of carotenoids is used routinely. Typically, the absorbance of the carotenoid solution is determined at λ_{max} in a 1-cm light path. Using the specific absorbance coefficient ($A^{1\%}_{1cm} = E^{1\%}_{1cm}$) for the carotenoid [33], then the absorption at

TABLE 2. Separation of Carotenoids by HPLC

Pigments	Experimental conditions: Column	Solvent	Flow rate (ml/min)	Ref.
Carotenes	Sperisorb ODS, 5 μm	ACN: isocratic	2.0	42
Carotenes	Nucleosil 10, C_{18}	MeOH/H_2O(2:1)40% THF/ACN(1:1)60% isocratic	1.5	43
Carotenes	Altex C_{18}	30% 2Pol/70%(90%) ACN to 55%2Pol/45% (90% ACN); sigmoidal gradient	0.67	44
Carotenes	Ultrasphere ODS, 5 μm	2Pol/ACN/H_2O (30:63:7); isocratic	0.67	45
Carotenoids	Lichrosorb RP8, 10 μm	MeOH/ACN(25:75)/ H_2O; gradient	1.7	46
Carotenoids	Spherisorb ODS, 5 μm	ACN/MeOH/2Pol (85:10:5); isocratic	0.8	47
Carotenoids	Chromosil C_{18}, 10 μm	Acetone/H_2O(9:1)and ACN/2Pol/H_2O (200:288:13); isocratic	1.5	48
Carotenoids	Spherisorb S5 ODS	ACN/THF 92:8	1.0	49
Carotenoids	Zorbax ODS, 5 μm	EtAc in ACN/H_2O (9:1); gradient	—	50
Xanthophylls	Zorbax C_{18}	Acetone/H_2O(75:25) and acetone/MeOH or (75:25); gradients	1.0	51

Abbreviations: ACN, acetonitrile; MeOH, methanol; H_2O water; 2Pol, 2-propanol; EtAc, ethyl acetate; THF, tetrahydrofuran.

that wavelength is related to x g of carotenoid in y ml of solution by the following equation:

$$x\ g = \frac{Ay}{A_{1cm}^{1\%} \cdot 100}$$

If $A_{1cm}^{1\%}$ is unknown, an arbitrary value of 2500 is usually taken as an approximation.

TABLE 3. Absorption Maxima in Petroleum Ether and Absorption Coefficients ($A^{1\%}_{1cm}$) of Carotenoids Found in Fungi

Carotenoid	Maxima (nm)[a]	$A^{1\%}_{1cm}$
Aleuriaxanthin	434, 460, 491	2600
Astaxanthin	— , 468, —	
Canthaxanthin	— , 466, —	2200
α-Carotene	422, 444, 473	2800
β-Carotene	(425),449, 476	2592
γ-Carotene	437, 462, 494	3100
δ-Carotene	378, 400, 425	2550
β-Cryptoxanthin	425, 449, 476	2386
3,4-Didehydrolycopene	465, 492, 527	3000
Echinenone	— , 458,(482)	2158
Lycopene	444, 470, 502	3450
Neurosporaxanthin	— , 477,(504)	1715
Neurosporene	414, 439, 467	2918[b]
Phillipsiaxanthin	488, 518, 554	—
Phytoene (15-*cis*)	276, 286, 297	1250
Phytofluene	331, 348, 367	1350
Plectianaxanthin	445, 471, 502	2505[c]
Rhodoxanthin	458, 490, 524	2500
Rubixanthin	434, 460, 490	2909[c]
Torularhodin	467, 501, 537	2040
Torulene	454, 480, 514	3240
β-Zeacarotene	406, 428, 454	2520
Zeaxanthin	424, 449, 476	2348

[a]Brackets indicate a shoulder in the spectrum.
[b]Hexane.
[c]Acetone.
Source: Data taken from Refs. 31 and 33.

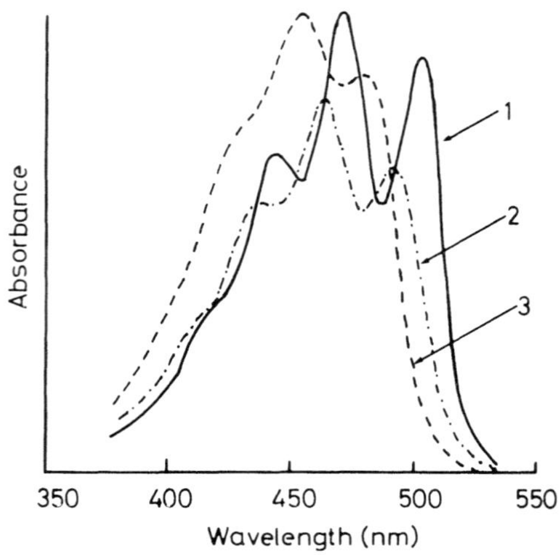

FIGURE 3. Light absorption spectra of lycopene (1), γ-carotene (2), and β-carotene (3) in petroleum ether.

IV. DISTRIBUTION OF CAROTENOIDS

A. Taxonomic Distribution

Carotenoids are not ubiquitous in fungi. They have been detected in some 200 species, but examination of another 130 has shown that they lack carotenoids (listed by Ref. 1) and certain species only produce carotenoids in the light (see Section VII.A).

The carotenes, especially β- and γ-carotene, are the most frequently found carotenoids in fungi. The Phycomycetes show the most limited distribution, with β-carotene predominating in all except the blastocladiales, where γ-carotene is the major pigment. In all the Ascomycetes, except for the Pezizales, the carotenoids are similar to those of the Phycomycetes, i.e., carotenes only. *Neurospora crassa* and *silophila*, however, also contain torulene, 3,4-didehydrolycopene, and the apocarotenoid neurosporaxanthin [60-62]. The Pezizales are of particular interest, since they contain several unique fungal carotenoids such as β,γ-carotene, phillipsiaxanthin, aleuriaxanthin, and plectaniaxanthin (Fig. 3) [63-66]. The Heterobasidiomycetes are characterized by the presence of carotenes rather than xanthophylls, as typified by *Ustilago* spp. [67]. When xanthophylls do occur, they contain hydroxyl groups at C-3, e.g., β-cryptoxanthin, rubixanthin, and zeaxanthin [68].

The order Hymenomycetes shows a different pattern since ketocarotenoids are major pigments, e.g., canthaxanthin in *Cantharellus* [69], astaxanthin and echinenone in *Pencophora* spp. [70], and *Sphaerobolus stellatus* [71]. In some *Cantharellus* species, however, neurosporene is accumulated, as the final dehydrogenation step to lycopene (see Fig. 7) is blocked [70].

The Deuteromycetes contain a relatively wide range of carotenoids, including carotenes with Δ^3 double bonds, and apocarotenoids such as neurosporaxanthin and torularhodin. It is interesting to note that the astaxanthin from the red yeast *Phaffia rhodozyma* has the 3R,3'R chirality

TABLE 4. Sources of Carotenoids from Fungi

Compound	Source	Amount: μg/g weight	Amount: % total carotenoids	Comments	Ref.
		(a) Carotenes			
Phytoene	C5 strain of *Phycomyces*	8,966	~100	*car*B mutant + dimethyl phthalate	72
Phytofluene	Wild-type *Phycomyces*	231	7		73
ζ-Carotene	S442 strain of *Phycomyces*	634	15	leaky *car*B	73
Neurosporene	*Cantharellus ianthinoxanthus*	?	74		64
Lycopene	C9 strain of *Phycomyces*	2,800	≥90	*car*R mutant	75
β-Carotene	C115 strain of *Phycomyces*	3,500	100	*car*S mutant	76
	S218*S219 strain of *Phycomyces*	15,317	≥90	Heterokaryon + trisporate B	77
	Blakeslea trispora	9,800	—	Mated cultures + soybean oil	78
	" "	12,960[a]	—	+ oil, detergent, and β-ionone	79
γ-Carotene	*Cronartium fusiforme*	311	64	—	80
	Rhizophlyctis rosea	69	46	In late stages of growth	81
β,γ-Carotene	*Caloscypha fulgens*	?	8	Only source in fungi	82
Torulene	*Gibberella fujikuroi* SG68	1,180	>90	Superproducing strain	83
		(b) Xanthophylls			
Astaxanthin	*Phaffia rhodozyma*	696	87	3R,3'R configuration	85
Plectaniaxanthin	*Plectania coccinea*	45	—	Unique to fungi	86
Torularhodin	*Rhodotorula glutinis* 48-23T	62	26	—	87
Neurosporaxanthin	SG22 strain of *Gibberella fujikuroi*	1,540	>90	Super-producing strain	83
Rhodoxanthin	*Epicoccum nigrum*	15	—	—	84

[a] μg/100 ml culture medium.

(Fig. 2), which is opposite to that of the lobster astaxanthin [71]. The retrocarotenoid rhodoxanthin, is also found in several Deuteromycetes [72].

B. Subcellular Localization

Carotenoids are found predominantly in lipid droplets in fungi, with only minor amounts associated with membranes. In the C115 mutant of *Phycomyces*, some 98% of the β-carotene was found in oil droplets, with the remainder located in an unidentified membrane fraction which has a buoyant density of 1.10 g/ml [73]. A later study with *N. crassa* came to similar conclusions [76]. In this ascomycete, synthesis and storage occur in different cell components, as the spherosomes function mainly for the storage of carotenoids which have been formed on the endoplasmic reticulum (ER). A similar pattern may occur in *P. blakesleeanus* [79]. In contrast, the cell wall of *B. trispora* contains carotenoids [80].

C. Sources of Fungal Carotenoids

Fungi can be useful or unique sources of carotenoids for both scientific and commercial applications. From a commercial viewpoint, the fungal carotenoids must be produced cheaply in large amounts or else the organism must be the sole source of that pigment. Scientific requirements may center on the relative abundance of a standard carotenoid from an easily cultured fungal strain. Some representative examples of yields and sources of carotenoids are shown in Table 4. It is possible, of course, that these values could be increased under different cultural conditions.

V. BIOSYNTHESIS

Extensive studies over the past 35 years have established the general features of carotenoid formation in higher plants, algae, bacteria, and fungi (reviewed by 1,4,87-91). The pathway from acetyl-CoA to the cyclic carotenes is well characterized, but the later steps leading to the formation of xanthophylls and apocarotenoids are less well understood.

A. Formation of Phytoene

The biosynthesis of phytoene can be divided into three stages: (a) the formation of isopentenyl pyrophosphate (IPP), (b) conversion of IPP to geranyl geranyl pyrophosphate (GGPP), and (c) the condensation of two molecules of GGPP to form phytoene.

1. Formation of IPP

IPP is the five-carbon compound from which all isoprenoids are derived. It is formed from acetyl-CoA via a sequence of six enzymic steps (Fig. 4). The enzymes that catalyze the conversion of acetyl-CoA to IPP are cytosolic, with the exception of 3-hydroxy-3-methylglutaryl coenzyme A (HMG-CoA) reductase, which is found in the mitochondria of yeast. All of these have been purified, although not necessarily from fungi, and HMG-CoA reductase is particularly well characterized (reviewed by 92). An alternative pathway to the HMG-CoA exists in *P. blakesleeanus* via the carboxylation of a metabolite of leucine [93,94].

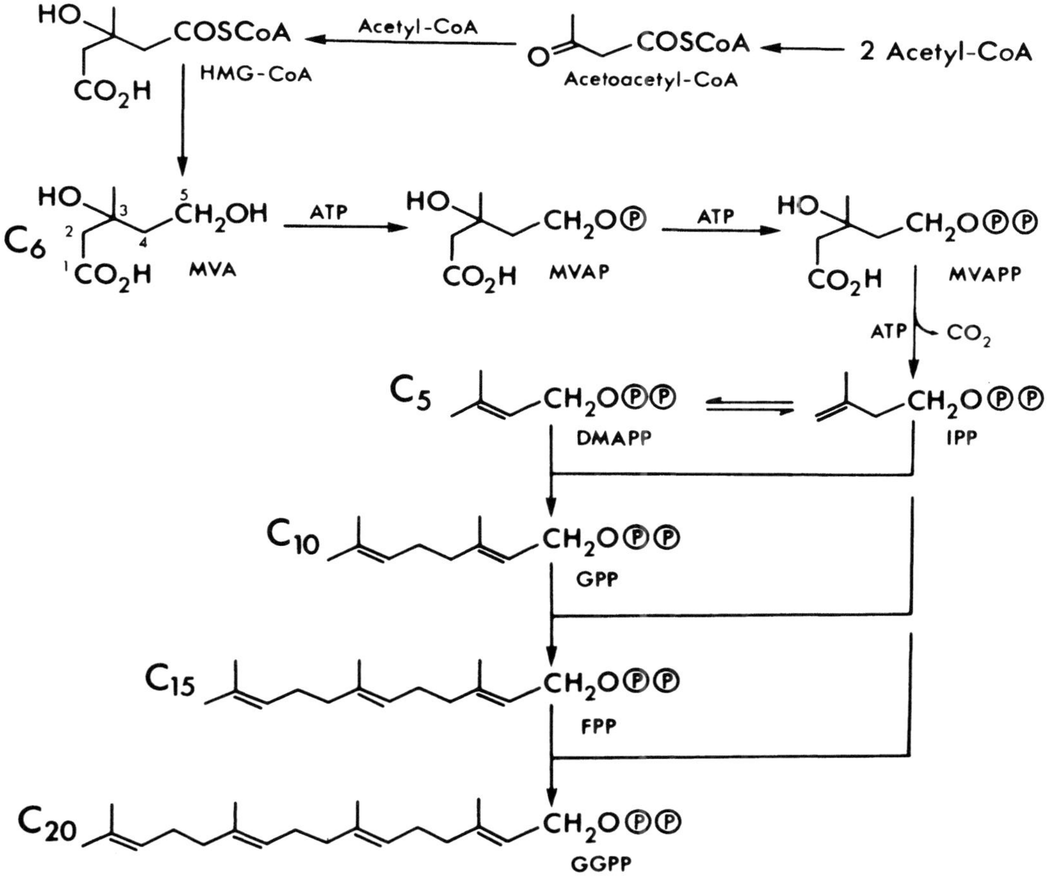

FIGURE 4. Pathway for the biosynthesis of GGPP from acetyl-CoA. (From Ref. 104, reproduced with permission.)

2. Conversion of IPP to GGPP

In the sequence of chain elongation from C_5 (IPP) to C_{20} (GGPP) the steps are catalyzed by prenyltransferases, the number and catalytic activities of which vary from organism to organism [95]. In *N. crassa* there are at least two soluble prenyltransferases. The major protein is farnesyl pyrophosphate (FPP) synthetase, which uses IPP plus geranyl pyrophosphate (GPP) as substrate, but not IPP plus FPP. The minor prenyltransferase (GGPP synthetase) can use either IPP plus GPP or IPP plus FPP as substrate (Harding, unpublished observations). Recently, the GGPP synthase from strain M1 of *P. blakesleeanus* was purified to homogeneity [96]. It has a subunit molecular weight of 30,000 and only accepts IPP and FPP as substrates for GGPP formation. Interestingly, the activity of the enzyme in the M1 strain (a *carS* mutant, see Section VIII.B), is fourfold higher than the wild-type activity.

3. Formation of Phytoene from GGPP

The first step that is unique to the biosynthesis of carotenoids is the head-to-head condensation of two molecules of GGPP to form phytoene (Fig. 5). It is now recognized that the only intermediate between GGPP and phytoene is prephytoene pyrophosphate (PPPP) (Fig. 5; reviewed by 1, 104). The proposed mechanism for the formation of PPPP is the same as that for presqualene pyrophosphate (PSPP), except that in its conversion to phytoene the cyclopropylcarbinyl ion is stabilized by the loss of a proton. Which proton is lost depends on whether 15-*cis*- or 15-*trans*-phytoene is formed. Samples of phytoene from fungi are predominantly 15-cis with only 1-2% in the all-trans configuration [97-99].

B. Desaturation of Phytoene

The desaturation of phytoene occurs in a stepwise sequence of reactions to form phytofluene, ζ-carotene, neurosporene, and lycopene, which have 5, 7, 9, and 11 conjugated double bonds, respectively (Fig. 6). In *N. crassa* [100] and *P. blakesleeanus* [101] the asymmetrical isomer of ζ-carotene, 7,8,11,12-tetrahydrolycopene, is also found.

The stereochemistry of hydrogen removal at C-7,8, C-11,12, C-7',8', and C-11',12' is by the loss of the 5-*pro*-R- and 2-*pro*-S hydrogens [102, 103]. The mechanism of hydrogen removal has yet to be elucidated, although the presence of hydroxyphytoenes in herbicide-treated plants may reflect the involvement of a hydration mechanism [50].

The existence of 3,4-dehydrolycopene but not an "isomeric neurosporene (i.e., 3,4-dehydro-7',8',11',12'-tetrahydrolycopene) in *N. crassa* and

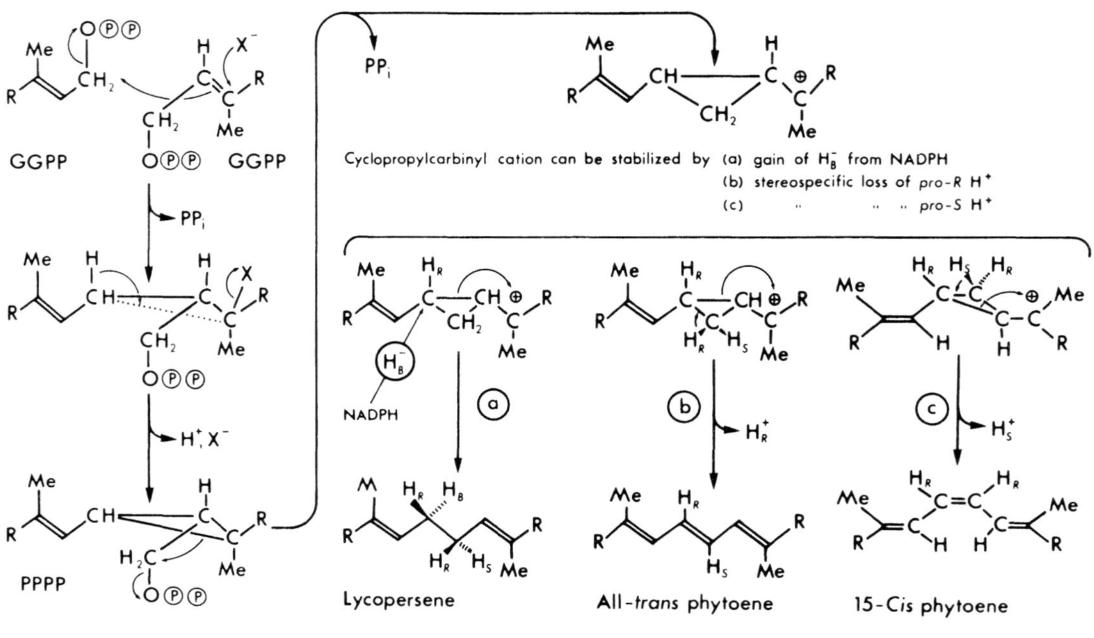

FIGURE 5. Formation of PPPP from GGPP and its conversions to (a) lycopersene, (b) all *trans*-phytoene, and (c), 15-*cis*-phytoene. See text for details. (Reproduced with permission from Ref. 104.)

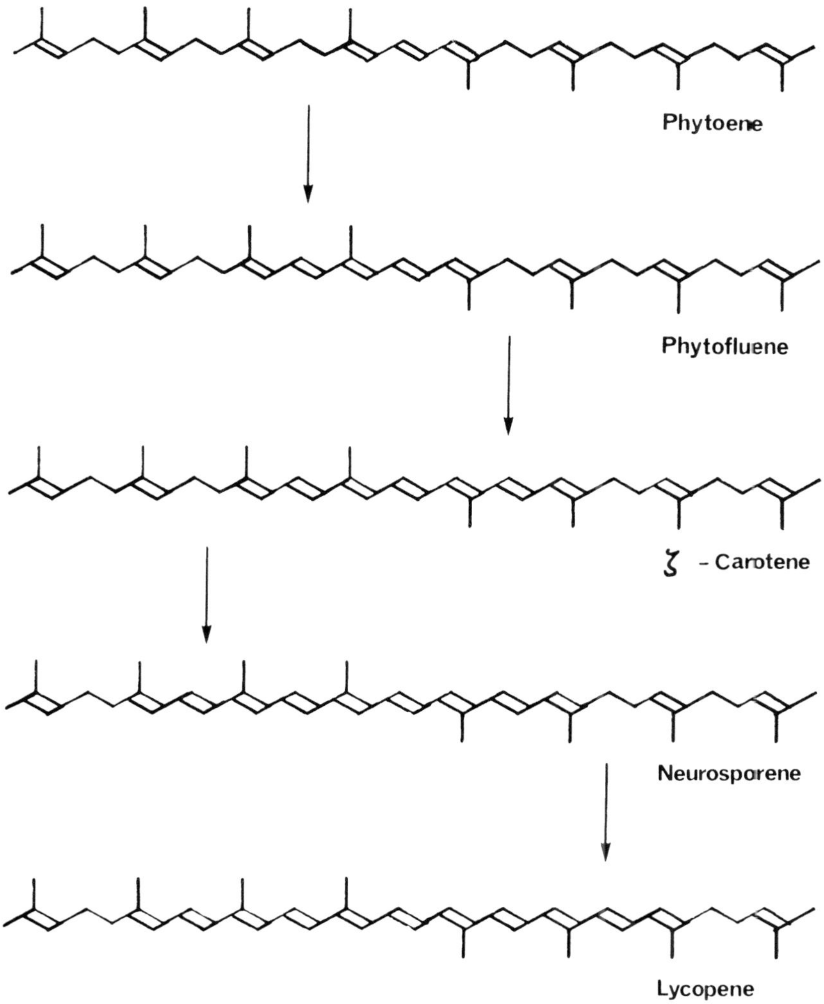

FIGURE 6. General pathway for the desaturation of phytoene into lycopene. For convenience, all the structures are drawn as all-trans isomers, although a cis-trans isomerization must occur at either phytoene or phytofluene. See Section V.B for details.

Gibberella fujikuroi indicates that there is a higher substrate specificity for the removal of the hydrogens from C-3 and C-4 than for other dehydrogenations; perhaps the enzyme which converts lycopene to its 3,4-dehydro derivative is different from that responsible for the other dehydrogenations in *Neurospora*. If this is correct, one would expect a mutant, accumulating lycopene, could be obtained from *N. crassa* (or *G. fujikuroi*). Such a strain has not been isolated (see Section VIII.A).

Although phytoene has the 15-cis configuration in all fungi examined (see Section V.A), the configuration of ζ-carotene, neurosporene, and lycopene is all-trans. Thus, in the conversion of phytoene to phytofluene,

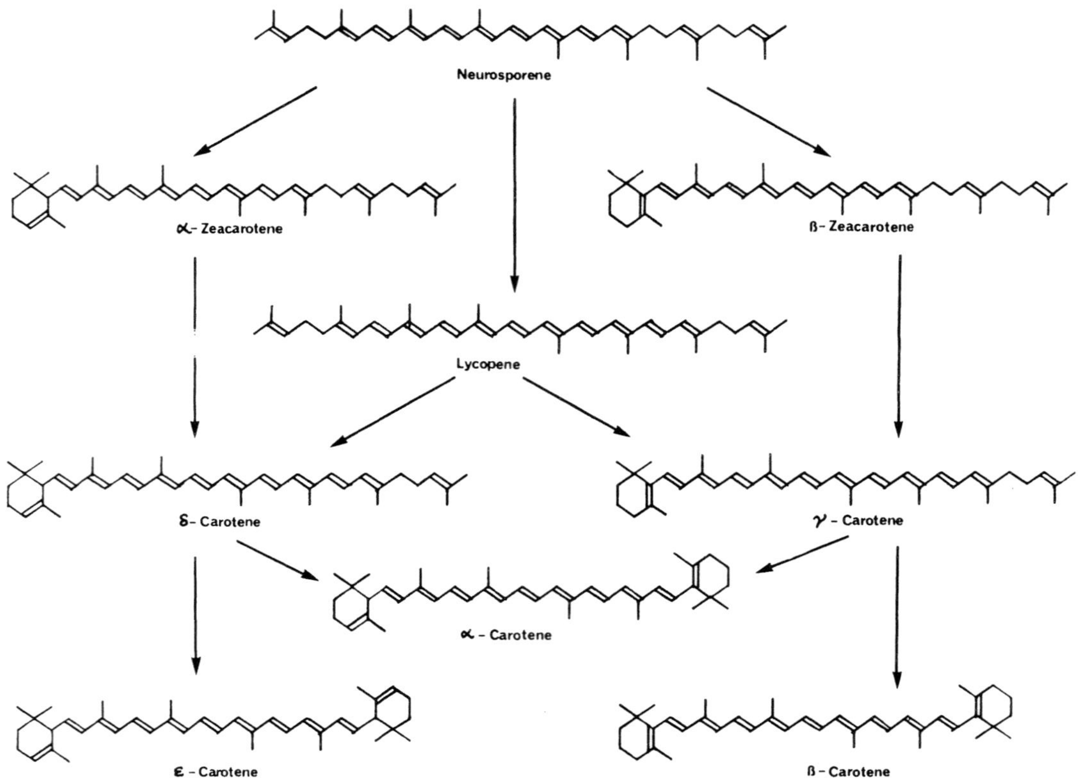

FIGURE 7. Overall scheme for the biosynthesis of cyclic carotenes.

an isomerization of the 15-15' double bond must occur. Studies on higher plants strongly suggest that the isomerization occurs at the phytofluene level [105,106], but the situation in fungi has yet to be thoroughly investigated. It may be that a strained phytoene molecule present in an enzymic complex is the active intermediate, as suggested by Goodwin [107].

C. Cyclization Reactions

Mono- and bicyclic carotenes are characteristically found in many fungi, particularly the Phycomycetes and Ascomycetes (see Section IV.A). The most common alicyclic end rings are either β- (e.g., β-carotene, γ-carotene) or ε- (e.g., α-carotene), whereas β,γ-carotene, found in *Caloscypha fulgens* (Fig. 2) [63,108], is a rare example of a carotene containing a γ ring.

There is substantial experimental evidence to show that the β and ε rings are formed independently, although probably from a common precursor [109,110]. Less is known about the biosynthesis of the α ring but the 6'S chirality of β,γ-carotene [108], which is opposite to that of carotenoids with ε rings, indicates that a different enzyme(s) is required for its formation. It is assumed that it is also formed independently of the β ring. Indeed, the accepted mechanism of cyclization to form all three ring systems involves proton attack at C-2, followed by stabilization of a common carbonium ion intermediate by the loss of a proton [111,112]. There are several routes for the formation of cyclic carotenes, depending on whether desaturation is completed before or after cyclization (Fig. 7).

D. Formation of Xanthophylls

The introduction of oxygen functions into carotenes occurs at various stages of desaturation and/or cyclization, and the large variety of xanthophylls found in fungi precludes the description of a general pathway. The majority of fungal xanthophylls contain cyclic end groups, although phillipsiaxanthin (Fig. 2) is an unusual exception. Its biosynthetic pathway is unknown, but may be similar to the formation of the structurally related ketocarotenoids of purple nonsulfur bacteria [86].

Pathways have been suggested for the formation of some fungal xanthophylls. The carotenoid acid torularhodin, characteristic of red yeasts, is probably formed from torulene by a series of oxidations via torularhodin alcohol and torularhodinaldehyde [87]. Neurosporaxanthin (Fig. 2) may be formed by the oxidation of torulene or γ-carotene, the latter being more likely on the basis of inhibitor studies [113].

A pathway for the formation of astaxanthin from β-carotene in *Phaffia rhodozyma* was proposed [85]. β-Carotene is first converted to echinenone, then hydroxylated to 3-hydroxyechinenone, which in turn is oxidized to phoenicoxanthin followed by hydroxylation at C-3. The 3R,3'R configuration of astaxanthin from *Phaffia*, which is opposite to that from other sources, indicates that the stereochemical specificity of the enzyme(s) responsible is different in this yeast.

The biosynthesis of echinenone and canthaxanthin presumably occurs by the oxidation of β-carotene, in the same fashion as that demonstrated in the marine invertebrate *Artemia salina* [114]. The formation of xanthophylls typically associated with higher plants, e.g., zeaxanthin, β-cryptoxanthin, rhodoxanthin, has not been investigated in fungi. They may be produced in an analogous way in both photosynthetic and nonphotosynthetic tissues.

E. Inhibitors of Carotenoid Biosynthesis

Many chemicals have been shown to inhibit fungal carotenogenesis. They can be broadly divided into two groups: (a) those that cause the accumulation of more saturated pigments, typically phytoene and phytofluene, and (b) those that inhibit cyclization (Table 5). The primary objective of studies with such inhibitors has been to aid in the elucidation of carotenoid biosynthesis. In this respect, it is important to note that they are not effective as inhibitors in all organisms and there is even variation in the response of fungi to the same compound. They can be used more pragmatically, however, as a way to produce unsaturated or acyclic carotenoids which may be in very low levels under normal growth conditions.

1. Inhibitors Which Cause the Accumulation of More Saturated Carotenes

This group of compounds is typified by the classic inhibitor of carotenogenesis, diphenylamine, which has been used for over 50 years. It is

TABLE 5. Inhibitors of Carotenoid Biosynthesis in Fungi

Compound	Organism	Ref.
(a) Accumulation of more saturated carotenoids		
Diphenylamine	*Allomyces* sp.	115
	Blakeslea trispora	116
	Epicoccum nigrum	117
	Neurospora crassa	118
	Phycomyces blakesleeanus[a]	74,119,120
	Rhodotorula gracilis	121
	Verticillium albo-atrium	122
CPTA	*Verticillium agaricinum*	123
9-Fluorenone	*Phycomyces blakesleeanus*	124
	Verticillium agaricinum	125
2-Methylheptanone	*Phycomyces blakesleeanus*	126
	Rhodotorula glutinis	127
	Verticillium agaricinum	125
	Verticillium albo-atrum	122
2-Hydroxybiphenyl	*Rhodotorula mucilaginosa*	128
β-Ionone	*Rhodotorula glutinis*	129
	Rhodotorula rubra	129
Fluridone, difunon, alkyl phenoxybenzamides	*Phycomyces blakesleeanus*	130
(b) Accumulation of lycopene		
CPTA	*Blakeslea trispora*	131
	Phycomyces blakesleeanus	132
	Rhodotorula sp.	133
Nicotine	*Phycomyces blakesleeanus*	135
	Verticillium agaricinum	123

[a]Also inhibitory in vitro [134].

effective with a range of fungi in vivo (Table 5a) and is also inhibitory in cell extracts of *Phycomyces* [134]. The in vivo accumulation of phytoene in *P. blakesleeanus* is greater than that predicted on the basis of the decrease in β-carotene [135] and this may be due to a loss of feedback inhibition by β-carotene on the step(s) before phytoene (see Section VIII). The in vitro effect is reversible, suggesting that diphenylamine inhibits the activity of phytoene dehydrogenase [134]. A range of structurally related biphenyl compounds are also inhibitory in vitro [130].

Several bleaching herbicides inhibit phytoene dehydrogenation in vitro but not in vivo (Table 5a), e.g., fluridone and a range of alkyl phenoxybenzamides cause the accumulation of phytoene, at the expense of β-carotene synthesis, in cell extracts of *Phycomyces* [130].

Finally, there are examples in this category of inhibitors which demonstrate the unpredictability of action on fungal carotenogenesis. In *Verticillium agaricinium* 2-(4-chlorophrenyl)trimethylamine (CPTA) inhibits phytoene metabolism [123], whereas in all other fungi it causes the accumulation of lycopene [132]. Also, β-ionone inhibits the formation of colored carotenoids in *Rhodotorula* spp. [129], yet it is a stimulator of carotenogenesis in the Mucorales (Table 8).

2. Cyclase Inhibitors

The best known inhibitors of cyclization are nicotine and CPTA (Table 5b). Typically, they cause the accumulation of lycopene, sometimes together with small amounts of monocyclic carotenes, e.g., β-zeacarotene. Compounds structurally related to CPTA have a similar effect in *B. trispora* [131], and a variety of other nitrogenous compounds are effective in *P. blakesleeanus* [132]. The presence of CPTA in the growth medium of a *carA* mutant of *Phycomyces* results in an increase in the lycopene and α-carotene content of this strain, which normally produces negligible quantities of carotenoid [136]. The function of the *carA* gene is discussed in Section VIII.B.

VI. PREPARATION AND PROPERTIES OF CAROTENOGENIC CELL EXTRACTS

Over 20 carotenogenic cell-free systems have been reported in the literature (reviewed by 89), but of these only five are from fungi (Table 6), and no fungal extracts capable of xanthophyll biosynthesis have been reported. Typically, preparations are crude extracts that convert ^{14}C-labeled acetate, mevalonic acid (MVA) or IPP into carotenes, although carotene-carotene conversions have been demonstrated with membrane preparations of *A. giganteus* [137,138] and *P. blakesleeanus* [147,148]. The early studies with cell-free systems were designed to confirm the biosynthetic pathways shown from in vivo investigations (see Section V), e.g., the incorporation of ^{14}C-labeled acyclic carotenes into β-carotene [105,147,148]. Later studies attempted to elucidate the properties of the enzymes themselves. Unfortunately, no carotenogenic enzyme has been purified to homogeneity from a fungus and so the properties of these proteins and the precise reactions they catalyze can only be surmised from the data obtained with crude or partially purified preparations. In fact, the only report of the purification of any carotenogenic enzyme is that of phytoene synthase from *Capsicum* chromoplasts [150].

A. Phytoene Synthesis

A variety of cofactors have been reported to stimulate phytoene biosynthesis, in vitro (Table 7). The best interpretation of the current information is probably that the use of impure enzyme preparations precludes accurate elucidation of the cofactor requirements for phytoene synthetase. There is some consensus, however, on the intracellular location of the enzyme. In preparations of *Phycomyces* [151], *Gibberella* [83] and *Aspergillus* [137], high salt concentrations remove the protein from its loose association with cell membranes, and so it is recovered in the cytosol. In contrast, phytoene synthetase of *N. crassa* can only be removed from its membrane support on treatment with detergents [142].

TABLE 6. Carotenogenic Activities of Fungal Cell Extracts

Organism	Method of preparation of cell extract	Carotenogenic activity: Substrate	Carotenogenic activity: Product	Ref.
Aspergillus giganteus	Lyophilization	MVA, IPP, PPP phytoene, lycopene	β-Carotene	137,138
Blakeslea trispora	Homogenization	Acetate, acetyl-CoA, malonyl-CoA	β-Carotene	139
Gibberella fujikuroi	Lyophilization	MVA	Acyclic carotenes + torulene	83
Neurospora crassa	Lyophilization	MVA IPP	Phytoene Phytoene	140 141
	Grinding with abrasives	MVA	Phytoene + other carotenes	142,143
Phycomyces blakesleeanus	Lyophilization	MVA	Carotenes	144-146
		Phytoene, neurosporene	β-Carotene	147,148
		GGPP	Phytoene	149

B. Phytoene Desaturation

There is no agreement on the cofactor requirements for the conversion of phytoene into lycopene (Table 7), but in all cases the enzyme(s) is membrane-bound. At present, the number of enzymes involved in the desaturation sequence is unknown, and in fungi containing 3,4-dehydrolycopene an extra dehydrogenation occurs, probably via a different enzyme (see Section V.B). In *Phycomyces* the phytoene-to-lycopene sequence is thought to be catalyzed by a multienzyme aggregate of four dehydrogenases which are multiple copies of the *carB* gene product (see Section VIII) [72,75,152]. This aggregate has been solubilized with retention of enzymic activity, from *Phycomyces* [153] and *Neurospora* [143].

C. Cyclization Reactions

Cyclization of lycopene, neurosporene, and γ-carotene into β-carotene has been demonstrated with a cell extract of *P. blakesleeanus* [147], as has the conversion of lycopene into β-carotene with an *A. giganteus* cell free preparation [137]. In both cases the enzyme(s) are membrane-bound. Although the preparation from *Phycomyces* was solubilized [153], purification of the enzymes was not reported. Genetic studies indicate that two cyclases are present in this fungus, both encoded by the *carR* gene (see Section VIII) [154].

TABLE 7. Properties of Carotenogenic Enzymes from Fungi

Organism	Cofactor requirements: Essential	Cofactor requirements: Stimulatory	Subcellular location	Photo-induction	Ref.
		A. Phytoene formation			
Aspergillus giganteus	—	ATP, DTT[a]	Cytosol	Yes	137
Gibberella fujikuroi	—	—	Cytosol	No	83
Neurospora crassa	Mg^{2+}	—	Membrane-associated	Yes	143
Phycomyces blakesleeanus	—	Mg^{2+},-SH reagent		—	144
	—	ATP	Cytosol	—	151
		B. Phytoene to lycopene			
A. giganteus	FAD	NAD(P), O_2, ATP, DTT	Membrane	Yes	137
G. fujikuroi	—	—	Membrane	—	83
N. crassa	Mg^{2+}	—	Membrane	Yes	143
P. blakesleeanus	—	NADP	Membrane	Yes	151
		C. Lycopene cyclization			
A. giganteus	FAD	ATP, O_2, DTT	Membrane	Yes	137

[a]DTT, dithiothreitol.

VII. REGULATION OF CAROTENOID BIOSYNTHESIS

Several environmental stimuli alter both the qualitative and quantitative profiles of carotenoids in fungi (Table 8). In addition, the isolation of regulatory mutants of filamentous fungi revealed a plethora of regulatory genes (see Section VIII.B). It is not yet possible, however, to explain at the molecular level the precise events that occur during these transitions. The following subsections summarize our current understanding of the regulation of carotenogenesis.

A. Light

Photoregulation of carotenoid biosynthesis has been reported in at least 15 species of fungi [166]. In some cases light enhances the content of colored carotenoids in the mycelia, e.g., *Phycomyces* [167,168]; in others, e.g., *N. crassa*, only small amounts of phytoene are formed in the dark

TABLE 8. Stimulators of Carotenogenesis in Fungi

Effector	Organism	Mode of action	Ref.
β-Ionone	Mucorales	See Section VII.C	155
Retinol + derivatives	Mucorales	See Section VII.C	156,157
Dimethyl phthalate	*Phycomyces*	See Section VII.D	158
Trisporic acids	Mucorales	See Section VII.C	77,159
Phenols	*Phycomyces*	See Section VII.D	158
Light	Various	See Section VII.A	
pH	*Phycomyces*	See Section VII.B	
Developmental processes	Various	See Section VII.B	
Citrus products	*Blakeslea trispora*	See Section VII.C	21
Inorganic phosphate	*Blakeslea trispora*	Decrease of phosphatases	160,161
F^-	*Blakeslea trispora*	Inhibition of phosphatases	162
Cu^{2+}	*Blakeslea trispora*	Derepression of trisporic acids; increase in MVA kinase	163
Penicillin	*Blakeslea trispora*	Increase in MVA kinase	164
Surfactants	Various	Unknown	165
Pyridine derivatives	*Blakeslea trispora*	Unknown	155

[169,170]; while a third group, typified by *A. giganteus* [138,171], exhibits strict photoregulation of carotene biosynthesis.

The kinetics of photoregulation in fungi follow two stages: a first light-dependent but temperature-independent phase is typically of short duration (the "light phase"), followed by a light-independent stage during which carotenoid biosynthesis takes place (the "dark reactions"). Between these two phases a lag period exists, and the duration of this also varies among species.

The identity of the photoreceptor that mediates carotenoid biosynthesis has been the subject of considerable debate and prolonged efforts for many years. The action spectra in fungi are generally very similar and resemble those of universally occurring blue light responses [173]. Reports that red light is effective in *Verticillium* [172] were recently discounted [174]. Discussions on the relative merits of flavoproteins and/or carotenoids as candidates for the photoreceptor are found in several reviews [4,166, 175,176]. There is the possibility that two separate photoreceptors exist in those fungi which exhibit a biphasic fluence response of photocarotenogenesis, e.g., *N. crassa* [177,178] and *P. blakesleeanus* [179]. Transduction of the light signal is thought to be mediated by redox reactions [see 166, 176].

Regardless of the nature of the photoreceptor, two questions can be asked regarding other aspects of photocarotenogenesis: (a) What enzymes are photoinducible? (b) At what metabolic level does photoregulation occur?

1. Induction of Carotenogenic Enzymes

Relatively few in vitro systems are available from photoregulated fungi (Table 7), but some information has been obtained from *N. crassa*, *A. giganteus*, *G. fujikuroi*, and *P. blakesleeanus*.

Using an in vitro preparation of *Neurospora*, it was demonstrated that the steps to phytoene are stimulated in light-grown cultures compared with dark-grown controls [181]. Earlier investigations had shown that phytoene formation and its dehydrogenation were independently regulated by light in *N. crassa* wild-type and the *al*-1 (phytoene-accumulating) mutant [182]. Light-grown cultures of *P. blakesleeanus* 1+ contained twice the activity of the dark-grown preparation [168]. However, an alternative hypothesis for the photoaccumulation of β-carotene in *P. blakesleeanus* suggests that blue light inhibits β-carotene metabolism rather than stimulating its formation [183,184].

In order to assay each carotenogenic enzyme specifically, ^{14}C-labeled GGPP, PPPP, phytoene, and lycopene were incubated with cell extracts from dark- and light-grown cultures of *A. giganteus*. These experiments showed that the enzymes catalyzing the conversions of PPPP to β-carotene are totally absent in dark-grown cells and are photoinduced upon illumination [138]. HMG-CoA reductase is photoregulated in *Rhodotorula minuta*, but the enzymes which convert phytoene to colored carotenoids are not photoinducible in this yeast [185].

2. Metabolic Level of Photoregulation

In vivo studies with inhibitors of transcription and translation have led to the conclusion that photoregulation affects transcription, e.g., *Fusarium aquaeductuum* [166], *A. giganteus* [138]. Evidence for the involvement of de novo mRNA synthesis was also obtained with *Fusarium* [180,186, 187]. The derepression of carotenoid gene transcription in *N. crassa* is controlled by the intracellular concentration of cAMP. Light causes a decrease in cAMP levels which in turn alleviates carotenoid gene repression [188-191].

B. Developmental Processes

There are numerous reports in the literature which show that changes in growth rate, age of culture, media constituents (especially carbon and nitrogen sources), aeration, etc., affect the carotenoid content of fungal cells. In many cases, any connection may only be a reflection of the metabolic state of the cell under certain cultural conditions.

Carotenoids often accumulate in fungi after cessation of growth, or where growth rates are limited by nutrient supplies, e.g., *P. blakesleeanus* [192], *Epicoccum nigrum* [84]. The final level of carotenoids in wild-type *Phycomyces* is dependent on the C/N ratio in the medium, while conidiation is associated with the appearance of carotenoids, e.g., *N. crassa* [175].

C. Bioinducers

The large increase in β-carotene production when two mating types [designated (+) and (-) of Mucorales (Zygomycetes)] are cultured together led to the observation that the trisporic acids act as stimulators of carotenogenesis [77,78,193,194]. They are formed from β-carotene [7] via sex-specific precursors, the prohormones P^+ and P^- [196]. The precise mechanism of control by trisporic acids is not yet understood. An increase in RNA synthesis in *Blakeslea* has been observed on addition of trisporic acids, and a concomitant increase in cAMP levels occurs in mated cultures. It is not clear, however, whether the effects of trisporic acids are directly on a cAMP-mediated system or are indirect. The enzymes that are induced following trisporic acid production are reported to be those responsible for the conversion of mevalonic acid 5-phosphate (MVAP) to dimethylallyl pyrophosphate (DMAPP) [195]. It should be noted that sexual stimulation of carotenogenesis in *Phycomyces* is additive to increases caused by light, retinol, dimethyl phthalate, and mutation of the *carS* gene (see Section VIII), but not β-ionone [76].

Naturally occurring compounds that are structurally related to trisporic acids are also stimulators of carotenoid formation in *Blakeslea* and *Phycomyces* [21,77,159]. Retinol is a particularly effective stimulator of carotenogenesis in the wild-type, *carA*, *carB*, and *carR* strains of *P. blakesleeanus*. The effects of both β-ionone and retinol are inhibited by cycloheximide [197], although the significance of this was recently challenged [198].

D. Miscellaneous Effectors

A range of widely differing effectors here have been reported to alter carotenoid biosynthesis in fungi (Table 8). In some cases no mode of action has been ascertained, and these effects may be of a very indirect nature. However, the stimulation of carotenoid formation in *Blakeslea* by phosphate, Cu^{2+}, fluoride, and penicillin is linked to an increase in the intracellular levels of prenyl pyrophosphates, either by the inhibition of phosphatases or by an increase in MVA kinase activity [162-164]. Whether or not the normal concentrations of these intermediates are rate limiting has not been demonstrated.

Several compounds mimic the photoinduction of carotenoid biosynthesis, e.g., *p*-chloromercuribenzoate [199], antimycin A and hydrogen peroxide [200]. The mode of action of these stimulators is unlikely to be the same as that of light, especially since, in *F. aquaeductuum* at least, the effect of chemoinduction is additive to that of photoinduction [201].

Finally, the demand for large amounts of naturally produced carotenoids has led to many attempts to supplement or alter the growth media of filamentous fungi in order to increase yields of carotenoids. The results of these efforts with the Mucorales were recently reviewed [21]. At present this approach is largely empirical, and so there is little understanding of the mechanisms that underly such increases of carotenes.

VIII. CAROTENOGENIC GENES

Although many fungal species synthesize a wide variety of carotenoid compounds (see Section IV and Table 4), only a limited number have been used in mutational studies. The major work has been carried out with *Gibberella fujikuroi*, *Neurospora crassa*, and *Phycomyces blakesleeanus*. Mutants affected in carotenoid metabolism are normally easily recognized by their differences in color compared to the wild type and result from alterations either of an enzyme directly involved in the carotenoid pathway (structural genes) or of another cellular component with regulatory interactions with carotenogenesis (regulatory genes). These two types of mutants can be distinguished by quantitative and qualitative analysis of their carotenoids, examination of their enzymes or cell-free systems, and by their responses to known regulatory effectors (light, chemicals, etc.). Complementation and mapping studies identify the number and distribution of the genes involved.

A. Structural Genes

Fungal mutants with lesions in specific carotenogenic enzymes are shown in Table 9. These mutants possess similar overall carotenoid levels to wild types but are characterized by changes in the patterns of carotenoids with an accumulation of intermediates prior to the blocked step. These mutants have helped elucidate the carotenoid biosynthetic pathways, the nature of the participating enzymes, and isolation of the encoding genes. Mutations in steps earlier than FPP likely to be lethal, the latter being a precursor for steroid as well as carotenoid biosynthesis (Fig. 5). Condensation of GGPP forms either 15-*cis*-phytoene or all-*trans*-phytoene, depending on the stereochemistry of hydrogen removal (206; see Section V). Both isomers are normally present in the same organism, but by far the predominant form found in fungi is the 15-cis isomer [88]. Since the later, more saturated carotenes are all-trans isomers, there must be a cis-trans isomerization step during the desaturation sequence. Experimental evidence suggests that this occurs at phytoene in *P. blakesleeanus* [97] and the existence of an isomerase enzyme is suggested by in vitro studies [146,147]. As yet, no mutant for isomerase activity has been identified.

The sequential desaturation of phytoene to lycopene involves a series of four dehydrogenations (see Fig. 7 and Section V) and raises the possibility of four discrete dehydrogenases, one specific to each step. However, from the mutants isolated so far (Table 9) it seems more likely that a single enzyme, phytoene dehydrogenase, is responsible for catalyzing all four steps (phytoene to lycopene) in fungi. *Ustilago* is a possible exception. The patterns of carotene accumulated by white and colored strains of *U. violacea* have been interpreted as resulting from lesions in three separate genes, each coding for different dehydrogenases: phytoene dehydrogenase (phytoene—neurosporene), neurosporene dehydrogenase (neurosporene—lycopene), and *cis*-β-zeacarotene dehydrogenase (β-zeacarotene—γ-carotene) [44]. As a cautionary note, Bejarano and coworkers [73] recently isolated a ζ-carotene-accumulating mutant (S442) of *P. blakesleeanus* which, by complementation analysis, mapped at the *carB* gene locus responsible for the single dehydrogenase.

Even fewer mutants have been reported blocked in cyclization (Table 9). In spite of extensive studies with *G. fujikuroi* and *N. crassa*, no easily recognizable lycopene-accumulating mutants have been identified. In both *P. blakesleeanus* [75,154,157] and *U. violacea* [44], only mutants accumulating lycopene and not γ-carotene have been observed. Therefore, it also seems likely that a single cyclase enzyme is responsible for the conversion

TABLE 9. Mutants with Altered Structural Genes

Organism	Mutant	Accumulated product	Defective enzyme	Ref.
Gibberella fujikuroi	SG4, SG75, SG76	PPPP	Phytoene synthetase	202-204
	SG43, SG78	Phytoene	Phytoene dehydrogenase	202-204
Neurospora crassa	*al*-1	Phytoene	Phytoene dehydrogenase	209,231
	al-2	PPPP	Phytoene synthetase	181
	al-3		GGPP synthetase	181,205
	ylo-1	Lycopene, α-carotene	3,4-Dehydrogenase	209
Phycomyces blakesleeanus	*carB*	Phytoene	Phytoene dehydrogenase	207
	carR (*carRA* complex locus)	Lycopene	Lycopene cyclase	75,154 157
Ustilago violacea	White (*wB*)	Phytoene	Phytoene dehydrogenase	4,44
	Yellow (*y*)	No lycopene	Neurosporene dehydrogenase	4,44
	AB278a-1 (double mutant)	*cis*-β-Zeacarotene	Neurosporene dehydrogenase and *cis*-β-zeacarotene dehydrogenase	4,44
	O/P complex locus	Lycopene (and some β-carotene)	Lycopene cyclase	44

of lycopene to β-carotene in fungi. Interestingly, in both organisms the cyclase gene is not simple but is associated with another function (*A*) in the *carRA* bifunctional gene in *P. blakesleeanus* and part of a complex locus (*O/P*) in *U. violacea*. Genetic analysis [154] showed that gene *carRA* contains two segments, the proximal segment, *carR* (5' end of mRNA), coding for lycopene cyclase [146,207], and the distal segment, *carA*, coding for *A* protein thought to be responsible for substrate transfer [208]. It is postulated that the *carRA* gene codes for a single polypeptide which is processed into the cyclase and A proteins. Therefore, nonsense or frame shift mutations in the *R* segment would result in the *carRA* phenotype and missense mutations in the *R* segment and any mutation in the *A* segment would produce the *carR* and *carA* phenotypes, respectively.

Cerda-Olmedo and coworkers suggest that the dehydrogenase (four copies of the *carB* gene product) and cyclase (two copies of the *carR* product) are sequentially arranged in a membrane-bound, multienzyme complex in *P. blakesleeanus*. This proposal is based on quantitative analyses of the nuclear proportions and carotenes produced in heterokaryons containing nuclei with the appropriate wild-type and mutant (*B* or *R*) alleles. Positive proof for such an enzyme complex requires its isolation and detailed characterization. Similarly, purified *carA* product is needed to determine its precise function. It is difficult to explain the observed absence of phytoene in *carA* mutants, both in vivo [207] and in vitro [146], simply by their lack of substrate transfer ability.

As shown in Table 9, two fungal mutants have been identified with lesions in enzymes involved in further modifications of carotenoids. These are the SG68 mutant of *G. fujikuroi* [202,203] and the *ylo*-1 mutant of *N. crassa* [209] defective in dioxygenase (?) and 3,4-dehydrogenase activities, respectively.

B. Regulatory Genes

Mutants defective in some regulatory function with respect to carotenogenesis produce either reduced or increased levels of carotenoids compared to the wild type or fail to respond to external effectors. By far the most studied system is that of *P. blakesleeanus*. Carotenoid biosynthesis in this organism is controlled by the end product, β-carotene, light, various chemicals, and sexual interactions (see Section VII), and mutants affected in these responses have been isolated. These mutants and some of their characteristics are summarized in Table 10. Studies with such mutants have been particularly helpful in trying to understand the regulatory mechanism involved (for review, see 22, 198).

As stated earlier, *carB* and *carR* are structural genes encoding the enzymes phytoene dehydrogenase and lycopene cyclase, respectively. When β-carotene synthesis is inhibited either genetically by *carB* or *carR* mutation [157] or chemically (by diphenylamine [120] or CPTA [210]), total carotene synthesis is greatly increased. Also, in partially inhibited situations, β-carotene levels remained constant even though the levels of intermediates were increased [72,208]. Such experiments support the conclusion that β-carotene regulates its own synthesis. Two groups of mutants, recessive to the *carS* or *carD* genes, accumulate large quantities of β-carotene compared to wild-type [76,211]. β-Carotene end product regulation is also thought to involve the *carS* gene product (pS) based on observations that introduction of the *carS* gene into *carB* mutants failed to stimulate phytoene synthesis [76] and the inability of chemical inhibitors to increase overall carotene production in *carS* mutants [74,134]. The *carD* mutants have carotene levels intermediate between wild-type and *carS* mutants. However, the *carDcarS* double mutants have similar levels of β-carotene to the *carS* mutants. It appears that the *carD* gene product (pD) is also implicated in β-carotene regulation but, by virtue of the epistatic nature of the *carS* over the *carD* mutation, is likely to function by increasing the amount or activity of pS [211]. Cerda-Olmedo believes that the *carA* gene product (pA) also plays a central role and suggests that *carA* mutants (i.e., those stimulated by retinol) have greatly reduced threshold levels for β-carotene inhibition [136,198]. Certainly, whatever its function, it is essential for operation of the carotenoid pathway since *carRA* mutants are unable to produce carotenes under any conditions [152].

TABLE 10. Carotenogenic Mutants for *Phycomyces blakesleeanus*[a]

Mutant gene(s)	Linkage[b] group	Proposed function	Major carotene	Dark levels (μg/g dry weight)	Response to[c]	
					Light	Retinol
Wild-type	—	—	β-Carotene	52	+	+
carB	IV	Phytoene dehydrogenase	Phytoene	1800	–	+
carR	IV	Lycopene cyclase	Lycopene	1600	–	+
carA	IV	Substrate transfer (pA)	β-Carotene	2	–	+
carRA	IV	R and A functions	Lycopene	2	–	–
carS	IV	End product control (pS)	β-Carotene	2700	–	+
carD	?	Control of pS	β-Carotene	958	+	+
carC	III	Photocarotenogenesis	β-Carotene	13	–	+
car1	?	Chemostimulation	β-Carotene	29	+	–
madA	II	Photoresponses	β-Carotene	56	–	+
madB	IX	Photoresponses	β-Carotene	54	–	+
picA	?	Photocarotenogenesis	β-Carotene	44	–	+
picB	?	Photocarotenogenesis	β-Carotene	46	–	+
carAcarS			β-Carotene	2-98	–	+

carBcarS	Phytoene	2153		+
carIcarS	β-Carotene	5595		–
carRAcarS	Lycopene	5		–
carBcarR	Phytoene	1430	–	
carAcarC	β-Carotene	3		
carBcarC	Phytoene	1900		
carRcarC	Lycopene	1650		
carScarC	β-Carotene	1460		

[a]Table compiled from Refs. 76, 77, 157, 194, 211-213.
[b]Mapping data from Ref. 232.
[c]+, Stimulation; -, no or partial stimulation.

Light stimulates carotenoid biosynthesis in *P. blakesleeanus* (see Section VII) and, as shown in Table 10, a number of mutants have been identified blocked in this effect. These include the single mutants *carB*, *carR*, *carA*, *carRA*, and *carS*, and the double mutants *carBcarR* and *carAcarS*, suggesting that this response also requires β-carotene and the gene products, pA and pS [157,212]. In addition, the *carC*, *picA*, *picB*, *madA*, and *madB* mutants all have reduced photocarotenogenesis [167,207, 212,213]. The *mad* mutants are also defective in other blue light responses, such as phototropism and photomorphogenesis [167,207], suggesting a common requirement for these gene products (for review see 214, 215). *carC* mutants produce lower dark levels of β-carotene than wild-type, but this negative effect is not expressed in the double mutants *carCcarB*, *carCcarR*, and *carCcarS* [213]. Studies with the heterokaryons, *carC***carS*, also showed that all three normal gene products were required for maximum photocarotenogenic response [216]. Thus, β-carotene, pA, and pS appear to be necessary for *carC* gene action. Typically, the β-carotene content of the wild type increases with light intensity above a certain threshold value. The intensity-response curves of the *mad* and *picA* mutants are similar to those of the wild-type but displaced to higher light intensities [183,212]. However, the *picB* and *carC* mutants produced β-carotene levels at all light intensities [212]. These types of behavior are believed to reflect lesions in differing stages in the photocarotenogenic pathway, the *mad* and *picA* mutants being defective in an "early" stage and the *picB* and *carC* mutants in a "late" stage [212].

Another light response, different from photocarotenogenesis, has been observed in the double mutant, *carScarE*. This mutant accumulates a red β-carotene complex in the dark and the yellow free form of β-carotene in the presence of light or retinol. Light has no effect in vitro. The *carE* gene behaves as an extranuclear genetic factor [217].

Carotenoid biosynthesis in *Phycomyces* is also stimulated by a variety of seemingly unrelated chemicals (see Section VII) and their effects on the different *car* mutants investigated. Retinol, dimethyl phthalate, and veronal were reported to increase carotene synthesis in wild types and mutants, *carB*, *carR*, *carA*, *carS*, and *carD* (retinol also in *carBcarS* and *carAcarS* double mutants) indicating that neither β-carotene nor the normal *carS* gene product is required for the chemostimulatory response [76,157, 158,211]. The *carRA* mutants remain unable to produce carotenes [152,157]. Recently, Cerda-Olmeda [198] reinterpreted the earlier experimental data [76,157] such that retinol affected only the wild-type and *carA* strains, the increased β-carotene synthesis resulting from an interaction between retinol and the *carA* gene product as well as counteracting the *carA* mutation. Dimethyl phthalate and veronal affect different strain ranges from retinol and are considered to have different modes of action [218]. Another class of mutants, *carI*, have been isolated which are normal with respect to β-carotene levels and photocarotenogenesis but fail to be stimulated by chemicals, including retinol [198,219]. The chemo- and photosensory pathways are linked since *picB* mutants are defective in both responses [212]. Therefore, chemoregulation requires, at least, functional *carA*, *carI*, and *picB* gene products.

Sexual interaction (or simulation by treatment of homokaryons with trisporates or intersexual heterokaryon formation) increases β-carotene production in wild-type, *carA*, *carS*, and *carD* mutants and is additive with the effects of light and retinol [76,77,157]. β-Carotene levels are important in sexual development since mutants either lacking or overproducing β-carotene are sexually incompetent [220]. Stimulation by trisporates also requires the presence of β-carotene in the receiving cells [77]. *carI* mutants are also insensitive to trisporates [77], thus linking the stimulatory pathways of trisporates and other chemicals.

Summarizing, a number of regulatory genes have been identified whose products interact with the carotenoid pathway mediating its regulation by β-carotene, light, chemicals, and sexual stimulation. In addition, certain of these gene products play a role in more than one stimulatory pathway. Cerda-Olmedo [198] proposed an interesting scheme which attempts to explain β-carotene regulation and its involvement in photo- and chemostimulation. The essential tenets are that formation of an inactive pA-β-carotene-pS complex determines the availability of pA and the flux of the pathway and any situation which reduces the levels of either β-carotene (*carB* and *carR* mutations) or pS (*carS* mutation) would increase carotenogenesis. He suggests that, in some way, light destabilizes the pA-β-carotene-pS complex permitting the pathway to operate and the mutant *carC* gene product counteracts the proposed light effect by stabilizing the complex. Retinol is thought to increase the flux of the carotenogenic pathway by combining with pA and therefore reducing the levels of the pA-β-carotene-pS complex.

However, this hypothesis fails to explain other experimental observations. For example, some double *carAcarS* mutants produce the same β-carotene levels as single *carA* mutants [76] although others produce more depending on the *carA* allele [212]. Protein synthesis inhibitors block the stimulation of β-carotene synthesis by light and chemicals [136,157]. Also, within the limits of the efficiencies of such systems, the characteristics of the in vitro carotene-synthesizing systems from the wild-type, *carA*, and *carS* cultures parallel those observed in vivo, suggesting control of the enzyme level rather than flux of the pathway [146]. A detailed study comparing carotene levels produced both in vivo and in vitro by different strains under different regulatory regimes confirms that increased enzyme activity plays a major role in distinguishing different mutants and regulation by light and retinol but, in certain instances (e.g., *carS*), other mechanisms may be involved (Bejarano and Cerda-Olmedo, personal communication).

Carotenoid biosynthesis is constitutive in *N. crassa* conidia but is induced by blue light in mycelia [170,221]. That photoinduction involves de novo synthesis of some enzymes is supported by the increase in phytoene synthetase and GGPP synthetase activities in cell-free extracts from cells after in vivo light treatment [181] and by protein synthesis inhibitor studies [222,223]. Indeed, using the *al*-3 gene as a probe, GGPP synthetase mRNA levels were shown to increase 15-fold in light-treated cultures [224]. A group of white collar (*wc*) mutants were isolated which remain white on illumination although, unlike the *al* mutants (discussed earlier), their conidia show normal constitutive carotenoid synthesis [225]. Also, the phytoene and GGPP synthetase activities are unchanged in a light-treated *wc* mutant [181]. Similarly, GGPP synthetase mRNA levels are unaffected by light in this mutant [224]. Therefore, the structural genes for carotenogenesis appear to be functional in these mutants but are blocked in photoinduction. All the wc mutants map at two loci, *wc*-1 in linkage group VII and *wc*-2 in linkage group I [226,227]. Observations that other blue light effects, such as phototropism of perithecal beaks and photoinduction of protoperitheca formation, are defective in *wc* mutants but not in *al* mutants help to substantiate the hypothesis that these *wc* genes regulate the initial photoresponse steps for all blue light effects in *Neurospora* [226,228]. In this sense, the *wc* genes appear to be comparable to the *madA* and *madB* genes in *Phycomyces*.

In another fungal species, *Fusarium aquaeductum*, carotenoid formation is normally strictly photoinduced [57,229]. *p*-Chloromercuribenzoate (PCMB) simulates photoinduction in that it induces carotenoid synthesis in the dark [200]. The PCMB effect is additive to that of light and, unlike photoinduction, is inhibited by thiols [201,230]. Both effects are blocked by protein and RNA synthesis inhibitors suggesting regulation at the transcriptional level [200]. A number of mutants have been described which lack photoregulation and which synthesize carotenoids in the dark [231,232]. However, these mutants still respond to PCMB [199,200]. It may be that two separate carotenogenic systems exist, one responsive to light and the other to PCMB [199,200].

Superproducer mutants of *G. fujikuroi* were isolated which accumulate carotenoids [202,204] and, at least under certain growth conditions, also gibberellins and sterols [203]. *Gibberella* differs from *Phycomyces* in that end product regulation appears not to operate [202].

IX. CONCLUSIONS

A thorough understanding of fungal carotenoid biosynthesis and its regulation must await the isolation and characterization of the proteins and genes responsible for carotenoid formation. To date, no fungal carotenogenic enzyme or regulatory protein has been purified and characterized. The enzymes are membrane-bound and therefore difficult but not impossible to purify. Some progress is being made, however, and we anticipate that the enzymes responsible for carotene biosynthesis will be purified in the near future, using modern protein purification techniques such as HPLC and affinity chromatography. The regulatory proteins pose different problems. They are likely to be present in very small quantities and there are no suitable assays for their identification.

One obvious route to carotenogenic gene isolation would be to screen appropriate cDNA or genomic DNA expression libraries with antibodies prepared to carotenogenic proteins. This task should be simplified by the relatively low haploid DNA levels in fungi and the possibility, at least in certain cases of isolating mRNA from induced (e.g., by light, chemicals, etc.) states. Another approach would be to identify either cDNA or genomic DNA clones which encode for the carotenogenic proteins by complementation of the different carotenoid mutants. A prerequisite for such an approach would be the development of highly efficient transformation systems and suitable cloning vectors for the fungi of interest. Such a system was developed for *Neurospora crassa* [233-236] and, not surprisingly, was used for the isolation of the *al*-1 [237,238], *al*-3 [224], and *ylo* genes [236,238]. It seems highly likely that the other *Neurospora* genes will be isolated by the same approach in the near future. An efficient transformation system for *Phycomyces* was also developed [239,240] and should soon be used successfully to identify the carotenoid genes. A transformation system for *G. fujikuroi* has also been recently reported [241]. In any case, their usefulness is limited to those with identifiable mutants. Obviously, as the different carotenogenic genes become available, it might be possible to use these as heterologous probes for the equivalent genes from other species. This approach is currently being used to identify the phytoene dehydrogenase genes of *Aspergillus giganteus* (this laboratory) and *P. blakesleeanus* (M. Jarayam, personal communication) and the lycopene cyclase of *P. blakesleeanus* (S. Torres-Martinez, personal communication). This is an exciting period in the application of modern recombinant DNA techniques to studies on carotenogenesis and its regulation, and hopefully will provide answers to many difficult questions.

A knowledge of the structure and properties of the carotenogenic enzymes will be of fundamental value to the agrochemical industry for the development of new commercial inhibitors (e.g., herbicides). Also, the isolated genes offer the potential for developing commercially acceptable sources for increased carotenoid production.

ABBREVIATIONS

acetyl-CoA	acetyl coenzyme A
AMO-1618	2-isopropyl-5-methyl-4-trimethylammonium chloride
CPTA	2-(4-chlorophenyl)triethylamine
CHAPS	3-(3-cholamidopropyl)dimethylammono-1-propanesulfonate
DMAPP	dimethylallyl pyrophosphate
ER	endoplasmic reticulum
FPP	farnesyl pyrophosphate
GGPP	geranylgeranyl pyrophosphate
GPP	geranyl pyrophosphate
HMG-CoA	3-hydroxy-3-methylglutaryl coenzyme A
HPLC	high-performance liquid chromatography
IPP	isopentenyl pyrophosphate
MVA	mevalonic acid
MVAP	mevalonic acid 5-phosphate
MVAPP	mevalonic acid 5-pyrophosphate
NMR	nuclear magnetic resonance
PCMB	p-chloromercuribenzoate
PPPP	prephytoene pyrophosphate
PSPP	presqualene pyrophosphate
TLC	thin-layer chromatography

REFERENCES

1. Goodwin, T. W. *The Biochemistry of the Carotenoids*, Vol. 1, 2nd ed. Chapman and Hall, London, 1980.
2. Castillo, R., Negre-Sadargues, G., and Level, R. General survey of the carotenoids in crustacea. In *Carotenoid Chemistry and Biochemistry* (G. Britton and T. W. Goodwin, eds.), Pergamon Press, Oxford, 1982, pp. 211-224.
3. Kayser, H. Carotenoids in insects. In *Carotenoid Chemistry and Biochemistry* (G. Britton and T. W. Goodwin, eds.), Pergamon Press, Oxford, 1982, pp. 195-210.
4. Ruddat, M. and Garber, E. D. Biochemistry, physiology and genetics of carotenogenesis in fungi. In *Secondary Metabolism and Differentiation in Fungi* (J. W. Bennett and A. Ciegler, eds.), Marcel Dekker, 1983, pp. 95-151.

5. Will, O. H. III, Newland, N. A., and Kovacs, A. Photoinactivation and long-term photosensitivity of seven species of the fungal genus Ustilago. *Bot. Gaz.*, *147*:461-465, 1986.
6. Will, O. H. III, and Scovel, C. A. Photoprotective functions of carotenoids. In *Carotenoids: Chemistry and Biology* (N. I. Krinsky, M. M. Mathews-Roth, and R. F. Taylor, eds.), Plenum Press, New York, 1990, pp. 229-236.
7. Austin, D. J., Bu'Luck, J. D., and Drake, D. The biosynthesis of trisporic acids from β-carotene via retinal and trisporal. *Experientia*, *26*:348-349, 1970.
8. Jerebzoff-Quintin, S., and Jerebzoff, S. Action de la diphénylamine sur la stabilisation de la période du rythme de zonation du Leptosphaeria michotii. Rôles comparés de la lumière et de l'antimycine A. *C.R. Acad. Sci. Ser. D.*, *278*:1239-1242, 1974.
9. Jayaram, M., Presti, D., and Delbruck, M. Light-induced carotene synthesis in Phycomyces. *Exp. Mycol.*, *3*:42-52, 1979.
10. Gooday, G. W., Fawcett, P., Green, D., and Shaw, G. The formation of fungal sporopollenin in the zygospore wall of Mucor mucedo: a role for the sexual carotenogenesis in the Mucorales. *J. Gen. Microbiol.*, *74*:233-239, 1973.
11. Klaui, H. Industrial and commercial uses of carotenoids. In *Carotenoid Chemistry and Biochemistry* (G. Britton and T. W. Goodwin, eds.), Pergamon Press, Oxford, 1982, pp. 309-328.
12. Peto, S. The marked differences between carotenoids and retinoids: methodological implications for biochemical epidemiology. *Cancer Surv.*, *2*:327-340, 1983.
13. Mathews-Roth, M. M. Carotenoids and cancer prevention—experimental and epidemiological studies. *Pure Appl. Chem.*, *57*:717-722, 1985.
14. Mathews-Roth, M. M. Photoprotection by carotenoids. *Fed. Proc.*, *46*:1890-1893, 1987.
15. Jansen, C. T. Placebo treatment of polymorphous light sensitivity. *Photodermatology*, *1*:201-202, 1984.
16. Jansen, C. T. Oral carotenoid treatment in polymorphous light eruption: a cross-over comparison with oxychloroquine and placebo. *Photodermatology*, *2*:166-169, 1986.
17. Raab, W. P., Tronnier, H., and Wiskemann, A. Photoprotection and skin colouring by oral carotenoids. *Dermatologica*, *171*:371-373, 1985.
18. Krinsky, N. I. Carotenoids in medicine. In *Carotenoids: Chemistry and Biology* (N. I. Krinsky, M. M. Mathews-Roth, and R. F. Taylor, eds.), Plenum Press, New York, 1990, pp. 279-291.
19. Ben-Amotz, A., and Avron, M. Accumulation of metabolites by halotolerant algae and its industrial potential. *Annu. Rev. Microbiol.*, *37*:95-119, 1983.
20. Ninet, L. and Renaut, J. Carotenoids. In *Microbial Technology*, Vol. 1, (H. J. Peppler and D. Perlman, eds.), New York, Academic Press, 1979, pp. 529-544.
21. Lampila, L. E., Wallen, S. E., and Bullerman, L. B. A review of factors affecting biosynthesis of carotenoids by the order Mucorales. *Mycopathogia*, *90*:65-80, 1985.
22. Cerda-Olmedo, E. Carotene mutants of Phycomyces. *Meth. Enzymol.*, *110*:220-243, 1985.
23. Commission on Biochemical Nomenclature. Tentative rules for the nomenclature of carotenoids. *Biochemistry*, *10*:4827-4837, 1971.

24. Commission on Biochemical Nomenclature. Nomenclature of carotenoids, amendments. *Biochemistry, 14*:1803-1804, 1975.
25. Straub, O. *Key to Carotenoids*. Birkhauser, Basel, 1987.
26. Britton, G. Carotenoids and polyterpenoids. *Terpenoids and Sterols, 11*:133-162, 1982.
27. Britton, G. Carotenoids and polyterpenoids. *Natural Prod. Rep., 3*: 591-614, 1980
28. Eugster, C. H. Characterization, chemistry and stereochemistry of carotenoids. *Pure Appl. Chem., 51*:463-506, 1979.
29. Britton, G., and Goodwin, T. W. (eds.), *Carotenoid Chemistry and Biochemistry*. Pergamon Press, Oxford, 1982.
30. Taylor, R. F., and Krinsky, N. I. (eds.), *Carotenoids: Chemistry and Biology*, Plenum Press, New York, 1990.
31. Britton, G. General carotenoid methods. *Meth. Enzymol., 111*:113-149, 1985.
32. Liaaen-Jensen, S., and Andrewes, A. g. Analysis of carotenoids and related polyene pigments. *Meth. Microbiol., 18*:235-255, 1985.
33. Davies, B. H. Carotenoids. In *Chemistry and Biochemistry of Plant Pigments*, Vol. 2 (T. W. Goodwin, ed.), Academic Press, New York, 1976, pp. 38-155.
34. Simpson, K. L., Chichester, C. O., and Phaff, H. J. Carotenoid pigments of yeasts. In *The Yeasts*, Academic Press, New York, 1971, pp. 493-515.
35. Muller, R. K. Berhard, K., Mayer, H., Ruttiman, A., and Vechi, M. Beitrag zur Analytik und synthese von 3-hydroxy-4-oxocarotinoiden. *Helv. Chim. Acta, 63*:1654-1664, 1980.
36. Jacobs, P. B., Le Boeuf, R. D., McCommas, S. A., and Taylor, J. D. The cleavage of carotenoid esters by cholesterol esterase. *Comp. Biochem. Physiol. B., B72*:157-160, 1982.
37. Britton, G., and Goodwin, T. W. Biosynthesis of carotenoids. *Meth. Enzymol., 18*:654-706, 1971.
38. Strain, H. H., and Sherma, J. Chloroplast pigments of higher plants, green algae, and brown algae and their influence upon the invention, modification and applications of Tswelt's chromatographic method. *J. Chromatogr., 73*:371-397, 1972.
39. Taylor, R. F., and Ikawa, M. Gas chromatography, gas chromatography—mass spectrometry and high-pressure liquid chromatography of carotenoids and retinoids. *Meth. Enzymol., 67*:2333-2361, 1980.
40. Schwartz, S. J., and Von Elbe, J. H. High performance liquid chromatography of plant pigments—review. *J. Liquid Chromatogr., 5*:43-73, 1982.
41. Taylor, R. F. Chromatography of carotenoids and retinoids. In *Advances in Chromatography*, Vol. 22 (J. C. Giddings, E. Grushka, J. Cazes, and P. R. Brown, eds.), Marcel Dekker, New York, 1983, pp. 157-213.
42. Pfander, H., Schurtenberger, H., and Meyer, V. R. Trennung von Carotinen und Diterpen mit Hochleistungflussig chromatographie. *Chimia, 34*:179-180, 1980.
43. Beyer, P., Kreuz, K., and Kleinig, H. β-carotene synthesis in isolated chromoplasts from Narcissus pseudonarcissus. *Planta, 150*:435-438, 1980.
44. Will, O. H. III., Ruddat, M., Garber, E. D., and Kezdy, F. J. Characterization of carotene accumulation in Ustilago violacea using high performance liquid chromatography. *Curr. Microbiol., 10*:57-64, 1984.
45. Ruddat, M., and Will, O. H. III. High performance liquid chromatography of carotenoids. *Meth. Enzymol., 111*:189-200, 1985.

46. Braumann, T., and Grimme, L. H. Reversed-phase HPLC of chlorophylls and carotenoids. *Biochim. Biophys. Acta, 637*:8-17, 1981.
47. Märki-Fischer, E., Marti, U., Buchecker, R., and Eugster, C. H. Das Carotinoidspektrum der Hagebutten von Rosa pumifera: Nachweis von (5Z)-Neursporin: Synthese von (3R, 15Z) - Rubixanthin. *Helv. Chim. Acta, 66*:494-513, 1983.
48. Rüedi, P. HPLC—a powerful tool in carotenoid research. *Pure Appl. Chem., 57*:793-800, 1985.
49. Daood, H. G., Biacs, P. A., Hoschke, A., Harkay-Vinkler, M., and Hajdu, F. Separation and identification of tomato fruit pigments by TLC and HPLC. *Acta Alimentaria, 16*:339-350, 1987.
50. Britton, G., Barry, P., and Young, A. J. The mode of action of diflufenican: its evaluation by HPLC. *Proc. Br. Crop. Protection Conf.—Weeds*, Vol. 3, 1987, pp. 1015-1022.
51. Fisher, C., and Kocis, J. A. Separation of paprika pigments by HPLC. *J. Agr. Food Chem., 35*:55-57, 1987.
52. Budziekiewicz, H. Mass spectra of carotenoids-labelling studies. In *Carotenoid Chemistry and Biochemistry* (G. Britton and T. W. Goodwin, eds.), Pergamon Press, Oxford, 1982, pp. 155-166.
53. Rose, M. E. Metastable ion techniques for the analysis of carotenoid isomers. In *Carotenoid Chemistry and Biochemistry* (G. Britton and T. W. Goodwin, eds.), Pergamon Press, Oxford, 1982, pp. 167-174.
54. Englert, G. N. M. R. of carotenoids—new experimental techniques. *Pure Appl. Chem., 54*:801-821, 1985.
55. Haxo, F. Studies on the carotenoid pigments of Neurospora I. Composition of the pigment. *Arch. Biochem., 20*:400-421, 1949.
56. Aasen, A. J., and Liaaen-Jensen, S. Fungal carotenoids. II. The structure of the carotenoic acid neurosporaxanthin. *Acta Chem. Scand., 19*:1843-1853, 1965.
57. Liaaen-Jensen, S. On fungal carotenoids and the natural distribution of spirilloxanthin. *Phytochemistry, 4*:925-931, 1965.
58. Arpin, N. Les carotinoides des discomycètes: Essai chimiotaxinomique. Doctoral thesis, Université de Lyon, Lyon, 1968.
59. Buckecker, R., Arpin, N., and Liaaen-Jensen, S. Absolute configuration of aleuriaxanthin. *Phytochemistry, 15*:1013-1014, 1976.
60. Valadon, L. R. G., Mummery, R. S., von Eijk, G. W., Roeymans, H. J., and Britton, G. Taxonomic implications of the carotenoids of Iodophanus carneus. *Trans. Br. Mycol. Soc., 74*:187-190, 1980.
61. Garber, I. D., Baird, M., and Chapman, D. Genetics of Ustilago violacea. I. Carotenoid mutants and carotenogenesis. *Bot. Gaz., 136*:341-346, 1975.
62. Czeczuyga, B. Investigations on carotenoids in fungi. VI. Representatives of the Helvellaceae and Morchellaceae. *Phyton. Horn. Austria, 19*:225-232, 1979.
63. Haxo, F. Carotenoids of the mushroom Cantharellus cinnobarinus. *Bot. Gaz., 112*:228-231, 1950.
64. Fiasson, J-L. Les carotenoides des Basidiomycetes: survol chimiotaxinomique. Doctoral thesis, Universite Lyon, 1968.
65. Alasoadura, S. O., and Visser, S. A. Pigment study of Sphaerobolus stellatus, *Mycopathol. Mycol. Appl., 47*:295-300, 1972.
66. Andrewes, A. G., and Starr, M. P. (3R,3'R)-Astaxanthin from the yeast Phaffia rhodozyma. *Phytochemistry, 15*:1009-1011, 1976.
67. Gribanovski-Sassu, O., and Foppen, F. H. The carotenoids of the fungus Epicoccum nigrum Link. *Phytochemistry, 6*:907-909, 1967.

68. Riley, G. J. P., and Bramley, P. M. The subcellular distribution of carotenoids in Phycomyces blakesleeanus C115 car42 mad-107(-). *Biochim. Biophys. Acta, 450*:429-440, 1976.
69. Mitzka-Schnabel, U., and Rau, W. The subcellular distribution of carotenoids in Neurospora crassa. *Phytochemistry, 19*:1409-1413, 1980.
70. Riley, G. J. P. and Bramley, P. M. The biosynthesis of carotenes in cell organelles of Phycomyces blakesleeanus C115 car42 mad-107 (-). *Cytobios, 34*:97-104, 1982.
71. Feofilova, E. P., and Pivovarova, T. M. Differentiation of Blakeslea trispora medium in relation to carotene production. *Mikrobiologiya, 45*:854-860, 1976.
72. Aragon, C. M. G., Murillo, F. J., de la Guardia, M. D., and Cerda-Olmedo, E. An enzyme complex for the dehydrogenation of phytoene in Phycomyces. *Eur. J. Biochem., 63*:71-75, 1976.
73. Bejarano, E. R., Govind, N. S., and Cerda-Olmedo, E. ζ-Carotene and other carotenes in a Phycomyces mutant. *Phytochemistry, 26*:2251-2254, 1987.
74. Lee, T-C., Rodriguez, D. B., Karasawa, I., Lee, T. H., Simpson, K. L., and Chichester, C. J. Chemical alteration of carotene biosynthesis in Phycomyces blakesleeanus and mutants. *Appl. Microbiol., 30*:988-993, 1975.
75. de la Guardia, M. D., Aragón, C. M. G., Murillo, F. J., and Cerda-Olmedo, E. A carotenogenic enzyme aggregate in Phycomyces: evidence from quantitative complementation. *Proc. Natl. Acad. Sci. USA, 68*:2012-2015, 1971.
76. Murillo, F. J., and Cerda-Olmedo, E. Regulation of carotene synthesis in Phycomyces. *Mol. Gen. Genet., 148*:19-24, 1976.
77. Govind, N. S., and Cerda-Olmedo, E. Sexual activation of carotenogenesis in Phycomyces blakesleeanus. *J. Gen. Microbiol., 132*:2775-2780, 1986.
78. Ciegler, A., Arnold, M., and Anderson, R. F. Microbiological production of carotenoids. IV. Effect of various grains on production of β-carotene by mated strains of Blakeslea trispora. *Appl. Microbiol., 7*:94-98, 1959.
79. Anderson, R. F., Arnold, M., Nelson, G. E. N., and Ciegler, A. Microbiological production of β-carotene in shaken flasks. *J. Agr. Food Chem., 6*:543-545, 1958.
80. Valadon, L. R. G., and Porter, D. The carotenoids of southern pine rust Cronartium fusiforme. *Phytochemistry, 13*:649-650, 1974.
81. Davies, B. H. The carotenoids of Rhizophylctis rosea. *Phytochemistry, 1*:25-29, 1961.
82. Arpin, N., Fiasson, J-L., Bouchez-Dangye-Caye, M. P., Francis, G. W., and Liaaen-Jensen, S. A new $C_{40}H_{56}$ carotene with a terminal methylene group. *Phytochemistry, 10*:1595-1601, 1971.
83. Avalos, J., Mackenzie, A., Nelki, D. S., and Bramley, P. M. Terpenoid biosynthesis in cell extracts of wild type and mutant strains of Gibberella fujikuroi. *Biochim. Biophys. Acta, 966*:257-265, 1988.
84. Foppen, F. H., and Gribanovski-Sassu, O. Lipids produced by Epicoccum nigrum in submerged culture. *Biochem. J., 106*:97-100, 1967.
85. Andrewes, A. G., Phaff, H. J., and Starr, M. P. Carotenoids of Phaffia rhodozyma, a red pigmented fermenting yeast. *Phytochemistry, 15*:1003-1007, 1976.
86. Arpin, N., and Jensen, S. L. Chemotaxonomic studies on fungi. Fungal carotenoids IV. New carotenoids isolated from Plectiana coccinea as tertiary esters. *Phytochemistry, 6*:995-1005, 1967.

87. Simpson, K. L., Nakayama, T. O. M., and Chichester, C. O. Biosynthesis of yeast carotenoids. *J. Bacteriol.*, *88*:1688-1694, 1964.
88. Spurgeon, S. L., and Porter, J. W. Biosynthesis of carotenoids. In *Biosynthesis of Isoprenoid Compounds*, Vol. 2 (J. W. Porter and S. L. Spurgeon, eds.), John Wiley and Son, New York, 1983, pp. 1-122.
89. Bramley, P. M. The in vitro biosynthesis of carotenoids. *Adv. Lipid Res.*, *21*:243-279, 1985.
90. Jones, B. L., and Porter, J. W. Biosynthesis of carotenes in higher plants. *CRC Crit. Rev. Plant Sci.*, *3*:295-324, 1986.
91. Harrison, D. M. The biosynthesis of carotenoids. *Natural Products Rep.*, *3*:205-215, 1986.
92. Dugan, R. E. Regulation of HMG-CoA reductase. In *Biosynthesis of Isoprenoid Compounds*, Vol. 1 (J. W. Porter and S. L. Spurgeon, eds.), John Wiley and Sons, New York, 1983, pp. 95-159.
93. Chichester, C. O. Nakayama, T., Mackinney, G., and Goodwin, T. W. On the incorporation of leucine-carbon into carotene by Phycomyces. *J. Biol. Chem.*, *214*:515-517, 1955.
94. Chichester, C. E., Yokoyama, H., Nakayama, T. O. M., Lukton, A., and Mackinney, G. Leucine metabolism and carotene biosynthesis. *J. Biol. Chem.*, *234*:598-602, 1959.
95. Poulter, C. D., and Rilling, H. C. Prenyl transferases and isomerase. In *Biosynthesis of Isoprenoid Compounds*, Vol. 1 (J. W. Porter and S. L. Spurgeon, eds.), John Wiley and Sons, New York, 1981, pp. 161-224.
96. Lütke-Brinkhaus, F., and Rilling, H. C. Purification of geranylgeranyl diphosphate synthase from Phycomyces blakesleeanus. *Arch. Biochem. Biophys.*, *266*:607-612, 1988.
97. Than, A., Bramley, P. M., and Davies, B. H. Stereochemistry of phytoene. *Phytochemistry*, *11*:3187-3192, 1972.
98. Herber, R., Maudinas, B., Villoutreix, J., and Granger, P. Formation de phytoene all-trans chez Mucor hiemalis. *Biochim. Biophys. Acta*, *280*:194-202, 1972.
99. Valadon, L. R. G., Herber, R., Villoutreix, J., and Maudinas, B. cis- and trans-Phytoene in Verticillium agaricinum. *Phytochemistry*, *12*:161-163, 1973.
100. Davies, B. H., Hallet, C. J., London, R. A., and Rees, A. F. The nature of θ-carotene. *Phytochemistry*, *13*:1209-1217, 1974.
101. Davies, B. H., and Rees, A. F. 7',8',11',12' - tetrahydro - γ-carotene: a novel carotene from Phycomyces blakesleeanus. *Phytochemistry*, *12*:2745-2750, 1973.
102. Williams, R. J. H., Charlton, J. M., Britton, G., and Goodwin, T. W. The stereospecific biosynthesis of phytoene and polyunsaturated carotenes. *Biochem. J.*, *104*:767-777, 1967.
103. McDermott, J. C. B., Britton, G., and Goodwin, T. W. Carotenoid biosynthesis in a Flavobacterium sp. Stereochemistry of hydrogen elimination in the desaturation of phytoene to lycopene, rubixanthin and zeaxanthin. *Biochem. J.*, *134*:1115-1117, 1973.
104. Davies, B. H. In *Lipids and Lipid Polymers in Higher Plants* (M. Tevini and H. K. Lichtenthaler, eds.), Springer-Verlag, Berlin and New York, 1977, pp. 197-217.
105. Qureshi, A. A., Andrewes, A. G., Qureshi, N., and Porter, J. W. The enzymatic conversion of cis- [^{14}C] phytofluene, trans- [^{14}C] phytofluene and trans-[^{14}C] ζ-carotene to more unsaturated acyclic, monocyclic and dicyclic carotenes by a cell-free preparation of red tomato fruits. *Arch. Biochem. Biophys.*, *162*:93-107, 1974.

106. Camara, B., and Moneger, R. Biosynthetic capabilities and localization of enzymatic activities in carotenoid metabolism of Capsicum annuum isolated chloroplasts. *Physiol. Veg.*, *20*:757-773, 1982.
107. Goodwin, T. W. Developments in carotenoid biochemistry over 40 years. *Biochem. Soc. Trans.*, *11*:473-483, 1983.
108. Hallenstet, M., Buchecher, R., Borch, G., and Liaaen-Jensen, S. Absolute configuration of β, γ-carotene and biosynthetic implications. *Phytochemistry*, *16*:583-585, 1977.
109. Goodwin, T. W., and Williams, R. J. H. A mechanism for the cylization of an acyclic precursor to form β-carotene. *Biochem. J.*, *94*:5C-7C, 1965.
110. Goodwin, T. W., and Williams, R. J. H. A mechanism for the biosynthesis of α-carotene. *Biochem. J.*, *97*:28C-32C, 1965.
111. Britton, G., Lockley, W. J. S., Powls, R., Goodwin, T. W., and Heyes, L. M. Carotenoid transformation during chloroplast development in Scenedsmus obliquus PG1 demonstrated by deuterium labelling. *Nature*, *268*:81-82, 1977.
112. Britton, G., Lockley, W. J. S., Patel, N. J., Goodwin, T. W. and Englert, G. Use of deuterium labelling to elucidate the stereochemistry of the initial step of the cyclization reaction in zeaxanthin biosynthesis in a Flavabacterium. *J. Chem. Soc. Chem. Commun.*, 656-656, 1977.
113. Valadon, L. R. G., and Mummery, R. S. Biosynthesis of neurosporaxanthin. *Microbios*, *1A*:3-8, 1969.
114. Davies, B. H., Hsu, W-J., and Chichester, C. O. The mechanism of the conversion of β-carotene into canthaxanthin by the brine shrimp, Artemia salina. *Comp. Biochem. Physiol.*, *33*:601-615, 1970.
115. Turian, G. Recherches sur l'action anticarotenogene de la diphenylamine et ses consequences sur la morphogenese reproductive chez Allomyces et Neurospora. *Physiol. Plant.*, *10*:667, 1957.
116. Thomas, D. M. and Goodwin, T. W. Studies on carotenogenesis in Blakeslea trisposa. I. General observations on synthesis in mated and unmated strains. *Phytochemistry*, *6*:355-360, 1967.
117. Foppen, F. H., and Gribanovski-Sassu, O. Carotenogenesis in diphenylamine-treated Epicoccum nigrum. Link. *Biochim. Biophys. Acta*, *176*:357-366, 1969.
118. Davies, B. H., Jones, D., and Goodwin, T. W. The problem of lycopersene in Neurospora crassa. *Biochem. J.*, *87*:326-329, 1963.
119. Goodwin, T. W., Jamikorn, M., and Willmer, J. S. Studies in carotenogenesis 8. Further observations concerning the action of diphenylamine in inhibiting the synthesis of β-carotene in Phycomyces blakesleeanus. *Biochem. J.*, *53*:531-538, 1953.
120. Olson, J. H., and Knizley, H. The effect of diphenylamine on carotenoid, sterol and fatty acid synthesis in Phycomyces blakesleeanus. *Arch. Biochem. Biophys.*, *97*:138-145, 1962.
121. Slechta, L., Gabriel, O., and Hoffmann-Ostenhof, O. Separation of carotenoid synthesis from fat synthesis in Rhodotorula gracilis by diphenylamine. *Nature*, *181*:268-269, 1958.
122. Valadon, L. R. G., and Mummery, R. S. Inhibition of carotenoid biosynthesis in a mutant of Verticillium albo-atrum. *J. Gen. Microbiol.*, *45*:531-540, 1966.
123. Valadon, L. R. G., and Mummery, R. S. Carotenogenesis in Verticillium agaricinum in response to nicotine and CPTA. *Microbios*, *10A*:97-104, 1974.

124. Valadon, L. R. G., Bramley, P. M., and Mummery, R. S. The effect of 9-fluorenone on carotenogenesis in Phycomyces blakesleeanus. *Microbiols Lett.*, *17*:81-86, 1981.
125. Valadon, L. R. G., and Mummery, R. S. Effect of certain inhibitors of carotenogenesis in Verticillium agaricinum. *Microbios*, *7*:173-180, 1973.
126. Mackinney, G., Nakayama, T., and Chichester, C. O. On the biosynthesis of carotenoids. *J. Biol. Chem.*, *220*:759-763, 1956.
127. Simpson, K. L., Nakayama, T. O. M., and Chichester, C. O. Biosynthesis of yeast carotenoids in Rhodotorula glutinis 48-23T. *J. Bacteriol.*, *88*:1688-1694, 1964.
128. Nusbaum-Cassuto, E., Villoutreix, J., and Malenge, J. P. Biosynthèse des carotenoids acycliques chez un mutant de Rhodotorula mucilaginosa. *Biochim. Biophys. Acta*, *136*:459-472, 1967.
129. Scharf, S. S., and Simpson, K. L. Attempts to detect lycopersene formation in yeast. *Biochem. J.*, *106*:311-315, 1968.
130. Bramley, P. M., Clarke, I. E., Sandmann, G., and Böger, P. Inhibition of carotene biosynthesis in cell extracts of Phycomyces blakesleeanus. *Z. Naturforsch.*, *39C*:460-463, 1984.
131. Hsu, W-J., Poling, S. M., and Yokoyama, H. Effect of amines on carotenogenesis of Blakeslea trisposa. *Phytochemistry*, *13*:415-419, 1974.
132. Elahi, M., Lee, T-C., Simpson, K. L., and Chichester, C. O. Effect of CPTA on the biosynthesis of carotenoids by Phycomyces blakesleeanus mutants. *Phytochemistry*, *12*:1633-1639, 1973.
133. Hayman, E. P., Yokoyama, H., Chichester, C. O., and Simpson, K. L. Carotenoid biosynthesis in Rhodotorula glutinis. *J. Bacteriol.*, *120*:1339-1343, 1974.
134. Clarke, I. E., de la Concha, A., Murillo, F. J., Sandmann, G., Skone, E. J., and Bramley, P. M. The effect of diphenylamine on carotenogenesis in Phycomyces blakesleeanus. *Phytochemistry*, *22*:435-439, 1983.
135. Davies, B. H. Carotene biosynthesis in fungi. *Pure Appl. Chem.*, *35*:1-28, 1973.
136. Murillo, F. J. Effect of CPTA on carotenogenesis by Phycomyces carA mutants. *Plant Sci. Lett.*, *17*:201-205, 1980.
137. El-Jack, M., Bramley, P. M., and Mackenzie, A. Carotene biosynthesis by a cell extract of Aspergillus giganteus mut. alba. *Phytochemistry*, *26*:2525-2529, 1987.
138. El-Jack, M., Mackenzie, A., and Bramley, P. M. The photoregulation of carotenoid biosynthesis in Aspergillus giganteus mut. alba. *Planta*, *174*:59-66, 1987.
139. Neujahr, H. Y., and Björk, L. Utilisation of malonyl-CoA for the biosynthesis of β-carotene and ergosterol in cell-free preparations from Blakeslea trispora. *Acta Chem. Scand.* *24*:2361-2365, 1970.
140. Bobowki, G. S., Barker, W. G., and Subden, R. E. The conversion of [2-^{14}C] mevalonic acid into triterpenes and tetraterpenes by cell-free extracts of a Neurospora crassa albino mutant. *Can. J. Bot.*, *55*:2137-2141, 1977.
141. Spurgeon, S. L., Turner, R. V., and Harding, R. W. Biosynthesis of phytoene from isopentenyl pyrophosphate by a Neurospora enzyme system. *Arch. Biochem. Biophys.*, *197*:23-29, 1979.

142. Mitzka-Schnabel, U., and Rau, W. Subcellular site of carotenoid biosynthesis in Neurospora crassa. *Phytochemistry, 20*:63-69, 1981.
143. Mitzka-Schnabl, U. Carotenogenic enzymes of Neurospora. *Pure Appl. Chem., 57*:667-669, 1985.
144. Lee, T-C., and Chichester, C. O. Geranylgeranyl pyrophosphate as the condensing unit for enzymatic synthesis of carotenes. *Phytochemistry, 8*:603-609, 1969.
145. Bramley, P. M., and Davies, B. H. Carotene biosynthesis by cell extracts of mutants of Phycomyces blakesleeanus. *Phytochemistry, 14*:463-469, 1975.
146. De la Concha, A., Murillo, F. J., Skone, E. J., and Bramley, P. M. In vitro carotenogenesis by wild type and mutants of Phycomyces blakesleeanus. *Phytochemistry, 22*:441-445, 1983.
147. Bramley, P. M., and Davies, B. H. β-Carotene biosynthesis by extracts of the C115 mutant of Phycomyces blakesleeanus. *Phytochemistry, 15*:1913-1916, 1976.
148. Bramley, P. M., Than, A., and Davies, B. H. Alternative pathways of carotene cyclization in Phycomyces blakesleeanus. *Phytochemistry, 16*:235-238, 1977.
149. Gregonis, D. E., and Rilling, H. C. The stereochemistry of trans-phytoene synthesis. Some observations on lycopersene as a carotene precursor and a mechanism for the synthesis of cis and trans-phytoene. *Biochemistry, 13*:1538-1542, 1974.
150. Dogbo, O., Laferriere, A., D'Harlingue, A., and Camara, B. Carotenoid biosynthesis: Isolation and characterization of a bifunctional enzyme catalyzing the synthesis of phytoene. *Proc. Natl. Acad. Sci. USA, 85*:7054-7058, 1988.
151. Bramley, P. M. Carotenogenic enzymes from Phycomyces. *Pure Appl. Chem., 54*:671-673, 1985.
152. Eslava, A. P., and Cerda-Olmedo, E. Genetic control of phytoene dehydrogenation in Phycomyces. *Plant. Sci. Lett., 2*:9-14, 1974.
153. Bramely, P. M., and Taylor, R. F. The solubilization of carotenogenic enzymes of Phycomyces blakesleeanus. *Biochim. Biophys. Acta, 839*:155-160, 1985.
154. Torres-Martinez, Z., Murillo, F. J., and Cerda-Olmedo, E. Genetics of lycopene cyclization and substrate transfer in β-carotene biosynthesis in Phycomyces. *Genet. Res., 36*:299-309, 1980.
155. Ninet, L., Renaut, J., and Tissier, R. Activation of the biosynthesis of carotenoids by Blakeslea trispora. *Biotechnol. Bioeng., 11*:1195-1210, 1969.
156. Dandekar, S., Modi, V. V., and Jani, K. Chemical regulators of carotenogenesis in Blakeslea trispora. *Phytochemistry, 19*:795-798, 1980.
157. Eslava, A. P., Alvarez, M. I., and Cerda-Olmedo, E. Regulation of carotene biosynthesis in Phycomyces by vitamin A and β-ionone. *Eur. J. Biochem., 48*:617-623, 1974.
158. Cerda-Olmedo, E., and Hüttermann, A. Förderung und Hemmung der carotinsynthese bei Phycomyces durch Aromaten. *Angewandte Botanik., 69*:59-70, 1986.
159. Gooday, G. W. Fungal sex hormones. *Annu. Rev. Biochem., 43*:35-49, 1974.
160. Dholakia, J. N., and Modi, V. V. Effect of phosphate concentration on carotenogenesis in Blakeslea trispora. *Biochem. Int., 7*:599-603, 1983.

161. Dholakia, J. N., and Modi, V. V. Regulation of carotenogenesis by inorganic phosphate in Blakeslea trispora. *J. Gen. Microbiol.*, *130*: 2043-2049, 1984.
162. Desai, H. G., Desai, J. D., and Modi, V. V. Regulation of carotenogenesis in Phycomyces blakesleeanus. *Indian J. Microbiol.*, *13*:57-62, 1973.
163. Govind, N. S., Amin, A. R., and Modi, V. V. Stimulation of carotenogenesis in Blakeslea trispora by cupric ions. *Phytochemistry*, *21*:1043-1044, 1982.
164. Desai, H. G., and Modi, V. V. Stimulation of carotenogenesis by penicillin in Blakeslea trispora. *Phytochemistry*, *16*:1373-1376, 1977.
165. Anderson, R. F., Arnold, N., Nelson, G. E. N., and Ciegler, A. Microbiological production of β-carotene in shaken flasks. *J. Agr. Food Chem.*, *6*:543-545, 1958.
166. Rau, W. Photoregulation of carotenoid biosynthesis. In *Biosynthesis of Isoprenoid Compounds*, Vol. 2 (J. W. Porter and S. L. Spurgeon, eds.), John Wiley and Sons, New York, 1983, pp. 123-157.
167. Bergman, K., Eslava, A. P., and Cerda-Olmedo, E. Mutants of Phycomyces with abnormal phototropism. *Mol. Gen. Genet.*, *123*:1-16, 1973.
168. Sandmann, G., and Hilgenberg, W. Förderung der β-Carotinsynthese durch Licht bei Phycomyces blakesleeanus Bgff. *Biochem. Physiol. Pflanz.*, *172*:401-407, 1978.
169. Zalokar, M. Studies on the biosynthesis of carotenoids in Neurospora crassa. *Arch. Biochem. Biophys.*, *50*:71-80, 1954.
170. Harding, R. W., Huang, P. C., and Mitchell, H. K. Photochemical studies of the carotenoid biosynthetic pathway in Neurospora crassa. *Arch. Biochem. Biophys.*, *129*:696-707, 1969.
171. Zurzycka, A. Aspergillus giganteus Wehm. mut. alba Zurz. *Acta Soc. Bot. Pol.*, *32*:715-718, 1963.
172. Osman, M., and Valadon, L. R. G. Effect of light quality on the photoinduction of carotenoid synthesis in Verticillium agaricinum. *Microbios*, *18*:229-234, 1978.
173. Presti, D., and Delbruck, M. Photoreceptors for biosynthesis, energy, storage and vision. *Plant Cell Environ.*, *1*:81-100, 1978.
174. Bjorn, G. S. Reinvestigation of the proposed red light effect on carotenogenesis in the fungus Verticillium agaricinum. *Physiol. Plant.*, *68*:648-652, 1986.
175. Harding, R. W., and Shropshire, W., Jr. Photocontrol of carotenoid biosynthesis. *Annu. Rev. Plant Physiol.*, *31*:217-238, 1980.
176. Rau, W. Mechanism of photoregulation of carotenoid biosynthesis in plants. *Pure Appl. Chem.*, *57*:777-784, 1985.
177. Schrott, E. L. Fluence response relationship of carotenogenesis in N. crassa. *Planta*, *150*:174-179, 1980.
178. Schrott, E. L. The biphasic fluence response of carotenogenesis in Neurospora crassa: temporary insensitivity of the photoreceptor system. *Planta*, *151*:371-374, 1981.
179. Jayaram, M., Presti, D., and Delbruck, M. Light induced carotene synthesis in Phycomyces. *Exp. Mycol.*, *3*, 42-52, 1979.
180. Mitzka-Schnabel, U., Warm, E., and Rau, W. Light-induced changes in the protein pattern translated in vivo and in vitro accompanying carotenogenesis in Neurospora crassa and Fusarium aquaeductuum. In *Blue Light Effects in Biological Systems* (H. Senger, ed.), Springer-Verlag, Berlin, 1984, pp. 264-269.

181. Harding, R. W., and Turner, R. V. Photoregulation of the carotenoid biosynthetic pathway in albino and white collar mutants of Neurospora crassa. *Plant Physiol.*, *68*:745-749, 1981.
182. Lansbergen, J. C., Renaud, R. L., and Subden, R. E. Phytoene photoinduction in a carotenoid mutation of Neurospora crassa. *Can. J. Bot.*, *54*:2445-2448, 1976.
183. Raugei, G., Dohrmann, U., Pohl, U., and Russo, V. E. A. Kinetics of photoaccumulation of β-carotene in Phycomyces blakesleeanus. *Planta*, *155*:296-300, 1982.
184. Pohl, U., Dohrmann, U., Raugei, G., and Russo, V. E. A. Influence of blue light on the accumulation of carotenoids in Phycomyces blakesleeanus. *Ber. Deutsch. Bot. Ges.*, *97*:327-333, 1984.
185. Tada, M. H., Shiroishi, M., Hasegaura, K. Suzuki, T., and Iwai, S. Mechanism of photoregulated carotenogenesis in Rhodotorula minuta IV. Effect of light on the production of ergosterol and phytoene and on the composition of carotenoid pigments. *Plant Cell Physiol.*, *23*:607-614, 1982.
186. Schrott, E. L. and Rau, W. Versuche zum direkten Nachweiss einer Beteiligung von mRNS an der Photoinduktion der Carotinoidsynthese. *Ber. Deutsch Bot. Ges.*, *88*:233-243, 1975.
187. Schrott, E. L., and Rau, W. Evidence for a photoinduced synthesis of poly(A)-containing mRNA in Fusarium aquaeductuum. *Planta*, *136*: 45-48, 1977.
188. Afansieva, T. P., Filippovich, S. Yu., Sokolovsky, V. Yu., and Kritsky, M. S. Developmental regulation of NAD^+ kinase in Neurospora crassa. *Arch. Microbiol.*, *133*:307-311, 1982.
189. Kritsky, M. S., Sokolovsky, V. Y., Belozerskaya, T. A., and Chernyshova, E. K. Relationship between cyclic AMP level and accumulation of carotenoid pigments in Neurospora crassa. *Arch. Microbiol.*, *133*:206-208, 1982.
190. Kritsky, M. S. Photoregulation of fungal cell metabolism. *FEMS Symp. Environ. Regul. Microb. Metab.*, *23*:137-144, 1985.
191. Chernysheva, E. K., and Kritskii, M. S. Nicotinamide coenzyme pool and regulation of asexual light cycle and carotenogenesis in N. crassa. *Pnkl. Biokhim. Mikrobiol.*, *22*:471-479, 1986.
192. Garton, G. A., Goodwin, T. W., and Lijinsky, W. Studies in carotenogenesis. I. General conditions governing β-carotene synthesis by the fungus Phycomyces blakesleeanus Burgff. *Biochem. J.*, *48*:154-163, 1951.
193. Sutter, R. P., Capage, D. A., Harrison, T. L., and Kren, W. A. Trisporic acid biosynthesis in separate (+) and (-) cultures of Blakeslea trispora: identification by Mucor assay of two mating-type specific components. *J. Bacteriol.*, *114*:1074-1082, 1974.
194. Murillo, F. J., Calderon, L., Lopez-Diaz, I., and Cerda-Olmedo, E. Carotene-superproducing strains of Phycomyces. *Appl. Environ. Microbiol.*, *36*:639-642, 1978.
195. Rao, S., and Modi, V. V. Carotenogenesis: possible mechanism of action of trisporic acid in Blakeslea trispora. *Experientia*, *33*:31-33, 1977.
196. Bu'Luck, J. D., Drake, D., and Winstanley, D. J. Specificity and transformation of the trisporic acid series of fungal sex hormones. *Phytochemistry*, *11*:2011-2018, 1972.

197. Murillo, F. J., and Cerda-Olmedo, E. Regulation of carotene synthesis in Phycomyces. *Planta, 150*:134-139, 1976.
198. Cerda-Olmedo, E. Carotene. In *Phycomyces* (E. Cerda-Olmedo and E. D. Lipson, eds.), Cold Spring Harbor Lab., New York, 1987, pp. 199-222.
199. Parn, P., and Seviour, R. J. Pigments induced by organomercurial compounds in Cephalosporium disopyros. *J. Gen. Microbiol., 85*:229-236, 1974.
200. Theimer, R. R., and Rau, W. Untersuchungen über die Lichtabhängige Carotinoidsynthese V. Aufhebung der lichtinduktion durch Reduktionsmittel und Ersatz des Lichts durch Wasserstoffperoxid. *Planta, 92*: 129-137, 1970.
201. Theimer, R. R., and Rau, W. Untersuchungen über die Lichtabhängige Carotinoidsynthese VIII. Die unterschiedlichen Wirkungsmechanismen von Licht und Mercuribenzoat. *Planta, 106*:331-343, 1972.
202. Avalos, J., and Cerda-Olmedo, E. Carotenoid mutants of Gibberella fujikuroi. *Curr. Genet., 11*:505-511, 1987.
203. Nelki, D. S. Terpenoid biosynthesis in Gibberella fujikuroi. Ph.D. thesis, University of London, 1987.
204. Avalos, J., and Cerda-Olmedo, E. Chemical modification of carotenogenesis in Gibberella fujikuroi. *Phytochemistry, 25*:1837-1841, 1986.
205. Kushwaha, S. C., Kates, M., Renaud, R. L., and Subden, R. E. The terpenyl pyrophosphates of wild type and tetraterpene mutants of Neurospora crassa. *Lipids, 13*:352-355, 1978.
206. Williams, R. J. H., Britton, G., Charlton, J. M., and Goodwin, T. W. The stereospecific biosynthesis of phytoene and polyunsaturated carotenes. *Biochem. J., 104*:767-777, 1967.
207. Ootaki, T., Lightly, A. C., Delbruck, M., and Hsu, W. J. Complementation between mutants of Phycomyces deficient with respect to carotenogenesis. *Mol. Gen. Genet., 121*:57-70, 1973.
208. Murillo, F. J., Torres-Martinez, S., Aragon, C. M. G., and Cerda-Olmedo, E. Substrate transfer in carotene biosynthesis in Phycomyces. *Eur. J. Biochem., 119*:511-516, 1981.
209. Goldie, A. H., and Subden, R. E. The neutral carotenoids of wild type and mutant strains of Neurospora crassa. *Biochem. Genet., 10*:275-284, 1973.
210. Hsu, W. J., Yokoyama, H., and Coggins, C. W. Carotenoid synthesis in Blakeslea trispora. *Phytochemistry, 11*:2985-2990, 1972.
211. Salgado, L. M., Bejarano, E. R., and Cerda-Olmedo, E. unpublished.
212. Lopez-Diaz, I., and Cerda-Olmedo, E. Relationship of photocarotenogenesis to other behavioural and regulatory responses in Phycomyces. *Planta, 150*:134-139, 1980.
213. Revuelta, J. L., and Eslava, A. P. A new gene (carC) involved in regulation of carotenogenesis in Phycomyces. *Mol. Gen. Genet., 192*:225-229, 1983.
214. Cerda-Olmedo, E. Behavioural genetics of Phycomyces. *Annu. Rev. Microbiol., 31*:535-547, 1977.
215. Lipson, E. D. Sensory transduction in Phycomyces photoresponses. In *The Blue Light Syndrome* (H. Senger, ed.), Springer-Verlag, Berlin, 1980, pp. 110-118.
216. Revuelta, J. L., and Eslava, A. P. Photoregulation of carotenogenesis in Phycomyces. *Curr. Gen., 8*:261-264, 1984.

217. De la Concha, A., and Murillo, F. J. Accumulation of a complex form of β-carotene by Phycomyces blakesleeanus cytoplasmic mutants. *Planta, 161*:233-239, 1984.
218. Bejarano, E. R., Parra, F., Murillo, F. J., and Cerda-Olmedo, E. End-product regulation of carotenogenesis in Phycomyces. *Arch. Microbiol., 150*:209-214, 1988.
219. Roncero, M. I. G., and Cerda-Olmedo, E. Genetics of carotene biosynthesis in Phycomyces. *Curr. Genet., 5*:5-8, 1982.
220. Sutter, R. P. Mutations affecting sexual development in Phycomyces blakesleeanus. *Proc. Natl. Acad. Sci. USA, 72*:127-130, 1975.
221. Zalokar, M. Biosynthesis of carotenoids in Neurospora. Action spectrum of photoactivation. *Arch. Biochem. Biophys., 51*:318-325, 1955.
222. Harding, R. W. and Mitchell, H. K. The effect of cycloheximide on carotenoid biosynthesis in Neurospora crassa. *Arch. Biochem. Biophys., 128*:814-818, 1968.
223. Rau, W., Lindemann, I., and Rau-Hund, A. Untersuchungen über die lichtabhangige Carotenoidsynthese. III. Die Farbstoffbildung von Neurospora crassa in Submerskultur. *Planta, 80*:309-316, 1968.
224. Nelson, M. A., Morelli, G., Carattoll, A., Romano, N., and Macino, G. Molecular cloning of a Neurospora crassa carotenoid biosynthetic gene (albino-3) regulated by blue light and the products of the white collar genes. *Mol. Cell. Biol., 9*:1271-1276, 1989.
225. Perkins, D. D., Glassey, M., and Bloom, B. A. New data on markers and rearrangements in Neurospora. *Can. J. Genet. Cytol., 4*:187-205, 1962.
226. Degli-Innocenti, F., and Russo, V. E. Isolation of new white collar mutants of Neurospora crassa and studies on their behavior in the blue light-induced formation of protoperithecia. *J. Bacteriol., 159*:757-761, 1984.
227. Perkins, D. D., Radford, A., Newmeyer, D., and Björkman, M. Chromosomal loci of Neurospora crassa. *Microbiol. Rev., 46*:426-570, 1982.
228. Harding, R. W., and Melles, S. Genetic analysis of phototropism of Neurospora crassa perithecial beaks using white collar and albino mutants. *Plant Physiol., 72*:996-1000, 1983.
229. Isler, O. Introduction. In *Carotenoids* (O. Isler, ed.), Birkhauser, Basel, 1972, pp. 11-28.
230. Lang, W., and Rau, W. Untersuchungen über die lichtabhängige carotinoidsynthese. IX Zum induktions-mechanismus der carotinoid-bildenden enzyme bei Fusarium aquaeductuum. *Planta, 106*:345-354, 1972.
231. Subden, R. E., and Goldie, A. H. Biochemical analyses of isoallelic series at the al-1 locus of Neurospora crassa. *Genetica, 44*:615-620, 1973.
232. Orejas, M., Alvarez, M. I., and Eslava, A. P. *Abstr. Phycomyces Meet.*, La Rabida, Huelva, Spain, 1986.
233. Vollmer, S. J., and Yanofsky, C. Efficient cloning of genes of Neurospora crassa. *Proc. Natl. Acad. Sci. USA, 83*:4869-4873, 1986.
234. Case, M. E., Schweizer, M., Kushner, S. R., and Giles, N. H. Efficient transformation of Neurospora crassa by utilizing hybrid plasma DNA. *Proc. Natl. Acad. Sci. USA, 76*:5259-5263, 1979.
235. Akins, R. A. and Lambowitz, A. M. General method for cloning Neurospora crassa nuclear genes by complementation of mutants. *Mol. Cell. Biol., 5*:2272-2278, 1985.

236. Orbach, M. J., Porro, E. B., and Yanofsky, C. Cloning and characterization of the gene for β-tubulin from a benomyl-resistant mutant of Neurospora crassa and its use as a dominant selectable marker. *Mol. Cell Biol.*, *6*:2452-2461, 1986.
237. Schmidhauser, T. J., Lauter, F. R., Russo, V. E. A., and Yanofsky, C. Cloning, sequence, and photoregulation of *al-1*, a carotenoid biosynthetic gene of Neurospora crassa. *Mol. Cell. Biol.*, *10*:5064-5070, 1990.
238. Schmidhauser, T. *Fungal Gen. Newslett.*, *34*:26, 1987.
239. Arnau, J., Murillo, F. J., and Torres-Martinez, S. Expression of Tn5-derived kanamycin resistance in the fungus Phycomyces blakesleeanus. *Mol. Gen. Genet.*, *212*:375-377, 1988.
240. Revuelta, J. L., and Jayaram, M. Transformation of Phycomyces blakesleeanus to G-418 resistance by an autonomously replicating plasmid. *Proc. Natl. Acad. Sci. USA*, *83*:7344-7347, 1986.
241. Sanchez-Fernandez, R., Campbell, E. I., Macro, J. A., Unkles, S. E., Cerda-Olmedo, E., and Kinghorn, J. R. Transformation of the filamentous fungus Gibberella fujikuroi using the A. niger *niaD* gene-encoding nitrate reductase. *Mol. Gen. Genet.* (in press).

17

NUCLEIC ACIDS AND NUCLEOTIDES

DAVID H. GRIFFIN *College of Environmental Science and Forestry, State University of New York, Syracuse, New York*

I. INTRODUCTION

Nucleic acids and nucleotides are important to the biotechnology of fungi for several reasons, directly as products of commercial interest, indirectly as facilitators of the genetic modification of fungi to obtain useful properties for other purposes, and as sources of useful genetic information that can be transferred to other organisms. Fungi are important to the fermentation industry for the production of a wide variety of products including antibiotics, chemicals, enzymes, and mushrooms. They are used for the modification of other materials, e.g., steroids for the drug industry, wood pulping for the paper industry, and lignin wastes and other recalcitrant molecules in the waste disposal industry. They have many potentially useful properties as symbionts, both pathogens and mutualists, in biological control of insect pests and weeds, and through their mycorrhyzal associations in enhancing the growth of plants for agriculture, forestry, and land reclamation. As recombinant organisms, fungi may be developed with specific functions tailored to an extensive array of applications. This discussion will deal with the properties of fungal nucleic acids and nucleotides for both their direct and indirect applications.

II. DNA

The DNA of fungi, like other eukaryotes, is found in several places within the organism. The principal genome is in the nucleus, but other genetic elements include the mitochondrial genome and plasmid DNA located both in nuclei and mitochondria. Considerable variation occurs among the fungi in all of these genetic elements. This variation is being used extensively to characterize species and strains of fungi genetically for taxonomic and population studies [1-9] and has been proposed as a means of positive identification specific enological strains of *Saccharomyces cerevisiae* [10].

A. Nuclear Genomic DNA

The genomes of fungi were among the smallest of eukaryotes (Table 1), generally occurring within the range of 3-11 times the size of *Escherichia*

TABLE 1. Genome Size and Organization

Fungus	Genome (Mbp)	Multiple of *E. coli*	Repetitive (%)	Reiteration frequency	Foldback (%)	Ref.
Chytridiomycetes						
Allomyces arbuscula	35	—	10-15	320	10	13,20
Oomycetes						
Achlya ambisexualis	44	11	17.6	91	2	21
Achlya bisexualis	46	—	2 14	2200 80	2	18
Saprolegnia	200	—	—	—	—	11
Zygomycotina						
Phycomyces blakesleeanus	31	6.7	20	100	10	14
Mucor azygospora	30-34	8.3	15-18	103	—	22,23
Entomophaga aulicae	8200	—	—	—	—	12
Absidia glauca	36	8.6	25	—	10	24
Ascomycotina						
Aspergillus nidulans	28	—	2-3	60	—	16
Fusarium graminearum	24	5.4	21-23	520	—	25
Gaeumannomyces graminis	29	—	0	0	—	17
Saccharomyces cerevisiae	13.8	3.3	11.4 1.7	—	—	19,26
Neurospora crassa	37	—	8	185	2	27,28
Neurospora spp.	32-33	7.8	10-15	100	—	22,23
Basidiomycotina						
Agaricus brunescens	34	8.0	16	50	1	29,30
Coprinus lagopus	35-38	8.8	10-18	112	—	22,23,31
Pleurotus ostreatus	21	—	23.4	17	—	30
Phaenerochaete chrysosporium	40-50	—	—	—	—	32
Schizophyllum commune	38	—	7	49	2	33,34
Stropharia spp.	29-37	—	—	—	—	35

coli, with two notable exceptions, *Saprolegnia* and *Entomophaga* [11,12]. Genome sizes of 70 species of smut fungi were estimated by microfluorimetry to range from 1.35 to 9.58 × 10^{10} D (12-88 megabase pairs, Mbp) [4]. This is a much wider range of genome sizes than for any of the other groups of fungi other than the *Saprolegnia* and *Entomophaga* estimates.

Reassociation analysis revealed three kinetic classes of DNA in most fungi: foldback or snapback DNA, repetitive DNA, and unique sequence DNA (Table 1). Foldback DNA, which reassociates immediately on placement of the DNA in reassociative conditions, commonly accounted for 1-2% of the genome, but two notable exceptions were reported: *Allomyces* [13] and *Phycomyces* [14] with 10% foldback. These sequences include inverted repeat sequences that can readily self-reassociate [15]. Repetitive sequences were the next most rapidly annealing fraction, and account for 0-23% of the genome with most reports in the range of 10-15%. These sequences include the ribosomal DNA cistrons (rDNA) and some other DNA of less well-defined function. Since the rDNA was uniformly found to be reiterated (see below) and were detectable even at the 2-3% level in *Aspergillus nidulans* [16], the inability to detect repetitive DNA in *Gaeumannomyces graminis* [17] was truly exceptional. While most fungi had only one kinetic class of repetitive DNA, *Achlya bisexualis* [18] and *S. cerevisiae* [19] had two classes. The reiteration frequencies commonly ranged from 50 to 150, but several exceptions occurred both lower and higher (Table 1). Several kinds of sequences were found in the repetitive fraction of nuclear DNA.

Alternating field gel electrophoresis has allowed the separation of chromosome sized DNA molecules of fungi and other lower eukaryotes [36-40]. *Neurospora crassa* chromosomal DNAs were estimated to range from 2.5 to 11 Mbp based on physical measurements of the genome size and cytological estimates of the sizes of the seven chromosomes. However, using contour-clamped homogeneous electric field electrophoresis (CHEF) and *Schizosaccharomyces pombe* chromosomal DNAs for comparison, the chromosomal DNAs of *N. crassa* were tentatively estimated to range from 4 to 12.6 Mbp [41]. The four resolved bands were identified by hybridization to genes from each of the seven linkage groups. The DNAs from the two largest chromosomes (I and V) comigrated, as did the DNAs from the three smallest chromosomes (II, VI, and VII). The two largest DNAs were later resolved. The rDNA was located on chromosome V [41]. Magee et al. [40] used orthogonal field alternating gel electrophoresis (OFAGE), field inversion gel electrophoresis (FIGE), and CHEF to resolve chromosomal size DNA molecules of *Candida albicans*, a diploid asexual yeast. Up to 11 bands were resolved ranging in size from about 1 to >3 Mbp. By probing Southern blots with cloned genes identified by their abilities to complement known mutations of *S. cerevisiae*, they were able to show that several of the bands represented separated homologs of four of the chromosomes. They concluded that *C. albicans* has seven chromosomes [40]. Chromosome length polymorphisms were observed in *C. albicans* and *Histoplasma capsulatum* [42,43]. The ability to separate intact chromosomal DNAs of fungi was a powerful tool for the analysis of genetically well-studied fungi as well as for the analysis of genetically undeveloped organisms.

1. Methylation of DNA

Modification of DNA by methylation of cytosine residues to 5-methylcytosine, mainly in CG doublets, was relatively common among eukaryotes [44].

While DNA methylation has been reported in several fungi, it was by no means a universal phenomenon, or even a common one [45-47]. A survey of 20 fungi representing 20 species in seven classes and 15 families found undetectable levels of methylation in most species [48]. Only *Sporotrichum dimorphosporum* and *Phycomyces blakesleeanus* contained 0.2 and 0.5% of total cytosine methylated, respectively. 5-Methylcytosine was clustered in long tracts of 10-30 kbp in *Coprinus cinereus* [49]. The tracts were variable among isolates of *C. cinereus*, and were inherited both mitotically and meiotically, but the lengths of the tracts were variable [50]. Similarly faithful inheritance of methylated tracts was observed in *N. crassa* [51]. Specificity of methylation on a smaller scale was observed in the rDNA genes of *N. crassa* [52], in which CG methylation was restricted to the nontranscribed spacer region of the rDNA repeat. The biological significance of DNA methylation among fungi is not known. The specificity of the methylation of CG doublets was like that of prokaryotic methylation in the DNA restriction modification process, but the irregularity of methylation and the lack of restriction enzymes suggests that some other process must be involved [53].

Other patterns of DNA methylation besides the faithful CG doublets have been observed in *N. crassa*. DNA introduced by transformation which had become stably amplified in tandem arrays was methylated but lacked the preferential methylation of CG pairs [54]. Perhaps related to this was the observation that DNA introduced into *N. crassa* was heavily methylated requiring the use of *E. coli* mutants deficient in methylation-sensitive restriction enzymes for recovering the transforming DNA [55]. On the other hand, DNA of the ζ-η region of *N. crassa* was properly methylated after introduction by transformation whether into its homologous site or into other regions in the genome, and further, adjacent regions of transforming DNA, whether bacterial or neurosporal in origin, were rarely methylated [56]. In a separate phenomenon, extensive rearrangements induced premeiotically (RIPping) occurred associated with DNA introduced by transformation of *N. crassa* which was associated with methylation of CG doublets [57]. There seem to be several different DNA methylation recognition processes, but the signals regulating them have not been discovered.

2. Ribosomal DNA

The ribosomal DNA (rDNA) cistrons were the most studied repetitive DNA [58]. The rDNA, middle repetitive with 40-240 copies per haploid genome, occurred as tandem repeats without interspersion of unique sequence DNA [18,58]. That rDNA was separable from the main nuclear DNA as a satellite band in CsCl density gradients in several fungi [59] was due to its different base composition from the main nuclear DNA and to the repetitive nature of the genes.

Cloning of the rDNA repeat and elucidation of the sequence organization of the rDNA genes within it has been accomplished for several fungi. Three of the four nuclear rDNA genes—the 17S, 5.8S, and 25S genes—were always present in the repeat in the same juxtaposition (Fig. 1). But there was variation among the several groups of fungi in the inclusion and arrangement of the 5S gene. Three patterns of organization have been found: Pattern I found in *S. cerevisiae* [60,61], *Torulopsis* [62], *Mucor racemosus* [63], and *Achlya ambisexualis* [64] had the 5S rRNA gene located within the nontranscribed spacer (NTS) sequences that separate the transcribed

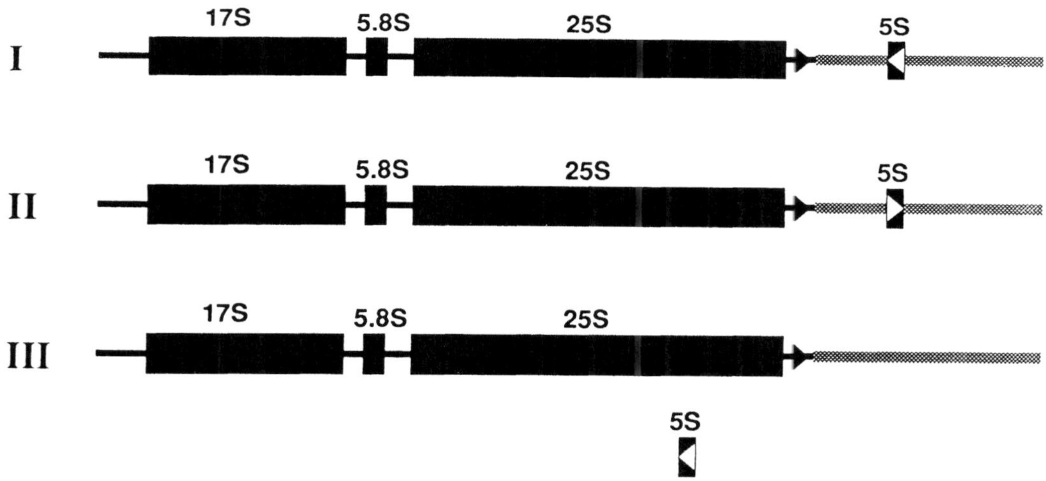

FIGURE 1. Patterns of ribosomal DNA repeats found among the fungi. Solid bars are transcribed in the formation of rRNA with arrow heads at the 3' termini indicating the directions of transcription of the two primary transcripts of rRNA. Wide solid bars represent sequences of the mature rRNAs, and narrow solid bars the transcribed spacer regions. Shaded bar represents the nontranscribed spacer (NTS). Patterns indicated by I, II, and III are as follows: I, *Saccharomyces*, *Torulopsis*, *Mucor*, and *Achlya*; II, various Basidiomycotina; III, *Neurospora*, *Aspergillus*, *Yarrowia*, and *Schizosaccharomyces*.

cluster (the 17S-nontranscribed spacer (NTS) sequences that separate the transcribed cluster (the 17S-5.8S-25S genes) in the tandemly repeated element, but in the opposite orientation. Pattern II, found among the Basidiomycotina, was like I except that the 5S gene was in the same orientation as the other rRNA genes [65-68]. Pattern III, found in *S. pombe* [69,70], *Yarrowia lipolytica* [71], *A. nidulans* [72], and *N. crassa* [73-75], had the 5S rDNA located elsewhere in the genome. The 100 copies of the 5S rDNA of *N. crassa* are dispersed rather than tandemly repeated. The *Neurospora* 5S includes 12 or more different sequences based on different primary genetic sequences [74,76-78]. The location of the 5S rRNA gene of *Cephalosporium acremonium* was not determined as the cloned *S. cerevisiae* fragment had insufficient homology to hybridize to the *C. acremonium* DNA [79]. While the 5S rDNA was in the major repeat of *Coprinus cinereus* and five other agarics, *Flammulina velutipes*, *Agaricus bisporus*, *C. micaceus*, *C. atramentarius* and *C. commatus*, all but *C. commatus* had the 5S gene in the same orientation as *C. cinereus*, i.e., transcribed from the same DNA strand as the 17S, 5.8S, and 25S genes. In *C. commatus* the 5S gene was inverted as has been found for pattern I fungi.

Variation in the rDNA repeats includes a range in size from 7.8 kbp for *A. nidulans* [80] to 11.8 kbp for *Phytophthora cryptogea* [81]. The Oomycetes had the largest repeating units of any of the fungi, ranging from 10.1 kbp for *Apodachlya pyrifera* to 11.8 kbp for *P. cryptogea* [81]. The largest repeat found outside the Oomycetes was 9.8 kbp for *Flammulina velutipes* [67]. Variation in the spacer region of the repeats has been observed among isolates of several fungi [1,67,71,82]. Evidence for within-strain variation of the spacer region has been obtained for *Yarrowia lipolytica* [71].

3. Other Repetitive DNA

While nearly all of the repeated sequence DNA could be accounted for by rDNA in *A. nidulans* [16], and no repetitive DNA was detectable in *G. graminis* [17], other fungi clearly contain repetitive DNA of other kinds. In *S. cerevisiae* only 2.4% of the genome was rDNA [83] while 13.1% of the genome was repetitive [19], in *A. bisexualis* the rDNA accounted for 0.65% of the genome [84] with the repetitive fraction accounting for 16% [18]. Nevertheless, the amount of repetitive DNA in fungi was considerably less than that of higher eukaryotes.

Several repetitive elements have been studied in some detail [85-88]. Non-rDNA repetitive sequences have been cloned in *S. cerevisiae* [87,89], *Mucor racemosus* [86], and *Absidia glauca* [24]. Ty, found through genetic studies [89], occurred as a middle repetitive element of 30-35 copies per haploid genome of *S. cerevisiae*. The sequences were all about 5.9 kbp and were composed of a 5.2-kbp unique region, called epsilon, flanked by two long terminal direct repeats of about 330 bp (LTR), called delta (Fig. 2). Ty resembled retrotransposons and had homology to conserved regions of retroviruses and retrotransposons of other eukaryotes [90]. Ty sequences were somewhat heterogeneous and were divided into three classes differing by base sequence in certain coding domains [90-92]. Within each class minor variations were recognized by restriction site polymorphisms. Several Ty elements have been sequenced revealing two open reading frames, TYA and TYB, that overlap by 38 or 44 bp in classes I and II, respectively (Fig. 2). Transcription forms mainly a full length RNA of 5.7 kbp beginning in the left delta and ending in the right delta giving a 50-bp terminal redundancy. Two minor RNA transcripts of 2.2 and 5.0 kbp were also formed. By attaching the LTR to the regulatable promoter from GAL1, viruslike particles (VLPs) appeared when yeast was stimulated to overproduce the 5.7-kbp transcript [93]. DNA-DNA interactions involving Ty can lead to genomic rearrangements including deletions, inversions, and translocations [94,95].

Other repetitive elements, sigma and tau, have been described in *S. cerevisiae*, which resembled the delta regions of Ty. Sigma elements were AT-rich, low-repetitive elements of 340 kbp with 8-bp inverted repeats at the ends. They were flanked by 5-bp direct repeats, and were found within 16-18 bp of the 5' ends of several tRNA genes [88]. The sequences did not contain any open reading frames. Tau was 371 bp with the same inverted repeat found at the ends of some Ty, sigma, copia, and spleen necrosis virus [87,96]. The tau inverted repeats, like sigma, were flanked by a 5-bp direct repeat. They occurred in multiple copies and tau^+ and tau^- allelic pairs were found. Tau, sigma, and delta had significant homology [87,96]. A subset of 25-30 sigma elements responded to the sex pheromone, α factor, and permitted expression of distal sequences [97]. These may play roles in the mating process or in transposition of the sigma sequence. Most sigma elements in *S. cerevisiae* were solo elements, but a few were clustered in pairs. One of these pairs was a direct repeat with 4.7 kbp between them [90]. While most delta elements were clustered in pairs forming the ends of Ty elements, they also occurred as solo elements [98].

The ability to regulate Ty function with engineered combinations led to the investigation of the use of Ty as a vehicle for a variety of applications

FIGURE 2. Ty element structure. Solid triangles represent the long terminal direct repeats, delta; the transcribed region represented by the shaded bars overlaps the delta at either end as shown, and contains two open reading frames TYA, the wider shaded bar, and TYB, the narrower shaded bar, which overlap each other as shown.

[85]. Fusion of a Ty element to the GAL1 promoter in a yeast plasmid allowed regulation of Ty activity by galactose, which resulted in the induction of transposition and a large increase in VLPs containing Ty RNA and reverse transcriptase activity [93]. This may be a key discovery to unlocking the mystery of the nearly ubiquitous VLPs of other fungi. While most fungal VLPs contained double-stranded RNA (see below), the nature of the RNA in the VLPs induced by Ty was not determined [93]. Fusion proteins comprised of most of the Ty protein p1 and part of the HIV protein gp120 can form Ty-VLPs creating particulate polyvalent antigens [99]. The human interferon (IFN) gene was fused in frame with the TYA gene yielding a p1-IFN fusion protein that was assembled into VLPs. They were easily purified and induced antibody response to interferon in rabbits [100].

B. Mitochondrial Genomic DNA

The genomes of fungal mitochondria show great diversity in size and organization in keeping with the evolutionary diversity of the fungi. All fungal mitochondrial genomic DNA (mtDNA) were covalently closed circular DNAs except for the yeast *Hansenula mrakii* [101], which had a linear genome. The size range of fungal mtDNA was intermediate between the 15- to 18-kbp range reported for animals and the 135- to 2400-kbp range reported for vascular plants [102-104]. The smallest fungal mtDNA reported was 18.9 kbp in *Torulopsis glabrata* [105] and the largest was 176 kbp in an isolate of *Agaricus bitorquis* [106]. Considerable size diversity occurred among closely related fungi, as the threefold range among members of the genus *Suillus* from 36 to 121 kbp [107] and the variation from 148 to 176 kbp among isolates of *A. bitorquis* [106]. In contrast to this, considerable uniformity in the complexities of mtDNA among 16 species representing four orders of Oomycete fungi [108]. While physical sizes ranged from 36.2 kbp in *Phytophthora infestans* to 73 kbp in *Pythium torulosum*, the presence of an inverted repeat of variable size resulted in genome complexities ranging only from 36 to 45 kbp. Heterogeneity of the inverted repeat also followed taxonomic groupings, ranging from 8.6 to 11.1 kbp among the Saprolegniales to 21.8 to 28.9 among *Pythium*, while members of the closely related genus *Phytophthora* lacked the inverted repeat. The size variation among *Suillus* was accounted for by insertions and deletions [108]. It was not determined whether the variable DNA among these genomes had functional significance.

Another source of variation in mtDNAs were the plasmidlike DNAs, defective mtDNAs, of the petite mutants of *S. cerevisiae* [109], stopper mutants of *N. crassa* [110,111], ragged mutants of *Aspergillus amstelodami* [112], some forms of senescence in *Podospora anserina* [113], and the plasmidlike small circular DNAs of *Cochliobolus heterostrophus* [114,115]. These were derived from the normal mtDNA by deletion and/or rearrangements that commonly resulted in the formation of multimeric circles [112-114, 116]. The senescence phenomenon of *P. anserina* involved the displacement of the normal mtDNA by the defective DNA [113], but in the ragged mutants of *A. amstelodami* the defective DNA coexisted with the normal [112,116]. The plasmidlike DNAs of *Cochliobolus heterostrophus* [114,115] did not cause any noticeable phenotype, but did contain all or part of the mitochondrial ATPase subunit 6 gene [115]. True mitochondrial plasmids, discussed below, may or may not integrate into the mtDNA but do not contain any recognizable genes for mitochondrial function.

C. Plasmid DNA

Because of their requirement for their hosts' replicative and transcriptional enzymes, plasmid DNAs must occur in either the nuclear or mitochondrial compartments for these functions. Movement outside of these compartments was not precluded, and movement between the two compartments has been suggested [117]. Proof of occurrence in the nucleus has been difficult [118]. The weight of evidence shows that the yeast 2-μm plasmid has a nuclear location [118], but it did not copurify with nuclei [119], and cytoplasmic inheritance patterns indicated that existence outside of the nucleus also occurred [120]. Copurification of plasmids with mitochondria has been taken as proof of mitochondrial localization [121-128]. Inability to demonstrate association with mitochondria suggested nuclear localization [129], but further evidence was needed. The maternal inheritance patterns of several of the mitochondrial plasmids of filamentous fungi was consistent with their mitochondrial location.

The plasmids of yeasts have been most extensively studied. Most laboratory strains of *S. cerevisiae* contained small covalently closed circular DNAs with contour lengths of about 2 μm [118]. Similar plasmids have been found in other yeast species in the genera *Saccharomyces* and *Zygosaccharomyces* [118,130], but they were rare elsewhere [130]. These share several common properties. They were all about 6 kbp, occurred in high copy number, 50-100 per cell [119,130], have a pair of inverted repeats separated by 1.7-3.2 kbp unique sequences that resulted in dumbbell-shaped molecules after melting and reannealing, and conferred no discovered phenotype on the host. The 2-μm plasmid of *S. cerevisiae* has been extensively used for gene cloning and other studies of the molecular biology of fungi [118].

Despite the shared properties of the several yeast 2-μm-like plasmids, there were great sequence differences among these plasmids. While the plasmids harbored in several closely related species to *S. cerevisiae* all shared significant sequence homologies, those from *Zygosaccharomyces* spp. had little homology either with each other or with the *S. cerevisiae* plasmids, not even amino acid sequence homology among the open reading frames within them [130]. However, homology to chromosomal DNA in *Endomyces monospora*, *Pichia membranaefaciens*, and *Z. bisporus* with one of the plasmids (pSB3) of *Z. rouxii* was found. Even so, these plasmids shared similar functions, autonomously replicating sequences, ARS, and the FLP enzyme that mediates an internal recombination event [130]. Many questions about the phylogenetic origins and significance of their structural organization remain.

Another plasmid, larger than the 2-μm plasmid, called the 3-μm plasmid was found in low copy number in strains of *S. cerevisiae*, *S. carlsbergensis*, *Schizosaccharomyces pombe*, and *Torulopsis glabrata*. These contained rDNA sequences that were indistinguishable from the rDNA repeats of their host strains. They may be produced by excision and circularization of the rDNA. Since rDNA contains an ARS sequence, these fragments could be propagated autonomously [118].

Scytalidium flavo-brunneum produces 15-azasterol and a pigment. These abilities were conferred on the fungus by the presence of a 9.1-kb plasmid obtained from wild-type cultures. This is the first demonstration of a correlation of a specific biochemical trait with the presence of a native filamentous fungal plasmid [131]. Another report suggested that mitochondrial plasmids of *Nectria haematococca* were involved in pathogenicity, but this proved to be an artifact [126,132].

Following the limitations proposed by Lambowitz et al. [133] we may define mitochondrial plasmids as those mitochondrial-replicating nucleic acids of independent origin to the normally functional mitochondrial genomic DNA. Mitochondrial plasmids may be DNA or RNA species present in mitochondria that were not derived from standard mtDNA. This distinguishes them from defective mtDNAs derived from the standard mtDNA by deletion and/or rearrangement of mtDNA [109-115]. The latter were defective in that they contained some, but not all, of the mitochondrial-borne information required for the formation of functional mitochondria. All mitochondrial plasmids described so far were DNA, but the occurrence of plasmidlike RNA sequences (discussed below) in the cytosol suggests that we should make allowance for the existence of such molecules in mitochondria.

Plasmids, in this restricted sense, have been found associated with mitochondria in a number of fungi [115,134]. Some of these in *N. crassa* and *N. intermedia* were multimeric circles [121,123], but most were linear [122,124-128]. All of these existed as independent molecules, but those of *Claviceps purpurea* were also found as integrated sequences in the mtDNA; integrated and independent sequences were not found to coexist [125,135].

The Mauriceville and Varkud mitochondrial plasmids of *Neurospora* spp. were closely related closed circular DNAs of 3.6 and 3.7 kbp whose nucleotide sequences were similar to mt introns and retroviruses. They occurred as single elements or as oligomers of tandemly repeated copies with a single open reading frame of 710-amino-acid codons with characteristics of retrovirus reverse transcriptase [136]. Mitochondrial plasmids appear to be widespread among the fungi, and many more are likely to be found as the search continues. The autonomously replicating nature of these plasmids indicates that they may be useful for cloning genes in fungi.

Two linear DNA plasmids with killer phenotypes have been found in *Kluyveromyces lactis* [137]. These were quite large plasmids, 8.9 and 13.4 kbp, that occurred in 50 to 100 copies per cell. These plasmids have similar killer function to the double-stranded RNA (dsRNA) killer plasmids of *Saccharomyces* and *Ustilago* discussed below [138]. They were stabily, cytoplasmically inherited, but their precise localization within the cell was not determined [137]. Another large linear DNA plasmid (14.2 kbp) with high copy number has been found in *S. kluyveri*. These plasmids are of considerable interest as potential cloning vehicles, and sources of useful sequences for chimeric plasmids, especially ARS and telomeres.

III. RNA

A. Ribosomal RNA

Ribosomes occur in the cytosol and in the mitochondria. These differ from each other in size, and in the kinds and sizes of RNA molecules within them. Although there were no quantitative data on the relative amounts of the two kinds of ribosomes, it was generally assumed that the cytosolic ribosomes, 80S ribosomes, comprise the bulk of the ribosomes of the organism. Indeed, one of the principal difficulties in studying mitochondrial ribosomes has been the contamination of mitochondrial fractions by cytoplasmic ribosomes [139]. Considerably more is known about cytosolic ribosomes than mitochondrial ribosomes.

The cytosolic ribosomes contain four species of RNA (rRNA), commonly referred to by their sedimentation coefficients as the 25S, 17S, 5.8S, and 5S rRNAs. The largest two species varied considerably in molecular size [140]; the 25S rRNA ranged from 1.2 MDa in *S. carlbergensis* and *Botryodiplodea theobromae* [141,142] to 1.45 MDa in *Smittium culicis* [143], while the 17S rRNA ranged from 0.67 MDa in *B. theobromae* and *Conidiobolus coronatus* [142,143] to 0.75 MDa in *S. culicis* [143]. However, these differences were obtained with acrylamide gel electrophoresis, which is sensitive to differences in the secondary structure of RNA molecules [144]. The variation in size of the 5S rRNA was proportionately considerably less, ranging from 118 to 121 bases [145]. No functional significance to these size variations has been identified.

The base composition of rRNAs of fungi fell in a relatively narrow range from 47.6 to 55.2% G+C [146]. In *S. cerevisiae* some bases were modified during processing of the primary rRNA transcript by methylation of some bases and conversion of uridine to pseudouridine as follows: 25S, 43 methyl groups, 32 pseudouridine; 18S, 24 methyl, 14 pseudouridine 5.8S, no methyl groups, 1 pseudouridine; 5S, no methyl, 0-1 pseudouridine [58]. Comparable studies in other fungi have not been performed.

The pathway of synthesis of rRNA involved the transcription by RNA polymerase I of a large segment of DNA containing the genes for the 25, 5.8, and 17S rRNA (Fig. 1) [146,147]. The DNA was transcribed as a unit forming an initial 35S precursor. This was processed by nucleases to a 32S piece, by discarding a portion, then split into a 20S piece containing the 18S rRNA and a 27S piece containing the 25 and 5.8S molecules. The 27S intermediate was cut to form the mature 25S moiety and a 7S intermediate that was further reduced to the 5.8S molecule. Methylation occurred prior to processing of the 35S precursor. The 5S rDNA was transcribed independently [147].

The 70S mitochondrial ribosomes were smaller than the 80S cytosolic ribosomes. They differed from the cytosolic ribosomes in having smaller RNAs and lacking both the 5S and 5.8S molecules [139,148]. The two RNA molecules of the mitochondrial ribosomes were generally smaller than their counterparts in the bacterial ribosome, 15S and 21S for *S. cerevisiae* [144, 149-151], 14S and 19S for *S. pombe* [152] compared to 16S and 23S for the bacterial ribosomes, and 17S and 25S for cytosolic ribosomes. The 15S molecule from *S. cerevisiae* showed 44% sequence homology to the *E. coli* 16S rRNA indicating a strong evolutionary relationship [150].

B. Messenger RNA

Messenger RNA (mRNA), a small but critical fraction of cellular RNA, comprises from 2 to 10% of the RNA as estimated in several fungi [153-157].

This fraction of RNA was heterodisperse in sucrose gradients, ranging in sedimentation from about 6S to greater than 30S [158-160]. The average sizes of mRNA molecules estimated from hybridization kinetics experiments was 1100-1200 b [155,161,162].

Like higher eukaryotes, some of the mRNA of fungi had polyadenylate (polyA) as tails on the 3' ends of the molecules and caps of 7-methylguanosine-5-triphosphoryl-5'-adenosine (or guanosine) on the 5' ends [159]. Other than the 5' cap and a few adjacent nucleotides, mRNA was largely unmethylated, which distinguished it from the ribosomal and transfer RNAs [154]. The presence of the polyA tails has been extensively used as handle to retrieve the mRNA population from the cell by affinity chromatography, but RNA that functions in vitro as mRNA may also lack the polyA tails [156]. Only 25% of the mRNA fraction of *Schizophyllum commune* contained polyA [157], and only 50% of the mRNA fraction from polysomes from *Botryodiplodia theobromae* contained polyA [163]. The nonadenylated mRNAs have been largely ignored.

The lengths of the polyA tails varied considerably; the longest average length, 100 b, was from ungerminated spores of *Rhizopus stolonifer* [161], while most reports ranged from 20 to 60 b [155,156,158,160,163-169]. Another frequent observation was that within a fungus different modal size classes of polyA tails occurred which varied with developmental stage and length of radioisotopes labeling period [153,158,161,164]. These differences have been attributed to a maturation process [156] but may also represent discrete classes of mRNAs with different synthetic and turnover rates [164].

Discrete frequency classes of mRNAs have been described for several fungi [155,170,171]. *S. cerevisiae* polyA containing mRNA occurred in three frequency classes with average copy numbers of 20, 400, and 2400 per cell [170], and *Achlya ambisexualis* had classes with average copy numbers of 5-10, 60-140, and 440-1000 [155]. Differential regulation of sequences within such frequency classes with development has been observed [172], but no correlation of these classes with other properties of mRNA have been noted. Most recent studies on mRNA have dealt with the regulation of specific genes and studies on processing of introns [173-175], which are beyond the purview of this paper.

C. Small Nuclear RNA

Small nuclear RNAs, snRNA, are a metabolically stable class of RNA found in the nuclei of eukaryotes [176]. In mammals these were implicated in intron processing, 3' editing of histone mRNA and pre-rRNA processing. They had a distinctive cap structure containing 2,2,7-trimethylguanosine (m_3G). This was similar to the 5' cap of mRNAs which had instead 7-methylguanosine. Six of the snRNAs of mammals (U1-U6) were highly abundant, 2×10^5 to 1×10^6 copies per cell. The others named U7 and greater were of lower abundance, but had similar physical properties. All were associated with protein to form small ribonucleoprotein particles (snRNPs) which shared a common antigen that reacted with the Sm antibodies associated with the autoimmune disease lupus. This property forms the basis of isolating and identifying similar snRNPs in other organisms. Techniques used for study include precipitation of the snRNPs with Sm antibodies followed by deproteinization and purification of the snRNA and precipitation of the snRNA with m_3G antibodies. Analysis of the snRNAs was usually done by two-dimensional acrylamide gel electrophoresis of the radioactively labeled RNA detected by autoradiography.

Sm antibodies precipitate similar snRNPs of fungi and plants [177]. The nuclear locations of these RNPs was not ascertained for fungi; however, most of these shared the unique 5' cap structure involving m_3G. The complexity of fungal snRNPs was greater than that of mammals. About 20 different snRNAs were detected in human cell extracts, budding yeasts (*Saccharomyces* and *Candida*) contained more than 30 detectable species, and *A. nidulans* and *S. pombe* contained more than 40 species of RNA precipitable by antibodies to m_3G. The size range of fungal presumptive snRNAs extends to a larger size, greater than 1000 b, than those from mammals, which varied from greater than 100 to about 300 b [177,178]. The greater diversity of snRNAs in fungi does not necessarily mean that there was a greater diversity or complexity in their function.

Some snRNAs were essential for viability and participated in the splicing of pre-rRNA [169,175] or in mRNA processing [178-180]. Mutants of snR20, snR7, and snR14, coding for snRNAs apparently analogous to the mammalian U2, U5, and U4 snRNAs, respectively, were lethal [179]. In another study two SNR genes called SNR128 and SNR190 because their RNA transcripts were 128 and 190 bases long, respectively, were mutated by gene disruption demonstrating that SNR128 was essential but SNR190 was not [169]. Both RNAs were implicated in ribosome biosynthesis; snR128 associated with deproteinized 27S pre-rRNA, and snR190 associated with the 20S intermediate. However, evidence that snR128 (also known as U14) is involved in processing of the 20S rRNA precursor from the 35S precursor has also been obtained [181]. Another subset of snRNAs coded by single-copy genes, SNR3, SNR4, SNR5, SNR8, SNR9, and SNR10, were not required for viability although SNR10 mutants showed conditionally impaired growth at low temperature or elevated osmotic strengths, but faster growth at 25°C [182,183]. Multiple mutants were constructed to test the hypothesis that the lack of necessity of these snRNAs was caused by overlapping roles allowing complementation among them. A quintuple mutant of SNR3, 4, 5, 8, and 9 was indistinguishable from wild-type and the sextuple mutant including these and the SNR10 mutant was indistinguishable from the single SNR10 mutant [182]. However, considering the large number or snRNAs, these results may simply mean that the correct combinations were not chosen. Further, the curious conditional response of the SNR10 mutant [183] suggests that the appropriate conditions to elicit mutant phenotypes may not have been used.

Cytoplasmically located small RNA molecules associated with ribonucleoprotein particles have also been described. The best characterized of these is the 7SL RNA, which is involved in signal peptide recognition in the protein secretory pathway [184]. RNAs with cytoplasmic location analogous to these have been shown to be functionally important in yeasts [185-187]. Other small cytoplasmic RNAs have been described from mammalian cells, but little is known of them in fungi.

D. Double-Stranded RNA

Double-stranded RNA molecules (dsRNA) were commonly found in fungi, primarily associated with viruslike particles of a benign nature [188]. However, the discovery of viruses in fungi did involve the study of a disease of mushrooms. The early history of "La France disease" of commercial mushrooms, *Agaricus bisporus* = *A. brunescens*, also called X disease, Brown disease, and watery stipe, from its first report by Sinden in 1948 was detailed by van Zaayen [189]. The disease was transmitted by hyphal anastomosis [190,191] and had a viral etiology [192]. VLPs of several forms, often in combination, have been associated with diseased mushrooms, including spherical or polyhedral particles from 19 to 50 nm, and bacilliform or rod-shaped particles from 19 × 50 nm to 17 × 150-300 nm [189]. Both single-stranded RNA (ssRNA) and dsRNA have been reported associated with the particles [193]. The sizes of the dsRNA molecules ranged from 1.0 to 3.4 kbp, with a multiplicity of sizes present in the preparations [193]. Subsequent purification of individual viruses from this mixture has been obtained [194-196]. Fungal viruses have been classified into several different types, and like these *Agaricus* viruses, they commonly have segmented dsRNA genomes [188].

A survey of 36 wild-type isolates of *N. crassa* and *N. intermedia* found seven strains containing dsRNA, which ranged in size from 500 bp to 18 kbp [197]. From one to three different dsRNA molecules occurred in each strain, and several cross-homologies were detected among them. A 9.0-kbp dsRNA found in three of the isolates had homology to genomic sequences in all 36 strains, suggesting genomic rather than viral origin. Since the genomic restriction fragments totaled less than 9 kbp, a simple direct relationship between the genomic sequences and the 9-kbp dsRNA was not indicated. Some of the dsRNA containing strains showed senescent or slow growth phenotypes, but a causal relationship to the dsRNAs was not shown [197].

The most detailed studies of the biology of fungal viruses were done with those associated with the killer phenomenon in yeasts. Killer yeasts secrete a toxin to which they were immune but which was lethal to sensitive yeasts [138,198]. First discovered in *S. cerevisiae*, the phenomenon has also been described from strains of *Candida*, *Cryptococcus*, *Debaryomyces*, *Hansenula*, *Kluyveromyces*, *Pichia*, *Torulopsis*, and the basidiomycete, *Ustilago maydis* [199,200]. In most *S. cerevisiae* killer strains, the toxin was produced from a mycoviral dsRNA genome, m-dsRNA. A second mycoviral genome was also present in these strains, L-dsRNA, which coded for the capsid protein of both viruses. The encapsidated isometric VLPs were 33-40 nm. The dsRNA viral system was demonstrated only in *Saccharomyces* and *U. maydis*. The genes for the killer system in *K. lactis* depended on two linear dsDNA plasmids (discussed above). The location of the killer genes in other systems was uncharacterized, but presumed to be nuclear. The mycoviral genomes of *S. cerevisiae* were dependent on nearly 30 chromosomal genes for their maintenance, which lack of autonomy was apparently different from other viruses. However, little has been done to study the genetic basis of viral maintenance in other fungi.

Molecular analysis of the dsRNAs of the *S. cerevisiae* killers revealed a range of sizes of dsRNA from 0.73 to 4.7 kbp [138]. Each strain contained a segmented genome consisting of M-dsRNA, 1.5-1.9 kbp, and L-dsRNA, 4.7 kbp, with other molecules of 2.7 kbp, 2.25 kbp, slightly greater than 4.7 kbp observed in some strains. The dsRNAs had no detectable sequence homology to genomic DNA [201-204]. The molecules were completely double-stranded except for single 3'-terminal adenylates, and an asymmetrically located AU-rich bubble of 200 bp in M-dsRNA. No repeated sequences were reported.

The dsRNA genomes of the *U. maydis* killer strains were also segmented, containing from three to seven different dsRNA molecules [205,206]. These have been placed in three size classes, H from 3.6 to 6.2 kbp, M from 1.1

to 1.7 kbp, and L, 0.36 kbp. Each killer strain contained at least one dsRNA moiety from each size class. The biology of the VLPs of this system has been reviewed in detail [200].

The dsRNAs associated with hypovirulence in *Endothia* (*Cryphonectria*) *parasitica* were even more complex, ranging from two to eight molecules per isolate [207,208]. These were not encapsilated as in typical fungal viruses, but were contained in vesicles in a European hypovirulent strain of the fungus, while an American isolate had no detectable particles [209, 210]. Perhaps these can be thought of as RNA plasmids, but nature will probably confound any attempts to maintain a precise, compact terminology for these various entities.

Structural organization of the dsRNAs from the European strain EP113, transmitted to an American strain of the fungus and labeled EP713, had certain similarities to the organization of the dsRNAs of an American strain, GH2 [211-213]. Each had a group of large molecules called major bands, ranging from 1 to 9 kbp and several minor bands less than 1 kbp. Within each strain these shared considerable sequence homologies, with the exception of the 1.0-kbp S-RNA moiety from GH2, but no homologies were detected among the dsRNAs between the two strains [213]. Smaller pieces apparently were derived from larger pieces by internal deletions. Each species of dsRNA had a 3' segment of polyadenylic acid paired to a 5' segment of polyuridylic acid on the opposite strand. Evidence suggested that the polyA-containing strand functioned as mRNA, and a protein corresponding to one of the three open reading frames found on the 9-kbp dsRNA of EP713 was detected in lysates of the hypovirulent strain [213]. This is the first gene product encoded by a hypovirulent-associated dsRNA to be detected.

IV. NUCLEOTIDES

From a biotechnology viewpoint, 5'-ribonucleotides of guanosine, inosine, and xanthosine are of interest as flavoring agents in the food industry [214]. These may be made directly by fermentation with bacterial mutants or indirectly by chemical modification of RNA prepared from yeast. Most of this work has been done in Japan, and no new information has become available since the review by Bigelas. Only a brief discussion of this work will be presented here.

A. RNA Nucleotides

Demain [215] reviewed the early work which dealt primarily with bacteria, but also with some yeast and molds. Evidence that nucleotide secretion was the result of breakdown of RNA was primarily from bacteria (*B. subtilis*). Consonant with this hypothesis were the observations that over 60% of the nucleotide material secreted by *S. cerevisiae* was poly- and oligonucleotides, other yeasts also secreted oligonucleotides [216], and secretion was stimulated by suspension of the cells in nonnutrient buffer. This was presumed to be primarily from the breakdown of rRNA via the latent nucleases bound to the ribosomes [215]. Contrary to this, however, were the observations that the presence of a nutrient glucose and nitrogen source may also stimulate secretion of nucleotides [217,218]; also, the observation that the ribosomal RNA content did not decrease during the early phases of excretion, but only during a later autolytic phase suggested an alternate source of nucleotides [219]. More specific secretory mechanisms may also be involved in some cases, e.g., only GMP was secreted from *Candida lipolytica* [220]. The observations that added nitrogen sources stimulated excretion [217,218] contrasted strikingly with the finding that growth stimulatory amino acids inhibited RNA secretion in *B. subtilis* [221], suggesting that the bacterial model may not be adequate to explain the fungal situations.

The RNA degradation model for production of nucleotides implies the presence of appropriate intracellular RNases, of which fungi indeed have an abundant supply [222,223]. Phosphodiesterases cleaving on one or the other side of the phosphate to yield 5'- and 3'-nucleoside monophosphates have been described [224]. A mutant of *Zygosaccharomyces soja* leaked 5'-nucleoside monophosphates and diphosphates suggesting that nucleotides were produced by degradation of RNA by polynucleotide phosphorylase [218]. The presence of these enzymes was necessary but insufficient to demonstrate their role in this process.

Production of nucleotides by direct fermentation has been studied most extensively with bacteria, but some genetic studies with fungi indicate some potential with these organisms as well. *N. crassa* and *S. pombe* adenine-requiring mutants blocked at IMP accumulated hypoxanthine [215]. High levels of XMP have been achieved using mutants blocked in the conversion of IMP and XMP to AMP and GMP. Provision of adenine and guanosine allows these mutants to grow. Accumulation of GMP by mutants blocked in GMP metabolism would not be feasible because of their lethality. The specific accumulation of GMP by *C. lipolytica* has not been explained [220].

B. Cyclic Nucleotides

The role of cyclic adenosine-3',5'-monophosphate (cAMP) as a secondary messenger in bacteria and animals has been extensively investigated, but there has been considerable doubt concerning the validity of similar roles for cAMP in the fungi. Stringent criteria for identification of cAMP and determining the role of cAMP in metabolism have been suggested [225]. Changes in cAMP levels may be rapid with a half-life of ~15 s in *Neurospora*. Reports of cAMP, adenylate cyclase, phosphodiesterase, cAMP-dependent protein kinase, and cAMP-binding proteins were widespread among fungi. A variety of functions have been proposed for control by cAMP including metabolism of carbon storage materials, mitochondrial functions, utilization of exogenous carbon sources, and diverse developmental phenomena. The concentrations of cAMP changed rapidly and transiently during spore germination of several zygomycete fungi and *S. cerevisiae*, triggering a protein phosphorylation cascade [226]. *Alternaria solani* and *Phymatotrichum omnivorum* produced substances identifiable as cAMP by several criteria [227]. *A. solani* secreted cAMP into the medium during growth, and the levels changed during growth in ways inconsistent with autolytic release.

V. SUMMARY AND CONCLUSIONS

There is a wealth of knowledge available on the molecular biology of the fungi that has been obtained to a great extent from studies of *Saccharomyces cerevisiae* and to a lesser extent from *Aspergillus nidulans* and *Neurospora crassa*. Other fungi have been studied to far lesser extents. While many broad patterns of similarity are apparent among the fungi and which are shared with other eukaryotes, it is also clear that there are great, ancient evolutionary diversities among the fungi which distinguish them from other eukaryotes and also from each other. The evidence for extensive diversity among the fungi suggests that we have only begun to learn of these differences and that considerable new insights into the nature of all living organisms will be obtained by broadening our study of molecular biology to include other organisms to provide a basis of comparison. The tools of molecular biology and recombinant DNA technology that have been developed now make this possible.

REFERENCES

1. Anderson, J. B., Bailey, S. S., and Pukkila, P. J. Variation in ribosomal DNA among biological species of Armillaria, a genus of root-infecting fungi. *Evolution, 43*:1652-1662, 1989.
2. Armstrong, J. L., Fowles, N. L., and Rygiewicz, P. T. Restriction fragment length polymorphisms distinguish ectomycorrhizal fungi. *Plant and Soil, 116*:1-7, 1989.
3. Bruns, T. D., Fogel, R., and Taylor, J. W. Amplification and sequencing of DNA from fungal herbarium specimens. *Mycologia, 82*:175-184, 1990.
4. Durán, R., and Gray, P. M. Nuclear DNA, an adjunct to morphology in fungal taxonomy. *Mycotaxon., 36*:205-219, 1989.
5. Goodwin, P. H., Kirkpatrick, B. C., and Duniway, J. M. Identification of Phytophthora citrophthora with cloned DNA probes. *Appl. Environ. Microbiol., 56*:669-674, 1990.
6. Kohn, L. M., Petsche, D. M., Bailey, S. R., Novak, L. A. and Anderson, J. B. Restriction fragment length polymorphisms in nuclear and mitochondrial DNA of Sclerotinia species. *Phytopathology, 78*:1047-1051, 1988.
7. Rygiewicz, P. T., Fowles, N. L., and Armstrong, J. L. DNA RFLPs distinguish ectomycorrhizal fungi. *Agriculture Ecosyst. Environ., 28*:431-435, 1990.
8. Smith, M. L., and Anderson, J. B. Restriction fragment length polymorphisms in mitochondrial DNAs of Armillaria—identification of North American biological species. *Mycol. Res., 93*:247-256, 1989.
9. Spitzer, E. C., Lasker, B. A., Travis, S. J., Kobayashi, G. S., and Medoff, G. Use of mitochondrial and ribosomal DNA polymorphisms to classify clinical and soil isolates of Histoplasma capsulatum. *Infection and Immunity, 57*:1409-1412, 1989.
10. Vezinhet, F., Blondin, B., and Hallet, J. N. Chromosomal DNA patterns and mitochondrial DNA polymorphism as tools for identification of enological strains of Saccharomyces cerevisiae. *Appl. Microbiol. Biotechnol., 32*:568-571, 1990.
11. Tanaka, K., Heath, I. B., and Moens, P. B. Karyotype, synaptonemal complexes and possible recombination modules of the oomycete fungus Saprolegnia. *Can. J. Genet. Cytol., 24*:385-396, 1982.
12. Murrin, F., Holtby, J., Nolan, R. A., and Davidson, W. C. The genome of Entomophaga aulicae (Entomophthorales, Zygomycetes): base composition and size. *Exp. Mycol., 10*:67-75, 1986.

13. Ojha, M., Turler, H., and Turian, G. Characterization of Allomyces genome. *Biochim. Biophys. Acta, 478*:377-391, 1977.
14. Dusenberry, R. L. Characterization of the genome of Phycomyces blakesleeanus. *Biochim. Biophys. Acta, 378*:363-377, 1975.
15. Pellegrini, M., Timberlake, W. E., and Goldberg, R. B. Electron microscopy of Achlya deoxyribonucleic acid sequence organization. *Mol. Cell. Biol., 1*:136-143, 1981.
16. Timberlake, W. E. Low repetitive DNA content in Aspergillus nidulans. *Science, 202*:973-975, 1978.
17. McFadden, J. J. P., and Buck, K. W. Sequence complexities of the nuclear and nitochondrial genomes of the take-all fungus, Gaeumannomyces graminis var. tritici. *J. Gen. Microbiol., 129*:3515-3517, 1983.
18. Hudspeth, M. E. S., Timberlake, W. E., and Goldberg, R. DNA sequence organization in the watermold Achlya. *Proc. Natl. Acad. Sci. USA, 74*:4332-4336, 1977.
19. Bicknell, J. N., and Douglas, H. C. Nucleic acid homologies among species of Saccharomyces. *J. Bacteriol., 101*:505-512, 1970.
20. Ojha, M., and Turian, G. Ribosomal RNA cistrons in Allomyces arbuscula. *Biochim. Biophys. A., 478*:392-406, 1977.
21. Dodd, I. G., Horgen, P. A., and Strauss, N. A. Isolation and characterization of the deoxyribonucleic acid in the water mould, Achlya ambisexualis. *Cytobios., 13*:31-36, 1975.
22. Dutta, S. K., and Ohja, M. Relatedness between major taxonomic groups of fungi based on measurement of DNA nucleotide sequence homology. *Mol. Gen. Genet., 114*:232-240, 1972.
23. Dutta, S. K. Repeated DNA sequences in fungi. *Nucl. Acids Res., 1*:1411-1419, 1974.
24. Wöstemeyer, J., and Burmester, A. Structural organization of the genome of the Zygomycete Absidia glauca: evidence for high repetitive DNA content. *Curr. Genet., 10*:903-907, 1986.
25. Szécsi, A. Kinetic complexity and repetitiveness of Fusarium graminearum nuclear DNA. *Exp. Mycol., 5*:323-329, 1981.
26. Lauer, G. D., Roberts, T. M., and Klotz, L. C. Determination of the nuclear DNA content of Saccharomyces cerevisiae and implications for the organization of DNA in yeast chromosomes. *J. Mol. Biol., 114*:507-526, 1977.
27. Krumlauf, R., and Marzluf, G. A. Characterization of the sequence complexity and organization of the Neurospora crassa genome. *Biochemistry, 18*:3705-3713, 1979.
28. Krumlauf, R., Marzluf, G. A. Genome organization and characterization of the repetitive and inverted repeat DNA sequences in Neurospora crassa. *J. Biol. Chem., 255*:1138-1145, 1980.
29. Arthur, R., Herr, F., Shaus, N., Anderson, J., and Horgen, P. Characterization of the genome of the cultivated mushroom Agaricus brunescens. *Exp. Mycol., 7*:127-132, 1982.
30. Horgen, P. A., Arthur, R., Davy, O., Moum, A., Herr, F., Strauss, N., and Anderson, J. The nucleotide sequence homologies of unique DNA's of some cultivated and wild mushrooms. *Can. J. Microbiol., 30*:587-593, 1984.
31. Dutta, S. K., Penn, S. R., Knight, A. R., and Ojha, M. Characterisation of DNAs from Coprinus lagopus and Mucor azygospora. *Experientia, 28*:582-584, 1972.

32. Raeder, U., and Broda, P. Comparison of the lignin-degrading white rot fungi Phanerochaete chrysosporium and Sporotrichum pulverulentum at the DNA level. *Curr. Genet.*, *8*:499-506, 1984.
33. Dons, J. J. M., De Vries, O. M. H., and Wessels, J. G. M. Characterization of the genome of the basidiomycete Schizophyllum commune. *Biochim. Biophys. Acta*, *563*:100-112, 1979.
34. Ullrich, R. C., Droms, K. A., Doyon, J. D., and Specht, C. A. Characterization of DNA from the basidiomycete Schizophyllum commune. *Exp. Mycol.*, *4*:123-134, 1980.
35. Jahnke, K. D., and Bahnweg, G. Assessing natural relationships in the Basidiomycetes by DNA analysis. *Trans. Br. Mycol. Soc.*, *87*:175-191, 1986.
36. Schwartz, D. C., and Cantor, C. R. Separation of yeast chromosome-sized DNAs by pulsed field gradient gel electrophoresis. *Cell*, *37*:67-75, 1984.
37. Smith, C. L., Matsumoto, T., Niwa, O., Kico, S., Fan, J.-B., Yanagida, M., and Cantor, C. R. An electrophoretic karyotype for Schizosaccharomyces pombe by pulsed field gel electrophoresis. *Nucl. Acids Res.*, *15*:4481-4489, 1987.
38. Snell, R. G., and Wilkins, R. J. Separation of chromosomal DNA molecules from C. albicans by pulsed field gel electrophoresis. *Nucl. Acids Res.*, *14*:4401-4406, 1986.
39. Vollrath, D., and Davis, R. W. Resolution of greater than 5 megabasepair DNA molecules by contour-clamped homogeneous electric fields. *Nucl. Acids Res.*, *15*:7865-7876, 1987.
40. Magee, G. G., Koltin, Y., Gorman, J. A., and Magee, P. T. Assignment of cloned genes to the seven electrophoretically separated Candida albicans chromosomes. *Mol. Cell. Biol.*, *8*:4721-4726, 1988.
41. Orbach, M. J., Vollrath, D., Davis, R. W., and Yanofsky, C. An electrophoretic karyotype of Neurospora crassa. *Mol. Cell. Biol.*, *8*:1469-1473, 1988.
42. Lasker, B. A., Carlel, G. F., Kobayashi, G. S., and Medoff, G. Comparison of the separation of Candida albicans chromosome-sized DNA by pulsed-field gel electrophoresis techniques. *Nucl. Acids Res.*, *17*:3783-3793, 1989.
43. Steele, P. E., Carle, G. F., Kobayashi, G. S., and Medoff, G. Electrophoretic analysis of Histoplasma capsulatum chromosomal DNA. *Mol. Cell. Biol.*, *9*:983-987, 1989.
44. Razin, A., Cedar, H., Riggs, A. D., Razin, A., Cedar, H., and Riggs, A. D. *DNA Methylation: Biochemistry and Biological Significance*. Springer-Verlag, New York, 1984, pp. 392.
45. Guseinov, V. A., Mazur, A. L., Vanyushin, B. F., and Belozersky, A. N. Pyrimidine units in the DNA of some fungi. *Biochemistry*, *37*:312-318, 1972.
46. Tamame, M., Antequera, F., Villanueva, J. R., and Santos, T. High frequency conversion to a "Fluffy" developmental phenotype in Aspergillus spp. by 5-azacytidine treatment: evidence for involvement of a single nuclear gene. *Mol. Cell. Biol.*, *3*:2287-2297, 1983.
47. Clutterbuck, A. J., Spathas, D. H. Genetic and environmental modification of gene expression in the brlA12 variegated position effect mutant of Aspergillus nidulans. *Genet. Res. Camb.*, *43*:123-138, 1984.
48. Antequera, F., Tamame, M., Villanueva, J. R., and Santos, T. DNA methylation in fungi. *J. Biol. Chem.*, *259*:8033-8036, 1984.

49. Zolan, M. E., and Pukkila, P. J. DNA methylation in Coprinus cinereus. In *Molecular Genetics of Filamentous Fungi* (W. E. Timberlake, ed.), Alan R. Liss, New York, 1985, pp. 333-344.
50. Zolan, M. E., and Pukkila, P. J. Inheritance of DNA methylation in Coprinus cinereus. *Mol. Cell. Biol.*, *6*:195-200, 1986.
51. Selker, E. U., and Stevens, J. N. Signal for DNA methylation associated with tandem duplication in Neurospora crassa. *Mol. Cell. Biol.*, *7*:1032-1038, 1987.
52. Russell, P. J., Rodland, K. D., Cutler, J. E., Rachlin, E. M., and McCloskey, J. A. DNA methylation in Neurospora: chromatographic and isoschizomer evidence for changes during development. In *Molecular Genetics of Filamentous Fungi* (W. E. Timberlake, ed.), Alan R. Liss, New York, 1985, pp. 321-332.
53. Adams, R. L. P., and Burdon, R. H. *Molecular Biology of DNA Methylation*. Springer-Verlag, New York, 1985, p. 247.
54. Bull, J. H., and Wootton, J. C. Heavily methylated amplified DNA in transformants of Neurospora crassa. *Nature*, *310*:701-704, 1984.
55. Orbach, M. J., Schneider, W. P., and Yanofsky, C. Cloning of methylated transforming DNA from Neurospora crassa in Escherichia coli. *Mol. Cell. Biol.*, *8*:2211-2213, 1988.
56. Selker, E. U., Jensen, B. C., and Richardson, G. A. A portable signal causing faithful DNA methylation de novo in Neurospora crassa. *Science*, *238*:48-53, 1987.
57. Selker, E. U., Cambareri, E. B., Jensen, B. C., and Haack, K. R. Rearrangement of duplicated DNA in specialized cells of Neurospora. *Cell*, *51*:741-752, 1987.
58. Bollon, A. P. Organization of fungal ribosomal RNA genes. In *The Cell Nucleus*, Vol. 10 (H. Busch and L. Rothblum, eds.), Academic Press, New York, 1982, pp. 67-125.
59. Garber, R. C., and Yoder, O. C. Isolation of DNA from filamentous fungi and separation into nuclear, mitochondrial, ribosomal and plasmid components. *Anal. Biochem.*, *135*:416-422, 1983.
60. Bell, G. I., DeGennaro, L. J., Gelfand, D. H., Bishop, R. J., Valenzuela, P., and Rutter, W. J. Ribosomal RNA genes of S. cerevisiae. I. Physical map of the repeating unit and location of the regions coding for 5S, 5.8S, 18S, and 25S ribosomal RNAs. *J. Biol. Chem.*, *252*:8118-8125, 1977.
61. Verbeet, M. P., Klootwijk, J., van Heerikhuizen, H., Fontijn, R., Vreugdenhil, E., and Planta, R. J. Molecular cloning of the rDNA of Saccharomyces rosei and comparison of its transcription initiation region with that of Saccharomyces carlsbergensis. *Gene*, *23*:53-63, 1983.
62. Tabata, S. Structure of the 5-S ribosomal RNA gene and its adjacent regions in Torulopsis utilis. *Eur. J. Biochem.*, *110*:107-114, 1980.
63. Cihlar, R. L., and Sypherd, P. S. The organization of the ribosomal RNA genes in the fungus Mucor racemosus. *Nucl. Acids Res.*, *8*:793-804, 1980.
64. Rozek, C. E., and Timberlake, W. E. Restriction endonuclease mapping by crossed contact hybridization: the ribosomal RNA genes of Achlya ambisexualis. *Nucl. Acid Res.*, *7*:1567-1578, 1979.
65. Cassidy, J. R., and Pukkila, P. J. Novel arrangement of the rRNA genes in the basidiomycete Coprinus cinereus. *J. Cell. Biol.*, *99*:16a, 1984.

66. Cassidy, J. R., Moore, D., Lu, B. C., and Pukkila, P. J. Usuaual organization and lack of recombination in the ribosomal RNA genes of Coprinus cinereus. *Curr. Genet.*, *8*:607-613, 1984.
67. Cassidy, J. R., and Pukkila, P. J. Inversion of 5S ribosomal RNA genes within the genus Coprinus. *Curr. Genet.*, *12*:33-36, 1987.
68. Buckner, B., Novotny, C. P., and Ullrich, R. C. Organization of the ribosomal RNA genes of Schizophyllum commune. *Curr. Genet.*, *13*:417-424, 1988.
69. Tabata, S. Nucleotide sequence of the 5S ribosomal RNA genes and their adjacent regions in Schizosaccharomyces pombe. *Nucl. Acids Res.*, *9*:6429-6437, 1981.
70. Mao, J., Appel, B., Schaack, J., Sharp, S., Yamada, H., and Söll, D. The 5S RNA genes of Schizosaccharomyces pombe. *Nucl. Acids Res.*, *10*:487-500, 1982.
71. Clare, J. J., Davidow, L. S., Gardner, D. C. J., and Oliver, S. G. Cloning and characterization of the ribosomal RNA genes of the dimorphic yeast, Yarrowia lipolytica. *Curr. Genet.*, *10*:449-452, 1986.
72. Bartnik, E., Strugala, K., and Stepien, P. Cloning and analysis of recombinant plasmids containing genes for Aspergillus nidulans 5S ribosomal RNA. *Curr. Genet.*, *4*:173-176, 1981.
73. Free, S. J., Rice, P. W., and Metzenberg, R. L. Arrangement of the genes coding for ribosomal ribonucleic acids in Neurospora crassa. *J. Bacteriol.*, *137*:1219-1226, 1979.
74. Selker, E. U., Yanofsky, C., Driftmuir, K., Metzenberg, R. L., Alzner-De Weerd, B., and Raj Bhanday, U. L. Dispersed 5S RNA genes in Neurospora crassa: structure, expression and evolution. *Cell*, *24*:819-828, 1981.
75. Morzycka-Wroblewska, E., Selker, E. U., Stevens, J. N., and Metzenberg, R. L. Concerted evolution of dispersed Neurospora crassa 5S RNA genes: pattern of sequence conservation between allelic and nonallelic genes. *Mol. Cell. Biol.*, *5*:46-51, 1985.
76. Metzenberg, R. L., Stevens, J. N., Selker, E. U., and Morzycka-Wroblewska, E. Identification and chromosomal distribution of 5S rRNA genes in Neurospora crassa. *Proc. Natl. Acad. Sci. USA*, *82*:2067-2071, 1985.
77. Selker, E. U., Stevens, J. N., and Metzenberg, R. L. Heterogeneity of 5S RNA in fungal ribosomes. *Science*, *227*:1340-1343, 1985.
78. Grayburn, W. S., and Selker, E. U. A natural case of RIP: degeneration of the DNA sequence in an ancestral tandem duplication. *Mol. Cell. Biol.*, *9*:4416-4421, 1989.
79. Carr, L. G., Skatrud, P. L., Ingolia, T. D., and Queener, S. W. Organization of the 5.8S, 16-18S, and 23-28S ribosomal RNA genes of Cephalosporium acremonium. *Curr. Genet.*, *12*:209-214, 1987.
80. Borsuk, P. A., Nagiec, M. M., Stepien, P. P., and Bartnik, E. Organization of the ribosomal RNA gene cluster in Aspergillus nidulans. *Gene*, *17*:147-152, 1982.
81. Klassen, G. R., McNabb, S. A., Dick, M. W. Comparison of physical maps of ribosomal DNA repeating units in Pythium, Phytophthora, and Apodachlya. *J. Gen. Microbiol.*, *133*:2953-2959, 1987.
82. Russell, P. J., Wagner, S., Rodland, K. D., Feinbaum, R. L. Russell, J. P., Bret-Harte, M. S., Free, S. J., and Metzenberg, R. L. Organization of the ribosomal ribonucleic acid genes in various wild-type and wild-collected strains of Neurospora. *Mol. Gen. Genet.*, *196*:275-282, 1984.

83. Schweizer, E., MacKechnie, C., and Halvorson, O. The redundancy of ribosomal and transfer RNA genes in Saccharomyces cerevisiae. *J. Mol. Biol.*, *40*:261-277, 1969.
84. Jaworski, A. J., Horgen, P. A. The ribosomal cistrons of the water mold Achlya bisexualis. *Arch. Biochem. Biophys.*, *157*:260-267, 1973.
85. Kingsman, A. J., Adams, S. E., Fulton, S. M., Malim, M. H., Rathjen, P. D., Wilson, W., and Kingsman, S. M. The yeast retrotransposon Ty and related elements. *Symp. Soc. Gen. Microbiol.*, *43*:223-246, 1988.
86. Dewar, R., Katayama, C., Sypherd, P. S., and Cihlar, R. L. Dispersed repetitive DNA sequence of Mucor racemosus. *J. Bacteriol.*, *162*:438-440, 1985.
87. Genbauffe, F. S., Chisholm, G. E., and Cooper, T. G. Tau, sigma, and delta a family of repeated elements in yeast. *J. Biol. Chem.*, *259*:10518-10525, 1984.
88. del Rey, F. J., Donahue, T. F., and Fink, G. R. Sigma, a repetitive element found adjacent to tRNA genes of yeast. *Proc. Natl. Acad. Sci. USA*, *79*:4138-4142, 1982.
89. Cameron, R. R., Loh, E. Y., and Davis, R. W. Evidence for transposition of dispersed repetitive DNA families in yeast. *Cell*, *16*:739-751, 1979.
90. Clark, D. J., Bilanchone, V. W., Haywood, L. J., Dildine, S. L., and Sandmeyer, S. B. A yeast sigma composite element, TY3, has properties of a retrotransposon. *J. Biol. Chem.*, *263*:1413-1423, 1988.
91. Kingsman, A. J., Gimlich, R., Clarke, L., Chinault, C., and Carbon, J. Sequence variation in dispersed repetitive sequences in Saccharomyces cerevisiae. *J. Mol. Biol.*, *145*:619-632, 1981.
92. Warmington, J. R., Waring, R. B., Newlong, C. S., Indge, K. J., and Oliver, S. G. Nucleotide sequence characterization of Ty1-17, a class II transposon from yeast. *Nucl. Acids Res.*, *13*:6679-6692, 1985.
93. Garfinkel, D. J., Boeke, J. D., and Fink, G. R. Ty element transposition: reverse transcriptase and virus-like particles. *Cell*, *42*:507-517, 1985.
94. Fink, G. R., Boeke, J. D., and Garfinkel, D. J. The mechanism of retrotransposition. *Trends Genet.*, *2*:118-123, 1986.
95. Kupiec, M., and Petes, T. D. Meiotic recombination between repeated transposable elements in Saccharomyces cerevisiae. *Mol. Cell. Biol.*, *8*:2942-2954, 1988.
96. Chisholm, G. E., Genbauffe, F. S., and Cooper, T. G. Tau, a repeated DNA sequence in yeast. *Proc. Natl. Acad. Sci. USA*, *81*:2965-2969, 1984.
97. van Arsdell, S. W., Stetler, G. L., and Thorner, J. The yeast repeated element sigma contains a hormone-inducible promoter. *Mol. Cell. Biol.*, *7*:749-759, 1987.
98. Williamson, V. M. Transposable elements in yeast. *Int. Rev. Cytol.*, *83*:1-25, 1983.
99. Adams, S. E., Dawson, K. M., Gull, K., Kingsman, S. M., and Kingsman, A. J. The expression of hybrid HIV:Ty virus-like particles in yeast. *Nature*, *329*:68-70, 1987.
100. Malin, M. H., Adams, S. E., Gull, K., Kingsman, A. J., and Kingsman, S. M. The production of hybrid Ty:IFN virus-like particles in yeast. *Nucl. Acids Res.*, *15*:7571-7580, 1987.

101. Weslowski, M., and Fukuhara, H. Linear mitochondrial deoxyribonucleic acid from the yeast Hansenula mrakii. *Mol. Cell. Biol., 1*:387-393, 1981.
102. Wallace, D. C. Structure and evolution of organelle genomes. *Microbiol. Rev., 46*:208-240, 1982.
103. Ward, B. L., Anderson, R. S., and Bendich, A. J. The mitochondrial genome is large and variable in a family of plants (Cucurbitaceae). *Cell, 35*:793-803, 1981.
104. Sederoff, R. R. Structural variation in mitochondrial DNA. *Adv. Genet., 22*:1-108, 1984.
105. Clark-Walker, G. D., Sriprakach, K. S., McArthur, C. R., and Azad, A. A. Mapping of mitochondrial DNA from Torulopsis glabrata: location of ribosomal and transfer RNA genes. *Curr. Genet., 1*:209-217, 1980.
106. Hintz, W. E., Mohan, M., Anderson, J. B., and Horgen, P. A. The mitochondrial DNAs of Agaricus: heterogeneity in A. bitorquis and homogeneity in A. brunnescens. *Curr. Genet., 9*:127-132, 1985.
107. Bruns, T. D., Palmer, J. D., Shumard, D. S., Grossman, L. I., and Hudspeth, M. E. S. Mitochondrial DNAs of Suillus: three fold size change in molecules that share a common gene order. *Curr. Genet., 13*:49-56, 1988.
108. McNabb, S. A., and Klassen, G. R. Uniformity of mitochondrial DNA complexity in Oomycetes and the evolution of the inverted repeat. *Exp. Mycol., 12*:233-242, 1988.
109. Locker, J., Lewin, A., and Rabinowitz, M. The structure and organization of mitochondrial DNA from petite yeast. *Plasmid, 2*:155-181, 1979.
110. Bertrand, H., Collins, R. A., Stohl, L. L., Goewert, R. R., and Lambowitz, A. M. Deletion mutants of Neurospora crassa mitochondrial DNA and their relationship to "stop-start" growth phenotype. *Proc. Natl. Acad. Sci. USA, 77*:6032-6036, 1980.
111. Gross, S. R., Hsieh, T. S., Levine, P. H. Intramolecular recombination as a source of mitochondrial chromosome heteromorphism in Neurospora. *Cell, 38*:233-239, 1984.
112. Lazarus, C. M., Earl, A. J., Turner, G., and Küntzel, H. Amplification of a mitochondrial DNA sequence in the cytoplasmically inherited "ragged" mutant of Aspergillus amstelodami. *Eur. J. Biochem., 106*:633-641, 1980.
113. Wright, R. M., Horrum, M. A., and Cummings, D. J. Are mitochondrial structural genes selectively amplified during senescence in Podospora anserina? *Curr. Genet., 7*:457-464, 1982.
114. Garber, R. C., Turgeon, B. G., and Yoder, O. C. A mitochondrial plasmid from the plant pathogenic fungus Cochliobolus heterostrophus. *Mol. Gen. Genet., 196*:301-310, 1984.
115. Garber, R. C., Lin, J. J., and Yoder, O. C. Mitochondrial plasmids in Cochliobolus heterostrophus. In *Extrachromosomal Elements in Lower Eukaryotes* (R. B. Wickner, A. Hinnebusch, A. M. Lambowitz, I. C. Gunsalus, and A. Hollaender, ed.), Plenum Press, New York, 1986, pp. 105-118.
116. Lazarus, C. M., and Küntzel, H. Anatomy of amplified mitochondrial DNA in "ragged" mutants of Aspergillus amstelodami: excision points within protein genes and a common 215 bp segment containing a possible origin of replication. *Curr. Genet., 4*:99-107, 1981.

117. Bertrand, H. The kalilo senescence factor of Neurospora intermedia: a mitochondrial IS-element derived from a nuclear plasmid. In *Extrachromosomal Elements in Lower Eukaryotes* (R. B. Wickner, A. Hinnebusch, I. C. Gunsalus, and A. Hollaender, eds.), Plenum Press, New York, 1986, pp. 93-103.
118. Gunge, N. Yeast DNA plasmids. *Ann. Rev. Microbiol.*, *37*:253-276, 1983.
119. Clark-Walker, G. D., and Miklos, G. L. G. Localization and quantification of circular DNA in yeast. *Eur. J. Biochem.*, *41*:359-365, 1974.
120. Livingston, D. M. Inheritance of the 2 μmDNA plasmid from Saccharomyces. *Genetics*, *86*:73-84, 1977.
121. Stohl, L. L., Collins, R. A., Cole, M. D., and Lambowitz, A. M. Characterization of two new plasmid DNAs found in mitochondria of wild-type Neurospora intermedia strains. *Nucl. Acids Res.*, *10*:1439-1458, 1982.
122. Mohan, M., Meyer, R. J., Anderson, J. B., and Horgen, P. A. Plasmid-like DNAs in the commercially important mushroom genus Agaricus. *Curr. Genet.*, *8*:615-619, 1984.
123. Bertrand, H., Chan, B. S.-S., Griffiths, A. J. F. Insertion of a foreign nucleotide sequence into mitochondrial DNA causes senescence in Neurospora intermedia. *Cell*, *41*:877-884, 1985.
124. Kistler, H. C., and Leong, S. A. Linear plasmid-like DNAs in the plant pathogenic fungus Fusarium oxysporum f. sp. conglutinans. *J. Bacteriol.*, *167*:587-593, 1986.
125. Tudzynski, P., Esser, K. Extrachromosomal genetics of Claviceps purpurea. II. Plasmids in various wild strains and integrated plasmid sequences in mitochondrial genomic DNA. *Curr. Genet.*, *10*:463-467, 1986.
126. Samac, D. A., and Leong, S. A. Two linear plasmids in mitochondria of Fusarium solani f. sp. cucurbitae. *Plasmid*, *19*:57-67, 1988.
127. Normand, P., Simonet, P., Glasson, L., Ravel-Chapins, P., Fortin, J. A., and Lalonde, M. Presence of a linear plasmid-like DNA molecule in the fungal pathogen Ceratocystis fimbriata Ell. and Halst. *Curr. Genet.*, *11*:335-338, 1987.
128. Hintz, W. E. A., Anderson, J. B., and Horgen, P. A. Physical mapping of the mitochondrial genome of the cultivated mushroom Agaricus brunnescens (= A. bisporus). *Curr. Genet.*, *14*:43-49, 1988.
129. Hashiba, T., Homma, Y., Hyakumachi, M., and Matsuda, T. Isolation of a DNA plasmid in the fungus Rhizoctonia solani. *J. Gen. Microbiol.*, *130*:2067-2070, 1984.
130. Toh-e, A., Utatsu, I., Utsunomiya, A., Sakamoto, S., and Imura, T. Two-micron DNA-like plasmids from non-saccharomyces yeasts. In *Extrachromosomal Elements in Lower Eukaryotes* (R. B. Wickner, A. Hinnebusch, A. M. Lambowitz, I. C. Gunsalus, and A. Hollander, eds.), Plenum Press, New York, 1986, pp. 425-437.
131. Caprioglio, D. R., Parks, L. W. Temporal expression of transcription and relative copy number of plasmid pSFB-1 in Scytalidium flavobrunneum. *J. Bacteriol.*, *171*:4876-4880, 1989.
132. Samac, D. A., and Leong, S. A. Lack of association of mitochondrial plasmids and pathogenicity in Nectria haematococca (Fusarium solani f. sp. cucurbitae). *Mol. Plant-Microbe Interact.*, *2*:128-131, 1989.
133. Lambowitz, A. M., Akins, R. A., Kelley, R. L., Pande, S., and Nargang, F. E. Mitochondrial plasmids of Neurospora and other filamentous fungi. In *Extrachromosomal Elements in Lower Eukaryotes*

(R. B. Wickner, A. Hinnebusch, A. M. Lambowitz, I. C. Gunsalus, and A. Hollaender, eds.), Plenum Press, New York, 1986, pp. 83-92.

134. Samac, D. A., and Leong, S. A. Mitochondrial plasmids in filamentous fungi. *Mol. Plant-Microbe Interact.*, *2*:155-159, 1989.
135. Tudzynski, P., Duvell, A., and Oeser, B. Linear plasmids in the phytopathogenic fungus Claviceps purpurea. In *Extrachromosomal Elements in Lower Eukaryotes* (R. B. Wickner, A. Hinnebusch, A. M. Lambowitz, I. C. Gunsalus, and A. Hollaender, eds.), Plenum Press, New York, 1986, pp. 119-127.
136. Akins, R. A., Kelley, R. L., and Lambowitz, A. M. Characterization of mutant mitochondrial plasmids of Neurospora spp. that have incorporated tRNAs by reverse transcription. *Mol. Cell. Biol.*, *9*:678-691, 1989.
137. Gunge, N., Tamaru, A., Ozawa, F., and Sakaguchi, K. Isolation and characterization of linear deoxyribonucleic acid plasmids from Kluyveromyces lactis and the plasmid-associated killer character. *J. Bacteriol.*, *145*:382-390, 1981.
138. Tipper, D. J., and Bostian, K. A. Double-stranded ribonucleic acid killer systems in yeasts. *Microbiol. Rev.*, *48*:125-156, 1984.
139. Piechulla, B., Hahn, U., McLaughlin, L. W., and Küntzel, H. Nucleotide sequence of 5S ribosomal RNA from Aspergillus nidulans and Neurospora crassa. *Nucl. Acid Res.*, *9*:1445-1450, 1981.
140. Lovett, J. S., and Haselby, J. A. Molecular weights of the ribonucleic acid of fungi. *Arch. Mikrobiol.*, *80*:191-204, 1971.
141. Planta, R. J., Retel, J., Klootwijk, J., Meyerink, J. H., De Jonge, P., Van Keulen, H., and Brand, R. C. Synthesis and processing of ribosomal ribonucleic acid in eukaryotes. *Biochem. Soc. Trans.*, *5*: 462-466, 1977.
142. Knight, R. H., and Van Etten, J. L. Characteristics of ribonucleic acids isolated from Botryodiplodia theobromae pycnidiospores. *Arch. Microbiol.*, *109*:45-50, 1976.
143. Porter, D., and Smiley, R. Ribosomal RNA molecular weights of Trichomycetes and Zygomycetes. *Exp. Mycol.*, *3*:188-193, 1979.
144. Faye, G., Kujawa, C., and Fukuhara, H. Physical and genetic organization of petite and grande yeast mitochondrial DNA IV. In vivo transcription products of mitochondrial DNA and localization of 23S ribosomal RNA in petite mutants of Saccharomyces cerevisiae. *J. Mol. Biol.*, *88*:185-203, 1974.
145. Chen, M., Anné, J., Volckaert, G., Huysmans, E., Vandendberghe, A., and De Wachter, R. The nucleotide sequences of the 5S rRNAs of seven molds and a yeast and their use in studying ascomycete phylogeny. *Nucl. Acids Res.*, *12*:4881-4892, 1984.
146. Russell, P. J., and Wilkerson, W. M. The structure and biosynthesis of fungal cytoplasmic ribosomes. *Exp. Mycol.*, *4*:281-337, 1980.
147. Udem, S. A., and Warner, J. R. Ribosomal RNA synthesis in Saccharomyces cerevisiae. *J. Mol. Biol.*, *65*:227-242, 1972.
148. Lizardi, P. M., and Luck, D. J. L. Absence of a 5S RNA component in the mitochondrial ribosomes of Neurospora crassa. *Nature (New Biol.)*, *229*:140-142, 1971.
149. Sor, F., and Fukuhara, H. Sequence nucleotidique du gene de l'ARN ribosomique 15S mitochondrial de la levure. *C.R. Acad. Sci. Paris Ser. D*, *191*:933-936, 1980.

150. Sor, F., and Fukuhara, H. Nucleotide sequence of the small ribosomal RNA gene from the mitochondria of Saccharomyces cerevisiae. In *Mitochondrial Genes* (P. Slonimski, P. Borst, and G. Attardi, eds.), Cold Spring Harbor Laboratory, New York, 1982, pp. 255-262.
151. Sor, F., and Fukuhara, H. Complete DNA sequence coding for the large ribosomal RNA of yeast mitochondria. *Nucl. Acids Res.*, *11*:339-348, 1983.
152. Lang, B. F., Ahne, F., Distler, S., Trinkl, H., Kaudewitz, F., and Wolf, K. Sequence of the mitochondrial DNA, arrangement of genes and processing of their transcripts in Schizosaccharomyces pombe. In *Mitochondria 1983* (R. J. Schweyen, K. Wolf, and F. Kaudewitz, ed.), de Gruyter, Berlin, 1983, pp. 313-329.
153. Johnson, S. A., Lovett, J. S., and Wilt, F. H. The polyadenylated RNA of zoospores and growth phase cells of the aquatic fungus Blastocladiella. *Dev. Biol.*, *56*:329-342, 1977.
154. Roheim, J. R., Knight, R. H., and Van Etten, J. L. Synthesis of ribonucleic acids during the germination of Rhizopus stolonifer sporangiospores. *Dev. Biol.*, *41*:137-145, 1974.
155. Rozek, C. E., Orr, W. C., and Timberlake, W. E. Diversity and abundance of polyadenylated RNA from Achlya ambisexualis. *Biochemistry*, *17*:716-722, 1978.
156. Van Etten, J. L., and Freer, S. N. Polyadenylate-containing RNA in dormant and germinating sporangiospores of Rhizopus stolonifer. *Exp. Mycol.*, *2*:301-312, 1978.
157. Zatinge, B., Dons, J. J. M., and Wessels, J. G. H. Comparison of poly(A)-containing RNAs in different cell types of the lower eukaryote Schizophyllum commune. *Eur. J. Biochem.*, *101*:251-260, 1979.
158. Groner, B., Hynes, N., and Phillips, S. Length heterogeneity in poly(adenylic acid) region of yeast messenger ribonucleic acid. *Biochemistry*, *13*:5378-5383, 1974.
159. Harper, J. F., Clancy, M. J., and Magee, P. T. Properties of polyadenylate-associated ribonucleic acid from Saccharomyces cerevisiae ascospores. *J. Bacteriol.*, *143*:958-965, 1980.
160. Jaworski, A. J. Synthesis of polyadenylic acid RNA during zoospore differentiation and germination in Blastocladiella emersonii. *Arch. Biochem. Biophys.*, *173*:201-209, 1976.
161. Freer, S. N., Mayama, M., and Van Etten, J. L. Synthesis of polyadenylate-containing RNA during the germination of Rhizopus stolonifer sporangiospores. *Exp. Mycol.*, *1*:116-127, 1977.
162. Timberlake, W. E., Schumard, D. S., and Goldberg, R. B. Relationship between nuclear and polysomal RNA populations of Achlya: a simple eucaryotic system. *Cell*, *10*:623-632, 1977.
163. Van Etten, J. L., and Rawn, C. D. Polyadenylate-containing RNA in dormant and germinating Botryodiplodia theobromae pycnidiospores. *Can. J. Microbiol.*, *25*:375-379, 1979.
164. Law, D. J., Rozek, C. E., and Timberlake, W. E. Polyadenylate metabolism in Achlya ambisexualis. *Exp. Mycol.*, *2*:198-210, 1978.
165. McLaughlin, C. S., Wamer, J. R., Edmonds, M., Nakazato, H., and Vaughn, M. H. Polyadenylic acid sequences in yeast messenger ribonucleic acid. *J. Biol. Chem.*, *248*:1466-1471, 1973.
166. Reed, J., and Wintersberger, E. Adenylic acid-rich sequences in messenger RNA from yeast polysomes. *FEBS Lett.*, *32*:213-217, 1973.

167. Rosen, D., Edelman, M., and Galun, E. Characterization of polyadenylate from the fungus Trichoderma viride. *J. Bacteriol.*, *123*:765-767, 1975.
168. Silver, J. C., and Horgen, P. A. Hormonal regulation of presumptive mRNA in the fungus Achlya ambisexualis. *Nature*, *249*:252-254, 1974.
169. Zagorski, J., Tollervey, D., and Fournier, M. J. Characterization of an SNR gene locus in Saccharomyces cerevisiae that specifies both dispensible and essential small nuclear RNAs. *Mol. Cell. Biol.*, *8*:3282-3290, 1988.
170. Hereford, L. M., and Rosbash, M. Number and distribution of polyadenylated RNA sequences in yeast. *Cell*, *10*:453-462, 1977.
171. Zatinge, B., Hoge, J. H. C., and Wessels, J. G. H. Frequency and diversity of RNA sequences in different cell types of the fungus Schizophyllum commune. *Eur. J. Biochem.*, *113*:381-389, 1981.
172. Timberlake, W. E., and Hamer, J. E. Regulation of gene activity during conidiophore development in Aspergillus nidulans. *Gen. Eng.*, *8*:1-29, 1986.
173. Langford, C. J., Nellen, J., Niessing, J., and Gallwitz, D. Yeast is unable to excise foreign intervening sequences from hybrid gene transcripts. *Proc. Natl. Acad. Sci. USA*, *80*:1496-1500, 1983.
174. Kaufer, N., Simanis, V., and Nurse, P. Fission yeast Schizosaccharomyces pombe correctly excises a mammalian RNA transcript intervening sequence. *Nature*, *318*:78-80, 1985.
175. Legrain, P., Seraphin, B., and Rosbash, M. Early commitment of yeast pre-mRNA to the spliceosome pathway. *Mol. Cell. Biol.*, *8*:3755-3760, 1988.
176. Busch, H., Reddy, R., Rothblum, L., C, C. Y. SnRNAs, SnRNPs, and RNA processing. *Annu. Rev. Biochem.*, *51*:617-653, 1982.
177. Tollervey, D. High level of complexity of small nuclear RNAs in fungi and plants. *J. Mol. Biol.*, *196*:355-361, 1987.
178. Ares, M., Jr. U2 RNA from yeast is unexpectedly large and contains homology to vertebrate U4, U5, and U6 small nuclear RNAs. *Cell*, *47*:49-59, 1986.
179. Siliciano, P. G., Brow, D. A., Roiha, H., and Guthrie, C. An essential snRNA from S. cerevisiae has properties predicted for U4, including interaction with a U6-like snRNA. *Cell*, *50*:585-592, 1987.
180. Kretzner, L. Saccharomyces cerevisiae U1 small nuclear RNA secondary structure contains both universal and yeast-specific domains. *Proc. Natl. Acad. Sci. USA*, *87*:851-855, 1990.
181. Li, H. V., Zagorski, J., and Fournier, M. J. Depletion of U14 small nuclear RNA (snR128) disrupts production of 18S rRNA in Saccharomyces cerevisiae. *Mol. Cell. Biol.*, *10*:1145-1152, 1990.
182. Parker, R., Simmons, T., Shuster, E. O., Siliciano, P. G., and Guthrie, C. Genetic analysis of small nuclear RNAs in Saccharomyces cerevisiae: viable sextuple mutant. *Mol. Cell. Biol.*, *8*:3150-3159, 1988.
183. Tollervey, D. Deletion of a yeast small nuclear RNA gene impairs growth. *EMBO J.*, *4*:3873-3878, 1985.
184. Walter, P., and Lingappa, V. R. Mechanism of protein translocation across the endoplasmic reticulum membrane. *Annu. Rev. Cell. Biol.*, *2*:499-516, 1986.
185. Brennwald, P., Liao, X., Holm, K., Porter, G., and Wise, J. A. Identification of an essential Schizosaccharomyces pombe RNA homologous to the 7SL component of signal recognition particle. *Mol. Cell. Biol.*, *8*:1580-1590, 1988.

186. Poritz, M. A., Siegel, V., Hansen, W., and Walter, P. Small ribonucleoprotein in Schizosaccharomyces pombe and Yarrowia lipolytica homologous to signal recognition particle. *Proc. Natl. Acad. Sci. USA*, *85*:4315-4319, 1988.
187. Felici, F., Cesareni, G., and Hughes, J. M. X. The most abundant small cytoplasmic RNA of Saccharomyces cerevisiae has an important function required for normal cell growth. *Mol. Cell. Biol.*, *9*:3260-3268, 1989.
188. Buck, K. W. Fungal virology: an overview. In *Fungal Virology* (K. W. Buck, ed.), CRC Press, Boca Raton, 1986, pp. 1-84.
189. van Zaayen, A. Mushroom viruses. In *Viruses and Plasmids in Fungi* (P. A. Lemke, ed.), Marcel Dekker, New York, 1979, pp. 239-324.
190. Gandy, D. G. "Watery stipe" of cultivated mushrooms. *Nature*, *185*:482, 1960.
191. Gandy, D. G. A transmissible disease of cultivated mushrooms ("watery stipe"). *Ann. Appl. Biol.*, *48*:427, 1960.
192. Hollings, M. Viruses associated with a die-back disease of cultivated mushroom. *Nature*, *196*:962-965, 1962.
193. Marino, R., Saksena, K. N., Schuler, M., Mayfield, J. E., and Lemke, P. A. Double stranded ribonucleic acid in Agaricus bisporus. *Appl. Environ. Microbiol.*, *31*:433-438, 1976.
194. Saksena, K. N. Isolation and large-scale purification of mushroom viruses. *Dev. Ind. Microbiol.*, *16*:134-144, 1975.
195. Barton, R. J., and Hollings, M. Purification and some properties of two viruses infecting the cultivated mushroom Agaricus bisporus. *J. Gen. Virol.*, *42*:231, 1979.
196. Tavantzis, S. M., Romaine, C. P., and Smith, S. H. Purification and partial characterization of a bacilliform virus from Agaricus bisporus: a single-stranded RNA mycovirus. *Virology*, *105*:94-102, 1980.
197. Myers, C. J., Griffiths, A. J. F., Kraus, S. R., and Martin, R. R. Double stranded RNA in natural isolates of Neurospora. *Curr. Genet.*, *13*:495-501, 1988.
198. Young, T. W. Killer Yeasts. In *The Yeasts*, Vol. 2 (A. H. Rose and J. S. Harrison, ed.), Academic Press, London, 1987, pp. 131-164.
199. Rosini, G. The occurrence of killer characteris in yeasts. *Can. J. Microbiol.*, *29*:1462-1464, 1983.
200. Koltin, Y. The killer systems of Ustilago maydis. In *Fungal Virology* (K. W. Buck, ed.), CRC Press, Boca Raton, 1986, pp. 109-142.
201. Bobek, L. A., Bruenn, J. A., Field, J., and Gross, K. W. Cloning of cDNA to a yeast viral double-stranded RNA and comparison of three viral RNAs *Gene*, *19*:225-234, 1982.
202. Bostian, K. A., Burn, V. E., Jayachandran, S., and Tipper, D. J. Yeast killer dsRNA plasmids are transcribed in vivo to produce full and partial-length plus-stranded RNAs. *Nucl. Acids Res.*, *11*:1077-1097, 1983.
203. Hastie, N., Brennan, V., and Bruenn, J. No homology between double-stranded RNA and nuclear DNA of yeast. *J. Virol.*, *28*:1002-1005, 1978.
204. Sommer, S. S., and Wickner, R. B. Yeast L dsRNA consists of at least three distinct RNAs; evidence that the non-Mendelian genes [HOK], [NEX] and [EXL] are on one of these dsRNAs. *Cell*, *31*:429-441, 1982.

205. Koltin, Y. Virus-like particles in Ustilago maydis: mutants with partial genomes. *Genetics, 86*:527-534, 1977.
206. Koltin, Y., and Kandel, J. S. Killer phenomenon in Ustilago maydis: the organization of the viral genome. *Genetics, 88*:267-276, 1978.
207. Van Alfen, N. K., and Hansen, D. R. Hypovirulence: a unique phenotypic expression of dsRNA in a pathogenic fungus. In *Double-stranded RNA Viruses* (R. W. Compans and D. H. L. Bishop, eds.), Elsevier, New York, 1983, pp. 463-468.
208. Van Alfen, N. K. Hypovirulence of Endothia (Cryphonectria) parasitica and Rhizoctonia solani. In *Fungal Virology* (K. W. Buck, ed.), CRC Press, Boca Raton, 1986, pp. 143-162.
209. Newhouse, J. R., Hoch, H. C., and MacDonald, W. L. The ultrastructure of Endothia parasitica. Comparison of a virulent with a hypovirulent isolate. *Can. J. Bot., 61*:389-399, 1983.
210. Newhouse, J. R., Macdonald, W. L., and Hock, H. C. Virus-like particles in hyphae and conidia of European hypovirulent (dsRNA-containing) strains of Cryphonectria parasitica. *Can. J. Bot., 68*:90-101, 1990.
211. Hiremath, S., L'Hostis, B., Ghabrial, S. A., and Rhoads, R. E. Terminal structure of hypovirulence-associated dsRNAs in the chestnut blight fungus Endothia parasitica. *Nucl. Acids Res., 14*:9877-9896, 1986.
212. Tartaglia, J., Paul, C. P., Fulbright, D. W., and Nuss, D. L. Structural properties of double-stranded RNAs associated with biological control of chestnut blight fungus. *Proc. Natl. Acad. Sci. USA, 83*: 9109-9113, 1986.
213. Rae, B. P., Hillman, B. I., Tartaglia, J., and Nuss, D. L. Characterization of double-stranded RNA genetic elements associated with biological control of chestnut blight: organization of terminal domains and identification of gene products. *EMBO J., 8*:657-663, 1989.
214. Bigelis, R. Primary metabolism and industrial fermentations. In *Gene Manipulations in Fungi* (J. W. Bennett and L. L. Lasure, eds.), Academic Press, Orlando, 1985, pp. 357-401.
215. Demain, A. L. Production of purine nucleotides by fermentation. *Progr. Ind. Microbiol., 8*:35-72, 1968.
216. Nakao, Y., Imada, A., Wada, T., and Ogata, K. Degradation of nucleic acids and their related compounds by microbial enzymes. Part XVI. Excretion of mononucleotides by yeasts. *Agr. Biol. Chem., 28*:151-159, 1964.
217. Lewis, M. J., and Phaff, H. J. Release of nitrogenous substances by brewers' yeast. 2. Effect of environmental conditions. *Am. Soc. Brew. Chem. Proc., 1963*:114-123, 1963.
218. Tsukada, Y., and Sugimori, T. Excretion of ultraviolet-absorbing materials by Zygosaccharomyces soja mutant. *Agr. Biol. Chem., 28*: 471-478, 1964.
219. Higuchi, M., and Uemura, T. Changes in subcellular components of yeast as related to nucleotide release from the cells incubated in citrate buffer. *Plant Cell Physiol., 3*:249-261, 1962.
220. Horitsu, H., Satake, T., Ogawa, K., and Tamoyeda, M. Leakage of 5' guanosine monophosphate from Candida lipolytica (IFO,0746) and its mutants. *Agr. Biol. Chem., 41*:1667-1672, 1977.
221. Demain, A. L. Amino acid control of ribonucleic acid excretion in Bacillus subtilis. *Biochem. Biophys. Res. Commun., 24*:39-43, 1966.

222. Ivanova, G. S., Valiukaite, R. V., Youngol, L. M., and Bezborodov, A. M. Multiple forms of intracellular and extracellular ribonucleases of some species of Aspergillus and Penicillium. In *Proc. 1st Symp. on Ribosomes and RNA Metabolism* (J. Zelinka and J. Balan, eds.), Publishing House of the Slovak Acad. Sci., Prague, 1974, pp. 485-498.
223. Ivanova, G. S. Ribonucleases in the differentiation of fungi. In *Biotechnology and Fungal Differentiation* (J. Meyrath and S. D. Bu'Lock, eds.), Academic Press, London, 1977, pp. 77-101.
224. Nakao, Y., Nogami, I., and Ogata, K. Degradation of nucleic acids and their related compounds by microbial enzymes Part IX. Degradation of ribonucleic acid and deoxyribonucleic acid by intracellular enzymes of Rhodotorula glutinis. *Agr. Biol. Chem.*, *27*:507-517, 1963.
225. Pall, M. L. Adenosine 3',5'-phosphate in fungi. *Microbiol. Rev.*, *45*:462-480, 1981.
226. Thevelein, J. M. Regulation of trehalase activity by phosphorylation-dephosphorylation during developmental transitions in fungi. *Exp. Mycol.*, *12*:1-12, 1988.
227. Jayaswal, R. K., Bressan, R. A., and Handa, A. K. Occurrence of cyclic adenosine 3',5'-monophosphate in the phytopathogenic fungi Alternaria solani and Phymatotrichum omnivorum. *Arch. Microbiol.*, *135*:125-129, 1983.

18

BIOSYNTHESIS AND PHYSIOLOGY OF ERGOT ALKALOIDS

HELENA SOČIČ and VLADIMIRA GABERC-POREKAR *Boris Kidrič Institute of Chemistry, Ljubljana, Slovenia, Yugoslavia*

I. INTRODUCTION

Very few fungi have such a fascinating history as *Claviceps*. These so-called ergot fungi grow on wild grasses and cereals where they form sclerotia, a latent form which contains pharmacologically active ergot alkaloids. Before recognizing their therapeutic use, ergot alkaloids were known only by their toxic effects, causing mass poisoning in animals and humans over the centuries.

The classical way of ergot alkaloid production is based on the extraction of alkaloids from the field-cultivated sclerotia but since the early 1950s *Claviceps* fungi have been successfully grown in vitro by means of a fermentation technology which represents the main commercial production today.

Because of the wide spectrum of ergot applicability a continuous and still growing interest in these fungal metabolites exists. Current studies are directed mainly toward optimization of the fermentation process, and this requires a better understanding of the processes involved, particularly the genetics and physiology of alkaloid formation. In the last 10 years, investigations in these directions have received much attention but much work remains to be done, especially in the domain of genetic and regulatory mechanisms governing alkaloid biosynthesis.

In this chapter we want to provide in outline an up-to-date survey on the formation of ergot alkaloids with a perspective for the future.

II. ERGOT ALKALOIDS

A. Chemistry

Fungi of the genus *Claviceps* are capable of producing various chemical types of ergot alkaloids. More than 40 different alkaloids have been isolated, and hundreds of chemical modifications of natural alkaloids have been prepared. Several authoritative reviews are available in the literature on the chemistry of these compounds and therefore only basic structures of different alkaloid types will be described [1-4].

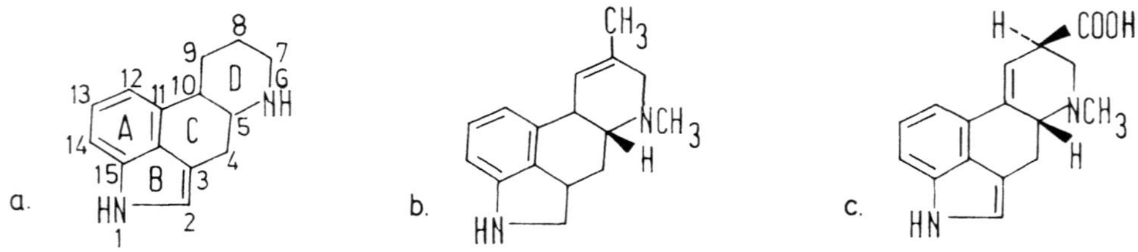

FIGURE 1. The ergoline ring system and some derivatives: (a) ergoline, (b) agroclavine, (c) lysergic acid.

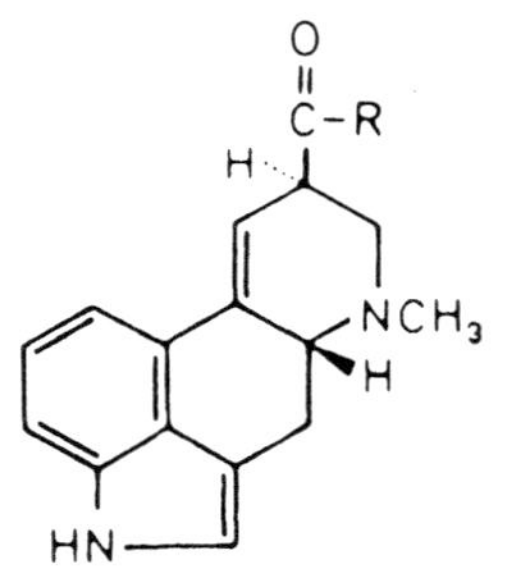

Name	R
Ergometrine (Ergobasine)	$-HN-C(CH_3)(H)-CH_2OH$
α-Hydroxyethyl-lysergamide	$-HN-CH(CH_3)-OH$
Lysergic acid	$-OH$
$\Delta^{8,9}$-Lysergic acid	$-OH$

FIGURE 2. Structure of simple lysergic acid derivatives.

The common structural characteristic of ergot alkaloids is the tetracyclic ergoline system (Fig. 1). Most naturally occurring alkaloids possess 6,8-dimethyl-$\Delta^{8,9}$ or $\Delta^{9,10}$-ergolene structure. On the basis of the substituent at C-8, ergot alkaloids are classified as lysergic acid derivatives and as clavine alkaloids. The "classical ergot" alkaloids are derivatives of lysergic acid or its stereoisomer isolysergic acid, which differ only in configuration at C-8 and occur as matching partners in natural series. The isolysergic acid derivatives are strongly dextrorotatory, whereas their lysergic acid counterparts are levorotatory or, occasionally, weakly dextrorotatory. While many of the compounds in the lysergic acid series exhibit pronounced pharmacological effects, the corresponding isolysergic acid derivatives are almost completely devoid of pharmacological activity. The substituent at C-8 can be a simple amide or a cyclol-structured tripeptide.

The simple derivatives of lysergic or isolysergic acid (Fig. 2) were first isolated from saprophytic cultures [5] and have been used as key substances for further chemical derivatization in obtaining pharmacologically active compounds.

Of special interest are naturally occurring, medicinally important peptide of alkaloids, where lysergic acid is combined with a tripeptide moiety containing α-hydroxy-α-amino acid forming a cyclol with the carboxyl group of L-proline (Fig. 3). This so-called ergopeptine group is found in most of the naturally occurring peptide ergot alkaloids. Some of them were discovered very recently [6] as analogs of ergocornine and α-ergocryptine having an additional methyl group in position 12, which is considered the starting point of the biosynthesis of the cyclol structure and may add to elucidation of this poorly known pathway. Besides the ergopeptine group,

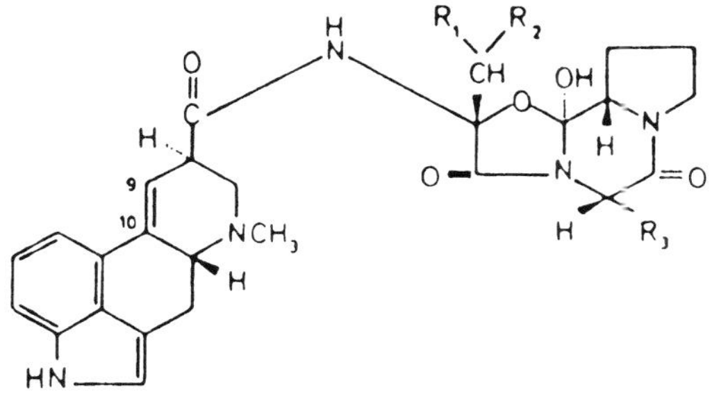

Name	R_1	R_2	R_3	Amino acid sequence of the tripeptide
Ergotamine	H	H	$CH_2-C_6H_5$	-L-Alanine-L-Phenylalanine-L-Proline
Ergosine	H	H	$CH_2CH(CH_3)_2$	-L-Alanine-L-Leucine-L-Proline
Ergocristine	CH_3	CH_3	$CH_2-C_6H_5$	-L-Valine-L-Phenylalanine-L-Proline
α-Ergokryptine	CH_3	CH_3	$CH_2CH(CH_3)_2$	-L-Valine-L-Leucine-L-Proline
β-Ergokryptine	CH_3	CH_3	$CH(CH_3)CH_2CH_3$	-L-Valine-L-Isoleucine-L-Proline
Ergocornine	CH_3	CH_3	$CH(CH_3)_2$	-L-Valine-L-Valine-L-Proline
Ergostine	H	CH_3	$CH_2-C_6H_5$	-L-α-aminobutyric acid-L-Phenylalanine--L-Proline

FIGURE 3. Naturally occurring peptide ergot alkaloids.

Stadler [7] described another, similarly structured group of peptide alkaloids, the ergopeptames, containing D-proline and the tripeptide chain in a noncyclol lactam form.

The group of simple structured ergoline derivatives, the clavine alkaloids, is considered to be of minor biological importance; however, their recently discovered pharmacological activities connected with antiprolactin action and antinidation are very promising. Some substances of this group are intermediates in lysergic acid biosynthesis (Fig. 10) and may also serve as key substances in the chemical synthesis of compounds with dopaminergic activity. More than 20 different clavines have been isolated and described in review articles [9-11].

B. Therapeutic Use

The complexity of chemical structure of the natural and semisynthetic alkaloids is reflected in the diversity of their pharmacological activity. The broad-action spectrum includes central, neurohumoral, and peripheral effects. Fortunately, structural modification allows the manipulation of these activities to selectively enhance some and suppress others, leading to a number of clinically useful agents [4]. The major therapeutic effects of ergot alkaloids have been reported as the following: uterotonic action; increase and decrease of blood pressure; treatment of migraine, other vascular headaches, and senile cerebral insufficiency. Within the last few years, new aspects of the pharmacology of ergot alkaloids have been recognized [12,13]. Some ergot alkaloid derivatives have been found to produce stimulation of dopaminergic receptors and can be used in the treatment of parkinsonism and, further, as potential therapeutic agents for the treatment of acromegaly, amenorrhea, prolactin-dependent galactorrhea, breast cancer, and possibly prostate cancer. A few naturally occurring alkaloids, and particularly the semisynthetic compound LSD (lysergic acid diethylamide), exhibit an extremely potent hallucinogenic activity.

III. MYCOLOGY OF *CLAVICEPS* STRAINS

A. Life Cycle

Claviceps is a genus within the large class of higher fungi known as the plant-parasitic Ascomycetes. The genus comprises several species of *Claviceps* which during flowering infect the host plant, usually rye or grasses. In nature, the biological cycle of *Claviceps* is quite complex, consisting of an asexual and a sexual stage. The asexual reproduction can occur either on a host plant or in saprophytic conditions, while sexual reproduction is possible only under parasitic conditions. The sequence of events in the natural life cycle of *Claviceps purpurea* is illustrated in Fig. 4, as presented by Spalla [14].

The basic life cycle of *Claviceps* fungi is from spore to mycelium, to sclerotium, and, following a meiotic division, back to spore again. Natural infection or parasitism starts with germination of ascospores in the flower of the host plant, in which a filamentous mycelium (sphacelia) develops, proliferating unicellular conidia. This haploid, vegetative stage represents the asexual stage of fungal development. The conidia, which are suspended in honeydew, are spread by rain, wind, and insects to other flowers of the host plant, giving rise to secondary infection.

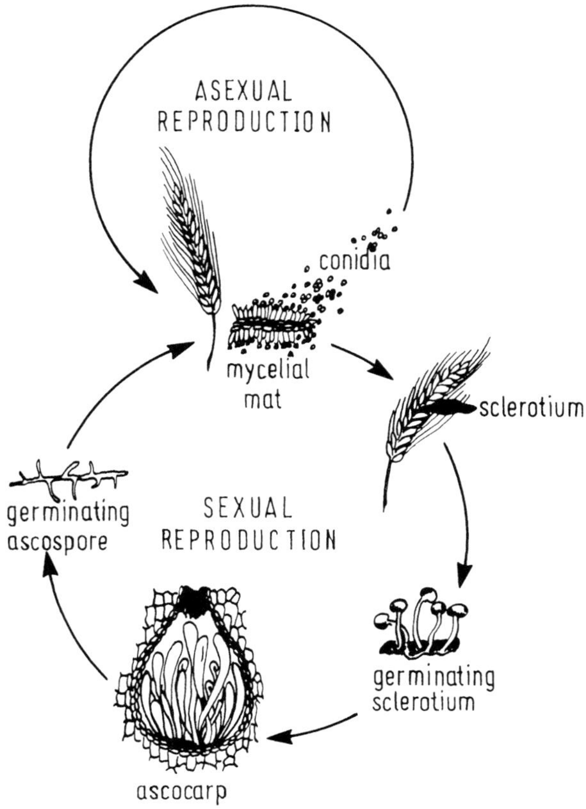

FIGURE 4. Life cycle of *C. purpurea*. (From Ref. 14.)

From conidia the developing mycelium grows rapidly, forming sclerotia which contain about 0.01-1% alkaloid. Sclerotium can be considered as the early stage of the sexual differentiation of *Claviceps*. The ripened sclerotium falls off and, after hibernation, starts to germinate, forming several stromata with ascoscarps and ascospores as the products of sexual reproduction. The ascospores, which are ejected during flowering of the host plant, infect the floret and conclude the cycle. For a clear illustration, see [15,16].

In saprophytic and particularly in submerged conditions, no sexual reproduction is possible, and therefore formation of true sclerotia never occurs, but cells similar to those of sclerotial cells can be observed.

B. Morphology of *Claviceps* Submerged Cultures

The developmental cycle of *Claviceps* in submerged cultures involves several stages which vary in their biochemical and morphological differentiation, where under the term differentiation a result of different expression of the genome under various conditions at different times is considered [17]. Submerged cultures can represent two types of development observed in nature: the sphacelial stage with conidiogenesis and the sclerotium stage with alkaloid synthesis.

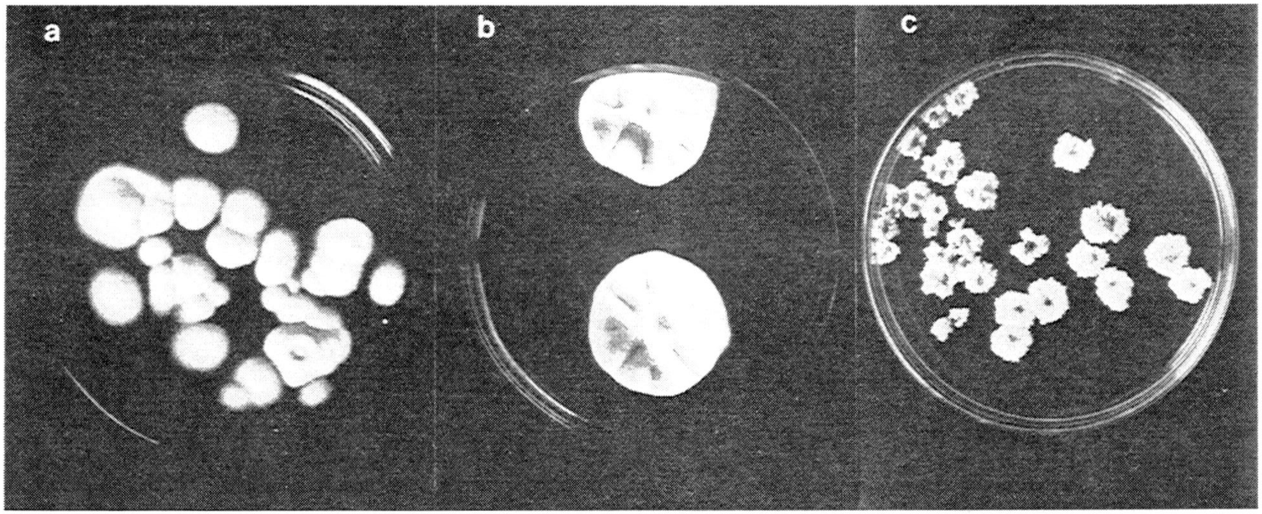

FIGURE 5. Morphology of various *Claviceps* strains on agar plates: (a) *C. paspali* producing simple lysergic acid derivatives; (b) ergotamine-producing *C. purpurea*; (c) ergotoxin-producing *C. purpurea*.

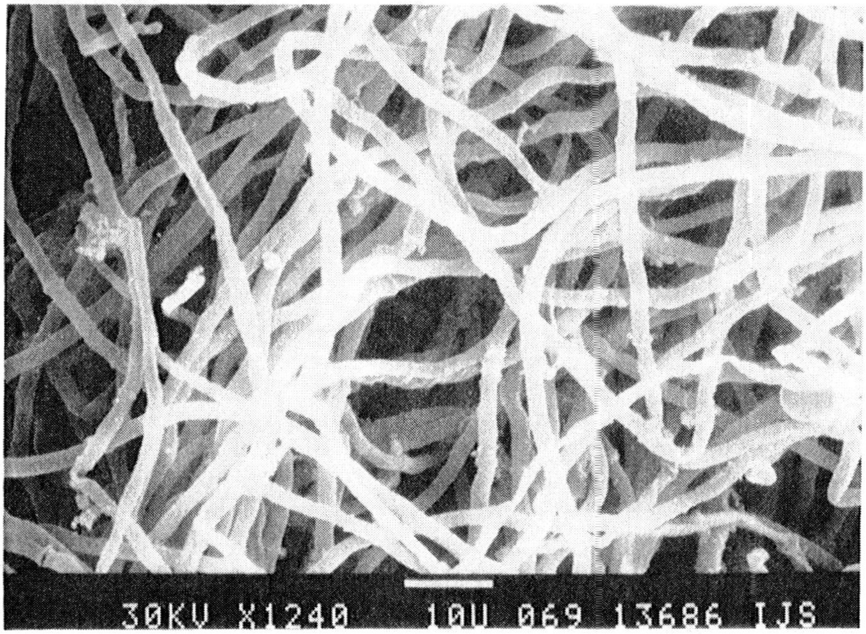

FIGURE 6. Filamentous growth of *C. paspali* in submerged cultures.

The development of *Claviceps* fungi for submerged fermentation starts with colonial growth on solid agar media. Colonies of various strains differ in their typical morphology (Fig. 5), but a common feature of good producers, particularly of peptide alkaloid producers, is sclerotial growth with more or less pronounced pigmentation [18-22]. White colonies with diffuse edges rarely give rise to submerged cultures with high alkaloid titers. At this stage, on solid agar, unicellular conidia or arthrospores may be formed, which represent efficient propagation units and serve as inocula for both parasitic and saprophytic production of alkaloids. Some strains used for submerged production do not form conidia, and in that case mycelial or hyphal fragments can be used for the liquid seed culture preparation.

In the liquid media conidia or arthrospores, as well as mycelial fragments, grow out into filamentous hyphae (Fig. 6) consisting mainly of thin-walled, elongated, multinucleate cells which are not capable of synthesizing alkaloids. Such hyphae represent vegetative growth, but low-producing and degenerated strains usually keep this characteristic feature throughout the fermentation process.

In further development to a productive form, some strains, like *Claviceps paspali*, retain their filamentous structure, but with thick-walled cells, while most of the strains, especially peptide alkaloid producers, differentiate toward ovoid cells (Fig. 7) which resemble the natural sclerotial cells. In nature, these isodiametric cells, which form plectenchymatic tissue, are typically thick-walled with a high content of lipids, and the same holds for the ovoid, sclerotialike cells in submerged cultivation [18,23-26]. Dickerson et al. [27] assigned to the sclerotial cells of *Claviceps fusiformis*

FIGURE 7. Sclerotialike cells of submerged *C. purpurea* cultures.

the dominant role of alkaloid producers, but other authors demonstrated the independence of morphology and alkaloid production [28,29].

Though morphological differentiation is not always accompanied by an enhanced alkaloid production, with numerous *Claviceps* strains a close parallelism between morphological differentiation toward sclerotialike cells and alkaloid-synthesizing ability was ascertained. For this reason several investigators tried to induce differentiation from sphacelial to sclerotial cells by changing the media composition with various additives [24,26,29], by monitoring the cultivating conditions [26], or by mutagenic treatment [24]. Respiration appears to be of major importance in the maintenance of fungal growth and differentiation as well. Though respiration rate is lowered during morphological differentiation into sclerotialike cells, it is a known fact that these cells show an increasing oxygen demand [26,30,31]. Our recent investigations support this statement, with the additional observation that the "diauxic" growth of the fungus *Claviceps purpurea* also depends on the ratio of dissolved and gaseous oxygen uptake rates. Thus, the respiration of sclerotial cells presumably proceeds with gaseous oxygen, while sphacelial cells require dissolved oxygen [32].

Some special cultivating conditions may induce differentiation of hyphae to the conidia or arthrospores which are generally not favorable for alkaloid production in submerged culture [14,33]. Most industrial strains which are potent producers of ergot alkaloids sporulate poorly in the liquid medium, and only a few strains are reported where conidia formation and alkaloid synthesis proceed simultaneously, such as *C. fusiformis* producing simple clavines [34] or clavine-producing *C. purpurea* [35].

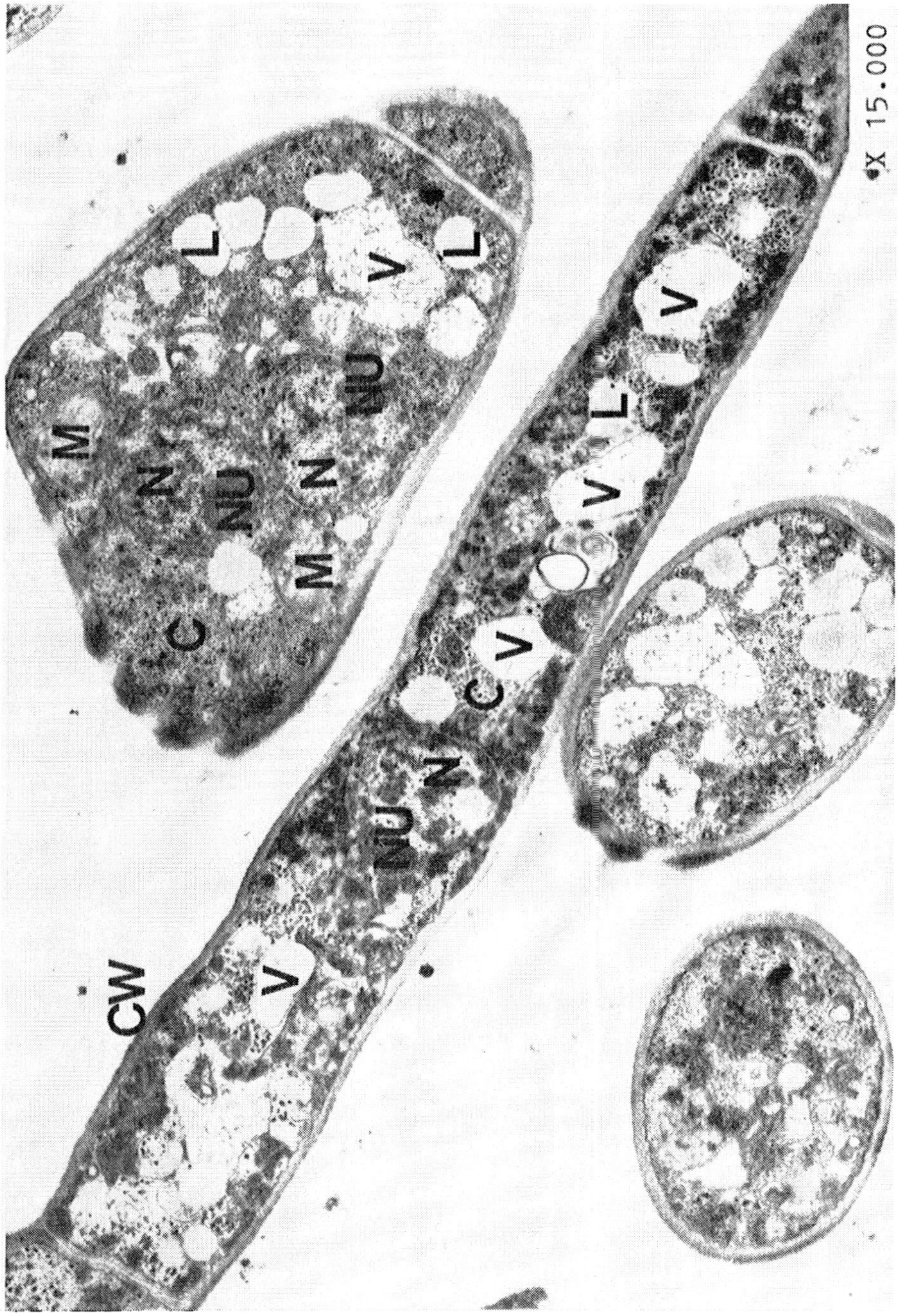

FIGURE 8. Ultrastructural characteristics of vegetative mycelium of a *C. purpurea* submerged culture. M, mitochondria; N, nucleus; NU, nucleolus; V, vacuole; L, lipid; CW, cell wall.

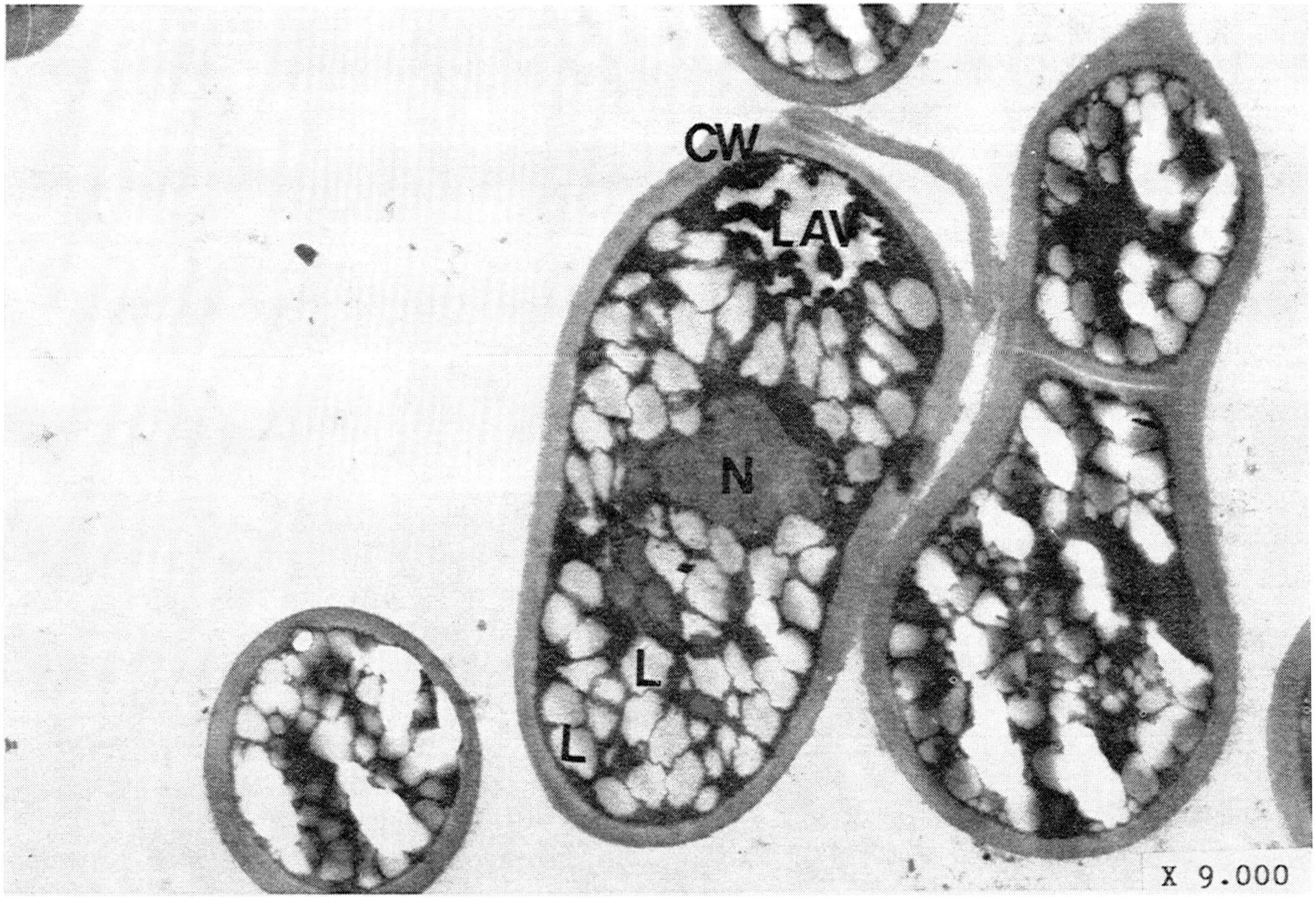

FIGURE 9. Ultrastructure of production mycelium of a *C. purpurea* submerged culture. L, lipid bodies; LAV, lipid-absorbing vacuoles; N, nucleus; CW, cell wall.

The above data clearly demonstrate that beside a particular biochemical differentiation, specific morphological differentiation is also required for alkaloid biosynthesis.

C. Structure and Fine Structure of Cells

To clarify the morphogenesis of submerged *Claviceps* strains in relation to alkaloid production, several investigations of their structure by electron microscopy have been performed.

The ultrastructure of vegetative mycelium reveals all the organelles typical of intensively growing and metabolically active cells. In the granular cytoplasm several nuclei with nucleoli, well-expressed mitochondria which are the site of major oxidative processes, vacuoles, and multivesicular bodies can be observed. At this stage of development only a few lipid bodies are formed [36-39] (Fig. 8). A similar cytoplasm structure was also reported for the sphacelial stage during sclerotium development [40].

As described by various authors [36-38], in saprophytic cultivation no significant changes in the ultrastructure of the mycelium in the early production phase could be observed. But a further developed productive mycelium is characterized by an increasing number of lipid droplets occupying the major part of the cells (Fig. 9). Such lipid-rich cells are particularly characteristic of those strains which accumulate alkaloids inside the cells in the same way as observed in sclerotia of the parasitically grown *C. purpurea* [41]. The cytoplasm of the producing mycelium usually exhibits clearly outlined nuclei, some mitochondria, and extended endoplasmic reticulum membranes which may form numerous vesicles. Interruptions or invaginations on the plasma membrane, as observed recently in a *C. paspali* strain capable of producing and excreting high amounts of simple lysergic acid derivatives into the medium, are very likely an indication of the secretory activity of producing cells [38]. Similar excretory structures were described on cytoplasmic membranes of cells of a developing sclerotium [42].

According to Vořišek and Řeháček [43], the endoplasmic reticulum is the location of alkaloid synthesis, since acetyl-CoA carboxylase, an enzyme catalyzing the synthesis of malonyl-CoA, is localized in the endoplasmic reticulum of sclerotialike cells. This finding was partly supported by Lösecke et al. [37], who also observed an extended endoplasmic reticulum forming a lot of small vesicles which might be related to alkaloid synthesis and transport in strains which excreted alkaloids into the medium. The strong connection between the alkaloid storage compartment (lipid droplets) and endoplasmic reticulum may serve as further evidence for the assumption that endoplasmic reticulum functions as the site of alkaloid synthesis. Ultracytochemical localization of malonyl-CoA formation was confirmed again in 1982 [44] but, as stated by Lösecke et al. [37], further experimental evidence is necessary to exactly localize the site of alkaloid biosynthesis.

IV. BIOSYNTHESIS OF ERGOT ALKALOIDS

Our present knowledge concerning the biosynthesis of ergot alkaloids is the result of many investigations performed during the past 25 years. Several extensive reviews on this subject [3,7,11,33,45] provide the biosynthetic pathway intermediates, some information on the enzymes involved, as well as the possible mechanisms of ergoline system formation. On the

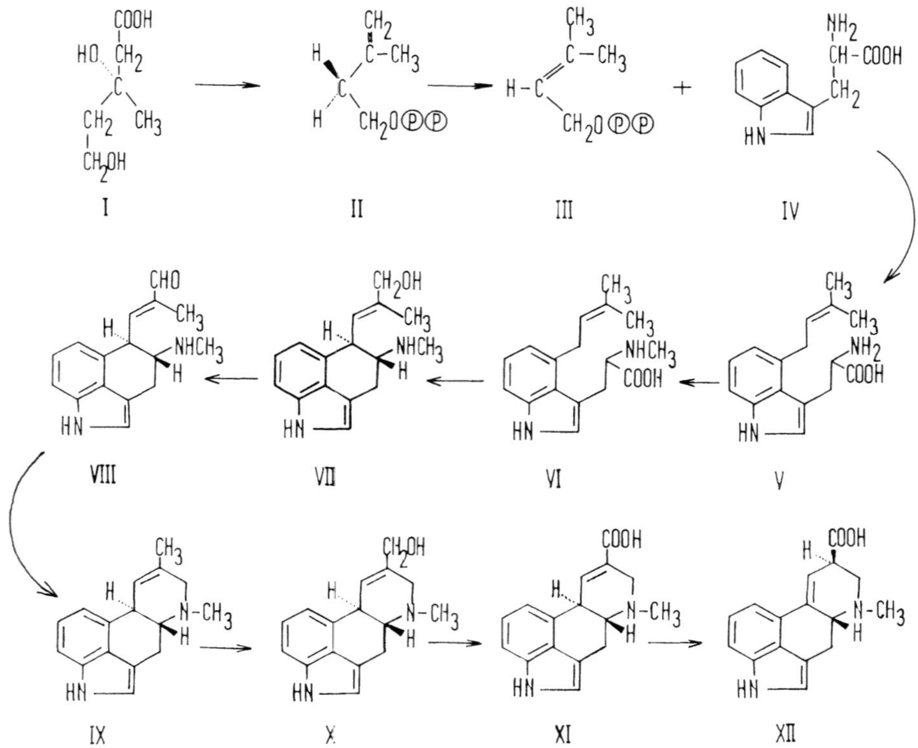

FIGURE 10. Biosynthesis of the ergoline ring system and lysergic acid. I, Mevalonate; II, isopentenylpyrophosphate; III, dimethylallylpyrophosphate; IV, tryptophan; V, 4-dimethylallyltryptophan; VI, N^{α}-methyldimethylallyl-tryptophan; VII, chanoclavine-I; VIII, chanoclavine-I aldehyde; IX, agroclavine; X, elymoclavine; XI, paspalic acid; XII, lysergic acid.

basis of these results a biosynthetic pathway (Fig. 10) for alkaloid formation was also proposed; however, because of the conflicting results obtained in various laboratories, several details remain to be clarified.

A. Precursors and Pathway Intermediates

Tryptophan, mevalonic acid, and the methyl group of methionine have been recognized as the main precursors to ergoline [46-48]. The first step in this biosynthetic pathway is the isoprenylation of L-tryptophan at C-4 by a dimethylallylpyrophosphate molecule to form 4-dimethylallyltryptophan. The next step represents the introduction of an *N*-methyl group in dimethylallyltryptophan [45]. The *N*-methyl group arises, via a transmethylation reaction, from *S*-adenosylmethionine [48,49]. According to Otsuka and coworkers [49], this methylation step of the amino group precedes the closure of ring C and the loss of the carboxyl group. As a matter of fact, Barrow and Quigley [50] had isolated N^{α}-methyl-4-γ,γ-dimethylallyltryptophan from *Claviceps* cultures incubated under unaerobic conditions. The same intermediate, fed to *Claviceps* cultures, was later shown to be directly converted to elymoclavine [49].

The next step in the biosynthetic pathway involves the formation of the intermediate chanoclavine-I with an open ring D. The ring closes after oxidation to the corresponding chanoclavin-I aldehyde, which was detected in a blocked mutant of *C. purpurea* no earlier than 1980 [51], though it had already been postulated as a natural intermediate in ergoline formation [45,52]. The first tetracyclic compound in the pathway is agroclavine, which is transformed by some oxidative steps to lysergic acid via elymoclavine. Some authors [45,53] report that transformation of elymoclavine into lysergic acid and its simple derivatives proceeds via $\Delta^{8,9}$-lysergic acid, the so-called paspalic acid which, in contrast to lysergic acid, occurs readily in nature and submerged cultivation, and easily isomerizes to lysergic acid. So, the true intermediate in the lysergic acid amide synthesis may be lysergic acid in the form of an activated lysergyl-CoA [54,55]. The amide components of ergonovine and lysergic acid-α-hydroxyethylamide are derived from L-alanine [56].

The biosynthesis of peptide alkaloids which consist of a moiety of lysergic acid linked to a cyclol peptide side chain is not as well understood as the formation of the ergoline ring system. Several hypotheses exist on the formation of the peptide chain, which may start from the lysergic acid or the proline end [3,45,57]. By feeding experiments it was possible to obtain incorporation of the appropriate amino acids as single compounds; however, only amino acids with nonpolar groups can be incorporated [58]. When di- or tripeptides are used in the feeding experiments, they have to be split enzymically to free amino acids before incorporation takes place [3]. The fact that addition of inhibitors of protein synthesis does not prevent the formation of the peptide alkaloids supports the idea of peptide chain formation being a nonribosomal process, similar to the biosynthesis of peptide antibiotics and presumably taking place on a multienzyme complex [57]. The proposed biosynthetic pathway, beginning with the linkage of L-proline to the enzyme, is depicted in Fig. 11.

B. Enzymes

Reactions involved in the biosynthetic pathway are catalyzed by specific enzymes or enzymic systems. The first reaction in the alkaloid biosynthesis, isoprenylation of L-tryptophan to 4-dimethylallyltryptophan, is catalyzed by L-tryptophan dimethylallyltransferase, commonly known as dimethylallyltryptophan (DMAT) synthetase. This enzyme is the best characterized and the first enzyme isolated by Heinstein et al. [59] from *Claviceps* strain SD58. The enzyme, which is activated by Fe^{2+}, Mg^{2+}, and particularly Ca^{2+}, consists of a single polypeptide chain with a molecular weight of 73,000 [60]. The enzyme is absent in growing cells and starts to appear at the end of the growth phase just preceding the appearance of alkaloids. Of special interest is the finding that this enzyme is inhibited by the end products agroclavine and elymoclavine [59] in an uncompetitive way with either substrate. This inhibition occurs at alkaloid concentrations which are reached in a typical fermentation [45,60] and may thus be of physiological significance in terminating alkaloid synthesis.

At this stage in the pathway, the enzymic hydroxylation of γ,γ-dimethylallyltryptophan, leading to 4-(*E*-4'-hydroxy-3'-methylbut-2'-enyl)-L-tryptophan, was also reported [61], but it seems that this reaction is not significant for the major ergoline pathway.

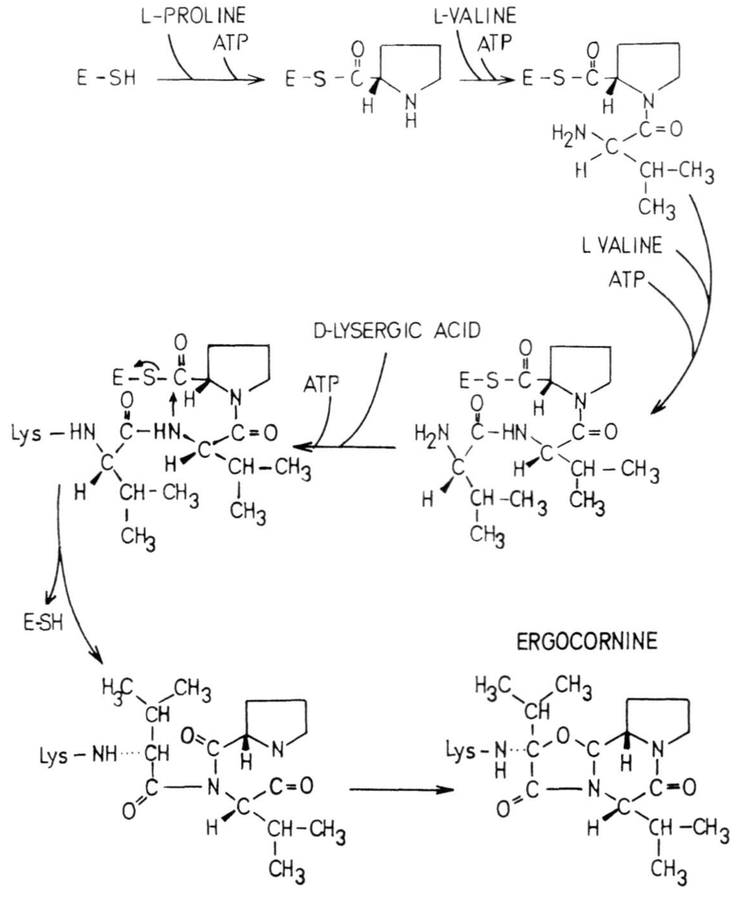

FIGURE 11. Hypothetical scheme for the peptide side chain of ergot alkaloid formation. (From Ref. 45.)

The next step, introduction of an *N*-methyl group in dimethylallyltryptophan, seems to be a reaction of high substrate specificity [45], which is not very often the case in secondary metabolite synthesis. This second pathway-specific step in ergoline biosynthesis is catalyzed by dimethylallyltryptophan-*N*-methyltransferase, an enzyme which is strongly inhibited by Zn^{2+} ion. The time course of this enzyme activity parallels that of alkaloid formation [49].

The enzyme involved in ring D cyclization, chanoclavine cyclase, requires ATP, Mg^{2+}, and NADPH for its action [62]. The time course of the appearance and decline of this enzyme in the cultures closely resembles that demonstrated for dimethylallyltryptophan synthetase and dimethylallyltryptophan-*N*-methyltransferase. The cyclase is inhibited by elymoclavine and lysergic acid, and presumably flavine coenzyme is needed in the reaction [45].

The conversions of agroclavine to elymoclavine and elymoclavine to paspalic and lysergic acid are catalyzed by microsomal oxygenases. The enzyme designated as agroclavine-17-monooxygenase or agroclavine hydroxylase has been frequently studied [63-65]. Agroclavine-17-monooxygenase and elymoclavine-17-monooxygenase, respectively, are dependent on NADPH and molecular oxygen; the enzyme activities are enhanced by NADH, and dependence on ATP has often been reported. Inhibition of both enzymes by carbon monoxide [64] and cytochrome C and the independence of NaCN strongly suggest that clavine monooxygenases are cytochrome P-450-dependent enzymes [65]. It is interesting to note that a mammalian enzyme system had also been found capable of converting agroclavine to noragroclavine and elymoclavine, which also depended on cytochrome P-450 [66].

From *C. purpurea* protoplasts and lyophilized mycelium a cell-free system possessing "cyclol synthetase" activity was also obtained capable of incorporating leucine, alanine, and proline into the peptide side chain of ergosine [67]. In 1985 an enzyme system from *C. purpurea* (Fr.) Tul. catalyzing the incorporation of phenylalanine into ergotamine—ergotamine synthetase—was isolated, but it could not be separated from phenylalanyl-tRNA synthetase by the purification procedure applied [68].

From *Claviceps* sp. SD58 cultures another enzyme, isopentenylpyrophosphate isomerase:dimethylallyl pyrophosphate isomerase, catalyzing one of the first reactions common to terpenoid and alkaloid biosynthesis, was isolated in purified form and characterized [69].

But to conclude, as far as concerns the enzymology of ergot alkaloid biosynthesis, a considerable amount of work still remains to be done if a clear picture of each step in the biosynthetic pathway is the ultimate aim.

C. Regulation Mechanisms

Biosynthesis of ergot alkaloids as secondary metabolites depends on the precursors and energy generated through the primary metabolism. Primary and secondary metabolism are intimately related in complex ways and even the regulation processes are governed by similar mechanisms, some of which will be mentioned shortly. We must admit that the regulation of ergot alkaloid biosynthesis remains a very complex and virtually unexplored field.

1. Trophophase-Idiophase Relationship

An interesting aspect of secondary metabolism is the growth (trophophase)-dissociated production of many secondary metabolites during the production phase-idiophase. But with most ergot alkaloid-producing strains, alkaloid production also appears to occur during the growth, especially when plotted against dry cell weight. Comparison with nucleic acid content shows a clearer dissociation of the growth (increase in nucleic acid) and production (increase in ergot alkaloids). As often in secondary metabolism, a certain region of low specific growth rates favors alkaloid production [70], indicating that the enzymes responsible for alkaloid biosynthesis might be repressed during rapid growth.

2. Induction

Early in the development of ergot fermentation technology it was noticed that alkaloid production could be increased by adding tryptophan to saprophytic cultures [71,72]. Though the stimulatory action of tryptophan was first considered to be a result of precursor activity, by 1964 the role of tryptophan as an inducer in ergoline biosynthesis was recognized [73].

It is known that tryptophan analogs which are not incorporated into the alkaloids are stimulatory for their production [74,75]. Tryptophan must be added during the growth phase of alkaloid fermentation to achieve stimulation, even though alkaloid synthesis occurs at a later time. Tryptophan intracellular accumulation reaches a peak just prior to alkaloid production, and activities cf several enzymes directly involved in ergoline synthesis, like DMAT synthetase, were shown to be induced by tryptophan or its analogs. All these experimental data certainly confirm the inducing role of tryptophan; in any case, the stimulatory effect of this amino acid on alkaloid production is not universal for all ergot strains [76]. Even a decrease in ergotamine synthesis upon tryptophan addition was reported, although the inhibition occurred at the peptide synthesis stage rather than the stage of ergoline synthesis [54].

3. Phosphate Effect

For many years it has been known that inorganic phosphate concentrations which are optimal for microbial growth often inhibit secondary metabolism, and ergot alkaloid biosynthesis has also been shown to be subject to this phosphate regulation. One of the most important cellular activities in which phosphate is involved is the storage of energy, and one of the possible explanations for phosphate regulation is its influence on the energy state of the cell or the concentration of particular phosphonucleotides. It is interesting to note that addition of tryptophan or its analogs can overcome the inhibition of alkaloid synthesis by phosphate, which suggests that phosphate inhibition might somehow be mediated through tryptophan [77].

4. Carbon and Nitrogen Catabolite Regulation

Glucose and other rapidly metabolized carbon sources, which are also excellent energy sources for fungal growth, have been found to suppress markedly the production of ergot alkaloids, the phenomenon being quite general in secondary metabolite production [78]. Nitrogen catabolite regulation functions in a similar way, ammonia being the rapidly utilized nitrogen source which usually is almost depleted from the medium until a high production of ergot alkaloids starts [79].

5. End Product Inhibition

Several enzymes involved in ergoline skeleton synthesis, like anthranilate synthetase [80], DMAT synthetase [81], and chanoclavine-I cyclase [82], were observed to be repressed upon addition of agroclavine or elymoclavine. In this way the first enzymes of alkaloid synthesis are feedback inhibited by the alkaloid end products.

It is very likely that still other kinds of regulation exist, especially between the secondary and primary metabolic pathways, but these interactions are quite complex and difficult to understand at the present state of knowledge.

D. Directed Biosynthesis

Biosynthesis of the peptide ergot alkaloids can be partially controlled by changing the relative amino acid concentrations in the internal cell pool

TABLE 1. Amino Acids in the Ergotoxine Alkaloid Side Chain

Alkaloid	Amino acid in position		
	1	2	3
α-Ergocryptine	Valine	Leucine	Proline
γ-Ergocryptine	Valine	Isoleucine	Proline
Ergocornine	Valine	Valine	Proline

and thus influencing the composition of the peptide moiety of the molecule. This direction of biosynthesis can be accomplished by feeding the culture with appropriate amino acids, by genetic manipulation, or by changing the environmental factors.

Kobel and Sanglier [31] described feeding experiments in parasitic and saprophytic conditions. By feeding solutions of the amino acids which are the normal constituents of the peptide moiety of ergot alkaloids to the host plant, or by addition of the appropriate amino acids to the fermentation medium, it was found possible to modify the alkaloid spectrum. As reported by several authors [31,83,84], by supplementing the appropriate amino acids, the amino acid in position 2 of the peptide side chain which characterizes the different ergotoxine alkaloids (Table 1) can be changed, resulting in a changed ratio of the individual alkaloids in the spectrum without major differences in the total amount of alkaloid produced.

Similar results were achieved with *C. purpurea* ergotamine, ergocryptine, and trace amounts of an ergostine-producing strain. By supplementation with aminobutyric acid [85], this amino acid was preferentially introduced in position 1 of the peptide side chain and the production of ergostine was stimulated.

Exogenously administered amino acids can obviously alter the relative concentrations of the amino acid pool in the cell, the incorporation of the particular amino acid depending on the relative concentrations of the competing amino acids. This competition seems to be restricted to structurally related nonpolar amino acids [31].

Directed biosynthesis of peptide alkaloids by incorporation of synthetic amino acid analogs was also described [86]; such experiments with mutasynthesis showed promise of obtaining new pharmacologically active alkaloids. Auxotrophs of an ergocristine-producing strain (Phe^-) and an ergocornine-ergocryptine-producing strain (Leu^-) were supplemented with phenylalanine analogs and leucine analogs, respectively, resulting in the formation of the appropriate alkaloid analogs. New, sulfur-containing peptide alkaloids were also obtained by feeding the proline analog L-thiazolidine-4-carboxylic acid to a *C. purpurea* strain [87].

Blocked mutants have become useful in directed biosynthesis as well. Such a mutant, blocked for agroclavine formation, and capable of producing large amounts of chanoclavine-I and chanoclavine-I aldehyde, was isolated from an ergotoxine-producing *C. purpurea* strain [51]. After supplementation with agroclavine, the mutant produced a further alkaloid, ergosine, which had never been present in the wild-type parent strain.

An internally controlled reversible change in the alkaloid spectrum under different environmental conditions was also reported [88]. The alkaloid spectrum was fundamentally changed when a *C. purpurea* ergocornine-ergocryptine-producing strain was transferred from saprophytic to parasitic cultivating conditions, where ergotamine and ergocristine were mainly produced. As observed in our laboratory, changes in growth velocities caused by different environmental conditions in submerged fermentation can also influence the qualitative alkaloid composition as well as the alkaloid intra- and extracellular distribution from a *C. purpurea* ergotoxine-producing strain [70].

V. PHYSIOLOGY OF ERGOT ALKALOID FORMATION

Ergot alkaloids are usually represented as fungal secondary metabolites in a manner similar to that of antibiotic formation, since their biochemical role in the culture cells is not yet fully understood. Ergot alkaloids are certainly not waste products or physiologically inert products of cell metabolism, and some authors consider them as kinds of detoxifying agents [33].

Division of fungal metabolism into primary and secondary metabolism is, anyway, quite artificial since alkaloids are formed from the basic intermediates of primary metabolism. The classic pathways of biogenesis, such as synthesis via acetyl-CoA or malonyl-CoA and the mexose monophosphate pathway are common and may produce secondary metabolites that differ greatly in their functions. Insignificant differences in the structure of the members of a family of secondary metabolites suggest that the enzyme systems involved in their formation are not highly specific. Even the enzyme systems of secondary metabolism are basically regulated by the same mechanisms as those of primary metabolism, e.g., by feedback regulation, catabolite repression, and enzyme induction [89].

In any case, in this chapter we refer to primary and secondary metabolism in terms of the metabolic changes which must be introduced to trigger the switch from pure growth to the stage of alkaloid biosynthesis. In that sense, the inoculum preparation or growth phase can be treated as a phase of primary metabolism, and alkaloid production corresponds to a phase of secondary metabolism.

A. Primary Metabolism

Though the conditions required for primary growth are not as highly specific as those for alkaloid production, the vegetative media are chosen so as to achieve rapid growth and to get an inoculum of suitable quantity and quality for the next stage—the alkaloid production phase.

Preculture of inoculum preparation proceeds in seed or "vegetative" liquid culture media which usually contain the same carbon, nitrogen, phosphate, carboxylic acid, and mineral sources as production media, but in different proportions: the carbon content in seed culture media is lowered at the expense of nitrogen, so the resulting C/N ratio is also lower. Frequently, the vegetative media comprise an additional nitrogen source, e.g., chick pea meal [72] or Proflo [90,91]. Such vegetative media composition enables rapid utilization of nutrients and supports high metabolic activity: an intensive synthesis of RNA, amino acids, and proteins. These syntheses are of great importance for promoting high levels of the enzymes responsible for primary metabolic pathways and also for the later synthesis of secondary metabolites. Little information is available in the literature on the development of *Claviceps* seed cultures since the saprophytic *Claviceps* strains were the ones most frequently investigated, including morphological and biochemical changes, in connection with alkaloid biosynthesis. However, since the inoculum represents a complex biochemical and regulatory system maintaining its effects throughout fermentation, it is very important to understand the factors influencing inoculum development.

By comparing the metabolic patterns of a mycelial *C. paspali* strain with conidial and mycelial *C. purpurea* strains during submerged inoculum development, we found that biosyntheses of macromolecular substances (RNA, DNA, and proteins) follow a similar course in all *Claviceps* strains, with an early accumulation of these substances in the cells [92,93]. These *Claviceps* strains differ basically in the alkaloids produced and in their morphological characteristics. Both *C. purpurea* strains are producers of peptide alkaloids and show considerable morphological differentiation during their developmental cycle. The *C. paspali* strain is a producer of simple lysergic acid derivatives, lacking any significant differentiation. The maximum and the subsequent fall in the relative RNA content of the cells in all three strains proved to be associated with major changes in the metabolism and the onset of secondary metabolic pathways. It can serve as a biochemical indicator for an optimally developed inoculum, as proved by checking the alkaloid-synthesizing activity of seed cultures at different developmental stages [92,93].

In contrast, the biosynthetic patterns for carbohydrates and lipids in the vegetative mycelia of *Claviceps* strains are not so uniform. The percentage of these "storage" compounds gradually increases throughout the development of conidial *C. purpurea* seed cultures, which undergo the whole development process from conidia through vegetative hyphae into sclerotialike cells. But in mycelial *Claviceps* strains, where the seed cultures start from biologically more differentiated mycelium fragments, the relative amounts of these storage compounds reach their maxima early in the growth phase and decline to the minimal values approximately at the time of the optimally developed inoculum. The percentage of mycelial lipids and carbohydrates starts to increase again with cellular and morphological differentiation, and accompanies the transition from primary to secondary metabolism [93].

1. Sporulation

Claviceps fungi are normally able to produce conidia which represent efficient propagation units and serve as inoculum for parasitic or saprophytic alkaloid production. In the natural cycle, sporulation takes place in the presence of a sticky liquid, called honeydew, which is very rich in various sugars. In saprophytic cultivation, sporulation can emerge on solid agar media or in shaken liquid media. Though the genetic information necessary for conidiogenesis and alkaloid formation exists simultaneously, the physiological conditions necessary for conidiation and germination of conidia are different from the conditions controlling alkaloid synthesis [94]. Conidial differentiation is linked to an intensive phosphate metabolism, the optimal concentration for conidiogenesis being higher than for simple growth [95]. Though a strong interaction between the concentration of the energy source and the phosphate concentration exists, the concentration of the carbon source used is lower than for alkaloid production. During sporulation the pentose phosphate pathway represents the major glycolytic route with most fungi, and it is also a known fact that conidiation is usually correlated with depletion of the nutrients from the medium. But with the *Claviceps* fungi effective sporulation in the liquid medium can be achieved by monitoring the respiration rate. Deficiency of dissolved oxygen can stop vegetative growth of *C. purpurea* and induce conidiogenesis [96].

Sporulating *Claviceps* strains are desirable because of their greater stability and suitability for genetic manipulation, but it is extremely difficult to induce *Claviceps paspali* strains to form asexual spores. Some authors [97,98] have succeeded in obtaining *C. paspali* capable of sporulating on agar plates, but few reports deal with sporulation in submerged cultures [99]. It was also shown that a conidia-induced mutant of *C. paspali*, isolated after γ-irradiation, showed repressed primary and secondary metabolic activities and a changed mycelial composition [79]. Generally, conidiation in a submerged culture and high alkaloid productive ability are mutually exclusive [100,101].

B. Secondary Metabolism

1. Nutritional Factors

In nature, ergot alkaloids are formed in sclerotia, while under saprophytic conditions alkaloid production proceeds in specific, mostly synthetic production media, under specific balanced environmental conditions.

Carbon source: Under its natural environmental conditions on rye ears, ergot sclerotium is a heterotrophic organism which depends entirely on the supply of nutrients from the host plant. The basic composition of the phloem sap of all vascular green plants is surprisingly uniform; besides sucrose, which is the most common transport form of saccharides and which commonly occurs in the sieve tubes at a 30% concentration, free amino acids (which are the main transport form of nitrogen) and organic acids of the citric acid cycle, especially citrate and malate, are the major components [102]. Hence, sucrose represents the principal carbon source for alkaloid synthesis in sclerotium.

In ergot submerged fermentations, the slowly metabolized carbon sources, such as mannitol, sorbitol, and sucrose, are preferentially used [72,91]. As often in secondary metabolism, carbon catabolite repression has been found in ergot alkaloid production, as already mentioned. Glucose, for example, represents an excellent substrate for *Claviceps* growth, but rapid glucose utilization and expression of secondary metabolism are mutually exclusive [78].

With clavine alkaloids, it was found that the magnitude of total production of alkaloids, as well as the proportion of individual clavines in the mixture, can be continuously affected by the carbon source [103]. Kren and Řeháček [104] reported a rare stimulating effect of extracellular glucose on elymoclavine synthesis and its inhibitory effect on the total alkaloid production. The phenomenon was explained by inhibition of phosphofructokinase through glucose or its metabolites, and thus a feed-forward regulation of the glycolytic pathway. In the presence of higher concentrations of glucose, the activity of the glycolytic pathway is suppressed, and glucose catabolism is shifted to the hexose-monophosphate pathway, which yields NADPH, necessary for elymoclavine formation.

Nitrogen source: The production of ergot by *C. purpurea* parasitic culture seems to be subject to nitrogen catabolite regulation as well [105]. All the changes in the composition of the developing sclerotium at the very beginning of the production phase are due to a crucial change in the nutrition provided by the host. Previous intake and digestion of water and nitrogen-rich inner flower organs is replaced by an intensive intake of assimilates, mainly glycides, which is reflected in an abrupt drop in total nitrogen content and a corresponding increase in the intensity of lipid biosynthesis. In submerged fermentation, a similar situation is reached by monitoring the composition of the production medium so as to achieve an exceptionally high C/N ratio. In analogy with parasitic growth, the production of alkaloids usually starts after exhausting the majority of ammonium nitrogen from the nutrient medium. For *Claviceps* strains, ammonium ion coupled with a carboxylic acid is a commonly used nitrogen source. Sometimes amino acids like asparagine and aspartic acid are used [106]. Addition of complex compounds of not exactly defined composition, such as peptone or caseinate, can enhance growth and alkaloid production. With our *C. paspali* and *C. fusiformis* strains a rich, meat peptone-containing medium [99,107] was used for selection, resulting in highly improved alkaloid yields.

Organic acids: Successful course of the alkaloid synthesis requires the presence of an organic acid, usually one of the tricarboxylic acid cycle acids or a closely related one (citric, succinic, malic, oxalic). As speculated by Spalla et al. [108], the organic acid supplies a readily available carbon source after the growth of the fungus has ceased and alkaloids are being produced. But in most *C. purpurea* fermentations, the growth and alkaloid production phase are not so strictly separated [109], and the phenomenon of diauxic growth does not coincide with the exhaustion of the primary carbon source. It seems more likely that the role of organic acids is not so much in providing an additional carbon source as in their metabolic activity. These acids enhance the conversion from the tricarboxylic acid cycle to the glyoxylate cycle by suppressing citrate synthetase and activating acetyl-CoA carboxylase, the key enzyme of fatty acid and alkaloid biosynthesis. During the alkaloid formation phase acetyl-CoA enters the tricarboxylic acid cycle only to a limited degree and takes an alternative path to the synthesis of lipids and alkaloids [26]. Citrate also represents a negative allosteric effector of phosphofructokinase which is the key glycolytic enzyme. An excess of citrate therefore leads to a preference for the pentose phosphate cycle, this being the most efficient route to the formation of NADPH, a direct source of energy for both alkaloid and lipid biosynthesis [102].

Phosphate: Another important medium component in submerged fermentation of ergot alkaloids is inorganic phosphate, which is readily captured from the medium and partially transformed into storage compounds such as polymetaphosphates and into permanent structures [26]. Transport of inorganic phosphate into the cell against the concentration gradient presumably involves a periplasmic binding protein [110]. The concentration of inorganic phosphate in the nutrient medium is quite decisive. Phosphorylated intermediates take part at all levels of primary and also secondary metabolism; on the other hand, the phosphorylated polymetaphosphates in the cells show a multiple inhibitory action which results in the inhibition of alkaloid synthesis. Several enzymes, directly or indirectly involved in alkaloid formation, are repressed by inorganic phosphate, e.g., dimethylallyltryptophan synthetase, the first enzyme of alkaloid biosynthesis [75], chanoclavine cyclase [82], tryptophan synthetase [111], and the key enzyme of the pentose phosphate cycle glucose-6-phosphate dehydrogenase [95].

It is possible that the mechanism of phosphate inhibition is mediated through its influence on the relative proportions of ATP, ADP, and AMP in the cell. An increased phosphate concentration in the medium, which is optimal for growth, might also increase intracellular ATP formation and lead to a high-energy charge [77]. This condition might be inhibitory to alkaloid formation since it was also shown that alkaloid synthesis started only after the intracellular phosphates and ATP had reached their maximal values and their content in the mycelium began to decline [109]. A high-phosphate medium was even shown to activate the alkaloid-degrading enzymes [112], indicating that alkaloid concentration is in a dynamic state, that alkaloid turnover exists and that alkaloids cannot be considered as end products of metabolism. It was also shown that high concentrations of phosphate simultaneously reduced alkaloid yields and the level of phosphatases, which suggests the involvement of phosphorylation-dephosphorylation reactions in alkaloid biosynthesis [113].

2. Environmental Conditions

Environmental conditions play a very important role in submerged *Claviceps* fermentations; it is also a known fact that secondary metabolism has narrower ranges of trace metal, pH, and temperature requirements than vegetative growth. With most strains, the metal ions Mg^{2+}, Fe^{2+}, and Zn^{2+} in trace amounts enhance alkaloid production [114], pH optima for growth and alkaloid synthesis are in the range of 5.5, and temperature optima are usually near 24°C. Induction of alkaloid biosynthesis requires osmotic pressures between 10 and 20 bar, attained by high sugar concentrations which can be partially replaced by sodium chloride [115]. The high osmotic pressure presumably inhibits conidiogenesis, induces differentiation to sclerotialike cells, and enhances the entrance of nutrients into the cells. Reduced osmolality of the production medium has a negative influence on peptide alkaloid formation. In contrast, we have found that in *C. paspali* fermentations (producing simple lysergic acid derivatives), the osmolality drop during the course of fermentation is quite significant and probably connected with alkaloid release into the surrounding medium (Gaberc-Porekar and Pertot, unpublished results).

3. Metabolic Pattern During Alkaloid Production

By following the course of fermentation of three highly productive *C. purpurea* strains, synthesizing mainly ergotamine, ergocristine, and ergocornine-ergocryptine types of alkaloids, it has been found that composition of the mycelia and cultivation media changes considerably during submerged cultivation [109].

The initial stage in the fermentation process is characterized by rapid exhaustion of phosphate and ammonium nitrogen (Fig. 13) from the nutrient medium and rapid synthesis of all the components of protoplasm, nucleic acids and proteins in particular. The respiration rate Q reaches its maximum value at this stage. Depletion of inorganic phosphate from the medium coincides approximately with the maximal values of intracellular phosphates, RNA, and proteins, and also with minimal pH values of the surrounding medium. Beyond this point major changes in the metabolism must occur since alkaloid biosynthesis starts, being accompanied by activated synthesis of storage compounds, carbohydrates, and particularly lipids. The relative RNA content, which reflects active growth, declines after this point, but biomass expressed as mycelium dry weight continues to increase, mostly at the expense of changed morphology to sclerotialike cells which have a high content of lipids and are surrounded by thick cellular walls. All three *C. purpurea* high-yielding strains show a decreasing protein-to-nonprotein compound ratio of mycelial components during the course of fermentation. This phenomenon, together with low growth rates, seems to be necessary for high peptide alkaloid production in *C. purpurea* batch fermentations [70,109].

In the mycelia of highly productive strains the total nitrogen and protein content of the cells decreases, though various authors have observed an intense proteosynthesis during alkaloid formation, however without a subsequent increase in the overall content of cell proteins [116], indicating a substantial protein turnover. Proteinases involved in such processes have been isolated from various *C. purpurea* strains and characterized [117,118]. Intracellular proteinases exhibit a maximum level on the sixth day of the production phase, whereas extracellular activity continues to increase throughout submerged fermentation. Obviously, amino acids freed by protein degradation pass into the metabolic pool and take part in various synthetic processes [119], including the building of the amino acid moiety of the peptide alkaloid molecule. Řeháček et al. [120] also demonstrated a positive relationship between alkaloid synthesis and the activity of glutamine synthetase, which converts most of the free cell ammonia into nontoxic glutamine, which takes part in the synthesis of tryptophan.

Lipid and fatty acid syntheses in the mycelia of *Claviceps* fungi have attracted much attention since lipid and alkaloid formation show a strict correlation. Parasitically or saprophytically cultivated, the sclerotial cells, which are the usual alkaloid producers, accumulate up to 50% of their weight as lipids. It was also recognized quite early that mature sclerotia or productive saprophytic cultures contained a special fatty acid—ricinoleic acid—which was recognized as a specific indicator of alkaloid-producing ability [106,121]. Recently, ricinoleic acid was found in different *Claviceps* species, indicating that it is neither a specific chemotaxonomic marker nor a specific indicator of peptide alkaloid synthesis. In any case, it is interesting to note that the composition of fatty acids is indicative of individual strains [122] and that even the relationship among strains can be deduced from the fatty acid composition.

The carbohydrate metabolism proceeds almost identically in productive and nonproductive *Claviceps* cultures up to the oxidative decarboxylation of pyruvate to acetyl-CoA [119]. The most effective component of the carbohydrate degradation is the tricarboxylic acid cycle, but in the alkaloid production phase the enzyme activities of this cycle are decreased [119], and acetyl-CoA, the key intermediate of primary and secondary metabolism, enters the tricarboxylic acid cycle to a lesser extent and serves preferentially for lipid and alkaloid synthesis.

Voříšek and Řeháček [43] found that acetyl-CoA carboxylase, an enzyme catalyzing the synthesis of malonyl-CoA, the first step in fatty acid synthesis, was localized in the endoplasmic reticulum, which is very likely also the site of alkaloid synthesis. Though mevalonic acid, a direct precursor of the ergot alkaloid skeleton, is synthesized from acetyl-CoA normally through another route, high concentrations of carboxylase always parallel alkaloid biosynthesis and its cofactor biotin also stimulates alkaloid yields [123]. Lipid and alkaloid biosynthesis do not compete only for acetyl-CoA but also for the same donor of reducing equivalents, NADPH, which is provided by the pentose phosphate pathway. This pathway, according to some authors [26,108,124], is supposed to be the major route of glucose catabolism during intensive alkaloid production.

Alkaloid synthesis always parallels lipid formation, but the inverse correlation is not always true. In our laboratory, an interesting *C. purpurea* mutant was isolated which does not produce alkaloids despite a remarkable lipogenesis (Gaberc-Porekar and Didek-Brumec, unpublished results). From the lipid/alkaloid ratio in the mycelia of effective alkaloid producers it must be obvious that biosynthesis of ergot alkaloids requires a substantially lower quantity of acetyl-CoA and NADPH; neither of these substrates can be a limiting factor for alkaloid synthesis so far.

VI. GENETICS OF *CLAVICEPS*

Despite the fact that biosynthesis of ergot alkaloids is a genetically regulated process, genetic studies of the *Claviceps* fungi have received limited attention compared to genetic studies of other industrially important microorganisms. This is probably due to the complex life cycle of *Claviceps* species and their genetic structure which makes genetic manipulations difficult.

A. Genetic Determination of Alkaloid Production

The genus *Claviceps*, as already mentioned, is characterized by its ability to synthesize various ergot alkaloids, the spectrum of alkaloids produced by a pure strain being genetically determined. For example, a strain which can synthesize ergotoxines in sclerotia will produce the same alkaloids in saprophytic cultures. Therefore, it is of primary importance to know the genetic factors that govern alkaloid production.

First investigations in this direction showed that the mycelium of a producing strain was characterized by plurinucleate cells. Such strains, when grown on solid nutrient media, formed colonies with distinct sectors, which led to the conclusion that producing strains were heterokaryotic [14,125]. But, as later confirmed in several laboratories, homokaryotic strains are also capable of producing alkaloids [95,126,127]. Esser and Tudzynsky [128] finally proved that heterokaryosis is not required for alkaloid synthesis by infecting rye plants with conidia originating from single ascospores; however, they stated that alkaloid production is correlated with distinct genotypes.

The synthesis of peptide alkaloids is apparently a nonribosomal process. It was found that genetic blocks in the biosynthesis of the peptide moiety do not influence the biosynthesis of the remaining part of the molecule [58]. Thus we may suppose that the alkaloid synthesis proceeds under the control of at least two independent operons.

The development of new technologies, like protoplast fusion and recombinant DNA technology, has greatly expanded the possibilities for basic genetic studies and genetic manipulation. Generally, there are three basic genetic approaches to improving the alkaloid titer in *Claviceps*: mutation, recombination, and gene cloning.

B. Mutation and Selection

Until recently genetic manipulations with *Claviceps* were performed mainly by classical methods of mutation and selection. The majority of high-yielding strains were obtained by mutagenesis performed either by UV and γ-irradiation or by chemical mutagens. A wide variety of chemical mutagenic agents were used, among which *N*-methyl-*N*'-nitro-*N*-nitrosoguanidine (MNNG), ethylmethanesulfonate (EMS), and nitrous acid proved to be the most effective. Very successful programs were carried out in successive steps by using various mutagens [22,31].

The ideal material for mutagenic treatment are mononucleate conidia, while mycelial fragments of nonsporulating strains consisting of multinucleate cells are less convenient. Therefore, high doses of mutagen are needed to allow the survival of one nucleus per unit. Such high doses of mutagens may cause undesirable cryptic mutations [31]; however, this problem can now be overcome by protoplast formation [20,129]. By using different lytic enzymes, protoplasts of *Claviceps* can be easily obtained [55,130,131], and, as recently found, protoplasts can also be stabilized for longer times by immobilization [132].

Several types of *Claviceps* mutants have been isolated with altered primary metabolic pathways (auxotrophs) or blocked in the secondary metabolic pathway. After isolation of the first auxotrophic *C. purpurea* mutant by Strnadova [133], more than 100 auxotrophs have been reported [20,129,134]. Some of them were used to show the influence of amino acids on ergot alkaloid production and some as markers in genetic recombination tests, but they were not further characterized. Recently, such a characterization was accomplished in our laboratory with five mutants of *C. purpurea* blocked in arginine biosynthesis [135].

Besides auxotrophic mutants, mutants resistant to various fungicides can also be used in genetic recombinations. For instance, resistance to benomyl has been found a suitable marker. Such mutants have been isolated from a parasitic *Claviceps* strain [134] and recently also from saprophytic cultures [136].

C. Genetic Recombination and Gene Cloning

Genetic recombination becomes an important adjunct to mutagenesis once several lineages of mutants have been established. It is promising not only in strain development but it also offers the possibility of elucidating genetic control and regulation of alkaloid biosynthesis. Considering the biology of *Claviceps*, two possibilities exist for genetic recombination studies. In nature, where a sexual cycle exists, meiotic recombination can be accomplished. In this way Tudzynsky and Esser [134] succeeded in exchanging the genetic material between two different mutant strains of *C. purpurea* with 50% of recombinants recovered. Some of the recombinants showed a higher alkaloid content than either of the parent strains. Another possibility for genetic recombination involves mitotic recombination through the parasexual cycle which was recently conclusively determined to function in *Claviceps* [137].

Generally, in recombination studies with axenic cultures, the protoplast fusion technique is beginning to find useful applications either in producing improved mutant strains by intraspecies crosses, or in formation of a novel spectrum of products by interspecies hybrids. There are only a few reports available on this approach to *Claviceps* strains.

The first intraspecific fusion experiments were performed by Spalla and Marnati [130], who succeeded in fusing a *C. purpurea* ergocristine producing strain with another *C. purpurea* strain producing high levels of ergocornine and ergocryptine. The fused strain produced all three alkaloids in significant amounts but was stable only on a minimal medium, while segregating back to the starting strains on a complete medium. The same phenomenon was observed by Robbers et al. [77] with the complemented strain obtained by fusion of different alkaloid producing SD58 strains. In our laboratory, strains complemented in alkaloid biosynthesis in nutritional requirements and resistance to different fungicides were obtained by protoplast fusion of a submerged inactive ergocristine strain with a submerged active ergocornine-ergocryptine strain [138].

Interspecific recombination has also been performed: protoplast fusion of the strains *C. paspali* (producing simple lysergic acid derivatives) and *C. purpurea* (producing low amounts of ergocristine) resulted in a fused strain which showed complementation in the nutritional requirements but produced only ergocristine. Better results were obtained by fusion of an ergotamine *C. purpurea* strain with a clavine producing *Claviceps* SD58 where the resulting hybrid strain produced both types of alkaloids [77]. Recombinant DNA technology holds the promise that we may be able to insert selected foreign genes into existing genotypes of high-yielding strains. But this approach may necessitate extensive basic research on genetics before its potential can be realized with *Claviceps*. In order to develop a system for gene cloning, Tudzynsky and Esser [139] initiated a molecular analysis of the mitochondrial DNA system. From the wild strains of *C. purpurea* they isolated several linear mitochondrial plasmids comparable to those of higher plants. Further, they established the existence of plasmid homologous sequences in genomic mtDNA of two nonrelated strains, which indicates possible transposonlike properties of the *C. purpurea* plasmids [140].

In conclusion, the application of new genetic technologies offers exciting prospects for the systematic genetic construction of improved *Claviceps* strains.

VII. INDUSTRIAL ASPECTS OF ALKALOID PRODUCTION

For the large-scale production of ergot alkaloids two methods are used: the classical parasitic cultivation of *Claviceps* strains on host plants, and the modern saprophytic cultivation which proceeds in bioreactors. Total chemical synthesis of ergot, like that of ergotamine [141], the ergotoxine group [142], and other peptide alkaloids, as well as of lysergic acid and its derivatives [2,143], was also accomplished but has been ruled out for economic reasons. In contrast, partial chemical synthesis of ergot derivatives starting from lysergic acid has become very important, particularly for the preparation of newly developed medicinally used ergoline derivatives [4].

A. Parasitic Ergot Alkaloid Production

Until recently, ergot alkaloids were obtained exclusively by extraction from field-cultivated sclerotia. In this process rye flowers are mechanically inoculated by spores of high productive strains of *C. purpurea*. This way about 200-500 kg sclerotia from 1 hectare (2.47 acres) of rye can be harvested. Isolation of alkaloids from sclerotia proceeds by extraction with organic solvents and further purification using the same methods as described for the isolation of alkaloids from a fermentation broth [144,145]. Such production of ergot alkaloids has been widely used in several European countries: Switzerland (Sandoz), Hungary (Richter Gedeon), Yugoslavia (LEK), Italy (Farmitalia), Czechoslovakia (Spofa). Despite technical and scientific advances, the agricultural production of ergot has several economic disadvantages: laboratory preparation and field application of inoculum are expensive, and the harvest of sclerotia depends on the weather and is possible only once a year. For these reasons, and because fermentative production proved to be more reproducible, field production began to decline. The last 10 years has seen the introduction of alkaloid production by the fermentation technique in most of the above cited countries.

B. Production in Bioreactors

Several advantages of the fermentative production of ergot alkaloids over field production can be expressed in economic terms. Alkaloid yields in fermentors as a percentage of the mycelial dry weight may exceed 10%, whereas sclerotia contain only 0.01-1.0%. Further, the production in bioreactors is not limited to seasonal variations and the whole manufacturing process can be performed under control and at the same site. Problems arise in scale-up of the fermentation process [31]. The production strains are very sensitive to mechanical stress, have a high oxygen demand, and require a well-balanced system of aeration and agitation. Very slight moderations of media and environmental conditions may considerably lower alkaloid yields, and the long cultivation periods (12-20 days) require very high standards of sterility. In addition, degeneration of high-yielding strains frequently occurs and a careful maintenance of strains is necessary.

Successful production of ergot alkaloids depends on the strain used and the conditions of its cultivation.

1. Producing Microorganism

The key to commercial production of ergot alkaloids is the high-yielding strains of the fungus *Claviceps*. The availability of such strains is rather limited, being mostly kept in the laboratories where they were or are being developed. Only a small number of strains, and not the high-yielding ones, may be obtained from the microbial culture collections.

The commonly used method for obtaining producing strains is through isolation from sclerotia, collected from rye or wild grasses. Parasitic strains are easily cultivated in the laboratory but only a small percentage is capable of synthesizing alkaloids. The most commonly used *Claviceps* species for submerged production of ergot alkaloids are listed in Table 2.

Difficulties in isolation of high-yielding *Claviceps* strains lie in their great variability, limited stability, and the degeneration phenomenon. The first step in finding a suitable strain is to screen a large number of sclerotia by determining their alkaloid content, by isolating monocultures on agar media, and by testing them in saprophytic cultures for alkaloid production. Such a selection of strain can be a laborious process, and therefore an appropriate preselection method prior to laboratory fermentation could be helpful. Among several methods for preselection, those involving applying a spray reagent for detecting alkaloids in the colonies seem to be most convenient. The intensity of the color [129] or UV fluorescence [146,147] gives an approximation to the amount of alkaloid formed.

TABLE 2. *Claviceps* Species Used for the Fermentative Production of Ergot Alkaloids

Species	Alkaloid
C. fusiformis	Agroclavine Elymoclavine
C. paspali	Lysergic acid amides $\Delta^{9,10}$-Lysergic acid $\Delta^{8,9}$-Lysergic acid Ergometrine
C. purpurea	Ergotamine α-Ergocryptine γ-Ergocryptine Ergocornine Ergocristine Ergosine

For the yield improvement of isolated strains, mutagenesis followed by selection still remains the basic technique. As shown in Fig. 12, considerable improvement can be achieved by a good selection-mutation program. In this way high-yielding *C. purpurea* strains, capable of producing ergotoxines and ergotamine, were developed in our laboratory [21,22]. The production capacity of the initial isolate was about 20 mg alkaloids/liter, while industrial production requires a yield of about 2 g/liter. The increase in productivity of both strains was accompanied by a significant change of their cell morphology, the sphacelial mycelium being transformed into sclerotial tissue. The composition of the alkaloids so far was not affected by mutagenic treatment; only the proportion of the two main alkaloids was slightly changed. As found in most cases, the quantity of alkaloids produced is readily altered by mutation while the spectrum of alkaloids is rather stable [95].

A frequently occurring problem with high-producing *Claviceps* strains is their degeneration, which can be observed in a decline of productivity on repeated transfers on agar media, and changed morphology [19,97]. Moreover, degenerated cultures are unable to utilize large amounts of nutrients. Therefore, the preservation and long-term storage of high-producing *Claviceps* strains are of prime importance for a successful commercial fermentation process. Freeze-drying seems to be the most convenient method [148,149]. Lyophilized cultures are easy to handle and can be kept at temperatures below 5°C. In most cases, the preserved strains retain good viability and productivity for many years. For example, *C. purpurea* and *C. paspali* strains isolated in our laboratory have retained their high-producing ability in the lyophilized state for more than 8 years. Some authors prefer cooling the cultures to -70°C and transferring them to a liquid nitrogen container [95].

2. Process of Alkaloid Production

As for the majority of secondary metabolites, ergot fermentation proceeds in two basic phases: inoculum preparation and production phase.

	Alkaloids (mg/L)
Sclerotium	
↓ *Primary cultures*	*none*
↓ *Submerged culture (3^d, generation)*	*trace*
↓ *MNNG^x*	*176*
↓ *UV light*	*187*
↓ *Selection*	*241*
↓ *Gamma rays*	*390*
↓ *Selection*	*540*
↓ *Selection*	*999*
↓ *Selection*	*1263*
↓ *Gamma rays*	*1386*
↓ *Selection*	*1762*
↓ *Selection*	*2180*

FIGURE 12. Isolation scheme of a high-yielding ergotoxine-producing *Claviceps purpurea* strain. MNNG^X = *N*-methyl-*N'*-nitro-*N*-nitrosoguanidine.

Inoculum preparation: Inoculum preparation is a very important phase since poorly prepared inocula lead to cultures which never perform well. Therefore, a good quality of inoculum, i.e., its age, genetic homogeneity, and metabolic activity, has to be assured to build up a high biomass in a short time. The development of inoculum starts from 14- to 28-day-old colonies, cultivated in the vegetative or seed culture media some of which are presented in Table 3. The cultivation proceeds in shaken flasks for 4-6 days. After a desired level of growth has been achieved, the contents are transferred into the seed tank and developed 3-4 days under operating conditions similar to those described for alkaloid production. Large-scale inoculum preparation requires two or more stages in the seed tank to obtain the required volume of cultures for inoculating the fermentation tank. It

TABLE 3. Composition of the Media

	Seed media		Production media	
Component (g)	[72]	[91]	[72]	[91]
Mannitol	40		100	
Sucrose		100		240
Succinic acid	10		30	
Chick pea meal	2			
Proflo		10		
Ammonium oxalate		3		9.6
Urea				1.7
KCl		0.12		0.31
KH_2PO_4	3	0.25	1	0.625
$MgSO_4 \cdot 7H_2O$	0.3	0.25	0.3	0.625
$FeSO_4 \cdot 7H_2O$		0.0166		0.025
$ZnSO_4 \cdot 7H_2O$		0.0068		0.010
Distilled water (ml)		1.000		1.000
Tap water (ml)	1.000		1.000	

has to be inoculated with at least 10% v/v of preculture, since the inoculum size affects growth and alkaloid production, as well as the extra- and intracellular alkaloid distribution [70].

Alkaloid production phase: The production phase is carried out in bioreactors of 30,000-liter volume which are generally equipped with instruments for continuous monitoring and recording of temperature, air flow, dissolved oxygen, and pH. A sufficient oxygen supply is usually secured by an aeration rate of 0.5-1.5 vvm, progressively increased, and agitation of 75-200 rpm. The optimal temperature lies between 24 and 27°C depending on the strain. Some media for alkaloid production are presented in Table 3. Cultivation in these media lasts for 12-16 days, with continuous control of pH, which proved to be optimal in the range of 5.2-6.0. The operating conditions, however, have to be adjusted to each individual strain since *Claviceps* fungi can show enormous differences not only from species to species but also from mutant to mutant. The progress of a typical fermentation is illustrated in Fig. 13 for an ergotoxin-producing *C. purpurea* strain.

Today several processes for production of therapeutically important peptide alkaloids are available with various *C. purpurea* strains, producing about 2 g/liter. Fermentative production of simple lysergic acid derivatives with *C. paspali* strains is also commercially accomplished, the production of $\Delta^{8,9}$-lysergic or paspalic acid being particularly interesting. Paspalic acid can be easily rearranged to lysergic acid, which is widely used for further chemical derivatization. Production of agroclavine by *C. fusiformis* with very high yields of about 6 g/liter is also accomplished. Table 4 lists commercially produced ergot alkaloids and their main manufacturers.

C. Optimization of the Fermentation Process

1. Mathematical Modeling

The growth of fungus and alkaloid formation in a submerged batch fermentation can be described by a mathematical model that can be used in automatic process control and optimization of an industrial process.

Submerged batch cultivation of *C. purpurea* was described by Grm et al. [150] with a growth model, based on the morphological features of a cell population during the fermentation process. Inorganic phosphate was determined as the limiting growth factor. Pažoutova et al. [151] proposed another model accentuating the antagonistic effects of phosphate on growth and alkaloid production. But a batch fermentation process cannot be optimized by merely adjusting the phosphate concentration (dosage) because other nutrients [103] as well as the cultivation conditions can influence the growth and alkaloid production. Another possibility for process optimization lies in feeding the culture and thus prolonging the idiophase of the fermentation process [70]. This procedure might be a step toward process optimization by the feed-batch cultivation technique that seems to be one of the perspectives in industrial alkaloid production.

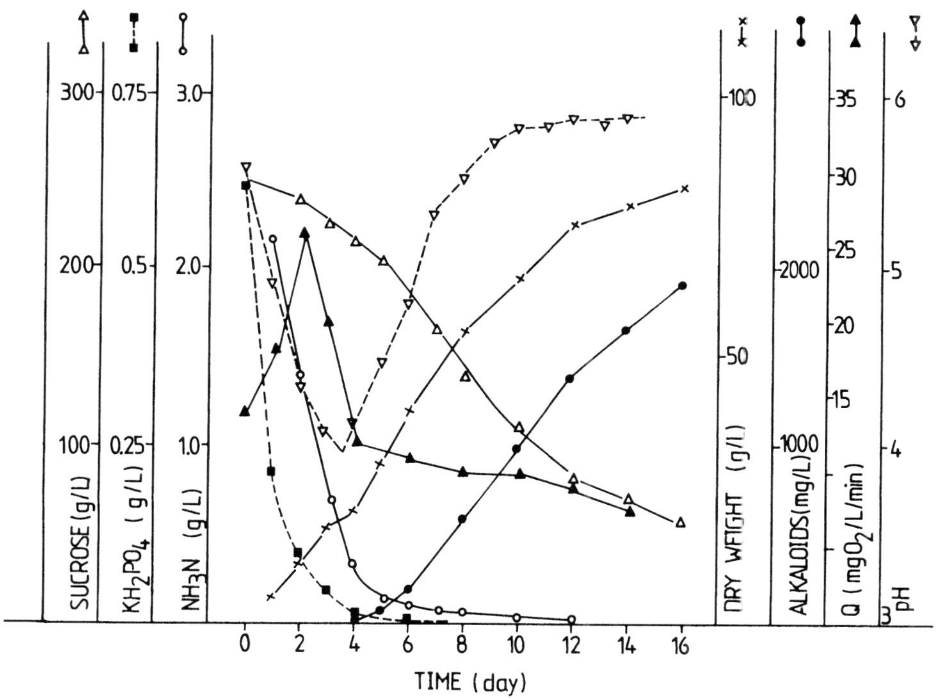

FIGURE 13. The fermentation course of ergotoxine alkaloid production with a *Claviceps purpurea* strain.

TABLE 4. Production of Ergot Alkaloids by Fermentation

Alkaloid	Manufacturers 1	2	3	4	5	6	7
Ergotamine	+	+	+	+	+	+	
Ergocryptine	+	+	+	+	+	+	+
Ergocornine	+	+	+	+	+	+	+
Ergocristine	+	+	+	+	+	+	+
Ergometrine	+			+			
Lysergic acid	+		+	+	+		+

Manufacturers: 1, Sandoz (Switzerland); 2, Gedeon Richter (Hungary); 3, Farmitalia (Italy); 4, Lek (Yugoslavia); 5, Spofa (Czechoslovakia); 6, Boehringer (Germany); 7, Poli (Italy).

Very recently, a kinetic model was proposed which indicates that a combination of fermentation and separation units in a closed loop would enable one to perform cultivation of *Claviceps* at the possible physiological maximum of specific alkaloid production rates [152].

2. Fermentation by Immobilized Cells

A promising method currently under investigation is the use of immobilized microbial cells in the production of secondary metabolites under conditions of semicontinuous or continuous fermentation.

Calcium alginate appears to be the most useful agent for the immobilization of *Claviceps* cells [153-155]. Kopp and Rehm [156] reported a semicontinuous production of ergometrine and clavines with immobilized mycelia of *C. purpurea* which retained their activity for 200 days. A 550-day semicontinuous cultivation was achieved with immobilized cells of a clavine-producing *C. fusiformis* strain [154]. In our laboratory, the prolonged productivity of the cells was also observed by following the alkaloid production of immobilized *C. paspali* cells during semicontinuous cultivation [155].

These findings support the view that overall alkaloid yields in saprophytic *Claviceps* cultures can be improved by offering the mycelium a stable support, calcium alginate, thus simulating the conditions of parasitically living cells [153,156]. The immobilized cell technique is particularly suitable for those *Claviceps* strains which produce extracellular alkaloids at high specific rates.

REFERENCES

1. Hofmann, A. *Die Mutterkornalkaloide*. F. Enke Verlag, Stuttgart, 1964, pp. 14-112.
2. Rutschmann, J., and Stadler, P. A. Chemical background (of alkaloids). In *Ergot Alkaloids and Related Compounds* (B. Berde and O. Schield, eds.), Springer-Verlag, Berlin, 1978, pp. 29-85.

3. Gröger, D. Ergot alkaloids—recent advances in chemistry and biochemistry. In *Antibiotics and Other Secondary Metabolites* (R. Hütter, T. Leisinger, J. Nüesch, and W. Wehrli, eds.), Academic Press, New York, 1978, pp. 201-217.
4. Stadler, P. A., and Giger, R. Ergot alkaloids and their derivatives, in medicinal chemistry and therapy. In *Natural Products and Drug Development* (P. Krogsgard-Larsen, C. H. Christensen, and H. Kofod, eds.), Munksgaard, Copenhagen, 1984, pp. 463-485.
5. Arcamone, F., Bonino, C., Chain, E. B., Ferreti, A., Pennella, P., Tonolo, A., and Vero, L. Production of lysergic acid derivatives by a strain of Claviceps paspali in submerged culture. *Nature, 187*:238-239, 1960.
6. Crespi-Perellino, N., Ballabio, M., Goia, B., and Minghetti, A. Two unusual ergopeptines produced by saprophytic culture of Claviceps purpurea. *J. Natural Prod., 50*:1065-1074, 1987.
7. Stadler, P. A. Neuere Ergebnisse der Mutterkorn-Alkaloid-Forschung. *Planta Medica, 46*:131-144, 1982.
8. Stütz, P. L., Stadler, P. A., Vigouret, J., and Jaton, A. Ergot alkaloids. New ergolines as selective dopaminergic stimulants. *J. Med. Chem., 21*:754-757, 1978.
9. Gröger, D. Ergot. In *Microbial Toxins* (S. Kadis, A. Ciegler, and S. J. Ajl, eds.), Academic Press, New York, 1972, pp. 321-373.
10. Winning, L. C., and Taber, W. A. The ergot alkaloids. In *Economic Microbiology*, Vol. 3 (A. H. Rose, ed.), Academic Press, New York, 1979, pp. 389-340.
11. Floss, H. G., and Anderson, J. A. Biosynthesis of ergotoxins. In *The Biosynthesis of Mycotoxins: A Study in Secondary Metabolism* (P. S. Steyn, ed.), Academic Press, New York, 1980, pp. 17-67.
12. Berde, B., and Stürmer, E. Introduction to the pharmacology of ergot alkaloids and related compounds. In *Ergot Alkaloids and Related Compounds*, Vol. 49 (B. Berde and O. Schild, eds.), Springer-Verlag, Berlin, 1978, pp. 1-28.
13. Rall, T. W., and Schleifer, L. S. Oxytocin, prostaglandins, ergot alkaloids and other agents. In *The Pharmacological Basis of Therapeutics* (A. G. Gilman, L. S. Goodmann, and A. Gilman, eds.), Macmillan, New York, 1980, pp. 935.
14. Spalla, C. Genetic problems of production of ergot alkaloids in saprophytic and parasitic conditions. In *Genetics of Industrial Microorganisms*, Vol. 2 (Z. Vanek, Z. Hoštalek, and J. Cudlin, eds.), Elsevier, Amsterdam, 1973, pp. 393-403.
15. Mantle, P. G. Industrial exploitation of ergot fungi. In *The Filamentous Fungi*, Vol. 1 (J. E. Smith and D. R. Berry, eds.), Edward Arnold, London, 1975, pp. 281-300.
16. Esser, K. *Kryptogamen*. Springer-Verlag, Berlin, 1976, pp. 396-399.
17. Turian, G. Fungal differentiation. In *Biotechnology and Fungal Differentiation* (J. Meyrath and J. D. Bu'Lock, eds.), Academic Press, London, 1977, pp. 1-15.
18. Kobel, H., and Sanglier, J. J., Qualitative change of alkaloid spectrum by mutation in Claviceps purpurea. In *Proceedings of the First International Symposium on the Genetics of Industrial Microorganisms*, Vol. 2 (Z. Vanek, Z. Hoštalek, and J. Cudlin, eds.), Elsevier, Amsterdam, 1973, pp. 421-425.

19. Segal, E., and Germanier, R., Degeneration phenomenon of alkaloid-producing strains of Claviceps paspali. *Z. Allg. Mikrobiol.*, *14*:145-151, 1974.
20. Keller, U. Highly efficient mutagenesis of Claviceps purpurea by using protoplasts. *Appl. Environ. Microbiol.*, *46*:580-584, 1983.
21. Puc, A., Didek-Brumec, M., and Sočič, H. Development of Claviceps purpurea high producing strains for the submerged production of ergot alkaloids. *Food Technol. Biotechnol. Rev.*, *23*:29-34, 1985.
22. Didek-Brumec, M., Puc, A., Sočič, H., and Alačević, M. Isolation and characterization of a high yielding ergotoxines producing Claviceps purpurea strains. *Food Technol. Biotechnol. Rev.*, *25*:103-109, 1987.
23. Mantle, P. G., and Tonolo, A. Relationship between the morphology of Claviceps purpurea and the production of alkaloids. *Trans. Br. Mycol. Soc.*, *51*:499-505, 1968.
24. Strnadova, K., and Kybal, J. Morphological changes of Claviceps purpurea induced by the mutagenic effect of nitrous acid. *Planta Medica*, *30*:395-398, 1976.
25. Pažoutova, S., Řeháček, Z., and Vorišek, J. Induction of sclerotia-like mycelium in axenic cultures of Claviceps purpurea producing clavine alkaloids. *Can. J. Microbiol.*, *26*:363-370, 1980.
26. Udvardy, E. Consideration of the development of an ergot alkaloid fermentation process. *Proc. Biochem.*, *15*:5-8, 1980.
27. Dickerson, A. G., Mantle, P. G., and Sczyrbak, C. A. Autolysis and extracellular glucans produced in vitro by a strain of Claviceps fusiformis. *J. Gen. Microbiol.*, *60*:403-415, 1970.
28. Esser, K., and Tudzynski, P. Genetics of the ergot fungus Claviceps purpurea. I. *Theor. Appl. Genet.*, *53*:145-149, 1978.
29. Schmauder, P., and Gröger, D. Über den Einfluss potentiell morphogener Substanzen auf Wachstum und Alkaloidbildung von Claviceps purpurea. *Pharmazie*, *34*:356-357, 1979.
30. Arcamone, F., Cassinelli, G., Ferni, G., Penco, S., and Pol, C. Ergotamine production and metabolism of Claviceps purpurea strain 275 FI in stirred fermentors. *Can. J. Microbiol.*, *16*:923-931, 1970.
31. Kobel, H., and Sanglier, J. J. Formation of ergotoxine alkaloids by fermentation and attempts to control their biosynthesis. In *Antibiotics and Other Secondary Metabolites* (R. Hütter, T. Leisinger, J. Nüesch, and W. Wehrli, eds.), Academic Press, New York, 1978, pp. 233-242.
32. Miličič, S., Kremser, M., Gaberc-Porekar, V., Didek-Brumec, M., and Sočič, H. The effect of aeration and agitation on Claviceps purpurea dimorphism and alkaloid synthesis during submerged fermentation. *Appl. Microbiol. Biotechnol.*, *31*:134-137, 1989.
33. Řeháček, Z. New trends in ergot alkaloid biosynthesis. *Proc. Biochem.*, 22-29, 1983.
34. Tonolo, A., and Udvardy-Nagy, E. Production of clavine alkaloids by Claviceps fusiformis in submerged culture. *Acta Microbiol. Acad. Sci. Hung.*, *15*:29-40, 1968.
35. Řeháček, Z., Kozova, J., Sajdl, P., and Vorišek, J. The physiology of conidial formation in submerged cultures of Claviceps purpurea (Fr.) Tul. producing alkaloids. *Can. J. Microbiol.*, *20*:1323-1329, 1974.
36. Vorišek, J., Ludvik, J., and Řeháček, Z. Morphogenesis and ultrastructure of Claviceps purpurea during submerged alkaloid formation. *J. Bacteriol.*, *120*:1401-1408, 1974.

37. Lösecke, W., Neumann, D., Schmauder, H. P., and Gröger, D. Changes in cytoplasmic ultrastructure during submerged cultivation of a peptide alkaloids-producing strain of Claviceps purpurea (Fr.) Tul. *Z. Allg. Mikrobiol.*, *22*:49-61, 1982.
38. Pertot, E., Jezernik, K., Didek-Brumec, M., and Sočič, H. Ultrastructural characteristics of a submerged active Claviceps paspali strain. *J. Basic Microbiol.*, *27*:369-376, 1987.
39. Didek-Brumec, M., Jezernik, K., Puc, A., and Sočič, H. Ultrastructural characteristics of Claviceps purpurea seed cultures. *J. Basic Microbiol.*, *28*:589-598, 1988.
40. Lösecke, W., Neumann, D., Gröger, D., and Schmauder, H. P. Changes of the cytoplasmic ultrastructure during development of sclerotia in Claviceps purpurea (Fr.) Tul. *Arch. Microbiol.*, *125*:251-257, 1980.
41. Neumann, D., Lösecke, W., Maier, W., and Gröger, D. Localization of alkaloids in sclerotia and suspension cultures of Claviceps purpurea (Fr.) Tul. *Biochem. Physiol. Pflanzen*, *174*:504-508, 1979.
42. Streiblova, E., Ryc, M., and Kybal, J. Fine structure of imbibed sclerotial cells of Claviceps purpurea (Fr.) Tul. revealed by freeze-etching. *Z. Allg. Mikrobiol.*, *18*:123-134, 1978.
43. Voříšek, K., and Řeháček, Z. Fine structural localization of alkaloid synthesis in endoplasmic reticulum of submerged Claviceps purpurea. *Arch. Microbiol.*, *117*:297-302, 1978.
44. Voříšek, J., and Sigler, K. Electron-cytochemical study of lipid deposit condensation and degradation in Ascomycetes: self-assembly of unit membranes. *Cell. Mol. Biol.*, *28*:221-233, 1982.
45. Floss, H. G. Biosynthesis of ergot alkaloids and related compounds. *Tetrahedron*, *32*:873-912, 1976.
46. Mothes, K., Weygand, F., Gröger, D., and Grisebach, H. Untersuchungen zur Biosynthese der Mutterkornalkaloide. *Z. Naturforsch.*, *13b*:41-44, 1958.
47. Gröger, D., Mothes, K., Simon, H., Floss, H. G., and Weygand, F. Über den Einbau von Mevalonsäure in das Ergolinsystem der Clavin-Alkaloide. *Z. Naturforsch.*, *15b*:141-143, 1960.
48. Baxter, R. M., Kandel, S. I., Okany, A., and Pyke, R. G. Biosynthesis of ergot alkaloids: The origin of N-methyl group. *Can. J. Chem.*, *42*:2936-2938, 1964.
49. Otsuka, H., Quigley, F. R., Gröger, D., Anderson, J. A., and Floss, H. G. In vivo and in vitro evidence for N-methylation as the second pathway-specific step in ergoline biosynthesis. *Planta Medica*, *40*:109-119, 1980.
50. Barrow, K. D., and Quigley, F. R. Ergot alkaloids III. The isolation of N-methyl-4-dimethylallyltryptophan from Claviceps fusiformis. *Tetrahedron Lett.*, 4269-4270, 1975.
51. Maier, W., Erge, D., Schmidt, J., and Gröger, D. A blocked mutant of Claviceps purpurea accumulating chanoclavine-I-aldehyde. *Experientia*, *36*:1353-1354, 1980.
52. Floss, H. G., Tcheng-Lin, M., Chang, C. J., Naidoo, B., Blair, G. E., Abou-chaar, C. I., and Cassady, J. M. Biosynthesis of ergot alkaloids: Studies on the mechanism of the conversion of chanoclavine-I into tetracyclic ergolenes. *J. Am. Chem. Soc.*, *96*:1898-1909, 1974.
53. Agurell, S. Biosynthesis of ergot alkaloids in Claviceps paspali. Part III. Incorporation of 6-methyl-$\Delta^{8,9}$-ergolene-8-carboxylic acid-^{14}C and lysergic acid-8-^{3}H. *Acta Pharm. Suecica*, *3*:65-70, 1963.

54. Keller, U., Zocher, R., and Kleinkauf, H. Biosynthesis of ergotamine in protoplasts of Claviceps purpurea. *J. Gen. Microbiol., 118*:485-494, 1980.
55. Maier, W., Schumann, B., Erge, D., and Gröger, D. Biosynthesis of ergot alkaloids by protoplasts of various Claviceps purpurea strains. *Biochem. Physiol. Pflanzen, 175*:815-819, 1980.
56. Gröger, D., Erge, D., and Floss, H. G. Über die Herkunft der Seitenkette im d-Lysergsäuremethylcarbinolamid. *Z. Naturforsch., 23b*:175-178, 1968.
57. Floss, H. G., Tcheng-Lin, M., Kobel, H., and Stadler, P. Biosynthesis of peptide ergot alkaloids. *Experientia, 30*:1369-1370, 1974.
58. Bianchi, M. L., Cattaneo, P. A., Crespi-Perellino, N., Guicciardi, A., Minghetti, A., and Spalla, C. Mechanisms of control in the qualitative biosynthesis of lysergic acid alkaloids. In *Overproduction of Microbial Products*, FEMS Symposium No. 13 (V. Krumphanzl, B. Sikyta, and Z. Vanek, eds.), Academic Press, London, 1982, pp. 235-242.
59. Heinstein, P. F., Lee, S. L., and Floss, H. G. Isolation of dimethylallylpyrophosphate: tryptophan dimethylallyl transferase from the ergot fungus (Claviceps sp.). *Biochem. Biophys. Res. Commun., 44*:1244-1251, 1971.
60. Lee, S. L., Floss, H. G., and Heinstein, P. Purification and properties of dimethylallylpyrophosphate: tryptophan dimethylallyl transferase, the first enzyme of ergot alkaloid biosynthesis in Claviceps sp. SD58. *Arch. Biochem. Biophys., 177*:84-94, 1976.
61. Petroski, R. J., and Kelleher, W. J. Biosynthesis of ergot alkaloids. Cell-free formation of 4-(E-4'-hydroxy-3'-methyl-but-2'-enyl)-L-tryptophan. *FEBS Lett., 82*:55-57, 1977.
62. Ogunlana, E. O., Wilson, B. J., Tyler, V. E., and Ramstad, E. Biosynthesis of ergot alkaloids. Enzymatic closure of ring of the ergolene nucleus. *Chem. Commun.*, 775-776, 1970.
63. Hsu, J. C., and Anderson, J. A. Agroclavine hydroxylase of Claviceps purpurea. *Biochim. Biophys. Acta, 230*:518-525, 1971.
64. Kim, S. U., Cho, Y. J., Floss, H. G., and Anderson, J. A. Conversion of elymoclavine to paspalic acid by a particular fraction from an ergotamine-producing strain of Claviceps sp. *Planta Medica, 48*:145-148, 1983.
65. Maier, W., Schumann, B., and Gröger, D. Microsomal oxygenases involved in ergoline alkaloid biosynthesis of various Claviceps strains. *J. Basic Microbiol., 28*:83-93, 1988.
66. Wilson, B. J., Ramstad, E., Jansson, E., and Orrenius, S. Conversion of agroclavine by mammalian cytochrome P-450. *Biochim. Biophys. Acta, 252*:348-356, 1971.
67. Maier, W., Erge, D., Schumann, B., and Gröger, D. Incorporation of L-(U-^{14}C) leucine into ergosine by cell-free extracts of Claviceps purpurea (Fr.) Tul. *Biochem. Biophys. Res. Commun., 99*:155-162, 1981.
68. Maier, W., Krauspe, R., and Gröger, D. Enzymatic biosynthesis of ergotamine and investigation on some aminoacyl-tRNA synthetases in Claviceps. *J. Biotechnol., 3*:155-166, 1985.
69. Bruenger, E., Chayet, L., and Rilling, H. C. Isopentenyl pyrophosphate isomerase: dimethylallyl pyrophosphate isomerase: isolation from Claviceps sp. SD 58 and comparison to the mammalian enzyme. *Arch. Biochem. Biophys., 248*:620-625, 1986.

70. Miličič, S., Kremser, M., Gaberc-Porekar, V., Didek-Brumec, M., and Sočič, H. Correlation between growth velocity and biosynthesis of ergot alkaloids in Claviceps purpurea batch fermentation. *Appl. Microbiol. Biotechnol.*, *27*:117-120, 1987.
71. Brady, L. R., and Tyler, V. E., Jr. The biosynthesis of clavine alkaloids. *Planta Medica*, *7*:225-233, 1959.
72. Arcamone, F., Chain, E. B., Ferretti, A., Minghetti, A., Pennella, P., Tonolo, A., and Vero, L. Production of a new lysergic acid derivative in submerged culture by a strain of Claviceps paspali Stevens and Hall. *Proc. R. Soc., Ser. B.*, *155*:26-54, 1961.
73. Floss, H. G., and Mothes, U. Über den Einfluss von Tryptophan und analogen Verbindungen auf die Biosynthese von Clavinalkaloiden in saprophytischer Kultur. *Arch. Mikrobiol.*, *48*:213-221, 1964.
74. Arcamone, F., Chain, E. B., Ferretti, A., Minghetti, A., Pennella, P., and Tonolo, A. The effect of tryptophan and methyltryptophans on submerged lysergic acid fermentation by Claviceps paspali Stevens and Hall. *Biochim. Biophys. Acta*, *57*:174-176, 1962.
75. Krupinski, V. M., Robbers, J. E., and Floss, H. G. Physiological study of ergot: Induction of alkaloid synthesis by tryptophan at the enzymatic level. *J. Bacteriol.*, *125*:158-165, 1976.
76. Gröger, D., and Tyler, V. E., Jr. Alkaloid production by Claviceps paspali in submerged culture. *Lloydia*, *26*:174-191, 1963.
77. Robbers, J. E. The fermentative production of ergot alkaloids. In *Advances in Biotechnological Processes*, Vol. 3 (A. Mizrahi and A. L. Wezel, eds.), Alan R. Liss, New York, 1984, pp. 197-239.
78. Drew, S., and Wallis, D. A. Regulation of secondary metabolism and keys to its manipulation. In *Secondary Metabolism and Differentiation in Fungi*, Vol. 5 (J. W. Bennet and A. Ciegler, eds.), Marcel Dekker, New York, 1983, pp. 35-54.
79. Gaberc-Porekar, V., Sočič, H., and Pertot, E. Metabolic changes in a conidia induced Claviceps paspali strain during submerged fermentation. *Can. J. Microbiol.*, *33*:602-606, 1987.
80. Mann, D. F., and Floss, H. G. Purification and properties of anthranilate synthetase from the ergot fungus, Claviceps spec., strain SD58. *Lloydia*, *40*:136-145, 1977.
81. Cheng, L. J., Robers, J. E., and Floss, H. G. End product regulation of ergot alkaloid formation in intact cells and protoplasts of Claviceps species, strain SD58. *J. Natural Prod.*, *43*:329-339, 1980.
82. Erge, D., Maier, W., and Gröger, D. Untersuchungen über die enzymatische Umwandlung von Chanoclavine-I. *Biochem. Physiol. Pflanzen*, *164*:234-247, 1973.
83. Spalla, C. Production of ergot alkaloids by fermentation. In *Indole and Biogenetically Related Alkaloids* (J. D. Phillipson and M. H. Zenk, eds.), Academic Press, New York, 1980, pp. 271-284.
84. Puc, A., Miličič, S., Kremser, M., and Socic, H. Regulation of ergotoxine biosynthesis in Claviceps purpurea submerged fermentation. *Appl. Microbiol. Biotechnol.*, *25*:449-452, 1987.
85. Crespi-Perellino, N., Guicciardi, A., Minghetti, A., and Spalla, C. Incorporation of α-aminobutyric acid into ergostine by Claviceps purpurea. *Experientia*, *37*:217-219, 1981.
86. Beacco, E., Bianchi, M. L., Minghetti, A., and Spalla, C. Directed synthesis of analogues of ergot peptide alkaloids with Claviceps purpurea. *Experientia*, *34*:1291-1293.

87. Baumert, A., Erge, D., and Gröger, D. Incorporation of thiazolidine-4-carboxylic acid into ergosine by Claviceps purpurea. *Planta Med.*, *44*:122-123, 1982.
88. Breuel, K., Braun, K., Dauth, Ch., and Gröger, D. Qualitative changes in peptide alkaloid biosynthesis of Claviceps purpurea by environmental conditions. *Planta Medica, 44*:121-122, 1982.
89. Demain, A. L. Overproduction of microbial metabolites and enzymes due to alteration of regulation. *Adv. Biochem. Eng.*, *1*:113-142, 1971.
90. Wack, G., Perényi, E., and Udvardy Nagy, E. Hungarian Patent 153,739, 1967.
91. Kobel, H., and Sanglier, J. J. French Patent 2,307,870, 1976.
92. Sočič, H., Gaberc-Porekar, V., and Didek-Brumec, M. Biochemical characterization of the inoculum of Claviceps purpurea for submerged production of ergot alkaloids. *Appl. Microbiol. Biotechnol.*, *21*:91-95, 1985.
93. Sočič, H., Gaberc-Porekar, V., Pertot, E., Puc, A., and Miličič, S. Developmental studies of Claviceps paspali seed cultures for the submerged production of lysergic acid derivatives. *J. Basic Microbiol.*, *26*:533-539, 1986.
94. Řeháček, Z. Physiology of ergot alkaloid production in submerged Claviceps cultures. In *Overproduction of Microbial Products*, FEMS Symp. No. 13 (V. Krumphanzl, B. Sikyta, and Z. Vanek, eds.), Academic Press, London, 1982, pp. 221-234.
95. Kobel, H., and Sanglier, J. J. Ergot alkaloids. In *Biotechnology*, Vol. 4 (H. J. Rehm and G. Reed, eds.), VCH Verlagsgesellschaft, Weinheim, 1986, pp. 569-609.
96. Milicic, S., Kremser, M., Povšič, Z., and Sočič, H. Growth and sporulation of the fungus Claviceps purpurea in batch fermentation. *Appl. Microbiol. Biotechnol.*, *20*:356-359, 1984.
97. Kobel, H. Degenerationsprobleme bei Produktionsstämmen von Claviceps. *Pathol. Microbiol.*, *34*:249-251, 1969.
98. Segal, E., and Germanier, R. Degeneration phenomenon of alkaloid producing strains of Claviceps paspali. *Z. Allg. Mikrobiol.*, *14*:145-151, 1974.
99. Pertot, E., Puc, A., and Sočič, H. Isolation and morphological characterization of a conidia-forming Claviceps paspali mutant. *Folia Microbiol.*, *31*:363-366, 1986.
100. Spalla, C., and Marnati, M. P. Genetic aspects of the formation of ergot alkaloids. In *Antibiotics and Other Secondary Metabolites*, FEMS Symposium No. 5 (R. Hütter, T. Leisinger, J. Nüesch, and W. Wehrli, eds.), Academic Press, London, 1978, pp. 219-232.
101. Řeháček, Z. Ergot alkaloid biosynthesis. In *Modern Biotechnology* (V. Krumphanzl and Z. Řeháček, eds.), Czechoslovak Acad. Sci.-UNESCO, Prague, 1984, pp. 855-895.
102. Kybal, J., Svoboda, E., Strnadova, K., and Kejzlar, M. Role of organic acid metabolism in the biosynthesis of peptide ergot alkaloids. *Folia Microbiol.*, *26*:112-119, 1981.
103. Kren, V., Mehta, P., Rylko, V., Flieger, M., Kozova, J., Sajdl, P., and Řeháček, Z. Substrate regulation of elymoclavine formation by some saccharides. *Zentralbl. Mikrobiol.*, *142*:71-85, 1987.
104. Kren, V., and Řeháček, Z. Feedforward regulation of phosphofructokinase in a submerged culture of Claviceps purpurea producing clavine alkaloids. *Spec. Sci. Technol.*, *7*:223-226, 1984.

105. Kybal, J., Kleinerova, E., and Bulant, V. Ergot alkaloids, VI. Nitrogen metabolism during the development of sclerotium of Claviceps purpurea. *Folia Microbiol.*, *21*:474-480, 1976.
106. Mantle, P. G., and Nisbet, L. J. Differentiation of Claviceps purpurea in axenic culture. *J. Gen. Microbiol.*, *93*:321-334, 1976.
107. Sočič, H., Pertot, E., and Didek-Brumec, M. The effect of media composition on the fermentative production of clavine ergot alkaloids. *Vestn. Slov. Kem. Drus.*, *29*:323-329, 1982.
108. Spalla, C., Filippini, S., and Grein, A. A hypothesis on the regulation mechanisms governing the biosynthesis of alkaloids in Claviceps. *Folia Microbiol.*, *23*:505-508, 1978.
109. Gaberc-Porekar, V., Puc, A., Didek-Brumec, M., and Sočič, H. Metabolic pattern of different Claviceps purpurea strains during submerged fermentation. 8th Regional Meeting of Biochemists, biophysicists and biotechnologists, Graz, 1987, Abstracts, p. P32.
110. Patel, H. C., and Desai, J. D. Phosphate transport in Claviceps sp. SD-58. *Experientia*, *41*:96-97, 1985.
111. Rao, K. K., and Gupta, A. K. Inhibition of tryptophan synthetase by phosphate in alkaloid producing strain of Aspergillus fumigatus. *Naturwiss.*, *62*:394-395, 1975.
112. Robbers, J. E., Eggert, W. W., and Floss, H. G. Physiological studies on ergot: time factor influence on the inhibitory effect of phosphate and the induction effect of tryptophan on alkaloid production. *Lloydia*, *41*:120-129, 1978.
113. Vaida, H. C., and Desai, J. D. Alkaloid production by Claviceps sp. SD-58; Involvement of phosphatase isozymes. *Folia Microbiol.*, *28*:12-16, 1983.
114. Mizrahi, A., and Miller, G. A defined medium for the production of lysergic acid alkaloids by Claviceps paspali. *Biotechnol. Bioeng.*, *12*:641-644, 1970.
115. Puc, A., and Sočič, H. Carbohydrate nutrition of Claviceps purpurea for alkaloid production related to the osmolality of media. *Eur. J. Appl. Microbiol.*, *4*:283-287, 1977.
116. Bu'Lock, J. D., and Barr, J. G. Regulation mechanism linking tryptophan uptake and synthesis with ergot alkaloid synthesis in Claviceps. *Lloydia*, *31*:342-354, 1968.
117. Kregar, I., Puc, A., and Turk, V. Extracellular proteinases of Claviceps purpurea. Isolation and characterization of an aspartic proteinase. *Eur. J. Appl. Microbiol. Biotechnol.*, *17*:129-132, 1983.
118. Kregar, I., Babnik, J., Turk, V., and Koleša, I. Proteinases of various Claviceps purpurea strains. *Biomed. Biochim. Acta*, *45*:1537-1540, 1986.
119. Řeháček, Z. Ergot alkaloids and their biosynthesis. *Adv. Biochem. Eng.*, *14*:33-60, 1980.
120. Řeháček, Z., Desai, J. D., Sajdl, P., and Pažoutova, S. The cellular role of nitrogen in the biosynthesis of alkaloids by submerged culture of Claviceps purpurea (Fr.) Tul. *Can. J. Microbiol.*, *23*:596-600, 1977.
121. Mantle, P. G., Morris, L. J., and Hall, S. W. Fatty acid composition of sphacelial and sclerotial growth forms of Claviceps purpurea in relation to the production of ergot alkaloids in culture. *Trans. Br. Mycol. Soc.*, *53*:441-447, 1969.

122. Kren, V., Rezanka, T., Sajdl, P., and Řeháček, Z. Identification of fatty acids in submerged cultures of Claviceps species. *FEMS Microbiol. Lett.*, *30*:359-363, 1985.
123. Desai, J. D., Desai, A. J., and Patel, H. C. Effect of biotin on alkaloid production during submerged cultivation of Claviceps sp. SD-58. *Appl. Environ. Microbiol.*, *54*:1694-1696, 1983.
124. Glund, K., Kirstan, C., and Schlee, D. Glucose catabolism during alkaloid production in a strain of Claviceps purpurea. *Folia Microbiol.*, *26*:398-402, 1981.
125. Amici, A. M., Scotti, T., Spalla, C., and Tognoli, L. Heterokaryosis and alkaloid production in Claviceps purpurea. *Appl. Microbiol.*, *15*:611-615, 1967.
126. Strnadova, K., and Kybal, J. Ergot alkaloids V. Homokaryosis of the sclerotia of Claviceps purpurea (Fr.) Tul. *Folia Microbiol.*, *19*:272-280, 1974.
127. Robbers, J. E., Jones, L. G., and Krupinsky, V. M. Ergot physiology: The production of ergot alkaloids in saprophytic culture by a homokaryotic strain of ergot. *Lloydia*, *37*:108-111, 1974.
128. Esser, K., and Tudzynsky, P. Genetics of the ergot fungus Claviceps purpurea 1. Proof of a monoecious life cycle and segregation patterns for mycelial morphology and alkaloid production. *Theor. Appl. Genet.*, *53*:145-149, 1978.
129. Srikrai, S., and Robbers, J. E. Methods for mutation and selection of the ergot fungus. *Appl. Environ. Microbiol.*, *45*:1165-1169, 1983.
130. Spalla, C., Guicciardi, A., Marnati, M. P., and Oddo, N. Protoplasts and production of alkaloids with a heterokaryotic strain of Claviceps purpurea. Abstracts. Fifth Int. Ferment. Symposium, Berlin, 1976, p. 192.
131. Stahl, C., Neumann, D., Schmauder, A., and Gröger, D. Preparation of protoplast from Claviceps. *Biochem. Physiol. Pflanzen*, *171*:363-368, 1977.
132. Komel, R., Rozman, D., Puc, A., and Sočič, H. Effect of immobilization on the stability of Claviceps purpurea protoplasts. *Appl. Microbiol. Biotechnol.*, *23*:106-109, 1985.
133. Strnadova, K. Methoden zur Isolierung und Ermittlung auxotropher Mutanten by Claviceps purpurea (Fr.) Tul. *Z. Pflanzenzüchtung*, *51*:167-169, 1964.
134. Tudzynski, P., Esser, K., and Gröschel, H. Genetics of the ergot fungus Claviceps purpurea. Part 2: Exchange of genetic material via meiotic recombination. *Theor. Appl. Genet.*, *61*:97-100, 1982.
135. Kus, B., Didek-Brumec, M., Sočič, H., Gaberc-Porekar, V., and Alačević, M. Determination of metabolic blocks in Claviceps purpurea arg mutants. *Food Technol. Biotechnol. Rev.*, *25*:127-131, 1987.
136. Kus, B., Didek-Brumec, M., Alačević, M., and Sočič, H. Benomyl resistant mutant of Claviceps purpurea strains. *Biol. Vestnik*, *34*:43-52, 1986.
137. Brauer, K. L., and Robbers, J. E. Induced parasexual processes in Claviceps sp. strain SD-58. *Appl. Environ. Microbiol.*, *53*:70-73, 1987.
138. Didek-Brumec, M., Kus, B., Sočič, H., Gaberc-Porekar, V., and Alačević, M. Influence of genetic recombinations on ergot alkaloid biosynthesis. Abstracts 18th FEBS Meeting, Ljubljana, Yugoslavia, 1987, p. 99.

139. Esser, K., Kück, U., Stahl, U., and Tudzynski, P. Cloning vectors of mitochondrial origin for eukaryotes. *Curr. Genet.*, *7*:239-243, 1983.
140. Tudzynski, P., and Esser, K. Extrachromosomal genetics of Claviceps purpurea. II. Plasmids in various wild strains and integrated plasmid sequences in mitochondrial genomic DNA. *Curr. Genet.*, *10*:463-467, 1986.
141. Hoffman, A., Frey, A. J., and Ott, H. Die Totalsynthese des Ergotamines. *Experientia*, *17*:206-210, 1961.
142. Stadler, P. A., Guttmann, S., Hauth, H., Huguenin, R. L., Sandrin, E. D., Versin, G., and Hoffmann, A. Die Synthese der Alkaloide des Ergotoxin-Gruppe. *Helv. Chim. Acta*, *52*:1549-1564, 1969.
143. Stadler, P. A., and Stütz, P. The ergot alkaloids. In *The Alkaloids*, Vol. 15 (R. H. F. Manske, ed.), Academic Press, New York, 1975, pp. 1-44.
144. Stoll, A., Brack, A., Hoffmann, A., and Kobel, H., U. S. Patent, 2,809,920, 1957.
145. Amici, A. M., Minghetti, A., Scotti, T., Spalla, C., and Tognolli, L., Production of peptide ergot alkaloids in submerged culture by three isolates of Claviceps purpurea. *Appl. Microbiol.*, *18*:464-468, 1969.
146. Gaberc-Porekar, V., Didek-Brumec, M., and Sočič, H. Direct selection of active Claviceps colonies on agar plates. *Z. Allg. Mikrobiol.*, *23*:95-98, 1983.
147. Durand, H., and Tibardy, G. In La genetique des Microorganismes Industrieles, Poster, Societe Francaise de Microbiologie, Paris, 1984.
148. Mizrahi, A., and Miller, G., Longterm preservation of a nonsporulating strain of Claviceps paspali. *Appl. Microbiol.*, *16*:1100-1101, 1968.
149. Pertot, E., Puc, A., and Kremser, M. Lyophilization of non-sporulating strains of the fungus Claviceps. *Eur. J. Appl. Microbiol.*, *4*:289-294, 1977.
150. Grm, B., Mele, M., and Kremser, M. Model of growth and ergot alkaloid production by Claviceps purpurea. *Biotechnol. Bioeng.*, *22*: 255-270, 1980.
151. Pažoutova, S., Votruba, J., and Řeháček, Z. A. A mathematical model of growth and alkaloid production in the submerged culture of Claviceps purpurea. *Biotechnol. Bioeng.*, *23*:2837-2849, 1981.
152. Flieger, M., Votruba, J., Kren, V., Pazoutova, S., Rylko, V., Sajdl, P., and Řeháček, Z. Physiological control and process kinetics of clavine alkaloid production by Claviceps purpurea. *Appl. Microbiol. Biotechnol.*, *29*:181-185, 1988.
153. Kopp, B., and Rehm, H. J. Alkaloid production by immobilized mycelia of Claviceps purpurea. *Eur. J. Appl. Microbiol. Biotechnol.*, *18*:257-263, 1983.
154. Kren, V., Ludvik, J., Kofronova, D., Kozova, J., and Řeháček, Z. Physiological activity of immobilized cells of Claviceps fusiformis during long term semicontinuous cultivation. *Appl. Microbiol. Biotechnol.*, *26*:219-226, 1987.
155. Pertot, E., Rozman, D., Miličić, S., and Sočič, H. Morphological differentiation of immobilized Claviceps paspali mycelium during semicontinuous cultivation. *Appl. Microbiol. Biotechnol.*, *28*:209-213, 1988.
156. Kopp, B., and Rehm, H. J. Semicontinuous cultivation of immobilized Claviceps purpurea. *Appl. Microbiol. Biotechnol.*, *19*:141-145, 1984.

19

PRODUCTION OF PENICILLINS AND CEPHALOSPORINS BY FUNGI

ABDEL-HALIM M. M. EL-SAYED* *Faculty of Pharmacy, Zagazig University, Zagazig, Egypt*

I. INTRODUCTION

Penicillin is the oldest recognized and the progenitor of the most significant chemotherapeutic agents developed during the past 50 years. A large amount of industrially relevant research has been carried out with the penicillin-producing fungus *Penicillium chrysogenum*, which makes it possible to give a coherent picture about the mode of biosynthesis of this antibiotic. Moreover, many of the principles derived from studies of *P. chrysogenum* and other penicillin-related antibiotic-producing fungi are equally applicable to organisms synthesizing other secondary metabolites [1]. These studies permitted the improvement of productivity of the industrially used fungus by more than 1000-fold that of the original *Penicillium* strains [2].

Penicillin proved not to be a single compound or even a unique series of compounds. The cephalosporin antibiotics were subsequently discovered and these proved to be structurally related to penicillin [3,4]. Penicillins and cephalosporins are unusual molecules due to a four-membered β-lactam ring systems (Fig. 1). The β-lactam antibiotics are natural products and industrial production of all β-lactam antibiotics depends on microbial fermentation. While total chemical synthesis of penicillin and cephalosporin has been successful [4-6], these efforts cannot compete with the microbial synthesis. The microbially synthesized penicillins and cephalosporins are called natural penicillins and natural cephalosporins, respectively (Table 1). These natural products can either be directly applied clinically or can be further modified through chemical or enzymatic methods to give the more valuable semisynthetic penicillins or cephalosporins (Tables 2 and 3) [2].

This chapter summarizes recent studies carried out to reveal the development in the field of production of natural penicillins and cephalosporins by fungi.

II. MICROORGANISMS

Penicillins and cephalosporins are derivatives of a bicyclic ring system. All, with exception of 6-aminopenicillanic acid (6-APA), have an acyl group

*Present affiliation: SymBiotech, Inc., Wallingford, Connecticut.

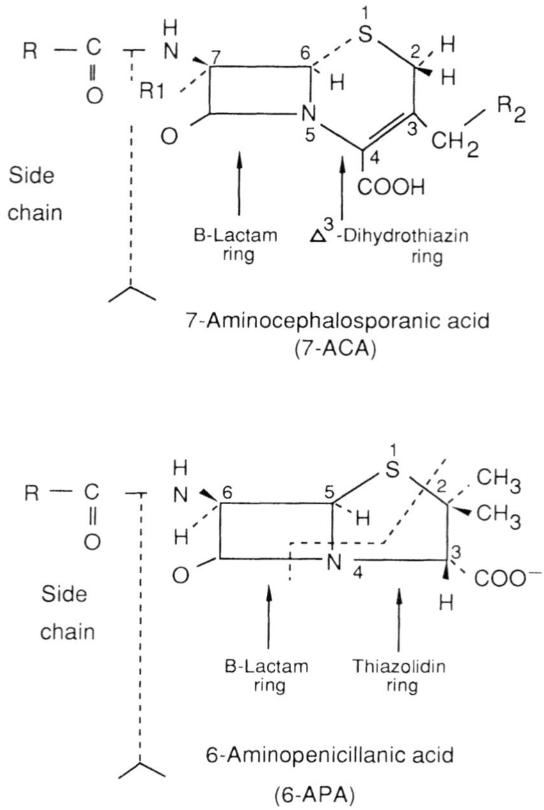

FIGURE 1. Chemical structure of penicillins and cephalosporins. (From Ref. 3.)

attached as a side chain to the amino group of a heterocyclic nucleus. The penicillin ring system (penam) contains a four-membered, β-lactam, ring fused with a five-membered thiazolidine ring. Cephalosporins are *N*-acylated derivatives of 7-aminocephalosporanic acid (7-ACA), which is derived from the cephem nucleus. A cephem is the bicyclic ring system obtained by the fusion of a β-lactam ring with dihydrothiazine ring. Natural penicillins can be classified into two types which are designated as *Penicillium*-type penicillins and *Cephalosporium*-type penicillins (Table 1).

The *Penicillium* type is characterized by microbial biosynthesis of an extensive series of penicillin compounds. These penicillins are variable according to the *N*-acyl side chain available in the fermentation medium in sufficient concentration. The *N*-acyl side chains are carboxylic acid derivatives with an unsubstituted methylene group adjacent to the carboxyl group [7-9]. The synthesis of a specific penicillin is a function of the available side chain precursor in the culture medium. In the absence of a sufficient side chain precursor, 6-APA is formed in this type of fermentation [10]. All microorganisms producing *Penicillium*-type penicillins are eukaryotic. They belong mainly to the *Penicillium* species, such as *Penicillium chrysogenum*, *P. notatum*, *P. turbatum*, *P. bacculatum*, *P. avellaneum*, *P. rubens*, *P. fluorescens*, and *P. citreoroseum* [3,4]. Other fungi which

TABLE 1. Biosynthetic Penicillins (A) and Cephalosporins (B)

A. Biosynthetic penicillins

Name	R
Benzylpenicillin, penicillin G	Ph-CH_2-CO-
p-Hydroxybenzylpenicillin, penicillin X	HO-Ph-CH_2-CO-
Phenoxymethylpenicillin, penicillin V	Ph-O-CH_2-CO-
Heptylpenicillin, penicillin K	$CH_3(CH_2)_6$-CO-
2-Pentenylpenicillin, penicillin F	$CH_3CH_2CH{=}CHCH_2$-CO-
Amylpenicillin, penicillin-dihydro-F	$CH_3(CH_2)_4$-CO-
Isopenicillin-N, penicillin M.	HO_2C-$CH(NH_2)(CH_2)_3$-CO-
Penicillin N, cephalosporin N, synnematin B	HO_2C-$CH(NH_2)(CH_2)_3$-CO-

B. Biosynthetic cephalosporin C[a]

	R_1	R_2
Cephalosporin C	H	$-OCOCH_3$
Deacetylcephalosporin C	H	-OH
Deacetoxycephalosporin C	H	-H
7-Methoxycephalosporin C	OCH_3	$-OCOCH_3$
Deacetyl-3-*O*-carbamoylcephalosporin C	H	$-OCONH_2$
Deacetyl-3-*O*-carbamoyl-7-methoxycephalosporin C	OCH_3	$-OCONH_2$

[a]R = D-2-aminoadipic acid.

can also produce this type of penicillin include *Aspergillus flavus*, *A. parasiticus*, *A. giganticus*, *A. niger*, *A. nidulans*, *A. flavipes*, *A. oryzae*, *Trichophyton mentagrophytes*, *Epidermophyton floccosum*, *Cephalosporium* species, *Emericellopsis* species, and *Paecilomyces parasiticus* [4,11-13]. On an industrial scale, highly selected mutants of *P. chrysogenum* are used to produce penicillins of this type through the addition of a suitable side chain precursor [3,4].

On the other hand, the *Cephalosporium*-type penicillins include the penicillin nucleus (6-APA) and a single penicillin with the side chain D-2-aminoadipic acid. The fermentation of this later type is insensitive to the addition of a side chain precursor and a little or no 6-APA is accumulated in the fermentation medium. The microorganisms producing this type include both eukaryotic *Cephalosporium* species such as *C. acremonium* or other fungi including *Paecilomyces persicinus* [14], *Emericellopsis salmosynnemata* and *Acremonium kiliense* [15], as well as prokaryotics from streptomycetes

TABLE 2. Some Important Semisynthetic Penicillins

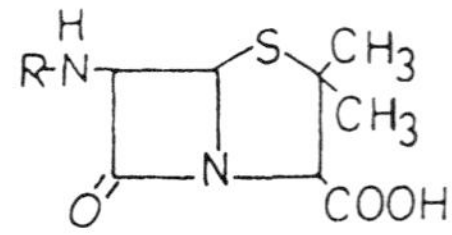

R (N-acyl side chain)	Name	Characteristics
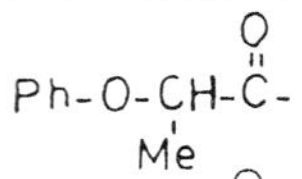 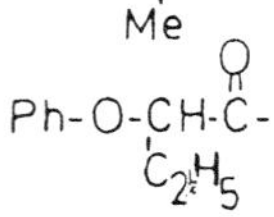 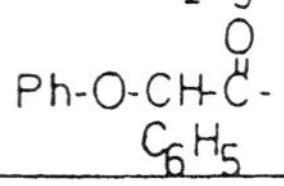	Phenoxyethylpenicillin (Phenethicillin) Phenoxypropylpenicillin (Propicillin) Phenoxybenzylpenicillin (Phenbencillin)	Effective against Gram`s positive bacteria, acid stable, senstive to penicillinases, absorbed orally
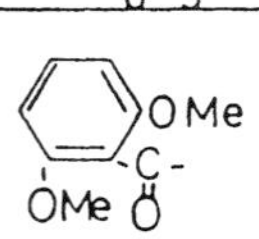	2,6-Dimethoxyphenylpenicillin (Methicillin)	Effective against Gram`s positive bacteria, acid labile, resists some penicillinases
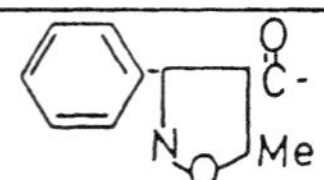 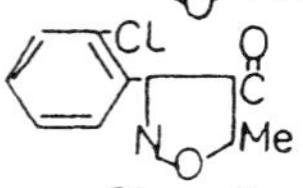 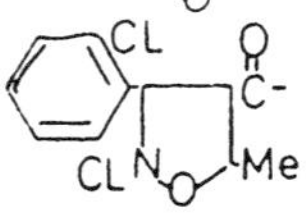 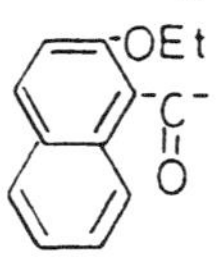	3-Phenyl-5-methyl-4-isooxazolylpenicillin (Oxacillin) 3-O-Chlorphenyl-5-methyl-4-isoxazolylpenicillin (Cloxacillin) 3-O,O-Dichlorophenyl-methyl-4-isoxazolylpenicillin (Dicloxacillin) 2-Ethoxy-1-naphthylpenicillin (Nafcillin)	Effective against Gram`s positive bacteria, acid stable, resist most beta-lactamases, strongly binds with serum protein, absorbed orally
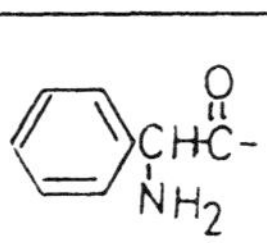	D-2-Aminobenzylpenicillin (Ampicillin)	Broad spectrum, very acid stable, sensitive to beta-lactamases, absorbed orally
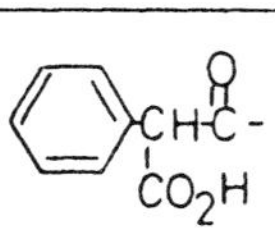	alfa-carboxybenzylpenicillin (Carbenicillin)	Broad spectrum, active against _Pseudomonas aeruginosa_, very acid stable, non effective orally

TABLE 3. Some Important Semisynthetic Cephalosporins

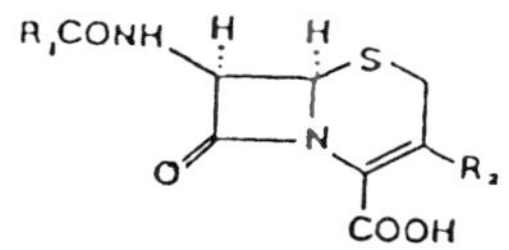

Name	R_1	R_2
1. Cephaloridine	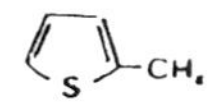	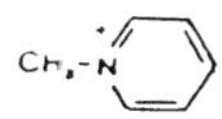
2. Cephalothin	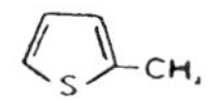	CH_2OCOCH_3
3. Cephalexin	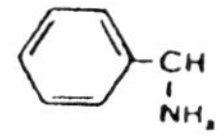	CH_3
4. Cephamandole	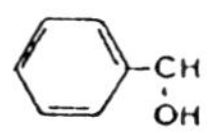	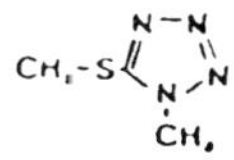
5. Celsulodin	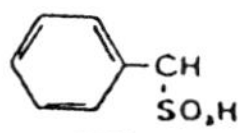	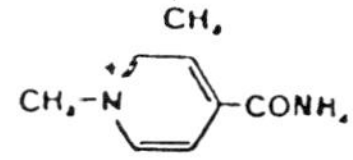
6. Cefoperazone	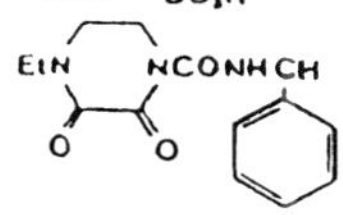	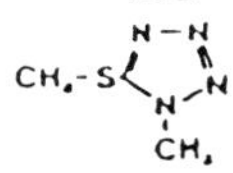
7. Cefotaxime	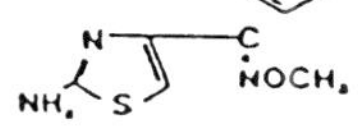	CH_2OCOCH_3
8. Ceftazidime	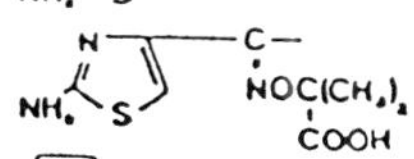	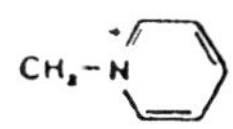
9. Cefuroxime	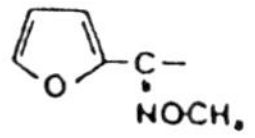	CH_2OCONH_2
10. Cefazolin	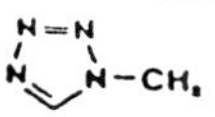	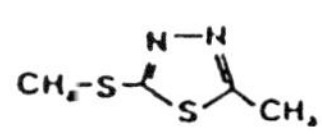
11. Ceforanide	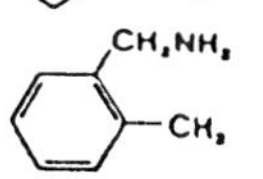	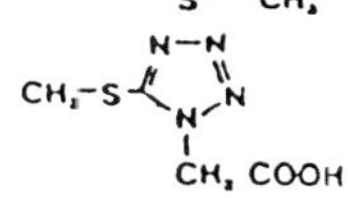
12. Cefotiam	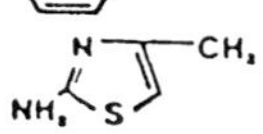	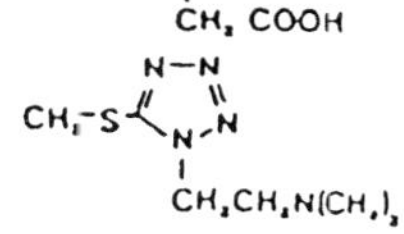
13. Cephaloglycin	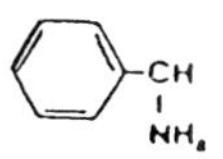	CH_2OCOCH_3
14. Cefatriazine	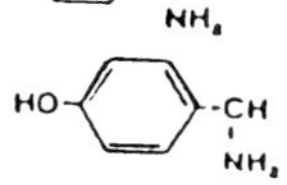	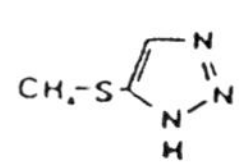

Source: Ref. 158.

such as *Streptomyces clavuligerus* and *S. lipmanii* [16]. These microorganisms can further utilize the naturally produced penicillin to form natural cephalosporin compounds (Table 1). With some exceptions, all cephalosporin compounds known to be formed biosynthetically possess the D-2-aminoadipyl side chain [17]. No 7-ACA is formed in these fermentations. For industrial cephalosporin fermentations, highly selected mutants from *Cephalosporium acremonium* (*Acremonium chrysogenum*) are used.

A. Screening, Selection, and Strain Improvement

Screening involves examination of all members of a population, e.g., a population of morphologically abnormal colonies. Selection describes isolation of a required type of variant under conditions that prevent recovery of other members of the population [1]. Strain improvement is the process carried out to improve the antibiotic titer under the required environmental conditions.

Generally, a good penicillin-yielding strain must have the following attributes [18]:

1. The strain should be a pure culture.
2. The strain must be genetically stable.
3. The strain must readily produce many vegetative cells and spores.
4. The strain should grow vigorously and rapidly after inoculation into seed tanks or other containers used to prepare large amounts of inoculum before an industrial fermentation in submerged cultures (the first strains produced penicillin only in surface cultures).
5. The strain should utilize highly complexed organic compounds with the ability to utilize phenylacetic acid as a side chain precursor.
6. The strain must not produce pigments which interfere with penicillin purification processes.
7. The strain should grow in the form of compact pellets which facilitate the separation of the biomass from fermentation medium at the end of production phase.
8. The strain should be readily maintained for a reasonably long period.
9. The strain should be amenable to change by mutagenic agents. A mutation program might be conducted to develop strains that give enhanced yields of the product.

The traditional screening method on agar plates has proved to be unsatisfactory for obtaining high-antibiotic-producing strains. Shake-flask fermentation still appears to be the dominant method of selection due to a greater correlation with results in production fermentors [2]. However, different environmental conditions can influence the performance of different strains. For example, Elander [19] reported that certain strains performed well in shake flasks and 5-liter vessels but not in 8000-gallon fermentors; others performed well in only 5-liter vessels and yet others on all scales except 5-liter vessels. From this example, the possibility of change in the activity of the strain under different environmental conditions can be expected. The main factor is that some correlation exists between performance in shake flasks and on larger scales. Any laboratory scale screening system must establish such a correlation in order to be industrially practical.

Other techniques have been used by many laboratories to screen and select high-antibiotic-producing strains. Among these techniques are the following:

1. Potency index method. In this method, *P. chrysogenum* colonies on agar medium are surrounded by a bacterial suspension and after incubation zones of bacterial lysis appear around the colonies. A potency index is given by zone plus colony diameter divided by the colony diameter. The potency index method allows a large number of colonies to be screened for penicillin production up to about 20 times faster than by the shake-flask method [1].

2. Recovery of morphological mutants. Backus and Stauffer [20] reported that the correlation between altered morphology and improved titer was not obligate because the strains with reduced vigor did not have improved yield. Other morphological characteristics of the penicillin-producing strains will be discussed later in Section II.B.

3. Recovery of auxotrophs. Demain [21] reported that auxotrophs influence antibiotic synthesis in many ways. He emphasized that auxotrophs having pathways unrelated to the main biosynthetic pathway can exert a regulatory influence on the latter by preventing the formation or accumulation of intermediates [22].

4. Detoxification techniques. High-penicillin-producing mutants can grow on concentrations of otherwise toxic substances that may be precursors of penicillins, such as phenylacetic acid or a toxic compound that is inactivated by reaction with the antibiotic [1].

5. Detection of different types of recombinant segregants. The selection of different types of segregants is normally accomplished first by using morphological criteria such as spore color, segregation of drug resistance markers, spore size, or DNA determination per nucleus which enable segregants to be classified as haploid or diploid [23].

B. Morphological, Physiological, and Genetic Studies

1. Morphological Studies

Macromorphological and micromorphological features of a penicillin-producing microorganism can mirror its penicillin productivity and applicability on an industrial scale. The macromorphological features of the fungus include its colony's shape and size on solid media, formation of hyphae, production of pigments, formation of pellets in submerged cultures, branching frequency, and conidia formation. The micromorphological features include the study of numbers, dimensions, and structure of the surface of the hyphae and mycelial-cell organelles. The site of accumulation of the penicillin within the microorganism is also a micromorphological feature.

Generally, high-yielding mutants produce weaker conidiophores with fewer, smoother spores arranged in short chains. The color of the basal mycelia is generally yellowish white or gray-greenish white with no diffusible pigments. The colonies are mostly compact with deep radial fissures. The radial growth become "clumpy," which may be due to rapid branching of the hyphae. The colonies of the high-yielding mutants are also smaller in diameter than the low-yielding mutants [2,6,23,24].

In submerged cultures, high-yielding mutants often tend to form compact pellets. The low-yielding strains grow mostly in the form of large pellets or long filaments. The rate of growth is faster in case of the low-yielding mutants. With fermentation, the pellets became lysed in the center to form a hollow core or the mycelia become completely segmented to form short filaments. This segmentation is faster with low-yielding mutants compared to high-yielding ones [4,23-29].

In a comparison between the surface of the mycelia of high- and low-yielding mutants of *P. chrysogenum* during penicillin biosynthesis, the surface of the high-yielding mutant was smooth after 24 h of fermentation. After 48 h, an amorphous material was observed. In the later stages of fermentation, the surface of the mycelium became covered with amorphous and tubular structures. In the low-yielding mutant, the surface of the mycelium remained smooth throughout the course of biosynthesis. Also, the cells of the high-yielding mutant increased in size with the course of biosynthesis, while the cells of the low-yielding mutant remained unchanged [24]. Kurylowicz et al. [24] also observed differences in the ultrastructure of a high-yielding and a low-yielding penicillin-producing mutant from *P. chrysogenum*. These differences can be summarized as follows:

1. In the high-yielding mutant, thickening of the cell wall was accompanied by intensification of biosynthesis from 70 nm after 24 h incubation to 500 nm after 120 h. In the low-yielding strain, thickening of the cell wall was not changed and remained at only about 10 nm.
2. Presence of giant vacuoles (20×8 μm) in the high-yielding mutant, which enlarge in the course of aging of the cell, while these vacuoles are smaller in size (8×1 μm) and fewer in number in the low-yielding strain. They reported that these vacuoles may play an important role in the accumulation of metabolites from the cytoplasm.
3. Presence of equal number of lipid bodies in the cytoplasm of both strains during the initial stages of penicillin biosynthesis. The number of these lipid bodies remained unchanged in the low-yielding mutant, while it was decreased by about 50 times in the high-yielding mutant.
4. Presence of structures similar to incomplete virus in the mycelial cells of the high-yielding mutant.
5. After 120 h incubation, some cells of the high-yielding mutant underwent lysis and were totally lacking in cytoplasm. At this stage of biosynthesis, many more cells from the low-yielding mutant underwent lysis.

Recently, El-Sayed [25,26] reported differences in the morphology of a high- and a low-yielding strain of *P. chrysogenum* immobilized in calcium alginate beads and used in penicillin fermentation as compared with the freely suspended mycelia. The immobilization of the conidia in alginate beads and their development in the immobilized form to produce an active biomass was accompanied by morphological changes, whereby an increase of the penicillin production activity was observed [25,30,31]. These changes depend on the immobilization and fermentation conditions of each *P. chrysogenum* strain (Figs. 2-10). Some of the factors improving the penicillin production activity in the application of immobilized *P. chrysogenum* mycelia are (1) increasing the biological active surface by formation of micropellets inside the alginate beads, (2) increasing the branching frequency of the growing mycelia in the immobilized form, (3) preventing the mycelia from sticking together by separation in alginate, (4) improving the mass transfer rate, and (5) slowing the growth rate of the immobilized mycelia during the penicillin production phase.

On the other hand, in the cephalosporin fermentation, four morphological forms of *C. acremonium* are recognized: hyphae, arthrospores, conidia, and germinated conidia [32]. After evaluation of these types to synthesize cephalosporins, it was found that arthrospores were 40% more efficient for antibiotic synthesis than all other types. The arthrospore form has been implicated as the cell type of *C. acremonium* most widely used on commercial production [4,32].

2. Physiological Studies

Many studies continue to show disagreement about the physiological factors affecting the antibiotic yield whether at a laboratory scale in shake flasks, or in small fermentors, or even on an industrial production scale. These factors include culture age, growth state and growth rate, aeration, temperature, pH, foam formation, and the adaptation of cell metabolism to the concentration of some key nutrient compounds such as sugars, nitrogen sources, sulfur, phosphorus, or iron.

Calam [33] reported that the biochemistry and morphology of a penicillin-producing culture is mainly established during the germination process. The germinated mycelia became active within 8-15 h and remain active for as long as 3 days. Consequently, there are many age models describing the activity of *P. chrysogenum* during penicillin fermentation either in feed-batch or chemostat fermentation [27,34-38]. These age models are mainly based on the growth rate and the rate-limiting biochemical reactions.

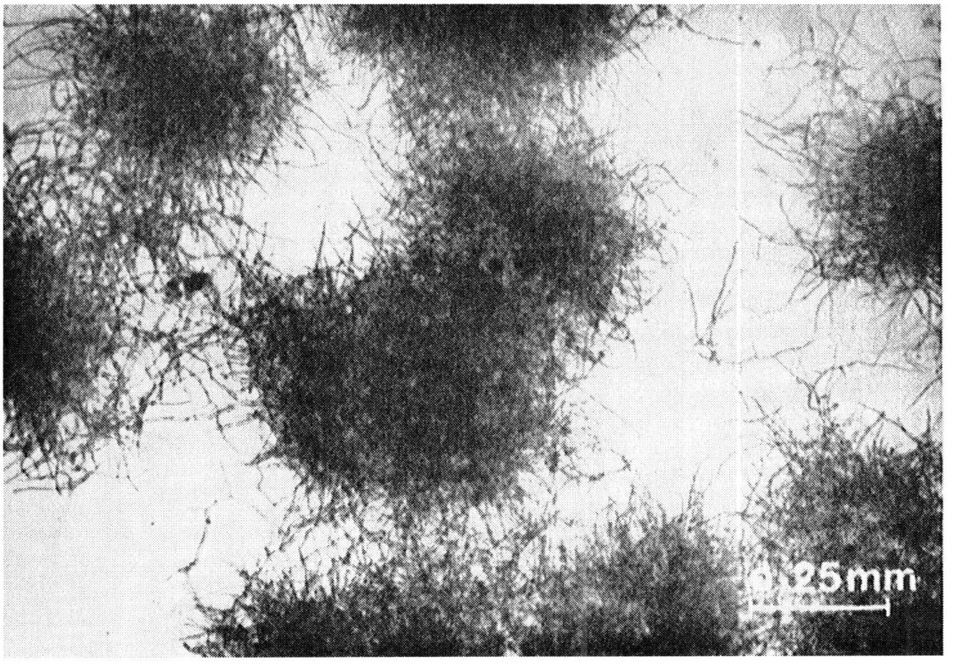

FIGURE 2. Light micrograph of free mycelia of *P. chrysogenum* ATCC 12690 after 96 h incubation in chemically defined growth medium (bar = 0.25 mm). (From Ref. 26.)

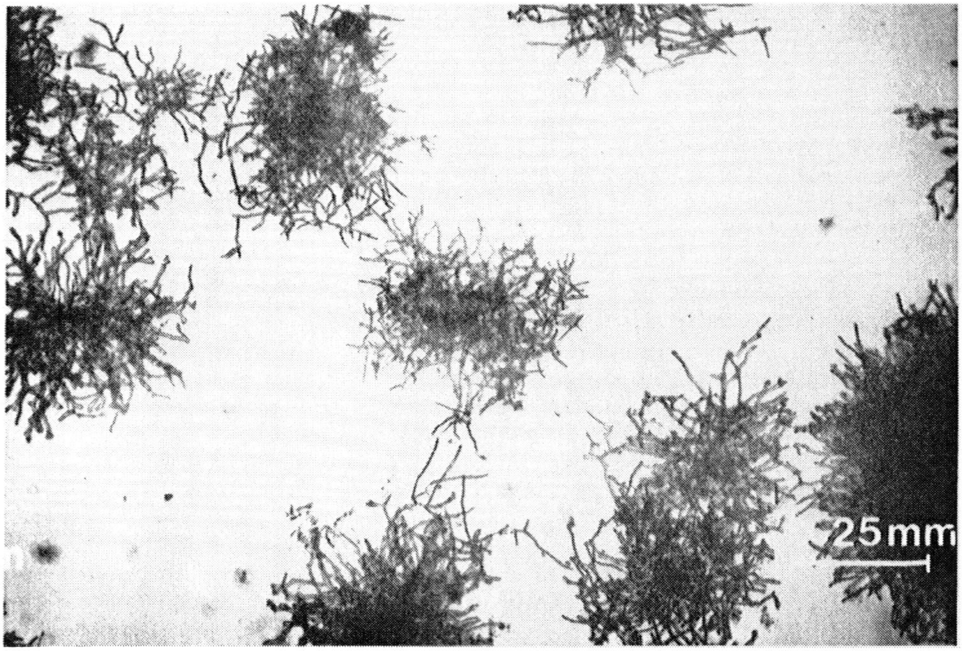

FIGURE 3. Light micrograph of free mycelia of *P. chrysogenum* S_1 after 96 h incubation in complex growth medium (bar = 0.25 mm). (From Ref. 26.)

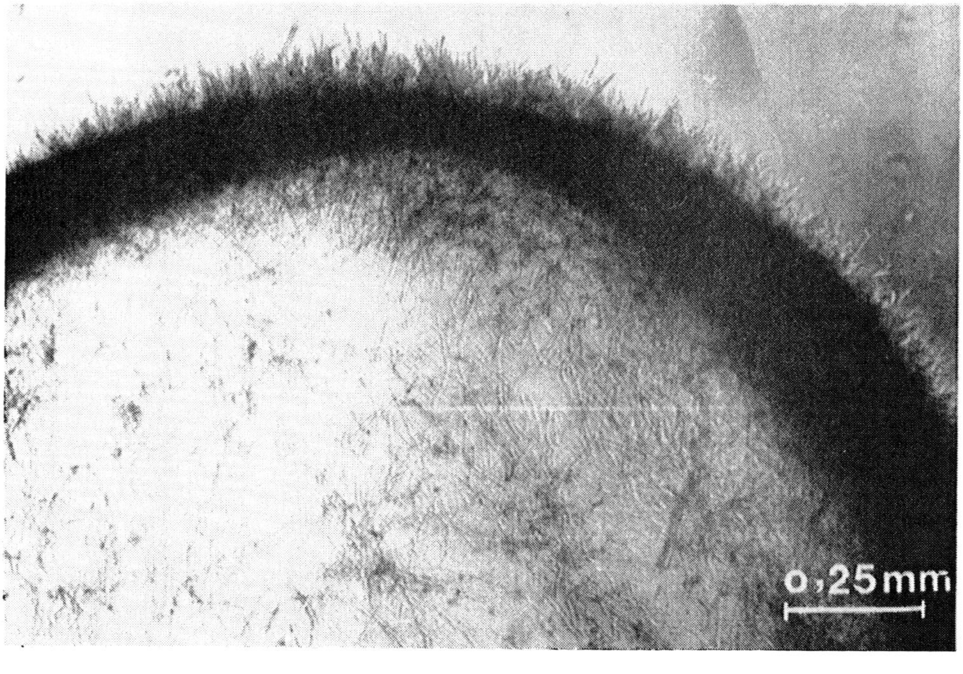

FIGURE 4. Light micrograph of immobilized *P. chrysogenum* ATCC 12690 after 96 h incubation in complex growth medium (bar = 0.25 mm). (From Ref. 26.)

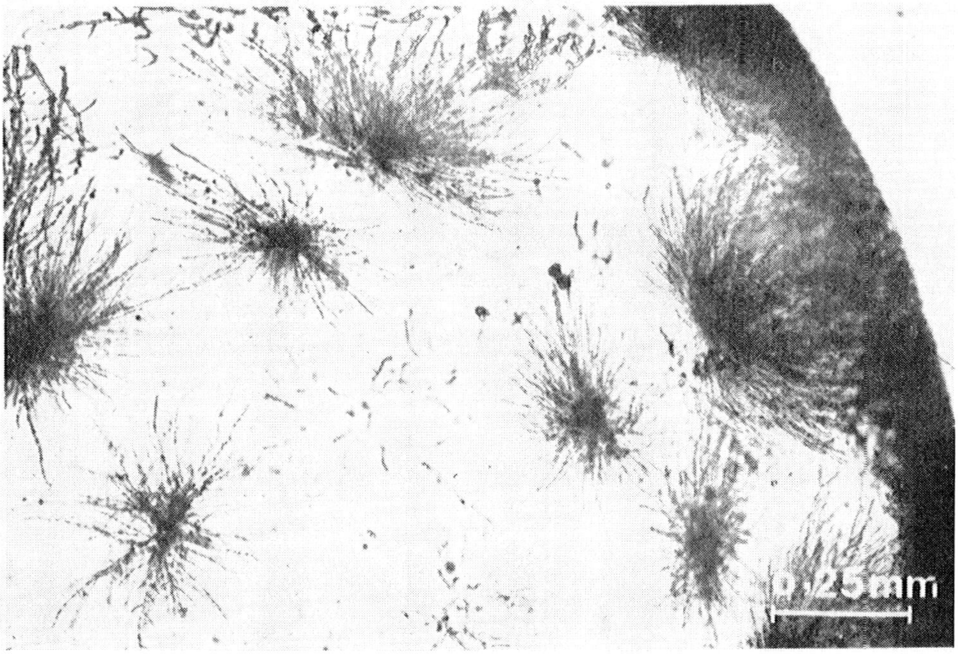

FIGURE 5. Light micrograph of immobilized mycelia of *P. chrysogenum* S_1 after 96 h incubation in complex growth medium (bar = 0.25 mm). (From Ref. 26.)

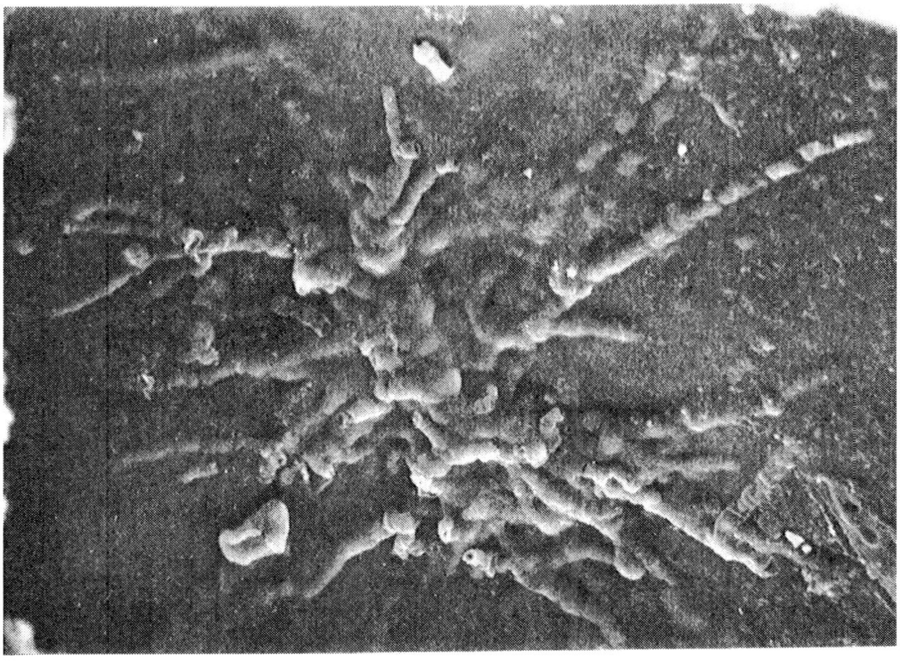

FIGURE 6. Scanning electron micrograph of a calcium alginate bead (central layers) with immobilized mycelia of *P. chrysogenum* S_1 after 96 h incubation in complex growth medium (×540). (From Ref. 26.)

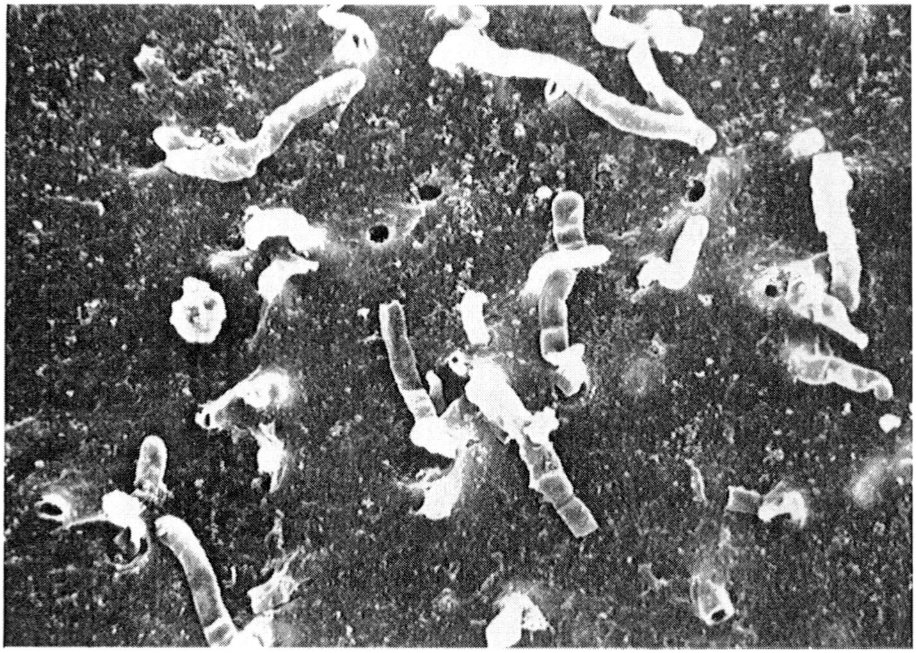

FIGURE 7. Scanning electron micrograph of a calcium alginate bead (central layers) with immobilized mycelia of *P. chrysogenum* S_1 after 96 h incubation in complex growth medium (×770). (From Ref. 26.)

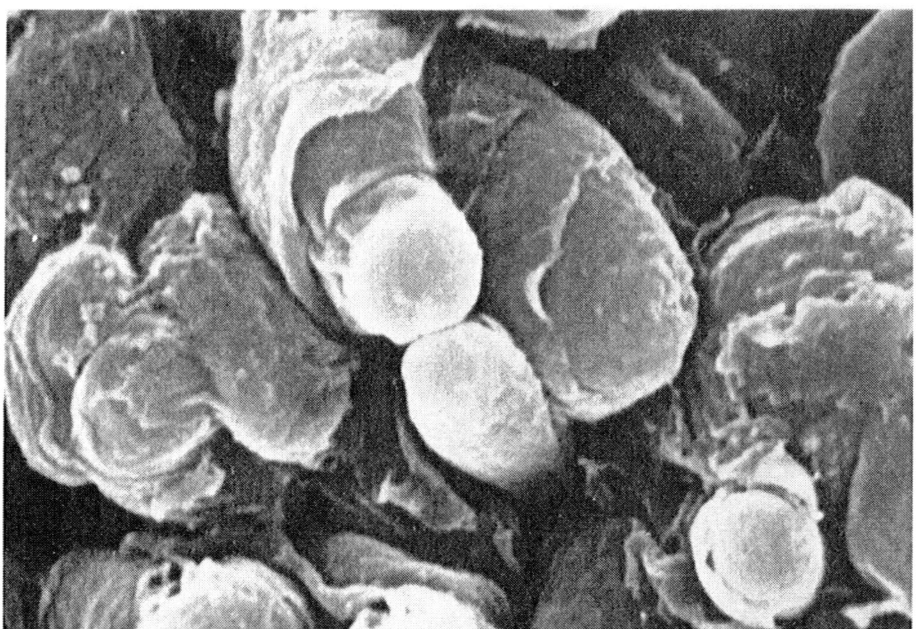

FIGURE 8. Scanning electron micrograph of a calcium alginate bead (outer surface) with immobilized mycelia of *P. chrysogenum* S_1 after 96 h incubation in complex growth medium (×7400). (From Ref. 26.)

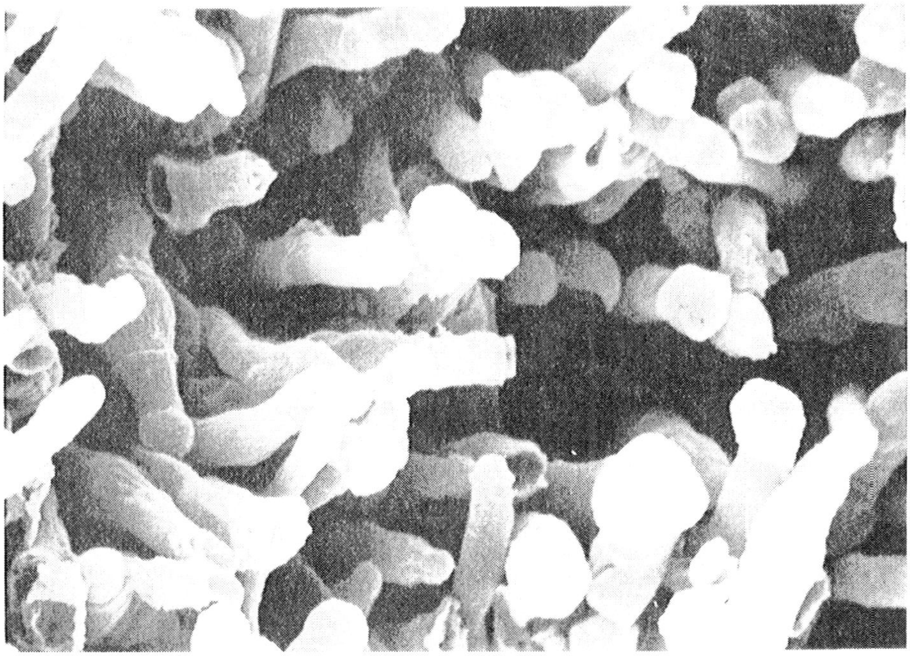

FIGURE 9. Scanning electron micrograph of a calcium alginate bead (outer surface) with immobilized mycelia of *P. chrysogenum* S_1 after 12 days incubation (4 days in growth medium and 8 days in production medium) in complex media (×2300). (From Ref. 26.)

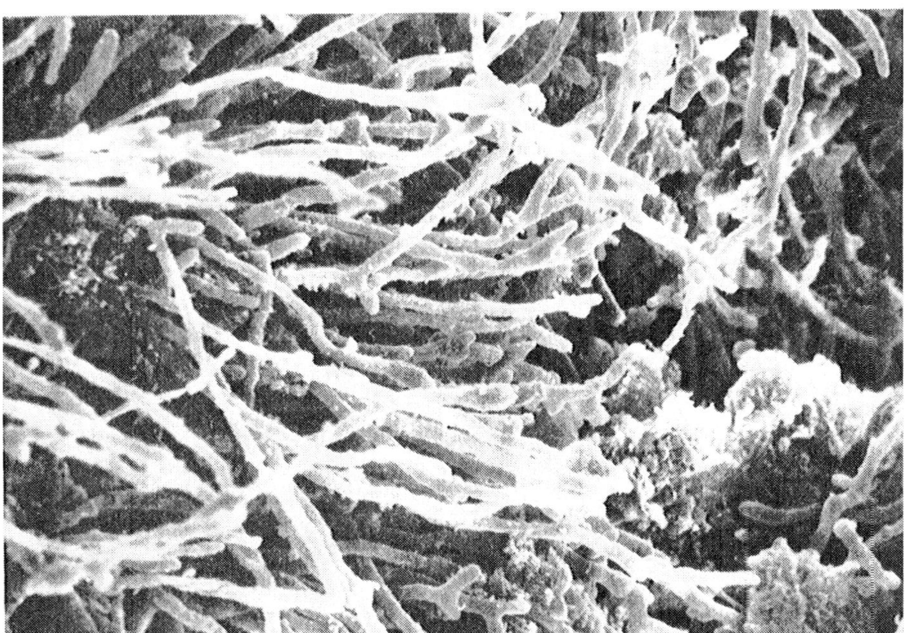

FIGURE 10. Scanning electron micrograph of free mycelia of *P. chrysogenum* S_1 after 96 h incubation in complex growth medium (×760). (From Ref. 26.)

Several workers [38-40] reported that a certain level of specific growth rate of the organism must be maintained in order to exploit the full capacity of biosynthetic activity of *P. chrysogenum*. Also, they determined the level of specific growth rate that has to be maintained during the production phase to be 0.015 h^{-1}. Furthermore, it was reported that the specific uptake rates determined under optimal conditions were 0.33 mM hexose/g cell/h, 1.6 mM oxygen/g cell/h, 2 mg NH_3-nitrogen/g cell/h, 2.8 mg SO_4-sulfur/g cell/h, and 1.8 mg phenylacetic acid/g cell/h [38].

Queener and Swartz [2] reported that the penicillin-synthesizing capacity of each fungal cell requires critical concentration of all nutrients essential for production of substrates and enzymes committed to penicillin synthesis. Fulfillment of this condition is largely a function of the correct medium composition and feed rates. In addition, it is necessary to ensure a metabolic state which allows production or assimilation of these substrates, e.g., 2-aminoadipic acid, cysteine, valine, and phenylacetic acid and synthesis of appropriate enzymes. This is largely a function of correct control of the oxidation of the carbohydrate constituent of the medium.

In the initial stages of fermentation, glucose is added as the main carbon source in a concentration higher than that required in order to achieve the desired growth rate. The presence of excess glucose at this stage causes a decrease in the pH and excessive biomass production. Too little glucose results in the use of organic nitrogen sources as a carbon source, which leads to an increase in the pH as well as delay in reaching the desired cell mass. In modern fermentations, glucose is fed throughout the fermentation on the basis of a pH value. Near the end of logarithmic growth phase where sufficient intermediate metabolites for penicillin biosynthesis are accumulated, the penicillin productivity of the fungal cell reaches a higher value. In this stage, the glucose must be critically controlled. Excess glucose leads to an increase in cell mass at the expense of the production of the intermediate metabolites for penicillin biosynthesis. Moreover, since oxygen demand increases and oxygen transfer decreases with increasing mycelium concentration, irreversible damage to the productivity of the mycelium can result [41-43]. On the other hand, in the absence of growth, glucose may directly inhibit the activity and/or synthesis of enzymes involved in penicillin biosynthesis [44]. With too little glucose, the growth rate is also decreased. When the growth rate is drastically decreased or stopped by interrupting the glucose feed, the mold autolyzes and pH increases; this leads not only to destruction of the penicillin productivity of the microorganism, but also to hydrolysis of the available penicillin in the fermentation broth. Autolysis can be prevented by supplying the fermentation medium with a maintenance ration of glucose (0.022 g glucose/g mycelial dry weight/h). If the glucose feed rate in chemostat cultures is decreased to the maintenance ration, mycelial differentiation can occur. Differentiation can involve increased hyphal vacuolation, a decreased degree of oxidation of glucose, breakdown and resynthesis of nucleic acids, and conidiation. The rates at which these changes occurred were inversely related to the growth rate prior to stopping the growth [40].

Many organic compounds such as corn-steep liquor, cottonseed flour (Pharmamedia), wheat bran, rice bran, and defated rice bran can be used as nitrogen source [2]. Lur'e et al. [45] demonstrated the importance of a continued supply of ammonia nitrogen. They found that it was essential to maintain a concentration of ammonia (by feeding ammonium sulfate) between 250 and 340 μg/ml for continued synthesis of penicillin. A lack of available ammonia nitrogen caused lysis of the mycelium and a drastic decrease in respiration relative to controls with a suitable nitrogen feed [2,45].

3. Genetic Studies

Mutation: In the early days of the penicillin-producing industry, yield improvement was carried out through spontaneous mutation or through mutation induced by physical or chemical mutagens [20]. Up to the mid-1960s, treatment with nitrogen mustard [methyl-bis-(β-chlorethyl)amine] and ultraviolet radiation were the most important methods for induction of mutations to improve penicillin yield in *P. chrysogenum* [19,46]. Subsequently, *N*-methyl-*N*'-nitro-*N*-nitrosoguanidine (NTG) has also been used and has been considered the preferred method by some workers for two reasons. First, if the NTG is used under the appropriate conditions, it produces an extremely high mutagenic but low killing effect relative to other mutagens. The second reason is that NTG can induce new mutations different from those induced by years of intensive use of ultraviolet irradiation and nitrogen mustard. Other mutagenic agents used in the development of the modern production strains of *P. chrysogenum* are listed in Table 4 [1,2].

Genetics related to parasexual cycle: Pontecorvo and Sermonti [47] discussed the parasexual cycle in *P. chrysogenum*. This was the stimulus for attempts to increase penicillin yield by breeding [1,19,48-52].

Breeding of *P. chrysogenum* strains begins with the selection of two parental strains for crossing [53]. The selected parents are then treated with a mutagen and auxotrophs which require different nutrients for growth, or resistant mutants which are tolerant to normally toxic concentration of certain chemicals are selected. The selected strains are then grown together on the same complete agar medium. This allows anastomosing of mycelial filaments from the two different strains with subsequent mixing of cytoplasm and transfer of nuclei. The produced filaments are called heterokaryons. The mixed colonies are then subcultured on a medium on which only heterokaryotic filaments can grow (simple minimal medium) and the heterokaryons

TABLE 4. List of Mutagens Used in the Development of High Penicillin Yield Mutants of *Penicillium chrysogenum*

Nitrogen mustard [methyl-bis-(β-chloroethyl)amine]
Ultraviolet light irradiation (275 and 253 nm)
N-Methyl-*N*'-nitro-*N*-nitrosoguanidine
Nitrous acid
Diepoxybutane
Ethylmethanesulfonate
X-Rays
γ-Rays
Ethylenimine

Source: Ref. 2.

are selected. The harvest of spores from mycelia containing heterokaryotic filaments may contain some diploid spores. The diploid spores are selected by plating on a medium which support only the growth of the diploid spores. Stable diploid strains are tested for antibiotic production in shake flasks, subjected to natural selection, mutation, and selection [1,2,4].

More details about the genetics of fungi can be found elsewhere in this book.

III. BIOSYNTHESIS

A. Theories of Biosynthesis

The biosynthesis of penicillins and cephalosporins is mainly related to the biosynthesis of the amino acids valine, cysteine, 2-aminoadipic acid, lysine, and glutamic acid by filamentous fungi. Also, sulfur metabolism related to the biosynthesis of cysteine must be considered (Fig. 11).

There are many studies, theories, and reviews that explain the aspects of penicillin and cephalosporin biosynthesis [3,4,10,17,54-64]:

1. The formation of the tripeptide 5-(L-2-aminoadipyl)-L-cysteinyl-D-valine
2. The cyclization of this tripeptide to form the penam and the 3-cephem bicyclic ring systems
3. The configuration of the amino acids involved at different steps of the biosynthetic pathway
4. The point of divergence from the tripeptide toward the synthesis of penam as well as 3-cephem antibiotics
5. The incorporation of different acyl radicals as side chains as well as the formation of 6-aminopenicillanic acid in *Penicillium*-type penicillins
6. The formation of penicillin-N instead of isopenicillin N through microorganisms of the *Cephalosporium* type

For long time, the Arnstein tripeptide theory (ATPT) for the biosynthesis of penicillins of the *Penicillium* type and the Oxford tripeptide theory (OTPT) for biosynthesis of penicillins of the *Cephalosporium* type were widely accepted. These theories explained the tripeptide formation but did not explain many other points in the biosynthesis pathways.

Isolation and characterization of many enzymes and intermediates involved in the different steps of the biosynthetic pathways and development of the ability to produce penicillin from both types by cell-free extracts in many laboratories has helped answer many questions regarding penicillin and cephalosporin biosynthesis [6,17,54,62,65-72].

Generally, the different steps for the biosynthesis of penicillins and cephalosporins can be summarized as follows:

1. Formation of the amino acids L-2-aminoadipic acid, L-cysteine, and L-valine through primary metabolism (Fig. 11).
2. Formation of the dipeptide L-2-aminoadipyl-L-cysteine.
3. Activation of the L-valine accompanied by change in configuration from L to D, as in the case of the biosynthesis of gramicidin-S where L-phenylalanine is changed to the D configuration.
4. Condensation of the active D-valine with the dipeptide to form the tripeptide 5-(L-2-aminoadipyl)-L-cysteinyl-D-valine (ACV) with the configuration LLD.

5. Cyclization of the tripeptide to form the bicyclic system through cyclase enzyme and formation of isopenicillin N.
6. In this latter step, the pathway bifurcates to produce both types of penicillins through the enzymes produced by different types of microorganisms.

1. Penicillium Type

In this group of microorganisms, epimerase or expandase enzymes cannot be detected. These microorganisms can produce different acyl-CoA transferases that are responsible for the replacement of L-2-aminoadipic acid by different activated side chain precursors to form different natural penicillins or by hydrogen ion to form 6-aminopenicillanic acid (6-APA). The free L-2-aminoadipic acid can be further utilized to produce another molecule of penicillin and so on (recycling).

2. Cephalosporium Type

This group of microorganisms can produce four other enzymes: (1) epimerase, which catalyzes the configurational change of the L-2-aminoadipic

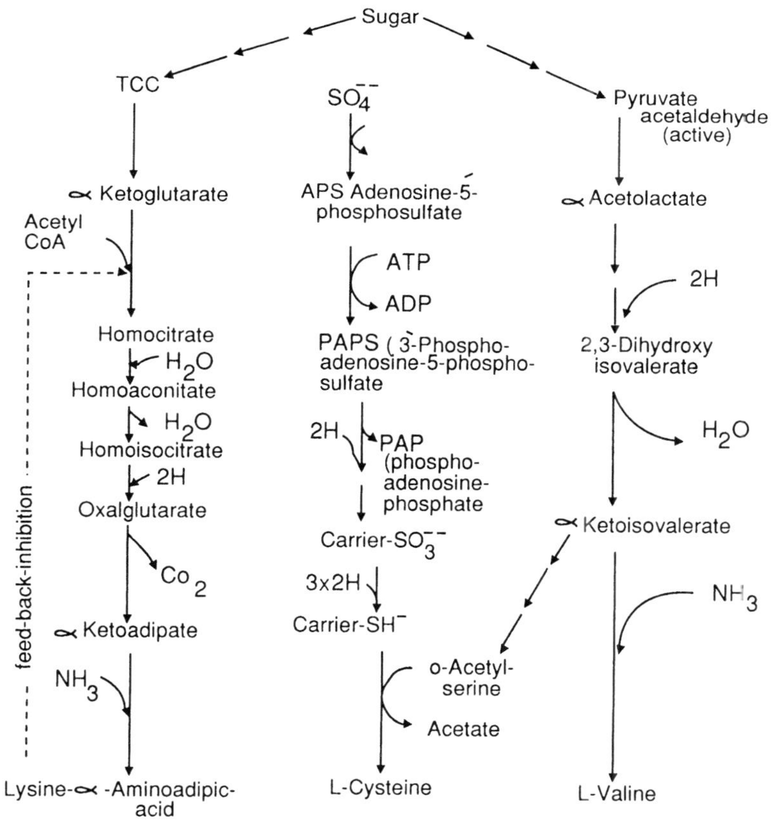

FIGURE 11. Possible primary metabolism pathways affecting the biosynthesis of penicillins and cephalosporins by fungi.

acid side chain of isopenicillin N to the D form, thereby producing penicillin N; (2) ring expandase, which catalyzes the expansion of the five-membered thiazolidine ring system of penicillin N to the six-membered dihydrothiazine ring system to produce deacetoxycephalosporin C (DOCSC); (3) oxidase, which is responsible for oxidation of the C_3-methyl group to produce the deacetylcephalosporin C (DCSC); (4) deacetylcephalosporin C-*o*-acetyltransferase which directs ester formation at C_3 and thus formation of cephalosporin C (CSC) from DCSC.

The different steps of these biosynthetic pathways can be represented as in Fig. 12.

B. Regulation of Biosynthesis

1. Primary Metabolism

The β-lactam antibiotics are secondary metabolites derived principally from compounds of primary metabolism. As previously discussed, isopenicillin N, penicillin N, and cephalosporin C can be formally divided into three amino acid residues: 2-aminoadipic acid, cysteine, and valine. Evidence that the β-lactam antibiotics derived from these amino acids come from experiments in which isotopic labels from precursors were incorporated into the antibiotics [4,17,64,67,68]. Further evidence comes from successful synthesis of penicillins and cephalosporins by cell-free extracts of these antibiotic-producing microorganisms from the tripeptide 5-[L-2-aminoadipyl)-L-cysteinyl]-D-valine as substrate, or by cell organelles (e.g., fungal vesicles from *P. chrysogenum*) with the same three amino acids as substrates for synthesis of penicillin [73].

2-Aminoadipic acid: 2-Aminoadipic acid is a common intermediate for the synthesis of lysine in higher fungi. Prokaryotes synthesize lysine by decarboxylation of another intermediate, 2,6-diaminopimelic acid [4].

Demain [74] observed that lysine inhibits the biosynthesis of penicillin by inhibition of homocitrate synthetase, an enzyme involved in the formation of 2-aminoadipic acid in *P. chrysogenum*. This inhibition could be reversed by the addition of 2-aminoadipic acid to the fermentation medium [75]. Thus, lysine inhibits penicillin biosynthesis by feedback regulation of the lysine pathway (Fig. 11), thereby limiting the endogenous supply of 2-aminoadipic acid [6,76]. On the other hand, D-2-aminoadipic acid represents the side chain of the *Cephalosporium*-type antibiotics. It has been shown that the L isomer of 2-aminoadipic acid is more efficiently incorporated into the side chain of these antibiotics than the corresponding D isomer [4,10]. It has also been demonstrated that lysine does not inhibit, and exogenous 2-aminoadipic acid does not stimulate, the antibiotic synthesis of *Cephalosporium*-type antibiotics [77-79].

Cysteine and sulfur metabolism: There are many papers describing the incorporation of labeled cysteine into benzyl penicillin [80-83] and in the C_3-cephem ring system of cephalosporins [79,84,85]. It was also found that the L form of cysteine was incorporated in both types of antibiotics [86-89].

In fungi, sulfur of cysteine can be derived from reduction of sulfate and can be further transferred to methionine. A reverse transsulfuration occurs from methionine to cysteine and is mediated by α-cystathionase and β-cystathionase. Methionine can repress reverse transsulfuration by inhibiting both α- and β-cystathionase [4].

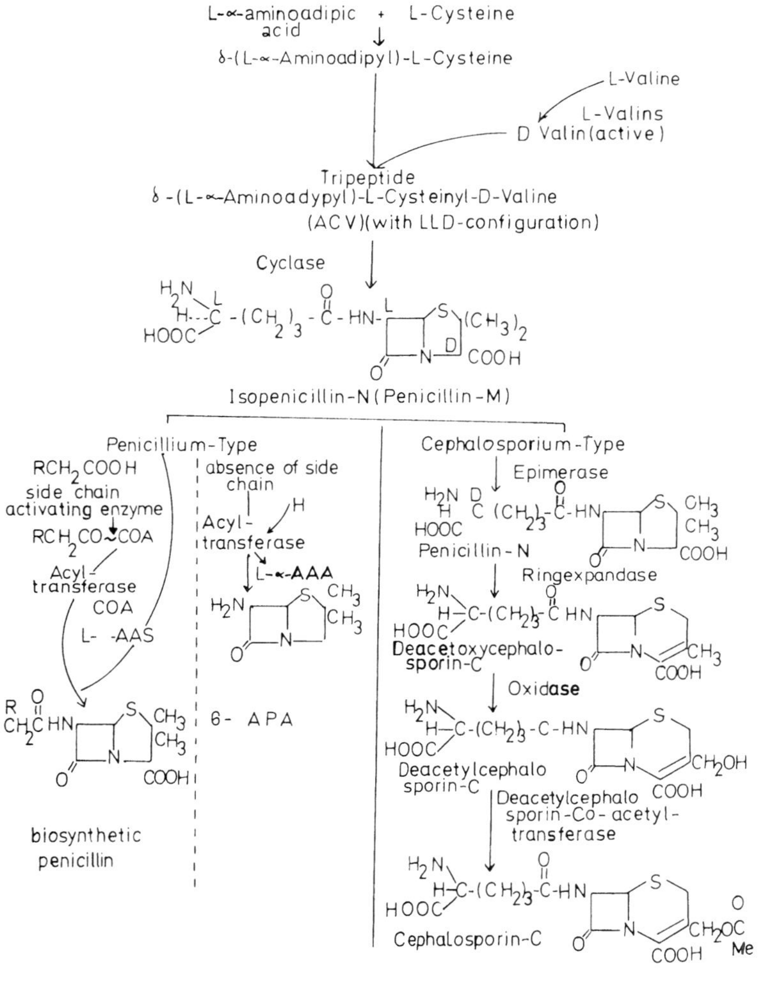

FIGURE 12. Biosynthesis of the two penicillin types by fungi.

The β-lactam antibiotics have a single sulfur atom derived from cysteine. In *Penicillium*, the sulfur atom is derived mainly via the sulfate reduction pathway from sulfate, but can be also derived via reverse transsulfuration from methionine [90,91]. In *Cephalosporium*, the sulfur for antibiotic synthesis is derived mainly from methionine through transsulfuration [4,92].

High-penicillin-producing strains are more efficient for sulfate anabolism than low producers [90,93]. Such strains have an increased ability to concentrate intermediates of the sulfate reduction pathway and have apparently relaxed control mechanisms regulating inorganic sulfate uptake and transfer to cysteine. Exogenous cysteine in fermentation medium containing inorganic sulfate does not stimulate penicillin production by *P. chrysogenum* [94]. A stimulatory effect by cysteine for penicillin productivity could only be obtained with starved mycelial suspensions [95]. Furthermore, although labeled methionine could be incorporated into penicillin, methionine does not stimulate *Penicillium*-type antibiotic biosynthesis [4,6,17,91].

In the production of *Cephalosporium*-type antibiotics, the role of sulfur-containing compounds is somewhat more complicated and may vary significantly from strain to strain. Sodium sulfate was found to inhibit cephalosporin production by some fungi and not to stimulate it significantly in any of the tested strains [4]. Methionine has multiple effects on the *Cephalosporium* fermentation. Synthesis of antibiotics by several *Cephalosporium*-type fungi is also enhanced by methionine, particularly the D isomer [4]. In studies with 42 cephalosporin-producing fungi, Kitano [96] found that methionine stimulated the production of cephalosporins by some strains of *Cephalosporium* and *Emericellopsis* but not by other strains of the two genera or in certain strains of *Paecilomyces*, *Arachnomyces*, *Anixiopsis*, and *Spiroidium*. Moreover, methionine inhibited the production of cephalosporins by *Paecilomyces persicinus* [97]. For the same strains, *S*-methylcysteine gave the same results as methionine [96]. Cysteine also stimulated the production of cephalosporins by some strains, inhibited it in others, and in yet others had no effect. There was no correspondence between the strains in which production was affected by cysteine or by methionine, respectively [17,96,98].

Valine: Valine is incorporated in the D configuration in *Penicillium*-type antibiotics as well as in penicillin N. It forms the α,β-unsaturated and substituted with acetoxy group at the γ-carbon fragment of the other cephalosporins [99-101].

In low-yielding strains of *P. chrysogenum* and *C. acremonium*, D-valine was found to be a poorer precursor than L-valine. With high-yielding strains of *P. chrysogenum*, both isomers were incorporated into penicillin to a similar extent [102]. Goulden and Chattaway [103] compared the end product control of acetohydroxy acid synthetase by valine in low- and high-penicillin-yielding strains of *P. chrysogenum*. They found that both strains were sensitive to valine as a major feedback inhibitor, but the high-penicillin-yielding strain was less sensitive [4,17,103].

2. Regulation of Enzymes of Secondary Metabolism

At the end of trophophase, sufficient primary metabolites are accumulated in the cells to be used in the formation of secondary products during the idiophase. The factor or factors that trigger secondary product formation at the end of the trophophase and during it are unknown. The triggering could be associated with catabolite regulation resulting from rapid sugar utilization, or may arise from induction of enzymes responsible for the formation of secondary products, in response to accumulation of the corresponding substrates in the primary metabolism during trophophase [18,104, 105]. During idiophase, some amino acids are also produced, but these amino acids are not utilized in the formation of cell protein and are mainly used in the synthesis of secondary metabolites [18]. Whatever the mechanism, there is no doubt that at the end of trophophase, marked changes occur in the enzymatic composition of the cells, and enzymes specifically related to formation of secondary products suddenly appear [17,18,55,105-111].

In penicillin and cephalosporin biosynthesis, the tripeptide 5-[L-(2-aminoadipyl)-L-cysteinyl]-D-valine (LLD-ACV) is first cyclized to form isopenicillin N [64,66,70-72,112-114]. Hollander et al. [114] isolated, purified, and characterized the isopenicillin N synthetase (cyclase) from *C. acremonium*. This enzyme has a molecular weight of 40,000-42,000. It was purified in 10% yield by a combination of protamine sulfate and ammonium sulfate precipitations, gel filtration, and ion exchange HPLC.

The purified enzyme could be stabilized with sucrose and stored at -20°C for several weeks without loss of activity. The cyclase enzyme converts LLD-ACV to isopenicillin N by oxidative cyclization in the presence of O_2, F^{2+}, and ascorbic acid. The synthesis of cyclase by *S. clavuligerus* was not affected by growth in high phosphate concentrations (100 mM phosphate), but its activity was inhibited by phosphate concentrations higher than 50 mM [111]. Also, cyclase formation was not affected by glucose catabolite repression in *C. acremonium* fermentation [62] or in resting cells of *C. acremonium* [115] or even in *S. clavuligerus* fermentation [67,68]. As the intracellular intermediate isopenicillin N forms, the biosynthetic pathway branches into either *Penicillium*-type or *Cephalosporium*-type antibiotics. The further utilization of isopenicillin N by *Penicillium*-type fungi to produce secondary metabolites of this type is under the control of: (1) the presence of sufficient concentration(s) of side chain precursor(s) [nonpolar acetic acid derivative(s)], side chain-activating enzyme, and energy source [116], (2) acyltransferase enzyme, which attaches the activated side chain precursor (Co-A ester) to the penicillin nucleus, 6-aminopenicillanic acid, to replace L-aminoadipic acid [116], (3) penicillin acylase, which is responsible for the conversion of isopenicillin N to 6-aminopenicillanic acid and L-aminoadipic acid, if no precursor is available [22,55,105,117]. The levels of these enzymes are said to be increased in high-penicillin-yielding mutants [17,55].

The cephalosporin branch involves still more enzymes and starts with the conversion of isopenicillin N to penicillin N by isopenicillin N epimerase [64] followed by subsequent reactions to form different natural cephalosporins. Jensen et al. [69] purified and characterized isopenicillin N epimerase activity from *S. clavuligerus*. As seen with other enzymes responsible for biosynthesis of cephalosporins, the epimerase enzyme in both prokaryotes and eukaryotes is the same and differs only in its greater stability in cell extracts from prokaryotes than those from eukaryotes [69,70,118]. Epimerase contains tightly associated, but not covalently bound, pyridoxal phosphate (PP). Purification of epimerase in the absence of PP could cause a gradual loss of PP and result in an irreversible loss of enzymatic activity. Moreover, addition of PP during purification of epimerase increases the resistance of the enzyme toward inhibitory substances such as hydroxylamine and iproniazid [69]. The production of epimerase is neither inhibited nor repressed by phosphate [111] or even glucose [62].

The second enzyme of this chain is the deacetoxycephalosporin C synthetase (expandase) which is responsible for the oxidative expansion of penicillin N to deacetoxycephalosporin C (DOCPC). Expandase is an α-ketoglutarate (α-KG)-linked dioxygenase requiring F^{2+}, a reducing agent (ascorbic acid), and α-KG as cofactors [65,71,119,120]. The activity of expandase is repressed by prior growth in high phosphate concentrations and inhibited by phosphate [111]. Also, expandase is subject to glucose catabolite repression which is responsible for suppressive effect of glucose on cephalosporin production by *C. acremonium* [62,115].

The next key enzyme in the production of cephalosporin C (CPC) is the deacetoxycephalosporin C hydroxylase which converts DOCPC to deacetylcephalosporin C (DCPC) in the presence of molecular oxygen. This enzyme is also a dioxygenase and requires α-KG, ascorbate, and F^{2+} for activity [17]. Mutants from *C. acremonium* [112,121-123] or from *S. lipmanni* and *S. clavuligerus* [124], which lack the ability to produce this enzyme, produce DOCPC in their fermentation medium and cannot produce DCPC or CPC [17,112].

The last enzyme in this chain is acetyl-CoA-deacetylcephalosporin C acetyltransferase, which converts DCPC to CPC. The failure of certain mutants to produce this enzyme leads to accumulation of DCPC in the fermentation medium [125]. Another cause of accumulation of DCPC in the fermentation medium instead of CPC, especially at the later stages of fermentation, may be cephalosporin C acetylhydrolase, which hydrolyzes CPC to DCPC and acetate [64,112,126].

IV. FERMENTATION

The earliest reports on penicillin production describe only the production of this antibiotic by surface cultures with a very low yield [94]. Thereafter, submerged culture fermentations with different parameters and control processes were developed. As fermentation technology continues to advance in parallel with increasing demand for innovative cost reduction, automatic control and optimization of fermentation processes is becoming an economic necessity [127]. Today virtually all industrial fermentation processes are carried out under automatic control of the overall process. As the complexity of the process control schemes and the multiplicity of the control variables increases, the use of a computer-based systems becomes an economic necessity for the control and monitoring of fermentation processes [28,29,36, 127-132].

The successful development of any fermentation from shake flask to large-scale production level depends to a high degree on the ability to measure and control the process variables. Although the shake flask is one of the most widely used cultivation devices, lack of detailed knowledge about all processes occurring therein makes scale-up from flask to large fermentor more difficult [109]. The use of aerated and agitated vessels facilitates the control and measurement of different variables during fermentation as well as scale-up of processes to production scale.

In this chapter, industrial penicillin and cephalosporin fermentations will be briefly discussed. As with other secondary metabolite fermentations, these fermentations are carried out in two stages. In the first stage, sporulation occurs and inoculum is developed from frozen or lyophilized stock culture, followed by inoculation of the production fermentor. The second stage occurs in the production fermentor from the moment of its inoculation to the separation of fungal mycelia from fermentation broth and the recovery of the antibiotic in a pure form. This second stage is characterized by three phases: the growth phase (trophophase), the production phase (idiophase), and the recovery phase (Fig. 13).

All steps of fermentation must be carried out under carefully controlled aseptic conditions. Contamination, especially with β-lactamase-producing microorganisms, can lead not only to destruction of the lot to be produced, but also to spoilage of other β-lactam antibiotic fermentations in the plant. Overcoming contamination problems can require weeks of elapsed time and many hours of labor.

A. Sporulation and Inoculum Development

A penicillin or cephalosporin fermentation begins with an aseptically freeze-dried spore stock culture (in the presence or absence of sterile soil) of a highly developed strain of *P. chrysogenum* or *C. acremonium*,

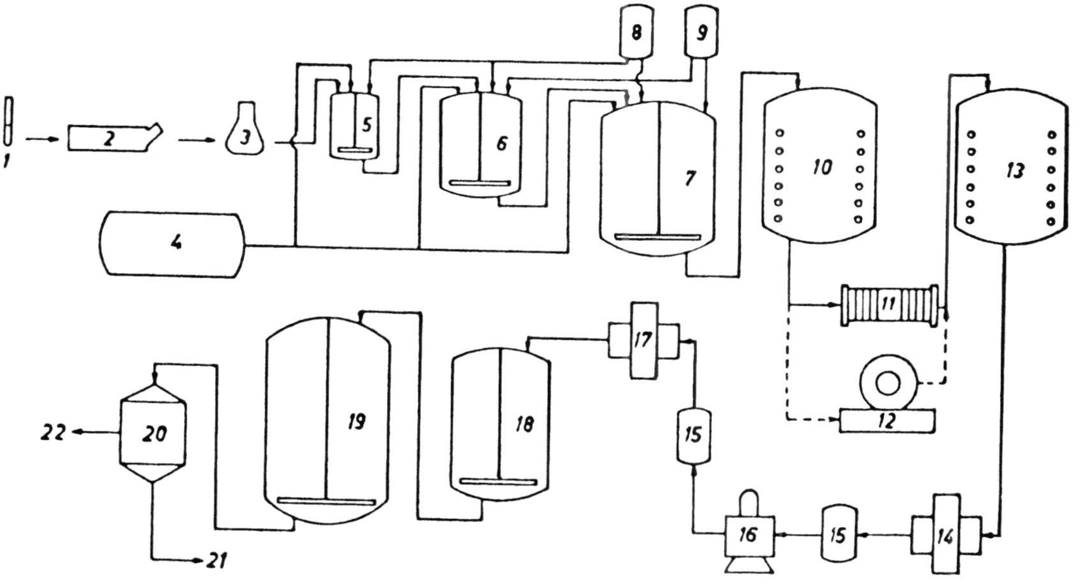

FIGURE 13. Scheme for different steps of penicillin fermentation; 1, 2, 3, 5, 6 = steps of inoculum preparation; 4 = subtrate feed tank; 7 = production fermentor; 8 = antifoam; 9 = side chain precursor; 10, 11 = cooling; 12 = rotatory vacuum filter; 13 = receiving vessel of culture filtrate; 14, 15, 16, 17 = extraction in different extractors; 18 = drying; 19 = precipitation vessel; 20 = separation vessel; 21 = recovery of organic solvent; 22 = drying and other purification steps. (From Ref. 3.)

respectively. The freeze-dried spores are suspended in sterile water or phosphate buffer (pH about 6.5), then used to inoculate the surface of sterile solid medium optimized for sporulation. Typically, such medium contains lactose, corn steep liquor (CSL), and/or peptone with inorganic salts and buffered at pH 5.5-5.8. Molasses and glycerol can be used as carbon sources. Sterile cooked rice can also be applied successfully as a sporulation medium for both *Penicillium* and *Cephalosporium*. These solid cultures (mostly slant cultures) are allowed to germinate, grow, and sporulate under rigorously controlled conditions of time, humidity, and light exposure. After sufficient spores are obtained (as indicated by the green color developed on the surface of the solid medium), the spores are washed off the solid medium with sterile water or phosphate buffer. About 1% (v/v) neutral diluent such as Tween 20 or Tween 80 is to be added to the water or the phosphate buffer followed by scraping with a pipette or with an inoculating needle or even by gentle sonication. The spore suspension is standardized (about 10^7-10^8 spores/ml) and used to inoculate liquid seed medium in shake flasks or further used to inoculate a larger solid medium surfaces in Blacke or Roux bottles. These are flat-sided bottles in which the solid medium is allowed to set with the bottle lying on one of its large sides. When the organism has grown and sporulated, the spores are harvested as before and used to inoculate shake flasks containing seed medium and then allowed to germinate [2,3,133,134].

The seed medium may contain soy meal, dextrin, soy oil, and water [24] or whey, lactose, CSL, KH_2PO_4 (anhydrous), $MgSO_4$, $CaCO_3$, peanut oil, and side chain precursor for the required penicillin [2]. Other possible ingredients are ammonium sulfate, $CaCO_3$, crude carbohydrate containing an equal amount of sucrose and starch, lard oil supplemented with 1% polyethylene glycol as antifoam [135], and other natural products such as cottonseed flour [43]. For cephalosporin fermentation, seed medium should be supplemented with about 0.5% (w/v) DL-methionine and contain sucrose instead of lactose [112].

About 0.5-1% (v/v) spore suspension is used to inoculate the seed medium and then incubated at carefully controlled conditions to allow the spores to germinate. The germination of spores is a critical step in the fermentation process [33,136]. Inoculum developed in the seed medium may be used to inoculate a secondary seed medium in order to produce sufficient inoculum for the first seed culture vessel, or may be prepared in larger quantities and divided into portions that are frozen (-70°C). Such frozen aliquots can then be used to inoculate secondary seed medium when needed [2,133]. From the first seed cultures, slants are prepared, incubated to produce spores, and then the spores are lyophilized and used as stock cultures for the next preparations.

In the first seed culture vessel, which is stirred and aerated, the temperature, pH, oxygen tension, and foam formation are controlled. The working volume of this vessel may range from 5 to 50 liters depending on the earlier stages of inoculum development.

This stage of fermentation is generally conducted batchwise in several sequential fermentations of increasing volume. The cells grown in one fermentation become the inoculum for the next fermentation of larger volume, until a volume of 40,000-200,000 liters is attained. Except in the case of the initial inoculation from a stock culture, the volume ratio of one fermentation to the next ranges from about 0.5 to 10%. Each step requires 36-48 h. During these development steps, little if any fermentation product accumulates because the cells are transferred to larger batches of media while they are still in their logarithmic growth phase [2,3,133]. The different steps of inoculum development are outlined in Fig. 13.

B. Production in Fermentors

Generally, the antibiotic yield per unit volume in a fermentor is dependent mainly on three parameters: X, the active cell concentration; Q, the specific rate of antibiotic biosynthesis; and t, the duration of the fermentation. The aim is to achieve high X and Q in a short time [2].

In penicillin and cephalosporin fermentations, most of the cell mass necessary to attain high antibiotic production is formed during the first 40 h of fermentation while the microorganism is growing at its maximum growth rate. After this, the growth rate tends to decrease and the antibiotic starts to accumulate in the fermentation medium. After the growth phase (60-72 h), an active cell mass sufficient to support a satisfactory antibiotic yield has been obtained. The growth must continue at a certain minimum rate if a high Q is to be obtained during the second production phase (idiophase) [38].

In the idiophase, the concentration of the antibiotic in the fermentation medium continues to increase until it reaches its maximum about 120-160 h after the beginning of the fermentation. The fermentation should be terminated when the value of Q declines to the point where the prolongation of the fermentation will lower the overall productivity (amount of antibiotic/volume of the fermentor/duration of the fermentation in hours). At this point, the antibiotic must be recovered before it is hydrolyzed by either microorganisms or medium constituents [2,3].

1. Growth Phase (Trophophase)

In commercial penicillin and cephalosporin fermentations, the organic nitrogen source is often CSL. Soy meal, peanut meal, yeast extract, caseine, peptone, or cottonseed flour can also be used. As carbon source, glucose, lactose, sucrose, starch, and/or vegetable oils are often used [2,3,42,43, 112,135]. Foam formation is controlled either mechanically and/or by addition of high molecular weight fatty alcohols, especially octadecanol, silicone, animal or vegetable oils [3]. In penicillin fermentation, about 1% (w/v) of side chain precursor is added at the beginning, followed by continuous or semicontinuous feeding of 10% (w/v) sodium or potassium salt solution of the precursor to give a final concentration of about 0.5-0.8% (w/v) of the entire fermentation contents [2]. Fermentation medium for cephalosporin production must include methionine (0.5%, w/v) [112]. Ammonium sulfate or other inorganic sulfate is used as a sulfur source for penicillin biosynthesis [4]. Phosphate concentration is a limiting factor for both cell growth and the activity of enzymes responsible for penicillin and cephalosporin production as mentioned before [2,111,137]. $CaCO_3$ (1%, w/v) is added to both fermentation media as a buffer. Mineral nutrition for both fungi is supplied by addition of a trace element solution in amount and concentration to increase growth rate and viability of the microorganism. Because iron can adversely affect the antibiotic productivity of the microorganisms, chemical or physical methods must be used to decrease the concentration of this element if its concentration in the medium is larger than required [104,138-140].

Dissolved oxygen concentration must be held higher than the critical oxygen concentration [41-43,112]. Both oxygen demand and viscosity increase with increase cell concentration. Higher viscosity causes the maximum oxygen transport rate of the fermentor to decline. Therefore, several companies have doubled the agitator power on existing fermentors to increase transfer rates. Some have also selected strains and culture conditions which decrease broth viscosity and thereby increase the oxygen transfer rate [2].

Optimum temperature for these fermentation ranges between 25 and 27°C. Above or below this range a decrease in the growth rate takes place preventing the biomass production from realizing the oxygen transfer capacity of a well-designed fermentor [139,141,142]. Optimum temperature can be maintained only by removal of excessive heat generated by the metabolic activity of the organism as well as the mechanical heat generated by the agitator [2].

The fermentation begins at a pH value of 5.6-5.8. Then pH increases slightly as inorganic and organic nitrogen sources are utilized with evolution of ammonia. Organic acids (lactate and acetate) available in the fermentation medium are also utilized in the first hours of fermentation and take part in the increase of the pH of the fermentation. After the enzymes responsible for sugar utilization have been induced, organic acids accumulate from sugar metabolism and this causes a decrease in pH. Near the end of the growth phase, the organic acids from sugar metabolism are utilized and this causes another increase in the pH to about 7-7.4. This latter pH value should then be maintained, almost at a constant, throughout the production phase.

The successful shift from a culture with a high Q value growing at an exponential rate to one with high Q growing at a low rate is a function of many variables [109]. Growth rate is limited mainly by restricting the amount of one or more nutrients (usually sugar) in the culture during this period [2].

2. Production Phase (Idiophase)

During this phase, the growth rate and oxygen consumption are lower in order that a satisfactory antibiotic production rate can be achieved. Growth rate is generally regulated by controlled feeding of glucose. Other factors such as oxygen or phosphate may be limited to lower growth rate [27,42,43,109,111,142]. Regulation of nitrogen metabolism is also an essential factor for both growth rate and β-lactam antibiotic production during this phase [45,106,107]. Also, temperature must be maintained at 25-27°C during the production phase.

In the penicillin fermentation, a feed-batch or continuous feeding of the fermentation with side chain precursor is an important factor for increasing the penicillin productivity of the fungus as well as for minimizing the toxic effects of the precursor (especially phenylacetate) on the fungus. The fermentation medium can also be supplemented with other ingredients which enhance penicillin biosynthesis such as 2-aminoadipic acid [2,3,22] or glutamic acid [3,108].

Optimum pH during the production phase ranges between 6.8 and 7.3 [112-135]. In recent fermentations pH values were used as a control signal in manipulating the feed rate of sugar or ammonium salts. If the pH falls too low, glucose feeding is stopped and an easily utilizable ammonium salt such as ammonium acetate is fed. The opposite occurs when pH increases. In general, changes in feeding in response to pH change are automatically controlled [25,30,31,135].

3. Recovery Phase

Recovery should begin before substantial autolysis of mycelia occurs. The onset of autolysis is marked by a decrease in dry weight and a parallel increase in the pH of the fermentation broth due to evolution of ammonia. As the result of the increase in the pH value, the available antibiotic may be destroyed.

These antibiotics are mainly extracellular and because their intracellular concentration can be neglected, the recovery begins by separation of mycelia and other solid materials from the antibiotic-containing broth. High-capacity rotary vacuum filters are used. Filter aids such as aluminum sulfate and $CaCl_2$ or tannic acid can be added prior to initial filtration, or prior to solvent extraction, to precipitate the soluble proteinaceous materials and prevent emulsion formation during the organic solvent extraction step. Addition of 1-2% formalin to the fermentation mixture during separation prevents or minimizes enzymatic degradation of the antibiotic either from the producing microorganism or from contaminants. Alternatively, the cell-free filtrate is cooled in a heat exchanger to 0-4°C, acidified to pH 2.5-3.0, and then extracted into a suitable organic solvent such as amylacetate or butyl acetate in a continuous countercurrent process. The antibiotic is back-extracted into an aqueous phase at pH 7.4 containing phosphate buffer. The process of extraction in organic solvent and back extraction into phosphate buffer can be repeated for purification. The salt of the antibiotic is then crystallized, washed, and dried [2,3].

V. FERMENTATION WITH IMMOBILIZED FUNGAL CELLS

In the last two decades, academic interest in immobilized biocatalysts has grown in parallel with their practical application [143-149]. Immobilized enzymes or bacterial cells have been investigated mainly to catalyze enzymatic single-step reactions. Some of these reactions have been scaled up to industrial production level, such as the use of immobilized penicillin acylase and glucose isomerase [145]. Practical application of immobilized fungi is still in its infancy, however. It appears that no such immobilized fungi have yet been applied even at the industrial pilot plant scale [150,151].

Few reports have been published on the utilization of immobilized fungal cells for the production of secondary metabolites like antibiotics or other products synthesized via multistep pathways [146,150-153]. Utilization of immobilized living cells of *P. chrysogenum* for total synthesis of penicillin has been attempted in some laboratories [25,26,30,31,154-156].

Morikawa et al. [154] were the first to report the production of penicillin G from glucose with immobilized mycelia of *P. chrysogenum* ATCC 12690. They entrapped mycelia harvested from a conventional batch fermentation at their maximum penicillin production rate in polyacrylamide (PAA), collagen membranes, and calcium alginate beads. Mycelia entrapped in calcium alginate beads showed the highest activity but the gel produced became too fragile to allow repeated use under fermentation conditions. The yield from calcium alginate-entrapped mycelia was only 45% that from washed mycelia within 5 h under the conditions used. PAA-entrapped mycelia showed activity (0.77 units/ml) only 15% that of washed mycelia within the same time period and under the same conditions. The kinetic half-life activity of the immobilized mycelia was 6 days compared with only 1 day for the washed suspended mycelia. Cells immobilized in collagen showed the lowest activity.

The second investigation was reported by Deo and Gaucher in 1984 [155]. They entrapped the conidia of the same strain (*P. chrysogenum* ATCC 12690) in Kappa carrageenan beads. After incubation in a growth medium, a penicillin-producing immobilized cell mass grew within the beads. This immobilized cell mass was used to produce penicillin from minimal production medium containing glucose as the sole carbon source in batch, semicontinuous, and continuous fermentations. They reported that immobilized mycelia exhibited a half-life activity for penicillin production activity ninefold greater than that of free mycelia. The continuous fermentation with immobilized mycelia was operated for 16 days with a half-life for penicillin productivity of more than 15 days.

In the third investigation, Wang et al. [156] produced penicillin by fermentation with confined fungal cells in a 200-liter tower fermentor. They began with the spores of a higher penicillin-producing strain (*P. chrysogenum* P_2) adsorbed on Celite microbeads (more than 180 μm in diameter). Compared to free mycelia in feed-batch fermentation, there was a significant improvement in the mass transfer coefficient when the cells were confined on the microbeads. Therefore, a higher production of penicillin was observed at a lower aeration rate with a reduction of the energy input required for the oxygen transfer. Furthermore, the confinement of the cells to the microcarrier allowed the intermittent harvesting of fermentation broth without reducing the cell mass in the fermentor.

Recently, we reported the production of penicillin by entrapped mycelia in calcium alginate beads [25,26,30,31,153]. The conidia of two strains with different penicillin production capabilities (*P. chrysogenum* ATCC 12690 and S_1) were entrapped in calcium alginate beads and incubated in different growth media to produce a penicillin-producing entrapped biomass. The growth phase was operated in two steps. The first step was a germination step for 48 h, after which the beads were filtered and washed with sterile water. The washed beads were resuspended in the same volume of freshly prepared growth medium supplemented with side chain precursor and reincubated for another 48 h (second step). Then the beads were filtered, washed, and transferred to the same volume of the production medium and reincubated at the same conditions.

The production of penicillin by entrapped mycelia was studied in batch, semicontinuous, and continuous fermentations using chemically defined and complex media. Fermentations were conducted in a shake flask, a bubble column, and a conical bubble bioreactor (Fig. 14). During the production phase, the pH was controlled by batchwise feeding of glucose, lactose, or ammonium acetate.

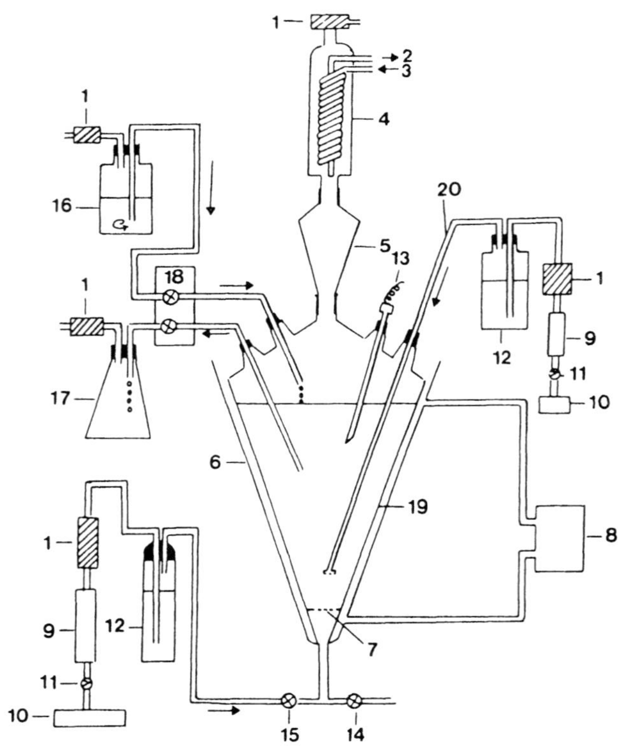

FIGURE 14. Continuous fermentation in conical bubble fermentor; 1 = air filter; 2 = cold water outlet; 3 = cold water inlet; 4 = reflux condensor; 5 = antifoam vessel; 6 = thermostating jacket; 7 = glass sieve, ϕ = 5 cm; 8 = thermostat; 9 = air flow meter; 10 = membrane pump; 11 = reducing valve; 12 = air humidifier; 13 = pH meter; 14 = medium outlet and sampling; 15 = air inlet valve; 16 = feed reservoir; 17 = product collection flask; 18 = peristaltic pump; 19 = bioreactor vessel; 20 = upper aeration tube. (From Ref. 31.)

The criteria of fermentation with *P. chrysogenum* strains are summarized as follows.

1. Low-Penicillin-Yielding Strain, ATCC 12690

This strain grows rapidly and this can limit the utilization of its immobilized form under favorable growth conditions. Such conditions include the use of complex fermentation media with high nutritional value, or fermentation at high aeration rate (as in a bubble column or a conical bubble fermentor) even when chemically defined fermentation media were used. Under these conditions, the alginate beads ruptured after about 24-48 h incubation in production medium and a very viscous fermentation mixture was obtained. Therefore, immobilized *P. chrysogenum* ATCC 12690 could only be used in shake-flask cultures using chemically defined media for batch or semicontinuous penicillin production.

As a comparison, both the immobilized and free mycelia were subjected to semicontinuous penicillin fermentation using chemically defined media in shake-flask cultures. Three production cycles, each for 6 days, were carried out. The cell growth of the free suspended mycelia was about 10% higher than that of the immobilized mycelia. The amount of penicillin produced by the immobilized form of this strain, however, was 102 units/mg DCW, about 2.5 times more than that of the free mycelia (Fig. 15) [25,30].

2. High-Penicillin-Yielding Strain, S_1

Because this *P. chrysogenum* strain grows at a lower rate, its immobilized form could be used successfully in batch, semicontinuous, and continuous fermentations using chemically defined or complex media in either shake-flask, bubble column, or conical bubble fermentors.

In shake-flask cultures using chemically defined media, the free mycelia of *P. chrysogenum* S_1 showed a total specific penicillin production of 331 units/mg CDW, compared to a value of about 1203 units of penicillin/mg immobilized CDW in three production cycles, each prolonged for 6 days (Fig. 16). Furthermore, the shaken cultures of this strain in either immobilized or free form could be used for semicontinuous penicillin production with complex media (Fig. 17). In the three production cycles, the immobilized mycelia produced about 1.1, 2.7, and 14.5 times more penicillin than the free mycelia in the first, second, and third cycles, respectively. Total specific penicillin production was 1650 units/mg CDW for immobilized mycelia and 846 units/mg CDW for free mycelia.

Even in the bubble column at a high aeration rate, *P. chrysogenum* S_1 grew slowly enough for its immobilized form to be suitable for semicontinuous penicillin production using either chemically defined or complex media. Using complex fermentation media, penicillin production by immobilized *P. chrysogenum* S_1 was 61% higher in a bubble column (Fig. 18) than in shaken cultures (Fig. 17). In bubble column cultures, about 173% more penicillin was produced using complex media as compared to chemically defined media. The total specific penicillin production with complex media using a bubble column was about 1845 units/mg CDW, about 2.6 times that obtained using chemically defined media in the same bioreactor.

Immobilized *P. chrysogenum* S_1 was also tested for continuous penicillin production in bubble column and conical bubble fermentor (Fig. 14). After batch fermentation in a complex production medium for 72 h, the fermentation was switched to continuous operation by adding new medium and withdrawing broth at the same flow rate using a peristaltic pump (Table 5).

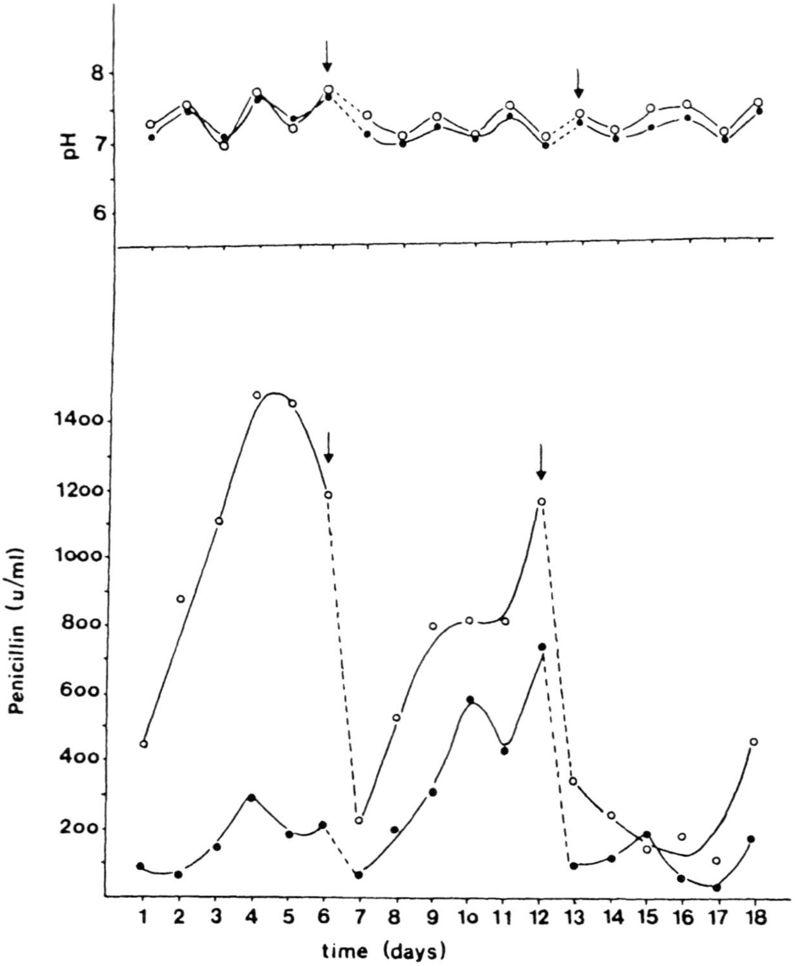

FIGURE 15. Semicontinuous penicillin production by immobilized and free cells of *P. chrysogenum* ATCC 12690 in shaken cultures using chemically defined media: (○—○) immobilized cells; (●—●) free cells; ↓ change of production medium. (From Ref. 30.)

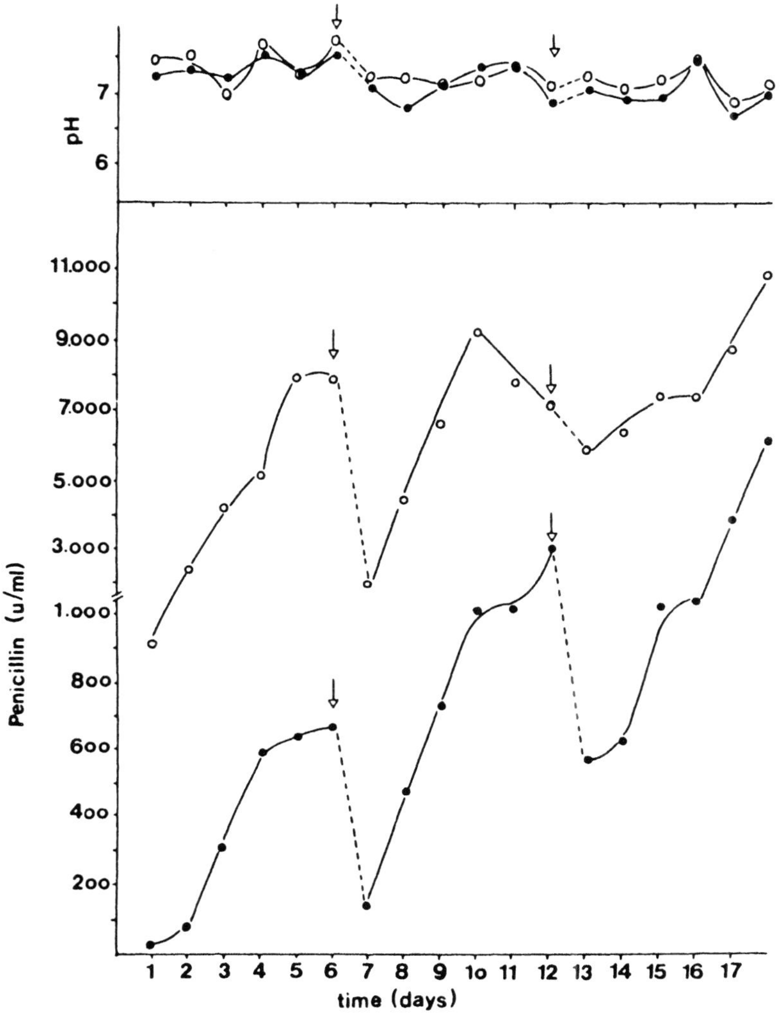

FIGURE 16. Semicontinuous penicillin production by immobilized and free cells of *P. chrysogenum* S_1 in shaken cultures using chemically defined media: (○—○) immobilized cells; (●—●) free cells; ↓ change of production medium. (From Ref. 30.)

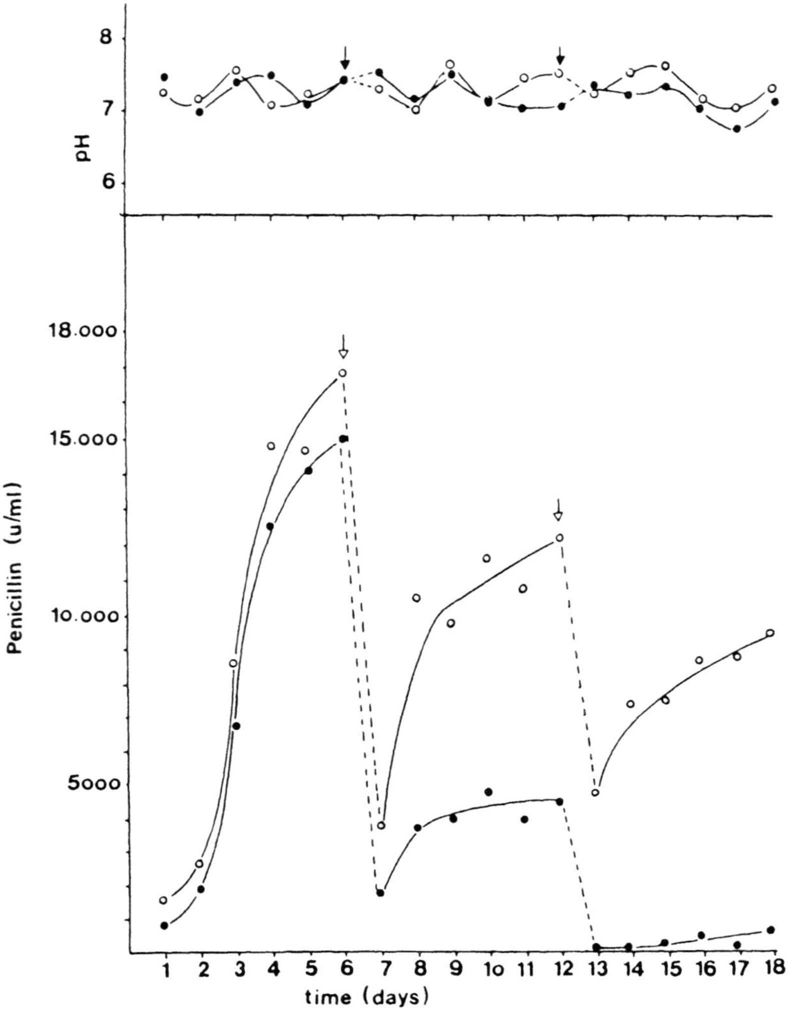

FIGURE 17. Semicontinuoue penicillin production by immobilized and free cells of *P. chrysogenum* S_1 in shaked cultures using complex media: (○—○) immobilized cells; (●—●) free cells; ↓ change of production medium. (From Ref. 30.)

TABLE 5. Operating Conditions in Different Continuous Fermentations with Immobilized *Penicillium chrysogenum* S_1

Fermentation run	Bioreactor	Sugar	Conc. (g/liter)	Flow rate (ml/h)	Dilution rate $(h^{-}\ )$	Mean residence time (h)
A	Bubble column	Lactose	20	7.50	0.0417	24
B	Bubble column	Glucose	20[a]	7.50	0.0417	24
C	Conical bubble fermentor	Lactose	40	18.75	0.0156	64
D	Conical bubble fermentor	Lactose	40	30.00	0.0250	40

[a] 2 g/liter ammonium acetate were added to the feed medium to prevent successive decrease in pH value.
Source: Ref. 31.

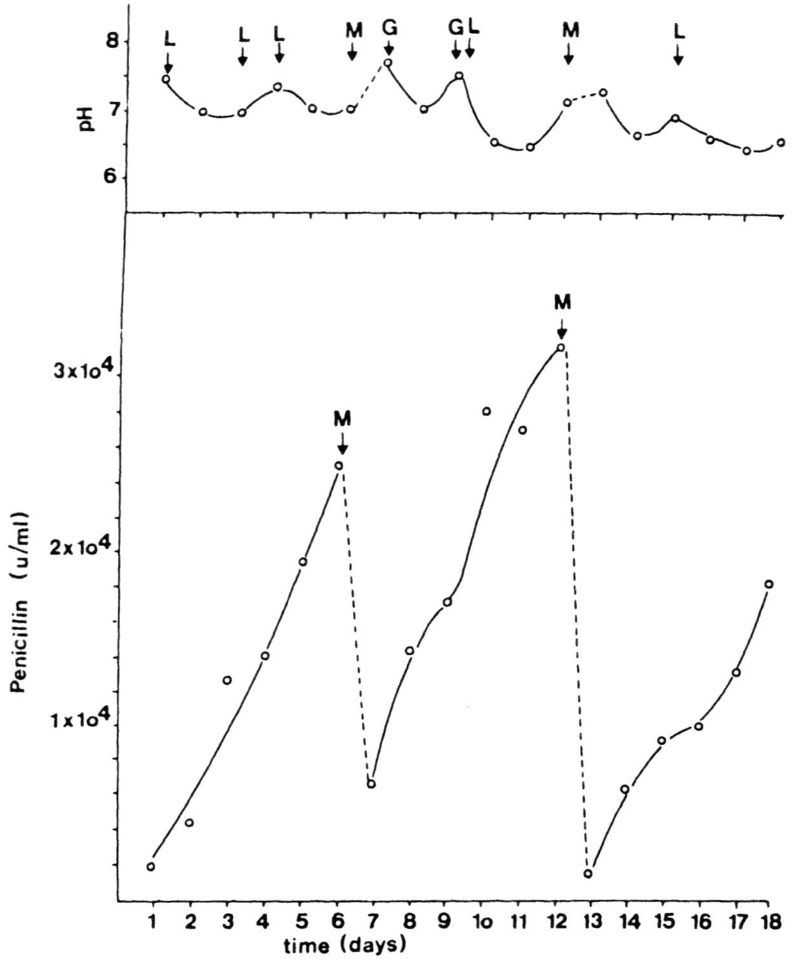

FIGURE 18. Semicontinuous penicillin production by immobilized cells of *P. chrysogenum* S_1 in bubble column using complex media. M↓ medium change; L↓ addition of 10 g/liter lactose; G↓ addition of 10 g/liter glucose. (From Ref. 30.)

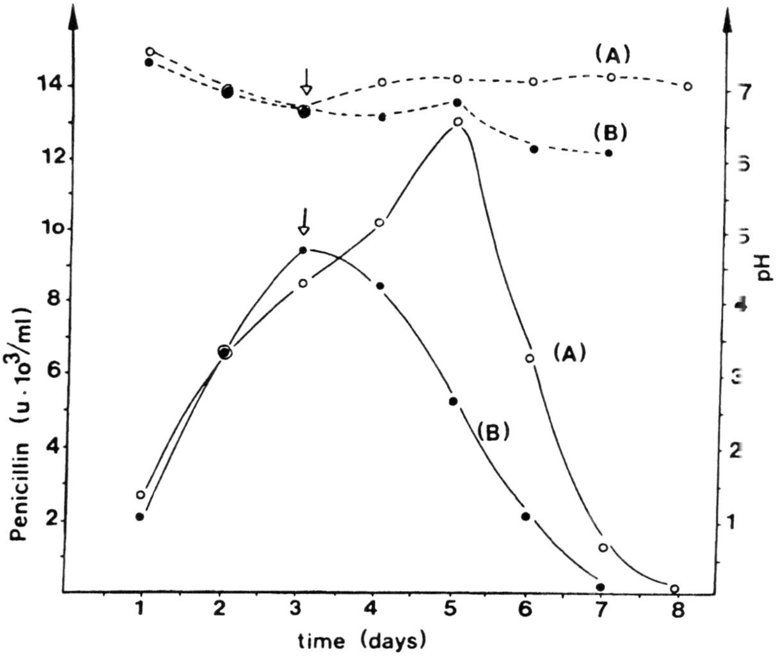

FIGURE 19. Penicillin concentration and pH of culture broth of two continuous penicillin fermentations with immobilized *P. chrysogenum* S_1 in bubble column. (o—o) Fermentation with lactose (A); (●—●) fermentation with glucose (B); (-----) pH value; ↓ start of continuous operation. (From Ref. 31.)

The courses of penicillin concentration and pH within the fermentation broth of fermentation A (with lactose) and fermentation B (with glucose) are shown in Fig. 19. The total amount of penicillin produced, the specific penicillin production (SPC), as well as the production rate (Q_{pen}) reached maximal level in the lactose fermentation (A) 48-72 h after continuous feeding was started. Similar maxima occured after 24-48 h in fermentation B [25,31].

In the conical bubble fermentor, fermentations were conducted with lactose (40 g/liter) at two different dilution rates, 0.015 h^{-1} (C) and 0.025 h^{-1} (D). At the lower dilution rate (C), the continuous fermentation ran effectively longer and penicillin concentration was stabilized at about 45% that which occurred at the beginning of the continuous feeding (Fig. 20). The total amounts of penicillin produced (T_p) increased more in the fermentation at the lower dilution rate (Fig. 21). The specific penicillin production rate (Q_{pen}) decreased gradually during the entire period of continuous fermentation (Fig. 22). Only 30-40% from the added lactose was utilized by the immobilized mold [25,31].

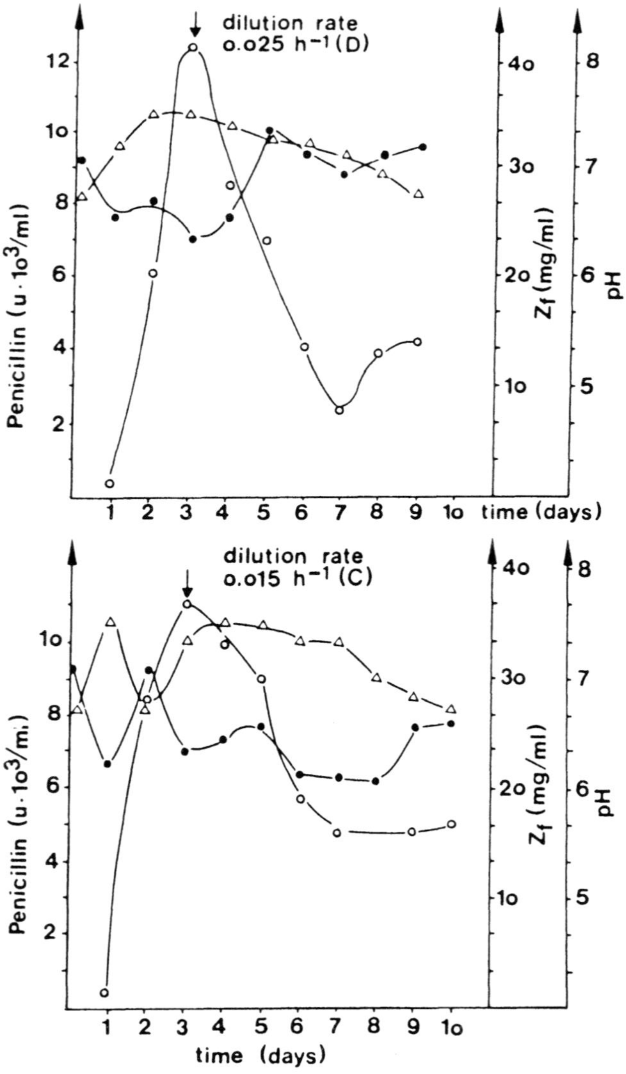

FIGURE 20. Penicillin concentration and reducing sugar concentration in culture broth of two continuous fermentations with immobilized *P. chrysogenum* S_1 in conical bubble fermentor. (o—o) Penicillin concentration (units/ml); (●—●) reducing sugar concentration (Z_f, mg/ml); (Δ—Δ) pH value; ↓ start of continuous operation. (From Ref. 31.)

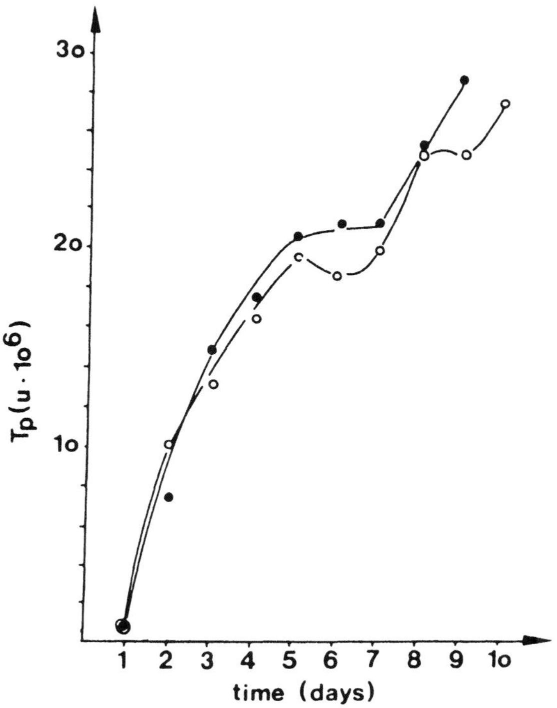

FIGURE 21. The total penicillin produced (T_p, units) in two continuous penicillin fermentations with immobilized *P. chrysogenum* S_1 in conical bubble fermentor. (o—o) Fermentation with dilution rate of 0.0156 h^{-1} (C); (●—●) fermentation with dilution rate of 0.0250 h^{-1} (D); ↓ start of continuous operation. (From Ref. 31.)

From these studies, the immobilization of *P. chrysogenum* appears to be a promising biotechnological method which can effectively improve the production of penicillin or other metabolites by fungi. However, further intensive studies concerning physicochemical changes and bioengineering of the process are needed.

No information could be found in the open literature regarding production of cephalosporin by immobilized whole cells of *C. acremonium* or by any other immobilized fungus. However, Freeman and Aharonowitz [157] reported the ability of immobilized *Streptomyces clavuligerus* in prepolymerized linear polyacrylamide gels to produce cephalosporins continuously for 96 h with yields similar to those from free resting cells.

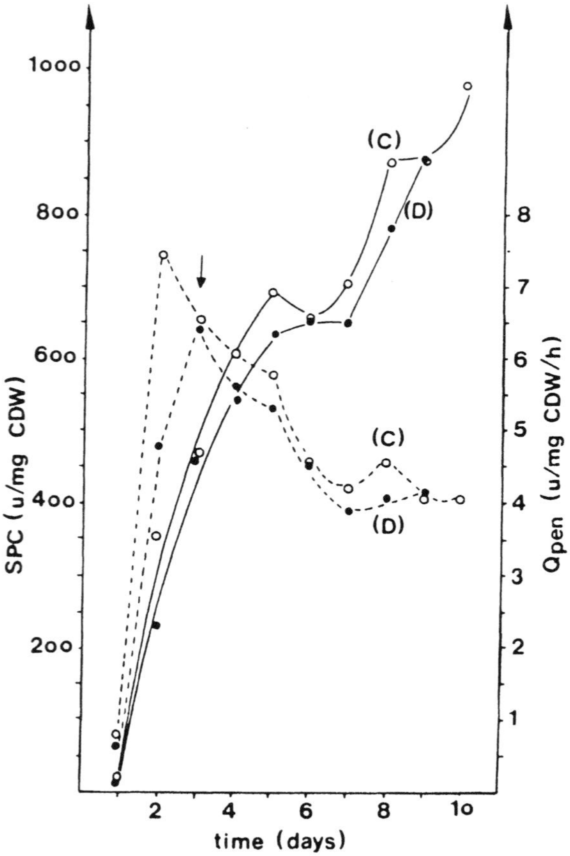

FIGURE 22. The specific penicillin production (SPC, units/mg CDW) and the specific penicillin production rate (Q_{pen}, units/mg CDW/h) of two continuous penicillin fermentations with immobilized *P. chrysogenum* S_1 in conical bubble fermentor. (○—○) Fermentation with dilution rate of 0.0156 h^{-1} (C); (●—●) fermentation with dilution rate of 0.0250 h^{-1} (D); (———) specific penicillin production (SPC, units/mg CDW); (-----) specific penicillin production rate (Q_{pen}, units/mg CDW/h); ↓ start of continuous operation. (From Ref. 31.)

VI. CONCLUSIONS

The efforts of many scientists over the past 50 years have contributed to the development of highly productive fungi for synthesizing β-lactam antibiotics. Careful management of the producing strains has been a key factor in this development. Strain development has been based on induced mutations to change the gene character in ways that increase the antibiotic production rate or overcome problems associated with production. Studies of morphology, biochemistry, and physiology in relation to environmental conditions are very important during the course of programs for screening and selecting highly productive strains.

Experiments with systems of cell organelles, especially with protoplasts, or with cell-free systems have clarified many points in the biosynthetic pathways of penicillin and cephalosporins by fungi. Such systems have been derived from highly active strains of *Penicillium, Cephalosporium,* and *Strpetomyces* used in the fermentation industries. The use of such systems in the mass production of penicillins and cephalosporins is still only a sought-after goal, however, because they require very expensive substrates and complex operating conditions.

Fermentation of penicillin or cephalosporin is conducted in sequential steps of increasing volume under highly controlled biological, chemical, and environmental conditions. The automatic control of different variables during fermentation is necessary to achieve high production rates that are cost-competitive.

Experiments with immobilized fungal cells offer the hope of further increase in the production capabilities of the producing fungi. The choice of the immobilization method and culturing conditions appears to be the most important factor governing the production of these products. However, the scale-up of fermentation with immobilized cells still requires cooperative efforts on the part of microbiologists, biochemists, and biochemical engineers.

REFERENCES

1. Ball, C. Genetics in the development of penicillin process. In *Antibiotics and Other Secondary Metabolites; Biosynthesis and Production*. FEMS Symp. No. 5 (R. Hütter, T. Leisinger, J. Nüesch, and W. Wehrli, eds.), Academic Press, London, 1987, pp. 165-176.
2. Queener, S., and Swartz, R. Penicillins, biosynthetic and semisynthetic. In *Economic Microbiology, Vol. 3, Secondary Products of Metabolism* (A. H. Rose, eds.), Academic Press, London, 1979, pp. 35-122.
3. Rehm, H.-J. Penicilline und cephalosporine. In *Industrielle Mikrobiologie* (ed.). Springer-Verlag, Berlin, 1980, pp. 453-483.
4. Lemke, P. A., and Brannon, D. R. Microbial synthesis of cephalosporin and penicillin compounds. In *Cephalosporins and Penicillins* (E. H. Flynn, ed.), Academic Press, London, 1972, pp. 370-437.
5. Abraham, E. B. Development in the chemistry and biochemistry of β-lactam antibiotics. In *Antibiotics and Other Secondary Metabolites, Biosynthesis and Production*. FEMS Symp. No. 5 (R. Hütter, T. Leisinger, J. Nüesch, and W. Wehrli, eds.), Academic Press, London, 1978, pp. 141-164.
6. Gorman, M., and Huber, F. β-lactam antibiotics. In *Annual Reports on Fermentation Process*, Vol. 1 (D. Perlman, ed.), Academic Press, London, 1977, pp. 327-346.
7. Behrens, O. K. In *The Chemistry of Penicillins* (H. Clarke, J. R. Johnson, and R. Robinson, eds.), Princeton University Press, 1949, pp. 657-679.
8. Behrens, O. K., Corse, J., Jones, R. G., Mann, M. J., Soper, Q. F., Van Abeele, R. F., and Chiang, M. C. Biosynthesis of penicillins. A. Biological precursors of benzylpenicillin (penicillin-G). *J. Biol. Chem.*, *175*:751-764, 1948.
9. Behrens, O. K., Corse, J., Jones, R. G., Kleiderer, E. C., Soper, Q. F., Van Abeele, F. R., Larson, L. M., Sylvester, J. C., Haines,

W. J., and Carter, H. E. Biosynthesis of penicillins. B. Utilization of deutrophenylacetyl-N^{15}-DL-valine in penicillin biosynthesis. *J. Biol. Chem.*, *175*:765-769, 1948.
10. Abraham, E. P., and Newton, G. G. F. Penicillins and cephalosporins. In *Antibiotics*, Vol. 2, *Biosynthesis* (G. Gottlieb and D. Shaw, eds.), Springer Verlag, Berlin, 1967, pp. 1-12.
11. Bibel, D. J., and Smiljanic, R. J. Interaction of Trichophyton mentagrophytes var-granulosum and Micrococci on skin cultures. *J. Invest. Dermatol.*, *72*:133-137, 1979.
12. Youssef, N., Wyborn, C. H. E., Holt, G., Noble, W. C., and Clayton, Y. M. Antibiotic production by dermatophyte fungi. *J. Gen. Microbiol.*, *105*:105-112, 1978.
13. Youssef, N., Wyborn, C. H. E., Holt, G., Noble, W. C., and Clayton, Y. M. Ecological effects of antibiotic production by dermatophyte fungi. *J. Hyg.*, *82*:301-308, 1979.
14. Papachariladu, E., and Pisano, M. A. Changes in the lipid composition of Paecilomyces persicinus P-10 M-1 during growth and cephalosporin-C production. *Appl. Environ. Microbiol.*, *48*:1084-1087, 1984.
15. Redhead, K., and Wright, S. J. L. Lysis of Cyanobacterium anabaena-flosaquae by antibiotic-producing fungi. *J. Gen. Microbiol.*, *119*:95-102, 1980.
16. Nagarajan, R., Boeck, L. D., Gorman, M., Hamil, R. L., Higgens, C. E., Hochn, M., Stark, W. M., and Whitney, J. G. β-lactam antibiotics from Streptomyces. *J. Am. Chem. Soc.*, *93*:2308-2310, 1971.
17. O'Sullivan, J., and Abraham, E. P. Biosynthesis of β-lactam antibiotics. In *Antibiotics, Vol. IV, Biosynthesis* (J. W. Corcoran, ed.), Springer-Verlag, Berlin, 1981, pp. 101-122.
18. Wolfgang, F. Sekundarmetabolite, Produkte aus Aminosauren. In *Biochemische Grundlagen der Mikrobiologie* (ed.), Fischer Verlag, Jena, 1978, pp. 150-159.
19. Elander, R. P. Enhanced penicillin biosynthesis in mutant and recombinant strains of Penicillium chrysogenum. *Abhd. Dtsch. Akad. Wiss Berlin*, Kl Med 2:403-423, 1967.
20. Backus, M. P., and Stauffer, J. F. The production and selection of a family of strains in Penicillium chrysogenum. *Mycologia*, *47*:429-463, 1955.
21. Demain, A. L. Mutation and the production of secondary metabolites. *Adv. Appl. Microbiol.*, *16*:177-202, 1973.
22. Demain, A. L. Biochemistry of penicillin and cephalosporin. *Lioydia*, *37*:147-167, 1974.
23. Cooney, C. L. Growth of microorganisms. In *Biotechnology*, Vol. 1 (H.-J. Rehm and G. Reed, eds.), Verlag Chemie, Weinheim, 1981, pp. 73-112.
24. Kurylowicz, W., Kurzalkowski, W., Woznicka, W., Polowiniak-Pracka, H., Paszkiewicz, A., Luba, J., and Piorunowski, J. *Atlas of Ultrastructure of Penicillium chrysogenum in Course of Biosynthesis of Penicillin.* Chemia Publishing Office, Warsaw, 1980.
25. El-Sayed, A.-H. M. M. Penicillin-Produktion bei immobilisierten Zellen von Penicillium chrysogenum. Dissertation, Fachbereich Biologie, Universität Münster, FRG, 1986.
26. El Sayed, A.-H. M. M., and Rehm, H.-J. Morphology of Penicillium chrysogenum strains immobilized in calcium alginate beads and used in penicillin fermentation. *Appl. Microbiol. Biotechnol.*, *26*:89-94, 1986.

27. Wittler, R., and Schügerl, K. Interrelation between penicillin productivity and growth rate. *Appl. Microbiol. Biotechnol.*, *21*:348-355, 1985.
28. König, B., Schügerl, K., and Seewald, C. Strategies for penicillin fermentation in tower loop reactor. *Biotechnol. Bioeng.*, *24*:259-280, 1982.
29. Wittler, R., Schügerl, K., Baumgärtl, H., and Lübbers, D. Oxygen transfer in Penicillium chrysogenum pellets. In *Reprints of the Third European Congress on Biotechnology, Munich*, Vol. 1. Verlag Chemie, Weinheim, 1984, pp. 513-520.
30. El-Sayed, A.-H. M. M., and Rehm, H.-J. Semicontinuous penicillin production by two penicillium chrysogenum strains immobilized in calcium alginate beads. *Appl. Microbiol. Biotechnol.*, *26*:211-214, 1987.
31. El-Sayed, A.-H. M. M., and Rehm, H. J. Continuous penicillin production by Penicillium chrysogenum immobilized in calcium alginate beads. *Appl. Microbiol. Biotechnol.*, *26*:215-218, 1987.
32. Nash, C. H., and Huber, F. M. Antibiotic synthesis and morphological differentiation of Cephalosporium acremonium. *Appl. Microbiol.*, *22*:6-10, 1971.
33. Calam, C. T. Factors governing the production of penicillin by Penicillium chrysogenum. In *FEMS Symp.* No. 13 (V. Krumphanzl, B. Sikyta, and Z. Vaněk, eds.), Academic Press, New York, 1982, pp. 89-95.
34. Shu, P. Mathematical models for the product accumulation in microbiological processes. *J. Microbiol. Technol. Eng.*, *3*:95-109, 1961.
35. Bajpai, R. K., and Reuss, M. A mechanistic model for penicillin production. *J. Chem. Technol. Biotechnol.*, *30*:332-344, 1980.
36. Bajpai, R. K., and Reuss, M. Evaluation of feeding strategies in carbon regulated metabolite production through mathematical modelling. *Biotechnol. Bioeng.*, *23*:717-738, 1981.
37. Pirt, S. J., and Righelato, R. C. Effect of growth rate on the synthesis of penicillin by Penicillium chrysogenum in batch and chemostat cultures. *Appl. Microbiol.*, *15*:1284-1290, 1967.
38. Ryu, D. D. Y., and Hospodka, J. Quantitative physiology of Penicillium chrysogenum in penicillin fermentation. *Biotechnol Bioeng.*, *22*:289-298, 1980.
39. Pirt, S. J. The theory of fed-batch culture with reference to penicillin fermentation. *J. Appl. Chem. Biotechnol.*, *24*:415-424, 1974.
40. Righelato, R. C., Trinci, A. P. J., Pirt, S. J., and Peat, A. The influence of maintenance energy and growth rate on the activity, morphology and conidiation of Penicillium chrysogenum. *J. Gen. Microbiol.*, *50*:399-412, 1968.
41. Feren, C. J., and Squires, R. W. The relationship between critical oxygen level and antibiotic synthesis of capreomycin and cephalosporin-C. *Biotechnol. Bioeng.*, *11*:583-592, 1969.
42. Larson, G., and Enfors, S. Q. Influence of oxygen starvation on the respiratory capacity of Penicillium chrysogenum. *Appl. Microbiol. Biotechnol.*, *21*:228-233, 1985.
43. Vardar, F., and Lilly, M. D. Effect of cycling dissolved oxygen concentration on product formation in penicillin fermentation. *Eur. J. Appl. Microbiol. Biotechnol.*, *14*:203-211, 1982.
44. Demain, A. L. Regulatory mechanisms and the industrial production of microbial metabolites. *Lioydia*, *31*:395-418, 1968.
45. Lur'e, L. M., Verhovtseva, T. P., Orlova, A. I., and Leritov, M. M. *Pharm. Chem. J.* (English translation), *10*:218, 1976. Reported in Queener and Swartz, 1979 [2].

46. Elander, R. P., Espenshade, M. A., Pathak, S. G., and Pan, C. H. The use of parasexual genetics in an industrial strain improvement program with Penicillium chrysogenum. In *Genetics of Industrial Microorganisms*, Vol. 2 (Z. Vaněk, Z. Hostalěk, and J. Cudlin, eds.), Elsevier, Amsterdam, 1973, pp. 239-253.
47. Pontecorvo, G., and Sermonti, G. Recombination without sexual reproduction in Penicillium chrysogenum. *Nature, 172*:126-127, 1953.
48. Anne, J. Comparison of penicillins produced by interspecies hybrids from Penicillium chrysogenum. *Eur. J. Appl. Microbiol. Biotechnol., 15*:41-46, 1982.
49. Malik, V. S. Genetics of secondary metabolism. In *Biotechnology*, Vol. 4 (H.-J. Rehm and G. Reed, eds.), Verlag Chemie, Weinheim, 1986, pp. 39-68.
50. Elander, R. P., and Demain, A. L. Genetics of microorganisms in relation to industrial requirements. In *Biotechnology*, Vol. 1 (H.-J. Rehm and G. Reed, eds.), Verlag Chemie, Weinheim, 1981, pp. 235-277.
51. Liang, P., and Liu, H. Interspecific somatic hybridization between Penicillium patulinum and Penicillium chrysogenum. 1. Protoplast fusion, hybrid development and induced segregation. *Acta Microbiol. Sin., 22*:248-256, 1982.
52. Carr, L. G., Skattrud, P. L., Ingolia, T. D., and Queener, S. W. Organization of the 5-8S, 16-18S and 22-28S ribosomal RNA genes of Cephalosporium acremonium. *Curr. Genet., 12*:209-214, 1987.
53. Macdonald, K. D., and Holt, G. Genetics of biosynthesis and overproduction of penicillin. *Sci. Progr. Oxford, 63*:547-573, 1976.
54. Bauer, K. Zur Biosynthese der Penicilline: Bildung von 5-(2-aminoadipyl)-Cysteinyl-Valin in Extrakten von Penicillium chrysogenum. *Z. Naturforsch., 25*:1125-1129, 1970.
55. Vandamme, E. J. Enzymes involved in β-lactam antibiotic biosynthesis. *Adv. Appl. Microbiol., 21*:89-123, 1977.
56. Abraham, E. P., and Newton, G. G. F. The structure of cephalosporin-C. *Biochem. J., 79*:377-393, 1961.
57. Arnstein, H. R. V., and Morris, D. Synthesis of optical isomers of cystinylvaline. *Biochem. J., 76*:318-323, 1960.
58. Arnstein, H. R. V., and Morris, D. The utilization of L-cystinyl-L-valine for penicillin biosynthesis. *Biochem. J., 76*:323-327, 1960.
59. Arnstein, H. R. V., and Morris, D. The structure of a peptide containing α-aminoadipic acid, cystine and valine, present in the mycelium of Penicillium chrysogenum. *Biochem. J., 76*:357-361, 1960.
60. Lodar, P. P., and Abraham, E. P. Isolation and nature of intracellular peptides from cephalosporin-C-producing Cephalosporium sp. *Biochem. J., 123*:471-476, 1971.
61. Arnstein, H. R. V., Morris, D., and Toms, E. J. Isolation of a tripeptide containing α-aminoadipic acid from the mycelium of Penicillium chrysogenum and its possible significance in penicillin biosynthesis. *Biochem. Biophys. Acta, 35*:361-362, 1959.
62. Heim, J., Shen, Y. Q., Wolfe, S., and Demain, A. L. Regulation of isopenicillin-N synthesis and deacetoxycephalosporin-C synthesis by carbon source during fermentation of Cephalosporium acremonium. *Appl. Microbiol. Biotechnol., 19*:232-236, 1984.
63. Hockenhull, D. J. D., Ramachandran, K., and Walker, T. K. The biosynthesis of penicillins. *Arch. Biochem., 23*:160-161, 1949.

64. Jensen, S. E., Westlake, D. W. S., and Wolfe, S. Production of penicillins and cephalosporins in an immobilized enzyme reactor. *Appl. Microbiol. Biotechnol., 20*:155-160, 1984.
65. Kohsaka, M., and Demain, A. L. Conversion of penicillin-N to cephalosporin(s) by cell-free extracts of Cephalosporium acremonium. *Biochem. Biophys. Res. Commun., 70*:464-473, 1976.
66. Fawcett, P. A., Usher, J. J., Huddleston, J. A., Bleany, R. C., Nisbet, J. J., and Abraham, E. P. Synthesis of δ(α-aminoadipyl) cysteinylvaline and its role in penicillin biosynthesis. *Biochem. J., 157*:651-660, 1976.
67. Jensen, S. E., Westlake, D. W. S., Bowers, R. J., and Wolfe, S. Cephalosporin formation by cell-free extracts from Streptomyces clavuligerus. *J. Antibiot., 35*:1351-1360, 1982.
68. Jensen, S. E., Westlake, D. W. S., and Wolfe, S. Cyclization of δ-(L-α-aminoadipyl)-L-cysteinyl-D-valine to penicillins by cell-free extracts of Streptomyces clavuligerus. *J. Antibiot., 35*:483-490, 1982.
69. Jensen, S. E., Westlake, D. W. S., and Wolfe, S. Partial purification and characterization of isopenicillin-N epimerase activity from Streptomyces clavuligerus. *Can. J. Microbiol., 29*:1526-1531, 1983.
70. Konomi, T., Hercken, S., Baldwin, J. E., Yoshida, M., Munt, N. A., and Demain, A. L. Cell-free conversion of δ-(L-α-aminoadipyl)-L-cysteinyl-D-valine into an antibiotic with the properties of isopenicillin-N in Cephalosporium acremonium. *Biochem. J., 184*:427-430, 1979.
71. Kupka, J., Shen, Y. Q., Wolfe, S., and Demain, A. L. Partial purification and the properties of the α-ketoglutarate-linked ring expansion of β-lactam biosynthesis of Cephalosporium acremonium. *FEMS Microbiol., 16*:1-6, 1983.
72. Kupka, J., Shen, Y. Q., Wolfe, S., and Demain, A. L. Studies on the ring expansion of β-lactam biosynthesis in Cephalosporium acremonium. *Can. J. Microbiol., 29*:488-496, 1983.
73. Kurzatkowski, W., Kurylowiçz, and Paszkiewiçz, A. Penicillin-G production by immobilized fungal vesicles. *Eur. J. Appl. Microbiol. Biotechnol., 15*:211-213, 1982.
74. Demain, A. L. Inhibition of penicillin formation by lysine. *Arch. Biochem. Biophys., 67*:244-245, 1957.
75. Somerson, N. L., Demain, A. L., and Tunheimer, T. D. Removal of lysine inhibition of penicillin production by α-aminoadipic acid. *Arch. Biochem. Biophys., 93*:238-241, 1961.
76. Goulden, S. A., and Chattaway, F. W. Lysine control of α-aminoadipate and penicillin synthesis in Penicillium chrysogenum. *Biochem. J., 110*: 55p-56p, 1968.
77. Demain, A. L., Newkirk, J. F., and Hendlin, D. Effect of methionine, norleucine and lysine derivatives on cephalosporin-C formation in chemically defined media. *J. Bacteriol., 85*:339-344, 1963.
78. Ott, S. L., Godezeski, C. W., Pavey, D., Farra, J. D., and Horton, D. R. Biosynthesis of cephalosporin-C. 1. Factors affecting the fermentation. *Appl. Microbiol., 10*:515-523, 1962.
79. Trown, P. W., Smith, B., and Abraham, E. P. Biosynthesis of cephalosporin-C from amino acids. *Biochem. J., 86*:284-291, 1963.
80. Arnstein, H. R. V., and Grant, P. T. The biosynthesis of penicillin. 1. The incorporation of some amino acids into penicillin. *Biochem. J., 57*:353-359, 1954.

81. Arnstein, H. R. V., and Crawhall, J. C. The biosynthesis of penicillin. 6. A study of the mechanism of the formation of the thiazolidine-β-lactam rings using tritium-labelled cysteine. *Biochem. J.*, *57*:180-187, 1957.
82. Bycroft, B. W., Wels, C. M., Corbett, K., and Lowe, D. A. Incorporation of (α-^{2}H)- and (α-^{3}H)-L-cystine into penicillin-G and the location of the label using isotope exchange and ^{2}H nuclear magnetic resonance. *J. Chem. Soc. Chem. Commun.*, 123-125, 1975.
83. Morecombe, D. J., and Young, D. W. Synthesis of chirally labelled cysteines and the steric origin of $C_{(5)}$ in penicillin biosynthesis. *J. Chem. Soc. Chem. Commun.*, 198-199, 1975.
84. Fujisawa, Y., Uchida, M., and Suzuki, M. S-sulfo-cysteine as a useful sulfur source for cephalosporin-C biosynthesis by Cephalosporium acremonium. *Agr. Biol. Chem.*, *24*:1519-1524, 1982.
85. Whitney, J. C., Brannon, D. R., Mabe, J. A., and Wicker, K. J. Incorporation of labelled precursors into A168 86B, a novel β-lactam antibiotic produced by Streptomyces clavuligerus. *Antimicrob. Ag. Chemother.*, *1*:247-251, 1972.
86. Aberhart, D. J., Lin, L. J., and Chu, JY.-R. Studies on the biosynthesis of β-lactam antibiotics. II. Synthesis and incorporation into penicillin-G of (2'-RS,3R,3'R)-(3.3'-$^3H_2$)-cystine. *J. Chem. Soc. Perkin*, *1*:2517-2523, 1975.
87. Huddleston, J. A., Abraham, E. P., Young, D. W., Morecombe, D. J., and Sen, P. K. The stereochemistry of β-lactam formation in cephalosporin biosynthesis. *Biochem. J.*, *169*:705-707, 1978.
88. Jeffery, J. D. A., Abraham, E. P., and Newton, G. G. F. Further degradation products of cephalosporin-C. Isolation and synthesis of 2-(4-amino-4-carboxybutyl)thiazole-4-carboxylic acid. *Biochem. J.*, *75*:216-223, 1960.
89. Young, D. W., Morecombe, D. J., and Sen, P. K. The stereochemistry of β-lactam formation in penicillin biosynthesis. *Eur. J. Biochem.*, *75*:133-147, 1977.
90. Segel, C. M., and Johnson, M. J. Intermediates in inorganic sulfate utilization by Penicillium chrysogenum. *Arch. Biochem. Biophys.*, *103*: 216-228, 1963.
91. Stevens, C. M., Vohre, P., Inamine, E., and Roholt, O. A. Utilization of sulfur compounds for the biosynthesis of penicillins. *J. Biol. Chem.*, *205*:1001-1006, 1953.
92. Caltrider, P. G., and Niss, H. F. Role of methionine in cephalosporin synthesis. *Appl. Microbiol.*, *14*:746-753, 1966.
93. Segel, I. H., and Johnson, M. J. Accumulation of intracellular inorganic sulfate by Penicillium chrysogenum. *J. Bacteriol.*, *81*:91-98, 1961.
94. Hockenhull, D. J. D. Studies in penicillin production by Penicillium chrysogenum in surface cultures. 2. Further studies in the metabolism of sulfur. *Biochem. J.*, *43*:498-504, 1948.
95. Demain, A. L. Inhibition of penicillin formation by amino acid analogs. *Arch. Biochem. Biophys.*, *64*:74-79, 1956.
96. Kitano, K. Studies on the production of β-lactam antibiotics. *J. Takeda Res. Lab.*, *36*:105-192, 1977.
97. D'Amato, R. F., and Pisano, M. A. A chemically defined medium for cephalosporin-C production by Paecilomyces persicinus. *Anatonie van Leeuwenhoek*, *42*:299-308, 1976.

98. Drew, S. W., and Demain, A. L. Production of cephalosporin-C by single and double sulfur auxotrophic mutants of Cephalosporium acremonium. *Antimicrob. Ag. Chemother.*, *8*:5-10, 1975.
99. Arnstein, H. R. V., and Clubb, M. E. The biosynthesis of penicillins. 5. Comparison of valine and hydroxyvaline as penicillin precursors. *Biochem. J.*, *65*:618-627, 1957.
100. Warren, S. C., Newton, G. G. F., and Abraham, E. P. The role of valine in the biosynthesis of penicillin-N and cephalosporin-C by a Cephalosporium sp. *Biochem. J.*, *103*:902-903, 1967.
101. Arnstein, H. R. V., and Magreiter, H. The biosynthesis of penicillin. 7. Further experiments on the utilization of L- and D-valine and the effect of cysteine analogues on penicillin biosynthesis. *Biochem. J.*, *68*:339-348, 1958.
102. Bycroft, B. W., Wels, C. M., Corbett, K., Maloney, A. P., and Lowe, D. A. Biosynthesis of penicillin-G from D- and L-(^{14}C)(α-^{3}H)-valine. *J. Chem. Soc. Chem. Commun.*, 923-924, 1975.
103. Goulden, S. A., and Chattaway, F. W. End-product control of acetohydroxy acid synthetase by valine in Penicillium chrysogenum Q_{176} and a high penicillin-yielding mutant. *J. Gen. Microbiol.*, *59*:111-118, 1969.
104. Jarvis, S. E., and Johnson, J. The mineral nutrition of Penicillium chrysogenum Q_{176}. *J. Bacteriol.*, *59*:51-60, 1950.
105. Demain, A. L. Overproduction of microbial metabolites and enzymes due to alteration of regulation. *Adv. Biochem. Eng.*, *1*:113-142, 1971.
106. Aharonowitz, Y. Nitrogen metabolite regulation of antibiotic biosynthesis. *Annu. Rev. Microbiol.*, *34*:209-233, 1980.
107. Aharonowitz, Y., and Demain, A. L. Nitrogen nutrition and regulation of cephalosporin production in Streptomyces clavuligerus. *Can. J. Microbiol.*, *25*:61-67, 1979.
108. Mateos, R. D. C. Effect of glutamine on penicillin formation in Penicillium chrysogenum. *Biotechnol. Lett.*, *6*:109-114, 1984.
109. Küenzi, M. T. Process design and control in antibiotic fermentation. In *Antibiotic and Other Secondary Metabolites: Biosynthesis and Production. FEMS Symp.* No. 5 (R. Hütter, T. Leisinger, J. Nüesch, and W. Wehrli, eds.), Academic Press, London, 1978, pp. 39-56.
110. Demain, A. L., Aharonowitz, Y., and Martin, J. F. Metabolic control of secondary biosynthetic pathways. In *Biochemistry and Genetic Regulation of Commercially Important Antibiotics* (L. C. Vining, ed.), Addison-Wesley, Reading, MA, 1983, pp. 49-72.
111. Lübbe, C., Wolfe, S., and Demain, A. L. Repression and inhibition of cephalosporin synthesis in Streptomyces clavuligerus by inorganic phosphate. *Arch. Microbiol.*, *140*:317-320, 1985.
112. Perez-Martinez, G., and Peberdy, J. F. Production of cephalosporin-C and its intermediates by raised titre strains of Acremonium chrysogenum. *Enz. Microbiol. Technol.*, *7*:389-394, 1985.
113. Fawcett, P. A., Lodar, P. B., Duncan, M. J., Beesley, T. J. and Abraham, E. P. Formation and properties of protoplasts from antibiotic-producing strains of Penicillium chrysogenum and Cephalosporium acremonium. *J. Gen. Microbiol.*, *79*:293-309, 1973.
114. Hollander, I. J., Shen, Y.-Q., Heim, J., Demain, A. L., and Wolfe, S. A pure enzyme catalyzing penicillin biosynthesis. *Science*, *224*:610-612, 1984.

115. Behmer, C. J., and Demain, A. L. Further studies on carbon catabolite regulation of β-lactam antibiotic in Cephalosporium acremonium. *Curr. Microbiol.*, *8*:107-114, 1983.
116. Brunner, R., and Röhr, M. Zur Biosynthese des Penicillins. Untersuchungen zur enzymatischen Aktivierung von Phenylessigsäure und Phenoxyessigsäure sowie zur Bildung von Penicillin aus 6-Aminopenicillansäure durch Mycelhomo-genate und zellfreie Extrakte von Penicillium chrysogenum. *Z. Physiol. Chem.*, *349*:95-103, 1968.
117. Demain, A. L. Biosynthesis of penicillins and cephalosporins. In *Biosynthesis of Antibiotics*, Vol. 1 (J. F. Snell, ed.), Academic Press, New York, 1966, pp. 29-94.
118. Sawada, Y., Baldwin, J. E., Singh, P. D., Solomon, N. A., and Demain, A. L. Cell-free cyclization of δ-(L-α-aminoadipyl)-L-cysteinyl-D-valine (ACV) to isopenicillin-N. *Antimicrob. Ag. Chemother.*, *18*: 465-470, 1980.
119. Hook, D. J., Chang, L. T., Elander, R. P., and Morin, R. B. Stimulation of the conversion of penicillin-N to cephalosporin by ascorbic acid, alpha-ketoglutarate and iron-II in cell-free extracts of strains of Cephalosporium acremonium. *Biochem. Biophys. Res. Commun.*, *87*:258-265, 1979.
120. Sawada, Y., Hunt, N. A., and Demain, A. L. Further studies on microbiological ring-expansion of penicillin-N. *J. Antibiot.*, *32*:1303-1310, 1979.
121. Fujisawa, Y., Kitano, K., and Kanzaki, T. Accumulation of deacetoxycephalosporin-C by mutant of Cephalosporium acremonium. *Agr. Biol. Chem.*, *39*:2049-2055, 1975.
122. Liersch, M., Nüesch, J., and Treichler, H. J. Final steps in the biosynthesis of cephalosporin-C. In *Second Int. Symp. Genet. Ind. Microbiol.* (K. D. Macdonald, ed.), Academic Press, London, 1976, pp. 179-195.
123. Queener, S. W., Capone, J. J., Radue, A. B., and Nagarajan, R. Synthesis of deacetoxycephalosporin-C by a mutant of Cephalosporium acremonium. *Antimicrob. Ag. Chemother.*, *6*:334-337, 1974.
124. Higgens, C. E., Hamil, R. L., Sands, T. H., Hoehn, M. M., Davis, N. E., Nagarajan, R., and Boeck, L. D. The occurrence of deacetoxycephalosporin-C in fungi and streptomyces. *J. Antibiot.*, *27*:298-300, 1974.
125. Fujisawa, Y., Shirafuji, H., Kida, M., Nara, K., Yoneda, M., and Kanzaki, T. Accumulation of deacetylcephalosporin-C by cephalosporin-C-negative mutants of Cephalosporium acremonium. *Agr. Biol. Chem.*, *39*:1295-1301, 1975.
126. Fujisawa, Y., Shirafuji, H., Kida, M., Nara, K., Yoneda, M., and Kanzaki, T. New findings on cephalosporin-C biosynthesis. *Nature*, *246*:154-155, 1973.
127. Ryu, D. D. Y., and Humphrey, A. E. Examples of computer-aided fermentation systems. *J. Appl. Chem. Biochem.*, *23*:283-295, 1973.
128. Van Breugel, J., and Van Ebergen, G. Ch. Computer control of bioreactors by indirect measurement of the process variables; some applications. In *Reprints of the Third European Congress on Biotechnology, Munich*, Vol. 3. Verlag Chemie, Weinheim, 1984, p. 593.
129. Kanzaki, T., and Fujisawa, Y. Recent progress in cephalosporin fermentation. *J. Takeda Res. Lab.*, *34*:324-349, 1975.

130. König, B., Seewald, C., and Schügerl, K. Process engineering investigations of penicillin production. *Eur. J. Appl. Microbiol. Biotechnol.*, *12*:205-211, 1981.
131. Lim, H. C., Tayeb, Y. J., Modak, J. M., and Bonte, P. Computational algorithms for optimal feed rates for a class of fed-batch fermentation. Numerical results for penicillin and cell mass production. *Biotechnol. Bioeng.*, *28*:1408-1420, 1986.
132. Luttman, R., Bellgardt, K. H., Bartzke, U., Nothnagel, J., Rapmundt, W., and Wonka, F. Process computer and microcomputer control of biotechnological process in a fermentor hall: concept and applications. In *Reprints of the Third European Congress on Biotechnology, Munich*, Vol. 3. Verlag Chemie, Weinheim, 1984, pp. 585-593.
133. Nakayama, K. Sources of industrial microorganisms. In *Biotechnology*, Vol. 1 (H.-J. Rehm and G. Reed, eds.), Verlag Chemie, Weinheim, 1981, pp. 355-410.
134. Rehm, H. J. *Einfuhrung in die Industrielle Mikrobiologie*. Springer-Verlag, Berlin, 1971.
135. Pan, C. H., Heppler, L., and Elander, R. P. Control of pH and carbohydrate addition in the penicillin fermentation. *Dev. Ind. Microbiol.*, *13*:103-112, 1972.
136. Calam, C. T., and Smith, G. M. Regulation of the biochemistry and morphology of Penicillium chrysogenum in relation to initial growth. *FEMS Microbiol. Lett.*, *10*:231-234, 1981.
137. Martin, J. F., and Demain, A. L. Control of antibiotic biosynthesis. *Microbiol. Rev.*, *44*:230-251, 1980.
138. Weinberg, E. Biosynthesis of secondary metabolites. Role of trace metals. In *Advances in Microbial Physiology*, Vol. 4 (A. H. Rose and F. Wilkinson, eds.), Academic Press, London, 1970, pp. 1-44.
139. Weinberg, E. D. Biosynthesis of microbial metabolites. Regulation by mineral elements and temperature. In *Overproduction of Microbial Metabolites*, FEMS Symp. No. 13 (V. Krumphanzl, B. Sikyta, and Z. Vaněk, eds.), Academic Press, London, 1982, pp. 181-194.
140. Pan, C. H., Heppler, L., and Elander, R. P. The effect of iron on a high-yielding industrial strain of Penicillium chrysogenum and production of penicillin-G. *J. Ferment. Technol.*, *53*:854-861, 1975.
141. McCann, E. P., and Calam, C. T. The metabolism of Penicillium chrysogenum and the production of penicillin using a high-yielding strain at different temperatures. *J. Appl. Chem. Biotechnol.*, *22*:1201-1208, 1972.
142. Giona, A. R., De Santis, R., Marrelli, L., and Toro, L. The influence of oxygen concentration and of specific rate of growth on the kinetics of penicillin production. *Biotechnol. Bioeng.*, *18*:493-512, 1976.
143. Abbott, B. J. Immobilized cells. In *Annual Reports on Fermentation Processes*, Vol. 1 (D. Perlman, ed.), Academic Press, London, 1977, pp. 205-233.
144. Abbott, B. J. Immobilized cells. In *Annual Reports on Fermentation Processes*, Vol. 2 (D. Perlman, ed.), Academic Press, London, 1978, pp. 91-123.
145. Poulsen, P. B. European and American trends in the industrial application of immobilized biocatalysts. *Enz. Microbiol. Technol.*, *3*:271-273, 1981.
146. Vandamme, E. J. Peptide antibiotics production through immobilized biocatalyst technology. *Enz. Microbiol. Technol.*, *5*:403-416, 1983.

147. Jack, T. R., and Zajic, J. E. The immobilization of whole cells. In *Advances in Biochemical Engineering*, Vol. 5 (T. K. Ghose, A. Fiechter, and N. Blakebrough, eds.), Springer-Verlag, Berlin, 1977, pp. 125-145.
148. Venkatasubramanian, K., and Vieth, W. R. Immobilized microbial cells. In *Progress in Industrial Microbiology*, Vol. 15 (J. Bull, ed.), American Chemical Society, Washington, DC, 1979, pp. 61-86.
149. Klein, J., Becke, J. W., Kressdorf, B., Steiner, H.-J., and Verlop, K. D. Matrix development for specific requirements. *Special Seminar of EFB-Working Parties, Seminar 4, "Immobilized Biocatalysts." Third European Congress on Biotechnology*, Munich, 1984.
150. Kopp, B., and Rehm, H.-J. Alkaloid production by immobilized mycelia of Claviceps purpurea. *Eur. J. Appl. Microbiol. Biotechnol.*, *18*:257-263, 1983.
151. Kopp, B., and Rehm, H.-J. Semicontinuous cultivation of immobilized Claviceps purpurea. *Eur. J. Appl. Microbiol. Biotechnol.*, *19*:141-145, 1984.
152. Eikmeier, H., and Rehm, H.-J. Production of citric acid with immobilized Aspergillus niger. *Appl. Microbiol. Biotechnol.*, *20*:365-370, 1984.
153. Kopp, B., El-Sayed, A.-H., Mahmoud, W., and Rehm, H.-J. Production of ergot alkaloids, penicillins and chlortetracycline by immobilized microorganisms. In *Reprints of the Third European Congress on Biotechnology*, Vol. 1, Munich. Verlag Chemie, Weinheim, 1984, pp. 281-286.
154. Morikawa, Y., Karuba, I., and Suzuki, S. Penicillin-G production by immobilized whole cells of Penicillium chrysogenum. *Biotechnol. Bioeng.*, *21*:261-270, 1979.
155. Deo, Y. M., and Gaucher, G. M. Semicontinuous and continuous production of penicillin-G by Penicillium chrysogenum immobilized in K-carrageenan beads. *Biotechnol. Bioeng.*, *26*:285-295, 1984.
156. Wang, D. I. C., Meier, J., and Yokoyamata, K. Penicillin fermentation in 200-liter tower fermentor using cells confined to microbeads. *Appl. Biochem. Biotechnol.*, *9*:105-116, 1984.
157. Freeman, A., and Aharonowitz, Y. Immobilization of microbial cells in crosslinked, prepolymerized linear polyacrylamide gels: antibiotic production by immobilized Streptomyces clavuligerus. *Biotechnol. Bioeng.*, *23*:2747-2759, 1981.
158. Brown, A. G., and Reading, C. The β-lactam antibiotics. In *Culture*, Vol. 4, Oxoid Ltd., Hampshire, England, September 1983, p. 2.

20

FUNGAL PLANT GROWTH REGULATORS

B. K. LONSANE and P. K. R. KUMAR* *Central Food Technological Research Institute, Mysore, India*

I. INTRODUCTION

Plant growth regulators (PGRs), the chemical compounds with the capability of coordinating the growth and morphogenesis of plants [1] are distinct from the plant nutrients in their ability to act in minute quantity [2] and to move in the plant body from the site of production or application to the site of action for evoking characteristic responses [1]. PGRs are sometimes divided into phytohormones and phytoinhibitors based on the action pattern [3], although the same PGR can promote or inhibit plant growth depending on the concentration [4]. The terms PGR, phytohormone, and phytoinhibitor are used in an inconsistent manner in some cases [5]. A large number of PGRs are known at present and newer compounds are continually being reported. The major groups include gibberellins (GAs), cytokinins (CK), auxins (AX), ethylene (ET), and abscisic acid (ABA). A miscellaneous group containing a horde of compounds is also known.

The use of GAs, ABA, ET, and AX has become a routine practice throughout the world in plants and crops of high commercial value [6-8]. The resulting gains, i.e., improved quality, increased yield, and desired changes in harvest time of the plant product, are of economic importance. For example, the defoliation and loss of fruit which occurs in grapes at about 50 and 30%, respectively, at the time of harvest are not only overcome by the application of GA_3; application also leads to increased berry size and higher cluster weight [9,10]. The PGRs are thus becoming the industrially important group of chemicals in agriculture and horticulture. Even at present the synthetic and biologically derived PGRs form a sizable agrochemical business though their sales volume and value are smaller than those of the pesticides. It is predicted that the rate of increase in dollar volume will be much faster for PGRs than for pesticides during the next several years.

It is interesting to note that the cost of GA_3 is lower than that of CK and ABA probably due to its commercial production by fungal fermentation. However, no serious attention has been paid by industrial mycologists to produce these other PGRs by fermentation processes. A bright prospect for achieving higher yields and economy in the fermentative production

---

Present affiliation: Institute für Technische Chemie, University of Hannover, Hannover, Germany.

of PGRs also needs to be exploited in light of the recent advances in successful extension of solid state fermentation (SSF) techniques to the production of GA_3 [11]. The present chapter is therefore an effort to critically analyze the available information on the production of PGRs by fungal fermentation.

II. OCCURRENCE

The PGRs occur, as a rule, in the plant kingdom, e.g., angiosperms, gymnosperms, pteridophytes, and bryophytes. The members of CK and AX are also present in the animal kingdom while ET is the only PGR which is also present in such sources as polluted air, smoke, and fumes. Most of the PGRs are also generally produced by different classes of microorganisms such as bacteria, yeasts, actinomycetes, fungi, and even algae. The chemical structures and physicochemical characteristics of the typical members of PGRs are presented in Fig. 1 and Table 1, respectively.

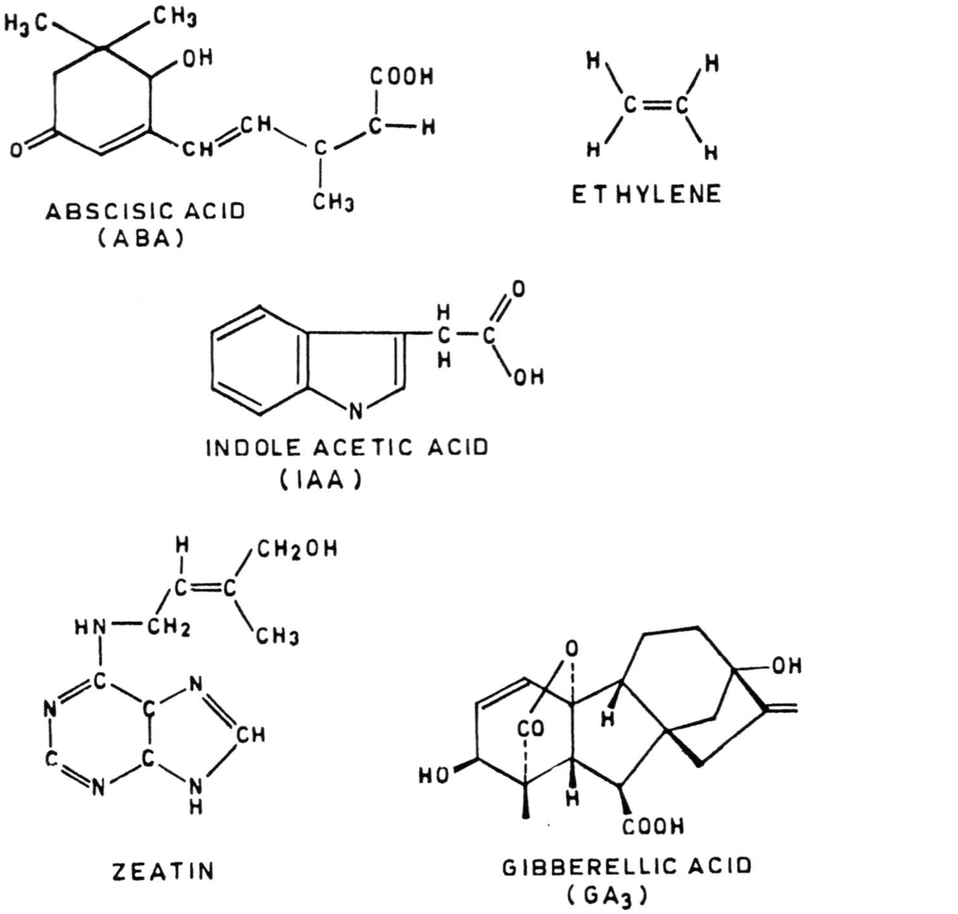

FIGURE 1. Chemical structures of zeatin, ethylene, GA_3, IAA, and ABA.

TABLE 1. Physicochemical Characteristics of Typical Members of Major Groups of PGRs

Common name	Chemical name	Molecular formula	Molecular weight	Melting point (°C)	Solubility in:
Auxin	Indole-3-acetic acid	$C_{10}H_9NO_2$	175.18	168-170	Ethanol, acetone, ether
Gibberellic acid	2,4α,7-Trihydroxy-1-methyl-8-methylenegibb-3-ene-1,10-dicarboxylic acid 1,4α-lactone	$C_{19}H_{22}O_6$	346.37	233-235	Methanol, ethanol, acetone, aqueous $NaHCO_3$, sodium acetate
Kinetin	*N*-(2-Furamylmethyl)-^{1}H-purine-6-amine-N^6-furfuryladenine	$C_{10}H_9N_5O$	215.21	266-267	Methanol, ethanol, cold water, aqueous solution of HCl or NaOH
Ethylene	Ethene	C_2H_4	28.05	-181[a]	Ether, slightly in acetone, ethanol, benzene
Abscisic acid	5-(1-Hydroxy-2,6-6-trimethyl-4-oxo-2-cyclohexen-1-yl)-3-methyl-2,4-pentadienoic acid	$C_{15}H_{20}O_4$	264.31	160-161	Chloroform, acetone, ethyl acetate, ether, aqueous $NaHCO_3$

[a]Freezing point.

III. CURRENT COMMERCIAL PRODUCTION PRACTICES

The routes employed for commercial production of PGRs (Table 2) are based primarily on the availability of the technology, technoeconomic feasibility, and the cost of production. Except for GA_3 and other GAs, the PGRs are mainly produced by chemical synthesis. Small quantities of some PGRs are also produced for use in research by extraction from plant materials.

IV. ADVANTAGES OF FERMENTATIVE PRODUCTION

Most of the PGRs have a complex molecular structure and consequently their production by chemical synthesis involves a series of steps, with the use of cost-intensive reactants/reagents in each step and stringent process controls. The yields of the intermediates and the final product are also usually lower. It is interesting to note that all attempts to chemically synthesize GA_3 at a production cost comparable to that by fermentation technology have not been successful thus far [11], despite the extensive efforts regarding chemical synthesis routes [12,13]. In contrast, the chemical synthesis routes are preferred due to the low cost of raw materials, ease of operation, and involvement of a few efficient synthetic steps in cases involving PGRs with simpler molecules.

Similarly, the extraction of natural PGRs from plant materials poses many problems. The level of GA_3 in most plants is 0.001-1.0 mg of GA_3 equivalent per kg fresh weight [14]. Similarly, 9 mg ABA or 1 mg CK can be obtained from 225 kg of immature cotton fruits [15] or 60 kg sweet corn kernels [16], respectively. The level of ABA in plants also changes rapidly in response to drought [17] and the product undergoes isomerization during extraction [18]. The nonspecific phosphatases present in plant tissues also hydrolyze CK nucleotides to their corresponding nucleosides [19]. The GA_3 is also mostly present in plants in bound form [20] which makes it highly polar as compared to free GA. These problems and limitations dictate that additional steps, stringent controls, and higher expenses for purification be expended to achieve efficient extraction. The availability of plants and plant tissues for extraction of PGRs is also governed by many factors such as the vagaries of nature, seasonal variation in the availability of plant materials and the level of PGRs, and the long cultivation time needed to reach a particular stage in the plant life. The plant material also tends to be limited in supply while labor- and land-intensive methods are required for cultivation and collection of plant materials. It is worthwhile to note that the extraction of GA_3 from plants on a commercial scale is universally accepted as economically nonfeasible as it cannot compete with fermentation technology [11].

The recent advances in plant cell and tissue cultures and the possibility of operation at the 20,000-liter level [21] also opens a potential avenue for their use in the production of PGRs as they offer many advantages [22]. However, this route has not yet been exploited commercially, probably due to poor economy as compared to fermentation technology, as well as several limitations [23,24]. The questionable feasibility of the production of PGRs through plant cell culture on a commercial level was also stressed by Spier and Fowler [24].

TABLE 2. Commercial Production Practices Employed Currently for the Production of PGRs

Group	Commercially available PGRs	Production scale	Production practice
Gibberellins	Gibberellic acid	Large	Fungal fermentation
	GA_1, GA_4, GA_7, GA_9	Small	Fungal fermentation
Cytokinins	cis, trans, and mixed isomers of zeatin; trans and mixed isomers of zeatin ribosides; kinetin and its ribosides	Medium	Chemical synthesis
	t-RNA	Small	From plants, animal, yeasts, and *E. coli*
Auxins	Ten different compounds	Large	Chemical synthesis
Abscisic acid	(+)-enantiomer, (-)-enantiomer, (+)-cis-trans isomer, mixed isomer	Small	Chemical synthesis
Ethylene	Ethylene	Large	Chemical synthesis

TABLE 3. Advantages of Fermentation Technique Over the Route Involving Extraction from Plant Materials

Easier to cultivate under near-ideal and closely controlled conditions
Increased yield of product by controlling cultivation parameters
Shorter generation time of the fungal culture
Brief fermentation batch time of 1-10 days vs. 2 months to several years in the case of plants
Needs comparatively smaller area
Easier and comparatively rapid process optimization
Highly amenable to genetic alterations for achieving higher productivity
Easier scale-up of the process and design of fermentation plant
Simpler expansion of plant capacity to meet increased demand
Adoptability of fermentation plant to switch from one PGR to another or to any other value-added biochemicals in the event of changing demand.

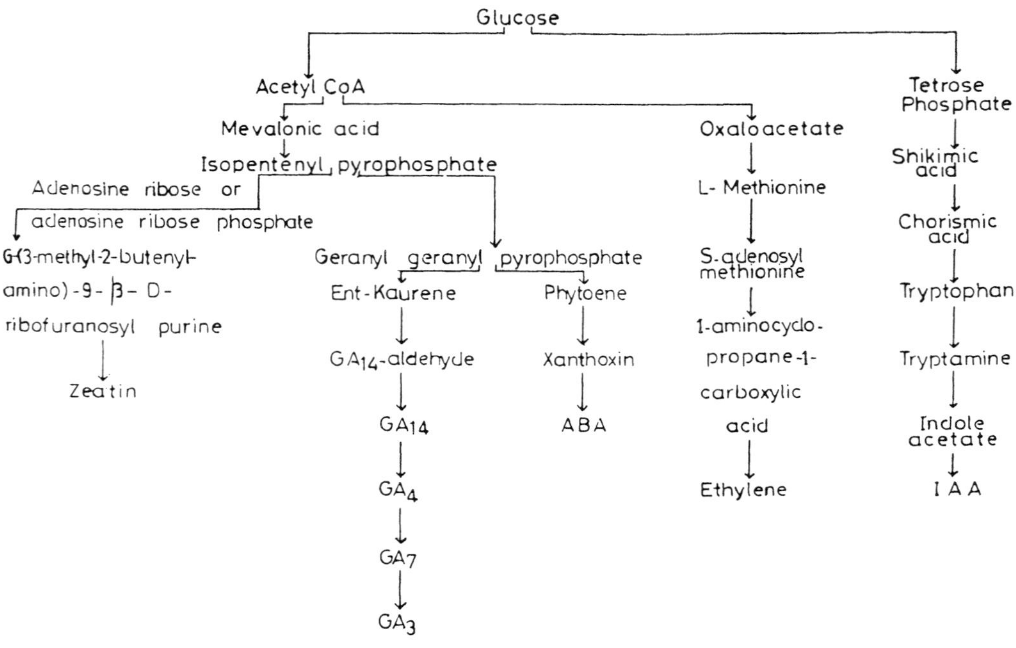

FIGURE 2. Broad steps involved in the biosynthesis of GA_3, ABA, ethylene, zeatin, and IAA.

As compared to the above three routes, the fermentation process technology for the production of PGRs offers a comparatively easier and largely amenable alternative with the capability to largely overcome the limitations associated with other routes. The general advantages of the fermentation route over processes based on extraction from plant materials are presented in Table 3. The basic steps involved in the biosynthesis of PGRs are depicted in Fig. 2.

V. FUNGI VS. OTHER MICROORGANISMS

The selection of a microorganism in any fermentation process is a critical feature that often governs the overall efficiency of the process. Remarkably few species of microorganisms are employed on an industrial scale as compared to the existence of vast numbers of microorganisms. Various criteria specified for selection of a strain in an industrial fermentation process [25] are usually followed. However, fungal organisms are often preferred as they offer many advantages over other types of microorganisms (Table 4). The well-established fermentation process for production of GA_3 also uses fungal cultures. To date only fungal strains are known to produce ABA.

TABLE 4. Advantages of Fungal Cultures Over Other Types of Microorganisms

Capability to grow in liquid surface, solid state, and submerged fermentation processes
Comparatively easier separation of cell biomass
Yields of the products superior and many times higher in some cases by solid state fermentation as compared to submerged fermentation
Metabolites mostly extracellular, thus eliminating the need for cell rupture in downstream processing
Extreme versatility in their nutrition and to grow in extreme environment
Formation of sexual and asexual spores in most fungi offers advantages in inoculum development
Capability of increasing the fermentation plant size indefinitely without altering the ratio between protoplasmic volume and surface area
Amenability to grow in filamentous, fragmented mycelial, and pellet forms
Highest suitability for growth in solid state fermentation
Comparatively less danger of the loss of fermentation batch due to phage contamination
Well-established use in many industrial fermentation processes

VI. GIBBERELLINS

A. Status

A total of 71 GAs are known to date and more are being discovered [11] owing to the possibilities of permutations in GA structure [26]. Of these, a total of 25 GAs are produced by microorganisms [27], but GA_3 has received greatest attention. A total of 43 fungal cultures belonging to 25 genera and 39 species, in addition to many bacteria, yeasts, and actinomycetes, are known to produce GA_3, other GAs, and GA-like compounds [11], *Gibberella fujikuroi* and *Fusarium monoliforme* are employed on an industrial scale because of their higher yields. The strains used are selected after extensive screening programs and are further improved by genetic manipulations [28,29]. The industrially employed strain also coproduces GA_1, GA_4, GA_7, and GA_9 in small quantities [11] and these have some advantages over GA_3 in specific plants and crops [30,31] but the difficulty of their separation poses problems [32]. Other cometabolites produced include products related to GA_3, degradation products of GA_3, plant growth repressors, PGRs other than GAs, and general metabolites [11,33]. A critical and exhaustive state of the art on fermentative production of GA_3 was recently published [11].

The submerged fermentation (SmF) process involving the use of the fermentors containing 5000-6000 gal medium [34] is used currently by the industry to produce GA_3 [35] though liquid surface fermentation was employed in the earlier period [36]. The yields achieved are 0.6-1.0 g GA_3/liter fermentation medium [11] depending on the use of precursors such as mevalonic acid, (-)-kaurene, (-)-isokaurene, (-)-kaurenol, (-)-13-epimanoyl oxide, and alkaline earth salts of senecicic acid [37,38]. The SmF process employed by the industry is of the most advanced quality and highest possible efficiency [11]. Extensive kinetic studies [39,40], optimization using statistical methods [41,42], and formulation of mathematical models [43] were also performed. Exploratory attempts have also been made at the laboratory/bench scale for evaluating some of the improved fermentor operating strategies such as two- or multistep fermentation [44-46], fed-batch culture [34,47], continuous fermentation [48,49], and the use of immobilized whole cells [50,51]. In spite of these attempts, GA_3 is a costly biochemical selling at U.S. $1000-3000/Kg depending on purity and potency. The analysis of fermentation data collected over the period of 18 years showed that the currently employed process offers negligible scope for further economy [52]. The technology used by the industry is a closely guarded secret and the fermentative production of GA_3 is a monopoly of few industrial concerns in the world. In most countries, its requirement is wholly met by import and a restraint exists on its extensive use for socioeconomic benefit.

B. Factors Affecting Production

Being a secondary metabolite, the fermentation involves trophophase and idiophase [53], which may be further subdivided into five to six phases [40,49,54]. The onset of idiophase is triggered by exhaustion of one of the essential nutrients, thereby implicating the importance of initial concentrations of essential nutrients such as nitrogen, phosphate, and carbohydrates [39,40,49,54]. The yield of GA_3 is also governed by the type and concentration of carbohydrates [11,39,40,55]. The range of various parameters employed (Table 5) shows that these are based on the strain used [11]. Other factors, such as the use of air supplemented with CO_2 [56], supplementation of the medium with other carbon and nitrogen sources [57,58], and inclusion of precursors [11], are also important. Foam control is obtained by use of octadecanol [35] or by allowing the foam to pass through a hole in the lid of the fermentor [54]. The economics of the use of precursors was doubted by Jefferys [33].

C. Downstream Processing

Downstream processing involves handling of a large volume of liquid for separating extremely low quantities of GA_3 [11], and thus is cost-intensive when compared to other fermentation products such as citric or gluconic acids. After separation of mycelial cells by filtration or centrifugation, GA_3 is either adsorbed on suitable resins/adsorbents [35,59] or extracted in appropriate solvents [20]. The resins and/or adsorbents used are charcoal, diatomaceous earth, strongly basic anionic exchange resins, Amberlite XAD-4, Amberlite IRA-401S, and Wofatite SBW/MB in the -OH form [11]. From the resins/adsorbents, GA_3 is eluted with ammonical methanol, acetone, aqueous solution of electrolytes, buffered solution of alkaline earth metal salts, or ether. Ethyl acetate is used mostly for extracting GA_3 from cell-free culture broth [11]. Further purification involves a series of operations such as repeated liquid-liquid partitioning, concentration under vacuum, and final processing to obtain either amorphous powder or crystals of GA_3. A total of 14 different downstream-processing strategies and some of the

TABLE 5. Physicochemical Factors Influencing Production of GA_3

Parameter	Range used	Remarks
Initial pH	3.5-5.8	Depends on strain, no control during fermentation
Temperature	25-34°C	Maintained during fermentation
Aeration	0.25-0.50 vvm	Biosynthesis of GA_3 progresses through compounds with increasing order of oxidation
Agitation	100-1400 rpm	
Light	Dark or different intensity	Contradictory reports
Glucose	Less than 1% at any given time	Controlled feeding
Ammonium nitrate	1.0-4.0 g/liter	Critically important concentration
C/N ratio	6:1 to 188:1	Most important factor
Mg^{2+}, K^+, P^{3+}	As used in Czapek-Dox or Raulin-Tom media	Leads to higher yield
Cu^{2+}, Fe^{2+}, Mn^{2+}, Zn^{2+}, Ca^{2+}, B^{3+}, Al^{3+}, MO^{4+}	As supplied by impurities present in other constituents of the medium	Essential for higher productivity
Inoculum ratio	2-10%	Successive development of active spore inoculum

additional steps for specific purposes were recently detailed [11]. It is worthwhile to stress that the strain used by the industry also produces fusaric acid [36,60], fusarinic acid [36], and another unidentified plant growth inhibitor [61] as concomitant products along with GA_3. These are not completely removed even by the efficient downstream-processing strategy and are present in low concentrations in some commercial products [11]. These compounds are diluted to nonphytotoxic levels when commercial GA_3 is solubilized to the desired concentration before application [33].

VII. CYTOKININS

A. Status

To date a total of 27 different CKs are chemically identified [19] in addition to a large number of cytokininlike compounds [5,62,63]. These promote the division and differentiations of plant cells and tissues as well as the control of senescence in plants [19,62]. Some CK also evokes response in animal and mammalian cells, a characteristic unique to this group of PGRs [62]. As compared to GA_3, the use of CK is on an extremely small

TABLE 6. Fungal Cultures Capable of Producing Free Cytokinins and Cytokininlike Compounds

Fungal cultures	Ref.
Agaricus bisporus, Amanita rubescens	66,67
Boletus cothurnatus, B. edulis var. *pinicolus*	68,69
Botrytis anthophyla, Coprinus micaceous	70,71
Cytospora sp., *Dictyostelium discoideum*	72,73
Erysiphe graminis var. *hordei, Heleboma* sp.	74,75
Exobasidium myrtilli, E. uvae-ursi	76
Helminthosporium avenae, H. turcicum	77,78
Monilinia fructicola, M. fructigena, M. laxa	72
Nectria galligena, Plasmodiophora brassicae	79,80
Pleurotus sojor-caju, Rhizopus roseolus	81,82
Rhizopogen ochracearubens, R. roseolus, R. luteolus	66,82,83
Suillus punctipes, Ustilago esculenta	84,85
Taphrina sp., *T. deformans, T. cerasi*	86,87
Taphrina amentorum, T. betulina	88
Uromyces phaseoli, U. fabae	89,90
Ectendotrophic, mildew and rust fungi	84,91,92

scale and is mainly confined to research laboratories and plant cell/tissue cultures. CK is also used in the manufacture of commercial growth regulator mixtures such as promalin, a combination of GA_3 and CK, which is commercially used for improving typiness in the Red Delicious apple [64]. The cost of CK is very high as compared to that of GA_3, ranging between 0.0076 and 12.27 U.S. \$/mg [31].

Diverse bacteria, fungi, yeasts, actinomycetes, and algae, isolated from rhizosphere regions of plants, have been reported to produce CK [5,62,65]. Among these, a total of 30 species of fungi, including rust and mildew pathogens capable of producing CK, are shown in Table 6. Some fungi such as *Uromyces phaseoli, Trichoderma viride, Taphrina betulina, T. deformans, T. cerasi,* and *T. amentorum* also contain cytokinin moieties in their tRNA constituents [5,62]. It is obvious that the obligate parasitic, mycorrhizal, ascomycetous, and basidiomycetous fungi have no potential in fermentative production of CK due to problems in cultivation and slow growth rate.

B. Factors Affecting Production

Studies on the production of CK by fungi have mostly been of an exploratory nature dealing with the ability of different species to elaborate various types of CK [5]. In fact, no serious attempt has been made thus far to formulate an efficient fermentation medium or the process for developing highly productive and cost-effective technology. Few reports, including high yield in several media, medium formulation based on workable combinations, and ease in purification of the product, have been made [82] while the growth parameters were selected arbitrarily [84]. Moreover, all reports are based on laboratory scale processes in flasks except for the two reports involving a 7.5-liter fermentor containing 4 liters of medium [93] and a 10- to 15-liter batch carried out in small carboys [82]. Consequently, information on factors affecting the production of CK by fungi is meager (Table 7).

C. Downstream Processing

The strategies employed do not give proper consideration to efficiency of the unit operations and cost factors in most reports. The methods are

TABLE 7. Parameters Employed in the Production of CK by Fungi

Parameter	Details[a]
Inoculum ratio	1.25% [93], liquid form [93], or chopped mycelial mat [82]
Fermentation medium	Hagem's [82], modified Hagem's [94], potato dextrose liquid [86], complex biological [95], Knop's solution [79], chemically defined [93,96]
Organic N_2 source	Critically important, also affect cell growth [79]
Incubation temp., °C	20 [39], 22 [95], 27 [82,96], ambient [84]
Initial pH	5.9 [93], 6.0-6.3 [95,97], 6.5 [79], not optimized, no maintenance during fermentation
Agitation	100-200 rpm [93], 200 cycles/min at 1.5-in. strokes [95]
Aeration	Done, rate not specified
Light	Incubation in dark [96], under incandescent light [93]
Foam control	Spraying of jar by Antifoam-A before autoclaving [93]
Fermentation time, days	4-5 [93], 12 [79], 6-27 [82], 21 [84]
Production rate	Influenced by asparagine/adenine [76], inhibition and stimulation by postculture liquids of Actinomycetes [83], varies at different stages of cap opening [98], higher during formation of fructification bodies [72].

[a]Figures in brackets are reference numbers.

TABLE 8. Downstream Processing Schemes Employed in the Production of Cytokinins

Miller [82], Johnston and Trione [93]

Mycelium-free culture broth
↓
Acidification to pH 2.5 with 1 N HCl
↓
Absorption on Dowex 50W-X8, H^+
↓
Elution with 5 or 6 N NH_4OH
↓
Evaporation in vacuo at 40°C for ammonia removal
↓ (Miller [82], Johnston and Trione [93]; → Crafts and Miller [84], see below)
pH adjustment to 8.0 and extraction with 1:1 volume of *n*-butanol
↓
Organic phase evaporation
↓
Crystallization

Crafts and Miller [84] (continuing from evaporation in vacuo at 40°C for ammonia removal)

Solubilization with 6 g K_2HPO_4 in 10 ml aqueous concentration
↓
Five extractions with ethyl acetate
↓
Evaporation and reextraction with water
↓
Evaporation and solubilization of the residue in little ethanol
↓
Separation on silica gel column
↓
Elution with water saturated with butanol
↓
Crystallization

Kovacs [98]

Mycelium-free culture broth
↓
Column chromatography using polyvinyl pyrrolidone at pH 6.4
↓
Elution with phosphate buffer
↓
Cytokininlike compound

based on selective adsorption of CK on specific ion exchange resins [82] or its selective extraction into appropriate solvents [87]. Further purification is achieved by repeated fractionation or partitioning with different solvents and chromatography. The isomers of the CK are separated on a reverse phase column using a gradient elution with increasing amount of methanol [99]. Some of the methods employed are presented in Table 8.

VIII. AUXINS

A. Status

As a generic term, auxins (AXs) include a variety of compounds that are generally acids with an unsaturated cyclic nucleus or their derivatives [69], which are grouped together on the basis of their ability to induce elongation of Avena coleoptiles. Some AXs also exert other plant growth regulating effects in the plants. To data, a total of 86 AXs are known that include a group of antiauxin compounds which either show inhibitory action or dual activity based on the dose employed. Indole-3-acetic acid (IAA) and indole-3-butyric acid (IBA) are some of the important members among a total of 38 AXs formed in higher plants and other natural sources. Their cost is cheaper than that of GA_3, ABA, and CK and ranges around U.S. $17.5-39.3/25 g [101]. It is emphasized that the use of IBA has become routine throughout the world [102].

AXs are produced by bacteria, fungi, yeasts, actinomycetes, and algae [3,5] under various conditions such as pathogenesis, rhizosphere-mycorrhizal association, and microbial interactions [103]. An exhaustive list of a total of 115 fungal genera comprising 343 species is available [5] based on the positive response for AX production by both bioassay and chromatography or chromatography alone. Obviously, the number will be much higher if the reports relying on bioassay alone are also considered. The major genera are shown in Table 9. Recently, the ability of *Balansia epichlöe* [104] and *Neurospora crassa* [105] to produce AX was also reported. All the reports on production of AX by fungi are confined to shake-flask level [104,105] except for a lone report involving production using large vessels to gather 140 liters of fermented medium in several lots [106]. Being a secondary metabolite, the fermentation phase involving *Rhizopus suinus* consists of rapid growth to reach maximum mycelial mass on the second day, low production of AX up to the second day, rapid increase in AX during the third day, and peak production on the fourth day [107].

B. Factors Affecting Production

There is limited information available on physicochemical and nutritional factors affecting the production of AX based on flask level experiments.

The initial pH of the medium has been shown to exert significant influence on the production of AX by different fungal species and strains [107]. The dependence of tryptophan uptake on the pH of the medium and enhanced production of AX in medium buffered with phosphate [108] give indirect evidence for the beneficial effects of pH control. The requirement of two different pH levels for optimum growth of the culture and maximum production of AX [109] and the pH-mediated shifts in the metabolic routes leading to bioconversion of indole acetaldehyde to IAA or tryptophol [110] were also reported. However, no information on its effect on AX

TABLE 9. Major Fungal Genera Capable of Producing Auxins

Genera	No. of species	Genera	No. of species
Aspergillus	22	*Alternaria*	8
Fusarium	27	*Boletus*	8
Helminthosporium	10	*Phytophthora*	8
Penicillium	15	*Rhizopus*	8
Ustilago	13	*Gloeosporium*	7
Taphrina	15	*Gibberella*	3
Exobasidium	13	*Cunninghamella*	5

production is available. The initial pH values employed by various workers were in the range of 4.5-7.0 [104,107,111]. The production of AX is also affected by incubation temperature [111], which is governed by interference with the uptake of tryptophan [112]. The incubation temperatures were in the range of 24-29°C [103,104,108,111,113]. The incubation period necessary for reaching peak in AX level was reported to be 3-12 days for most of the cultures [79,103,114-118]. In case of some slow-growing cultures such as *Boletus* spp., *B. felleus*, *Amanita frostiana*, and *Balansia epichlöe*, the incubation period required was several months, i.e., up to 20 months [104,113], and their ability to produce AX thus must be viewed cautiously [5]. The increase in AX production with the increase in incubation period as reported by some workers [102,113,119] is probably related to the secondary metabolite nature of AX and the C/N ratio employed in the medium [107].

The decomposition of IAA was to the extent of 10-15% and 40% at 1500 and 2800 lux units of fluroescent light, respectively. No decomposition after several hours exposure at 1000 lux [120] was reported. Moreover, the decomposition after exposure for 7 days to 3200 lux units of red and blue light was found to be 50 and 90%, respectively [120]. In spite of such a drastic effect, the culture flasks were not protected from light in many cases [103-106] except for the experiments involving total dark incubation [108,113,114,121]. The incubation of the culture flasks either on shakers [76,105,121] or in stationary state [103,104,108,113,114] was generally employed, appearing to be determined by the earlier belief that IAA is decomposed by the oxygen present in air or the fact that no decomposition occurs even after aeration at a rate of 1 vol air/vol medium/min in a 3-liter airlift fermentor or in 500-ml-capacity aluminum foil-capped flasks shaken at 140 rpm for 12 days [120]. The higher production of IAA under aerated-agitated as compared to stationary or anaerobic cultures [76,107,122,123] or due to aeration at 25 liters air/h [106] were also reported. In one case, a two-stage fermentation process involving shaking of a sporulation medium for 10 days followed by stationary culture in TM medium for 4 weeks was used to obtain higher yields [104].

Although no systematic studies have been carried out on the effect of carbon sources on auxin production, the available information indicates that the efficiency of the synthesis of IAA, especially in the absence of tryptophan is governed by the carbohydrate source [103]. In general, 2-3% glucose [78,106,108] or 3-4% sucrose [104,113,114] was employed. Other carbon compounds such as sorbitol, succinic acid, citric acid, and tartarate were also added to the medium [104,115,124], although no information was given on their role in AX production. The addition of phenols such as chlorogenic acid, gallic acid, morin, rutine, and phloridzin to tryptophan-containing media was shown to result in enhanced IAA production [125,125].

Both inorganic and organic nitrogenous compounds, including amino acids, were reported to be effective although the response varied in different fungal cultures [76,107,108,127-133]. Although no chemical specificity of the nitrogen source was stressed [107], the production of AX was reported to be 10 times higher when inorganic nitrogen source was replaced by organic nitrogen compounds [76,106,107]. The use of a comparatively higher concentration of inorganic nitrogen for better growth of the culture and enhanced production of AX was shown to result in adverse effects on AX titers due to the increased osmotic value of the medium [107]. Among the organic nitrogen sources evaluated, tryptophan, uric acid, and tyrosine were found to be effective [115,129] based on the strain employed. Other organic nitrogen sources used were peptone [106], hemoglobin [134], and egg albumin [135]. It is interesting to note that the production of AX by *Aspergillus flavus* was 1.45 and 4.57 times higher in mineral-egg albumin medium as compared to the same medium without egg albumin and glucose-peptone medium, respectively [135]. The copious production of AX in casein-supplemented medium by *Helminthosporium oryzae* was reported [136] and the role played by proteinase produced by the culture in partial release of AX from protein-auxin complex as well as the presence of tryptophan in casein were stressed.

The essential nature of a high molecular weight amino acid for IAA formation by fungi and its identification as tryptophan are the important landmarks in AX research [106,115]. The presence of tryptophan in peptone [106] and the consequent higher yields of IAA with the use of peptone or tryptophan-containing media were subsequently confirmed by many workers [106,107,130,137]. In general, the media were supplemented with 0.10-0.24% DL- or L-tryptophan [104,113,114] as the concentration of tryptophan varies in different brands and batches of the peptone [5]. The AX was not formed by some cultures in the absence of tryptophan [104,113,114,132]. In contrast, some cultures formed AX even in the absence of any added tryptophan [108,115,129,133] because they are able to form it in appreciable amounts as a metabolic byproduct under proper experimental conditions. In another set of cultures, a different pathway for biosynthesis of IAA was reported that does not involve tryptophan [138]. Similarly, the different yields of IAA from tryptophan by some fungal cultures [5] could be explained by its diversion to the production of a number of indole derivatives such as indoleacetylnitrite, indolepyruvic acid, tryptamine, and tryptophol [137]. It is interesting to note that the tryptophan is not required for the growth of the culture [106].

Increased production of IAA due to Mg^{2+}, Zn^{2+}, phosphate, and pyridoxine and marked reduction by 0.2 M KCl were reported [76,107,131]. Interestingly, Mg^{2+} was shown to have no effect on growth of *Rhizopus suinus* [107] in contrast to that of *Exobasidium myrtolli* [5]. Vitamins such as thiamine hydrochloride and nicotinic acid have been incorporated in the synthetic media [104] but no information is available on their requirement.

TABLE 10. Typical Schemes Employed in Downstream Processing of Auxins Produced by Fermentation

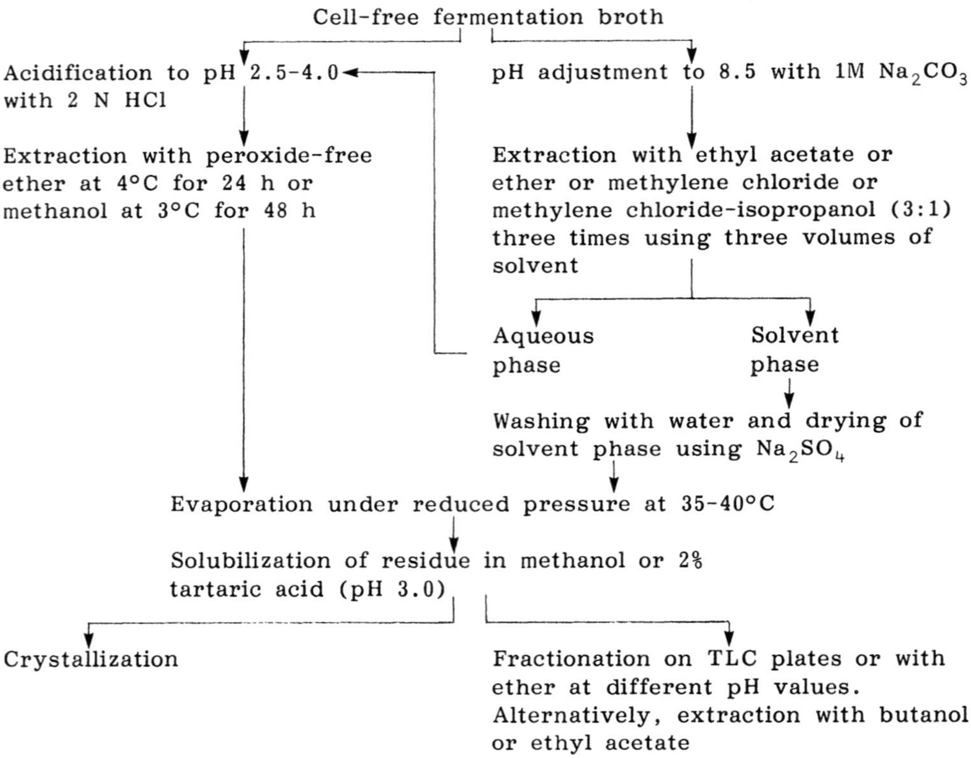

Source: Schemes are from Refs. 104,108,113,114,116, and 121.

C. Downstream Processing

At the end of fermentation, the AX from cell-free media is usually extracted in appropriate solvents and concentrated at reduced pressure. In some cases, the AX from the solvent phase was adsorbed on suitable resins, eluted, and fractionated before concentration to obtain a pure, amorphous or crystalline product. Detailed data on purification parameters have also been reported in the case of IAA production using *Rhizopus suinus* [106]. Some of the typical downstream-processing strategies employed at a laboratory scale are presented in Table 10.

IX. ABSCISIC ACID

A. Status

Abscisic acid (ABA), a stress hormone found in plants and some fungi [17], is a lone member of its class of PGRs except for the presence of its precursors, conjugates, and metabolites in natural sources [139]. The natural ABA consists entirely of (+)-enantiomer while the synthetic (-)-enantomer has almost similar physiological effects though at lower magnitude

[139]. ABA is the most expensive of all the major groups of PGRs. For example, (+)-cis-trans isomer of about 98% purity is available at U.S.$ 422.65 for 5 mg packing [31]. It is interesting to note that the physiological effects of ABA in plants are opposite to those of GA_3, although mevalonate is a common precursor [140]. Similarly, it is produced by fungi in relatively higher amounts as compared to plants [17], although it is a secondary metabolite and not essential for fungal growth [141].

To date, only four phytopathogenic fungal species belonging to three genera, i.e., *Cercospora rosicola* [141], *C. cruenta* [142], *Diplocarpon rosae* [143], and *Botrytis cinerea* [144] have been reported to produce ABA. The titers of ABA reported are 10 and 52 mg/liter in 9 and 7 days, respectively, by *C. cruenta* [142] and *B. cineria* [144]. In addition, ABA was also detected in *Penicillium italicum* collected from infected oranges but it was not able to produce ABA on Czapek-Dox agar. Thus, it was concluded that the fungus possibly absorbed ABA from orange tissue [145]. The conclusion is probably not valid as an inability of *C. rosicola* to produce ABA in Czapek-Dox agar, even after enrichment with thiamine, and an ability to produce it in other agar media are well documented [141,146]. Surprisingly, no report on the ability of bacteria, yeasts, and actinomycetes to produce ABA is available though the production of lunalaric acid, i.e., ABA-like compounds produced by algae is known [139].

B. Factors Affecting Production

In spite of the relatively recent discovery of ABA, a great deal of information on *C. rosicola* is available. However, it is based on the use of various liquid and agar media [141-144, 146-148] at laboratory scale. Fermentors have not been used in spite of the recognition of the potential of the fermentation route [146].

The initial pH of the medium employed was in the range of 6.5-6.8 [141,147,149] while it was not even specified in some cases [142-144]. Changes in the pH of the medium after autoclaving were also given in selecting stable nitrogen source [146]. The increased production of ABA due to pH control in potato dextrose medium enriched with narcissus extract was also reported [147]. The use of $CaCO_3$ or $Mg_2(PO_4)_2$ for pH control was found to have varied effects such as reduced production of ABA by $Mg_2(PO_4)_2$, inhibition of cell growth and ABA production by $CaCO_3$. Improved production of ABA due to higher aeration of the medium resulting from adherence of the cells to $CaCO_3$ particles rather than to walls of the culture tube and effective control of pH at 7.0-8.2 throughout the period of 20 days incubation was also reported [148]. The ability of 0.1% corn steep liquor and the use of a higher inoculum ratio to overcome the inhibition of cell growth and ABA production caused by $CaCO_3$ and $Mg_2(PO_4)_2$ was also reported [148].

The effect of incubation temperature and aeration/agitation on ABA production appears to be negligible. The selection of proper incubation temperature is probably governed by the optimum growth temperature of the culture. It ranged between 22 and 28°C in all reported studies [140-144,146-150]. The incubation time employed was from 7 days to over 1 month based on the culture and physicochemical as well as nutritional parameters employed. In general, the flasks were incubated on rotary or reciprocating shakers except for the use of stationary culture in a few cases [141,143]. The increased production of ABA due to improved aeration in $CaCO_3$-containing media also indicates the necessity of aeration [147]. The inoculum ratio employed was 2.0-2.5% (v/v) in most cases [146, 147,150]. It was usually grown on agar media and blended in a Waring blender with sterile water for 30 s to 10 min [140,147]. A synthetic medium was also used [150].

The increase in the production of ABA by *C. rosicola*, to an extent of 2-5 times in media incubated under light as compared to those in dark, has been reported [144,146,147]. However, the cultures incubated under light contained 5-40% more 2-*trans*-ABA (t-ABA) than those incubated in dark [146]. A light-catalyzed isomerization of ABA to t-ABA is indicated by several observations such as (a) the presence of about 35% t-ABA and one-twentieth of ABA in medium incubated below black light blue lamp as compared to a value of 2% t-ABA when incubated below blue light [144], and (b) the decrease in the ratio of ABA/t-ABA with time [146]. In contrast, the product from *Botrytis cineria* contained 2.0 and 2.4% t-ABA when cultured under blue light or in the dark, respectively [144]. It is interesting to note that the incubation was invariably carried out under light except for one report involving *C. rosicola* incubated in the dark [147], whereas information on this aspect was not given in case of *Diplocarpon rosai* [143].

Glucose has been a major carbon source employed in synthetic and biological media used for production of ABA by *C. rosicola*. Its substitution by other carbon or carbohydrate sources leads to either the same or lower yields [146,148,150] as well as reduction in cell growth [148]. Contradictory results, i.e., complete inhibition or enhanced production of ABA, have been reported regarding the effect of ammonium salts as well as amino and organic acids when used individually to supplement potato dextrose or synthetic media [146,148]. Other significant trends include about 77% inhibition of ABA formation by leucine [146], a major component of peptone [141], and enhanced production of ABA with the use of combination of two or three specific amino acids as compared to a single amino acid of highest efficiency [146]. The data on the effect of inorganic nutrients on ABA production are too scanty as most of the workers probably depended on the presence of these compounds as impurities in medium constituents.

The requirement of vitamins and growth factors for improved cell growth and ABA production have been indicated and these include thiamine [146] and corn steep liquor [148]. Certain biological products such as yeast or malt extract, oat meal, and potato are usually included in the media as sources of growth factors [141]. The enhancement of ABA production by plant extracts from narcissus, carnation, rose, stock, and chrysanthemum in contrast to the inhibition by those from fressia, tiger lily, and iris has been reported [147]. Narcissus extract is most effective, increasing the production of ABA by 36 times while its inhibitory action on cell growth could be overcome by adding it to the medium on the fifth day of fermentation instead of at the start. Detailed descriptions of various phases of the fermentative process for ABA production in liquid media with and without pH control as well as solid medium have been reported [146-148]. It is interesting to note that the production of ABA could be increased from 75 to 319 μg/ml of the medium by feeding additional potato dextrose broth to the systems on the third and fourth days [146].

C. Downstream Processing

ABA is an extracellular product [147] and no measurable activity was found in ethyl acetate extracts of the mycelial pellets [146]. The excretion of

TABLE 11. Typical Schemes for Downstream Processing of ABA

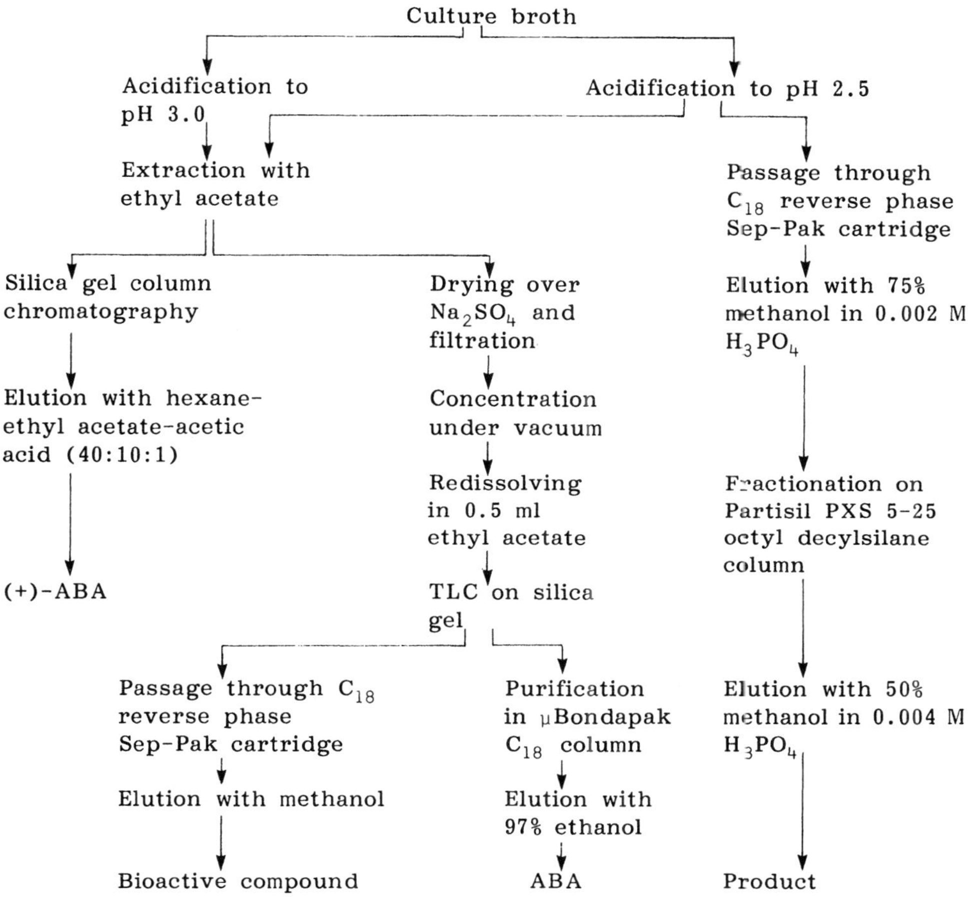

Source: Schemes are from Refs. 142,146,147, and 150.

more than 90% ABA formed during the course of fermentation and 70% extraction efficiency of ABA from the medium to which it was added at a concentration of 100 μg/ml has been reported [147]. It is also necessary to carry out extraction in either dark or dim light to avoid isomerization of the product. Some of the typical downstream-processing strategies employed at the laboratory scale are presented in Table 11.

X. ETHYLENE

A. Status

Ethylene (ET), a simple organic compound that has been recognized as a PGR [151], is also a stress or ripening hormone as it is induced in plants in response to parasitic infection [5] and hastens ripening in certain plants [62]. It is also unique in existing in the gaseous state [62] and is the cheapest of all the PGRs. It has practical application in accelerating the ripening of sugarcane, enhancing the flow of the latex in the rubber

TABLE 12. Fungal Cultures Not Included in the Extensive List of Ethylene Producers by Mahadevan [5] and Also Those Reported in Subsequent Years

Culture	Ref.
Bipolaris sorokiniana	152
Botrytis cinerea	153
Collectotrichum dematium var. *fruncatum*	154
Fusarium oxysporum f. sp. *vasinfectum*	154
Penicillium chrysogenum	155
P. italicum	156
P. urticae	157
Pleurotus osteratus	158
Pythium ultimum	159
Rhizoctonia solani	160
Septoria musiva	161
Trichomonas vaginalis	162
Verticillium dahliae	154

tree, and stimulating root formation as well as promoting flowering in pineapple [7]. Its importance can be established by the fact that the regulation of plant growth by all natural and synthetic PGRs is exerted by their action on ET production and/or activity [7]. Due to the difficulties in the handling and application of gaseous ET, it is used in the form of 2-chloroethylphosphonic acid, an ET-generating compound. An extensive list of bacteria, yeasts, fungi, and actinomycetes, with the capability of producing ET, is available and includes 51 species of 26 fungal genera [5]. The fungal cultures that are not included in this list and also those reported in the subsequent years are cited in Table 12.

B. Factors Affecting Production

In spite of the recognition of the promises of the fermentation processes, especially from the viewpoint of the global shortage of crude oil deposits [155,157], the work on the production of ET to date has been confined to laboratory scale. This might be due to the need to use culture vessels for retaining gaseous ET or carrying it to a collection vessel as ET is not soluble in aqueous media [155]. Though ET production and cell growth were stated to be directly related in earlier years [163,164], it was subsequently established that the growth precedes ET formation [165-167]. It is further confirmed by the detailed work on various phases involved in the production of ET by fungal cultures [166-168].

In general, media with an initial pH in the range of 3.0-6.8 were employed [155,157,166,167] without pH control except in one report [166]. The cultures were incubated at 22-28°C for a period of 3-8 days [155,157,

166,169] and enhanced production was reported under light [154]. The existence of varying metabolic routes in fermentations carried out under static or shaken culture and the essential nature of oxygen in the process have been indicated as in the following: (a) linear relationship in ET production vs. time as long as O_2 was present in the bioreactor [157]; (b) inhibition of ET production by low levels of O_2 and high concentrations of CO_2 [167]; ability of methionine to act as a inducer in shaken culture but not in the static state [170]; differences in phosphate-mediated control of ET production in static vs. shaken cultures [171]; and higher yields in static culture [166]. In most cases, the cultures were incubated on shakers [157,169] and even humidified air was passed through the culture vessel at a rate of 250 ml/min [172].

Though various carbohydrates can support the growth of *Septoria musiva*, it was shown to produce ET only in glucose-containing media [161, 173]. However, glucose itself was not efficient in shake culture unless used in combination with methionine in *Penicillium digitatum* [170], whereas *Rhizoctonia solani* produced ET in glucose-containing media with reduced oxygen tension [160]. The ability of glucose, mannose, mannitol, glycerol, alanine, and citric acid to promote ET production; reduction in its production with the use of sucrose; and higher effectiveness of sucrose as compared to fructose have been reported [163,174]. Of 24 amino acids evaluated, methionine and L-cysteine were the most effective in stimulating ET production by *S. musiva* [173]. Other substrates such as methionine, ethionine, glucose, sucrose, starch, acetate, and succinate [159] as well as casamino acids and L-arginine were reported to be effective for ET production by fungi [165].

Enhanced ET production with ferrous sulfate in *P. digitatum* [165] and with zinc, boron, iron, and cupric sulfate in *S. musiva* [161] and inhibited production with zinc, cobalt, and nickel in the former and nickel chloride in the latter culture were reported. The phosphate-mediated control of ET production in *P. digitatum* was indicated including 200- to 250-fold increase in ET production in low-phosphate media. The reversion of the effect by addition of phosphate in the media and markedly lower intracellular acid-soluble phosphate levels in ET producer vs. non-ET producer strains were reported [171,175]. Vitamins and growth factors such as riboflavin, biotin, complete vitamin mixture, and yeast extract were reported for their stimulatory action on growth as well as ET production by *P. digitatum* and *Verticillium dahliae* [154,155,167]. The autoclaving of the yeast extract was also shown to result in reduced production of ET by 10-30% [167].

The compounds reported as precursors in the biosynthesis of ET in fungi include methionine, acrylic acid, glutamic acid, 2-ketoglutaric acid, acetate, and citrate [158,166,176,177]. However, methionine was effective only in shaken cultures of *P. digitatum* [170]. Its action was shown to be concentration dependent and involved induction of a specific enzyme system. Moreover, some of these precursors also acted as substrates [154, 170]. Some compounds such as NaN, KCN, catalase, and monofluoroacetate were reported to be inhibitory for ET biosynthesis [154,166].

It is interesting to note the autocatalytic effect of ET on its own production and the influence of gaseous atmospheric conditions on ET production as indicated by significantly higher production in cotton-stoppered flasks compared to sealed flasks [167]. Consequently, the replacement of the gaseous atmosphere in the culture vessel by flushing at least once every 24 h was stressed to maintain maximum growth rate of the culture and enhanced production of ET [167]. Similarly, the medium renewal in static culture, accomplished by removing spent medium present below the fungal mat and its replacement with fresh medium, was shown to lead to a significant increase in the rate of ET production [166]. The higher tonicity of the medium was also reported to result in enhanced ET production [178].

The inoculum was generally developed in modified Pratt's medium using a small moist chamber [167] or by preculture in a fermentation medium to which 3% sodium-carboxymethylcellulose was added [157]. Alternatively, the spore suspension from potato dextrose agar slant was used [166,170]. In one case, the mycelial cells from the preculture were filtered and used directly or after slicing of large pellets/compacted mycelium into 1- to 2-mm^3 sections [169]. The single-celled fungi grown in preculture were also centrifuged before use as inoculum [169]. The inoculum ratio used was 5% v/v [165,170] or 0.1-0.2 g dry weight of the cells per 10 ml glass syringe, a culture vessel [169].

XI. MISCELLANEOUS REGULATORS

In addition to the PGRs belonging to the five major groups described above, a large number of fungal metabolites with capabilities to regulate plant growth have been reported. Some of these are recent reports. All of these are mostly neglected. These regulatory compounds are listed in Table 13.

XII. POTENTIAL OF SOLID STATE FERMENTATION TECHNIQUE

Solid state fermentation (SSF) technique, involving fermentation of moist, water-insoluble solid substrate such as wheat bran by microorganisms in the absence of free water [201], offers numerous advantages over submerged fermentation (SmF) process. These include superior productivity, lower capital investment, reduced energy requirement, dispensation with rigorous control of some fermentation parameters, easier product recovery and scale-up, compactness of the fermentor, lower wastewater outout, elimination of foam problems, and simpler growth media, plant, machinery, fermentor control system, and inoculum build-up procedure [201-203]. With the realization of the product being in too dilute an SmF broth and the consequent high cost in downstream processing as well as disposal of large volumes of waste effluent [204,205], more attention is currently being paid to SSF technique [206]. In spite of the suitability of SSF technique to fungal strains, the information available on its extension to the production of PGRs is negligible.

The exploratory efforts on the extension of SSF technique to GA_3 production were not successful due to instability of a strain of *Gibberella fujikuroi* FT-2 [11]. Subsequently, a laboratory scale process was developed after selection of a stable strain and extensive studies on all aspects of the development of an efficient and economic fermentation process [11]. The preliminary cost calculation and its comparison with presently used SmF process revealed several gains. For example, the SSF process yields 1.05 and 1.54 g GA_3/kg dry moldy bran (DMB) under batch and fed-batch processes [207-209]. Consequently, the expenses on fermentation medium ingredients at these yields works out to be 71 and 49%, respectively, as

TABLE 13. Some Miscellaneous Plant Growth Regulators

PGR	Fungal culture capable of production	Ref.
Aspertin	*Aspergillus nidulans*	179
3-Carboxy-2,4-diphenylbut-2-enoic anhydride	*Aspergillus nidulans*	180
Cladosporium A and B	*Cladosporium cladosporioides*	181
Cotylenin A and B	*Cladosporium* spp.	182
Coumarin, herniarin, umbelliferon	*Phytophthora infestans*	183
Fusicoccin	*Fusicoccini amygdali* Del	184
Helminthosporal	*Helminthosporium sativum*	185
N-Jasmonyl- and *N*-dihydroxyjasmonoylisoleucine	*Gibberella fujikuroi*	186
Malformin	*Aspergillus niger*	187
Nagilactone	*Podocarpus nagi*	188
Neovasinone	*Neocosmophora vasinfecta*	189
Nigeragin A and B	*Aspergillus niger*	190
Pestolotin	*Pestalotina cryptomariaecola*	191
Pevalic acid	*Penicillium variabile*	192
L-β-Phenyllactic acid	*Exobasidium* spp.	193
2-pyruvoylaminobenzamide	*Collectotrichum lagenarium*	194
Radiclonic acid	Unidentified fungus	195
Sclerin	*Sclerotinia sclentiorum*	196
Terrein	*Aspergillus terreus*, *Penicillium raistrichii*	197,198
Victorin	*Helminthosporium victoriae*	199
Viridicatin	*Penicillium crustosum*	200

compared to SmF process giving 0.6 g GA_3/liter [210]. Similarly, about 57 and 39% volume of the extract would be necessary to be subjected to downstream-processing operations at these yields, respectively, under SSF as compared to SmF processes. These calculations are based on the recovery of GA_3 from 1 kg DMB in 1 liter extract by applying multiple-contact countercurrent leaching technique [211] in which vacuum concentration of extract is not needed. This constitutes a tremendous gain in economy as the ratio of enzyme recovery to the fermentation cost in SmF processes is as high as 2.0 [212] and even the separation of the biomass from fermented medium by centrifugation was shown to involve 48% of the total production cost [205].

Considerable savings in effluent disposal can also occur due to higher concentration of the product in the SSF process [204,206,213]. It is also universally accepted that the expenditure on effluent treatment is a most significant cost element in fermentation processes [214]. The economic gains in other aspects can also be judged from the data on the comparative economics in the production of amyloglucosidase under both SmF and SSF processes [215].

Likewise, the economic benefits will also be available in production of other PGRs under SSF technique. However, limited information is available on the potential of SSF technique for the production of AX and ET while indirect indications exist in case of CK and ABA. For example, the essential requirement of solid substrates in AX production by *Absidia ramosa* and *Rhizopus suinus* were stressed in earlier years [127,216]. In addition, marked increase in the production of AX by *R. suinus* due to increase in surface area of the culture from 16.6 to 575 cm^2 was also reported [107], although it was stressed that the solid substrate apparently has no significant effect on AX production. Unfortunately, all auxinologists subsequently concentrated the efforts on the use of liquid media in AX production while SSF technique has been ignored primarily due to a report on the production of AX by *Aspergillus niger* both in solid and liquid media [134].

The production of ET in specific stages of the growth of commercial mushrooms on moist solid media under SSF technique is well documented [158,172]. The culture conditions, similar to those operative under SSF technique, were also reported to lead to enhanced production of ET. For example, ET production was about 20 times higher in still culture as compared to shaken cultures, although the growth in standing culture was about one-fourth that in shaken cultures [167]. Some other observations such as (a) a submerged pelleted growth of the culture in shaken flask which was unable to reverse the trend of lower ET production even when further incubated in stationary culture, (b) the direct relationship between increased production of ET in stationary culture with the surface growth of the culture, and (c) reduction in ET production due to accidental submersion of the surface mat in stationary culture collectively indicate the potential of SSF technique in ET production.

It is interesting to note that the species of *Taphrina* and *Exobasidium* formed yeastlike cells and spores in surface layers on the parasitized tissues of the host plant [93]. When grown by SmF process in synthetic media, the CK production by these fungi was too small to cause extensive morphological changes that occur naturally in the host plant in spite of the growth in SmF medium in the form of yeastlike cells [93]. It was therefore stressed that the production of CK by pathogenic hyphal cells of *Taphrina* growing on host tissues was undoubtedly different in quantities [93]. The close resemblance of the growth of fungi under SSF technique to the above type of growth of the fungi on the host tissues is well known [201,203].

The indications showing the potential of SSF technique in the production of ABA include (a) higher (fivefold) [147] production of ABA, in solid media vs. liquid broth [141], (b) erratic and unpredictable growth of the fungus in shake culture involving the use of complex natural extracts [149] and, (c) increased production of ABA in media containing certain plant extracts [147]. The preparation of plant extract involves homogenization of the material by cost-intensive centrifugation for the separation of the residue. On the other hand, the use of these plant materials in the SSF process will not involve drastic homogenization and thus the centrifugation can be eliminated.

XIII. R&D NEEDS

Further R&D efforts in GA_3 production by the SmF process may not be promising due to saturation in terms of yield and economy [11]. Several avenues, however, should be explored in order to further exploit the advantages of the SSF process:

1. Scale-up studies, design of efficient fermentation processes, determination of commercial viability, and comparison of economics with the best SmF process to stimulate industrial interest.
2. Formulation of an efficient and economic process for purification of the product, as the methodology employed at laboratory scale [217] is cost-intensive and largely unsuitable for industrial scale applications.
3. Development of kinetic and other mathematical models for various relationships in the SSF process as well as engineering studies on transport phenomena for efficient plant design and optimization.
4. Investigations on separation of brassinosteroidlike compounds, concomitantly produced along with GA_3 in SSF process [217], for obtaining costly byproduct or parameter standardization to optimizing its yield, and development of alternate technology to its extraction from plants or chemical synthesis.
5. Identification and characterization of other concomitant products, such as novel sterols, with capability to disturb defensive mechanisms in plants against microbial attack. A polyketide pigment with antimicrobial properties also deserves attention [217].

Based on the state of the art of the fermentative production of other PGRs such as CK, AX, ABA, ET, and miscellaneous compounds, the R&D needs that emerge are as follows:

1. Extensive screening of fungal cultures for selecting best producing strains;
2. Standardization of physicochemical and nutritional factors to achieve increased production;
3. Development of efficient fermentation process and cost-effective downstream processing strategies which are practicable for use at an industrial scale;
4. Comparison of optimized fermentation process with the currently employed chemical synthesis routes/extraction from plant materials for comparative economics, and
5. Extension of SSF technique and determination of its comparative economics against the SmF process

In addition, the economic importance of other aspects may also be investigated. For example, it is well recognized that the development of the mycorrhizosphere on most plants by fungi is a chance encounter in agricultural and economic plants. The inoculation of the plant system with suitable fungal inoculants may prove useful in overcoming these uncertainties and also in improving the development of the plants due to the beneficial effects of the PGRs produced by them. Investigations on these aspects and studies on the feasibility of commercial production of appropriate inoculants are therefore of significance.

Similarly, the ability of filamentous fungi to penetrate deeply into intracellular and intercellular spaces in the SSF process is well known

[201,218]. The moldy substrate obtained from SSF process is thus a firmly bound complex, very similar to immobilized growing cells [11], and was used earlier in the continuous production of cellulases in a packed column reactor [219]. The use of such complexes of solid substrate and PGR-producing mycorrhizal or rhizosphere fungi in agriculture, horticulture, and forestry as well as on economic plants may have a longer-lasting effect than spraying of the PGR. Such application may also become economical since it will eliminate major expenses on cost-intensive downstream processing operations.

Intensive R&D efforts on some of the miscellaneous plant growth regulators may also lead to useful and potent PGRs for use in specific commercial applications.

XIV. CONCLUSIONS

The PGRs produced by fungi form a industrially important group of fungal metabolites. Among them, the fermentative production of GA_3 is the only example of the commercial exploitation of fungal biotechnology. The other major and miscellaneous PGRs are more or less neglected by industrial biotechnologists, and information on these is widely scattered. An effort to provide a condensed but reasonably specific state-of-the-art capsule on fungal PGRs with a view to capturing the attention of fungal biotechnologists and stimulating the interest of the fermentation industry in these commercially valuable fungal metabolites has been attempted. The potentials of fermentation techniques for the production of the PGRs was also analyzed and extension of the SSF technique is recommended.

ACKNOWLEDGMENTS

The authors thank Dr. M. S. Prasad for critical comments on the manuscript and several useful remarks. Thanks are also due to Dr. N. G. Karanth and Dr. B. L. Amla for the interest in the work. P. K. R. Kumar is thankful to the Council of Scientific and Industrial Research, New Delhi, for the research fellowship.

REFERENCES

1. Letham, D. S., Higgins, T. J. V., Goodwin, P. B., and Jacobsen, J. V. Phytohormones in retrospect. In *Phytohormones and Related Compounds: A Comprehensive Treatise, Vol. 1, The Biotechemistry of Phytohormones and Related Compounds* (D. S. Letham, P. B. Goodwin, and T. J. V. Higgins, eds.), Elsevier, Amsterdam, 1978, pp. 1-27.
2. Anonymous. Nomenclature of Plant Regulators. *Plant Physiol.*, *29*:307-308, 1954.
3. Robbins, W. J. The expanding concepts of plant growth regulation. In *Plant Growth Regulation: Fourth International Conference on Plant Growth Regulation*, Iowa State University Press, Ames, 1961, pp. 13-21.
4. Thimann, K. V. The natural plant hormones. In *Plant Physiology, Vol. 6B: Physiology of Development—The Hormones* (F. C. Steward, ed.), Academic Press, New York, 1972, pp. 3-365.

5. Mahadevan, A. *Growth Regulators, Microorganisms and Diseased Plants.* IBH, New Delhi, 1984, pp. 1-468.
6. Nickell, L. G. Plant Growth Regulators. *Chem. Eng. News, 56(41)*:18-34, 1978.
7. Nickell, L. G. *Plant Growth Regulators: Agricultural Uses*. Springer-Verlag, Berlin, 1982, pp. 1-173.
8. Nickell, L. G. *Plant Growth Regulating Chemicals*, Vols. 1 and 2. CRC Press, Boca Raton, 1983, pp. 1-280 and 1-256.
9. Weaver, R. J. Effect of gibberellic acid on fruit set and berry enlargement in seedless grapes of Vitis vinifera. *Nature, 181*:851-852, 1958.
10. Sidahmed, O. A., and Kliewer, W. M. Effects of defoliation, gibberellic acid and 4-chlorophenoxy acetic acid on growth and composition of Thompson seedless grape berries. *Am. J. Enol. Vitic., 31*:149-153, 1980.
11. Kumar, P. K. R., and Lonsane, B. K. Microbial production of gibberellic acid: State of the art. In *Advances in Applied Microbiology*, Vol. 34 (S. L. Neidleman, ed.), Academic Press, New York, 1989, pp. 29-139.
12. Corey, E. J., Danheiser, R. L., Chandrasekaran, S., Siret, P., Keck, G. E., and Grass, J. L. Stereospecific total synthesis of gibberellic acid. *J. Am. Chem. Soc., 100*:8031-8034, 1978.
13. Hook, J. M., Mander, L. N., and Rudolf, U. Total synthesis of gibberellic acid. The hydrofluorene route. *J. Am. Chem. Soc., 102*:6628-6629, 1980.
14. Graebe, J. E., and Ropers, H. J. Gibberellins. In *Phytohormones and Related Compounds* (D. S. Letham, P. B. Goodwin, and T. J. V. Higgins, eds.), Elsevier, Amsterdam, 1978, pp. 107-204.
15. Ohkuma, K., Lyon, J. L., Addicott, F. T., and Smith, O. E. Abscisin II, an abscission-accelerating substance from young cotton fruit. *Science, 142*:1592-1593, 1963.
16. Letham, D. S. Zeatin, a factor inducing cell division from Zea mays. *Life Sci., 2*:569-573, 1963.
17. Hirai, N. Abscisic acid. In *Chemistry of Plant Hormones* (N. Takahashi, ed.), CRC Press, Boca Raton, 1986, pp. 201-247.
18. Lenton, J. R., Perry, V. M., and Saunders, P. F. The identification and quantitative analysis of abscisic acid in plant extracts by gas-liquid chromatography. *Planta, 96*:271-280, 1971.
19. Koshimizu, K., and Iwamura, H. Cytokinins. In *Chemistry of Plant Hormones* (N. Takahashi, ed.), CRC Press, Boca Raton, 1986, pp. 154-199.
20. Russell, S. Extraction, purification and chemistry of gibberellins. In *Gibberellins and Plant Growth* (H. N. Krishnamoorthy, ed.), Wiley Eastern Ltd., New Delhi, 1975, pp. 1-34.
21. Fowler, M. W. The large scale cultivation of plant cells. *Prog. Ind. Microbiol., 16*:207-229, 1982.
22. Millis, N. F. The organisms of biotechnology. In *Comprehensive Biotechnology, Vol. 1, The Principles of Biotechnology: Scientific Fundamentals* (A. T. Bull and H. Dalton, eds.), Pergamon Press, Oxford, 1985, pp. 7-19.
23. Fowler, M. W. Commercial application and economic aspects of mass plant cell cultures. In *Plant Biotechnology* (S. H. Mantell and H. Smith, eds.), Cambridge University Press, Cambridge, 1983, pp. 3-37.
24. Spier, R. E., and Fowler, M. W. Animal and plant cell cultures. In *Comprehensive Biotechnology, Vol. 1. The Principles of Biotechnology:*

Scientific Fundamentals (A. T. Bull and H. Dalton, eds.), Pergamon Press, Oxford, 1985, pp. 301-329.

25. Bull, A. T., Ellwood, D. C., and Ratledge, C. The changing scene in microbial biotechnology. In *Microbial Technology* (A. T. Bull, D. C. Ellwood, and C. Ratledge, eds.), Cambridge University Press, Cambridge, 1979, pp. 1-27.
26. Crozier, A. Aspects of the metabolism and physiology of gibberellins. In *Advances in Botanical Research* (H. W. Woolhouse, ed.), Academic Press, London, 1981, pp. 33-149.
27. Takahashi, N., Yamaguchi, I., and Yamane, H. Gibberellins. In *Chemistry of Plant Hormones* (N. Takahashi, ed.), CRC Press, Boca Raton, 1986, pp. 57-151.
28. Erokhina, L. I. Gibberellin production by some mutants of Fusarium monoliforme on media containing a plant oil as a carbon source. *Genetika, 7*:77-82, 1967.
29. Imshenetskii, A. A., and Ul'yanova, O. M. Biochemical activity of Gibberella fujikuroi mutants. *Mikrobiologia, 31*:832-837, 1962.
30. Taylor, B. K. Reduction of apple skin russeting by gibberellins A_{4+7}. *J. Hort. Sci., 50*:169-172, 1975.
31. Sigma Chemical Co. *Biochemicals and Organic Compounds for Research and Diagnostic Clinical Reagents*. Sigma Chem. Co., St. Louis, 1988, pp. 1-1864.
32. Martin, G. C. Commercial uses of gibberelline. In *The Biochemistry and Physiology of Gibberellins*, Vol. 2 (A. Crozier, ed.), Praeger, New York, 1983, pp. 395-444.
33. Jefferys, E. G. The Gibberellin fermentation. In *Advances in Applied Microbiology*, Vol. 13 (D. Perlman, ed.), Academic Press, New York, 1970, pp. 283-316.
34. Bergman, D. E., Denison Jr., J. E., and Friedland, W. O. Culture process for gibberellic acid. U.S. Patent 3,021,261, 1962.
35. Stodola, F. H., Raper, K. B., Fennell, D. I., Conway, H. F., Johns, V. E., Langford, C. T., and Jackson, R. W. The microbiological production of gibberellin A and X. *Arch. Biochem. Biophys., 54*:240-245, 1955.
36. Yabuta, T., Kambe, K., and Hayashi, T. Biochemistry of the bakane fungus of rice. Part I. Fusaric acid, a new product of the bakane fungus. *J. Agr. Chem. Soc., 10*:1059-1068, 1934.
37. Imperial Chemical Industries Ltd. Gibberellic acid. Br. Patent 957,634, 1964.
38. Redemann, C. T. Gibberellins. U.S. Patent 2,918,413, 1959.
39. Borrow, A., Brown, S., Jefferys, E. G., Kessell, R. H. J., Lloyd, E. C., Lloyd, P. B., Rothwell, A., Rothwell, B., and Swait, J. C. The kinetics of metabolism of Gibberella fujikuroi in stirred culture. *Can. J. Microbiol., 10*:407-444, 1964.
40. Borrow, A., Brown, S., Jefferys, E. G., Kessell, R. H. J., Lloyd, E. C., Lloyd, P. B., Rothwell, A., Rothwell, B., and Swait, J. C. The effect of varied temperature on the kinetics of metabolism of Gibberella fujikuroi in stirred culture. *Can. J. Microbiol., 10*:445-466, 1964.
41. Maddox, I. S., and Richert, S. H. Use of response surface methodology for the rapid optimization of microbiological media. *J. Appl. Bacteriol., 43*:197-204, 1977.
42. Ganchev, K., Krachanova, M., and Popova, R. Studies on the isolation of gibberellic acid. *Acta Microbiol. Bulg., 15*:43-49, 1984.

43. Gohlwar, C. S., Sethi, R. P., Marwaha, S. S., Seghal, V. K., and Kennedy, J. F. Gibberellic acid biosynthesis from whey and simulation of cultural parameters. *Enz. Microb. Technol.*, *6*:312-316, 1984.
44. Birch, A. J., Grove, J. F., and Nixon, I. S. Gibberellic acid. Br. Patent 844, 341, 1959.
45. Borrow, A., Jefferys, E. G., and Nixon, I. S. Gibberellic acid. U.S. Patent 2,906,670, 1959.
46. Borrow, A., Jefferys, E. G., and Nixon, I. S. Gibberellic acid. U.S. Patent 2,906, 673, 1959.
47. Abbott Laboratories, Br. Patent 919,186, 1963 as cited by Jefferys [25].
48. Holme, T., and Zacharias, B. Gibberellic acid formation in continuous culture. *Biotechnol. Bioeng.*, *7*:405-415, 1965.
49. Bu'Lock, J. D., Detroy, R. W., Hošťálek, Z., and Munim-Al Shakarchi, A. Regulation of secondary biosynthesis in Gibberella fujikuroi. *Trans. Br. Mycol. Soc.*, *62*:377-389, 1974.
50. Kahlon, S. S., and Malhotra, S. Production of gibberellic acid by fungal mycelium immobilized in sodium alginate. *Enz. Microb. Technol.*, *8*:613-616, 1986.
51. Kumar, P. K. R., and Lonsane, B. K. Immobilized growing cells of Gibberella fujikuroi P-3 for production of gibberellic acid and pigment in batch and semicontinuous cultures. *J. Appl. Microbiol. Biotechnol.*, *28*:537-542, 1988.
52. Vass, R. C., and Jefferys, E. G. Gibberellic acid. In *Economic Microbiology*, Vol. 3, *Secondary Products of Metabolism* (A. H. Rose, ed.), Academic Press, London, 1979, pp. 421-433.
53. Bu'Lock, J. D., Hamilton, D., Hulme, M. A., Powell, A. J., Smalley, H. M., Shepherd, D., and Smith, G. N. Metabolic development and secondary biosynthesis in Penicillium urticae. *Can. J. Microbiol.*, *11*: 765-778, 1965.
54. Borrow, A., Jefferys, E. G., Kessell, R. H. J., Lloyd, E. C., Lloyd, P. B., and Nixon, I. S. The metabolism of Gibberella fujikuroi in stirred culture. *Can. J. Microbiol.*, *7*:227-276, 1961.
55. Darken, M. A., Jensen, A. L., and Shu, P. Production of gibberellic acid by fermentation. *Appl. Microbiol.*, *7*:301-303, 1959.
56. Borrow, A., Jefferys, E. G., and Nixon, I. S. Gibberellic acid from Gibberella fujikuroi. Br. Patent 803,591, 1958.
57. Sanchez-Marroquin, A. Microbiol production of gibberellic acid in glucose media. *Appl. Microbiol.*, *11*:523-528, 1963.
58. Maddox, I. S., and Richert, S. H. Production of gibberellic acid using a dairy waste as the basal medium. *Appl. Environ. Microbiol.*, *33*:301-302, 1977.
59. Borrow, A., Brian, P. W., Chester, V. E., Curtis, P. J., Hemming, H. G., Henehan, C., Jefferys, E. G., Lloyd, P. B., Nixon, I. S., Morris, G. L. F., and Radley, M. Gibberellic acid, a metabolic product of the fungus Gibberella fujikuroi: some observations on its production and isolation. *J. Sci. Food Agr.*, *6*:340-348, 1955.
60. Kurosawa, E. Concerning the results of transplanting rice seedlings affected with bakane disease. *Ann. Phytopathol. Soc.*, *4*:33-34, 1934.
61. Thakur, M. S., and Vyas, K. M. Production of plant growth regulators by some Fusarium spp. *Folia Microbiol.*, *28*:124-129, 1983.
62. Letham, D. S., Goodwin, P. B., and Higgins, T. J. V. (eds.). *Phytohormones and Related Compounds: A Comprehensive Treatise*,

Vol. 1, The Biochemistry of Phytohormones and Related Compounds; and Vol. 2, Phytohormones and the Development of Higher Plants. Elsevier, Amsterdam, 1978, pp. 1-641 and 1-648.
63. Letham, D. S., and Miller, C. O. Identity of kinetin-like factors from Zea mays. *Plant Cell. Physiol.*, *6*:355-359, 1965.
64. Greenhalgh, W. J., and Godley, G. L. *Aust. J. Exp. Agr. Animal Husb.*, *16*:592, 1976 as cited by Nickell [7].
65. Greene, E. M. Cytokinin production by microorganisms. *Bot. Rev.*, *46*:25-74, 1980.
66. Miller, C. O. Ribosyl-trans-zeatin, a major cytokinin produced by crown gall tumour tissue. *Proc. Natl. Acad. Sci. USA*, *71*:334-338, 1974.
67. Miller, C. O. Naturally occurring cytokinins. In *Biochemistry and Physiology of Plant Growth Substances, Proc. 6th Int. Conf. Plant Growth Substances* (F. Wrightman and F. Setterfield, eds.), Runge Press, Ottawa, 1968, pp. 33-45.
68. Miller, C. O. Cytokinin production by mycorrhizal fungi. USDA Misc. Publ. No. 1189, 168-174, 1971.
69. Gogola, N. Effect of natural cytokinins of Pinus sylvestris and other growth substances on the mycelial growth of Boletus edulis var. pinicolus Osterr. *Bot. Z.*, *118*:321-333, 1970.
70. Mazin, V. V., Ivanova, O. A., Churikova, V. V., and Talieva, M. N. Cytokinin activity of culture medium of Botrytis anthophyla Bond, the pathogen of red clover anther mold. *Mikol. Fitopatol.*, *14*:385-388, 1980.
71. Szabo, L., Pozsar, D. I., and Kota, M. Cytokinin produced in relation to the development of pea root nodules. *Can. J. Bot.*, *54*:2155-2162, 1970.
72. Vizárová, G. Contribution to the study of cytokinin production by phytopathogenic fungi. *Biol. Plant.*, *17*:380-382, 1975.
73. Nomura, T., Tanaka, Y., Abe, H., and Uchiyama, M. Cytokinin activity of discadenine: a spore germination inhibitor of Dictyostelium discoideum. *Phytochemistry*, *16*:1819-1820, 1977.
74. Vizárová, G. Determination of cytokinine in barley after an attack by powdery mildew. *Biol. Bratisl.*, *26*:29-56, 1971.
75. Karapetyam, O. A. Elaboration of physiologically active kinin compounds by rhizosphere microorganisms. *Biol. Zh. Arm.*, *26*:72-76, 1973.
76. Norberg, S. O. Studies in the production of auxins and other growth stimulating substances by Exobasidium. *Symbol. Bot. Upsal.*, *19*:1-117, 1968.
77. Mandahar, C. L., and Arora, R. K. Infection of oat (Avena sativa L) leaves by Helminthosporium avenae Eiderm II. Secretion of cytokinins by the pathogen in culture. *Phytopathol. Medit.*, *17*:52-54, 1978.
78. Mandahar, C. L., and Arora, A. K. Secretion of cytokinins by germinating conidia of Helminthosporium turcicum. *Natl. Acad. Sci. Lett.*, *1*:245-246, 1978.
79. Taris, B., and Clemencet, Y. Extraction and study of growth substances produced in pure culture by Nectria galligena, an agent of European canker in fruit trees. *C.R. Acad. Sci. Ser. D.*, *271*(1):78-81, 1970.
80. Ingram, D. S. Abnormal growth of tissues infected with Plasmodiophora brassicae. *J. Gen. Microbiol.*, *56*:56-57, 1969.

81. Dua, I. S., and Jandaik, C. L. Cytokinins in two cultivated edible mushrooms. *Sci. Hort.*, *10*:301-304, 1979.
82. Miller, C. O. Zeatin and zeatin-riboside from a mycorrhizal fungus. *Science*, *157*:1055-1056, 1967.
83. Edmunds, S., Kempest, M., and Michalski, L. Production of cytokinin like substances by mycorrhizal fungi of pine (Pinus sylvestris) in cultures with and without metabolites of actinomycetes. *Acta Microbiol. Pol.*, *34*:177-186, 1985.
84. Craft, C. B., and Miller, C. O. Detection and identification of cytokinins produced by mycorrhizal fungi. *Plant Physiol.*, *54*:586-588, 1974.
85. Chan, Y. S., and Thrower, L. B. The host-parasite relationship between Zizania caduciflora Turcz. and Ustilago esculenta P. Henn. IV. Growth substances in the host parasite combination. *New Phytol.*, *85*:225-233, 1980.
86. Sommer, N. F. Production by Taphrina deformans of substances stimulating cell elongation and division. *Physiol. Planta*, *14*:460-469, 1961.
87. Barthe, P., and Bulard, C. Identification of one cytokinin from pure culture of Taphrina cerasi by gas chromatography. *Can. J. Bot.*, *52*:1515-1518, 1974.
88. Kern, H., and Naef-Roth, S. Zur bildung von auximen und cytokinen durch Taphrina arten. *Phytopathol. Z.*, *83*:193-222, 1975.
89. Dekhuijzen, H. M., and Staples, R. C. Mobilization factors in uredospores and bean leaves infected with bean rust fungus. *Contrib. Boyce Thompson Inst.*, *24*:39-52, 1968.
90. Kiraly, Z., Pozsar, B., and Hammady, M. El. Cytokinin activity in rust-infected plants: juvenility and senescence in diseased leaf tissue. *Acta Phytopathol.*, *1*:29-37, 1966.
91. McNew, G. L. The broader concepts of plant growth regulation. In *Plant Growth Regulation, Proceedings of the Fourth International Conference on Plant Growth Regulation*. Iowa State University Press, Ames, 1961, pp. 3-11.
92. Engelbrecht, L. Cytokinin in den "Grunen Inseln" des herbstlabes. *Flora*, *159*:208-214, 1968.
93. Johnston, J. C., and Trione, E. J. Cytokinin production by the fungi Taphrina cerasi and Taphrina deformans. *Can. J. Bot.*, *52*:1583-1589, 1974.
94. Modess, O. *Symbolae Bot. Upsal.*, *5*:1, 1941, as cited by Miller [49].
95. Abe, H., Uchiyama, M., Tanaka, Y., and Saito, H. Structure of discadenine, a spore germination inhibitor from the cellular slime mold Dictyostelium discoideum. *Tetrahedron Lett.*, 3807-3810, 1976.
96. Laloue, M., and Hall, R. H. Cytokinins in Rhizopogon roseolus: secretion of N-(9-B-D-ribofuranosyl-9H) purin-6yl carbamyl threonine into the culture medium. *Plant Physiol.*, *51*:559-562, 1973.
97. Tanaka, Y., Abe, H., Uchiyama, M., Taya, Y., and Nishimura, S. Isopentenyladenine from Dictyostelium discoideum. *Phytochemistry*, *17*:543, 1978.
98. Kovacs, E. Cytokinin activity of Agaricus bisporus. *Acta. Aliment.*, *11*:169-178, 1982.
99. Peter, M., and Willy, H. Isomers of zeatin and zeatin riboside in clubroot tissue: Evidence for trans-zeatin biosynthesis by Plasmodiophora brassicae. *Physiol. Planta*, *66*:240-250, 1986.

100. Marumo, S. Auxins. In *Chemistry of Plant Hormones* (N. Takahashi, ed.), CRC Press, Boca Raton, 1986, pp. 9-56.
101. Aldrich Chemical Company. *Catalog Handbook of Fine Chemicals*, Milwaukee, 1988, pp. 1-2212.
102. Thimann, K. V. Fifty years of plant hormone research. *Plant Physiol.*, *54*:450-453, 1974.
103. Singaracharya, M. A., Reddy, S. R., and Reddy, S. M. Influence of carbon sources on IAA synthesis by two seed borne fungi. *Curr. Sci.*, *48*:781-782, 1979.
104. Porter, J. K., Bacon, C. W., Cutler, H. G., Arrendale, R. F., and Robbins, J. D. In vitro auxin production by Balansia epichlöe. *Phytochemistry*, *24*:1429-1431, 1985.
105. Tomita, K., Kitsuwa, T., Murayama, T., and Nakamura, T. Identification of indole-3-acetic acid in Neurospora crassa. *Agr. Biol. Chem.*, *51*:2633-2634, 1987.
106. Thimann, K. V. On the plant growth hormone produced by Rhizopus suinus. *J. Biol. Chem.*, *109*:279-291, 1935.
107. Bonner, J. The production of growth substance by Rhizopus suinus. *Biol. Z.*, *52*:565-582, 1932.
108. Fenn, P., Durbin, R. D., and Kuntz, J. E. Synthesis of tryptophol and O-acetyl-tryptophol from tryptophan by Ceratocystis fragacearum. *Phytochemistry*, *16*:899-901, 1977.
109. Nielsen, N., and Hartelius, V. Growth of Aspergillus niger at different hydrogen-ion concentration with and without the addition of growth promoting substance B. *Compt. Rend. Trav. Lab. Carlsberg*, *19*(15): 1-22, 1933.
110. Marchal, P., and Rigaud, J. Sur la presence et le metabolisme de l'indolyl-3-acetaldehyde chez Nectria galligena Bres. var. major Wr. *Compt. Rend. Acad. Soc.*, *271D*:479-481, 1970.
111. Hirata, S. C. Phytohormones in the deformed parts of diseased plants. VI. Production of auxin by Fusarium and the growth promotion of rice seedlings by their culture filtrates. *Bull. Faculty Agr. Univ. Miyazaki*, *4*:119-127, 1958.
112. Marlow, J. L., and Kosuge, T. Tryptophan and indole acetic acid transport in the olive and oleander knot organism Pseudomonas savastanoi (E. F. Smith) Stevens. *J. Gen. Microbiol.*, *72*:211-219, 1972.
113. Mahadevan, A. In vitro production of indole acetic acid by Fusarium vasinfectum Atk. *Experientia*, *27*:433, 1965.
114. Mahadevan, A. Synthesis of indole acetic acid by Fusarium oxysporum f. lycopersici. *Sci. Cult.*, *33*:132-133, 1967.
115. Boysen-Jensen, P. Über die bildung and biologische bedeutung des wachstumm regulators bei Aspergillus niger. *Biochem. Z.*, *250*:270-280, 1932.
116. Barthe, P. Recherche de substances de type cytokinine synthetises par Nectria galligena Bres. var major Wr. en culture pure. *Compt. Rend. Acad. Sci.*, *272D*:2881-2883, 1971.
117. Torev, A. K. The higher hymenomycetes as a source of physiologically active substances for plants. *Dokl. Mosk. Selskokhoz. Akad.*, *70*:179-187, 1961.
118. Ulrich, J. M. Auxin production by mycorrhizal fungi. *Physiol. Planta*, *13*:429-443, 1960.
119. Narania, K., and Reddy, S. M. In vitro production of IAA by five hypomycetes. *Geobios*, *6*:260-262, 1979.

120. Yamakawa, T., Kurahashi, O., Ishida, K., Kato, S., Kodamai, T., and Minoda, Y. Stability of indole-3-acetic acid to autoclaving, aeration and light illumination. *Agr. Biol. Chem.*, *43*:879-880, 1979.
121. Sahai Srivastava, B. I. S., and Shaw, M. The biosynthesis of indole acetic acid in Melampsora lini (Pers.). Lev. *Can. J. Bot.*, *40*:309-315, 1962.
122. Crady, E. E., and Wolf, F. T. The production of indole acetic acid by Taphirina deformans and Dibotryon morbosum. *Physiol. Planta*, *12*:526-533, 1959.
123. Beltra, R., Ballesteros, A. M., and Lahoz, R. Production of growth factors by Nectria galligena. *Microbiol. Espan.*, *22*:41-53, 1969.
124. Dolk, H. E., and Thimann, K. V. Studies on the growth hormones of plants. *Proc. Natl. Acad. Sci. USA*, *18*:30-46, 1932.
125. Drakina, T. I. Effect of phenolic compounds on the synthesis of indoles by Taphrina fungi. Fenol, yue *Soedin. Ikh. Biol. Funkt, Mater, Vses. Simp 1st* 1966, pp. 259-261.
126. Nair, K. J., and Sridhar, R. Phenol-tryptophan interaction and IAA formation by Piricularia oryzae and Helminthosporium oryzae. *Phytopathol. Z.*, *80*:76-78, 1974.
127. Nielsen, N. *Jahrb. Wissensch. Bot.*, *73*:125, 1930, as cited by Gruen [195].
128. Nielsen, N. Über wuches-toffe der hefe. *Biochem. Z.*, *237*:244-246, 1931.
129. Reddy, S. M., Pradeep Kumar, B., and Acharya, M. A. S. Influence of different nitrogen sources on the indole acetic acid synthesis by two fruit rot fungi. *Indian Phytopathol.*, *33*:508-510, 1980.
130. Gentile, A. C., and Klein, R. M. The apparent necessity of indole acetic acid for the growth of Diplodia (fungi, Imperfecti). *Physiol. Planta*, *8*:291-299, 1955.
131. Alghisi, P., Ambra, V. D., and Rude, G. Produzione di auxina da triphfano ad opera Ustilago maydis influenza della vitamin B_6. *Riv. Pat. Veg. Pavia Ser.*, *34*:197-207, 1964.
132. Wolf, F. T. The production of indole acetic acid by Ustilago zeae and its possible significance in tumour formation. *Proc. Natl. Acad. Sci. USA*, *38*:106-111, 1952.
133. Matsuyama, N., and Misawa, T. The hypertrophic disease caused by Taphrina species. IV. Production of auxins and hemiauxin by pathogens. *Tohoku J. Agr. Res.*, *17*:37-49, 1966.
134. Boysen-Jensen, P. Formation of a growth regulator by Aspergillus niger. *Biochem. Z.*, *239*:243-249, 1931.
135. Reddy, M. M., Swami, U. B. S., and Reddy, S. M. Synthesis of IAA by two species of Aspergillus. *Geobios*, *8*:178-179, 1981.
136. Mahadevan, A., and Chandramohan, D. Auxin production by the rice leaf blight fungus Helminthosporium oryzae and other species of Helminthosporium. In *Symposium on Disease of Rice, Maize, Sorghum and Millets*, New Delhi, Indian Phytopathol. Soc., 1966, pp. 91-96.
137. Schwin, F. J. Unter suchungen zum wirtparasit-verhaltnis bei der kragenfaule des apfelbaumes Phytophthora cactorum und zld ihrer bekampfung. II. Zur rolle der indol wuchsstoffe für den erreger. *Phytopathol. Z.*, *54*:162-184, 1965.
138. Lecea de, J. R., Rosa La De, C., and Beltra, R. Growth substances in Agrobacterium tumefaciens and its fixed L-forms: influence of aseptic tumour induction. *Phyton*, *29*:119-126, 1972.

139. Milborrow, B. V. Abscisic acid. In *Phytohormones and Related Compounds: A Comprehensive Treatise*, Vol. 1 (D. S. Letham, P. B. Goodwin, and T. J. V. Higgins, eds.), Elsevier, Amsterdam, 1978, pp. 295-347.
140. Norman, S. M., Poling, S. M., Maier, V. P., and Orme, K. D. Inhibition of abscisic acid biosynthesis in Cercospora rosicola by inhibitors of gibberellin biosynthesis and plant growth retardants. *Plant Physiol.*, *71*:15-18, 1983.
141. Assante, G., Merlini, L., and Nasini, G. (+)-Abscisic acid, a metabolite of the fungus Cercospora rosicola. *Experientia, 33*:1556-1557, 1977.
142. Oritani, T., Ichimura, M., and Yamashita, K. The metabolism of analogs of abscisic acid in Cercospora cruenta. *Agr. Biol. Chem.*, *46*:1959-1960, 1982.
143. Wani, S. P., Rai, P. V., and Patil, R. B. Abscisic acid production by Diplocarpon rosae and its role in disease production. *Curr. Sci.*, *49*:238-239, 1982.
144. Marumo, S., Katayama, M., Komori, E., Ozaki, Y., Natsume, M., and Koudo, S. Microbiol production of abscisic acid by Botrytis cinerea. *Agr. Biol. Chem.*, *46*:1967-1968, 1982.
145. Rudnicki, R., Borecka, H., and Pieniazek, J. Abscisic acid in Penicillium italicum. *Planta, 86*:195-196, 1969.
146. Norman, S. M., Maier, V. P., and Echols, L. C. Influence of nitrogen source, thiamine and light on biosynthesis of abscisic acid by Cercospora rosicola Passerini. *Appl. Environ. Microbiol.*, *41*:981-985, 1981.
147. Takayama, T., Yoshida, H., Araki, K., and Nakayama, K. Microbial production of abscisic acid with Cercospora rosicola. I. Stimulation of abscisic acid accumulation by plant extracts. *Biotechnol. Lett.*, *5*:55-58, 1983.
148. Takayama, T., Yoshida, H., Araki, K., and Nakayama, N. Microbial production of abscisic acid with Cercospora rosicola. II. Effect of pH control and medium composition. *Biotechnol. Lett.*, *5*:59-62, 1983.
149. Norman, S. M., Maiyer, V. P. and Echols, L. C. Development of a defined medium for growth of Cercospora rosicola Passerini. *Appl. Environ. Microbiol.*, *41*:334-336, 1981.
150. Bennett, R. D., Norman, S. M., and Maier, V. P. Biosynthesis of abscisic acid from $1,2^{13}$-C_2) acetate in Cercospora rosicola. *Plant Physiol.*, *20*:2343-2344, 1981.
151. Imaseki, H. Ethylene. In *Chemistry of Plant Hormones* (N. Takahashi, ed.), CRC Press, Boca Raton, 1986, pp. 250-264.
152. Coleman, L. W., and Hodges, C. F. The effect of methionine on ethylene and 1-aminocyclopropane-1-carboxylic acid production by Bipolaris sorokiniana. *Phytopathology, 76*:851-855, 1985.
153. Williamson, C. E. Ethylene, a metabolic product of diseased or injured plants. *Phytopathology, 40*:205-208, 1950.
154. Tzeng, D. D., and Devay, J. E. Ethylene production and toxigenicity of methionine and its derivatives with riboflavin in cultures of Verticillium, Fusarium, and Collectotricum spp. exposed to light. *Physiol. Planta, 62*:545-552, 1984.
155. Fujii, T., Ogawa, T., and Fukuda, H. A screening system for microbes which produce olefin hydrocarbons. *Agr. Biol. Chem.*, *49*:651-657, 1985.

156. Biale, J. B. Effect of emanations from several species of fungi on respiration and colour development of citrus fruits. *Science, 91*:458-459, 1940.
157. Fukuda, H., Fujii, T., and Ogawa, T. Microbial production of C_2-hydrocarbons, ethane, ethylene and acetylene. *Agr. Biol. Chem., 48*:1363-1365, 1984.
158. Yamanaka, K. Ethylene production during the mycelial development and fruit body formation of Pleurotus ostreatus. *Nara-ken Ringyo Shikenjo Kenkyu Hokoku, 11*:1-6, 1982.
159. El-Sharouny, H. M. Screening of ethylene producing root infecting fungi in Egyptian soil. *Mycopathologia, 85*:13-15, 1984.
160. Brown-Skrobot, S. K., and Brown, L. R. The production of carbon monoxide and ethylene by Rhizoctonia solani. *Dev. Ind. Microbiol., 22*:725-731, 1981.
161. Brown-Skrobot, S. K., and Brown, L. R. Ethylene and carbon monoxide production by Septoria musiva. *Dev. Ind. Microbiol., 25*:749-755, 1984.
162. Ishigura, T. Gas chromatographic mass spectrometric analysis of gas produced by Trichomonas vaginalis. *Nippon Sanka Fujinka Gakkai Zasshi, 37*:1097-1102, 1985.
163. Fergus, C. L. The production of ethylene by Penicillium digitatum. *Mycologia, 46*:543-555, 1954.
164. Young, R. E., Pratt, H. K., and Biale, J. B. Identification of ethylene as a volatile product of the fungus Penicillium digitatum. *Plant Physiol., 26*:304-310, 1951.
165. Fukuda, H., Fujii, T., and Ogawa, T. Preparation of a cell-free ethylene forming system from Penicillium digitatum. *Agr. Biol. Chem., 50*:977-981, 1906.
166. Chou, T. W., and Yang, S. F. The biogenesis of ethylene in Penicillium digitatum. *Arch. Biochem. Biophys., 157*:73-82, 1973.
167. Spalding, D. H., and Lieberman, M. Factors affecting the production of ethylene by Penicillium digitatum. *Plant Physiol., 49*:645-648, 1965.
168. Pozout, J., Pazoutova, S., and Vancura, V., Effects of light, phosphate and oxygen on ethylene formation and conidiation in surface culture of Penicillium cyclopium. *Curr. Microbiol., 7*:133-136, 1982.
169. Ilag, L., and Curtis, R. W. Production of ethylene by fungi. *Science, 159*:1357-1358, 1968.
170. Chalutz, Z., Lieberman, M., and Sisler, H. D. Methionine-induced ethylene production by Penicillium digitatum. *Plant Physiol., 60*:402-406, 1977.
171. Mattoo, A. K., Chalutz, E., Anderson, J. D., and Lieberman, M. Characterization of the phosphate-mediated control of ethylene production by Penicillium digitatum. *Plant Physiol., 64*:55-60, 1979.
172. Wood, D. A., and Hammond, J. B. W. Ethylene production by axenic fruiting cultures of Agaricus biosporus. *Appl. Environ. Microbiol., 34*:228-229, 1977.
173. Brown-Skrobot, S. K., Brown, L. R., and Filter Jr., T. H. Mechanism of ethylene and carbon monoxide production by Septoria musiva. *Dev. Ind. Microbiol., 26*:567-573, 1985.
174. Phan-Chon-Ton, C. R. Contribution a "etude" de la production de l'ethylene par le Penicillium digitatum Sacc. *Rev. Gen. Bot., 69*:505-543, 1962.

175. Kapulnik, E., Mattoo, A. K., Chalutz, E., and Chet, I. The relationship between the production of ethylene by certain fungi, their enzyme constitution and their ability to accumulate intracellular phosphate. *Z. Pflanzenphysiol.*, *109*:347-354, 1983.
176. Jacobsen, D. W., and Wang, C. H. The biogenesis of ethylene in Penicillium digitatum. *Plant Physiol.*, *43*:1959-1966, 1968.
177. Ketring, D. L., Young, R. E., and Baile, J. B. Effect of monofluoroacetate on Penicillium digitatum metabolism and on ethylene biosynthesis. *Plant Cell Physiol.*, *9*:617-631, 1966.
178. Burg, S. P. The physiology of ethylene formation. *Annu. Rev. Plant Physiol.*, *13*:265-302, 1962.
179. Argoudelis, A. D., and Zieserl, J. F. Structure of U-13, 933, a new antibiotic. *Tetrahedron Lett.*, 1969-1973, 1966.
180. Hamasaki, T., Nakajima, H., Yokota, T., and Kimura, Y. A new metabolite 3-carboxy-2, 4-diphenyl-but-2-enoic anhydride produced by Aspergillus nidulans. *Agr. Biol. Chem.*, *47*:891-892, 1983.
181. Hirota, A., Sakai, H., and Isogai, A. New plant growth regulators, cladospolide A and B, macrolides produced by Cladosporium cladosporioides. *Agr. Biol. Chem.*, *49*:731-735, 1985.
182. Sassa, T. Cotylenins, leaf growth substances produced by a fungus. Part I. Isolation and characterization of cotylenins A and B. *Agr. Biol. Chem.*, *35*:1415-1418, 1971.
183. Austin, D. J., and Clarke, D. D. Production of coumarins by Phytophthora infestans. *Nature*, *210*:1165-1166, 1966.
184. Ballio, A., Chain, E. B., De Leo, P., Erlanger, B. F., Mauri, M., and Touolo, A. Fusicoccin: A new wilting toxin produced by Fusicoccin amygdali Del. *Nature*, *203*:297, 1964.
185. Tamura, S., Sakurai, A., Kainuma, K., and Takai, M. Isolation of helminthosporol as a natural plant growth regulator and its chemical structure. *Agr. Biol. Chem.*, *27*:738-739, 1963.
186. Cross, B. E., and Webster, G. R. B. New metabolites of Gibberella fujikuroi. Part XV. N-Jasmonoyl and N-dihydrojasmonoyl-isoleucine. *J. Chem. Soc.*, 1839-1842, 1970.
187. Takahashi, N., and Curtis, R. W. Isolation and characterization of malformin. *Plant Physiol.*, *36*:30-36, 1961.
188. Hayashi, Y., Takahashi, S., Ona, H., and Sakan, T. Structure of nagilactone A, B, C and D, novel nor- and dinor diterpenoides. *Tetrahedron Lett.*, 2071-2076, 1968.
189. Nakajima, H., Nishimura, K., Hamasaki, T., Kimura, Y., Yakota, T., and Udagawa, S. Structure of neovasinone, a new α-pyrone plant growth regulator produced by the fungus Neocosmopora vasinfecta E. F. Smith. *Agr. Biol. Chem.*, *51*:1212-1224, 1987.
190. Iwamoto, T., Hirota, A., Shima, S., Sakai, H., and Isogai, A. Nigerazine A, an isomer of nigerazine B from Aspergillus niger. *Agr. Biol. Chem.*, *49*:3323-3325, 1985.
191. Kumura, Y., Katagiri, K., Inoue, T., and Tamura, S. Isolation and biological activity of pestalotin, a gibberellin synergist from Pestalotia cryptomeriaecola. *Agr. Biol. Chem.*, *35*:1313-1314, 1971.
192. Nakajima, H., Hamasaki, T., Shikata, Y., and Kimura, Y. Structure of pelvalic acid, a new metabolite from Penicillium valiabile Sopp., and its biological activity. *Agr. Biol. Chem.*, *50*:753-755, 1986.
193. Tamura, S., and Chang, C. F. Isolation of L-β-phenyllactic acid as a plant growth regulator produced by Exobasidium. *Agr. Biol. Chem.*, *29*:1061-1062, 1965.

194. Kimura, Y., Inoue, T., and Tamura, S. Isolation of 2-pyruvoylamino-benzamide as an antiauxin from Colletotrichum lagenarium. *Agr. Biol. Chem.*, *37*:2213-2214, 1973.
195. Sassa, T., Tomizuku, K., Ikeda, M., and Muira, Y. Isolation of a new root growth stimulating substance from fungus. *Agr. Biol. Chem.*, *37*:1212-1222, 1973.
196. Satomura, Y., and Sato, A. Isolation and physiological activity of sclerin, a metabolite of Sclerotinia fungus. *Agr. Biol. Chem.*, *29*:337-344, 1965.
197. Grove, J. F. The structure for terrein. *J. Chem. Soc.*, 4693-4694, 1954.
198. Barton, D. H. R., and Miller, E. The constitution and stereochemistry of terrein. *J. Chem. Soc.*, 1028-1029, 1955.
199. Pringle, R. B., and Braun, A. C. The isolation of the toxin of Helminthosporium victoriae. *Phytopathology*, *47*:369-371, 1957.
200. Taniguchi, M., and Satomura, Y. Isolation of viridicatin from Penicillium crustosum and physiological activity of viridicatin and its 30-carboxymethylene derivative on microorganisms and plants. *Agr. Biol. Chem.*, *34*:506-510, 1970.
201. Lonsane, B. K., Ghildyal, N. P., Budiatman, S., and Ramakrishna, S. V. Engineering aspects of solid state fermentation. *Enz. Microb. Technol.*, *7*:257-265, 1985.
202. Hesseltine, C. W. Biotechnology report. Solid state fermentations. *Biotechnol. Bioeng.*, *14*:517-532, 1972.
203. Lonsane, B. K., Ghildyal, N. P., and Murthy, V. S. Solid state fermentation processes and their challenges. In *Technical Brochure: Symp. on Fermented Foods, Food Contaminants, Biofertilizers and Bioenergy*. Association of Microbiologists of India, Mysore, pp. 12-18, 1982.
204. Hahn-Hägerdal, B. Water activity: a possible external regulator in biotechnical processes. *Enz. Microb. Technol.*, *8*:322-327, 1986.
205. Datar, R. Economics of primary separation steps in relation to fermentation and genetic engineering. *Proc. Biochem.*, *21*:19-26, 1986.
206. Steinkraus, K. H. *Acta Biotechnol.*, *4*:83-88, 1984 as cited by Hahn-Hägerdal [204].
207. Kumar, P. K. R., and Lonsane, B. K. Gibberellic acid by solid state fermentation: consistent and improved yields. *Biotechnol. Bioeng.*, *30*:267-271, 1987.
208. Kumar, P. K. R. and Lonsane, B. K. Potential of fed-batch culture in solid state fermentation for production of gibberellic acid. *Biotechnol. Lett.*, *9*:179-182, 1987.
209. Kumar, P. K. R., and Lonsane, B. K. Batch and fed-batch solid state fermentation: kinetics of cell growth, hydrolytic enzymes and gibberellic acid production. *Proc. Biochem.*, *23*:43-47, 1988.
210. Somal, B. S., Shastri, K. S. N., Thakur, R. N., Pandotra, V. R., and Atal, C. K. Prospects of production of gibberellic acid in India by submerged fermentation. In *Proceedings, Industrial Fermentation Symposium*, Jammu, Regional Research Laboratory, 1978, pp. 100-106.
211. Kumar, P. K. R., and Lonsane, B. K. Extraction of gibberellic acid from dry mouldy bran produced under solid state fermentation. *Proc. Biochem.*, *22*:139-143, 1987.
212. Rosen, C.-G., and Datar, R. Primary separation steps in fermentation processes. In *Biotechnology 1983, Proceedings of First International*

Conference on Commercial Applications and Implications of Biotechnology, Online, Northwood, 1983, pp. 201-224.
213. Arima, K. Microbial enzyme production. In *Global Impacts of Applied Microbiology* (M. P. Starr, ed.), John Wiley and Sons, New York, 1964, pp. 277-294.
214. Hepner, L. Workshop 3. Economics, Report. In *Third European Congress on Biotechnology: Proceedings*, Vol. 4, VCH, Munchen, 1985, pp. 465-469.
215. Ghildyal, N. P., Lonsane, B. K., Sreekantiah, K. R., and Murthy, V. S. Economics of submerged and solid state fermentations for the production of amyloglucosidase. *J. Food Sci. Technol.*, *22*:171-176, 1985.
216. Nielsen, N. Untersuchungen über stoffe die das wachstum der avena-coleoptile beschleunigen. *Planta*, *6*:376-378, 1928.
217. Kumar, P. K. R. Microbial production of gibberellic acid. Ph.D. Thesis, University of Mysore, 1987, pp. 1-426.
218. Chahal, D. S. Growth characteristics of microorganisms in solid state fermentation for upgrading of protein values of lignocelluloses and cellulase production. In *Foundation of Biochemical Engineering, Kinetics and Thermodynamics in Biological Systems, ACS Symp. Ser. No. 207* (H. W. Blanch, E. T. Papoutsakis, and G. Stephanopoulos, eds.), American Chemical Society, Washington, D.C., 1983, pp. 421-442.
219. Zetelaki-Horvth, K. Cellulase production by Phanerochete chrysoporium on a solid substrate column. *Proc. Biochem.*, *19*:65-66, 1984.

21

PRODUCTION OF ETHANOL BY YEAST

TAHÍA BENÍTEZ and ENRIQUE MARTÍNEZ-FORCE *University of Seville, Seville, Spain*

JUAN JIMÉNEZ* *University of Oxford, Oxford, England*

ANDRÉS AGUILERA† *New York University, New York, New York*

I. INTRODUCTION

Ethanol is an ideal fuel and the basis for countless chemical transformations. Optimal conversion of carbohydrates to ethanol requires cells that are tolerant to high concentrations of both substrate and product and are able to efficiently produce ethanol at relatively high temperatures [1].

Saccharomyces species are the most ethanol-tolerant of eukaryotic organisms [2]. It is widely believed that the maximum level of ethanol produced by each *Saccharomyces* strain reflects its intrinsic ability to tolerate ethanol. Thus, sake yeasts, which produce wines of 20% ethanol, are much more ethanol-tolerant than brewer's yeasts, which normally make beers of 4-5% ethanol [2]. However, there are a number of different ways by which ethanol production can be improved by a strain, e.g., increasing the range of substrates which can be used as feedstock, improving the efficiency of substrate conversion to ethanol, raising the fermentation temperature, or improving tolerances to ethanol and/or osmotic pressure, etc. As part of this trend some attention has also been given to the effects of ethanol on yeasts other than *Saccharomyces* which are capable of fermenting polysaccharides and sugars not accessible to the former such as inulin, starch, lactose, cellobiose, and xylose [3-6]. The development of interest in other species with the ability, for example, to convert xylose to ethanol or to ferment at high temperatures indicates that no existing strain of *Saccharomyces* meets the specifications for all current and future uses [7-9].

Ethanol is the main end product of glycolysis in *Saccharomyces* and is known to inhibit sugar fermentation and cause other unfavorable effects in yeast cells [10]. However, the molecular mechanism of ethanol inhibition of fermentation is not clear, and little has been reported on its action on individual enzymes of the glycolytic pathway [11,12]. In addition, ethanol has been described as a noncompetitive inhibitor of other cellular functions such as growth rate [13-16]. Ethanol has also been found to strongly

---

Present affiliations:
*University of Malaga, Malaga, Spain.
†University of Seville, Seville, Spain.

inhibit the transport of sugar and amino acids and numerous other biological processes, many of which are associated with membrane lipids [17,18]. The plasma membrane is the first sensitive organelle to make contact with ethanol. Since ethanol is an amphipathic compound, the lipid composition of the plasma membrane may be important for ethanol tolerance [19]. When cells are grown in the presence of ethanol, they respond to membrane lipid-ethanol interactions by modifying their membrane lipid composition [18,20]. This is consistent with the structural changes observed in the cell membranes of microorganisms tolerant to high concentrations of ethanol [21]. However, trials to obtain more tolerant strains by growing cultures in media supplemented with ethanol have been unsuccessful, and no increase in tolerance has been attained [22-24]. *Saccharomyces* might adapt their membranes to ethanol but this adaptation might not result in an enhancement of parameters such as growth and fermentation because the kinetics of the inhibition of these processes by ethanol probably depends on mechanisms that are not membrane-bound [7,14,25].

Ethanol tolerance is also strongly influenced by environmental and nutritional conditions [7,21]. The production of high concentrations of ethanol is limited by (a) the inhibitory effect on the fermenting microorganism exerted by the ethanol produced [26], (b) the substrate concentration which affects osmotic pressure [27], and (c) the synergistic effect of ethanol and substrate [28]. Ethanol production can also be limited by noninhibitory mechanisms. There are reports on the favorable effects of the addition of invertase [29] and of supplements containing nitrogenous constituents [30] and the direct increase of the microbial biomass achieved by cell recycle [31]. However, under fixed conditions different strains differ in their ability to tolerate ethanol and within each strain ethanol tolerance is a reproducible characteristic, implying that it is genetically controlled.

Since ethanol tolerance in *Saccharomyces* species is controlled by a large number of genes [18], it is very difficult to obtain highly ethanol-tolerant yeast mutants, and the isolation of such mutants usually requires long-term selection techniques in continuous culture [32,33]. In contrast, ethanol-sensitive mutants have been obtained in such high frequency and with so many defined complementation groups that it has been suggested that almost every gene of *Saccharomyces* is able to reduce the ethanol tolerance of a given strain [34]. Genetic analysis of ethanol-tolerant strains has confirmed that many genes limit cell proliferation in the presence of ethanol [33]. These genes are different in nonisogenic strains, indicating that nonisogenic hybridization (genetic complementation) might generate yeast hybrids more tolerant than their parental strains [22,23]. In fact, hybridization has been the usual method of generating ethanol-tolerant strains [35]. A further complication is that genes that limit growth are different at different ethanol concentrations [18]. Therefore, the selection of ethanol-tolerant strains should be carried out in a wide range of ethanol concentrations since selection at any fixed concentration does not necessarily lead to an increase in tolerance at other levels [33].

The mitochondrial genome may also play an important role in the ethanol tolerance of yeast cells. Ethanol itself is a powerful mutagen of the yeast mitochondrial genome [36]. *rho*$^-$ cells cannot regenerate their respiratory ability since such mutants are the result of a deletion in the mitochondrial genome [36]. Therefore, there must be a selective advantage to *rho*$^+$ cells, especially in the presence of ethanol. In fact, *rho*$^-$ mutants are more ethanol-sensitive than their *rho*$^+$ parental strains [37,38].

A relationship between the effects of high temperatures and ethanol concentrations on yeast cells has already been described: the presence of ethanol enhances the lethal effects of high temperatures and decreases the maximal temperature permitting growth [39]. Also, the ability of a yeast strain to resist the inhibitory effect of ethanol decreases as the temperature increases [39]. Exposure of *Saccharomyces* cells to high temperatures also results in an increased rate of *rho*$^-$ mutations, the mutagenic effect of ethanol being enhanced at high temperatures [39]. The relationship between temperature, ethanol, and the *rho*$^-$ mutation frequency in *Saccharomyces* species leads to the hypothesis that it is the mitochondrial membrane that is sensitive to the toxic effects of high temperatures [39].

This chapter summarizes current views of the genetic basis of ethanol tolerance and production in yeast. The relationships between nuclear genes, the mitochondrial genome, and ethanol tolerance are described and recent developments in the genetic improvement of ethanol-producing yeasts are discussed.

II. YEAST CHARACTERIZATION FOR ETHANOL PRODUCTION: PHYSIOLOGICAL CHARACTERIZATION

A. Relationship Between Growth and Fermentation

Productivity depends on strain and on culture conditions [40]. To improve the productivity of a given strain in the formation of energy metabolism end products such as ethanol, it is important to establish whether or not the product formation rate depends on other metabolic processes such as the growth rate μ [41,42]. Three different situations have been dedescribed with regard to growth rate and product formation rate [41]: those in which both processes are totally linked, those in which they are not linked, and those in which the product formation rate is partly linked to growth rate. Although the relationship can be established by measuring the metabolic coefficient q_p [42], in practice it is sometimes difficult to differentiate growth-linked processes from partly growth-linked product formation when the product formation rate not linked to growth has a very low value [44]. A more sensitive method of establishing the degree of linkage between these two metabolic processes involves determining the yield, Ypx (product/biomass yield). For independent processes q_p has a constant value K independent of the dilution rate D, and Ypx = K/D (Fig. 1A); when a metabolic process is totally linked to growth q_p = K'D and Ypx has a constant value K' (Fig. 1B); finally, when a metabolic process is partly growth-linked, q_p correlates with D and has a value K when D is nil. In this case Ypx varies in inverse proportion to D but will have a value K' when D >> K (Fig. 1C) [44]. It had previously been observed that ethanol production was partly linked to growth because cell growth stops at low ethanol concentrations while fermentation stops at relatively higher ones [1] and fermentation continues when growth stops [45-47]. These observations were confirmed by the isolation of ethanol-sensitive mutants unable to grow at over 0.5 M ethanol but which ferment and respire at ethanol concentrations as high as 1.5 M [34]. The measurement of Ypx coupled with these observations of partial linkage allows D to be adjusted in particular situations [48]. For instance, when the substrate is expensive, the yield Ypx is more important than the production rate q_p and the dilution rate D can be maintained at very low values that maximize Ypx [49]. When inexpensive substrates are used, the production rate is of primary interest and D can be maintained at its maximal value (Fig. 1C).

An interesting point is that although only 3-5% of total glucose is used in respiration [50,51], respiration causes a considerable increase in the

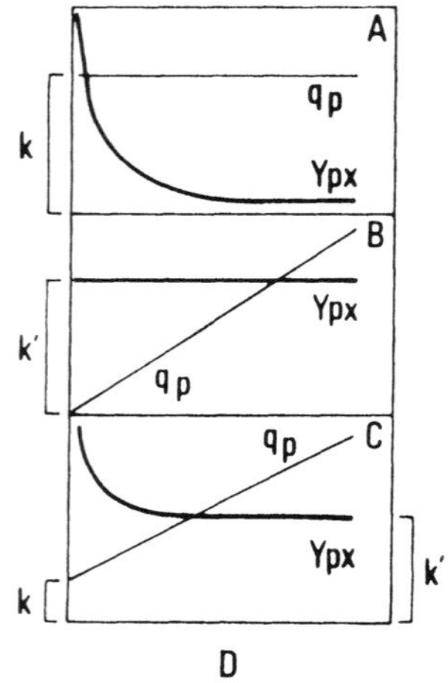

FIGURE 1. Variations in the rate of product formation q_p and the yield product/biomass Ypx with the dilution rate D when the formed metabolic product is (A) independent from growth, (B) totally growth-linked, or (C) partly growth linked. In (A), q_p = K and Ypx = 0 when D >> K; in (B), Ypx = K' and q_p = 0 when D = 0; in (C), q_p = K when D = 0 and Ypx = K' when D >> K. (From Ref. 44.)

growth rate and, in consequence, Yxs (biomass/substrate) is higher in grande than in petite mutants [44]. This lower Yxs is probably one of the reasons for the higher sensitivity of the petite strains to ethanol [52]. Similar results have been described by other authors [52,53] who found that the maximum growth rate under anaerobic conditions was always less than under aerobic conditions in both batch and continuous cultures.

B. Inhibition of Growth and Fermentation by Substrate

The production of high concentrations of ethanol is frequently limited by the inhibitory effect on the growth and productivity of the fermenting microorganism exerted by the substrate, the concentration of which affects osmotic pressure [27]. Substrate inhibition of growth rate usually begins at much lower concentrations than inhibition of fermentation [2]. In general, direct substrate inhibition of fermentative ability becomes significant somewhere between the range of 15-25% sugar concentrations [54,55].

The nature of the sugar is also important, as substrate inhibition of fermentation by glucose is greater than that by equal concentrations of sucrose [56]. Values of specific ethanol production rate and specific sugar uptake have been found to decrease almost linearly with the increase in sugar concentration [26]. In *Zymomonas mobilis*, ethanol inhibition of sucrose metabolism occurred at relatively low ethanol concentrations compared to those inhibiting glucose metabolism [57]. In *Candida pseudotropicalis* the effect of ethanol was formulated as an exponential function of ethanol concentration but the empirical constant is different when glucose or lactose are used as a substrate [58]. Also, in the production of ethanol by *Kluyveromyces fragilis*, the medium containing glucose gave the lowest yield of ethanol, that containing galactose gave the highest, and lactose medium produced intermediate yields [5]. Panchal and Stewart [27] found that when *Saccharomyces uvarum* was grown in minimal medium containing sucrose and a nonmetabolized sugar, sorbitol, significant levels of intracellular ethanol were obtained. This intracellular ethanol concentration, which decreased as the osmotic pressure of the medium was lowered, caused a reduction in cell viability. Since it is known that ethanol permeates yeast membranes by passive diffusion, these results have to be reexamined in light of the development of recent techniques to measure intracellular ethanol [59].

In order to maintain cell viability, it is necessary to reduce the osmotic pressure of the medium. An effective way to do this is to alter the way substrate is introduced during fermentation. When the substrate is introduced into the fermentation media in several batches (feed-batch), ethanol yields are higher than when substrate is provided only at the beginning [60,61]. Cell viability when the substrate is slowly infused is close to 95% compared to 40% for a single-batch run [62]. Addition of corn syrup extract at various times during the fermentation process also results in a higher final ethanol concentration. These improvements are attributed to increased yeast growth [2].

However, one must take into account that when measuring μ in media supplemented with increasing glucose concentrations, different behaviors have been observed. In some *Saccharomyces* strains μ is increasingly inhibited by glucose concentrations over 10%; this inhibition ranges from 25 to 40% in 25% glucose [15]. Similar inhibitory effects of high substrate concentrations were widely shown in strains of *Candida* [28]. "Flor" yeasts, however, show low growth rate values, μ, at low substrate concentrations but these values are maintained at up to 25% glucose. Over this concentration all strains are equally inhibited [15] (Fig. 2). As a consequence, strains which are inhibited by a glucose concentration above 10-12% show significant improvement in their growth (μ) and fermentation (ν) rate values when a total amount of 25% glucose is added in steps compared to the results obtained with a 25% initial glucose concentration. However, this is not the case with flor yeasts. Therefore, the efficient enhancement of μ by stepwise addition of sugar appears to be strongly dependent on the inhibitory kinetics of μ by substrate [15]. In other *Saccharomyces* strains [63] higher ethanol production rates and final ethanol concentrations are obtained when the substrate is added stepwise during the fermentation process.

It has been reported that in *Saccharomyces* species acclimated to higher concentrations of glucose (increased osmotolerance), ethanol tolerance levels decreased [2]. However, recent studies indicate that some sugar-tolerant yeasts are also alcohol-tolerant, growing at 15% added ethanol and fermenting at 18%, and fermenting the carbohydrate of cultures supplemented with 50% sucrose to give final ethanol concentrations of over 14% [64]. This does not imply, though, that these two features, ethanol and sugar tolerance, are necessarily related. When attempting to improve the osmotolerance of *Saccharomyces* or other yeast strains one should bear in mind that increasing production of glycerol could represent a substantial loss of efficiency in converting substrate to ethanol. For instance, the production of polyalcohols by osmophilic yeasts can represent 40-60% of the carbohydrate consumed [65]. In addition, when ethanol or sugar tolerance are measured, other physiological factors which can alter a cell's susceptibility to substrate and/or ethanol are ignored (i.e., nutrients, pH, etc.) [66].

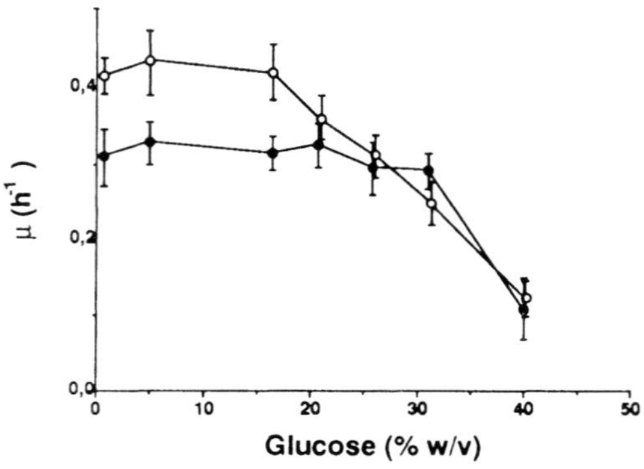

FIGURE 2. Inhibition of μ by increasing glucose concentrations shown by "flor" yeasts FJF206 and FJF414 (●), as compared to the other strains ACA4, ACA21, ACA174, IFI82, IFI256, D5A, and DS8 (○). Results are the means and standard deviations of three or four experiments carried out with each strain. (From Ref. 15.)

Some authors have reported that the factor limiting the high yield of ethanol by brewing yeasts is a nutritional deficiency rather than ethanol toxicity [60]. Supporting this approach, several groups reported improvements in alcoholic fermentation and final ethanol concentration by nutrient supplementation with additives that include proteins, vitamins, and lipids.

Casey et al. [67] demonstrated that the factor limiting the production of high levels of ethanol by brewing yeasts was actually a nutritional deficiency. When a nitrogen source was added to worts it was possible to produce beers of up to 16.2% ethanol, mainly because yeast viability remained high.

Other authors studying ethanol production by *K. fragilis* found that using mash, juice, or the pulp of Jerusalem artichoke without prior extraction as substrate led to the production of the highest concentration of ethanol (12.8%). These high concentrations are explained as due to protein supplementation mechanisms [68,69]. Supplementation of a simple medium with 2-4% soy flour also increased the final concentration of ethanol and the rate of fermentation by *Saccharomyces bayanus* [60]. Soy flour is an abundant source of proteins and lipids. This improvement is not attributed to an increase in the ethanol tolerance of the yeast but to the correction of nutritional deficiencies.

Dombek and Ingram [70,71] found that the addition of yeast extract, peptone, or ashed preparations of both restored the fermentative ability of cells which had previously begun to decline in the early stage of batch fermentation before high concentrations of ethanol had accumulated. Magnesium was identified as the active component.

The maximum growth yield of *Hensenula polymorpha* and *Beneckea natriegens* on glucose-supplemented media was achieved by using formate as an additional energy source [40]. In addition, many yeasts require growth factors to grow in molasses [72-74]. It was found that with wine yeasts growth in molasses without supplements is very poor and growth is highly stimulated by the addition of a nitrogen source or biotin [15]. The effect of adding each supplement is independent and additive in both growth and fermentation. Some of the wine-forming strains which in laboratory rich medium show a glucose consumption of about 4 g/liters/h were unable to ferment molasses. This inability is due to the lack of invertase activity [15]. Other investigators reported that in fermentation of cane molasses into ethanol by *Saccharomyces* strains, ethanol production is considerably increased by the addition of fungal invertase [29], indicating that sucrose hydrolysis could be a limiting step. In wine-forming strains, though, with the exception of some strains which do not possess invertase, this characteristic is not the rate-limiting step for molasses fermentation, but rather the lack of biotin or a nitrogen source [15].

C. Inhibition of Growth and Fermentation by Ethanol

Concentrations of ethanol above 15% result in immediate inactivation of most organisms [21]. Most bacteria exhibit a dose-dependent inhibition of growth over the range 1-10% ethanol, and few grow at concentrations above 10% [21]. Among eukaryotes, *Saccharomyces* appears to be the most alcohol-resistant organism. It is able to grow in concentrations of 8-12% ethanol and to survive in concentrations of up to 15% and to ferment sugars to produce ethanol in concentrations between 12 and 20% [21].

Many authors reported on the different and separable effects of ethanol on the specific rate of growth of a yeast population, its viability, and its specific rate of fermentation [57]. Different inhibition constants are derived for the action of ethanol on each of these parameters [57]. Inhibition of cell growth and viability is observed to increase with increasing ethanol concentrations, whereas high fermentative capacity is only inhibited at high ethanol concentrations [60].

Growth and fermentation are partly linked parameters [44], so that the fermentation rate increases proportionally to a given increase in the growth rate. Thus, any increase in ethanol-tolerant growth of a strain gives rise to a proportional increase in the rate of ethanol production [33]. In addition, yeast cells are usually able to continue producing ethanol after growth has stopped. This fermentative capacity not linked to growth varies from one strain to another [64]. Those strains with a long fermentative phase not linked to growth are probably of most interest in industrial applications (Table 1).

With regard to viability, Day et al. [47] examined the survival pattern of yeasts in aqueous ethanol as a possible means of predicting alcohol tolerance in high-gravity fermentation. For a wide range of yeasts under standardized conditions there is a correlation between survival rate (the time to a chosen % survival) in aqueous ethanol and the levels of ethanol tolerance in a synthetic growth medium (Fig. 3). These authors propose that survival of cells in aqueous ethanol be used as a method of determining ethanol tolerance. As will be seen later, this correlation can be disrupted in ethanol-sensitive mutants because some strains are as viable as the wild type or ferment as well, but stop growing at much lower concentrations of ethanol than the latter [14].

1. Kinetics of Ethanol Inhibition

Although cells stop fermenting at higher concentrations than that which completely inhibits growth, there is no striking difference between the response of growth and ethanol fermentation. The growth of microorganisms is a very complex phenomenon and the specific growth rate μ is usually

TABLE 1. Minimum Concentration of Ethanol (% w/v) Required Before Any Inhibition of Growth (μ) or Fermentation (ν) Rates Is Observed (threshold concentrations), Lowest Concentration of Ethanol Which Prevents Growth or Fermentation, and Concentration of Ethanol Which Reduces Growth or Fermentation by 50% (μ, ν = 50% μ_0, ν_0) for Different *Saccharomyces* Yeasts

Strain	Threshold concentrations		Concentration at which		Concentration at which		Ref.
	($\mu < \mu_0$)	($\nu < \nu_0$)	(μ = 50% μ_0)	(ν = 50% ν_0)	(μ = 0)	(ν = 0)	
Saccharomyces baker's yeast	—[a]	—	—	—	7.0	—	2
"	—	—	—	10.5	7.6	—	2
"	—	—	—	—	9.5	—	2
"	—	—	—	—	11.3	—	2
S. cerevisiae var. *ellipsoideus*	2.6	—	—	—	6.9	—	2
S. uvarum ATCC 26602	—	—	—	—	9.5	—	2
" NCYC 1324	2.0	—	—	9.1	9.4	—	2
S. sake IFO 2347	3.2	0.8	6.4	10.8	10.3	20.8	2
S. cerevisiae ATCC 4126	2.5	—	—	—	9.3	—	2
" NRRL Y-132	—	—	—	—	8.7	—	2
" UQM 70Y	—	—	—	—	10.0	—	2
" UG5	2.0	2.0	2.7	2.8	9.6	—	16

"	NCYC 366	6.3	—	—	—	9.5	—	2
"	NCYC 366	3.5	—	—	—			2
"	NCYC 366	—	—	—	10.8	5.4	—	77
"	NCYC 431	—	—	—	—	10.1	—	77
"	NCYC 478	—	—	—	—	9.3	—	77
"	NCYC 479	—	—	—	—	10.1	—	77
"	CBS 1198	—	—	—	—	9.3	—	77
S. cerevisiae	S288C (30°C)	1.2	2.3	4.6	4.6	9.2	11.5	14
"	BE288-1 (30°C)	0.5	0.2	3.5	3.5	11.5	11.5	14
S. cerevisiae	S288C (37°C)	1.2	—	3.7	—	6.9	—	24
	BE288-1 (37°C)	0.5	—	3.3	—	6.7	—	24
Saccharomyces	"flor" yeast	4.8	—	8.0	—	10.0	—	15
"	wine "	3.2	—	6.4	—	9.0	—	15
"	wine "	3.2	—	5.6	—	10.0	—	15

[a]—, not determined.

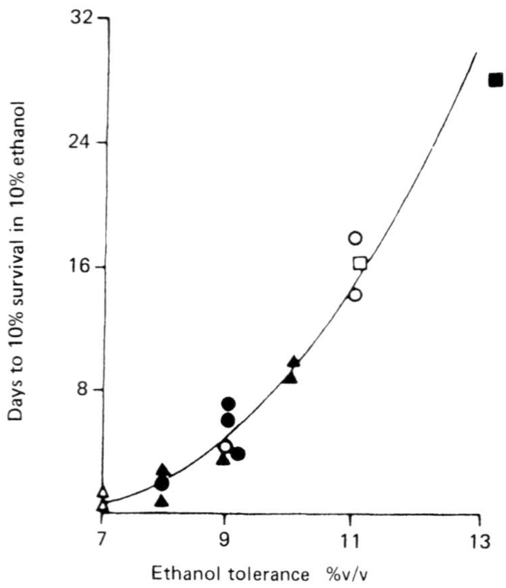

FIGURE 3. Relationship in yeasts between ethanol tolerance and survival in ethanol. (●) *Saccharomyces cerevisiae*; (○) *S. carlsbergensis*; (▲) *S. rouxii*; (■) *S. bailii*; (□) *S. diastaticus*. (From Ref. 47.)

expressed as a function of the limiting substrate concentration s, so that $\mu = \mu_0 \cdot (s/K_s + s)$, where μ_0 and K_s are the maximum specific growth rate and the Monod constant [75], respectively [76].

During alcohol fermentation, ethanol accumulates in the medium to such an extent that the presence of ethanol decreases the value of the specific growth rate. Thus, the previous equation must be extended to include the ethanol concentration p: $\mu = f \cdot (s, p)$. Since the product (ethanol) inhibition affects only μ but not K_s, the growth rate of microorganisms appears to be inhibited by ethanol in a noncompetitive manner [77]. The specific growth rate can then be represented by the equation $\mu = \mu_i \cdot (s/K_s + s)$, where μ_i is the value of the maximum specific growth rate in the presence of ethanol.

Many investigators have attempted to establish a correlation between μ/μ_0 and the ethanol concentration. Four types of dependence of μ on the ethanol concentration p are reported in the literature: (a) a *linear relationship*: $\mu = \mu_0 \cdot (1 - p/p_m)$. This equation was derived for bacteria by Hinshelwood [78]. Holzberg et al. [79] later confirmed the equation and further suggested that there was a threshold concentration of ethanol below which there was no inhibition and above which a linear relationship was exhibited, (b) an *exponential relationship*: from experiments with a respiratory-deficient mutant of *Saccharomyces cerevisiae*, Aiba et al. [13] proposed the following exponential model: $\mu = \mu_0 \cdot e^{-kp}$; (c) a *hyperbolic relationship*: Aiba and Shoda [80] proposed a hyperbolic relation to correlate μ and p in *Saccharomyces* bakers' yeast: $\mu = \mu_0 \cdot (1/1 + (p/k))$. Novak et al. [53] applied this equation to correlate their experimental data, also with *S. cerevisiae*. Exponential and hyperbolic relationships, however, predict a continuous decrease of the specific growth rate as the concentration of ethanol rises, but also imply that the cells are capable of growing and producing ethanol indefinitely; (d) a *parabolic relationship*: other authors [81] suggested the parabolic equation $\mu = \mu_0 \cdot (1 - p/p_m)^{0.5}$ to describe the observed kinetics of product inhibition in the strain *S. cerevisiae*.

Another generalized nonlinear equation has been proposed to account for the influence of ethanol product [76]: $\mu = \mu_0 \cdot (1 - p/p_m)^n$. By using this equation one can approach the previous equations in various ways: linearly (n = 1), with a rapid initial drop in rate followed by a slow approach to p_m (n > 1) or vice versa (n < 1). Similarly, the equations $\mu/\mu_0 = 1 - (p/p_m)^\alpha$ (for growth) and $\nu/\nu_0 = 1 - (p/p'_m)^\beta$ (for fermentation) approach linear, when μ and ν values are 1 and nonlinear relationships when μ and ν values are $\gtrless 1$ [76]. These equations describe the kinetic pattern of ethanol inhibition on the specific rates of growth and ethanol fermentation. When the presence of ethanol plays no role, μ or $\nu = \mu_0$ or ν_0; when, the growth or fermentation of microorganisms is completely inhibited by the presence of ethanol, μ or $\nu = 0$ (p_m and p'_m are the ethanol concentrations at which μ or $\nu = 0$).

Chattaway et al. [16] also proposed a kinetic model of ethanol fermentation conducted under a variety of conditions in a continuous reactor. The expressions for specific growth and product formation rates are: $\mu = \mu_0 \cdot \exp(-k_1 p)(1 - x/x_1)$ and $\nu = \nu_0 \cdot \exp(-k_2 p)(1 - x/x_2)$. According to their results, parameters were identified by nonlinear programming and shown to fit data correctly for steady states of seven different experiments, although the same model also fitted batch fermentation data (Fig. 4).

It was suggested that the differences in the reported experimental results are due to the physiological conditions of the microorganism and the medium and environmental conditions used [76]. Aiba et al. [13] using a respiration-deficient mutant of baker's yeast found an exponential relationship; Brown et al. [82] using a strain of *S. uvarum* and Ghose and Tyagi [83] using *S. cerevisiae* with cellulose hydrolyzate as the carbon source found a linear relationship. Similarly, Wasungu and Simard [84] found that the specific growth rate of baker's yeast decreased linearly as the ethanol concentration increased in batch and continuous cultures. Ritcher [85] confirmed the linear relationship between growth rate and ethanol with a threshold ethanol inhibitory concentration of 42.5 g/liter. From experimental data, a general model for the inhibition of growth and alcohol production caused by ethanol is deduced: $\mu = \mu_0 - a'(p - p'_1)^{b'}$ and $\nu = \nu_0 - a(p - p_1)^b$, so that if the inhibitory effects are linear, b and b' become 1.

Jiménez and Benítez [15] observed three main types of behavior displayed by different strains in relation to μ-inhibition kinetics by ethanol: a linear inhibition shown by laboratory and some industrial strains, a parabolic inhibition shown by distiller and wine-forming strains and, finally, a lack of inhibition at up to 6% ethanol shown by flor strains. Above this concentration μ was linearly inhibited with increasing ethanol concentrations.

In addition, Aguilera and Benítez [14] found that the inhibitory effect of added ethanol on μ and ν of *S. cerevisiae* differed according to whether respiratory processes occurred or not. The relationship of μ and ν to ethanol concentration varied exponentially for both a petite strain and a grande strain grown under anaerobic conditions. This relationship was linear in a grande strain in aerobiosis. Grande strains of *S. cerevisiae* [14], *Bacterium lactis aerogenes* [78], *Kluyveromyces marxianus* [86],

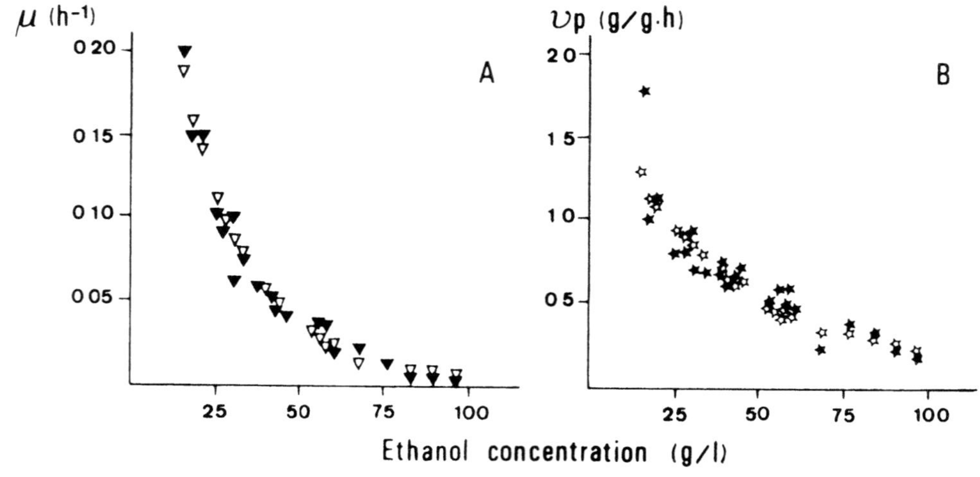

FIGURE 4. Experimental (full symbols) and calculated (open symbols) specific growth and ethanol-production rates plotted vs. corresponding ethanol concentrations. (From Ref. 16.)

and other strains of *S. cerevisiae* [79] showed a linear relationship when grown in aerobiosis. In contrast petite mutants and a grande strain of *S. cerevisiae* grown in anaerobiosis [14], petite mutants of *S. sake* [13] and *C. pseudotropicalis* grown in anaerobiosis [58] all showed an exponential relationship. The dependency of μ and ν on ethanol varies according to whether or not the cells respire.

2. Added and Produced Ethanol

Most authors have studied the effect of ethanol concentration on growth and fermentation rates and/or viability using added ethanol [89]. In general, little inhibition is seen below 2-3% ethanol [85] (Table 1). However, it seems that ethanol inhibition of growth and fermentation rates occurs at lower concentrations of ethanol if the yeast produces it endogenously. At higher ethanol concentrations, however, this does not seem to be so. For grande and petite strains of *S. cerevisiae* the final ethanol concentration produced when using 26% glucose was the same as the level that could be tolerated by fermentation when the media was supplemented with ethanol (2.4 M) [24].

However, with wine-forming yeasts in laboratory media, strains tolerate ethanol levels that are higher than those normally produced in fermentation [64]. Similar results have been reported for brewing yeasts [47]. The increased toxicity of produced vs. added ethanol was thought to be a result of its accumulating intracellularly to concentrations higher than those of the external medium, and effects such as cell concentration have been related to the different inhibitory effects of autogenously produced vs. extracellular added ethanol [87].

Growth inhibition of *Saccharomyces carlsbergensis* by ethanol has also been related to the retention of ethanol inside the cells: yeast cells stop dividing when the intracellular ethanol concentration reaches a critical value. A linear relationship was described between external and internal ethanol concentrations as determined with washed and centrifuged cells [88]. Higher intracellular ethanol concentrations and stronger inhibition were found at higher temperatures [10]. Novak et al. [53] studied the inhibitory effect of ethanol during alcoholic fermentation in anaerobiosis and demonstrated that ethanol produced during batch fermentation is more inhibitory than added ethanol in the range of 0-72.6 g/liter. They claimed that the measurement of the intracellular alcohol concentration explained the dual inhibitory effects of ethanol.

Nagodawithana and Steinkraus [87] proposed that, due to an imbalance between the rates of production and the net outflux of ethanol, there would be an intracellular accumulation of ethanol which in turn would explain the apparently greater inhibitory potency of endogenously produced ethanol present in the medium, a hypothesis that was supported by the findings of several authors [53]. However, from dialysis experiments Loureiro and Ferreira [89] reached the conclusion that the cell membranes are highly permeable to ethanol and that it seemed unlikely that any substantial binding to cell components took place. Recently, Guijarro and Lagunas [59] found that cells do not accumulate ethanol against a concentration gradient and that ethanol permeates the yeast cell plasma membrane by simple diffusion. More accurate experiments led to the conclusion that ethanol permeates the yeast cell plasma membrane without involvement of any carrier. Also, the outflow rate of ethanol was greater than the ability of the organism to produce ethanol, indicating that intracellular accumulation of ethanol is not possible. The intracellular concentration of ethanol found was similar to the concentration in culture media in all tested conditions (Fig. 5). A possible explanation for these alterations in the kinetic pattern when yeasts produce their own ethanol as compared to added ethanol is that of an osmotic effect due to high substrate concentrations, since ethanol and substrate show synergistic effects.

D. Synergistic Effects of Substrate and Ethanol

Some authors observed that the growth kinetics of several yeast strains as affected by ethanol varies with the initial substrate concentration. In *C. pseudotropicalis* the effects of both ethanol and substrate have been formulated and it was demonstrated that, when lactose and glucose were present in the medium with a small amount of alcohol, a synergistic effect on the rate of fermentation appeared, considerably limiting the rate of fermentation [58]. In addition, ethanol seemed to be more inhibitory when lactose was the substrate rather than glucose. Inhibition by glucose was only observed at high sugar concentrations (Fig. 6). The same authors [28] found that the inhibitory effects of alcohol and substrate were always less with a respiratory-deficient mutant of *C. pseudotropicalis* than with the parent strain. Similar synergistic effects were found by Mota et al. [90] in several *Saccharomyces* and *Kloeckera* strains, and Brown et al. [37] in a *S. uvarum* strain and its respiratory-deficient mutant when comparing ethanol inhibition in media with 2 and 15% glucose, respectively. In this case, however, the respiratory-deficient mutant showed increased inhibition by ethanol primarily at high glucose concentrations (Table 2).

Aguilera [91] measured ethanol production from sucrose concentrations varying from 0 to 70%. He found that *S. cerevisiae* did not grow at sucrose concentrations of 60% and growth was nonexponential at sucrose concentrations

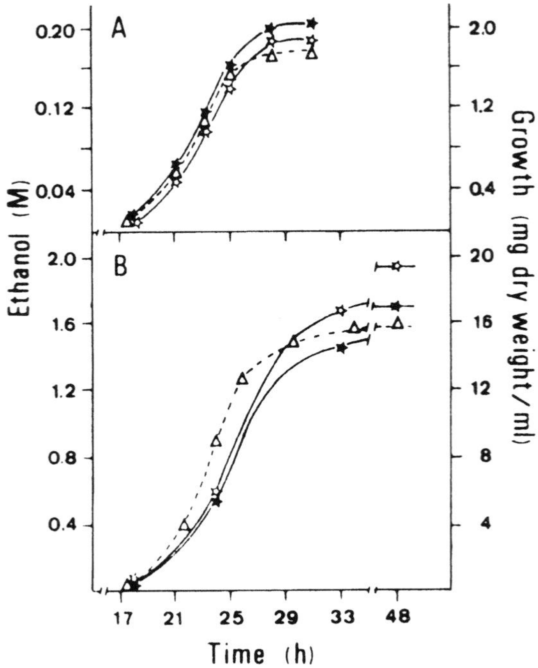

FIGURE 5. Intracellular and extracellular ethanol content at different stages of growth. (A) *S. cerevisiae* ATCC 42407 and (B) *S. cerevisiae* ACA174 were grown with 2 and 20% glucose, respectively. At the indicated times, samples of the culture were taken to measure growth (△), intracellular ethanol content (★), and ethanol concentration in the medium (☆). Ethanol was measured by enzymatic analysis, and its intracellular concentration was calculated with an assumed homogeneous distribution of this compound in the interstitial, periplasmic, and intramembranous space of yeast cells. Mean values of the results of two experiments are shown. (From Ref. 59.)

of over 40% [64], although fermentation was detectable until near 70% sucrose [91]. In addition, due to the synergistic effects of sucrose and ethanol, 30-35% sucrose was the substrate concentration which produces the maximal amount of ethanol (about 90% of the substrate was consumed producing about 13-14% v/v ethanol) [91]. From low concentrations and up to 30-35% sucrose there was an increase in final ethanol production, although at a slower rate. Above this sucrose concentration the rate of ethanol production was even lower and the final concentration decreased, being 0 at 70% sucrose (Fig. 7). When using glucose at concentrations ranging between 0 and 40% [24], 26% was the concentration at which maximum ethanol production occurred (2.4 M).

Different approaches have been taken to avoid the problems caused by these synergistic effects. Some researchers have added the substrate in stages to reduce the osmotic pressure of the medium [61], while others have isolated mutants tolerant to either high ethanol [46] or high sugar concentrations (osmotolerant mutants) [92].

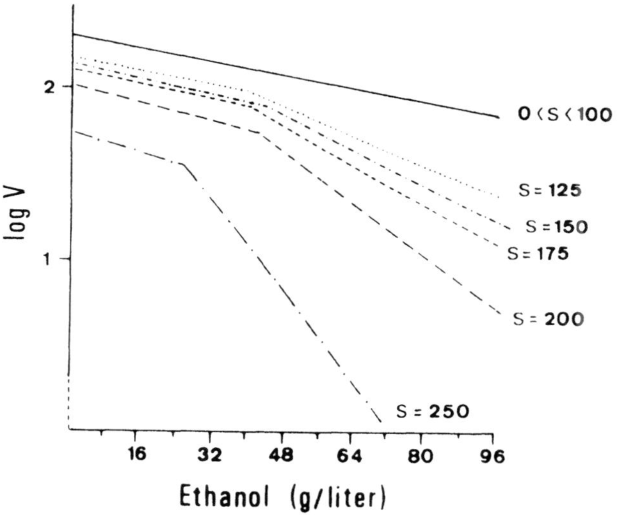

FIGURE 6. Specific rate of ethanol production (expressed by log V) plotted against ethanol concentration (g/liter) for different concentrations of glucose as substrate, s (g/liter). (From Ref. 58.)

TABLE 2. Growth Inhibition of *Saccharomyces uvarum* 5D-cyc by Ethanol

	Grande		Petite	
	Growth rate $\mu(h^{-1})$	Inhibition (%)	Growth rate $\mu(h^{-1})$	Inhibition (%)
2% (w/v) glucose				
Control (- ethanol)	0.312	—	0.254	—
+ 3% (w/v) ethanol	0.245	21.5	0.156	38.6
+ 6% (w/v) ethanol	0.123	60.6	0.096	62.2
15% (w/v) glucose				
Control (- ethanol)	0.285	—	0.220	—
+ 3% (w/v) ethanol	0.211	26.0	0.116	47.3
+ 6% (w/v) ethanol	0.074	74.0	0.035	84.1

Source: Ref. 37.

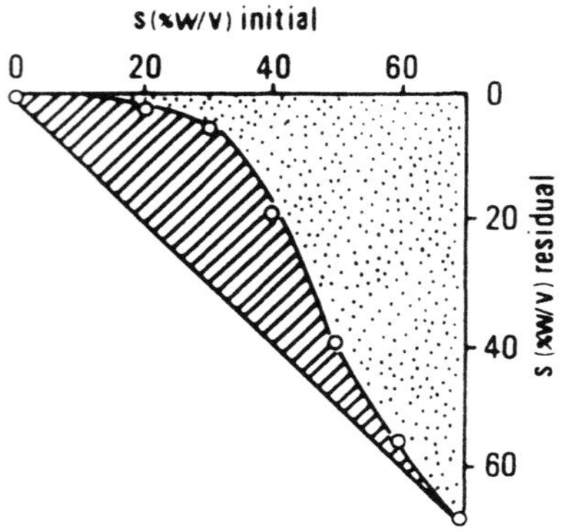

FIGURE 7. Consumed (▨) and residual (⬚) sucrose for strain S288C of *S. cerevisiae* in YEPS medium supplemented with different concentrations of sucrose at 30°C. (From Ref. 91.)

E. Effect of Ethanol on Yeast Enzymes

The observed inhibition of growth and fermentation rates by ethanol leads one to predict that the kinetic patterns for these inhibitions may reflect the effect of ethanol on specific enzymes, mainly those related to nutrient transport across the plasma membrane and those involved in the glycolytic pathway [7].

1. Inhibition by Ethanol of Transport Systems

Alcohols have been shown to inhibit transport systems in eukaryotic microorganisms. Using D-xylose as a nonmetabolizable sugar analog, Leao and van Uden [93] studied the effects of ethanol on glucose transport. Their results showed that ethanol does not affect the affinity of the transport system for D-xylose (noncompetitive inhibition) but lowers the rate of sugar uptake. Similar results have been found by Loureiro-Dias and Peinado [94] who examined the effects of ethanol on maltose transport, and by Sá-Correira and van Uden [95] when studying the fructose transport system of *K. fragilis*. The results were recently confirmed by Pascual et al. [11] who studied glucose transport in *S. cerevisiae* (Fig. 8A).

Thomas et al. [19] reported that ethanol depressed the rates at which glucose and several amino acids were taken up by a *Saccharomyces* strain, suggesting that ethanol inhibition of glucose uptake and fermentation may largely be due to interference with transport mechanisms. Using a nonmetabolizable analog of ammonium, van Uden [7] also found that ethanol has a more powerful effect on ammonium transport in *Saccharomyces* than on the transport of sugars. At ethanol concentrations at which the transport system for both glucose and ammonium were inhibited, the inhibition of the latter was much stronger than that of the former. The results suggested that fermentation may still proceed at ethanol concentrations at which growth in glucose-ammonium medium is already completely inhibited [2,7].

In *Candida wickerhamii* grown in cellobiose, cells transported cellobiose as glucose by a glucose-proton symport after previous hydrolysis of the disaccharide by an exocellular β-glucosidase [96]. Ethanol reversibly inhibited the two glucose transport systems (facilitated diffusion and glucose-proton symport) with exponential noncompetitive kinetics. Also, in experiments with *S. cerevisiae*, ethanol was reported to noncompetitively inhibit ATPase activity in membranes [97]. However, although an obvious candidate for the ethanol effect is the proton ATPase of the yeast plasma membrane, there are indications that this enzyme is unaffected by relatively low concentrations of ethanol [11].

2. Inhibition by Ethanol of Glycolytic Enzymes

Although ethanol is a major end product of anaerobic glycolysis in *S. cerevisiae*, the amount of available information on feedback effects of the alcohol on the activity and synthesis of glycolytic enzymes is not very extensive. Nagodawithana et al. [12] reported noncompetitive inhibition of the activities of hexokinase and α-glycerophosphate dehydrogenase in extracts of *S. cerevisiae* (Fig. 8B). With the two enzymes a reversible inhibition-enzyme complex is formed. Further studies have confirmed the inhibitory effect of ethanol on hexoquinase from *S. cerevisiae* [98]. In addition, a critical role has been given to pyruvate decarboxylase and alcohol dehydrogenase in the control of ethanol production by *S. cerevisiae* [99]. By deriving strains with altered levels of these two enzymes, these authors found that a high level of both enzymes was necessary for faster ethanol production. In line with these results, *S. cerevisiae* strains identified as being either slow or fast ethanol producers [100] had variable levels of four key enzymes—invertase, pyruvate decarboxylase, alcohol dehydrogenase, and aldehyde dehydrogenase—during fermentation. A good

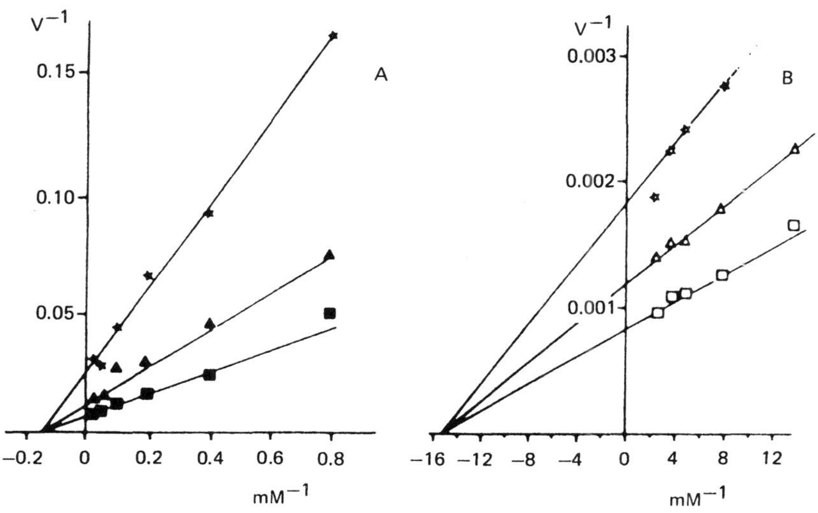

FIGURE 8. (A) Double-reciprocal plots of uptake rates of D-glucose without ethanol (■), with 5.2 M ethanol (▲), and with 10.3 M ethanol (★). Glucose uptake rates V measured at 30°C and plotted as reciprocals of μmol/min/g dry wt. (From Ref. 11.) (B) Lineweaver-Burk plot for hexokinase. (☆), 15% ethanol addition; (△), 10% ethanol addition; (□), no ethanol. (From Ref. 12.)

TABLE 3. Ethanol (EtOH) Inactivation of Glycolytic Enzymes in Yeasts with Different Alcohol Resistance

Strain		Enzyme activity (μmol/mg prot./min)					
		Hexokinase	Phospho-fructokinase	Pyruvate kinase	Alcohol de-hydrogenase	Aldolase	Glucose-6-P dehydroge-nase
Saccharomyces aviformis							
Source: Ref. 98	- EtOH	8.0	0.44	37	5.7	—	—
	+ 10% EtOH	4.2	0.5	35	4.5	—	—
Kloeckera apiculata							
Source: Ref. 98	- EtOH	3.7	0.44	3.5	3.0	—	—
	+ 10% EtOH	0.1	0.1	0.3	0.1	—	—
Saccharomyces uvarum							
Source: Ref. 12	- EtOH	11.76	0.54	—	—	0.15	0.93
	+ 15% EtOH	5.32	0.1	—	—	0.15	0.59
Saccharomyces cerevisiae							
Source: Ref. 11	- EtOH	0.34	0.36	3.55	—	0.64	0.09
	+ 10% EtOH	0.29	0.30	2.24	—	0.73	0.12
Saccharomyces Baker's yeasts MK270							
Source: Ref. 206	- EtOH	1.3	—	—	3.1	—	—
	+ 15% EtOH	0.615	—	—	id	—	—
Saccharomyces Baker's yeasts MK388							
Source: Ref. 206	- EtOH	1.9	—	—	2.2	—	—
	+ 15% EtOH	0.76	—	—	2.1	—	—
Ingledew 235[a]							
Source: Ref. 2	- EtOH	1	1	1	—	1	—
	+ 15% EtOH	0.9	0.9	0.9	—	0.9	—

[a]Relative values.

correlation between alcohol dehydrogenase activity and ethanol productivity has been established in two ethanol-tolerant *S. cerevisiae* yeasts. They were compared for their invertase and alcohol dehydrogenase activity and ethanol productivity and the strains, which showed significantly higher ethanol production compared to other yeast strains tested, also had 40-100% greater alcohol dehydrogenase activity. This was not true, however, for invertase activity [101].

There are also reports of a general resistance of glycolytic enzymes to ethanol denaturation in ethanol-tolerant strains at concentrations that inactivate the corresponding enzymes in sensitive strains [98]. In addition, ethanol concentrations which denaturate yeast glycolytic enzymes and those which inhibit fermentation [11,34] have been described to be higher than those which inhibit growth. This indicates that glycolytic enzymes are more ethanol-tolerant than other enzymes involved in cell growth (Table 3). Accordingly, Millar et al. [102] demonstrated that concentrations of ethanol below 12% do not denature glycolytic enzymes of *S. cerevisiae* or cause appreciable inhibition of activity in vitro under substrate-saturating conditions (Table 3). Similar conclusions have been obtained by Dombek and Ingram [10]: the activities of glycolytic and alcohologenic enzymes measured in vitro remained high in the presence of ethanol and did not appear to be limiting. Furthermore, the specific activities of many of these enzymes continued to increase even after increases in total cell protein had ended, suggesting that they may be preferentially synthetized [10].

Recently, Pascual et al. [11] studied the effect of ethanol on the activities of the key enzymes of the glycolytic pathway and on two membrane functions related to fermentation—the glucose uptake system and proton extrusion rate. The results indicate that ethanol at up to 2 M does not cause any change in the glucose uptake rate or any substantial change in the key glycolytic enzyme activities even though the fermentation rate is reduced by about 50%. Almost no glycolytic enzyme was denatured at ethanol concentrations up to 3 M. However, transmembrane proton flow was found to be the most ethanol-sensitive of the functions tested, so that it could represent the first target of ethanol action on fermentation.

The investigations demonstrate that, in addition to *Saccharomyces*, other strains with appropriate hydrolases like *Candida*, *Hansenula*, etc., can be used for ethanol production in special processes like corn mash fermentation [103]. Their efficiencies are at the standard level at the first phase of fermentation. However, amylases were shown in *Schwanniomyces castelli* to be inhibited at ethanol concentrations which are often obtained in fermentation. This inhibition fit a pattern of noncompetitive kinetics and was reversible [104]. In addition, in *C. wickerhamii* ethanol affected the cellobiase in a complex way, a major effect being deviation from Michaelis-Menten kinetics for ethanol concentrations higher than 4% [96].

Transfer of a heterologous function, like a certain hydrolase activity, from an ethanol-sensitive strain to a tolerant one would give rise to a lack of activity in the heterologous gene product under conditions inhibitory to the sensitive parental. However, induced cellobiase activity in *Candida* is extremely resistant to ethanol denaturation ($K_i \simeq 12\%$ ethanol) whereas K_i for cellobiase in cells grown in ethanol is 2.25% [105]. The latter result may reflect indirect inhibition of cellobiase, i.e., inhibition of other enzymes related to cellobiase synthesis. Further support for this suggestion is that when hybrids between *Candida* and *Saccharomyces* strains were formed by protoplast fusion there was an increase in the K_i values over 5% v/v ethanol in some of the hybrids (Fig. 9).

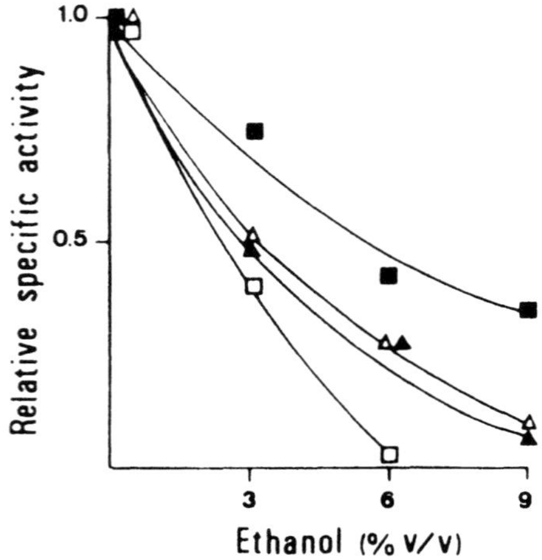

FIGURE 9. Inhibition by ethanol (% v/v) of induced cellobiase activity (grown in SCE medium) of Candida lusitaniae CLM1 (□) and *Candida/Saccharomyces* hybrids SCH1 (△), SCH2 (▲), and SCH3 (■). (From Ref. 105.)

III. YEAST CHARACTERIZATION FOR ETHANOL PRODUCTION. GENETIC CHARACTERIZATION: ROLE OF NUCLEAR GENES

A. Genetic Characterization of Ethanol-Producing Yeasts

Yeast strains used in commercial alcohol fermentations are usually polyploid or aneuploid, nonmating, and unable to sporulate [65]. Rarely, some strains do yield viable meiotic segregants which can be subjected to genetic analysis or developed by sexual hybridization [65].

B. Brewing Yeasts

1. Genetic Characterization

With regard to brewing yeasts, the challenges are first to control changes in the genome of these brewing strains (lack of markers makes it extremely difficult to readily distinguish a mutant from a contaminating strain [106]), and second, to manipulate strains to fit brewing needs and to use genetics as an approach to understand and better utilize yeast behaviors. Sexual production of recombinants by involving the haploid/diploid life cycle has proved to be a difficult and unfruitful task with brewing strains of *Saccharomyces* [106]. Brewing strains are polyploid or aneuploid and usually heterozygous for mating type. In addition, strains sporulate poorly. If sporulation does occur, many spores are not viable and the isolation of viable spores with mating activity is rare. The low spore viability, uneven segregation of mating type, and occurrence of aneuploid spores can all be due either to polyploidy/aneuploidy or to reduced recombination in meiosis leading to nondisjunction of chromosomes or even to recessive lethal genes or genes for homothallism [107].

For the brewer, breeding of strains with appropriate flocculation, improved carbohydrate utilization, and lower levels of undesirable compounds are of special interest [108]. Introduction of characters from nonbrewing yeasts into brewing yeasts have been carried out by rare-mating protoplast fusion, and procedures of isolating mating clones from brewer's strains have also been described [108]. Mutation, genetic transformation, and single-chromosome transfer have the advantage that minor parts of a genome are altered or added. One can therefore introduce well-defined traits into a brewer's yeast with lower risk of affecting other qualities [107]. Therefore, the genetic difficulties are overcome by conventional and rare-mating techniques, the latter consisting of isolating on appropriate selective media hybrids between respiratory-deficient mutants of nonmaters with an auxotrophic strain showing $\underline{a}$ or α mating type. The hybrids appeared with a frequency of 10^{-6} to 10^{-8} and their formation was explained by mitotic recombination occurring at the mating type locus in the nonmater strain [106]. Spencer et al. [109, 110] used these techniques to produce hybrids of poorly sporulating brewing strains with $\underline{a/a}$ or α/α auxotrophic diploids. Some of these hybrids (presumed tetraploids) sporulated freely yielding viable spores.

The transfer of a single chromosome from one nuclear genotype to another is a technique that has been used to genetically characterize brewing strains [111,112]. This technique makes use of the *kar1* mutation which impedes nuclear fusion [113], so that a mating involving this defect results in a transient heterokaryon budding off mainly haploid cells of the two parental genotypes [114]. At low frequency they also yield progeny that have the nuclear genotype of one parent plus an extra chromosome of the other [112].

Auxotrophic mutants have also been induced by UV irradiation and chemical treatment in meiotic segregants of a distiller's production strain of yeast. Tetrad analysis of hybrids between spore-derived clones of the distiller's strain and haploid *S. cerevisiae* strains revealed that the spores derived were frequently disomic. In addition, a high degree of polymorphism was detected using DNA restriction analysis [115] and several copies of chromosome III, site of the sexual locus, were frequently found, which explains the low frequency of maters and poor spore viability of these strains.

2. Genetic Improvement

Cross-breeding is a common means of producing strains of increased brewing performance [47,111]. Sexual hybridization has been used successfully and it is feasible to select ethanol-tolerant hybrids of brewing yeasts that still retain desirable brewing characteristics. Day et al. [47] constructed a series of hybrids between an ethanol-tolerant strain of *S. diastaticus* and poorly ethanol-tolerant brewer's yeast. The hybrids were more ethanol-tolerant and attenuated better than the poorly ethanol-tolerant parental, but they were less flocculent and less head forming than either parent. Some of the hybrids had a faster fermentation rate than either parent and maintained a viability approaching that of the ethanol-tolerant parent throughout the fermentation.

Somatic fusion has also been used to select hybrids between the commercial baker's yeast *S. cerevisiae* and a petite *S. diastaticus* brewer's strain using a selective medium with starch and glycerol. Respiratory-competent fusion products were capable of using dextrins and starch as carbon sources [116]. In addition, when interspecific and intergeneric yeast hybrids were obtained by protoplast fusion of *S. diastaticus* and other strains, in most of the hybrids the *S. diastaticus* parent contributed most of the genome and only a single chromosome or a fragment of a chromosome appeared to be transferred to the *S. diastaticus* nucleus to form the genome of the fusion product [110].

Stable spheroplast fusion products of the polyploid brewing yeast *S. uvarum* and *S. diastaticus* had enhanced ethanol-producing capability in defined media and in whole corn mash, produced ethanol at a very rapid rate, outperforming the parental strains, and had increased thermotolerance and enhanced osmotolerance [60,92]. Protoplast fusion techniques were also used to construct novel yeast strains from polyploid brewing yeasts to select for high-performing strains capable of producing elevated ethanol levels [92]. Fusion products between an industrial strain of *S. cerevisiae* and *S. mellis*, an osmotolerant yeast, were also obtained and examined with regard to their fermentative capacity. Whereas at low substrate concentrations they were indistinguishable from the parental strains, at 35% glucose the fusion products were far superior with a higher fermentation efficiency [60].

Although mutations are probably induced with the same frequency in diploids or polyploids as they are in haploids, they do not usually reveal themselves, unless they are dominant, because of the presence of non-mutated alleles. Secondary techniques are normally used that can lead to the expression of a recessive mutant phenotype in a diploid or polyploid yeast such as chromosome loss or mitotic recombination [117,118] and additional mutation or sporulation [118]. Continuous culture is a very powerful technique for selecting mutants, particularly those with an altered affinity for substrate or any selective advantage. Within a few generations, the mutant population completely replaces the parent strain [106].

Genetic transformation is the change of the genetic setup of an organism by the introduction of purified DNA. So far the only type of DNA that has been used with genes of specific brewing interest is native yeast DNA. A gene for flocculence has been transferred by this kind of transformation as has a gene for dextrin utilization [119,120].

C. Wine-Forming Yeasts

1. Genetic Characterization

Most wine-forming strains appear to be homothallic and to contain a great deal of heterozygosis for lethal genes or genes affecting growth rate [121]. Some may be polyploid or aneuploid. A number of auxotrophic mutants have been induced in some wine-forming strains and some of these have been used to produce tetraploid strains through cell-mating or protoplast fusion techniques.

In a study carried out by Benítez et al. [64], 632 wine-forming yeasts were subjected to screening and genetic analysis. Most of the strains were nonsporulating and nonmaters. The majority of the strains able to sporulate were homothallic and most of their meiotic products were also homothallic. Tetrads with either two or four viable spores per asci with about equal frequency indicate the presence of aneuploidy and/or heterozygosis for a recessive lethal mutation located in one extra chromosome [18]. Hybrids between spores of homothallic strains and a cycloheximide-resistant laboratory strain were obtained by the rare-mating selective method described for brewing yeasts with approximate frequencies of 1×10^{-4}. Tetrad analyses of some of the hybrids indicate that the strains were true hybrids between the laboratory and wine-forming yeasts and that the wine-forming yeasts yielded nearly haploid spores after meiosis.

Homothallism in wine-forming yeast: Most wine-forming yeasts are homothallic. During a yeast cycle an isolated spore germinates and divides mitotically. However, after a few generations there is a change in the mating type in one of the cells. As a consequence a further conjugation process between the cell with the changed mating type and any cell of the population takes place, originating self-diploid clones which are isogenic. These clones are homozygous for all genes except for the sexual locus, which is heterozygous and which allows the cell to sporulate but makes it unable to conjugate [122]. Frequently, however, wine-forming yeast strains are isolated which are homothallic and nevertheless heterozygous for numerous alleles [18]. Heterozygosis for auxotrophic alleles can be explained either by mating between spores of the same ascus or spontaneous mutation. However, heterozygosis for lethal alleles can only be explained by spontaneous mutation [102].

The fact that most wine-forming yeasts are homothallic makes it particularly difficult to genetically analyze them and to attempt to improve their desirable characteristics—a statement applicable to most industrial strains, including brewing yeasts. Spontaneous heterothallic spores have been isolated [122] and, additionally, under appropriate selective conditions, homothallic spores can be conjugated with heterothallic strains in a fashion similar to the rare-mating techniques described for brewer's yeasts.

"Flor" yeasts: Usually under conditions of high ethanol concentrations a film of yeasts ("flor"), mainly *Saccharomyces* strains, growing on the surface of the medium is developed [123]. Because of the relationship between flor and ethanol-tolerance, special attention has been given to the type of genetic control involved in the process of flor formation [123].

Santa María and Vidal [123] using tetrad analysis of six different *Saccharomyces* species isolated from sherry wine and able to form flor concluded that the six species monogenically segregated for homothallism/heterothallism and for flor/flor absence. Each characteristic segregated independently from the other (Table 4). Other authors [38] found spore viability in flor yeasts to be low as compared to laboratory control strains. The systematic survival of only two spores from each tetrad of some of the strains suggests that these strains might be either aneuploid (2n - 1) or heterozygous for a recessive lethal mutation. Similarly, the systematic survival of no more than two spores per asci in other strains might indicate either a heterozygous state for two recessive lethal mutations or the presence of complex aneuploidies. These observations suggest that the sporulation rate is lower than the mutation rate in the natural environment and are in agreement with findings of Sancho et al. [124] who suggested the existence of sexual isolation in yeast populations during wine production.

When flor formation was analyzed in flor strains and their meiotic products [38], all the meiotic products showed this feature, indicating that either the strains are homozygous for flor or that this characteristic is governed by several redundant genes. When hybrids formed between meiotic products of flor yeasts and laboratory strains were studied, analysis

TABLE 4. Segregation of Characters in "flor" *Saccharomyces*

Organism	A[a]	Segregation in asci 2:0:2	2:1:1	2:2:0	"flor" : no "flor"
S. gaditensis SBY 1668	3	1	2		2:2
S. hispanica SBY 294	3	1	1	1	2:2
S. oxidans SBY 2592	1	1			2:2
S. cordubensis SBY 1672	7		5	2	2:2

[a]Number of asci yielding four viable ascospores per ascus.
Source: Ref. 123.

of the genetic feature flor rendered a digenic segregation for this characteristic, and tetrad segregation indicates that each of the dominant alleles of the two genes involved has to be present for the phenotype to be expressed.

All meiotic products of flor yeasts were able to mate with laboratory strains of either α or a mating type, indicating that the strains are heterothallic [38]. Further studies, however, indicated that some of the supposed heterothallic meiotic products were homothallic, but mating type change only showed after a variable number of generations so that single-cell spores are able to conjugate [125].

Flocculating yeasts: Many wine-forming yeasts are able to flocculate [64]. As with the flor feature, flocculation has been observed to be a dominant trait displayed in constructed wine laboratory hybrids [38]. Tetrad analysis of these hybrids also rendered a digenic segregation so that each of the dominant alleles of the two nuclear genes involved has to be present for the phenotype to be expressed [126]. Other authors [127] reported that the mechanism of flocculation may be governed by the expression of the single dominant gene *FLO*1, although the phenotypic expression of *FLO*1 may be modified by a number of other genes not linked to it [127]. Thornton [128] also proposes that flocculation in yeasts is genetically controlled and that a single nuclear gene can confer the flocculent characteristic. He also described the modification of a wine-producing yeast's properties by the use of genetic techniques in which the property of flocculation was introduced into a powdery wine-producing yeast and undesirable characters were eliminated by backbreeding. Suzzi et al. [31] found that flocculation of some strains showed a degree of instability in the expression of the character which is typical of cytoplasmic inheritance.

Other genetic features: Undesirable strains of a yeast sometimes spoil the fermentation. Double-stranded RNA plasmids which confer the superkiller phenotype have been transferred to wine-forming yeasts by cytoduction using the protoplast fusion technique. The killer wine-forming yeast constructed completely suppressed the growth of killer-sensitive strains of *Saccharomyces* [129].

Hybridization has also been used to develop new strains of pure culture wine-producing yeasts and undesirable characters like foaming, poor bouquet, or poor fermentation have been readily eliminated by this technique. Some hybrids produce wines comparable to those obtained by proven strains [128]. Thornton [130] found that sulfur dioxide tolerance was under the control of dominant polymeric genes. The hybrid strains had wine-making qualities comparable to those of the parent strains but had an increased sugar conversion efficiency.

2. Genetic Basis of Ethanol Tolerance in Wine-Producing Yeasts

Because ethanol tolerance in *Saccharomyces* species is known to be genetically controlled by a large number of genes [23], it is extremely difficult to isolate ethanol-tolerant mutants by conventional screening and selection techniques. Although mutants of *S. uvarum* whose fermentation rate in ethanol is higher than that of the wild type have been obtained in continuous culture [32], no data are available to explain the genetic basis of this increase in tolerance.

Ethanol-tolerant alleles: Ismail and Ali [23] carried out crosses between haploid products of *S. cerevisiae* showing different levels of tolerance and studied their meiotic segregation at ethanol concentrations very close to that able to completely suppress growth. These workers provided the first direct evidence of a polygenic character of ethanol tolerance in the strains studied. In addition, crosses between different haploids yielded diploids with a wide range of ethanol tolerance which in some cases exceeded that of the parental strains. Jiménez and Benítez [38] carried out genetic analysis to determine those genes which limit growth of *Saccharomyces* at high ethanol concentrations and those which limit growth at lower ethanol concentrations. Due to the enormous difficulty in isolating highly ethanol-tolerant yeast mutants, the study was carried out with ethanol-tolerant wine strains already selected [64]. Growth of the meiotic products of these strains was either equal to or slower than that of the parental strains but it was never faster than the latter. This variability prevented the use of classical genetic methods to study ethanol tolerance of these strains since appropriate selection of homothallic-heterothallic hybrids could not be assured and thus the actual ethanol tolerance of the hybridized wine spores could not be observed. Two consecutive generations of homothallic strains were analyzed as an alternative procedure [18].

The simplest interpretation given to the meiotic segregation obtained is the existence of recessive alleles which in homozygosis reduce the ethanol tolerance of the strains. However, segregation was also observed in the second generation. This phenomenon cannot be explained solely by heterozygosis but requires the simultaneous presence of aneuploidy and heterozygosis. The consequence of the homothallic nature of these strains is that after meiosis the subsequent generations have an even number of extra chromosomes and a tendency to euploidy which in turn gives rise to more viable products [18].

Generation time as a measurement of ethanol tolerance in medium supplemented with ethanol has also been measured in flor strains, highly ethanol-tolerant and heterothallic, and their meiotic products. Ethanol tolerance of these meiotic products varied greatly, the spores in every case being less tolerant than the parental (Fig. 10A). The observed segregation for growth in ethanol suggests, as before, that ethanol tolerance is subjected

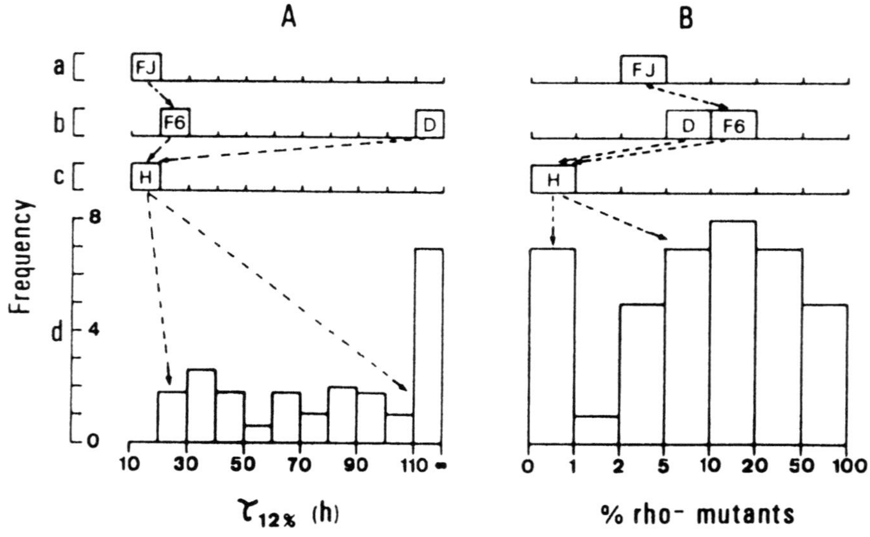

FIGURE 10. (A) Range of variation in the generation time (h) in YPDE (12% ethanol) and (B) range of variation in the frequency of spontaneous (YPD) petite mutant formation of (a) the "flor" yeast FJF414 (FJ); (b) the meiotic product of FJF414, FJF414/6 (F6), and the laboratory strain D517-4B (D); (c) the wine laboratory hybrid FDH1 (H), formed between the laboratory strain D517-4B and the wine meiotic product FSP414/6; (d) the meiotic products of the wine laboratory hybrid FDH1. About $1\text{-}3 \times 10^3$ colonies per strain were counted. (Part A from Ref. 18; Part B from Ref. 131.)

to polygenic control [23] and that there exist a great number of genes able to spontaneously mutate to alleles of sensitivity to ethanol in heterothallic strains. Hybrids between flor spores and laboratory strains were more tolerant to ethanol than the parental strains (Fig. 10A), indicating as before that genes of laboratory strains are able to complement recessive alleles of the wine spores and that genes involved in ethanol tolerance are different in the different strains [38].

To determine how many genes differentiate the ethanol tolerance of a given ethanol-tolerant flor strain and a certain laboratory strain, tetrad analysis was carried out in some flor laboratory hybrids after sporulation [131]. In the simplest situation, i.e., with no gene interactions, independent gene segregation, and additive gene effects, Fig. 10A indicates that the observed differences in ethanol tolerance were due to three to four diallelic genes.

Meiotic products of flor yeasts also segregated for growth in the absence of ethanol [38]. However, a lack of relationship has been found between genes which limit growth in the absence of ethanol and those which do so in the presence of this compound (Fig. 11). Neither has a relationship been found between genes which limit growth at 0, 5, 8, and 12% ethanol. From these observations it can be concluded that growth inhibition by ethanol is the result of the inhibition of different cellular functions with increasing ethanol concentrations. These results are supported by experiments indicating that different enzymes of the same strain are differently inhibited by ethanol [11] and explain why the kinetics of growth inhibition by ethanol is characteristic of the particular strain [7].

As a practical approach, the reported results suggest that nonisogenic hybridization might provide the most useful method to improve ethanol tolerance. This was shown empirically by other authors [92] as will be extensively described further. In addition, the selection of the most ethanol-tolerant hybrids has to be carried out in a wide range of ethanol concentrations [18].

Ethanol-sensitive alleles: To determine the genetic basis of ethanol tolerance in yeasts, ethanol-sensitive mutants were isolated and characterized [34]. These ethanol-sensitive mutants were unable to grow at ethanol concentrations at which a wild-type strain of *Saccharomyces*, haploid and heterothallic, was able to grow. Mutants showed phenotypic alterations of their temperature sensitivity, viability in ethanol, cell morphology, and other physiological features (Table 5). Each mutant was affected at a different gene, which indicates in all probability that the affected genes codify for proteins involved in the general metabolism of the yeast cell [34]. About one-third of the ethanol-sensitive mutants also showed thermosensitivity [34]. It therefore could be that those changes which convert a normal protein to a thermosensitive one might frequently make it ethanol-sensitive as well. Results described show that ethanol and temperature produce similar effects on membrane fluidity [20].

The ethanol inhibition constant for growth in the mutants varied from 0.35 to 0.8 M, suggesting not only that different mutations may affect gene expression with different intensity, but also that some gene products

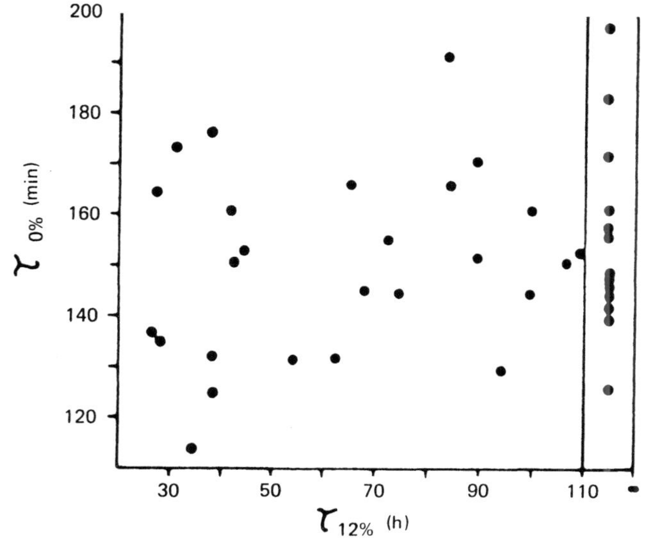

FIGURE 11. Relationship between generation time (min) (τ) of the meiotic products of the wine genetic hybrid FDH1 in YPD (τ_0) vs. YPDE (12% ethanol) ($\tau_{12\%}$). (From Ref. 18.)

TABLE 5. Ethanol Inhibition Constant for Growth (K_i), Growth Rate (μ) in YPD Medium, and Viability in Phosphate Buffer (2.5 M ethanol) of Ethanol-Sensitive Mutants and Wild-Type S288C[a]

K_i (M)	Strains	μ (h^{-1}) in YPD medium	Viability (%) in ethanol (2.5 M)
0.35	MSE1	0.30	2
	MSE2	0.30	0.002
	MSE3	0.37	100
	MSE4	0.40	0.0002
	MSE5	0.37	50
	MSE7	0.34	0.0002
0.45	MSE6	0.30	25
	MSE9	0.38	1
0.50	PSA1	0.38	100
	MSE8	0.30	0.0002
	MSE14	0.40	5
	MSE16	0.40	100
0.70	MSE10	0.36	1
	MSE11	0.36	1
	MSE12	0.35	10
	MSE13	0.40	5
	MSE15	0.38	1
	MSE17	0.36	100
	MSE18	0.40	10
0.80	MSE19	0.40	1
	MSE20	0.40	20
1.05	S288C	0.37	100

[a]The viability of the wild type was considered as the unit (100%).
Source: Ref. 34.

are more important than others with regard to ethanol tolerance. In most ethanol growth-sensitive mutants the glycolytic pathway was not affected [34]. The results demonstrated that growth can, at least partly, be regulated by genes not related to those which control fermentation [44]. Those mutants which are as viable and as good fermenters as the wild type but which possess the lowest K_i for growth (Table 5) are particularly interesting [34] and would be very useful in industries in which substrates are expensive.

When effects on ethanol production of increases in gene dosage in the yeasts *Candida* [132,133] and *Saccharomyces* [134] were compared for an isogenic ploidy series ranging from haploid to tetraploid, the per-cell rate of ethanol accumulation increased with cell ploidy (Fig. 12). This increase in the tetraploid strain was as great as seven times the level of accumulation in the haploid, i.e., the efficiency of ethanol production per unit cell mass was greater in cells of higher ploidy.

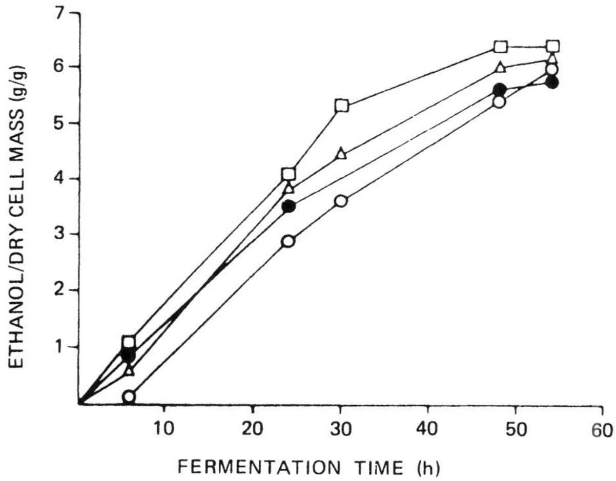

FIGURE 12. Influence of the ploidy of *C. shehatae* CBS2779 on the rate of ethanol production from 9% w/v D-xylose. (○) Haploid strain; (●) diploid strain; (△) triploid strain; (□) tetraploid strain. (From Ref. 133.)

D. Role of Genes Related to Cell Membrane Composition

Most of the reported physiological results implicate cell membranes as being important determinants of alcohol resistance [21], so it could be that those genes involved in the different tolerance between, for instance, highly ethanol-tolerant wine yeasts and laboratory strains [38] code mainly for cell membrane functions.

It has been reported that lipid composition in yeasts is of great importance for the improvement of ethanol tolerance [19]. During growth in ethanol *Saccharomyces* strains synthesize lipids enriched in $C_{18:1}$ fatty acyl residues to compensate for a decrease in palmityl residues [19]. These changes, together with the reports that supplementation with unsaturated fatty acids favors alcohol tolerance by lowering membrane leakage [17], led to the conclusion that the cell membrane is the main target for alcohol inhibition [21] and imply that membrane fatty acyl residues are important determinants of resistance to alcohol [135].

Ingram [135] reported that ethanol induces cell lysis in *Escherichia coli*. This effect appears to result from the weakening of hydrophobic interactions by ethanol rather than from intercalation of ethanol into membranes. Ingram [20,136] suggested that cells of *E. coli* radically change their fatty acid composition when grown in the presence of alcohols and that these changes represent an adaptive membrane alteration compensating for the direct interaction of alcohols with the membrane. Since alcohols directly affect membrane fluidity, Ingram proposed that changes in the degree of membrane fluidity were the primary signal eliciting an adaptive response in fatty acid composition. He confirmed in *E. coli* that ethanol acted at the level of fatty acid biosynthesis and altered lipid composition by decreasing the proportion of saturated acyl chains available for the synthesis of phospholipids [137].

Jones and Greenfield [138] working with *S. cerevisiae* described a novel and accurate permeation technique to measure the effects of ethanol

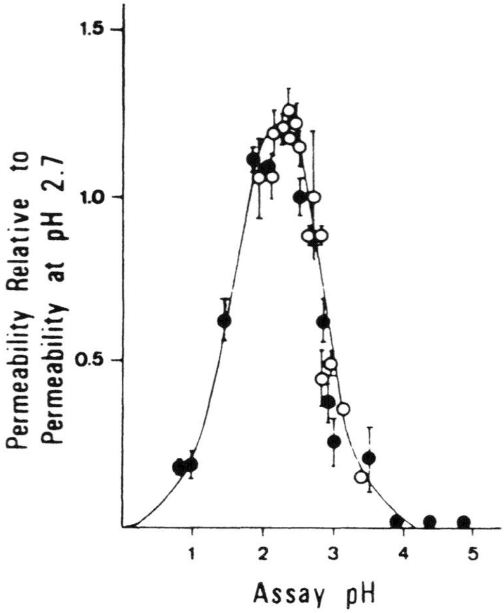

FIGURE 13. Effect of assay pH on membrane permeability. (○), Dry wt/OD = 24; (●), dry wt/OD = 16. (From Ref. 138.)

concentration on membrane fluidity. As Fig. 13 shows, the system works at an optimal pH around 2.6-2.8 and the different cell population sizes have no effect on the nature of the response [138]. They presented evidence for an exponential increase in yeast plasma membrane fluidity with ethanol concentration. These authors argue that the role of adaptation of a yeast strain to ethanol is reflected in the existence of a threshold concentration before the appearance of a fluidizing effect and that the sensitivity of a cell membrane to fluidization by ethanol is influenced by the physiological state of the yeast cell. Therefore, the concept that cell membranes increased fluidity as an adaptive response to ethanol, thereby conferring ethanol tolerance, is disputed by these authors. Rather, they think that the observed increase in fluidity is the result of a number of more fundamental changes. If fluidity were of adaptive benefit to yeast cells in the presence of ethanol, the more highly fluid batch yeast cells should exhibit a higher threshold to ethanol than that exhibited by the continuously grown adapted cells. The fact that these cells exhibit a higher fluidity is more likely an expression of poor membrane state than any mechanism of ethanol tolerance.

Accordingly, it was found in *Saccharomyces* strains [24] that whatever changes occur in lipid composition in the cells as a consequence of the presence of ethanol in the media, these cells do not improve their growth or fermentation rates. Furthermore, the inhibition rate of both parameters in grande strains in ethanol did not change with time whereas in petite mutants this inhibition did increase. When fatty acid and ergosterol were analyzed in both grande and petite strains of *Saccharomyces*, no differences were found between them [24]. In the presence of ethanol, both strains lengthened their fatty acid chains whereas the degree of unsaturation remained unaltered. It could be that these strains compensate for the effect of ethanol on membrane fluidity by lengthening their fatty acid chains instead of increasing the degree of unsaturation [20] because *Saccharomyces* already possesses a high degree of unsaturated fatty acids.

The differences in ethanol sensitivity displayed by grande and petite strains of *Saccharomyces* [24] cannot be explained by differences in their lipid composition. Those changes seem to be the consequence of the inhibition exerted by the ethanol on the lipid biosynthetic pathway, as already reported in *E. coli* [137]. In addition, the lipid composition of ethanol-sensitive mutants [14] grown both with and without ethanol has also been found to be very similar to that of the wild type even though the lipid composition of yeasts has been related to ethanol tolerance [140-142]. In all these cases it is concluded that, although the genes connected with the lipid biosynthetic pathway are important, there are other genes involved in other cellular functions which are of great importance in ethanol tolerance.

The explanation of the adverse effects of ethanol has centered on the interaction of ethanol with plasma membrane lipids. Pascual et al. [11] measured the effects of ethanol on two membrane functions related to fermentation: the glucose uptake system and proton extrusion rate. Transmembrane proton flow was found to be the most sensitive to ethanol of the functions tested. There are also reports [142] on the effect of chemicals which inhibit the proton pump of yeast cells. It was recently found using the extracellular acidification technique [25], which has proven to be a reliable measurement of the degree of ethanol interference with cell membrane [7], that the ethanol tolerance of the cell plasma membrane of a highly ethanol-tolerant wine strain is very different from that of less tolerant laboratory strains [143], and that these differences correlated highly with their growth ethanol tolerance [25]. Increase in sensitivity to ethanol in yeasts can be associated with an increase in passive diffusion of protons through the cell membrane. It has been suggested that changes observed in the membrane tolerance to ethanol reflect changes in the protein composition of the cell membrane [143]. In addition, it has been found that the fatty acids/dry weight ratio of cells incubated in media with ethanol falls as the concentration of ethanol rises [139]. These results, together with the suggestion [144] that a decrease in the lipid/protein ratio could be essential for ethanol tolerance, indicate that the protein component of cell membranes may play an important role in the ethanol tolerance of yeast strains and that the different fatty acids/dry weight ratios of grande and petite cells could partly account for their different ethanol sensitivities [14].

Ethanol and other alkanols inhibited transport in a noncompetitive, exponential way; the exponential inhibition constant being positively correlated with the lipid/buffer partition coefficients of the alkanols (Fig. 14). This indicates that hydrophobic membrane sites are the primary targets of alkanol inhibition of transport [145]. When yeast cells are suspended in water containing glucose, protons are extruded and the extracellular pH drops to a final value following a characteristic acidification curve [25]. The alkanols affected the acidification curves of energized yeast suspensions in such a way that the final pH values were linear functions of the alkanol concentrations. These results were consistent with a balance between active and passive proton movements at the final pH. The exponential constant of ethanol-enhanced passive proton diffusion across the plasma

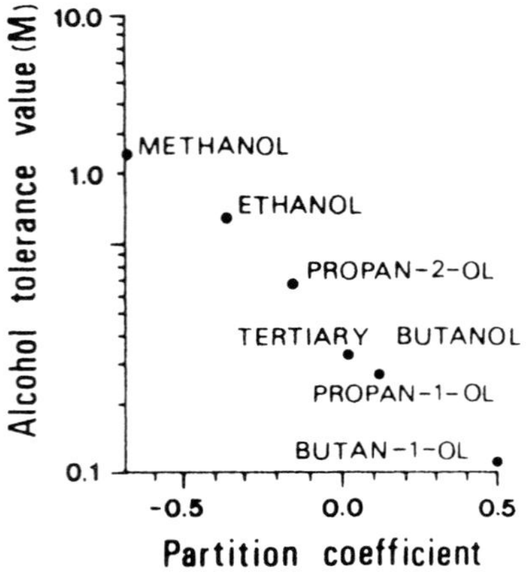

FIGURE 14. Inhibition by alcohols of fermentation in *S. cerevisiae*. Alcohols causing equivalent levels of inhibition of fermentation (glucose utilization) are plotted as a function of the membrane-water partition coefficients for each alcohol. There is an excellent correlation between hydrophobicity (partition coefficients) and the potency of these alcohols as inhibitors of fermentation. (From Ref. 21.)

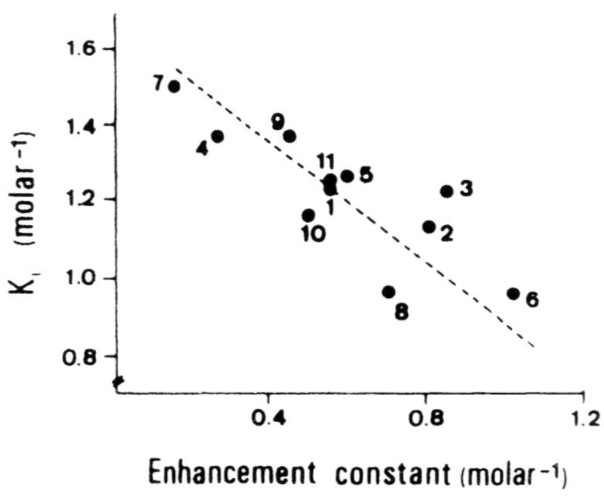

FIGURE 15. Relations between K_i (concentration of added ethanol that reduced the specific growth rate of yeast growth by 50%) and k (exponential constant of ethanol-enhanced proton diffusion across the yeast plasma membrane). Numbers refer to yeast strains. (From Ref. 25.)

membrane of yeast can be estimated rapidly and easily [145], so that it might constitute a rapid method for the preliminary evaluation of ethanol tolerance in yeasts [25], especially useful when large numbers of mutants and other strains have to be screened. This method has been tested by using a number of strains to establish whether a correlation existed between the values of the exponential constant of ethanol-enhanced proton influx $\underline{k}$ and the ethanol concentration that reduced the specific growth rate by 50% (K_i). In 10 out of 11 strains the relation between final pH and ethanol concentration was linear and estimates for $\underline{k}$ were calculated from the slopes. The relationship between K_i and $\underline{k}$ in the 11 strains was also linear (Fig. 15).

IV. YEAST CHARACTERIZATION FOR ETHANOL PRODUCTION. GENETIC CHARACTERIZATION: ROLE OF MITOCHONDRIAL GENES

A. Relationship Between Tolerance to Ethanol and Resistance to *rho*$^-$ Induction

Ethanol inhibits many biological processes, some of which have been associated with functions of the cell membrane lipids [14]. In addition, the observed differences in cell lipid composition have been attributed to mitochondrial functions [14]. Much research has been carried out to investigate the possible relationship between mitochondrial functions and ethanol tolerance.

Bacila and Horii [146] found that the amount of growth measured as protein production was greater in *rho*$^+$ strains of *S. uvarum*, but the yield of ethanol produced by the *rho*$^-$ cells was nearly twice as much. Alexander and Detroy [147] used different types of mitochondrial mutants, including *rho*^{0} strains which completely lack mitocondrial DNA, and 2% initial sugar concentration and found that the number of doublings, the time stationary phase was reached, and the rate of ethanol production for the first 24 h were about the same for all the mitochondrial mutants. However, after 24 h incubation, the amount of ethanol present in the wild-type cultures dropped continuously, probably because by then the glucose had become limiting and ethanol was being utilized by the wild-type but not the *rho*- mutants [147]. This may also be the explanation for the increase in ethanol production shown by the *rho*- mutants [146].

Recent reports indicate that, in general, petite mutants of *S. cerevisiae* with either nuclear or mitochondrial mutations, involving either the complete mitochondrial genome or just part, but with total inability to respire, have a lower tolerance to ethanol [60,131]. The reasons for the discrepancy found between these and previous results may be the low glucose concentrations used in some of the investigations [146,147], so that the final concentration of ethanol was not high enough to exert any inhibitory effect on yeast cells of either *rho*$^+$ or *rho*$^-$ phenotype, or the lack of glucose, which induced only the *rho*$^+$ but not the *rho*$^-$ strain to utilize ethanol. This may have led to misinterpretation of the results.

In contrast to other eukaryotic cells, the mitochondrial genome in *Saccharomyces* is not essential. Respiratory-deficient mutants (*rho*$^-$) of *Saccharomyces* arise spontaneously at very high frequencies [148] and such mutants grow and survive under optimal culture conditions almost as well as the *rho*$^+$ strains [33].

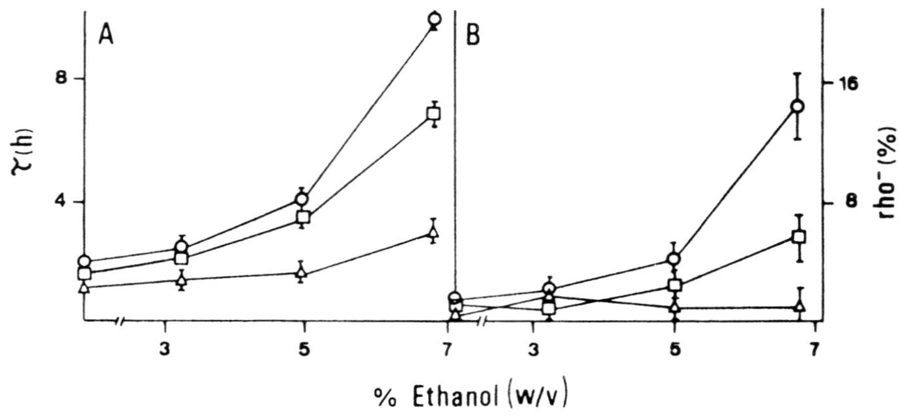

FIGURE 16. (A) Generation time (τ) and (B) percentage of *rho*⁻ mutants vs. ethanol concentration for laboratory strain ABQ21 (○), wine strain ACA21 (△), and heteroplasmon ABQ21-H1 (□) at 30°C. Results are the average and standard deviation of five trials. (From Ref. 33.)

When the frequency of *rho*⁻ mutants was determined in wine strains described as highly ethanol-tolerant [64] and in laboratory strains, in both the presence and the absence of 10% ethanol [33], the spontaneous frequency of *rho*⁻ mutants was higher in the laboratory than in the wine strains. The results suggest that ethanol growth tolerance correlates with a low frequency of spontaneous as well as ethanol-induced *rho*⁻ mutants (Fig. 16) [33].

However, in addition to the mitochondrial genomes, many nuclear genes affect the stability of the *rho*⁺ phenotype, as shown by the meiotic segregation of wine laboratory hybrids with regard to *rho*⁻ mutant induction (Fig. 10B) [131]. This segregation is similar to that shown by the same wine-genetic hybrids with regard to growth ethanol tolerance (Fig. 10A) [18]. The influence of nuclear genes is above all deduced from the high stability of the *rho*⁺ phenotype (Fig. 10B) [131] and the increase in growth ethanol tolerance [18] (Fig. 10A) of wine laboratory hybrids, indicating that nuclear genes from nonisogenic strains are able to complement their functions.

Some of the wine yeasts used in the experiments described [33] were flor strains responsible for the production of sherry wine having a high alcohol content. *rho*⁻ mutants of these flor strains were unable to form "velum" (flor characteristic). The flor feature may have arisen in these yeasts whose natural environments are musts with an alcohol content of over 15% [123] as an adaptation permitting growth on a liquid surface where oxidative metabolism is possible and where the oxygen uptake is not restricted.

B. Ethanol Tolerance of Isogenic *rho*⁺ and *rho*⁻ Strains

Aguilera and Benítez [14] reported that the relationship established between inhibition of growth and fermentation rates and ethanol concentration was adjusted to a linear equation for a *rho*⁺ strain of *S. cerevisiae* (S288C) but to an exponential one for the isogenic *rho*$^-$ strain [14]. When comparing a laboratory strain (ABQ21) possessing functional mitochondria (*rho*$^+$ phenotype) and a highly ethanol-tolerant wine yeast (ACA7) [64] with their corresponding isogenic *rho*$^-$ mutants, a stronger growth inhibition effect by ethanol (measured as generation time (τ) in media with increasing ethanol concentrations (0-12% ethanol) was observed in the *rho*$^-$ mutants of both strains (Fig. 17) [126].

To see whether the greater tolerance to ethanol shown by strains with respiratory metabolism was exclusively due to the presence of functional mitochondria, and whether these characteristics followed either a nuclear or a cytoplasmic inheritance, functional mitochondria were transferred from a *rho*$^+$ laboratory strain (S288C) to its *rho*$^-$ isogenic strain (BE288-1) by a cytoduction process using a strain possessing a *kar* mutation, described before, which prevents nuclear fusion after conjugation [113]. Heteroplasmons possessing the nucleus of the *rho*$^-$ strain but functional mitochondria were isolated on appropriate selective media and afterward the relationship between the presence of functional mitochondria in the recipient strain (BE288-SC) and ethanol tolerance of this yeast was studied.

Once heteroplasmons with functional mitochondria were selected, a spontaneous *rho*$^-$ mutant was newly isolated which lacked respiratory activity (BE288-SP). As Table 6 shows, the heteroplasmon strain which had received functional mitochondria (BE288-SC) had considerably increased its ethanol tolerance, which was similar to the *rho*$^+$ original strain S288C.

The spontaneous BE288-SP *rho*$^-$ mutant, obtained from the heteroplasmon, lost this increase in ethanol tolerance, showing an ethanol sensitivity for growth and fermentation similar to the original isogenic *rho*$^-$ strain BE288-1. Results demonstrated that mitochondria (but not nuclear genes codifying for mitochondrial functions) are responsible for the increase in tolerance to ethanol shown by the *rho*$^+$ strains S288C and BE288-SC with regard to the *rho*$^-$ yeasts BE288-1 and BE288-SP [14].

C. Ethanol Tolerance of Nonisogenic *rho*$^+$ and *rho*$^-$ Strains

Previous experiments on the role of mitochondria in ethanol tolerance of yeast strains [14] have been supported by experiments of mitochondria transfer between nonisogenic strains, using highly ethanol-tolerant wine yeasts as mitochondria donors [64] and laboratory yeasts, more sensitive to ethanol and with a *rho*$^-$ phenotype, as recipient strains [33]. As a control, laboratory strains isogenic to the recipient strains but with functional mitochondria (*rho*$^+$ phenotype) were used as mitochondria donors.

Heteroplasmons possessing the nucleus of the laboratory strain but mitochondria coming from the highly tolerant wine yeasts showed a significant decrease of the frequency of ethanol-induced *rho*- mutants together with a considerable increase of their ethanol tolerance (Fig. 16) [33].

Ethanol is a powerful mutagenic agent of yeast mitochondria [36]. In addition, the presence of functional mitochondria seems to be important to the maintenance of low levels of ethanol inhibition with regard to growth and fermentation rates in yeasts. These facts suggest that highly ethanol-tolerant wine yeasts possess mitochondria especially adapted to tolerate ethanol. This would explain why, when wine yeasts were isolated from wines with a high ethanol content (sherry wines), all the isolated strains possessed a *rho*$^+$ phenotype [64]. It also explains the low percentage of *rho*$^-$ mutants that can be induced by ethanol in these wine strains (Figs. 16

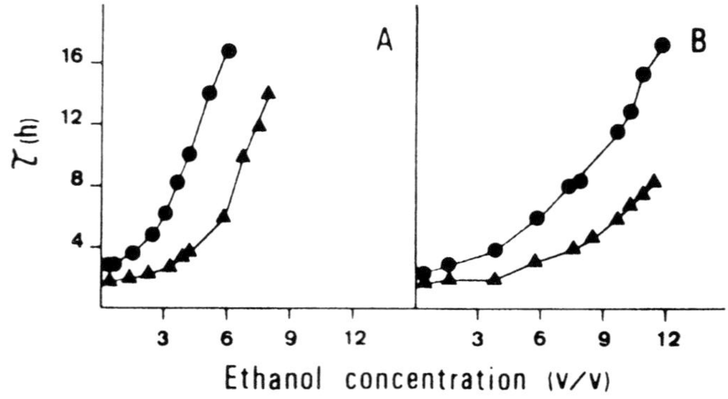

FIGURE 17. Generation time (τ) as a function of ethanol concentration (% v/v) of (A) the laboratory strain ABQ21 (*rho*$^+$) and its *rho*$^-$ mutant and (B) the wine yeast ACA7 (*rho*$^+$) and its *rho*$^-$ mutant. (From Ref. 126.)

TABLE 6. Growth (μ) (h^{-1}) and Fermentation (ν) (μl $CO_2/10^6$ cell/h) Rates of the Grande Strains S288C and BE288-SC and of the Petite Mutants BE288-1 and BE288-SP, Grown in YPD and YPDE (1 M ethanol)

	μ			ν		
Strains	YPD	YPDE	%i	YPD	YPDE	%i
S288C (RHO$^+$)	0.390	0.185	53.6	8.6	4.3	50.0
BE288-1 (RHO$^-$)	0.345	0.095	72.5	8.4	2.6	68.5
BE288-SC (RHO$^+$)	0.380	0.190	50.0	6.8	4.8	42.5
BE288-SP (RHO$^-$)	0.330	0.098	70.3	6.7	2.5	61.7

[a]%i: % inhibition of μ and ν in YPDE with respect to YPD. Results represent the average of three to four experiments.
Source: Ref. 14.

and 17) or in heteroplasmons which possess the nucleus of a laboratory strain but mitochondria transferred from highly ethanol-tolerant wine yeasts (Fig. 16) [33].

In addition, the greater stability in high ethanol concentrations of the *rho*$^+$ phenotype conferred by the wine yeast mitochondrial genome (Fig. 16) [33] would be the main factor responsible for the greater viability displayed by yeast strains possessing these mitochondria because their *rho*$^+$ stability permits respiration in conditions where other mitochondria are not functional. Although only 3-5% of the total glucose is used in respiration in a *rho*$^+$ strain, this is reflected in a substantial increase in the growth rate [24,44] and in the yield Ypx, which is higher in the *rho*$^+$ than in the *rho*$^-$ mutant [44]. This inefficiency of the *rho*$^-$ strains with regard to the yield Ypx could partly account for their greater sensitivity to ethanol.

D. Induction of *rho*⁻ Mutants by Membrane-Active Agents

Ethanol, isopropanol, propanol, and butanol enhanced thermal death in *S. cerevisiae* by increasing the entropy of activation of thermal death [149] and through nonspecific action on membrane structure. The specific death rates at constant temperature were an exponential function of the alkanol concentration [149]. Ethanol, *n*-butanol, and isoamylalcohol have been observed to cause a decrease in the viability of *S. cerevisiae* cells, with butanol and isoamyl alcohol acting in a synergistic manner with each other and with ethanol in causing cell death in suspensions of nongrowing cultures [150].

Though the similarity in the values of the exponential constants [7] and the correlation with the lipid-buffer partition coefficients of the alkanols [21] points to alkanol-induced changes in the membranes as the ultimate cause of their toxic effects on *S. cerevisiae*, there may be differences in the fine mechanisms of the alkanol-membrane interactions.

Ingram [135] found that the presence of alkanols in cultures of *E. coli* led to changes in the composition of the plasma membrane. Changes in membrane lipid composition induced by ethanol and effects on lipid metabolism were analogous with those caused by chaotropic agents and by a decrease in growth temperature: ethanol, chaotropic salts, and decrease in temperature all weaken hydrophobic interactions, which is consistent with a common mechanism for the effects of these agents on fatty acid synthesis [137]. Changes in the strength of hydrophobic interactions are thought to alter enzyme structure, substrate structure, or the equilibrium between the soluble enzymes of fatty acid synthesis and their respective acyl carrier protein substrates [137].

Sterols have been identified as important components of biological membranes influencing critical membrane functions such as transport and the activity of membrane-bound enzymes [151]. The sterol composition of *S. cerevisiae* can be manipulated as the organism cannot synthesize unsaturated fatty acids or sterols under anaerobic conditions but can incorporate into its membrane lipids a wide variety of added fatty acids and sterols [152].

Watson [153] manipulated the membrane lipid composition of *S. cerevisiae* by growing cells anaerobically with or without ergosterol and unsaturated fatty acids, and found that cells low in ergosterol but with high values of unsaturated fatty acid residues in their membrane phospholipids produced high concentrations of ethanol (13-15%). Other authors [154] also grew *S. cerevisiae* anaerobically in the absence of oleic acid and ergosterol successfully, although anaerobic growth was stimulated by the addition of Tween 80 and ergosterol to the anaerobic medium.

There also have been many suggestions that sterol synthesis and respiratory adaptation are closely linked [155]. Adaptation of anaerobically grown yeast cultures to oxidative conditions was found to produce a rapid burst of ergosterol synthesis with the concomitant formation of mitochondria.

Reports on the effect of alkanols point to the inner mitochondrial membrane as the target region of thermal death and its enhancement by alkanols [39]. Furthermore, it was suggested [7] that interference of alkanols with the inner membrane mitochondria of *S. cerevisiae* may underlie the effects of temperature on growth and cell viability, whereas the inhibition of growth and fermentation by alkanols at any permissive temperature is probably due to interference with nutrient transport across the plasma membrane or with glycolytic enzymes. In fact, processes identified as underlying alkanol sensitivity in *S. cerevisiae* include, as indicated before, growth, fermentation, viability [131], thermal death [93], passive proton influx across the plasma membrane [7,150], sugar and ammonium transport [93, 150], etc.

It is well known that some mutagenic agents such as ethidium bromide interact directly with the mitochondrial DNA, so that it is the DNA rather than the mitochondrial membrane that is involved in the *rho*⁻ formation process [36]. Other mutagenic agents such as nitrosoguanidine also induced *rho*⁻ formation by base shifts in DNA [33]. However, ethanol at 24% has been observed to induce respiratory-deficient mutants with no specific nuclear effects [36], so that it has been suggested [43] that the mutagenic effect of ethanol involves damage to the mitochondrial membranes [36]. This suggestion is supported by reports on the isolation of petite mutants induced by a variety of chemical agents, many of which are known to act as membrane solvents [146].

These chemical agents which act at the membrane level and that affect hydrophobic interactions [146] directly transform *rho*⁺ cells into *rho*⁻ mutants under both growth or nonproliferating conditions [131]. Furthermore, isopropanol, *tert*-butanol, and sodium dodecyl sulfate (SDS) have been observed to be more powerful mutagenic agents than ethanol [131].

It has been observed that constitutive alterations in the membrane composition (i.e., *erg* mutants, unable to synthesize ergosterol) also lead to an increase in the frequency of petite mutants [33]. This observation supports the idea that mitochondrial membrane alterations are involved in the mechanism of petite mutant formation.

Wine yeasts were reported to be generally more resistant to spontaneous or organic solvent-induced petite mutant formation than laboratory strains (Fig. 18) [131]. When mitochondria were transferred from wine strains to laboratory yeasts, the heteroplasmons obtained displayed an increase in resistance to spontaneous and induced petite mutation, indicating that the mitochondrial DNA was also directly involved in the degree of resistance to petite formation shown by a particular strain. As was reported in regard to ethanol, in addition to the genes that give rise to altered sterol composition, many nuclear genes affect the stability of the grande phenotype as shown by the high stability of the *rho*⁺ phenotype (Fig. 18) of the wine laboratory hybrids.

A correlation has been observed between the mutagenic effect of a chemical agent and its effect on cell growth [156]. Petite mutant induction has been observed even when the strains were maintained in distilled water supplemented with different solvents, supporting reports on the utilization of ethanol and other compounds [146] as petite mutant inducers under nonproliferating conditions. The mutagenic effect of the organic solvent has been correlated with cell death. Both mutation rate and death rate have been described to be closely associated parameters and both are transferable via mitochondria [33].

It has been suggested that the stability of the *rho*⁺ genotype depends on the mitochondrial membrane and on the mitochondrial genome, which points to the existence of a mitochondrial membrane-DNA complex [36]. This complex may be affected by membrane composition alterations and by membrane solvents [136] giving rise to a high proportion of petite mutations. There are also many other nuclear genes involved in the stability of the grande phenotype in the absence as well as in the presence of membrane solvents.

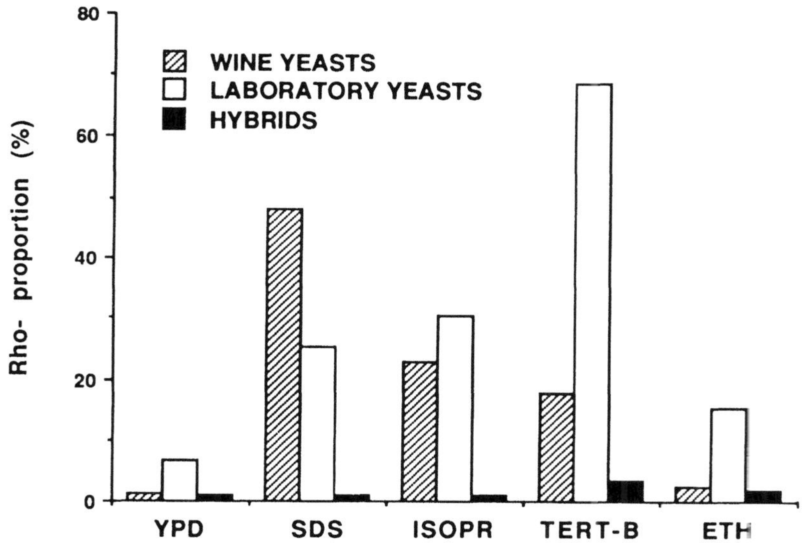

FIGURE 18. Average frequency of petite mutants, either spontaneous (HPD) or alcohol- or SDS-induced of laboratory yeasts (□), wine strains (■), and wine laboratory hybrids (▨). (From Ref. 131.)

E. Synergistic Effects of Ethanol and Temperature

The specific growth rate (μ_g) of cell populations generally falls off sharply between the optimum (T_{op}) and the maximum (T_{max}) temperature for growth. This behavior can be suitably represented by Arrhenius plots. The sharp decline of μ_g values observed between the optimum (T_{op}) and the maximum (T_{max}) temperatures for growth has been explained in terms of irreversible and reversible thermal denaturation of enzymes [78]. van Uden and Madeira-Lopes [157] and Oliveira-Baptista and van Uden [158] suggested additional mechanisms that contributed to the kinetics of growth in the supraoptimal temperature range: (a) correlation in a number of yeast species between exponential death at supermaximal temperatures and the T_{max} of growth, and (b) concurrence of exponential population death with exponential population growth within the supraoptimal temperature range.

In addition to cell growth (Fig. 19A) [151] and death rates, the effect of temperatures from 15 to 40°C on the highest amount of dry weight resulting from the steady-state continuous culture growth of *S. cerevisiae* was obtained at approximately 30°C (Fig. 19B), although growth remained vigorous between 20 and 35°C [84].

After an initial period of deathless growth, the duration of which is a function of temperature, yeast exponential death sets in [157]. During the second period, the specific rate of population change (μ) is the difference between the specific growth rate (μ_g) and the specific death rate (μ_d) which decreases the apparent specific growth rate (μ). Above a maximal temperature (T_{max}) the specific death rate (μ_d) is greater than the specific growth rate so that the population suffers net exponential death and eventually dies out [157]. At T_{max}, μ_g was equal to μ_d and net growth

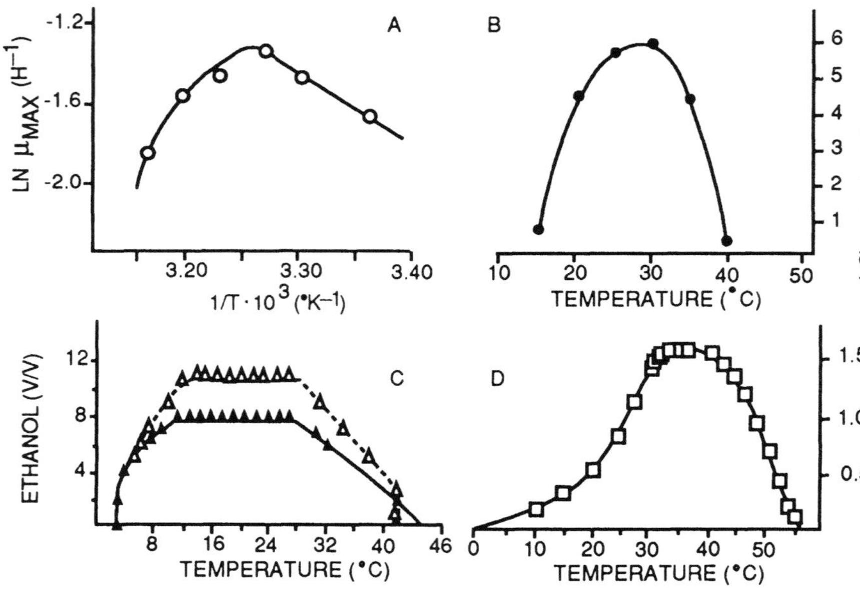

FIGURE 19. (A) Arrhenius plot of ln μ_m against 1/T. (From Ref. 151.) (B) Influence of temperature on the growth of *S. cerevisiae* in continuous culture. (From Ref. 84.) (C) Temperature profiles of maximum ethanol tolerance of *S. cerevisiae* (○) and *Kluyveromyces fragilis* (●). Experimental points indicate the concentrations of ethanol above which growth could not be detected. (From Ref. 91.) (D) Temperature profile curve of ethanol formation of the strain *S. cerevisiae* Sc5. (From Ref. 85.)

no longer occurred [157,158]. Based on kinetic evidence, it was postulated [159] that yeasts have thermosensitive target sites. If these sites are irreversibly destroyed, their inactivation implies loss of cell viability. The sites ("death sites") determine the maximum temperature for growth (T_{max}).

Using a strain of *S. cerevisiae* grown at supraoptimal temperatures, Simoes-Mendes et al. [160] found that after an initial period of exponential growth, a second period followed during which exponential death and net exponential petite mutation concurred with exponential growth. The specific rates of the three exponential processes were of the same order of magnitude and varied with temperature. Net exponential petite mutation did not occur during the deathless first period of growth. These results suggest that mitochondria are possible targets of thermal death in yeasts.

Ethanol at concentrations above 3% decreased the maximum temperature for growth of *S. cerevisiae*. At concentrations of 6%, ethanol depressed the optimum temperature for growth from 37 to 25°C and the maximum temperature for growth from 40-44°C to 33-36°C [161]. When the ethanol concentration becomes high enough, the yeast population becomes increasingly subject to ethanol-enhanced thermal death [162]. Ethanol affected the temperature profile of industrial wine yeasts much as it did that of *S. cerevisiae* [163]. With other yeasts like *K. fragilis* [164], ethanol enhanced thermal death by increasing the entropy of activation of thermal death [7]

(Fig. 19C), which is what occurred in *Saccharomyces*. It was therefore observed that the effects of ethanol on ethanol production [162] and specific growth and death rates of *S. cerevisiae* were strongly temperature-dependent (Fig. 19D) and, in turn, the *S. cerevisiae* temperature profile was a function of ethanol concentration.

Net growth of *S. uvarum* was found to be most resistant to inhibition by ethanol at temperatures (25-30°C) just below the growth optimum (35°C) [46]. Other authors found that the optimal temperature for growth was 34°C for the same strain (*S. uvarum*) (Fig. 19A), while the specific ethanol production rate was maximal at 37-43°C [151]. Up to 37°C, the inhibitory effects of ethanol on growth and the specific ethanol production rate were unaffected by temperature, although above this temperature ethanol inhibition increased significantly. Temperatures as high as 36 and 40°C did not negatively affect the ethanol productivity of Jerusalem artichoke juice batch fermentation by *K. marxianus*, and the final concentrations of ethanol were close to those produced at lower temperatures [69]. The improved fermentation [69] was attributed to the juice complex composition.

In an attempt to diminish the energy available for growth (increase of the yield product/substrate) because of the increasing maintenance requirement of the yeast cells at higher temperatures [7], yeast cells were brought to stationary phase by raising the temperature, so that ethanol was produced but cells did not grow. Raising the temperature from 30 to 39°C resulted in less growth but higher ethanol productivity by *S. cerevisiae* strains [165]. At temperatures above 39°C, however, cell death predominated.

A similar enhancement of thermal death in *S. cerevisiae* by ethanol was also observed when other alkanols were used [164]. As in the case of ethanol [161], it was possible to express the maximum temperature for growth of the yeast as a function of alkanol-enhanced thermal death [163]. The fact that the enhancement of thermal death in *S. cerevisiae* by different alkanols correlates with their lipid-buffer partition coefficients suggests that the thermal death sites of this yeast are associated with a membrane system [150]. In addition and as indicated before, the elevated growth temperatures at which exponential death concurred with exponential growth dramatically increased the spontaneous mutation rate to respiratory deficiency (petite mutation) in *S. cerevisiae* and other yeasts [7].

These observations of the effects of ethanol on T_{max}, on thermal death, and on petite mutation in *S. cerevisiae* suggested that the T_{max} sites, the thermal death sites, as well as the primary target sites of ethanol-enhanced petite mutation are located on a mitochondrial membrane.

F. Relationship Between Ethanol Tolerance and Thermotolerance

In yeasts a temperature shift from 23 to 36°C induces the transient synthesis of a set of proteins termed heat shock proteins (*hsp*) [166], which seems to confer protection against a subsequent and otherwise lethal rise in temperature. Although the production of *hsp* has been correlated with the acquisition of thermotolerance, ethanol has also been identified as an inducer of *hsp* [166]. Furthermore, the concentration of ethanol required for *hsp* induction confers thermotolerance [166], suggesting that these *hsp* or ethanol-induced proteins probably function in a general cellular response to stress.

At temperatures under 37°C, the K_i for growth and fermentation rates in a grande strain of *S. cerevisiae* is always higher than the K_i for its isogenic respiratory-deficient mutant (*rho*$^-$) [24]. However, at 37°C these differences disappear and both *rho*$^+$ and *rho*$^-$ strains are similarly inhibited by ethanol [24]. Grande cells are more inhibited by ethanol at 37°C than at lower temperatures because at 37°C mitochondria become nonfunctional due to synergistic effect that ethanol and temperature exert on these mitochondria [33]. In consequence, at 37°C grande strains are as inhibited by ethanol as petite mutants. However, when the growth rate or inhibition constant for growth (K_i) was determined at 37°C in the presence of increasing ethanol concentrations in a laboratory *S. cerevisiae* strain, a considerable increase in both parameters was observed, either when the cells had previously been grown in glycerol instead of glucose as carbon source, so that only oxidative metabolism was allowed, or when mitochondria had been transferred from highly ethanol-tolerant wine yeasts to *rho*$^-$ mutant of the laboratory strain [24]. The greater stability of the *rho*$^+$ phenotype under conditions of high temperatures and high ethanol concentrations was conferred by the wine yeast mitochondrial genome and this genome is the main factor responsible for the higher viability displayed by yeast strains possessing these mitochondria [131], since their *rho*$^+$ stability permits respiration under conditions where other mitochondria are not functional.

Under extreme conditions of ethanol and temperatures (38°C, 4-8% ethanol) the death rate μ_d is proportional to the *rho*$^-$ induction rate λ at any ethanol concentration. μ_d and λ are parameters characteristic of each particular strain [31,40]. When mitochondria are transferred from highly ethanol-tolerant wine strains to laboratory strains, μ_d and λ values in the heteroplasmons possessing the nucleus of the laboratory strain but wine yeast mitochondria are equal to μ_d and λ values shown by the corresponding wine yeast which acted as donor [33]. Results indicated that mitochondrial functions and survival are linked parameters and that the mitochondrial genome (but not mitochondrial functions coded for by the nuclear genome) is responsible for cell survival under extreme conditions of high ethanol concentration and/or high temperatures.

Cell proliferation under conditions of high temperatures and high ethanol concentrations depends on cell growth, but it is also influenced by cell death. This influence is especially significant at temperatures above 30°C and ethanol concentrations above 5% where the higher viability of heteroplasmons leads to a greater growth ethanol tolerance than the laboratory strains (Fig. 16) [131]. The influence of cell viability mediated by the mitochondrial genome is even higher. The maximal temperatures at which there were viable cells (T_{max}) were higher for the wine strains than for the laboratory strains (Table 7) [33], and heteroplasmons possessing mitochondria from the wine strains improved their T_{max} (Table 7).

It therefore seems that nuclear genes are responsible for tolerance to ethanol when its concentrations are low enough to allow growth and cell division, since cell growth and division are basically dependent on nuclear genes. Under these conditions of ethanol concentrations, cells do grow and divide and mortality is almost imperceptible. However, under extreme conditions of high ethanol concentrations and/or high temperatures the mitochondrial genome is responsible for ethanol tolerance since, under such conditions, cells do not grow or divide, and the parameter which determines ethanol tolerance is cell viability.

TABLE 7. Maximal Temperatures for Growth ($Tmax_g$) and Survival ($Tmax_s$) (°C) of Strains ABQ21 and ACA21, and Heteroplasmons ABQ21-H1 and ABQ21-N, in YPD and YPDE Media at Indicated Ethanol Concentrations[a]

	Media			
	YPD (no ethanol)		YPDE (10% ethanol)	
Strain	$Tmax_g$	$Tmax_s$	$Tmax_g$	$Tmax_s$
ABQ21	38.0	42.0	25.0	27.2
ABQ21-H1	40.5	46.0	25.5	27.7
ABQ21-N	38.0	42.0	25.0	27.2
ACA21	40.5	46.0	28.7	ND

[a]Results are the average of three trials. ND = nondetermined.
Source: Ref. 33.

V. IMPROVEMENT OF ETHANOL PRODUCTION

A. Improvement of the Environmental Conditions

1. Effects of Supplements

Ethanol production is known to be limited by the inhibitory effects of ethanol [167], substrate [27], and the synergistic effects of both [58] on the microorganisms which carry out the fermentative process. These effects are so obvious that sometimes low ethanol production has mistakenly been attributed to ethanol when it is due to effects other than ethanol or substrate inhibition, which can also limit growth and fermentation.

The factor limiting the production of high levels of ethanol by brewing yeasts was shown to be a nutritional deficiency [168], so that when a nitrogen source, ergosterol, or oleic acid was added, it was possible to produce beers of up to 16.2% ethanol (Fig. 20).

Proline is present in high concentration in wine media, although it is not usually utilized in fermentation as a nitrogen source. For this reason research has been directed toward the identification of conditions under which proline could be utilized as a yeast food [66]. Also with wine yeasts, growth and fermentation in beet molasses is very poor. The addition of an ammonium salt and/or biotin is highly stimulating for both growth and fermentation rates [15].

Ethanol tolerance of *Saccharomyces* strains is generally lower in a defined medium than in a complex medium. In the presence of ethanol, increased levels of pantothenate improve yeast viability and, in consequence, ethanol production [47].

S. cerevisiae grown in a minimal medium produces more ethanol when the concentration of the ammonium salt which is the sole source of nitrogen is raised [30]. The controlled addition of such ammonium salts to *S. cerevisiae* fermenting cells, which results in an accelerated glycolytic rate, shortens the fermentation time independent of the final ethanol produced [30].

Ethanol production from Jerusalem artichoke by *K. fragilis* has been described to be 12.8% with the consumption of 95% of initial sugars. These

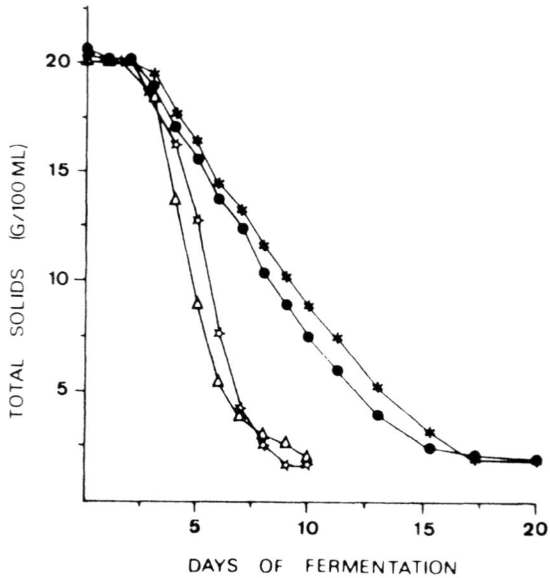

FIGURE 20. Sugar utilization in control and test fermentations to which yeast foods have been added. (△), control; (*), glycine; (☆), added urea; (●) addition of yeast extract. (From Ref. 66.)

high concentrations of ethanol are explained by protein supplementation mechanisms [68]. In addition, the low yield of ethanol production from Sudanese molasses by *K. fragilis* has been attributed to a lack of growth factors in the molasses. Higher yields are obtained when vitamins and salts are added to the medium [72]. Ethanol-induced inhibition of the specific growth and fermentation rates of *K. fragilis* and ethanol-induced stimulation of the specific death rate are also found to be much more drastic in simple medium than in Jerusalem artichoke juice [69].

Dombek and Ingram [144] reported that nutrient limitation was the major factor responsible for the apparent toxicity of low levels of ethanol on *S. cerevisiae* Cultures which showed a decline in fermentative activity during the early stages of fermentation. Supplementing fermentation media with 0.5 mM magnesium resulted in increased yeast cell mass and reduced the decline in fermentative activity. These two effects reduced the time required for the conversion of 20% glucose into ethanol by one-third [71].

When cane molasses was fermented by *S. cerevisiae* yeasts, ethanol production was also increased by the addition of invertase [29], indicating that the hydrolytic activity of yeast can be a limiting step. Similarly, treatment of beet pulp with the enzyme system of a culture filtrate of *Talaromyces emersonii* considerably improved beet pulp conversion to ethanol [169].

In general, the trace elements have only a minor effect on cell yield Yx/s but a major effect on the long-term viability of cells and on the fermentation rate. There exist a number of cations which at varied concentrations act as inhibitors of growth and fermentation [170]. Molasses used as substrates for ethanol production have low levels of phosphorus and iron, and sometimes zinc is absent. Copper, however, could be toxic to some yeasts at the level observed [170]. When the supply of phosphate is restricted in a glucose-limited chemostat culture of baker's yeast to an extent that residual phosphate in the medium can hardly be observed, the value of the critical dilution rate is apparently enhanced [171]. In addition, the specific ethanol productivity with *S. cerevisiae* grown aerobically in a chemostat was found to increase from zero to 13 mmol/g cell dry matter/h when the potassium content in the substrate used was decreased to 0.05 mol/kg glucose [170].

Acetate, a byproduct of ethanol fermentation by *S. cerevisiae*, was shown to inhibit cell growth if present in high concentrations. In an immobilized cell reactor, when steady state is not achieved and cell overgrowth is a problem, acetate is added to batch and plug flow systems. As a consequence the productivity increases by as much as 50% while cell growth is greatly reduced [49].

Summarizing, together with the reports on the favorable effects of supplements containing mainly nitrogenous constituents, there are also unfavorable effects due to the inhibition of cells by compounds other than ethanol and fermentable substrate. To characterize and compare ethanol production of fermenting yeasts, the standardization and further optimization of the medium composition must be taken into consideration.

2. Effects of Biomass Recycle

Ethanol production can also be enhanced by direct increase of the microbial biomass achieved by cell recycle, either by use of flocculent strains [31] or by cell immobilization [172].

When using molasses at concentrations above 15% sucrose, growth is almost nil and fermentation impractical [29]. To accelerate the fermentation process the same grown biomass is used in consecutive fermentations [15] with a constant biomass increase when fermenting molasses with 12% sucrose, which allows a higher rate of sucrose consumption, producing the same final ethanol concentration [15]. When molasses with 20% sucrose was used in the last fermentation, higher ethanol concentrations were obtained at the same rate as when 12% sucrose was used. Cell recycle was achieved either using flocculent strains or entrapping powdery, nonflocculent strains in alginate [15]. Cultures of *K. marxianus* immobilized in alginate beads and used in a packed-bed bioreactor system for the continuous production of ethanol from extract of Jerusalem artichoke have a volumetric ethanol productivity 15 times higher than that of an ordinary continuous stirred tank bioreactor using free cells [167]. The increase in fermentation rate during the recycle process is probably due to an increase in cell concentration caused by a reduction of ethanol inhibition and growth stimulation by possible supplements absent in the previous fermentation media.

Fermentation systems need to be improved to provide higher rates and alcohol levels to make alcohol production cheaper [55]. Maia et al. [173] reported on alternative techniques for alcoholic fermentation by *S. cerevisiae* yeasts and concluded that batch process leads to high yields but the productivity was very low, whereas in the feed-batch process inhibition by the substrate was reduced which permitted an increase in ethanol concentration as well as an improvement in productivity, the largest increase in productivity being observed in continuous techniques. Continuous ethanol production from glucose, sucrose, cane molasses, corn wet-milling waste, etc., by cells of *S. cerevisiae* immobilized on rotating

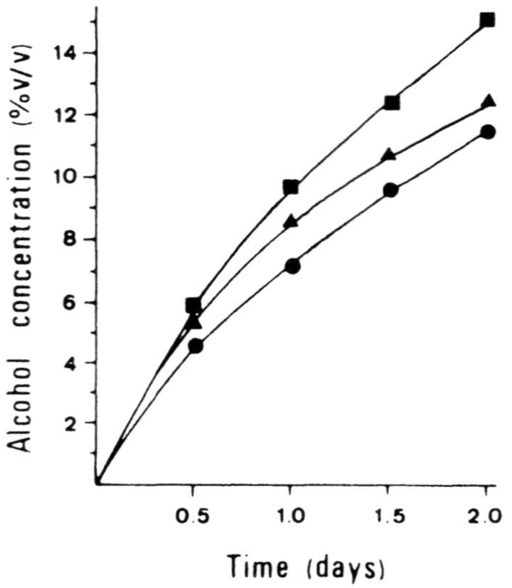

FIGURE 21. Average alcohol concentrations of different stages of cocurrent (■), cell recycle (▲), and countercurrent fermentations (●) (30°C, 9.8 g/liter cells, 30% (w/v) glucose). (From Ref. 55.)

fiber disks by calcium alginate gel in a continuous dynamic immobilized cell bioreactor have also been described as showing high ethanol productivity and very long half-life [174].

To establish the optimal conditions for alcohol production, experiments have been conducted with *S. cerevisiae* cells and various feeding procedures. Among the cell-feeding procedures a cocurrent design (Fig. 21) was found to be better than recycle and countercurrent designs. The specific alcohol productivity of the cocurrent system was the highest. This high productivity has been explained as due to the ability of the cells to generate ATP before producing inhibitory levels of alcohol, thus allowing them to adapt or maintain their cellular structure [55].

Continuous ethanol production with internal separation of yeast cells by filtration, either as a rotating microporous filter (as frequently used with molasses) or as filtration through a thin layer of diatomaceous earth [73] has been described. The maximum rate and ethanol productivity was obtained in cultures of baker's yeast (26 g/liter/h) when 10% glucose was supplied.

Physical characteristics such as flock density function, flock size distribution, and relative flock strength of a number of flocculent strains have been measured with regard to ethanol production [128]. When a mutant of *S. cerevisiae* which forms large multicellular flocks in liquid cultures was used, media containing high concentrations of glucose was rapidly fermented in a continuous nonaerated tower fermentor which operated continuously for 7 months [175].

Fluidized bed bioreactor systems coupled to a small secondary reactor in which the biomass particles were recycled have been used to convert glucose to ethanol using particles of a flocculating strain of *Z. mobilis* [176], since this species appears to give a much higher production rate than the more common yeast *Saccharomyces* [177,178]. Flocculent strains of *Zymomonas* isolated from sugarcane juice have therefore been used to produce ethanol in upflow floc continuous reactors [179]. Finally, another aspect to be considered in the process of ethanol production is the volume of liquid to be transformed. The required volume of fermentation broth is much smaller in the case of the continuous method using immobilized cells than in the case of conventional batch fermentation methods [180].

3. Effects of the Addition of Lipids

A series of studies has been carried out to explore the influence of the lipid composition of the cell membrane and of lipid addition, which in turn modifies lipid composition [152] (Table 8), on ethanol tolerance. When membranes were enriched with linoleyl residues (18:2) rather than oleyl residues (18:1), the decline in cell viability of *S. cerevisiae* in the presence of ethanol was found to be less [19]. This protection by polyunsaturated fatty acid residues was further increased by incorporation of sterols with unsaturated side chains (ergosterol) rather than ones with saturated side chains. Similarly, the increase in generation time in cultures of *S. cerevisiae* grown in ethanol enriched with ergosterol and linoleic acid was observed to be less than when ergosterol and oleic acid were added [19]. Furthermore, supplementation of molasses fermentation medium with oils and fatty acids increased fermentation and the tolerance of distiller's yeast to higher temperatures and alcohol concentrations [181].

TABLE 8. Effect of Different Supplements of Ergosterol on the Ergosterol Content of Whole Cells and Isolated Mitochondrial Precursor Structures of *S. cerevisiae*[a]

Ergosterol supplement (mg/liter medium)	Yield of cells (g dry wt/liter)	Ergosterol: Cellular (mg/g dry wt)	Ergosterol: Mitochondrial (mg/g protein)
0	0.95	0.74	7.3
5	0.95	1.15	35.8
10	1.25	2.20	22.1
10	1.25	0.53	39.5
10	1.34	1.80	47.0
15	1.34	2.20	105.0
20	1.45	1.95	32.6
40	1.34	2.20	60.0
40	1.45	3.25	40.7
40	1.45	3.10	47.0

[a]Results are for 10 individual experiments. Cellular fatty acids of all preparations contain 70-80% unsaturated fatty acids (C16:1 + C18:1).
Source: Ref. 152.

This increased fermentation appears to be the result of both the increased cell concentration due to growth stimulation and the higher percentage of active cells in the population.

Synthetic medium containing a subinhibitory level of ethanol (6%) supplemented with oleate improved *S. cerevisiae* yeast viability [47]. In grape must of high sugar concentration, yeast growth viability and fermentation activity were stimulated when oleanolic acid was added. However, the stimulation was mainly the result of the yeast cell sterol content which increased when certain sterols (ergosterols) were exogenously supplied [141]. Other authors also observed that when minimal medium was supplemented with ergosterol and Tween 80 increased *S. cerevisiae* growth resulted [30]. After the addition of ergosterol and oleic acid to brewing yeast, cell viability also remains high. The yeasts can be repitched at least five times and it is possible to produce beers of up to 16.2% ethanol [67].

Some authors claim that unsaturated fatty acids but not ergosterol are essential for high ethanol production in *Saccharomyces* [153]. While a supplement of ergosterol in the fermentation of brewer's yeast improved attenuation and cell yield, there was very little improvement in final cell viability [47]. Furthermore, when cells were grown anaerobically with or without ergosterol and unsaturated fatty acids, those cells low in ergosterol but enriched in unsaturated fatty acid produced concentrations of ethanol of 15.5%—higher than those produced by the nonsupplemented unsaturated fatty acid control. The fermentation activity of *S. cerevisiae* supplemented with oleic acid and ergosterol is also less affected by different concentrations of ethanol than some cells supplemented with linoleic acid and ergosterol, the viability of these latter cells being more affected [182].

The fermentation ability of a strain of *K. fragilis* already selected for rapid lactose-fermenting capability improved dramatically with the addition of unsaturated fatty acids and ergosterol to the medium. The fermentation time decreased by two-thirds [140] and the effects of added lipids were noted on specific growth rate and biomass production (Table 9).

E. coli also needs a minimum amount of unsaturated fatty acid which is 15-20% of the total phospholipid fatty acid. When *E. coli* cells were grown in media supplemented with either vaccenic or palmitic acid, several cellular parameters associated with membrane functions were less inhibited by ethanol in those cells grown in vaccenic acid [183]. Metabolite leakage was proposed as the primary event associated with bacterial inactivation in *E. coli* in

TABLE 9. Influence of Ethanol Concentration on Maximum Growth Rate (μ_m) and Maximum Biomass Concentration (X_m) in Anaerobic Culture

Ethanol conc. (%, v/v)	μ_m (h^{-1})		X_m (g/liter)	
	Control	+ Lipids	Control	+ Lipids
0	0.17	0.22	3.82	4.10
2	0.15	0.20	1.86	3.60
4	0.08	0.10	1.42	2.02
6	0.02	0.02	0.56	0.54

Source: Ref. 140.

buffered solutions of ethanol, and the increase in acyl chain length as the beneficial aspect of vaccenic acid (18:1) incorporation rather than the increase in membrane unsaturation [183].

Clearly, enrichment with unsaturated fatty acyl residues conferred increased tolerance to ethanol, resulting in improved viability and increased ethanol production. Since increased unsaturation of the cell membranes would presumably result in increased membrane fluidity, the effect of unsaturated fatty acid addition compensated for changes in membrane fluidity induced by ethanol [2]. In addition, a functionally important part of the adaptive changes to the presence of ethanol is the addition of lipids, which provoke a tendency toward increased chain length [139]. In support of this point, extremely ethanol-tolerant strains of *Lactobacillus* capable of spoiling sake wines containing over 20% ethanol have been found to contain high levels of unusually long-chain fatty acids (C_{20}-C_{30}) in their phospholipids [184].

Proteolipids from mycelia of *Aspergillus oryzae* act as a high-concentration alcohol-producing factor and caused the accumulation of more than 20% alcohol by *Saccharomyces sake* yeasts, when added to a synthetic medium [17].

In addition to *S. sake*, the presence of *A. oryzae* proteolipid at the early stage of fermentation was essential to obtaining yeast cells capable of producing high concentrations of alcohol in strains of *S. cerevisiae*, *S. uvarum*, and other yeast strains [61]. The final ethanol production was over 20% under these conditions in some of the strains. Furthermore, a new alcohol-tolerant *S. uvarum* yeast which had the ability to ferment a high variety of substrates including inulin produced up to 22.4% ethanol after the addition of fungal mycelia into the synthetic medium at 15°C. The concentration dropped to 11.6% at 40°C (Fig. 22) [63].

4. Stepwise Addition of the Substrate

When the glucose concentration is decreased to near exhaustion, respiration and alcohol production are naturally slowed. Besides, glucose levels of 5% or higher cause a shift to fermentation only even in the presence of oxygen [55]. However, higher glucose concentrations affect the alcohol production rate significantly, although this inhibition is much less pronounced than that caused by alcohol concentrations. In the case of continuous or feed-batch fermentations the glucose concentration could be kept between two inhibition levels to avoid any effect from the substrate [55]. The range where both inhibitions can be avoided has been determined in *S. sake* [55], specific alcohol productivity staying constant between 2.5 and 10% glucose. If the initial glucose concentration is 30 or 40%, the initial specific alcohol productivity will be only 75 or 50% of the specific productivity without substrate inhibition.

Maia et al. [173] reviewed the parameters and techniques of alcoholic fermentation with *S. cerevisiae* and found that in the feed-batch process inhibition by the substrate is much reduced, permitting a considerable increase in the ethanol concentration in the fermentation medium as well as an improvement in the productivity. Similarly, Ciftci et al. [55] conducted experiments to establish the optimal conditions for alcohol production with cell feeding at various stages and found that the glucose concentration should be kept under 10% for maximum specific alcohol productivity. Additionally, when growth and fermentation rates of brewer's, distiller's, and wine yeasts [61] are measured in media supplemented with increasing substrate

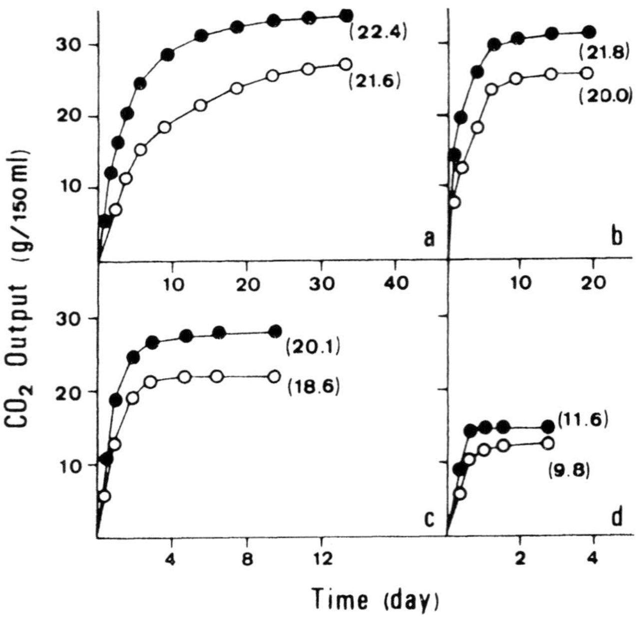

FIGURE 22. Time course of ethanol production by *Saccharomyces uvarum* var. *inulyticus*. The basal synthetic medium (150 ml) was supplemented with (●) or did not contain (○) 10 g of fungal mycelia. It was then inoculated with actively growing cells to the extent of 1×10^8 cells/ml of medium. Fermentation tests were carried out at (a) 15°C, (b) 20°C, (c) 30°C, and (d) 40°C. Final ethanol concentration (%v/v) is shown in parentheses in the figure. (From Ref. 63.)

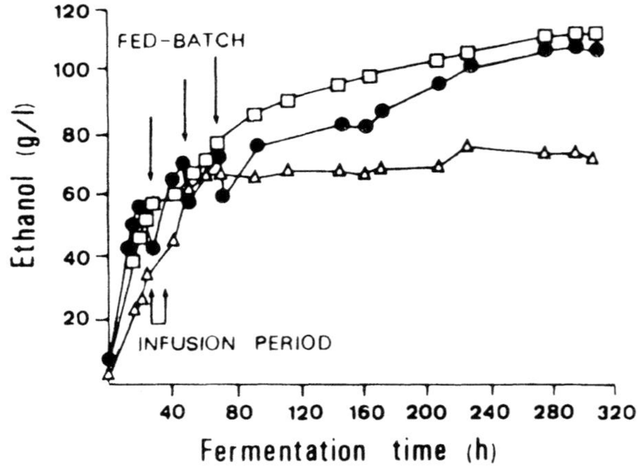

FIGURE 23. Production of ethanol by a flocculent strain of *S. cerevisiae* under various conditions of substrate addition. (△) Control; (●) stepwise addition of the substrate; (□) infusion addition of the substrate. (From Ref. 62.)

concentrations, i.e., glucose, these rates are normally increasingly inhibited with increasing sugar concentration over 10%, with this inhibition being between 25 and 50% in 25% glucose [15]. This substrate inhibition can be avoided by the stepwise addition of the fermentable sugar [61].

When the substrate is introduced to the fermentation media of *S. cerevisiae* in four batches (feed-batch), the levels of ethanol obtained in the cell-free media were much higher than those with a single-batch operation (Fig. 23) [62]. When part of the substrate was slowly infused over an 8-h period the concentration of ethanol achieved was even higher (Fig. 23) [62].

The formation of concentrations of alcohol up to 20-21% was accomplished by *S. sake* at 20°C with a gradual addition of sucrose [185]. This led to an increased rate of ethanol production [185] and higher final ethanol concentrations [61]. Results obtained when the fermentation tests were carried out with the stepwise addition of sucrose have been extended to yeasts other than *S. sake* [61]. When *S. uvarum, S. cerevisiae, Saccharomyces formosensis, S. cerevisiae* var. *ellipsoideus*, and *Schizosaccharomyces pombe* were tested, they all produced over 15% ethanol at 20°C. *S. uvarum* and *S. cerevisiae* produced about 20% ethanol.

Highly ethanol-tolerant wine strains have also been found to show a significant improvement of their growth and fermentation rate values when 25% glucose was added stepwise compared with the results obtained with a 25% initial glucose concentration [15], although no differences in the maximal amount of ethanol produced were found. When the production of ethanol by *Z. mobilis* was studied in feed-batch cultures, glucose inhibition of growth rate also occurred at concentrations greater than 8%. At 20% glucose there was no growth and almost no ethanol production. This inhibition could be effectively overcome with a feed-batch system of culture while maintaining good yields of ethanol from glucose and volumetric productivity [179]. The results thus indicate the effectiveness of modified feeding conditions on ethanol production.

5. Other Improvements Due to Environmental Conditions

Oxygen has a profound influence on the activity of yeast and particularly on yeast growth. Certain yeast enzymes react only with oxygen which cannot be replaced by other hydrogen acceptors [186]. This applies to the enzymes involved in the synthesis of unsaturated fatty acids and sterols which are vital elements of the cell membranes [186]. Furthermore, the presence of oxygen is required for aerobic respiration with its very high-energy yield and growth potential [51]. The oxygen requirement by brewing strains of *S. uvarum* varies with the strain [186]. The strains having a high demand for oxygen are the most sensitive. In fermentation the typical consequences of oxygen deficiency include restricted yeast growth, reduced yeast viability, and slow and incomplete fermentation [186].

It was described above that the addition of unsaturated fatty acids, ergosterol, or proteolipids to the fermentation medium improved yeast survival in the presence of ethanol and contributed to increased ethanol production. The regulation of unsaturated fatty acid metabolism can be influenced by the availability of oxygen [47]. To overcome this problem, fermentations with controlled oxygen levels [47] were found to be satisfactory for reproducing brewery behavior. Only by supplying oxygen were viabilities maintained at high levels [47]. In fermenting molasses, however, there were only minor differences in dissolved oxygen concentration and no correlation was observed between the rate or extent of fermentation and dissolved oxygen concentration [181].

One key objective of new processes to produce ethanol by fermentation must be to reduce the energy requirements through improved distillation systems, nontraditional methods of separation, and thermophilic fermentation operation or other appropriate techniques. Under anaerobic conditions ethanol is a primary metabolite and it is impossible to uncouple growth and ethanol production, but the energy available for growth can be diminished at higher temperatures [165]. Raising the temperature from 30 to 39°C in a culture of *S. cerevisiae* led to less cell growth and higher ethanol productivity [165].

In addition, the optimum temperature for alcoholic fermentation by distiller's yeast is about 30°C, but in tropical and subtropical regions the temperature of fermenting media increases above 40°C resulting in the inactivation of the yeast and inefficient fermentation. Supplementation of the fermentation medium (molasses) with oils and fatty acids increased the tolerance of distiller's yeast to higher temperatures and alcohol concentrations [181]. It is likely that supplementation with oils and fatty acids altered the membrane composition in a favorable manner, thereby reducing the harmful effects of higher temperatures.

Ethanol batch fermentation of *S. cerevisiae* revealed that lower culture temperature caused slower growth and slower ethanol production but the final cell mass and ethanol concentrations reached levels which were higher than those for higher culture temperatures [187]. Alcohol tolerance and production decreases markedly with a rise in temperature. *S. sake* produced more than 20% alcohol at 10 or 20°C. At 30°C, fermentation was most rapid but it ceased at a stage where the concentration of alcohol reached about 17% [61]. *S. uvarum* also produced 22.4, 21.8, 20.1, and 11.6% ethanol at 15, 20, 30, and 40°C, respectively [63] (Fig. 22).

The specific alcohol productivity of *S. cerevisiae* increased with increasing temperature, at temperatures between 20 and 40°C [55]. Furthermore, when cells were exposed to temperatures between 30 and 60°C to estimate the specific alcohol productivity, an optimum between 40 and 42.5°C was observed [55]. Alcohol productivity increased drastically between 30 and 35°C and there was a clear decrease above 45°C. The maximum specific ethanol formation rate has been found by other authors [188] to be reached within the temperature limits of 32-36°C and the maximum temperature for ethanol formation (T_{max}) was 57°C [188]. The results suggest that when the separation of the growth and the alcohol production stages is possible, the fermentor should be operated at temperatures between 40 and 42.5°C if high specific alcohol productivity is needed.

Temperature studies on immobilized cells or enzymes are also useful in a number of ways [189]. For instance, with free cells of *K. marxianus* grown on Jerusalem artichoke extracts [189], the ethanol and biomass yields were relatively constant over the temperature range 25-35°C but dropped sharply beyond 35°C. However, with immobilized cells, ethanol yield remained almost constant in the temperature range 25-45°C, and the specific ethanol production rate and specific total sugar uptake rate attained their maximum values at 40°C.

S. cerevisiae and other yeasts acidify the medium during growth on fermentable carbon sources [48] so that fermentation often stops due to the low pH values rather than an inhibitory effect of the ethanol produced. This effect can be avoided in continuous culture by the incorporation of a pH control system [38,48] or can be minimized under certain conditions such as cell immobilization in alginate. For instance, for the pH range between 3 and 7, the free cell optimum for growth and product formation of *K. marxianus* grown on Jerusalem artichoke extracts was found to be pH 5. At values higher or lower than pH 5 a sharp decrease in specific ethanol production rate was observed. In comparison, immobilized cells showed a broad optimum pH profile. The best ethanol production rates were observed between pH 4 and 6 [189].

B. Improvement of Industrial Strains

1. Introduction of New Properties

Genetic improvement of industrial yeasts frequently requires the transference of new properties, absent in the original strains, from organisms not always taxonomically related [65]. This transfer is carried out by conjugation if the two strains are genetically compatible or by genetic engineering or protoplast fusion when this is not so. Lactose is one of the least expensive carbon sources for ethanol production. Although attempts have been made to use lactose for alcohol production, the major problem has been that *S. cerevisiae* is unable to ferment lactose [9]. The relatively small number of other microorganisms which are capable of fermenting lactose usually produce a low yield of ethanol. Therefore, genetic engineering techniques have been used to introduce the β-galactosidase gene from *Kluyveromyces lactis* into *S. cerevisiae* [190]. Protoplast fusion has also been used to construct hybrids between *K. fragilis* capable of fermenting lactose and highly ethanol-tolerant *S. cerevisiae* [9]. The fusants obtained were capable of assimilating lactose and producing ethanol in excess of 13% and their DNA content per cell suggested that multiple fusions had taken place.

The most important substrate used for ethanol production is cellulose, since it is the most abundant and the easiest to obtain. Hydrolysis of cellulose with fungal cellulase complexes such as those obtained from *Trichoderma reesei* results in the liberation of glucose and cellobiose [191]. The inability of industrial glucose-fermenting yeasts to ferment cellobiose results in the incomplete conversion of the cellulose hydrolyzate to ethanol. This has given rise to an interest in yeasts that are capable of fermenting cellobiose [192,193]. Furthermore, those yeasts which are able to utilize cellobiose such as strains of the genera *Candida, Zygosaccharomyces, Pichia, Kluyveromyces, Brethanomyces*, etc., are not as efficient ethanol producers as *Saccharomyces* strains [64]. Hybrids have been obtained by protoplast fusion either between *Zygosaccharomyces* and *Saccharomyces* strains [132] or between *Candida* and *Saccharomyces* [105]. These hybrids combine the best characteristics of both types of yeast species, namely, ethanol production and a capacity for cellobiose utilization. Even in intraspecific protoplast fusion, mitotic instabilities, ploidy levels which deviate from the expected, and chromosomal rearrangements arise during hybrid formation [117]. In these cases genetic engineering techniques are used to introduce the β-glucosidase gene from *Candida* [3] or the carboxymethylcellulase (1,4-β-glucanase) gene from *Cellulomonas* [142] into *Saccharomyces*. Although there is not always a satisfactory expression of these gene products due to additional problems related to substrate transport or regulation of such heterologous genes [3], yeasts transformed with a recombinant plasmid secreted carboxymethylcellulase activity [142]. Secretion of active

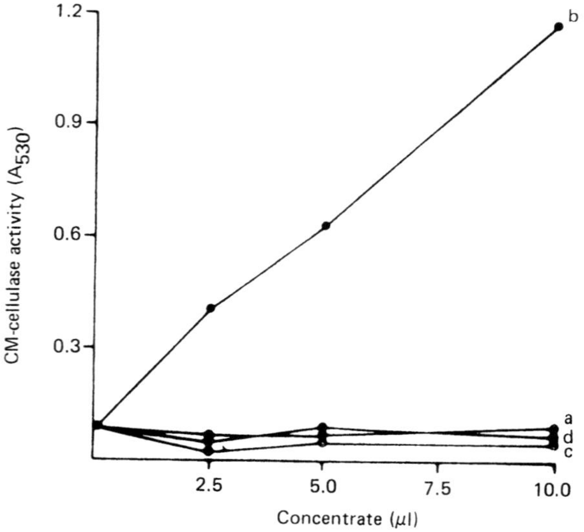

FIGURE 24. Production of reducing sugar by yeast extracellular carboxymethylcellulase. (a) Untransformed control; (b) culture containing pK 2.4 plasmid; (c) culture with pK 1.3 plasmid; (d) culture with pL 5.19 plasmid. Plasmid pL 5.19 carries a yeast leader peptide incorporated into the cloned genes which allows the carboxymethylcellulase to be excreted. (From Ref. 192.)

enzyme was greatly increased when the leader of a secreted yeast protein, the K1 toxin, was inserted in the bacterial cellulase sequence (Fig. 24). This is, up to now, the first step in constructing a functional cellulase complex in *Saccharomyces* strains.

The production of industrial and fuel ethanol from starchy biomass, one of the most common substrates for ethanol production, involves a three-step process: liquefaction of starch by an endoamylase, saccharification to produce glucose from dextrins, and fermentation of glucose. A variety of *S. diastaticus* strains have been collected and used to prepare starch-fermenting strains and a series of hybrids constructed for efficient direct starch fermentation [116]. In addition, the α-amylase gene from *Bacillus* [194] has been transferred to *S. cerevisiae* and the heterologous gene product expressed under the desired conditions.

Xylan is a major component of hemicellulose and contains substantial amounts of D-xylose. There might be an advantage if yeasts were available that could accomplish both hydrolysis of xylan and subsequent bioconversion of the monosaccharides to ethanol. Yeasts have been identified that are xylanolytic but not cellulolytic [6]. Some of these yeasts (*Pichia stipitis*) are able to convert xylan to ethanol with an appreciable yield [6]. Selected *K. marxianus* strains are also able to directly ferment D-xylose to ethanol under aerobic conditions [26]. Other yeast species have been found to produce ethanol from xylose, *Candida tenuis* being a good fermenter of both xylose and cellobiose under the test conditions used [8]. However, these strains produced over 1 g of ethanol per liter in the fermentation test [8] so that genetic engineering has been also used to transfer the xylose-isomerase gene from *E. coli* to a highly ethanol-tolerant *Saccharomyces* strain [195]. Protoplast fusion has also been used to construct presumptive polyploids of the D-xylose-fermenting yeast *Candida shehatae* [133]. The increased level of ploidy was found to have an effect on the rate of ethanol production from D-xylose (Fig. 12) [133].

2. Increase of the Efficiency of Substrate Conversion

Yeasts in general utilize glucose preferentially in mixtures of glucose and other carbon sources [196]. Glucose prevents the uptake of other carbon sources by catabolite regulation of the uptake system, resulting in the sequential utilization of the two substrates [196]. The repression of the utilization of the other substrate as a result of the presence of glucose is a major factor limiting the industrial use of those strains that are able to utilize several carbohydrates such as lactose, starch, or cellobiose [132]. Carbon-limited growth frequently stimulates the production of catabolite-repressed enzymes [197,198]. Another route to improvement involves genetic deregulation and selection of constitutive mutants or mutants insensitive to catabolite repression [198].

There are several reports on the isolation of yeast mutants with a number of enzymes that are no longer affected by glucose [199-201]. In some cases mutants have been isolated that are resistant to glucose analogs [202]. Some of the mutants appear to affect the repression on only one enzyme [202] while most of them have a pleiotropic effect [202].

The search for a satisfactory method of disposal of the unutilized whey produced in the manufacture of cheese remains an area of intense concern. One alternative is to ferment the lactose into ethanol for use as a fuel or chemical feedstock [202]. Mutants of industrial strains of *Saccharomyces* which rapidly and completely fermented mixtures of glucose and galactose to ethanol have been isolated that are resistant to 2-deoxyglucose, a toxic glucose analog. One class apparently specifically affects the utilization of galactose and allows sequential utilization of first glucose and then galactose. The second class was resistant to general catabolite repression and produced maltase, invertase, and galactokinase in the presence of repressive levels of glucose [202]. A third class of mutant was characteristically able to utilize both sugars simultaneously. These mutants were capable of fermenting a mixture of 10% glucose and 10% galactose to completion in less than 37 h (Fig. 25) [202].

Mutants that do not repress enzymes subject to catabolite repression have been isolated that are resistant to the glucose analog 2-deoxyglucose in the presence of raffinose, starch, cellobiose, and cellulose [197] or that are able to grow in maltose, lactate, and other carbon sources in the presence of glucosamine [197] or by the procedure of growth on a noninducing substrate [198]. By using these procedures glucose-resistant mutants have been isolated able to ferment starch, lactose, cellobiose, and other substrates in the presence of glucose [203].

3. Highly Ethanol-Productive Yeast Strains

An important development in current ethanol production technology is the fermentation of several carbohydrates with maximal efficiency. This requires a yeast that is alcohol-tolerant, viable when fermentation ends, and suitable for reuse in subsequent fermentations. Nevertheless, little work has been published on genetic aspects of ethanol tolerance in yeast

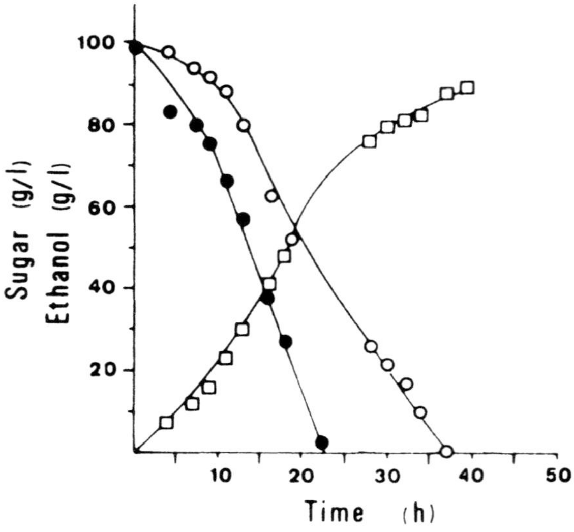

FIGURE 25. Fermentation kinetics of mutant strain SR. A mixture of 10% glucose plus 10% galactose was inoculated (10% v/v) from a culture of strain SR grown in YP medium plus 5% dextrose. (●), glucose consumption; (○), galactose consumption; (□) ethanol production. (From Ref. 202.)

or the genetic improvement of ethanol-producing yeasts [47]. Failure to select for ethanol tolerance during subculture [24] and segregation of ethanol tolerance in wine and brewing yeast meiotic products [131] is compatible with the proposal that ethanol tolerance in yeast is under complex genetic control [47]. For this reason, strain improvement by hybridization has been used successfully and it is feasible to select ethanol-tolerant yeast hybrids which gather together desirable characteristics distributed in various nonisogenic strains [22]. With such hybrids it is possible to extend the upper limits of ethanol production which are determined by yeast survival and crop viability. In addition, as an alternate way to improve ethanol production, strains from several yeast species have been examined for their ability to grow and survive in a range of ethanol concentrations and to ferment several carbohydrates [64], and the fastest and most tolerant yeast strains have been appropriately selected and improved [23].

Selection: With regard to yeast selection, polyploid yeast strain construction has been described [121] as an obligatory technique in the improvement of ethanol production. Polyploid strains make it possible to manipulate gene dosage, and the ability to do this may be important in yeast-breeding programs because differences in gene dosage can lead to differences in the enzymatic content of cells [204].

Screening and selection techniques have frequently been used after isolating highly ethanol-tolerant yeast strains from their natural environments (sake, wine with high alcohol content, etc.) [63].

A new variant of *S. uvarum* was isolated from fermented rice wine [63] which displays an extraordinary ability to ferment inulin into ethanol and a hypertolerance to high concentration of ethanol at a wide range of temperatures (Fig. 22). In contrast to other *S. uvarum* strains described

[63], the new variant grew well at 40°C and did not require any vitamin or growth factor for normal growth.

Type strains of 200 species of yeasts able to ferment glucose and grow on xylose were screened for fermentation of D-xylose. Of these, several produced more than 1 g ethanol per liter in the fermentation test with 2% xylose [8]. Maximum ethanol concentrations of 5.6 g/liter were obtained after screening strains of *K. marxianus* able to ferment D-xylose directly to ethanol under aerobic conditions [26].

Protoplast fusion techniques are used to obtain polyploids though the fusion products may be of unknown ploidy level and genetically unstable [121]. For this reason, measurements of DNA content, cell volumes, resistance to mutant induction or survival in the face of several mutagenic treatments, as well as the segregation of genetic markers when available [205] are normally used to determine ploidy level in isolated fusants. This protoplast fusion technique was employed in the preparation of presumptive diploid, triploid, and tetraploid strains of the D-xylose-fermenting yeast *C. shahetae*, and the hybrid nature of the fusant strains confirmed by analyzing cell size, cell volume, DNA content, and induced mitotic segregation. The most interesting fact was that increased ploidy level had an effect on the ethanol yield from D-xylose [133] (Fig. 12). Increased ploidy level has also been found to have a favorable effect on ethanol production by *Saccharomyces* strains [134], since the efficiency of ethanol production per unit cell mass was greater in cells of higher ploidy.

Mutation: The complexity of the effects of ethanol on yeast cells means that attempts to isolate ethanol-tolerant mutants by conventional screening and selection techniques are unlikely to be successful. Therefore, in order to isolate such mutants a continuous selection system has been developed [46] in which the intensity of selection is dictated by the culture itself via a feedback control circuit. This has permitted the isolation of viable mutants of *S. uvarum* which showed increased fermentation rates at high concentrations of ethanol. This method of continuous selection uses a feedback control system based on the rate of CO_2 production by the culture and has yielded yeast mutants with considerably improved ethanol tolerance (Table 10) [46].

Korhola [206] also developed more ethanol-tolerant strains by feeding in exogenous ethanol during continuous cultivations of industrial *Saccharomyces* yeasts: these more ethanol-tolerant strains showed less inhibition of growth and fermentation by ethanol as compared to the starting strain, whereas both grew equally well in the absence of ethanol.

Possible genetic markers for ethanol tolerance have been identified following different types of phenotypic observations. To improve the conversion of hexoses and pentoses to ethanol, a pyruvate-negative mutant of *Clostridium saccharolyticum* having lower acetate kinase activity was obtained [207] which uses more substrate (glucose or xylose) and produces more biomass and ethanol, but less acetic acid. This improved production of ethanol was also accompanied by an increased tolerance to ethanol (K_i = 6.5%) in the mutant as compared to 3.5% for the parent strain [207].

Transformation: An application of transformation in *Saccharomyces* is in the breeding of industrial strains, especially those that have little or no sex. In such applications, it may be of interest to carry out the transformation with plasmid DNA in the construction of which only *Saccharomyces* DNA has been involved, i.e., to have a yeast transformation vector

TABLE 10. Fermentation Performance of *Saccharomyces cerevisiae* Wild-Type (5D-cyc) and Five Ethanol-Tolerant Mutants Which Were Isolated from a Continuous Fermentor 1133 h Into the Selection Experiment

Strain	Control Q_{CO2}	% control in presence of ethanol		
		10% (w/v)	20% (w/v)	35% (w/v)
Wild-type	0.822	74.6	71.9	66.0
154	0.823	80.5	71.1	77.7
155	0.793	78.8	70.7	77.7
158	0.993	80.7	76.8	73.8
159	1.33	93.4	80.7	75.8
160	1.74	89.4	83.5	74.6

Source: Ref. 32.

system without the involvement of foreign DNA. Attempts to develop such a system are based on the use of 2-μm DNA which is a naturally occurring circular plasmid in *S. cerevisiae* [120].

In fact, a pure yeast plasmid consisting of 2-μm DNA with an insertion of a chromosomal sequence including the *HIS*4 gene was constructed [120]. The composite plasmid can be transferred during *kar*1-mediated cytoduction of rho^+ cells to rho^- recipient strains [112]. By using circular molecules of the yeast plasmid 2-μm DNA, a strain of *S. cerevisiae* carrying a deletion in the *his*4 locus was transformed to histidine prototrophy. The transformant contained a 2-μm DNA in which a chromosomal DNA region containing the *HIS*4 gene had been inserted [120].

In addition, when tetrad analysis of strains used for ethanol production is impossible due to a low yield of viable spores and/or conjugation is not possible due to the fact that the strains are nonmaters, single chromosome transfer is used for genetic analysis and/or transfer of new properties among industrial yeasts [111].

Hybridization: Many genes limit cell proliferation in the presence of ethanol [18]. Since these genes are different in nonisogenic strains, nonisogenic hybridization generates yeast hybrids more tolerant than their parental strains [47], so that hybridization has been the usual method of generating ethanol-tolerant yeasts [18]. Day et al. [47] compared the ability to grow and survive in fairly high ethanol concentrations (up to 15%) and the ability to ferment concentrated substrates in several sake yeasts (*S. cerevisiae*), osmophilic strains (*Saccharomyces rouxii* and *Saccharomyces bailey*), *S. diastaticus* and brewing strains of *S. cerevisiae* and *S. carlsbergensis*. Strain improvement was attempted by hybridization of highly ethanol-tolerant *Saccharomyces* strains with production strains of *S. cerevisiae*. The resulting hybrids showed a range in their ethanol tolerance levels but some of the hybrids had a faster fermentation rate than either parental [47].

Selective hybridization over three generations produced a significant increase in fermentation efficiency, i.e., from 84 to 93%. Wine-making characteristics were monitored during the hybridization program, and selected hybrid strains of *Saccharomyces* wine yeasts had wine-making qualities comparable to those of the parent strains but with increased sugar conversion efficiency [130].

Hybrids of brewing strains of *S. carlbergensis* were constructed by mating meiotic segregants [108]. One of the hybrids produced by the mating of two presumably haploid clones was equal to the parental strain in brewing performance on a pilot scale. Thus the separation of nuclear genomes of brewer's yeast into individual cells and their successful mating opens the possibility of introducing genetic alterations in brewing strains by mutagenesis and genetic transformation of the haploid lines [119].

Haploid progeny produced from a rapidly fermenting strain of *S. diastaticus* were used for hybridization with other *S. diastaticus* and *S. cerevisiae* haploids. In this case, although the dosage genes (starch or dextrin fermentations) may enhance ethanol production, epistatic effects in certain strain combinations caused a decrease in starch-fermenting activity [193]. However, results demonstrated that direct fermentation of starch by *S. diastaticus* can be accomplished with considerable efficiency through a combination of strain selection, hybridization, and systematic optimization of fermentation conditions. Furthermore, among the hybrids investigated, some of them showed additive effects and one behaved synergistically with regard to gene dosage and the rate of ethanol production [193].

To improve ethanol production, Seki et al. [35] hybridized a flocculent yeast of *S. cerevisiae* with other high-alcohol-producing strains by protoplast fusion. Fusant strains produced 12.4% ethanol even at 40°C, in a continuous alcohol fermentation with a cell-recycling system. The fusant showed an average productivity of 13 g/liters/h in a repeated batch culture with cell reuse at 37°C (Fig. 26A) [35].

Protoplast fusion has been used to construct hybrids between auxotrophic strains of *S. cerevisiae* having high ethanol tolerance and auxotrophic strains of lactose-fermenting *K. fragilis*. Some of the hybrids isolated were prototrophic and capable of assimilating lactose and, above all, producing ethanol in excess of 13% (Fig. 26B) as compared to the ethanol-tolerant *K. fragilis* parental which only produced about 10% ethanol under the same conditions [9].

Hybridization and spheroplast fusion have also been employed to construct novel yeast strains from polyploid brewing yeasts to select for high-performing strains capable of producing elevated ethanol levels [92]. When polyploid yeasts and genetically manipulated yeast strains were subjected to anaerobic fermentation, the polyploid strains exhibited superior performance in whole corn mash, while the genetically manipulated strains did so in defined media [92]. There was a 10-30% improvement in ethanol production by the hybrid strains over that of the polyploid strain (Fig. 27).

Since genes that limit growth are different at different ethanol concentrations [18], the selection of ethanol-tolerant strains should be carried out in a wide range of ethanol concentrations. This consideration may be particularly relevant to ethanol production where the fermenting strains need to grow and ferment under constantly increasing concentrations of ethanol rather than at any particular set value. Perhaps the most promising procedure for the selection of a more ethanol-tolerant yeast is that employing continuous culture conditions after hybridization. In fact, hybrids between naturally occurring wine yeast strains and laboratory strains have been formed and subjected to competition experiments under continuous

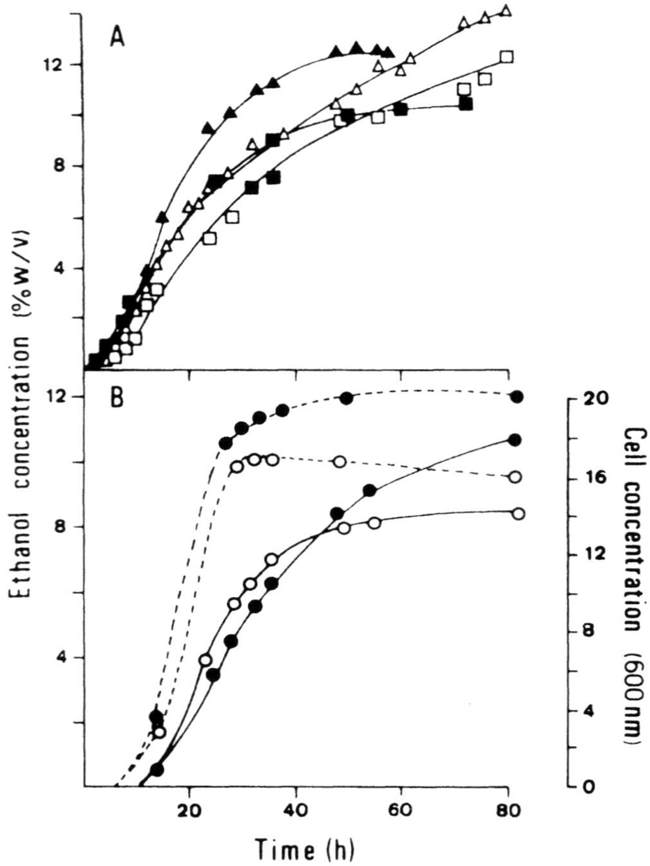

FIGURE 26. (A) Time courses of ethanol production by fusants KL411 (▲) and KL501 (△) and by the parental strains TJ1 (□) and K11 (■) in batch cultures. (From Ref. 35.) (B) Fermentation profiles for both *K. fragilis* (○) and the fusant (●) grown on lactose as a sole carbon source at 30°C. Solid lines, ethanol concentration; broken lines cell concentration. (From Ref. 9.)

culture controlled by pH with increasing ethanol concentrations over a wide range to select the fastest growing strain at any concentration of ethanol [38]. The continuous culture system was obtained by controlling the dilution rate of a chemostat connected to a pH meter (Fig. 28A). Under these conditions when the medium was supplemented with ethanol (Fig. 28B), the ethanol concentration of the culture increased with each pulse of dilution. A hybrid strain was selected by this procedure that was more tolerant than any of the highly ethanol-tolerant wine yeast strains at any concentration of ethanol and was able to grow in up to 16% ethanol. This improvement in ethanol tolerance led to an increase in both the ethanol production rate and the total amount of ethanol produced [33,38].

Finally, breeding of ethanol-producing yeast by hybridization involves genetic information of both nuclear and cytoplasmic (principally mitochondrial) origin [208]. As described before, there are significant differences between the fermentation behavior of some ethanol-producing yeasts and their isogenic petite derivatives. This means that breeding exclusively for mitochondria could have a practical significance. When brewing heteroplasmons are constructed and compared with their parental brewing yeasts, minor but significant differences have been found both in the speed of fermentation and in the formation of some flavor compounds [208]. These results, together with those described in which a laboratory strain substantially improved its ethanol and thermotolerance when provided mitochondria from highly ethanol-tolerant strains, indicate that mitochondria transfer have to be taken into account when planning genetic improvement experiments. This is especially true when the external conditions are such that cell growth and division are minimal and ethanol production depends mainly on cell viability which in turn depends mainly on mitochondria.

4. Other Characteristics

Yeast strains of interest for ethanol production harbor a great deal of genetic variability. These differences in yeast strains include ethanol-producing ability, ethanol and sulfur dioxide tolerance, tannin resistance,

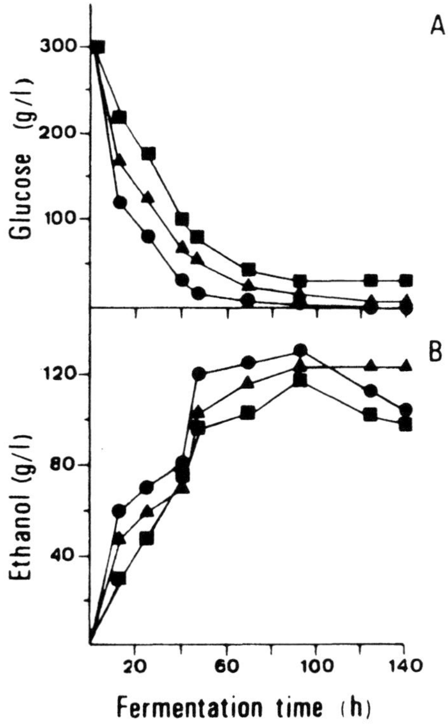

FIGURE 27. Fermentation characteristics of *Saccharomyces* sp. containing DEX genes in defined media (30% glucose-peptone-yeast extract). (■) Parental strain 62; (▲) hybrid strain 1384; (●) hybrid strain 1400. (A) Glucose uptake; (B) Ethanol production. (From Ref. 92.)

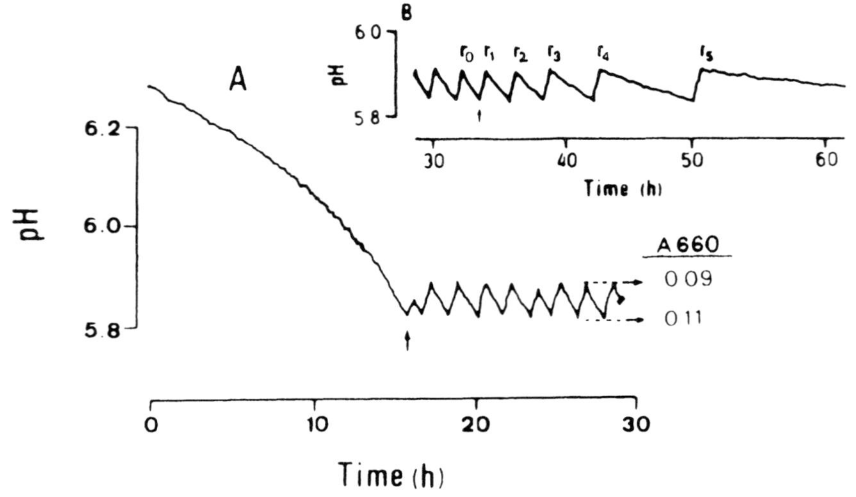

FIGURE 28. (A) Time course traces of pH in a culture of the strain FJ414 growing in YPDE (10% ethanol), subjected to turbidostatic control by pH. In (B), arrow indicates the time when the medium supply was supplemented to 18% ethanol. The r_n values indicate the ethanol concentration after n pulses of dilution. (From Ref. 38.)

ability to ferment efficiently at unusually high or low temperatures, volatile acid and fuel oil production, ability to synthesize diacetyl and hydrogen sulfide, altered degrees of flocculence, greater rates of wort fermentation, better strains of baker's yeast, nonfoaming strains, and a number of other characteristics of importance in the process of ethanol production [121], only a few of which will be discussed in this chapter.

In breweries, an early start of cell aggregation causes weak fermentation of the wort, but the absence of flocculation impedes the clarification of beer [209]. Batch fermentation is the most common process used in the biological production of ethanol. The necessary separation of the yeast cells from the liquid requires yeasts which flocculate under controlled conditions, which considerably reduces the energy costs of the separation processes [209]. The utilization of flocculent yeast in continuous fermentation is even more essential than in the classical batch fermentation. The productivity of a continuous procedure increases when the cells from the outflow are washed and then returned to the bioreactor thus keeping the concentration of the cells fairly constant [209]. Flocculating yeasts are better for recycling than nonflocculating yeasts [55].

In biomass production from liquids containing low amounts of utilizable carbohydrates, flocculating yeasts may increase productivity by reducing the size of the fermentors. This is because of the selective retainment of biomass in the reactor and the recovery of cells from external sedimentation tanks [209].

Flocculation properties have been introduced into powdery wine yeast strains by mating spores of the wine strains with haploid vegetative cells of a laboratory yeast strain which carried a dominant gene for flocculation. The genetic input of the laboratory yeast was reduced to less than 2%

by five generations of back breeding. When the new flocculent hybrids were finally isolated and their wine-making properties evaluated, somehybrids strains had similar fermentation patterns and produced wines similar to those of the original wine strains [128].

Induction of yeast cell aggregates in a column reactor has been initiated by packing yeast cell paste of *S. uvarum* into the column. Thereafter yeast cells aggregated in the column within 48 h without a supply of oxygen. When this yeast-aggregate column reactor was used for continuous ethanol production, a final ethanol concentration of 10.8% was obtained and the column reactor was maintained at a steady state for more than a month [210].

Finally, mitochondrial petites flocculate considerably less than the corresponding wild strains. Petites of a weak flocculent show no cell aggregation at all, while petite cells from a strong flocculent show a reduction in aggregation ability [209]. Thus, ethanol-tolerant yeast mitochondria confer an increase in the ability to flocculate together with an increase in ethanol tolerance, thermotolerance, and cell viability [209].

Ethanol tolerance is strongly dependent on growth temperature [33]. In biological ethanol production processes, the fermentation techniques do not normally require elaborate instrumentation or control, but it is desirable that the process operate at a temperature of at least 40°C in batch cultures. Yeasts have been isolated from different stages of the sugar-refining process from molasses in an attempt to isolate thermotolerant strains which would grow and produce ethanol at high temperatures. Several strains which gave acceptable yields at 40°C have been isolated and identified. Many of these yeasts were capable of growing at 40°C and producing high yields of biomass. The marked differences in viability observed at 55°C suggest that this would be the appropriate temperature to use in any rapid selection technique for isolation of thermotolerant yeasts [72].

Close examination of selected ethanol-producing yeasts reveals that they still have some undesirable qualities such as excessive sulfur dioxide or hydrogen sulfide formation and foaming [130]. These properties can be eliminated over a period of time by systematic strain selection or by mutant selection techniques. Undesirable properties which have a simple genetic basis, i.e., foaming, have been eliminated more precisely and quickly by selective hybridization [130].

Sulfur dioxide is frequently added to musts to inhibit wild yeasts and bacteria, aid clarification and settling, and prevent undesirable enzyme reactions [130]. Wine yeasts are tolerant of relatively high levels of $SO_2 \cdot SO_2$ tolerance is under the control of several dominant genes [130]. Therefore different combinations of genes can give rise to different tolerance levels. This suggestion is supported by the inheritance pattern in several hybrids in which the tolerance level was greater than that of either of the wine yeast parents [130].

With regard to osmotolerance, hybrids between *S. diastaticus*, an industrial yeast strain, and *S. rouxii*, an osmotolerant species, have been obtained by protoplast fusion [109]. These were similar to the *S. diastaticus* parent in fermentation pattern but were tolerant of sugar concentrations higher than those tolerated by the industrial parental strain [110]. Furthermore, segregants from several asci formed by these hybrids had a greater fermentative capacity than either the original hybrid or even a commercial baker's yeast grown under similar conditions [110].

Some of the fusion products of a brewing strain of *S. uvarum* and *S. diastaticus* possessed enhanced ethanol-producing capability. In medium containing 30% glucose a hybrid outperformed all other strains of *Saccharomyces* investigated [92], producing in excess of 12% ethanol. Some of the fusion products were also more tolerant to sorbitol- and mannitol-induced osmotic pressures than either of the fusion partners [92]. This efficient ethanol-producing hybrid was tolerant to elevated ethanol concentrations as well as osmotolerant to the high concentrations of sugar present in the medium, producing high concentrations of ethanol [92].

In the process of ethanol production a starter culture of a specific strain is frequently inoculated to reduce fermentation times, prevent contamination, and assure the desired product quality [130]. Contamination by a killer yeast can be a serious problem if the starter strain is sensitive to the killer toxin. To prevent contamination by undesirable yeast strains, killer yeasts suitable for ethanol production have been constructed by introducing killer and superkiller plasmids using cytoduction [129] and a killer strain for ethanol production was constructed without altering the nuclear genome of the producing strain [211]. The strains harboring the killer factor inhibited the growth of killer-sensitive cells when they were grown in fermentation medium [211].

VI. CONCLUSIONS

1. Growth and fermentation are partly linked parameters, so that the fermentation rate (ν) increases proportionally to a given increase in the growth rate (μ), and any increase in ethanol-tolerant growth gives rise to a proportional increase in the rate of ethanol production. Yeast cells are able to continue producing ethanol after growth has stopped. Ethanol inhibition of μ and ν occurs at lower concentrations of ethanol if the ethanol is produced endogenously, probably because of an osmotic effect due to high substrate concentrations since ethanol and substrate show synergistic effects. The kinetic patterns for the observed inhibition of μ and ν by ethanol may reflect the effect of ethanol on specific enzymes.

2. Yeast strains used in commercial alcohol fermentation are usually polyploid or aneuploid, nonmating, and unable to sporulate. They are homothallic and contain a great deal of heterozygosis for lethal genes or genes affecting growth rate. Breeding of strains with appropriate flocculation, osmotolerance, improved carbohydrate utilization, and lower levels of undesirable compounds are of special interest. Introduction of these characters into ethanol-producing yeasts has been carried out by rare-mating protoplast fusion, isolation of mating clones from industrial strains, mutation, genetic transformation, and single-chromosome transfers.

3. Although ethanol tolerance is strongly influenced by environmental and nutritional conditions, under fixed conditions different strains differ in their ability to tolerate ethanol and within each strain ethanol tolerance is a reproducible characteristic, implying that it is genetically determined. Ethanol tolerance is subject to polygenic control and there are a great number of genes able to spontaneously mutate to alleles of sensitivity. Genes are able to complement recessive alleles of ethanol sensitivity and those involved in ethanol tolerance are different in nonisogenic strains. Growth inhibition by ethanol is the result of the inhibition of different cellular functions with increasing ethanol concentrations. Therefore non-

isogenic hybridization provides the most useful method to improve ethanol tolerance and the selection of the most ethanol-tolerant hybrid has to be carried out in a wide range of ethanol concentrations.

4. Most of the reported physiological results indicate that cell membranes are important determinants of alcohol resistance. An increase in sensitivity to ethanol in yeasts can be associated with an increase in passive diffusion of protons through the cell membrane. The exponential constant of ethanol-enhanced passive proton diffusion across the plasma membrane might constitute a rapid method for the preliminary evaluation of ethanol tolerance in yeasts that would be especially useful when large numbers of mutants and other strains have to be screened.

5. Although only a small percentage of mitochondrial functions are coded for by mitochondrial genes, the mitochondrial genome is directly involved in the ethanol tolerance of the cell. The lack of the *rho*$^+$ phenotype leads to a decrease in both growth and viability in ethanol. When ethanol concentration becomes high enough, the yeast population becomes increasingly subject to ethanol-enhanced thermal death. The fact that the enhancement of thermal death in yeasts by different alkanols correlates with their lipid-buffer partition coefficients suggests that the thermal death sites of these yeasts are associated with a membrane system. In addition, the elevated growth temperatures at which exponential death concurred with exponential growth dramatically increase the spontaneous mutation rate to respiratory deficiency in yeasts, suggesting that the T_{max} sites, the thermal death sites, as well as the primary target sites of ethanol-enhanced petite mutation are located on a mitochondrial membrane. The mitochondrial genome is responsible for cell survival under extreme conditions of high ethanol concentration and/or high temperatures.

6. Sometimes low ethanol production due to effects other than ethanol or substrate inhibition has been mistakenly attributed to ethanol. Frequently, the addition of an ammonium salt and/or biotin or other supplements such as enrichment with unsaturated fatty acyl residues is highly stimulating for both growth and fermentation rates and increases the tolerance of yeasts to higher temperatures and alcohol concentrations. Ethanol production can be enhanced by direct increase of the microbial biomass achieved by cell recycle either by the use of flocculent strains or by cell immobilization. When the substrate is introduced to the fermentation media of yeasts in several batches (feed-batch) the levels of ethanol obtained are much higher than those with a single-batch operation. In addition, fermentation may also stop due to oxygen deficiency, low pH values, pressure, etc. When cultures are fermented with controlled oxygen, pressure, and pH, these conditions led to reproducible yeast behavior.

7. Genetic improvement of industrial yeasts frequently requires the transfer via conjugation or protoplast fusion of new properties absent in the original strains. In intergeneric protoplast fusion many of the hybrids are unstable. In these cases genetic engineering techniques are used. Genes have been transferred to yeasts and the heterologous gene product expressed under the desired conditions. There are also several reports on the isolation of yeast mutants with a number of enzymes that are no longer repressed by glucose and are capable of fermenting mixtures of different sugars, with a corresponding increase in substrate conversion efficiency. Ethanol production technology requires a yeast that is alcohol-tolerant, viable at the end of fermentation, and suitable for reuse in subsequent fermentations. Screening and selection techniques have been used

after isolating highly ethanol-tolerant yeast strains from their natural environments. Alternatively, a continuous selection system has been described which has yielded yeast mutants with considerably improved ethanol tolerance. Hybridization has been the usual method of generating ethanol-tolerant yeast. Another procedure for the selection of a more ethanol-tolerant yeast is that of employing continuous culture conditions after hybridization when hybrids are subjected to competition experiments under continuous culture controlled by pH with increasing ethanol concentrations. Hybrids have been selected that are more tolerant than any of the highly ethanol-tolerant yeast strains at any concentration of ethanol, which led to an increase in both the ethanol production rate and the total amount of ethanol produced. Finally, breeding of ethanol-producing yeast by hybridization involves genetic information of both nuclear and mitochondrial origin. Mitochondria transfer has therefore to be taken into account when planning genetic improvement procedures.

ACKNOWLEDGMENTS

We thank M. Guinea, R. Maldonado, and C. Zamora for skillful technical assistance; B. Cubero, I. L. Calderón, and J. Casadesús for helpful discussions; and R. Rhett for correcting the manuscript.

REFERENCES

1. Stokes, S. L. Influence of temperature on the growth and metabolism of yeasts. In *The Yeasts*, Vol. 2 (A. H. Rose and J. S. Harrison, eds.), Academic Press, London, 1970, pp. 119-134.
2. Casey, G. P., and Ingledew, W. M. Ethanol tolerance in yeasts. *CRC Crit. Rev. Microbiol.*, *13*:219-280, 1986.
3. Kohchi, C., and Toh-e, A. Cloning of Candida pelliculosa B-glucosidase gene and its expression in Saccharomyces cerevisiae. *Mol. Gen. Genet.*, *203*:89-94, 1986.
4. Spencer-Martins, I. Extracellular isoamylase produced by the yeast Lipomyces kononenkoae. *Appl. Environ. Microbiol.*, *44*:1253-1257, 1982.
5. Duvnjak, Z., Houle, C., and Mok, K. L. Production of ethanol and biomass from various carbohydrates by Kluyveromyces fragilis. *Biotechnol. Lett.*, *9*:343-346, 1987.
6. Lee, H., Biely, P., Latta, R. K., Barbosa, M. F. S., and Schneider, H. Utilization of xylan by yeasts and its conversion to ethanol by Pichia stipitis strains. *Appl. Environ. Microbiol.*, *52*:320-324, 1986.
7. van Uden, N. Ethanol toxicity and ethanol tolerance in yeasts. *Annu. Rep. Ferment. Proc.*, *8*:11-58, 1985.
8. Toivola, A., Yarrow, D., van der Bosch, E., Dijken, J. P., and Scheffers, W. A. Alcoholic fermentation of D-xylose by yeasts. *Appl. Environ. Microbiol.*, *47*:1221-1223, 1984.
9. Farahnak, F., Seki, T., Ryu, D. D. Y., and Ogrydziak, D. Construction of lactose-assimilating and high-ethanol-producing yeasts by protoplast fusion. *Appl. Environ. Microbiol.*, *51*:362-367, 1986.
10. Dombek, K. M., and Ingram, L. O. Ethanol production during batch fermentation with Saccharomyces cerevisiae. Changes in glycolytic enzymes and internal pH. *Appl. Environ. Microbiol.*, *53*:1286-1291, 1987.

11. Pascual, C., Alonso, A., García, I., Romay, C., and Kotyk, A. Effect of ethanol on glucose transport, key glycolytic enzymes and proton extrusion in Saccharomyces cerevisiae. *Biotechnol. Bioeng.*, *32*:374-378, 1988.
12. Nagodawithana, T. W., Whitt, J. T., and Cutaia, A. J. Study of the feedback effect of ethanol on selected enzymes of the glycolytic pathway. *ASBC J.*, *35*:179-183, 1977.
13. Alba, S., Shoda, M., and Najatami, M. Kinetics of product inhibition in alcohol fermentation. *Biotechnol. Bioeng.*, *10*:845-865, 1968.
14. Aguilera, A., and Benítez, T. Role of mitochondria in ethanol tolerance of Saccharomyces cerevisiae. *Arch. Microbiol.*, *142*:389-392, 1985.
15. Jiménez, J., and Benítez, T. Characterization of wine yeasts for ethanol production. *Appl. Microbiol. Biotechnol.*, *25*:150-154, 1986.
16. Chattaway, T., Goma, G., and Renaud, P. Y. Modelling ethanol and secondary inhibitions of ethanol fermentation in a multistate reactor. *Biotechnol. Bioeng.*, *32*:271-276, 1988.
17. Hayashida, S., Feng, D. D., Ohta, K., Chaitiumvong, S., and Hongo, M. Compositions and a role by Aspergillus oryzae proteolipids as a high concentration alcohol-producing factor. *Agr. Biol. Chem.*, *40*:73-78, 1976.
18. Jiménez, J., and Benítez, T. Genetic analysis of highly ethanol-tolerant wine yeasts. *Curr. Genet.*, *12*:421-428, 1987.
19. Thomas, D. S., Hossack, J. A., and Rose, A. H. Plasma-membrane lipid composition and ethanol tolerance in Saccharomyces cerevisiae. *Arch. Microbiol.*, *117*:239-245, 1978.
20. Ingram, L. O. Adaptation membrane lipids to alcohols. *J. Bacteriol.*, *125*:670-678, 1976.
21. Ingram, L. O., and Buttke, T. M. Effects of alcohols on microorganisms. *Adv. Microb. Physiol.*, *25*:253-300, 1984.
22. Ismail, A. A., and Ali, M. M. Selection of high ethanol yielding Saccharomyces. I. Ethanol tolerance and the effect of training in Saccharomyces cerevisiae Hansen. *Folia Microbiol.*, *16*:346-349, 1971.
23. Ismail, A. A., and Ali, M. M. Selection of high ethanol yielding Saccharomyces. II. Genetics of ethanol tolerance. *Folia Microbiol.*, *16*:300-354, 1971.
24. Aguilera, A., and Benítez, T. Synergistic effects of ethanol and temperature on yeast mitochondria. *Curr. Microbiol.*, *18*:179-188, 1989.
25. Jiménez, J., and van Uden, N. Use of extracellular acidification for the rapid testing of ethanol tolerance in yeasts. *Biotechnol. Bioeng.*, *27*:1596-1598, 1985.
26. Margaritis, A., and Bajpai, P. Direct fermentation of D-xylose to ethanol by Kluyveromyces marxianus strains. *Appl. Environ. Microbiol.*, *44*:1039-1041, 1982.
27. Panchal, C. J., and Stewart, G. G. The effect of osmotic pressure on the production and excretion of ethanol and glycerol by a brewing yeast strains. *J. Inst. Brew.*, *86*:207-210, 1980.
28. Moulin, G., Boze, P., and Galzy, P. A comparative study of the inhibitory effect of ethanol and substrate on the fermentation rate of the parent and a respiratory deficient mutant. *Biotechnol. Lett.*, *3*:351-356, 1981.
29. Park, Y. K., and Sato, H. P. Fungal invertase as an aid for fermentation of cane molasses into ethanol. *Appl. Environ. Microbiol.*, *44*:988-989, 1982.

30. Harding, P. J., Obi, E. I., and Slaughter, J. C. The potential action of NH_4 as stimulate of ethanol production by Saccharomyces cerevisiae. *FEMS Microb. Lett.*, *21*:185-187, 1984.
31. Suzzi, G., Romano, P., and Zambonelli, C. Flocculation of wine yeasts: frequency differences and stability of the character. *Can. J. Microbiol.*, *30*:36-39, 1984.
32. Brown, S. W., and Oliver, S. G. Isolation of ethanol-tolerant mutants of yeast by continuous selection. *Eur. J. Appl. Microbiol. Biotechnol.*, *16*:116-122, 1983.
33. Jiménez, J., and Benítez, T. Yeast cell viability under conditions of high temperature and ethanol concentrations depends on the mitochondrial genome. *Curr. Genet.*, *13*:461-469, 1988.
34. Aguilera, A., and Benítez, T. Ethanol-sensitive mutants of Saccharomyces cerevisiae. *Arch. Microbiol.*, *143*:337-344, 1986.
35. Seki, T., Myuga, S., Limtong, S., Vedono, S., Kumnuanda, J., and Taguchi, M. Genetic construction of yeast strains for high ethanol production. *Biotechnol. Lett.*, *5*:351-356, 1983.
36. Bandas, E. L., and Zakharov, I. A. Induction of *rho*$^-$ mutations in yeast Saccharomyces cerevisiae by ethanol. *Mutat. Res.*, *71*:193-199, 1980.
37. Brown, S. W., Sugden, D. A., and Oliver, S. G. Ethanol production and tolerance in grande and petite yeasts. *J. Chem. Technol. Biotechnol.*, *34B*:116-120, 1984.
38. Jiménez, J., and Benítez, T. Selection of ethanol-tolerant yeast hybrids in pH regulated continuous culture. *Appl. Environ. Microbiol.*, *54*:917-922, 1988.
39. Cabeca-Silva, C., Madeira-Lopes, A., and van Uden, N. Temperature relations of ethanol-enhanced petite mutation in Saccharomyces cerevisiae mitochondria as targets of thermal death. *FEMS Microbiol. Lett.*, *15*:149-151, 1982.
40. Babel, W., Müller, R. H., and Markuske, K. D. Improvement of growth yield of yeasts on glucose to the maximum by using an additional energy source. *Arch. Microbiol.*, *136*:203-208, 1983.
41. Pirt, S. T. *Principles of Microbe and Cell Cultivation*. Blackwell Scientific, Oxford, 1975.
42. Andrews, G. F. Parameter estimation from batch culture data. *Biotechnol. Bioeng.*, *26*:824-825, 1984.
43. Goma, G., Moletta, R., and Novack, M. Comments on the "maintenance coefficient" changes during alcohol fermentation. *Biotechnol. Lett.*, *1*:415-420, 1979.
44. Aguilera, A., and Benítez, T. Relationship between growth, fermentation and respiration rates in Saccharomyces cerevisiae: a study based on the analysis of the yield Ypx. *Biotechnol. Bioeng.*, *32*:240-244, 1988.
45. Moulin, G., Boze, H., and Galzy, P. Utilization of a respiratory deficient mutant for alcohol production. *J. Ferment. Technol.*, *60*:25-29, 1982.
46. Brown, S. W., and Oliver, S. G. The effect of temperature on the ethanol tolerance of the yeast Saccharomyces uvarum. *Biotechnol. Lett.*, *4*:269-274, 1982.
47. Day, A., Anderson, E., and Martin, P. A. Ethanol tolerance of brewing yeasts. In *15th Eur. Brew. Conv. Proc. Congr.*, IRL Press, Oxford, pp. 377-391, 1975.
48. Watson, T. G. The present states and future prospects of the turbidostat. *J. Appl. Chem. Biotechnol.*, *22*:229-243, 1972.

49. Vega, J. L., Clansen, E. C., and Gaddy, J. L. Acetate addition to an immobilized yeast column for ethanol production. *Biotechnol. Bioeng.*, *29*:429-435, 1987.
50. Lagunas, R. Energy metabolism of Saccharomyces cerevisiae discrepancy between ATP balance and known metabolic functions. *Biochim. Biophys. Acta*, *440*:661-674, 1976.
51. Lagunas, R. Energetic irrelevance of aerobiosis for Saccharomyces cerevisiae growing on sugars. *Mol. Cel. Biochem.*, *27*:139-146, 1979.
52. Walker-Caprioglio, H. M., and Parks, L. W. Autoconditioning factor relieves ethanol-induced growth inhibition of Saccharomyces cerevisiae. *Appl. Environ. Microbiol.*, *53*:33-35, 1987.
53. Novack, M., Strehaiano, P., and Morena, M. Alcoholic fermentation: on the inhibitory effect of ethanol. *Biotechnol. Bioeng.*, *23*:201-211, 1981.
54. Grant, S. J., and McGhee, J. E. Continuous ethanol production by immobilized Saccharomyces cerevisiae: effects of cell concentration and glucose flow rate. *Trans. Ill. State Acad. Sci.*, *79*:117-128, 1986.
55. Ciftci, T., Constantimides, A., and Wang, S. S. Optimization of conditions and cell feeding procedures for alcohol fermentation. *Biotechnol. Bioeng.*, *25*:2007-2023, 1983.
56. Lillie, S. H., and Pringle, J. R. Reserve carbohydrate metabolism in Saccharomyces cerevisiae. Responses to nutrient limitation. *J. Bacteriol.*, *143*:1384-1394, 1980.
57. Lee, K. J., Skotnicki, M. L., Tribe, D. E., and Rogers, P. L. The kinetics of ethanol production by Zymomonas mobilis on fructose and sucrose media. *Biotechnol. Lett.*, *3*:207-212, 1981.
58. Moulin, G., Boze, H., and Galzy, P. Inhibition of alcoholic fermentation by substrate and ethanol. *Biotechnol. Bioeng.*, *22*:2375-2381, 1980.
59. Guijarro, J. M., and Lagunas, R. Saccharomyces cerevisiae does not accumulate ethanol against a concentration gradient. *J. Bacteriol.*, *160*:874-878, 1984.
60. D'Amore, T., and Stewart, G. G. Ethanol tolerance of yeast. *Enz. Microb. Technol.*, *9*:321-330, 1987.
61. Hayashida, S., and Ohta, K. Formulation of high concentrations of alcohol by various eyasts. *J. Inst. Brew.*, *87*:42-44, 1981.
62. Panchal, C. J., and Stewart, G. G. Ethanol production by a highly flocculent brewing yeast strain. *Dev. Ind. Microbiol.*, *22*:711-717, 1981.
63. Flor, P. Q., and Hayashida, S. Saccharomyces uvarum inulyticus var. nor., a new high-concentration ethanol tolerant yeast from rice wine. *Eur. J. Appl. Microbiol. Biotechnol.*, *18*:148-152, 1983.
64. Benítez, T., del Castillo, L., Aguilera, A., Conde, J., and Cerdá-Olmedo, E. Selection of wine yeasts for growth and fermentation in the presence of ethanol and sucrose. *Appl. Environ. Microbiol.*, *45*:1429-1436, 1983.
65. Tubb, R. S. Genetics of ethanol-producing microorganisms. *CRC Crit. Rev. Biotechnol.*, *1*:241-261, 1984.
66. Ingledew, W. M., Magnus, C. A., and Sosulski, F. W. Influence of oxygen on proline utilization during the wine fermentation. *Am. J. Enol. Vitic.*, *38*:246-248, 1987.
67. Casey, G. P., Magnus, C. A., and Ingledew, W. M. High gravity brewing: effects of nutrition on yeast composition, fermentative ability and alcohol production. *Appl. Environ. Microbiol.*, *48*:639-646, 1984.

68. Rosa, M. F., Vieira, A. M., Cartolomeu, M. L., Sá-Correira, I., Cabral, J. M. S., and Novais, J. M. Production of high concentration of ethanol from mash, juice and pulp of Jerusalem artichoke tubers by Kluyveromyces fragilis. *Enz. Microb. Technol.*, *8*:673-676, 1986.
69. Rosa, M. F., Sá-Correira, I., and Novais, J. M. Production of ethanol at high temperatures in the fermentation of the Jerusalem artichoke juice and a simple medium by Kluyveromyces marxianus. *Biotechnol. Lett.*, *9*:441-444, 1987.
70. Dombek, K. M., and Ingram, L. O. Magnesium limitation and its role in apparent toxicity of ethanol during yeast fermentation. *Appl. Environ. Microbiol.*, *52*:975-981, 1986.
71. Dombek, K. M., and Ingram, L. O. Nutrient limitation as a basis for the apparent toxicity of low levels of ethanol during fermentation. *J. Ind. Microbiol.*, *1*:219-226, 1986.
72. Idris, E. T. A., and Berry, D. R. Selection of thermotolerant yeast strain for biomass production from sudanese molasses. *Biotechnol. Lett.*, *2*:61-66, 1980.
73. Haraldson, A., and Rosen, C. G. Studies on continuous ethanol fermentation of sugar cane molasses. I. A system for continuous fermentation. *Eur. J. Appl. Microbiol. Biotechnol.*, *14*:216-219, 1982.
74. Haraldson, A., and Rosen, C. G. Studies on continuous ethanol fermentation of sugar cane molasses. II. Continuous alcohol fermentations and product removal in a laboratory scale plant. *Eur. J. Appl. Microbiol. Biotechnol.*, *14*:220-224, 1982.
75. Monod, J. *Recherches sur la croissance des cultures bacteriennes*. Paris, Hermann, 1942.
76. Luong, J. H. T. Kinetics of ethanol inhibition in alcohol fermentations. *Biotechnol. Bioeng.*, *27*:280-285, 1985.
77. Rose, A. H., Beavan, M. J., and Charpentier, C. Physiological basis for enhanced ethanol production by Saccharomyces cerevisiae. In *Overproduction of Microbiae Products* (V. Krumphanzel and D. W. Tempest, eds.), Academic Press, London, 1982, pp. 211-219.
78. Hinshelwood, C. M. *Chemical kinetics of bacterial cell*. London, Oxford University Press, 1946.
79. Holzberg, I., Finn, R. K., and Steinkraus, K. H. A kinetic study of the alcoholic fermentation of grape juice. *Biotechnol. Bioeng.*, *9*: 413-418, 1967.
80. Aiba, S., and Shoda, M. Reassessment of the product inhibition of alcohol fermentation. *J. Ferment. Technol.*, *47*:790-798, 1969.
81. Bazua, C. D., and Wilke, C. R. Ethanol effects on the kinetics of a continuous fermentation with Saccharomyces cerevisiae. *Biotechnol. Bioeng. Symp.*, *7*:105-108, 1977.
82. Brown, S. W., Oliver, S. G., Harrison, D. E. F., and Righelato, R. C. Ethanol inhibition of yeast growth and fermentation: differences in the magnitude and complexity of the effect. *Eur. J. Appl. Microbiol. Biotechnol.*, *11*:151-157, 1981.
83. Ghose, T. K., and Tyagi, R. D. Rapid ethanol fermentation of cellulose hydrolysate. II. Product and substrate inhibition and optimization of fermentor design. *Biotechnol. Bioeng.*, *21*:1401-1408, 1979.
84. Wasungu, K. M., and Simard, R. E. Growth characteristics of bakers' yeast in ethanol. *Biotechnol. Bioeng.*, *24*:1125-1134, 1982.
85. Richter, K. Inhibitory effects of ethanol in alcoholic fermentation. *Acta Biotechnol.*, *6*:234-243, 1986.

86. Bajpai, P., and Margaritis, A. Ethanol inhibition kinetics of Kluyveromyces marxianus grown on Jerusalem artichoke juice. *Appl. Environ. Microbiol.*, *44*:1325-1329, 1982.
87. Nagodawitana, T. W., and Steinkraus, K. H. Influence of the rate of ethanol production and accumulation on the viability of Saccharomyces cerevisiae in "rapid fermentation." *Appl. Environ. Microbiol.*, *31*:158-162, 1976.
88. Navarro, J. M. Fermentation alcoolique: influence des conditions de culture sur l'inhibition par l'ethanol. *Cell Mol. Biol.*, *26*:241-246, 1979.
89. Loureiro, V., and Ferreira, H. G. On the intracellular accumulation of ethanol in yeast. *Biotechnol. Bioeng.*, *25*:2263-2269, 1983.
90. Mota, M., Strehaiano, P., and Goma, G. Studies on the conjugate effects of substrate (glucose) and product (ethanol) on cell growth kinetics during fermentation of different yeast strains. *J. Inst. Brew.*, *90*:360-365, 1984.
91. Aguilera, A. Fisiología y genética de la tolerancia a etanol en Saccharomyces cerevisiae. Ph.D. thesis, University of Seville, Spain, 1983.
92. Panchal, C. J., Peacock, L., and Stewart, G. G. Increased osmotolerance of genetically modified ethanol producing strains of Saccharomyces sp. *Biotechnol. Lett.*, *4*:639-643, 1982.
93. Leao, C., and van Uden, N. Effects of ethanol and other alkanols on the glucose transport system of Saccharomyces cerevisiae. *Biotechnol. Bioeng.*, *24*:2601-2604, 1982.
94. Loureiro-Dias, M. C., and Peinado, J. M. Effects of ethanol and other alkanols on the maltose transport system of Saccharomyces cerevisiae. *Biotechnol. Lett.*, *4*:721-724, 1982.
95. Sá-Correira, I., and van Uden, N. Effect of ethanol on the fructose transport system of Kluyveromyces fragilis. *Biotechnol. Lett.*, *5*:413-418, 1983.
96. Spencer-Martins, I., and van Uden, N. Substrate and endproduct regulation of cellobiose uptake by Candida wickerhamii. *Appl. Microbiol. Biotechnol.*, *25*:272-276, 1987.
97. Cartwright, C. P., Veazey, F. J., and Rose, A. H. Effect of ethanol on activity of the plasma-membrane ATPase in, and accumulation of glycine by Saccharomyces cerevisiae. *J. Gen. Microbiol.*, *133*:857-866, 1987.
98. Llorente, P., and Sols, A. Ethanol inactivation of glycolytic enzymes in yeasts with different alcohol resistance. *Psc. 6th FEBS Meeting*, Madrid, 1969, p. 123.
99. Sharma, S., and Tauro, P. Control of ethanol production by yeast: role of pyruvate decarboxylase and alcohol dehydrogenase. *Biotechnol. Lett.*, *8*:735-738, 1986.
100. Sharma, S., and Tauro, P. Detection and biochemical characterization of fast and slow ethanol-producing yeasts. *J. Appl. Microbiol. Biotechnol.*, *3*:155-160, 1987.
101. Gokhale, D. V., Rao, B. S., and Sivaramakrishnam, S. Alcohol dehydrogenase and invertase activities in ethanol tolerant yeasts. *Enzyme Microbiol. Technol.*, *8*:623-626, 1987.
102. Millar, D. G., Griffiths-Smith, K., Algar, E., and Scopes, R. I. C. Activity and stability of glycolytic enzymes in the presence of ethanol. *Biotechnol. Lett.*, *9*:601-606, 1982.

103. Laube, K., Wesenberg, J., and Lietz, P. Selection of distiller's yeast with particular respect to non Saccharomyces strains. *Acta Biotechnol.*, *7*:111-118, 1987.
104. Malfait, M. H., Moulin, G., and Galzy, P. Ethanol inhibition of growth, fermentation and starch hydrolysis in Schwanniomyces castelli. *J. Ferment. Technol.*, *64*:279-284, 1986.
105. Martín-Rendón, E., Jiménez, J., and Benítez, T. Ethanol inhibition of Saccharomyces and Candida enzymes. *Curr. Genet.*, *15*:7-16, 1989.
106. Tubb, R. S. Applying yeast functions in brewing. A current assessment. *J. Inst. Brew.*, *85*:286-289, 1979.
107. Kielland-Brandt, M. C. The genetics of brewers yeast. *EBC Congr.*, pp. 263-276, 1981.
108. Gjermamsen, C., and Sigsgaard, P. Construction of a hybrid brewing strain of Saccharomyces carlsbergensis by mating of meiotic segregants. *Carlsberg Res. Commun.*, *46*:1-11, 1981.
109. Spencer, J. F. T., Brizeau, C., Reynolds, N., and Spencer, D. M. The use of mitochondrial mutants in hybridization of industrial yeast strains. VI. Characterization of the hybrid Saccharomyces diastaticus × Saccharomyces rouxii obtained by protoplast fusion and its behavior in simulated dought raising tests. *Curr. Genet.*, *9*:649-652, 1985.
110. Spencer, J. F. T., Spencer, D. M., Brizeau, C., Martini, A. V., and Martini, A. The use of mitochondrial mutants in hybridization of industrial yeast strains. V. Relative parental contributions to the genomes of interspecific and intergeneric yeast hybrids obtained by protoplast fusion as determined by DNA reassociation. *Curr. Genet.*, *9*:623-625, 1985.
111. Nilsson-Tillgren, T. Genetic differences between Saccharomyces carlsbergensis and Saccharomyces cerevisiae. Analysis of chromosome III by simple chromosome transfer. *Carlsberg Res. Commun.*, *46*:65-76, 1981.
112. Kielland-Brandt, M. C., Nilsson-Tillgren, T., Petersen, J. G. L., and Holmberg, S. Transformation in yeast without the involvement of bacterial plasmids. In *Molecular Genetics in Yeast* (D. von Welstein et al., 3ds.), Munksgaard, Copenhagen, pp. 369-382, 1981.
113. Conde, J., and Fink, G. R. A mutant of Saccharomyces cerevisiae defective for nuclear fusion. *Proc. Natl. Acad. Sci. USA*, *73*:3651-3655, 1976.
114. Benítez, T., Del Castillo, L., Aguilera, A., and Conde, J. Instability of Saccharomyces cerevisiae heterokaryons. *Curr. Genet.*, *8*:345-352, 1984.
115. Keiding, A. K. Genetic and molecular characterization of a distiller's yeast. *Carlsberg. Res. Commun.*, *50*:95-125, 1985.
116. Figueroa, L. I., Richard, M. F., and Broock, M. R. Interspecific protoplast fusion of the baker's yeast Saccharomyces cerevisiae and Saccharomyces diastaticus. *Biotechnol. Lett.*, *6*:269-274, 1984.
117. Klinner, V., and Bottcher, F. Chromosomal rearrangements after protoplast fusion in the yeast Candida maltosa. *Curr. Genet.*, *9*:619-621, 1985.
118. Shepherd, M. G., Poulter, R. T. M., and Sullivan, P. A. Candida albicans: biology, genetics and pathogenicity. *Ann. Rev. Microbiol.*, *39*:579-614, 1985.

119. Petersen, J. G. L., and Piñon, R. In vivo labelling of proteins associated with folded chromosomes of yeast. *Carlsberg Res. Commun.*, *44*:395-402, 1979.
120. Kielland-Brandt, M. C., Nilsson-Tillgren, T., Holmberg, S., Petersen, J. G. L., and Svenningsen, B. A. Transformation of yeast without the use of foreign DNA. *Carlsberg Res. Commun.*, *44*:77-87, 1979.
121. Snow, R. Toward genetic improvement of wine yeast. *Am. J. Enol. Vitic.*, *30*:33-37, 1979.
122. Fujimura, H., and Yanagishima, N. Mating-type-specific cell cycle arrest and shmoo formation by pheromone of Saccharomyces cerevisiae in Hansenula wingei. *Arch. Microbiol.*, *136*:79-80, 1983.
123. Santa María, J., and Vidal, D. Genetic control of "flor" formation by Saccharomyces. *J. Bacteriol.*, *113*:1078-1080, 1973.
124. Sancho, E. D., Hernández, E., and Rodriguez-Navarro, A. Presumed sexual isolation in yeast populations during production of sherry like wine. *Appl. Environ. Microbiol.*, *51*:395-397, 1986.
125. Jiménez, J., and Benítez, T. Hibridación de capas vinicas homotalicas con cepas heterotalicas. *Microbiologia SEM*, *4*:117-124, 1988.
126. Jiménez, J. Caracterización genética de cepas de levaduras para la producción industrial de etanol. Ph.D. thesis, University of Seville, Spain, 1986.
127. Miki, B. L. A., Poon, N. H., Jams, A. P., and Seligy, V. L. Possible mechanism for flocculation interactions governed by gene FLO1 in Saccharomyces cerevisiae. *J. Bacteriol.*, *150*:878-889, 1982.
128. Thornton, R. J. The introduction of flocculation into a homothallic wine yeast. A practical example of the modification of wine-mating properties by the use of genetic techniques. *Am. J. Enol. Vitic.*, *36*:47-49, 1985.
129. Seki, T., Choi, E. H., and Ryn, D. Construction of killer wine yeast strain. *Appl. Environ. Microbiol.*, *49*:1211-1215, 1985.
130. Thornton, R. J. Selective hybridization of pure culture wine yeasts. II. Improvement of fermentation efficiency and inheritance of SO_2 tolerance. *Appl. Environ. Microbiol.*, *14*:159-164, 1982.
131. Jiménez, J., Longo, E., and Benítez, T. Induction of petite yeast mutants by membrane active agents. *Appl. Environ. Microbiol.*, *54*: 3126-3132, 1988.
132. Pina, A., Calderón, I. L., and Benítez, T. Intergeneric hybrids of Saccharomyces cerevisiae and Zygosaccharomyces fermentati obtained by protoplast fusion. *Appl. Environ. Microbiol.*, *51*:995-1003, 1986.
133. Johannsen, E., Eagle, L., and Bredenham, G. Protoplast fusion used for the construction of presumptive polyploids of the D-xylose fermenting yeast Candida shehatae. *Curr. Genet.*, *9*:313-319, 1985.
134. Dilorio, A. A., Weathers, P. J., and Campbell, D. A. Comparative enzyme and ethanol production in an isogenic yeast ploidy series. *Curr. Genet.*, *12*:9-14, 1987.
135. Ingram, L. O. Mechanism of lysis of Escherichia coli by ethanol and other chaotropic agents. *J. Bacteriol.*, *146*:331-336, 1981.
136. Ingram, L. O., and Vreeland, N. S. Differential effects of ethanol and hexanol on the Escherichia coli cell envelope. *J. Bacteriol.*, *144*: 481-488, 1980.
137. Ingram, L. O. Regulation of fatty acid composition in Escherichia coli: a proposed common mechanism for changes induced by ethanol, chaotropic agents, and a reduction of growth temperature. *J. Bacteriol.*, *149*:166-172, 1982.

138. Jones, R. P., and Greenfield, P. F. Ethanol and the fluidity of the yeast plasma membrane, *Yeast*, *3*:223-232, 1987.
139. Casey, V. C., and Ingram, L. O. Lipid composition of Zymomonas mobilis: effects of ethanol and glucose. *J. Bacteriol.*, *154*:1291-1300, 1983.
140. Janssens, J. H., Burris, N., Woodward, A., and Bailey, R. Lipid-enhanced ethanol production by Kluyveromyces fragilis. *Appl. Environ. Microbiol.*, *45*:598-602, 1983.
141. Larme, F., Lafon-Lafourcade, S., and Riderean-Gayon, P. Relationship between the sterol content of the yeast cells and their fermentation activity in grape must. *Appl. Environ. Microbiol.*, *39*:808-811, 1980.
142. Serrano, R. Effect of ATPase inhibition on the proton pump of respiratory-deficient yeast. *Eur. J. Biochem.*, *105*:419-424, 1980.
143. Jiménez, J., and Benítez, T. Adaption of yeast cell membranes to ethanol. *Appl. Environ. Microbiol.*, *53*:1196-1198, 1987.
144. Dombeck, K. M., and Ingram, L. O. Effects of ethanol on the Escherichia coli plasma membrane. *J. Bacteriol.*, *157*:233-239, 1984.
145. Leao, C., and van Uden, N. Effects of ethanol and other alkanols on passive proton influx in the yeast Saccharomyces cerevisiae. *Biochim. Biophys. Acta*, *747*:43-48, 1984.
146. Bacila, M., and Horii, J. Improving the efficiency of alcohol production with respiration-deficient yeast mutants. *Trends Biol. Sci.*, *March*: 59-61, 1979.
147. Alexander, N. J., and Detroy, R. W. Alcohol production by mitochondrial mutants and wild type. *Biotechnol. Lett.*, *5*:165-168, 1983.
148. Evans, R. J., Oakly, K. M., an d Clark-Walker, G. D. Elevated levels of petite formation in strains of Saccharomyces cerevisiae restored to respiratory competence. I. Association of both high and moderate frequences of petite mutant formation with the presence of aberrant mitochondrial DNA. *Genetics*, *111*:389-402, 1985.
149. Leao, C., and van Uden, N. Effects of ethanol and other alkanols on the kinetics and the activation parameters of thermal death in Saccharomyces cerevisiae. *Biotechnol. Bioeng.*, *24*:1581-1590, 1982.
150. Okolo, B., Johnston, J. R., and Berry, D. R. Toxicity of ethanol, n-butanol and isoamyl alcohol in Saccharomyces cerevisiae when supplied separately and in mixtures. *Biotechnol. Lett.*, *9*:431-434, 1987.
151. Lees, N. D., Lofton, S. L., Woods, R. A., and Bard, M. The effects of varied energy source and detergent on the growth of sterol mutant of Saccharomyces cerevisiae. *J. Gen. Microbiol.*, *118*:209-214, 1980.
152. Cobon, G. S., and Haslam, J. M. The effect of altered membrane sterol composition on the temperature dependence of yeast mitochondrial ATPase. *Biochim. Biophys. Res. Commun.*, *52*:320-326, 1973.
153. Watson, K. Unsaturated fatty acid but not ergosterol is essential for high ethanol production in Saccharomyces. *Biotechnol. Lett.*, *4*:397-402, 1982.
154. Macy, J. M., and Miller, M. W. Anaerobic growth of Saccharomyces cerevisiae in the absence of oleic acid and ergosterol? *Arch. Microbiol.*, *134*:64-67, 1983.
155. Nes, W. R., Sekula, B. C., Nes, W. D., and Adler, J. H. The function importance of structural features of ergosterol in yeast. *J. Biol. Chem.*, *253*:6218-6225, 1978.
156. Takahata, N., and Muruyama, T. A mathematical model of extranuclear genes and the genetic variability maintained in a finite population. *Genet. Res.*, *37*:291-302, 1981.

157. van Uden, N., and Madeira-Lopes, A. Concurrent exponential growth and death of cell populations of Saccharomyces cerevisiae at supra-optimal growth temperatures. *Z. Allg. Mikrobiol.*, *10*:515-526, 1970.
158. Oliveira-Baptista, A., and van Uden, N. Occurrence of two maximum temperatures for growth in yeasts. *Z. Allg. Mikrobiol.*, *11*:59-61, 1971.
159. Madeira-Lopes, and van Uden, N. Thermal association and dissociation in thermosensitive mutants of Saccharomyces cerevisiae. *Z. Allg. Mikrobiol.*, *19*:303-305, 1979.
160. Simoes-Mendes, B., Madeira-Lopes, A., and van Uden, N. Kinetics of petite mutation and thermal death in Saccharomyces cerevisiae growing at superoptimal temperatures. *Z. Allg. Mikrobiol.*, *18*:275-279, 1978.
161. van Uden, N., and da Cruz Duarte, H. Effects of ethanol on the temperature profile of Saccharomyces cerevisiae. *Z. Allg. Mikrobiol.*, *21*:743-750, 1981.
162. Richter, K. Effects of temperature on microbial ethanol production. II. A thermodynamic interpretation of the temperature profile curves of ethanol production. *Acta Biotechnol.*, *7*:127-140, 1987.
163. Loureiro, V., and van Uden, N. Effects of ethanol on the maximum temperature for growth of Saccharomyces cerevisiae. A model. *Biotechnol. Bioeng.*, *24*:1881-1884, 1982.
164. Sá-Correira, I., and van Uden, N. Effects of ethanol on thermal death and on the maximum temperature for growth on the yeast Kluyveromyces fragilis. *Biotechnol. Lett.*, *4*:805-808, 1982.
165. Krouwel, P. G., and Braber, L. Ethanol production by yeast at supra-optimal temperatures. *Biotechnol. Lett.*, *1*:403-408, 1979.
166. Plesset, J., Palm, C., and McLaughlin, C. S. Induction of heat shock proteins and thermotolerance by ethanol in Saccharomyces cerevisiae. *Biochem. Biophys. Res. Commun.*, *108*:1340-1345, 1982.
167. Margaritis, A., and Bajpai, P. Continuous ethanol production from Jerusalem artichoke tubers. I. Use of free cells of Kluyveromyces marxianus. *Biotechnol. Bioeng.*, *24*:1473-1482, 1982.
168. Ingledew, W. M., Magnus, C. A., and Patterson, J. R. Yeast foods and ethyl carbamate formation in wine. *J. Enol. Vitic.*, *38*:332-335, 1987.
169. Moloney, A. P., O'Prorke, A., Considine, P. J., and Coughlan, M. P. Enzymic saccharification of sugar beet pulp. *Biotechnol. Bioeng.*, *26*:714-718, 1984.
170. Jones, R. P., and Greenfield, P. F. A review of yeast ionic nutrition. Part I: Growth and fermentation requirements. *Proc. Biochem.*, *19*: 48-60, 1984.
171. Ogawa, T., Inoue, T., Sato, S., Shimoda, M., and Aiba, S. Phosphate mediation of the aerobic fermentation in baker's yeast. *Biotechnol. Lett.*, *2*:89-94, 1980.
172. Bajpai, P., and Margaritis, A. Optimization of studies for ethanol production from Jerusalem artichoke extract using immobilized Kluyveromyces marxianus cells. *J. Microb. Biotechnol.*, *1*:57-65, 1986.
173. Maia, A., Strehaiano, P., and Goma, G. Alternative techniques for alcoholic fermentation by Saccharomyces cerevisiae. *Rev. Latino-Am. Microbiol.*, *29*:175-182, 1987.

174. Parekh, S. R., and Waymam, M. Performance of a novel continuous dynamic immobilized cell bioreactor in ethanolic fermentation. *Enz. Microb. Technol.*, *9*:406-410, 1987.
175. Jones, S. T., Korus, R. A., Admassu, W., and Heimsch, R. C. Ethanol fermentation in a continuous tower fermentor. *Biotechnol. Bioeng.*, *26*:742-747, 1984.
176. Scott, C. D. Ethanol production in a fludized-bed bioreactor utilizing flocculating Zymomonas mobilis with biomass recycle. *Biotechnol. Bioeng. Symp.*, *13*:287-298, 1983.
177. Amin, G., van den Eynde, E., and Verachtert, H. Determination of by-products formed during the ethanolic fermentation, using batch and immobilized cell systems of Zymomonas mobilis and Saccharomyces bayanus. *Eur. J. Appl. Microbiol. Biotechnol.*, *18*:1-5, 1983.
178. Arcuri, E. J. Continuous ethanol production and cell growth in an immobilized-cell bioreactor employing Zymomonas mobilis. *Biotechnol. Bioeng.*, *24*:595-604, 1982.
179. Rodríguez, E., and Callieri, D. A. S. Conversion of sucrose to ethanol by a flocculent Zymomonas sp. strain in a continuous upflow floc reactor. *Eur. J. Appl. Microbiol. Biotechnol.*, *18*:186-188, 1983.
180. Chibata, I., and Tosa, T. Immobilized microbial cells and their applications. *Trends Biol. Sci.*, *April*:88-90, 1980.
181. Saigal, D., and Viswanathan, L. Effects of oils and fatty acids in molasses fermentation by distillers' yeast. *Enz. Microb. Technol.*, *6*:78-80, 1984.
182. Morais, S. M. F., Aquarone, E., Rose, A. H., and Beavan, M. Ethanol effect on fermentative activity of Saccharomyces cerevisiae. *Rev. Microbiol.*, *17*:371-375, 1986.
183. Eaton, L. C., Tedder, T. F., and Ingram, L. O. Effects of fatty acid composition on the sensitivity of membrane functions to ethanol in Escherichia coli. *Substance and Alcohol Actions*, *3*:77-87, 1982.
184. Uchida, K. Occurrence of saturated and mono-unsaturated fatty acids with unusually long chains C_{20}-C_{30} in Lactobacillus heterohiochii alcoholphilic bacterium. *Biochim. Biophys. Acta*, *348*:86-90, 1974.
185. Hayashida, S., Feng, D. D., and Hongo, M. Function of the high concentration alcohol-producing factor. *Agr. Biol. Chem.*, *10*:2001-2006, 1974.
186. Jakobsen, M., and Thorne, R. S. W. Oxygen requirements of brewing strains of Saccharomyces uvarum (carlsbergensis) bottom fermentation yeast. *J. Inst. Brew.*, *86*:284-287, 1980.
187. Akira, N., Nishizaur, Y., Tsuchiya, Y., and Nagai, S. Kinetic analysis for batch ethanol fermentation of Saccharomyces cerevisiae. *J. Ferment. Technol.*, *65*:277-284, 1987.
188. Richter, K., and Becker, U. Effects of temperature on microbial ethanol production. I. The temperature profile curve of ethanol production of the yeast strain Saccharomyces cerevisiae Se5. *Acta Biotechnol.*, *7*:87-92, 1987.
189. Bajpai, P., and Margaritis, A. The effect of temperature and pH on ethanol production by free and immobilized cells of Kluyveromyces marxianus grown on Jerusalem artichoke extract. *Biotechnol. Bioeng.*, *30*:306-313, 1987.
190. Dickson, R. C. Expression of a foreign eukaryotic gene in Saccharomyces cerevisiae: β-galactosidase from Kluyveromyces lactis. *Gene*, *10*:347-352, 1980.

191. Kilian, S. G., Prior, B. A., and Lategan, P. M. Diauxic utilization of glucose-cellobiose mixtures by Candida wickerhamii. *Eur. J. Appl. Microbiol. Biotechnol.*, *18*:369-373, 1983.
192. Skipper, N., Sutherland, M., Davies, R. W., Kilburn, D., Miller, R. C., Warren, A., and Wong, R. Secretion of bacterial cellulase by yeast. *Science*, *230*:958-960, 1985.
193. Laluce, C., and Mattoon, J. R. Development of rapdily fermenting strains of Saccharomyces diastaticus for direct conversion of starch and dextrins to ethanol. *Appl. Environ. Microbiol.*, *48*:17-25, 1984.
194. Baulcombe, D., and Buffard, D. Gibberellic acid regulated expression of α-amylase and six other genes in wheat aleurone layers. *Planta*, *157*:493-498, 1983.
195. Maleszka, R., Wang, P. Y., and Schneider, H. A col E_1 hybrid plasmid containing Escherichia coli genes complementing D-xylose negative mutants of Escherichia coli and Salmonella Typhimurium. *Can. J. Biochem.*, *60*:144-148, 1982.
196. Bajpai, R. K., and Ghose, T. K. An induction-repression model for growth of yeasts on glucose-cellobiase mixtures. *Biotechnol. Bioeng.*, *20*:927-935, 1978.
197. Gancedo, J. M., and Gancedo, C. Catabolite repression mutants of yeast. *FEM Microbiol. Rev.*, *32*:179-187, 1986.
198. Schaefer, E. J., and Cooney, C. L. Production of maltase by wild-type and a constitutive mutant of Saccharomyces italicus. *Appl. Environ. Microbiol.*, *43*:75-80, 1982.
199. Ciriacy, M. A yeast mutant with glucose-resistant formation of mitochondrial enzymes. *Mol. Gen. Genet.*, *129*:329-335, 1978.
200. Ciriacy, M. Isolation and characterization of further cis- and trans-acting regulatory elements involved in the synthesis of glucose-repressible alcohol dehydrogenase (ADH II) in Saccharomyces cerevisiae. *Mol. Gen. Genet.*, *176*:427-431, 1979.
201. Zimmermann, F. K., and Scheel, I. Mutants of Saccharomyces cerevisiae resistant to carbon catabolite repression. *Mol. Gen. Genet.*, *154*:75-82, 1977.
202. Bailey, R. B., Benítez, T., and Woodward, A. Saccharomyces cerevisiae mutants resistant to catabolite repression: use in cheese whey hydrolysate fermentation. *Appl. Environ. Microbiol.*, *44*:631-634, 1982.
203. McCann, A. K., and Barnett, J. A. Starch utilization by yeasts: mutants resistant to carbon catabolite repression. *Curr. Genet.*, *8*: 525-530, 1984.
204. Yutaka, O., Endo, A., Iwasaki, H., Ito, Y., and Goto, S. Isolation and properties of ethanol-using yeasts with acid and ethanol tolerance. *J. Ferment. Technol.*, *64*:379-382, 1986.
205. Olaiya, A. F., and Sogin, S. J. Ploidy determination of Candida albicans. *J. Bacteriol.*, *140*:1043-1049, 1979.
206. Korhola, M. Ethanol tolerant yeast. In *Proceedings of the International Conference on Molecular Yeast Genetics*. Montpellier, France, 1982.
207. Murray, W. D., Wemyss, K. B., and Khan, A. W. Increased ethanol production and tolerance by a pyruvate-negative mutant of Clostridium saccharolyticum. *Eur. J. Appl. Microbiol. Biotechnol.*, *18*:71-74, 1983.
208. Conde, J., and Mascort, J. L. Effect of the mitochondrial genome on the fermentation behaviour of brewing yeast. *Eur. Brew. Con. Congress*, pp. 177-186, 1981.

209. Esser, K., and Kües, K. Flocculation and its implication for biotechnoloty. *Proc. Biochem., Dec.*:21-23, 1983.
210. Lifu, C., and Gong, C. S. Continuous ethanol production using induced yeast aggregates. *Appl. Microbiol. Biotechnol., 25*:208-212, 1986.
211. Ouchi, K., Kawase, N., Nakano, S., and Akiyama, H. Stabilization of yeast killer factor by glycerol. *Agr. Biol. Chem., 42*:1-5, 1978.

22

FUNGAL ENZYMES

DAVID A. LOWE *Industrial Division, Bristol-Myers Squibb Company, Syracuse, New York*

I. INTRODUCTION

The industrial use of enzymes is now an integral part of a wide variety of commercial processes and applications ranging from their use in the complex large-scale manufacture of chemicals and processed foods to their simple inclusion in many commodity supermarket products. The diverse properties of enzymes have allowed them to be applied to a wide variety of processes and products. Enzymes are now manufactured in a variety of formulations and specific activities. The cost of the product is usually inversely proportional to the volume manufactured and variety of application. At one extreme are the specialty enzymes produced for research needs, which are typically of high purity, high cost, and available in many varieties and forms. At the other extreme are the bulk technical, detergent, and food grade enzymes, typically proteases and carbohydrate-hydrolyzing enzymes. These are at low cost and available in less than two dozen general types. Pharmaceutical grade enzymes are of a higher quality than food grade and normally have had some degree of purification. Analytical grade enzymes are of high purity and are similar to those of research grade. In most cases, enzyme costs are greater than $1/g for research quality materials and less than $10/kg for industrial quality.

Economics of enzyme production are governed by the demand for bulk technical and food grade enzymes as these are the largest category in terms of volume and are generally of high quality. In large-scale enzyme manufacture, the current tendency is to limit the range of enzymes produced to as few as possible but to attain as wide a use as possible for their application. In this manner, the economics of scale can maintain low manufacturing costs for these products.

Historically, the enzyme industry was developed through the use of plant and animal materials as sources of enzymes, although notable exceptions exist, e.g., the use of fungi in koji, tofu, tempeh, and other oriental foods [1]. With the development of stirred-tank fermentations and the ease of genetic manipulation of microbial cultures for byproduct overproduction, microorganisms are now the major source of enzymes both in volume and variety. In many cases, microbial enzymes have been specifically selected and developed to replace existing plant and animal enzymes. The use of microbes as enzyme sources has reduced the major risks in their manufacture that used to be associated with the availability of animal wastes and the climatic and political growth of agricultural crops. The present trend toward the use of microorganisms will continue. Genetic engineering technology will accelerate the development and production of enzymes in both bacteria and fungi. Bacteria have the advantage of fast growth and short fermentation cycles. However, fungi are able to produce large quantities of secretable enzymes and process protein with carbohydrate components. Presently, fungal sources do provide the largest variety of bulk commercial enzymes and have the widest variety of application.

The enzyme industry is very competitive and current information on actual manufacturing processes is difficult to obtain directly. However, some relevant information can be collected from a variety of indirect sources. Godfrey and Reichert [2] listed approximately 260 sources of commercial enzymes produced by 27 manufacturers throughout the world. From this total, 97 were from fungi, 66 from facteria, 23 from plants, 25 from animals, 9 from yeast, 7 from actinomycetes, and the remainder not categorized. However, in terms of quantity of enzymes produced, fungi only accounted for 25% of the total, compared to 40% from bacteria, 25% from plant, and 10% from animal origin. In 1982, Hepner and Male [3] identified only 16 enzymes with markets of over $0.5 million worldwide. A wider range of enzyme types exists for research quality products as is indicated from commercial catalogs such as Sigma [4] 740, Boehringer [5] 280, and Calbiochem [6] 180. Less than 10% of these enzymes are of fungal origin. This may reflect the historical development of these biochemical enzymes with the major academic emphasis on animal and bacterial sources.

The American Type Culture Collection [7] lists over 462 fungal cultures that have been deposited with them as sources of patentable enzymes. These cultures produce 106 different enzymes (Table 1). Several enzyme types have large numbers of different deposited producing cultures, e.g., proteases have 62 different deposited cultures (Table 2) and urate oxidase has 47 (Table 3). The popularity of urate oxidase patents reflects the future potential for this enzyme rather than its current commercial use.

Among fungal sources the genus *Aspergillus* provides the greatest variety of enzymes. From the 97 commercial sources of fungal enzymes listed by Godfrey and Reichert [2] (see Table 4), 61 were from aspergilli with *A. niger* providing 39 enzyme products and 16 different enzyme types. Further versatility of *Aspergillus* can be seen by examining the patent-related enzymes held by the American Type Culture Collection (Table 5). This list contains some variance when compared to other data [2], and it appears that one cannot always equate patented expectations with commercial reality.

II. SAFETY AND HANDLING

Humans, through the historical use of enzymes in food and in other processes such as leather tanning, have had prolonged contact with and exposure to a variety of enzymes. Most of these enzymes were used in their crude form and by a process of trial and error were found to be active and beneficial. However, with the introduction of more active, highly purified enzymes, empirical approaches could not be used. Currently, governments often maintain tight control over the testing and release of all new enzymes [8]. This change occurred over the last 25 years during which time there were three major factors that influenced this present environment. First,

TABLE 1. Enzymes Produced from Patented Fungal Cultures Deposited with the American Type Culture Collection

Enzyme type	Number of cultures
Acetamidase	1
β-*N*-Acetylglucosaminidase	1
cis-Aconitic acid decarboxylase	1
Aconitic hydrase	1
Adenylate cyclase	2
Agroclavine hydroxylase	1
Alcohol dehydrogenase	2
Alcohol oxidase	2
Aldehyde reductase	1
Aminopeptidase	2
Amylase	22
Aniline oxidase	1
Blasticidin S deaminase	2
Byssochlamyopepidase A	1
Carbohydrase	2
Carboxypeptidase acid	4
Catalase	1
Cellobiohydrolase	1
Cellobiose dehydrogenase	2
Cellulase	29
Cephalosporin acetylesterase	1
Chitin synthetase	4
Chitinase	7
Chitosanase	1
Collagenase	1
Cutinase	3
Decarboxylases	4
Dextranase	6
o-Dianisidine peroxidase	1
Dihydrogeodin oxidase	1
endo-β-1,4-Galactanase	1
endo-Polygalacturonase	6

TABLE 1. (Continued)

Enzyme type	Number of cultures
Esterase	4
Ferrichrome reductase	1
Formamidase	1
Fructosyltransferase	1
Galactofuranoside	1
Galactose oxidase	1
α-Galactosidase	9
β-Galactosidase	4
Glucamylase	1
Glucan-1,4-α-glucosidase	1
Glucanase	19
Glucoamylase	2
Glucose dehydrogenase	1
Glucose isomerase	11
Glucose oxidase	3
α-Glucosidase	14
β-Glucosidase	13
β-D-glucosidase cellobiase	6
Glucuronosyltransferase	1
Glutamate decarboxylase	1
Glutaminase	1
Hydroxycinnamate ester hydrolase	2
11-α-Hydroxylase	1
Invertase	3
Isoamylase	1
Isocitrate lyase	2
Isopen N synthetase	1
Keratinase	7
Kynureninase	3
Laccase	19
Lactate dehydrogenase	2
Ligninase	1
Lipase	16

Enzyme type	Number of cultures
L-Lysine oxidase	1
Maltase	1
α-Mannosidase	2
Mycodextranase	1
Myrosinase	1
Neuraminidase	2
Nitrate peroxidase	1
Nitrate reductase	1
Pectate lyase	3
Pectin esterase	3
Pectin lyase	4
Penicillin amidase	1
Peptidase	1
Peroxidase	3
Phenoloxidase	1
Phosphodiesterase	2
6-Phosphogluconate dehydrogenase	2
Phospholipase	4
Polyamine oxidase	2
Pectinase	7
Protease	62
Pterin deaminase	4
Purine β-ribonucleosidase	9
Pyruvate carboxylase	1
Pyruvate decarboxylase	1
Quercetinase	1
Ribonuclease	3
Ribonucleotide reductase	1
Salicylate hydroxylase	1
Sulfate adenylyltransferase	2
Superoxide dismutase	1
Tannase	1
Threonine synthetase	1

TABLE 1. (Continued)

Enzyme type	Number of cultures
Tryptophan aminopeptidase	1
Tryptophan synthetase	1
Tryrosinase	2
Urate oxidase	47
Vanillate hydroxylase	1
Xylanase	5
B-D-Xylosidase	1
D-Xylulose reductase	1

TABLE 2. Fungal Cultures Patented for Protease Production and Deposited with the American Type Culture Collection

Acremonium chrysogenum	*Fusarium* spp.
Acremonium kiliense	*Gibberella saubinetii*
Actinomucor elegans	*Haloprus cystisinus*
Alternaria tenuis	*Irpex lacteus*
Arthrographis kalrai	*Lenzites styracina*
Aspergillus awamori	*Metarrhizium anisopliae*
Aspergillus flavus	*Monilinia fructigena*
Aspergillus flavis var. *columnaris*	*Mycelia sterilia*
Aspergillus foetidus	*Neurospora crassa*
Aspergillus foetidus var. *acidus*	*Neurospora sitophila*
Aspergillus fumigatus	*Paecilomyces farinosus*
Aspergillus niger var. *macrosporus*	*Penicillium citreonigrum*
Aspergillus oryzae	*Penicillium glabrum*
Aspergillus oryzae var. *oryzae*	*Penicillium roquefortii*
Aspergillus oryzae var. *viridis*	*Penicillium spinulosum*
Aspergillus phoenicis	*Phanerochaete chrysosporium*
Aspergillus sojae	*Polyporus ostreiformis*
Beauveria bassiana	*Polyporus palustris*
Cladosporium spp.	*Pycnoporus cinnabarinus*
Conidiobolus heterosporus	*Pycnoporus sangguineus*
Conidiobolus khandalensis	*Rhizomucor pusillus*
Conidiobolus lamprauges	*Rhizopus oligosporus*

TABLE 2. (Continued)

Coprinus cinereus	*Scytalidium lignicolum*
Coprinus radians	*Talaromyces thermophilus*
Fomitopsis castanea	*Verticillium lecanii*

TABLE 3. Fungal Cultures Patented for Urate Oxidase Production and Patented with the American Type Culture Collection

Absidia glauca	*Monosporium apiospermum*
Alternaria alternata	*Mucor hiemalis*
Aspergillus awamori	*Mucor mucedo*
Aspergillus carbonarius	*Mucor racemosus*
Aspergillus clavatus	*Paecilomyces lilacinus*
Aspergillus flavus	*Penicillium canescens*
Aspergillus nidulans	*Penicillium expansum*
Aspergillus niger	*Penicillium funiculosum*
Aspergillus ochraceus	*Penicillium glabrum*
Aspergillus oryzae	*Penicillium granulatum*
Aspergillus sydowi	*Penicillium griseum*
Aspergillus tamarii	*Penicillium purpurescens*
Aspergillus terreus	*Penicillium purpurogenum*
Aspergillus terricola	*Penicillium raistrichii*
Aspergillus ustus	*Penicillium spinulosum*
Aspergillus wentii	*Penicillium thomii*
Cephalosporium spp.	*Penicillium velutinum*
Fusarium coeruleum	*Penicillium waksmani*
Fusarium moniliforme	*Rhizopus oryzae*
Fusarium orthoceras	*Scopulariopsis brevicaulis*
Fusarium oxysporum	*Stemphylium macrosporoideum*
Fusarium solani	*Talaromyces spiculisporus*

TABLE 4. Sources of Industrial Fungal Enzymes

Common name	EC IUB No.	Culture source	Industrial supplier
α-Amylase (diastase)	3.2.1.1	*Aspergillus niger*	Hazyme[1]
		Aspergillus oryzae	Amylozyme,[2] Amylasep,[1] Clarase,[3] Takamyl,[4] Fungamyl,[5] Starzyme HM[6]
		Endomyces spp.	
γ-Amylase Amyloglucosidase	3.2.1.1	*Aspergillus niger*	Ambazyme,[2] Diase,[1] Agidex,[7] Optidex L,[3] Diazyme,[4] AMG,[5] StarzymeAG[6]
		Aspergillus oryzae	
		Aspergillus phoenicis	
		Rhizopus delemar	
		Rhizopus niveus	AllcoholaseII[8]
Catalase	1.11.1.6	*Aspergillus niger*	Fermcolase,[9] CatalaseLA3[6]
		Penicillium spp.	
Cellulase	3.2.1.4	*Aspergillus niger*	BiocellulaseA,[10] Cellulase,[9] CellulaseP4000,[3] CellulaseCA[6]
		Aspergillus sojae	
		Penicillium funiculosum	CellulaseCP[6]
		Trichoderma viride	BiocellulaseT,[10] MKC Cellulase,[3] Celluclast,[5] Cellulase CT[6]
Dextranase	3.2.1.11	*Penicillium* spp.	Dextranase[5]
α-Galactosidase (melibiase)	3.2.1.22	*Aspergillus niger*	
β-Galactosidase (lactase)	3.2.1.23	*Aspergillus niger*	
		Aspergillus oryzae	Takamine Fungal Lactase[4]
		Penicillium emersonii	

β-Glucanase	3.2.1.6	*Aspergillus niger* *Aspergillus oryzea* *Penicillium emersonii*	Cellobiase,[5] Finizym[5] Beta-Glucanase,[2] Barley-B, Glucanase[7]
Glucose oxidase	1.1.3.4	*Aspergillus niger*	Ovazyme,[9] Maxazyme GO,[1] MKC Glucose Oxidase,[3] Glucox[6]
		Penicillium spp.	
Hemicellulase	3.2.1.78	*Aspergillus niger*	Hemicellulase,[4] Gamanase[5]
		Aspergillus oryzae *Aspergillus saitoi*	
Inulinase	3.2.1.7	*Aspergillus* spp.	Novozym 230[5]
Invertase	3.2.1.26	*Aspergillus niger* *Aspergillus oryzae*	
Lipase (esterase)	3.1.1.3	*Aspergillus niger* *Aspergillus oryzae* *Mucor miehei* *Rhizopus arrhizus* *Rhizopus delemar*	Novozym206,[5] Lipase A[6] Piccantase[1] Lipase[11]
Penicillin amidase	3.5.1.11	*Basidiomycetes* spp. *Fusarium oxysporum*	
Peroxidase	1.11.1.7	*Aspergillus* spp. *Penicillium* spp.	
Pectinase	3.2.1.15	*Aspergillus niger*	Pectinol[12] Extractase,[9] Klerzyme,[1] Rapidase CPE,[1] Optclar,[3] Clarex,[3] Spakl,[3] Pectinex,[5] Panzym[6]

TABLE 4. (Continued)

Common name	EC IUB No.	Culture source	Industrial supplier
		Aspergillus ochraceus	
		Aspergillus oryzea	
(anthocyanase)		*Aspergillus niger*	Pectinol DL[12]
(naringinase)		*Aspergillus niger*	
Protease			
(alkaline)	3.4.21.14	*Aspergillus flavus*	Rhozyme P-11[12]
		Aspergillus oryzae	Rhozyme 41[12]
(acid)	3.4.23.6	*Aspergillus niger*	
		Aspergillus oryzae	Acid Fungal Protease P[3]
		Aspergillus spp.	Fungal Proteinase S[13]
		Endothia parasitica	
		Mucor miehei	
		Mucor pusillus	
(neutral)	3.4.23.4	*Aspergillus oryzae*	Panazyme 77,[2] Fungal Proteinase P[13]
Rennets	3.4.23.6	*Endothia parasitica*	
		Mucor miehei	Fromase,[1] Marzyme,[4] Rennilase[5]
		Mucor pusillus	
Tannase	3.1.1.20	*Aspergillus niger*	
		Aspergillus oryzae	
Xylanase	3.2.1.32	*Aspergillus niger*	

1, Gist Brocades NV; 2, ABMC Food Division; 3, Miles Kali-Chemie GmbH; 4, Miles Laboratories; 5, Novo Industri A/S; 6, J. and E. Sturge Ltd.; 7, Glaxo Operations; 8, Alltech Inc.; 9, Fermco Biochemicals Inc.; 10, Biocon Ltd.; 11, Hughes and Hughes (Enz.) Ltd.; 12, Corning; 13, Rohm GmbH.

TABLE 5. Enzymes Produced by *Aspergillus niger* Cultures Deposited with the American Type Culture Collection

Aconitic hydrase	Kyureninase
Adenylate cyclase	Maltase
Amylase	Myrosinase
Cellulase	Phosphodiesterase
Glucamylase	Protease
B-Glucanase	Ribonuclease
Glucoamylase	Urate oxidase
Keratinase	

the rapid and widespread introduction of enzyme detergents with the active agent in a powdered form highlighted the sensitivity of large numbers of people to allergic reactions to protein products of this type. Second, the subsequent collapse of the bacterial enzyme market left many companies with excess fermentation capacity primarily designed for crude microbial enzyme production. This excess capacity was soon filled by the introduction of new enzymes made from microorganisms that both competed with existing enzymes from the more conventional animal and plant sources, and opened new areas for enzyme application, e.g., replacement of chemical hydrolytic reactions. This enabled expansion in the production of a wide variety of enzymes in a variety of forms, particularly as immobilized enzymes. Finally, a third stage resulted from applications of genetic engineering where novel enzymes can now be produced in large amounts from the propagation of highly efficient, genetically engineered hosts such as *Escherichia coli* and *Aspergillus* spp. Thus, it is now possible to produce a large range of enzymes from nontraditional sources such as mammalian cells, viruses, thermophilic, pathogenic, and fastiduous microorganisms.

Such widespread manufacture and use of these new enzymes has necessitated careful handling and toxicity and safety testing of these products. Enzyme products have the potential for a variety of adverse reactions due to (a) their catalytic properties, e.g., damage to mucous membranes by proteases; (b) allergic response as foreign antigens both as the enzyme product and as associated products of manufacture, e.g., impurities from the producing culture and additives introduced in manufacture; and (c) the presence of toxic products from the producing organism, e.g., aflatoxins and other mycotoxins. Fortunately, such toxic compounds are produced as secondary metabolites at a later stage than that which is usually optimal for enzyme production, usually in the fast-growing primary fermentation stage.

Generally, enzymes derived from animal and plant sources consumed by humans are not required to have toxicological tests provided they meet the required purity and minimal microbiological content. Certain enzymes produced from fungal cultures that have historically been used by man have similar standings. In the United States, these are termed "generally regarded as safe" (GRAS); see Table 6. However, new enzymes produced from these fungal cultures and enzymes from other microorganisms not covered by GRAS must be subjected to toxicity testing. Long-term toxicity studies are not required for enzymes when used as food-processing aids as they are present at very low concentrations and daily intakes are extremely low.

TABLE 6. Enzymes Permitted for Use in Food Products

Aspergillus niger	GRAS	α-Amylase Cellulase Glucoamylase Lactase	Pectinase Glucose oxidase Lipase Catalase
A. oryzae	GRAS	α-Amylase Lactase	Glucoamylase Lipase
Rhizopus oryzae	CFR	α-Amylase Pectinase	Glucoamylase
R. niveus	CFR	Glucoamylase	
Morteirella vinaceae	CFR	α-Galactosidase	
Endothia parasitica	CFR	Rennet	
Mucor miehei	CFR	Rennet	
M. pusillus	GRASP	Rennet	

Abbreviations: CFR, Code of the Federal Register; GRAS, generally regarded as safe; GRASP, GRAS petition.

For ease of use and minimization of hazardous conditions, bulk enzymes are supplied either in the concentrated liquid form or as granular solids. The latter can be soluble and insoluble immobilized enzymes, and as viscous solutions, both splashing and aerosol formation can be minimized. High enzyme concentrations reduce the need for preservatives although some antimicrobial agents may still be required, thereby enabling the product to be stored and handled under ambient conditions. Concentrated forms can also reduce shipping charges. Autodigestion can occur with proteases and in such cases enzymes are made in pelleted or encapsulated forms. Today the use of powdered forms is avoided. In some industrial applications, immobilized forms are preferred [9-11]. Immobilized enzymes do have a higher initial production cost. However, this initial expense is more than offset by the enhanced stability of these products, ease of reuse, and separation of product from the biological catalyst, giving them distinct process advantages. For example, immobilized enzymes in column reactors do not require any handling and are easily separated from the reaction products that are not contaminated by enzyme protein.

In the liquid form, the enzyme protein often represents only a minor portion, depending on the required activity of the preparation. The bulk is composed of stabilizers such as corn starch, lactose, dextrins, glucose, NaCl, or KCl. Caramel or annato can be used for coloring. The most common preservatives are benzoic acid or its esters or alkyl parabens.

The activity of most bulk commercial enzymes is expressed in arbitrary units derived from historical use or the analytical method used to assay activity. The systematic use of international enzyme units (IU), the conversion of 1 μmol of substrate to product per minute, has not been adopted although many of the arbitrary units can be converted to these standardized units [12].

Commercial enzymes are supplied with relevant technical information to guide the user on their optimum use. Such information often includes a general description of what the enzyme does, a range of substrates, and a description of the physical properties of the enzyme.

III. PRODUCTION

A. Cultures

Each manufacturer of a particular enzyme will have its own high-producing culture, one that has been selected and improved for high enzyme production through strain selection, media development, process development, and scale-up programs. Some of the early strains in the improvement plan may be deposited for patent purposes, and the strains are usually available from the ATTC or other collections for research purposes and as a patent requirement. The enzyme activities and properties of these cultures are only stable representatives of the original isolate. Superior and current production cultures are not deposited or available outside the company in order to maintain the proprietary nature of these cultures and their associated processes.

Most commercial bulk enzymes are produced from the genus *Aspergillus*. This is a highly variable genus with many closely related species. Mutation programs have certainly increased this diversity into many subspecies. Fortunately, aspergilli are mainly haploid in their vegetative stage and are prolific producers of conidia which facilitate mutation and selection programs. For effective mutation and strain improvement, it is important to have conidia with single nuclei in order to minimize genetic variability and subsequent culture rundown.

B. Fermentation

All fungi are aerobic and can be cultivated in two forms for enzyme production. Their natural form is growth on the surface of a substrate. However, they can easily be adapted to submerged growth in deep fermentors provided they are aerated. Both forms are used for enzyme manufacture.

1. Surface Culture

Solid state enzyme fermentations have evolved from the oriental koji-type processes for soybean and wheat fermentation [1]. Methods have been described for the production of amylogycosidase from *Rhizopus* [13], proteases from *A. oryzae* [14] and *A. niger* [15], cellulase from *Trichoderma* [16], and pectinase from *A. niger* [17].

There are several advantages to the use of submerged fermentation culture which include (a) low energy input as air compression is not required, intermittent agitation and sterilization possibly not required; (b) less space required per unit of substrate due to limited use of water; (c) product more concentrated and needing less workup to the final product; and (d) minimal effluent problems. In some cases surface cultures can be more cost-effective than submerged fermentations. However, the process is still considered low technology and has limitations for further process improvement as culture conditions are difficult to control and measure. Effective scale-up may present a major problem due to the two-dimensional nature of the processes.

2. Submerged Culture

Stirred-tank fermentations have been successfully used to produce enzymes in a manner similar to their use in antibiotic manufacture. Some fungal strains may require special treatments or mutation/selection to obtain the correct growth morphology in the stirred tank, namely, the avoidance of the formation of large pellets and excessive surface and wall growth.

Enzyme fermentations are generally of shorter duration compared to fungal antibiotic cycles as no secondary stage is required and the processes are optimized for rapid cell growth. Most enzymes are produced extracellularly. Vessel sizes can range from 1000 to 50,000 gal [18]. The tanks are built with stainless steel walls and components and can withstand the pressures of two atmospheres required for sterilization and the in-cycle use of high pressure to maintain sufficient dissolved oxygen levels during periods of peak metabolic demand. Agitation is carried out by single or multiple Rushton-type turbines, and sterile air is introduced through spargers usually at the bottom of the tank. Due to the fast growth and high metabolic rates attained for enzyme production, sufficient surface area has to be available inside the tank for cooling. This is accomplished by exterior cooling jackets in small tanks and by internal cooling coils or panels in larger tanks.

Media formulations are strictly proprietary and have been developed through many years of experimentation. However, as a basis they are usually composed of inexpensive raw materials, typically agricultural byproducts [19]. In the production of enzymes it is important to avoid any catabolyte repression by the presence of glucose. Therefore, glucose-containing substrates have to be fed in-cycle or as slowly utilized carbohydrates such as lactose, glycerol, or sucrose. Often an inducer is required for optimum enzyme production, which normally is the main substrate that the enzyme utilizes. In this manner, cellulose should be present for cellulase production and starches for amylase production. In some cases, products of the enzyme reaction can be used, e.g., phenoxyacetic acid can induce the production of penicillin V acylase in *Fusarium* [20].

Complex agricultural products can serve as both carbohydrate and nitrogen sources. Soybean meal or flour is a relatively inexpensive source of protein (45% by weight) and has a varied carbohydrate content (starch, cellulose, xylans). Other meals include cottonseed, rape, peanut, sunflower, fish, and corn gluten. Corn steep liquor is an inexpensive source of soluble nitrogen, vitamins, and growth factors and is often used in the cultivation of the seed or inoculum tank. It can be used in combination with meals for large-tank fermentations. However, in some circumstances the presence of reducing sugars can cause enzyme repression.

Most fungal enzyme fermentations are run in a batch or feed-batch mode. Continuous fermentations are not successful as they tend to produce dilute enzyme-containing broths which require more detailed extraction. In addition, highly mutated cultures often revert to lower or even nonproducing strains.

C. Recovery and Purification

With the exception of oxidase enzymes and penicillin amidases the majority of enzymes are extracellular. In these cases, the fermentations have been designed for optimal production of culture filtrate high in enzyme activity and relatively low in other products that may interfere with extraction and recovery. In the simplest case, the broth filtrate is concentrated by evaporation or spray-dried without further treatment. These crude technical enzyme grades are often suitable for some process uses. However, to produce enzymes for specific purposes, some degree of fractionation and purification may be required.

Enzyme purification procedures are similar for both intracellular and extracellular enzymes and can be considered together. In the harvesting of both enzyme types, it is important to chill the fermentation broth before downstream processing to avoid the growth of contaminants and suppress enzyme inactivation due to its own instability or from the action of coproduced proteases. Both centrifugation and filtration are used to separate the cell/solid mass from the solution [21,22]. The choice depends on the physical form of the whole broth and whether the enzyme activity resides in the solid or liquid phase. Filtration methods can include the use of string filters if the mycelial mat is sufficiently fibrous or use of filter aids on a drum filter. More recently, the use of ultrafiltration has become popular due to flexibility in the choide of membrane sizes [23].

Intracellular enzymes have to be released by cell breakage, which is commonly performed by Gaulin or ball mill homogenization. The broken cell debris is then removed by further filtration or centrifugation, often with the addition of flocculating or nucleoprotein-precipitating agents such as polyethyleneimine to aid processing.

To minimize expense in the purification of these bulk products, enzyme fractionations are limited mainly to the use of salt and solvent precipitation. Ammonium sulfate precipitation is often used with solvent-sensitive enzymes. However, when applicable, the use of water-miscible solvents such as acetone and isopropanol are preferred, as both of these solvents are inexpensive, easily recovered, easily removed from the enzyme, and easy to handle. The use of ion exchange resins is somewhat restrictive in the manufacture of bulk enzymes due to their high purchase cost and high operational cost. For further details on enzyme purification, see Refs. 18, 24, and 25.

IV. ENZYMES AND THEIR USES

A. Carbohydrate-Hydrolyzing Enzymes

1. Amyloglucosidase (Glucoamylase)

This enzyme is responsible for the direct hydrolysis of starch and hydrolyzed starch intermediates to glucose. It is usually used in combination with other amylolytic enzymes for the preparation of specific sugar products. Catalytically it is an *exo*-α-amylase which hydrolyzes glucose units from the terminal ends of the α-1-4 amylose chains. In combination with fungal α-amylase, it can produce high-glucose syrups. It is produced primarily from strains of *A. niger* and *A. awamori*, as well as from *A. oryzae*, *A. phoenicis*, and *Rhizopus delemar* [14,26]. *A. niger* strains are known to produce an interfering transglucosidase enzyme which can form unfermentable sugars from glucose and maltose. This enzyme can be selectively removed by the use of polyacrylic acids such as Carbopol [27]. As an example of using the most appropriate substrate for maximum enzyme production, high amyloglucosidase activities have been claimed by using ground corn which has been hydrolyzed to a more usable and better inducible substrate by the addition of bacterial α-amylase [28].

2. α-Amylase

Fungal α-amylases are preferred over bacterial and cereal enzymes in the baking industry due to their high content of proteolytic activity. Hydrolytic activity can be regulated in the baking process by heat inactivation. This enzyme is primarily produced from *A. oryzae* [29] but can be produced from strains of *A. niger* and *P. expansum*. Enzyme activity from *P. expansum* has been purified 1500-fold by solvent (methanol and isopropanol) and ammonium sulfate precipitation. Such active preparations can be used in the textile industry for desizing [30]. α-Amylase is an endo-α-amylase which hydrolyzes internal 1-4 linkages for maltose units and is mainly used for the production of low-glucose, high-maltose syrups which are preferred for making hard confections.

Considerable quantities of α-amylase are used in paper manufacture in the controlled hydrolysis of starch to dextrins. The addition of dextrins to papermaking can impart a wide variety of properties on the paper. Typically, the dextrin content of paper is between 5 and 20%. Starches are used throughout the world as adhesive sizes in the weaving of fabrics. This size has to be removed before the dyeing and texturizing stages. Sizes can be removed by the use of hot water alone. However, a preliminary enzyme hydrolysis is more efficient. Fungal amylases are used for desizing. However, they are often not thermostable and are inactivated above 55°C. Bacterial thermostable amylases are now favored by the industry as they can operate at temperatures up to 100°C. Fungal amylases are used as supplements to barley amylases in the manufacture of bread and beer. In the latter case, their presence improves fermentability and can produce low-calorie beer [31].

3. Pullulanase

Amylopectin, a highly branched starch component, has 1-6 glycosidic linkages at the branching points. These 1-6 linkages are not hydrolyzed by amylases or amyloglucosidases and can lead to incomplete starch hydrolysis. Pullulanase from *Cladosporium resinae*, for example, can hydrolyze pullulan, a 1-6-glycoside-linked carbohydrate; when supplemented to starch-hydrolyzing systems, it can significantly increase glucose yields [32]. A highly stable enzyme from *Thermus aquaticus* was recently described [33].

4. Cellulases

The cost-effective enzymatic hydrolysis of cellulose to glucose has yet to be realized; however, the large quantities of cellulose available in nature make cellulase-catalyzed sugar production an interesting and inexpensive alternative to conventional processes. Major limitations to the ready use of celluloses have been the unavailability of cellulase enzymes and the poor substrate affinity of cellulose in its natural microcrystalline form. Cellulase is produced primarily from *Trichoderma reesei* [34], *Thielavia terrestris* [35], and *Sporotrichum cellulophilum* [36].

All of these cellulases are heat-tolerant. However, they produce cellobiose as an end product which has an inhibitory feedback on the enzyme activity. *Trichoderma reesei* cellulase productivity has been improved more than fourfold by strain improvement programs [37]. A cellulase from *P. funiculosum* is less inhibited by cellobiose; however, this enzyme is not heat-tolerant [38].

Cellulase enzyme preparations are effective as additives to silage fermentations, where they can increase the release of fermentable sugars and nitrogen. Cellulases have uses in the preparation of a variety of plant products where cellulosic contaminants have to be removed, as in protein purification. They are also used as additives to aid maceration of plant material to increase extraction efficiencies, e.g., olive oil recovery and coffee concentrates [17]. Cellulase preparations from *A. niger* have been used to reduce waste disposal problems associated with the residues from apple, plum, and pear brandy distillations [39].

5. α-Galactosidase (Melibiase)

This enzyme is able to hydrolyze raffinose, a trisaccharide of glucose, fructose, and galactose in beet sugar juice, which prevents sucrose crystallization. Raffinose and stachyose are also the flatulent sugars that occur in legumes. α-Galactosidase is produced from both *Absidia griseola* and *Circinella muscae* [40]. Sucrose manufacture from beet requires preparations low in invertase [41]. Production of α-galactosidase from *P. duponti* entails the use of aqueous waste from the blanching of legumes as the major fermentation substrate, as this material is rich in galactose dimers [42].

6. β-Glucanase

β-Glucans are present in barley and can give rise to viscous solutions which can cause problems in beer production. The addition of fungal glucanases to malted barley mash increases wort separation and filtration rates and removes haze from the beer [43]. β-Glucans also occur in some wine processes involving the use of botrytized grapes; however, this glycan is made from 1-3 linkages instead of the 1-4 linkages that occur in barley glucans. Bacterial 1-3-glucanases have been developed to hydrolyze these products. Feeding barley with unhydrolyzed glucans to chickens results in the production of sticky bird droppings. A thermostable β-glucanase is produced by *P. emersonii* [44] and an *endo*-β-glucanase by *A. phoenicis* [45].

7. β-Galactosidase (Lactase)

The sugar lactose is the major constituent of whole milk, skim milk, whey, and yogurt. This disaccharide has several properties that adversely affect its usefulness as a sugar source. The nutritional value is limited as large populations are unable to hydrolyze it due to a lack of intestinal lactase. In addition, its low solubility prevents its concentration to syrups, and as a sugar it is not very sweet. These problems are all overcome by the conversion of the sugar to its monosaccharide components, glucose and galactose. This can be done by acid or enzyme hydrolysis. With the availability of lactase, enzymatic hydrolysis is the favored route as it avoids any color formation or the need for demineralization. There are many process advantages to this treatment. Hydrolysis of liquid milk increases its digestibility and sweetness. Enzyme-hydrolyzed milk powders can be tolerated by infants with lactase deficiencies. Milk products can be concentrated without the precipitation of lactose, e.g., sweet condensed milk. Enzyme-treated milk speeds up the manufacturing time of some cheeses and yogurts. Enzyme-treated whey products can be concentrated to syrups and used in many food products.

β-Galactosidases are produced from yeasts, *Kluyveromyces* spp., or *A. niger* and *A. oryzae*. The yeast enzymes have a pH range of 6-7 and work well on whole-milk products. However, the fungal enzymes have pH optima at pH 4-5 and are commonly used with acid whey derived from cottage cheese manufacture [46].

β-Galactosidases are used commercially in both the soluble form and the immobilized form, depending on the requirements of the process. Batch hydrolysis of whole-milk products uses a soluble enzyme as this gives a greater flexibility to the process, provided that there are no disadvantages of having the added enzyme in the final product. In the hydrolysis of whey, particularly the large-scale production of hydrolyzates for feed input for other fermentations, the use of immobilized enzymes are preferred as they allow for a simple continuous process whereby the initial cost of the enzyme can be spread out over many cycles of use. In addition, the product is free of lactase. Commercial scale immobilized enzyme processes have been developed using controlled-pore glass [47], ion exchange resins [48], and polyvinylchloride silica sheets [49,50].

Some improved β-galactosides enzymes have been reported. An enzyme with improved acidic stability from *A. oryzae* was shown to give higher conversion yields with acid wheys [51]. A thermostable, acid-stable enzyme was isolated from *Mucor pusillus* [52].

8. Pectinases

Pectin is a water-soluble polysaccharide occurring widely in higher plants. Its colorlessness and tastelessness make it ideal for the setting of jams and as thickeners and emulsifiers. The polymer is made of partially esterified D-galacturonic acid units linked by α-1-4 bonds. However, in many manufacturing processes the presence of pectins in juice and wine products is not desirable. Pectinases are commercially produced from *A. niger* and *Rhizopus oryzae* [53]. Addition of this enzyme to processed fruit extracts can remove any unwanted haze, e.g., in citrus fruits and wines [54]. Pectinase treatments are used in both red and white wine processes. When included in the pulped grapes, the total fermentation time is decreased by action of the pectinase in aiding the release of fermentable sugars and color from skins in the case of red wines. Similar treatments increase the amount of pressed juice from the white grapes prior to fermentation and from the fermented juices from red grapes. When used in combination with cellulases, pectinases can help increase the recovery of oil from macerated plant oil sources such as palm oil, olive oil, rape seed, and cottonseed. They are useful processing aids in the extraction of coffee flavors, spices, and other natural plant flavorings and colors. Pectinases aid the concentration of citrus juices by reducing the viscosity of associated pectin gums.

Pectin esterase enzymes have been described from *Aspergillus*, *Penicillium*, and *Fusarium* which can hydrolyze the methyl ester on the pectin, thus converting high-methoxyl pectins to low-methoxyl pectins [55]. Low-methoxyl pectins have the improved property of forming gels in the absence of sugars and acids and are used in low-calorie jams and related food products

9. Dextranase

α-1-6-Linked glucose dextran polymers are mainly produced by *Leuconostoc* species. Such cultures can infect damaged sugar cane and the resulting dextrans can create high viscosities during sugar evaporation. Their specific removal can occur by the use of dextranases made from *P. lilacinum* [2]. The enzyme has a pH optimum of 5-6 and a temperature optimum of 50°C and is used prior to the evaporation stage. Dextranases have also been reported to be produced from *Fusarium fusarioides* [56]. α-1-3-Linked dextrans are produced by oral streptococci and are a major contributor to dental caries. These dextrans can be hydrolyzed with a glucanase from *Cladosporium resinae* [57].

B. Protease and Peptidases

1. Milk Coagulation Enzymes

Milk has to be coagulated by a specific protease to produce the curd necessary for cheese manufacture. Historically, the most suitable enzyme came from the fourth stomach of the calf and was an acidic protease called renin or chymosin. This enzyme specifically cleaves the κ-casein between phenylalanine and methionine, and the resulting *para*-κ-casein is precipitated as a calcium complex. This curd then undergoes a variety of proteolytic changes during the curing stage, the diversity of which led to the unique flavor and texture of the cheese. Many alternative protein-hydrolyzing enzymes have been sought to replace the use of calf rennet. Procine and bovine pepsins can coagulate milk; however, their action is slow and there is an excess of proteolytic activity which leads to undesirable textural changes in the final product. Bacterial coagulants have similar properties; however, fungal enzymes have been found to be more suitable as rennet replacements. From screening programs, peptidase activity from *Mucor miehei* was found to have the same specific cleavage site as rennet [58]. Now over one-half of cheese manufactured worldwide uses fungal coagulants from *M. miehei*, *M. pusillus*, or *Endothia parasitica*. Thermolabile *Mucor* enzymes have been developed to enable the thermal destruction of the enzyme in the resulting whey, which then enables the whey to be formulated into other dairy products such as baby food and dietary aids. Other refinements on enzyme properties include the inactivation of cellulase activity which, if left active, can disintegrate the cheese cloth used to hold the curd together [59].

2. Other Fungal Proteases

Although bacterial proteases tend to dominate the market, especially in detergents, there are several specific uses for fungal proteases. Acid proteases are produced from *Rhizopus rhizopodiformis* [60] and *A. saitoi*, and neutral and alkaline proteases are produced from *A. oryzae* and *P. lilacinum* [61]. A thermostable alkaline protease, suitable for detergent use, was isolated from a thermophilic Eumycetes [62]. Proteases have uses as aids in the manufacture of fish and meat proteins, membrane cleaning, viscosity reduction in gelatin-containing products, increased efficiency in starch and oil recovery, and in leather manufacture. Specific acid proteases from *Aspergillus* species are used as flavor developers in processed cheeses. Fungal proteases can also be used to make enzyme-modified cheese, where the complete process can accelerated and final flavors modified to be similar to cheddar, Swiss, romano, and edam [2].

C. Lipases

Fungal lipases are produced from *Rhizopus* spp. and *A. niger*, the latter being more tolerant of acid conditions [2]. A lipase with esterase activity was reported from *Mucor miehei*, which has the ability to hydrolyze a wide range of fats with fatty acids from 4 to 20 carbons [63]. Together with pancreatic lipases they are used in the ripening of Italian-style cheeses and some blue cheeses, where their enzymatic actions increase the production of the highly flavored, shorter chain fatty acids. Lipases are also used to modify butterfat to releases the butyric, caproic, caprylic, and capric acids associated with butter flavor. These preparations are used in the manufacture of butter-flavored toffees and margarines and in cheese-flavored dips and sauces [2].

Lipase from *Rhizopus delemar* was used in the reverse direction to synthesize specific medium-chain triglycerides from fatty acids and glycerol [64].

D. Catalase

Fungal catalase is produced by strains of *A. niger* [2]. Its main use is to destroy the excess of hydrogen peroxide that is added to sterilize raw milk, as the presence of excess peroxide can have detrimental effects of the resultant cheese processing. Use of peroxide is mainly used in countries where refrigeration and pasteurization equipment is lacking. Catalase is also used to destroy excess hydrogen peroxide in the sterilization of soft lenses. Catalase is often used in combination with glucose oxidase.

E. Glucose Oxidase

Fungal glucose oxidase is produced primarily from *A. niger* and *P. notatum* [2,65]. Preparations from *P. amagasakaiense* are reported to be free of galactose and mannose oxidase activity [66]. Optimization of *A. niger* culture conditions were recently described [67]. This intracellular enzyme is available commercially in a variety of forms and at varying purities depending on use. Typically, crude forms have catalase present which can be beneficial for its intended use. However, in analytical preparations of the enzyme, all catalase activity has to be removed. Glucose oxidase reacts with glucose in the presence of molecular oxygen to form gluconic acid and hydrogen peroxide. It is used in a variety of food products to remove the oxygen to prevent changes in color and flavor [68]. It is routinely used in soft drinks, canned foods, beer, and wine and in mayonnaise and similar dressings. In combination with catalase it is used to remove glucose from prepared foods. This is particularly important in the use of egg and egg white preparations where it prevents browning in the egg powders. Hydrogen peroxide is added to provide an excess of oxygen (from the action of catalase). In combination with ascorbic acid, catalase-free preparations are used for oxidizing flours. Glucose oxidase is also used for analytical and diagnostic agents for the measurement of glucose in body fluids. It has been used in many forms from test strips to immobilized electrodes and membranes. The production of oxygen is proportional to the glucose concentration, and the released oxygen can be measured chemically by linkage to receptor dyes [66], or by potentiometric (polarographic) or amperometric (galvanic) electrodes [69].

F. Penicillin V Amidases

Penicillin antibiotics are semisynthetic side chain modifications of the active penicillin nucleus 6-aminopenicillanic acid (6-APA). This nucleus is made from the enzymatic hydrolysis of the fermented penicillins benzylpenicillin (penicillin G) and phenoxymethylpenicillin (penicillin V). Fermentation of these two forms is dependent on the precursor used: phenylacetic acid for penicillin G and phenoxyacetic acid for penicillin V. Bacterial enzymes are used to hydrolyze penicillin G. However, penicillin V is only effectively hydrolyzed by fungal enzymes. Major producers of penicillin V include Bristol-Myers Squibb and Biochemie.

Penicillin V amidase has been well described from the basidiomycete *Bovista plumbea*, now identified as *Pleurotus ostreatus* [70]. The enzyme has been purified over 220-fold to a specific activity of 23 IU/mg protein, and the activity of the amidase was not significantly inhibited by the presence of the hydrolysis products 6-aminopenicillanic acid and phenoxyacetic acid [71]. Whole-cell and concentrated enzyme extract preparations of this basidiomycete have been immobilized with polyacrylic resin [72] and with chitosan/polyphosphate/glutaraldehyde treatments [73].

Penicillin V amidase activity from *Fusarium oxysporum* was partially purified to a specific activity of 4.8 IU/mg protein [20]. Both of these enzymes appear to have properties suitable for the commercial hydrolysis of penicillin V to 6-aminopenicillanic acid. An immobilized crude cell preparation from an unknown fungal source was described [74] and developed into a commercial product (Novozym 217) by Novo [75]. The kinetics of this immobilized enzyme product has been investigated in recycled single-stage and multistage fixed-bed reactors [76,77].

V. FUTURE PROSPECTS

Industrial enzyme manufacture is now a mature business as most markets and applications have been saturated, leading to an excess of production capacity. This has lead to competitive pricing by many manufacturers. In bulk enzyme manufacture, future trends will be toward the lowering of manufacturing costs of existing enzyme products. Lower costs will be realized by increasing the productivity of the fermentations by the selection and use of superior strains and the optimization of the fermentation and recovery processes. Further cost reductions can occur through the use of less expensive raw materials in the fermentation and simplified extraction procedures.

Opportunities will arise from the use of genetic engineering to produce these conventional enzymes with improved properties. For example, many thermotolerant enzymes have been isolated from thermophiles, e.g., cellulases and amylases from *Talaromyces emersonii* [78] and xylanases from *Thermoascus aurantiacus* [79].

In situations where the isolated culture is a low producer, it can be more cost-effective in terms of research and development resources to isolate and purify the enzyme for sequence analysis in order to make a DNA probe to recover the full complement protein DNA sequence from the producing culture for insertion into a more suitable host. Hybridization by protoplast fusion between closely related species was recently reported [80], indicating the potential of conventional breeding techniques. Recombination into fungal hosts has been well documented [81,82]. High enzyme yields have been reported; for example, a protein from *Rhizomucor miehei* was produced at 3 g/liter in *A. oryzae* [83]. In other applications, active calf chymosin was integrated and expressed in *Trichoderma reesei* [84]. Opportunities also exist for protein engineering by site-directed mutagenesis to improve the stability properties and expand the substrate spectrum of conventional enzymes [85]. In a similar manner, genetic recombination could insert a variety of novel enzyme-producing DNAs, primarily from nonmicrobial sources, into *Aspergillus* to be used as protein-manufacturing cultures for the production of low-cost enzyme.

REFERENCES

1. Wood, B. J. B., and Min, Y. F. Oriental food fermentations. In *The Filamentous Fungi*, Vol. 2 (J. E. Smith and D. R. Berry, eds.), John Wiley and Sons, New York, 1975, pp. 265-280.
2. Godfrey, T., and Reichelt, J. In *Industrial Enzymology: The Application of Enzymes in Industry*. Nature Press, New York, 1983.
3. Hepner, L., and Male, C. Industrial enzymes: present status and opportunities. In *Enzyme Technology* (R. M. Lafferty, ed.), Springer-Verlag, New York, 1983, pp. 7-8.
4. *Biochemicals: Organic Compounds for Research and Diagnostic Reagents*. Sigma, St. Louis, 1988.
5. Boehringer Mannheim Biochemicals, Indianapolis, 1987-1988.
6. Calbiochem Biochemical/Immunochemical Catalog, Boehringer Diagnostics, San Diego, 1987.
7. *Catalogue of Fungi/Yeasts*, 17th ed. American Type Culture Collection, Rockville, 1987.
8. *Food Chemicals Codex*, 2nd ed., 1st Suppl. National Academy of Sciences, Washington, D.C., 1974.
9. Vandamme, E. J. Immobilized enzyme and cell technology to produce peptide antibiotics. In *Enzyme Technology* (R. M. Lafferty, ed.), Springer-Verlag, New York, 1983, pp. 237-270.
10. Oestergaard, J., and Knudsen, S. L. Use of Sweetzyme in industrial continuous isomerization. *Die Starke, 28*:350-356, 1976.
11. Klibanov, A. M. Immobilized enzymes and cells as practical catalysts. *Science, 219*:722-727, 1983.
12. Dixon, M., and Webb, E. C. Enzyme techniques. In *Enzymes*, 3rd ed. (M. Dixon and E. C. Webb, eds.), Academic Press, New York, 1979, pp. 7-22.
13. Arima, K. Microbial enzyme production. In *Global Impacts of Applied Microbiology* (M. P. Starr, ed.), John Wiley and Sons, New York, 1964, pp. 277-294.
14. Yoshida, F., and Ichishima, E. Method of producing a proteolytic enzyme by use of black Aspergillus type molds. U.S. Patent 3,149,051, 1964.
15. Underkofler, L. A. Development of a commercial enzyme process: glucoamylase. In *Cellulases and Their Applications* (R. F. Gould, ed.), American Chemical Society, Washington, D.C., 1969, pp. 343-358.
16. Toyama, N. Feasibility of sugar production from agriculture and urban cellulosic waste with Trichoderma viride cellulase. *Biotechnol. Bioeng. Symp., 6*:207-219, 1976.

17. Beckhorn, E. J., Labbee, M. D., and Underkofler, L. A. Production and use of microbial enzymes for food processing. *J. Agr. Fd. Chem.*, *13*:30-34, 1965.
18. Lambert, P. W., and Meers, J. L. The production of industrial enzymes. *Phil. Trans. Roy. Soc. Lond.*, *B300*:263-282, 1983.
19. Zabriskie, D. W., Arminger, W. B., Phillips, D. H., and Albano, P. A. *Traders' Guide to Fermentation Media Formulation*, Traders Oil Mill Co., 1980.
20. Lowe, D. A., Romancik, G., and Elander, R. P. Enzymatic hydrolysis of penicillin V to 6-aminopenicillanic acid by Fusarium oxysporum. *Biotechnol. Lett.*, *8*:151-156, 1986.
21. Boss, F. C. Filtration. In *Fermentation and Biochemical Engineering Handbook* (H. C. Vogel, ed.), Noyes, New Jersey, 1983, pp. 163-174.
22. Boss, F. C. Centrifugation. In *Fermentation and Biochemical Engineering Handbook* (H. C. Vogel, ed.), Noyes, New Jersey, 1983, pp. 296-316.
23. Drioli, E. Membrane processes in the separation, purification, and concentration of bioactive compounds from fermentation broths. In *Separation, Recovery and Purification in Biotechnology* (J. A. Asenjo and J. Hong, eds.), ACS Symposium Series 314, American Chemical Society, Washington, D.C., 1986, pp. 52-66.
24. Street, G. Large scale industrial enzyme production. *Crit. Rev. Biotechnol.*, *1*:59-85, 1983.
25. Volesky, B., and Luong, J. H. T. Microbial enzymes: production, purification and isolation. *Crit. Rev. Biotechnol.*, *2*:119-146, 1985.
26. Aschengreen, N. H. Microbial enzymes for alcohol production. *Proc. Biochem.*, *4*:23-25, 1969.
27. Strenberg, M. Removal of transflucosidase from amyloglucosidase. U.S. Patent 3,917,513, 1975.
28. Smith, J. A., and Frankiewicz, J. R. Process for preparing glucoamylase. U.S. Patent 4,169,013, 1979.
29. Beckhorn, E. J. Production of microbial enzymes. In *Microbial Technology* (H. J. Peppler, ed.), Reinhold, New York, 1967, pp. 366-380.
30. Belloc, A., Florent, J., Mancy, D., and Verrier, J. New amylase and its preparation by fermentation of a Penicillium. U.S. Patent 3,906,113,1975.
31. Harrison, J. G., and Rowsell, J. E. Glucoamylase in modern brewing. *Proc. Biochem.*, *5*:37-39, 1970.
32. Marshall, J. J. Starch-degrading enzymes derived from Cladosporium resinae. U.S. Patent 4,211,842, 1980.
33. Plant, A. R., Morgan, H. W., and Daniel, R. M. A highly stable pullulanase from Thermus aquaticus YT-1. *Enz. Microb. Technol.*, *8*:668-672, 1986.
34. Marsden, W. L., and Gray, P. P. Enzymatic hydrolysis of cellulose in lignocellulosic materials. *Crit. Rev. Biotechnol.*, *3*:235-276, 1986.
35. Skinner, W. A., and Tokuyama, F. Production of cellulase by a thermophilic Thielavia terrestris. U.S. Patent 4,081,328, 1978.
36. Komura, I., Awao, T., and Yamada, K. Thermostable cellulase and a method for producing the same. U.S. Patent 4,106,989, 1978.
37. Durand, H., Clanet, M., and Tiraby, G. Genetic improvement of Trichoderma reesei for large scale cellulase production. *Enz. Microb. Technol.*, *10*:341-346, 1988.

38. Wood, T. M., and McRae, S. I. Purification and some properties of a 1,4-beta-D-glucan glucohydrolase associated with the cellulase from the fungus Penicillium funiculosum. *Carbohydr. Res.*, *110*:291-303, 1982.
39. Gunde-Cimerman, N., and Cimerman, A. Aspergillus niger mutants for bioconversion of apple distillery wastes. *Enz. Microb. Technol.*, *8*:166-170, 1986.
40. Watanabe, M., Narita, S., Kagaya, I., Miura, M., Suzuki, Y., Tanaka, S., Sasaki, M., Miyatani, S., and Terayama, K. Method for production of alpha-galactosidase. U.S. Patent 3,894,913, 1975.
41. Narita, S., Naganishi, H., Yokouchi, A., and Kagaya, I. Method for manufacture of alpha-galactosidase by microorganism. U.S. Patent 3,957,578, 1976.
42. Arnaud, N., and Bush, D. A. Alpha-galactosidase production. U.S. Patent 3,966,555, 1976.
43. Enkelund, J. Externally added beta-glucanase. *Proc. Biochem.*, *7*:27-29, 1972.
44. James, A. E., Fare, G., Sagar, B. F., Lucas, F., and Degray, M. I. Beta-1,4/beta-1,3 glucanase. U.S. Patent 3,880,742, 1975.
45. Hjortshoj, K., and Aunstrup, K. Beta-glucanase. U.S. Patent 4,110,163, 1978.
46. Scott, R. "Rennets" and cheese. In *Topics in Enzyme and Fermentation Biotechnology*, Vol. 3 (A. Wiseman, ed.), John Wiley and Sons, New York, 1979, pp. 109-169.
47. Dohan, L. A., Baret, J. L., Pain, S., and Delalande, P. Lactose hydrolysis by immobilized lactase: semi-industrial experience. In *Enzyme Engineering*, Vol. 5 (H. H. Weetall and G. P. Royer, eds.), Plenum Press, New York, 1979, pp. 279-293.
48. Prenosil, J. E., Stuker, E., Hediger, T., and Bourne, J. R. Enzymatic whey hydrolysis in the pilot plant "Lactohyd." *Biotechnology*, *2*:441-444, 1984.
49. Goldberg, B. S. Method of immobilizing proteinaceous substance. U.S. Patent 4,169,014.
50. Amerace Corporation. APH Pilot Reactor for Hydrolysis of Whey Permeate. New Product Bulletin 022283, Amerace Corporation, New Jersey, 1985.
51. Kiuchi, A., and Tanaka, Y. Process for preparing beta-galactosidase. U.S. Patent 3,919,049, 1975.
52. Crisan, E. V., and Sorensen, S. G. Production and utilization of lactase. U.S. Patent 4,007,283, 1977.
53. Fogarty, W. M., and Ward, O. P. Pectic substances and pectinolytic enzymes. *Proc. Biochem.*, *7*:13-17, 1972.
54. Robbins, R. H. Clarification of fruit juices. *Proc. Biochem.*, *3*:38-40, 1968.
55. Ishii, S., Kiho, K., Sugiyama, S., and Sugimoto, H. Novel pectin esterase, process for ots production, and process for producing demethoxylated pectin by the use of said pectin esterase. U.S. Patent 4,200,694, 1980.
56. Glasziou, K. T., and Slack, C. R. Biological products. U.S. Patent 3,875,009, 1975.
57. Reese, E. T., Magaire, A., and Parrish, F. W. Alpha-1,3-glucanases of fungi and their relationship to mycodextranase. *Ferment. Technol. Today*, 735-742, 1972.
58. Aunstrup, K. Preparation of a milk-coagulation enzyme. U.S. Patent 3,988,207, 1976.

59. Kokusho, Y., Higashi, S., Machida, H., and Iwasaki, S. Process for improving the quality of microbial rennet. U.S. Patent 3,950,221, 1976.
60. Lehmann, R., Pfeiffer, H. F., Schindler, J., and Schreiber, W. Process of producing acid stable protease. U.S. Patent 4,062,732, 1977.
61. Nakao, Y., Nara, K., Katamoto, K., Ito, K., and Ikeda, S. Preparation of a dustless, powdery, protease-containing composition. U.S. Patent 3,846,240, 1974.
62. Craveri, R., Manachini, P. L., Monzese, C., and Aragozzini, F. Process for the manufacture of proteases. U.S. Patent 3,813,319, 1974.
63. Feldman, L. I., and Dooley, J. G. Lipolytic enzyme flavoring system. U.S. Patent 4,065,580, 1977.
64. Tahoun, M. K., and Ali, H. A. Specificity and glyceride synthesis by mycelial lipases of Rhizopus delemar. *Enz. Microb. Technol.*, *8*:429-432, 1986.
65. Blain, J. A. Industrial enzyme production. In *The Filamentous Fungi*, Vol. 1 (J. E. Smith and D. R. Berry, eds.), John Wiley and Sons, New York, 1975, pp. 193-211.
66. Rosner, L., and Foster, R. O. Method for detecting the presence of glucose in cervical mucus. U.S. Patent 3,116,223, 1963.
67. Rogalski, J., Fiedurek, J., Szczordrak, J., Kapusta, K., and Leonowicz, A. Optimization of glucose oxidase synthesis in submerged cultures of Aspergillus niger G-13 mutant. *Enz. Microb. Technol.*, *10*:508-511, 1988.
68. Reed, G. *Enzymes in Food Processing*. Academic Press, New York, 1966.
69. Lee, H. Y., and Tsao, G. T. Oxygen electrodes. *Adv. Biochem. Eng.*, *13*:35-86, 1979.
70. Stoppok, E., Wagner, F., and Zadrazil, F. Identification of a penicillin V acylase processing fungus. *Eur. J. Appl. Microb. Biotechnol.*, *13*: 60-61, 1981.
71. Schneider, W. J., and Roehr, M. Purification and properties of penicillin acylase of Bovista plumbea. *Biochem. Biophys. Acta*, *452*:177-185, 1976.
72. Brandl, E., Kleiber, W., and Knauseder, F. Process for the production of 6-aminopenicillanic acid. U.S. Patent 3,737,375, 1973.
73. Kluge, M., Klein, J., and Wagner, F. Production of 6-aminopenicillanic acid by immobilized Pleurotus ostreatus. *Biotechnol. Lett.*, *4*:293-296, 1982.
74. Gestrelius, S. Immobilized penicillin V acylase. Development of an industrial catalyst. *Appl. Biochem. Biotechnol.*, *7*:19-21, 1982.
75. Novo Enzyme Information No. 236: Novozym 217, immobilized penicillin V acylase. Novo Industri A/S, Denmark, 1982.
76. Haagensen, P., Karlsen, L. G., Petersen, J., and Villadsen, J. The kinetics of penicillin V deacylation on an immobilized enzyme. *Biotechnol. Bioeng.*, *25*:1873-1895, 1983.
77. Karlsen, L., and Villadsen, J. Optimization of a reactor assembly for the production of 6-APA from penicillin V. *Biotechnol. Bioeng.*, *26*:1485-1491, 1984.
78. Bunni, L., McHale, L., and McHale, A. P. Production, isolation and partial characterization of an amylase system produced by Talaromycese emersonii CBS 814.70. *Enz. Microb. Technol.*, *11*:370-375, 1989.

79. Yu, E. K. C., Tan, L. U. L., Deschatelets, M. K.-H. C. L., and Saddler, J. N. Production of thermostable xylanase by a thermophilic fungus. Thermoascus aurantiacus. *Enz. Microb. Technol.*, *9*:16-24, 1987.
80. Ogawa, K., Ohara, H., and Toyama, N. Interspecific hybridization of Aspergillus awamori var. kawachi and Aspergillus oryzae by protoplast fusion. *Agr. Biol. Chem.*, *52*:1985-1991, 1988.
81. Upshall, A. Filamentous fungi in biotechnology. *Biotechniques*, *4*:158-166, 1986.
82. Esser, K., and Mohr, G. Integrative transformation of filamentous fungi with respect to biotechnological applications. *Proc. Biochem.*, *21*:153-159, 1986.
83. Christensen, T., Woeldike, H., Boel, E., Mortensen, S. B., Hjortshoej, K., Thim, L., and Hansen, M. T. High level expression of recombinant genes in Aspergillus oryzae. *Biotechnology*, *6*:1419-1422, 1988.
84. Harkki, A., Uusitalo, J., Bailey, M., Pettia, M., and Knowles, J. K. C. A novel fungal expression system: secretion of active calf chymosin from the filamentous fungus Trichoderma reesei. *Biotechnology*, *7*:596-603, 1989.
85. Tatsumi, H., Ohsawa, M., Tsuji, R. F., Murakami, S., Nakano, E., Motai, H., Masaki, A., Ishida, Y., and Murakami, K. Cloning and the sequencing of alkaline protease from Aspergillus oryzae. *Agr. Biol. Chem.*, *52*:1887-1888, 1988.

23

THE ROLE OF THERMOPHILIC FUNGI IN AGRICULTURE

H. S. SHEKHAR SHARMA *Department of Agriculture for Northern Ireland, Belfast, Northern Ireland*

B. N. JOHRI *G. B. Pant University of Agriculture and Technology, Pantnagar, India*

I. INTRODUCTION

Temperature plays an important role in influencing the growth, development, metabolic activity, and distribution of living organisms on the face of the earth. The fungi, like all other groups of microorganisms, have during their evolution become adapted to survive in a wide range of temperatures, which by no means is as wide as for the prokaryotes. Fungi can be classified as psychrophiles, mesophiles, and thermophiles, but if temperature characteristics for growth, i.e., maximum, optimum, and minimum, are listed, the profiles will fall on the uninterrupted cline (Fig. 1). By far the majority of fungi are mesophiles growing between 5 and 37°C and psychrophiles extend below that range of temperature. Many definitions of thermophilic fungi have been attempted but the one given by Cooney and Emerson [1] has been used quite extensively. According to these workers, a thermophilous fungus is one that has a maximum temperature for growth at or above 50°C and a minimum for growth at or above 20°C. This definition is based on the fact that there are a few fungi that grow at temperatures above 50°C and that some of these are unable to grow at 20°C. However, definition based on optimum temperature for growth at or above 40°C has been used by Crisan [2,3]. The thermophilic fungi owe their ubiquity and common occurrence to this inherent ability to occupy a temperature niche which most other fungi cannot exploit [4]. There is a small group of species described thus far that has become especially active at temperatures where competition from mesophiles would appear to be minimal.

II. HISTORY

As in many a pioneering study, the first two thermophilic fungi to be identified were chance contaminants on organic substrates incubated at high temperatures. Lindt [5] reported *Mucor pusillus* on bread whereas Tsiklinskaya [6] found *Humicola lanuginosa* on a potato disk which was inoculated with soil particles. Later Miehe's [7] investigation of self-heating

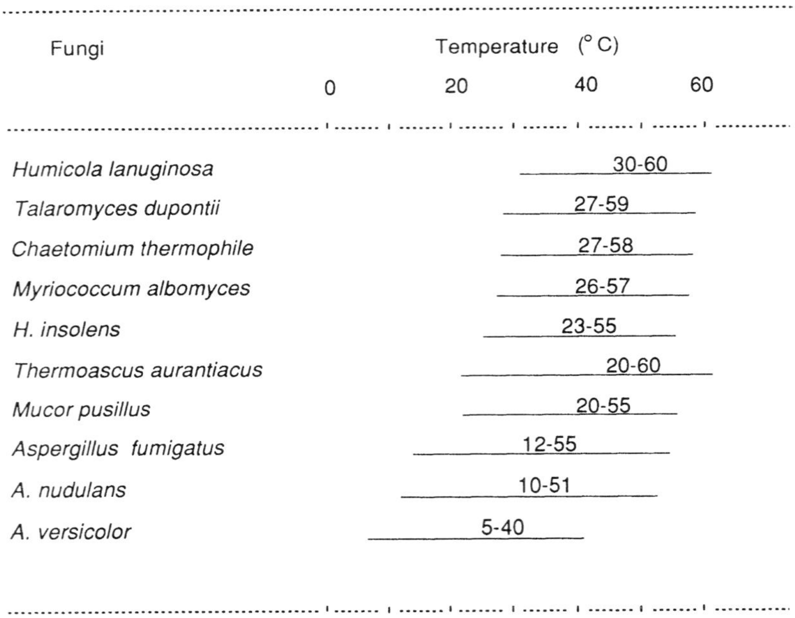

FIGURE 1. Cardinal temperatures for growth of some thermophilic and thermotolerant fungi.

hay produced the first extensive report on thermophily in fungi. Until the publication of their monograph in 1964 by Cooney and Emerson [1], the book *Die Selbsterhitzung des Heus* was the only authoritative publication on the biology of thermophilic fungi. Miehe [7-9] isolated and studied a range of thermophilic fungi, which among others included *Thermoascus aurantiacus* and *Malbranchea pulchela* var. *sulfurea*. Griffon and Maublanc [10] were the first to describe *Talaromyces dupontii*, which until recently remained the only thermophilic member of the genus *Penicillium*. However, the complicated nature of thermophily in fungi was only beginning to be realized [11,12]. La Touche [13] reported a new species of thermophilic fungus *Chaetomium thermophile*, which generated much interest due to its cellulolytic nature and ushered in a new period of intensive research on different aspects of fungal thermophily. Cooney and Emerson [1] reported on the morphology, taxonomy, biology, and economic importance of the thermophilic fungi. Other aspects of the thermophilic fungi were later reviewed by Emerson [4], Crisan [14], and Johri [15]. An updated list of thermophilic fungi is given in Table 1. Thermophilic yeasts have also been reported but their growth temperatures do not fall within the generally accepted range of thermophilic fungi. Also, extensive new reports on the application of thermophilic fungi in agriculture appeared which have prompted a review to provide a concise overview on the role of thermophilic fungi in agriculture.

TABLE 1. Thermophilic Fungi, Their Cardinal Temperatures and Substrates

Fungi	Cardinal temp. of growth (°C)			Substrate	Source
	Min.	Opt.	Max.		
Zygomycetes					
Absidia blakesleeana	~25	~37	~50	PK	134
A. ramosa	~25	~37	~50	PK	134
Mucor miehei	24-25	35-45	55-57	HG, AD	1,2,135,136,137
M. pusillus	20-27	35-55	55-60	CT, B, P, TL	1,2,133,135
Rhizopus sp.	28	48-59	60		137
Rhizopus sp.	25	45-52	60	CT	135,136
Ascomycetes					
Allescheria terrestris	22	42-45	55	CT, AS	136,139
Byssochlamys sp.	30	50	60	GC	1
Chaetomium thermophile	25-27	50	58-60	MC, RC	1,13,33
C. thermophile var. *coprophile*	25-28	45-55	58-60	GC, S, CT	1,2,136
C. thermophile var. *dissitum*	25-28	45-50	58-60	HG, AD, BN	1,2,136
C. fusisporale	~25	~37	~50	PK	134
Corynascus thermophilus	ND	45	ND	RC	122
Dactylomyces crustaceus	20	37	55	CT	136
Myriococcum albomyces	25-30	40-45	55-60	MC, CT	1,2,135
Talaromyces Byssochlamydoides	25	40-45	50		140

TABLE 1. (Continued)

Fungi	Cardinal temp. of growth (°C)			Substrate	Source
	Min.	Opt.	Max.		
T. emersonii	25-30	40-50	57-60	B, CT	2,136,141
T. thermophilus	25-30	45-50	57-60	CT, MC, GN	2,136,141
T. duponti	25	45-50	60	TS, GN	1,118
Thermoascus aurantiacus	20-35	40-46	55-65	HG, GC, P, CT	1,2,136,141,142
T. crustaceus	~20	37	55	CT, SM, GN	136,141,142,143,144
Thielavia australiensis	ND	45	ND	RC	123
T. thermophila	25	43	55	MC	45,145,146
Basidiomycetes					
Coprinus sp.	~20	~45	55	MC	147
Deuteromycetes					
Aspergillus fumigatus	12-20	37-43	52-55	CT, P, MC, RC	1,135,136,138
Acremonium alabamensis	ND	45	ND	TS	123
Calcarisporiella thermophilia	ND	45	ND	TS	123
Chrysosporium sp.	25	40-45	55	CT	135,136
Humicola grisea var. *thermoidea*	20-24	38-46	55-56	GC, AD, CT, GN	1,2,136
H. insolens	20-23	35-45	55	MC, AD, CT	1,2,136
H. lanuginosa	28-30	45-55	60	HG, GC, AD, RC	1,2,33

H. stellata	24	40	50	HG	1,148
Malbranchea pulchella var. *sulfurea*	25-30	45-46	53-57	HG, AD	1,2,136
Melanocarpus albomyces	ND	~45	ND	RH, PE	129
Myceliophthora thermophilum	ND	45	ND	TS, RC, RH	129
Paecilomyces sp.	30	45-50	55-60	P, PK	134,138
Penicillium sp.	20	42	55	CT	135,136
Scytalidium thermophilum	ND	45	ND	B, RC, PE	17,123
S. indonesiacum	18-21	45	55	TS	149
Sporotrichum thermophile	18-24	40-50	55	AD, AS, CT	2,136,139,146
Stilbella thermophila	~25	35-50	~55	MC	20,42
Thermomyces ibadanensis	31-35	42-47	60-61	PK, RC	123,150,151
T. lanuginosus	20-25	47	52	MC, TS	42,118
Thermoidium sulphureum	25	45	55	CT	136
Torula thermophila	23	35-45	58	MC, AD, BN, RC	3,33,42
Mycelia sterila					
Papulospora thermophila	29	~45	52	MC	152

Abbreviations: AD = animal dung; AS = alluvial soil; B = bagasse; BN = bird-nesting materials; CP = cocoa pods; CT = coal spoil tips; GC = garden compost; GN = ground nut; HG = hay and grass; MC = mushroom compost; ND = not determined but other data indicate possible thermophilism; P = peat; PK = palm kernals and oil; PE = palm oil effluent; RC = refuse compost; RH = rice heads; RM = rotten maize; TS = tropical soil; TL = tobacco leaf; S = straw; SM = salt marshes.

III. COLONIZATION AND HEATING OF AGRICULTURAL PRODUCE

A wide range of fungi, bacteria, and actinomycetes are usually present on agricultural produce. The rate of their growth depends on the conditions of storage, particularly the water activity of the substrate, aeration, ambient temperature, and nutritional factors.

If the energy released during the respiration of plant and microbial cells cannot escape as quickly as it is produced, the substrate heats up and conditions within may favor growth of thermophilic organisms. Detailed studies have been carried out with baled hay [16], baled sugarcane bagasse [17], grains [18,19], mushroom compost [20-22], retting guayule [1] and wood chip piles [23] with respect to colonization by thermophilic fungi.

But again, fungal colonization is affected by the water activity, aeration, temperature, nutrition, pH, and type of substrate [24]. These variables are considered below in some detail.

A. Water Activity

The water activity (a_w) of a material is the ratio of the vapor pressure of water over the substance to the vapor pressure over pure water at the same temperature [25]. For any particular substrate, there is a relationship between its water content and the activity of its water [26]. Numerically a_w is the same as equilibrium relative humidity but it is expressed as a proportion of one rather than as a percentage. As changes in temperature affect the vapor pressure over the substance and over water approximately proportionally, a_w over the substance does not change very much with change in temperature. The changes which do occur are usually in the direction of increases in a_w with increasing temperature at constant moisture content.

The extent of microbial activity a substrate can support, is determined by the water activity and consequently how it influences the growth of a particular fungal species and the rate of energy release and heating. Most species of fungi require at least $0.75a_w$ while no fungal growth occurs with less than $0.61a_w$ [27]. A material of water activity $0.75a_w$, for example, is in equilibrium with air of relative humidity 70%.

B. Aeration

Aerobic respiration releases energy from microorganisms and plant cells and consequently heats up the substrate. Experiments carried out with different substrates have shown that heating ceases as soon as O_2 is exhausted [28-31]. The requirement for oxygen increases with temperature to a maximum at about 40°C but does not decrease greatly until near the lethal growth at a temperature of 65°C. The heat generated above 65°C results from exothermic chemical reactions [28,32]. Henssen [22] concluded that thermophilic fungi can undergo growth as facultative aerobes on the basis of the growth of the test fungi in sealed containers together with alkaline pyrogallol. The growth of fungi was stopped in the presence of 100% nitrogen and the fungi started sporulating only in the presence of oxygen [33,34]; most fungi require at least 0.2% for trace growth, 0.7-1.05 for moderate growth, and 1.0% for sporulation.

C. Temperature

Ayerst [35] reported that temperature and water activity interact to determine the growth range of individual species of mesophilic isolates of *Aspergillus* spp. However, there are few or no comparable data for thermophilic fungi. Hussain [36] showed that hay (moisture content 50%) inoculated with thermophilic bacteria and actinomycetes would increase in temperature from 30 to 55-70°C. However, hay inoculated with the bacteria alone continued to heat to 260°C only after autocatalytic chemical process. Heating of substrate by the activity of the microbes is limited to a maximum of 65-70°C and, unless spontaneous ignition occurs, the maximum temperature of heating is closely related to the water content [17].

D. Nutrition

Rapid growth of thermophilic microorganisms which leads to self-heating depends on the presence of soluble organic substances in the substrate. Available substrates such as starch, cellulose, and lignocellulose may be utilized by thermophilic microorganisms including fungi [16,17,37-42].

Festenstein [43] studied the changes in the carbohydrate and nitrogen content of hay at 40% moisture during a 9-day incubation period. Total nitrogen appeared to increase as a result of loss in dry matter, but the proportion that was soluble decreased steadily during the breakdown. However, the proportion of volatile nitrogen increases as a result of proteolysis and deamination of amino acids. The lipid content usually decreases as hay breaks down [17,43,44].

Thermophilic fungi are tolerant of a range of pH with most species growing between pH 4 and 8. The pH of hay at baling is about 6.5 and primary colonists like *Mucor miehei* and *M. Pusillus* both have pH optima close to the pH of hay at baling; secondary colonists, e.g., *Humicola lanuginosa*, *Malbranchea pulchella* var. *sulfurea*, and *Talaromyces thermophilus* have pH optima close to neutrality and occur late in succession.

Allescheria terrestris, *Chrysosporium pruinosum*, and *Talaromyces emersonii* can grow in acidic environment (pH 3.4-6.1) of bagasse and have a pH optima of approximately 3 [45].

Other factors such as speed, degree, and persistence of heating may be influenced by the size and density of the mass of material [46].

IV. COLONIZATION OF SUBSTRATES

Agricultural produce like grain and hay are colonized by storage fungi. The storage fungi do not invade grains to any appreciable degree before harvest [47,48]. This is well established from tests with different batches of cereal grains from different areas over a long period of time.

A. Grain

Since the pioneering farmers of Neolithic times, human society and security have depended on the storage of food products in time of plenty for use in time of need, and man has been struggling to protect his supplies of food and fodder from the inevitable attacks of pests of one type or another [4]. Moldy corns and hay were well known in Roman times and directions for preventing spontaneous heating and combustion of agricultural products are to be found in the classical treatises of the first century A.D. [49]. The cause of deterioration and spoilage of corn and corn meal was also exhaustively studied by McHargue [50]. Darsie et al. [51] determined the amount of heat produced by germinating seeds, which have higher moisture contents and rates of respiration than stored seeds. They concluded that heat produced from germinating seeds in Dewar flasks raised the temperature within the bottle by only 1-3°C, whereas the samples of seed overgrown with fungi increased 10°C in temperature within 4-5 days. Ramstad and Geddes [52] proved that heating of stored grains was caused by fungal colonization and not respiration of moist grain itself; this fact was later confirmed by several other workers [28,29,53-57]. Thermophilic fungi such as *Absidia* sp., *Aspergillus* spp., *Thermomyces lanuginosus*, *M. pusillus*, and *Rhizomucor* spp. were isolated from grains [55,58,59]. Okafor [58] also isolated thermophiles *T. lanugenosus*, *M. pusillus*, and *Rhizomucor* sp. from rotting maize cobs.

Walker [60] was the first to determine the important role of water in spontaneous combustion of grains. The heating process in storage bins undergoes the following sequence of events: damp-stored grain can become colonized by *Aspergillus restrictus* and other fungi which grow slowly, probably without affecting the temperature or moisture. *A. glaucus* can grow rapidly, provided the increase in temperature reaches 35-40°C; this is associated with a rise in moisture content of the colonized grains. When the moisture content exceeds 15-15.5%, *A. candidus* takes over and under the right conditions can increase the moisture content and temperature of the grain rapidly. When the moisture present in the seeds reaches 18.5%, *A. flavus* can grow. Both *A. candidus* and *A. flavus* can push the temperature of the grain to 55°C and maintain that temperature for weeks. If this condition is maintained depending on the increase in moisture content, the next stage of heating begins involving thermophilic bacteria and fungi. By this stage free water may be available for the microorganisms and may carry the temperature to 75°C, after which autocatalytic process may take over and increase the temperature to the point of ignition. The final stage of spoilage of grain can be detected when the organism involved can be detected as mustiness or caking or total decay.

The problem of storage of moist barley grain may be resolved to some extent by formic acid or other fungicide treatments. The treated samples produced much reduced population of thermophilic or other types of mold [61]. The respiration and heating of damp wheat seeds could be controlled by reducing the moisture content of the seed from about 20 to 12% which in turn would reduce the colonization of grains by thermophilic fungi [62].

Spoilage of grains can also be prevented by storing moist grain in sealed stores [53,63]. Because of the absence of O_2 and the high concentration of CO_2, moldiness and self-heating is reduced to a minimum; the stored moist grain retains a perfect appearance for several months. However, in large-scale trials the moist grain in large, sealed commercial storage showed general increases in yeast population and some deterioration of grains in the surface layers [53,63]. This method of storing grains would be better in a perfectly sealed grain silo [64].

B. Hay

It has been well known for many centuries that damp hay may set a barn on fire and there has been continuing controversy during this time concerning the cause of spontaneous combustion of stored plant materials. The spontaneous heating and ignition of hay led to the first detailed study of thermophilous fungi by Miehe [7]. Many theories have been put forward by various workers to explain this phenomena: friction of materials, vital heat, chemical theory, low-temperature carbonization, heat-resistant enzymes, catalytic oxidation, and pyrophoric iron. Browne [49] reported in an extensive review all the different theories. Cohn [65,66], was the first to draw a distinction between primary production of heat (thermogenesis) by the living cells and secondary production of heat by saprophytic microorganisms (microbial thermogenesis). Later Miehe [7,8] made a detailed investigation to elucidate self-heating of hay and to differentiate between the role of mesophilic and thermophilic microorganisms. He reported that damp, packed hay or leaves self-heated to 55 or 60°C within one to several days.

Similar hay that was steamed, reducing the microbial population to a minimum, did not heat. However, steamed hay reinoculated with a soil suspension or washings from untreated hay did heat as also did hay inoculated with various pure cultures of microorganisms prevalent on self-heated hay. These experiments convinced Miehe that heating of moist plant materials to 60-70°C was caused by microorganisms and he isolated a range of thermophilic fungi, bacteria, and actinomycetes. Miehe [28] further demonstrated that mesophiles such as *Rhizopus nigricans* or *Penicillium glaucum* could only raise the temperature to 35 or 40°C whereas thermophilic fungi like *Malbranchea pulchella* var. *sulfura* (*thermoidium*) or *Humicola lanuginosa* (*thermomyces*) raised the temperatures to 60°C or above. This was later verified by Christensen and Gordon [55].

Thus the series of events leading to self-heating and ultimate spontaneous combustion of stored plant materials is now generally agreed on [57,67-70]. The temperature rises quickly after the initial buildup of heat from the metabolism of the mixed saprophytic microbial flora, often in a matter of hours depending on the size of the material and availability of nutrients, and soon passes the optimum for mesophilic forms.

The thermophilic fungi and actinomycetes take over, multiply rapidly, and raise the temperature to the peak (70°C or slightly higher) that can be reached by microbial activity. At this stage the level of soluble sugars decreases rapidly and the pH value rises to 7 or above [16]. Autocatalytic chemical process takes over and causes further heating leading to ignition if the conditions are maintained [1]. The microbiological evidence from samples has been taken into account in legal cases where criminal incendiary action is suspected [71]. However, samples from the suspected fire need to be taken very quickly on site; otherwise changes in the microbial population of thermophiles could be affected. The only satisfactory way to cut losses from overheating or combustion is to store hay or straw at moistures below 19% or less [68].

C. Other Substrates

A wide range of agricultural produce, e.g., groundnuts, wood chips, bagasse, wool, cotton, hemp, flax fibers, and straw, undergo self-heating when they are moist and improperly stored. The colonization of these substrates associated with self-heating follows the same pattern as grains or hay, although the species of microorganisms involved may be different. This may be due to differences in pH and nutrients available in different substrates; bagasse has acid pH and high sugar content while hay comes near neutral pH and high carbohydrate content.

The microbial colonization and heating of stored materials is usually associated with loss in quality and nutritional value. It can also lead to potential health hazards to workers handling such materials and to persons and animals consuming them. Spoiled grain for seed or malting may have low viability while baking qualities are adversely affected by heating above 40°C [72] and the consequent reduction of dry matter and major, minor, and trace nutrients together with loss of digestibility of proteins [73,74].

Other possible health hazards include infection, allergy, and mycotoxicosis. A number of thermophilic fungi have been reported as causing diseases to warm-blooded animals including man. *Mucor pusillus* has been associated with measles in cattle and with a wide variety of mycoses. *Aspergillus fumigatus* is well known as causing aspergillosis in birds, especially poultry, but also in gulls and ostriches. Aspergillosis is a respiratory disease and, when severe, the lumen of the trachea and the alveoli become completely lined with bluish green conidial heads. Spores of *A. fumigatus* can cause asthma and other thermophilic actinomycetes have been implicated in allergic alveolitis [46,75,76].

V. COMPOSTING OF PLANT AND ANIMAL WASTES

In nonintensive farming the animal acts as its own disposal system distributing the waste about the land on which it grazes. Treatment is brought about in the soil and nutrients recycled via the food chain. The land does not become enriched and no pollution problems exist. However, with the introduction in the late 1960s of intensive production of cereals and animals, large quantities of wastes are concentrated in the countryside. Probably the largest agricultural waste is that from cereal production [77]. In 1972, about a third of the total crop of 9 million tons was surplus straw and a substantial part of this was burned in the field [78].

Methods for producing compost were worked out in detail long before the scientific explanation of the process was elaborated. Composting is a process involving conversion of organic residues into lignoprotein complexes (humus) via thermophilic organisms under optimum moisture and aeration. In the process, CO_2 is evolved and the temperature of the pile may reach 60-80°C. The physicochemical and microbiological principles of the process have been extensively studied and reviewed [79-81].

The lignocellulosics are largely deficient in nitrogen (Table 2) and microorganisms are better able to colonize substrates where nitrogen is not deficient. A cheap source of nitrogen is usually required as a supplement and animal waste (Table 3) not only provides this but enables a satisfactory means for disposal of these waste products of agriculture. Microorganisms and in particular the fungi offer many advantages requiring very simple conditions for composting plant and animal wastes [82]. Fungi are able to grow under semisolid conditions and ramify through the solid substrate, colonizing it very quickly [83]. In the process fungal mycelium can penetrate the straw and translocate the nutrients [84].

Early studies by Waksman and Gerretsen [85] on the process of composting indicated that lignin and cellulose were degraded at high rates by mixed microbial populations as incubation temperatures were raised. Rosenburg [42] screened a wide range of thermotolerant and thermophilic

TABLE 2. Composition of Lignocellulosic Materials (%)

Raw	Cellu-lose	Hemi-cellulose	Lignin	Extrac-tives	Ash	Ref.
Materials						
Spruce wood	43.0	27.0	28.6	1.8	0.4	153
Pine wood	44.0	26.0	27.8	5.3	0.4	153
Birch wood	40.0	39.0	19.5	3.1	0.4	153
Pine kraft pulp	77.0	18.0	5.0	0.2	0.4	153
Bagasse	33.4	30.0	18.9	6.0	2.4	154
Wheat straw	30.5	28.4	18.0	3.5	11.0	154
Rice straw	32.1	24.0	12.5	4.6	17.5	154
Bamboo	ND	19.6	20.1	1.2	3.3	154
Cotton	80-95	5-20	ND	ND	ND	155
Municipal refuse	76	ND	ND	ND	ND	156

ND = no data available.
Source: Data from Refs. 153-156.

TABLE 3. Composition of Animal Wastes (%)

Component	Cattle manure	Swine waste	Broiler litter
Total dry matter	91.6	ND	84
Crude protein	13.5	20	30
True protein	ND	15.6	15
Crude fiber	31.0	14.8	17
Lignin	9.4	—	—
Cellulose	30.3	—	—
Hemicellulose	26.1	—	—
Ether extract	2.8	7.0	3.4

ND = not determined.
Source: Ref. 82.

fungi for their ability to degrade cellulose and lignocellulose. Composting a mixture of agricultural banana plantation waste and chicken litter and forestry residues (pine needles) was carried out on the Canary Islands [86]. During the initial phase of composting, the microorganisms present were mostly bacteria with some fungi. This was followed by an advanced stage of breakdown whereby the population of thermophilic fungi peaked followed by the fungal population receding gradually. These authors did not identify the thermophilic fungi present in the compost pile.

Seal and Eggins [87] investigated composting of a mixture of straw and pig slurry and reported that temperatures of about 50°C in a slightly alkaline pH increased the rate of biodegradation and suppressed fungal pathogens such as *A. fumigatus* and *M. pusillus*. Addition of a mixture of a soil inoculum and 25% of the recycled end product further enhanced biodegradation of straw and slurry. At the end of the process, the products were reduced in bulk and odor and were easier to handle. In the early stages, the temperature inside the compost pile may rise to 80°C. The speed of decomposition by the thermophilic fungi and other microorganisms depended on the nature of the material, water, air, nitrogen, and minerals such as potassium and phosphorus.

A. Heat Production by Composting Organic Matter

The biochemical process of microbial respiration during composting of organic material liberates heat which normally results in a rise in temperature; this is dependent on the sufficient insulation provided by the composting mass. However, this energy is low grade since the temperature of a composting medium may seldom exceed 70°C. Gallardo-Lara and Nogales [79] reviewed the composition of compost produced in different parts of the world and concluded that the components were variable and heterogeneous. Therefore, rates of heat production are different. Energy production by composting pulverized municipal refuse, oat straw, and wool was reported by Wiley [88], Carlyle and Norman [67], and Walker and Harrison [89], respectively. The energy may be useful for such operations as heating greenhouses and brooding chickens.

Mote and Griffis [90] evaluated the energy released from different composting materials and concluded that heat produced by composting gin trash could be effectively used for brooding broiler chickens during winter. In nature, microbial heat released during composting may be used for controlling the temperature of nests during incubation of eggs of mallee fowls or alligators [91,92]. The heat released may help to counteract the cooler nighttime temperature of the swamps.

B. Mode of Substrate Degradation

The efficiency of degradation of plant residues by a particular fungus depends on its capacity to degrade cellulose and lignin since these are not readily degraded. The weight loss of the substrates after degradation ranges from 20 to 30%; the C/N ratio is also decreased by these fungi [93,94]. Chang and Hudson [95] studied the succession of thermotolerant and thermophilic fungi during degradation of wheat straw compost and reported that soluble substrates were utilized followed by cellulose and hemicellulose. The ability of thermophilic fungi to use complex carbon sources and the ability to thrive at high temperature are the two important characteristics of successful colonizers of compost [96]. Tansey [23] studied growth, cellulose decomposition, and cellulase activity of thermophilic fungi and observed that some grew and decomposed cellulose rapidly but their culture filtrates had low cellulase activity against insoluble cellulose. Of the 21 species of thermophilic fungi tested by Rosenburg [42], *Chrysosporium pruinosum* and *Sporotrichum pulverulentum* were found to degrade ligno-cellulose extensively. The role of other enzymes, i.e., proteases and lipases produced by thermophilic fungi, was also reported by Johri et al. [44]. Jain et al. [93] reported that thermophilic fungi degraded both cellulose and lignin present in wheat straw although the extent of degradation varied. Martin and Haider [97] observed that some fungi which degrade lignin actively may not be efficient in humus formation. On the other hand, those which degrade lignin and cellulose may also form humus efficiently [97]. Deacon [98] studied the succession of thermophilic fungi on a modified system of filter paper cellulose and concluded that noncellulolytic fungi such as *T. lanuginosus* could enhance cellulose breakdown when coinoculated with cellulolytic fungus such as *C. thermophile* by using freely available sugars that would otherwise slow down the rate of degradation. He further added that secondary fungal colonizers which become most active later in a succession must be able to tolerate growth-inhibiting substances produced by the primary colonists.

C. Soil Modification with Compost

Microorganisms may contribute to the formation of humus either by extra-cellular transformation of plant and animal constituents into humic compounds or by synthesis of humic substances within the cells. There has been a general increase in humic substances in the final degraded material by all fungi, although some such as *A. fumigatus* and *H. insolens* formed more humic substances than other thermophilic fungi [93]. The end product of composting is usually of variable composition depending on several factors such as climate, type of nutrients, season of the year, quality of composting, and time of maturation.

The beneficial effects of humus in improving soil structure are reflected in increased aeration, reduction in power requirement for ploughing heavy soils, rapid germination of seeds, and reduced volume weight [99]. However, incorporation of some type of compost in the soil may increase the levels of organic pollutants and electrical conductivity of soils [100]. Some humus components may serve as chelates for improving cation uptake by plant roots or may even show hormonelike activity [101].

The use of compost as a fertilizer has resulted in many investigations aimed at assessing the yields of a wide range of crops. The applications of these materials increased yields of sorghum [102], corn [103], potatoes [104,105], tobacco [106], tomatoes [107], beet [108], and ryegrass [109, 110]. The incorporation of organic matter prepared by composting agricultural wastes and low-grade rock phosphate in the soil improved the water-holding capacity and yield of cereal crops [111]. It was also established that crop response appears to be conditioned by the innate soil fertility in such a way that positive increases were obtained with poor soils but not fertile ones [112].

Compost application to soil had a positive effect on the microbial population and rhizosphere microorganisms, and also contributed to the reduction of nematode populations in plants [113-115].

VI. COLONIZATION IN NATURAL HABITATS

Thermophilic fungi are widely present in a varied range of natural habitats such as soils from both cool and hot climates, hot water springs, warm soil, and air.

Earlier work reported thermophilic fungi to be rarely present in soil [2]. However, recent studies suggested that thermophilic fungi may be commonly present in soil [116,117]. These investigations suggest that the fungal propagules are able to survive in a wide range of soil types. Eggins and Malik [116] isolated *A. fumigatus*, *C. thermophile*, *H. grisea*, *H. insolens*, *H. lanuginosa*, *Malbranchea pulchella*, *M. pusillus*, *Myriococcum albomyces*, *Sporotrichum thermophile*, *T. dupontii*, and *T. aurantiacus* from soil samples collected from a site in England, suggesting wide distribution for these fungi. Thermophilic fungi were also encountered in loamy garden soil irrespective of pH. Garden soil rich in humus and animal manure evidently served as a suitable substrate providing, through microbial decomposition, the temperature ranges required for the growth of thermophilic fungi [118]. Most of the thermophilic fungi isolated from soil have been found to be cellulolytic [116,119,120].

Thermophilous molds were also reported from soils of the Mount Yangan-Tau, the burning mountain in the southern Urals [121]. *Dactylaria gallopava* was isolated from acid thermal soils and effluents of acid hot springs of the Yellowstone National Park in the United States [122]. Hedger [117] found much higher population of thermophilous fungi from mud and plant litter from pools in hot springs (40-50°C) in western Java than in soil (20°C) from the surrounding dense tropical rainforest; he attributed this to the marked differences between the sites. The population of thermophiles in the forest soils were no higher than in temperate soils. However, in eastern Java, which has a drier monsoon climate, the populations of thermophiles in soils of the more open teak forests were higher. Hedger proved that under conditions of high insolation growth of thermophilous fungi may be stimulated in tropical countries. Ogbonna and Pugh [123] isolated a maximum number of fungi from soil during the hot season in Nigeria compared to the winter months. A wide range of thermophilic fungi were also isolated from the soils of other tropical countries such as India [124], Malaysia [125], and the Bahamas [126].

The occurrence of a wide range of thermophilic and thermotolerant fungi in the tropics and the role of insolation in the development of thermophiles indicates the potential for causing biodeterioration of particularly root crops in the tropics.

Studies on the incidence of thermophilic fungi in air [124,127,128] showed that *A. fumigatus*, *M. pusillus*, and *T. lanuginosa* were the most commonly occurring fungi. *A. fumigatus* was found to be abundant in the atmosphere, on living leaves and in freshly harvested rice grains [128-131]. In the tropical countries, the common thermophilic fungi were detected throughout the winter and summer months [124]. The common species encountered in air were also frequently isolated from sun-heated soils [132, 133].

Fungal spores present on the surface of freshly harvested rice seeds may proliferate when the seeds are stored, but whether thermophilous fungi are present as a contaminant or whether the fungi are capable of active growth in tropical conditions in the field on seed surfaces needs to be investigated [129]. Preliminary investigations suggest that spores of thermophilic fungi are capable of germinating on green leaves of tomato at 30°C (unpublished data, A. J. Kuthubutheen).

Most thermophilic and thermotolerant fungi are dispersed by airborne spores, which are the major source of inoculum. The ecological investigations may help in understanding the cycle of colonization of soil or plant materials by thermophilic fungi and consequently dispersion of the fungal spores in the air [157].

VII. CONCLUSION

Aerobic thermophiles have a higher rate of digestion of cellulose and possibly lignin, and can grow at temperatures which can kill plant and human pathogens in compost. The evidence seems overwhelming that thermophilic fungi play an important role in the aerobic decomposition and self-heating of composting plant materials. In this chapter, studies dealing particularly with fungi are emphasized even though thermophilic bacteria and actinomycetes also contributed largely to the microbial thermogenesis of plant materials.

REFERENCES

1. Cooney, D. G., and Emerson, R. *Thermophilic Fungi: An Account of Their Biology, Activities and Classification*. W. H. Freeman, San Francisco, 1964, pp. 1-188.
2. Crisan, E. V. The isolation and identification of thermophilic fungi. M.S. thesis, Purdue Univ., 1959, pp. 102-107.
3. Crisan, E. V. Isolation and culture of thermophilic fungi. *Contr. Boyce Thompson Institute Plant Res.*, *22*:291-301, 1964.
4. Emerson, R. 1968. Thermophiles. In *The Fungi: An Advanced Treatise*, Vol. 3 (G. C. Ainsworth and A. S. Sussman, eds.), Academic Press, New York, 1968, pp. 105-128.
5. Lindt, W. Mittheilungen über einige neue pathogene Schimmelpilze. *Arch. Exp. Pathol. Pharmakol.*, *21*:269-297, 1886.
6. Tsiklinskaya, P. Sur les mucedinees thermophiles. *Ann. Inst. Pasteur*, *13*:500-505, 1899.
7. Miehe, H. Über die Selbsterhitzung des Heues. *Arb. Deutsch. Landwirtsch. Gesellsch.*, *111*:76-91, 1905. (Anhang zu: Falke, Friedrich. Die Braunheubereitung zugleich eine Schilderung der gebräuchlichsten Heubereitungsarten. zweite Aufl. Ibid., 111:1-75, 1905.
8. Miehe, H. *Die Selbsterhitzung des Heus Eine biologische. Studie*. Gustav Fischer Jena., 1907, pp. 1-127.
9. Miehe, H. Thermoidium sulfureum n.g.n. sp., ein neuer Wärrmpilz. *Ber. Deutsch. Bot. Gesl.*, *25*:510-515, 1907.
10. Griffon, E. and Maublanc, A. Deux moisissures thermophiles. *Bull. Soc. Mycol. France*, *27*:68-74, 1911.
11. Noack, K. Beiträge zur Biologie der thermophilen Organismen. *Jahrb. Wissensch. Bot.*, *51*:598-648, 1912.
12. Noack, K. Der Betriebstoffwechsel der thermophilen Pilze. *Jahrb. Wissensch. Bot.*, *59*:413-468, 1920.
13. La Touche, C. J. On a thermophilic species of Chaetomium. *Trans. Br. Mycol. Soc.*, *33*:94-104, 1950.

14. Crisan, E. V. Current concepts of thermophilism and the thermophilic fungi. *Mycologia, 65*:1171-1198, 1973.
15. Johri, B. N. Biology of thermophilic fungi. In *Recent Advances in Biology of Microorganisms* (K. S. Bilgrami and K. M. Vyas, eds.), Bishen Singh, Dehra Dun, India, 1980, pp. 265-278.
16. Gregory, P. H., Lacey, M. E., Festenstein, G. N., and Skinner, F. A. Microbial and bio-chemical changes during the moulding of hay. *J. Gen. Microbiol., 33*:147-174, 1963.
17. Lacey, J. Moulding of sugar-cane bagasse and its prevention. *Ann. Appl. Biol., 76*:63-76, 1974.
18. Sinha, R. N., and Wallace, H. A. H. Ecology of a fungus-induced hot spot in stored grains. *Can. J. Plant Sci., 45*:48-59, 1965.
19. Lacey, J. The microbiology of moist barley stored in unsealed silos. *Ann. Appl. Biol., 69*:187-212, 1971.
20. Fergus, C. L. Thermophilic and thermo-tolerant moulds and actinomycetes of mushroom compost during peak heating. *Mycologia, 56*:267-284, 1964.
21. Henssen, A. Beiträge zur Morphologie und Systematik der thermophilien Actinomyceten. *Arch. Mikrobiol., 26*:373-414, 1957.
22. Henssen, A. Über die Bedeutung der thermophilen Mikroorganismen fur die Zerset zung der Stallmistes. *Arch. Mikrobiol., 27*:63-81, 1957.
23. Tansey, M. R. Isolation of thermophilic fungi from self-heated industrial wood chip piles. *Mycologia, 65*:537-547, 1971.
24. Johri, B. N., and Satyanarayan, T. Ecology of thermophilic fungi. In *Progress in Microbial Ecology* (K. G. Mukherji, V. P. Singh, and R. P. Singh, eds.), Print House, Lucknow, India, 1984, pp. 349-361.
25. Ayerst, G. Water activity: its measurement and significance in biology. *Int. Biodeter. Bull., 1*:12-13, 1965.
26. Snow, D., Crichton, M. H. G., and Wright, N. C. Mould deterioration of feeding stuffs in relationship to humidity of storage. II. The water uptake of feeding stuffs at different humidities. *Ann. Appl. Biol., 31*:111-116, 1944.
27. Christensen, C. M., and Kaufman, H. H. Microflora. In *Storage of Cereal Grains and Their Products* (C. M. Christensen, ed.), Am. Assoc. Cereal Chemists, Minnesota, 1974, pp. 158-192.
28. Miehe, H. Die Wärmebildung von Reinkulturen im Hinblick aut die Äleologie der Selbsterhizung Pflanzlicher Stoffe. *Arch. Mikrobiol., 1*:78-118, 1930.
29. Milner, M., and Geddes, W. F. Grain storage studies. 2. The effect of aeration, temperature and time of respiration of soyabeans containing excessive moisture. *Cereal Chem., 22*:484-501, 1945.
30. Carter, E. P. and Young, G. Y. Role of fungi in the heating of moist wheat. U.S. Dept. Agr. Circ. No. 838, 1950.
31. Festenstein, G. N., Lacey, J., Skinner, F. A., Jenkins, P. A., and Pepys, J. Self-heating of hay and grain in Dewar flasks and the development of farmer's lung antigens. *J. Gen. Microbiol., 41*:389-407, 1965.
32. Currie, J. A., and Festenstein, G. N. Factors defining spontaneous heating and ignition of hay. *J. Sci. Food Agr., 22*:223-320, 1971.
33. Kane, B. E., and Mullins, J. T. Thermophilic fungi in a municipal waste compost system. *Mycologia, 65*:1087-1100, 1973.
34. Deploey, J. J., and Fergus, C. L. Growth and sporulation of thermophilic fungi and actinomycetes in O_2-N_2 atmospheres. *Mycologia, 67*:780-797, 1975.

35. Ayerst, G. The effects of moisture and temperature on growth and spore germination in some fungi. *J. Stored Prod. Res.*, *5*:127-141, 1969.
36. Hussain, H. M. Ökologische Untersuchungen über die bedeutung thermophiler Mikroorganismen für die Selbsterhitzung von Heu. *Zeit. Allgemeine Mikrobiol.*, *13*:323-334, 1973.
37. Kuo, M. J. and Hartman, P. A. Isolation of amylolytic strains of Thermoactinomyces vulgaris and production of thermophilic actinomycete amylases. *J. Bacteriol.*, *92*:723-726, 1966.
38. Fergus, C. L. The production of amylase by some thermophilic fungi. *Mycologia*, *61*:1171-1175, 1969.
39. Fergus, C. L. The cellulolytic activity of thermophilic fungi and actinomycetes. *Mycologia*, *61*:120-129, 1969.
40. Hedger, J. N. and Hudson, H. J. Nutritional studies of Thermomyces lanuginosus from wheat straw compost. *Trans. Br. Mycol. Soc.*, *62*:129-143, 1974.
41. Oso, B. A. The production of cellulase by Talaromyces emersonii. *Mycologia*, *70*:577-585, 1978.
42. Rosenburg, S. L. Cellulose and lignocellulose degradation by thermophilic and thermotolerant fungi. *Mycologia*, *70*:1-13, 1978.
43. Festenstein, G. N. Biochemical changes during moulding of self-heated hay in Dewar flasks. *J. Sci. Food Agr.*, *17*:130-133, 1966.
44. Johri, B. N., Jain, S., and Chouhan, S. Enzymes from thermophilic fungi: proteases and lipases. *Proc. Ind. Acad. Soc.* (Plant Sci.), *94*:175-196, 1988.
45. Rosenburg, S. L. Temperature and pH optimum for 21 species of thermophilic and thermotolerant fungi. *Can. J. Microbiol.*, *21*:1535-1540, 1975.
46. Lacey, J. Colonization of damp organic substrates and spontaneous heating. In *Microbial Growth and Survival in Extremes of Environments* (G. W. Gould and J. E. L. Correy, ed.), Academic Press, London, 1980, pp. 53-68.
47. Tuite, J. F. Low incidence of storage moulds in freshly harvested seed of soft red winter wheat. *Plant Dis. Reptr.*, *43*:470, 1961.
48. Tuite, J. F., and Christensen, C. M. Grain storage studies. 23. Time of invasion of wheat seed by various species of Aspergillus responsible for deteriorationof stored grain, and source of inoculum of these fungi. *Phytopathology*, *47*:265-268, 1957.
49. Browne, C. A. The spontaneous combustion of hay. U.S. Dept. Agr. Tech. Bull. No. 141, 1929, pp. 1-38.
50. McHargue, J. S. The cause of deterioration and spoiling of corn and corn-meal. *J. Ind. Eng. Chem.*, *12*:257-262, 1920.
51. Darsie, M. L., Elliott, C., and Pierce, G. J. Study of the germinating power of seeds. *Botan. Gaz.*, *58*:101-136, 1914.
52. Ramstad, P. E., and Geddes, W. F. The respiration and storage behavior of soyabeans. *Minn. Agr. Expt. St. Tech. Bull.*, *156*:54, 1943.
53. Burmeister, H. R., Hartman, P. A., and Saul, R. A. Microbiology of ensiled high moisture grain. *Appl. Microbiol.*, *14*:31-34, 1966.
54. Carter, E. P. The role of fungi in the heating of moist wheat. U.S. Dept. Agr. Circ. No. 838, 1950, p. 26.
55. Christensen, C. M., and Gordon, D. R. The mold flora of stored wheat and corn and its relation to heating of moist grain. *Cereal Chem.*, *25*:42-51, 1948.

56. Milner, M., and Geddes, W. F. Grain storage. 3. The relation between moisture content, mould growth, and respiration of soybeans. *Cereal Chem.*, *23*:225-247, 1946.
57. Norman, A. G., Richards, L. A., and Carlyle, R. E. Microbial thermogenesis in decomposition of plant materials. I. An adiabatic fermentation apparatus. *J. Bacteriol.*, *41*:689-697, 1941.
58. Okafor, N. Thermophilic micro-organisms from rotting maize. *Nature*, *210*:2209-2221, 1966.
59. Gilman, J. C., and Barron, D. H. Effect of molds on temperature of stored grain. *Pl. Physiol.*, *5*:565-573, 1930.
60. Walker, I. K. The role of water in the spontaneous combustion of solids. *Fire Res. Abst. Rev.*, *9*:5-22, 1967.
61. Mulinge, S. K., and Apinis, A. E. Occurrence of thermophilous fungi in stored moist barley grain. *Trans. Br. Mycol. Soc.*, *53*:361-370, 1969.
62. Larmour, R. K., Clayton, J. S., and Wrenshall, J. L. A study of the respiration of damp wheat. *Can. J. Res.* *12*:627-645, 1935.
63. Harrison, J. Some observations on the storage of high moisture barley in unsealed silos. *Rep. Agr. Res. Council*, *117*, 1965.
64. Nicholas, A. A., and Leaver, C. W. Methods of examining damp grain at harvest and after sealed and open storage. Changes in the microflora of damp grain during sealed storage. *J. Appl. Bacteriol.*, *29*: 566-581, 1966.
65. Cohn, F. Über thermogene Wirkung von Pilzen. *Jahresber Schles. Geselsch. vaterl Kult*, *66*:150-156, 1889.
66. Cohn, F. Ueber Wärme-Erzeugung durch Schimmelpilze und Bakterien. *Jahresber. Schles. Geselsch. vaterl. Kult.*, *68*:23-29, 1890.
67. Carlyle, R. E., and Norman, A. G. Microbial thermogenesis in the decomposition of plant materials. *J. Bacteriol.*, *41*:699-724, 1941.
68. Glathe, H. Selbsterhitzung und Selbstenzündung von Erntestoffen und ihre Verhütung. *Ergeb. Landwistsch. Forsch.*, *3*:83-98, 1960.
69. Niese, G. Mikrobiologische Untersuchungen zur Frage der Selbsterhitzung organischer Stoffe. *Arch. Mikrobiol.*, *34*:285-318, 1959.
70. Pallmann, H., Hüni, K., and Dönzo, O. Der Wärmehaushalt und die Kontrolle überhitzer Futterstöcke. Zur Diagnose der Selbstenzündung von Dürrfutter und Getreidestöcken. *Schweiz. Lansdwertsch. Monatshefte.*, *23*:223-243, 1945.
71. Glathe, H. Die mikrobiologische Analysey im Dienste der Brandursachenermittlung. *Kriminalistik: Zeitschr. gesam. Kriminal. Wessensch. Praxis*, *14*:121-123, 1960.
72. Westermarch-Rosendahl, C. Spontaneous heating in newly harvested wheat and rye. II. Effects on the technological quality of flour. *Acta Agr. Scand.*, *28*:159-168, 1978.
73. Watson, S. J., and Nash, M. J. *The Conservation of Grass and Forage Crops*. Oliver & Boyd, Edinburgh, 1960, pp. 5-44.
74. Lacey, J. Potential hazards to animals and man from microorganisms in fodders and grain. *Trans. Br. Mycol. Soc.*, *65*:171-184, 1975.
75. Lacey, J. Occupational and environmental factors in allergy. In *Allergy* (M. A. Ganderton and A. W. Frankland, eds.), Pittman Medical, London, 1975, pp. 303-319.
76. Lacey, J., Pepys, J., and Cross, T. Actinomycete and fungus spores in air as respiratory allergens. In *Safety in Microbiology* (D. A. Shapton and R. G. Board, eds.), Society of Applied Bacteriology Technical Series No. 6, Academic Press, London, 1972, pp. 151-184.

77. O'Callaghan, J. R. The waste products of agriculture. *J. Roy. Soc. Arts*, *123*:139-149, 1975.
78. Anon. Report on the use and disposal of straw, National Farmers' Union, Working party on straw disposal, 1972.
79. Gallardo-Lara, F., and Nogales, R. Effect of the application of town refuse compost on the soil-plant system: A review. *Biolog. Wastes*, *19*:35-62, 1987.
80. Gotaas, H. B. *Composting, Sanitary Disposal and Reclamation of Organic Wastes*. World Health Organization, Geneva, Monograph Series No. 31, 1956.
81. Gray, K. R., and Biddlestone, A. J. Decomposition of urban waste. In *Biology of Plant Litter Decomposition*, Vol. 2, C. H. Dickinson and G. J. F. Pugh (eds.), Academic Press, London, 1974, pp. 743-775.
82. Shuler, M. L. Utilization of farm wastes for food. In *Utilization and Recycle of Agricultural Waste and Residues* (M. L. Shuler, ed.), CRC Press, Boca Raton, 1980, pp. 67-133.
83. Hudson, H. J. *Fungal Biology*. Edward Arnold, London, 1986, pp. 56-63.
84. Eggins, H. O. W., and Allsopp, D. Biodeterioration and biodegradation by fungi. In *The Filamentous Fungi*, Vol. 1 (J. E. Smith and D. R. B. Berry, eds.), Edward Arnold, London.
85. Waksman, S. A., and Gerretsen, F. C. Influence of temperature and moisture upon the nature and extent of decomposition of plant residues by microorganisms. *Ecology*, *12*:33-60, 1931.
86. Corominas, E., Perestelo, F., Perez, M. L., and Falcon, M. A. Microorganisms and environmental factors in composting of agricultural waste of the Canary Islands. In *Compost Production, Quality and Use* (M. D. E. Bertoldi, M. P. Ferranti, P. L'Hermite, and F. Zucconi, eds.), Elsevier, London, 1987, pp. 127-138.
87. Seal, K. J., and Eggins, H. O. W. The use of thermophilic fungi in the biodeterioration of pig waste. *Proc. 3rd Int. Biodeg. Symp.* 1975, pp. 687-692.
88. Wiley, J. S. Progress report on high rate composting studies. *Proc. 12th Indus. Waste Conf.* Purdue Univ. Series 42, 1953, pp. 596-603.
89. Walker, I. K., and Harrison, W. J. The self-heating of wet wool. *N. Zeal. J. Agr. Res.*, *3*:861-895, 1960.
90. Mote, C. R., and Griffis, C. L. Heat production by composting organic matter. *Agr. Wastes*, *4*:65-73, 1982.
91. Frith, H. J. *The Mallee-Fowl: The Bird That Builds an Incubator*. Angus and Robertson, Sydney, 1962, pp. 136.
92. Cadman, W. H. Records of temperature humidity and carbon-dioxide under the sitting hen. *Harper Adams Util. Poult. J.*, *8*:390-398, 1923.
93. Jain, M. K., Kapoor, K. K., and Mishra, M. M. Cellulase activity, degradation of cellulose and lignin and humus formation by thermophilic fungi. *Trans. Br. Mycol. Soc.*, *73*:85-89, 1973.
94. Stoller, B. B., Smith, F. B., and Brown, P. E. A mechanical apparatus for the rapid, high temperature microbial decomposition of fibrous, cellulosic materials in the preparation of composts for mushroom cultivation. *J. Am. Soc. Agron.*, *29*:717-723, 1937.
95. Chang, Y., and Hudson, H. J. The fungi of wheat straw compost I. Ecological studies. *Trans. Br. Mycol. Soc.*, *50*:649-666, 1967.
96. Chang, Y. The fungi of wheat straw compost. II. Biochemical and physiological studies. *Trans. Br. Mycol. Soc.*, *50*:667-677, 1967.

97. Martin, J. P., and Haider, K. Microbial activity in relation to soil humus formation. *Soil Sci., 11*:54-63, 1971.
98. Deacon, J. W. Decomposition of filter paper cellulose by thermophilic fungi acting singly, in combination and in sequence. *Trans. Br. Mycol. Soc., 85*:663-669, 1985.
99. Banse, H. J. Beeinflussung der physikalishen bodeneigenschaften durch kompostgaben. *Internationale. Arbeitsgemeinschaft für Müllforschung Informatiosblatt, 13*:30-34, 1961.
100. Chanyasak, V., Hirai, M., and Kubota, H. Changes of chemical components and nitrogen transformation in water extracts during composting of garbage. *J. Ferment. Technol., 60*:439-446, 1982.
101. Larsen, S. Soil phosphorus. *Adv. Agr., 19*:151-210, 1967.
102. Hortenstine, C. C., and Rothwell, D. F. Pelletized municipal refuse compost as a soil amendment and nutrient source for sorghum. *J. Environ. Qual., 2*:343-345, 1973.
103. Zobac, J., and Vana, J. The agronomical effectiveness of industrially produced compost in dependence of the technology of production. *Rostlinna Vyroba, 20*:931-939, 1974.
104. Purvis, D., and McKenzie, E. J. Effects of applications of municipal compost on uptake of copper, zinc and boron by garden vegetables. *Plant Soil, 39*:361-371, 1973.
105. Trenel, M. Zur Bodenverbesserung durch torf verrateten müll und Klarschlamm. *Albrecht-Thaer-Arch., 5*:199-255, 1961.
106. Duggan, J. Utilization of municipal refuse compost. I. Field scale compost demonstration. *Compost Sci., 14*:1, 1973.
107. Vlamis, J., and Williams, D. E. Utilization of municipal wastes as agricultural fertilizers. *California Agr., 25*:7-9, 1971.
108. Schneider, F. Wirkung unter schiedlich gerotteter Stadtkomposte auf die Pflanzenerträge bei periodischer anwendung als organischer. *Dünger Arch Acker Pflanzenbau Bodenk, 18*:747-758, 1974
109. Fioromanti, S., and Marty, J. R. Contribution a l'etude du compost obtenupar fermentation des orchures ménagères de la ville de Tolouse *Bull. Sci. Sol*, 18-43, 1966.
110. Nogales, R., Zamora, M. A., Gomez, M., and Gallardo-Lara, F. Evaluación del potencial fertilizante de un compost de basura urbana. Effecto sobre gereminación y rendimientos secuenciales de cosecha. *Anal. Edaf. Agrobiol., 43*:183-194, 1984.
111. Singh, R. D., and Yadav, D. Y. Evaluation of a low grade rock phosphate composted with some agricultural wastes for use in a crop rotation. *Agr. Wastes, 18*:73-79, 1986.
112. Stephen, R. C., and Linn, Y. C. The value of Bankok compost addition to Hong Kong soils on the yield and quality of Chinese white cabbage. *Agr. Sci. Hong Kong, 1*:161-174, 1970.
113. Chopra, P., and Magu, S. P. Effect of selected herbicides and city compost on the rhizospheric microflora of wheat and maize. *Indian J. Agron., 30*:5-9, 1985.
114. Patrick, Z. A., Sayre, R. M., and Thorpe, H. J. Nematicidal substances selective for plant-parasitic nematodes in extracts of decomposing rye. *Phytopathology, 55*:702-704, 1965.
115. Pera, A., Vallini, G., Sireno, I., Bianchin, M. L., and DeBertoldi, M. Effect of organic matter on rhizosphere micro-organisms and root developments of sorghum plants in two different soils. *Plant Soil, 74*:3-18, 1983.

116. Eggins, H. O. W., and Malik, K. A. The occurrence of thermophilic cellulolytic fungi in a pasture land soil. *Antonie van Leeuwenhoek*, *35*:178-184, 1969.
117. Hedger, J. W. The ecology of thermophiles in Indonesia. In *Biodeterioration et Humification* (G. Kilburn, O. Reisinger, A. Mourney, and J. Cancela da Fonseca, eds.), Sarreguemines. Ed. Pierron. 1975, pp. 59-65.
118. Kuthubutheen, A. J. Thermophilous fungi from Malaysia. *Trans. Br. Mycol. Soc.*, *79*:548-552, 1982.
119. Pugh, G. J. F., and Kuthubutheen, A. J. Effect of fungicides on cellulose and starch hydrolysis by thermophilous and mesophilic fungi. *J. Thermal Biol.*, *4*:23-28, 1979.
120. Tansey, M. R. 1967. Agar-diffusion assay of cellulolytic ability of thermophilic fungi. *Arch Mikrobiol.*, *77*:1-11, 1967.
121. Loginova, L. G., Kosmachev, A. E., Golovacheva, R. S., and Sergina, L. M. A study of thermophilic microflora of Mount Yangan-Tau in the southern Urals. *Microbiologiya*, *31*:1082-1086 (*Microbiology*, *31*: 877-880, 1962).
122. Tansey, M. R., and Brock, T. D. Isolation of thermophilic fungi from hot spring effluents and thermal soils of the Yellow Stone National Park. *Bacteriol. Proc.*, abstract, 6, 1971.
123. Ogbonna, C. K., and Pugh, G. J. F. Thermophilic and thermotolerant fungi from Nigeria. *Int. Biodet. Bull.*, *19*:69-75, 1983.
124. Sandhu, D. K., and Singh, S. Airborne thermophilous fungi at Amritsar, India. *Trans. Br. Mycol. Soc.*, *84*:41-45, 1985.
125. Kuthubutheen, A. J. Thermophilous fungi associated with freshly harvested rice seeds. *Trans. Br. Mycol. Soc.*, *23*:357-359, 1972.
126. Gochenaur, S. E. Distributional patterns of mesophilous and thermophilous micro-fungi in two Bahamian soils. *Mycopathologia*, *57*:155-164, 1975.
127. Evans, H. C. Thermophilous fungi isolated from the air. *Trans. Br. Mycol. Soc.*, *59*:516-519, 1972.
128. Hudson, H. J. Thermophilous and thermo-tolerant fungi in the aerospora at Cambridge. *Trans. Br. Mycol. Soc.*, *61*:596-598, 1973.
129. Kuthubutheen, A. J. Thermophilous fungi associated with freshly harvested rice seeds. *Trans. Br. Mycol. Soc.*, *73*:357-359, 1979.
130. Kuthubutheen, A. J. and Pugh, G. J. F. The effects of fungicide on the occurrence of thermophilous fungi on leaf surface. *Mycopathol. Mycol. Appl.*, *62*:131-141, 1977.
131. Thomas, A. R., and Hill, E. C. Aspergillus fumigatus and supersonic aviation. I. Growth of Aspergillous fumigatus. *Int. Biodeter. Bull.*, *12*:87-94, 1976.
132. Singh, S. Studies on thermophilous fungi in Indian soils. Ph.D. thesis, Guru Nanak Dev University, Amritsar, India, 1983.
133. Tansey, M. R., and Jack, M. A. Thermophilic fungi in sun-heated soils. *Mycologia*, *68*:1061-1075, 1976.
134. Kuku, F. O., and Adeniji, M. O. The effects of moulds on the quality of Nigerian palm kernals. *Int. Biodeter. Bull.*, *12*:37-41, 1976.
135. Evans, H. C. Thermophilous fungi of coal spoil tips. I. Taxonomy. *Trans. Br. Mycol. Soc.*, *57*:241-254, 1971.
136. Evans, H. C. Thermophilous fungi of coal spoil tips. II. Occurrence, distribution and temperature relationships. *Trans. Br. Mycol. Soc.*, *57*:255-266,1971.

137. Summer, J. L., and Evans, H. C. The effects of growth temperatures on the fatty acid composition of fungi of the order Mucorales. *Can. J. Microbiol.*, *15*:515-520, 1969.
138. Küster, E., and Locci, R. Studies on peat and peat microorganisms. II. Occurrence of thermophilic fungi in peat. *Arch. Mikrobiol.*, *48*: 319-324, 1964.
139. Apinis, A. E. Occurrence of thermophilous microfungi in certain alluvial soils near Nottingham. *Nova Hedwigia*, *47*:559-570, 1963.
140. Brock, T. D. *Thermophilic Micro-organisms and Life at High Temperature.* Springer-Verlag, New York, 1978, pp. 39-66.
141. Stolk, A. C. Thermophilic species of Talaromyces Benjamine and Thermoascus Miehe. *Antonie van Leeuwenhock*, *32*:262-276, 1965.
142. Apinis, A. E. Dactylomyces and Thermoascus. *Trans. Br. Mycol. Soc.*, *50*:573-582, 1967.
143. Apinis, A. E., and Chesters, C. G. C. Ascomycetes of some salt marshes and sand dunes. *Trans. Br. Mycol. Soc.*, *47*:419-435, 1964.
144. Brock, T. D., and Darland, G. K. Limits of microbial existence: Temperature and pH. *Science*, *169*:1316-1318, 1970.
145. Fergus, C. L., and Sinden, J. W. A new thermo-tolerant fungus from mushroom compost: Thielavia thermophila sp. nov. *Can. J. Bot.*, *47*:1635-1637, 1969.
146. Hedger, J. N., and Hudson, H. J. Thielavia thermophila and Sporotrichum thermophile. *Trans. Br. Mycol. Soc.*, *54*:497-500, 1970.
147. Reese, E. T. Aerobic decomposition of cellulose by micro-organisms at temperatures above 40°C. Ph.D. thesis, Pennsylvania State College, 1946, pp. 100.
148. Bunce, M. E. Humicola stellatus sp. nov., a thermophilic mould from hay. *Trans. Br. Mycol. Soc.*, *44*:372-376, 1961.
149. Hedger, J. N., Samson, R. A., and Basuki, T. Scytalidium indodesiacum, a new thermophilous hyphomycete from Indonesia. *Trans. Br. Mycol. Soc.*, *78*:364-366, 1982.
150. Apinis, A. E.,and Eggins, H. O. W. Thermomyces ibadanensis sp nov from oil palm kernal stacks in Nigeria. *Trans. Br. Mycol. Soc.*, *49*:629-632, 1966.
151. Eggins, H. O. W., and Coursey, D. G. Thermophilic fungi associated with Nigerian oil palm produce. *Nature*, *203*:1083-1084, 1964.
152. Fergus, C. L. The temperature relationships and thermal resistance of a new thermophilic Papulasporia from mushroom compost during peak heating. *Mycologia*, *63*:426-431, 1971.
153. Rydholm, S. A. *Pulping Process.* Wiley-Interscience, New York, 1965, pp. 5-60.
154. Clark, T. F. Annual crop fibres and the bamboos. In *Pulp and Paper Manufacture*, Vol. 2 (R. G. McDonald and J. N. Franklin, eds.), McGraw-Hill, New York, 1969, pp. 1-74.
155. Cowling, E. B., and Kirk, T. K. Properties of cellulose and lignocellulosic materials as substitutes for enzymatic conversion process. *Biotech. Bioeng. Symp.*, *6*:95-123, 1976.
156. Dunlap, C. E. Comparative evaluation of cellulose resource. In *Proc. Bioconversion Bioeng. Symp.*, Ind. Inst. Technol., New Delhi, 1981, pp. 397-412.
157. Sharma, H. S. S. Economic importance of thermophilous fungi. *Appl. Microbiol. Biotechnol.*, *31*:1-10, 1989.

24

BIOTECHNOLOGICAL POTENTIAL OF THERMOPHILIC FUNGI

T. SATYANARAYANA *University of Delhi, Delhi, India*

B. N. JOHRI *G. B. Pant University of Agriculture and Technology, Pantnagar, India*

JOACHIM KLEIN *Gesellschaft für Biotechnologische Forschung mbH, Braunschweig, Germany*

I. INTRODUCTION

Temperature is one of the extremely important environmental variables that plays a key and decisive role in the survival, growth, distribution, and diversity of microorganisms on the surface of the earth. The majority of microorganisms grow in the temperature range 10-45°C and are termed mesophiles; some are psychrophiles (0-20°C); a few thermophiles that are capable of growth in the temperature range 25-80°C and a very few such as *Pyrodictium furiosum* and *P. brockii* are extreme thermophiles which grow optimally at 90-105°C. Fungi are the only eukaryotic organisms with a maximum growth temperature of 60-62°C [1]. Among 100,000 known species of fungi, very few are thermophiles. According to Crisan [2], thermophilic molds are those with an optimum temperature requirement for growth at or above 40°C.

A comprehensive monograph on thermophilic fungi published in 1964 [3] aroused considerable interest that led to investigations of their occurrence, ecology, nutrition, physiology, biochemistry, and secretion of enzymes and antibiotics. The extensive literature on these aspects has been elegantly reviewed from time to time [4-10]. The present chapter is aimed at highlighting the biotechnological potential of thermophilic molds.

Thermophilic molds have been reported from various types of soils, and in habitats where decomposition of plant material takes place. These include compost and wood chip piles, nesting materials of birds and animals, and municipal refuse. In these habitats thermophiles may occur either as resting propagules or as active mycelia depending on the availability of nutrients and favorable environment. Soils in tropical countries do not appear to have a higher population of thermophilic fungi than soils in temperate countries as presumed earlier. Their widespread occurrence could well be due to the dissemination of propagules from self-heating masses of organic material [11]. Since they thrive in the complex organic materials by degrading various components, thermophilic molds have been presumed to possess a battery of extracellular enzymes for the hydrolysis of polymers such as starch, protein, pectin, hemicellulose, and cellulose. Investigations with pure cultures have clearly demonstrated their ability to degrade various components of organic matter and their ability to produce a large number of extracellular enzymes.

The biotechnological potential of these molds stems from their ability to efficiently degrade organic matter acting as biodegradants and natural savengers; to produce a long array of industrially important extracellular as well as intracellular enzymes, antibiotics, amino acids, phenolic compounds, and polysaccharides; from their role in the production of SCP and nutritionally enriched feeds; and from their suitability as agents of bioconversions and experimental systems for genetic manipulations. Due to their multifarious potentialities, they appear to be nature-borne biotechnologists.

II. CHARACTERISTICS ENDOWED ON THERMOPHILIC MOLDS

A. Simple Nutritional Requirements

Most thermophilic fungi grow in simple media containing carbon and nitrogen sources and a few mineral salts, suggesting that they are auxotrophic to vitamins [3,12-15]. Simple nitrogen sources like sodium and potassium nitrates and asparagine support good growth of these molds [14]; nitrites of sodium and potassium, nitrate and sulfate of ammonium are poor nitrogen sources [14,16-17]. Among 11 fungal species, *Acremonium alabamensis*, *Humicola lanuginosa*, and *Rhizopus microsporus* were able to utilize urea and grow on starch, cellulose, hemicellulose, lignocellulose, and pectins [14,18,19]. *Humicola lanuginosa* utilized glucose, sucrose, maltose, mannose, galactose, xylose, starch, and lignin, while *H. insolens* utilized glucose, maltose, xylose, and starch [20]. With the exception of *Torula thermophila*, all others were capable of growth on C-1 compounds, with methanol and sodium formate as the source of carbon and energy [21]. Thermophilic fungi were observed to require succinate in addition to glucose for optimal growth [22,23]. Gupta and Maheswari [24] unequivocally demonstrated that the improved growth in the presence of organic acid was a result of its buffering action; the organic acid was not required as a source of carbon. Oxygen availability is also an important factor for their growth [25,26]. The pH optima varies widely; from a low of 3.4-5.4 for *Talaromyces emersonii* and it reaches a value of 7.2-8.1 for *T. thermophilus* [15]. During growth on lignocellulose, thermophiles are capable of solubilizing tricalcium phosphate and rock phosphate [27]. Cooney and Emerson [3] reported that *T. dupontii* developed imperfect conidial stage under aerobic conditions, and ascocarps under anaerobic conditions. Mehrotra and Mehrotra [28] and Joshi [29] showed the induction of zygospores in *Mucor miehei* and *Thermomucor indicae-seudaticae* by aspartic acid and phenylalanine.

B. Wider Extracellular Enzymatic Machinery

Thermophilic molds grow in plant organic matter by utilizing the soluble products of hydrolysis of complex polymers such as starch, protein, pectin, hemicellulose, cellulose, and lignin [4,8,34]. They produce a number of

TABLE 1. Extracellular (EC) and Intracellular (IC) Enzymes of Thermophilic Molds

Enzyme	Producing organisms	EC/IC	Ref.
Acid phosphatase	*H. insolens, S. thermophile, T. emersonii*	EC, IC	35,116
Alkaline phosphatase	*Malbranchea pulchella, T. emersonii, Chaetomium thermophile, Thermoascus aurantiacus, S. thermophile*	EC, IC	35
Aminopeptidases	*S. pulverulentum, T. emersonii, Acremonium alabamensis, S. thermophile*	EC, IC	35
Carboxypeptidases	*Talaromyces duponti*	EC, IC	34
Amylases and glucoamylase	*Mucor miehei, M. pusillus, H. lanuginosa, Papulospora thermophila, Torula thermophila, A. fumigatus, T. duponti*	EC, IC	36-40
	C. cellulolyticum, C. thermophile, H. insolens, Myceliophthora thermophila, S. thermophile, T. emersonii, T. aurantiacus	EC	41-46
α-Galactosidase	*S. thermophile, H. insolens, Thermomucor indicae-seudaticae*	EC, IC	35
β-Glucosaminidase	*T. emersonii, H. insolens, C. thermophile, T. aurantiacus*	EC, IC	35
Invertase	*Thermomyces lanuginosus*	IC	47
Lactase	*H. insolens, A. alabamensis, S. thermophile, T. emersonii*	EC, IC	48,35
Lipases	*H. lanuginosa, M. miehei, M. pusillus, S. thermophile, H. insolens*	EC	35,40,49-52
Proteases	*H. lanuginosa, M. pulchella, M. pusillus, M. miehei, S. thermophile, T. thermophilus, P. thermophila*	EC	53-60
Trehalase	*H. lanuginosa*	IC	61
Xylanases	*Myceliophthora thermophila, T. aurantiacus, S. thermophile, C. cellulolyticum, M. pulchella, Melanocarpus albomyces*	EC	63-70

extracellular enzymes that hydrolyze complex biopolymers (Table 1). Elaboration of intracellular enzymes like lactase, invertase, amino- and carboxypeptidases, β-glucosaminidase, etc., is also known. Some thermostable enzymes possess extraordinary stability in the presence of detergents and aqueous organic solvents [30-32] and show lesser loss of activity during storage and purification [10].

C. Higher Growth Rates

Three distinct patterns of mycelial biomass production have been recorded in coal mine strains of thermophilic fungi [6,71,72]. In the first, the lag phase was fairly long and gradually merged with the log phase of growth. Consequently, dry mycelial production was slow and autolysis did not occur until the fifteenth day of incubation, e.g., *Absidia corymbifera* and *Torula thermophila*. The second pattern consisted of an extended lag and log phase with sharp rise in mycelial production between 9 and 12 days, e.g., *Humicola grisea* and *Thielavia minor*. In the third, lag and log phases combined together did not exceed 3 days and a decline in dry mycelial production commenced after the sixth day. As a result, there was a very sharp rise in dry weight followed by a significant decline, e.g., *Rhizopus rhizopodiformis*, *Thermomyces lanuginosus*, *Sporotrichum* sp., and *Thermoascus aurantiacus*. An abrupt drop in mycelial biomass production was reported in *H. lanuginosa* above the optimum temperature which also affected lag period and growth rate [4]. A systematic investigation has thus far been undertaken to clearly express growth rates of these molds under strictly controlled conditions because most reported studies are based on shake or static cultures. Thermophiles which are encountered in decomposing plant organic residues and composts at higher frequency, namely, *S. thermophile*, *H. lanuginosa*, *H. insolens*, *T. aurantiacus*, *T. emersonii*, and *C. thermophile* [3,36,73,77], are fast-growing and exhibit high growth rates [88].

D. Higher Temperature Requirements

Thermotolerant molds like *Aspergillus fumigatus* grow at or below 20°C, while thermophilic mold species such as *H. lanuginosa*, *C. thermophile*, *T. aurantiacus*, and *T. emersonii* show little or no growth at 20-30°C [14,15,78,81]. Therefore, thermophiles have varied optimum temperature requirements (Table 2). Several require higher temperature for growth than their mesophilic counterparts (25-37°C). Temperature has a definite effect on the formation of sterile or fertile ascocarps in ascomycetous thermophiles. *Chaetomium thermophile* and *T. aurantiacus* generally produce fertile ascocarps at 40-45°C but at higher temperatures they form either sterile ascocarps or vegetative mycelia [4,36,78,80,84]. Thermostable proteins and membranes (due to more saturated lipids) and polyamines might play an important role at elevated temperatures [1,85,86]. Their failure to grow at low temperatures could be due to their inability to synthesize a sufficient quantity of saturated fatty acids to maintain membrane integrity for proper functioning [85,87].

E. Thermostability of the Macromolecules

Thermophilic bacteria produce several unusually thermostable enzymes [89-91] and certain of their structural proteins also exhibit an intrinsic

TABLE 2. Optimum Temperature Requirements of Thermophilic Molds

Organism	Optimum temperature (°C)	Ref.
Aspergillus fumigatus	40	3,15,36,81
Chaetomium thermophile var. *coprophile*	50-55	3,15
C. thermophile var. *dissitum*	45-50	3,15,36
Humicola grisea var. *thermoidea*	38-46	3,15
H. insolens	40-45	3,15
H. lanuginosa	45-50	3,15,36
Malbranchea pulchella var. *sulfurea*	45	3,15
Mucor miehei	35-45	3,15
M. pusillus	35-45	3,14,15,36
Myriococcum albomyces	37-42	3,15
Sporotrichum thermophile	40-50	3,15,36
Talaromyces emersonii	40-45	15,81
T. duponti (synonym *T. thermophilum*)	40-50	3,15
Thermoascus aurantiacus	40-50	3,14,15,36
Thielavia thermophila	45	15,83
T. minor	40	14,36
Torula thermophila	40-45	3,15,36

thermostability [92,93]. However, the DNA and RNA of mesophiles and thermophiles do not appear to differ significantly in thermostability [94,95]. A comparable study by De Bertoldi et al. [96] with mesophilic and thermophilic *Humicola* species has shown that there is no significant difference in their GC content. Nageswara Rao and Cherayil [97] observed a higher number of ribose methylations in the RNA of *Thermomyces lanuginosus* than any other nonvertebrate, but its significance in thermophily is not clearly understood.

Based on electrophoretic analyses of cellular extracts of *Penicillium dupontii* and *H. lanuginosa*, Crisan [98] concluded that the production of thermostable proteins in thermophilic fungi was an exception rather than the rule. The proteins of *P. dupontii* do not appear unusually thermostable. However, the organism produced a greater number of thermostable proteins at higher incubation temperatures. Only one protein showed a direct relationship between the amount of protein produced and thermostability on increased growth temperatures. Thermophilic molds are known to produce thermostable proteases, lipases, glucoamylases, cellulases, ribonucleases, aminopeptidases, and carboxypeptidases [10,34,51,99-103].

Thermostable glucose-6-phosphate dehydrogenase [115], malate dehydrogenase, isocitrate dehydrogenase, and succinate dehydrogenase [104] have also been detected in thermophilic molds. Many of these molds do not appear to synthesize heat-stable malate dehydrogenase; some factors in vivo appear to impart thermostability to enzymes which are otherwise thermolabile. Salts of sodium, potassium, and ammonium and citrate could protect malate dehydrogenase of *H. lanuginosa*, *C. thermophile*, and *M. pusillus* against heat inactivation.

Proteases of thermophilic molds have optima at 45-55°C [53,101,102,105] and lipases at 45-50°C [50-52,101]. The average optimum temperature of thermophilic ribonuclease was 60.5°C [106], and for amylases, 45-50°C [36-40]. The xylanases and cellulases possess optima at 50-70°C [65,103, 107-111]; these are glycoproteins associated with varied amounts [13-68%) of carbohydrate [107,111,112]. Partial removal of carbohydrate renders such enzymes less heat-stable. The carbohydrate moiety appears to play a key role in stabilizing the conformational structure of such enzymes at high temperatures [111,112]. Pazure et al. [113] felt that the prolonged periodate oxidation may affect certain amino acids and thereby influence activity.

F. Stress Resistance

An isolate of thermophilic *Humicola* sp. was recovered from a sand dune sample of Thar desert, Rajasthan (India), where the organisms are exposed to desiccation stress [114]. This mold was able tc grow in a medium consisting of 10-50% sucrose with a minimum of 20% and an optimum of 40-50% (xerophilic or osmophilic). When mycelium was exposed to different levels of relative humidity, an increase in glucose concentration was noted (84.8% over control); this lowered the phospholipid level, but there was a marked increase in proline (114-fold higher than the control) and sterols. These changes appear to be adaptive mechanisms to desiccation stress.

III. POTENTIAL BIOTECHNOLOGIES

There are certain advantages to using thermophilic microbes and their enzymes [91]. Reduced viscosity of media enhances the quality of mixing and allows easier liquid-solid separation. Productivity would increase as the reaction rates of organisms and enzymes increase. Mass cultivation of thermophilic microbes would be cheaper than the mesophiles due to reduction in the investment needed for heat exchange equipment, the lack of necessity of cooling, and reduced contamination problems at high temperatures. Cost effectiveness would be enhanced as a result of longer lifetimes of organisms and/or enzymes, particularly for immobilized systems. Higher enzyme yields would result due to greater enzyme stability, which is important in enzyme recovery and purification. Isolation and purification of thermophilic enzymes could be carried out at ambient temperatures since the activity of thermostable enzymes is reduced at low temperatures. Thermostable enzymes appear to be more resistant to denaturing effects of detergents and organic solvents. Therefore, it is possible to use a high concentration of reactants in non- or partially aqueous reaction systems, and to be able to purge the systems efficiently with organic solvents. Volatilization of products may be valuable as a recovery step or as a means of removing potentially inhibitory products. There is also a highly reduced chance of pathogenicity with thermophilic microbes. Because of these advantages, thermophiles appear to be suitable for use in biotechnological processes.

A. Extracellular Enzymes

Thermophilic molds recovered from various sources have been found to produce a number of extracellular enzymes of biotechnological interest. The information on these enzymes is discussed below.

1. Amylases

Several thermophilic molds, i.e., *H. lanuginosa*, *M. pulchella*, *M. miehei*, *M. pusillus*, *T. thermophilus*, *S. thermophile*, and *T. thermophila*, produce extracellular amylolytic enzymes under both surface and submerged conditions [36-40,117-125]. Amylase production in *H. lanuginosa* was 8-12 times higher than that in *H. grisea* var. *thermoidea*; in the former maximum amylase production was achieved after autolysis and in the latter before autolysis. Increase in the amount of starch in the medium resulted in the enhancement of amylase, whereas there was no effect of increase in yeast extract on amylase secretion [118]. Barnett and Fergus [118] reported *H. lanuginosa* to be an excellent producer of amylase. Purified amylolytic enzyme preparation of *H. lanuginosa* consisted of α-amylase (MW 54-57 kDa) and glucoamylase (MW 70-76 kDa); isoelectric points of these enzymes were pH 3.4 and 4.0, respectively. An α-glucosidase (MW 40-48 kDa) is also produced by this organism. The higher yield and activity was recorded when low molecular weight dextrans were used as carbon source [119]. Rao et al. [37,38] reported that this mold produced a glucoamylase (52.1 U/mg protein) in static culture with very high specific activity. Taylor et al. [123] observed two glucoamylases from shaken cultures of *H. Lanuginosa* in a complex medium. They were designated as glucoamylase I and II; on further analysis one fraction was found to contain glucoamylase I and α-glucosidase, whereas the other contained only glucoamylase (II). *Rhizomucor pusillus* [124], *T. thermophila*, *M. miehei*, and *S. thermophile* [39,126] were also found to produce amyloglucosidase. *Rhizomucor pusillus* produced only one glucosidase with optimum temperature for gelatinized starch at 65°C and cornstarch at 50°C. This enzyme was a 68-kDa protein which could be modified by subtilisin to a 60-kDa molecule. The modified enzyme was unable to either digest or get adsorbed to raw starch. It could, however, still hydrolyze gelatinized potato starch and glycogen and compared well with the unmodified protein. Kanlayakrit et al. [124] suggested that the raw starch affinity site could be different from the active site of glucoamylase.

2. Proteases

Proteases of microorganisms have wide-ranging applications in the brewery, dairy, textile, and leather industries. Another important use of microbial proteases is in the baking industry for developing unique and variable properties of doughs. Fungal proteases have also found use in the food industry as a substitute to papain, rennin, and pepsin. The acid proteases of *M. pusillus* and *M. miehei* are used in the preparation of oriental foods like tempeh and koji, and in cheese making as a substitute for rennets. Thermophilic molds produce acid as well as alkaline proteases. *Mucor pusillus*, *M. miehei*, *Penicillium duponti*, *C. thermophile*, and *S.*

thermophile produce acid proteases [33,57-59,128,129], while *T. thermophila* [130], *M. pulchella* var. *sulfurea* [56], and a thermophilic eumycete Y-15 [102] produce alkaline proteases. *Humicola lanuginosa* produces thiol protease, which is also an alkaline protease [60]. *Mucor miehei* and *M. pusillus* also produce milk-clotting acid proteases [131-133]. Recently, Bailey and Siika-aho [134] reported *Rhizomucor pusillus* VTT-82193 to be the best source of microbial rennin.

The cultural conditions, methodology of purification, and properties and kinetics of proteases of thermophilic molds were reviewed by Johri et al. [10]. The level of extracellular proteases in thermophilic molds is generally greater than the level of intracellular proteases [55,105]. The extent of protease production depends on a variety of factors including the composition of the medium. In *M. pulchella* var. *sulfurea*, enzyme production was stimulated by 2% casein, while 1% casamino acids or 0.75% yeast extract completely inhibited enzyme formation [56]. Addition of ammonium salts to wheat bran medium increased protease production in *P. duponti* K-1014 whereas nitrate was detrimental [135]. An enhancement in the production of extracellular protease was shown by the addition of casein, gelatin, glucose, and maltose [53]. The favorable pH for protease secretion ranged from 4 to 9 and temperature from 35 to 50°C depending on the organism. The optimum temperature for proteolytic activity was reported to range from 45 to 70°C [10]. In general, proteases of thermophilic molds are more thermostable when compared to those from mesophiles [53,55,101,127,136]. Thermomycolase of *M. pulchella* var. *sulfurea* exhibited greater thermostability in the presence of calcium ions.

Bailey and Siika-aho [134] found milk-coagulating activity of *R. miehei* VTT-82193 (ATCC 16457) to compare favorably with two commercial rennins (Supranen, Pfizer, from *Endothia parasitica*; Rennilase, Novo Industri A/S, from *Mucor miehei*). The costs of industrial scale production were estimated to be well below those of commercial microbial coagulants.

3. Lipases

Lipolytic enzymes are long-chain fatty acid ester hydrolases which have applications in the fields of nutrition, food technology, clinical medicine, detergents, and analytical and preparative chemistry. Lipase in microbial protease preparations has usefulness in the acceleration of flavor development in some types of cheese [51]. Lipases have been applied in the degradation of sewage, in bread dough for the uniform production of monoglycerides that improve its resistance to staling, in digestive aids, and in detergents [10]. The use of lipases in the resolution of racemic α-halopropionic acids and racemic alkanoic acid intermediates for optically active adrenergic blocking agents (β blockers) has been demonstrated. Lipolytic thermophilic molds have been implicated in the spoilage of oil seeds under storage since they can cause a decrease in oil content accompanied by a rapid increase in free fatty acids resulting in seed deterioration [138-143]. Thermostable lipases on *H. lanuginosa* and others could be a useful tool in the determination of triacylglycerol structures since they differ in sites of attack from the better known lipases of animal origin [144].

Thermophilic molds that are known to produce extracellular lipases include *M. pusillus*, *M. miehei*, *T. thermophilus*, *T. emersonii*, *H. lanuginosa*, *T. crustaceus*, *A. alabamensis*, and *C. thermophile* [49-52,141, 142,145-152]. Arima et al. [145] obtained high production of lipase at

45°C in a medium containing olive oil; its optimum activity was found on polyvinyl alcohol-emulsified oil at pH 8 and at 60°C. Lipases of *M. pusillus* and *T. thermophilus* exhibited optimum activity at 45°C, while those of *T. crustaceus* and *H. grisea* var. *thermoidea* were active at 50°C [146]. The optimum temperature for synthesis and activity of lipase of *T. emersonii* was 40-45°C [141]. *T. emersonii* exhibited induced synthesis of lipase when various carbohydrates were used as carbon source in the presence of yeast extract. Lipases were synthesized by *H. grisea* var. *thermoidea*, *M. pusillus*, *T. thermophilus*, and *T. crustaceus* in a medium devoid of lipids; however, their addition enhanced lipase production [146]. In general, thermophilic fungal lipases can hydrolyze both natural and synthetic triglycerides [51,146]. To prevent hydrolytic rancidity in cheese, Somkuti [153] was able to resolve lipase and protease by gel filtration of enzyme mixtures of *M. pusillus* on Sephadex G-75 and Sephadex G-100.

Sulfhydryl reducing agents like dithiothreitol were not able to alter the enzyme activity of *H. lanuginosa* but rendered it more susceptible to heat inactivation [152]. Reducing compounds (cysteine, 2-mercaptoethanol), chelating agents (EDTA, *o*-phenanthroline), and thiol group poisons (pCMB, monoiodoacetate) did not have a detectable effect on glycerol ester hydrolase or esterase activity of *M. pusillus*; this suggested that the lipase was not a metalloenzyme and did not require either free -SH group(s) or an intact S-S bridge for its activity [147]. Lipase of *M. pusillus* lost 10, 27, and 50% of activity when exposed to 55 and 58°C for 10, 24, and 45 min, respectively [51]; however, the enzyme from *H. lanuginosa* S-38 was highly thermostable as it retained 100% activity after exposure to 60°C for 2 h [50]. The lipase activity was inhibited by Co^{2+}, Cu^{2+}, Ni^{2+}, Hg^{2+}, and Sn^{2+} [149]. The k_m for trilaurin was 14.2 mg/ml. This enzyme had specificity for substrates containing C_{12} fatty acids and exhibited 1,3 positional specifity [149]. When absorbed onto Amberlite XAD-7, this lipase could be repeatedly used in batch replacements up to 300 h [150].

4. Xylanases

The potential uses of xylanases center on their ability to degrade xylans [154]. Xylanases are used to liquefy coffee mucilage in the manufacture of liquid coffee. They are used in wine manufacture to change organoleptic and rheological properties of must and wines and for the isolation of protoplasts from potato leaves. *Streptomyces* E86 xylanase system has been used to produce xylobiose from a commercial hardwood xylan. Endoxylanase has been used for the ultrastructural localizations in the cell wall of graminae fibers. The recent interest in xylanase stems from its ability to digest xylan, a major constituent of woody angiosperms, gymnosperms, and stems of monocots, to xylose which can be used for fuel and chemical production using appropriate fermentations. In order to use xylanase commercially for the production of xylose from biomass, it must possess high stability and be recoverable for reuse.

A number of thermophilic molds, namely, *Absidia ramosa*, *A. alabamensis*, *A. fumigatus*, *C. cellulolyticum*, *H. lanuginosa*, *Melanocarpus albomyces*, *M. pusillus*, *S. thermophile*, *T. aurantiacus*, and *Thielavia terrestris* have been reported to produce xylanolytic enzymes [63-70,112,155-165]. Xylanolytic enzymes were produced by growing thermophilic molds in liquid media containing pure xylan or substrates consisting of hemicelluloses such as rice straw, wheat straw, sugarcane bagasse, and wheat bran under submerged/stationary conditions for a desired period at 40-50°C. Yoshioka et al. [162] and Kitpreechavanich et al. [155] obtained xylanases by solid state fermentation of molds on wheat bran and extracted it with tap/deionized water. The culture fluids were filtered/centrifuged for the separation of fungal biomass, and cell-free culture fluid has been used as a crude preparation of xylan-degrading enzymes. Xylanase was reported in *S. thermophile* and *A. alabamensis* [63] and *M. albomyces* [69,270]. It was necessary to incorporate xylose/xylose-containing substrate in the medium for the induction of xylanases. When arabinogalactan was used as a substrate, xylanase was not secreted by *A. alabamensis* and *S. thermophile* [63]. Surfactants like Tween 80 enhanced xylanase production in *C. cellulolyticum* [70]. Almost all xylanase synthesized was excreted into the medium (extracellular); the intracellular fraction contained very little enzyme. The xylanolytic system of *A. fumigatus* [163], *H. lanuginosa* [155,164], *M. pulchella* var. *sulfurea* [109,156-159], *T. byssochlamydoides* [162], and *T. aurantiacus* comprised *exo*-1,4-β-xylosidase and/or *endo*-1,4-β-xylosidase, and β-xylosidase, which are required for the total hydrolysis of xylan to xylose. There are two isoenzymes of xylanase in *M. albomyces*; the activity of xylanase preparation was enhanced when an exogenous β-glucosidase was added [69]. The amino acid composition of xylanase of *S. thermophile* [64] resembled that of *T. byssochlamydoides* [162] and *Schizophyllum commune* [166]. Xylanase of *S. thermophile* consisted of more acidic residues and amides than the basic residues, with high proportion of aspartic acid, threonine, serine, glutamic acid, glycine, and alanine [64].

Humicola lanuginosa xylanase was stable at pH 4-12 with an optimum at 5-6 [155], while xylanase and β-xylosidase of *M. pusillus* had optimal activity at pH 6-6.5 [109,159]. The xylanase of *T. byssochlamydoides* was stable in the pH range 3.0-9.0 with an optimum at 5.5 [162]. Xylanase of these molds has been reported to have an optimum activity in the temperature range 50-70°C [155,157,160,162,165]. The xylanase of *T. byssochlamydoides* was more thermostable than that of *M. pulchella* var. *sulfurea* [162]; the former retained 50% activity at 95°C for 5 min while optimum at 70-75°C; the latter was completely inactivated when heated for 5 min at 70°C, with an optimum at 55°C. The xylanase of *A. fumigatus* was found to contain 46-68% of carbohydrate [112]. When 37% of carbohydrate was removed by periodate oxidation, the specific activity remained the same, but resulted in only 3% of maximum activity after pretreatment at 65°C for 40 min as compared to that of 53% for the native enzyme. Based on this observation, Flannigan and Sellars [112] concluded that the carbohydrate moiety has a role in stabilizing the conformational structure of enzymes at high temperatures.

Malbranchea pulchella var. *sulfurea*, *T. byssochlamydoides*, and *M. albomyces* produced significantly large amounts of xylanolytic enzymes capable of hydrolyzing xylan to the extent of 90% or above as xylose. The extent of hydrolysis by the mixture of xylanases from *A. fumigatus* and *H. lanuginosa* was comparable with acid hydrolysis by using 0.8 N H_2So_4 at 121°C for 15 min [155]. The volumetric productivity of xylanase by *Thielavia terrestris* was 1044 IU/liter/h, which was claimed to be superior to many mesophilic xylanolytic microorganisms [66].

5. Cellulases

Several thermophilic molds, namely, *Acremonium cellulophilum*, *A. fumigatus*, *C. thermophile*, *C. cellulolyticum*, *H. insolens*, *S. thermophile*, *Myceliophthora thermophila*, *Thielavia terrestris*, *T. emersonii*, and *T.*

aurantiacus, have been reported to degrade cellulose and produce extracellular cellulolytic enzymes [36,41-46,76,107,108,110,111,168-206,213,228]. Some molds such as *Rhizopus rhizopodiformis, M. pusillus, M. miehei*, and *H. lanuginosa*, which were observed to be noncellulolytic [76,178], were later found to break down cellulose by producing cellulases [201,209, 210]. To assay cellulolytic ability and to obtain cellulolytic enzymes, these molds were grown on liquid media containing either insoluble cellulosic substrates, i.e., cereal straws (rice, wheat, etc.), filter paper, Whatman cellulose powder, solka floc (purified ball-milled spruce cellulose), Avicel (with or without pretreatments), wheat bran, or soluble carboxymethylcellulose (CMC). The cultures were grown at 40-50°C and pH 5-8.2 under stationary/submerged conditions for a desired period of time. The crude enzyme preparations were obtained by filtering/centrifuging the cultures for separation of biomass and residual solids. Nishio et al. [175] cultivated *Talaromyces* sp. on wheat bran in solid state and extracted cellulases with deionized water while Jain et al. [42] obtained cellulases by extracting mold-grown wheat straw with sodium citrate buffer. Grajek [181] reported greater production of cellulases in solid state fermentation (SSF) than in liquid fermentation (LF); enzyme productivity in *T. aurantiacus* and *S. thermophile* per unit volume was higher in SSF than in LF. Tween 80 has been shown to be beneficial in the production and recovery of cellulase [212]; however, this was detrimental to the degradation of cellulose by *S. thermophile* [191]. Romanelli et al. [191] felt that this detrimental effect could be due to its action as a detergent to prevent the adsorption of the extracellular cellulase on cellulose. The cell-associated cellulolytic activity reached a maximum by 3 days, while extracellular activity reached a peak by 6 days in *Thielavia terrestris* [189].

The spores of *S. thermophile* immobilized in agar/polyacrylamide/alginate gels have been shown to grow and produce cellulases [207]. The mycelia of *T. emersonii* immobilized in calcium alginate beads along with cellulose produced twice the maximum cellulase production by the nonimmobilized system [206]. The immobilized *S. cellulophilum* gave the same filter paper activity as the free mycelium, but the lag for cellulase production in the immobilized fungus was longer [229].

The cellulases are generally inducible enzymes. In spite of the fact that cellulose, cellobiose, sophorose, and lactose have been known to be good inducers [214], the precise character of the natural inducer is not understood. As cellulose is insoluble, the true inducer must be some compound derived from cellulose hydrolysis. Cellulase production in a medium containing cellobiose as inducer and carbon course was as high as that in cellulose as carbon source. Therefore, it is likely that cellobiose is the natural inducer of cellulolytic enzymes. Gong and Tsao [214] reported that *T. reesei* has a level of basal synthesis of cellulolytic enzymes that is independent of carbon source. They are of the opinion that these constitutive enzymes are sufficient for the induction of cellulases. Similar results were observed by Satyanarayana [36]; *C. thermophile, S. thermophile*, and *T. aurantiacus* produced a low level of cellulolytic enzymes in a medium devoid of cellulose or other known inducers.

Glucose represses the synthesis of cellobiohydrolase and endoglucanase in the presence of cellulose and other inducers. Mixtures of inducer and repressor produce a delayed induction. In *S. pulverulentum*, cellobiose is a better inducer than sophorose, while the reverse is true in *T. reesei* [216]. Glucose catabolism was a prerequisite for the induction of carboxymethylcellulase (CMCase) synthesis in *S. thermophile* [45]. The extracellular β-glucosidase is repressed by glucose and induced by β-glucosides like methyl-β-D-glucoside, amygdalin, salicin, and cellulose, but not by cellobiose [217]. However, Lusis and Becker [202] reported induction of cellobiase by cellulose and cellobiose; its synthesis was inhibited by the inhibitors of protein synthesis (cycloheximide) and metabolism (sodium azide, dinitrophenol).

Efforts have been made to develop hyperproductive cellulase mutants of thermophilic molds [180,185,187,218]. Hyperproductive cellulase mutants were obtained less often with UV than with nitrosoguanidine; a 1.5- to 2.2-fold increase in CMCase was found in mutants [218]. Since the increase was not proportional in all components of the cellulase system, Sen and Chakrabarty [218] concluded that genes controlling the production of the cellulase complex might not be under coordinate control in *M. thermophila*. Hayashida and Mo [180] obtained a protease-negative mutant no. 140 of *H. grisea* var. *thermoidea* with UV and nitrosoguanidine which exhibited enhanced endoglucanase activity. Moloney et al. [185] develoepd mutants of *T. emersonii* DBS 814.70 by using nitrosoguanidine with an improved capacity to produce different components of the cellulase system. Another mutant UV7 obtained with UV produced an enhanced amount of cellulase during growth on cellulose, lactose, and glucose-containing media.

The cellulolytic enzymes of *C. cellulolyticum* [170], *C. thermophile* [172,202], *H. grisea* var. *thermoidea* [180], *H. insolens* [107,171,196], *M. thermophila* [168,194,204,205], *S. thermophile* [169,170,213], *T. emersonii* [43,176], *T. aurantiacus* [44,174,192,193], and *Thielavia terrestris* [189,200] were fractionated by standard protein separation techniques such as solvent precipitation/salt precipitation, ion exchange chromatography, gel filtration, and isoelectric focusing. The cellulase systems of these molds contained all the three components, i.e., *exo*-β-1,4-glucanase (cellobiohydrolase), *endo*-β-1,4-glucanase, and β-glucosidase; the synergistic action of these enzymes is required for hydrolysis of cellulose [41]. Multiplicity in these components has also been observed in different molds. In the absence of O_2, cellulose breakdown was reduced by one-half in *S. pulverulentum* [35]; the two oxidizing systems were cellobiose oxidase and cellobiose-quinone oxidoreductase. An oxidative system did not appear to be involved in the degradation of cellulose by *T. emersonii* [203]. The *Myceliophthora thermophila* cellulase system consisted of eight components [205], while that of *T. emersonii* comprised five forms of exoglucanase, at least two forms of endoglucanase, and three forms of β-glucosidase [76]. Intracellular β-glucosidases were found in *S. thermophile* [225], *T. aurantiacus* [193], and *T. terrestris* [189]. In *T. aurantiacus* [193] intracellular β-glucosidase was an acid, dimeric glycoprotein that could be dissociated with 6 M urea to still active monomers. Its activity was inhibited by glucono-δ-lactone, suggesting the presence of at least one carboxylate ion together with a histidine residue in protonated form in the active site. The cellulases were more acid-stable. Multiple components of cellulases appear to be common even in thermophiles [9]. Cellobiose dehydrogenase was found to occur in *S. thermophile* [169]. Its induction was dependent on pH and cellobiose concentration. Based on their observations, Hayashida and Mo [180] reported the occurrence of two types of endoglucanases in *H. grisea* var. *thermoidea*, Avicel disintegrating, and Avicel nondisintegrating. The most common range of molecular weights of endonuclease in the above was 20-87 kDa. However, β-glucosidase of

H. insolens [171] and *S. pulverulentum* [220] appear to be larger with molecular weights of 250 and 165-182 kDa, respectively. The isoelectric point for most of the cellulase components is below 6.0 [9]. Temperature and pH optima for cellulolytic activity are in the range 50-70°C and 4-5, respectively. However, Durand et al. [200], Goksoyr and Eriksen [221], and Wojtczak et al. [231] remarked that there is little difference in thermal stability between the cellulases of thermophilic and mesophilic molds.

Glycosylation may be necessary for secretion of cellulases by *Talaromyces emersonii* since all of the components in culture filtrates were glycosylated [176,219]. By using tunicamycin to block *N*-glycosylation (attachment of carbohydrate to asparagine side chain) and 2-deoxyglucose to impair *N*- and *o*-glycosylation (attachment of carbohydrate to serine and threonine side chains), Montenecourt's group [222] postulated that *N*-glycosylation is required for thermal and pH stability of secreted cellulases and for protection against proteolysis. Similar observations were made in *H. insolens* and *M. meihei* cellulases [107,111,171,223]. Removal of carbohydrate either partially or completely rendered cellulases less heat- and pH-stable.

The antibodies of two enzyme antigens (β-glucosidase and endocellulase) of *H. insolens* exhibited complete cross-reactivity, suggesting the presence of common epitope on the two enzyme proteins [196]. However, the two enzymes differed in their amino acid composition and substrate specificity. Kinoshita et al. [190] reported continuous hydrolysis of cellulose by using crude enzymes of *Sporotrichum cellulophilum* in an ultrafilter membrane reactor with the intermittent addition of cellulose. During 5 days of operations, the product concentration declined from 30 to 16 g^{-1} at a flow rate of 3-30 ml/h. The inactivation of the system appeared to be due to a protease present in the cellulase preparation; the activity of protease was remarkably high at higher temperature.

The potential application of cellulases are in increasing the tensile strength of high α-sulfite pulp; recovering agar and the production of sea weed jelly; processing fibrous materials as ruminant feed from delignified bark; saccharifying the delignified cellulosic wastes; treating rice grains for sake brewing; removing soybean seed coat; extracting green tea components; isolating soybean and coconut proteins, sweet potato, and corn starch; modifying rice tissue, carrot, or other foodstuffs; producing fruit vinegar or pectin from citrus pulp; improving the solubility of raw materials in brewing; digesting excreta in septic tanks; supplementing enzymes in feed for poultry and pigs; and preparing protoplasts from higher plants and fungi [167,215]. These applications of cellulases are nonprofitable at present due to economic cost and their limited efficiency [9]. Cellulases have been employed for saccharification of wastes such as saw dust, sugarcane bagasse, straw, bark, etc., thus achieving pollution abatement. Vandamme et al. [110] showed the possibility of using cellulases of *A. fumigatus* in the preparation of protoplasts from barley callus tissue.

6. Other Enzymes

Malbranchea pulchella var. *sulfurea* was reported to produce extracellular enzymes such as polygalacturonase, pectinase, and chitinase [226]. This mold secretes a good level of polygalacturonase in a medium containing glucose as the carbon source. Polygalacturonase showed optimum activity at pH 4-5 and at 50°C, with K_m and V_{max} values 0.17% (w/v) and 37.03 μg/ml, respectively. Several other thermophilic molds were noted to produce pectinolytic enzymes [234].

Craveri and Colla [106] showed that *P. dupontii*, *H. lanuginosa*, *M. pulchella*, and *Thermoascus* sp. produce extracellular ribonucleases. Thermophilic and mesophilic fungal ribonucleases were almost alike in temperature and pH optima for activity. *Chaetomium thermophile* was reported to excrete polyphenol oxidase [230,269]. A patent was issued to Miles [232] for preparing an enzyme transglucosidase from *T. dupontii*. This enzyme is highly thermostable since it retains at least 40% activity after incubation at 70°C for 60 min at pH 5.5. This can be used for the conversion of maltose to panose, isomaltose, and glucose if the reaction is performed at 50-70°C for 1-6 days. β-Glucosaminidase was detected in the extracellular enzyme preparations of *H. insolens*, *T. emersonii*, and *T. aurantiacus* by Apizym technique, while phosphoamidase was produced by *H. insolens*, *S. thermophile*, and *T. emersonii* [35]. Several thermophilic molds have been observed to excrete thermostable lactases [48].

Thermomyces lanuginosa synthesizes constitutively an intracellular glucosidase, trehalase (α,α-glucoside-1-glucohydrolase) in glucose asparagine midum at 50°C [61]. A high level of this enzyme was found in nonsporulating mycelium as compared to that in sporulating mycelium. The trehalases of mesophilic (*Neurospora crassa*) and thermophilic (*T. lanuginosa*) isolates had similar temperature and pH optima for activity, but differed in molecular weight, electrophoretic mobility, and k_m [62]. Proteinaceous factors in the mycelium, casein, and bovine serum albumin provided protection against heat inactivation to trehalase of the thermophilic mold, but not that of the mesophile. The protective factors were sensitive to proteolytic enzymes [235]. The purified enzyme was less thermostable than the crude cell-free extract.

B. Exopolysaccharides

The production of expolysaccharides by these molds has not received adequate attention thus far. Chavant and his coworkers in France (personal communication) separated an exopolysaccharide from the culture filtrates of *M. pulchella* grown in Emerson's YpSs medium at 45°C for 3-6 days under stationary conditions. This polysaccharide showed antitumor activity.

C. Bioconversion of Sterols and Steroids, and Binding of Sterols

Thermophilic molds lack the enzymatic machinery to break down the aliphatic side chain of sterols [236]. However, *A. alabamensis* and *T. emersonii* transformed cholesterol to cholestenone, and stigmasterol and sitosterol to stigmastadienone. All the thermophilic molds tested exhibited avidity for binding sterols to the mycelium and metabolized sterols. Colombie et al. [237] demonstrated that *A. alabamensis* strongly absorbs progesterone using ^{14}C-labeled progesterone. This mold transforms progesterone to 6β-hydroxyprogesterone and 11 α-progesterone by the intracellular enzymes and these steroids are excreted into the medium.

D. Upgradation of Animal Feeds and SCP

The cellulolytic thermophiles have certain advantages over mesophiles, namely, high rates of cellulose breakdown, good sources of protein (being low in nucleic acid levels), activity over a wide range of temperatures

920-950°C, and higher specific growth rates [88]. *Chaetomium cellulolyticum* [46,238-241,248-251] and *S. pulverulentum* [242-246] have been used to upgrade animal feeds and to produce single-cell protein. *C. cellulolyticum* has been reported to be a good system [239] for the conversion of hemicellulose and cellulose to SCP. This organism was found to be fast-growing on 1% cellulose with a protein synthesis rate of 0.09 h^{-1}; this rate increased to 0.3 h^{-1} when 1% cellulose was added after the exhaustion of the first batch of substrate [250]. In liquid medium the growth was rapid with a specific growth rate of 0.14 h^{-1} and a doubling time of about 5 h [249]. As compared to *T. viride*, this mold has 50-100% faster growth rates and over 80% more final biomass [248]. The amino acid composition of *C. cellulolyticum* was better than that of *T. viride* and compared favorably with FAO reference protein, alfalfa, and soya meal. Acid pretreatment appeared to be better than alkali pretreatment if the product is to be used as a ruminant feed in terms of rate and the extent of protein production in the insoluble biomass product. Feeding trials suggested that the SCP products are suitably nutritious, digestible, and nontoxic in animal feed protein rations. The composition and properties of SCP prepared by the Waterloo process are given in Table 3 [240]. The average protein production by the parent strain of this mold was 0.006 g/liter, which gives a product with 5% protein content, while with a mutant 7S/7 was 0.01 g/liter/h, which gave a product with 13.3% crude protein [238]. Alkali (1% NaOh)-treated hardwood sawdust after fermentation for 48 h contained 12.5% crude protein and its in vitro digestibility was 30% [241].

Ghai et al. [252] used the thermotolerant fungi *C. cellulolyticum* and *Actinomucor* for producing SCP on canning industry wastes. Solid substrate fermentation of newsprint for 6 days with *S. thermophile* increased protein content from 0 to 6.5% of the total dry weight [253,254].

Leafy waste material was converted to brown-colored product with increased N, P, and K contents without foul smell by using *Sporotrichum* sp. D-14, *Mucor* sp. C-16, and *Humicola* sp. H-37 in 90 days [255]. The protein content did not increase beyond 10% in the *S. pulverulentum* fermented ball-milled and alkali-pretreated rice hulls, but the produced biomass contained 30-40% protein with a favorable amino acid composition [245]. Wastewater from pulp and fiber board mills consisting of water-soluble mono- and oligosaccharides and lignocellulosic material was fermented in an economically feasible way due to the combined effect of protein production and water purification [243]. The degree of purification was dependent on residence time. The residence time of 15 h resulted in a 60% reduction in BOD. The investigation on the nutritional aspects of the biomass produced with sheep gave more satisfactory results than with monogastric species (rats), with a digestibility of 82% crude protein and metabolizable energy of 70% of soya bean oil meal [246].

Among the thermophilic molds used, *T. aurantiacus* and *S. thermophile* produced a twofold increase in protein content of sugar beet pulp in 48 h [247]. Protein productivity per cubic meter working volume of the fermentor was 0.2 kg/h for *T. aurantiacus* and 0.162 kg/h for *S. thermophile*. Grajek [247] felt that the use of thermophiles could simplify the technology of processing by eliminating the need for heat removal and reducing contamination hazards.

Thermophilic *A. fumigatus* G-2-6 decolorized 75% melanoidin of molasses solution when this strain was cultivated at 45°C for 3 days in glycerol peptone medium [251]. Molasses melanoidin-decolorizing activity was found

TABLE 3. Composition and Properties of Waterloo Fungal SCP Product Prepared by Using *Chaetomium cellulolyticum*

Composition	
Protein	45% (dry matter)
Carbohydrate	35%
Fat	10%
Nucleic acid	5%
Vitamins	High
Minerals, ash	5%
Properties	
Color	Gray
Texture	Granular
Odor	Mushroomy
Digestibility	73%
Moisture	8%
Biomass content	73%
Lignin	12%
Ash	12%

Source: Ref. 240.

TABLE 4. Antibiotics Produced by Thermophilic Molds

Mold	Antibiotic	Antimicrobial spectrum	Refs.
Malbranchea pulchella	Penicillin	Antibacterial	258,261,262
	Tf-26 VX	-do-	263
	Malbranchin A	-do-	264
	Malbranchin B	-do-	264
Malbranchea 38-1	6-Aminopenicil-lanic acid	-do-	266
Malbranchea 34-n-1	-do-	-do-	266
Thermoascus sp.	-do-	-do-	266
Myriococcum albomyces	Myriocin	Antifungal	265
Mucor pusillus	Sillucin	Antibacterial	267
M. miehei	Miehein	-do-	267
Sterile mycelia	Thermozymocidin	Antifungal	268

superior to that by Basidiomycetes. Decolorization of kraft bleach plant effluent by *S. pulverulentum* was reported by Sundman et al. [256].

E. Antibiotics and Volatile Sporostatic Factors

Thermophilic molds have been reported to produce a number of antibacterial and antifungal antibiotics (Table 4). These molds also produce volatile inhibitory metabolites [257]. Mycelial growth of *H. lanuginosa* and *T. indicae-seudaticae* was inhibited by the volatiles produced by *M. pulchella* [258].

F. Phenolics, Organic Acids, and Amino Acids

Thermophilic molds such as *Mucor* sp., *Talaromyces* sp., and others have been reported to produce extracellular phenolic compounds and organic acids (lactic and citric acids) [259]. Subrahmanyam [260] observed some thermophilic molds to excrete different amino acids like glutamic acid, lysine, and alanine depending on the cultural conditions.

G. Thermophilic Mold Enzymes in Processing of Oil Seeds

During the recent study, Johri et al. (unpublished) observed better yield of oil from seeds contaminated with the thermophilic molds in storage. They obtained 5% higher yields in oil from cotton, sunflower, soybean, and castor using crude enzyme preparations of *A. fumigatus*, *H. lanuginosa*, and *S. thermophile*. Pure cellulase and hemicellulase from commercial sources were not as effective as the crude enzyme preparations. This is contrary to the known role of these molds in deteriorating oil quality.

H. Experimental Systems for Genetic Manipulation

As mentioned earlier, several attempts have been made to enhance cellulolytic enzymes by genetic manipulation of thermophilic molds by induced mutations using both chemical and physical mutagens. In order to isolate catabolite-resistant mutants of *T. emersonii* CBS 814.70, Moloney et al. [185] obtained a morphological mutant UCG42 by treating the parent strain with nitrosoguanidine; this mutant had an improved filter paper hydrolyzing activity. Ultraviolet irradiation of UCG42 enabled them to isolate more mutants with further enhancements in cellulase activity. All UCG mutants, unlike the wild-type CBS 814.70, effected clarification of cellulose-containing agar in the presence of D-glucose. However, in liquid cultures, neither mutants nor the wild-type strain produced cellulase in the presence of D-glucose. Addition of D-glucose led to the development of acidity with pH values as low as 1.9. This led to cessation of cellulase production or the rapid inactivation of cellulases. Moloney et al. [185] were of the opinion that by preventing pH from falling too low, the mutant strains might show greater enzyme activity than the wild type in the presence of D-glucose. This is an instance of suitability of thermophilic molds as experimental systems for genetic manipulations for achieving the desired objectives.

Fanrich and Irrgang [238] isolated five variants of *C. cellulolyticum* which differed in morphological and physiological characteristics. They obtained a strain 7S/7 from the progeny of a cross between a variant 4S and another 7S. The strain 7S/7 produced more CMCase than the parent strain. Sexual strains of thermophiles may offer a chance for genetic manipulations through breeding. Another possibility of genetic manipulation for achieving genetic recombination via protoplast fusion was shown by Johri and his coworkers (unpublished). Amino acid-requiring mutants were developed by using mutagens, and protoplast fusion was established in the presence of polyethylene glycol. Protoplasts were prepared by the method described by Sharma et al. [271] employing Novozyme SP249 and cellulase CP.

IV. CONCLUSIONS

Since thermophilic molds are economically very important, an understanding of their genetics is essential. The knowledge of genetics of the useful thermophilic fungal strains may help in generating hyperproductive strains and improving production of metabolites, potent agents of bioconversion and biodegradation. The techniques of genetic engineering such as gene amplification have opened up new vistas in the microbial production of the desired products.

Polyamines have been shown to play a crucial role in growth and differentiation, stabilization of nucleic acids, and protein synthetic organelles to function at high temperatures in thermophilic bacteria. Similar investigations are necessary to gain the knowledge of polyamines of these molds and the occurrence of any novel polyamines specific to this group. Information on the pathway of polyamine biosynthesis will help in employing polyamine biosynthesis inhibitors in controlling diseases caused by these molds and biodeterioration.

In the domain of biotechnology, immobilized systems have already found wide-ranging applications since they bypass many of the shortcomings which frequently make soluble enzymes commercially impractical. The superiority of immobilized systems stems from cheaper cost due to reuse, improved stability, and from the possibility of influencing the efficacy of the biocatalyst by modifying its microenvironment. However, very few attempts have been undertaken to investigate the feasibility of thermophilic fungal strains to achieve efficient new biocatalytic systems. Rao et al. [121] reported thermal stabilization of glucoamylase of *T. lanuginosus* by physically trapping it in polyacrylamide gels. McHale [206] observed twofold enhancement in cellulase production in immobilized *T. emersonii* CBS 814.70 as compared to that in nonimmobilized system. Tamada et al. [122] noted a high diffusion of air and cell growth in *S. cellulophilum* immobilized in a polymer made of 5% monomer polyethylene glycol No. 1000 demethacrylate (23G). Spores of *S. thermophile* immobilized in alginate beads could grow and produce extracellular cellulase [207]. Recently, Singh et al. [273] noted that immobilized dry cells of *Malbranchea* retained penicillin acylase activity up to 90 days. Immobilized cells could be maintained for 40 days with good reuse efficiency until the tenth cycle. These observations suggest a need for greater efforts to apply cell and enzyme immobilization technology to thermophilic fungal systems. There is also a need to look for novel biocatalytic properties of thermophilic molds.

REFERENCES

1. Tansey, M. R., and Brock, T. D. The upper temperature limit for eukaryotic organisms. *Proc. Natl. Acad. Sci. USA*, *69*:2426-2428, 1972.

2. Crisan, E. V. Isolation and culture of thermophilic fungi. *Contr. Boyce Inst. Pl. Res.*, *22*:291-301, 1964.
3. Cooney, D. G., and Emergon, R. *Thermophilic Fungi: An Account of Their Biology, Activities, and Classification*. W. H. Freeman, San Francisco, 1964, p. 188.
4. Emerson, R. Thermophiles. In *The Fungi*, Vol. 3 (G. C. Ainsworth and A. S. Sussman, eds.), Academic Press, New York, 1968, pp. 105-128.
5. Tansey, M. R., and Brock, T. D. Microbial life at high temperatures: ecological aspects. In *Microbial Life in Extreme Environments* (D. J. Kushner, ed.), Academic Press, London, 1978, pp. 131-216.
6. Johri, B. N. Biology of thermophilic fungi. In *Recent Advances in Biology of Microorganisms* (K. S. Bilgrami and K. M. Vyas, eds.), Mahendrapal Singh, Dehradun, 1980, pp. 265-277.
7. Johri, B. N., and Satyanarayana, T. Ecology of thermophilic fungi. In *Progress in Microbial Ecology* (K. G. Mukherji, V. P. Agnihotri, and R. P. Singh, eds.), Print House, Lucknow, 1984, pp. 349-361.
8. Johri, B. N., and Satyanarayana, T. Thermophilic moulds: perspectives in basic and applied research. *Indian Rev. Life Sci.*, *6*:75-100, 1986.
9. Satyanarayana, T., Jain, S., and Johri, B. N. Cellulases and xylanases of thermophilic moulds. In *Perspectives in Mycology and Plant Pathology* (Professor B. S. Mehrotra Festschrift), (V. P. Agnihotri, A. K. Sarbhoy, and D. Kumar, eds.), Malhotra Publishing House, New Delhi, 1988, pp. 24-60.
10. Johri, B. N., Jain, S., and Chauhan, S. Enzymes from thermophilic fungi: proteases and lipases. *Proc. Indian Acad. Sci. (Plant Sci.)*, *94*:175-196, 1985.
11. Maheswari, R., Kamalam, P. T., and Balasubramanyam, P. V. The biogeography of thermophilic fungi. *Curr. Sci.*, *56*:151-155, 1987.
12. Deploey, J. J. Carbohydrate nutrition of Mucor miehei and Mucor pusillus. *Mycologia*, *68*:190-194, 1976.
13. Sahm, D. F., and Chapman, E. S. Effects of carbon sources on Papulaspora thermophila at various temperatures. *Mycologia*, *68*:168-174, 1976.
14. Satyanarayana, T., and Johri, B. N. Thermophilic fungi of paddy straw compost: growth, nutrition and temperature relationships. *J. Indian Bot. Soc.*, *63*:164-170, 1984.
15. Rosenberg, S. L. Temperature and pH optima for 21 species of thermophilic and thermotolerant fungi. *Can. J. Microbiol.*, *21*:1535-1540, 1974.
16. Subrahmanyam, A. Nutritional requirements of Torula thermophila Cooney and Emerson at two different temperatures. *Nova Hedwigia*, *19*:85-89, 1977.
17. Subrahmanyam, A. Studies on Thermoascus aurantiacus. *Acta Mycologia*, *16*:121-131, 1980.
18. Basu, M. Utilization of hemicellulose by thermophilic microorganisms. *Natl. Acad. Sci. Lett.*, *3*:143-144, 1980.
19. Deploey, J. J. Pectin utilization by thermophilic fungi. *Proc. Pennsylvania Acad. Sci.*, *50*:179-181, 1976.
20. Reese, E. T. Arobic decomposition of cellulose by organisms at temperatures above 40°C. Ph.D. thesis, Pennsylvania State College, 1946.
21. Chauhan, S., Prakash, A., Satyanarayana, T., and Johri, B. N. Utilization of C-1 compounds by thermophilic fungi. *Natl. Acad. Sci. Lett.*, *8*:167-169, 1985.

22. Wali, A. S., Mattoo, A. K., and Modi, V. V. Stimulation of growth and glucose catabolite enzymes by succinate in some thermophilic fungi. *Arch. Microbiol.*, *118*:49-53, 1978.
23. Asundi, S., Mattoo, A. K., and Modi, V. V. Studies on a thermophilic Talaromyces sp.: mitochondrial fatty acids and requirement for a C_4 compound. *Indian J. Microbiol.*, *14*:75-80, 1974.
24. Gupta, S. D. and Maheswari, R. Is organic acid required for nutrition of thermophilic fungi? *Arch. Microbiol.*, *141*:164-169, 1985.
25. Prasad, A. R. S., Kurup, C. K., and Maheswari, R. Effect of temperature on respiration of a mesophilic and a thermophilic fungus. *Plant Physiol.*, *64*:347-348, 1979.
26. Deploey, J. J., and Fergus, C. L. Growth and sporulation of thermophilic fungi and actinomycetes in O_2-N_2 atmospheres. *Mycologia*, *67*:780-797, 1975.
27. Singh, C. P., Mishra, M. M., and Yadav, K. S. Solubilization of insoluble phosphates by thermophilic fungi. *Ann. Microbiol. (Inst. Pasteur)*, *131B*:289-296, 1980.
28. Mehrotra, B. S. and Mehrotra, B. M. Induction of zygospores at high temperatures in the thermophilic species Mucor miehei with aspartic acid and phenylalanine. *Experientia*, *36*:53-54, 1980.
29. Joshi, M. C. Studies on the mucoraceous fungi of Kumaun region. Ph.D. thesis, Kumaun University, Nainital (U.P.), 1982, p. 145.
30. Endo, T. Studies on protease produced by thermophilic bacteria. *J. Ferment. Technol.*, *40*:346-353, 1962.
31. Ohta, Y. Thermostable protease from thermophilic bacteria. *J. Biol. Chem.*, *242*:509-515, 1967.
32. McHale, A., and Coughlan, M. P. The cellulolytic system of Talaromyces emersonii. Purification and characterization of the extracellular and intracellular β-glucosidases. *Biochem. Biophys. Acta*, *662*:152-159, 1981.
33. Mehrotra, B. S. Thermophilic fungi: biological enigma and tools for the biotechnologist and biologist. *Indian Phytopathol.*, *38*:211-229, 1985.
34. Zuber, H. Sequence analysis of peptides with citrus carboxy peptidase and with a thermophilic carboxypeptidase from Talaromyces dupontii. Abst., 7th Int. Congr. Biochem., Tokyo, 3:541-542, 1967.
35. Satyanarayana, T., Chavant, L., and Montant, C. Applicability of API zym for screening enzyme activity of thermophilic moulds. *Trans. Br. Mycol. Soc.*, *85*:727-730, 1985.
36. Satyanarayana, T. *Thermophilic Microorganisms and Their Role in Composting Process*. Ph.D. thesis, Sagar University, India, 1978, p. 213.
37. Rao, V. B., Maheswari, R., Sastri, N. V. S., and Rao, P. V. S. Thermostable glucoamylase from thermophilic fungus Thermomyces lanuginosa. *Curr. Sci.*, *48*:113-115, 1979.
38. Rao, V. B., Sastri, N. V. S., and Rao, P. V. S. Purification and characterization of a thermostable glucoamylase from thermophilic fungus Thermomyces lanuginosa. *Biochem. J.*, *193*:379-387, 1981.
39. Subrahmanyam, A., Mangallam, S., and Gopalakrishnan, K. S. Amyloglucosidase production by Torula thermophila. *Indian J. Exp. Biol.*, *15*:495-496, 1977.
40. Adams, P. R., and Deploey, J. J. Amylase production by Mucor pusillus and Mucor miehei. *Mycologia*, *68*:934-938, 1976.

41. Eriksen, J. J. and Goksoyr, J. Cellulases from Chaetomium thermophile var. dissitum. *Eur. J. Biochem.*, *77*:445-450, 1977.
42. Jain, M. K., Kapoor, K. K., and Mishra, M. M. Cellulase activity, degradation of cellulose and lignin by thermophilic fungi. *Trans. Br. Mycol. Soc.*, *73*:85-89, 1979.
43. Coughlan, M. P., and McHale, A. The cellulolytic system of Talaromyces emersonii: identification of various components produced during growth on cellulosic media. *Biochem. Biophys. Acta*, *662*:145-151, 1981.
44. Tong, C. C., Cole, A. L., and Shepherd, M. G. Purification and properties of the cellulases from the thermophilic fungus Thermoascus aurantiacus. *Biochem. J.*, *191*:83-94, 1980.
45. Canevascini, G., Coudray, M. R., Rey, J. P., Southgate, R. J. G., and Meir, H. Induction and catabolite repression of cellulase synthesis in the thermophilic fungus Sporotrichum thermophile. *J. Gen. Microbiol.*, *110*:291-303, 1979.
46. Chahal, D. S., Swan, J. E., and Young, M. M. Protein and cellulase production by Chaetomium cellulolyticum grown on wheat straw. *Dev. Ind. Microbiol.*, *18*:433-442, 1977.
47. Palanivelu, P., Balasubramanyam, P. V., and Maheswari, R. Coinduction of sucrose transport and invertase activities in thermophilic fungus Thermomyces lanuginosus. *Arch. Microbiol.*, *139*:44-47, 1984.
48. Sorensen, S. Y., and Crisan, E. Y. Thermostable lactase from thermophilic fungi. *J. Food Sci.*, *39*:1184-1187, 1974.
49. Satyanarayana, T., and Johri, B. N. Lypolytic activity of thermophilic fungi of paddy straw compost. *Curr. Sci.*, *50*:680-682, 1981.
50. Liu, W. H., Beppu, T., and Arima, K. Cultural conditions and some properties of lipase of Humicola lanuginosa S-38. *Agr. Biol. Chem.*, *36*:1919-1924, 1974.
51. Somkuti, G. A., and Babel, F. J. Lipase activity of Mucor pusillus. *Appl. Microbiol.*, *16*:617-619, 1968.
52. Ogundero, V. W. Lipase activity of thermophilic fungi from mouldy ground nuts in Nigeria. *Mycologia*, *72*:118-126, 1980.
53. Satyanarayana, T., and Johri, B. N. Extracellular protease of thermophilic fungi isolated from decomposing paddy straw. *Trop. Plant Sci. Res.*, *1*:137-140, 1983.
54. Ramabadran, R. Production of amylase and protease by thermophilic fungi. *Annamalai Univ. Agr. Res. Ann.*, *2*:45-47, 1970.
55. Ong, P. S. and Gaucher, G. M. Protease production by thermophilic fungi. *Can. J. Microbiol.*, *19*:129-133, 1973.
56. Ong, P. S. and Gaucher, G. M. Production and characterization of thermomycolase, the extracellular serine protease of the thermophilic fungus Malbranchea pulchella var. sulfurea. *Can. J. Microbiol.*, *22*:165-176, 1976.
57. Khan, M. R., Blain, J. A., and Patterson, J. D. E. Extracellular proteases of Mucor pusillus. *Appl. Environ. Microbiol.*, *37*:719-724, 1983.
58. Khan, M. R., Blain, J. A., and Patterson, J. D. E. Partial purification of Mucor pusillus intracellular proteases. *Appl. Environ. Microbiol.*, *45*:94-96, 1979.
59. Somkuti, G. A., and Babel, F. J. Conditions influencing the synthesis of acid protease by Mucor pusillus Lindt. *Appl. Microbiol.*, 1309-1312, 1967.

60. Shenolikar, S., and Stevenson, K. J. Purification of thiol proteinase from the thermophilic fungus Humicola lanuginosa. *Biochem. J., 205*: 147-152, 1982.
61. Prasad, A. R. S., and Maheswari, R. Growth and trehalase activity in the thermophilic fungus Thermomyces lanuginosa. *Proc. Indian Acad. Sci., 87B*:231-241, 1978.
62. Prasad, A. R. S., and Maheswari, R. Temperature response of trehalase from a mesophilic (Neurospora crassa), and a thermophilic (Thermomyces lanuginosus) fungus. *Arch. Microbiol., 122*:257-280, 1979.
63. Satyanarayana, T., and Johri, B. N. Variation in xylanolytic activity of thermophilic fungi. *Bionature, 3*:39-41, 1983.
64. Margaritis, A., Merchant, R., and Yaguchi, M. Xylanase, CM-cellulase and avicelase production by the thermophilic fungus Sporotrichum thermophile. *Biotechnol. Lett., 5*:265-270, 1983.
65. Yoshioka, H., Chavanich, S., Nilubol, N., and Hayashida, S. Production and characterization of thermostable xylanase from Talaromyces byssochlamydoides YH-50. *Agr. Biol. Chem., 45*:579-586, 1987.
66. Merchant, R., Merchant, F., and Margaritis, A. Production of xylanase by the thermophilic fungus Thielavia terrestris. *Biotechnol. Lett., 10*:513-516, 1988.
67. Tan, L. U. L., Mayers, P., and Saddler, J. N. Purification and characterization of a thermostable xylanase from a thermophilic fungus Thermoascus aurantiacus. *Can. J. Microbiol., 33*:689-692, 1987.
68. Dubey, A. K., and Johri, B. N. Xylanolytic activity of thermophilic Sporotrichum sp. and Myceliophthora thermophilum. *Proc. Indian Acad. Sci. (Plant Sci.), 97*:247-255, 1987.
69. Maheswari, R., and Kamalam, P. T. Isolation and culture of a thermophilic fungus Melanocarpus albomyces, and factors influencing the production and activity of xylanase. *J. Gen. Microbiol., 131*:3017-3027, 1985.
70. Dubeau, H., Chahal, D. S., and Ishaque, M. Xylanase of Chaetomium cellulolyticum: its nature of production and hydrolytic potential. *Biotechnol. Lett., 9*:275-280, 1987.
71. Thakre, R. P. Ecological and physiological studies of soil fungi (thermophilic) from coal-mine localities in Madhya Pradesh. Ph.D. thesis, Sagar University, India, 1975, p. 163.
72. Thakre, R. P., and Johri, B. N. Occurrence of thermophilic fungi in coal-mine soils of Madhya Pradesh. *Curr. Sci., 45*:271-273, 1976.
73. Chang, Y., and Hudson, H. J. The fungi of wheat straw compost. I. Ecological studies. *Trans. Br. Mycol. Soc., 50*:649-666, 1967.
74. Kane, B. E., and Mullins, J. T. Thermophilic fungi in municipal waste compost system. *Mycologia, 65*:1087-1100, 1973.
75. Sandhu, D. K., and Sidhu, M. S. The fungal succession on decomposing sugar cane bagasse. *Trans. Br. Mycol. Soc., 75*:281-286, 1980.
76. Tansey, M. R. Isolation of thermophilic fungi from self-heated industrial wood chip piles. *Mycologia, 63*:537-547, 1971.
77. Satyanarayana, T., Johri, B. N., and Saksena, S. B. Seasonal variation in mycoflora of nesting materials of birds with special reference to thermophilic fungi. *Trans. Br. Mycol. Soc., 68*:307-309, 1977.
78. Singh, P. Thermophilous microfungi of soil and air and their enzymatic activity. Ph.D. thesis, Bhopal University, India.
79. Singh, S., and Sandhu, D. K. Growth response of some thermophilic fungi at different incubation temperatures. *Proc. Indian Acad. Sci. (Plant Sci.), 91*:153-158, 1980.

80. Millner, P. D. Radial growth responses to temperature by 58 Chaetomium species, and some taxonomic relationships. *Mycologia, 69*:492-502, 1977.
81. Quereshi, A. R., and Johri, B. N. Temperature relationship of some thermophilic fungi. *Bull. Bot. Soc. Univ. Sagar, 19*:29-30, 1972.
82. Stolk, A. C. Thermophilic species of Talaromyces Benjamin and Thermoascus Miehei. *Antonie von Leeuwenhoek, 31*:262-276, 1965.
83. Fergus, C. L., and Sinden, J. W. A new thermophilic fungus from mushroom compost: Thielavia thermophila sp. nov. *Can. J. Bot., 47*: 1635-1637, 1969.
84. Tansey, M. R. Effect of temperature on growth rate and development of the thermophilic fungus Chaetomium thermophile. *Mycologia, 64*: 1290-1299, 1972.
85. Crisan, E. V. Current concepts of thermophilism and the thermophilic fungi. *Mycologia, 65*:1171-1198, 1973.
86. Oshima, T. Unusual polyamines in an extreme thermophile, Thermus thermophilus. *Adv. Polyamine Res., 4*:479-487, 1983.
87. Chang, S. B., and Matson, R. S. Membrane stability (thermal) and nature of fatty acids in yeast cells. *Biochem. Biophys. Res. Commun., 46*:1529-1535, 1972.
88. Seal, K. J., and Eggins, H. O. W. The upgrading of agricultural wastes by thermophilic fungi. In *Food from Wastes* (G. G. Birch, K. J. Parker, and J. T. Worgan, eds.), Applied Sciences, London, 1976, pp. 58-78.
89. Amelunxen, R., and Lins, M. Comparative thermostability of enzymes from Bacillus stearothermophilus and Bacillus cereus. *Arch. Biochem. Biophys., 125*:765-769, 1968.
90. Koffler, H. Protoplasmic differences between mesophiles and thermophiles. *Bacteriol. Rev., 21*:227-240, 1957.
91. Sonnleitner, B. Biotechnology of thermophilic bacteria: growth, products, and application. *Adv. Biochem. Eng. Biotechnol., 28*:69-138, 1983.
92. Koffler, H., Mallett, G. E., and Adye, J. Molecular basis of biological stability to high temperatures. *Proc. Natl. Acad. Sci USA, 43*:464-477, 1957.
93. Smith, R. W., and Koffler, H. Bacterial flagella. *Adv. Microb. Physiol., 6*:219-339, 1971.
94. Friedman, S. M. Protein synthesizing machinery of thermophilic bacteria. *Bacteriol. Rev., 32*:27-38, 1968.
95. Stenesh, J., and Yang, C. Characterization and stability of ribosomes from mesophilic and thermophilic bacteria. *J. Bacteriol., 93*:930-936, 1967.
96. De Bertoldi, M., Lepidi, A. A., and Nuti, M. P. Significance of DNA base composition in classification of Humicola and related genera. *Trans. Br. Mycol. Soc., 60*:77-85, 1973.
97. Nageswara Rao, J. S., and Cherayil, J. D. Minor nucleotides in the ribosomal RNA of Thermomyces lanuginosus. *Curr. Sci., 48*:1-5, 1979.
98. Crisan, E. V. The proteins of thermophilic fungi. In *Current Topics in Plant Sciences* (J. E. Gunckel, ed.), Academic Press, New York, 1969, pp. 32-33.
99. Somkuti, G. A., Babel, F. J., and Somkuti, A. C. Cellulolysis by Mucor pusillus. *Appl. Microbiol., 17*:888-892, 1969.
100. Chapnis, R., and Zuber, H. Thermophilic aminopeptidases: API from Talaromyces duponti. *Meth. Enzymol., 19*:552-556, 1970.

101. Craveri, R., Manachini, P. L., and Aragozini, F. Thermostable and alkaline lipolytic and proteolytic enzymes produced by a thermophilic mold. *Mycopathol. Mycol. Appl.*, *54*:193-204, 1974.
102. Chauhan, S. Extracellular protease of thermophilic fungi, Ph.D. thesis, Bhopal University, India.
103. Loginova, L. G., and Tashpulatov. Zh. Multicomponent cellulolytic enzymes of thermotolerant and mesophilic fungi, closely related to Aspergillus fumigatus. *Microbiology*, *36*:828-831, 1967.
104. Wali, S., Mattoo, A. K., and Modi, V. V. Comparative temperature stability properties of malate dehydrogenases from some thermophilic fungi. *Int. J. Peptide Protein Res.*, *14*:19-106, 1979.
105. Zakirov, M. Z., Schrlokova, S. S., and Karavaeva, N. N. Torula thermophila UzPT-1 thermophilic fungus producing proteolytic enzymes. *Appl. Biochem. Microbiol.*, *21*:608-612, 1975.
106. Craveri, R., and Colla, C. Ribonuclease activity of mesophilic and thermophilic molds. *Ann. Microbiol.*, *16*:97-99, 1966.
107. Hayashida, S., and Yoshioka, H. Thermostable cellulases from Humicola insolens YH-8. *Agr. Biol. Chem.*, *44*:481-487, 1980.
108. Margaritis, A., and Creese, E. Thermal stability characteristics of cellulase enzymes of Sporotrichum thermophile. *Biotechnol. Lett.*, *3*:471-476, 1981.
109. Matsuo, M., Yasui, T., and Kobayashi, T. Purification and some properties of β-glucosidase from Malbranchea pulchella var. sulfurea no. 48. *Agric. Biol. Chem.*, *41*:1593-1599, 1977.
110. Vandamme, E. J., Logghe, J. M., and Geeraerts, H. A. M. Cellulase activity of a thermophilic Aspergillus fumigatus (Fresinius). *J. Chem. Technol. Biotechnol.*, *32*:968-974, 1982.
111. Yoshioka, H., and Hayashida, S. Relationship between carbohydrate moiety and thermostability of β-glucosidase from Mucor miehei YH-10. *Agr. Biol. Chem.*, *45*:571-577, 1981.
112. Flannigan, B., and Sellars, P. N. Production of xylanolytic enzymes by Aspergillus fumigatus. *Trans. Br. Mycol. Soc.*, *71*:353-358, 1978.
113. Pazure, J. H., Knull, H. R., and Simpson, L. L. Glycoenzymes: a note on the role for the carbohydrate moieties. *Biochem. Biophys. Res. Commun.*, *40*:110-116, 1970.
114. Mahajan, M. K., Johri, B. N., and Gupta, R. K. Influence of desiccation stress in a xerophilic thermophile Humicola sp. *Curr. Sci.*, *55*: 928-930, 1986.
115. Broad, T. E., and Shepherd, M. G. Purification and properties of glucose-6-phosphate dehydrogenase from the thermophilic fungus Penicillium duponti. *Biochim. Biophys. Acta*, *198*:407-414, 1971.
116. Arnold, W. N., Garrison, R. G., Mann, L. C., and Wallace, D. P. The acid phosphatases of Thermoascus crustaceus, a thermophilic fungus. *Microbios*, *54*:101, 1988.
117. Fergus, C. L. The production of amylase by some thermophilic fungi. *Mycologia*, *61*:1171-1175, 1969.
118. Barnett, E. A., and Fergus, C. L. The relation of extracellular amylase, mycelium and time in some thermophilic and mesophilic Humicola species. *Mycopathol. Mycologia Appl.*, *44*:131-141, 1971.
119. Jensen, B., Olsen, J., and Allermann, K. Effect of media on the production of extracellular amylase from the thermophilic fungus Thermomyces lanuginosus. *Biotechnol. Lett.*, *9*:313-316, 1987.

120. Jensen, B., Olsen, J., and Allermann, K. Purification and extracellular amylolytic enzymes from the thermophilic fungus Thermomyces lanuginosus. *Can. J. Microbiol.*, *34*:218-223, 1988.
121. Rao, B. V., Sastri, N. V. S., and Rao, P. V. S. Thermal stabilization of glucose oxidase glucoamylase by physical entrapment. *Biochem. J.*, *193*:389-394, 1981.
122. Tamada, M., Kasai, N., and Kaetsu, I. Effect of structure of polymer support on the growth of Sporotrichum cellulophilum immobilized by polymerization induced by gamma rays. *Biotechnol. Bioeng.*, *32*:386-390, 1988.
123. Taylor, P. M., Napier, E. J., and Fleming, I. D. Some properties of glucoamylase produced by thermophilic fungus Humicola lanuginosa. *Carbohydr. Res.*, *61*:301-308, 1978.
124. Kanlayakrit, W., Ishimatsu, K., Nakao, M., and Hayashida, S. Characteristics of raw starch-degrading glucoamylase from thermophilic Rhizomucor pusillus. *J. Ferment. Technol.*, *65*:379-385, 1987.
125. Deploey, J. J., Nasta, M., and Adams, P. R. *Mycologia*, *74*:847, 1982.
126. Mangallam, S., Subrahmanyam, A., and Gopalakrishnan, K. S. Preliminary observations on the production of amyloglucosidase by thermophilic fungi. *Curr. Sci.*, *46*:16, 1977.
127. Somkuti, G. A., and Babel, F. J. Purification and properties of Mucor pusillus acid protease. *J. Bacteriol.*, *95*:1407-1414, 1968.
128. Ottesen, M., and Rickert, W. The acid protease of Mucor miehei. *Meth. Enzymol.*, *19*:459-460, 1970.
129. Emi, S., Meyers, D. V., and Iacobucci, G. A. Purification and properties of the thermostable acid protease of Penicillium duponti. *Biochem. J.*, *15*:842-848, 1976.
130. Karavaeva, N. N., Zakirov, M. N., and Mukhiddinova, N. G. Partial purification and some properties of protease from Torula thermophila. *Biokhimiya*, *47*:1625-1630, 1975.
131. Arima, K., Iwasaki, K., and Tamura, G. Milk-clotting enzyme from microorganisms. V. Purification and crystallization of Mucor rennin from Mucor pusillus Lindt. *Appl. Microbiol.*, *16*:1727-2733, 1968.
132. Oka, T., Ishino, K., Tsuzuki, K., Morihara, K., and Arima, K. Specificity of a rennin like enzyme from Mucor pusillus. *Agr. Biol. Biol. Chem.*, *37*:1177-1184, 1973.
133. Sternberg, M. Bond specificity, active site and milk-clotting mechanism of the Mucor miehei protease. *Biochem. Biophys. Acta*, *285*:383-392, 1972.
134. Bailey, M. J., and Siika-aho, M. Production of microbial rennin. *Biotechnol. Lett.*, *10*:161-166, 1988.
135. Hashimoto, H., Iwaasa, T., and Yokotsuka, T. Thermostable acid protease produced by Penicillium duponti K-1014, a true thermophilic fungus isolated from compost. *Appl. Microbiol.*, *24*:936-992, 1972.
136. Hashimoto, H., Iwaasa, T., and Yokotsuka, T. Some properties of acid protease from the thermophilic fungus, Penicillium duponti K-1014. *Appl. Microbiol.*, *25*:578-582, 1973.
137. Kloosterman, M., Elferink, V. H. M., Van Lersel, J., Roskam, J., Meijer, E. M., Hulshof, L. A., and Sheldon, R. A. Lipases in the preparation of β-blockers. *Trends Biotechnol.*, *6*:251-256, 1988.
138. Eggins, H. O. W., and Coursey, D. G. Thermophilic fungi associated with Nigerian oil palm produce. *Nature*, *203*:1083-1084, 1964.

139. Apinis, A. E., and Eggins, H. O. W. Thermomyces ibadensis sp. nov. from oil palm kernel stalks in Nigeria. *Trans. Br. Mycol. Soc.*, *49*:629-632, 1966.
140. Oso, B. A. Thermophilic fungi from stalks of oil palm kernels in Nigeria. *Z. Allg. Mikrobiol.*, *14*:713-717, 1974.
141. Oso, B. A. The lipase activity of Talaromyces emersonii. *Can. J. Bot.*, *56*:1840-1843, 1978.
142. Ogundero, V. W. Isolation of thermophilic and thermotolerant fungi from stored groundnuts in Nigeria and determination of their lipolytic activity. *Int. Biodeter. Bull.*, *17*:51-56, 1981.
143. Ogundero, V. W. Thermophilic fungi from Nigerian palm produce. *Mycologia*, *73*:198-202, 1981.
144. Alford, J. A., and Smith, J. L. *J. Am. Chem. Soc.*, *42*:1083-1040, 1965.
145. Arima, K., Liu, W. H., and Beppu, T. Studies on the lipase production of thermophilic fungus Humicola lanuginosa. *Agr. Biol. Chem.*, *36*:893-895, 1972.
146. Ogundero, V. W. Lipase activity of thermophilic fungi from mouldy ground-nuts in Nigeria. *Mycologia*, *72*:118-126, 1980.
147. Somkuti, G. A., Babel, F. J., and Somkuti, A. C. Lipase of Mucor pusillis. *Appl. Microbiol.*, *17*:606-610, 1969.
148. Morinaga, T., Kanda, S., and Nomi, R. Lipase production of a new thermophilic fungus Humicola lanuginosa var. catenulata. *J. Ferment. Technol.*, *64*:451-453, 1986.
149. Omar, I. C., Hayashi, M., and Nagai, S. Purification and some properties of a thermostable lipase from Humicola lanuginosa No. 3. *Agr. Biol. Chem.*, *51*:37-45, 1987.
150. Ibrahim, C. O., Mishio, N., and Nagai, S. Acidolysis reaction by a thermostable lipase from Humicola lanuginosa. *Agr. Biol. Chem.*, *52*:2923-2925, 1988.
151. Liu, W. H., Beppu, T., and Arima, K. Effect of various inhibitors on lipase action of thermophilic fungus Humicola lanuginosa S-38. *Agr. Biol. Chem.*, *37*:2487-2492, 1973.
152. Liu, W. H., Beppu, T., and Arima, K. Physical and chemical properties of the lipase of thermophilic fungus Humicola lanuginosa S-38. *Agr. Biol. Chem.*, *37*:2493-2499, 1973.
153. Somkuti, G. A. Simple method for removal of lipase from Mucor rennet. *J. Dairy Sci.*, *57*:591, 1974 (Abstract).
154. Woodward, J. Xylanases: functions, properties and applications. In *Topics in Enzyme and Fermentation Biotechnology* (A. Wiseman, ed.), Ellis Horwood Ltd., Chichester, 1984, pp. 9-30.
155. Kitpreechavanich, V., Hayashi, M., and Nagai, S. Production of xylan degrading enzymes by thermophilic fungi, Aspergillus fumigatus, and Humicola lanuginosa. *J. Ferment. Technol.*, *62*:63-69, 1984.
156. Matsuo, M., Yasui, T., and Kobayashi, T. Production and saccharifying action for xylan of xylanase from Malbranchea pulchella var. sulfurea no. 48. *Nippon Nogaikagaku Kaishi*, *49*:263-270, 1975.
157. Matsuo, M., Yasui, T., and Kobayashi, T. Enzymatic properties of β-xylosidase from Malbranchea pulchella var. sulfurea no. 48. *Agr. Biol. Chem.*, *41*:1601-1606, 1977.
158. Matsuo, M., and Yasui, T. Hydrolysis of p-substituted phenyl-β-D-xylosides by β-xylosidase from Malbranchea pulchella var. sulfurea no. 48. *Agr. Biol. Chem.*, *45*:1603-1608, 1981.

159. Matsuo, M., and Yasui, T. Properties of xylanase of Malbranchea pulchella var. sulfurea no. 48. *Agr. Biol. Chem.*, *49*:839-841, 1985.
160. Yu, E. K. C., Tan, L. U. L., and Chan, M. K. H. Production of thermostable xylanase by a thermophilic fungus Thermoascus aurantiacus. *Enz. Microb. Technol.*, *9*:16-24, 1987.
161. Matsuo, M., and Yasui, T. Xylanases of Malbranchea pulchella var. sulfurea. *Meth. Enzymol.*, *160*:671-674, 1988.
162. Yoshioka, H., Nagato, N., Chavanich, S., Nilubol, N., and Hayashida, T. Purification and properties of thermostable xylanase from Talaromyces byssochlamydoides YH-50. *Agr. Biol. Chem.*, *45*:2425-2432, 1981.
163. Steward, J. C., Lester, A., Milburn, B., and Heptinstall, J. Xylanase enzymes of Aspergillus fumigatus Fresinius. *Biotechnol. Lett.*, *7*:581-584, 1985.
164. Grajek, W. Production of D-xylanases by thermophilic fungi using different methods of culture. *Biotechnol. Lett.*, *9*:353-356, 1987.
165. Hayashida, S., Ohta, K., and Mo, K. Xylanase of Talaromyces byssochlamydoides. *Meth. Enzymol.*, *160*:675-678, 1988.
166. Paice, M. G., Jurasek, L., Carpenter, M. R., and Smillie, L. B. *Appl. Environ. Microbiol.*, *36*:802-808, 1978.
167. Ghose, T. K., and Pathak, A. N. Cellulases. 2. Applications. *Proc. Biochem.*, *8*:20-21, 24, 1973.
168. Klyosov, A. A., and Ermolova, O. V. A thermostable endo-1,4-β-glucanase from Myceliophthora thermophila. *Biotechnol. Lett.*, *10*:351-354, 1988.
169. Coudray, M. R., Canevascini, G., and Meier, H. Characterization of a cellobiose dehydrogenase in the cellulolytic fungus Sporotrichum (Chrysosporium) thermophile. *Biochem. J.*, *203*:277-284, 1982.
170. Fanrich, P., and Irrgang, K. Conversion of cellulose to sugars and cellobionic acid by the extracellular enzyme system of Chaetomium cellulolyticum. *Biotechnol. Lett.*, *4*:775-780, 1982.
171. Yoshioka, H., and Hayashida, S. Purification and properties of β-glucosidase from Humicola insolens YH-8. *Agr. Biol. Chem.*, *44*:1729-1735, 1980.
172. Eriksen, J., and Goksoyr, J. The effect of temperature on growth and cellulase (β-1,4-endoglucanase) production in the compost fungus Chaetomium thermophile var. dissitum. *Arch. Microbiol.*, *110*:233-238, 1976.
173. Ogundero, V. W. Amylase and cellulase activities of thermophilic fungi causing deterioration of tobacco products in Nigeria. *Mycopathology*, *69*:131-135, 1979.
174. Shepherd, M. G., Tong, C. C., and Cole, A. L. Substrate specificity and mode of action of cellulases from the thermophilic fungus Thermoascus aurantiacus. *Biochem. J.*, *193*:67-74, 1981.
175. Nishio, N., Krisu, H., and Nagai, S. Thermophilic cellulase production by Talaromyces sp. in solid state cultivation. *J. Ferment. Technol.*, *59*:407-410, 1981.
176. McHale, A., and Coughlan, M. P. The cellulolytic system of Talaromyces emersonii. Identification of the various components produced during growth on cellulosic media. *Biochem. Biophys. Acta*, *662*:145-151, 1981.
177. Srivastava, R. G., Narain, R., and Mehrotra, B. S. Cellulolytic ability of mesophilic and thermophilic fungi isolated in India from manure and plant refuse. *Indian J. Mycol. Plant Pathol.*, *11*:66-72, 1981.

178. Fergus, C. L. The celluloytic activity of thermophilic fungi and actinomycetes. *Mycologia, 61*:120-129, 1969.
179. Thakre, R. P., and Johri, B. N. Degradation of natural cellulose by thermophilic and thermotolerant fungi. *Acta Bot. Ind., 8*:45-49, 1980.
180. Hayashida, H., and Mo, K. Production and characteristics of Avicel-disintegrating endoglucanase from a protease-negative Humicola grisea var. thermoidea mutant. *Appl. Environ. Microbiol., 51*:1041-1046, 1986.
181. Grajek, W. Comparative studies on the production of cellulases by thermophilic fungi in submerged and solid-state fermentation. *Appl. Microbiol. Biotechnol., 26*:126-129, 1987.
182. Kowamori, M., Takayama, K., and Takasawa, S. Production of cellulases by a thermophilic fungus, Thermoascus aurantiacus A-131. *Agr. Biol. Chem., 51*:647-654, 1987.
183. Deacon, J. W. Decomposition of filter paper cellulose by thermophilic fungi acting singly and in sequence. *Trans. Br. Mycol. Soc., 85*: 663-669, 1985.
184. Chapman, E. S., Evans, E., Jacobelli, M. C., and Logan, A. A. The cellulolytic and amylolytic activity of Papulaspora thermophila. *Mycologia, 67*:608-615, 1975.
185. Moloney, A. P., Hackett, T. J., Considine, P. J., and Coughlan, M. P. Isolation of mutants of Talaromyces emersonii CBS 814.70 with enhanced cellulase activity. *Enz. Microb. Technol., 5*:260-264, 1983.
186. Kvesitadze, G., Kvachadze, L., Aleksidze, T., and Chartishvili, D. K. Exogenous cellulases of thermophilic micromycetes. Part I. Selection of producers. *Acta Biotechnol., 1*:101-106, 1986.
187. Morrison, J., McCarthy, U., and McHale, A. P. Cellulase production by Talaromyces emersonii CBS 814.70 and a mutant UV7 during growth on cellulose, lactose, and glucose containing media. *Enz. Microb. Technol., 9*:422-425, 1987.
188. McHale, A. P., and Morrison, J. Cellulase production during growth of Talaromyces emersonii CBS 814.70 on lactose containing media. *Enz. Microb. Technol., 8*:749-754, 1986.
189. Breuil, C., Wojtczak, G., and Saddler, J. N. Production and localization of cellulases and β-glucosidase from the thermophilic fungus Thielavia terrestris. *Biotechnol. Lett., 8*:673-676, 1986.
190. Kinoshita, S., Chua, J. W., Kato, N., Yoshida, T., and Taguchi, H. Hydrolysis of cellulose by cellulases of Sporotrichum cellulophilum in an ultrafilter membrane reactor. *Enz. Microb. Technol., 8*:691-695, 1986.
191. Romanelli, R. A., Houston, C. W., and Barnett, S. M. Studies on thermophilic cellulolytic fungi. *Appl. Microbiol., 30*:276-281, 1975.
192. Feldman, K. A., Lovett, J. S., and Tsao, G. T. Isolation of the cellulase enzymes from the fungus Thermoascus aurantiacus and regulation of enzyme production. *Enz. Microb. Technol., 10*:262, 1988.
193. Bedino, S., Testore, G., and Obert, F. Kinetic properties and mechanism of action of an intracellular β-glucosidase from Thermoascus aurantiacus Miehe. *Italian J. Biochem.*, 207-220, 1986.
194. Roy, S. K., Raha, S. K., Dey, S. K., and Chakrabarthy, S. L. Induction and catabolite repression of β-glucosidase synthesis in myceliophthora thermophila D-14 (ATCC 48104). *Appl. Environ. Microbiol., 54*:2152-2153, 1988.

195. Collings, A., Davis, B., and Mills, J. Endo-β-1,4-glucanase, exo-β-1,4-glucanase, β-glucosidase and related enzyme activity in culture filtrates of thermophilic, thermotolerant and mesophilic fungi. *Microbios*, *56*:131-148, 1988.
196. Rao, V. S., and Murthy, S. K. Purification and characterization of β-glucosidase and endocellulase from Humicola insolens. *Indian J. Biochem. Biophys.*, *25*:687-694, 1988.
197. Ramabadran, R. The cellulase of Humicola insolens. *Indian J. Exp. Biol.*, *7*:186-187, 1969.
198. Araujo, E. F., Barrosa, E. G., Caldesh, R. A., and Silva, D. U. β-Glucosidase activity of a thermophilic cellulolytic fungus Humicola sp. *Biotechnol. Lett.*, *5*:781-784, 1983.
199. Coutts, A. D., and Smith, R. E. Factors influencing the production of cellulases by Sporotrichum thermophile. *Appl. Environ. Microbiol.*, *31*:819-825, 1976.
200. Durand, H., Soucaille, P., and Tiraby, G. Comparative study of cellulases and hemicellulases from four fungi: mesophiles Trichoderma viride and Penicillium sp. and thermophiles Thielavia terrestris and Sporotrichum cellulophilum. *Enz. Microb. Technol.*, *6*:175-180, 1984.
201. Johri, B. N., and Pandey, A. Cellulose breakdown by Rhizopus rhizopodiformis (Cohn) Zopf., a thermophilic mould. *Indian J. Microbiol.*, *22*:76-78, 1982.
202. Lusis, A. J., and Becker, R. R. The β-glucosidase system of the thermophilic fungus Chaetomium thermophile var. coprophile. *Biochim. Biophys. Acta*, *329*:5-16, 1973.
203. McHale, A., and Coughlan, M. P. Synergistic hydrolysis of cellulose by components of the extracellular system of Talaromyces emersonii. *FEBS Lett.*, *117*:319-322, 1980.
204. Sen, S., Abraham, T. K., and Chakrabarthy, S. L. Cellulolytic activity of Myceliophthora thermophila D-14. *Curr. Sci.*, *50*:598-600, 1981.
205. Sen, S., Abraham, T. K., and Chakrabarthy, S. L. Characteristics of the cellulase produced by Myceliophthora thermophila D-14. *Can. J. Microbiol.*, *28*:271-277, 1982.
206. McHale, A. P. Cellulase production by Talaromyces emersonii CBS 814.70 co-immobilized with cellulose in calcium alginate. *Biotechnol. Lett.*, *10*:361-364, 1988.
207. Singh, A., Goel, R., and Johri, B. N. Immobilized Sporotrichum thermophile produces cellulase. *Curr. Sci.*, *58*:207-209, 1989.
208. Olutiola, P. O. Characterization of cellulase from Humicola lanuginosa. *Experientia*, *38*:1332-1333, 1982.
209. Mishra, M. M., Kapoor, K. K., Jain, M. K., and Singh, C. P. Cellulose degradation by Humicola lanuginosa. *Trans. Br. Mycol. Soc.*, *76*:159-160, 1981.
210. Somkuti, G. A. Synthesis of cellulase by Mucor pusillus and Mucor miehei. *J. Gen. Microbiol.*, *81*:1-6, 1974.
211. Pugh, G. J. F., and Kuthubutheen, A. J. The effects of fungicides on cellulose and starch hydrolysis by thermophilous and mesophilic fungi. *J. Therm. Biol.*, *4*:23-28, 1979.
212. Reese, E. T. Extracellular purine B-ribosidases from fungi. *Can. J. Microbiol.*, *14*:377-383, 1968.
213. Fracheboud, D., and Canevascini, G. Isolation, purification and properties of the exocellulase from Sporotrichum (Chrysosporium) thermophile. *Enz. Microb. Technol.*, *11*(4):220-229, 1989.

214. Gong, C. S., and Tsao, G. T. Cellulase and biosynthetic regulation. *Ann. Rep. Ferment. Proc.*, *3*:111-140, 1979.
215. Doughlan, M. P. The properties of fungal and bacterial cellulases with comment on their production and application. *Biotechnol. Genet. Eng. Rev.*, *3*:39-109, 1985.
216. Eriksson, K. E. Biosynthesis of polysaccharases. In *Microbial Polysaccharides and Polysaccharases* (C. W. Berkeley, G. W. Gooday, and D. C. Ellwood, eds.), Academic Press, London, 1979, pp. 285-293.
217. Enari, T. M. Microbial Cellulases. In *Microbial Enzymes and Biotechnology* (W. M. Fogarty, ed.), Applied Science, London, 1983, pp. 183-223.
218. Sen, S., and Chakrabarty, S. L. Isolation of cellulase hyperproductive mutants of Myceliophthora thermophila. *Proc. 7th Int. Biotechnol. Symp.* (New Delhi), Vol 2, 1984, pp. 604-605.
219. Moloney, A. P., McCrae, S. T., Wood, T. M., and Coughlan, M. P. Isolation and characterization of the 1,4-β-D-glucan glucanohydrolases of Talaromyces emersonii. *Biochem. J.*, *225*:365-374, 1985.
220. Deshpande, V., Eriksson, K. E., and Pettersson, B. Production, purification and partial characterization of 1,4-β-glucosidase enzymes from Sporotrichum pulverulentum. *Eur. J. Biochem.*, *90*:191-198, 1978.
221. Goksoyr, J., and Eriksen, J. Cellulases. In *Economic Microbiology* (A. H. Rose, ed.), Academic Press, London, 1980, pp. 283-330.
222. Verivuori, H., Siegler, K. M., Sands, J. A., and Montenecourt, B. S. Regulation of cellulase biosynthesis and secretion in fungi. *Biochem. Soc. Trans.*, *13*:411-414, 1985.
223. Yoshioka, H., and Hayashida, S. Production and purification of thermostable β-glucosidase from Mucor miehei YH-10. *Agr. Biol. Chem.*, *44*:2817-2824, 1980.
224. Eriksson, K. E. Enzymatic mechanisms involved in cellulose hydrolysis by the rot fungus Sporotrichum pulverulentum. *Biotechnol. Bioeng.*, *20*:317-322, 1978.
225. Meyer, H. P., and Canevascini, G. Separation and some properties of two intracellular B-glucosidases of Sporotrichum (Chrysosporium) thermophile. *Appl. Environ. Microbiol.*, *41*:924-931, 1981.
226. Virk, S., Singh, S. P., and Johri, B. N. Extracellular enzymes from thermophilic Malbranchea. 29th Annual Conference of Association of Microbiologists of India held at Hissar in Feb. (9-11), 1989, India, p. 59 (Abstract).
227. Cossar, D., and Canevascini, G. Cellulase enzyme production during continuous culture growth of Sporotrichum (Chrysosporium) thermophile. *Appl. Microbiol. Biotechnol.*, *24*:306-310, 1986.
228. Bhat, K. M. and Maheswari, R. Sporotrichum thermophile: growth, cellulose degradation and cellulase activity. *Appl. Environ. Microbiol.*, *53*:2175-2182, 1987.
229. Tamada, M., Kasai, N., Kamakura, M., and Kaetsu, I. Periodical batch culture of the immobilized growing fungus Sporotrichum cellulophilum producing cellulase in the non-woven materials. *Biotechnol. Bioeng.*, *28*:1227-1232, 1986.
230. Ishigami, T., and Yamada, Y. Purification and properties of polyphenol oxidase from Chaetomium thermophile, a thermophile fungus. *J. Gen. Appl. Microbiol.*, *32*:293-302, 1986.

231. Wojtczak, G., Breuil, C., Yamada, J., and Saddler, J. N. A comparison of the thermostability of cellulases from various thermophilic fungi. *Appl. Microbiol. Biotechnol.*, *27*:82, 1987.
232. Miles. Method of preparing novel thermostable transglucosidase by cultivation of Talaromyces duponti. U.S. patent 779,051, 1987.
233. Yamanobe, T., Mitshuishi, Y., and Takasaki, Y. Method for the manufacture of cellulase. U.S. patent 4,562,150, 1985.
234. Craveri, R., Craveri, A., and Guicciardi, A. Research on the properties and activities of enzymes of eumycete thermophilic isolates of soil. *Anh. Microbiol. Enzymol.*, *17*:1-30, 1967.
235. Prasad, A. R. S., and Maheswari, R. Purification and properties of trehalase from the thermophilic fungus Humicola lanuginosa. *Biochim. Biophys. Acta*, *525*:162-170, 1978.
236. Satyanarayana, T., and Chavant, L. Bioconversion and binding of sterols by thermophilic fungi. *Folia Microbiologica*, *32*:353-359, 1987.
237. Colombie, F., Prome-Patouraux, D., Sancholle, M., Chavant, L., and Montant, Ch. Bioconversion of progesterone by a thermophilic fungus. Annual Conference of the Network of Mycology, Paul Sabatier University, 1985, p. 85 (Abstract). *Biotechnol. Bioeng.*, *22*
238. Fanrich, R., and Irrgang, K. Cellulase and protein production by Chaetomium cellulolyticum strains grown on cellulosic substrates. *Biotechnol. Lett.*, *3*:201-206, 1981.
239. Moo-Young, M., Chahal, D. S., and Vlach, D. SCP production from variously chemically pre-treated substrates using Chaetomium cellulolyticum. *Proc. Bioconversion Symp.*, IIT, Delhi, 1977, pp. 457-467.
240. Moo-Young, M., Daugulis, A. J., Chahal, D. S., and MacDonald, D. G. The Waterloo process for the production of SCP from wastes. *Proc. Biochem.*, *14*:38-40, 1979.
241. Pamment, N., Moo-Young, M., Hsieh, F. H., and Robinson, C. W. Growth of Chaetomium cellulolyticum on alkali-pretreated hard-wood saw dust solids and pretreatment liquor. *Appl. Environ. Microbiol.*, *36*:284-290, 1978.
242. Ek, M., and Eriksson, K. E. Conversion of waste fibres into protein. *Proc. Bioconversion Symp* IIT, Delhi, 1977, pp. 449-454.
243. Ek, M., and Eriksson, K. E. Utilization of the white-rot fungus Sporotrichum pulverulentum for water purification and protein production onmixed lignocellulosic waste water. *Biotechnol. Bioeng.*, *22*:2273-2284, 1980.
244. Eriksson, K. E., and Larsson, K. Fermentation of waste mechanical fibres from a news print mill by the white-rot fungus Sporotrichum pulverulentum. *Biotechnol. Bioeng.*, *17*:327-348, 1975.
245. Janus, I. J. M. Production of microbial protein from rice hulls. *Bioconversion Symp.*, IIT, Delhi, 1977, pp. 469-478.
246. Thomke, S., Rundgren, M., and Eriksson, K. E. Nutritional evaluation of the white-rot fungus Sporotricum pulverulentum as a feed stuff to rats, pigs, and sheep. *Biotechnol. Bioeng.*, *22*:2285-2303, 1980.
247. Grajek, W. Production of protein by thermophilic fungi from sugar beet pulp in solid-state fermentation. *Biotechnol. Bioeng.*, *32*:225-260, 1988.
248. Moo-Young, M., Chahal, D. S., Swan, J. E., and Robinson, C. W. SCP production by Chaetomium cellulolyticum, a thermotolerant cellulolytic fungus. *Biotechnol. Bioeng.*, *19*:527-528, 1977.

249. Chahal, D. S., and Hawksworth, D. L. Chaetomium cellulolyticum, a new thermotolerant and cellulolytic Chaetomium. I. Isolation, description, and growth rate. *Mycologia, 68*:600-610, 1976.
250. Chahal, D. S., and Wang, D. I. C. Chaetomium cellulolyticum, growth, behaviour on cellulose and protein production. *Mycologia, 70*:160-170, 1978.
251. Ohmomo, S., Kaneko, Y., Sirianuntapiboon, Somchai, P., and Atthasampunna, P. Decolorization of molasses waste by a thermophilic strain Aspergillus fumigatus G-2-6. *Agr. Biol. Chem., 51*:3339-3346, 1987.
252. Ghai, S. K., Kahlon, S. S., and Chahal, D. S. Single cell protein from canning industry waste: sag waste as substrate for thermotolerant fungi. *Indian J. Exp. Biol., 17*:789-791, 1979.
253. Barnes, J. G., Eggins, H. O. W., and Smith, E. L. Preliminary stages in the development of a process for the microbial upgrading of waste paper. *Int. Biodeter. Bull., 8*:112-116, 1972.
254. Zadrazil, F., and Brennert, H. Solid-state fermentation of lignocellulose containing plant residues with Sporotrichum pulverulentum Nov. and Dichomitus squalens (Karst.) Reid. *Eur. J. Appl. Microbiol. Biotechnol., 16*:45-51, 1982.
255. Sen, S., Abraham, T. K., and Chakrabarty, S. L. Biodeterioration and utilization of city wastes by thermophilic microorganisms. *Indian J. Exp. Biol., 17*:1284-1285, 1979.
256. Sundman, G., Kirk, T. K., and Chang, H. Fungal decolorization of kraft bleach plant effluent. *Tappi, 64*:145-148, 1981.
257. Satyanarayana, T., and Johri, B. N. Volatile sporostatic factors of thermophilic fungal strains of paddy straw compost. *Curr. Sci., 50*:1763-1766, 1981.
258. Prakash, A. Antagonistic attributes of thermophilic fungi and thermophilism. Ph.D. thesis, Bhopal University, India, 1984.
259. Renuka, B. R., Subrahmanyam, A., Vinayakumar, M., Phanikumar, K., and Rama Rao, P. Biochemical activities of some thermophilic fungi. *Curr. Sci., 55*:60-361, 1986.
260. Subrahmanyam, A. Studies on morphology and biochemical activities of some thermophilic fungi. D.Sc. thesis, Kumaun University, Nainital, U.P., India, 1985, p. 336.
261. Rode, L. J., Foster, J. W., and Schuhardt, V. T. Penicillin production by a thermophilic fungus. *J. Bacteriol., 53*:565-566, 1947.
262. Premabai, M., and Narasimha Rao, P. L. Thermophilic microorganisms. II. Fermentation characteristics of Malbranchea pulchella. *Indian J. Biochem., 3*:172-178, 1966.
263. Saito, M., Matsuura, I., and Okazaki, H. TF-26 VX, an antibiotic produced by a thermophilic fungus. *J. Antibiot., 32*:1210-1212, 1979.
264. Premabai, M., and Narasimha Rao, P. L. Thermophilic microorganisms. IV. Elaboration of Malbranchins A&B by Malbranchea pulchella. *Indian J. Biochem., 3*:187-190, 1966.
265. Kluepfel, D., Bagh, J., Baker, H., Charest, M. P., Kudelski, A., Sehgal, S. N., and Vezina, C. Myriocin, a new antibiotic from Myriococcum albomyces. *J. Antibiot., 25*:109-115, 1972.
266. Kitano, K., Kintaka, K., Katamoto, K., Nana, K., and Nakao, Y. Occurrence of 6-aminopenicillanic acid in culture broths of strains belonging to genera Thermoascus, Gymnoascus, Polypaecium and Malbranchea. *J. Ferment. Technol., 62*:63-69, 1975.

267. Somkuti, G. A., and Greenberg, R. Antimicrobial peptidase of thermophilic fungi. *Dev. Ind. Microbiol.*, *20*:661-672, 1979.
268. Craveri, R., Manachini, P. L., and Aragozzini, F. Thermozymocidin—a new antifungal antibiotic from a thermophilic eumycete. *Experientia*, *28*:867-868, 1972.
269. Ishigami, T., Hirose, Y., and Yamada, Y. Characterization of polyphenol oxidase from Chaetomium thermophile. *J. Gen. Appl. Microbiol.*, *34*:401-408, 1988.
270. Chaudhuri, S., Thakur, I. S., Goel, R., and Johri, B. N. Purification and characterization of two thermostable xylanases from Melanocarpus albomyces. *Biochem. Int.*, *17*:563-575, 1988.
271. Sharma, V. K., Goel, R., and Johari, B. N. Isolation, purification and regeneration of protoplasts from Sporotrichum thermophile conidiospores. *Biochem. Int.*, *17*:895-906, 1988.
272. Esser, K., Kuck, U., Lang-Hinrich, C., Lemke, P., Osiewacz, H. C., Stahl, U., and Tudzynski, P. *Plasmids of Eukaryotes: Fundamentals and Applications*. Springer-Verlag, Berlin, 1986, p. 124.
273. Singh, D., Goel, R., and Johri, B. N. Deacylation of penicillins by immobilized mycelia of the thermophile Malbranchea. *J. Gen. Appl. Microbiol.*, *34*:333-339, 1988.

25

FUNGAL DEGRADATION OF LIGNIN: BIOTECHNOLOGICAL APPLICATIONS

K. BOOMINATHAN and C. ADINARAYANA REDDY *Michigan State University, East Lansing, Michigan*

I. INTRODUCTION

Lignin (Latin *lignum* = wood) is the most abundant and widely distributed renewable aromatic polymer in the biosphere. Lignin, cellulose, and hemicellulose are the major structural components of woody tissues. The lignin component comprises about 20-30% of the dry mass of woody plants [1-5]. A number of important functional roles have been attributed to lignin [1,2,5]. These include conferring strength and rigidity with elasticity to plant tissues and minimization of water permeation across cell walls of xylem tissue. Because of its recalcitrance to biodegradation, lignin undoubtedly plays a role in conferring protection against infection by plant pathogenic microbes. About 70-90% of the lignin in wood fiber occurs in the primary and secondary cell wall layers. The middle lamella contains the rest with 10-30% of the lignin. In the plant cell wall, lignin is intimately interspersed with hemicellulose and forms a matrix around the cellulose microfibrils. Lignin also functions as an adhesive cement binding cells together.

There has been a great proliferation of research on lignin biodegradation in recent years for several reasons [3-10]. Lignin, because of its recalcitrance and close association with cellulose and hemicellulose, is an important factor limiting the access of these plant polymers to polysaccharide-digesting enzymes (i.e., cellulases and hemicellulases). Consequently, degradation or modification of the lignin barrier is important for the efficient utilization of cellulose and hemicellulose in woody plant biomass as ruminant feed and as fermentation substrates for the production of fuels and organic chemicals. Utilization of novel biotechnological means to remove lignin will therefore ensure efficient utilization of the billions of tons of lignocellulosic biomass produced annually in the biosphere. Lignin degradation is also very important in global carbon cycling because of the great abundance of lignin in the biosphere and because lignin is an important factor limiting the rate of degradation of cellulose and other plant polysaccharides. An efficient microbiological process for lignin removal would be extremely useful for pulp and paper industries as this will greatly reduce the cost (in both energy and chemicals) presently required for delignification to produce paper pulps. Recent studies indicate that lignin-degrading fungi can play an important role in biopulping, biobleaching of pulps, and in decolorizing and detoxifying the vast amounts of bleach plant effluents generated by pulp and paper industries worldwide [6]. Furthermore, numerous reports published in recent years indicate that lignin-degrading fungi are capable of degrading highly toxic environmental pollutants such as dioxins, benzo(a)pyrenes, and a variety of other xenobiotics that pose a serious health hazard to humans and animals worldwide [7-9].

In this chapter we present a brief summary of recent advances in lignin biodegradation, different strategies to maximize enzyme production, and the potential biotechnological applications of fungal lignin degradation systems. The reader is referred to a number of excellent reviews and books that have appeared in the literature covering lignin structure, biosynthesis, degradation, and the practical applications of lignin degradation systems [1-13].

II. LIGNIN SUBSTRATE

A. Structure

Lignin is a phenylpropanoid polymer that arises from three precursor cinnamyl alcohol derivatives: coumaryl alcohol, coniferyl alcohol (the guaiacyl units), and sinapyl alcohol (the syringyl units) (Fig. 1A) [1,4,5]. At the site of lignification in the cell wall, the phenolic hydroxyl groups of these cinnamyl alcohol derivatives undergo a peroxidase-mediated one-electron oxidation producing phenoxy radicals. Random coupling of the various

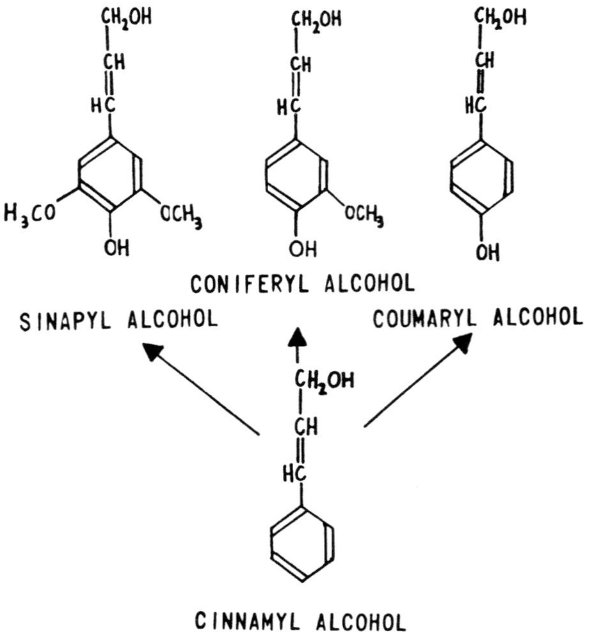

(A)

FIGURE 1. (A) Immediate precursors for lignin biosynthesis. (B) Schematic formula for a representative portion of a spruce lignin.

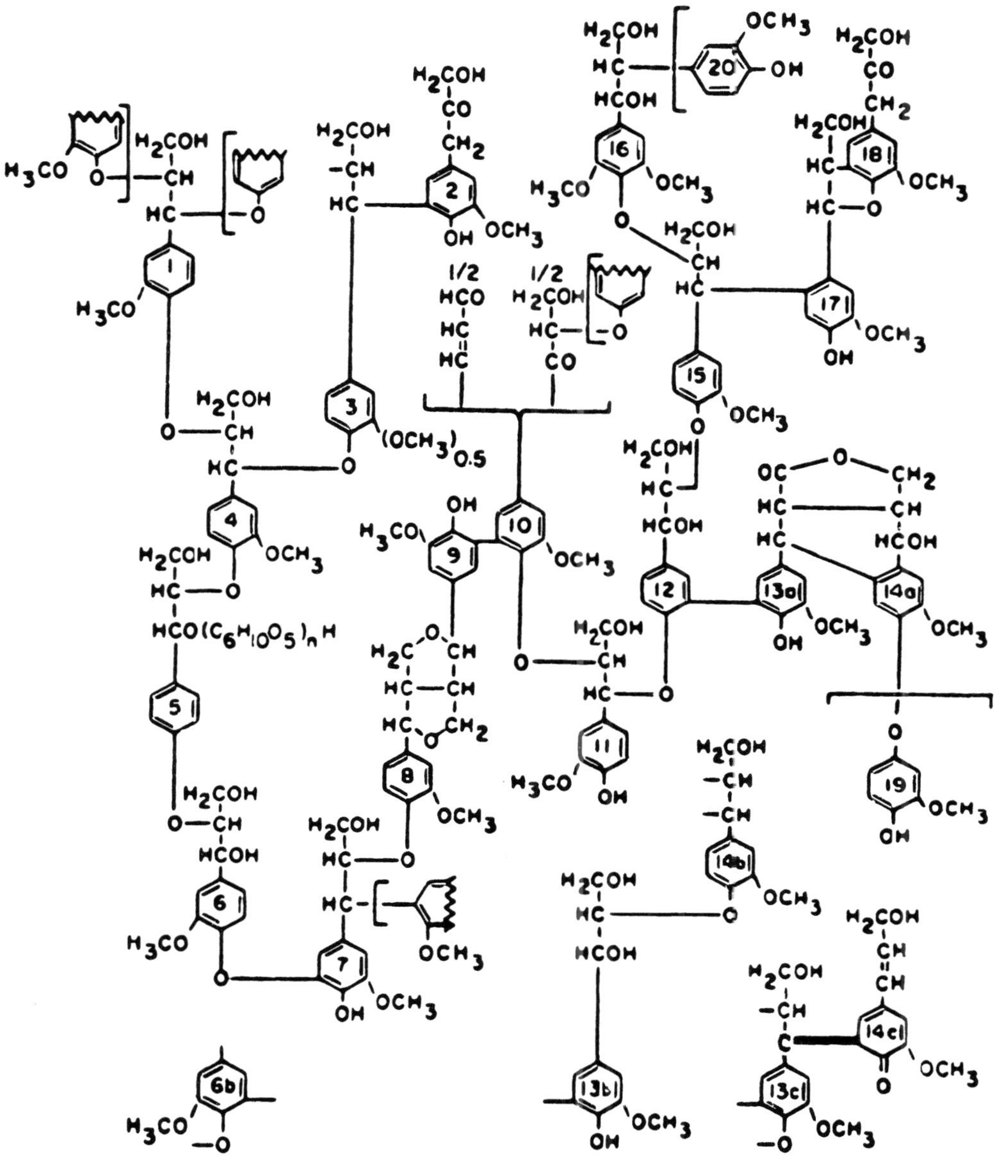

(B)

resonance forms of these phenoxy radicals with each other and also with the radicals in the growing lignin polymer results in the formation of a highly crosslinked, stereochemically complex, three-dimensional, amorphous, heterogeneous polymer that contains a variety of intermonomer linkages (Fig. 1B). Of these linkages, six to eight linkages are more important. The single most predominant linkage is the β-O-4 type between units 1 and 4, 5 and 4, 5 and 6, 7 and 8, 13b and 14b, and 15 and 16 (in Fig. 1B) and constitutes 40-60% of the total linkages in different lignins. The other linkages are relatively less numerous and include phenylcoumaran (between units 17 and 18; 6-12%), biphenyl (units 12 and 13a; 5-10%), diarylpropane (between units 16 and 20; 5-10%), diphenyl ether (between units 6 and 7; 4-6%); pinoresinol (between units 8 and 9; < 5%) and α-aryl ether linkages (between units 11 and 12; 6-8%). Lignins of hardwoods (angiosperms) are composed of approximately equal amounts of guaiacyl and syringyl units, while the lignins of softwoods (gymnosperms) contain mostly guaiacyl units. Grass lignins are composed of syringyl, guaiacyl, and coumaryl units, and *p*-coumaric acid esterified to the terminal hydroxyl groups of some of the propyl side chains of the lignins.

Lignin is a unique biopolymer. Unlike other naturally occurring biopolymers such as starch, cellulose, proteins, and nucleic acids, it does not contain identical, readily hydrolyzable, repeating linkages that occur at regular intervals [4]. Instead, as mentioned above, lignin is a highly irregular, three-dimensional polymer that has no precise chemical structure but contains a series of substructures which occur at random and are different in lignins from various sources. These structural features, together with the large size of the lignin polymer (mol wt 10^5-10^6 kDa), dictate that lignin-degrading biocatalysts be extracellular and relatively nonspecific. This indeed turns out to be the case (see below).

B. Methodology

One of the major bottlenecks in lignin degradation research until about 10 years ago was the lack of simple, sensitive, and accurate assays for measuring lignin decomposition. The advent of sensitive radioisotopic procedures involving the bioconversion of ^{14}C synthetic lignins (labeled in the ring, side chain, or methoxyl groups) to $^{14}CO_2$ represented a major advance for the study of lignin decomposition by microorganisms [15]. Lignocelluloses specifically labeled in the lignin component and labeled industrial-type lignins (e.g., kraft lignin) have also been employed by various investigators (see Refs. 8-10,16). However, there are important limitations to these methods. First, these procedures do not take into account the degradation of lignin into water-soluble intermediates. Second, the radiolabeled synthetic lignins (DHPs, for "dehydrogenative polymerizates") are still relatively expensive and not readily available. This makes DHPs inaccessible to laboratories in a majority of the developing countries. Furthermore, care should be taken to ensure that the synthetic lignin prepared is polymeric, which is a potential problem with DHPs. Polymeric dyes such as poly R, poly B, remazol blue, and others have been used as indicators of lignin degradation [17], but these are qualitative at best and any results obtained must be verified by studying the degradation of [^{14}C]DHPs, milled wood lignin, or other suitable substrates. Nonetheless, polymeric dyes have been extensively used in successfully isolating different classes of biochemical mutants by us as well as other investigators [18-21].

Numerous studies involving the use of lignin model compounds (mostly monomeric and dimeric compounds and some trimers) have been published. These model compounds have proven invaluable for the elucidation of the key steps involved in the degradation of the lignin polymer and for studying the mechanism of action of enzymes implicated in lignin degradation [4,5,8,10,11a,b,12]. Several studies have documented that results of research with model compounds are equally valid for the lignin polymer itself [11a,b]. In these model studies, β-aryl ether (β-O-4)-type model compounds are used extensively, as the β-O-4 substructure is the most prevalent intermonomer linkage in the lignin polymer (Fig. 1B).

III. LIGNIN DEGRADATION BY WHITE ROT FUNGI

Wood-rotting fungi are classified into three major categories based on the type of wood decay caused by these organisms: white rot fungi, brown rot fungi, and soft rot fungi. Of these three groups, white rot fungi, which primarily comprise a heterogeneous collection of hundreds of species of basidiomycetes, are the most important lignin degraders and have been the most extensively studied group. The reader is referred to a number of excellent reviews that recently appeared on wood decay by different groups of wood-rotting fungi, with a special focus on the white rot basidiomycetes [4,5,8,9,10]. We will present here only a brief summary of the physiology of lignin degradation by white rot fungi, especially strains of *Phanerochaete chrysosporium*,* which has become the most commonly used model organism in lignin biodegradation studies. The reasons for the popularity of the organism for lignin degradation studies include rapid growth and metabolism of lignin; ability to grow optimally at relatively high temperatures of 40°C; ability to produce conidia (asexual spores) and basidiospores (sexual spores)—an important advantage for studying genetics and molecular biology; ability to grow on chemically defined media; existence of an already considerable knowledge base on its ecology, physiology, biochemistry, molecular biology, and genetics; and low phenol oxidase activity [9,10].

A. Physiology

1. Lignin Is Not a Growth Substrate

Despite the fact that lignin is rich in carbon, it is not a growth substrate for microorganisms which are reported to degrade lignin. Lignin-degrading fungi metabolize various lignin preparations only in the presence of an alternate energy/carbon source [8-10]. Thus, lignin degradation can be truly considered "cometabolism" along with utilization of energy/carbon substrates such as cellulose, hemicellulose, simple carbohydrates, and glycerol. Apparently, the oxidative degradation of lignin by white rot fungi such as *P. chrysosporium* does not provide sufficient energy for growth [1,5,9,22]. Reports of growth of white rot fungi on lignin [23] have not been substantiated [24-26]. Furthermore, degradation of only lignin during wood decay has not been observed except in a single study with *P. chrysosporium* in which glucose was provided as a cosubstrate [27]. In another study, a cellulase-negative mutant of *Sporotrichum pulverulentum* (an anamorph of *P. chrysosporium*) degraded lignin in wood without attacking the cellulose, but it simultaneously degraded hemicelluloses [28,29]. Clearly, lignin does not serve as a growth substrate and strains degrading lignin require an easily metabolizable growth substrate. The basis for a cosubstrate requirement is not known at this time, but Kirk and Farrell [10]

---

*Abbreviations for genera throughout the text are as follows: *L*, *Lentinus*; *Pl*, *Pleurotus*; *Ph*, *Phlebia*; *C*, *Coriolus*; *Po*, *Polyporus*; *T*, *Trametes*; *M*, *Merulius*; *Ga*, *Ganoderma*; *S*, *Sporotrichum*; *P*, *Phanerochaete*.

speculate that the basis for this probably lies in the balance between energy-producing and energy-consuming reactions. It has been hypothesized from earlier studies that the normal mode for white rot fungi to degrade wood is by a simultaneous attack on polysaccharides and lignin. Degradation of cellulose provides glucose for the fungus and when the flow of sugar ceases, the fungus starves and thereby transfers from a primary metabolism to a secondary metabolism mode. Therefore, to degrade lignin, the fungus requires easily metabolizable nutrients such as sugars from the polysaccharides of wood. These sugars are also necessary for the production of hydrogen peroxide, which is an absolute prerequisite for the organism to degrade lignin [30,31].

2. Lignin Degradation is a Secondary Metabolic Process

Lignin is degraded only during secondary metabolism. The shift to secondary metabolism follows the cessation of growth due to the depletion of certain nutrients. Several ligninolytic fungi, if not all, apparently exhibit this feature [1,2,8-10,32-35]. Secondary metabolism in *P. chrysosporium* is triggered by nitrogen, carbon, or sulfur limitation but not by phosphorus limitation [35]. Lignin degradation stops when primary growth is restored by adding excess carbon or nitrogen to cultures limited for those nutrients [35]. In nitrogen-limited cultures, ligninolytic activity is associated with the formation of new secondary mycelia [25]. Since nitrogen is limiting in wood and is the natural substrate for wood-rotting fungi [36], expression of ligninolytic capability under nitrogen-limited conditions is advantageous for the organism so that sustained lignin degradation occurs.

White rot fungi such as *Lentinus edodes, Pleurotus ostreatus* [34], and an unidentified basidiomycete isolated from cattle dung [32] degraded radiolabeled lignin preparations in both nitrogen-rich and nitrogen-limited conditions. It is not clear whether such fungi exhibit a clear separation between primary and secondary metabolism or if lignin degradation is regulated differently in these organisms [10].

B. Enzymology

1. Lignin Peroxidases

An intensive search for the extracellular enzymes associated with lignin degradation followed the report of Forney et al. [37] who showed that a temporal relationship exists between ligninolytic activity and H_2O_2 production, and that H_2O_2 plays an integral role in lignin degradation (Fig. 2). This idea was further supported by the fact that catalase inhibited lignin degradation [38]. These results quickly led to the discovery of extracellular peroxidases in *P. chrysosporium* [9,10,39-45]. When cultivated under ligninolytic conditions, this organism produces two types of extracellular glycosylated heme proteins: lignin peroxidases and manganese-dependent peroxidases [39-51]. The lignin peroxidase (LIP) family contains multiple isozymes [41-43] with pI values ranging from 3.3 to 4.7 [43,44,49]. LIPs contain 1 mol of iron protoporphyrin IX per mole of enzyme and have a molecular mass ranging between 38,000 and 43,000. The isozymes are purified by anion exchange chromatography or a combination of anion exchange chromatography and gel filtration [41,43,46-49]. The isozymes were arbitrarily designated as H1, H2, ..., H10 (H refers to heme peak). The number of isozymes reported varied from as low as 2 to as high as 15

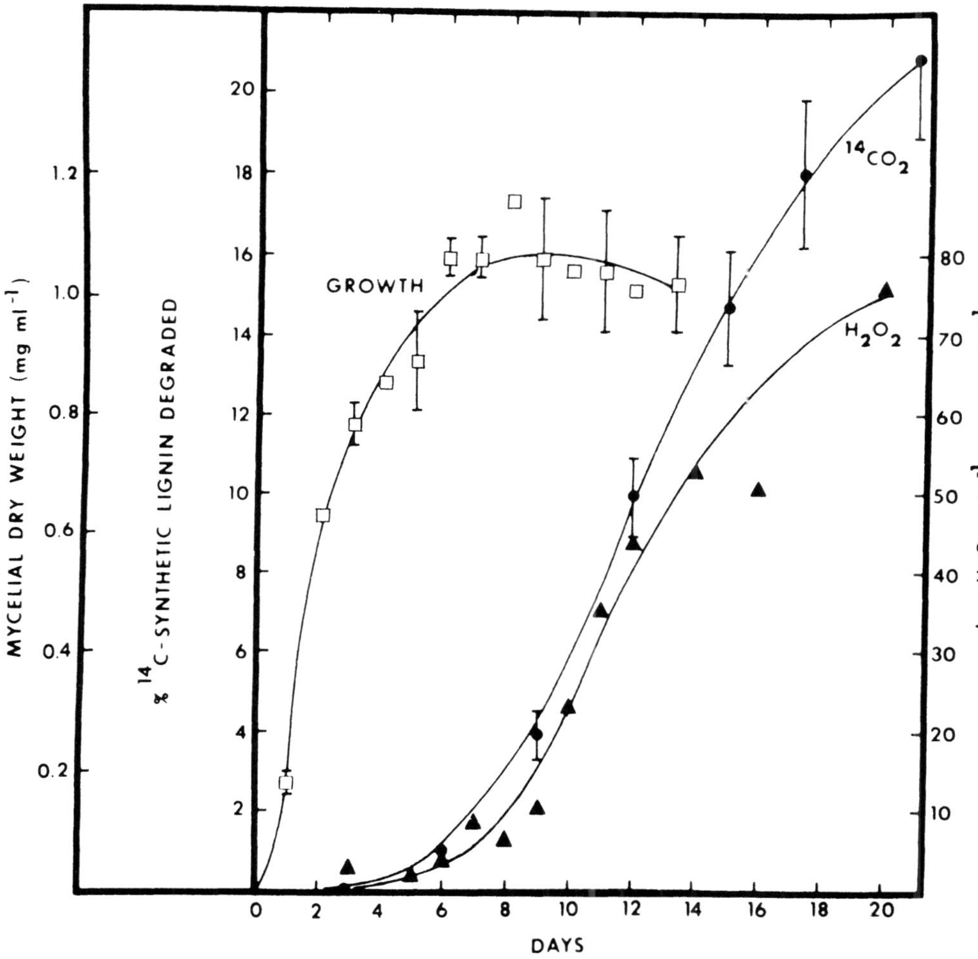

FIGURE 2. Relationship between the specific activity for H_2O_2 production in cell extracts (▲), the metabolism of 2'-^{14}C-labeled synthetic lignin to $^{14}CO_2$ (●), and mycelial dry weight (□) [37].

[41,43,47,49]. These reported differences in the number of isozymes reflect differences in strain, culture conditions, and purification/fractionation techniques employed by different investigators [10]. For example, a comparison of the fast-protein liquid chromatography (FPLC) profiles of the extracellular LIPs in *P. chrysosporium* strain BKM-F, grown in acetate-buffered static cultures, revealed major amounts of LIP isozymes H2 and H6 and moderate amounts of H8 and H10 whereas in identical cultures buffered with dimethylsuccinate (DMS) instead of acetate, major amounts of H8 and H10 proteins, moderate amounts of H2, and relatively minor amounts of H6 were observed. Similarly, a comparison of the FPLC profile of the extracellular fluid of strain ME446 with that of strain BKM-F revealed appreciable differences. Strain ME446 in DMS-buffered cultures generally produces three major LIP protein peaks whereas strain BKM-F produces six readily identifiable LIP peaks of which H2 and H8 are the predominant ones [20,42,46,47]. Recently, Odier and Delattre [48] reported separation of nine proteins with LIP activity from cultures of *P. chrysosporium* INA-12 grown in a medium containing glycerol as the carbon source and nonlimiting levels of nitrogen. These results show that by appropriate changes in the strains or culture conditions employed, the levels of various extracellular peroxidases can be selectively manipulated.

One of the LIP isozymes (H8) has been most extensively characterized [7] and crystallized [50]. The biochemical characteristics of the LIPs have been reviewed in detail elsewhere [7-12,51]. The essential substrate oxidation mechanism of LIP enzymes appears to be peroxidative. Oxidation of veratryl alcohol (3,4-dimethoxybenzyl alcohol) by the LIP isozyme H8 has been studied in detail [7,10,51]. LIP(H8) catalyzes stoichiometric oxidation of 1 mol of veratryl alcohol to 1 mol of veratraldehyde per mol of H_2O_2 consumed. A turnover number of 7.8 sec^{-1}, K_m of 29 μM for H_2O_2 and 72 μM for veratryl alcohol have been reported. Steady-state kinetic results indicated a "ping-pong" mechanism in which the LIP enzyme is first oxidized by H_2O_2, and the oxidized enzyme then reacts with veratryl alcohol to form veratraldehyde. The latter can be converted to native enzyme in the presence of excess veratryl alcohol. Since veratryl alcohol is a normal secondary metabolite produced by lignin-degrading culture, it has been suggested that LIP catalyzes the oxidation of veratryl alcohol to a cation radical and the latter acts as a diffusible one-electron oxidant to interact with lignin and lignin model compounds; however, the results of others do not appear to be consistent with this hypothesis [10, 55,62].

LIP(H8) has been reported to catalyze the H_2O_2-dependent oxidation of a variety of nonphenolic lignin model compounds including, but not limited to, oxidation of benzylic alcohols and cleavage of Cα-Cβ bonds, β-aryl ether bonds, and aromatic rings [9,10,51-62]. These reactions are consistent with a mechanism involving the initial one-electron oxidation of a variety of aromatic substrates by the oxidized LIP enzyme intermediates to form aryl cation radicals followed by a series of nonenzymatic reactions to yield a multiplicity of final products. A unified general scheme has been proposed for the reactions catalyzed by LIPs (Fig. 3) [7,10,51]. These enzymes, like other peroxidases, first react with H_2O_2 to form a two-electron oxidized intermediate (compound I). Compound I oxidizes the one-electron oxidation of lignin substrate or a variety of lignin model compounds (RH) to yield the one-electron oxidized intermediate compound II, and a lignin substrate cation radical. The substrate radical then undergoes a variety of reactions. The compound II returns to resting enzyme by oxidizing a second substrate.

LIPs oxidize a spectrum of substrates. Although the stoichiometry and products formed by most of the reactions catalyzed by LIPs are complex, knowledge obtained has been adequate to elucidate the mechanisms of these enzymes [7]. The LIP-mediated oxidation of some of the substrates and the formation of products are shown in Fig. 4. The reactions can be classified into five categories such as benzylic alcohol oxidation (A); carbon-carbon bond cleavage (B); hydroxylation (C); phenol dimerization or polymerization (D); and demethylation (E).

Studies on fungus-degraded lignin isolated from decayed wood suggested the presence of an aromatic cleavage system. The existence of such a cleavage system was reported by Umezawa and Higuhchi [57]. These

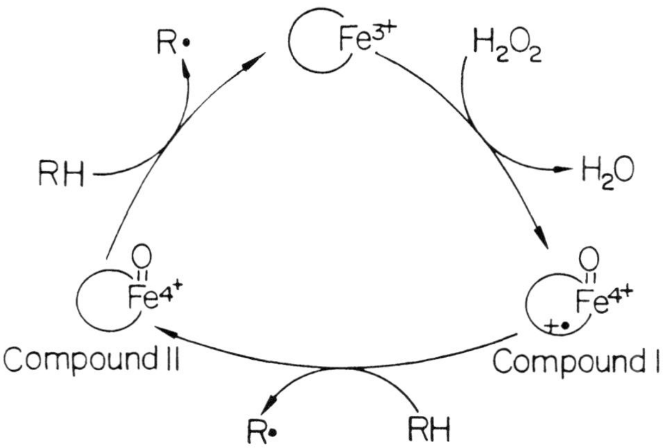

FIGURE 3. Catalytic cycle of lignin peroxidase indicating various reduction states of the heme. Resting enzyme reacts with H_2O_2 to produce the two-electron oxidized intermediate, compound I, which in turn oxidizes lignin substrate (RH) to yield the one-electron oxidized intermediate, compound II, and a substrate radical. Compound II returns to resting enzyme by oxidizing a second substrate molecule. The free radical (R) can undergo a variety of reactions [7].

investigators identified the formation of the ring cleavage products, β-γ-cyclic carbonates of aryl glycerol, from the culture filtrate of *P. chrysosporium* with β-O-4 model compound as substrate. Subsequently, LIP-mediated formation of the ring cleavage products was shown [58]. Umezawa and Higuhchi [59] proposed a scheme for the aromatic ring cleavage of β-O-4 lignin substructure model compounds (Fig. 5). The mechanism involves one-electron oxidation of the aromatic ring to the corresponding cation radical [substrate to (a)] followed by attack of nucleophile [H_2O (a) to (b) or the hydroxyl groups of Ca and Cr positions of propyl side chain] [(a) to (g)] and coupling with oxygen [(b) to (c), (e) to (f), (g) to (h)], with intramolecular addition of peroxy radicals to double bonds to form cyclic peroxide [(c) to (e)] and subsequent coupling with oxygen. Thus the aromatic ring cleavage reactions catalyzed by LIPs appear to be different from that of ring cleavage of aromatic compounds catalyzed by dioxygenases.

Recently, Tuisel et al. [60] made a detailed investigation of the effect of pH and temperature on the stability of another LIP isozyme, H2. They reported that the stability of this enzyme improved at pH 5.5, although the optimal pH for activity was 2.5. The spectral characteristics of H2 isozyme were reported to be similar to the LIP(H8) isozyme [7].

A major unsolved question in lignin degradation today is, does each LIP have a specific role in lignin degradation? If so, what is it? All the isozymes catalyze the oxidation of veratryl alcohol to veratraldehyde which, as described above, is the basis for the standard assay for this enzyme [44,45]. They have similar molecular weight (38-43 kDa) for glycosylated enzymes and approximately 37 kDa for the nonglycosylated proteins, the sequences of which were deduced from their cDNA sequences. They also have similar extinction coefficients of 109 nM [49]. Polyclonal antibody raised against H8 cross-reacted with all of the isozymes [10]. Subsequent

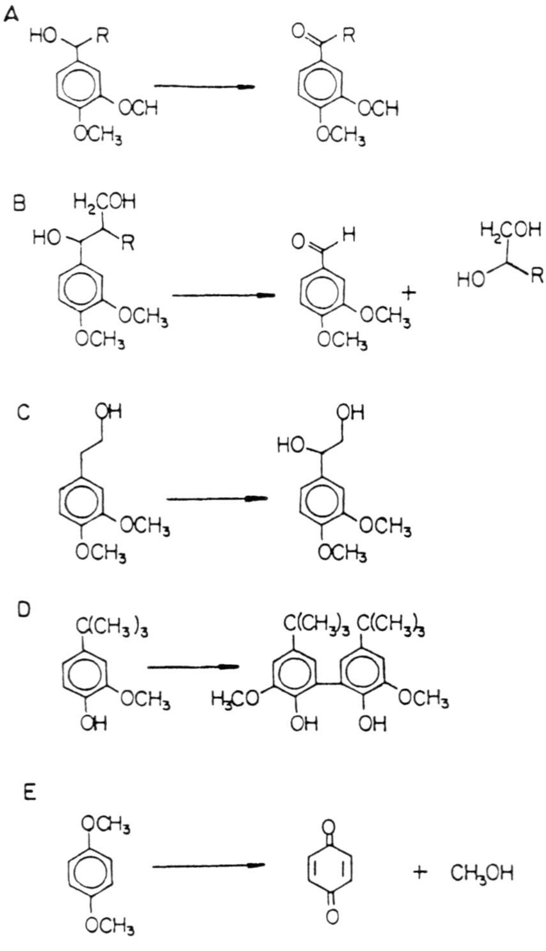

FIGURE 4. The reactions catalyzed by lignin peroxidase of *P. chrysosporium*. Most of these reactions can be accounted for by cation radical formations [7].

study by peptide-mapping analysis using V8 proteolysis showed the presence of four shared peptide fragments, indicating that the LIP isozymes are related to each other [43,49]. Furthermore, their peptide homology studies also suggested that isozymes H1 and H2 form one class; H6, H7, and H8 belong to a second class; and isozyme H10 belongs to a third class [49].

Recent studies showed the presence of the LIP enzyme system in other white rot fungi such as *Trametes versicolor* [63-67], *Pl. ostreatus* [63], *Bjerkandra adusta* [63], *Ph. radiata* [68-70], *Lentinus edodes* [71], *Chrysosporium pruinosum* [63], *Panus tigrinus* [72], and *Merulius tremellosus* [64,65].

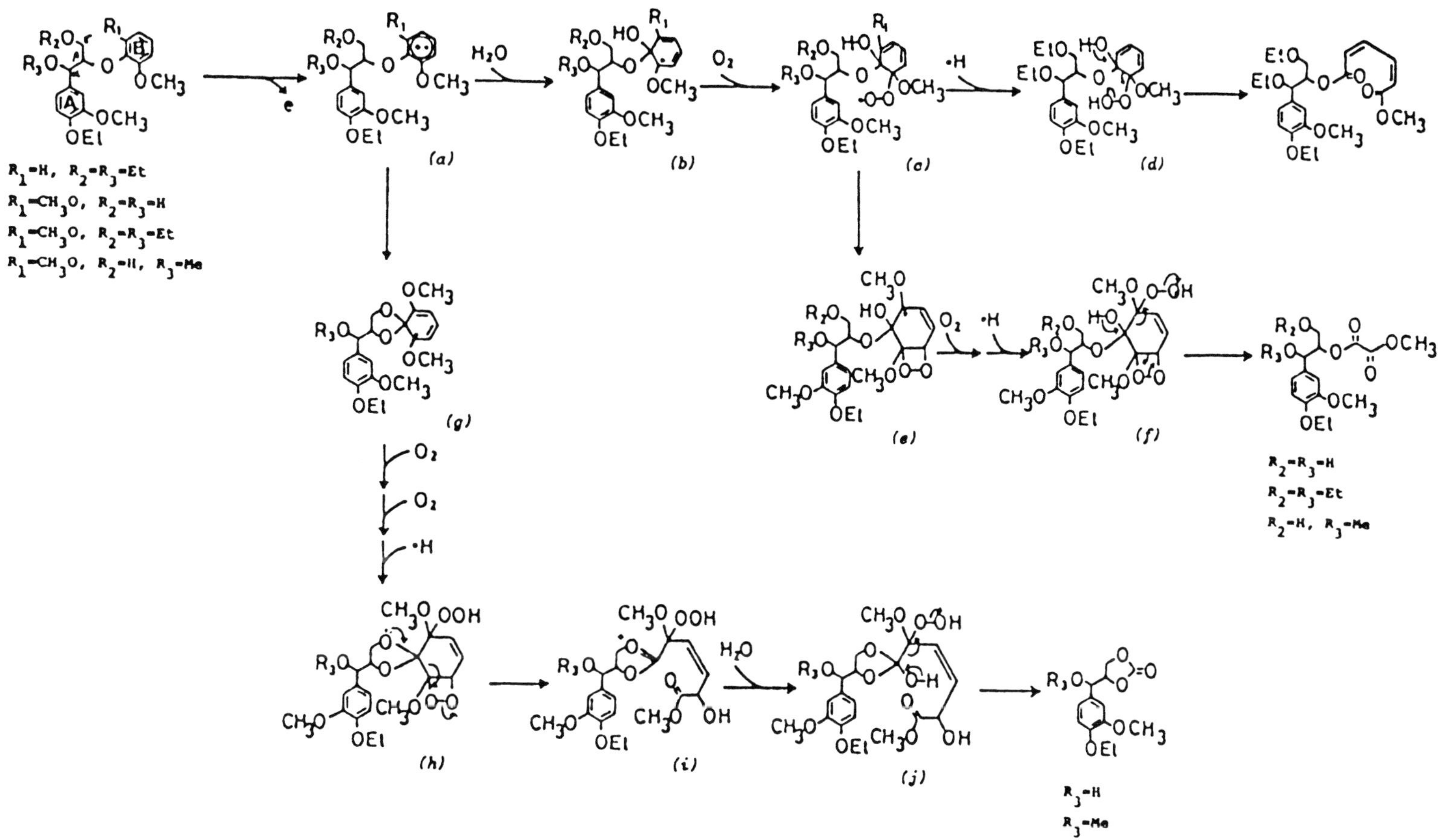

FIGURE 5. Mechanism for the formation of aromatic ring cleavage products from β-O-4 lignin model compounds. Formation of the α,β-cyclic carbonate can be explained in the same way as for β,γ-cyclic carbonate except for nucleophilic attack of CαOH of (a) instead of CγOH [11a].

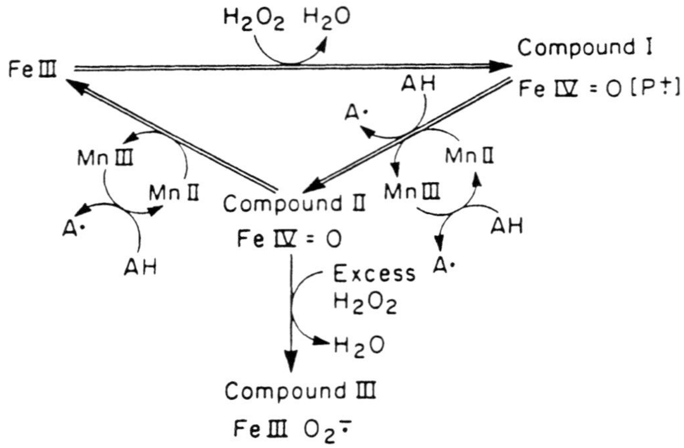

FIGURE 6. Catalytic cycle of manganese-dependent peroxidase of *P. chrysosporium* [51].

2. Manganese-Dependent Peroxidases

The second family of extracellular heme proteins produced by *P. chrysosporium* includes four to six or more manganese-dependent peroxidases (MNPs) with PI values ranging from 4.2 to 4.9 [10,43,51,73-79]. MNPs, similar to LIPs, are glycoproteins, contain 1 mol of iron protoporphyrin IX per mol of enzyme, and have a molecular mass of approximately 46 kDa [74-77]. MNPs oxidize Mn(II) to Mn(III) [76-78] and can be activated by lactate which probably acts by chelating Mn(III) to form stable complexes with a high redox potential [78,79]. The Mn(III) in turn oxidizes phenols to phenoxy radicals.

MNPs have essentially the same catalytic cycle as LIPs except that Mn(II) is obligatorily required for oxidation of phenols and other substrates. As shown in Fig. 6, the native enzyme first reacts with H_2O_2 and is oxidized to compound I, which is rapidly converted to compound II by reacting with one equivalent of Mn(II) resulting in the formation of Mn(III). A second unit of Mn(II) reduces compound II to the native ferric enzyme. The Mn(III) formed is involved in the oxidation of various phenolic substrates [51]. These enzymes could not efficiently complete the catalytic cycle in the absence of Mn(II), suggesting the absolute requirement of MnII [79]. The MNPs apparently utilize freely diffusible Mn(II)/Mn(III) as an obligatory redox couple to oxidize terminal phenolic substrates such as lignin and various model compounds. About 10-15% of the phenolic group in lignin is free and it is likely that the Mn(III) can diffuse into the polymeric matrix and oxidize the polymeric phenolic groups. That MNPs can catalyze oxidation of a free phenolic β-1 lignin model compound was also demonstrated [51]. The ability of these enzymes to oxidize phenols to phenoxy radicals led Higuchi [80] to propose that it has the potential to effect limited degradation of lignin. Indeed, we have demonstrated with the aid of an LIP-negative mutant that strains producing only MNPs and not LIPs can depolymerize and degrade lignin albeit to a limited extent [20]. A correlation between LIP production and ^{14}C-ring-labeled synthetic lignin degradation was also observed [81]. Paszczynski et al. [77] showed that MNPs can oxidize reduced substrates, including glutathione, reduced NADPH, and dihydroxymaleic acid, with coupled reduction of O_2 to H_2O_2. Therefore, these enzymes in addition to their peroxidative activity may also play a role in generating extracellular H_2O_2 in lignin-degrading cultures.

3. Nomenclature of the Extracellular Peroxidases

The extracellular peroxidases produced by the ligninolytic cultures of *P. chrysosporium* that oxidize veratryl alcohol to veratraldehyde have been variously referred to as "ligninases," "diaryl propane oxygenases," and "lignin peroxidases." However, the latter more accurate name is being widely used as it reflects the true function and the absolute dependence of LIPs on hydrogen peroxide. Similarly, the MNPs were referred to as "peroxidases," "manganese peroxidases," and "manganese-dependent peroxidases." We suggest that these peroxidases be referred to as "manganese-dependent peroxidases" since this designation clarifies their manganese dependence as the use of the term "manganese peroxidase" implies that manganese is an integral part of the enzyme.

Since the peroxidase enzyme family consists of the two subfamilies LIPs and MNPs, and in turn each subfamily has four or more isozymes, the individual isozymes are arbitrarily identified according to their FPLC elution profile such as H1, H2, ..., H10 (H refers to the heme peak) or by their isoelectric point. Recently, Farrell et al. [49] proposed a nomenclature for LIP and MNP isozymes. These investigators suggested that H8 protein, the first LIP isozyme isolated and characterized and having a pI of 3.5, should be referred to as LiP1 as this was the most abundant isozyme present in their medium. They also suggested that the second most abundant LIP isozyme with a pI of 4.4, previously referred to as H2, be designated LiP2, and so on. While the proposal of Farrell et al. [49] brings some order to the nomenclature, several disadvantages remain. For instance, the LIP isozyme H8 is not the predominant protein under all the culture conditions. Cultivation of this fungus with shaking in a medium buffered with sodium acetate as an alternate to the expensive and commonly used dimethylsuccinate buffer results in the production of a great abundance of MNP isozymes (especially H4) and the LIP isozyme H8 is produced in much lower amounts than the LIP isozyme H6 [47] (Fig. 7). It is also significant that H2 is a major LIP isozyme produced by *P. chrysosporium* under a variety of culture conditions. Tuisel et al. [60] reported that *P. chrysosporium* produces H8 and H2 isozymes in equal abundance. Therefore, the nomenclature LiP1 for H8 based on its reported abundance on some media is not entirely appropriate. It would be much less confusing and more logical if the first LIP isozyme on FPLC (H1) is designated LIP1, the second one (H2) designated LIP2, and so on. We propose the nomenclature scheme shown in Table 1 based on the following reasons:

1. Since the original designation of H1, H2, etc., is fairly entrenched in the minds of readers for several years, it is easier to switch from H series to LIP series in a serial manner as shown in Table 1 (i.e., LIP1 for H1, LIP2 for H2, etc.).
2. We suggest *lip* designation for a lignin peroxidase gene, as opposed to the *LiP* gene designation proposed by Farrell et al. [49], because according to the standard genetic nomenclature of lower eukaryotes,

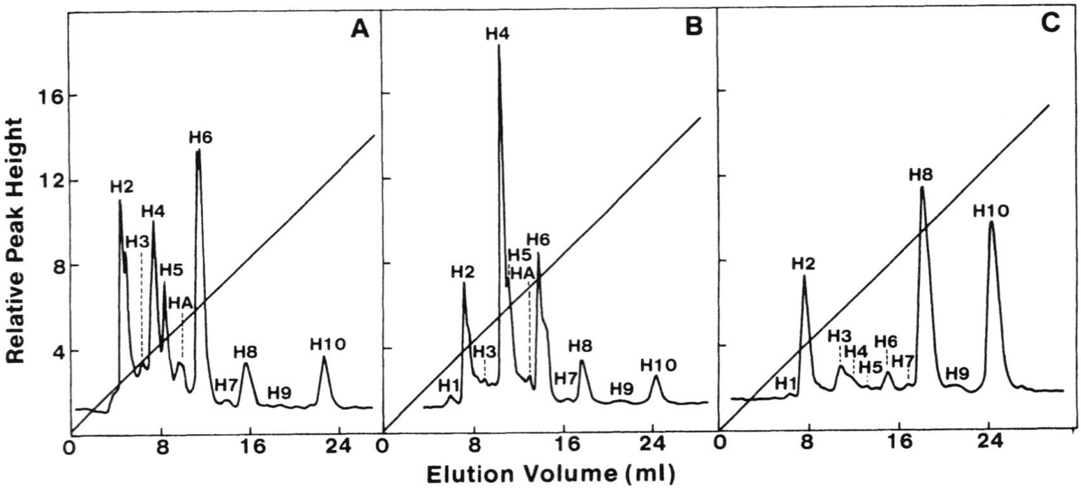

FIGURE 7. FPLC profiles of extracellular heme proteins from 6-day-old acetate or dimethylsuccinate-buffered cultures of *P. chrysosporium*. The peaks are designated as described by Tien and Kirk [42]. H1, H2, HA, H6, H7, H8, and H10 are lignin peroxidases; H3, H4, H5, and H9 are manganese-dependent peroxidases. The diagonal line is the sodium acetate gradient (10 mM to 1 M). Relative peak height represents absorbance at 405 nm. (A) Stationary acetate-buffered culture. (B) Shaken acetate-buffered culture. (C) Shaken dimethylsuccinate-buffered culture [47].

especially yeasts, dominant and recessive genes, respectively, are denoted by upper and lower case three-letter codes. Examples are *SUP*4 and *arg*2. However, in fungi, gene nomenclature is not as well formalized and the lower case designation is used most often (e.g., *ura*3, and *ade*2 genes). Hence, we propose the *lip* designation for the lignin peroxidase genes of *P. chrysosporium* and suggest dropping the *LIP*, *lpo*, *lgp*, and *LIG* designations that have been used by us and other investigators in the past. Note that the three-letter designation is underlined or italicized and generally all three letters are given in upper case or lower case and not a combination as in LiP. The lignin peroxidase mutants will simply be designated as *lip* mutant [20] or will be referred to as *lip*1 mutant, *lip*2 mutant, etc., in referring to mutants lacking a specific LIP isozyme.

3. Based on the above rationale, it is reasonable to designate LIP proteins as LIP1, LIP2, ..., LIP6, etc., and the corresponding genes as *lip*1, *lip*2, ..., *lip*6. Furthermore, designating lignin peroxidase proteins as LiP1, LiP2, and so on, apparently in analogy with MnP1, MnP2, etc., proposed for MNPs is misleading to the novice reader as "Li" may be interpreted by the reader as lithium the same way "Mn" stands for manganese. LIP designations would avoid all this confusion. The new designation for LIP proteins and genes is used in the rest of this review.

In keeping with the above discussion, we propose designation of *mnp*1, *mnp*2 etc., for manganese-dependent peroxidase genes and MNP1, MNP2,

TABLE 1. Proposed Nomenclature for Extracellular Peroxidases of *P. chrysosporium* BKM-F and the Genes Encoding These Proteins

Peroxidase[a]	Designation[b]	
	Gene	Protein
Lignin peroxidase		
H1	*lip*1	LIP1
H2	*lip*2	LIP2
H6	*lip*3	LIP3
H7	*lip*4	LIP4
H8	*lip*5	LIP5
H10	*lip*6	LIP6
Manganese-dependent peroxidase		
H3	*mnp*1	MNP1
H4	*mnp*2	MNP2
H5	*mnp*3	MNP3
H9	*mnp*4	MNP4

[a]From Ref. 42.
[b]Designations for various genes and the corresponding peroxidase encoded by these genes are given.

etc., for the proteins corresponding to these genes. The mutants would be referred to as *mnp*1 mutant, *mnp*2 mutant, and so on.

4. Hydrogen Peroxide-Producing Enzymes

Hydrogen peroxide production was shown to be temporally correlated with lignin degradation (^{14}C-synthetic lignin → $^{14}CO_2$) in this fungus [37]. The addition of catalase to ligninolytic cultures was shown to inhibit lignin degradation [38]. Production of H_2O_2, similar to the lignin-degrading activity, was shown to be triggered in response to nitrogen or carbohydrate starvation [31,83,84]. Exogenous addition of NH_4 or glutamate to ligninolytic cultures grown in low nitrogen media (2.4 mM nitrogen concentration) resulted in the inhibition of H_2O_2 production, lignin degradation, and LIP and MNP activity [19,20,84]. A strong positive correlation was observed between H_2O_2 production and the lignin degradation activity when *P. chrysosporium* was grown with different growth substrates [84]. Unique H_2O_2-producing, periplasmically located, microbody-type structures are observed only in lignin-degrading cultures grown in nitrogen-limited medium but not in non-lignin-degrading cultures grown in high-nitrogen (24 mM) media [83]. The strongest evidence for the importance of H_2O_2 in lignin degradation is the discovery of H_2O_2-dependent LIPs and MNPs which are believed to be important in lignin mineralization [7,10,20]. Glucose oxidase was shown to be the primary source of H_2O_2 production in lignin-degrading cultures [31,84]. Mutants (*gox*) that lacked glucose oxidase activity failed to degrade lignin and the glucose oxidase protein band was not seen on native polyacrylamide gel electrophoresis (PAGE) of extracts of these mutants. In contrast, the *gox* revertants not only exhibited ligninolytic activity but also showed a glucose oxidase protein band in cell extracts [85]. However, further studies showed that the *gox* mutants are regulatory mutants and are pleiotropic for a set of secondary metabolic characteristics including LIPs [86]. It is also of interest that *gox* mutants did not lose all secondary metabolic activities and that they did exhibit the ability to produce veratryl alcohol, a secondary metabolite produced by *P. chrysosporium*, and the ability to sporulate, i.e., form conidia [86]. The above results collectively indicate that glucose oxidase is an important source of H_2O_2 in lignin-degrading cultures of *P. chrysosporium*. The glucose-1-oxidase of *P. chrysosporium* has been purified to homogeneity and characterized [87,88].

A number of other H_2O_2-producing enzymes have also been demonstrated in ligninolytic cultures of *P. chrysosporium*. Ericksson et al. [30] purified and characterized glucose-2-oxidase and suggested that it contributes to hydrogen peroxide production in lignin-degrading cultures of *P. chrysosporium*. Kersten and Kirk [89] reported that the extracellular, idiophasic enzyme glyoxal oxidase is important as a source of hydrogen peroxide in lignin-degrading cultures. The enzyme has been purified and characterized [90]. Various other enzymes including MNPs have also been reported to contribute to hydrogen peroxide. Obviously, *P. chrysosporium* has several enzymes that are capable of generating hydrogen peroxide. The relative contribution of each enzyme to overall lignin degradation is currently being investigated in our laboratory with the aid of mutants defective in the production of one or more enzymes.

5. Ultrastructural Localization of Lignin Peroxidases

Several laboratories have focused their research on the cellular localization of H_2O_2 production and the extracellular peroxidases. Forney et al. [83] reported that hydrogen peroxide-producing enzymes are located in the periplasmic space of *P. chrysosporium* as evidenced by the deposits of oxidized diaminobenzidine only in ligninolytic cultures but not in nonligninolytic cultures (Fig. 8A). For example, the periplasmic microbody-like structures which are the sites of H_2O_2 production in *P. chrysosporium* were demonstrable in cultures grown in low-nitrogen media but not in high-nitrogen media. Even in low-nitrogen media, these microbody-like structures are not produced during primary metabolism but only during secondary metabolism; peak production of these structures is temporally correlated with the peak of lignin degradation activity [83].

The discovery of extracellular LIPs combined with recent advances in immunocytochemical techniques enabled researchers to localize the LIPs in the fungal cell. Using an anti-lignin peroxidase antiserum-protein A complex, Garcia et al. [91] presented evidence to indicate that LIPs are located very close to the cell membrane in white rot fungi such as *P. chrysosporium, Coriolus versicolor, and Dichomitus squalens* colonizing sawdust (under laboratory conditions) or in naturally decaying wood. These results also indicate that other white rot fungi are also able to produce LIP-like enzymes which are immunologically reactive with the antibodies raised against LIP enzymes of *P. chrysosporium*. Subsequent studies by Srebotnik et al. [92] also confirmed the intracellular localization of LIPs in the plasmamembrane region (Fig. 8B). Several investigators observed that lignins are tightly bound to fungal mycelia during lignin degradation [93,94]. However, some electron microscopic studies demonstrated that lignin is degraded at some distance from the fungal hyphae [95], suggesting that in nature LIPs do not necessarily have to be in close contact to catalyze degradation of the latter substrate. Further studies are needed to solve these seemingly conflicting results.

6. Other Enzymes Involved in Lignin Degradation

Studies with wood partially degraded by white rot fungi but still containing polymeric lignins have established the involvement of reactions other than Cα-Cβ bond cleavage in side chains, although Cα-Cβ cleavage is clearly a prominent reaction. Several additional reactions such as demethylation of arylmethoxyl groups, hydroxylation of aromatic rings ortho to methoxyl or hydroxyl groups, degradation of ring cleavage fragments, and possibly various side chain reductions take place. These results suggest the involvement of a host of other enzymes in lignin biodegradation. Kirk and Farrell [10] and Buswell and Odier [8] discussed the other enzymes such as laccase associated with lignin degradation in detail and the reader is referred to these excellent reviews for further information.

C. Molecular Biology

The discovery of LIPs of *P. chrysosporium* truly moved lignin biodegradation research into the realm of not only biochemistry but also molecular biology. As described above, LIPs are a family of extracellular, glycosylated heme proteins [7,49]. As many as 15 LIP isozymes have been resolved by anion exchange chromatography depending on the separation efficiency, extent of glycosylation, age of culture, composition of the culture medium, and differences in strains [43,47,49]. At the time these LIP isozymes were discovered it was not known whether they were distinct but related proteins or if they represented different posttranslational modifications of the protein products of a small number of structural genes. As soon as the LIPs were discovered, several laboratories immediately started on the trail of cloning the cDNAs/genes for LIPs because cloning and overproduction in suitable expression systems is important for the initial evaluation of LIPs in various biotechnological applications as well as for the eventual commercialization of these processes. Rapid advances have been made in cloning and sequencing the cDNA/genes encoding several of the LIPs as well as two of the MNPs of *P. chrysosporium*. Cloning of the *LIP* genes of other white rot fungi was also reported recently. A detailed discussion of the molecular biology of lignin degradation enzymes is beyond the scope of this chapter, but because of the importance of this area to eventual biotechnological application of lignin biodegradation systems of fungi, a brief overview of this rapidly evolving area is presented here.

1. Lignin Peroxidase cDNAs

Zhang et al. [96] were the first to report the isolation of two different types of LIP cDNAs from *P. chrysosporium* strain BKM-F. This was soon followed by the complete sequencing of these two types of cDNAs, *CLG4* and *CLG5*, and deduction of the amino acid (aa) sequences of the proteins encoded by these two cDNAs [97,98]. These data showed that cDNA *CLG4*

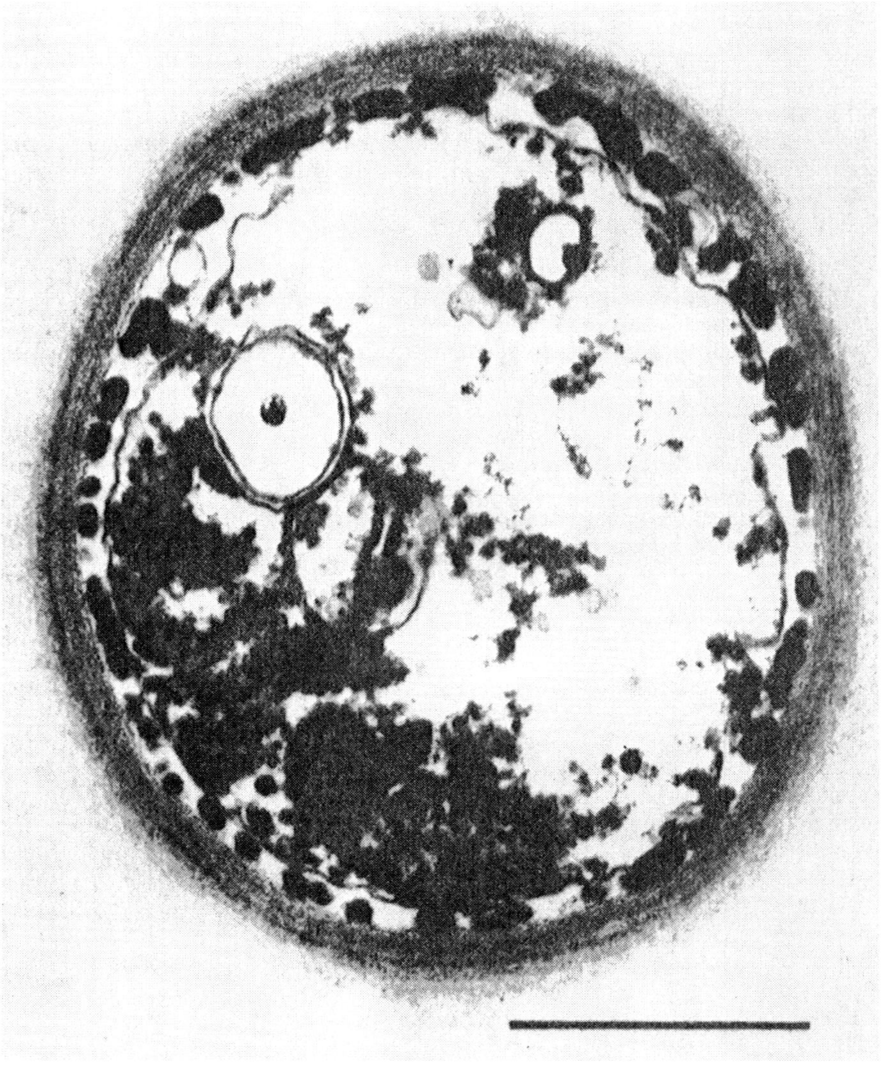

(A)

FIGURE 8. (A) Transmission electron micrograph of a transverse section of a cell of *P. chrysosporium*. Note the fairly even distribution of the deposits of oxidized diaminobenzidine in the periplasmic space indicating sites of H_2O_2 production [83]. Magnification, 21,500x; bar = 1 μm. (B) Immunogold labeling of *P. chrysosporium* grown on pine (shown on page 781). Small pieces of wood colonized by the fungus were fixed with 2% glutaraldehyde. Ultra-thin section of this fungus-colonized wood was subjected to immunocytochemical labeling [92]. Lignin peroxidase (indicated by dark particles) is located in the periplasmic space. HY, hypha; WW, wood cell wall. Bar = 1 μm. Note the striking similarity in the intracellular distribution of LIPs and H_2O_2 production sites in Figs. 8B and A, respectively.

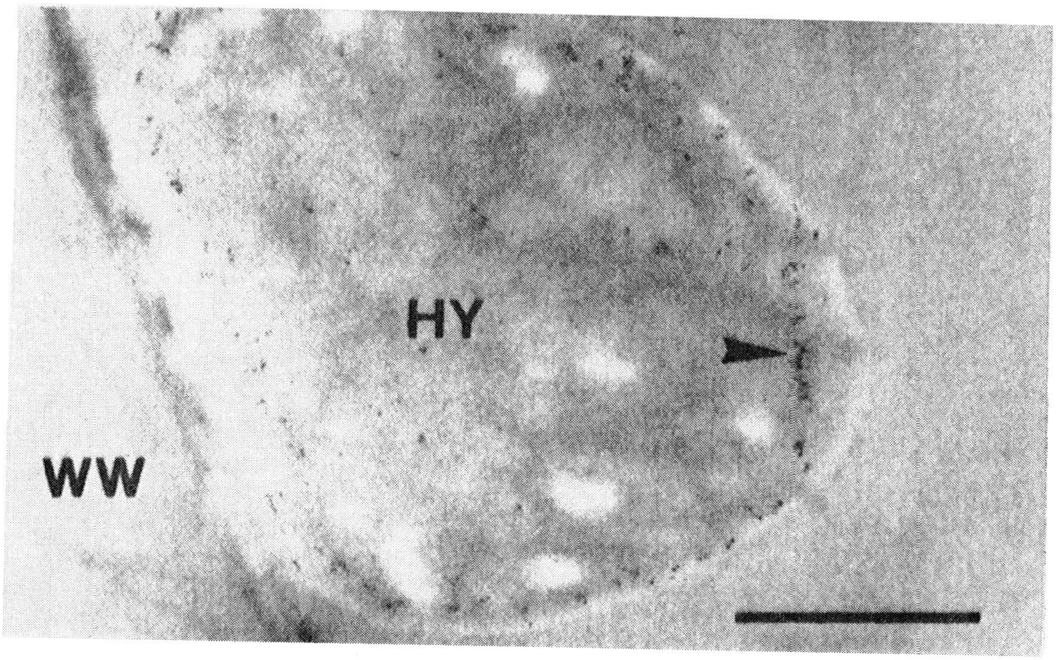

(B)

and *CLG5* encode LIP proteins LIP2 (H2) and LIP6 (H10), respectively. Mature LIP2 and LIP6 contained 344 amino acids each and have an M_r of slightly less than 37 kDa. Mature LIP2 and LIP6 proteins were preceded by a leader peptide containing 28 and 27 amino acids, respectively, and contained one and two *N*-glycosylation sites, respectively, with the general sequence Asn-X-Thr/Ser. Each contained numerous *O*-glycosylation sites. This is consistent with the reports that purified LIPs exhibit both *N*-glycosylation and *O*-glycosylation [49]. Nucleotide sequence (nt) homology between the coding regions of LIP2 and LIP6 encoding cDNAs was 71.5% and the aa homology between the corresponding proteins was 68.5%. In parallel with these studies, Tien and Tu [99] independently cloned and sequenced *ML1* cDNA that apparently encodes LIP5 (H8) protein. These data indicated that the LIP5 encoded by *ML1* contained 345 amino acids and was preceded by a leader peptide of 28 amino acids. The leader peptide for the three LIPs (i.e., LIP2, LIP5, and LIP6) end in a Lys-Arg proteolytic cleavage site. The nt homologies between *CLG4* and *ML1* and *CLG5* and *ML1* were 74 and 81%, respectively, whereas the aa homologies were 69 and 79%, respectively. These results indicated a high degree of relatedness among the cDNAs encoding LIP2, LIP5, and LIP6 and also showed that LIP5 and LIP6 are somewhat more related to each other than they are to LIP2-encoding cDNA. Northern blotting studies using each of the three cDNAs as probes showed that LIP production is controlled at the level of mRNA production [98,99]. Farrell et al. [100] also reported on the cloning and expression of cDNA encoding LIP5. The cDNA data suggested that different LIP proteins are encoded by different structural genes, that the LIP proteins are structurally related as independently reported by other investigators [43,49], that each LIP protein contains an almost identical number of amino acids and thus pointed to the fact that the differences in molecular weights of individual LIPs are due to differences in glycosylation of these proteins.

Recently, Saloheimo et al. [101] isolated and sequenced an *lip* cDNA from *Ph. radiata*. This gene encoded a mature protein that contained 337 amino acids (35,983) and was preceded by a signal peptide of 24 amino acids which ended in a putative proteolytic cleavage site: Arg-Arg. This protein contained one putative *N*-glycosylation site. This *lgp*3 cDNA reportedly has about 60% homology at both the aa and nt levels to those of *P. chrysosporium*.

2. Manganese-Dependent Peroxidase cDNAs

Two cDNAs encoding two different MNPs of *P. chrysosporium* have also been cloned and sequenced [102,103]. The deduced MNP proteins encoded by *MnP1* and *λMP1*, respectively, contained 378 and 382 amino acids including leader sequences of 21-24 amino acids; one and four putative glycosylation sites were found in these two MNP proteins. Unlike the leader sequences of LIP proteins which end in dibasic residues, Lys-Arg (the presumed signal peptide cleavage site) the leader sequence of one MNP ends in Pro-Thr and the other ends in Glu-Ser. The aa homology of these two MNPs to the LIP proteins varied from 50 to 60%. The expression of *mnp* cDNA, similar to the *lip* cDNAs of *P. chrysosporium*, is controlled by nutrient nitrogen at the level of transcription [102,103].

3. Lignin Peroxidase Genes

The isolation and characterization of *lip* cDNAs paved the way for the isolation and study of the structural organization of *lip* genes. Zhang [98] in our laboratory probed the *P. chrysosporium* BKM-F genomic library with labeled LIP cDNAs *CLG4* and *CLG5* and isolated six distinct genomic fragments designated *GLG1*, *GLG2*, ..., *GLG6*. He subsequently determined the complete sequence of *GLG2* and showed that its sequence corresponds to cDNA *CLG5* and that the aa sequence of the protein encoded by *GLG2* is identical to that of LIP6 protein encoded by *CLG5*. The 5' sequence of this gene (hereafter referred to as the *lip6* gene) contained archetypal TATA box and CAAT boxlike sequences in the 5' noncoding region located, respectively, 51 and 97 nt upstream of the cDNA starting point. A comparison of the sequence of the *lip6* gene [98] with its corresponding cDNA *CLG5* [97] showed that the gene contains nine small introns which range in size from 50 to 62 base pairs (bp). These introns contained consensus splice junction sequences similar to those reported in other fungal and yeast introns, suggesting that the splicing mechanism in *P. chrysosporium lip* genes is probably similar to that of other eukaryotic organisms [98]. The proximal and distal histidine residues and one arginine residue known to be required for the catalytic activity of all peroxidase proteins examined to date are conserved in the *lip6* gene [98], as well as in LIP2 and LIP5 proteins deduced from their corresponding cDNAs [97,99]. A comparison of the codon usage between the *lip6* gene of *P. chrysosporium* and the genes of *Saccharomyces cerevisiae* showed that the codon usage in the *lip* gene is extremely biased in favor of codons ending in G or C. However, codon usage in the *lip6* gene of *P. chrysosporium* was relatively similar to that in *Neurospora* genes and to a lesser extent in *Aspergillus* genes as well [98].

In parallel with the above work of Zhang [98], several different laboratories [104-106,108,109] reported the sequence of a gene that is believed to encode LIP5 (H8) or a close homolog of it because the sequences reported were extremely similar (~95% similarity). The features of this gene were very similar to the *lip5* cDNA(λ*ML1* cDNA) previously reported and to the general features of the *lip6* gene except that the *lip5* gene contained only eight introns. Furthermore, the introns were very similar in size to the *lip6* introns, although slight differences were seen between the position of introns of *lip5* and *lip6* genes. Brown et al. [107] and Asada et al. [108] independently reported the sequence of two *lip* genes from *P. chrysosporium* ME446 which appear very similar to the *lip5* gene of *P. chrysosporium* BKM-F 1767 described above by the other laboratories.

Recently, Padmavathy and Reddy [110] sequenced LIP genomic fragment *GLG1* (see above) and by comparing this sequence with the *CLG4* sequence demonstrated that this fragment contains the *lip2* gene. These data also showed that the *lip2* coding sequence is interrupted by eight introns which range in size from 50 to 63 bp. A comparison of the *lip2* gene sequence to those of *lip6* [98] and *lip5* [104,105] showed that *lip2* is much less related to the latter genes than they are to each other. This is in keeping with the previous studies on LIP proteins [47,49] showing that LIP2 forms a distinct subclass within the LIP protein family.

It has been well established that LIP production by *P. chrysosporium* occurs only during secondary metabolism and is triggered in response to nitrogen or carbon starvation. Holzbaur and Tien [106] determined the relative levels of expression of two major LIP isozymes LIP2 and LIP5 under conditions of nitrogen and carbon starvation by analyzing RNA and protein levels. Their results indicated that under nitrogen limitation in the medium employed [116], the fungus produces major amounts of LIP2 and LIP5 proteins and the corresponding mRNAs. However, carbon-limited cultures showed a major peak for LIP2 only. These results plus the recent results of Dass and Reddy [47] and others [10,116] suggest that different LIP isozymes are differentially regulated in response to environmental stress.

Recently, we isolated and sequenced a *lip* gene from *T. versicolor*, another widely studied white rot fungus (A. Black and C. A. Reddy, unpublished data). This gene encodes a mature LIP protein that is 339 amino acids long (M_r = 36,607) and is preceded by a signal sequence of 25 amino acids. As compared to the *lip* genes of *P. chrysosporium*, which contain eight to nine introns, the *T. versicolor* gene contains six small introns which vary in size from 49 to 64 bp. The LIP protein encoded by this gene had an aa homology of 55-60% to the LIPs of *P. chrysosporium*, similar to that exhibited by the *Ph. radiata* LIP protein [101] described above.

The data presented above clearly show that rapid progress is being made in understanding the structural organization of the LIP genes of various white rot fungi. However, a great deal more work is needed to understand the structural organization and regulation of expression of LIP genes of white rot fungi.

4. Other Studies

Several other areas of molecular biology research on *P. chrysosporium* are under active pursuit. Our laboratory and that of Dr. Cullen and coworkers at the University of Wisconsin have independently studied genome organization of this organism. Preliminary studies from our laboratory indicate that there are at least 7 chromosomes in *P. chrysosporium* and that the *lip* genes appear to be distributed on two to three chromosomes.

A DNA transformation system is critical for molecular genetic analyses of any microorganism. Hence, this was and remains an active area of research in many laboratories. Alic et al. [111] recently described an integrative transformation system for *P. chrysosporium*, whereas Randall et al. [112] from our laboratory described a nonintegrative transformation system. These studies represent a modest progress at best and more intensive studies are needed to improve these transformation systems. A critical need is a system by which one can conduct reliable disruption studies to enable determination of the functional roles of various *lip* and *mnp* genes and perhaps other genes involved in the lignin degradation system of *P. chrysosporium*.

IV. STRATEGIES FOR MAXIMIZING ENZYME PRODUCTION

To fully exploit the capabilities of white rot fungi in various biotechnological processes, optimal conditions for LIP production on a large scale need to be explored. There are three basic approaches to maximize enzyme production. First is optimization of the culture conditions including nutrition of the organism, environmental factors, medium additives, and culture size. Second is strain improvement, including selection of superior wild-type strains and development of improved strains through mtuagenesis. The third approach is cloning and high level expression of the LIP genes employing recombinant DNA technology.

A. Manipulation of Culture Conditions

Research over the past 10 years has contributed a great deal to our understanding of lignin degradation as affected by various culture conditions [8-10]. The level of lignin degradation and LIP activity was low in non-agitated cultures in which the LIP activity in the extracellular fluid was first detected. Furthermore, any agitation of the cultures or even disturbance of the cultures during incubation severely affected lignin degradation and LIP activity. This was a limitation to their use in large-scale production and in their use in biotechnological processes. However, recent developments largely helped to overcome this problem. Jager et al. [46] showed that addition of nonionic surfactants such as sorbitol polyoxyethylene monooleate (Tween 80), sorbitol polyoxyethylene monolaurate (Tween 20), or 3-(3-cholamidopropyl)dimethylammonio-1-propanesulfonate to the culture medium afforded LIP production with agitation. It was shown that wild-type *P. chrysosporium* cultures and mutants can be activated to produce LIP activity under agitation [113,114]. Reid et al. [115] reported that lignin is degraded as rapidly in agitated as in nonagitated systems by mycelial pellets of *P. chrysosporium*. Thus the ability to produce LIPs under agitated conditions represented another milestone in the advancement of lignin biodegradation research and also in the large-scale production of LIPs and MNPs. A number of subsequent studies have focused their research on the manipulation of other culture conditions to increase the enzyme yield.

Kirk et al. [116] showed that addition of veratryl alcohol, a secondary metabolite of *P. chrysosporium*, and excess trace elements improved LIP activity. Leisola et al. [117] reported strategies for high levels of LIP production by manipulating carbon and nitrogen levels in agitated cultures supplemented with veratryl alcohol. Addition of solid manganese(IV) oxide to cultures of *P. chrysosporium* at the beginning of ligninolytic activity not only improved production and enzymatic activity but also improved the stability of LIP isozymes produced [118]. It has been suggested that MnO_2 addition protects LIPs against inactivation and damage by hydrogen peroxide via catalytic decomposition of H_2O_2 by MnO_2. Roch et al. [119] investigated LIP production by several strains of *P. chrysosporium* grown on glycerol under conditions of nitrogen sufficiency. Strains which grew most poorly on glycerol produced the highest titers of LIPs whereas enzyme levels were much lower when greater biomass values were obtained. In the case of *P. chrysosporium* strain INA-12, the nature of the nitrogen source had a pronounced effect on both growth and enzyme production. The highest biomass values were obtained when L-glutamate or L-glutamine served as the major nitrogen source but enzyme synthesis was repressed completely. LIP activity was maximal in this strain when the initial pH of the culture medium was adjusted to 5 [119]. The results of another study with strain INA-12 showed that a shift in growth temperature elevates the LIP titers and that there are two temperature optima connected with LIP production: one at 37°C which is optimal for the mycelial growth and the other at 30°C which is optimal for the LIP production [122a]. A summary of the levels of LIP produced by strains of *P. chrysosporium* grown under different culture conditions is presented in Table 2. It is apparent from the data that the yield of LIP using various strains of *P. chrysosporium* has improved substantially as compared to the activities obtained when the LIP was first discovered. Although improvements on the enzyme production by various additives to the culture medium is evident from these studies, their application in commercial large-scale production seems to be restricted.

LIPs were shown to be localized around the plasmamembrane using immunogold-labeled antibodies [92]. At this time it is not clear as to what percentage of LIPs produced by *P. chrysosporium* is released into the growth medium. Attempts to change the membrane permeability to allow release of most of the LIP synthesized inside the cell may prove rewarding. Tween 80 supplementation to the cultures of *P. chrysosporium* INA-12 resulted in a 1.5-fold enhancement of LIP activity when compared to non-supplemented cultures [123]. Supplementing cultures with phospholipids and fatty acids also increased the LIP production by 1.5-fold followed by changes in fatty acid composition of *P. chrysosporium* cells [122b]. However, the relation between the observed change in fatty acid and phospholipid composition of *P. chrysosporium* cell to LIP production is not known.

Recent studies show that LIP or MNP production can be selectively affected by regulating manganese levels in the culture medium [124,125]. Addition of 40 ppm of manganese in the form of $MnCl_2$ to the medium resulted in no LIP production but a great enhancement in MNP production. In Mn(II)-free cultures, on the other hand, LIP activity appeared earlier and was 2.5-fold higher than that in the cultures grown with 11.15 ppm Mn(II). Brown et al. [124] showed that the presence of Mn(II) in the medium is absolutely required for MNP production by *P. chrysosporium*. These investigators demonstrated that the MNP mRNA was not detectable in *P. chrysosporium* cultures grown in the absence of manganese. In cultures grown in the presence of 180 μM manganese, MNP mRNA was detected on day 4 and reached a maximum on day 5, indicating the regulation of *mnp* gene expression by manganese at the transcriptional level. The above results demonstrate convincingly that manganese is a critical requirement for the synthesis of MNP and that levels of manganese in the medium have a dramatic effect on the LIP production as well.

Studies on the kinetics of production of LIPs in media that are nitrogen-limited but contain nonlimiting levels of carbon indicate that the activity levels peak around the sixth day followed by a relatively sharp decline in LIP levels, regardless of the carbon substrate, i.e., glucose or glycerol [8,119,122b], culture conditions [46,82,120], or the strains used [20,102, 119,120]. A temporal relationship between a decrease of LIP activity following day 6 and an increase in extracellular protease activity during the late secondary metabolism has been reported [126]. Subsequently, Dosoretz et al. [127] demonstrated protease-mediated degradation of LIPs in liquid cultures of *P. chrysosporium* during late secondary metabolism. The protease production in idiophase appears to be related to carbon starvation because the protease levels could be lowered by daily addition of glucose after day 6. This decrease in protease level was in turn reflected by the high levels of extracellular LIPs observed under such conditions [127]. This study also showed that the idiophasic protease is produced de novo in response to glucose starvation as the addition of cyclohexamide to the

TABLE 2. Effect of Culture Parameters on Lignin Peroxidase (LIP) Activity of Different Strains of *P. chrysosporium*

Culture condition	Strain	Culture volume (ml)	LIP (U/liter)	Ref.
Stationary cultures				
Standard[a] medium (S)	ME446	10	5	82
	Cel^-	10	1-21	120
	BKM-F	10	23	45,120
	SC26	10	34	116
S + 1-2 mM VA	BKM-F	10	60	117
S + 0.4 mM VA, 7X TM	BKM-F	10	100	120
S + 0.4 mM VA, 7X TM	der8-5	10	460	19
S + pine wood	BKM-F	10	140	117
S + high N + 7X TM	der8-5	10	380	19
S + high N + 7X TM	PSBL-1	10	1000	136
Glycerol,[b] high N	INA-12	10	350	137

Agitated Cultures				
S + 0.4 mM VA, 7x TM	BKM-F	30	< 5	46
S + 0.4 mM VA + 7X TM + 0.05% Tween 80	BKM-F	30	125	46
S + 2mM VA; pellets concentrated[c]	BKM-F	50	120	117
High N, low C[d] + 2 mM VA; pellets concentrated	BKM-F	50	670	117
S + 7X TM + 3 mM VA + 0.1% Tween 80	BKM-F	45	350-650	121
S + 7X TM + 8 mM benzyl alcohol + 0.1% Tween 80	BKM-F	600	650-800	121
S + 7X TM + 6 mM benzyl alcohol + 0.1% Tween 80	SC26	600	750-1675	121

[a]Standard medium was the same as the nitrogen-limited (2.2 mM) medium described by Kirk et al. [116]. High-nitrogen medium contained 22 mM nitrogen. Unless specified otherwise, excess glucose was the carbon source.
[b]Glucose was replaced with glycerol (1%).
[c]Pellet concentration was several fold higher than in normal cultures.
[d]Medium contained 0.15% glucose.
Abbreviations: S = standard medium; VA = veratryl alcohol; TM = trace minerals, N = nitrogen; C = carbon.
Source: Adapted from Ref. 121.

cultures on day 6 was followed by severe inhibition of protease levels and in turn high levels of LIP activity [127].

Addition of veratryl alcohol, trace minerals, and Tween 80 seem to enhance peroxidase production but is not essential to the process. Lewandowski et al. [128] investigated the effect of various nutrient media and bioreactor configurations to grow the fungus to induce its extracellular enzyme system, develop kinetic data for the degradation of 2-chlorophenol, and use chemical engineering analysis to design an effective reactor. In their study they omitted veratryl alcohol and Tween 80 from the medium but were still able to improve the chlorophenol biodegradation rate by 40-fold when the fungus was immobilized. Such studies open up the possibility of producing enzymes utilizing simplified media in order to reduce the cost of enzyme production.

B. Use of Lignin Peroxidase Overproducers

One logical approach to improving LIP yields by *P. chrysosporium* is the isolation of LIP-overproducing mutants. Gold and Cheng [129] developed a method which facilitated mutant analysis by inducing colonial growth on solid agar plates and developing replica-plating techniques. Methods have also been developed for the selection of mutants using asexual spores [19,20,86,133] as well as the production of sexual spores for genetic recombination [134]. Tien et al. [135] devised a procedure whereby amino acids are covalently linked to lignin model compounds. The result is that LIP action on the models frees amino acids essential for growth in a nitrogen-deficient medium. While this approach appears to be suitable for the isolation of mutants affected in their response to nitrogen levels in the medium [136], simpler methods [18-20] appear to be equally effective. Ultraviolet mutagenesis of *P. chrysosporium* strain BKM-F and selection for enhanced decolorization of "E1 effluent," the effluent from the first alkali extraction of a kraft pulp bleach plant, yielded a mutant SC26 [116]. Although its hyphae have a greater polysaccharide sheath as revealed by phosphotungstic acid staining, the growth requirements are the same as for the wild type and it produced higher levels of LIP activity [116]. The qualitative analysis of the extracellular protein profile showed its resemblance to the wild-type strain, and the increased level of activity was attributed to the greater accumulation of heme proteins in the culture medium. Buswell et al. [137] reported a strain INA-12 that produced much higher levels of enzyme than the well-studied strains ME446 and BKM-F. Strain INA-12 differed from BKM-F and ME446 not only in its capacity to produce higher levels of LIPs but also in degrading lignin under conditions of nitrogen deficiency as well as sufficiency when glycerol served as the carbon source. The protein profiles of the extracellular fluid showed dissimilarities from those produced by strain BKM-F. Strain INA-12 produced LIPs but not MNPs under these conditions. Whether the increase in LIP activity is due to an increase in the production of specific LIP isozymes or all the LIP isozymes has not been determined. Mutant strains of *P. chrysosporium* selected on the basis of cellulase overproduction enabled Galliano et al. [138] to isolate LIP-overproducing strains. One of these strains showed enhanced ligninolytic activity. This strain is of particular interest for research on delignification of lignocellulosic materials to enhance their value as feedstock for ruminant animals and in designing efficient biopulping processes.

The use of nitrogen-deregulated mutants (*der* mutants) would be more desirable for various biotechnological processes as such mutants will be able to produce extracellular peroxidases and degrade lignin under both nitrogen-deficient and nitrogen-sufficient conditions. Furthermore, cell biomass of the wild type is restricted by the low amount of nitrogen present in the usual media used for the production of LIPs by *P. chrysosporium*. Use of nitrogen-deregulated mutants would allow for better growth and perhaps better LIP production. In our laboratory, we isolated two nitrogen-deregulated mutants of *P. chrysosporium* ME446, *der*8-2 and *der*8-5, which are no longer subject to nitrogen regulation for ligninolytic activity, by subjecting the wild-type conidia to γ-irradiation [19]. The *der* mutants showed higher levels of LIP, MNP, glucose oxidase, and ligninolytic activities than the wild-type strain under nitrogen-limited conditions. The mutants also showed significant amounts of these activities in nitrogen-rich media. Further analyses by FPLC of the concentrated extracellular fluids of the *der* mutants indicated the presence of at least one new lignin peroxidase protein (designated LN) which had the highest specific activity as compared to the other LIP isozymes produced. Another nitrogen-deregulated mutant of *P. chrysosporium* was isolated and characterized [21].

TABLE 3. List of Biochemical Mutants of *P. chrysosporium*

Strain no.	Characteristic(s)[a]	Ref.
44	Cel^-; Xyl^-; Pox^-	28
44-2	Cel^-; Xyl^-; Pox^+	130
63-2	Cel^-; Xyl^-; Pox^+	130
31	Cel^+; LIG^+	131
3113	Cel^-; LIG^+	131
85118	Cel^-; LIG^+	131
SC 26	LIG - enhanced	116
LMT 26	Pox^-; LIG-enhanced	132
gox$^-$	Gox^-; LIG^-	85
*lip*5	LIP^-; MNP^+	20
*der*8-2	Der(N): LIG-enhanced	19
*der*8-5	Der(N); LIG-enhanced	19
MA 4	Der(N): LIP and MNP-enhanced	21
PSBL-1	Der(N): LIG - enhanced	136

[a]Cel = cellulase; Xyl = xylan degradation; Pox = phenol oxidase; LIG = lignin degradation; + = positive; - = negative; *der*(N) = nitrogen deregulated; LIP = lignin peroxidase; MNP = manganese-dependent peroxidase.
Source: Partly adapted from Ref. 10.

Similar to the above mentioned *der* mutants, this mutant also produced higher levels of LIPs in both nitrogen-rich and nitrogen-limited media and produced a new LIP protein [21]. Tien and Myer [136] reported the characterization of a nutrient-deregulated mutant of *P. chrysosporium* BKM-F which produced about 1000 U/liter LIP under nitrogen-rich growth conditions. Table 3 summarizes the various classes of biochemical mutants isolated to date. Further research on the production of strains with enhanced LIP production, as well as those specifically deregulated for other nutrients such as carbon, will be useful from the standpoint of both basic research and biotechnological applications.

The production of extracellular peroxidases in *P. chrysosporium* is seen only during secondary metabolism. Culture periods of 3 and 5 days are required for the appearance of MNPs and LIPs, respectively. Therefore, the third attractive approach to enhance LIP and MNP production will be to generate constitutive mutants that can produce LIPs and MNPs even during primary growth, which will substantially reduce the culture incubation period. However, whether or not this is an attainable objective remains to be determined.

C. Expression of Lignin Peroxidase Genes in a Heterologous System

Another approach used to maximize enzyme production is to clone and express *LIP* genes in different host systems by recombinant DNA techniques. It should be possible to manipulate the *lip* genes or cDNAs appropriately and thus engineer the LIP proteins to increase either enzyme efficiency or enzyme stability. This approach will allow for an increase in LIP enzyme production through altered gene regulation, thus increasing enzyme production in larger scale.

Several laboratories have cloned and sequenced *LIP* genes. All these genes display a considerable degree of homology at both the nucleotide and the amino acid level. One of the genes encoding the isozyme H8 (*LIP*5) has reportedly been expressed [139]. The recombinant LIP (rH8) has been reported to catalyze the aromatic demethoxylation and C_α oxidation of the substrates tetramethoxybenzene and veratryl alcohol, respectively. The specific activity of the rH8 is comparable to that of *P. chrysosporium* LIP5. Furthermore, the rH8 exhibited substrate-dependent peroxide uptake in the presence of kraft lignin or milled wood lignin and the rate and extent of this reaction was comparable to that of *P. chrysosporium* H8. Recently, Salohemio et al. [101] cloned and expressed a *LIP* cDNA from *Ph. radiata* in *Trichoderma reesei*. *LIP* transcripts were seen in the transformants. However, the LIP protein was not detected either in the extracellular culture fluid or inside the mycelium. These investigators have suggested that in the absence of heme, the LIP protein is incorrectly folded and rapidly degraded by proteases. Alternatively, the structure of the novel RNA produced by the expression construct interferes with its translation. Studies are in progress in our laboratory and others to express different *LIP* genes encoding different LIP isozymes.

D. Large-Scale Production

The extracellular enzymes (i.e., LIPs and MNPs) associated with lignin degradation were first isolated from shallow stationary cultures grown in 125-ml flasks. Initial attempts to scale up production to larger flasks or to large, shallow trays resulted in low or no activity. However, Kirk et al. [116] showed LIP production using SC26, a mutant of *P. chrysosporium* which, unlike the wild-type strain, adhered to plastic disks in the fermentor. Activities equivalent to the levels seen in small stationary cultures were obtained in 3-liter fermenters. Paczcynski et al. [77] scaled the production of lignin peroxidases using a rotating plastic drum apparatus in which the mycelium was immobilized to the interior wall. Linko et al. [141] demonstrated continuous and repeated batch production of LIPs using immobilized mycelia in agar or in agarose gels. The biocatalyst was more stable in agarose gel and produced about 245 U/liter of LIP in a 70-h batch. Subsequently, LIP was produced continuously by immobilizing *P. chrysosporium* with nylon web or polyurethane in a modified Biostat E bioreactor, agitated with either air or oxygen flow alone, or in combination with a mechanical stirrer. Under these conditions, LIP production started rapidly, and activities about 600 U/liter were obtained in about 3 days. A total activity yield of 10,000 U was obtained during 1 week of continuous production with the working volume of 7.6 liters. The maximum activity was about 750 U/liter [142]. *P. chrysosporium* immobilized on polyurethane foam could be refrigerated up to 8 weeks or more and then reactivated for producing LIPs. *P. chrysosporium* strain INA-12 immobilized with polyurethane foam produced high LIP levels in 24 h in repeated batch cultures when biomass was partially regenerated on day 5. The enzyme synthesis continued at a high level for eight batches, with only 7% activity lost every 24 h. Immobilization of *P. chrysosporium* on different carriers is being carried out in a number of laboratories worldwide in an attempt to develop a system for the continuous production and to maximize the enzyme yield [143,144].

The ability to grow *P. chrysosporium* in the form of pellets in shake culture enabled the use of this organism in fungal pellet reactor systems. Previous studies more than a decade back showed that mycelial pellets are formed by the aggregation of growing hyphae and their formation is greatly influenced by the rate of agitation, media composition, strain, inoculum concentration, and other factors [145]. Fungal pellets provide a convenient means of immobilizing the biomass within a reactor. Once formed pellets can serve as biocatalysts for the production of secondary metabolites. Fungal pellets have been used in reactor systems for the production of LIPs [140,141,146]. Michel et al. [146] carefully studied the mechanism of pellet formation and growth of pellets in continuously stirred bioreactors used for LIP production. Their studies indicate that each pellet on the average is an aggregate of 2×10^2 to 6×10^3 hyphal fragments and that aggregation of the mycelium into pellets ceases in about 3 h postinoculation after which the number of pellets remains relatively constant. A 1.5-liter airlift reactor was developed which produced LIPs from mycelial pellets but the enzyme yield was low [140]. A 78-ml column reactor using preformed mycelial pellets was also used for LIP production which produced LIPs continuously [141]. Another study investigated the possibility of production of LIPs in a submerged stirred-tank reactor [147]. Various factors such as inoculum, pellet size, and certain organic compounds (i.e., polypropylene glycol or polyethylene glycol) were investigated. For example, a rich inoculum resulted in the formation of small pellets, fast depletion of glucose, and no production of enzyme whereas reducing the inoculum size prolonged the development of the culture followed by a secondary metabolic phase. However, the enzyme activities were very low. Linko and Zhong [143] compared different methods of immobilization of *P. chrysosporium* for enzyme production and found that nylon served as the best carrier for immobilization with about 340 U/liter LIPs being obtained. A continuously stirred-tank reactor (CSTR) system for the production of LIPs using preformed mycelial pellets was also described recently [146]. A maximum of 180 U/liter was obtained in this system. Attempts to produce LIP-producing mycelial pellets in the CSTR itself were not successful. A combination of selection for LIP-overproducing strains together with further improvements in optimization of culture parameters for producing LIPs in reactor systems would be very useful for large-scale production of these enzymes and their use in biotechnological applications.

V. BIOTECHNOLOGICAL APPLICATIONS

Within the last decade a number of major research advances moved the field of lignin biodegradation a quantum leap forward. These include the discovery of LIPs and MNPs, an understanding of culture conditions which facilitate the selective production of LIPs or MNPs, the development of reactor methods which hold promise for large-scale production of LIPs/MNPs, the generation of LIP overproducing mutants, and the recent major advances in molecular biology including cloning and characterization of several *lip* genes/cDNAs (see above). These advances in turn greatly accelerated the pace of research designed for the application of white rot fungi or their enzymes to various biotechnological processes. Some of these developments are reviewed here.

A. Pulp and Paper Industry

1. Biopulping

A major and perhaps the most obvious application of white rot fungi and their products is in connection with the pulp and paper industry. In the manufacture of pulp, efficient and selective removal of lignin with minimal destruction of cellulose and hemicellulose is a major objective. Currently, delignification of wood for producing paper pulp is accomplished by the kraft pulping process in which the wood chips are cooked in an alkaline environment at a high temperature or by sulfite pulping in which the wood chips are cooked in the presence of sodium sulfite. Thus, the processes in use today employ an excessive amount of energy, a high volume of chemicals, and generate vast amounts of effluents that contain toxic chlorolignols and pose a serious environmental hazard. Use of white rot fungi, especially mutants lacking the ability to degrade cellulose or their enzymes for biopulping, offer many potential advantages. (a) Fungal processes with a high degree of selectivity for lignin removal as compared to the existing chemical processes can be designed. This in turn should result in a high-quality pulp. (b) Fungal treatment processes would be relatively milder and require much lower levels of chemicals. (c) The high energy demands of the paper industry today would be greatly minimized by the use of biological pulping processes. Considering the high cost and uncertainty of energy supplies, especially in the developing world, an efficient biopulping process would be greatly advantageous from the standpoints of both economy and ensuring continuous availability of good-quality pulp. (d) Biopulping would essentially eliminate the pollution hazards associated with the chemical pulping processes. Hence, it is not surprising that the pulp and paper industries worldwide are very interested in investigating the possibilities of using microbes and their products for pulping. In practice, however, biological delignification is generally too slow to compete with chemical pulping.

The production of thermomechanical pulps involves relatively low energy input, but the subsequent secondary refining requires substantial energy for development of desired pulp properties. Such high-energy input can be reduced by employing lignin-degrading microorganisms as exemplified by white rot fungi. In fact, most of the research to date on biological pulping has focused on the combination of biological treatment and mechanical pulping, often referred to as biomechanical pulping.

The possible areas in the pulp and paper industry for using lignin-degrading fungi include primary pulp manufacture, pulp modification, waste treatment, recycled fiber treatment, and byproduct conversion [6]. *P. chrysosporium* is an attractive candidate for such applications owing to its high lignin-degrading ability and production of relatively nonspecific peroxidases implicated in lignin degradation. A totally specific attack on the lignin component, e.g., by *P. chrysosporium*, does not occur with the whole fungus as it first degrades cellulose to provide the energy for growth, metabolism, and partly to produce hydrogen peroxide, which is essential for lignin degradation [30,37,85]. If the capacity of *P. chrysosporium* to degrade lignin could be made more specific, it would be possible to save a great deal of energy and chemical cost in pulping processes. To achieve this, various approaches have been employed by different investigators. In one approach, Eriksson and Vallander [148] used *cel*$^-$-strains that are defective in the production of cellulase enzymes and in cellulose degradation but retain the capacity to degrade lignin. The strains employed were *cel*44 (a *cel*$^-$-mutant of *S. pulverulentum* which is now redesignated as *P. chrysosporium*) and *cel*26 (a *cel* mutant of *P. gigantea*). While the use of these mutants resulted in energy savings in mechanical pulping, some disadvantages such as discoloration of the pulp and a very slow rate of delignification were observed. In another approach, Eriksson et al. [149] impregnated wood chips with glucose before inoculation with *S. pulverulentum* and this was shown to promote rapid colonization of the wood by the fungus. Glucose impregnated in the wood chips apparently served as an easily metabolizable growth substrate because lignin does not serve as growth substrate and the white rot fungi essentially require a growth substrate such as glucose or cellobiose for efficient lignin degradation. Therefore, the cel$^-$-mutants preferentially utilized added glucose leaving behind the hemicelluloses during the delignification process. However, it is probably not economically feasible to impregnate wood chips with sugars. Microbial delignification of coarse thermomechanical pulp prior to the refining stage resulted in energy savings but the pretreatments resulted in poor strength properties of the final pulp [150]. Bar-Lev et al. [151] showed that the pretreatment of a coarse thermomechanical pulp with *P. chrysosporium* substantially reduced the energy requirement without diminishing the strength properties. Most recent studies (I. D. Reid and T. K. Kirk, personal communication) show that pretreatment of aspen wood chips with *Ph. brevispora* decreased the energy requirement by 47% compared to untreated chips. *P. chrysosporium*, *Ph. subserialis*, and *Ph. tremellosa* also caused 30-40% energy savings. Furthermore, these fungi also significantly increased the tensile strength of the pulp. The process economics appear promising, but further research is needed to develop to commercial scale. Improved strains of *P. chrysosporium* that have greater specificity of attack on the lignin component and that substantially reduce the time required for delignification need to be developed to make the process economically feasible. More detailed studies are required on the physiological requirements during lignin degradation, enzyme mechanisms, and particularly more about primary attack on the lignin polymer by the fungal peroxidases to be able to successfully develop and run a large-scale biotechnological process using fungal pretreatment of wood chips for energy reduction in mechanical pulping [152]. An examination of pulp qualities and the energy savings combined with the economic evaluations suggest that the microbial pretreatment of wood chips for pulp production may have a viable future. It is easier to integrate a delignification step, using inexpensively produced LIPs, into existing mechanical pulping processes. Although the technical feasibility for this approach is uncertain today, this approach has the greatest potential for success in the coming years. This should lead to a lowered energy requirement and greatly increased selectivity yielding a high-quality pulp product.

2. Biobleaching

The kraft pulping process involves the treatment of wood chips with $Na_2S/NaOH$ at about 170°C for about 2 h to degrade and solubilize lignin [153-155]. The lignin present in the wood chips is subjected to a variety of chemical reactions resulting in partial depolymerization, formation of phenolic hydroxyl groups, and dissolution of lignin in alkaline pulping liquor [156-159]. The kraft pulping process removes about 90% of the lignin. The remaining 10% or so in the pulp consists of various conjugated structures, including quinones, complexed catechols, chalcones, and stilbenes [157,160] which are mainly responsible for the brown color characteristic of kraft pulp. Chlorination of this pulp followed by alkaline extraction is considered an efficient way to degrade and remove most of the residual lignins and lignin-derived products [165]. Some chlorine becomes bound to lignin and other pulp components and the effluents resulting from such treatments contain various toxic chlorinated organic compounds which are of environmental concern. Many of the different compounds released during chemical bleaching processes are toxic, mutagenic, and/or carcinogenic [161,162]. More than 80% of organic chlorinated material in spent bleach liquors contains high molecular weight lignins. Therefore, such effluents must be pretreated before they are disposed off into the environment. Although alternative methods employing either oxygen or ozone have been developed to reduce the use of chlorine and chlorine oxides in bleaching and the associated environmental concerns, these processes do not appear to be as effective in bleaching as the chlorine-based treatments.

A number of studies focused on replacing the chlorination step with biological bleaching. This process ideally should remove as much residual lignin as possible without damaging the polysaccharides. Hiroi and Eriksson [163] and Lindquist et al. [164] showed that white rot fungi are effective in removing residual lignin and causing bleaching.

Kirk and Yang [166] showed that *T. versicolor* or *P. chrysosporium* removes residual lignin from pine kraft pulp. Experiments with shallow stationary cultures grown in pulp-amended medium containing nutrients required for fungal growth demonstrated that about 50-70% decrease of the kappa number (a measure of lignin content) occurs in 6-8 days. The removal of lignin occurred in three phases: a lag phase in which the culture was established, followed by a rapid phase of delignification and then a slower phase. Their studies also suggested that under the conditions employed, *P. chrysosporium* is more effective in the bleaching process than the other white rot fungi tested such as *T. versicolor, Gloeoporus dichorus*, and *L. edodes*. *P. chrysosporium* was also shown to degrade 50% of the high molecular weight chlorinated lignins in bleached pulp to CO_2 in 50 days [3]. Although the fungus was shown to deplete up to 50% of the pulp cellulose, this cellulose utilization could be prevented by the addition of glucose (or malt extract, cane molasses, corn syrup, starch, etc.) to repress the cellulose degradation system. Alternatively, the use of cellulase-negative mutant strains described above should be useful. Cocultivation of white rot fungi, *Peniophora gigantea, P. chrysosporium*, and *Heterobasidion annosum*, with brown rot fungi or an unidentified fungus, D37, on filter paper with a birch lignin preparation showed stimulation in ligninolysis resulting from the interaction between fungi. *Pe. gigantea* + D37 and *P. chrysosporium* + D37 combination was also accompanied by inhibition of cellulose degradation, an advantage in biopulping [167].

The biological bleaching of kraft pulp using white rot fungi in aerated agitated cultures was shown to significantly increase the pulp brightness and reduce the lignin content, but the resulting pulp contains fungal biomass (up to 10% w/w) and separation of the latter from the pulp is a major problem. Immobilization of the fungi, *P. chrysosporium* [144] and *T. versicolor* [168a,b], during biological bleaching results in the production of biologically bleached pulp-free mycelium. For example, *T. versicolor* immobilized on polyurethane foam was shown to effectively delignify both softwood and hardwood kraft pulp. This approach allowed separation of the pulp from the fungus after treatment resulting in a pulp that is free of the fungus. This allows reuse of the immobilized fungus to delignify more pulp. This process also resulted in an increase in the brightness of the pulp. Reid et al. [169] recently showed that delignification with *T. versicolor* followed by an alkali extraction was able to remove up to two-thirds of the residual lignin from softwood kraft pulp. This delignification increased the bleachability of the pulp with ClO_2, the pulp yield, and the pulp sheet strength. These investigators, in confirming earlier results, added sufficient glucose to pulp-containing cultures to repress cellulases and hemicellulases.

While these and other studies [170] suggest the potential for application of *P. chrysosporium* and other lignin-degrading fungi in biopulping and biobleaching processes, several technical problems will need to be addressed before commercial scale biopulping and biobleaching operations become a reality. These include environmental control requirements during pretreatment, sterility considerations, and the exposure times [171], but these problems do not appear insurmountable.

3. Treatment of Waste Effluents

The pulp and paper industry generates annually over 700 billion gallons of toxic and intensely colored waste effluents which primarily contain high molecular weight, modified, and chlorinated lignins [172,173]. The effluents are treated by biological oxidation in aerated lagoons and/or activated sludge systems before being discharged into streams. Such treatments reduce the BOD (biological oxygen demand) and COD (chemical oxygen demand) of the effluents but the color of the effluent persists. Therefore effective treatment of the colored pulp mill bleach plant effluents is both difficult and expensive. Furthermore, conventional water treatment processes are relatively ineffective at removing the toxic high molecular weight chlorolignols [174]. Several investigators demonstrated that kraft mill effluents can be decolorized by cultures of the white rot fungi, in particular *P. chrysosporium* [172-177]. Decolorization of pine kraft black liquor by *Po. versicolor* has also been demonstrated [178]. Although part of the lignin was metabolized by this fungus, physical absorption has been reported to be the main mechanism of color removal, rendering the process less attractive from a technological point of view. Fukuzumi et al. [175] demonstrated removal of 99% of the color of a kraft bleach plant effluent in 4 days by *Tintoporia borbonica*. Up to 60% decolorization of E1 state liquor (first alkali extraction after chlorination) within 4 days was achieved by *P. chrysosporium* [177]. Later studies using a thin-film reactor showed about 80% color removal [143]. Unlike *Po. versicolor, P. chrysosporium* decolorized the E1 effluent by degradation of chlorinated phenols involving methylation, oxidation, and reduction [172]. *P. chrysosporium* has also been employed in a continuous process for decolorizing bleach plant effluents which has been referred to as the MyCoR (mycelial color removal) process. This process involves the immobilization of the fungus in a rotating biological contactor (RBC) reactor and is reported to be the first wastewater treatment system to be developed based on a pure fungal culture. Supplemental nutrients such as glucose or cellulose may be added to the reactor to provide the energy and carbon required by the fungus for effective decolorization. This process not only reduced the color of the effluent significantly but also reduced the COD making the process ideal for effluent treatment technology for the pulp and paper industry. Both high and low molecular weight chlorinated organic compounds were removed by this process. A preliminary economic evaluation suggested that the rate of decolorization and the life span of the fungus are the most critical factors. Yin et al. [179] suggested that treatment of the E1 effluent by ultrafiltration before RBC treatment would be economically attractive. Their study also suggested that combining ultrafiltration and the MyCoR system might maximize the efficiency of the MyCoR process and reduce the treatment cost, thereby further making the process more economically feasible for industrial use. When supplemented an exogenous carbohydrate source, daily decolorization rates of 50-80% of the bleach plant effluents were achieved in the MyCoR and MyCoPOR (a process in which the fungus is immobilized on a porous carrier) processes [180,181].

Several environmental parameters have been reported to influence color removal. Campbell et al. [176] evaluated two bench scale MyCoR processes using a RBC reactor. The fastest color removal of 200 color units/liter/day (CU/LD) occurred using an oxygen-enriched atmosphere with the effluent at an elevated temperature (40°C) and pH 4.5. A color removal rate of 600 CU/LD was obtained using an air atmosphere. The problems encountered with the operation of MyCoR RBC processes were short fungal lifetime (5-7 days), high oxygen concentration required, and biomass wasting. Recently, Prouty [182] examined the use of an aerated reactor for color removal from bleach plant effluents. The aerated reactor reportedly has two advantages over the MyCoR RBC process. (a) The reactor uses air instead of pure oxygen (and therefore may be less costly and less hazardous to operate); and (b) the fungal lifetime in the aerated reactor is longer than in the MyCoR RBC process. Although the aerated reactor color removal rate is lower than that of the MyCoR RBC process utilizing pure oxygen, the removal rate is significantly higher than that seen with the MyCoR RBC process using air atmosphere. The results obtained in this bench scale process provide a reasonable basis for evaluation of an industrial fungal color reduction process.

The extracellular LIPs and MNPs of *P. chrysosporium* and other white rot fungi have been implicated in the decolorization of kraft bleach effluents but their roles have not been critically examined. A basic understanding of the enzymology will be invaluable to the development of large-scale effluent biotreatment processes and achieve reductions in effluent treatment costs. In our laboratory, we investigated the role of MNPs and LIPs of *P. chrysosporium* in the decolorization of kraft bleach plant effluents with the aid of biochemical mutants and by varying culture conditions to manipulate the levels of LIPs and MNPs (F. C. Michel, Jr., S. B. Dass, E. A. Grulke, and C. A. Reddy, Abstr. Annu. Mtg. Am. Soc. Microbiol. #060, 1990). Rate of effluent decolorization in relation to the LIP and MNP production levels was monitored. In agreement with earlier reports [20,43], MNP activity was first detected in the culture fluid in 2 or 3 days and increased to a maximum value between the fourth and fifth day and declined thereafter (Fig. 9). LIP activity, on the other hand, first appeared between 4 and 5 days, and reached a maximum around day 6 to 7. Rapid decolorization of the effluent was observed when it was added to nitrogen-limited cultures on days 2-7 (Fig. 10). Little or no decolorization was observed in high-nitrogen culture in which the production of LIPs and MNPs was blocked indicating that these enzymes play a key role in decolorization of the effluents. An observation of particular interest is the fact that decolorization occurred on days 2, 3, and 4 before the appearance of LIPs. The rate of decolorization also reached a maximum level on day 4 when MNP activity reached its highest level and did not increase significantly when lignin peroxidase activity reached its maximum on days 6 and 7. High rates of decolorization were observed with LIP-negative (*lip*) mutant of *P. chrysosporium* which does not produce LIPs but does produce a normal complement of MNPs. Furthermore, high levels (about 90%) of decolorization were also observed when LIP production was completely suppressed by adding high levels of manganese (100 ppm) to the cultures (Fig. 11). Finally, adding only minimal levels of manganese to cultures which greatly suppress MNP activity resulted in a severe decrease in decolorization. These results indicated for the first time that MNPs play a predominant role in decolorizing bleach plant effluents and LIPs appear to play only a minor role in this process.

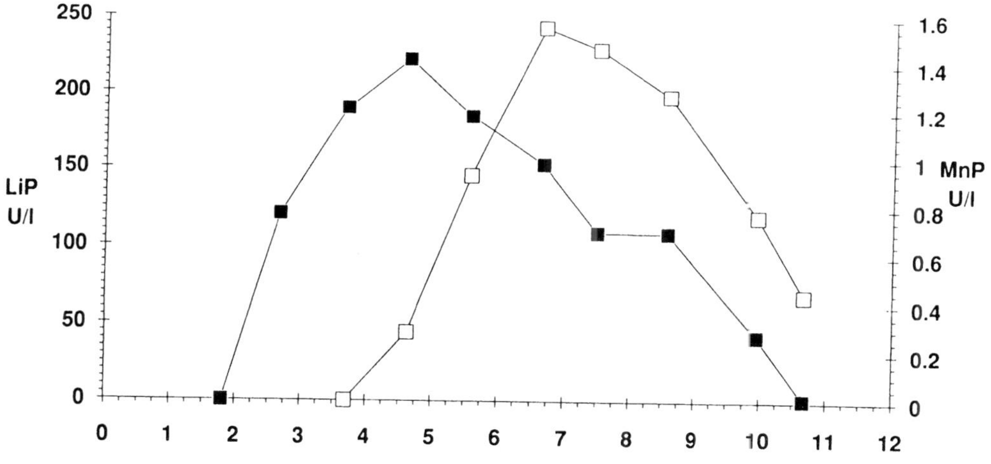

FIGURE 9. Kinetics of lignin peroxidase and manganese-dependent peroxidase production by *P. chrysosporium*. Acetate-buffered media in rubber stoppered 250-ml Erlenmeyer flasks (85 ml culture volume) were inoculated with a 10% (v/v) mycelial inoculum. The cultures were agitated at 173 rpm with daily oxygenation. Lignin peroxidase activity was measured as described [45] and manganese-dependent peroxidase activity was measured using the procedure of Kuwahara et al. [74]. A unit of manganese-dependent peroxidase activity is defined as 1 mmol of phenol red oxidized per minute, using an extinction coefficient of $4660M^{-1}$. ■, MNP; □, LIP.

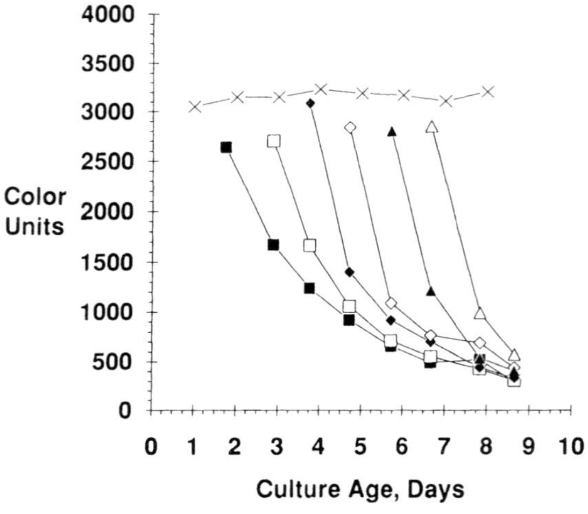

FIGURE 10. Decolorization of kraft bleach plant effluent (BPE) by nitrogen-limited (2.4 mM), agitated cultures of *P. chrysosporium* and an uninoculated control. BPE was added on different days during incubation. Values represent duplicate cultures. ■, day 2; □, day 3; ◆, day 4; ◇, day 5; ▲, day 6; △, day 7; -x-, no inoculum.

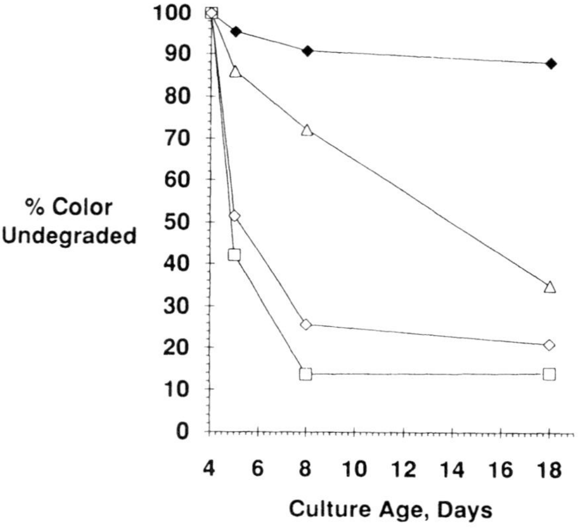

FIGURE 11. The effect of Mn(II) and nitrogen levels on kraft bleach plant effluent decolorization in agitated cultures of *P. chrysosporium*. Color remaining represents the total undegraded color in a single culture flask. The effluent was added 4 days after the culture inoculation. ◆, N-sufficient; △, N-limited, 0 ppm Mn(II); □, N-limited, 12 ppm Mn(II); ◇, N-limited, 100 ppm Mn(II).

B. Lignocellulose as a Renewable Raw Material

The rapidly increasing costs and uncertain petroleum supplies have brought a sense of urgency to the need for developing new technologies for the use of alternative raw materials as petroleum substitutes. Plant biomass has become the likely choice since, unlike petroleum, it is renewable and enormous quantities are available annually on a global scale. Upon eventual depletion of the finite fossil fuel reserves in the world, renewable resources such as plant biomass will be invaluable in meeting our needs for energy and particularly for carbon-based chemicals.

Wood is composed of 70-80% polymeric carbohydrates and 20-30% lignin. Wood and other lignocellulosics can be transformed by various microbes to produce many of the industrially important organic chemicals produced by the chemical industry. Many of these chemicals can be recovered as byproducts of chemical pulping processes. As early as 1952, Wise and Jahn [183] recognized the great potential of lignins, specifically those from the pulping industry, as one of the chief sources of organic materials. Large quantities of lignin are potentially available for chemical conversion in spent pulping liquors.

Cellulose in lignocellulosic materials constitutes the most abundant organic polymer in the biosphere and therefore an inexhaustible renewable source for sugars and fermentation feedstocks to produce feeds, fuels, and organic chemicals [184]. However, cellulose occurs in close association with lignin in plant cell walls and only a portion of this cellulose is accessible to microbial enzymes. Hence, the lignin-cellulose association must be disrupted by a pretreatment process before cellulases can optimally gain access to the cellulose in these materials. Steam treatment increases the susceptibility of lignocellulosic material to enzymes [185] but this process generates inhibitors of hydrolysis and fermentation [186]. The inhibitors can be removed by water extraction but this step might also remove hemicellulose (pentose) sugars which are solubilized during steaming [187]. An interesting pretreatment alternative is biological delignification with fungi or microbial enzymes which leave behind the carbohydrate. Such a process has the potential advantages of low capital cost, low energy input, greater specificity, and high yield without generating polluting byproducts.

LIPs catalyze the cleavage of the propane side chain between the two chiral $C\alpha$-$C\beta$ carbons in dimeric model compounds of the β-O-4 aryl ether and β-1 arylpropane types, the reactions observed to occur in ligninolytic cultures of *P. chrysosporium* [188-191]. This together with the fact that β-ether linkages (about 48%) and 1,2-diarylpropane structures (about 7%) account for over one-half the total intermonomeric bonds in lignin [1] and cleavages of these two linkages itself will lead to the extensive depolymerization of lignin makes the *P. chrysosporium* and its LIPs suitable sources for biotechnological processes. Attempts to use ligninolytic enzymes for the depolymerization of native lignin have not been successful so far. Until now, only processes based on whole organism have been shown to be effective. Ritter et al. [193] developed a MyCoPOR process for the polymerization and depolymerization of lignocellulose using *P. chrysosporium* immobilized on foam and found significant transformation. Other fungi such as *Po. dichomus* [194], *Schizophyllum commune*, and *C. versicolor* [195] have also been used. Transformation of lignocellulosic substances using three enzymes—cellulase, glucose oxidase, and fungal peroxidases—from *T. versicolor* and *Inonotus radiatus* immobilized with silanized porous glass beads has also been demonstrated [192]. Twelve phenolic compounds including the three monomeric compounds caffeic, sinapic, and coumaric acids were identified as products of this biotransformation process. While some of these earlier results were mildly encouraging, a great deal of research needs to be done before an objective evaluation of the feasibility of lignin/lignocellulose as a resource for feedstock sugars, chemicals, and fuels can be made. A reasonable prediction is that it will be many years before a commercial scale application becomes a reality.

The extensive coal deposits in the United States constitute an important source of energy and chemicals. In an attempt to utilize this resource, there is a growing interest in the potential use of microorganisms in the solubilization of coal and related substances to chemicals and fuels. Diverse microorganisms including *Streptomyces*, *Pseudomonas* [192a,b], and lignin-degrading fungi such as *Po. versicolor* and *Poria monticoloa* [192c] have been shown to cause coal solubilization. It was suggested that wood-degrading systems of the fungi were also responsible for coal solubilization. Subsequent studies [192d,e] suggested that extracellular enzymes play a role in coal solubilization by *Po. versicolor*. However, Fredrickson et al. [192f] presented data to indicate that coal solubilization is mediated by siderophore-like substances and not by extracellular enzymes based on the fact that gel permeation chromatography fractions with coal-solubilizing activity retained activity even after boiling at 100°C for 40 min. Whether or not coal solubilization by *Po. versicolor* cultures is enzymatic or non-enzymatic, the ability of this fungus to solubilize coal is of particular interest in biotechnological applications. These studies further open up avenues for screening fungi with high lignin-degrading potential, such as *P. chrysosporium* for coal solubilization. While these preliminary studies are encouraging, much needs to be done before such fungi are employed in biotechnological processes for converting coal to organic chemicals.

C. Conversion of Lignocellulosic Materials into Feedstock

In view of the vast amounts of lignocellulosic byproducts produced by the agricultural and forest industries, considerable efforts have been made in recent years to develop systems for upgrading their nutritive value. It has been estimated that several billion tons of recyclable cellulose is present in the agroindustrial and forestry residues generated each year on a global scale. Examples of such residues include hardwoods, softwoods, grasses, wheat and rice straw, sugarcane bagasse, cornstalks, animal manures, waste fibers from chemical pulping processes, and urban refuse. However, the nutritional value of lignocellulosic substrates is limited by the fact that low polysaccharide degradation is achieved in the gastrointestinal tract owing to the presence of lignin which limits the accessibility of the cellulose and hemicellulose to microbial hydrolytic enzymes. The digestibility of various woods to enzymatic hydrolysis by a mixture of microbial polysaccharidases and to digestion by rumen fluid was inversely proportional to the lignin content [196]. Microbial pretreatment of lignocellulosic materials for delignification increases the nutritional quality by increasing the digestibility of the lignocellulosic materials in the rumen. Digestion of kraft pulps prepared from hard- and softwoods increased rapidly with delignification, although this relationship was seen with softwood pulps only after an initial phase during which extensive delignification occurred with only minor changes in digestibility [197].

Three major criteria, namely, increase in the digestibility of lignocellulose for ruminants, increase in protein content, and improvement in the palatability of dry roughages, are being considered to increase the nutritive value of lignocellulosic materials [198]. Various approaches, including physical and chemical methods, have been used for delignification [199,200]. But with the identification of microorganisms, especially white rot fungi belonging to the basidiomycetes group, which have the ability to degrade lignin, use of biological methods in delignification was considered [201-206]. The ability of white rot fungi to convert straw and other plant wastes into feed and the various parameters affecting the progress of solid state fermentation have been investigated [204,207-209].

Fungal delignification of wood has been observed under natural conditions. The final product of delignification, referred to as "palo podrido," is a white decomposed wood which is used as animal feed [210]. The wood degraded to palo podrido is usually caused by fungi such as *Ganoderma applanatum* or *Armillaria* spp. Zadrazil and Brunnert [209] found that lower temperatures increase the digestibility of fermented product derived from wheat straw. Relatively higher lignin degradation rates are obtainable using straw as substrate as compared to wood [211]. Nevertheless, biological delignification of wood preparations may also offer possibilities for the production of ruminant feedstuff [199,200,212,213]. Solid-state fermentation of Aspen wood with *Ph. tremellosus* removed 52% of the lignin but only 12% of total wood weight and increased cellulose digestibility from 18 to 53% after 8 weeks of treatment. Zadrazil [214] screened over 200 fungi for substrate degradation, lignin decomposition, and increase in digestibility of wheat straw at different temperatures and over different periods of fermentation. The results showed that white rot fungi such as *Abortioporus biennis, Ga. lucidum, Hymenochaete tabacina, L. edodes, Pl. eryngii, Pl. sajor caju, Pl. ostreatus, Pl. serotinus, Po. arculais, Po. galactinus, Poria expansa, S. pulverulentum,* and *T. hirsuta* have a high capacity to increase in the vitro digestibility of wheat straw.

Successful biological delignification to increase the feed value of lignocellulosic materials demands a fungus with a high selectivity for lignin degradation over carbohydrate consumption. Several species of white rot fungi have been screened on a variety of substrates [199]. Some but not all fungi delignified or increased the digestibility of their substrates. A comparative study on the delignification by several species of *Ganoderma* indicated that *Ga. colossum* and *Ga. oregonense* caused extensive delignification. *Ga. colossum*, being an efficient lignin-degrading thermophilic fungus, will be of use for delignification purposes [215]. Most of the above studies are preliminary because the fermentation conditions have been optimized or even explored for only a few fungal species or substrates [199,200]. Much remains to be done to identify and optimize fungal pretreatment processes that selectively remove most or all of the lignin in lignocellulosic materials with only a minimal loss (utilization) of the polysaccharide component, so that maximum nutritive value can be realized from the substrate that is being pretreated.

An important and economically attractive lignocellulose bioconversion process for the utilization of lignocellulosic residues is the cultivation of edible fungi, a process that has been recognized for centuries. Fungi such as *L. edodes* (shiitake mushroom), *Tremella fusiformis*, and *Pleurotus* spp. have been successfully cultivated on timber and tree stumps. *Volveriella volvaceae* (paddy straw mushroom) has been successfully grown on rice straw in China for centuries via preparation of straw beds 60-75 cm high, wetting, and spawning. Productivity of this mushroom has been improved by cultivating it on partially composted rice straw and cotton waste supplemented with additional nitrogen sources such as chicken manure. Large-scale commercial cultivation of *Pleurotus* spp. on shredded straw supplemented with corn cob, soya bean flour, or other supplement is also widely practiced. A composting process is usually used for cultivating *Agaricus bisporus*, the most popular edible mushroom species accounting for 75% of the world production.

TABLE 4. List of Toxic Xenobiotics and Other Chemicals Known To Be Degraded by *P. chrysosporium*

Compound	Involvement of LDS[a]	Ref.
DDT 1,1,1-trichloro-2,2-bis-4-chlorophenylethane	?	220
2,4,5-Trichloroacetic acid	Yes	221
Fluorene	Yes	222
Anthracene, phenanthrene	?	233
Thianthrene	Yes	227
Pentachlorophenol	Yes	223
Crystal violet	Yes	224
Veratrylglycerol β-2,4-dichlorophenyl ether	?	225
Benzo(a)pyrene	Yes	217,218
Chloroaniline	?	226
Pyrene dibenzo(*p*)dioxin, 2-chlorodibenzo(*p*)dioxin, 2,3,8,8-tetrachlorodibenzo-*p*-dioxin	Yes	216,217
3,4,3',4'-Tetrachlorobiphenyl, 2,4,5,2'4',5'-hexachlorobiphenyl, 2,3,7,8-tetrachlorodibenzo-*p*-dioxin	Yes	216
2,4,6-Trinitrotoluene (TNT), 2,4-dinitrotoluene	?	220
Lindane, chlordane, aldrin, dielrin, heptachlor, mirex	?	216,228

[a]Although the lignin-degrading system (LDS) is believed to be involved in the degradation of these xenobiotics, unequivocal data in support of this are not available in many instances.

D. Detoxification of Environmental Pollutants

P. chrysosporium and its lignin-degrading system has generated a great flurry of interest in academia, industry, and government because of the recent discovery that *P. chrysosporium* degrades and detoxifies a broad spectrum of toxic xenobiotic substances [216-219]. Biological degradation of aromatic hydrocarbons, e.g., dioxins and benzo(a)pyrenes, in the environment is of increasing interest since some of these compounds have been shown to be highly carcinogenic or otherwise harmful to humans and animals. A list of toxic organic compounds degraded by *P. chrysosporium* is presented in Table 4.

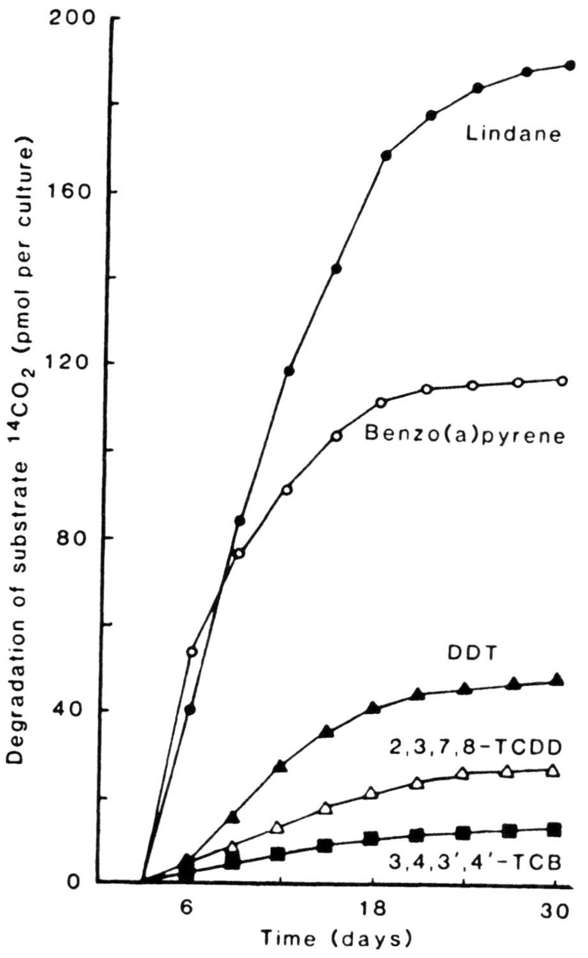

FIGURE 12. Rate of oxidation of selected environmentally persistent organopollutants to CO_2 by *P. chrysosporium* [216].

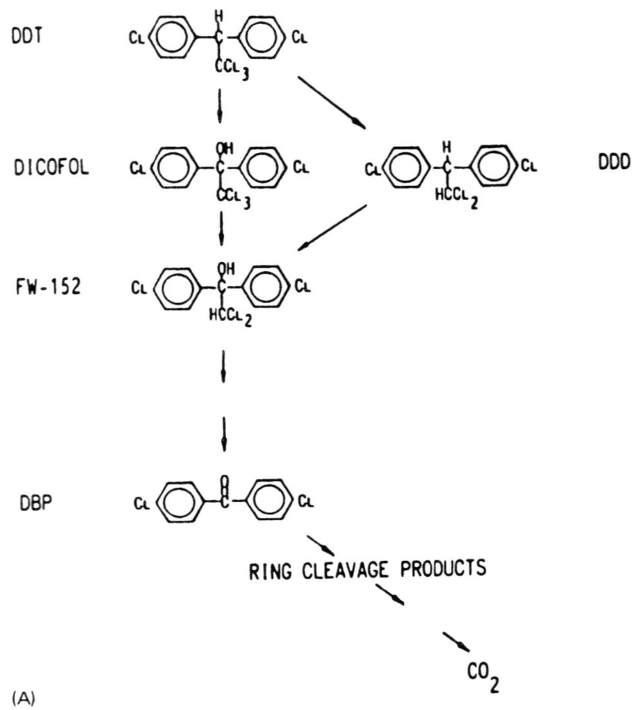

(A)

FIGURE 13. (A) Proposed pathway for the degradation of DDT [229]. (B) Proposed scheme for aerobic and anaerobic oxidation of benzo(a)pyrene [B(a)P] by lignin peroxidase of *P. chrysosporium*. Pathway to B(a)P-1,6-quinone is shown. Other quinones would be formed analogously [217].

One of the primary motivations for the investigators' looking for such metabolic activities on the part of *P. chrysosporium* was the fact that carbon skeletons similar to those of many of the pollutants are found within the lignin polymer and that the lignin-degrading enzymes (e.g., LIPs and MNPs) are known to be relatively nonspecific [1]. Bumpus et al. [216] reported that radiolabeled (^{14}C at the ring carbon) organopollutants such as DDT, 2,4,5,2',4'5'-hexachlorobiphenyl, 3,4,3',4'-tetrachlorobiphenyl, 2,3,7,8-tetrachlorodibenzo-*p*-dioxin, lindane, and benzo(a)pyrene were oxidized to $^{14}CO_2$ by nitrogen-limited, ligninolytic cultures of *P. chrysosporium* (Fig. 12). They suggested that the lignin-degrading enzyme system may be responsible for such activities. This suggestion was based on the fact there was a direct positive correlation between the appearance of lignin-degrading system and the onset of mineralization of xenobiotic pollutants.

Recent studies of Bumpus and Aust [229] indicated the formation of dicofol, DDD, FW-152, and DBP as the intermediates from DDT degradation (see Fig. 13A). Based on these results, a dissimilatory pathway for DDT degradation by *P. chrysosporium* was proposed [229]. Although the involvement of the ligninolytic system in the degradation of DDT has been suggested, a recent study indicated that the LIPs at least have no role in DDT degradation based on the fact DDT was degraded even under nitrogen-rich culture

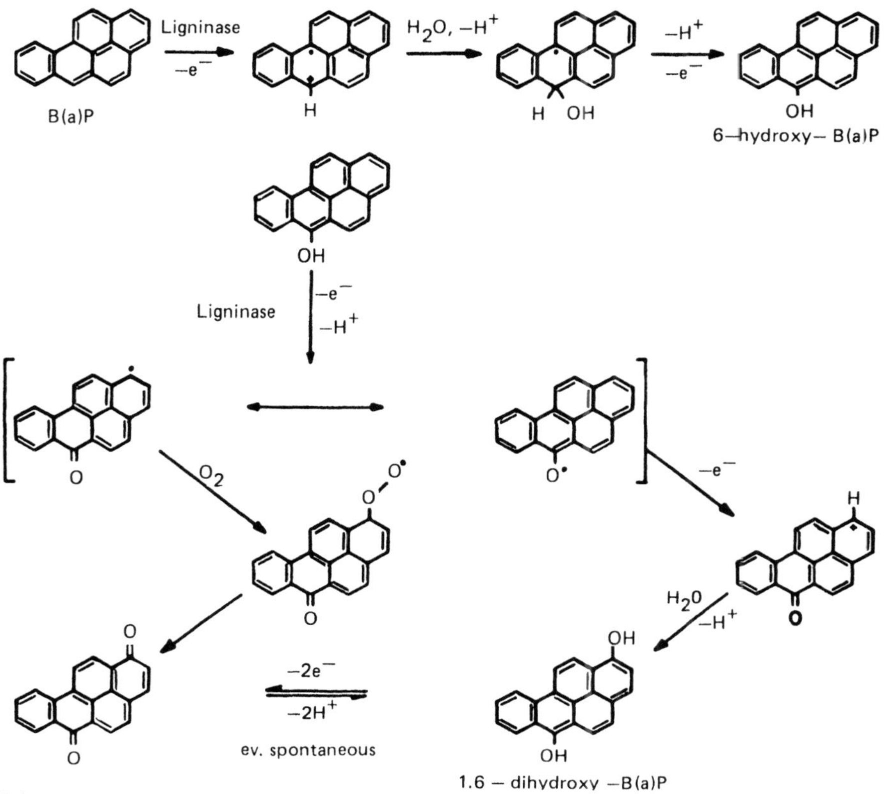

(B)

conditions which are known to repress LIP production [230]. Furthermore, these investigators noted lack of temporal correlation between the onset of DDT degradation and LIP production even under ligninolytic conditions, in that liberation of $^{14}CO_2$ from ^{14}C-labeled DDT was seen even before the LIP activity was demonstrable in the extracellular fluid. However, the possibility that DDT would be taken into the fungal cell and metabolized before the secretion of the LIP enzymes into the extracellular milieu occurs cannot be ruled out. Furthermore, the MNPs are known to be produced much earlier than LIPs [20,43] and it is possible that MNPs may account for the degradation of DDT. Therefore, further studies are needed to better understand the enzyme system involved in the degradation of DDT by *P. chrysosporium*. Whatever enzyme system is involved, the fungus is capable of mineralizing DDT, raising the possibility of its use in decontamination of the sites contaminated with DDT. Eaton [219] reported that *P. chrysosporium* was able to degrade polychlorinated biphenyls in cultures in the secondary metabolic phase when the extracellular peroxidases and other components of the lignin-degrading system are produced.

Polycyclic aromatic hydrocarbons are a group of hazardous substances accumulating in aquatic and terrestrial environments. Because of the toxic nature of these substances many studies have been initiated to understand their biological degradation by various microbes. Benzo(a)pyrene, a typical and perhaps the most investigated representative of polycyclic aromatic hydrocarbons, has been reported to be degraded by many biological systems. When grown in nigrogen-limited medium with agitation, *P. chrysosporium* was shown to effectively degrade ^{14}C-labeled benzo(a)pyrene to $^{14}CO_2$ during secondary metabolism, suggesting the involvement of the ligninolytic enzyme system [218]. Detailed analyses indicated that LIPs play a critical role in the initial oxidation of benzo(a)pyrene. Crude and purified preparations of LIPs oxidized benzo(a)pyrene to benzo(a)quinones. In keeping with the earlier proposal that veratryl alcohol acts as a mediator in oxidation reactions catalyzed by LIPs, Haemmerli et al. [217] suggest that benzo(a)pyrene oxidation is mediated by veratryl alcohol. The oxidation of benzo(a)pyrene was also shown to be enhanced in the presence of veratryl alcohol. These investigators suggested that veratryl alcohol cation radicals function as one-electron oxidants and these radicals react with benzo(a)pyrene resulting in the benzo(a)pyrene cation radical and veratryl alcohol (Fig. 13B). This suggestion was further supported by the fact that LIPs are known to attack different types of aromatic compounds generating cation radicals [39,40,44,45,74,231,232]. A similar cation-mediated oxidation and dibenzo(p)dioxins, 2-chlorodibenzo(p)dioxins, and pyrenes by LIPs has been proposed [217]. Polycyclic aromatic hydrocarbons present in anthracene oil, a distillation product obtained from coal tar, were shown to be degraded by *P. chrysosporium* [233]. [^{14}C]Phenanthrene was shown to be degraded to $^{14}CO_2$ to a substantial extent. About 70-100% disappearance was seen during 27 days of incubation. The above studies indicate that LIP-catalyzed one-electron oxidation applies not only to lignin and lignin-related substructures but to certain polycyclic aromatic and heteroaromatic pollutants as well, making the lignin-degrading white rot fungi in general and *P. chrysosporium* in particular attractive candidates for waste treatment systems.

Successful use of *P. chrysosporium* in the bioremediation of soils contaminated with xenobiotic substances depends on its ability to degrade them in the soil environment. There is a considerable interest in developing in situ bioremediation systems to clean up waste sites using *P. chrysosporium*. A preliminary study by Lamar et al. [234] indicated that *P. chrysosporium* transformed about 98% of the pentachlorophenol from the soil microcosm. The highest rate of transformation was noticed during the first week. The nature of the products (i.e., whether extractable or soil-bound) was greatly influenced by the soil type. Aust [235] reported promising results with biodegradation of agrochemicals such as DDT, pentachlorophenol, lindane, 2,4,5-trichloroacetic acid, chloroanilines, chlordane, and heptachlor in soil and aqueous systems. Lewandowski et al. [128] investigated the possibility of constructing a reactor system for use in 2-chlorophenol degradation by *P. chrysosporium* employing simplified media in order to reduce the cost for commercial applications. While studies on the practical feasibility are still in preliminary stages, the metabolic capability of *P. chrysosporium* appears to hold promise in processes designed for detoxification of various environmental pollutants.

Biological applications of lignin-degrading fungi have taken a giant step forward with the discovery of LIPs. One of the more successful applications of these enzymes may be in the detoxification of selected xenobiotic compounds. This goal is within the reach as rapid advances are being made in large-scale cultivation of LIP-overproducing mutants of *P. chrysosporium*. The fact that a number of *LIP* genes have been cloned and sequenced not only from *P. chrysosporium* but also from other fungi such as *Ph. radiata* and *T. versicolor* makes overproduction of recombinant LIPs in suitable expression systems and the use of the recombinant enzymes for biodegradation of xenobiotics a reasonable future goal.

VI. CONCLUDING REMARKS

The discovery of LIPs in 1982 greatly accelerated the pace of lignin biodegradation research. Major advances have been made in our understanding of the biochemistry of LIPs as well as the second closely related family of isozymes, MNPs, although some details of the catalytic mechanism still need to be researched. Enzymes other than LIPs and MNPs (e.g., laccase) also seem to play a role in lignin degradation in some organisms such as *T. versicolor*. Recent studies suggest that other white rot fungi have generally similar enzyme systems although there may be differences in details. Future research will have to unravel the specific roles of individual LIPs and MNPs in lignin degradation. Also, more research is needed to determine the contribution of different hydrogen peroxide-producing enzymes to the overall lignin degradation process.

Great progress has been made in cloning and sequencing cDNAs and genes which encode various LIPs and MNPs in *P. chrysosporium* as well as in other white rot fungi. Both integrative and nonintegrative transformation systems have been developed that should facilitate rapid advances in the molecular genetic analysis of the lignin-degrading enzyme system of *P. chrysosporium* and perhaps other lignin-degrading fungi. Progress is also being made in understanding the regulation of the expression of genes encoding various extracellular peroxidases, but much remains to be done. Furthermore, a major emphasis needs to be placed in the next few years on accomplishing overexpression of active LIPs, which is essential for assessing the potential industrial applications of lignin-degrading enzymes. This is not a trivial problem and little success has been achieved to date, although a number of laboratories are working on this aspect, and one can hope that enough has been learned about the system to ensure success of these efforts. This in turn will accelerate further research on potential applications of LIPs and perhaps other enzymes of the lignin-degrading systems of fungi.

Biotechnological applications of white rot fungi and their enzymes is an intensive area of research. The use of whole *P. chrysosporium* and to some extent LIPs appears to hold promise in processes designed for the detoxification of various environmental pollutants as well. Some success has been accomplished in the use of whole *P. chrysosporium* and *T. versicolor* in decolorizing bleach plant effluents. Successful use of LIPs in biopulping processes will have a great favorable impact on the world economy, sparing scarce energy sources, and alleviating the current environmental problems with waste effluents generated by pulp and paper industry. Use of efficient biomimetic catalysts is another potential future application in the pulp and paper industry. The upgrading of wood and agricultural wastes as a feed resource is a very important area of research which, if successful, should yield rich dividends for the efficient production of milk and meat while simultaneously sparing soybeans and grains for use by needy humans, especially in developing countries. Use of wood and lignocellulosic agricultural wastes for the production of edible mushrooms is already a major industry in some countries and is likely to spread to other countries as well.

ACKNOWLEDGMENTS

The research from our laboratory reported here was supported in part by grant DE-FG02-85 ER 13369 from the U.S. Department of Energy, Public Health Service grant R01 GM 39032-01 from the National Institutes of Health, Research Excellence Funds from the State of Michigan, National Science Foundation Center for Microbial Ecology, and the Michigan State University Agriculture Experiment Station. We also acknowledge a number of colleagues who allowed us to use the figures from their publications.

REFERENCES

1. Crawford, R. L. *Lignin Biodegradation and Transformation*. Wiley-Interscience, New York, 1981.
2. Reddy, C. A. Physiology and biochemistry of lignin degradation. In *Current Perspectives in Microbial Ecology* (M. J. Klug and C. A. Reddy, eds.), American Society for Microbiology, Washington, D.C., 1984, pp. 558-571.
3. Eriksson, K. E., and Kirk, T. K. Biopulping, biobleaching and treatment of kraft bleaching effluents with white-rot fungi. In *Comprehensive Biotechnology*, Vol. 3 (C. L. Cooney and A. E. Humphrey, eds.), Pergamon Press, 1985, pp. 271-294.
4. Higuchi, T., Chang, H.-M., and Kirk, T. K. (eds.). *Recent Advances in Lignin Biodegradation Research*. Tokyo Univ., 1983.
5. Kirk, T. K., Higuchi, T., Chang, H.-M. (eds.). *Lignin Biodegradation: Microbiology, Chemistry, and Potential Applications*, Vols. 1 and 2. CRC Press, Boca Raton, 1978.
6. Kirk, T. K. Advances in biotechnology in pulp and paper manufacture: Overview of the 1989 international conference. *Tappi*: 33-43, 1989.
7. Tien, M. Properties of ligninase from Phanerochaete chryosporium and their applications. *CRC Crit. Rev. Microbiol.*, *15*:141-168, 1987.
8. Buswell, J. A., and Odier, E. Lignin biodegradation. *CRC Crit. Rev. Biotechnol.*, *6*:1-60, 1987.
9. Buswell, J. A. Fungal degradation of lignin. In *Handbook of Applied Mycology, Vol. 1: Soil and Plants* (D. K. Arora, Bharat Rai, K. G. Mukerji, and G. Knudsen (eds.), Marcel Dekker, New York, 1991.
10. Kirk, T. K., and Farrell, R. L. Enzymatic "combustion": The microbial degradation of lignin. *Annu. Rev. Microbiol.*, *41*:465-505, 1987.
11. (a) Higuchi, T. Biodegradation of lignin and its potential applications. In *Bioprocess Engineering* (T. K. Ghose, ed.), Ellis Harwood Limited, London, 1985, pp. 39-58. (b) Higuchi, T. Lignin biochemistry: Biosynthesis and biodegradation. *Wood Sci. Technol.*, *24*:23-63, 1990.
12. Umezawa, T. Mechanisms for chemical reactions involved in lignin biodegradation by Phanerochaete chrysosporium. *Wood Res.*, *75*:21-79, 1988.
13. Lewis, N. G., and Yamamoto, E. Lignin: Occurrence, biogenesis and biodegradation. *Annu. Rev. Plant Physiol. Plant Mol. Biol.*, *41*:455-496, 1990.
14. Sarkanen, K. V., and Ludwig, C. H. (eds.). *Lignins: Occurrence, Formation, Structure, and Reactions*. Interscience, New York, 1971.
15. Kirk, T. K., and Brunow, G. Synthetic ^{14}C-labeled lignins. *Meth. Enzymol.*, *161*:65-73, 1988.

16. Crawford, R. L., and Crawford, D. L. (^{14}C)Lignin-labeled lignocelluloses and ^{14}C-labeled milled wood lignins: preparation, characterization, and uses. *Meth. Enzymol.*, *161*:18-30, 1988.
17. Gold, M. H., Glenn, J. K., and Alic, M. Use of polymeric dyes in lignin biodegradation assays. *Meth. Enzymol.*, *161*:74-78, 1988.
18. Kelley, R. L., Ramasamy, K., and Reddy, C. A. Characterization of glucose oxidase-negative mutants of a lignin-degrading basidiomycete Phanerochaete chrysosporium. *Arch. Microbiol.*, *144*:254-257, 1986.
19. Boominathan, K., Dass, S. B., Randall, T. a., and Reddy, C. A. Nitrogen-deregulated mutants of Phanerochaete chrysosporium, a lignin-degrading basidiomycete. *Arch. Microbiol.*, *153*:521-527, 1990.
20. Boominathan, K., Dass, S. B., Randall, T. A., Kelley, R. L., and Reddy, C. A. Lignin peroxidase-negative mutant of the white-rot basidiomycete Phanerochaete chrysosporium. *J. Bacteriol.*, *172*:260-265, 1990.
21. Kuwahara, M., and Asada, Y. Production of ligninases, peroxidases and alcohol oxidase by Phanerochaete chrysosporium and its mutant. In *Lignin Enzymic and Microbial Degradation* (E. Odier, ed.), INRA, Paris, 1987, pp. 171-177.
22. Cowling, E. B. Comparative biochemistry of decay of sweetgum sapwood by white-rot and brown-rot fungi. U.S. Dept. of Agriculture Technical Bulletin, Washington, D.C., 1961, p. 1258.
23. Pelczar, Jr., M. J., Gottleib, S., and Day, W. C. Growth of Polyporus versicolor in a medium with lignin as the sole carbon source. *Arch. Biochem.*, *25*:449-451, 1950.
24. Kirk, T. K., Connors, W. J., and Zeikus, J. G. Requirement of a growth substrate during lignin degradation by two wood rotting fungi. *Appl. Environ. Microbiol.*, *32*:192-196, 1976.
25. Leisola, M., Ulmer, D., Haltmeir, T., and Fiechter, A. Rapid solubilization and depolymerization of purified kraft lignin by thin layers of Phanerochaete chrysosporium. *Appl. Microbiol. Biotechnol.*, *17*:117-120, 1983.
26. Ulmer, D. C., Leisola, M. S. A., Schmidt, B. H., and Feichter, A. Rapid degradation of isolated lignins by Phanerochaete chrysosporium. *Appl. Environ. Microbiol.*, *45*:1795-1801, 1983.
27. Yang, H. H., Effland, M., and Kirk, T. K. Factors influencing fungal decomposition of lignin in a representative lignocellulosic, thermomechanical pulp. *Biotechnol. Bioeng.*, *22*:65-77, 1980.
28. Ander, P., and Eriksson, K.-E. The importance of phenol oxidase activity in lignin degradation by the white-rot fungus Sporotrichum pulverulentum. *Arch. Microbiol.*, *109*:1-8, 1976.
29. Ander, P., and Eriksson, K.-E. Methanol formation during lignin degradation by Phanerochaete chrysosporium. *Appl. Microbiol. Biotechnol.*, *21*:96-102, 1985.
30. Eriksson, K. E., Patterson, B., Volc, J., and Musilek, V. Formation and partial characterization of glucose-2-oxidase, a H_2O_2 producing enzyme in Phanerochaete chrysosporium. *Appl. Microbiol. Biotechnol.*, *23*:257-262, 1986.
31. Reddy, C. A., and Kelley, R. L. The central role of hydrogen peroxide in lignin biodegradation by Phanerochaete chrysosporium. In *Biodeterioration* (S. Barry, D. R. Houghton, G. C. Llewellyn, and C. E. O'Rear, eds.), CAB International Mycological Institutes, London, 1986, pp. 535-542.

32. Freer, S. N., and Detroy, R. W. Biological delignification of ^{14}C-labelled lignocelluloses by basidiomycetes: Degradation and solubilization of lignin and cellulose components. *Mycologia, 74*:943-951, 1982.
33. Hatakka, A. I., Buswell, J. A., Pirhonen, T. I., and Uusi-Rauva, A. K. Degradation of ^{14}C-labelled lignins by white-rot fungi. See Ref. 4, p. 279.
34. Leatham, G. F., and Kirk, T. K. Regulation of ligninolytic activity by nutrient nitrogen in white-rot basidiomycetes. *FEMS Microbiol. Lett., 16*:65-67, 1983.
35. Jeffries, T. W., Choi, S., and Kirk, T. K. Nutritional regulation of lignin degradation by Phanerochaete chrysosporium. *Appl. Environ. Microbiol., 42*:290-296, 1981.
36. Cowling, E. B., and Merrill, W. Nitrogen in wood and its role in wood deterioration. *Can. J. Bot., 44*:1539-1554, 1966.
37. Forney, L. J., Reddy, C. A., Tien, M., and Aust, S. D. The involvement of hydroxyl radical derived from hydrogen peroxide in lignin degradation by the white-rot fungus Phanerochaete chrysosporium. *J. Biol. Chem., 257*:11455-11462, 1982.
38. Kutsuki, H., and Gold, M. H. Generation of hydroxyl radical and its involvement in lignin degradation by the white-rot fungus Phanerochaete chrysosporium. *Biochem. Biophys. Res. Commun., 109*:320-325, 1982.
39. Tien, M., and Kirk, T. K. Lignin-degrading enzyme from the hymenomycete Phanerochaete chrysosporium Burds. *Science, 221*:661-663, 1983.
40. Glenn, J. K., Morgan, M. A., Mayfield, M. B., Kuwahara, M., and Gold, M. H. An extracellular H_2O_2-requiring enzyme preparation involved in lignin biodegradation by the white-rot basidiomycete Phanerochaete chrysosporium. *Biochem. Biophys. Res. Commun., 114*: 1077-1083, 1983.
41. Renganathan, V., Miki, K., and Gold, M. H. Multiple molecular forms of diarylpropane oxygenase, an H_2O_2-requiring, lignin-degrading enzyme from Phanerochaete chrysosporium. *Arch. Biochem. Biophys., 342*: 304-314, 1985.
42. Tien, M., and Kirk, T. K. Lignin peroxidase of Phanerochaete chrysosporium. *Meth. Enzymol., 161*:238-249, 1988.
43. Leisola, M. S. A., Kozulic, B., Meussdorffer, F., and Fiechter, A. Homology among multiple extracellular peroxidases from Phanerochaete chrysosporium. *J. Biol. Chem., 262*:419-424, 1987.
44. Gold, M. H., Kuwahara, M., Chiu, A. A., and Glenn, J. K. Purification and characterization of an extracellular H_2O_2-requiring diarylpropane oxygenase from the white-rot basidiomycete Phanerochaete chrysosporium. *Arch. Biochem. Biophys., 234*:353-362, 1984.
45. Tien, M., and Kirk, T. K. Lignin-degrading enzyme from Phanerochaete chrysosporium. Purification, characterization and catalytic properties of a unique H_2O_2-requiring oxygenase. *Proc. Natl. Acad. Sci. USA, 81*:2280-2284, 1984.
46. Jäger, A., Croan, C., and Kirk, T. K. Production of ligninases and degradation of lignin in agitated submerged cultures of Phanerochaete chrysosporium. *Appl. Environ. Microbiol., 50*:1274-1278, 1985.
47. Dass, S. B., and Reddy, C. A. Characterization of extracellular peroxidases produced by acetate-buffered cultures of the lignin-degrading basidiomycete Phanerochaete chrysosporium. *FEMS Microbiol. Lett., 69*:221-224, 1990.

48. Odier, E., and Delattre, M. Multiple lignin peroxidases of Phanerochaete chrysosporium INA-12. *Enz. Microb. Technol.*, *12*:447-452, 1990.
49. Farrell, R. L., Murtagh, K. E., Tien, M., Mozuch, M. D., and Kirk, T. K. Physical and enzymatic properties of lignin peroxidase isoenzymes from Phanerochaete chrysosporium. *Enz. Microb. Technol.*, *11*:322-328, 1989.
50. Troller, J., Smith, J. G., Leisola, S. A., Kallen, K., Winterhalter, K. H., and Fiechter, A. Crystallization of a lignin peroxidase from the white-rot fungus Phanerochaete chrysosporium. *Bio/Technology*, *6*:571-573, 1988.
51. Gold, M. H., Wariishi, H., and Valli, K. Extracellular peroxidases involved in lignin degradation by the white-rot basidiomycete Phanerochaete chrysosporium. In *Biocatalysis in Agricultural Biotechnology* (J. Whitaker and P. E. Sonnet, eds.), American Chemical Society, Washington, D.C., 1989, pp. 127-140.
52. Miki, K., Renganathan, V., and Gold, M. H. Novel aryl ether rearrangement catalyzed by the lignin peroxidase of Phanerochaete chrysosporium. *FEBS Lett.*, *203*:235-238, 1986.
53. Miki, K., Renganathan, V., and Gold, M. H. Mechanism of β-aryl ether dimeric lignin model compound oxidation by lignin peroxidase of Phanerochaete chrysosporium. *Biochemistry*, *25*:4790-4796, 1986.
54. Renganathan, V., and Gold, M. H. Spectral characterization of the oxidized states of lignin peroxidase, an extracellular enzyme from the white-rot basidiomycete Phanerochaete chrysosporium. *Arch. Biochem. Biophys.*, *432*:304-314, 1986.
55. Schoemaker, H. E., Harvey, P. J., Bowen, R. M., and Palmer, J. M. On the mechanism of enzymatic lignin breakdown. *FEBS Lett.*, *183*: 7-12, 1985.
56. Kersten, P. J., Tien, M., Kalyanaraman, B., and Kirk, T. K. The ligninase of Phanerochaete chrysosporium generates cation radicals from methoxy benzenes. *J. Biol. Chem.*, *260*:2609-2612, 1985.
57. Umezawa, T., and Higuchi, T. Aromatic ring cleavage in degradation of β-O-4 lignin substructure by Phanerochaete chrysosporium. *FEBS Lett.*, *182*:257-259, 1985.
58. Umezawa, T., Shimada, M., Higuchi, T., and Kusai, K. Aromatic ring cleavage of β-O-4 lignin substructure model dimers by lignin peroxidase of Phanerochaete chrysosporium. *FEBS Lett.*, *205*:287-292, 1986.
59. Umezawa, T., and Higuchi, T. Mechanism of aromatic ring cleavage of β-O-4 lignin substructure models by lignin peroxidase. *FEBS Lett.*, *218*:255-260, 1987.
60. Tuisel, H., Sinclair, R., Bumpus, J. A., Ashbaugh, W., Brock, B. J., and Aust, S. D. Lignin peroxidase H2 from Phanerochaete chrysosporium: Purification, characterization and stability to temperature and pH. *Arch. Biochem. Biophys.*, *279*:158-166, 1990.
61. Hammel, K. E., Kalyanaraman, B., and Kirk, T. K. Substrate free radicals are intermediates in ligninase catalysis. *Proc. Natl. Acad. Sci. USA*, *83*:3708-3712, 1986.
62. Harvey, P. J., Palmer, J. N., Schoemaker, H. E., Dekker, H. L., and Wever, R. Pre-steady-state kinetic study on the formation of compound I and II of ligninase. *Biochim. Biophys. Acta*, *994*:59-63, 1989.
63. Waldner, R., Leisola, M., and Fiechter, A. Production of extracellular ligninolytic enzymes by different white-rot fungi. *Proc. 3rd Int. Conf.*

Biotechnology in the Pulp and Paper Industry, Stockholm, 1986, pp. 17-19.
64. Biswas-Hawkes, D., Dodson, A. P. J., Harvey, P. J., and Palmer, J. M. Ligninase from white-rot fungi. *Proc. Int. Symp. Lignin Enzymic and Microbial Degradation*, INRA Publications, Inc., Paris, 1987, pp. 171-176.
65. Dodson, P. J., Evans, C. S., Harvey, P. J., and Palmer, J. M. Production and properties of an extracellular peroxidase from Coriolus versicolor which catalyzes Cα-Cβ cleavage in a lignin model compound. *FEMS Microbiol. Lett.*, *42*:17-22, 1987.
66. Jonsson, L., Johansson, T., Sjostrom, K., and Nyman, P. O. Purification of ligninase isozymes from the white-rot fungus Trametes versicolor. *Acta Chem. Scand.*, *B41*:766-769, 1987.
67. Lobarzewski, J. The characteristics and functions of the peroxidases from Trametes versicolor in lignin biotransformation. *J. Biotechnol.*, *13*:111-117, 1990.
68. Hatakka, A., Kantelinen, A., Tervila-Wilo, and Viikari, L. Production of ligninases by Phlebia radiata in agitated cultures. *Proc. Int. Symp. Lignin Enzymic and Microbial Degradation*, INRA Publications, Inc., Paris, 1987, pp. 185-189.
69. Niku-Paavola, M. L. Ligninolytic enzymes of the white-rot fungus Phlebia radiata. *Proc. Int. Symp. Lignin Enzymic and Microbial Degradation, Paris*, 1987, pp. 119-123.
70. Karhunen, E., Kantelinen, A., and Niku-Paavola, M. L. Mn-dependent peroxidases from the lignin-degrading white-rot fungus Phlebia radiata. *Arch. Biochem. Biophys.*, *279*:25-31, 1990.
71. Oki, T., Shinmoto, M., and Ishikawa, H. Enzymatic degradation of guaiacylglycerol-β-guaiacyl ether. *Mokuzai Gakkaishi*, *32*:448-456, 1986.
72. Maltseva, O. V., Myasoedowa, N. M., Leotievski, A. A., and Golovleva, L. A. Characteristics of the ligninolytic system of Panus tigrinus. *Proc. Sov. Finn. Sem. Micro. Degradation Lignocellul. Raw Mater*, Tbilisi, Georgia, USSR, 1986, pp. 74-82.
73. Huynh, V.-B., and Crawford, R. L. Novel extracellular enzymes (ligninases) of Phanerochaete chrysosporium. *FEMS Lett.*, *28*:119-123, 1985.
74. Kuwahara, M., Glenn, J. K., Morgan, M. A., and Gold, M. H. Separation and characterization of two extracellular H_2O_2-dependent oxidases from ligninolytic cultures of Phanerochaete chrysosporium. *FEBS Lett.*, *169*:247-250, 1984.
75. Paszczynski, A., Huynh, V.-B., and Crawford, R. L. Enzymatic activities of an extracellular manganese dependent peroxidase from Phanerochaete chrysosporium. *FEMS Microbiol. Lett.*, *29*:37-41, 1985.
76. Glenn, J. K., and Gold, M. H. Purification and characterization of an extracellular Mn(II)-dependent peroxidase from the lignin-degrading basidiomycete Phanerochaete chrysosporium. *Arch. Biochem. Biophys.*, *242*:329-341, 1985.
77. Paszczynski, A., Huynh, V.-B., and Crawford, R. L. Comparison of lignin peroxidase-I and peroxidase-M2 from the white-rot fungus Phanerochaete chrysosporium. *Arch. Biochem. Biophys.*, *244*:750-765, 1986.
78. Glenn, J. K., Akileswaran, L., and Gold, M. H. Mn(II) oxidation is the principal function of extracellular Mn-peroxidase from Phanerochaete chrysosporium. *Arch. Biochem. Biophys.*, *251*:688-696, 1986.
79. Wariishi, H., Akileswaran, L., and Gold, M. H. Manganese peroxidase from the basidiomycete, Phanerochaete chrysosporium: Spectral

characterization of the oxidized states and the catalytic cycle. *Biochemistry*, *27*:5365-5370, 1988.
80. Higuchi, T. Catabolic pathways and role of ligninase for the degradation of lignin substructure model compounds by white-rot fungi. *Wood Res.*, *73*:58-81, 1986.
81. Perez, J., and Jeffries, T. W. Mineralization of ^{14}C-ring-labeled synthetic lignin correlates with the production of lignin peroxidase, not of manganese peroxidase or laccase. *Appl. Environ. Microbiol.*, *56*:1806-1812, 1990.
82. Faison, B. D., and Kirk, T. K. Factors involved in the regulation of ligninase activity in Phanerochaete chrysosporium. *Appl. Environ. Microbiol.*, *49*:299-304, 1985.
83. Forney, L. J., Reddy, C. A., and Pankratz, H. S. Ultrastructural localization of hydrogen peroxide production in ligninolytic cultures of Phanerochaete chrysosporium. *Appl. Environ. Microbiol.*, *44*:732-736, 1982.
84. Kelley, R. L., and Reddy, C. A. Identification of glucose oxidase activity as the primary source of hydrogen peroxide production in ligninolytic cultures of Phanerochaete chrysosporium. *Arch. Microbiol.*, *144*:248-253, 1986.
85. Ramasamy, K., Kelley, R. L., and Reddy, C. A. Lack of lignin degradation by glucose oxidase negative mutants of Phanerochaete chrysosporium. *Biochem. Biophys. Res. Commun.*, *131*:436-441, 1985.
86. Kelley, R. L., Ramasamy, K., and Reddy, C. A. Characterization of glucose oxidase-negative mutants of a lignin degrading basidiomycete Phanerochaete chrysosporium. *Arch. Microbiol.*, *114*:254-257, 1986.
87. Kelley, R. L., and Reddy, C. A. Purification and characterization of glucose oxidase from ligninolytic cultures of Phanerochaete chrysosporium. *J. Bacteriol.*, *166*:269-274, 1986.
88. Kelley, R. L., and Reddy, C. A. Glucose oxidase of Phanerochaete chrysosporium. *Meth. Enzymol.* *161*:307-316.
89. Kersten, P. J., and Kirk, T. K. Involvement of a new enzyme, glyoxal oxidase in extracellular H_2O_2 production by Phanerochaete chrysosporium. *J. Bacteriol.*, *169*:2195-2202, 1987.
90. Kersten, P. J. Glyoxal oxidase of Phanerochaete chrysosporium: Its characterization and activation by lignin peroxidase. *Proc. Natl. Acad. Sci. USA*, *87*:2936-2940, 1990.
91. Garcia, S., Latge, M. C., Prevost, M. C. and Leisola, M. Cytochemical study of wood degradation by Phanerochaete chrysosporium. *Proc. Int. Symp. Lignin Enzymic and Microbial Degradation*, INRA Publications, Inc., Paris, 1987, pp. 277-282.
92. Srebotnik, E., Messner, K., Foisner, R., and Petterson, B. Ultrastructural localization of ligninase of Phanerochaete chrysosporium by immunogold labeling. *Curr. Microbiol.*, *16*:221-227, 1989.
93. Janshekar, H., Brown, C., Haltmeir, T., Leisola, M., and Fiechter, A. Bioalteration of kraft pine lignin by Phanerochaete chrysosporium. *Arch. Microbiol.*, *132*:14-21, 1982.
94. Chua, M. G. S., Choi, S., and Kirk, T. K. Mycelium binding and depolymerization of synthetic ^{14}C-labeled lignin during decomposition by Phanerochaete chrysosporium. *Holzforschung*, *37*:55-61, 1983.
95. Blanchette, R. A., and Reid, I. D. Ultrastructural aspects of wood delignification by Phlebia(Merulius) tremellosus. *Appl. Environ. Microbiol.*, *52*:239-245, 1986.

96. Zhang, Y. Z., Zylstra, G. J., Olsen, R. H., and Reddy, C. A. Identification of cDNA clones for ligninase from Phanerochaete chrysosporium using synthetic oligonucleotide probes. *Biochem. Biophys. Res. Commun.*, *137*:649-654, 1986.

97. de Boer, H. A., Zhang, Y. Z., Collins, C., and Reddy, C. A. Analysis of nucleotide sequences of two ligninase cDNAs from a white-rot filamentous fungus, Phanerochaete chrysosporium. *Gene*, *60*:93-102, 1987; *69*:369, 1988.

98. Zhang, Y. Z. Molecular cloning of lignin peroxidase cDNAs and genes from a white-rot basidiomycete fungus Phanerochaete chrysosporium. Ph.D. thesis, Michigan State University, 1987.

99. Tien, M., and Tu, C-PD. Cloning and sequencing of a cDNA for a ligninase from Phanerochaete chrysosporium. *Nature*, *326*:520-523, 1987; corrigendum *238*:742.

100. Farrell, R. L., Gelep, P., Algis, A., Kashayar, J., Malone, T. E., James, R., Bruce, A., and Jennifer, A. Recombinant ligninase. Eur. Patent Application No. 87810516.2 (1988).

101. Saloheimo, M., Barajas, V., Niku-Paavola, L. N., and Knowles, J. K. C. A lignin peroxidase-encoding cDNA from the white rot fungus Phlebia radiata: Characterization and expression in Trichoderma reesei. *Gene*, *85*:343-351, 1989.

102. Pease, E. A., Andrawis, A., and Tien, M. Manganese-dependent peroxidase from Phanerochaete chrysosporium. Primary structure deduced from cDNA sequence. *J. Biol. Chem.*, *264*:13531-13535, 1989.

103. Pribnow, D., Mayfield, M. B., Nipper, V. J., Brown, J. A., and Gold, M. H. Characterization of a cDNA encoding a manganese peroxidase from the lignin-degrading basidiomycete Phanerochaete chrysosporium. *J. Biol. Chem.*, *264*:5036-5040, 1989.

104. Smith, T. L., Schalch, J., Gaskell, S., Covert, S., and Cullen, D. Nucleotide sequence of a ligninase gene from Phanerochaete chrysosporium. *Nucl. Acids Res.*, *16*:1219, 1988.

105. Walther, I., Kalin, M., Reiser, J., Suter, F., Fritsche, B., Saloheimo, M., Leisola, M., Teeri, T., Knowles, J. K. C., and Fiechter, A. Molecular analysis of a Phanerochaete chrysosporium lignin peroxidase gene. *Gene*, *70*:127-137, 1988.

106. Holzbaur, E. L. F., and Tien, M. Structure and regulation of a lignin peroxidase gene from Phanerochaete chrysosporium. *Biochem. Biophys. Res. Commun.*, *155*:626-633, 1988.

107. Brown, A., Sims, P. F. G., Raeder, U., and Broda, P. Multiple ligninase related genes from Phanerochaete chrysosporium. *Gene*, *73*:77-85, 1988.

108. Asada, Y., Kimura, Y., Kuwahara, M., Tsukamoto, A., Koide, K., Oka, A., and Takanami, M. Cloning and sequencing of a ligninase gene from a lignin-degrading basidiomycete Phanerochaete chrysosporium. *Appl. Microbiol. Biotechnol.*, *29*:469-473, 1988.

109. Schalch, H., Gaskell, G., Smith, T. L., and Cullen, D. Molecular cloning and sequences of lignin peroxidase genes of Phanerochaete chrysosporium. *Mol. Cell. Biol.*, *9*:2743-2747, 1989.

110. Padmavathy, S. N., and Reddy, C. A. Analysis of nucleotide sequence of a lignin peroxidase gene from a white-rot basidiomycete Phanerochaete chrysosporium BKM-F 1767. Abstr. Annu. Meet. Am. Soc. Microbiol., 1989, p. 181.

111. Alic, M., Kornegay, J. R., Pribnow, D., and Gold, M. H. Transformation by complementation of an adenine auxotroph of the lignin-degrading basidiomycete Phanerochaete chrysosporium. *Appl. Environ. Microbiol.*, *55*:406-411, 1989.
112. Randall, T. A., Reddy, C. A., and Boominathan, K. A novel extrachromosomally maintained transformation vector for the lignin-degrading basidiomycete Phanerochaete chrysosporium. *J. Bacteriol.*, *173*:776-782.
113. Leisola, M. S. A., and Feichter, A. New trends in lignin biodegradation. *Adv. Biotechnol. Processes*, *5*:59-89, 1985.
114. Gold, M. H., Kuwahara, M., Chiu, A., and Glenn, J. K. Purification and characterization of an extracellular H_2O_2-requiring diarylpropane oxygenase from the white rot basidiomycete, Phanerochaete chrysosporium. *Arch. Biochem. Biophys.*, *234*:353-362, 1984.
115. Reid, I. D., Chao, E. E., and Dawson, S. S. Lignin degradation by Phanerochaete chrysosporium in agitated cultures. *Can. J. Microbiol.*, *31*:88-90, 1985.
116. Kirk, T. K., Croan, S., Tien, M., Murtagh, K. E., and Farrell, R. L. Production of multiple ligninases by Phanerochaete chrysosporium: Effect of selected growth conditions and use of a mutant strain. *Enz. Microb. Technol.*, *8*:27-32, 1986.
117. Leisola, M. S. A., Thanei-Wyss, U., and Feichter, A. Strategies for production of high ligninase activities by Phanerochaete chrysosporium. *J. Biotechnol.*, *3*:97-107, 1985.
118. Kern, H. W. Improvement in the production of extracellular lignin peroxidases by Phanerochaete chrysosporium: Effect of solid manganese-IV oxide. *Appl. Microbiol. Biotechnol.*, *32*:223-234, 1989.
119. Roch, P., Buswell, J. A., Cain, R. B., and Odier, E. Lignin peroxidase production by strains of Phanerochaete chrysosporium grown on glycerol. *Appl. Microbiol. Biotechnol.*, *31*:587-591, 1989.
120. Kirk, T. K., Tien, M., Johnsrud, S. C., and Eriksson, K.-E. Lignin-degrading activity of Phanerochaete chrysosporium Burds: Comparison of cellulase-negative and other strains. *Enz. Microb. Technol.*, *8*:75-80, 1986.
121. Kirk, T. K., Tien, M., Croan, C., McDonagh, T., and Farrell, R. L. Production of ligninases by Phanerochaete chrysosporium. *Proc. 3rd Int. Congress on Biotechnology in Pulp and Paper Industry, Stockholm*, 1986, pp. 5-6.
122. (a) Asther, M., Capdevilla, C., and Corrieu, G. Control of lignin peroxidase production by Phanerochaete chrysosporium INA-12 by temperature shifting. *Appl. Environ. Microbiol.*, *54*:3194-3196, 1988.
(b) Asther, M., Lesage, L., Drapon, R., Corrieu, G., and Odier, E. Phospholipid and fatty acid enrichment of Phanerochaete chrysosporium INA-12 in relation to ligninase production. *Appl. Microbiol. Biotechnol.*, *27*:393-398, 1988.
123. Asther, M., Corrieu, G., Drapon, R., and Odier, E. Effect of Tween 80 and oleic acid on ligninase production by Phanerochaete chrysosporium INA-12. *Enz. Microb. Technol.*, *9*:245-249, 1987.
124. Brown, J. A., Glenn, J. K., and Gold, M. H. Manganese regulates expression of manganese peroxidase by Phanerochaete chrysosporium. *J. Bacteriol.*, *172*:3125-3130, 1990.
125. Bonnarme, P., and Jeffries, T. W. Mn(II) regulation of lignin peroxidases and manganese-dependent peroxidases from lignin-degrading white-rot fungi. *Appl. Environ. Microbiol.*, *56*:210-217, 1990.

126. Dosoretz, C. D., Chen, H. C., and Grethlein, H. E. Effect of environmental conditions on the extracellular protease activity in ligninolytic cultures of Phanerochaete chrysosporium. *Appl. Environ. Microbiol.*, *56*:395-400, 1990.
127. Dosoretz, C. G., Dass, S. B., Reddy, C. A., and Grethlein, H. E. Protease mediated degradation of lignin peroxidase in liquid cultures of Phanerochaete chrysosporium. *Appl. Environ. Microbiol.*, *56*:3429-3434.
128. Lewandowski, G. A., Armenante, P. M., and Pak, D. Reactor design for hazardous waste treatment using a white-rot fungus. *Water Res.*, *24*:75-82, 1990.
129. Gold, M. H., and Cheng, T. M. Induction of colonial growth and replica plating of the white-rot basidiomycete Phanerochaete chrysosporium. *Appl. Environ. Microbiol.*, *35*:1223-1225, 1978.
130. Eriksson, K.-E., Johnsrud, S. C., and Valander, L. Degradation of lignin and lignin model compounds by various mutants of the white-rot fungus, Sporotrichum pulverulentum. *Arch. Microbiol.*, *135*:161-168, 1983.
131. Johnsrud, S. C., and Eriksson, K.-E., Cross-breeding of selected and mutated homokaryotic strains of Phanerochaete chrysosporium K-3: New cellulase-deficient strains with increased ability to degrade lignin. *Appl. Microbiol. Biotechnol.*, *21*:320-327, 1985.
132. Liwicki, R., Paterson, A., MacDonald, M. J., and Broda, P. Phenotypic classes of phenol oxidase-negative mutants of the lignin-degrading fungus, Phanerochaete chrysosporium. *J. Bacteriol.*, *162*:641-644, 1985.
133. Gold, M. H., Cheng, T. M., and Mayfield, M. B. Isolation and complementation studies of auxotrophic mutants of the lignin-degrading basidiomycete Phanerochaete chrysosporium. *Appl. Environ. Microbiol.*, *44*:996-1000, 1982.
134. Alic, M., and Gold, M. H. Genetic recombination in the lignin-degrading basidiomycete Phanerochaete chrysosporium. *Appl. Environ. Microbiol.*, *50*:27-30, 1985.
135. Tien, M., Kersten, P. J., and Kirk, T. K. Selection and improvement of lignin-degrading microorganisms: Potential strategy based on lignin model amino acid adducts. *Appl. Environ. Microbiol.*, *53*:242-245, 1987.
136. Tien, M., and Myer, S. Selection and characterization of mutants of Phanerochaete chrysosporium exhibiting ligninolytic activity under nutrient-rich conditions. *Appl. Environ. Microbiol.*, *56*:2540-2544, 1990.
137. Buswell, J. A., Mollet, B., and Odier, E. Ligninolytic enzyme production by Phanerochaete chrysosporium under conditions of nitrogen sufficiency. *FEMS Microbiol. Lett.*, *25*:295-299, 1984.
138. Galliano, H., Gas, G., and Durand, H. Lignocellulose biodegradation by mutant strains of Phanerochaete chrysosporium hyper producing cellulases. *Biotechnol. Lett.*, *10*:655-660, 1988.
139. Maione, T. E., Javahaerian, K., Belew, M. A., Gomez, L. E., and Farrell, R. L. Activities of recombinant ligninase. *Proc. Int. Symp. Lignin Enzymic and Microbial Degradation*, INRA Publications, Inc., Paris, 1987, pp. 177-183.
140. Leisola, M., Troller, J., Fiechter, A., and Linko, Y. Y. Production and characterization of ligninolytic enzymes of Phanerochaete chrysosporium. *Proc. Biotech. Pulp and Paper Industry*, 1986, pp. 16-19.

141. Linko, Y.-Y., Leisola, M., Lindholm, N., Troller, J., Linko, P., and Fiechter, A. Continuous production of lignin peroxidase by Phanerochaete chrysosporium. *J. Biotechnol.*, *4*:282-289, 1986.
142. Linko, S. Production and characterization of extracellular lignin peroxidase from immobilized Phanerochaete chrysosporium in a 10-l bioreactor. *Enz. Microb. Technol.*, *10*:410-417, 1988.
143. Linko, S., and Zhong, L. C. Comparison of different methods of immobilization for lignin peroxidase production by Phanerochaete chrysosporium. *Biotechnol. Tech.*, *1*:251-254, 1987.
144. Kirkpatrick, N., and Palmer, J. M. Semicontinuous ligninase production using foam-immobilized Phanerochaete chrysosporium. *Appl. Microbiol. Biotechnol.*, *27*:129-133, 1987.
145. Metz, B. and Kossen, N. W. F. The growth of molds in the form of pellets. Literature Review. *Biotechnol. Bioeng.*, *19*:781-799, 1977.
146. Michel, F. C. Jr. M., Grulke, E. A., and Reddy, C. A. Development of a stirred tank reactor system for the production of lignin peroxidases (ligninases) by Phanerochaete chrysosporium BKM-F-1767. *J. Ind. Microbiol.*, *5*:103-112, 1990.
147. Janshekar, H., and Fiechter, A. Cultivation of Phanerochaete chrysosporium and production of lignin peroxidases in submerged stirred tank reactors. *J. Biotechnol.*, *8*:97-112, 1988.
148. Eriksson, K.-E., and Vallender, L. Properties of pulps from thermomechanical pulping of chips pretreated with fungi. *Sven. Papperstidn.*, R33-R38, 1982.
149. Eriksson, K.-E., Grunewald, A., and Kringstad, K. Studies on growth conditions in wood for three white-rot fungi and their cellulase-less mutants. *Biotechnol. Bioeng.*, *22*:363-376, 1980.
150. Samuelsson, L., Mjoberg, P. J., Hartler, N., Vallander, L., and Eriksson, K.-E. Influence of fungal treatment on the strength versus energy relationship in mechanical pulping. *Sven. Papperstidn.*, *83*:221-225, 1980.
151. Bar-Lev, S. S., Chang, H. M., and Kirk, T. K. Evidence that fungal treatment can reduce the energy requirement for secondary refining of thermomechanical pulp. *Tappi*, *65*:111-113, 1982.
152. Kirk, T. K., and Eriksson, K.-E. Role of biotechnology in manufacture. In *Wood Pulp and Paper Technology* (F. Roberts, ed.), *International Review for the Pulp and Paper Industry*, Sterling, London, 1989, pp. 23-28.
153. Rydholm, A. A. *Pulping Processes*. Interscience, New York, 1965.
154. Sanyer, N., and Chidester, C. H. Manufacture of wood pulp. In *The Chemistry of Wood* (B. L. Browning, ed.), Interscience, New York, 1963, pp. 441-534.
155. Sjostrom, E. *Wood Chemistry: Fundamentals and Applications*. Academic Press, New York, 1981.
156. Gierer, J. The reactions of lignin during pulping: A description and comparison of conventional pulping processes. *Sven. Papperstidn.*, *73*:571-576, 1970.
157. Gierer, J. Chemical aspects of delignification. In *The Ekman-Days International Symposium on Wood and Pulping Chemistry*, Swedish Forest Products Laboratory, Stockholm, Sweden, Vol. 2, 1981, pp. 12-17.
158. Ljunggren, S. The significance of aryl ether cleavage in kraft delignification of soft wood. *Sven. Papperstidn.*, *83*:363-369, 1980.

159. Marton, J. Reactions of lignin in alkaline pulping. See Ref. 14, pp. 639-694.
160. Falkehag, I., Marton, J., and Adler, E. Chromophores in kraft lignin. *Adv. Chem. Ser.*, *59*:75-89, 1966.
161. Ander, P., Eriksson, K. E., Kolar, M. C., Kringstad, K., Rannug, U., and Ramel, C. Studies on the mutagenic properties of bleaching effluents. 1. *Sven. Papperstidn.*, *80*:454-459, 1977.
162. Eriksson, K. E., Kolar, M. C., and Kringstad, K. Studies on the mutagenic properties of bleaching effluents. 2. *Sven. Papperstidn.*, *82*:95-104, 1979.
163. Hiroi, T., and Eriksson, K. E. Microbiological degradation of lignin. 1. Influence of cellulose on the degradation of lignins by the white-rot fungus Pleurotus ostreatus. *Sven. Papperstidn.*, *79*:157-161, 1976.
164. Lundquist, K., Kirk, T. K., and Connors, W. J. Fungal degradation of kraft lignin and lignin sulfonates prepared from synthetic ^{14}C-lignins. *Arch. Microbiol.*, *112*:291-296, 1977.
165. Loras, V. Bleaching of chemical pulps. In *Pulp and Paper Chemistry and Chemical Technology*, Vol. 1, 3rd ed. (J. P. Casey, ed.), John Wiley and Sons, New York, pp. 663.
166. Kirk, T. K., and Yang, H. H. Partial delignification of unbleached kraft pulp with ligninolytic fungi. *Biotechnol. Lett.*, *1*:347-352, 1979.
167. Lundborh, A. Increased degradation of steam exploded birch lignin during interaction between white-rot fungi and other fungi. *Mater. Org. (Berl).*, *23*:37-48, 1988.
168. (a) Kirkpatrick, N., Reid, I., Ziomek, E., Ho, C., and Paice, M. G. Relationship between fungal biomass production and the brightening of hardwood kraft pulp by Coriolus versicolor. *Appl. Environ. Microbiol.*, *55*:1147-1152, 1989. (b) Kirkpatrick, N., Reid, I., Ziomek, E., and Paice, M. G. Biological bleaching of hardwood kraft pulp using Trametes versicolor immobilized in polyurethane foam. *Appl. Microbiol. Biotechnol.*, *33*:105-108, 1990.
169. Reid, I. D., Paice, M. G., Ho, C., Jurasek, L. Biological bleaching of softwood kraft pulp with the fungus Trametes(Coriolus) versicolor. *Tappi. J.*, *73*:149-153, 1990.
170. Pellinen, J., Abuhasan, J., Joyce, T. W., and Chang, H.-M. Biological delignification of pulp by Phanerochaete chrysosporium. *J. Biotechnol.*, *10*:161, 1989.
171. Kirk, T. K., and Chang, H.-M. Potential applications of bioligninolytic systems. *Enz. Microb. Technol.*, *3*:189-196, 1981.
172. Huynh, V.-B., Chang, H. M., Joyce, T. W., and Kirk, T. K. Dechlorination of chloroorganics by a white-rot fungus. *Tappi. J.*, *68*:98-102, 1985.
173. Sundmann, G., Kirk, T. K., and Chang, H. M. Fungal decolorization of kraft bleach plant effluent. Fate of chromophoric material. *Tappi. J.*, *64*:145-148, 1981.
174. Eriksson, K.-E., and Kolar, K. C. Microbial degradation of chlorolignins. *Environ. Sci. Technol.*, *19*:1086-1089, 1985.
175. Fukuzumi, T., Nishida, A., Aoshima, K., and Minami, K. Decolorization of kraft waste liquor with white-rot fungi. I. Screening of fungi and culturing condition for decolorization of kraft waste liquor. *Mokuzai Gakkaishi*, *23*:290-298, 1977.
176. Campbell, A. G. A bench scale evaluation of the MyCOR process for decolorization of bleach plant effluent using the white-rot fungus Phanerochaete chrysosporium. Ph.D. thesis, North Carolina State University, 1983.

177. Eaton, D. C., Chang, H.-M., and T. K. Kirk. Fungal decolorization of kraft bleach effluents. *Tappi. J., 63*:103-106, 1985.
178. Marton, J., Stern, A. M., and Marton, T. Decolorization of kraft liquor with Polyporus versicolor, a white-rot fungus. *Tappi. J., 52*:1975-1979, 1969.
179. Yin, C. F., Joyce, T. W., and Chang, H.-M. Kinetics of bleach plant effluent decolorization by Phanerochaete chrysosporium. *J. Biotechnol., 10*:67-76, 1989.
180. Chang, H.-M., Joyce, T. W., Matsumoto, Y., Yin, C. F., Vasudevan, B., Coecht, C. A., and Kirk, T. K. Decolorization of bleach plant effluents by a white-rot fungus. In *Proc. 3rd International Symposium on Biotechnology in Pulp and Paper Industry*, Swedish Forest Products Laboratory, Stockholm, 1986, pp. 120-123.
181. Messner, K., Ertler, G., and Farcher, J., Treatment of bleach plant effluents by the MyCOPOR system. In *Proc. 4th International Symposium on Biotechnology in Paper and Pulp Industry, Raleigh, North Carolina*, 1989, pp. 67-68.
182. Prouty, A. L. A bench-scale development and evaluation of a fungal bioreactor for color removal from bleach effluents. *Appl. Microbiol. Biotechnol., 32*:490-493, 1990.
183. Wise, L. E., and Jahn, E. C. *Wood Chemistry*, 2nd ed. Reinhold, New York, 1952, p. 412.
184. Linden, B. E., and Dale, J. C. Fermentation substrates and economics. *Ann. Rep. Ferm. Proc., 7*:107-134, 1984.
185. Saddler, J. N., and Brownell, H. H. Pretreatment of wood cellulosics to enhance enzymatic hydrolysis to glucose. *Proc. International Symposium on Ethanol from Biomass*, Winnipeg, Canada, 1982.
186. Mes-Hartree, M., and Saddler, J. N. The nature of inhibitory materials present in pretreated lignocellulosic substrates which inhibit the enzymatic hydrolysis of cellulose. *Biotechnol. Lett., 5*:531-536, 1983.
187. Mes-Hartree, M., Yu, E. K. C., Reid, I. D., and Saddler, J. N. Suitability of aspenwood biologically delignified with Phlebia tremellosus for fermentation to ethanol or butanediol. *Appl. Microbiol. Biotechnol., 26*:120-125, 1987.
188. Umezawa, T., Higuchi, T., and Nakatsubo, F. Difference in $^{18}O_2$ incorporation in oxygenative degradation of β-O-4 and β-1 lignin substructures by Phanerochaete chrysosporium. *Agr. Biol. Chem., 47*:2945-2948, 1983.
189. Enoki, A., Goldsby, G. P., and Gold, M. H. Metabolism of the lignin model compounds veratrylglycerol β-guaiacylether and 4-ethoxy-3-methoxyphenylglycerol-β-guaiacylether by Phanerochaete chrysosporium. *Arch. Microbiol., 125*:227-232, 1980.
190. Enoki, A., and Gold, M. H. Degradation of the diaryl propane lignin model compound 1-(3',4'-diethoxyphenyl)-1,3-dihydroxy-2-(4''-methoxyphenyl)-propane and derivatives by the basidiomycete Phanerochaete chrysosporium. *Arch. Microbiol., 132*:123-130, 1982.
191. Kirk, T. K., and Nakatsubo, F. Chemical mechanism of an important cleavage reaction in the fungal degradation of lignin. *Biochim. Biophys. Acta, 756*:376-384, 1983.
192. Lobarzewski, J., and Paszczynski, A. Lignocellulose biotransformation with immobilized cellulase, D-glucose and fungal peroxidases. *Enz. Microb. Technol., 7*:564-566, 1985.

192. (a) Strandberg, G. W., and Lewis, S. N. The solubilization of coal by an extracellular product from Streptomyces setonii 75 Vi2. *J. Ind. Microbiol.*, *1*:371-375, 1987. (b) Teintze, M., Hossain, M. B., Barnes, C. L., Leong, J., and van der Helm, D. Structure of ferric pseudobactin, a siderophore from a plant growth promoting Pseudomonas. *Biochemistry*, *20*:6446-6456, 1981. (c) Cohen, M. S., and Gabriele, P. D. Degradation of coal by the fungi Polyporus versicolor and Poria monticola. *Appl. Environ. Microbiol.*, *44*:23-27, 1982. (d) Pyne, J. W., Stewart, D. L., Fredrickson, J., and Wilson, B. W. Solubilization of leonardite by an extracellular fraction from Coriolus versicolor. *Appl. Environ. Microbiol.*, *53*:2844-2848, 1987. (e) Cohen, M. S., Bowers, W. C., Aronson, H., and Gray, Jr., E. T. Cell-free solubilization of coal by Polyporus versicolor. *Appl. Environ. Microbiol.*, *53*:2840-2843, 1987. (f) Fredrickson, J. K., Stewart, D. L., Campbell, J. A., Powell, M. A., McMulloch, M., Pyne, J. W., and Bean, R. M. Biosolubilization of low-rank coal by a Trametes versicolor siderophore-like product and other complexing agents. *J. Ind. Microbiol.*, *5*:401-406, 1990.
193. Ritter, D., Farcher, S. J., Messner, K., and Stachelberger, H. Polymerization of lignosulfonate by Phanerochaete chrysosporium immobilized on foam. *J. Biotechnol.*, *13*:229-241, 1990.
194. Selin, J. F., Sundman, V., and Raika, M. Utilization and polymerization of lignosulfonates by wood-rotting fungi. *Arch. Microbiol.*, *103*:63-70, 1975.
195. Bergbauer, M., and Kraepelin, G. Polymerization and depolymerization of waste water lignins by white-rot fungi. In *Proc. of 4th International Symposium on Biotechnology in Pulp and Paper Industry, Raleigh, North Carolina*, 1989, p. 79.
196. Kirk, T. K. and Moore, W. E. Removing lignin from wood with white-rot fungi and digestibility of resulting wood. *Wood Fiber*, *4*:72-79, 1972.
197. Baker, A. J. Effect of lignin on the in vitro digestibility of wood pulp. *J. Anim. Sci.*, *36*:768, 1973.
198. Kamara, D. N., Zadrazil, F. Microbiological improvement of lignocellulosics in animal feed production. In *Treatment of Lignocellulosics with White-Rot Fungi* (F. Zadrazil and P. Reiniger, eds.), Elsevier, New York, 1986, pp. 56-61.
199. Reid, I. D. Solid-state fermentations for biological delignification. *Enz. Microb. Technol.*, *11*:786-803, 1989.
200. Reid, I. D. Optimization of solid-state fermentation for selective delignification of aspen wood with Phlebia tremellosa. *Enz. Microb. Technol.*, *11*:804-809, 1989.
201. Hartley, R. D., Jones, E. C., King, N. J., Smith, J. A. Modified wood waste and straw as potential components of animal feeds. *J. Sci. Food Agr.*, *25*:431-437, 1974.
202. Kaneshiro, T. Lignocellulosic agricultural wastes degraded by Pleurotus ostreatus. *Dev. Ind. Microbiol.*, *18*:591-597, 1977.
203. Lindenfelser, L. A., Detroy, R. W., Ramstack, J. M., and Worden, K. A. Biological modification of the lignin and cellulose components of wheat straw by Pleurotus ostreatus. *Dev. Ind. Microbiol.*, *20*:541-551, 1978.
204. Zadrazil, Z. Conversion of different plant waste into feed by basidiomycetes. *Eur. J. Appl. Microbiol. Biotechnol.*, *9*:31-35, 1980.

205. Hattaka, A. I. Pretreatment of wheat straw by white-rot fungi for enzymic saccharification of cellulose. *Eur. J. Appl. Microbiol. Biotechnol.*, *18*:350-357, 1983.
206. Midgett, R. E., and Paradis, A. J. Solid-state fermentation of natural birch lignin by Phanerochaete chrysosporium. *Enz. Microb. Technol.*, *7*:150-154, 1985.
207. Zadrazil, F. The conversion of straw into feed by basidiomycetes. *Eur. J. Appl. Microbiol. Biotechnol.*, *4*:291-294, 1977.
208. Zadrazil, F., and Brunnert, H. The influence of ammonium nitrate supplementation on degradation and in vitro digestibility of straw colonized by higher fungi. *Eur. J. Appl. Microbiol. Biotechnol.*, *9*:37-44, 1980.
209. Zadrazil, F., and Brunnert, H. Investigation of physical parameters important for the solid state fermentation of straw by white-rot fungi. *Eur. J. Appl. Microbiol. Biotechnol.*, *11*:183-188, 1981.
210. Tangol, R. *Diccionario etimologico Chiloe*, Chile, 1976, referenced in *Treatment of Lignocellulosics by White-Rot Fungi* (F. Zadrazil and P. Reiniger, eds.), Elsevier, New York, 1986.
211. Agosin, E., Tollier, M. T., Brillout, J. M., Thivend, P., and Odier, E. Fungal pretreatment of wheat straw: Effects on the biodegradability of cell walls, structural polysaccharides, lignin and phenolic acids by rumen microorganisms. *J. Sci. Food Agr.*, *37*:97-106, 1986.
212. Reade, A. E., and McQueen, R. E. Investigation of white-rot fungi for the conversion of poplar into a potential feedstuff for ruminants. *Can. J. Microbiol.*, *29*:457-463, 1983.
213. Reid, I. D. Biological delignification of aspen wood by solid state fermentation with white-rot fungus Merulius tremellosus. *Appl. Environ. Microbiol.*, *50*:133-139, 1985.
214. Zadrazil, F. Screening of fungi for lignin decomposition and conversion of straw into feed. *Angew. Botanik.*, *59*:433-452, 1985.
215. Adaskaveg, J. E., Gilbertson, R. L., and Blanchette, R. A. Comparative studies of delignification caused by Ganoderma species. *Appl. Environ. Microbiol.*, *56*:1932-1943, 1990.
216. Bumpus, J. A., Tien, M., Wright, D., and Aust, S. D. Oxidation of persistent environmental pollutants by a white-rot fungus. *Science*, *228*:1434-1436, 1985.
217. Haemmerli, S. D., Leisola, M. S. A., Sanglard, D., and Feichter, A. Oxidation of benzo(a)pyrene by extracellular ligninases of Phanerochaete chrysosporium. *J. Biol. Chem.*, *261*:6900-6903, 1986.
218. Sanglard, D., Leisola, M. S. A., and Feichter, A. Role of extracellular ligninases in biodegradation of benzo(a)pyrene by Phanerochaete chrysosporium. *Enz. Microb. Technol.*, *8*:209-212, 1986.
219. Eaton, D. C. Mineralization of polychlorinated biphenyls by Phanerochaete chrysosporium: A ligninolytic fungus. *Enz. Microb. Technol.*, *7*:194-196, 1985.
220. Fernando, T., Aust, S. B., and Bumpus, J. A. Effects of cuture parameters on DDT-1,1,1-trichloro-2-2-bis-4-chlorophenylethane biodegradation by Phanerochaete chrysosporium. *Chemosphere*, *19*:1387-1398, 1989.
221. Ryan, T. P., and Bumpus, J. A. Biodegradation of 2,4,5-T in liquid culture and in soil by the white rot fungus Phanerochaete chrysosporium. *Appl. Microbiol. Biotechnol.*, *31*:302-307, 1989.
222. George, E. J., and Neufeld, R. D. Degradation of fluorene in soil by the fungus Phanerochaete chrysosporium. *Biotechnol. Bioeng.*, *33*:1306-1310, 1989.

223. Mileski, G. J., Bumpus, J. A., Jurek, M. A., and Aust, S. D. Biodegradation of pentachlorophenol by the white rot fungus Phanerochaete chrysosporium. *Appl. Environ. Microbiol.*, *54*:2885-2889, 1988.
224. Bumpus, J. A., and Brock, B. J. Biodegradation of crystal violet by the white rot fungus Phanerochaete chrysosporium. *Appl. Environ. Microbiol.*, *54*:1143-1150, 1988.
225. Zitzelsberger, W., Zielger, W., and Wallnoefer, P. R. Stereospecificity of the degradation of veratrylglycerol beta-2,4-dichlorophenylether, a model compound for lignin bound xenobiotic residues by Phanerochaete chrysosporium. *Chemosphere*, *16*:1137-1142, 1987.
226. Arjmand, M., and Sandermann, H. Plant biochemistry of xenobiotics mineralization of chloroaniline-lignin metabolites from wheat by the white rot fungus Phanerochaete chrysosporium. *Z. Naturforsch. Sect. C. Biosci.*, *41*:206-214, 1986.
227. Schreiner, R., Stewans, S. E., and Tien, M. Oxidation of thianthrene by ligninase of Phanerochaete chrysosporium. *Appl. Environ. Microbiol.*, *54*:1858-1860, 1988.
228. Kennedy, D. W., Aust, S. D., and Bumpus, J. A. Comparative biodegradation of alkyl halide insecticides by the white-rot fungus Phanerochaete chrysosporium. *Appl. Environ. Microbiol.*, *56*:2347-2353, 1990.
229. Bumpus, J. A., and Aust, S. D. Biodegradation of DDT [1,1,1-trichloro-2,2-bis(4-chlorophenyl)ethane] by the white-rot fungus Phanerochaete chrysosporium. *Appl. Environ. Microbiol.*, *53*:2001-2008, 1987.
230. Koehler, A., Jaeger, A., Willershauen, H., and Graf, H. Extracellular ligninase of Phanerochaete chrysosporium Burdsall has no role in the degradation of DDT. *Appl. Microbiol. Biotechnol.*, *29*:618-620, 1988.
231. Hammel, K. E., Tien, M., Kalyanaraman, B., and Kirk, T. K. Mechanism of oxidative $C\alpha$-$C\beta$ cleavage of a lignin model dimer by Phanerochaete chrysosporium ligninase. *J. Biol. Chem.*, *260*:8348-8353, 1985.
232. Leisola, M. S. A., Schmidt, B., Thanei-Wyss, U., and Fiechter, A. Aromatic ring cleavage of veratryl alcohol by Phanerochaete chrysosporium. *FEBS Lett.*, *189*:267-270, 1985.
233. Bumpus, J. A. Biodegradation of polycyclic aromatic hydrocarbons by Phanerochaete chrysosporium. *Appl. Environ. Microbiol.*, *55*:154-158, 1989.
234. Lamar, R. T., Glaser, J. A., and Kirk, T. K. Fate of pentachlorophenol (PCP) in sterile soils inoculated with the white-rot basidiomycete Phanerochaete chrysosporium: Mineralization, volatilization and depletion of PCP. *Soil Biol. Biochem.*, *22*:433-440, 1990.
235. Aust, S. D. Biodegradation of agrochemicals by white-rot fungi. Abst. 5th International Symposium on Microbial Ecology, Kyoto, Japan, S20-2, 1989.

26

VA MYCORRHIZAL FUNGI: CHALLENGES FOR COMMERCIALIZATION

TIM WOOD *NPI, Salt Lake City, Utah*

I. INTRODUCTION

Vesicular-arbuscular (VA) mycorrhizal fungi are beneficial soil microorganisms that form symbiotic associations (mycorrhizae) with the fine roots of plants. The fungi colonize root cortical tissues and extend their hyphae into surrounding soils forming an important linkage between the plant and its below-ground environment (Fig. 1). The fungus obtains photosynthate and other growth factors from its host. In turn it extends and increases the absorbing surface area of the root system allowing the plant to more fully tap soils for nutrients and water. Mycorrhizal plants frequently demonstrate greater uptake of phosphorus and trace metals particularly when those nutrients are sparingly soluble in soils [1]. They can also show improved resistance to drought, environmental stresses, and some root pathogens [2-4], effects that may result from improved mineral nutrition [5]. As a result, mycorrhizal plants frequently exhibit increased growth, yield, and survival over nonmycorrhizal controls [6-7].

VA mycorrhizal fungi have extremely broad host ranges. They colonize the roots of approximately 90% of all plant taxa [8] including most commercially important agronomic and horticultural crops (e.g., corn, cotton, soybean, citrus, grape, geranium, etc.). Given this fact and the above functional attributes, VA mycorrhizal fungi hold considerable potential for use as inoculants to improve plant production in agriculture and horticulture [7,9]. Commercially important benefits that can be derived from their use include increases in plant growth and yield [6,7], improved crop uniformity [10], reductions in phosphorus and trace metal fertilizer requirements [11], reduced losses due to environmental stresses and root disease [2,12-14], reduced stunting on fumigated soils [15-17], improved transplant establishment [18-20], and reduced cropping times [21]. All of these benefits translate to increased profits for farmers and nurserymen through improved production efficiencies and increased crop values.

Several groups including private companies and government agencies have worked to develop commercial VAM inoculants. Superior strains of fungi have been identified for given applications [22,23]. Culture systems for these fungi have been developed, and some have been patented [24-26]. Several inoculant formulations have been developed [7,27,28], and techniques for applying inoculants to nursery and agricultural soils have been

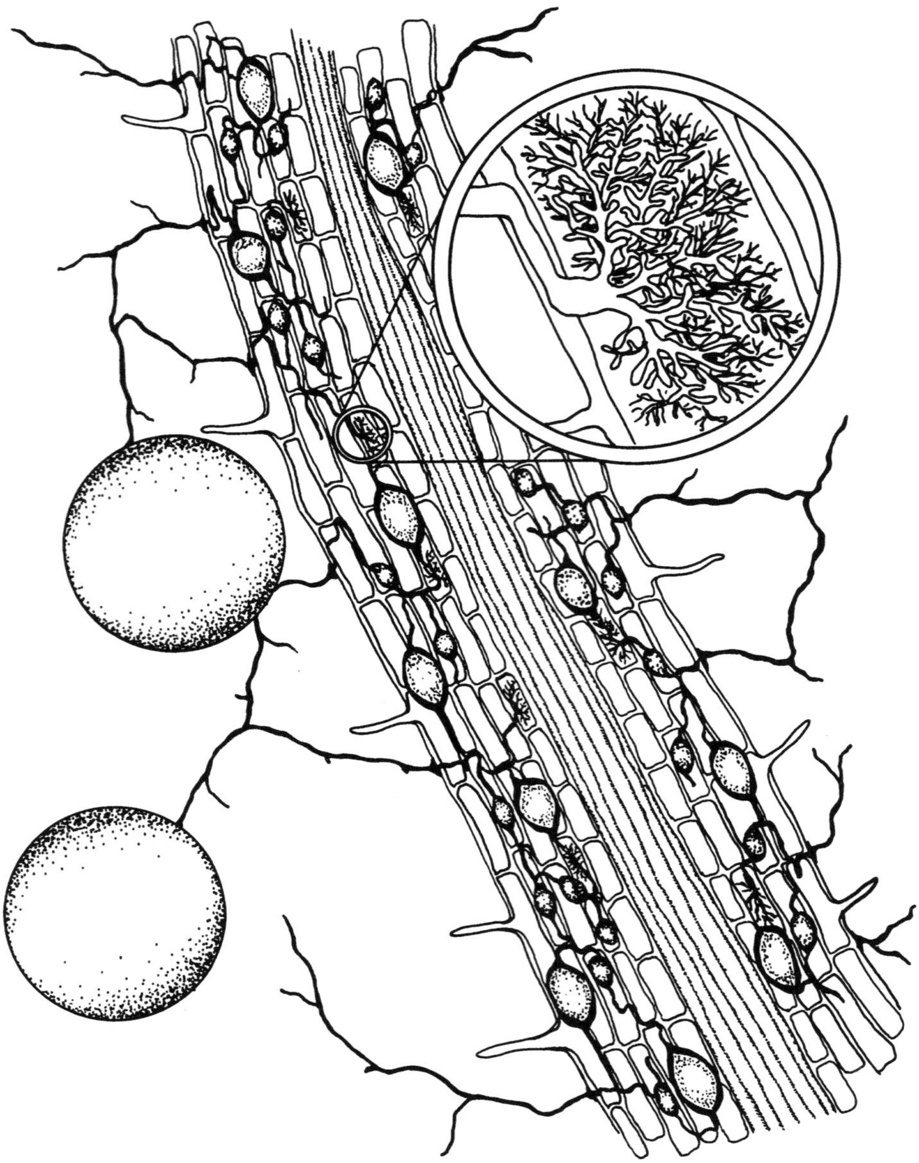

FIGURE 1. Diagram of a typical VA mycorrhizal association. This longitudinal section through a mycorrhizal root shows fungal arbuscules (inset), vesicles, external hyphae, and spores. Arbuscules form within cortical cells and are thought to mediate exchanges between host and fungus. Vesicles are thought to be fungal storage organs.

designed and tested [29,30]. In short, the technologies for producing and applying commercial mycorrhizal inoculants are in place.

Despite these advances, VA mycorrhizal fungi have not been widely commercialized. This chapter discusses limits that the biology of VA mycorrhizal fungi and mycorrhizal associations place on the efficacy of inoculants. It further describes the challenges that these limits pose for commercialization of VAM technologies.

II. MARKETS FOR VA MYCORRHIZAL INOCULANTS

Three principal markets for VA mycorrhizal inoculants exist. The first and potentially the largest segment comprises agronomic crops. Most field crops including corn, wheat, soybean, cotton, peanut and many vegetables are VA mycorrhizal and could potentially benefit from inoculation. In fact, 10-20% yield increases have been realized from VA mycorrhizal inoculation of such crops in the field [31-34]. Agronomic crop production occurs on approximately 100 million ha annually in the United States [35], and if VA mycorrhizal inoculants applied in furrow or as seed treatments were to prove generally effective, markets for inoculum could be substantial.

Horticultural (nursery-grown) crops present additional opportunities for inoculation. Two broad market segments exist: horticultural crops grown on fumigated field soils, and horticultural crops grown in containers of artificial soil mix.

Approximately 20,000-30,000 ha of land are fumigated with methyl bromide and related compounds in the United States each year. The fumigants are used to kill soilborne pathogens and to ensure disease-free production of such high-value crops as strawberries, peppers, onions, bulbs, woody ornamentals, shade trees, fruit trees, and forestry planting stock. VA mycorrhizal fungi, being sensitive to methyl bromide, are also killed in the process [36], and this loss can lead to stunted plant growth, particularly when crops are dependent on mycorrhizae and when the soils contain low levels of available phosphorus and/or trace elements. Reinoculation with appropriate strains of mycorrhizal fungi has been shown to correct stunting in citrus [15-17], apple [37,38], grape [39], peach [40], bell pepper [41,42], shade trees, and forestry seedlings [43,44]. As these benefits are relatively consistent and recognized within the industry, crops grown on fumigated soils offer perhaps the most immediate markets for VA mycorrhizal inoculants.

Horticultural crops grown in containers constitute a large but diverse market for mycorrhizal inoculants. Approximately 3 billion containerized plants are grown annually in the United States [45]. Crops include woody landscaping ornamentals, bedding plants, potted florals, vegetable transplants, forestry seedlings, and reclamation transplants. Literally thousands of plant varieties are produced, and again the majority form mycorrhizae. Cultural conditions employed in production (i.e., artificial soil mix compositions, fertilization regimes, pesticide applications, cropping times, etc.) are equally diverse. As such the industry is fragmented in an operational sense. No two growers produce plants in the same way, and this poses challenges for developing broadly effective inoculants. Nevertheless, benefits of mycorrhizae have been documented for many containerized crops [10,46,47], and this segment offers significant market opportunities for VA mycorrhizal inoculum.

The benefits of inoculating horticultural crops are not limited to growth responses in the nursery. Mycorrhizal planting stock can also show improved establishment when transplanted in the field [18-20]. During the establishment period, transplant root systems are confined to relatively small soil volumes, and mycorrhizal associations are thought to help plants more fully tap those limited volumes for nutrients and water, thus reducing drought and infertility stresses. Evidence supporting this benefit is limited, and additional research is needed to clarify how frequently and under what circumstances it is commercially important. At present the primary applications may be in forestry and reclamation where establishment problems can be severe and where a premium exists on technologies for improving transplant success.

III. CONSTRAINTS TO VAM EFFICACY

Given the breadth of potential applications for VA mycorrhizae in agriculture and horticulture, one might assume that markets for inoculum could be as broad as markets for many of the major agricultural chemicals and that sales could total hundreds of millions of dollars annually in the United States alone. For such widespread and routine use to be adopted, however, mycorrhizal inoculants would have to prove effective over a wide range of commercially important crops grown in diverse soils and under diverse cultural regimes. They would have to work reliably time after time for individual growers. The benefits realized would have to be commercially important (e.g., improvements in yield, plant quality, cropping time), and increased revenues derived from inoculation would have to be large enough to justify the costs of inoculation.

VA mycorrhizal inoculants have yielded consistent and commercially significant results in some applications. For example, inoculation of citrus seedlings grown in fumigated, low-phosphorus soils is generally beneficial [48]. Unfortunately, such applications are limited, and alone they constitute only small, niche markets for inoculum. More broad-scale efficacy with major agronomic and horticultural crops has been difficult to achieve, and major markets for inoculum have not developed.

Six factors act to constrain the potential efficacy of VA mycorrhizal inoculants. These are:

1. Variation between plant genotypes in mycorrhizal dependency
2. Variation in soil nutrient status
3. Competition with indigenous populations of mycorrhizal fungi
4. Phosphorus inhibition of mycorrhizal development
5. Slow growth of mycorrhizal fungi and slow development of mycorrhizal associations, and
6. Incompatibilities with established cultural practices

The remainder of this section examines each limitation in turn, first describing the biological nature of the constraint and then discussing the challenges each poses for commercialization.

A. Variation Between Plant Genotypes in Mycorrhizal Dependency

Mycorrhizal dependency has been defined as the degree to which a given plant genotype requires mycorrhizal associations to attain maximum growth in a given soil at a given level of soil fertility [49]. It has been measured numerically as the ratio of the dry weight of a mycorrhizal plant divided by the dry weight of a nonmycorrhizal plant grown under identical conditions [48]. Such indexes of mycorrhizal dependency typically range from values of 1.0 (no dependency) to 3.0 or more (high dependency).

Unfortunately, there are no comprehensive studies giving definitive comparisons of mycorrhizal dependency for large numbers of agricultural and horticultural crops. The values reported in the literature come from a wide variety of experiments employing diverse methods. Because growth responses to inoculation, and hence mycorrhizal dependencies, vary markedly with soil fertility and other cultural conditions, it is difficult to compare results between studies and derive consistent ratings among the plants that have been tested. Establishment of standard protocols is needed.

Nevertheless, results in the literature as a whole allow qualitative ranking of many economically important crops. On a broad scale, plants such as citrus, grape, cotton, and leek frequently respond to mycorrhizal inoculation with growth increases of 100% or more over noninoculated controls. As a result they appear to have high mycorrhizal dependencies. Corn, wheat, alfalfa, and clover typically show intermediate responses to inoculation and more moderate mycorrhizal dependencies, while other plants, like tomato, typically show insignificant growth responses to inoculation and low mycorrhizal dependencies. Still other crops, most notably those in the Crucifereae and Chenopodiaceae, do not readily form mycorrhizae of any sort and have no mycorrhizal dependency. Such crops include radish, rape, spinach, sugar beet, cabbage, and broccoli.

On a finer scale, it is also known that varieties within given crop species differ in the degree to which they form and depend on VA mycorrhizae. Heckman and Angle [50] noted significant differences between 15 soybean cultivars in the rates at which they formed VA mycorrhizae. Similar genotypic differences have been found for pearl millet [51], wheat [52], and corn [53]. Varietal differences in mycorrhizal dependencies (growth responses to inoculation) have been documented for alfalfa [54], citrus [55,56], and wheat [52].

Factors underlying these differences are not fully understood. Variation in rates of mycorrhizal formation may be related to the permeability of root membranes and the supply of sugar and amino acid exudates available for fungal growth and development [52,57,58]. Differences in mycorrhizal dependency are thought to be related to root morphology. Baylis [59] suggested that plants such as tomato and ryegrass with finely divided, fibrous root systems and with many long root hairs do not require mycorrhizal fungi to a significant degree. Their extensive root systems fully tap soils for nutrients and water without the aid of mycorrhizal fungi. At the other extreme, plants like citrus, peanut, and leek with thick, fleshy roots and few root hairs are more dependent. Genotypic variation in root fineness has been noted between varieties of several crop species including corn [60] and wheat [61], and it has been correlated with differences in phosphorus uptake kinetics. Such variation may in part underlie intraspecific variation in mycorrhizal dependency.

Variation in mycorrhizal dependency limits the potentially broad efficacy of inoculants. Mycorrhizal fungi will not prove effective with all commercial crops because some crops have little or no mycorrhizal dependency in typical agricultural and horticultural soils. Identifying those commercial plants with high mycorrhizal dependencies and thus identifying priority markets for inoculum remains an important challenge for commercialization.

B. Variation in Soil Nutrient Status

VA mycorrhizal fungi can dramatically improve plant uptake of phosphorus and trace elements when those nutrients are present but sparingly soluble in soils. Such conditions frequently occur at extremes in soil pH. Under acid conditions (below pH 5.5) phosphorus can be stripped from soil solutions and chemically fixed by a variety of iron and aluminum compounds [62]. Under basic conditions (above pH 7.0), phosphorus can be chemically bound by calcium minerals [62], and trace elements such as zinc, copper, manganese, and iron become insoluble [63]. In such cases, equilibrium concentrations of these nutrients in soil solutions are reduced to very low levels, and their diffusion through soils to plant roots is limited. As such, plants absorb phosphorus and trace metals from only the narrow cylinder of soil surrounding their roots [64,65]. Mycorrhizal fungi are thought to improve nutrient uptake in these instances by extending the absorptive surface area of the root system and allowing plants to tap larger soil volumes for nutrients. Some have speculated that VA mycorrhizae may improve nutrient uptake by a second mechanism, enhanced solubilization of phosphorus and trace elements in the rhizosphere. Most evidence suggests, however, that mycorrhizal and nonmycorrhizal roots absorb P from the same labile pool in the soil and that solubilization effects, if present, are minor [66].

VA mycorrhizae can improve plant uptake of phosphorus from nonphosphorus-fixing soils (e.g., from very sandy soils or from artificial peat-based soil mixes), but significant effects are generally found only when solution phosphorus concentrations are very low [67]. Mycorrhizae tend to have little impact on plant growth in soils where phosphorus and trace elements are abundant and soluble [1]. Under such circumstances plant roots themselves absorb sufficient quantities of nutrients without the assistance of fungal symbionts.

These generalizations suggest the importance of developing models that use measurements of soil fertility to predict plant growth responses to inoculation. Such models can help in targeting inoculum usage to soils in which growth responses are most likely to occur. Three approaches to developing predictors can be taken.

The first and most rigorous involves developing specific models for individual crop cultivars with given mycorrhizal dependencies. Menge et al. [48] developed such a model for predicting growth responses of Troyer citrange to mycorrhizal inoculation in fumigated field soils and artificial horticultural mixes. They found that there was a high probability of achieving significant growth responses to inoculation when soils contained less than 34 ppm bicarbonate-extractable phosphorus, less than 27 ppm DTPA-extractable Mn, less than 12 ppm DTPA-extractable Zn, or less than 3% organic matter. In related work, Ojala et al. [68] found that phosphorus availability indexes differed in their predictive power and that those based on water extractions (i.e., saturation extract P, 1:10 soil-water P, and anion exchange resin P) provided better predictors than those based on stronger chemical extractants (i.e., sodium bicarbonate-extractable P and ammonium fluoride-extractable P). This refinement and others allowed Menge and coworkers [48] to develop multiple-regression

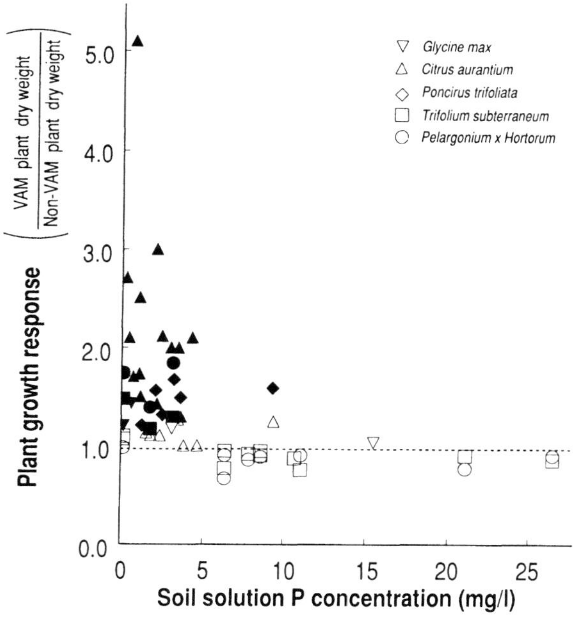

FIGURE 2. Composite results from studies in the literature show that VAM inoculation of container-grown plants frequently improves plant growth when soil solution phosphorus concentrations are maintained below 5 mg P/liter. Closed symbols indicate statistically significant increases in shoot growth over noninoculated controls.

equations for predicting the magnitude of growth responses for Troyer citrange with accuracy. Similar models could be developed individually for many crops, but the work involved would be enormous.

Alternatively, measures of fertility can be used to develop general rules of thumb for predicting whether or not growth responses to mycorrhizal inoculation in a given soil are likely for plants in general. This represents a less rigorous and far simpler approach to the problem.

As an example, Figure 2 plots composite data from studies in the literature in which both plant growth responses to inoculation and levels of soil solution phosphorus were measured [69-71]. All were greenhouse pot experiments, although host genotypes, VA mycorrhizal fungal strains, soil types, and cultural conditions varied markedly between studies. Nevertheless, a generally useful pattern emerges. Significant growth responses to VA mycorrhizal inoculation (closed symbols) typically occur when soil solution phosphorus concentrations are below 5 mg P/liter. Above this level, responses to inoculation are largely insignificant (open symbols).

Wood and Jedrzejek (unpublished observations) developed a similar predictor for growth responses on fumigated field soils. They analyzed methyl bromide-fumigated soils from a variety of commercial nurseries on which plants were responsive vs. nonresponsive to VA mycorrhizal inoculation. Soils ranged from acid clay loams to basic sandy loams. Crops included Brazilian sour orange, peach rootstock, Russian olive, giant sequoia, *Pyrus betulaefolia*, grape, and others. Of 11 soil characteristics measured,

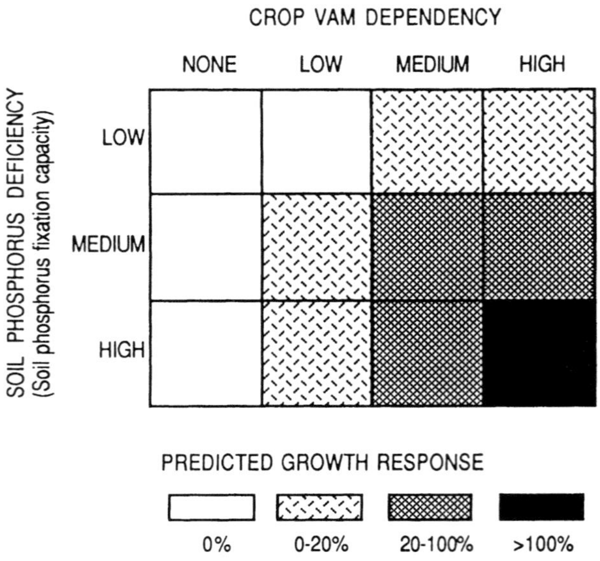

FIGURE 3. Qualitative model for predicting plant growth and yield responses to VAM inoculation on fumigated field soils. Estimates, expressed as percent increases in growth or yield over noninoculated controls, vary with host VAM dependency and degree of soil phosphorus deficiency. They are based on general trends in the literature.

two showed consistent and significant differences between responsive and nonresponsive soils. On the average, nonresponsive soils contained about 10 times more 1:10 soil-water-extractable phosphorus than did responsive soils (7.8 vs. 0.9 μg P/g). Nonresponsive soils also showed on the average only one-tenth of the phosphorus fixation capacity that responsive soils did (37 vs. 370 μg P/g). The data when taken as a whole suggest that for fumigated field soils, growth responses to mycorrhizal inoculation are likely for plants in general when levels of 1:10 soil-water-extractable phosphorus are below 5 μg/g soil and/or when significant phosphorus fixation capacities exist.

The rule-of-thumb models discussed here carry advantages associated with simplicity and disadvantages associated with poor definition and accuracy. Most importantly, they ignore differences in host mycorrhizal dependency. Figure 3 illustrates a third approach to modeling that addresses this problem. It presents a qualitative model for predicting plant response to mycorrhizal inoculation on fumigated soils as a function of host mycorrhizal dependency and soil nutrient status. It shows approximate growth responses (% growth increase over noninoculated controls) for crops with no, low, medium, and high mycorrhizal dependencies grown on soils with low, medium, and high phosphorus deficiencies (or phosphorus fixation capacities). Again, there are no standard and established techniques for quantifying the absolute VA mycorrhizal dependency of given crop genotypes, and one must rely on trends in the literature for assigning crops to given dependency ratings (see previous sections). Phosphorus deficiency categories can be quantified based on results such as those of Ojala et al.

[68] and Wood and Jedrzejek (unpublished observations). Experience suggests that high deficiency coincides with levels of 0-2 μg P/g soil, medium deficiency with levels of 2-5 μg P/g, and low deficiency with levels above 5 μg P/g (all 1:10 soil-water-extractable phosphorus). Plant growth responses to inoculation ranging from 0 to 100% and more were assigned to various combinations of mycorrhizal dependency and soil phosphorus deficiency based on collective results in the literature. These assignments represent best-guess estimates of growth responses that might be expected. This model again is a generalization, but it serves as a partially refined tool for prioritizing research and targeting markets for inoculum. A similar model will be presented in the next section for nonfumigated field soils.

The three modeling approaches illustrated above differ in rigor and predictive power, but they generate a common conclusion. Marked growth responses to mycorrhizal colonization, in general, occur at low and fairly restricted levels of available phosphorus. For fumigated field soils, levels of 0-5 μg P/g soil (1:10 soil-water-extractable phosphorus) are conducive to growth responses. Above this range, mycorrhizal inoculation is less likely to improve plant growth. It is difficult to know what percentage of the total acreage of fumigated crop production falls in this fertility category, but it may be small. Revenues from sales of high-value horticultural crops justify the use of fertile land and sufficient levels of fertilizers. As such, growth responses to inoculation may be generally restricted to production of crops with relatively high mycorrhizal dependencies grown on phosphorus and trace element-fixing soils (where fertilization is inefficient).

Similarly, mycorrhizal inoculation of container crops is most likely to provide significant increases in plant growth when levels of soil solution phosphorus fall below 5 mg P/liter. These levels are below the 20-100 mg P/liter that are typically maintained by recommended applications of slow release and liquid feed fertilizers [72]. As a result, growth responses to inoculation again may be generally restricted to production of crops that have fairly high mycorrhizal dependencies and/or when soil mixes demonstrate significant phosphorus fixation capacities.

This does not mean that use of mycorrhizal inoculation will be restricted to such situations, as the advantages associated with improved transplant success should be realized for inoculated crops regardless of whether a growth response in the nursery is realized. However, transplanting benefits will require establishing good mycorrhizal colonization in the nursery. This factor will be discussed in a later section.

C. Competition with Native Populations of Mycorrhizal Fungi

Mycorrhizal inoculation of crops grown on nonfumigated field soils is subject to an additional constraint to efficacy: competition from indigenous mycorrhizal fungi. Published estimates of native VA mycorrhizal fungal spore populations in agricultural soils range widely from 0.05 spores/g soil to 100 spores/g soil [73]. Much of this variation may arise from differences in spore extraction techniques [74], and the higher end of the published range (10-100 spores/g soil) may best reflect true densities. Percent viability of native spores has not been assessed, but live spores together with mycorrhizal roots and root fragments in the soil constitute natural inoculum against which commercial inocula must compete.

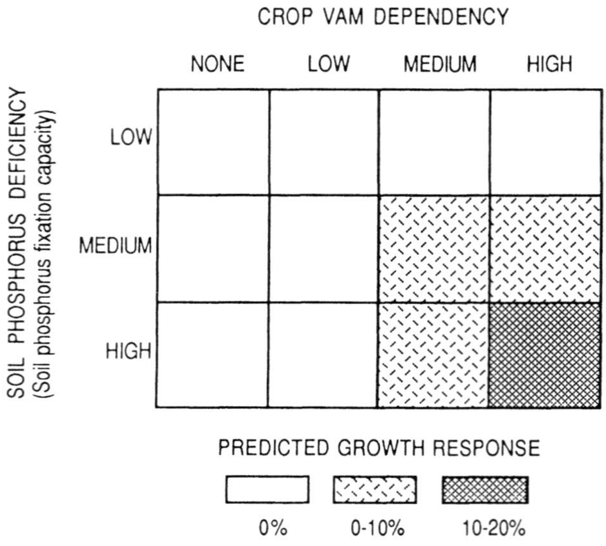

FIGURE 4. Qualitative model for predicting plant growth and yield responses to VAM inoculation on nonfumigated field soils. Estimates, expressed as percent increases in growth or yield over noninoculated controls, vary with host VAM dependency and degree of soil phosphorus deficiency. They are based on general trends in the literature.

Native inocula reduce the efficacy of commercial inocula in two ways. First, if soils are conducive, native VAM mycorrhizal fungi form effective mycorrhizas and produce a background of improved plant growth that can obscure effects of applied fungi. Second, native VAM mycorrhizal fungi will consume root exudates and other required growth factors and compete with applied strains for colonization of the root. This in turn slows the spread of inoculant organisms and reduces their efficacy.

Figure 4 presents a model for predicting mycorrhizal growth responses of agronomic crops grown in nonfumigated field soils. It is analogous to Fig. 3 and was developed in the same way from literature on mycorrhizal inoculation of crops grown on intact field soils. It is provided here to make one point. Competition with native mycorrhizal fungi in nonfumigated soils generally reduces the expected magnitude of growth responses to inoculation for any given combination of host mycorrhizal dependency and soil fertility status. In turn, the likelihood of finding a significant increase in growth under a given set of conditions is also reduced. These estimates suggest that significant (10-20%) growth and yield responses to inoculation are likely to occur only when crop mycorrhizal dependencies are medium to high and when soil phosphorus deficiencies are medium to high. We do not know what percentage of U.S. agricultural crop production falls within these conditions, and hence on what acreage inoculation is likely to prove effective. Extensive field testing is required to address this question and to test the model in general.

Populations of native mycorrhizal fungi are known to fluctuate in agricultural soils with soil type, fertilization regime, cropping history, pesticide treatments, tillage, and season [73,75-77]. The effects that such fluctuations might have on the efficacy of applied inoculants are unknown at this time and need to be studied.

D. Phosphorus Inhibition of Mycorrhizal Development

High levels of available phosphorus and high-phosphorus fertilization regimes inhibit mycorrhizal development. VA mycorrhizal fungi can colonize the roots of phosphorus-deficient plants relatively rapidly. They can also infect roots of plants that are phosphorus-sufficient, but spread of colonization through the root cortex is inhibited [78]. Current understanding of the biochemical and molecular mechanisms that underlay this inhibition is limited. Ratnayake et al. and Graham et al. [57,58] proposed that under conditions of phosphorus deficiency, plasmalemmas of cortical cells are permeable and leaky, and the supply of exudates required by VA mycorrhizal fungi for growth and development is high. In contrast, P sufficiency reduces membrane permeability, and thus reduces exudation and suppresses VA mycorrhizal development.

Plant defenses against fungal invasion may also limit VA mycorrhizal development. Limited evidence suggests that when VA mycorrhizal fungi colonize cortical tissues they trigger host defense mechanisms (e.g., production of phytoalexins and phenolic compounds), which may in turn slow further mycorrhizal development [79,80]. These mechanisms are also thought to be involved in part in the cross-protection against some root diseases that results from mycorrhizal colonization. The relationship between host defense mechanisms and host phosphorus nutrition has not been defined.

Levels of soil solution phosphorus at which inhibition of mycorrhizal development occurs will vary between plant genotypes and cultural conditions. However, Fig. 5 indicates a rule of thumb for plants in general. It represents a composite of results from greenhouse pot studies in which both levels of VA mycorrhizal colonization and levels of soil solution phosphorus were measured [69-70,81]. Host genotypes, VA mycorrhizal fungal strains, soil types, cropping periods, and cultural conditions varied markedly between studies, but collectively the results provide a useful conclusion. In greenhouse culture of containerized plants, levels of soil solution phosphorus should be maintained at less than 10 mg/liter if relatively high levels and rapid rates of VA mycorrhizal development are to be achieved.

Most fertilizers used in production of containerized crops contain high levels of soluble phosphorus. Typical elemental N/P ratios fall in the range of 7:1 to 2:1, and when used at recommended rates they supply phosphorus at concentrations of 20-100 mg P/liter [72]. These concentrations are far in excess of those conducive to vigorous mycorrhizal colonization (Fig. 5). As a result, high levels of VA mycorrhizal development are often difficult to achieve on a routine basis in greenhouse production of containerized plants.

This problem can be solved in part by reducing fertilization rates, but in most instances this will reduce plant growth. Figure 6 illustrates the challenge. It shows shoot growth and percent VA mycorrhizal colonization for miniature rose variety Royal Ruby as functions of increasing levels of a preplant, slow-release fertilizer with a 4.3:1.0:2.7 elemental N/P/K ratio. In this example, plant growth increases with increasing fertilizer applications, but VA mycorrhizal colonization peaks early and then declines. A portion of the reduction in percent colonization is due to a dilution effect.

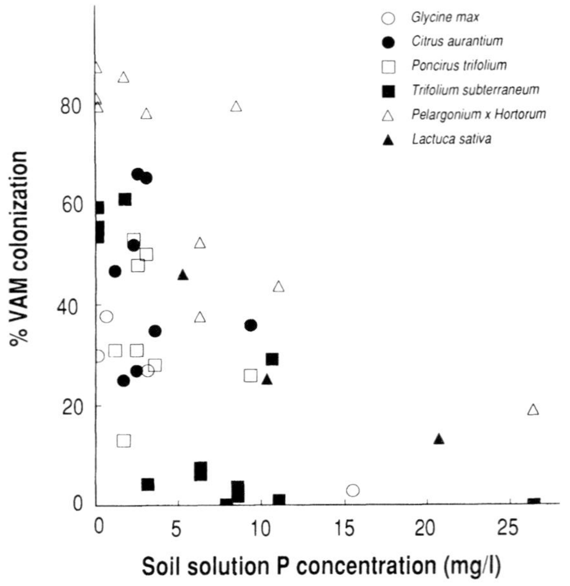

FIGURE 5. Composite results from studies in the literature suggest that high levels of VAM development can be achieved on roots of container-grown plants when levels of soil solution phosphorus are maintained below 10 mg P/liter. Above this level, phosphorus inhibition of VAM development appears marked.

Increases in P fertilization promote faster root growth, and roots can actually outgrow VA mycorrhizal fungal spread. The remainder of the decline results from phosphorus suppression of mycorrhizal development. Importantly, however, in Fig. 6 VA mycorrhizal development is low at levels of acceptable plant growth.

Reducing nitrogen-to-phosphorus ratios to supply sufficient nitrogen for plant growth and low levels of phosphorus conducive to full mycorrhizal development may solve this problem in part. Figure 7 shows shoot growth and percent mycorrhizal colonization of *Trifolium subterraneum* as functions of the phosphorus concentration of liquid feed fertilizers applied at 10-day intervals. In this instance, fertilizers with different elemental N/P/K ratios were applied, but all were diluted to supply 500 mg N/liter. Results show that a 9:1:11 formulation gave both good shoot growth and VA mycorrhizal development. More work is needed to confirm this result with a wide variety of horticultural crops, but the approach looks promising.

Graham and Timmer [71] showed that using relatively insoluble forms of phosphorus, such as rock phosphate, as substitutes for soluble phosphorus can yield good plant growth and good mycorrhizal colonization.

E. Slow Growth of Mycorrhizal Fungi and Slow Development of Mycorrhizal Associations

VA mycorrhizal fungi are relatively slow growing. Culture periods, measured as the time required for mature spores to germinate, colonize a plant, and produce new mature spores, are typically measured in months. New spores can begin forming at 3-4 weeks following inoculation, but large numbers of mature spores are formed only after 3-4 months. This factor alone places constraints on commercial inoculum production and makes VA mycorrhizal inoculants relatively expensive.

Slow development of VA mycorrhizal associations in plant roots places additional constraints on efficacy. Figure 8 shows rates of VA mycorrhizal colonization of cotton, onion, and tomato roots under conditions of moderate phosphorus supply. Here *Glomus intraradices*, a fairly aggressive colonizer, requires 4-8 weeks to colonize more than 50% of the root systems of cotton and onion, both of which are VA mycorrhizal-dependent hosts. Colonization

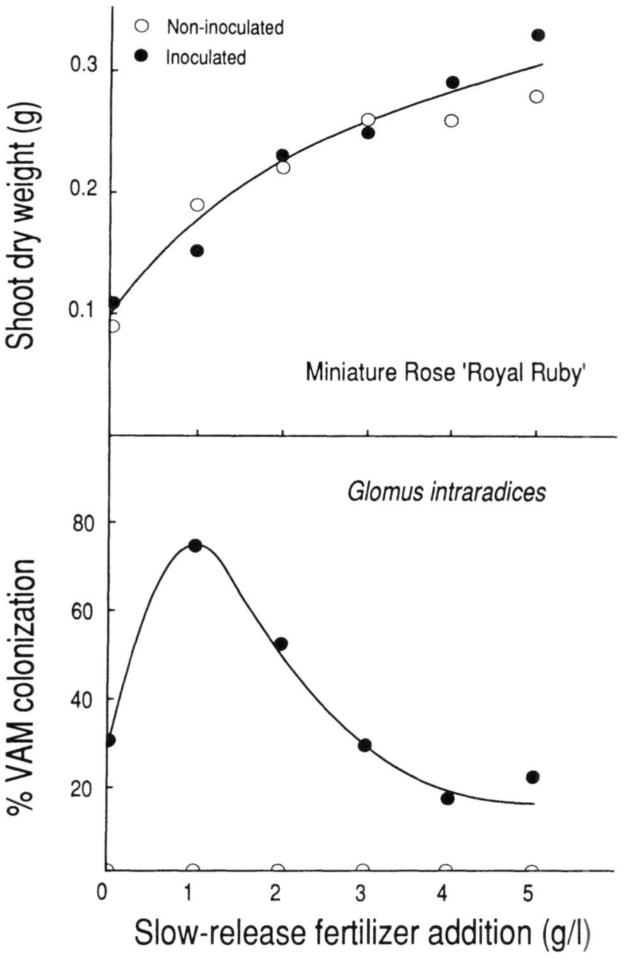

FIGURE 6. Shoot growth for the miniature rose "Royal Ruby" (upper panel) increased with increasing applications of a preplant, slow-release fertilizer with a 4.3:1.0:2.7 elemental N/P/K ratio. In contrast, percent VA mycorrhizal colonization peaked at low levels of fertilization and then declined (lower panel).

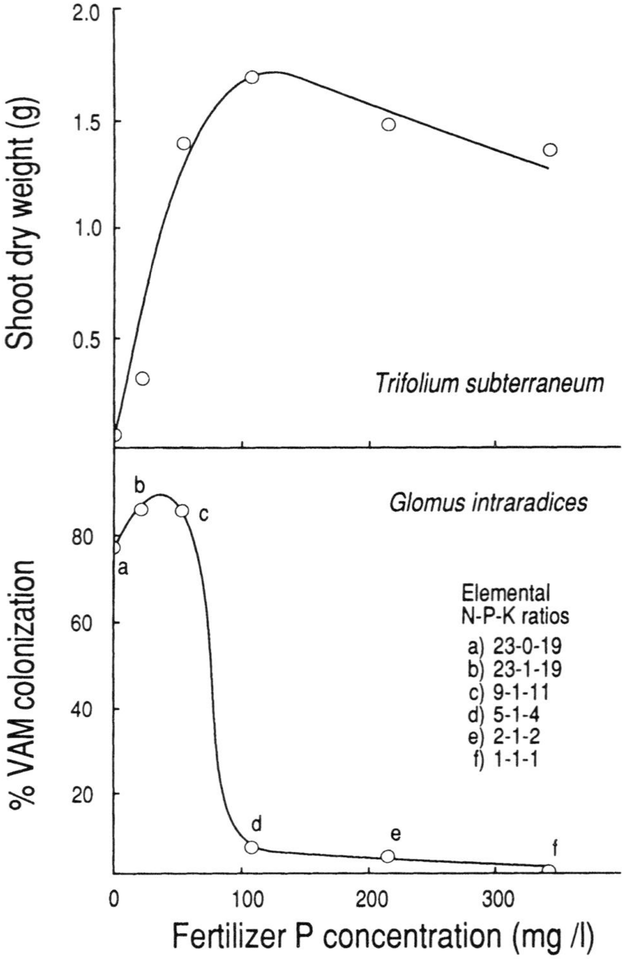

FIGURE 7. High levels of shoot growth (upper panel) and VA mycorrhizal colonization (lower panel) were found on *Trifolium subterraneum* seedlings fertilized with a complete but relatively low-phosphorus nutrient solution with a 9:1:11 elemental N/P/K ratio.

of tomato, a host with low VA dependency, occurs even more slowly. Such rates can be increased with increasing inoculum dosage [82], but high dosages bring higher inoculation costs.

Slow VA mycorrhizal colonization limits potential efficacy in several ways. Annual agronomic crops may benefit most from mycorrhizae early in development when seed reserves of nutrients are depleted but when root systems are not fully developed. At this time mycorrhizal associations could theoretically play their most significant role in increasing nutrient uptake and promoting growth. If VA mycorrhizal development is slow, however, this opportunity may be lost.

Similar opportunities for suppression of root diseases may be compromised by slow VA mycorrhizal development. It is believed that VA mycorrhizal

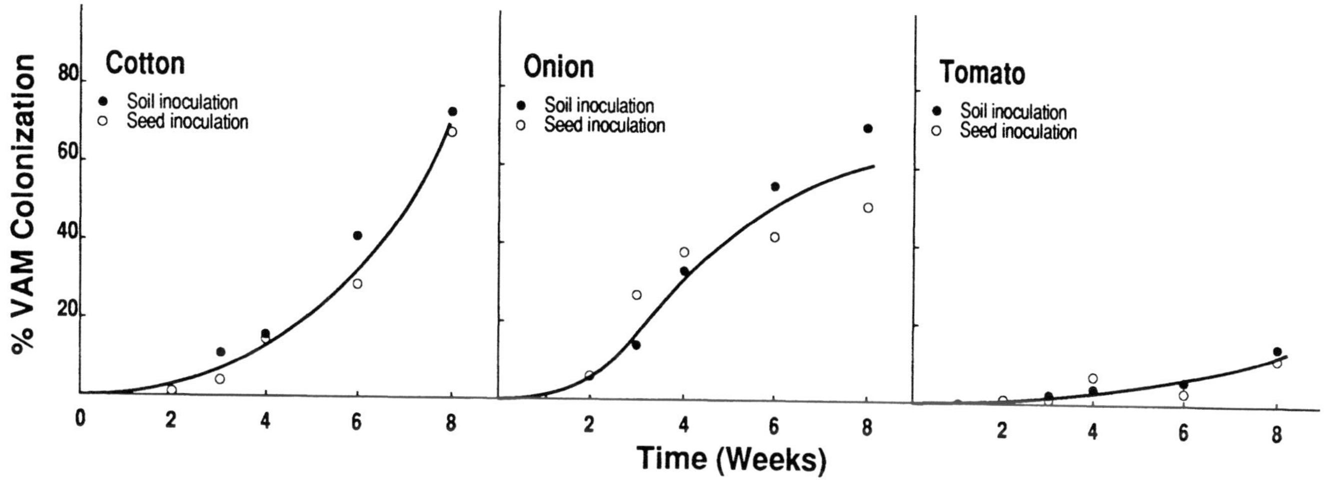

FIGURE 8. Time course for VA mycorrhizal colonization of cotton, onion, and tomato plants grown in containers under conditions of moderate phosphorus supply.

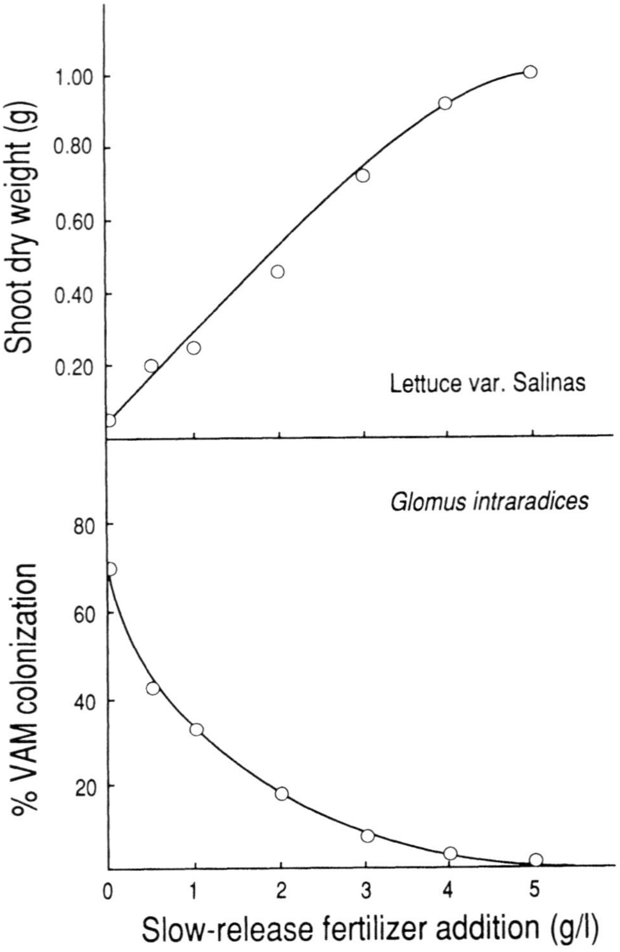

FIGURE 9. Low levels of VA mycorrhizal colonization were found on greenhouse-grown lettuce transplants supplied with the amounts of phosphorus fertilizer required to give rapid shoot growth.

fungi can reduce development of some root diseases through competition with pathogens for root exudates and/or by cross-protection. It is also believed, however, that for control to be realized, VA mycorrhizal associations must be established on the root system to a significant degree prior to challenge by the pathogen [83,84]. Again, slow VA mycorrhizal development may preclude significant protection, particularly of seedling diseases.

Given these problems, short cropping times pose a challenge for using VA mycorrhizal inoculants with containerized crops such as bedding plants and vegetable transplants, industry segments in which a premium is placed on high turnover to maximize the number of crops that can be produced per year. Such seedlings are typically grown in the nursery for 4-6 weeks before being sold and transplanted. Because VA mycorrhizal associations are slow to develop, it is difficult to establish high levels of colonization on plants within this time, particularly when high levels of phosphorus fertilizers, which suppress VA mycorrhizal development, are used.

VA mycorrhizal inoculation of lettuce offers a case in point. Lettuce production in California and Florida is hampered by corky root disease caused by a gram-negative bacterium [85]. Preliminary evidence indicates that VA mycorrhizal colonization of lettuce roots may suppress disease development. This in turn suggests that inoculation of transplants prior to installation in the field may offer some disease control. Lettuce transplants, however, are grown in 4-week rotations, and establishing the high levels of colonization necessary to promote disease protection remains problematic. Figure 9 shows low levels of VA mycorrhizal colonization on lettuce seedlings at levels of fertilization necessary to give rapid seedling growth.

Promoting high levels of mycorrhizal development on young, rapidly growing plants remains an important challenge in commercial use of mycorrhizal inoculants. Potential solutions to this problem will be discussed in a following section.

F. Incompatibilities with Established Cultural Practices

Several horticultural and agronomic practices are incompatible with mycorrhizal inoculation. The effects of high-phosphorus fertilization regimes on mycorrhizal development and efficacy have already been discussed. Two additional incompatibilities deserve mention.

Tree bark is frequently used as a component of artificial soil mixes employed in containerized plant production. A wide variety of materials differing in source and degree of composting are used. Many composted tree barks have been shown to have fungicidal and fungistatic properties capable of suppressing such plant pathogens as *Phythophthora, Pythium, Rhizoctonia, Fusarium,* and *Thielaviopsis* [86]. Two mechanisms are thought to be involved. Bark composts support high levels of antagonistic microbes and phagous organisms that suppress fungal populations. Also some barks appear to contain suppressive compounds of their own. Hardwood bark composts seem to have the strongest fungistatic properties. Pine barks appear to be less active.

Mycorrhizal associations are sometimes difficult to establish in soil mixes containing large quantities of bark. Many failures are related to the excessive acidity of poorly composted barks and can be corrected by pH adjustment. Other problems may be associated with fungicidal and fungistatic properties of the composts. Unfortunately, nurserymen frequently have little control over the quality and degree of composting of bark materials they receive from suppliers, and inconsistencies here can lead to inconsistencies in plant quality and mycorrhizal performance. This important problem has received little attention and deserves study.

Chemical fungicides also threaten the efficacy of VA mycorrhizal inoculants. This topic has been extensively reviewed by Menge [36]. It will suffice to say here that major problems in the use of mycorrhizal inoculants have been encountered when fungicides used to control pathogens have inadvertently killed mycorrhizal inoculants and suppressed VA mycorrhizal development. Not all fungicides harm VA mycorrhizae. Those specific for the pythiaceous fungi (e.g., Metalaxyl and Aliette) are compatible with inoculation. Other compounds (e.g., PCNB and Benlate) are toxic to VA mycorrhizal fungi, and their use must be avoided when mycorrhizal development is a priority.

IV. CHALLENGES FOR COMMERCIALIZATION

The constraints to efficacy discussed in the previous section pose challenges for research and development. Four research priorities are discussed here. Three focus on technological development: identification of superior fungal strains, development of improved culture systems for VA mycorrhizal fungi, and definition of integrated crop cultural systems that are compatible with mycorrhizal inoculation. The fourth area involves the need for extensive, commercially relevant field testing of inoculants.

A. Identification of Superior Fungal Strains

As discussed in the previous section, mycorrhizal fungi fail to show broad commercial efficacy, in part because (a) mycorrhizal development is a slow process; (b) high levels of phosphorus fertilization both interfere with mycorrhizal development and mask potential benefits; and (c) other established cultural practices are incompatible with inoculation. In theory, these problems could be reduced by selecting or developing strains of fungi that were rapid and aggressive root colonizers and that showed tolerances to pesticides, bark composts, and/or high levels of phosphorus fertilizers.

Selection of mycorrhizal fungi for rapid colonization involves several facets. Rapid and synchronous spore germination, absence of spore dormancy, and vigorous germ tube growth are important characteristics for assuring high inoculum potential and early root infection. As such these traits should be considered in the selection process.

Selection of mycorrhizal fungi for rapid spread within the root system is also important. This problem is related to selection for tolerance to high levels of P fertilization, as host phosphorus status is a major factor governing colonization rate. Strains of VA mycorrhizal fungi differ in their sensitivity and tolerance to applied P. Jasper et al. [87] found that isolates derived from unfertilized virgin soils of low-phosphorus status were more sensitive to increasing P supply than fungi derived from adjacent fertilized agricultural soils. Sylvia and Schenck [88] compared six strains of VA micorrhizal fungi for their ability to colonize bahiagrass in a high-P soil containing 199 ppm double-acid-extractable P. Three isolates were capable of forming mycorrhizae, and the others were not. As such, selection for P-tolerant, and hence rapidly colonizing, strains appears feasible, and isolation from phosphorus-rich agricultural soils offers a promising strategy.

Importantly, there may be limits to selecting VA mycorrhizal fungi for aggressive colonization. Slow development and phosphorus sensitivity may be fundamentally important characteristics of VA mycorrhizal fungi that work to maintain a beneficial symbiotic balance in mycorrhizae. In other words, these qualities may be important in distinguishing VA mycorrhizal fungi from root pathogens. If this is true, it may prove difficult to find markedly aggressive strains of VA mycorrhizal fungi that prove generally beneficial in commercial use.

The feasibility of selecting mycorrhizal fungi for tolerance to bark composts is uncertain. The mechanisms of suppression are not fully understood, and it is difficult to know what specific traits need to be selected for. Nevertheless, the problem can be approached empirically by screening large numbers of strains for efficacy in bark mixes known to be suppressive to pathogens.

Development of VA mycorrhizal fungi that are tolerant to given fungicides represents an important challenge to commercialization. Isolation of strains from soils with prolonged histories of fungicide application offers one avenue for research. Genetic engineering presents a second, but progress here will be difficult as VA mycorrizal fungi have not yet been grown in pure culture and their genetics is poorly understood.

B. Development of Improved Culture Systems for VAM Fungi

VA mycorrhizal fungi are obligate symbionts and are grown in association with living root tissues. Several culture techniques based on this constraint are applicable for the commercial scale production of inoculum. The most widely used is pot culture. Here, VA mycorrhizal fungi are produced in association with the roots of container-grown plants in a greenhouse [24]. Related approaches include aeroponic culture [89], nutrient film techniques (B. Mosse, personal communication), and axenic root organ culture techniques [26]. All are two-member (plant-plus-fungus) systems. All are technically feasible and hold commercial potential. But importantly, all involve extended culture periods of several months, making VA mycorrhizal inoculants relatively expensive to produce. As such development of rapid and more efficient culture systems remains an important challenge for commercialization. Pure (single-member) culture systems for VA mycorrhizal fungi have not been reported. Success here would represent a significant scientific breakthrough and might yield more rapid fungal growth and development.

C. Definition of Integrated Horticultural and Agricultura Systems Compatible with VA Mycorrhizal Inoculation

Fertilizers, fungicides, incompatible soil mixes, and short cropping times pose individual constraints to VA mycorrhizal efficacy. Any one of these factors alone can undermine an inoculation program, and solving one problem at a time while ignoring others invites failure. If VA mycorrhizal inoculants are to be widely adopted in agriculture and horticulture, integrated cultural systems compatible with VA mycorrhizal inoculation must be developed and marketed. Such systems must cooptimize fertilization, fungicide use, soil mix compositions, VA mycorrhizal fungal strain characteristics, inoculum dosage rates, and plant cropping schedules to provide excellent plant growth and high levels of mycorrhizal colonization.

Progress toward this goal is being made through development of specialty soil mixes formulated for compatibility with VA mycorrhizal inoculation and marketed as such. Similar work is underway toward development and marketing of VA mycorrhizal-compatible fertilizer formulations. Fungicides could be clearly labeled as being compatible or incompatible with mycorrhizal inoculation, and alternative pest control measures could be encouraged. In short, many technical elements of the solution are in place or under development.

D. Commercially Relevant Field Testing of Inoculants

The scientific literature is replete with examples of plant growth, yield, and survival being dramatically improved through VA mycorrhizal inoculation. However, when many of these studies are examined in detail, their results have little significance to commercial agriculture and horticulture. Greenhouse pot studies showing growth responses of soybean, wheat, and corn to VA mycorrhizal inoculation have little relevance to what might happen in field agriculture. Similarly, studies conducted at low levels of fertilization can yield significant growth responses, but results will not be commercially relevant unless the growth of inoculated plants is acceptable to industry.

The need for commercially relevant field and greenhouse testing of VA mycorrhizal fungi remains an important challenge for inoculum development. In short, more field data are needed before the commercial potential of inoculants can be fully assessed. Whenever possible, studies should be conducted in conjunction with commercial growers and preferably in commercial facilities or fields. Commercially relevant fertilization regimes, fungicide applications, soil mixes, cropping times, inoculum dosages, and inoculation techniques should be employed. And plant quality at the conclusion of the experiment must be evaluated in terms of commercial standards. It is only through such research that the true commercial potential of VA mycorrhizal inoculants can be evaluated.

V. CONCLUSIONS

Commercial use of VA mycorrhizal inoculants is currently associated with high-value, VA mycorrhizal-dependent crops grown on fumigated, low-phosphorus soils. It is here that the benefits of inoculation are most dramatic, and it is here that efficacy has been most routinely demonstrated. Prospects for expanding these markets to include larger segments of agriculture and horticulture are good, but the bounds of that expansion remain uncertain.

The efficacy of VA mycorrhizal inoculants is limited in part by aspects of VA mycorrhizal biology. VA mycorrhizal fungi, as noted, are slow growing, and VA mycorrhizal associations are both slow to develop and sensitive to host phosphorus sufficiency. Selection for aggressive VA mycorrhizal fungal strains may reduce these constraints.

The efficacy of VA mycorrhizal inoculants is also limited in part by the background of chemical fertilizers and pesticides currently used in agriculture and horticulture. Phosphorus fertilizers and fungicides both inhibit mycorrhizal development and substitute for benefits that VA mycorrhizal inocula can provide. Current concerns for the threats that agricultural chemicals pose for environmental quality and human health presage efforts to reduce chemical inputs and identify alternatives. Such efforts may open opportunities for VA mycorrhizal inocula, as mycorrhizal fungi offer an environmentally sound, biological alternative to chemistry for maintaining plant quality and productivity in agriculture and horticulture.

REFERENCES

1. Abbott, L. K., and Robson, A. D. The effect of mycorrhizae on plant growth. In *VA Mycorrhiza* (C. L. Powell and D. J. Bagyaraj, eds.), CRC Press, Boca Raton, 1984, pp. 113-130.
2. Nelson, C. E., and Safir, G. R. Increased drought tolerance of mycorrhizal onion plants caused by improved phosphorus nutrition. *Planta, 154*:407-413, 1982.

3. Allen, M. F., and Boosalls, M. G. Effects of two species of VA mycorrhizal fungi on drought tolerance of winter wheat. *New Phytol.*, *93*:67-76, 1983.
4. Bagyaraj, D. J. Biological interactions with VA mycorrhizal fungi. In *VA Mycorrhiza* (C. L. Powell and D. J. Bagyaraj, eds.), CRC Press, Boca Raton, 1984, pp. 131-154.
5. Graham, J. H. Citrus mycorrhizae: potential benefits and interactions with pathogens. *Hort. Sci.*, *21*:1302-1306, 1986.
6. Menge, J. A. Mycorrhiza agriculture technologies. In *Background Papers for Innovative Biological Technologies for Lesser Developed Countries*, U.S. Congress, Office of Technology Assessment, Washington, D.C., U.S. Government Printing Office, 1981, pp. 383-424.
7. Powell, C. L. Field inoculation with VA mycorrhizal fungi. In *VA Mycorrhiza* (C. L. Powell and D. J. Bagyaraj, eds.), CRC Press, Boca Raton, 1984, pp. 205-234.
8. Gerdemann, J. W. Vesicular-arbuscular mycorrhiza and plant growth. *Annu. Rev. Phytopathol.*, *6*:397-418, 1968.
9. Ferguson, J. J. *Applications of Mycorrhizal Fungi in Crop Production*. Gainesville, Florida Cooperative Extension Service, University of Florida, 1984.
10. Biermann, B., and Linderman, R. G. Increased geranium growth using preplant inoculation with a mycorrhizal fungus. *J. Am. Soc. Hort. Sci.*, *108*:972-976, 1983.
11. Menge, J. A., Labanauskas, C. K., Johnson, E. L. V., and Platt, R. G. Partial substitution of mycorrhizal fungi for phosphorus fertilization in the greenhouse culture of citrus. *Soil Sci. Soc. Am. J.*, *42*:926-930, 1978.
12. Ojala, J. C., Jarrell, W. M., Menge, J. A., and Johnson, E. L. V. Influence of mycorrhizal fungi on the mineral nutrition and yield of onion in saline soil. *Agron. J.*, *72*:255-259, 1983.
13. Hussey, R. S., and Roncadori, R. W. Vesicular-arbuscular mycorrhizae may limit nematode activity and improve plant growth. *Plant Dis.*, *66*:9-14, 1982.
14. Bartschi, H., Gianinazzi-Pearson, V., and Vegh, I. Vesicular-arbuscular mycorrhiza formation and root rot disease (Phytophthora cinnamomi), development in Chamaecyparis lawsoniana. *Phytopathol. Z.*, *102*:213-218, 1981.
15. Menge, J. A., Lembright, H., and Johnson, E. L. V. Utilization of mycorrhizal fungi in citrus nurseries. *Proc. Int. Soc. Citric*, *1*:129, 1977.
16. Timmer, L. W., and Leyden, R. F. Stunting of citrus seedlings in fumigated soils in Texas and its correction by phosphorus fertilization and inoculation with mycorrhizal fungi. *J. Am. Soc. Hort. Sci.*, *103*:533-537, 1978.
17. Kleinschmidt, G. D., and Gerdemann, J. W. Stunting of citrus seedlings in fumigated nursery soils related to the absence of endomycorrhizae. *Phytopathology*, *62*:1447-1453, 1972.
18. Johnson, C. R., and Crews, C. E. Survival of mycorrhizal plants in the landscape. *Am. Nurseryman*, *150*:15, 1979.
19. Menge, J. A., Davis, R. M., Johnson, E. L. V., and Zentmyer, G. A. Mycorrhizal fungi increase growth and reduce transplant injury in avocados. *Calif. Agr.*, *32*:6-7, 1978.

20. Schenck, N. C., and Tucker, D. P. H. Endomycorrhizal fungi and the development of citrus seedlings in Florida fumigated soils. *J. Am. Soc. Hort. Sci.*, *99*:284, 1974.
21. Sasa, M., Zahka, G., and Jakobsen, I. The effect of pretransplant inoculation with VA mycorrhizal fungi on the subsequent growth of leeks in the field. *Plant Soil*, *97*:279-283, 1987.
22. Govinda Rao, Y. S., Bagyaraj, D. J., and Rai, P. V. Selection of an efficient VA mycorrhizal fungus for finger millet. I. Glasshouse screening. *Zentralbl. Microbiol.*, *138*:409, 1983.
23. Powell, C. L., Clark, G. E., and Verberne, N. J. Growth response of four onion cultivars to several isolates of VA mycorrhizal fungi. *N. Zealand J. Agr. Res.*, *25*:465, 1982.
24. Menge, J. A. Inoculum production. In *VA Mycorrhiza* (C. L. Powell and D. J. Bagyaraj, eds.), CRC Press, Boca Raton, 1984, pp. 187-203.
25. Dehne, H. W., and Backhaus, G. F. The use of vesicular-arbuscular mycorrhizal fungi in plant production. 1. Inoculum production. *Z. Pflanzenkrankheiten Pflanzenschutz*, *93*:415-424, 1986.
26. Mugnier, J., and Mosse, B. Vesicular-arbuscular mycorrhizal infection in transformed root-inducing T-DNA roots grown axenically. *Phytopathology*, *77*:1045-1050, 1987.
27. Hall, I. R. Soil pellets to introduce vesicular-arbuscular mycorrhizal fungi into soils. *Soil Biol. Biochem.*, *11*:85-86, 1979.
28. Warner, A. Mycorrhizal seed pellets. U.S. Patent 4,551,165.
29. Hayman, D. S., Morris, E. J., and Page, R. J. Methods for inoculating field crops with mycorrhizal fungi. *Ann. Appl. Biol.*, *99*:247, 1981.
30. Ferguson, J. J., and Menge, J. A. Response of citrus seedlings to various field inoculation methods with Glomus deserticola in fumigated nursery soils. *J. Am. Soc. Hort. Sci.*, *111*:288-292, 1986.
31. Ross, J. P., and Harper, J. A. Effect of Endogone mycorrhiza on soybean yields. *Phytopathology*, *60*:1552-1556, 1970.
32. Kuo, C. G., and Huang, R. S. Effect of vesicular-arbuscular mycorrhizae on the growth and yield of rice stubble cultured soybeans. *Plant Soil*, *64*:325, 1982.
33. Powell, C. L., and Bagyaraj, D. J. VA mycorrhizal inoculation of field crops. *Proc. N. Zealand Agron. Soc.*, *12*:85, 1982.
34. Black, R. L. B., and Tinker, P. B. Interaction between effects of vesicular-arbuscular mycorrhiza and fertilizer phosphorus on yields of potatoes in the field. *Nature*, *267*:510.
35. U.S. Dept. Agriculture. *Agricultural Statistics 1984*. U.S. Government Printing Office, Washington, D.C., 1984.
36. Menge, J. A. Effect of soil fumigants and fungicides on vesicular-arbuscular fungi. *Phytopathology*, *72*:1125-1132, 1982.
37. Koch, B. L., Covey, R. P., and Larsen, H. J. Response of apple seedlings in fumigated soil to phosphorus and vesicular-arbuscular mycorrhiza. *Hort. Sci.*, *17*:232-233, 1982.
38. Hoepfner, E. F., Koch, B. L., and Covey, R. P. Enhancement of growth and phosphorus concentrations in apple seedlings by vesicular-arbuscular mycorrhizae. *J. Am. Soc. Hort. Sci.*, *108*:207-209, 1983.
39. Menge, J. A., Raski, D. J., Lider, L. A., Johnson, E. L. V., Jones, N. O., Kissler, J. J., and Hemstreet, C. L. Interactions between mycorrhizal fungi, soil fumigation and growth of grapes in California. *Am. J. Enol. Viticulture*, *34*:117-121, 1983.

40. Strobel, N. E., Hussey, R. S., and Roncadori, R. W. Interactions of vesicular-arbuscular mycorrhizal fungi, Meloidogyne incognita, and soil fertility on peach. *Phytopathology, 72*:690-694, 1982.
41. Haas, J. H., and Krikun, J. Efficacy of endomycorrhizal-fungus isolates tested at standardized inoculum potential levels and the inoculum quantities required for growth response of bell pepper. *New Phytol., 100*: 613-621, 1985.
42. Haas, J. H., Bar-Tal, A., Bar-Yosef, B., and Krikun, J. Nutrient availability effects on vesicular-arbuscular mycorrhizal bell pepper (Capsicum annum L.) seedlings and transplants. *Ann. Appl. Biol., 108*:171-179, 1986.
43. Schultz, R. C., Kormanik, P. P., and Bryan, W. C. Effects of fertilization and vesicular-arbuscular mycorrhizal inoculation on growth of hardwood seedlings. *Soil Sci. Soc. Am. J., 45*:961-965, 1981.
44. Kormanik, P. P., Schultz, R. C., and Bryan, W. C. The influence of vesicular-arbuscular mycorrhizae on the growth and development of eight hardwood tree species. *For. Sci., 28*:531-539, 1982.
45. U.S. Bureau of the Census. *1978 Census of Agriculture, Vol. 5, Part 7, 1979 Census of Horticultural Specialties*. Washington, D.C., U.S. Government Printing Office, 1980.
46. Johnson, C. R., Joiner, J. N., and Crews, C. E. Effects of N, K, and Mg on growth and leaf nutrient composition of three container grown woody ornamentals inoculated with mycorrhizae. *J. Am. Soc. Hort. Sci., 105*:286, 1980.
47. Maronek, D. M., Hendrix, J. W., and Kiernan, J. Differential growth response to the mycorrhizal fungus Glomus fasciculatus of Southern Magnolia and Bar Harbor Juniper grown in containers in composted hardwood bark-shale. *J. Am. Soc. Hort. Sci., 105*:206-208, 1980.
48. Menge, J. A., Jarrell, W. M., Labanauskas, C. K., Osala, J. C., Huszar, C., Johnson, E. L. V., and Sibert, D. Predicting mycorrhizal dependency of Troyer citrange on Glomus fasciculatus in California citrus soils and nursery mixes. *Soil Sci. Soc. Am. J., 46*:762-768, 1982.
49. Gerdemann, J. W. Vesicular-arbuscular mycorrhizae. In *The Development and Function of Roots* (J. G. Torrey and D. T. Clarkson, eds.), Academic Press, London, 1975, pp. 575-591.
50. Heckman, J. R., and Angle, J. S. Variation between soybean cultivars in vesicular-arbuscular mycorrhiza fungi colonization. *Agron. J., 79*: 428-430, 1987.
51. Krishna, K. R., Shetty, K. G., Dart, P. J., and Andrews, D. J. Genotype dependent variation in mycorrhizal colonization and response to inoculation of pearl millet. *Plant Soil, 86*:113-125, 1985.
52. Azcon, R., and Ocampo, J. A. Factors affecting the vesicular-arbuscular infection and mycorrhizal dependency of thirteen wheat cultivars. *New Phytol., 87*:677-685, 1981.
53. Toth, R., Page, T., and Castleberry, R. Differences in mycorrhizal colonization of maize selections for high and low ear leaf phosphorus. *Crop Sci., 24*:994-996, 1984.
54. Satterlee, L., Melton, B., McCaslin, B., and Miller, D. Mycorrhizae effects on plant growth, phosphorus uptake, and N_2 (C_2H_2) fixation in two alfalfa populations. *Agron. J., 75*:715-716, 1983.
55. Menge, J. A., Johnson, E. L. V., and Platt, R. G. Mycorrhizal dependency of several citrus cultivars under three nutrient regimes. *New Phytol., 81*:553, 1978.

56. Mehraveran, H. Mycorrhizal dependency of six citrus cultivars. Ph.D. thesis, Univ. of Illinois, Champaign-Urbana, 1977.
57. Ratnayake, M., Leonard, R. T., and Menge, J. A. Root exudation in relation to supply of phosphorus and its possible relevance to mycorrhizal formation. *New Phytol.*, *81*:543, 1978.
58. Graham, J. H., Leonard, R. T., and Menge, J. A. Membrane-mediated decrease in root exudation responsible for phosphorus inhibition of vesicular-arbuscular mycorrhiza formation. *Plant Physiol.*, *68*:548, 1981.
59. Baylis, G. T. S. The magnolioid mycorrhiza and mycotrophy in root systems derived from it. In *Endomycorrhizas* (F. E. Sanders, B. Mosse, and P. B. Tinker, eds.), Academic Press, London, 1975, pp. 373-389.
60. Baligar, V. C., and Barber, S. A. Genotypic differences of corn in ion uptake. *Agron. J.*, *71*:870-873, 1979.
61. Romer, W., Augustin, J., and Schilling, G. The relationship between phosphate absorption and root length in nine wheat cultivars. *Plant Soil*, *111*:199-202, 1988.
62. Lindsay, W. L. *Phosphates: Chemical Equilibria in Soils*. Wiley-Interscience, New York, 1979, pp. 162-205.
63. Brady, N. C. *The Nature and Properties of Soils*. Macmillan, New York, 1974.
64. Bhat, K. K. S., and Nye, P. H. Diffusion of phosphate to plant roots in soil. I. Quantitative autoradiography of the depletion zone. *Plant Soil*, *38*:161-175, 1973.
65. Bhat, K. K. S., and Nye, P. H. Diffusion of phosphate to plant roots in soil. III. Depletion around onion roots without root hairs. *Plant Soil*, *41*:383-394, 1974.
66. Sanders, F. E., and Tinker, P. B. Mechanism of absorption of phosphate from soil by endogone mycorrhiza. *Nature*, *233*:278-279, 1971.
67. Gemma, J. G. Physiology and ecology of vesicular-arbuscular mycorrhizal fungi in a sand dune system. Ph.D. thesis, Univ. of Rhode Island, Kingston, 1987.
68. Ojala, J. C., Jarrell, W. M., Menge, J. A., and Johnson, E. L. V. Comparison of soil phosphorus extractants as predictors of mycorrhizal dependency. *Soil Sci. Soc. Am. J.*, *47*:958-962, 1983.
69. Biermann, B., and Linderman, R. G. Effect of container plant growth medium and fertilizer phosphorus on establishment and host growth response to vesicular-arbuscular mycorrhizae. *J. Am. Soc. Hort. Sci.*, *108*:962-971, 1983.
70. Bethlenfalvay, G. J., and Yoder, J. F. The Glycine-Glomus-Rhizobium symbiosis. I. Phosphorus effect on nitrogen fixation and mycorrhizal infection. *Physiol. Plant*, *52*:141-145, 1981.
71. Graham, J. H., and Timmer, L. W. Rock phosphate as a source of phosphorus for vesicular-arbuscular mycorrhizal development and growth of citrus in soilless medium. *J. Am. Soc. Hort. Sci.*, *110*:489-492, 1985.
72. Ball, V. *Ball Redbook, Greenhouse Growing*, Reston, Reston, VA, 1985.
73. Sutton, J. C., and Barron, G. L. Population dynamics of Endogone spores in soil. *Can. J. Bot.*, *50*:1909-1914, 1972.
74. Ianson, D. C., and Allen, M. F. The effects of soil texture on extraction of vesicular-arbuscular mycorrhizal fungal spores from arid sites. *Mycologia*, *78*:164-168, 1986.

75. Hayman, D. S. Endogone spore numbers in soil and vesicular-arbuscular mycorrhiza in wheat as influenced by season and soil treatment. *Trans. Br. Mycol. Soc.*, *54*:53-63, 1970.
76. Kruckelmann, H. W. Effects of fertilizers, soils, soil tillage, and plant species on the frequency of endogone chlamydospores and mycorrhizal infection in arable soils. In *Endomycorrhizas* (F. E. Sanders, B. Mosse, and P. B. Tinker, eds.), Academic Press, London, 1975, pp. 511-525.
77. Smith, T. F. The effect of season and crop rotation on the abundance of spores of vesicular-arbuscular (VA) mycorrhizal endophytes. *Plant Soil*, *57*:475-479, 1980.
78. Schwab, S. M., Menge, J. A., and Leonard, R. T. Comparison of stages of vesicular-arbuscular mycorrhiza formation in sudangrass grown at two levels of phosphorus nutrition. *Am. J. Bot.*, *70*:1225-1232, 1983.
79. Morandi, D., Bailey, J. A., and Gianinazzi-Pearson, V. Isoflavonoid accumulation in soybean roots infected with vesicular-arbuscular mycorrhizal fungi. *Physiol. Plant Pathol.*, *24*:357-364, 1984.
80. Krishna, K. R., and Bagyaraj, D. J. Phenols in mycorrhizal roots of Arachis hypogaea. *Experimentia*, *40*:85-86, 1984.
81. Hepper, C. M. The effect of nitrate and phosphate on the vesicular-arbuscular mycorrhizal infection of lettuce. *New Phytol.*, *92*:389-399, 1983.
82. Carling, D. E., Brown, M. F., and Brown, R. A. Colonization rates and growth responses of soybean Glycine max plants infected by vesicular-arbuscular mycorrhizal fungi. *Can. J. Bot.*, *57*:1769, 1979.
83. Dehne, H. W., and Schoenbeck, F. The influence of the endotrophic mycorrhiza on the fusarial wilt of tomato. *Z. Pflanzenkr. Pflanzenschutz*, *82*:630, 1975.
84. Stewart, E. L., and Pfleger, F. L. Development of Poinsettia as influenced by endomycorrhizae, fertilizer and root rot pathogens Pythium ultimum and Rhizoctonia solani. *Florist Rev.*, *159*:37, 1977.
85. van Bruggen, A. H. C., Grogan, R. G., Bogdanoff, C. P., and Water, C. M. Corky root of lettuce in California caused by a gram-negative bacterium. *Phytopathology*, *78*:1139-1144, 1988.
86. Hoitink, H. A. J. Composted bark, a lightweight growth medium with fungicidal properties. *Plant Dis.*, *64*:142-147, 1980.
87. Jasper, D. A., Robson, A. D., and Abbott, L. K. Phosphorus and the formation of vesicular-arbuscular mycorrhizas. *Soil Boil. Biochem.*, *11*:501-505, 1979.
88. Sylvia, D. M., and Schenck, N. C. Application of superphosphate to mycorrhizal plants stimulates sporulation of P-tolerant vesicular-arbuscular mycorrhizal fungi. *New Phytol.*, *95*:655, 1983.
89. Hung, L. L., and Silvia, D. M. VAM inoculum production in aeroponic culture. In *Mycorrhizae in the Next Decade* (D. M. Silvia, L. L. Hung, and J. H. Graham, eds.), IFAS, University of Florida, Gainesville, 1987, p. 272.

27

THEORETICAL CONSIDERATIONS IN THE COMMERCIAL UTILIZATION OF MYCORRHIZAL FUNGI

R. WHETTEN and ANNE J. ANDERSON *Utah State University, Logan, Utah*

I. INTRODUCTION

Commercial utilization of mycorrhizal inoculum has increasing potential as planting practices tend toward a less chemical and more biological approach. The current problems in generation of profitable and active vesicular-arbuscular (VA) mycorrhizal inoculum have been reviewed in Chapter 26. In this chapter we take a more theoretical approach, stressing the importance of the ecology of the systems in which mycorrhizal fungi may be beneficial. We also examine environmental factors which may limit the efficacy of potentially successful inoculum. We speculate on the role of molecular engineering of mycorrhizal systems. These discussions cover areas which will remain as challenging after the basic problems described in the previous chapter are resolved. We hope that inclusion of these considerations in this chapter will stimulate novel research directions that will promote the efficacy of mycorrhizal commercialization.

We initiate our discussion by consideration of potential markets for mycorrhizal inocula in addition to those discussed in Chapter 26. At present some of these areas are not cost-effective but they may become so, especially if inoculum can be cheaply and easily produced. We extend our discussion to examine factors that are involved in the compatibility of the mycorrhizal inoculum in the agricultural environment. These discussions augment the coverage provided in Chapter 26 for the present commercial markets. The section is completed by speculation on ecological considerations of commercial release and the prospects for the genetic engineering of fungi with improved traits.

II. POTENTIAL BENEFITS FROM COMMERCIAL UTILIZATION OF MYCORRHIZAL FUNGI

Plants grown under environmental stress, especially phosphate deficiency, may have growth performances improved by conversion to mycorrhizal status. The benefit of mycorrhizal formation in low phosphate soils is discussed in detail in Chapter 26. Other stress situations include soils which have been exposed to conditions which deplete indigenous flora, including mycorrhizal fungi. Growth in sterilized soil in nursery fumigated agricultural land, or stored mine soils are examples. Mycorrhizal status may also be advantageous in nonsterile soils because of suppression of root pathogens.

Additional situations which benefit from mycorrhizal inoculation involve growth under water-deficient and toxic conditions. Drought may occur naturally, as testified by the dearth of rainfall in many areas of the United States in the spring and summer of 1988, or as the result of inadequate irrigation. The recirculation of irrigation water is rendering increasing salinity in large acreages of land; a condition that is unacceptable for many crop species presently cultivated. Mining and heavy industries are causing the accumulation of heavy metals in the soil as well as the intensification of such gaseous pollutants as ozone and sulfur dioxide and precipitation of acid rain. In each of these situations, cultivation of mycorrhizal plants may offer improved growth performance. Literature which supports the protective effects of mycorrhiza in various adverse conditions is discussed.

A. Nursery and Containerized Plantings

Common nursery practices (the use of nonsoil potting mixture or sterilized soil, growth under enclosed conditions, treatment with biocides, and high doses of fertilizers) are not commensurate with the establishment of a mycorrhizal root system. The inoculum may be absent or, if present, the infection potential of the mycorrhizal fungus may be suppressed. However, establishment of mycorrhizae at the seedling stage has been demonstrated to enhance growth and promote survival on transplantation. As discussed by Linderman [1], the addition of suitable inoculum to containerized plants could be beneficial in "alleviating drought stress, aiding in transplanting and enhancing rooting of currings." Indeed, an increasing utilization of mycorrhizal inoculum in nurseries may arise from legislation regulating the quality of water effluents from such industries where minimum levels of phosphate and nitrogen are expected. These concerns may enforce limitations on heavy fertilization schemes.

Beneficial effects of added VA mycorrhizal inoculum to the nursery and containerized industry have been discussed elsewhere in this book (see Chapter 26). Ectomycorrhizal inoculum may also be of commercial significance. Molina [2] and Graham and Linderman [3] determined that mycorrhizal formation in conifer hosts would occur consistently with *Laccaria laccata* and *Cenococcum geophilum* under containerized conditions. Compatibility with fertilization schedules was observed with *L. laccata* [4]. Parke et al. [5] demonstrated that Douglas fir containerized seedlings colonized with the ectomycorrhizal fungus *Rhizopogon vinicolor* were more tolerant to drought than nonmycorrhizal plants.

Studies by Fortin et al. [6,7] also illustrate how fertilization schedules and choice of inoculum can be implemented to maximize results of survival and growth improvement in nursery soil cultivation. Their data confirm that fungal isolates differ in their competitive abilities and consequently they emphasize the importance of selection of correct isolates for amendments. McAfee and Fortin [7] further indicate that other soil biota in addition to mycorrhizal fungi may influence, either positively or negatively, the formation of mycorrhizae and consequently the potential of mycorrhizal amendments in nursery practices.

B. Protection Against Pathogens

Certain mycorrhizal associations may benefit plant health by suppression of pathogens. The beneficial effect of VA mycorrhizal inoculum in the protection of lettuce against a bacterial agent of corky root disease is discussed in Chapter 26. Ectomycorrhizae help in the suppression of root pathogens, like *Phytophthora*, *Pythium*, *Rhizoctonia*, and *Fusarium oxysporum* [8,9]. Nutrient competition or a barrier effect caused by the sheathing of the root in the fungal mantle are two suggested mechanisms. The production of antibiotics by the mycorrhizal fungus effective against the pathogens is another proposed beneficial trait. For example, Sylvia and Sinclair [10] demonstrated that antibiotics produced by *L. laccata* were able to impede development of *F. oxysporum*. Recent studies by Duschesne et al. [11] on resistance of *Pinus resinosa* to *Fusarium oxysporum* mediated by the ectomycorrhizal fungus *Paxillus involutus* indicate that the synthesis of the antifungal compound is stimulated by root exudate factors.

Host root metabolism may be involved in another way with resistance. Sylvia [12] suggested that activation of host defense mechanisms in mycorrhizal roots could participate in disease suppression. Sylvia and Sinclair [10] demonstrated that invasion by *L. laccata* promoted phenolic deposition in the cortical cells of the host root. They propose these phenolics may inhibit the infection potential of subsequent invasions by *F. oxysporum*. These preliminary data demonstrate that factors from both the host and the fungus are responsible for disease suppression. Identification of these parameters may permit genetic manipulation of the mycorrhizal fungus and of the host to maximize disease resistance potential.

C. Growth Under Conditions of Drought and Salinity

Mycorrhiza formation may, with certain plants, enhance growth under drought and saline growth conditions. Beneficial effects of VA mycorrhizal colonization under drought conditions were reported for *Bouteloua* by Allen [13] and for wheat by Allen and Boosalis [14]. Dakessian et al. [15] indicate a correlation between soil water at the wilting potential and plant growth enhancement for soybean. Another legume, *Hedysarum boreale*, demonstrated enhanced plant growth after inoculation with VA mycorrhizal fungi and *Rhizobium* under nutrient- and water-stress conditions [17,18]. Beneficial effects are also observed for mycorrhizal plants compared with nonmycorrhizal plants growing under saline conditions [17,18].

Allen [13] proposed that an important feature in drought tolerance is the ability of the hyphae extending from a mycorrhizal root to tap additional water sources unavailable to the root—a property of more significance to plants growing in arid rather than mesic conditions. Other studies suggest that beneficial effects of mycorrhizal formation in alleviation of drought or salinity stress may only be apparent in growth under phosphate limitation because the addition of phosphate to nonmycorrhizal plants boosted their performance under drought and salinity. This effect was demonstrated in onion exposed to salt or drought [17] and big blue stem [19] grown under drought. However, studies of citrange suggest that the beneficial effects of VA mycorrhizal inoculation under saline conditions are not fully explained by improved phosphate nutrition [20]. Similarly, Dakessian et al. [15] indicate that phosphate nutrition does not explain the improved water relations of VA mycorrhizal-inoculated soybean.

Clearly, commercial inoculum cannot be expected to provide universal relief from drought and saline stress but in selected examples it could well substitute for phosphate fertilization. Selection of the fungal isolate may be an important factor in commercial procedures. Allen [21] reported that an isolate of *G. fasciculatum* from a native soil high in salinity could establish mycorrhizae at higher salt concentrations than an isolate obtained from cultivated soils. Spores of *G. fasciculatum* are more drought-tolerant than the larger spored *G. mosseae* [21]. These observations suggest that there are genetic parameters in the fungi which could be identified and utilized to maximize the efficiency of mycorrhizal fungi to tolerate these adverse plant growth conditions. In addition, molecular knowledge of the mechanisms by which mycorrhizal colonization affects water transportation in the plant (leaf conductance, root hydraulic conductivity, and accumulation of plant osmolarity regulators, e.g., betaine [22]) may be of value in future genetic manipulation of host and of fungus.

D. Improved Resistance to Pollutants

Plant growth in the presence of pollutants (heavy metals, acid rain, ozone) may be boosted by mycorrhiza formation. Consequently, inoculation of species outplanted into such areas where these pollutants are accumulated would seem valuable. Loblolly pine seedlings were protected against exposures to ozone and sulfur dioxide, singly and in combination [23], by ectomycorrhizal formation with *Pisolithus tinctorius*. Ectomycorrhizae have been associated with heavy-metal tolerance. Ectomycorrhizal colonization of *Pinus* and *Picea* by *Suillus luteus* was also demonstrated to be beneficial for growth in soils with Cd, Cu, Ni, Pb, and Zn amendments [24]. Threshold levels of metal concentrations for the beneficial effects were observed [24]. These levels may correlate with decreasing ability of the fungus to form mycorrhizae as the heavy metal concentration increases. Birch is one of the few tree species which survive in soils polluted with copper and nickel and its survival is correlated to its association with ectomycorrhizal fungi. Studies by Jones and Hutchinson [25,26] indicate that performance of birch infected with *Scleroderma flavidum* exceeded that of other mycorrhizal associations and nonmycorrhizal plants. Thus, choice of inoculum is significant in commercialization. Of interest is the observation of Gildon and Tinker [27] that certain VA mycorrhizal isolates are more tolerant than other isolates to heavy metals. Heavy-metal resistance in other organisms has been correlated to discrete genetically based detoxification mechanisms. Consequently, molecular examination of the heavy-metal-tolerant isolates may allow characterization of genes that could be subsequently used in genetic exchange systems to improve the performance of mycorrhizal fungi otherwise less adapted to environments rich in heavy metals. The molecular basis of the induced tolerance may also reside in interactions between the plant root and the fungus, perhaps involving an elevated phosphate level within the root [26].

E. Reclamation of Mine Soils

Reclamation of mined soils presents a variety of problems which may have synergistic influences upon mycorrhiza formation in seeded plants. Frequently, the top soil has been removed and stored. Prolonged storage causes a reduction of microflora, including indigenous mycorrhizal inoculum, a decrease in organic materials such as humus and phytin, and altered texture and moisture levels. The disturbed site may also have extremes of pH and accumulations of salt, heavy metals, or oil products. For example, Call and McKell [28] state that "processed oil shale is a highly saline, highly alkaline, biologically sterile, nutrient deficient material with no structure."

Reclamation of the soil requires reconstruction not only of the above-ground flora but also of the below-ground flora [29]. Reestablishment of mycorrhizal flora is essential for the process of succession in plant establishment to take its natural course. Mycorrhizal inoculum would naturally be deposited on the reclaimed land with time. However, transplantation of preinoculated plants appears to have strong merit, by reducing the time factor and limiting invasion of the land by weedy plants that typically do not form mycorrhizae [21].

Aldon [30,31] indicated that endomycorrhizae enhanced the survival and growth of shrubs on mine spoils. Other pioneering studies by Marx [32] demonstrated the greater survival of pine mycorrhizal with *Pisolithus tinctorius* than nonmycorrhizal transplants in strip-mined land. Two characteristics of *P. tinctorius* may have been contributing features to its success: (a) the sterility of the soil limited the exposure of *P. tinctorius* to other competitors and underplayed the relatively slow growth rate of this fungus, and (b) the high soil temperatures were compatible to growth of *P. tinctorius* which in culture grows well up to 34°C. Call and McKell [28] observed that endomycorrhizae enhanced the growth of native shrub species in native soil and processed oil shale. Consequently, the use of VA mycorrhizal or ectomycorrhizal inoculum, precolonized plants, and better management of top soils to avoid storage may enhance our ability to restore mined land to a vegetated state [21].

III. COMPATIBILITY WITH THE AGRICULTURAL ENVIRONMENT

The benefit of added inoculum on a commercial level is highly dependent on the conditions under which the crop is raised—conditions which must maximize colonization potential. These factors may impose a limit to efficacy even though the host and the inoculum have been selected for maximum effects. Consequently, several features of agricultural practices or environmental parameters are considered.

A. Environmental Parameters

1. Growth Conditions

Soil conditions are of paramount importance and the effects of one significant parameter, pH, are discussed in Chapter 26. Other conditions may also impose limits on the efficacy of the inoculum in the field. Soil moisture level may effect the degree and efficacy of mycorrhizal colonization. Reduction in VA mycorrhizal colonization has been correlated with increasing water table and beneficial effects of mycorrhiza formation may be absent [33,34]. Restricted water availability through drought and salinity can affect plant development. As indicated in the previous chapter, mycorrhiza formation offers protection, in comparison to nonmycorrhizal plants, against drought and salinity provided these factors are below a threshold level.

Adequate light supply is required to permit growth stimulation to occur. Impaired light may be especially significant during seedling establishment when the energy drain toward mycorrhiza formation may have major effects on the stress level of the plant [35]. A prolonged period of cloudy conditions in the field or inadequate lighting in greenhouse culture may reduce the expected beneficial effects of added VA mycorrhizal inoculum. Removal of aerial portions of the plant by herbivory may also be detrimental to VA mycorrhiza formation. Bethlenfalvay and Dakessian [36] demonstrated that grazing reduced the extent of VA mycorrhizal fungal colonization in range grasses. This situation could be significant in cases where renewal of inoculum is dependent on spread from established plantings that become exposed to grazing.

2. Effects of Temperature

Soil temperature may strongly influence the growth potential of the fungus, the host, and the mycorrhizae; each may possess a different optima. Trappe [37] suggested that monitoring temperature requirements for growth of ectomycorrhizal fungi might promote the matching of inoculum with the conditions at the amended site. Cline et al. [38] monitored the growth of multiple isolates of *Pisolithus tinctorius*, *Cenococcum geophilum*, *Thelophora terrestris*, and *Suillus granulatus* over a wide temperature range. Each fungal species showed a high degree of intraspecies variation with little correlation to geographic origin. The data did support the observation that *P. tinctorius* is able to survive at high soil temperatures as demonstrated by its successful application in revegetation of mine spoils in the southern United States [32]. Other studies of temperature optima on VA mycorrhiza formation with cotton, which tolerates high soil temperature, illustrated that growth stimulation was maximum at 30°C with plants inoculated with *G. margarita* and *G. intraradices* [39]. A third isolate, *G. ambisporum*, was more proficient at 36°C.

Parke et al. [40] suggested that the ability of indigenous mycorrhizal fungi to function over a temperature range is an important attribute. They stated that growth of the fungi even at low rates at submaximal temperatures could be highly significant in seedlings established in areas where the early spring is cool. A revaluation of the development of VA mycorrhizae in winter cereals by Dodd and Jeffries [41] indicated substantial infection early in seedling growth under cooler soil temperatures in November and December. Previous studies had indicated that VA mycorrhizal fungal infection was not significant until late spring or early summer. Dodd and Jeffries [41] suggested that a different subfraction of VA mycorrhizal fungal flora may be responsible for colonization in these winter months. Correlation of VA mycorrhizal colonization with seasonal root development was also observed by Allen [42]. He observed that *Atriplex gardneri* progressed from a mycorrhizal status at snowmelt to a nonmycorrhizal form in the summer, even though spores were available in the soil throughout the seasons. Low to medium soil temperatures and soil moisture were considered key factors in this seasonal pattern. These data indicate that the timing of application of mycorrhizal inoculum can be critical.

3. Environmental Pollutants

Pollutants in the environment may alter the efficacy of added inoculum. Studies on pine species indicated that acid precipitation decreases mycorrhizal association and carpophore formation [43-45]. The causes of this response may be diverse: fungal growth inhibition at low pH, enhanced solubilization by acidic pH of toxic metals, and decreases in mycorrhiza formation by increased nitrogen mineralization in the soil. Prolonged exposure to ozone also decreased mycorrhiza formation [45].

Pollution by high concentrations of heavy metals and arsenic is an additional deleterious factor in mycorrhiza formation [23,46]. Decreased mycorrhiza formation in old apple orchard sites has been correlated to arsenic levels in the soil by accumulation of lead arsenate used as a pesticide [47]. Thus consideration of prior cropping procedures may be pertinent in evaluating the benefits from mycorrhizal inoculum.

B. Agricultural Practices

1. Fertilization

The addition of inorganic and organic fertilizers may not be compatible with the commercial utilization of mycorrhizal fungi. The significance of phosphate level in VA mycorrhizal fungal colonization is discussed in detail in Chapter 26. Here we again emphasize that increasing levels of phosphate have been demonstrated in numerous systems to deter the beneficial effect of mycorrhiza formation. Mycorrhizal fungus colonization declines as the plant utilizes the more readily available fertilizer [48,49]. However, some mycorrhizal fungal isolates have been shown to be more tolerant to high phosphate levels [48]. It is proposed that the extent of mycorrhiza formation may then be dependent on the phosphate level within the plant root. Clearly it may be advantageous to use tolerant isolates if some degree of P fertilization is to be maintained. The application of nitrogenous fertilizers also is negatively correlated with mycorrhiza formation and growth benefit. Even the use of green manure has been shown to be detrimental in monoculture situations, although little effect was observed in crop rotation systems [50].

2. Pesticide Utilization

Beneficial consequences of commercially applied mycorrhizal fungi must be compatible with programs for chemical pest control measures. This was already discussed in Chapter 26 for specific growth conditions such as the container-grown plant industry. This section emphasizes again the complexity of managing an agroecosystem involving added biologicals. Variability in effects of fungicides and biocides on mycorrhizal fungi and mycorrhizae has been widely investigated and reviewed [51]. The selectivity of certain fungicides for different taxa dictates that they should be compatible with nontarget mycorrhizal fungi. For example, as the thiazoles are more effective on *Zygomoycotina* than *Basidio* or *Ascomycotina*, they have little effect on ericoid and ectomycorrhizal development but markedly suppress spore formation by Endogonaceae and mycorrhiza formation by zygomycetous symbionts. Such specificity suggests that the use of thiazoles would be appropriate for control of pathogens in nursery-raised *Pinaceae* inoculated with ectomycorrhizal fungi. Trappe et al. [51] also stated that herbicides, nematicides, and insecticides may have nontarget effects extending to mycorrhizal potential. They indicate certain interactions in which insecticides, herbicides, and nematicides have deleterious effects on mycorrhizal function. Kough et al. [52] confirmed that benomyl and captan significantly reduced growth of VA mycorrhizal onion but not of non-VA mycorrhizal plants. Growth reduction in the VA mycorrhizal plants was correlated with decrease in succinate dehydrogenase in root extracts. Consequently, they suggested that monitoring succinate dehydrogenase in root extracts could be a useful method for prediction of deleterious effects of fungicides (or other biocides) on mycorrhizal function. Deleterious effects of benomyl on VA mycorrhizae was also reported by Fitter [52] and Buttery et al. [54].

In contrast to the deleterious effect of pesticides on mycorrhiza formation, beneficial results are also reported. Metalaxyl was demonstrated by Afek and Menge (personal communication) to promote growth of VA mycorrhizal onions. According to them, the pesticide causes suppression of *Pythium* as a root pathogen and permits growth of healthier mycorrhizal roots. Another interesting observation is the ability of simazine to promote mycorrhizae in a nonhost, *Chenopodium quinona*, perhaps through enhanced secretion of sugars and amino acids from the roots [55].

The use of microbes as biocontrol agents was questioned by Garbaye and Bowen [56] because of side effects on mycorrhiza formation. Studies by Bowen et al. [56,57] revealed antagonism between many soil bacteria and ectomycorrhizal fungi. Garbaye and Bowen [56] thus suggest that released microorganisms which suppress fungal pathogens may impair the potential for mycorrhizal formation through their actions on these nontarget fungal species.

3. Cropping History

The cropping history of a field may prejudice the efficacy of added inoculum. Precropping with a nonhost plant did not decrease the potential of VA mycorrhizal inoculum naturally present in the soil upon rotation to a mycorrhizal host plant [58,59]. These data were obtained with a range of nonhost species (radish, cabbage, sugarbeet, and oil seed rape). However, Baltruschat and Dehne [50] indicate that the use of nonhost crops as green manure were significant in the decline of VA mycorrhizal inoculum potential in monoculture of wheat. These workers also observed that the negative effect of green manure on VA mycorrhizal inoculum potential was accentuated by nitrogen fertilization. Schenck and Kinlock [60] revealed a decline in VA mycorrhizal colonization upon monoculture in a short-term (3 years) growth period. These studies suggest that supplementation of mycorrhizae could be beneficial under monoculture conditions.

C. Effects of Indigenous Soil Microorganisms

1. Competition with Other Mycorrhizal Inoculum

The effectiveness of added inoculum may be influenced strongly by competition with naturally occurring isolates with mycorrhizal potential. This possibility is described in Chapter 26 as one of the major factors in constraining efficacy of inocula and is further discussed here.

VA mycorrhizal fungi are demonstrated to differ in their aggressiveness in colonization. Wilson [61,62] showed that an isolate of *Gigaspora decipiens*, although weaker in its infection potential than two *Glomus* species, was more aggressive in limiting secondary infections of the root by other VA mycorrhizal fungal species. Consequently, the success of added inoculum may depend on the aggressive potential of the indigenous flora.

Lopez-Aguillon and Mosse [63] attempted to create artificially the scenario of competition between indigenous and added inoculum by placing inocula of different endomycorrhizal fungi in defined positions relative to roots. Field soil was used in the studies. They observed a dominance pattern for three VA mycorrhizal fungi which was the same for two hosts, clover and sorghum. Competition, but not apparent antagonism, was noted between the three isolates. The extent of root occupancy was highly related to invasiveness, a measure of the successful spread of the organism. However, root occupancy as measured by percentage of root colonization was not correlated with stimulation in growth of the hosts. These data illustrate that it may be difficult to predict the overall effectiveness of an isolate under field conditions even though the individual characteristics of that isolate are known.

Studies with seedlings of eucalyptus, *Alnus*, *Helianthemum*, and conifers indicated that VA mycorrhizal and ectomycorrhizal fungi may be competitors for colonization of the same roots. Kough et al. [64] demonstrated that four species of conifers were responsive in terms of seedling uniformity and size to VA mycorrhizal inoculation. Studies of eucalyptus, *Alnus*, and *Helianthus* showed that the VA mycorrhizal fungi were superseded by the ectomycorrhizal fungi, which seemed to limit additional VA mycorrhizal colonization [65]. Chilvers et al. [65] suggest that for eucalypts the initial effects of VA mycorrhizal colonization may help promote establishment in low-nutrient or calcareous soils.

Other studies on eucalypts also suggest a possible succession in ectomycorrhizal establishment as roots penetrate through different soil layers. Malujczuk et al. [66] observed that *Hysterangium* and *Cortinarius* species would form ectomycorrhizae of a superficial nature with eucalypts. The abundance of *Cortinarius* spp. in litter and *Hysterangium* spp. in the underlying soil-litter interface suggests these interactions could be significant in early seedling establishment in nature.

These studies suggest that multiple types of mycorrhizal inoculum may be desirable to promote host survival, especially for some trees and shrubs, which can form both endo- and ectomycorrhizal associations.

2. Effects of Nonmycorrhizal Microorganisms

The potential for mycorrhizal fungi to stimulate plant growth may be influenced by the nature of other soil organisms [67]. Several studies have demonstrated that mycorrhiza formation causes reduction in rhizosphere colonization by certain microorganisms but may enhance rhizosphere colonization by other species. Species reported to be enhanced in the rhizosphere of mycorrhizal roots include a gram-negative nonmotile rod [68], and *Azotobacter* species [69], and *Pseudomonas* species [70,71]. Linderman and coworkers [72-74] also indicated that equilibrium in the rhizosphere and rhizoplane of corn or clover was changed by mycorrhiza formation; facultative aerobes increased and fluorescent pseudomonads and actinomycetes decreased in the rhizosphere of the mycorrhizal plants. However, analysis of population numbers may not reveal the whole story. The mycosphere organisms may be antagonistic, neutral, or stimulatory to the function of the mycorrhizal root. Certain microorganisms are reported to act synergistically with the mycorrhizal fungi and promote plant growth. Phosphate-solubilizing bacteria promoted growth of mycorrhizal alfalfa in sand-vermiculite amended with rock phosphate [75]. An *Azotobacter*, which could assimilate molecular nitrogen, plus *Glomus mosseae* increased significantly the dry mass of tall fescue [76]. Meyer and Linderman [72] revealed that coinoculation of clover with VA mycorrhizae and *Pseudomonas putida* synergistically enhanced shoot dry weight, more rapid colonization by the VA mycorrhizal fungi, and nodulation by *Rhizobium*.

However, beneficial effects may be extremely isolate-specific. Linderman (personal communication) demonstrated great variability in the ability of different VA mycorrhizal fungi to stimulate the numbers of nodules formed by a *Rhizobium* strain on a legume. The effect was not related to improved phosphate nutrition caused by the functioning mycorrhizae. Variability in isolate interaction is a dominant observation in studies of *Rhizobium*-mycorrhizal symbiosis [77]. Occasional pairings of a *Rhizobium* isolate and of a VA mycorrhizal fungal isolate were found to promote the functions of each symbiont and enhancement of soybean growth. More generally, the interactions showed antagonism in the extent of VA mycorrhizal colonization and nodule formation and function [78]. These results were generated using control plants that had N and P amendments to balance the P effect of VA mycorrhizae. Previous studies by Bethlenfalvay and Yoder [79] had illustrated that under phosphate limitation, VA mycorrhiza formation was essential to provide the additional P required for *Rhizobium*-mediated nitrogen fixation. Bayne and Bethlenfalvay [80] stress that screening for "desirable symbiosis parameters should be part of any selection process to identify desirable agricultural VAM isolates."

Studies of Bowen and Theodorou [57] generally revealed antagonism between ectomycorrhizal fungi and soil pseudomonads or *Bacillus* isolates. Antagonism and occasional enhancement was observed in a later study [56] in ectomycorrhiza formation on *Pinus radiata*. Antagonism of mycorrhizal fungi by soil microorganisms was suggested by Ross [81] to contribute to the suppression of mycorrhizal growth response that has been observed in comparison of growth in nonsterile to sterile soils. Ross [81] observed reduction in spore production and spore size upon addition of raw field soil into pot cultures of soybean and *Glomus*. He proposed that metabolites from the mycosphere organisms acted as suppressants of sporulation. Wilson et al. [82] from studies of VA mycorrhizal fungal effects on big blue stem using raw soil amendment suggested another explanation that the additional mycosphere organisms competed with the mycorrhizal fungus for available nutrients.

Direct parasitism of mycorrhizal fungi may be another cause for reduced efficacy of mycorrhizal inoculum. Fungi which can parasitize spores of VA mycorrhizal fungi have been identified [83,84]. Paulitz and Menge [85] demonstrated that at low inoculum density of *Glomus deserticola* the mycoparasite *Anguillospora pseudolongissima* reduced the extent of root colonization and growth response of onions. Paulitz and Menge [85] observed variability in the occurrence of the mycoparasite in field sites. These observations indicate that there are factors which regulate the level of the mycoparasite and these constitute other potential variables in the commercialization of mycorrhizal fungi. We do not have the knowledge to predict whether release of VA mycorrhizal fungal inoculum would boost the production of their mycoparasites in the soil and counteract their expected benefits.

D. Selection of Host and Isolate

Beneficial effects of mycorrhizal fungi are controlled by genetic factors within the plant and the fungus as well as by the environmental parameters previously discussed. Presumably there are factors in the plant and in the microbe which can be manipulated to maximize the positive effects of their interaction in a mycorrhiza. Studies of the relative ability of mycorrhizal fungi to effectively form mycorrhizae with one host compared to others may promote identification of these genes. Specificity is observed in both ectomycorrhiza and VA mycorrhiza formation.

Certain ectomycorrhizal fungi display clear preferences in hosts with which complete functional mycorrhizae and fruiting structures are formed. The incompatible hosts appear to mount effective defense mechanisms against the challenging ectomycorrhizal fungus. Incompatibility also exists toward VA mycorrhizal fungi. Although VA mycorrhiza formation is widespread and the VA mycorrhizal fungi are often referred to as generalists, there are certain plants which grow naturally in a nonmycorrhizal state. Many of the nonmycorrhizal species belong to the families Chenopodiaceae and Cruciferae [86,87]. The biochemistry of incompatibility in these plants is not resolved [87]. However, the existence of the nonmycorrhizal species means that certain gene products from the plant and the fungus will condition an incompatible response. Minor levels of incompatibility also occur in VA mycorrhizal fungi-compatible hosts. Many evidences demonstrate certain VA mycorrhizal fungal isolates outperform others. Identification of the genes involved in compatibility and incompatibility will permit maximization of the commercial utilization of mycorrhizal fungi through genetic engineering. At present we can learn from these studies which fungal isolate may be the better performer in field and nursery conditions.

1. Specificity in Ectomycorrhizal Formation

The choice of ectomycorrhizal inoculum for trees and shrubs is relevant to nursery cultivation of seedlings that will be outplanted in reforestation and revegetation programs. Recent studies spearheaded by the research group of Molina and Trappe [88-90] have demonstrated clear incompatibilities in ectomycorrhizal associations. High specificity in production of mycorrhizae and sporocarps was observed with conifers and alders. For example, *Rhizopogon vinicolor* was compatible with Douglas fir but incompatible with lodgepole pine and other conifers, whereas *Rhizopogon occidentales* was only compatible with lodgepole pine. Likewise alder produced mycorrhizae with only 4 out of 28 fungal isolates of diverse genera which had mycorrhizal potential on other hosts. Other fungi were less specific in their hosts, forming mycorrhizae and sporocarps with wider host ranges. For example, *L. laccata* induced mycorrhizae with eucalypts, conifers, and ericaceous shrubs. The industrial value of *L. laccata* was demonstrated by several studies of container-grown conifers and shrubs. However, as discussed by Molina and Trappe [88], the value of the specific symbionts in nature is unknown, and we are unable to predict the effects of addition of monocultures, selected by man, to the environment.

Variability between isolates of a selected ectomycorrhizal species was reported [89,91] in studies of ectomycorrhiza development in loblolly pine inoculated with three isolates of *P. tinctorius*. Dixon et al. [89,91] support the views of previous researchers in suggesting that both the host and fungal genotypes are influential in symbiosis.

2. Genotype Effects in Ectomycorrhizal Fungi

Natural ectomycorrhizae involve dikaryotic mycelia in which two or more genetically different nuclei coexist. Debaud et al. [90,92] demonstrated that homokaryotic mycelia derived from specific matings of *Hebeloma cylindrosporum* have morphological, biochemical, and ultrastructural characteristics identical to those of wild-type dikaryons in interactions with *Pinus pinaster*. These data indicate that nuclei which are genetically distinct at least in their incompatibility groups are not a prerequisite for ectomycorrhiza formation by *Hebeloma* isolates. Previous studies by Gay and Debaud [91,93] illustrated that the ability of isolates of *Hebeloma cylindrosporum* to synthesize indole acetic acid (IAA) from tryptophan was as varied between homokaryons as it was between dikaryons prepared from control matings and naturally isolated dikaryotic mycelia. The synthesis of IAA was investigated because of the potential effect of this plant growth regulator on modifying the structure of the host root to perhaps promote mycorrhiza formation. These data indicate a lack of correlation between the level of IAA synthesis and mycorrhizal potential. However, studies of a key enzyme, glutamate dehydrogenase (GDH), involved in nitrogen assimilation by ectomycorrhizal fungi, revealed differences between the mono- and dikaryons [94]. The activity was depressed in the monokaryon and frequently elevated in the dikaryons. Whether this relationship is just an effect of gene dosage is unknown.

Studies by Kropp et al. [95,96] with *Laccaria bicolor* revealed a lesser degree of effectiveness of certain monokaryons in ectomycorrhiza formation. Although one isolated monokaryon had mycorrhizal potential, others which were initially able to form mycorrhizae lost their potential for colonization with time in culture. Another key observation was that certain dikaryons formed by anastomosis of selective monokaryons displayed reduced mycorrhizal potential with time. These data suggest that different types of nuclei are required for mycorrhiza formation in this fungus. Similar conclusions are also apparent from studies of *Suillus granulatus*, for which monospore isolates were demonstrated to be incapable of mycorrhiza formation in *P. pinaster* [97]. These data suggest a subtle but strong role for incompatibility loci in nature; presumably they permit the formation of dikaryons which have greater mycorrhizal potential.

The reduced mycorrhizal ability of certain monokaryons to produce mycorrhizae will limit the usefulness of these species in mutational analyses of the genes essential for symbiosis. However, the ability to regenerate an active dikaryon from monokaryons may permit one of the partners to be genetically engineered and to contribute this trait to a stable dikaryotic isolate suitable for agricultural release.

3. Specificity in VAM Symbiosis

Effects of the host: Plant species show wide variability in their responsiveness to dependency on VA mycorrhizae and this area is elegantly discussed in Chapter 26, especially in relation to models for efficacy. Research in this area is further discussed here to stress further the influence of the genetic potential of the host in the symbiosis. For example, Hepper [98] in studies of infection of clover by *G. mosseae* revealed one clover plant that was incompatible through prevention of fungal entry. These studies also demonstrated that the extent of colonization was variable with age and thus presumably with plant gene expression. Hepper observed decreased infection as clover aged, whereas leek upon aging retained compatibility. The concept of the root possessing only discrete tissue zones that are receptive to mycorrhizal fungus colonization is well discussed in Harley and Smith [99]. These data substantiate that plant cell differentiation, and thus host gene expression, is related to mycorrhizal potential.

Consequently, timing of application of inoculum may be crucial especially with certain plants that generate resistance to mycorrhizal fungi with aging.

In recent studies one cultivar of pea was found to have growth promoted by either VA mycorrhizal fungal inoculation or phosphate treatments whereas these treatments with another cultivar were ineffective [100]. A third cultivar was responsive only to phosphate but not to *Glomus* infections. Equal mycorrhizal fungus colonization occurred with all three cultivars. Cultivars of forage range grasses (*Agropyron*) showed different responses to VA mycorrhizal fungi; no correlation was observed between infection level or sporulation efficiency with VA mycorrhizal fungi and the different ploidy levels of the cultivars [101].

A differential host response to inoculum from a single source was apparent at the half-sib level in alfalfa [102]. Differences were observed in the mean percent root colonization, and thus resemble the variability observed between cultivars of other hosts [100]. The analysis of Lackie et al. [102] suggested that the heritability of the trait for VA mycorrhizal association was sufficiently strong to permit active selection. Indeed co-selection of mycorrhizal colonization with other traits may be occurring. In 1984, Toth et al. demonstrated that field-grown maize, which displayed a heritable trait for high P in the ear leaf at silking, was more mycorrhizal than maize cultivars with a low ear leaf P [103].

Loss of ability to form mycorrhizae may also be inbred. Crops selected for high yields in the modern agricultural setting of fertilization and chemical pesticide controls may demonstrate reduced potential to become mycorrhizal when raised under more nutrient-sufficient conditions. A survey of 15 soybean cultivars bread for a variety of traits including resistance to *Phytophthora megasperma* (Drechs.) var. *sojae* and *Rhizobium* nodulation demonstrated ranges in VA mycorrhizal colonization potential from 57 to 80% using VA mycorrhizal inoculum present in field soil [104]. Studies by Di [101] demonstrated that three cultivars of *Agropyron* which have been selected by plant breeders displayed little beneficial responses upon VA mycorrhizal colonization. In contrast, two wild *Agropyron* cultivars showed positive responses to colonization. These data suggest that traits for beneficial effects of mycorrhizal interactions could be maximized in field crops by breeding or genetic engineering.

Fungal effect of fungi: Several studies indicate differences between VA mycorrhizal fungal isolates in their effectiveness with a common host. Schubert and Hayman [48] working with onion revealed that different species of *Glomus* and *Gigaspora* varied in rate and extent of root colonization differentially with the level of phosphate amendment. Inoculation of apple with five species of *Glomus* indicated that two species, *G. intraradices* and *G. spigaeum*, were most effective in causing increases in dry weight and P uptake [105]. These observations correlated with the rapid ability of these more successful species to attain high levels of infection. Similarly with green ash one isolate, *G. etunicatum*, which colonized at a lower rate and to a lower extent than another *Glomus* species and a *Gigaspora* species, enhanced P uptake to a lesser degree [106]. However, little difference in effectiveness was observed between the three VA mycorrhizal isolates when examining height, root collar dimension, and root and shoot dry weights of the ash seedlings [106]. These data emphasize the importance of measuring the desired phenotype in comparative analysis of inoculum potential; measurement of phosphate level or colonization parameters alone could be misleading. Rate of colonization remains a highly important feature when considering the value of the isolate in commercial utilization, as discussed in Chapter 26.

Daniels-Hetrick and Bloom [107] indicate that the host plant may also influence the degree of sporulation of a VA mycorrhizal fungus. Sporulation by *Glomus fasciculatus* was influenced by the host plant, but *G. pecrocarpum* and *G. mosseae* were not. These data suggest that choice of inoculum may be critical if the effects of commercial application of inoculum are to depend on generation of spores within the treated crop.

Studies by Graham et al. [108] suggest that another variable in effectiveness between VA mycorrhizal fungal isolates may be the ability to produce external hyphae. The *Glomus* species, which were more effective in stimulating growth of Troyer citrange, colonized to the same extent as lesser effective isolates but produced greater amounts of external hyphae.

Examination of additional parameters may be needed in situations involving other organisms which colonize the rhizosphere or root. Three isolates of *G. mosseae* were shown to have a fourfold difference in growth promotion and to have variable effects on nitrogen fixation efficiency in inoculated soybean [109]. These data strengthen the earlier statement by Bethenfalvay et al. that *Rhizobium* and VA mycorrhizal strains need careful selection for maximum compatibility [80].

IV. ECOLOGICAL CONSIDERATIONS OF COMMERCIAL RELEASE OF INOCULUM

One primary concern of ecological consequences of commercial releases of inoculum is whether the additional supply of mycorrhizal fungus inoculum added to soils would have unwanted effects connected with nontarget hosts. Additional questions of effects on nontarget hosts are raised when considering the release of genetically modified fungi. These fungi may contain inserts of DNA or RNA which encode beneficial traits. There the movement of the modified genes may be distinct from the spread and development of the original vector, the mycorrhizal fungus. These modified genes may be transferred to nontarget hosts, where they may exert effects and be subjected to different factors which regulate their survival and spread.

These considerations are discussed here by considering the ecological effects of mycorrhizal fungi as organisms and then examining our present knowledge of mechanisms by which fungal genes can be exchanged and altered. The organismal studies we have selected center on the effects of mycorrhizal fungi in nutrient cycling and mechanisms for spread of added inoculum. Additional discussions in this area have been presented by Harley and Smith [99] and Daniels-Hetrick [110]. The discussion of genetic distribution examines known mechanisms of fungal genetic exchange systems, including the importance of extrachromosomal materials.

A. Nutrient Distribution

The interconnection of the roots of plants, even of different species, by mycorrhizal fungal hyphae is well documented. A plant may be colonized by more than one fungus and hyphae from one inoculated root may ramify to roots of many plants. Thus the hyphae of mycorrhizal fungi may constitute channels for underground exchange of nutrients. These hyphal connections mediate transport of phosphate, carbon, nitrogen, sulfur, and water between the colonized plants [111-113]. Competition between plants for soil reserves mediated through the mycorrhizal connections is documented. One species of a mycorrhizal *Agropyron* grass was more competitive than another in obtaining phosphate from soil spaces shared with sagebrush which was also mycorrhizal [120]. Caldwell et al. [121] demonstrated rapid shifts in phosphate acquisition between neighboring plants which were mycorrhizal, dependent on stresses induced in one of the plants involved in the system. It is likely that nontarget plants would be enveloped in the web of mycorrhizal connections emanating from plants with mycorrhizae developed from added inoculum. Studies by Pellet and Sieverding [122] indicated that weed species are highly mycorrhizal and could consequently act as unnecessary sinks for the soil nutrients. Indeed the nontarget plants may be better competitors for the nutrients than the target plants. However, an alternative view suggested by Trappe [123] is that potential nontarget plants could be utilized as persistent reservoirs for new inoculum throughout time. Whatever the effect of nontarget colonization might be, field preparation and choice of inoculum are important considerations.

B. Inoculum Dispersal

The spread of added inoculum is intimately associated with effects on nontarget plants. Spread by hyphal contact or by an infected root merging with a nonmycorrhizal root will occur over limited distances. Other evidence suggests that plant root pieces containing mycorrhizae may be transferred by animal and insect vectors and act as potent inoculum sources. More widespread distribution by spore dispersal would also occur.

Studies of the spreading mechanisms of different mycorrhizal fungi are required to provide models for prediction. Studies by Scheltema et al. [124] indicate that *Gigaspora calospora* and *Glomus fasciculatum* have quite distinct mechanisms of spread from a localized inoculum. Wilson [61] made similar conclusions in studying other *Gigaspora* and *Glomus* species. Spores, hyphae, and inoculated plant parts may also be transported by vectors. Allen et al. [125] indicate that gopher or ant activity played key roles in exposing mycorrhizal inoculum buried by accumulation of tephra and ash from the eruption of Mount St. Helens. They also showed that erosion participated in other sites to uncover the inoculum-bearing soil layers. In a strip mine area, Warner et al. [126] suggested that wind could account for the movement of large numbers of spores up to 2 km. Mammals and insects acted as vectors for lesser distances. Colonization of nontarget weed or hedgerow plants that are perennial could act as continual inoculum sources after the initial commercial application.

C. Genetic Redistribution in Mycorrhizal Fungi

The efficacy of added mycorrhizal inoculum over a prolonged time and the contribution of genetic material to the indigenous flora may depend on the potential for genetic exchange. The processes of genetic exchange may result in a dilution of genes of a derived phenotype from the added mycorrhizal inoculum. Furthermore genes that may have been genetically engineered into an isolate may be dispersed into other organisms and hence their patterns of spread and survival made distinct from those inherent in the initial vector.

D. Recombination Strategies

Genetic exchange between fungi may occur as a result of sexual and asexual processes which are governed by incompatibility factors. Sexual incompatibility prevents meiosis and the completion of a sexual cycle. Vegetative incompatibility blocks exchange that occurs following the hyphal fusion termed anastomosis. Parasexuality, a process in which hyphal anastomosis is followed by nuclear fusion, recombination, and resolution of the nuclei without meiosis, may occur at low frequency in mycelia which are sexually or vegetatively incompatible.

We are in our infancy in terms of our understanding the role of asexual and sexual gene exchanges in mycorrhizal fungi. Laboratory studies with ectomycorrhizal fungi of the species *Laccaria*, *Suillus*, and *Hebeloma* have been initiated [92,95,96,112]. Homokaryons have been generated from the dikaryons isolated from nature and selected dikaryons have been reconstructed. The mycorrhizal potential and other genetic parameters of these designated strains are being examined as discussed earlier.

Fries [127] observed that isolates of *Suillus luteus* taken from different loci within a test plot were vegetatively and sexually incompatible. Eleven different vegetative incompatibility groups were detected among 25 isolates at nine sampling sites. Only in two cases were isolates of different incompatibility groups detected at the same site. He proposed that colonization occurs as a result of a single infection and spreads from this focus to form a localized homogeneous population. This situation for *Suillus* resembles the data obtained with a wood decay basidiomycete, *Heterobasidion annosum*, which is believed to colonize from an initial site through root-to-root contact [128]. These data suggest that in nature the probability for genetic exchange between mycorrhizal basidiomycetes may be low. However, with commercial utilization of mycorrhizal fungi the spacing between isolates is negated by placing the inoculum directly into the soils which possess indigenous mycorrhizal species.

Nonsexual genetic exchange mechanisms may be particularly important for transfer of extrachromosomal elements between isolates. Viruslike particles (generally containing dsRNA) and plasmids have been detected in fungi, although their existence in mycorrhizal fungi has not been ascertained [129]. Plasmids from fungal mitochondria are classified into two types: those that have homology with the mitochondrial genome and those that do not. Whether some of the plasmids are able to integrate into the nuclear genome is being actively pursued. Studies with a fruiting basidiomycete, *Coprinus cinereus*, revealed that insertions into functional mitochondrial genes were expressed as an optional intron [130]. Crosses between different isolates of *C. cinereus* indicated that rearrangement of the mitochondrial insertion sites occurred [130]. Optional introns generate genes of different sizes [131]. Conceivably, these insertion, deletion, and recombination events could produce altered products with different properties. The changes may have beneficial consequences under certain conditions, e.g., temperature stress. These possibilities resemble the genetic variability observed in plants by the presence of transposons [132]. A comparison of the mitochondrial insertions to transposonlike elements has been made for the Kalilo-DNA of *Neurosospora intermedia*, which causes senescence by integration into mitochondrial DNA, and for the plant pathogen *Claviceps purpurea* [133] where the mitochondrial insertions are cryptic.

Cryptic plasmids have been identified in several other fungi possessing plant pathogenic potential, including the mitochondria of *Cochliobolus hetero-*

strophus [134], *Ceratocystis fimbriata* [135], and *Gaeumannomyces graminis* [136]. The *Claviceps* plasmids had sequence homology to plasmids in plants, including its host, maize [132]. Linear plasmids have been identified from the pathogen *Tilletia controversa* [137].

Other observations of extrachromosomal elements in plant pathogenic fungi suggest that they may participate in modulating the interaction of the fungus with the plant. Kistler et al. [138,139] demonstrated different plasmids in races of *F. oxysporum* f. sp. *conglutinans*. A plasmid isolated from a race 2 isolate showed homology with other race 2 plasmids but not race 1 and 5 plasmids. Race 1 and 5 elements possessed common DNA. These homologies are interesting in that these races have different patterns of pathogenicity; race 1 and 5 cause disease in cabbage and race 2 disease in radish. Samac and Leong [140] indicated that loss of linear mitochondrial plasmids from *F. solani* f. sp. *cucurbitae* race 1 reduced virulence on a susceptible host.

Cytoplasmic factors have also been correlated with altered virulence patterns. Finkler et al. [141] demonstrated that the presence of dsRNA within *Rhizoctonia solani* was associated with an increased virulence. In contrast, studies on the chestnut blight fungus [142] observed a relationship with the presence of dsRNA and loss of virulence. Both these viruslike dsRNAs are transmitted through anastomosis of vegetatively compatible hypha. Extrapolations of these findings with plant pathogenic fungi to mycorrhizal fungi suggest that extrachromosomal elements could be associated with regulation of the mycorrhizal potential of a fungus through modulation of host specificity and infection potential.

V. PROSPECTS FOR GENETIC ENGINEERING

The release of genetically engineered mycorrhizal fungi is a potential goal. The fungi may be modified by inclusion of genes that improve their mycorrhizal performance. Several targets for genetic engineering of VA mycorrhizal fungi are discussed in Chapter 26. More rapid colonization rates would be a major goal, but genes for enhanced resistance to pollutants or salinity may offer adaption of fungi to specific environments. Alternatively, the fungi may act as vectors for other beneficial traits, such as genes for the toxins from *Bacillus thuringiensis* which have antiinsect properties. Currently, these genes are being engineered into soil saprophytic pseudomonads to enhance the longevity of the chemical in the environment [143,144]. A third reason for genetically engineering a released fungus would be to provide markers so that the survival of the organism, and of the genetic trait, in the environment could be monitored. These types of studies are being utilized by the Environmental Protection Agency in their decisions on whether to permit commercial release of genetically engineered microorganisms [145].

A. Transformation

Transformation of a wide variety of fungi has been achieved through a process which involves DNA treatment of fungal protoplasts and regeneration of mycelia from the transformed cells. The protoplasts are derived from enzymatic treatment of fungal hyphae. This process has been successful with ascomycetes and basidiomycetes, including plant pathogens. It seems likely that mycorrhizal fungi would be amenable to this procedure. Indeed, Kropp and Fortin [146] published that viable *L. bicolor* protoplasts could be formed and would regenerate to give hyphae with mycorrhizal potential. Protoplasts and regenerated mycelia were also obtained by Hébraud and Fèvre [147] for three species of *Hebeloma* and *Suillus bellini*. Fungicide-resistant mutants were isolated after UV treatment of protoplasts derived from monokaryotic *H. cylindrosporum* [147].

Other techniques for transformation may be available. Dickman [148] reported that intact cells of the fungal pathogen *Colletotrichum trifolii* were transformed using lithium acetate treatments. The use of DNA-coated particles [149] and electroporation [150] is reported to transform plant cells; the former process may offer a route for transformation of VA mycorrhizal fungi by treatment of intact mycorrhizal roots. Transformation has been used in novel situations to study complementation of mutants in yeasts and recently in the basidiomycete *Schizophyllum commune* [151].

Extrachromosomal elements may also constitute the basis of vector systems [152,153]. The 2μ plasmids of yeasts have been utilized as vectors [154]. Powell and Kistler (personal communication) indicate that they may be able to utilize an extrachromosomal element from *Fusarium oxysporum* as a vector. Consequently, studies of extrachromosomal elements in mycorrhizal fungi may be profitable.

Current transformations have involved transfer of genes conferring resistance to chemicals. Different degrees of stability have been obtained. Keller et al. [155] indicate instability in some hygromycin-resistant isolates of *Cochliobolus heterostrophus*. In contrast, transformed isolates of *Glomerella cingulata* f. sp. *phaseoli* have shown excellent stability [156]. Recent transformation studies using the protoplast method demonstrated cotransformation with a second, nonselected gene [157-161]. Constructs of the *Escherichia coli lacZ* gene have been used as the second transformed gene and may provide a convenient selection technique based on the color-generating reactions catalyzed by β-galactosidase.

Transformants of the fungal plant pathogens *Septoria nodonum* [161] and *Glomerella cingulata* f. sp. *phaseoli* [156] were demonstrated to have retained virulence on their susceptible host. Transformants recovered from infected plants retained the transformed gene. In some fungi the transformation is reported to involve integration of the donated DNA into the genome [161] whereas in others the gene remains in the plasmid vector which may be present in single or multiple copies [156,157].

Expression of the gene within the transformed cell may require construction of an appropriate promoter. The procedures described by Turgeon [163] for the construction of a promoter library from the ascomycete *Cochliobolus heterostrophus* may be used to isolate promoters from mycorrhizal fungi. However, because subsequent studies indicated that these promoters were active in a variety of heterogeneous fungal species, suitable promoters may already be available. Another degree of sophistication may be to utilize a promoter for the engineered gene which is regulated by plant factors. These constructions would minimize gene expression outside the plant environment. Plant-regulated promoters have been characterized from the root-colonizing bacteria *Agrobacterium* and *Rhizobium* [164,165]. The regulated genes govern the expression of other genes which are involved in the colonization process. Recent studies by Osborn et al. [166] demonstrated that plant-activated promoters were present in the bacterial pathogen *Xanthomonas campestris*. Genes which are regulated by plant factors are also found in plant fungal pathogens. The synthesis by fungal pathogens of enzymes which degrade plant cell walls and cuticular structures is controlled through the availability of breakdown products of these polymers [167,168]. Consequently, it seems likely that plant factors also regulate fungal genes concerned with mycorrhiza formation, and that isolation of plant-activated promoters should be feasible.

B. Identification of Fungal Isolate

Issues of the survival and spread of commercial inoculum and the potential for gene exchange between the introduced and native fungi demand methods for their detection. These problems encountered in mycorrhizal detection have much in common with detection of disease-causing fungi. Two excellent reviews, one in the area of plant pathogenic fungi [169] and the other for medical fungi [170], summarize current technologies, problems, and successes.

Detection depends on unique properties of the isolate. Previous methodologies predominantly used by taxonomists to categorize mycorrhizal fungi were based on morphological characteristics, largely of spores and spore-bearing structures. Clearly these procedures have limitations. Molecular procedures can offer greater sensitivity, even to the isolate level, and speed, as well as permitting examination of a variety of tissue types (spores, hyphae, mycorrhizae). The molecular techniques employ studies of protein character, antibody specificities, and nucleic acid hybridization.

Isozyme analysis of easily detectable enzyme activities has been reported by Hepper et al. [171] to permit differentiation of three *Glomus* species. They examined extracts of roots colonized by the various species for isozymes of the enzymes esterase, glutamate oxaloacetate transaminase, and peptidase. Although the enzymes were identified as being of fungal origin by examination of extracts of noninfected roots, the possibility that they were plant products induced by VA mycorrhizal colonization has not been excluded. Ho and Trappe [172] examined isozyme production for acid and alkaline phosphatase and nitrate reductase from species of the ectomycorrhizal fungus *Rhizopogon*. They suggested that the isozyme patterns could distinguish between species but not between isolates of the same species. Studies by Stenlid [128] indicated that isozyme analysis of different isolates of the pathogenic basidiomycete *Heterobasidion annosum* correlated with characterization of the isolates by vegetative incompatibility groups.

Different immunological responses may be another technique used for isolate detection. Kough et al. [173] raised polyclonal antibodies against *Glomus* cell walls and demonstrated recognition of *Glomus* but not other VA mycorrhizal fungal species or soil- and root-inhabiting fungi. However, their recognition system for antibody-antigen interaction, which was dependent on fluorescence, was inadequate in peat soils because of autofluorescence [174]. Other methodologies for observing antibody-antigen recognition could be more appropriate. Greater specificity may arise by using selected monoclonal antibodies instead of polyclonal preparations. This technique is providing rapid and convenient identification of medical fungal pathogens [170].

The specificity of nucleic acid hybridization offers another independent strategy for fungal identification. The specificity of the probe depends on unique restriction fragments, frequently derived from the mitochondria or from ribosomal genes. Examination of isolates of *Phytophthora* by RFLP analysis of mitochondrial DNA supported their separation into distinct species [175]. Correlations between nuclear rDNA and mitochondrial DNA RFLPs with designation of forma speciales were observed by Kistler et al. [138,139] for strains of *F. oxysporum* pathogenic on crucifers. RFLP analysis of the plant pathogenic hymenomycete *Armillaria mellea* using mitochondrial and nuclear probes indicated greater variability between rather than within species [176]. Analysis of mtDNA by RFLP analysis demonstrated variability between species of the fruiting basidiomycete *Agaricus* as well as between isolates of species *A. brunnescens* [177]. Variability in plasmid size and composition was observed between isolates of the ascomycete *Ascobolus immersus* which were obtained from nature [153]. Recent studies by Armstrong et al. [178] utilized RFLP to distinguish isolates of *L. bicolor* and *L. laccata* collected from diverse geographic sites. They suggest that the RFLP method offers a useful tool for classification of ectomycorrhizal fungi.

VI. CONCLUSION

Utilization of mycorrhizal fungi in agriculture, sylviculture, and horticulture has significant potential. The information we are gaining from the initial field uses is providing guidance to the most promising targets and indicating the direction for future prospects. The developing techniques of molecular biology offer promise to reduce the problems that are inherent in the biology of mycorrhiza formation that currently limit commercialization.

REFERENCES

1. Linderman, R. G. Managing rhizosphere microorganisms in the production of horticultural crops. *HortScience, 21*:1299-1302, 1986.
2. Molina, R. *Ectomycorrhizal Inoculation of Containerized Western Conifer Seedlings*. U.S. Dept. of Agriculture Forest Service, Pacific Northwest Forest and Range Experiment Station, 1980, pp. 1-10.
3. Graham, J. H., and Linderman, R. G. Inoculation of containerized Douglas-fir with the ectomycorrhizal fungus Cenococcum geophilum. *Forest Sci., 27*:27-31, 1981.
4. Molina, R., and Chamard, J. Use of the ectomycorrhizal fungus Laccaria laccata in forestry. II. Effects of fertilizer forms and levels on ectomycorrhizal development and growth of container-grown Douglas-fir and ponderosa pine seedlings. *Can. J. For. Res., 13*:89-95, 1983.
5. Parke, J. L., Linderman, R. G., and Black, C. H. The role of ectomycorrhizas in drought tolerance of Douglas-fir seedlings. *New Phytol., 95*:83-95, 1983.
6. Gagnon, J., Langlois, C. G., and Fortin, J. A. Growth and ectomycorrhiza formation of containerized black spruce seedlings as affected by nitrogen fertilization, inoculum type, and symbiont. *Can. J. For. Res., 18*:922-929, 1988.
7. McAfee, B. J., and Fortin, J. A. Comparative effects of the soil microflora on ectomycorrhizal inoculation of conifer seedlings. *New Phytol., 108*:443-449, 1988.
8. Marx, D. H. Ectomycorrhizae as biological deterrents to pathogenic root infections. *Annu. Rev. Phytopathol., 10*:429-454, 1972.

9. Duchesne, L. C., Peterson, R. L., and Ellis, B. E. Interaction between the ectomycorrhizal fungi Paxillus involutus and Pinus resinosa induces resistance to Fusarium oxysporum. *Can. J. Bot.*, *66*:558-562, 1987.
10. Sylvia, D. M., and Sinclair, W. A. Suppressive influence of Laccaria laccata on Fusarium oxysporum and on Douglas-fir seedlings. *Phytopathology*, *73*:384-389, 1983.
11. Duchesne, L. C., Peterson, R. L., and Ellis, B. E. Pine root exudate stimulates the synthesis of antifungal compounds by the ectomycorrhizal fungus Paxillus involutus. *New Phytol.*, *108*:471-476, 1988.
12. Sylvia, D. M. Role of Laccaria laccata in protecting primary roots of Douglas-fir from root rot. *Plant Soil*, *71*:299-302, 1983.
13. Allen, M. F. Influence of vesicular-arbuscular mycorrhizae on water movement through Bouteloua gracilis (H.B.K.) lag ex steud. *New Phytol.*, *91*:191-196, 1982.
14. Allen, M. F., and Boosalis, M. C. Effects of two species of VA mycorrhizal fungi on drought tolerance of wheat. *New Phytol.*, *93*:61-76, 1983.
15. Dakessian, S., Brown, M. S., and Bethlenfalvay, G. J. Relationship of mycorrhizal growth enhancement and plant growth with soil water and texture. *Plant Soil*, *94*:439-443, 1986.
16. Carpenter, A. T., and Allen, M. F. Responses of Hedysarum aboreale Nutt. to mycorrhizae and Rhizobium: plant and soil nutrient changes in a disturbed shrub-steppe. *New Phytol.*, *109*:125-132, 1988.
17. Nelson, C. E., and Safir, G. R. Increased drought tolerance of mycorrhizal onion plants caused by improved phosphorus nutrition. *Planta*, *154*:407-413, 1982.
18. Pfeiffer, C. M., and Bloss, H. E. Growth and nutrition of guayule Parthenium argentatum in a saline soil as influenced by vesicular-arbuscular mycorrhiza and phosphorus fertilization. *New Phytol.*, *108*: 315-321, 1988.
19. Daniels-Hetrick, B. A., Gerschefske Kitt, D., and Thompson Wilson, G. Effects of drought stress on growth response in corn, sudan grass, and big bluestem to Glomus etunicatum. *New Phytol.*, *105*:403-410, 1987.
20. Graham, J. H., Syvertsen, J. P., and Smith, M. L. Water relations of mycorrhizal and phosphorus-fertilized non-mycorrhizal citrus under drought stress. *New Phytol.*, *105*:411-419, 1987.
21. Skujins, J., and Allen, M. F. Use of mycorrhizae for land rehabilitation. *Mircen J.*, *2*:161-176, 1986.
22. Duke, E. R., Johnson, C. R., and Koch, K. E. Accumulation of phosphorus, dry matter and betaine during NaCl stress of split-root citrus seedlings colonized with vesicular-arbuscular mycorrhizal fungi on zero, one or two halves. *New Phytol.*, *104*:583-590, 1986.
23. Mahoney, M. J., Chevone, B. I., Skelly, J. M., and Moore, L. D. Influence of mycorrhizae on the growth of loblolly pine seedlings exposed to ozone and sulfur dioxide. *Phytopathology*, *76*:679-682, 1985.
24. Dixon, R. K., and Buschena, C. A. Response of ectomycorrhizal Pinus banksiana and Picea glauca to heavy metals in soil. *Plant Soil*, *105*: 265-271, 1988.
25. Jones, M. D., and Hutchinson, T. C. Nickel toxicity in mycorrhizal birch seedlings infected with Lactarius rufus or Scleroderma flavidum. I. Effects on growth, photosynthesis, respiration and transpiration. *New Phytol.*, *108*:451-459, 1988.

26. Gildon, A., and Tinker, P. B. A heavy metal tolerant strain of a mycorrhizal fungus. *Trans. Br. Mycol. Soc.*, *77*:648-649, 1981.
27. Jones, M. D., and Hutchinson, T. C. Nickel toxicity in mycorrhizal birch seedlings infected with Lactarius rufus or Scleroderma flavidum II. Uptake of nickel, calcium, magnesium phosphorus and iron. *New Phytol.*, *108*:461-470, 1988.
28. Call, C. A., and McKell, C. M. Endomycorrhizae enhance growth of shrub species in processed oil shale and disturbed native soil. *J. Range Management*, *38*:258-261, 1985.
29. MacMahon, J. A. *Disturbed Lands and Ecological Theory: An Essay About a Mutualistic Association*. Restoration Ecology (W. R. Jordon, M. E. Gilpin, and J. D. Aber, eds.), Cambridge University Press, Cambridge, 1987, pp. 221-237.
30. Aldon, E. F. *Endomycorrhizae Enhance Survival and Growth of Four-wing Saltbrush on Coal Mine Spoils*. U.S. Dept. of Agriculture Forest Service Research Note RM-294, 1975.
31. Aldon, E. F. Endomycorrhizae enhance shrub growth and survival on mine spoils. In *The Reclamation of Disturbed Arid Lands* (R. A. Wright, ed.), University of New Mexico Press, Albuquerque, 1978, pp. 174-170.
32. Marx, D. H. Mycorrhizae and establishment of trees on strip-mined land. *Ohio J. Sci.*, *75*:288-297, 1975.
33. Hartmond, U., Schaesberg, N. V., Graham, J. H., and Syvertsen, J. P. Salinity and flooding stress effects on mycorrhizal and non-mycorrhizal citrus rootstock seedlings. *Plant Soil*, *104*:37-43, 1987.
34. Anderson, R. C., Ebbers, B. C., and Liberta, A. E. Soil moisture influences colonization of prairie cordgrass (Spartina pectinata Lind.) by vesicular-arbuscular mycorrhizal fungi. *New Phytol.*, *102*:523-527, 1986.
35. Tester, M., Smith, S. E., Smith, F. A., and Walker, N. A. Effects of photon irradiance on the growth of shoots and roots, on the rate of initiation of mycorrhizal infection and on the growth of infection units in Trifolium subterraneum L. *New Phytol.*, *103*:375-390, 1986.
36. Bethlenfalvay, G. J., and Dakessian, S. Grazing effects on mycorrhizal colonization and floristic composition of the vegetation on a semiarid range in northern Nevada. *J. Range Management*, *37*:312-316, 1984.
37. Trappe, J. M. Selection of fungi for ectomycorrhizal inoculations in nurseries. *Annu. Rev. Phytopathol.*, *15*:203-222, 1977.
38. Cline, M. L., France, R. C., and Reid, C. P. P. Intraspecific and interspecific growth variation of ectomycorrhizal fungi at different temperatures. *Can. J. Bot.*, *65*:869-875, 1987.
39. Smith, G. S., and Roncadori, R. W. Responses of three vesicular-arbuscular mycorrhizal fungi at four soil temperatures and their effects on cotton growth. *New Phytol.*, *104*:89-95, 1986.
40. Parke, J. L., Linderman, R. G., and Trappe, J. M. Effect of root zone temperature on ectomycorrhiza and vesicular-arbuscular mycorrhiza formation in disturbed and undisturbed forest soils of southwest Oregon. *Can. J. For. Res.*, *13*:657-665, 1983.
41. Dodd, J. C., and Jeffries, P. Early development of vesicular-arbuscular mycorrhizas in autumn-sown cereals. *Soil Biol. Biochem.*, *18*:149-154, 1986.
42. Allen, M. F. Formation of vesicular-arbuscular mycorrhizae in Atriplex gardneri (Chenopodiaceae): seasonal response in a cold desert. *Mycologia*, *75*:773-776, 1983.

43. Dighton, J., and Skeffington, R. A. Effects of artificial acid precipitation on the mycorrhizae of scots pine seedlings. *New Phytol.*, *107*:191-202, 1987.
44. Termorshuizen, A. J., and Schaffers, A. P. Occurrence of carpophores of ectomycorrhizal fungi in selected stands of Pinus sylvestris in the Netherlands in relation to stand vitality and air pollution. *Plant Soil*, *104*:209-217, 1987.
45. Stroo, H. F., Reich, P. B., Schoettle, A. W., and Amundson, R. G. Effects of ozone and acid rain on white pine (Pinus strobus) seedlings grown in five soils. II. Mycorrhizal infection. *Can. J. Bot.*, *66*:1510-1516, 1988.
46. Bell, B., Evans, C. S., and Roberts, E. R. Decreased incidence of mycorrhizal root tips associated with soil heavy-metal enrichment. *Plant Soil*, *106*:143-145, 1988.
47. Trappe, J. M., Stahly, E. A., Benson, N. R., and Duff, D. M. Mycorrhizal deficiency of apple trees in high arsenic soils. *HortScience*, *8*(1):52-53, 1973.
48. Schubert, A., and Hayman, D. S. Plant growth responses to vesicular-arbuscular mycorrhiza. XVI. Effectiveness of different endophytes at different levels of soil phosphate. *New Phytol.*, *103*:79-90, 1986.
49. Thomson, B. D., Robson, A. D., and Abbott, L. K. Effects of phosphorus on the formation of mycorrhiza by Gigaspora calospora and Glomus fasciculatum in relation to root carbohydrates. *New Phytol.*, *103*:751-765, 1986.
50. Baltruschat, H., and Dehne, H. W. The occurrence of vesicular-arbuscular mycorrhiza in agro-ecosystems. Influence of nitrogen fertilization and green manure in continuous monoculture and in crop rotation on the inoculum potential of winter wheat. *Plant Soil*, *107*:279-284, 1988.
51. Trappe, J. M., Molina, R., and Castellano, M. Reactions of mycorrhizal fungi and mycorrhiza formation to pesticides. *Annu. Rev. Phytopathol.*, *22*:331-359, 1984.
52. Kough, J. L., Gianinazzi-Pearson, V., and Gianinazzi, S. Depressed metabolic activity of vesicular-arbuscular mycorrhizal fungi after fungicide applications. *New Phytol.*, *106*:707-715, 1987.
53. Fitter, A. H. Effect of benomyl on leaf phosphorous concentration in alpine grasslands: a test of mycorrhizal benefit. 1986. *New Phytol.*, *103*:767-776, 1986.
54. Buttery, B. R., Park, S. J., Findlay, W. I., and Dhanvantari, B. N. Effects of fumigation and fertilizer on growth, yield, chemical composition, and mycorrhizae in white bean and soybean. *Can. J. Plant Sci.*, *68*:677-686, 1988.
55. Schwab, S. M., Johnson, E. L. V., and Menge, J. A. Influence of simazine on formation of vesicular-arbuscular mycorrhizae in Chenopodium quiona wild. *Plant Soil*, *64*:283-287, 1982.
56. Garbaye, J., and Bowen, G. D. Effect of different microflora on the success of ectomycorrhizal inoculation of Pinus radiata. *Can. J. For. Res.*, *17*:941-943, 1987.
57. Bowen, G. D., and Theodorou, C. Interactions between bacteria and ectomycorrhial fungi. *Soil Biol. Biochem.*, *2*:119-126, 1979.
58. Ocampo, J. A. Effect of crop rotations involving host and non-host plants on vesicular-arbuscular mycorrhizal infection of host plants. *Plant Soil*, *56*:283-291, 1980.

59. Ocampo, J. A., and Hayman, D. S. Influence of plant interactions on vesicular-arbuscular mycorrhizal infections. *New Phytol.*, *87*:333-343, 1981.
60. Schenck, N. C., and Kinloch, R. A. Incidence of mycorrhizal fungi on six field crops in monoculture on a newly cleared woodland site. *Mycologia*, *72*:445-456, 1980.
61. Wilson, J. M. Comparative development of infection by three vesicular-arbuscular mycorrhizal fungi. *New Phytol.*, *97*:413-426, 1984.
62. Wilson, J. M. Competition for infection between vesicular-arbuscular mycorrhizal fungi. *New Phytol.*, *97*:427-435, 1984.
63. Lopez-Aguillon, R., and Mosse, B. Experiments on competitiveness of three endomycorrhizal fungi. *Plant Soil*, *97*:155-170, 1987.
64. Kough, J. L., Molina, R., and Linderman, R. G. Mycorrhizal responsiveness of Thuja, Calocedrus, Sequoia, and Sequoiadendron species of western North America. *Can. J. Bot.*, *15*:1049-1054, 1985.
65. Chilvers, G. A., Lapeyrie, F. F., and Horan, D. P. Ectomycorrhizal vs endomycorrhizal fungi within the same root system. *New Phytol.*, *107*:441-448, 1987.
66. Malajczuk, N., Dell, B., and Bougher, N. L. Ectomycorrhiza formation in eucalyptus. III. Superficial ectomycorrhizas initiated by Hysterangium and Cortinarius species. *New Phytol.*, *105*:421-428, 1987.
67. Bagyaraj, D. J. Biological interaction with VA mycorrhizal fungi. In *VA Mycorrhiza* (C. L. I. Powell and D. J. Bagyariaj, eds.), CRC Press, Boca Raton, 1984, pp. 132-153.
68. Ames, R. N., Reid, C. P. P., and Ingham, E. R. Rhizosphere bacteria population responses to root colonization by a vesicular-arbuscular mycorrhizal fungus. *New Phytol.*, *96*:555-563, 1984.
69. Bagyaraj, D. J., and Menge, J. A. Interaction between a VA mycorrhiza and Azotobacter and their effects on rhizosphere microflora and plant growth. *New Phytol.*, *80*:567-573, 1978.
70. Barea, J. M., Azcón, R., and Hayman, D. S. Possible synergistic interactions between Endogone and phosphate-solubilizing bacteria in low-phosphate soils. In *Endomycorrhizas* (F. E. Sanders, B. Mosse, and P. B. Tinker, eds.), Academic Press, San Francisco, 1975, pp. 409-417.
71. Azcón, R., Barea, J. M., and Hayman, D. S. Utilization of rock phosphate in alkaline soils by plants inoculated with mycorrhizal fungi and phosphate-solubilizing bacteria. *Soil Biol. Biochem.*, *8*:135-138, 1976.
72. Meyer, J. R., and Linderman, R. G. Response of subterranean clover to dual inoculation with vesicular-arbuscular mycorrhizal fungi and a plant growth-promoting bacterium, Pseudomonas putida. *Soil Biol. Biochem.*, *18*:185-190, 1986.
73. Meyer, J. R. and Linderman, R. G. Selective influence on populations of rhizosphere or rhizoplane bacteria and actinomycetes by mycorrhizas formed by Glomus fasciculatum. *Soil Biol. Biochem.*, *18*:191-196, 1986.
74. Linderman, R. G. Mycorrhizal interactions with the rhizosphere microflora: the mycorrhizosphere effect. *Phytopathology*, *78*:366-371, 1988.
75. Piccini, D., and Azcón, R. Effect of phosphate-solubilizing bacteria and vesicular-arbuscular mycorrhizal fungi on the utilization of Bayovar phosphate by alfalfa plants using a sand-vermiculite medium. *Plant Soil*, *101*:45-50, 1987.

76. Ho, I. Interaction between VA-mycorrhizal fungus and Azotobacter and their combined effects on growth of tall fescue. *Plant Soil, 105*: 291-293, 1988.
77. Ames, R. N., and Bethlenfalvay, G. J. Localized increase in nodule activity but no competitive interaction of cowpea rhizobia due to pre-establishment of vesicular-arbuscular mycorrhiza. *New Phytol., 106*: 207-215, 1987.
78. Bethlenfalvay, G. J., Brown, M. S., and Stafford, A. E. Glycine-Glomus-Rhizobium symbiosis. *Plant Physiol., 79*:1054-1058, 1985.
79. Bethlenfalvay, G. J. and Yoder, J. F. The Glycine-Glomus-Rhizobium symbiosis. I. Phosphorus effect on nitrogen fixation and mycorrhizal infection. *Physiologia Plantarum, 52*:141-145, 1981.
80. Bayne, H. G. and G. J. Bethlenfalvay. The Glycine-Glomus-Rhizobium symbiosis. IV. Interactions between the mycorrhizal and nitrogen-fixing endophytes. *Plant. Cell Environ., 10*:607-612, 1987.
81. Ross, J. P. Effect of nontreated field soil on sporulation of vesicular-arbuscular mycorrhizal fungi associated with soybean. *Phytopathology, 70*:1200-1205, 1980.
82. Wilson, G. W. T., Daniels Hetrick, B. A., and Gerschefski Kitt, D. Suppression of mycorrhizal growth response of big bluestem by non-sterile soil. *Mycologia, 80*:338-343, 1988.
83. Ross, J. P., and Ruttencutter, R. Population dynamics of two vesicular-arbuscular endomycorrhizal fungi and the role of hyperparasitic fungi. *Phytopathology, 67*:490-496, 1977.
84. Daniels, B. A., and Menge, J. A. Hyperparasitism of vesicular-arbuscular mycorrhizal fungi. *Phytopathology, 70*:584-588, 1980.
85. Paulitz, T. C., and Menge, J. A. The effects of a mycoparasite on the mycorrhizal fungus, Glomus deserticola. *Phytopathology, 76*:351-354, 1985.
86. Hirrel, M. C., Mehravaran, H., and Gerdemann, J. W. Vesicular-arbuscular mycorrhizae in the Chenopodiaceae and Cruciferae: do they occur? *Can. J. Bot., 56*:2813-2817, 1978.
87. Tester, M., Smith, S. E., and Smith, F. A. The phenomenon of "nonmycorrhizal" plants. *Can. J. Bot., 65*:419-431, 1987.
88. Molina, R., and Trappe, J. M. Patterns of ectomycorrhizal host specificity and potential among Pacific Northwest conifers and fungi. *Forest Sci., 28*:423-458, 1982.
89. Molina, R. Pure culture synthesis and host specificity of red alder mycorrhizae. *Can. J. Bot., 57*:1223-1228, 1979.
90. Malajczuk, N., Molina, R., and Trappe, J. M. Ectomycorrhiza formation in eucalyptus. I. Pure culture synthesis, host specificity and mycorrhizal compatibility with Pinus radiata. *New Phytol., 91*:1-16, 1982.
91. Dixon, R. K., Garrett, H. E., and Stelzer, H. E. Growth and ectomycorrhizal development of Loblolly Pine progenies inoculated with three isolates of Pisolithus tinctorius. *Silvae Genetica, 36*:240-244, 1987.
92. Debaud, J. C., Gay, G., Prevost, A., Lei, J., and Dexheimer, J. Ectomycorrhizal ability of genetically different homokaryotic and dikaryotic mycelia of Hebeloma cylindrosporum. *New Phytol., 108*:323-328, 1988.
93. Gay, G., and Debaud, J. C. Genetic study on indole-3-acetic acid production by ectomycorrhizal Hebeloma species: inter- and intraspecific variability in homo- and dikaryotic mycelia. *Appl. Microbiol. Biotechnol., 26*:141-146, 1987.

94. Wagner, F., Gay, G., and Debaud, J. C. Genetical variability of glutamate dehydrogenase activity in monokaryotic and dikaryotic mycelia of the ectomycorrhizal fungus Hebeloma cylindrosporum. *Appl. Microbiol. Biotechnol.*, *28*:566-571, 1988.
95. Kropp, B. R., McAfee, B. J., and Fortin, J. A. Variable loss of ectomycorrhizal ability in monokaryotic and dikaryotic cultures Laccaria bicolor. *Can. J. Bot.*, *65*:500-504, 1987.
96. Kropp, B. R., and Fortin, J. A. The incompatibility system and relative ectomycorrhizal performance of monokaryons and reconstituted dikaryons of Laccaria bicolor. *Can. J. Bot.*, *66*:289-294, 1988.
97. Ducamp, M., Poitou, N., and Oliver, J. M. Cytological and biochemical comparisons of monospore cultures and fruit-body isolates of Suillus granulatus (fr. ex L.)/Kuntze. In *Physiological and Genetical Aspects of Mycorrhizae* (V. Gianinazzi-Pearson and S. Gianinazzi, eds.), NRA, Paris, 1986.
98. Hepper, C. M. Influence of age of roots on the pattern of vesicular-arbuscular mycorrhizal infection in leek and clover. *New Phytol.*, *101*:685-693, 1985.
99. Harley, J. L., and Smith, S. E. *Mycorrhizal Symbiosis*. Academic Press, New York, 1983.
100. Estaun, V., Calvet, C., and Haymay, D. S. Influence of plant genotype on mycorrhizal infection: response of three pea cultivars. *Plant Soil*, *103*:295-298, 1987.
101. Di, J. J. Occurrence of VAM in degraded rangelands and physiological responses of six Agropyron cultivars to infection. MS thesis. Utah State University, Logan, 1988.
102. Lackie, S. M., Bowley, S. R., and Peterson, R. L. Comparison of colonization among half-sib families of Medicago sativa L. by Glomus versiforme (Daniels and Trappe) Berch. *New Phytol.*, *108*:477-482, 1988.
103. Toth, R., Page, T., and Castleberry, R. Differences in mycorrhizal colonization of maize selections for high and low ear leaf phosphorus. *Crop Sci.*, *24*:994-996, 1984.
104. Heckman, J. R., and Angle, J. S. Variation between soybean cultivars in vesicular-arbuscular mycorrhiza fungi colonization. *Agron. J.*, *79*:428-430, 1987.
105. Reich, L. Rates of infection and effects of five vesicular-arbuscular mycorrhizal fungi on apple. *Can. J. Plant Sci.*, *68*:233-239, 1988.
106. Lamar, R. T., and Davey, C. B. Comparative effectivity of three Fraxinus pennsylvanica marsh. vesicular-arbuscular mycorrhizal fungi in a high-phosphorus nursery soil. *New Phytol.*, *109*:171-181, 1988.
107. Daniels-Hetrick, B. A., and Bloom, J. The influence of host plant on production and colonization of vesicular-arbuscular mycorrhizal spores. *Mycologia*, *78*:32-36, 1986.
108. Graham, J. H., Linderman, R. G., and Menge, J. A. Development of external hyphae by different isolates of mycorrhizal Glomus spp. in relation to root colonization and growth of Troyer citrange. *New Phytol.*, *91*:183-189, 1982.
109. Bethlenfalvay, G. J., Franson, R. L., Mihara, K. L., and Brown, M. S. (personal communication).
110. Daniels-Hetrick, B. A. Ecology of VA mycorrhizal fungi. In *VA Mycorrhiza* (C. L. I. Powell and D. J. Bagyaraj, eds.), CRC Press, Boca Raton, 1984, pp. 57-94.

111. Abuzinadah, R. A. and Read, D. J. The role of proteins in the nitrogen nutrition of ectomycorrhizal plants. I. Utilization of peptides and proteins by ectomycorrhizal fungi. *New Phytol.*, *103*:481-493, 1986.
112. Abuzinadah, R. A., and Read, D. J. The role of proteins in the nitrogen nutrition of ectomycorrhizal plants. III. Protein utilization by Betula, Picea and Pinus in mycorrhizal association with Hebeloma crustuliniforme. *New Phytol.*, *103*:507-514, 1986.
113. Abuzinadah, R. A., Finlay, R. D., and Read, D. J. The role of proteins in the nitrogen nutrition of ectomycorrhizal plants. II. Utilization of protein by mycorrhizal plants of Pinus contorta. *New Phytol.*, *103*:495-506, 1986.
114. Duddridge, J. A., Malibari, A., and Read, D. J. Structure and function of mycorrhizal rhizomorphs with special reference to their role in water transport. *Nature*, *287*:834-836, 1980.
115. Finlay, R. D., and Read, D. J. The structure and function of the vegetative mycelium of ectomycorrhizal plants. I. Translocations of ^{14}C-labelled carbon between plants interconnected by a common mycelium. *New Phytol.*, *103*:143-156, 1986.
116. Finlay, R. D., and Read, D. J. The structure and function of the vegetative mycelium of ectomycorrhizal plants. II. The uptake and distribution of phosphorus by mycelial strands interconnecting host plants. *New Phytol.*, *103*:157-165, 1986.
117. Haystead, P., Malajczuk, N., and Grove, T. S. Underground transfer of nitrogen between mature plants infected with vesicular-arbuscular mycorrhizal fungi. *New Phytol.*, *108*:417-423, 1988.
118. Newman, E. I., and Ritz, K. Evidence on the pathways of phosphorus transfer between vesicular-arbuscular mycorrhizal plants. *New Phytol.*, *104*:77-87, 1986.
119. Soderstrom, B., Finlay, R. D., and Read, D. J. The structure and function of the vegetative mycelium of ectomycorrhizal plants. IV. Qualitative analysis of carbohydrate contents of mycelium interconnecting host plants. *New Phytol.*, *109*:163-166, 1988.
120. Caldwell, M. M., Eissenstat, D. M., Richards, J. H., and Allen, M. F. Competition for phosphorus: differential uptake from dual-isotope-labeled soil interspaces between shrub and grass. *Science*, *229*:384-386, 1985.
121. Caldwell, M. M., Richards, J. H., Manwaring, J. H., and Eissenstat, D. M. Rapid shifts in phosphate acquisition show direct competition between neighboring plants. *Nature*, *327*:615-616, 1987.
122. Pellet, D., and Sieverding, E. Host preferential multiplication of fungal species of the Endogenaceae in the field, demonstrated with weeds. In *Physiological and Genetical Aspects of Mycorrhizae* (V. Gianinazzi-Pearson and S. Gianinazzi, eds.), NRA, Paris, 1986, pp. 555-558,
123. Trappe, J. M. *Mycorrhizae and Productivity of Arid and Semiarid Rangelands*. Oregon State University, 1981, pp. 581-599.
124. Scheltema, M. A., Abbott, L. K., Robson, A. D., and De'ath, G. The spread of mycorrhizal infection by Gigaspora calospora from a localized inoculum. *New Phytol.*, *106*:727-734, 1987.
125. Allen, M. F., MacMahon, J. A., and Anderson, D. Reestablishment of endogonaceae on Mount St. Helens: survival of residuals. *Mycologia*, *76*(6):1031-1038, 1984.

126. Warner, N. J., Allen, M. F., and MacMahon, J. A. Dispersal agents of vesicular-arbuscular mycorrhizal fungi in a disturbed arid ecosystem. *Mycologia, 79*:721-730, 1987.
127. Fries, N. Somatic incompatibility and field distribution of the ectomycorrhizal fungus Suillus luteus (Boletaceae). *New Phytol., 107*: 735-739, 1987.
128. Stenlid, J. Population structure of Heterobasidion annonsum as determined by somatic incompatibility, sexual incompatibility, and isoenzyme patterns. *Can. J. Bot., 63*:2268-2273, 1985.
129. Koltin, Y., and Leibowitz, M. L. *Viruses of Fungi and Simple Eukaryotes*. Marcel Dekker, New York, 1988, p. 434.
130. Economou, A., Lees, V., Pukkila, P. J., Zolan, M. E., and Casselton, L. A. Biased inheritance of optional insertions following mitochondrial genome recombination in the basidiomycete fungus Coprinus cinereus. *Curr. Genet., 11*:513-519, 1987.
131. Dujon, B., Colleaux, L., Jacquier, A., Michel, F., and Monteilhet, C. Mitochondrial introns on mobile genetic elements: the role of intron-encoded proteins. In *Extrachromosomal Elements in Lower Eukaryotes* (R. B. Wickner, A. Hinnebusch, A. M. Lambowitz, I. C. Gunsalus, and A. Hollaender, eds.), Plenum Press, New York, 1986, pp. 5-27.
132. Freeling, M. Plant transposable elements and insertion sequences. *Annu. Rev. Plant Physiol., 35*:277-298, 1984.
133. Tudzynski, P., and Esser, K. Extrachromosomal genetics of Claviceps purpurea. *Curr. Genet., 10*:463-467, 1986.
134. Garber, R. C., Lin, J. J., and Yoder, O. C. Mitochondrial plasmids in Cochliobolius heterostrophus. In *Extrachromosomal Elements in Lower Eukaryotes* (R. B. Wickner, A. Hinnebusch, A. M. Lambowitz, I. C. Gunsalus, and A. Hollaender, eds.), Plenum Press, New York, 1986, pp. 49-57.
135. Normand, P., Simonet, P., Giasson, L., Ravel-Chapius, P., Fortin, J. A., and Lalonde, M. Presence of a linear plasmid-like DNA molecule in the fungal pathogen Ceratocystis fimbriata Ell. & Halst. *Curr. Genet., 11*:335-338, 1987.
136. Honeyman, A. L., and Currier, T. C. Isolation and characterization of linear DNA elements from the mitochondria of Gaeumannomyces graminis. *Appl. Environ. Microbiol., 52*:924-929, 1986.
137. Kim, W. K., McNabb, S. A., and Klassen, G. R. A linear plasmid in Tilletia controversa, a fungal pathogen of wheat. *Can. J. Bot., 66*:1098-1100, 1988.
138. Kistler, H. C., Bosland, P. W., Benny, U., Leong, S., and Williams, P. H. Relatedness of strains of Fusarium oxysporum from crucifers measured by examination of mitochondrial and ribosomal DNA. *Phytopathology, 77*:1289-1293, 1987.
139. Kistler, H. C., and Leong, S. A. Linear plasmidlike DNA in the plant pathogenic fungus Fusarium oxysporum f. sp. conglutinans. *J. Bacteriol., 167*:587-593, 1986.
140. Samac, D. A., and Leong, S. A. Two linear plasmids in mitochondria of Fusarium solani f. sp. cucurbitae. *Plasmid, 19*:57-67, 1988.
141. Finkler, A., Ben-Zvi, B.-S., and Koltin, Y. dsRNA virus of Rhizotonia solani. In *Viruses of Fungi and Simple Eukaryotes* (Y. Koltin and M. J. Leibowitz, eds.), Marcel Dekker, New York, 1988, pp. 387-410.

142. Van Alfen, N. K. Viruses of Endothia parasitica. In *Viruses of Fungi and Single Eukaryotes* (Y. Koltin and M. J. Leibowitz, eds.), Marcel Dekker, New York, 1988, pp. 371-386.
143. Obukowicz, M. G., Perlak, F. J., Kusano-Kretzmer, K., Mayer, E. J., Bolten, S. L., and Watrud, L. S. Tn5-mediated integration of the delta-endotoxin gene from Bacillus thurigiensis into the chromosome of root-colonizing Pseudomonads. *J. Bacteriol.*, *168*:982-989, 1986.
144. Obukowicz, M. G., Perlak, F. J., Kusano-Kretzmer, K., Mayer, E. J., Bolten, S. L., and Watrud, L. S. IS50L as a non-self transposable vector used to integrate the Bacillus thurigiensis delta-endotoxin gene into the chromosome of root-colonizing Pseudomonads. *Gene*, *51*:91-96, 1987.
145. Levin, M. A., Seidler, R., Borquin, A. W., Fowle, J. R., III, and Barkey, T. EPA developing methods to assess environmental release. *Biotechnology*, *6*:38-45, 1988.
146. Kropp, B. R., and Fortin, J. A. Formation and regeneration of protoplasts from the ectomycorrhizal basidiomycete Laccaria bicolor. *Can. J. Bot.*, *64*:1224-1226, 1986.
147. Hébraud, M., and Fèvre, M. Protoplast production and regeneration from mycorrhizal fungi and their use for isolation of mutants. *Can. J. Microbiol.*, *34*:157-161, 1987.
148. Dickman, M. B. Whole cell transformation of the alfalfa fungal pathogen Colletotrichum trifolii. *Curr. Genet.*, *14*:241-246, 1988.
149. Klein, T. M., Wolf, E. D., Wu, R., and Sanford, J. C. High-velocity microprojectiles for delivering nucleic acids into living cells. *Nature*, *327*:70-72, 1987.
150. Shillito, R. D., Saul, M. W., Paszkowski, J., Muller, M., and Potrykus, I. High efficiency direct gene transfer to plants. *Biotechnology*, *3*:1099-1103, 1985.
151. Froeliger, E. H., Munoz-Rivas, A. M., Specht, C. A., Ullrich, R. C., and Novotny, C. P. The isolation of specific genes from the basidiomycete Schizophyllum commune. *Curr. Genet.* *12*:547-554, 1987.
152. Mohan, M., Meyer, R. J., Anderson, J. B., and Horgen, P. A. Plasmid-like DNAs in the commercially important mushroom genus Agaricus. *Curr. Genet.*, *8*:615-619, 1984.
153. Francou, F-X., Randscholt, N., Decaris, B., and Gregoire, A. Presence of several plasmids in the fungus Ascobolus immersus and homologies with the linear plasmid pA1. *J. Gen. Microbiol.*, *133*:311-316, 1987.
154. Andrews, B. J., Beatty, L. G., and Sadowski, P. D. Site specific recombination of the yeast plasmid 2 μ circle; intermediates in the bindary process. In *Extrachromosomal Elements in Lower Eukaryotes* (R. B. Wickner, A. Hinnebusch, A. M. Lambowitz, I. C. Gonsalus, and A. Hollaender, eds.), Plenum Press, New York, 1986, pp. 407-424.
155. Keller, N. P., Bergstrom, G. C., and Yoder, O. C. Stability of foreign DNA in the Cochliobolus heterostrophus genome. *J. Cell. Biochem.*, *12C*:282, Abstr. Y318, 1988.
156. Rodriguez, R. J., and Yoder, O. C. Selectable genes for transformation of the fungal plant pathogen Glomerella cingulata f. sp. phaseoli (Colletotrichum lindemuthianum). *Gene*, *54*:73-81, 1987.

157. Davis, M. A., Cobbett, C. S., and Hynes, M. J. An amdS-lacZ fusion for studying gene regulation in Aspergillus. *Gene, 63*:199-212, 1988.
158. Penttila, M., Nevalainen, H., Ratto, M., Salminen, E., and Knowles, J. A versatile transformation system for the cellulolytic filamentous fungus Trichoderma reesei. *Gene, 61*:155-174, 1987.
159. Kolar, M., Punt, P. J., van den Hondel, C. A. M. J. J., and Schwab, H. Transformation of Penicillium chrysogenum using dominant selection markers and expression of an Escherichia coli lacZ fusion gene. *Gene, 62*:127-134, 1988.
160. Wernars, K., Goosen, T., Wennekes, B. M. J., Swart, K., van den Hondel, C. A. M. J. J., and van den Broek, H. W. J. Cotransformation of Aspergillus nidulans: a tool for replacing fungal genes. *Mol. Gen. Genet., 209*:71-77, 1987.
161. Cooley, R. N., Shaw, R. K., Franklin, F. C. H., and Caten, C. E. Transformation of the phytopathogenic fungus Septoria nodorum to hygromycin β resistance. *Curr. Genet., 13*:383-389, 1988.
162. Kistler, H. C., and Benny, U. K. Genetic transformation of the fungal plant wilt pathogen, Fusarium oxysporum. *Curr. Genet., 13*:145-149, 1988.
163. Turgeon, G., Garber, R. C., and Yoder, O. C. Development of a fungal transformation system based on selection of sequences with promoter activity. *Mol. Cell. Biol., 7*:3297-3305, 1987.
164. Bolton, G. W., Nester, E. W., and Gordon, M. P. Plant phenolic compounds induce expression of the Agrobacterium tumefaciens loci needed for virulence. *Science, 232*:983-985, 1986.
165. Horvath, B., Bachem, C. W. B., Schell, J., and Kondorosi, A. Host-specific regulation of nodulation genes in Rhizobium is mediated by a plant-signal, interacting with the nodD gene product. *EMBO J., 6*:841-848, 1987.
166. Osbourn, A. E., Barber, C. E., and Daniels, M. J. Identification of plant-induced genes of the bacterial pathogen Xanthomonas campestris pathovar campestris using a promoter-probe plasmid. *EMBO J., 6*:23-28, 1987.
167. Woloshuk, C. P., and Kolattukudy, P. E. Mechanism by which contact with plant cuticle triggers cutinase gene expression in the spores of Fusarium solani f. sp. pisi. *Proc. Natl. Acad. Sci. USA, 83*:1704-1708, 1986.
168. Crawford, M. S., and Kolattukudy, P. E. Pectic lyase from Fusarium solani L. sp. pisi. Purification, characterization, in vitro translation of the mRNA and involvement in pathogenicity. *Arch. Biochem. Biophys., 258*:196-205, 1987.
169. Michelmore, R. W., and Hulbert, S. H. Molecular markers for genetic analysis of phytopathogenic fungi. *Annu. Rev. Phytopathol., 25*:383-404, 1987.
170. Kaufman, L., and Standard, P. G. 1987. Specific and rapid identification of medically important fungi by exoantigen detection. *Annu. Rev. Microbiol., 41*:209-225, 1987.
171. Hepper, C. M., Sen, R., and Maskall, C. S. Identification of vesicular-arbuscular mycorrhizal fungi in roots of leek (Allium Porrum L.) and maize (Zea mays L.) on the basis of enzyme mobility during polyacrylamide gel electrophoresis. *New Phytol., 102*:403-410, 1986.
172. Ho, I., and Trappe, J. M. Enzymes and growth substances of Rhizopogon species in relation to mycorrhizal hosts and infrageneric taxonomy. *Mycologia, 79*(4):553-558, 1987.

173. Kough, N., Malajczuk, N., and Linderman, R. G. Use of the indirect immunofluorescent technique to study the vesicular-arbuscular fungus Glomus epigaeum and other Glomus species. *New Phytol.*, *94*:57-62, 1983.
174. Kough, J. L., and Linderman, R. G. Monitoring extra-matrical hyphae of a vesicular-arbuscular mycorrhizal fungus with an immunofluorescence assay and the soil aggregation technique. *Soil Biol. Biochem.*, *18*:309-313, 1986.
175. Forester, H., Kinscherf, T. G., Leong, S. A., and Maxwell, D. P. Estimation of relatedness between Phytophthora species by analysis of mitochondrial DNA. *Mycologia*, *80*:466-478, 1988.
176. Anderson, J. B., Petsche, D. M., and Smith, M. L. Restriction fragment polymorphisms in biological species of Armillaria mellea. *Mycologia*, *79*:69-76, 1987.
177. Hintz, W. E., Mohan, M., Anderson, J. B., and Horgen, P. A. The mitochondrial DNAs of Agaricus: heterogeneity in A. bitorquis and homogeneity in A. brunnescens. *Curr. Genet.*, *9*:127-132, 1985.
178. Armstrong, J. L., Fowles, N. L., and Rygiewicz, P. T. Restriction fragment length polymorphisms distinguish ectomycorrhizal fungi. *Plant Soil*, *116*:1-7, 1989.

Part III

DECOMPOSITION OF BIOLOGICAL AND CHEMICAL WASTE

28

METABOLISM OF FOREIGN COMPOUNDS BY FUNGI

H. G. RAJ, MANJU SAXENA,* and ABDELAMIR ALLAMEH† *Vallabhbhai Patel Chest Institute, University of Delhi, Delhi, India*

K. G. MUKERJI *University of Delhi, Delhi, India*

I. INTRODUCTION

Man, animals, plants, and microorganisms live in a hostile environment with their existence being constantly threatened or modified by a variety of biological, chemical, and physical agents. They are exposed to a steadily increasing number of synthetic chemicals, insecticides, and other environmental pollutants produced by man. Some of these compounds are known as xenobiotics, which are considered to be nonnutritive, generally lipophilic compounds that interact with organisms to display toxic and sometimes mutagenic effects, e.g., many aromatic hydrocarbons. If such chemicals prove recalcitrant to microbial decomposition they will accumulate in the soil leading to serious ecological changes. There is a possibility that eventually these compounds could localize in animal tissues. In order to prevent such an occurrence, it is vital to understand how microorganisms degrade and detoxify the natural and synthetic chemicals. When an organic chemical is introduced into a terrestrial or aquatic ecosystem, it may be subjected to nonenzymatic or enzymatic reactions brought about by that environment. Most of the studies concerning xenobiotic metabolism are on animal tissues with special emphasis on liver. Some of these chemicals upon entry into the mammalian body are readily excreted in urine or feces. However, nonpolar or water-insoluble compounds are excreted very slowly. These compounds are metabolized by the action of oxidative enzymes called drug-metabolizing enzymes before being excreted in bile, urine, or through air [1-5]. The foreign compound metabolism can be divided into two phases [6]. The oxidation, reductions, and hydrolyses are in phase I; the second phase comprises synthetic reactions. Phase I reactions are generally responsible for the introduction of functional groups like hydroxyl (-OH), carboxyl (-COOH), and amino ($-NH_2$) into the xenobiotic molecule, which then undergoes synthetic reactions of the second phase. These reactions involved the union or coupling of a foreign compound or its phase I metabolites,

Present affiliations:
*University of Texas Medical Branch, Galveston, Texas.
†Faculty of Medical Sciences, Tarbiat Modaress University, Tehran, Iran.

with a readily available endogenous conjugating agent, such as glutathione, glucuronic acid, or a sulfur component [4,7].

Although not all the enzymes of mammalian drug-metabolizing systems have been detected in fungi, it appears that there is a close similarity between the two systems [8]. Some reviews are available on the microbial degradation of chemicals of environmental concern [9-13]; however, the reactions involved in xenobiotic metabolism by fungi have not been reviewed. Here an effort is made to review the reactions and enzymes involved in phase I and phase II reactions, in general, with special emphasis on fungi.

II. PHASE I REACTIONS

A. Microsomal Mixed-Function Oxidations

The endoplasmic reticulum of most of the cells contain flavoproteins and cytochromes that function in electron transport reactions requiring oxygen and reduced pyridine nucleotides. This system, which has cytochrome P-450 as the terminal oxidase, has been studied in different tissues of various organisms. Mixed-function oxidations are those in which one atom of molecular oxygen is reduced to water while the other is incorporated into the substrate. The term "monooxygenases" has also been used for the enzymes that catalyze the incorporation of one atom of molecular oxygen into the substrate. The overall oxidation reaction can be written as follows:

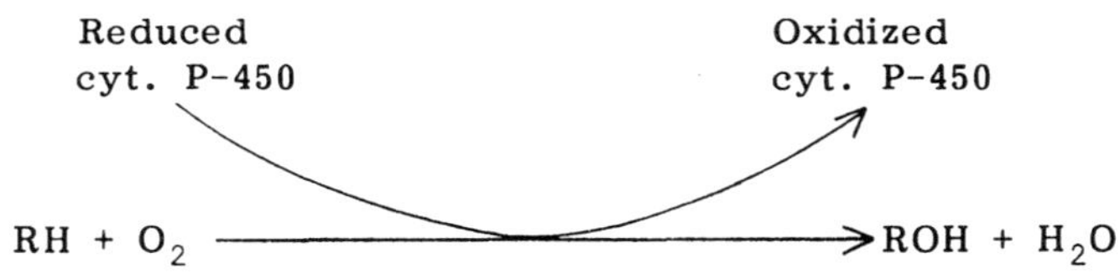

where RH is the substrate.

On the basis of available experimental evidence it is now possible to present a scheme for the pathways of electron flow and mechanism of oxygen-supported hydroxylation reactions (Fig. 1). The mechanism of hydroxylation reactions catalyzed by the cytochrome P-450 system has been comprehensively reviewed [14-16].

First, the oxidized form of cytochrome (Fe^{3+}) combines with the substrate (R). This reaction occurs rapidly. In the second step, the iron atom is reduced to ferrous state by the uptake of one electron. This is the reaction where phosphatidylcholine is thought to function to facilitate electron transfer from NADPH by NADPH cytochrome C reductase. Reduced cytochrome P-450 substrate complex combines with the oxygen in the next step to give a ternary complex. This complex may undergo two electron reductions to yield O_2^{2-}, the dianion of hydrogen peroxide.

Microorganisms can cause the transformation of foreign compounds in several ways [18-22]. It is well documented that bacteria degrade aromatic compounds principally by the action of dioxygenases [13,23,24] and monooxygenase enzyme systems [25]. These latter systems involve cytochrome P-450. The presence of cytochrome P-450 has been reported in many fungi like *Claviceps purpurea* [26,27], *Rhizopus nigricans* [28], yeasts [29], *Trichosporon cutaneum* [30], *Candida tropicalis* [31], *Cunninghamella bainieri* [32], *Aspergillus parasiticus* [33], *Fusarium oxysporum* [34,35], etc. The metabolism of xenobiotic compounds by fungal monooxygenase enzymes has

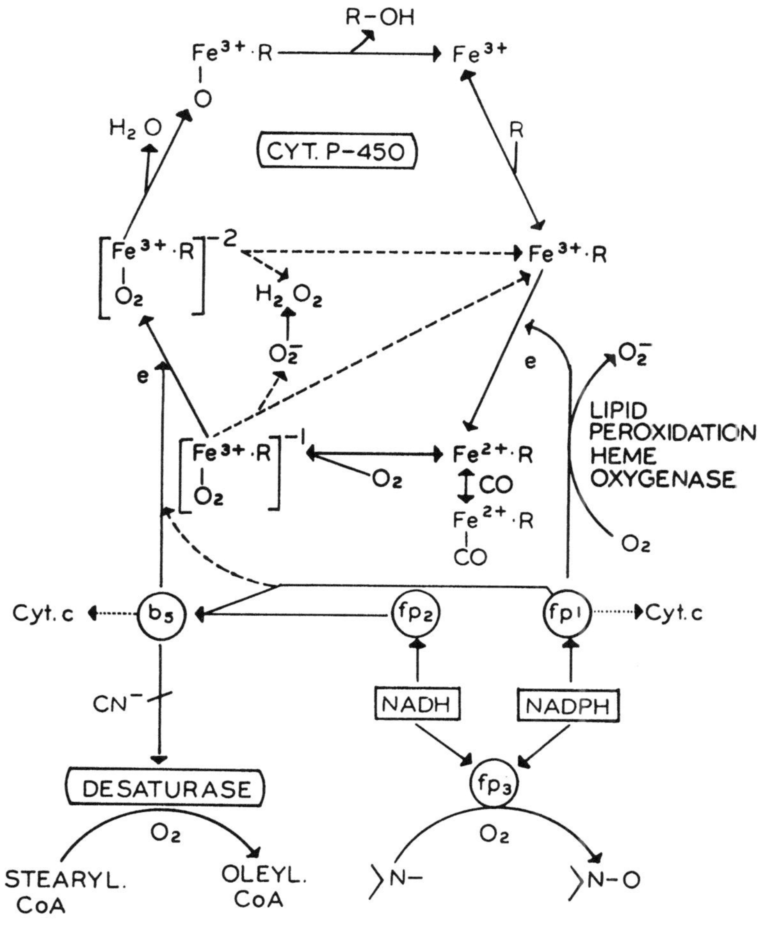

FIGURE 1. A schematic representation of the pattern of electron transport reactions occurring in microsomes. R, substrate; R-OH, hydroxylated substrate; Fp, flavoprotein. (From Ref. 17.)

been found to be very much similar to that of liver microsomal systems [36].

A positive correlation has been reported between the amount of cytochrome P-450 and the alkaloid synthesis by the fungus *C. purpurea* and the interconversion of cytochrome P-450 to cytochrome P-420 [27]. The interconversion of cytochrome P-450 and cytochrome P-420 represents a unique property of the cytochrome P-450 because after treatment with detergents or sulfhydryl reagents cytochrome P-420 can be converted back to cytochrome P-450 by treatment with polyols or reduced glutathione [37]. Similar changes have been observed in fungus *C. purpurea* [27] when incubated anaerobically with snake venom or sodium cholate, where the CO-binding component is converted to a soluble form exhibiting the spectral characteristics of cytochrome P-420.

The ability of microorganisms to metabolize 13 model compounds including acetanilide acronycine aniline, anisole, benzene, benzoic acid, biphenyl, chlorobenzene, coumarin, naphthalene, nitrobenzene, *trans*-stilbene, and toluene were tested and compared with that of mammalian metabolism of xenobiotics [38]. The following organisms were used in these studies: *Aspergillus niger*, *Penicillium chrysogenum*, *Cunninghamella blakesleeana*, *Aspergillus ochraceous*, *Gliocladium deliquescens*, *Streptomyces* sp., *Rhizopus stolonifer*, *Curvularia lunata*, *Cunninghamella bainieri*, and *Helicostylum piriforme*. In most instances the microbial model system yielded patterns of phenolic metabolites similar to those reported for cytochrome P-450 monooxygenases of hepatic microsomes.

Breskvar and Hudnik-Plevnik [28] showed that cytochrome P-450 is located in the membrane fraction sedimenting at 105,000 *g* in the case of *Rhizopus nigricans* by using the reduced CO difference spectra. Hydroxylation of progesterone by the cell-free extracts of *R. nigricans* [39] and *Aspergillus ochraceus* [40] are predominantly at the 11α position. The aryl hydrocarbon hydroxylase (AHH) enzyme was characterized in *Cunninghamella bainieri* [32] and found to be due to a cytochrome P-450 present in the 100,000 *g* pellet. This enzyme is NADPH-dependent and exhibits a pH optimum near 7.8. In particular this fungus catalyzes the NIH shift [41] and the oxidation of organic compounds in the same way as hepatic cytochrome P-450 [36]. The presence of polysubstrate monooxygenase (PMSO) activity, namely, cytochrome P-450 and NADPH cytochrome P-450 reductase, in the microsomal fraction of *Aspergillus parasiticus* suggests the involvement of these enzymes in the biosynthesis of aflatoxins [33]. This fungus also has the ability to metabolize benzephetamine [33], aminopyrine, and *p*-nitroanisole (M. Saxena, A. Allameh, H. G. Raj, and K. G. Mukerji, unpublished results) involving the cytochrome P-450 enzyme systems.

Mixed-function oxidases in fungi have been reported to be induced by known inducers of these enzymes in animal tissues. The level of cytochrome P-450 increased when the fungi were grown in presence of inducers like phenobarbitone [33], polycyclic hydrocarbons, e.g., 3-methylcholanthrene (3-MC) [27], soybean flour [42], and soy oil [34,35]. Enhanced production of secondary metabolites such as trisporic acids, alkaloids, and aflatoxins by phenobarbitone treatment in *Blakeslea trispora* [43], *C. purpurea* [27], and *A. parasiticus* [33], respectively, is attributed to increased mixed-function oxygenase activities in fungi. Ethionine and KCN, which decrease alkaloid content in *C. purpurea*, caused a parallel decrease in cytochrome P-450 [27]. Recently, Hamid and Smith [44] contemplated the role of this enzyme system in the degradation of aflatoxin B_1 and G_1 by intact mycelium and cell-free extracts of *A. flavus*.

Among yeasts, cytochrome P-450 has been found in *Candida tropicalis*, *C. guilliermondii*, *Endomycopsis lipolytica*, *S. cerevisiae*, *S. carlsbergensis*, and *Schizosaccharomyces pombe*. Hydroxylation of foreign chemicals by this enzyme system was found in *C. tropicalis* [45] and *S. cerevisiae* [46,47]. Lebeault et al. [31] isolated a soluble cytochrome P-450 from *C. tropicalis* which catalyzes the hydroxylation of hydrocarbons, drugs, and fatty acids. The growth of this fungus on tetradecane caused the induction of a cytochrome P-450-containing enzyme system [45]. Some strains of *S. cerevisiae* produce a detectable level of cytochrome P-450 during the logarithmic growth phase in 14-20% glucose [48-50]. The enzymatic system of *S. cerevisiae* exhibits oxidations, hydroxylations, and demethylation activities for many compounds [51-53], which was similar to the mammalian system [54]. Depending on the inducer used, more than two forms of cytochrome P-450 were shown to be present in yeast [55-57]. The level of cytochrome P-450 in yeast cells of the strain D7 was affected by ethanol produced by glycolysis during growth [58] and was induced by sodium phenobarbital [57,59,60].

In the case of yeasts, inhibitors of cytochrome P-450 caused a block of metabolic detoxification of the 4-nitroquinoline-1-oxide (4NQO) [61]. 4NQO is proposed to be metabolically reduced to 4-hydroxyaminoquinoline-1-oxide, which is then enzymatically acylated to form 4-aminoacyloxyamine-quinoline-1-oxide, believed to be an ultimate carcinogen for the animal cells [62,63]. Detoxification by cytochrome P-450 is more efficient in cells from the logarithmic growth phase than in cells from the stationary phase [61].

B. Epoxidation and Aromatic Hydroxylation

Epoxidation is an extremely important microsomal reaction as it can produce both stable epoxides and arene oxides—the epoxides of aromatic rings which are intermediates in aromatic hydroxylations. In the case of polycyclic hydrocarbons these reactive intermediates are known to be involved in carcinogenesis.

Literature regarding the mechanism of aromatic hydroxylation operating in fungi was reviewed by Boyd and coworkers [64]. Arene oxides as intermediates in monooxygenase-catalyzed hydroxylation of natural aromatic substrates have been reported for animals [65], plants [66], and the microorganism *Penicillium cyclopium* [67]. NIH shift in a series of fungi like *Aspergillus niger*, *A. foetidus*, *Rhizopus arrhizus*, *Mucor parasiticus*, *Cunninghamella elegans*, *Helicostylum piriforme*, *Curvularia falcata*, and *Trichothecium roseum* was studied [64]. The migration and retention of deuterium during 2,4-hydroxylation of anisole in fungi was similar to that observed with liver microsomal systems. However, the deuterium retention for the ortho or para hydroxylation of 3,5-dideuterioanisole for *A. niger* [68] was considerably less than that for liver microsomes. All of the fungi studied produced much more 2-hydroxyanisole than 4-hydroxyanisole, whereas 4-hydroxylation greatly exceeds 2-hydroxylation of anisole in *C. bainieri* and in liver microsomes [41]. Recently, Archelos et al. [69] studied the hydroxylation of various mono- or polycyclic β-lactam derivatives by the fungus *Beauveria sulfurescens* (ATCC 7159). Most of the mono-hydroxylated products obtained from these transformations were highly regiospecific.

C. Dealkylation

1. *O*-Dealkylation

A good example of *O*-dealkylation is the demethylation of *p*-nitroanisole. Due to the ease with which the product, *p*-nitrophenol, can be measured, this substrate is frequently used for the demonstration of mixed-function oxidase activity. The reaction is believed to proceed via an unstable methyl derivative. Other substrates of *O*-dealkylation include the drugs codeine and phenacetin and the insecticide methoxychlor. The *O*-dealkylation of organophosphorous triesters differs from the above reaction in that it involves the dealkylation of an ester rather than an ether. This reaction was first described for the insecticide chlorfenvinphos but it is known to occur with a variety of vinyl, phenyl, phenylvinyl, and naphthyl phosphates and the thionophosphate triesters. From the experiments performed by Boyd and coworkers [64] it is known that *O*-demethylation occurred in preference to aromatic hydroxylation during the early stages (spores and germlings) of fungal growth. Their results indicate that *O*-demethylation occurs during the metabolism of anisole by *Aspergillus niger*, involving the oxygen atom transfer enzymes. *O*-Demethylation of anisole was observed in other fungi, chiefly *Cunninghamella bainieri* [36], *C. blakesleeana*, *Aspergillus ochraceous*, *Rhizopus stolonifer*, *Helicostylum piriforme* [64].

2. *N*-Dealkylation

This is one of the common reactions in the metabolism of drugs, insecticides, and other xenobiotics. Both *N*- and *N*,*N*-dialkyl carbamate insecticides are readily dealkylated and in some cases the methyl intermediates are stable enough to be isolated. Another example is the insecticide carbaryl, which undergoes several different microsomal oxidation reactions including an attack on the *N*-methyl group. The oxidative demethylation of aminopyrine is frequently used to screen a series of fungi for nonspecific monooxygenases. For this *Cunninghamella bainieri* was found to be particularly efficient and the same metabolic products (monomethylaminopyrine and three other unidentified compounds) were produced in comparable amounts by the fungus and mouse liver microsomes [36]. Several soil fungi attack atrazine and 2-chloro-4-ethylamino-6-isopropylamino-S-triazine (widely used herbicides) by *N*-dealkylation, a process that does not necessarily ensure detoxication [70]. Isolates of *Paecilomyces*, *Beauveria*, *Penicillium*, and *Verticillium* species were shown to metabolize *n*-dodecylbenzene [71] and *n*-alkyltetrahydrothiophenes found in petroleum [72]. The ability of the fungi to metabolize *n*-alkylbenzene was dependent on the length of the alkyl side chain.

3. *S*-Dealkylation

This reaction is known to occur with a number of thioethers such as methylmercaptan, 6-methylthiopurine, etc. No such reaction has been reported in fungi.

D. Oxidations

1. *N*-Oxidation

This can occur in a number of ways including hydroxylamine formation, oxime formation, and *N*-oxide formation. Among these, *N*-oxide formation is primarily dependent on a flavoprotein enzyme localized in the microsomal fraction [73]. This oxidative reaction was reported in *Streptomyces griseus* [74-77] in addition to aromatic and aliphatic hydroxylations, *O*- and *N*-dealkylation, and C-C fission. All these oxidative transformations have been observed with alkaloids [78,79], coumarins [75], retenoids [76], and other complex xenobiotics [74,77].

2. *S*-Oxidation

This reaction is commonly observed among insecticides such as carbamates and organophosphates. Thioethers are generally oxidized by microsomal mixed-function oxidases to the sulfoxides, some of which are further metabolized to sulfones. These oxidative biotransformations were observed in few fungi, chiefly *Aspergillus niger* [80], *A. foetidus*, *Helminthosporium* sp., and *Mortierella isabellina* [81], and *Corynebacterium equi* [82]. These fungi were found to be effective in sulfoxidation reactions of 1,3-dithiolanes [83].

3. *P*-Oxidation

This reaction is a typical cytochrome P-450-mediated mixed-function oxidation. It involves the conversion of trisubstituted phosphine to phosphine oxide. Diphenylmethylphosphine and 3-dimethylaminopropyldiphenylphosphine are known substrates for this enzyme.

E. Desulfuration and Ester Cleavage

Insecticides such as phosphothionate and phosphorodithioate are converted to compounds which are relatively inactive toward cholinesterase into potent cholinesterase inhibitors by an oxidative reaction in which the P=S group is converted to a P=O group. Hydrolysis of the phosphorous ester bonds in organophosphorous insecticides is due to the oxidative dearylation, a typical cytochrome P-450-dependent oxidation reaction which requires NADPH and oxygen and is inhibited by CO.

F. Nonmicrosomal Oxidations

A number of other enzymes are involved in the oxidation of foreign compounds. These oxidoreductases are located in either the mitochondrial fraction or the cytosolic fraction of the tissue homogenates.

1. Alcohol Dehydrogenases

This class of enzyme catalyzes the conversion of alcohols to aldehydes or ketones. The reaction is reversible and reduces carbonyl compounds to alcohols. This enzyme is probably the most important dehydrogenase involved in the metabolism of foreign alcohols and carbonyl compounds. The enzyme is found in the soluble fraction of liver, kidney, and lung. It requires NAD or NADP as a coenzyme. The reaction proceeds at a slower rate with NADP. The oxidation of aldehydes to acids by aldehyde oxidase is a vital detoxification reaction since aldehydes are usually toxic and, because of their lipid solubility, are not readily excreted.

Alcohol dehydrogenase activity was detected in 24 species of Mucorales, e.g., *Rhizopus* sp., *Mucor* sp., *Thamnidium* sp., *Choanophora* sp., and *Helicostylum* sp. [84]. A number of dehydrogenases described from various sources play an important role in steroid, lipid, and carbohydrate metabolism.

2. Aldehyde Dehydrogenase

The oxidations of aliphatic and aromatic aldehydes to their corresponding acids allow the acids to be excreted or conjugated in phase II reactions. Aldehyde dehydrogenases have been isolated from a variety of sources, the most important of which is the liver. Dehydrogenation of the higher aldehyde dodecanol to lauric acid is accomplished by the membrane-bound aldehyde dehydrogenase of the yeast *Candida maltosa* [85]. This reaction proceeds under both aerobic and anaerobic conditions in the presence of exogenous NAD^+ and is concomitant with NAD^+ reduction.

3. Amine Oxidases

These involve the oxidation of biogenic amines formed during normal biological processes. Two groups of amine oxidases are involved in the oxidative deamination of naturally occurring amines as well as foreign compounds, i.e., monoamine oxidases and diamine oxidases. To our knowledge, the presence of amine oxidases is not documented in fungi.

4. Oxidoreductase

This is a soluble enzyme which catalyzes the NADPH-dependent reductions. An NADPH:biochanin A oxidoreductase was purified to apparent homogeneity from the fungus *Fusarium javanicum* [86]. This enzyme acts as a putative starting enzyme in the degradation of isoflavone, particularly biochanin A, thereby yielding isoflavonone.

G. Reduction Reactions

A number of functional groups such as nitro, diazo, carbonyl, disulfide, sulfoxide, and alkene are susceptible to reduction. In many cases it is difficult to determine whether these reactions proceed nonenzymatically by the action of biological reducing agents such as NADPH, NADH, FAD, etc., or through the mediation of functional enzyme systems. Cytoplasmic aldoketoreductases belong to a class of drug-metabolizing enzymes which are widely distributed in the tissues. They all have similar physicochemical characteristics, require NAD as cofactor for the reduction, and have broad substrate specificity. These reductases define a system of detoxification for aldehyde and ketone groups. They have been reported to be present in some fungi, e.g., in mushrooms [87,88].

H. Hydrolysis

Foreign compounds bearing functionalities such as esters, amides, or substituted phosphates are susceptible to hydrolysis. Hydrolytic reactions are the only phase I reactions that do not utilize energy. The acid and alcohol or amine formed on hydrolysis can be eliminated directly or conjugated by phase II reactions.

A variety of foreign compounds such as phthalic acid esters (plasticizers), phenoxyacetic acid and picolinic acid esters (herbicides), and pyrethroids and their derivatives (insecticides), as well as a variety of ester and amide derivatives of drugs, are detoxified by hydrolases. Sharabi and Bordeleau [89] isolated a *Penicillium* sp. from the soil which has the ability to hydrolyze karsil, a herbicide. They partially purified the karsil acrylamidase (EC 3.5.1.9) (arylacylamineamidohydrolase) and tested its ability to hydrolyze 25 related compounds.

I. Epoxide Hydration

Epoxide hydrases bring about hydration of the epoxide ring of alkene and arene compounds to form a corresponding *trans*-dihydrodiol. Epoxide hydrases play an important role similar to that of glutathione epoxide *S*-transferases in deactivating certain labile epoxides which may be carcinogenic. The enzyme(s) that mediated the hydration are located in the microsomal fraction of mammalian liver. The activity is optimal at alkaline pH and probably involves a nucleophilic attack of the -OH on the oxirane carbon. The enzyme epoxide hydrase was reported to be present in mammals [90,91], protozoa [92], and fungi [93]. Wackett and Gibson [93] used *p*-nitrostyrene oxide as substrate for epoxide hydrolase from a 100,000 *g* pellet obtained from disrupted mycelia of the filamentous fungus *Cunninghamella elegans*. Epoxide hydrolase activity in *C. elegans* was membrane-bound and strongly inhibited by 1,2-epoxy-3,3,3-trichloropropane exhibiting the same properties as liver microsomal epoxide hydrolase [90]. Soluble epoxide hydrolase activity with *p*-nitrostyrene oxide as the substrate was not detected in *C. elegans*. However, a soluble epoxide hydrolase that is active with *trans*-β-methylstyrene oxide but not styrene oxide was reported [94].

III. PHASE II REACTIONS

Foreign compounds containing functional groups such as hydroxyl, amino, carboxyl, epoxide, or halogen, or metabolites of phase I reactions as well as many naturally occurring compounds can undergo biosynthetic reactions referred to as conjugations. In this case natural or foreign compounds or its phase I metabolite is coupled with a readily available endogenous conjugating agent derived from a carbohydrate, a protein source, or a sulfur component. According to Williams [95], conjugation reactions are of two types. Type I reactions include the formation of glycosides, sulfates, and methylated and acetylated conjugates whereas type II reactions result in peptide conjugations.

Type I reaction

Activated conjugating agent (intermediate) + substrate → conjugated product

A. Glycoside Conjugation

For both glucoside and glucuronide conjugation, two initial steps are required. The first is the formation of an activated intermediate, uridine diphosphate glucose (UDPG), and uridine diphosphate glucuronic acid (UDPGA). The second step of glucuronic acid conjugation involves reaction of the active intermediate UDPGA with the aglycone mediated by glucuronyltransferases depending on the structure of the aglycone. The reaction results from nucleophilic displacement of the functional groups of the substrate with Walden inversion. UDPGA is in the α configuration and the glucuronide formed due to inversion is in the β configuration.

1. *O*-Glucuronides

1. Ether type. This type of glucuronide occurs with primary, secondary, and tertiary alcohols and with phenols.
2. Ester type. This type of conjugate is formed from carboxylic acids. Recently, Puls et al. [96] discovered a new fungal enzyme, α-glucuronidase. This enzyme acts synergistically with xylanase and liberates 4-*O*-methylglucuronic acid from 4-*O*-methylglucuronic acid-

substituted xylooligomones. α-Glucuronidase is readily produced by *Agaricus bisporus* and *Pleurotus ostreatus*; however, they are poor producers of xylanases and β-xylosidases.

3. Hydroxylamines. In this reaction the amino group of the substrate is hydroxylated by phase I reactions at the nitrogen group and then conjugated via phase II reactions. A typical example is *N*-hydroxylation of 2-acetylaminofluorane (AAF), when the amino group is hydroxylated by mixed-function oxidase system and later conjugated with UDPGA.

2. *N*-Glucuronides

Several types of *N*-glucuronides have been identified in which the glucuronyl moiety is attached to the nitrogen. They include the following functional groups: aromatic amino, carboxyl, and heterocyclic nitrogen.

3. *S*-Glucuronides

Conjugation of glucuronic acid with a variety of thiol compounds has been reported [97]. This conjugation involves the activity of UDP-glucuronyltransferases, a group of closely related enzymes with similar properties. They are difficult to solubilize and generally quite unstable after partial purification. Therefore, data on substrate specificity have been obtained using crude homogenates [97].

The activity of the enzyme UDP-glucuronosyltransferase in the fungus *Cunninghamella elegans* was entirely soluble. The product formed from soluble UDP-glucuronic acid and 3-hydroxybenzo(a)pyrene gave a fluorescence spectrum that was identical with that given by 3-benzo(a)pyrenyl-glucopyranosiduronic acid [93]. The properties of the glucuronosyltransferase were investigated in detail in the dimorphic fungus *Mucor rouxii* [98]. This enzyme was involved in the transfer of D-glucuronic acid (the major sugar of polyuronides) from uridine-5-diphosphoglucuronic acid to endogenous acceptors. Most of the glucuronosyltransferase activity present in cell free extracts from this fungi was found to be associated with the membrane fraction [99].

B. Sulfur Conjugation

Sulfate ester formation is known to occur with primary, secondary, and tertiary alcohols, phenols, and arylamines. Sulfate esters in terms of biological conjugation are in reality half-esters, i.e., $ROSO_3$, which are completely ionized, water-soluble, and quickly eliminated from the organism. Conjugation via the sulfate route requires a great deal of energy. The activation of the sulfate ion by ATP kinase has an absolute requirement for Mg^{2+}.

Adenosine-3'-phosphate-5'-phosphosulfate (PAPS) is the active sulfating agent sulfotransferase and forms sulfate conjugates, aryl sulfates, alkyl sulfates, and sulfamates. Phenol sulfotransferases, alcohol sulfotransferases, and arylamine sulfotransferases appear to be the most important enzymes involved in xenobiotic detoxification. Phenol sulfotransferase is found in the soluble fraction of mammalian liver, kidney, and intestinal mucosa, whereas some alcohol sulfotransferases have only been found in mammalian livers. A number of other sulfotransferases such as steroid sulfotransferase, choline sulfotransferase, mucopolysaccharide sulfotransferase, and cerebroside sulfotransferase are also known. However, of these, probably only the steroid sulfotransferase plays a role in detoxification of xenobiotics.

Enzymatic hydrolysis of water-soluble conjugates produced from benzo(a)pyrene demonstrates the presence of sulfotransferase activity in the fungus *Cunninghamella elegans* [100]. In addition, sulfate conjugates of naphthalene and biphenyl have been isolated from culture filtrates of *C. elegans* [101]. The formation of sulfate esters from phenols may be a significant step in the fungal metabolism of aromatic hydrocarbons [93].

C. Methyl Transferases

Methylation is a reaction in which a methyl group from methylating coenzyme is transferred to amines, phenols, and thiol compounds to form *N*-, *O*-, and *S*-methyl conjugates. The coenzymes that provide methyl groups for transfer are *S*-adenosylmethionine, N^6-methyltetrahydrofolate derivatives, and vitamin B_{12} derivatives. *S*-Adenosylmethionine is known to be the most important coenzyme for methylation of xenobiotics. *S*-Adenosylmethionine is formed from a methionine-activating enzyme as follows [102]:

$$\text{Methionine} + \text{AR-p-p-p} \rightarrow S\text{-adenosylmethionine} + \text{pyrophosphate} + \text{Pi}$$

Depending on the enzyme, methyltransferases can transfer the methyl group of *S*-adenosylmethionine to oxygen or nitrogen to form a methylated product and homocysteine. Although products formed as a result of methylation are less water-soluble, these reactions are considered to be detoxification reactions.

1. *N*-Methylation

Many natural and foreign amines are catalyzed by this reaction. For example, histamine is readily converted to *N*-methylhistamine with a highly specific enzyme, imidazone-*N*-methyltransferase [103]. Other enzymes studied in mammalian tissue are phenylethanolamine-*N*-methyltransferase (PNMT) and *N*-methyltransferases [104]. PNMT is the enzyme involved in the last step in the metabolism of catecholamines; it *N*-methylates noradrenaline to adrenaline, first demonstrated in minced adrenal gland [105].

2. *O*-Methylation

Methylation of epinephrine, norepinephrine, and other catechol derivatives are methylated by catechol-*O*-methyltransferase (COMT). COMT is widely distributed in mammalian tissue [106,107]. It is present mainly in the soluble fraction of the cell and serves an important role in the activation of circulating catecholamine. These enzymes are also reported in plants. In fruit plants an *O*-methylating enzyme is present that methylates catechols mainly on the meta position [108]. Methyltransferases are also reported from many fungi [109,110] and plants [111]. Methyltransferases purified from *Alternaria tenius* catalyze the methylation of alternariol, a phenolic mycotoxin [110]. Recently, a methyltransferase was purified from *Aspergillus parasiticus* to virtual homogeneity and its physicochemical properties were characterized. This enzyme converts sterigmatocystin to *O*-methylsterigmaticystin in the aflatoxin biosynthetic pathway [109].

3. *S*-Methylation

In this reaction methyl groups of certain xenobiotics are transferred to thiol groups. Conversion of mercaptoethanol to *S*-methylthioethanol by *S*-methyltransferase is a good example of this reaction.

4. Biomethylation of Elements

Environmental toxic elements like lead, mercury, tin, platinum, etc., as well as metalloids like arsenic, selenium, and sulfur are methylated. Methylation reactions and enzymes involved have been reported in animal tissues, but there is no report available in literature concerning their detection in fungi.

D. Glutathione *S*-Transferases (GSH *S*-Transferases)

A large number of foreign compounds are metabolized into mercapturic acids, conjugates of L-acetylcysteine, and eliminated via biliary excretion. In this reaction the cysteine portion of the conjugate is derived from reduced glutathione (GSH). The overall reaction involves the conjugation of foreign compounds possessing electrophilic centers for GSH followed by transfer of a glutamate group, loss of glycine, and, finally, acetylation. A number of enzymes are involved in the mercapturic acid pathway. The initial reaction, i.e., conjugation, does not require the use of a high-energy intermediate. However, the synthesis of GSH from various amino acids and the *N*-acetylation of the cysteine conjugate does require ATP. GSH *S*-transferases are known to involve the conjugation of GSH with certain xenobiotics and protect critical nucleophiles such as proteins and nucleic acids by facilitating the conjugation of these toxic electrophiles with endogenous GSH.

Information available on glutathione *S*-transferase in microorganisms is limited. Extremely low specific activities of glutathione S-transferases were reported in yeasts, algae, fungi, bacteria, and protozoa [112-114], when compared with other sources like mammalian liver, lung, kidney, etc. Glutathione S-transferase activity was also reported in various fungi, chiefly in *Rhizoctonia solani, Alternaria alternata, Aspergillus niger, Chaetomium* sp., *Penicillium digitatum, Rhizopus* sp., *Pythium* sp. [115], *Cunninghamella elegans* [83], *Fusarium solani* [115], *Aspergillus flavus* [113], *Mucor javanicus* [116], and yeast [117]. This enzyme activity in fungi was assessed using 1-chloro-2,4-dinitrobenzene (CDNB) and *p*-nitrobenzyl chloride (NBC), both of which are excellent substrates for liver glutathione *S*-transferases. Fungal glutathione *S*-transferase shares several properties including localization in cytoplasm, enzyme multiplicity, and stabilization with glycerol and a thiol reagent with the enzymes found in mammalian liver [118]. Degradation of pesticides in soil showed that *Mucor javanicus* metabolizes several dichloronitrobenzenes to form dichlorobenzamines and chloromethylthionitrobenzenes. Methylthio compounds were formed through glutathione conjugates as products of glutathione transferases [119,120]. *M. javanicus* glutathione *S*-transferase catalyzed the formation of glutathione conjugates of several dichloronitrobenzenes [116]. A positive correlation exists between aflatoxin production and glutathione *S*-transferase activity. Higher enzyme activity was found to be associated with aflatoxigenic strains as compared to nonaflatoxigenic strains [113]. Much evidence was given in support of *A. flavus* glutathione *S*-transferase induction by endogenous aflatoxins (M. Saxena, A. Allameh, H. G. Raj, and K. G. Mukerji, unpublished observations). Phenobarbitone, a known inducer of aflatoxin formation, increases the glutathione *S*-transferase activity in aflatoxigenic strain alone. Tolnaftate-treated *A. flavus* mycelium was found to have significantly diminished glutathione *S*-transferase activity coupled with reduced level of aflatoxin (M. Saxena, A. Allameh, H. G. Raj, and K. G. Mukerji, unpublished observations). Cytosolic glutathione *S*-transferase from *Fusarium oxysporum* [115] and *M. javanicus* [116] were purified. The molecular weight, estimated by SDS-polyacrylamide gel electrophoresis, of the purified enzyme was found to be similar to that of glutathione *S*-transferases subunits purified from insects, mammals, and plant enzymes [121,122].

E. Acylation

Foreign compounds, including many different amines and carboxylic acids, undergo biological acylation in mammalian species and in the process are converted to amide conjugates prior to excretion. Although this reaction renders the compound less water-soluble, it generally decreases the toxicity. Acylation reactions are catalyzed by enzymes (acyltransferases) which are specific for the foreign compound and involve coenzyme A (CoA-SH) derivatives. These reactions are of two general types. The first type involves an activated conjugating intermediate (type I reaction) and the second involves amino acid conjugation (type II reaction).

1. Acetylation (Type I Acylation)

Foreign amines are acylated by acetyl-*S*-CoA to form acetylated derivative. This reaction involves the enzyme *N*-acetyltransferase, which has been found in the soluble fraction of liver and associated with the reticuloendothelial cells of liver, spleen, and lung as well as with the intestinal mucosa. More than one *N*-acetyltransferase is known based on the acetylating capacity of mammalian tissues. Acetylation of amino, hydroxy, and sulfur compounds occurs quite readily in vivo. Acetylation of -OH groups has been demonstrated with choline while the acetylation of -SH groups is known to occur in the formation of acetyl-CoA. At present the acetylation of hydroxy or thiol groups present in foreign compounds has not been reported. Aromatic amines, sulfonamides, and certain amino acids are *N*-acetylated while aliphatic and phenyl-substituted aliphatic amines are not acetylated.

The ability to acetylate foreign compounds is influenced by developmental and genetic factors in addition to species differences. There are rapid and slow acetylations in both human and rabbit. Acetylation of foreign compounds has been reported in fungi by Smith and Rosazza [38]. They suggested that aniline is actively converted by *Cunninghamella blakesleeana*, *C. bainieri*, *Aspergillus ochraceous*, and *Rhizopus stolonifer* to 2-hydroxyacetanilide, 4-hydroxyaniline, and two unknown phenolic products. The reaction involved is *N*-acetylation of aniline which represents a common mammalian biotransformation observed particularly with aromatic amines [5,123].

2. Amino Acid Conjugation (Type II Acylation)

The reaction involves the activation of foreign carboxylic acid in the presence of ATP and CoA to form acyl *S*-CoA derivatives that then acylate the α-amino group of certain amino acids to form peptide conjugates. Peptide conjugation has been reported to occur with glycine and glutamine in mammals and certain primates. In other organisms amino acids which are utilized in peptide conjugation are different. The role of the proteases and peptidases in cellular nutrition, protein turnover, and protection against toxic peptides has been studied in bacteria [124], fungi [125], and yeast [126]. Further, peptidases have been purified and characterized in *C. albicans* and *S. cerevisiae* [127,128].

3. Deacetylation

Enzymes that deacetylate *N*-acetylated xenobiotics have been found in the liver and kidney of many species. The aromatic deacetylase is known as arylacylamidase, which cleaves the conjugate and forms the starting compounds. From the literature available only three fungi—*Penicillium chrysogenum*, *Aspergillus ochraceous*, and *Rhizopus stolonifer*—have been shown to deacetylate the acetanilide which is principally converted to 2-hydroxyacetanilide and aniline [123].

F. Phosphate Conjugation

The biosynthesis of phosphate esters is a general phenomenon found in all intermediary metabolism. However, phosphate conjugates of xenobiotics are rarely encountered. At present it appears that insects are the only species that utilize this mechanism as a phase II detoxification reaction. The enzyme requires ATP and Mg^{2+} for phosphorylation of 1-naphthol and *p*-nitrophenol.

IV. DEGRADATION OF ENVIRONMENTAL POLLUTANTS BY FUNGI

Generally, there are three possibilities by which some members of the soil microbial community acquire the ability to degrade an environmental pollutant such as a herbicide, pesticide, or synthetic polymer: (a) Some organisms require adaptation before attacking a pollutant having degradative enzymes of a constitutive type. (b) Some organisms acquire this ability through random mutation. (c) Some organisms may have adaptive enzymes which are induced in the presence of a particular pollutant.

Among microorganisms, *Nocardia* spp. have been reported to degrade phenoxyalkanoic acids or phenoxyalkylcarboxylic acids which comprise an aliphatic acid joined by ether linkage to a benzene ring. Chloro and methyl substitution of the aromatic ring and variation of the aliphatic chain length gives a number of herbicidal compounds. Mechanisms by which these herbicides are metabolized include β-oxidation of compounds with long fatty acid moieties, cleavage of the ether link between the side chain and the aromatic ring, and ring hydroxylation prior to attack on the side chain. *Aspergillus niger* both β-oxidized the side chain and hydroxylated the ring of some higher phenoxyalkanoates, an example being the conversion of phenoxybutyrate to 4-hydroxyphenoxyacetates [129]. Bartha and Pramer [130] reported the cleavage of propanil, 3,4-dichloropropionilide (DCPA), into 3,4-dichloroaniline (DCA) and propionic acid by microbial acylamidase in soil. Similar observations were recorded for the related herbicides dicryl and karsil [131]. However, pure culture studies [134] with *Fusarium solani* showed that though the organism used propanil as a sole carbon and energy source DCA accumulated in the medium without undergoing condensation to form an azo compound. A specific acylamidase hydrolyzing propanil was extracted from *Fusarium* [132,133]. Sharabi and Bordeleau [89] obtained fungi which hydrolyzed karsil, forming DCA as a major product which accumulates without forming the compound.

Aspergillus fumigatus have been shown to metabolize ethylamino side chains of Simazine leaving the less phytotoxic chlorine-substituted triazine ring intact [70,134]. Several soil fungi attacked atrazine by *N*-dealkylation.

Involvement of the lignin-degrading system in the biodegradation of environmental pollutants by the white rot fungus *Phanerochaete chrysosporium* was recently reviewed [10,135]. This fungus has the ability to degrade a wide variety of structurally diverse organic compounds including a number of environmentally persistent organopollutants such as DDT, polychlorinated biphenyls, and benzo(a)pyrene. The precise enzymological details of xenobiotic degradation in *P. chrysosporium* have yet to be established. There is considerable evidence to suggest that the degradation of a wide variety of organic compounds is mediated by aryl cation radicals [136-138]. Recently, Hammerli and coworkers [139] presented evidence demonstrating participation of the extracellular ligninase(s) in benzo(a)pyrene degradation by a lignin-degrading basidiomycete in *P. chrysosporium*. Multiple forms of diarylpropane oxygenases including ligninases and peroxidases have been purified from *P. chrysosporium* [140]. In addition to oxidation of low molecular weight phenols to form phenoxy radicals, these peroxidases also catalyze hydrogen peroxide-independent oxidations of molecules like reduced glutathione and NADPH. Glenn and Gold [141] suggested that the ligninolytic degradation activity of the fungus was also responsible for the decolorization of dyes like poly 8-411 and poly R-481.

V. CONCLUSION

Accumulation of a variety of synthetic chemicals and several other environmental pollutants in the soil can lead to serious ecological changes. Some of these chemicals are toxic and/or mutagenic to man. Many of these hazardous chemicals are degraded and metabolized by microorganisms in nature. The role played by microorganisms in the biodegradation and metabolism of these compounds makes them potentially useful in the treatment of xenobiotics. No adequate attention has been paid to the study of mechanisms involved in the xenobiotic metabolism by microorganisms, especially fungi. However, some reports indicate that fungi possess the machinery to handle chemicals foreign to the organism in a manner similar to animal tissues. Mixed-function oxidase systems are well implicated in most of the biotransformations reported in fungi. Cytochrome P-450 and its related enzymes have received considerable attention in many fungi, chiefly *Claviceps purpureus*, *Rhizopus nigricans*, *Cunninghamella bainieri*, yeasts, and *Aspergillus parasiticus*. Fungal enzyme-assisted phase II reactions are also quite common, i.e., acetylation, sulfation, and peptide conjugation. Some of the oxidases and conjugating enzymes have been purified and characterized in fungi, e.g., glutathione S-transferases from *Fusarium solani* [115] and *Mucor javanicus* [117]. Fungal glutathione S-transferases share several properties such as localization in cytosol, enzyme multiplicity, and stabilization with glycerol and thiol.

Phenobarbitone, a known inducer of liver cytochrome P-450, was shown to be an effective inducer of this enzyme in yeasts [60] and *Aspergillus parasiticus* [33]. Glutathione S-transferase in fungi can be induced by known inducers of this enzyme in liver, such as phenobarbitone. Also certain endogenous chemicals produced by fungi seem to exert a similar inductive effect, e.g., aflatoxin induced activation of *A. parasiticus* glutathione S-transferase. The existence of a positive correlation between the glutathione S-transferase activity and aflatoxin production in *A. parasiticus* could support the inductive effect of aflatoxin on glutathione S-transferase [113].

Apart from the mixed-function oxidase system, other enzyme systems such as the lignin degradation system also participate in the degradation of certain chemicals of environmental concern. There is a need to probe the utility of such enzymatic machinery of fungi to bring about biotransformation of chemicals. Detailed studies on inducers and inhibitors of these systems may be helpful in understanding the metabolism of the foreign chemicals by fungi. Manipulation of the drug-metabolizing enzymes may help in removing or converting harmful chemicals to less toxic chemicals by fungi.

ACKNOWLEDGMENT

One of us (M.S.) is the recipient of a Senior Research Fellowship from the Council of Scientific and Industrial Research.

REFERENCES

1. Chasseaud, L. F. The nature and distribution of enzyme catalysing the conjugation of glutathione with foreign compounds. *Drug. Metab. Reb.*, *2*:185-220, 1973.
2. Gillette, J. R., Davis, D. C., and Sesame, H. A. Cytochrome P-450 and its role in drug metabolism. *Annu. Rev. Pharmacol.*, *12*:57-64, 1972.
3. Kuntzman, R. Drugs and enzyme induction. *Annu. Rev. Pharmacol.*, *9*:21-36, 1969.
4. Parke, D. V. *The Biochemistry of Foreign Compounds*. Pergamon Press, New York, 1968, pp. 117-136.
5. Parke, D. V. *The Biochemistry of Foreign Compounds*. Pergamon Press, New York, 1968, pp. 224-226.
6. Williams, R. T. *Detoxification Mechanisms*, 2nd ed., Chapman and Hall, London, 1959.
7. Florkin, M., and Stotz, E. H. In *Comprehensive Biochemistry*, Vol. 13. Elsevier, New York, 1973.
8. Ainsworth, G. C., and Sussaman, A. S. *The Fungi: An Advanced Treatise*, Vol. 1. Academic Press, New York, 1956.
9. Alexander, M. Biodegradation of chemicals of environmental concern. *Science*, *211*:132-138, 1981.
10. Bumpus, J. A., and Aust, S. D. Biodegradation of environmental pollutant by the white rot fungus Phanerochaete chrysosporium: involvement of the lignin degrading system. *Bioassays*, *6*:166-170, 1987.

11. Dagley, S. A biochemical approach to some problems of environmental pollution. *Essays Biochem.*, *11*:81-138, 1975.
12. Dagley, S. Microbial degradation of organic compounds in the biosphere. *Surv. Prog. Chem.*, *8*:121-170, 1977.
13. Gibson, D. T. Microbial degradation of aromatic compounds. *Science*, *161*:1093-1098, 1968.
14. Coon, M. J., Strobel, H. W., Autor, A. P., Heidama, J., and Duppel, W. Functional components and mechanisms of action of the resolved liver microsomal enzyme system catalyzing fatty acid, hydrocarbon and drug hydroxylation. In *Biological Hydroxylation Mechanisms* (G. S. Boyd and R. M. S. Smellie, eds.), Academic Press, New York, 1972, p. 45.
15. Coon, M. J., White, R. E., Nordblam, G. D., Ballon, D. P., and Guengerich, F. P. Highly purified liver microsomal cytochrome P-450 properties and catalytic mechanism. *Cro. Chem. Acta*, *49*:163-167, 1977.
16. White, R. E., and Coon, M. J. Oxygen activation by cytochrome P-450. *Annu. Rev. Biochem.*, *49*:315-356, 1980.
17. Estabrook, R. W. Microsomal electron transport reactions. An overview. *Meth. Enzymol.*, *52*:43-47, 1973.
18. Charney, W., and Herzog, H. L. *Microbial Transformations of Steroids*. Academic Press, New York, 1967.
19. Fonken, G. S., and Johnson, R. A. *Chemical Oxidations with Microorganisms*. Marcel Dekker, New York, 1972.
20. Izuka, H., and Naito, A. *Microbial Transformations of Steroids and Alkaloids*. University of Tokyo Press, Tokyo, 1967.
21. Rosazza, J. P., Juhl, R., and Davis, P. Tryptophol formation by Zigosaccharomyces priorianus. *Appl. Microbiol.*, *26*:98-105, 1973.
22. Sih, C. J., Foss, P., Rosazza, J. P., and Lamberger, M. Microbiological synthesis of L-3,4-Dihydroxyphenylalanine. *J. Am. Chem. Soc.*, *91*:6204, 1969.
23. Dagley, S. In *Advances in Microbial Physiology*, Vol. 6 (A. H. Rose and J. F. Wilkinson, eds.), Academic Press, New York, 1971, pp.1-42.
24. Hayaishi, O. Crystalline oxygenases of Pseudomonads. *Bacteriol. Rev.*, *30*:720-731, 1966.
25. Yu, C. A., and Gunsalus, I. C. Crystalline cytochrome P-450. *Biochem. Biophys. Res. Commun.*, *40*:1431-1436, 1970.
26. Ambike, S. H., and Baxter, R. M. Cytochrome P-450 and b_5 in Claviceps purpurea: interconversion of P-450 and P-420. *Phytochemistry*, *9*:1959-1962, 1970.
27. Ambike, S. H., Baxter, R. M., and Zahid, N. D. The relationship of cytochrome P-450 levels and alkaloid synthesis in Claviceps purpurea. *Phytochemistry*, *9*:1953-1958, 1970.
28. Breskvar, K., and Hudnik-Plevnik, T. A possible role of cytochrome P-450 in hydroxylation of progesterone by Rhizopus nigricans. *Biochem. Biophys. Res. Commun.*, *74*:1192-1198, 1977.
29. Karenlampi, S. O., Marin, E., and Hanninen, O. O. P. Short communication: occurrence of cytochrome P-450 in yeasts. *J. Gen. Microbiol.*, *120*:529-533, 1980.
30. Laurila, H., Kappeli, O., and Fiechter, A. The cytochrome P-450 containing monooxygenase of Trichosporon cutaneum occurrence and properties. *Arch. Microbiol.*, *140*:257-259, 1984.

31. Lebeault, J. M., Lode, E. T., and Coon, M. J. Fatty acid and hydrocarbon hydroxylation in yeast: role of cytochrome P-450 in Candida tropicalis. *Biochem. Biophys. Res. Commun.*, *42*:413-419, 1971.
32. Ferris, J. P., MacDonald, L. H., Patrie, M. A., and Martin, M. A. Aryl hydrocarbon hydroxylase activity in the fungus Cunninghamella bainieri: evidence for the presence of cytochrome P-450. *Arch. Biochem. Biophys.*, *173*:443-452, 1976.
33. Bhatnagar, R. K., Ahmad, S., Kohli, K. K., Mukerji, K. G., and Venkitasubramanian, T. A. Induction of polysubstrate monooxygenase and aflatoxin production by phenobarbitone in Aspergillus parasiticus NRRL 3240. *Biochem. Biophys. Res. Commun.*, *104*:1287-1292, 1982.
34. Shoun, H., Sudo, Y., Sato, Y., and Beppu, T. Purification and properties of cytochrome P-450 of a fungus Fusarium oxysporum. *J. Biochem.*, *94*:1219-1229, 1983.
35. Shoun, H., Sudo, Y., and Beppu, T. Subterminal hydroxylation of fatty acids by a cytochrome P-450 dependent enzyme system from a fungus, Fusarium oxysporum. *J. Biochem.*, *97*:755-763, 1985.
36. Ferris, J. P., Fasco, K. J., Stylianopoulon, F. L., Jerina, D. M., Daly, J. W., and Jeffrey, A. M. Monooxygenase activity for a fungal system similar to liver microsomes. *Arch. Biochem. Biophys.*, *156*:97-103, 1973.
37. Ichikawa, Y., and Yamaro, T. Reconversion of detergent and sulfhydryl reagents produced P-420 to P-450 by polyols and glutathione. *Biochem. Biophys. Acta*, *131*:490-497, 1967.
38. Smith, R. V., and Rosazza, J. P. Microbial models of mammalian metabolism: Aromatic hydroxylation. *Arch. Biochem. Biophys.*, *161*:551-558, 1974.
39. Sallam, L. A. R., El Refai, A. H., and El Kady, I. A. The in vitro transformation of progesterons by Rhizopus nigricans REF 129. *Z. Allg. Mikrobiol.*, *11*:325-330, 1971.
40. Jayanthi, C. R., Madyastha, P., and Madyastha, K. M. Microsomal 11 α-hydroxylation of progesterone in Aspergillus ochraceus. I. Characterization of the hydroxylase system. *Biochem. Biophys. Res. Commun.*, *106*:1262-1268, 1982.
41. Daly, J. W., Jerina, D. M., and Witkop, B. Arene oxides and the NIH shift: The metabolism, toxicity and carcinogenicity of aromatic compounds. *Experientia*, *23*:1129-1264, 1972.
42. Sariaslani, F. S., and Kunz, D. A. Induction of cytochrome P-450 in Streptomyces griseus by soybean flour. *Biochem. Biophys. Res. Commun.*, *141*:405-410, 1986.
43. Bulock, J. D., and Winstanley, D. J. Trisporic acid production by Blakeslea trispora and its promotion by barbiturate. *J. Gen. Microbiol.*, *69*:391-394, 1971.
44. Hamid, A. B., and Smith, J. E. Degradation of aflatoxin by Aspergillus flavus. *J. Gen. Microbiol.*, *133*:2023-2029, 1987.
45. Duppel, W., Lebeault, J. M., and Coon, M. J. Properties of a yeast cytochrome P-450 containing enzyme system which catalyzes the hydroxylation of fatty alkanes and drugs. *Eur. J. Biochem.*, *36*:583-592, 1973.
46. Wiseman, A., Gondal, J. A., and Sims, P. 4'-Hydroxylation of biphenyl by yeast containing cytochrome P-450:radiation and thermal stability comparisons with liver enzyme (oxidized and reduced forms). *Biochem. Soc. Trans.*, *3*:278-281, 1975.

47. Woods, L. F. J., and Wiseman, A. Metabolism of benzo(a)pyrene by the cytochrome P-450/P-448, of Saccharomyces cerevisiae. *Biochem. Soc. Trans.*, *7*:124-127, 1979.
48. Del-Carratore, R., Bronzetti, G., Bauer, C., Corsi, C., Nieri, R., Paolini, M., and Giagoni, P. Cytochrome P-450 factors determining synthesis in strain D7 Saccharomyces cerevisiae. *Mutat. Res.*, *121*: 117-123, 1983.
49. Lindenmayer, A., and Smith, L. Cytochromes and other pigments of bakers yeast grown aerobically and anaerobically. *Biochem. Biophys. Acta*, *93*:445-461, 1964.
50. Wiseman, A., Lin, T. K., and Mcloud, C. Relationship of cytochrome P-450 to growth phase of brewer's yeast in 1% or 20% glucose medium. *Biochem. Soc. Trans.*, *3*:276-278, 1975.
51. Callen, D. F., and Philpot, R. M. Cytochrome P-450 and the activation of premutagens in Saccharomyces cerevisiae. *Mutat. Res.*, *45*:305-324, 1977.
52. Callen, D. F., Wolf, C., and Philpot, R. M. Cytochrome P-450 mediated genetic activity of seven halogenated aliphatic hydrocarbons in Saccharomyces cerevisiae. *Mutat. Res.*, *77*:55-68, 1980.
53. Sato, R., and Omura, T. *Cytochrome P-450*. Kodonisha, Tokyo, and Academic Press, New York, 1978.
54. Yoshida, Y., Kumaoka, H., and Sato, R. Studies on the microsomal electron transport system of anaerobically grown yeast. I. Intracellular localization and characterization. *J. Biochem.*, *75*:1201-1210, 1974.
55. Kelly, D. E., and Passy, J. M. Metabolic activation of cytochrome P-450/P-448 in yeast Saccharomyces cerevisiae. *Mutat. Res.*, *108*:147-159, 1983.
56. Kelly, S. L., Kelly, D. E., King, D. J., and Wiseman, A. Interaction between yeast cytochrome P-450 and chemical carcinogens. *Carcinogenesis*, *6*:1321-1325, 1985.
57. King, D. J., Azari, M. R., and Wiseman, A. The induction of cytochrome P-450 dependent benzo(a)pyrene hydroxylase in Saccharomyces cerevisiae. *Biochem. Biophys. Res. Commun.*, *105*:1115-1121, 1982.
58. Del-Carratore, R., Morganti, C., Galli, A., and Brozetti, G. Cytochrome P-450 inducibility by ethanol and 7-ethoxycoumarin O-deethylation in Saccharomyces cerevisiae. *Biochem. Biophys. Res. Commun.*, *123*: 186-192, 1984.
59. Wiseman, A., and Lin, T. K. Induction of cytochrome P-450 in yeast by phenobarbital. *Biochem. Soc. Trans.*, *3*:974-977, 1975.
60. Wiseman, A., and Woods, L. F. S. Metabolism of benzo(a)pyrene by cytochrome P-450/P-448 of Saccharomyces cerevisiae. *J. Chem. Technol. Biotechnol.*, *29*:320-324, 1979.
61. Del-Carratore, R., Cundari, E., Vellosi, R., Galli, A., and Bronzetti, G. Specific inhibitors of the monooxygenase system of Saccharomyces cerevisiae modified the mutagenic effect of 4-nitroquinoline 1-oxide and the deethylation activity of the yeast. *Carcinogenesis*, *7*:1127-1130, 1986.
62. Tada, M., and Tada, M. Seryl/tRNA synthetase and activation of the carcinogen 4-nitroquinoline 1-oxide. *Nature*, *255*:510-512, 1975.
63. Takahashi, K., Huang Khoda, K., and Kawazae, Y. Studies on chemical carcinogens. XXIV. Mutagenic potency of alkylated 4-nitroquinoline-1-oxide and its dependence on the rate of metabolic activation. *Chem. Pharm. Bull.*, *31*:959-965, 1983.

64. Boyd, D. R., Campbell, R. M., Craig, H. C., Watson, C. G., Daly, J. W., and Jerina, D. M. Mechanism of aromatic hydroxylation in fungi. Evidence for the formation of Arene oxide. *J. Chem. Soc. Perkin. I.*, *1976*:2438-2443, 1971.
65. Guroff, G., Daly, J. W., Jerina, D. M., Renson, J., Witkop, B., and Udenfriend, S. Hydroxylation-induced migration: the NIH shift. *Science*, *157*:1524-1530, 1967.
66. Zenk, M. H. Occurrence of the "NIH Shift" during in vivo hydroxylation of cinnamic acid in higher plants. *Z. Pflanzenphysiol.*, *57*:477-478, 1967.
67. Nover, L., and Luckner, M. Mixed functional oxygenation during the biosynthesis of cyclopenin and cyclophenol, benzodiazepine, alkaloids of Penicillium cyclopium Westling. *FEBS Lett.*, *3*:292-296, 1963.
68. Bocks, S. M., Lindsay-Smith, J. R., and Norman, R. O. C. Hydroxylation of phenoxyacetic acid and anisole by Aspergillus niger (Van Tiegh). *Nature*, *201*:398, 1964.
69. Archelos, A., Fourneron, J. D., and Furstoss, R. Microbial transformations 11. Regioselective hydroxylation of β-lactams by the fungus Beauveria sulfurescens. *Tetrahedron Lett.*, *29*:6611-6614, 1988.
70. Kaufman, D. D., and Blake, J. Degradation of atrazine by soil fungi. *Soil Biol. Biochem.*, *2*:73-80, 1970.
71. Fedorak, P. M., and Westlake, D. W. S. Fungal metabolism of n-alkylbenzenes. *Appl. Environ. Microbiol.*, *51*:435-437, 1986.
72. Fedorak, P. M., Payzant, J. D., Montgomery, D. S., and Westlake, D. W. S. Microbial degradation of n-alkyl tetrahydrothiophenes found in Petroleum. *Appl. Environ. Microbiol.*, *54*:1243-1248, 1988.
73. Zeigler, D. M., and Pettit, F. H. Microsomal oxidases. I. The isolation and dialkylarylamine oxygenase activity of pork liver microsomes. *Biochemistry*, *5*:2932-2938, 1966.
74. Chien, M., and Rosazza, J. P. Microbial transformation of natural antitumor agents: metabolism of bruseantin by Streptomyces griseus. *J. Chem. Soc. Perkin Trans.*, *1*:1352-1356, 1981.
75. Sariaslani, F. S., and Rosazza, J. P. Novel biotransformations of 7-ethoxycoumarin by Streptomyces griseus. *Appl. Environ. Microbiol.*, *46*:468-474, 1983.
76. Sariaslani, F. S., and Rosazza, J. P. Biotransformations of 1',2'-dihydrorotenone by Streptomyces griseus. *Appl. Environ. Microbiol.*, *49*:451-452, 1983.
77. Sariaslani, F. S., Beale, J. M. Jr., and Rosazza, J. P. Oxidation of rotenone by Polyporus anceps laccase. *J. Nat. Prod.*, *47*:692-697, 1984.
78. Sariaslani, F. S., Eckenrode, F. M., Beale, J. M. Jr., and Rosazza, J. P. Formation of a reactive Iminium derivative by enzymatic and chemical oxidations of 16-O-Acetylvindoline. *J. Med. Chem.*, *27*:749-754, 1984.
79. Kunz, D. A., Reddy, G. S., and Vatvars, A. Identification of transformation products arising from bacterial oxidation of codeine by Streptomyces griseus. *Appl. Environ. Microbiol.*, *50*:831-836, 1985.
80. Poje, M., Mota, O., and Balenovic, K. Stereoselective oxidation of geni-disulphides with Aspergillus niger. *Tetrahedron*, *36*:1895-1897, 1980.
81. Auret, B. J., Boyd, D. R., Cassidy, E. S., Hamilton, R., Turley, F., and Drake, A. F. Structure and absolute stereochemistry of thioacetal

sulphoxides obtained by fungal metabolism of 2-alkyl-1,3-Dithianes. *J. Chem.Soc. Perkin Trans.*, *1*:1547-1552, 1985.
82. Okamoto, Y., Ohta, H., and Tsuchihashi, G. I. Studies on enzymatic oxidation of sulfides. 6. Asymmetric microbial oxidation of formaldehyde dithioacetals. *Chem. Lett.*, *12*:2049-2052, 1986.
83. Auret, B. J., Boyd, D. R., and Dunlop, R. Stereoselectivity during fungal sulphoxidations of 1,3-dithiolanes. *J. Chem. Soc. Perkin Trans.*, *1*:2827-2829, 1988.
84. Gleason, F. H. Alcohol dehydrogenase in Mucorales. *Mycologia*, *63*:906-910, 1971.
85. Krauzora, V. I., Kuvichkina, T. N., Sharyshev, A. A., Romanova, I. B., and Lozinov, A. B. Lauric acid and NADH synthesis during dodecanal and dodecanal oxidation by membrane fractions of the yeast Candida maltosa grown on mexadecane. *Biokhimiya*, *51*:23-27, 1986.
86. Schlieper, D., and Barz, W. NADPH: biochanin A oxidoreductase from the fungus Fusarium javanicum. *Phytochemistry*, *26*:2495-2498, 1987.
87. Pollock, S. H. Dihydromorphinine ketone reductases. *Life Sci.*, *27*: 465-475, 1975.
88. Roeriag, S., Fujimoto, J. M., Wang, R. J. H., Pollock, S. H., and Lange, D. Preliminary characterization of enzymes for reduction of naloxone and naltrexone in rabbit and chicken liver. *Drug Metab. Dispos.*, *4*:53-58, 1975.
89. Sharabi, N. E., and Bordeleau, L. M. Biochemical decomposition of the herbicide N-(3,4-dichlorophenyl)-2-methylpentanamide and related compounds. *Appl. Microbiol.*, *18*:369-375, 1969.
90. Lu, A. Y. H., and Miwa, G. T. Molecular properties and biological functions of microsomal epoxide hydrase. *Annu. Rev. Pharmacol. Toxicol.*, *20*:513-531, 1980.
91. Oesch, F. In *Progress in Drug Metabolism*, Vol. 3 (J. W. Bridges and L. F. Chasseaud, eds.), John Wiley and Sons, Chichester, 1979, pp. 253-301.
92. Yawetz, A., and Agosin, M. Epoxide hydrase in Trypanosoma cruzi epimastigotes. *Biochim. Biophys. Acta*, *585*:210-219, 1979.
93. Wackett, L. P., and Gibson, D. R. Metabolism of xenobiotic compounds by enzymes in cell extracts of the fungus Cunninghamella elegans. *Biochem. J.*, *205*:117-122, 1982.
94. Ota, K., and Hammock, B. D. Cytosolic and microsomal epoxide hydrolases: differential properties in mammalian liver. *Science*, *207*: 1479-1480, 1980.
95. Williams, R. T. The biogenesis of conjugation and detoxification products. In *Biogenesis of National Compounds*, 2nd ed. (P. Bernfeld, ed.), Oxford, Pergamon Press, 1967, pp. 589-639.
96. Puls, J., Schmidt, O., and Granzow, C. α-Glucuronidase in two microbial xylanolytic systems. *Enz. Microb. Technol.*, *9*:83-88, 1987.
97. Dutton, G. J. Glucuronide forming enzymes. In *Handbook of Experimental Pharmacology, Vol. 28, Part 2, Concepts in Biochemical Pharmacology* (B. B. Brodies, J. R. Gillette, and H. S. Ackerman, eds.), Springer-Verlag, Berlin, 1971, p. 378.
98. Dow, J. M., Olona, P. M., and Villa, V. D. Glucuronosyl transferase from the dimorphic fungus Mucor rouxii. *Exp. Mycol.*, *6*:329-334, 1982.
99. Flores-Carreon, A., Balcajar, R., and Ruiz-Herrera, J. Characterization of glucuronosyl transferase from Mucor rouxii: requirement for polyuronide acceptors. *Exp. Mycol.*, *9*:294-301, 1985.

100. Cerniglia, C. E., and Gibson, D. T. Oxidation of benzo(a)pyrene by the filamentous fungus Cunninghamella elegans. *J. Biol. Chem.*, *245*:12174-12180, 1979.
101. Cerniglia, C. E., and Gibson, D. T. Metabolism of Cunninghamella elegans. *Appl. Environ. Microbiol.*, *34*:363-370, 1977.
102. Cantoni, G. L., and Durell, J. Activation of methionine for transmethylation. I. The methionine activating enzyme: studies on the mechanism of the reactions. *J. Biol. Chem.*, *225*:1033-1048, 1957.
103. Brown, D. D., Tomchick, R., and Axelrod, J. Distribution and properties of a histamine-methylating enzyme. *J. Biol. Chem.*, *234*: 2948-2950, 1959.
104. Axelrod, J., and Vessel, E. S. Heterogenicity of N- and O-methyl transferases. *Mol. Pharmacol.*, *6*:78-84, 1970.
105. Bulbring, E. Methylation of noradrenaline by minced suprarenal tissue. *Br. J. Pharmacol.*, *4*:234-244, 1949.
106. Axelrod, J., and Tomchick, R. Enzymatic O-methylation of epinephrine and other catechols. *J. Biol. Chem.*, *233*:702-705, 1958.
107. Bamshad, J., Lerner, A. B., and Mcguire, J. S. Catechol-O-methyl transferase in skin. *J. Invest. Dermatol.*, *43*:111-113, 1964.
108. Finkle, B. J., and Nelson, R. F. Enzyme reaction with phenolic compounds by a meta O-methyl transferase in plants. *Biochim. Biophys. Acta*, *78*:747-749, 1963.
109. Bhatnagar, D., Ullah, A. H. J., and Cleveland, T. E. Purification and characterization of a methyl transferase from Aspergillus parasiticus SRRC 163 involved in aflatoxin biosynthetic pathway. *Prep. Biochem.*, *18*:321-349, 1988.
110. Stinson, E. E., and Moreau, R. A. Partial purification and some properties of an Alternariol-O-methyl transferases from Alternaria tenius. *Phytochemistry*, *25*:2721-2724, 1986.
111. Poulton, J. E. In *The Biochemistry of Plants*, Vol. 7 (E. E. Conn, ed.), Academic Press, New York, 1981, pp. 471-474.
112. Lau, E. P., Niswander, L., Watson, D., and Fall, R. R. Detection of microbial glutathione S-transferases. *Fed. Proc.*, *39*:1696, 1980.
113. Saxena, M., Mukerji, K. G., and Raj, H. G. Positive correlation exists between glutathione S-transferase activity and aflatoxin formation in Aspergillus flavus. *Biochem. J.*, *254*:567-570, 1988.
114. Yawetz, A., and Agosin, M. Glutathione S-transferase and drug metabolism in Trypanosoma cruzi. In vivo and in vitro formation of thioethers. *Comp. Biochem. Physiol. C.*, *66*:265-267, 1980.
115. Cohen, E., Gamliel, A., and Katan, J. Glutathione and glutathione S-transferase in fungi. Effect of pentachloro nitrobenzene and 1-chloro-2,4-dinitrobenzene: purification and characterization of the transferase from Fusarium. *Pest. Biochem. Physiol.*, *26*:1-9, 1986.
116. Kumagai, H., Tamaki, H., Koshino, Y., Suzuki, H., and Tochikura, T. Distribution, formation and stabilization of yeast glutathione S-transferase. *Agr. Biol. Chem.*, *52*:1377-1382, 1988.
117. Ando, K., Honma, M., Chiba, S., Tahara, S., and Mizutani, J. Glutathione transferase from Mucor javanicus. *Agr. Biol. Chem.*, *52*:135-139, 1988.
118. Jakoby, W. B. *Enzymatic Basis of Detoxification*, Vol. 1. Academic Press, New York, 1980, pp. 1-6.
119. Hafsah, Z., Tahara, S., and Mizutane, J. Metabolic pathways of 2,4-dichloro-1-nitrobenzene in Mucor javanicus. *J. Pest. Sci.*, *12*: 617-625, 1987.

120. Tahara, S., Hafsah, Z., and Mizutani, J. A glutathione conjugate of 2,4-dichloro-1-nitrobenzene: its detection as a metabolic intermediate and further metabolism in Mucor javanicus. *J. Pest. Sci.*, *12*:607-616, 1987.
121. Motoyama, N., and Dauterman, W. C. Purification and properties of housefly glutathione S-transferase. *Insect Biochem.*, *7*:361-367, 1977.
122. Mozer, T. J., Tiemeir, D. C., and Jaworsky, E. G. Purification and characterization of corn glutathione S-transferase. *Biochemistry*, *22*:1068-1072, 1983.
123. Brodie, B. B., and Axelrod, J. The fate of acetanilide in man. *J. Pharm. Exp. Ther.*, *94*:29-38, 1948.
124. Hermsdorf, C. L., and Simmonds, S. Role of peptidases in utilization and transport of peptides by bacteria. In *Microorganisms and Nitrogen Sources* (J. W. Payne, ed.), John Wiley and Sons, New York, 1980, pp. 301-334.
125. Wolf, D. H., and Holzer, H. Proteolysis in yeast. In *Microorganisms and Nitrogen Sources* (J. W. Payne, ed.), John Wiley and Sons, New York, 1980, pp. 431-458.
126. Rose, B., Becker, J. M., and Naider, F. Peptidase activities in Saccharomyces cerevisiae. *J. Bacteriol.*, *139*:220-224, 1979.
127. Logan, D. A., Becker, J. M., and Naider, F. Peptide transport in Candida albicans. *J. Gen. Microbiol.*, *114*:179-186, 1979.
128. Logan, D. A., Naider, F., and Becker, J. M. Peptidases of yeast and filamentous forms of Candida albicans. *Exp. Mycol.*, *7*:116-126, 1983.
129. Byrade, R. J. W., and Woodcock, D. The metabolism of some phenoxy-n-alkylcarboxylic acids by Aspergillus niger. *Biochem. J.*, *65*:682-686, 1957.
130. Bartha, R., and Pramer, D. Pesticide transformation to aniline and azo compounds in soil. *Science*, *156*:1617-1618, 1967.
131. Bartha, R., Linke, H. A., and Pramer, D. Pesticide transformations: production of chloroazobenzene from chloroanilines. *Science*, *161*:582-583, 1968.
132. Lanzilotta, R. P., and Pramer, D. Herbicide transformation. I. Studies with whole cells of Fusarium solani. *Appl. Microbiol.*, *19*:301-306, 1970.
133. Lanzilotta, R. P., and Pramer, D. Herbicide Transformation. II. Studies with an Acylamidase of Fusarium solani. *Appl. Microbiol.*, *19*:307-313, 1970.
134. Kaarney, P. C., Kaufman, D. D., Alexander, M. Biochemistry of herbicide decomposition in soils. In *Soil Biochemistry* (A. D. Mclaren and G. H. Patterson, eds.), Marcel Dekker, New York, 1967.
135. Bumpus, J. A., Tien, M., Wright, D., and Aust, S. D. Oxidation of persistent environmental pollutants by a white rot fungus. *Science*, *228*:1434-1436, 1985.
136. Eaton, D. S. Mineralization of polychlorinated biphenyls by Phanerochaete chrysosporium: a ligninolytic fungus. *Enz. Microb. Technol.*, *7*:194-196, 1985.
137. Funderburk, H. H., and Bozarth, G. A. Review of the metabolism and decomposition of diquat and paraquat. *J. Agr. Food Chem.*, *15*:563-567, 1967.

138. Leatham, G. F., Crawford, R. L., and Kirk, T. K. Degradation of phenolic compounds and ring cleavage of catechol by Phanerochaete chrysosporium. *Appl. Environ. Microbiol.*, *46*:191-197, 1983.
139. Haemmerli, S. D., Leisola, M. S. A. Sanglard, D., and Fiecheter, A. Oxidation of benzo(a)pyrene by extracellular ligninases of Phanerochaete chrysosporium: veratryl alcohol and stability of ligninase. *J. Biol. Chem.*, *26*:6900-6903, 1986.
140. Paszczynski, A., Huynh, V. B., and Crawford, R. L. Comparison of ligninase-1 and peroxidase M2 from the white rot fungus Phanerochaete chrysosporium. *Arch. Biochem. Biophys.*, *244*:750-765, 1985.
141. Glenn, J. K., and Gold, M. H. Decolorization of several polymeric dyes by the lignin-degrading basidiomycete Phanerochaete chrysosporium. *Appl. Environ. Microbiol.*, *45*:1741-1747, 1983.

29

HYDROCARBON-DEGRADING YEASTS AND FILAMENTOUS FUNGI OF BIOTECHNOLOGICAL IMPORTANCE

NICHOLAS D. LINDLEY *Institut National des Sciences Appliquées, Toulouse, France*

I. INTRODUCTION

Hydrocarbons represent an enormous energy resource without which our modern lifestyle would be impossible. Principally exploited as fossilized fuels, hydrocarbons must be considered as a finite resource although methane and lignin (essentially a complex hydrocarbon) are certainly renewable. As well as being the predominant energy source in most countries, hydrocarbons are also an important feedstock for the chemical industry. Their potential impact in biotechnology is enormous, but as yet their exploitation has been limited.

The chemical nature of hydrocarbons is extremely diverse, varying from simple saturated aliphatic alkanes to complex polycyclic aromatic compounds. Such a range of carbon substrates can support the growth of many microorganisms using diverse and often badly understood metabolic sequences. The scope for innovative industrial application is therefore high. In part, the absence of a basic understanding of the biochemical nature of hydrocarbon metabolism has slowed down and in some cases halted many research projects. In addition, a variety of problems associated with the development of high-productivity fermentation strategies for insoluble high-energy status substrates necessitate an approach somewhat different from sugar-based fermentation technology. These problems ought to provide the impetus for research projects rather than the reverse, and the real reasons for the lack of hydrocarbon-based fermentation processes must be sought elsewhere.

The sharp rise in oil prices during the 1970s can be seen as a contributory factor, but viewed over a long term, oil prices have certainly been no less stable than a variety of agricultural feedstocks [1]. The more recent fall in oil prices ought to stimulate renewed interest in hydrocarbon feedstocks. Perhaps a better explanation for the lack of success of hydrocarbons within biotechnology can be attributed to a complex tissue of sociopolitical propaganda which mobilized public opinion against such uses. The power of farming pressure groups throughout the western world has effectively oriented biotechnology toward government subsidized agriculture-based substrates, if only to justify continued overproduction of cereals, e.g., witness the vast investment in alcohol production despite a questionable economic reality. Seen in this light, what future exists for hydrocarbons as substrates for biotechnological processes? A clear-cut answer is difficult since political influences are notoriously subject to abrupt changes of direction. It is, however, clear that any hydrocarbon-based process will need to produce higher priced products than the unicellular proteins extensively investigated in the decade prior to the oil crisis. Hydrocarbons most easily transformed by microorganisms are in strong demand as fuels and therefore likely to remain relatively expensive. Work on fungi capable of consuming renewable lignin hydrocarbons is promising since species able to consume cyclic compounds are being characterized [2]. Such compounds are more available but generally recalcitrant, although they do offer tremendous potential for partial transformation processes, which may prove to be the direction most likely to succeed.

Biotechnological use of hydrocarbons can to some extent be said to suffer nowadays from the huge investment in single-cell protein (SCP) production. A limited number of species were extensively studied, but the selection criteria were such that strains likely to overproduce metabolites were generally excluded. Metabolite overproduction is found in microorganisms in which carbon and/or energy flow is for some reason subject to imbalance. Yeasts suitable for SCP production are by their very nature species in which the regulation of metabolism is extremely efficient. Screening procedures of the type regularly undertaken within the pharmaceutical industry may well throw up new hydrocarbon-utilizing species more suitable for added-value product synthesis. The alternative is, of course, to genetically modify strains whose growth characteristics have been partially documented. While such an approach has already proven useful, the phenotypic variation (particularly among filamentous fungi) represents a huge reservoir of potential that should not be overlooked.

Although funding for hydrocarbon research is currently at a lull at both public and industrial levels, it should be remembered that a wide range of useful products can already be derived from hydrocarbon-grown microorganisms [3]. These products include the biomass itself (SCP), amino acids, organic acids, carbohydrates, nucleic acids, lipids, vitamins, coenzymes, and antibiotics. These applications use principally *n*-alkanes; aromatic degradation has to date been largely oriented toward depollution strategies. In addition, hydrocarbon-degrading fungi have a well-documented history of fuel tank contamination [4], necessitating economically expensive fuel treatment strategies. In this chapter the information obtained from the development of such studies will be presented, together with possible directions for future research.

II. BIOCHEMICAL MECHANISMS OF HYDROCARBON METABOLISM

Although hydrocarbon-degrading microorganisms were first reported over a century ago, much of our understanding of this type of metabolism has come only recently. The development of modern chromatographic techniques coupled more recently to mass spectrometry has made possible the study of the separate mechanisms involved in the catabolic breakdown of hydrocarbons. The information is far from complete and is restricted to a relatively small number of species. With a view to the biochemical mechanisms

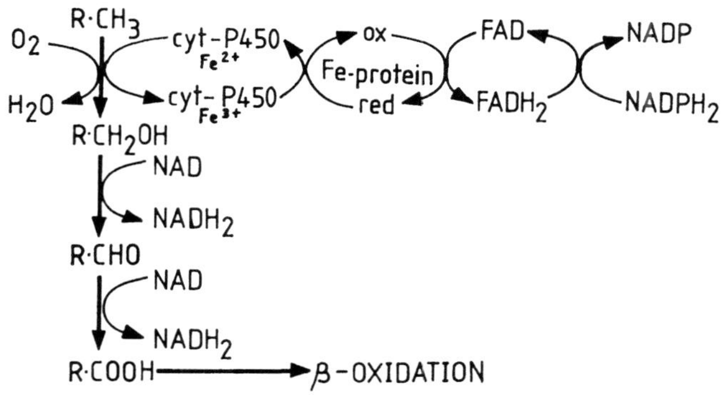

FIGURE 1. Mechanism of *n*-alkane oxidation most commonly encountered in yeasts and filamentous fungi.

involved it is convenient to subdivide the hydrocarbons into aliphatics and aromatics, although the degradation of cyclic aliphatics falls somewhere in between.

A. Aliphatic Hydrocarbons

Alkanes and in particular those of intermediate size (C_{10}-C_{20}) are the most rapidly metabolized hydrocarbons and were the substrate chosen for the SCP projects of the 1960s. As a result, the mechanism by which *n*-alkanes are metabolized has been fairly well documented. Several possible biochemical pathways are involved, although the initial mode of attack is most commonly at the site of the terminal methyl group. Various mechanisms have been proposed involving the insertion of molecular oxygen. The most frequent makes use of a mixed-function alkane monooxygenase located in the microsomes [5]. This enzyme is a typical monooxygenase enzyme producing an alkanol and involving an elaborate electron transfer system (see Fig. 1). This array involves cytochrome P450 as the terminal oxidase and an NADPH-cytochrome P450 reductase as the electron transfer component [6]. Following the oxidation of the alkane to primary alcohol, further oxidation to fatty acids via aldehyde occurs. These reactions involve NAD(P)H yielding dehydrogenases specific for long-chain alcohols and aldehydes. At the level of fatty acids, the normal route taken for further catabolism is via β-oxidation involving the initial activation of the fatty acid to the corresponding acyl-CoA ester. The enzyme involved is an acyl-CoA synthetase believed to be present in several forms. An alternative to β-oxidation is α-oxidation in which the first intermediate is an α-hydroxy fatty acid which is oxidatively decarboxylated. However, few microorganisms have been shown to use this mechanism. The acetyl-CoA resulting from such pathways can either serve as an anabolic precursor or may be further catabolized to carbon dioxide via the tricarboxylic acid cycle.

Although the mechanism described above is that most frequently encountered, a number of alternatives do occur, some of which have recently been exploited industrially. Two such alternatives have been reported [7]: subterminal oxidation and diterminal oxidation (see Fig. 2). This latter

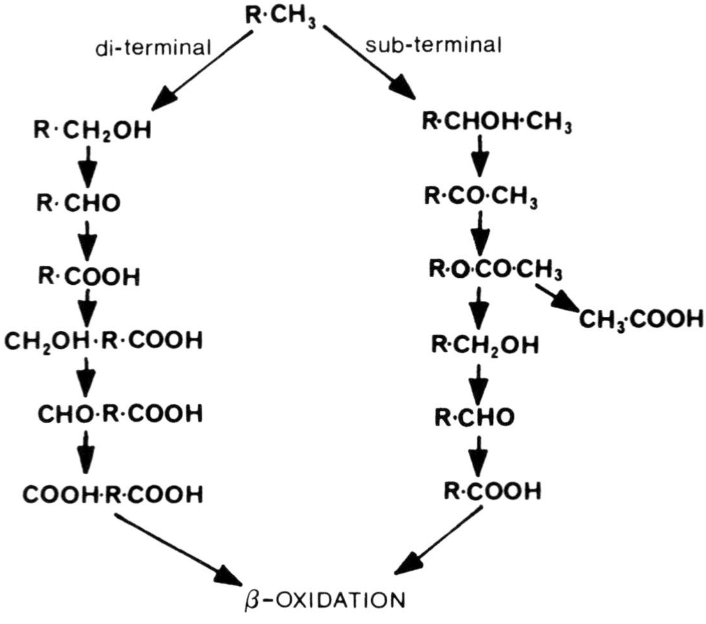

FIGURE 2. Subterminal and diterminal oxidation mechanisms of alkane degradation.

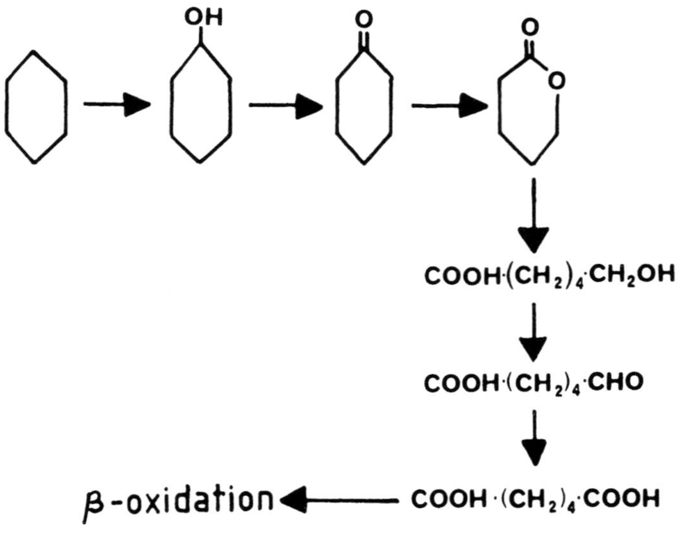

FIGURE 3. Microbial oxidation of cyclohexane. Hydroxylation by a mixed-function oxidase leads to an alicyclic alcohol, which is dehydrogenated to a ketone. Further oxidation involves the formation of a lactone and sequential diterminal attack to yield a dicarboxylic acid.

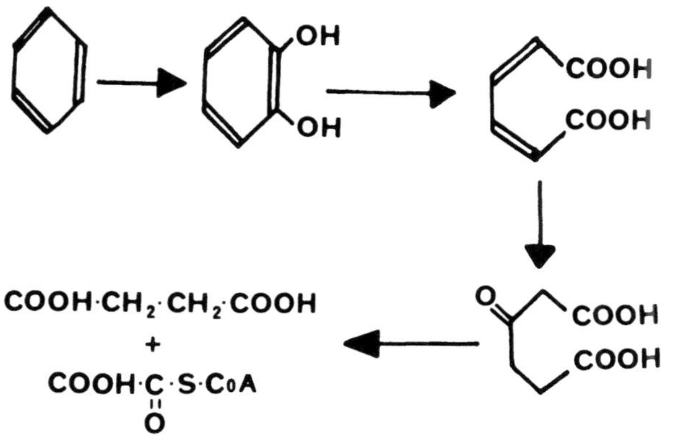

FIGURE 4. Mechanism of aromatic ring cleavage. Following oxidation to catechol, ortho cleavage results in *cis,cis*-muconic acid which is further metabolized to β-ketoadipic acid and then oxidatively cleaved to the tricarboxylic acid intermediates succinate and acetyl-CoA.

mechanism has been exploited to overproduce dicarboxylic acids [3] and seems to involve a two-stage attack whereby the second methyl group is oxidized following the formation of a fatty acid as previously outlined.

Alicyclic hydrocarbons such as cyclohexane have no terminal methyl groups and are utilized by a mechanism similar to subterminal oxidation [8]. This mechanism (Fig. 3) involves the formation of a dicarboxylic acid which is further metabolized by β-oxidation. Microorganisms capable of growth on cyclic aliphatics have rarely been isolated; more commonly, commensalism and cometabolism are involved.

B. Aromatic Hydrocarbons

The only eukaryote so far subject to intensive study as regards aromatic hydrocarbon degradation is *Trichosporon cutaneum* [9]. This organism has a metabolic versatility equal to, though somewhat different from, that of the more widely studied pseudomonads. The pathway of aromatic hydrocarbon metabolism used by fungi is summarized in Fig. 4, though it is likely that other mechanisms also exist. Condensed aromatics about which even less is known are also degraded by a similar strategy. Dihydroxylation and opening of one of the rings enables that ring to be degraded to pyruvate and carbon dioxide before the next aromatic ring is attacked in the same way.

More recently, research has focused on the various extracellular enzymes (possibly peroxidases) which degrade lignin aromatics [2]. These enzymes, most studied in white-rot fungi such as *Phanerochaete chrysosporium*, may prove also to be important in biotechnological use of aromatic hydrocarbons. Their mode of action appears to be relatively random involving the formation of highly reactive free radicals, but as yet no definite study of these various enzymes exists.

C. Cooxidation of Recalcitrant Hydrocarbons

Many complex hydrocarbons can be broken down or at least modified by microbial action, although growth does not take place. In the presence of an alternative substrate providing both the carbon and a major energy source for the microorganism, some degree of partial conversion of recalcitrant hydrocarbons can take place [10]. Breakdown rates are typically slow, but for certain halogenated aromatics cooxidation can offer some promise for depollution scientists. Coupling of mixed-substrate strategies with defined mixed populations is currently being examined in many laboratories. An alternative use of cooxidation potential would be to carry out controlled modification of pharmaceutical products, although turnover rates would need to be improved before such an application could realistically be developed.

D. Intracellular Distribution of Metabolic Pathways

In eukaryotic microorganisms, hydrocarbon catabolism is specifically associated with ultrastructural features [11]. The most striking and widely studied alkane-induced change concerns the appearance of microbodies. These microbodies, found in yeasts and filamentous fungi, are peroxisomes of a crystalline nature. They are clearly of importance although the alkane monooxygenase is not associated with such structures, being localized in the microsomes. The peroxisomes are believed to contain the alcohol and aldehyde dehydrogenases and the enzymes of β-oxidation. It would appear that certain enzyme sequences are duplicated in the various organelles, although the advantages to the cell are not clear. An example of this can be seen in the β-oxidation pathway found in both the peroxisomes and the mitochondria. However, the enzymes involved differ in chain length specificity and energetic efficiency [12], and must be assumed to endow the microorganisms with a degree of regulatory control, the exact nature of which is at present obscure. In addition to these metabolism-housing organelles, hydrocarbon-grown filamentous fungi have been shown to contain membrane-bound, electron-dense inclusions [13], absent during growth on glucose. As yet these microbodies have not been properly identified although they have been speculated to function as intracellular stores of nonmetabolized alkanes. This hypothesis is almost certainly correct; the appearance of such structures is coincident with recoverable quantities of unaltered hydrocarbons from within the cell. A possible explanation of such structures can be proposed based on formation of alkane-containing membrane vesicles at the cell membrane of *Candida tropicalis* by the process of pinocytosis [14]. This hypothesis will be examined later in relation to hydrocarbon uptake mechanisms.

III. PROBLEMS ASSOCIATED WITH GROWTH ON INSOLUBLE SUBSTRATES

Hydrocarbons exist as gases, immiscible liquids, and waxy solids, depending on their composition. The common feature of all these various forms is that they are virtually insoluble in water. This clearly poses a problem as regards growth on hydrocarbons. Certain generalized observations can be made regarding the suitability of the various types of hydrocarbons for microbial utilization:

1. Aliphatic hydrocarbons are degraded and assimilated by a wide range of microorganisms while cyclic and aromatic compounds may be partially oxidized by many (often in the presence of an alternative carbon substrate) but are assimilated by a few species, mostly bacteria.
2. *n*-Alkanes smaller than octane are not assimilated by most hydrocarbon-degrading microbes, though they can often be catabclized, thus serving as an additional energy source.
3. Unsaturated compounds are more difficult to degrade than saturated hydrocarbons.
4. Branching of the hydrocarbon chain increases its resistance to degradation.

Bearing these observations in mind, it is not surprising that the most commonly used hydrocarbon feedstock is the cut including *n*-alkanes of intermediary chain length. These compounds are liquid at the normally employed growth temperatures (25-35°C), particularly if a mixture is used. In general, the biomass yield increases with the molecular weight of the substrate, but some controversy exists as regards the relative rates of consumption. While it is true that the overall culture time necessary for complete oxidation of alkanes increases with the molecular weight, this is not sufficient justification to conclude that longer alkanes (e.g., C_{16}-C_{20}) are metabolized more slowly than their smaller counterparts (C_{10}-C_{15}). Studies with a range of yeasts and filamentous fungi have clearly shown that such a hypothesis is incorrect. Longer chain alkanes are in fact often metabolized more rapidly, but only after a prolonged period of adaptation [15], i.e., long lag phase (see below).

A. Growth Kinetics

The growth kinetics of both yeasts and filamentous fungi on insoluble hydrocarbons is somewhat unusual. The pattern obtained is a short-term exponential growth phase followed by a linear growth pattern (see Fig. 5). This profile can be explained simply by the following concept: an immiscible substrate can be present in excess concentrations within a culture although the availability of this substrate to the microorganism is restricted. In other words, the surface area of the substrate and hence the potential access for the microorganism can become limiting. The rate of this limitation will depend on the fermentation: presence of a surfactant will favor the formation of an emulsion and hence the contact surface will be greater. As a result, the transition to linear growth phases in surfactant-containing broths will occur much later than for cultures in which little or no surfactant is produced or added. Large-scale cultures of hydrocarbon-utilizing microorganisms therefore necessitate careful attention to mixing and the creation of microemulsions if rapid growth to industrially useful cell densities is to be achieved. Such cultures may also pose potential oxygen transfer problems since the oxygen demand for growth on hydrocarbons is extremely high. This is directly related to the degree of reductance of the substrate, which is exceptionally high. However, this problem can be minimized because oxygen is extremely soluble in hydrocarbons, and the carbon substrate can function as an oxygen vector [16], particularly if coupled to an aeration jet device [17]. In other respects nutritional requirements are not particularly different from those necessary for exploitation of other carbon substrates.

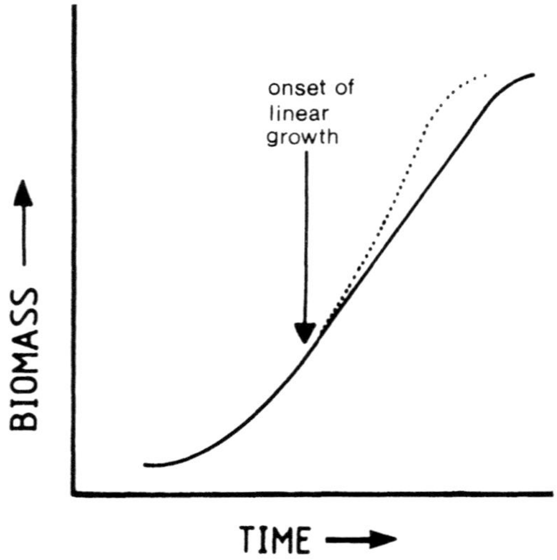

FIGURE 5. Typical growth curve for a hydrocarbon-utilizing yeast.

A realistic appraisal of substrate costs leads us to believe that microorganisms able to consume mixtures of hydrocarbons will be advantageous. It is therefore important to understand the way in which mixtures of hydrocarbons are utilized. There is a definite preference for aliphatic hydrocarbons and, more specifically, for *n*-alkanes. Furthermore, shorter chain alkanes such as decane are metabolized before longer chain alkanes but at slower specific consumption rates and less efficiently, i.e., lower biomass yields. This observation, first reported for *Candida lipolytica* and attributed to alkane-specific solubility effects [18], has now been extended to a wider range of yeasts and fungi [19], although in this later study no evidence for sequential production of extracellular solubilizing agents specific to individual alkanes was obtained. It would appear that some modification to the microorganism's physiology is necessary to enable hydrocarbons to be metabolized and that the ensuing lag phase is directly proportional to the alkanes chain length [15].

In a review of the potential of various eukaryotes to degrade diesel fuel hydrocarbons it was observed that in almost all cases *n*-alkanes were first degraded in order of size [19] followed by the methyl alkanes at reduced growth rates (unpublished observations). Alicyclic and aromatic hydrocarbons were not as a general rule removed from the complex mixture. An exception was observed for the yeast *Candida tenuis* which, although incapable of growth on aromatic compounds when presented as sole carbon sources, degraded such hydrocarbons at high rates from whole-fuel mixtures. This yeast is therefore fundamentally different from the more commonly studied microorganisms and merits further investigation. Such species may well prove to have a more promising biotechnological future than other species which grow faster on the more easily metabolized *n*-alkane fraction. Before passing to discussions of the growth-promoting activity of extracellular products it is perhaps useful to compare the growth potential of yeasts and filamentous fungi. The yeasts that have been studied extensively are mostly from the genera *Candida* (though some have been reclassified) and grow at specific rates significantly higher than the filamentous species studied to date. Furthermore, the biomass yields obtained with filamentous species are surprisingly low taking into consideration the energetic value of the hydrocarbons. As yet no convincing explanation for this phenomenon exists: biochemical sequences seem to be similar, though very few detailed studies exist for filamentous fungi. A possible explanation is that filamentous fungi are "leaky" as regards important intermediates; certainly a variety of organic acids are said to be found in spent medium [20]. If this is so, then their potential for metabolite overproduction is high. Examination of the growth kinetics of such species coupled with thorough physiological studies may be profitable.

B. Role of Surfactants

The production of surface-active emulsifying agents is a vital step in the successful assimilation of hydrocarbons by many microbial species. It has often been demonstrated that when active hydrocarbon assimilation occurs, a microemulsion of very small droplets is formed [21]. These range from 0.01 to 0.5 μm in diameter and are formed due to the presence of surfactants. In the case of *Candida lipolytica* this emulsion can be made up of droplets so fine that pseudosolubility has been discussed [18], though it is unlikely that true solubilization of the alkane into the aqueous phase at a distance from the cell surface is a common mechanism.

The major effect of these hydrocarbon-induced surfactants is to increase the available surface area of the hydrocarbon phase allowing more cells to achieve the essential cell-hydrocarbon contact. Surfactants having "solubilizing" properties are more likely to be found at the cell surface where they may function as transporters in hydrocarbon uptake mechanisms [22].

The diversity of surfactants produced by yeasts and fungi while growing on hydrocarbons ranges from simple fatty acids to complex polymers. In the case of yeasts, where the surfactants are often complicated lipid-sugar-peptide polymers [23], production is usually specific to growth on hydrocarbons (or rather the presence of hydrocarbons in the medium). Filamentous fungi, on the other hand, do not produce such active surfactants, and the metabolites playing this role are often simple fatty acids or phospholipids [20], which are usually produced, albeit to a lesser extent, on substrates other than hydrocarbons. The fortuitous function of such lipidic metabolites cannot be compared with the highly specific polymeric surfactants associated with yeasts grown on alkanes.

C. Specific Uptake Mechanisms for Hydrocarbons

To date no evidence has been published suggesting that hydrocarbons can be attacked by extracellular enzymes (other than nonspecific lignin peroxidases) and therefore some mechanism of substrate uptake is needed. Modification of the outer cell surface to increase the contact area and the hydrophobicity are common adaptations during hydrocarbon metabolism. This chemical change in cell structure facilitates the passive adherence of the substrate to the cell, thus providing both a point of anchorage at the droplet surface and direct substrate-cell contact. Such passive attraction does not, however, explain how the insoluble hydrocarbon enters the cell. After its immobilization at the cell surface hydrocarbon moves toward the plasmalemma, and this may be facilitated by the presence of pores and channels as are often reported [14]. If the initial enzymes were plasmalemma-bound, activation of the hydrocarbon might occur enabling aldehydes or fatty acids to enter the cytoplasm, but cytochemical location of the hydrocarbon-oxidizing enzymes (see above) does not agree with such a hypothesis. Instead, in many species unaltered hydrocarbons can be recovered from within the cytoplasm (though sometimes in membrane-bound vesicles). The kinetics of uptake suggests that the transfer mechanism is not a simple diffusion process [24,25], and the inhibitory effect of uncouplers such as dinitrophenol might be taken as evidence that active, energy-dependent transport occurs. The work to date has been inconclusive and while active transport may be involved, definite evidence of intracellular accumulation against a concentration gradient has never been obtained. Work with *Cladosporium resinae* [25] clearly showed that hydrocarbons were present at much higher concentrations within the cell than the medium, but that this may have been possible by facilitated diffusion bearing in mind the high cell wall-associated levels. One possibility is that the alkane is taken into the cell via pinocytosis of the plasmalemma, thus explaining the requirement for energy and avoiding hydrocarbon-water solubility problems since the hydrocarbon need not pass directly into the cytoplasm. Such vesicles have been reported in fungi [13] but conclusive support of such a process would necessitate electron microscopy evidence. If the nature of the uptake process cannot be adequately portrayed, the kinetics do enable sequential uptake of hydrocarbons to be related to the affinity of the microorganisms for various alkanes. Saturation constants for *n*-alkanes are related to the substrate chain length: microorganisms have more affinity for shorter alkanes which outcompete the longer substrates for uptake [26]. Again, details of the biochemical mechanism are absent, but some degree of alkane-specific modification to the physicochemical nature of the cell wall seems to be involved.

D. Fermentation Strategies

Clearly the fermentation strategy employed for a potential biotechnological process will depend on the nature of the product, i.e., biomass production will impose different requirements to metabolite overproduction. It is not within the scope of this chapter of hydrocarbon-utilizing yeasts and filamentous fungi to detail all possible fermentation strategies, which would in any case be somewhat independent of the substrate. Hydrocarbons by their insoluble nature pose certain problems for the microorganisms (see above) which biotechnologists need to take into account when planning how best to grow the organism. Clearly, the fermentation strategy will differ from those employed with soluble carbon feedstocks, e.g., sugars, methanol, etc., or solid wastes such as lignocellulose. The major consideration to be taken into account is the absolute necessity to maximize substrate dispersion within the fermentor, thus providing the maximum access to microorganism-hydrocarbon contact essential for substrate uptake. Provided adequate mixing is used, the production of highly active surfactants by many hydrocarbon-assimilating yeasts should ensure that a microemulsion is formed. Recovery of the surfactant and addition to future cultures (providing the same substrate is used) will often diminish the time of the culture since induction and biosynthesis lag times will be suppressed. However, care should be taken if a biosurfactant is added to cultures of nonrelated microorganisms, since these compounds when not self-produced can perturb cell metabolism and hence counteract the beneficial influence of improved substrate dispersion.

A separate problem needs also to be taken into account: the highly reduced nature of hydrocarbon substrates. While this characteristic of hydrocarbon feedstocks ensures high biomass yields and can be easily manipulated for metabolite overproduction, optimized growth on such reduced substrates necessitates a fermentation strategy with high oxygen transfer capacity. Considerable oxygen is demanded to assimilate hydrocarbons, both in the initial alkane monooxygenase step and in the process of oxidative phosphorylation via the cytochromes fed by the reduced coenzymes resulting from alcohol and aldehyde dehydrogenases, β-oxidation, and other catabolic pathways. To facilitate such a high oxygen demand it might be desirable to use oxygen jet devices whereby an oxygen-enriched hydrocarbon-water emulsion is injected into the fermentor. This technique would be particularly appropriate for filamentous fungi since these microorganisms do not produce active emulsifying agents. In such cultures substrate dispersion and oxygen supply could be overcome by a combined approach with a periodic reinjection of the hydrocarbon phase no longer emulsified. With yeasts this process may well prove to be inappropriate since the presence of surfactants may lead to serious foam problems.

Directly related to this high oxygen demand is the production of metabolic heat, necessitating an extremely efficient cooling device. Heat production associated with the incomplete recovery of biochemical energy can be seen as both a problem and a potential advantage. If the excess heat can be reused, e.g., in greenhouse style market gardening facilities adjacent to the fermentation unit, then product credits can accrue; but if not the cost of cooling can be high. The problem is generally more intense in laboratory scale units, being fairly simple to overcome during scale-up to pilot plants. However, since dense cell cultures (>50 g dry weight/liter) are increasingly used to attain maximum productivities, heat transfer may, together with oxygen supply, fix the upper level of biomass that can be efficiently exploited.

So far we have emphasized problems associated with maintaining hydrocarbon-metabolizing microorganisms active under biotechnologically feasible fermentation strategies. This daunting task does, however, have certain inherent advantages related to the nature of the hydrocarbons. Since the energetic content of the substrate is high, the biochemical energy yield associated with precursor metabolite formation satisfies the majority of the energy requirement for anabolism. As such the tricarboxylic acid cycle is partially inactivated as a catabolic pathway and often functions primarily as a source of metabolites, i.e., as an anabolic sequence. The net result of this is a relatively low content of carbon dioxide in the effluent gas and a diminished influence of this respiratory product on growth inhibition. For many substrates, carbon dioxide-associated growth inhibition can have a profoundly negative influence on growth rates [27], particularly at high biomass concentrations.

A possible advantage to hydrocarbons is that many filamentous fungi which grow as pellets with carbohydrate feedstocks grow as homogeneous cultures on hydrocarbons (Lindley, unpublished). In this respect certain of the nutrient transfer problems traditionally associated with fungal technology can be avoided.

All biotechnological processes are to some extent dependent on efficient downstream processing and complete use of the carbon feedstock (up to 60% of the process costs). Since one of the characteristics of fermentation is that the product will generally be present as a fairly dilute solution together with residual nutrients, an insoluble substrate can present certain advantages. Recovery and subsequent reuse of residual hydrocarbon can improve process economics, but in some cases two-phase systems can allow partitioning of product into the hydrocarbon phase, thus simplifying tertiary purification [28]. This could be applied to organic acid production, for example, so long as the pH optimum for growth is sufficiently close to (or below) the pK of the product to enable the establishment of concentration gradients of the free acid. In addition to reducing downstream processing costs, this type of liquid-liquid, extraction-reextraction technology can remove potentially inhibitory effects of accumulated metabolites.

Development of a complete fermentation strategy adapted for hydrocarbon-utilizing microorganisms need not pose more problems than for other feedstocks although certain scale-up procedures would need to be modified. However, a careful assessment of the entire process rather than treatment of a series of individual characteristics can often exploit the nature of hydrocarbons to arrive at a highly efficient industrial technology. This global approach does require a multidisciplinary approach often lacking in both academic laboratory workers and industrial biotechnologists. In this respect hydrocarbon-based processes are perhaps at a disadvantage at the present time, although such a statement should be taken as a call for specialists other than biologists to take an active interest rather than as a reason to abandon hydrocarbons as a feedstock.

IV. HYDROCARBONS AS BIOTECHNOLOGICAL FEEDSTOCKS

As was seen from the preceding sections of this chapter, the intensive study of hydrocarbon metabolism has been limited to relatively few yeasts and hardly any filamentous fungi (perhaps only *Cladosporium resinae*). It is perhaps all the more surprising in view of the superficial understanding of hydrocarbon utilization that such a range of potential biotechnological applications have been proposed in the literature [3]. Much of the early work was oriented toward producing single-cell protein and the problems proved to be not so much of a biological nature but rather those associated with gaining market approval. Since many of the objections raised as regards SCP are equally valid for other hydrocarbon-based products destined for human or animal consumption either as food additives or pharmaceuticals, the validity of these arguments needs to be objectively assessed. Thus although SCP from hydrocarbons now seems unlikely ever to become widespread, it is necessary to discuss this initial application.

A. Single-Cell Protein

For approximately a decade many industrialists in the petrochemical industry supported ambitious research projects to convert cheap alkane mixtures to yeast-based protein [30]. Some of these research ventures led to industrial development using *Candida lipolytica*, e.g., Toprina (British Petroleum Ltd.) and Liquipron (Liquichimica Spa.), but these were rapidly abandoned in the 1970s when oil prices increased dramatically. However, the difficulty associated with obtaining market clearance was a large contributory factor to the final decision to cut production.

By the early 1970s both Toprina [31] and Liquipron [32] had met all requirements for the initiation of clinical trials although doubts still existed on a number of key points:

1. Utilization and safety of fatty acids of uneven numbers of C atoms
2. Significance and fate of residual paraffins
3. Strength of the evidence for lack of toxicity or carcinogenicity of SCP proposed for animal feeds

The fatty acid content of hydrocarbon-grown yeasts reflects the alkane mixture used and since pure alkanes were not an economically feasible approach, high levels of odd-chain fatty acids were unavoidable. More recently, it was demonstrated that the distribution of fatty acids can be altered if *C. lipolytica* mutants are used which lack the acyl-CoA synthetase enabling intact incorporation or chain elongation. In such strains lipids are uniquely formed by de novo synthesis (see Fig. 6) and even-chain fatty acids account for at least 90% of total fatty acids even when fed on odd-chain alkanes [33]. This work, published in the late 1970s, was not available when SCP commercialization was being attempted and the potential problems arising from odd-chain fatty acids were never satisfactorily resolved.

Of more general interest was the fate of residual paraffins since these hydrocarbons are known to accumulate within adipose tissues. Metabolic studies showed that alkanes were to a large extent degraded, with only 7% being stored unchanged. It was, however, the fate of trace amounts of polycyclic aromatic hydrocarbons, and in particular 1,2,5,6-dibenzanthracene, 3,4-benzopyrene, and 1,12-benzoperylene, which might be expected to be present in gas-oil fractions that caused most concern. From a regulatory viewpoint, SCP derived from hydrocarbons was subject to far more detailed toxicological evaluation. Questions were posed as to how such polycyclic residues could be removed. Solvent extraction was used in certain processes to remove such compounds along with residual

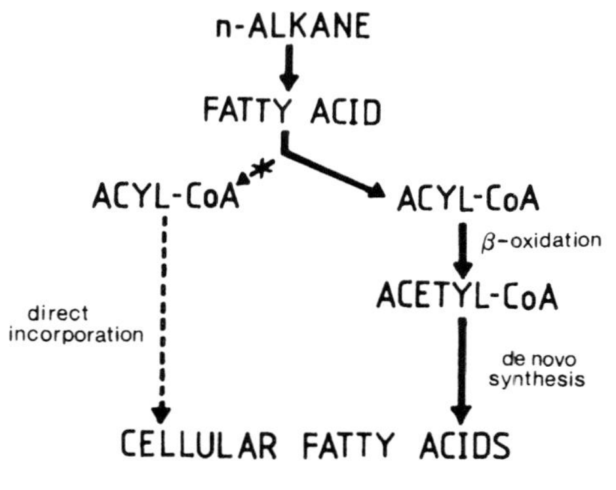

FIGURE 6. Fatty acid metabolism in mutants of *Candida lipolytica* lacking acyl-CoA synthetase I.

alkanes, and studies showed that the polycyclic aromatic content of solvent-extracted gas-oil-grown yeasts was lower than in various baker's yeasts [39]. It is, however, always difficult to demonstrate the absence of potentially carcinogenic compounds, particularly when the level of such aromatics is likely to vary depending on the origin of the hydrocarbon feedstock. This problem remains for future hydrocarbon-based products and the need for quasi-total removal of trace concentrations will have an obvious effect on the already high costs of downstream processing. With respect to SCP production polycyclic aromatics were at worst present only to the same levels as other sources of crude protein.

Hydrocarbon-derived SCP suitably treated was generally considered to be a good source of protein though fermentation and processing required special care to avoid desirable side effects [29]. Again nothing unusual; all SCP production varies as regards dietary tolerance if even minor modifications are made to the process. Obtaining the go-ahead to sell SCP grown on alkanes was difficult but finding a suitable way of incorporating such products into human diets was virtually impossible.

The product is a dry powder with a flavor ranging from bland to bitter. The principle use was not as a source of protein but rather as a minor vitamin-rich or flavoring agent (meaty or nutty taste of some yeasts). The amount of SCP used is at best small and widespread use necessitates its incorporation into food with an acceptable flavor such as already exists for its major competitor, soya flour. The acceptability of SCP might well be increased if filamentous fungi rather than yeasts were used. The texture of the final product is fibrous and therefore more closely akin to meat, while their mushroomlike flavor is generally acceptable. Fungal SCP from other feedstocks has been developed to commercial levels, but such a process for hydrocarbons seems unlikely. The growths rate and biomass yields of filamentous fungi are not nearly as good as for yeasts and the process costs would be unlikely to allow a competitive sale price to be fixed. Little doubt remains that acceptable protein-rich foods can be derived from hydrocarbons, but the cost is generally too high to justify production. Agricultural surpluses and the resulting economic considerations are such that a large-scale production of hydrocarbon-derived SCP is unlikely in the forseeable future.

B. Oil Spillage Treatment

At present pollution of both land and sea by hydrocarbons is reaching critical levels, particularly with respect to the marine environment [35]. Everyone is aware of major tanker disasters such as the Torey Canyon and Amoco Cadiz in the English Channel and the enormous environmental repercussions; but numerous minor "accidents" occur wherever oil tankers circulate. Since certain microorganisms are particularly adapted to using hydrocarbons, the possible use of such microbes in an oil spillage treatment strategy has been examined in various laboratories. In order to assess this type of strategy nutritional studies have been undertaken since a high capacity to assimilate hydrocarbons would be of little practical use if other elements are severely limiting. This type of study in marine environments has shown that first nitrogen and then phosphorus are often limiting nutrients in seawater when carbon substrate is present in excess [36]. It is therefore important to add such compounds to the oil spillage before seeding with microorganisms. It might also be argued that the addition of such nutrients would be sufficient to stimulate the growth of naturally occurring hydrocarbon-degrading microorganisms. Of course both strategies would be slow and unlikely to completely degrade crude oil. Test experiments have shown slightly improved depollution rates, but the ensuing eutrophication with food-web concentration of aromatic residues cannot be ruled out.

A more realistic use of microorganisms might be possible if the spillage could be contained or removed for separate treatment in deep-shaft pollution treatment facilities. The lack of nutrient loss (inevitable in the marine environment due to mixing) would be beneficial as would the improved growth environment (e.g., higher temperature, adequate phase dispersion), although the high costs of such a technology would have to be competitive with oil recovery techniques. If microorganisms were to be used as an integral part of an oil spillage treatment process, bacteria, being more metabolically versatile (or at least better understood as regards recalcitrant hydrocarbon oxidation) and more rapid in their rates of substrate turnover, would likely be more suitable than eukaryotes.

C. Surfactant Production

Again, both filamentous and unicellular fungi have been less widely exploited than bacteria, although many fungi, especially yeasts, produce highly active surfactants during growth on hydrocarbons [23]. Before describing some of these products it is necessary to discuss potential markets for emulsifying agents. Perhaps the most widespread use of biopolymers is by the food industry to stabilize emulsified sauces and related products. Bearing in mind some of the objections raised regarding SCP, it would take a brave industrialist to try and obtain clearance for use of a hydrocarbon-derived surfactant in food products. This is not to say that some products would not receive the blessing of the regulatory committees, but obtaining such an accord would take considerably longer than for a similar product derived from other substrates.

Such objections as might arise regarding food industry applications would not influence the use of surfactants within the petrochemical industry. Biosurfactants have a role to play in oil spillage treatment as well as recovery of otherwise noneconomic oil reserves. In this latter application, surfactant or the microorganism that will produce the surfactant is injected into the residual oil-bearing strata in such a way that a petroleum-water emulsion is formed which can then be pumped out and subjected to phase separation. Normally up to two-thirds of the oil will be considered uneconomical to extract [37], so that the use of simple technology to extract the remaining petroleum would be increasingly important as finite end-point dates are set for known oil reserves. Surfactants have a variety of alternative uses such as within the paint industry. New applications are regularly necessary and a more widespread analysis of the type of surfactants employed by microorganisms during growth on hydrocarbons would be useful.

The range of surfactants produced by fungi and yeasts is enormous, varying from simple fatty acids and phospholipids among filamentous species to complex polymers by various *Candida*-type yeasts. Quite often the surfactant activity has been arbitrarily measured and details of the chemical structure of biosurfactants are frequently inadequate. While a great many scientific reports exist, many of these have not the rigorous attention to fine detail and standardized assay conditions that would enable a meaningful comparison of the potential applications of each surfactant.

Among filamentous fungi, the extracellular metabolites of *Cladosporium resinae* have been the most intensively studied [20]. The phospholipids were the same as those produced intracellularly, although simple loss via lysis could be excluded since the fatty acid substituents were of a different nature. Extracellular phospholipids were predominantly dodecanoic acid-substituted [38]: this fatty acid was also present in significant amounts as a free acid within the broth. Various evolutionary reasons can be given for this presence including bacteriostatic effects and optimized emulsifying activity of C_{12}-fatty acids. Other filamentous fungi seem to have a similar range of fatty acids and phospholipids within the extracellular liquid, although in all cases such metabolites were also produced on nonhydrocarbon substrates. Highly active surfactant during hydrocarbon exploitation by *Aspergillus versicolor* was observed in the author's laboratory which was seemingly growth environment-dependent (nonpublished results); to date this product has not been characterized. With this possible exception biotechnological applications of extracellular biosurfactants would seem to be unlikely from filamentous species, unless short-chain phospholipids are of some economic value making downstream recovery costs worthwhile.

By contrast, the surfactants produced by yeasts are of high potential due to their highly active emulsifying properties. Produced at a given point in the growth phase and induced specifically by the presence of hydrocarbons, these biosurfactants are potent emulsifiers. Only a relatively small fraction of the carbon substrate (judged on carbon mass balances) is transformed into these various metabolites yet extremely rapid emulsification to microdroplet levels is achieved with oil-water mixtures. The chemical nature of these extracellular metabolites varies and undoubtedly further discoveries would occur if a pragmatic screening procedure were practiced. At the moment only species of *Torulopsis* (e.g., *magnoliae* and *gropengiesseri*) and *Candida* (e.g., *lipolytica*, *tropicalis*, and *petrophilum*) have been intensively investigated. The genus *Torulopsis* produces glycolipids that have similarities to bacterial rhamnolipids. These yeast sophorose lipids [39] contain an acetyl-substituted sophorose disaccharide attached to the hydroxyl function of a hydroxycarboxylic acid. In some species this sophorose lipid is modified in such a way that both the hydroxyl group and the carboxyl functions of the fatty acid are attached to the dissaccharide to form a lactone. The structure can to some extent be modified with respect to the fatty acid chain length by using carboxylic acids, methyl esters, hydrocarbons, or glycerides in the broth, thus enabling potentially interesting characteristics to be developed. Furthermore surfactant yields are greatly improved by such cosubstrate feeding strategies, i.e., yields of 17 g/liter [40].

Candida yeasts produce somewhat different surfactants which appear to be polysaccharide-lipid or lipopeptide complexes either produced within the infrastructure of the cell wall (e.g., *C. tropicalis*) or inducibly released into the medium. The surfactant of *C. lipolytica* seems to be somewhat unusual in that improved "pseudosolubility" can be obtained for the specific alkane metabolized [41]. Presumably some structural change to the chemical integrity of the surfactant takes place in response to specific alkane substrates, though whether or not this is merely fortuitous modification cannot be determined since identification of the polymer is lacking in detail. It may be a surfactant somewhat akin to that reported in *C. petrophilum*, which contains a peptide and one or more fatty acids [42]. The structure of the peptide fraction is similar to that of the bacterial lipopeptide, surfactin—the most effective biosurfactant yet reported. Clearly such a compound is of potential biotechnological value. The inherent properties of hydrocarbon-utilizing fungi for biosurfactant production offer one of the most promising areas for biotechnological exploitation. To date, the products associated with well-studied species grown on simply degraded alkanes show potential applications for the petrochemical industry. Modification of surfactant by careful use of lipidic cosubstrates would seem to offer scope for structural and possibly functional modification of the emulsifying properties. As was remarked in a recent review of surfactants [23], many of the lipidic components of microorganisms have emulsifying potential though few have been rationally examined. As such there remains an excellent potential for developing families of compounds with graduated surfactant properties.

D. Fuel Tank Contamination

Microbiological contamination is by no means a recent phenomenon, having first been reported in the 1940s for aircraft fuel tanks. Initially the major contamination was due to sulfate-reducing bacteria but a changeover from petrol to kerosene-type fuels in the 1950s saw a shift in the type of microorganisms reported. Buildup of fungal mycelia and ultimately blocking of fuel pumps led to operational difficulties resulting in several aircraft crashes in the early days of jet aircraft [43]. A variety of filamentous fungi and yeasts were isolated from contaminated fuel systems and the fungus *Cladosporium resinae* identified as the dominant organism in such populations. It is of course important to understand the context of such investigations: yeasts may well be major species within the population but they do not generally block the fuel lines as do mycelial fungi. This accounts for the research concentrated on studies of *C. resinae* with the formulation of fungicide strategies. Such an approach was never entirely successful since wide-spectrum fungicides are not easy to find, particularly for two-phase systems. A more successful although time-consuming method was to scrupulously clean all fuel tanks at regular intervals, paying particular attention to the removal of the accumulated water within the tanks. Such water enters from traces present in the fuel and from condensation. If such pockets of water can be eliminated, microbial growth will not occur and fungicide treatment will become largely superfluous, acting as a secondary backup system. Providing the fuel tanks are regularly cleaned and the fuel line filters properly monitored, aircraft fuel contamination is unlikely to develop within the tanks. Of course, such precautions accomplish nothing if the fuel used is of low quality, i.e., high water content or already contaminated in the fuel supply reservoir. It is ironic that extensive cleaning of the aircraft fuel systems now occurs while much less rigorous procedures are applied to fuel stocks. Growth of microorganisms in the fuel is not itself dangerous; rather, it is the blocking of fuel lines by accumulated mycelial mats which leads to problems.

Aircraft fuel tank contamination is generally considered no longer to be an important problem, but a recent development in ship engines has highlighted a similar problem for the modern marine industry [44]. Such gas turbine engines use diesel fuel similar to traditional diesel-powered motors but require a fuel supply of improved purity. A ship's fuel tanks make up an important percentage of the ship's weight, and in order to maintain a constant ballast the fuel is displaced with seawater. The turbine blades of modern engines are highly susceptible to the corrosive effect of saltwater so that an elaborate filter/coalescer assembly is necessary to protect the engine. These filters are prone to blockage if the inevitable fungal growth within the primary fuel tanks enters the fuel line. Since the manner used to combat fungal contamination in aircraft fuel systems cannot be applied in ships, two complementary approaches would seem to be necessary, i.e., an engineering approach in which fungal matter is removed by centrifugal pumps, reinforced by a fungicide strategy. Fungicide strategies would necessarily be expensive since huge volumes of both fuel hydrocarbons and seawater would be involved, and possibly periodic sterilization of empty tanks would be essential.

Though filter blocking has received a great deal of attention with a predominant role of filamentous fungi, a detrimental effect on coalescer performance has also been observed due to yeast growth or, more precisely, their surfactants which result in microemulsion formation. The effect of this is the arrival of water-contaminated fuel at the engines resulting in reduced power output and increased corrosion. Corrosion may also occur within fuel tanks due to the acidic metabolites of hydrocarbon degradation, although such corrosion is not nearly as important as that associated with sulfate-reducing bacteria.

The economic cost of hydrocarbon-degrading microorganisms in terms of contamination problems has never been realistically assessed but is sure to be high if similar problems associated with crude oil transport are also taken into account. This problem is a serious cause for concern within the petrol industry and is far from being resolved.

E. Metabolite Overproduction

A wide variety of biochemical intermediates of potential industrial value can be produced by yeasts and molds during growth on hydrocarbons (Table 1). Such products can be subdivided into metabolites that are only associated with hydrocarbon metabolism, such as long-chain dicarboxylic acids, and intermediates that are common to growth of an organism on a variety of substrates, such as organic acids. In some cases this latter type of compound may be produced at significantly higher concentrations during growth on hydrocarbons, while for others it is merely a cost performance advantage that favors the use of hydrocarbons. Bearing in mind the world surplus of agricultural carbohydrates, fermentations need to have high yields of a given product before hydrocarbons can offer much advantage. This is clearly not the case for products specifically associated with hydrocarbon degradation where competition is more likely to arise from the chemical synthesis industry.

1. Organic Acids

Of the big market products, amino acids have received considerable attention, but bacterial strains seem more suitable for this purpose. Work relating to glutamate production has shown that carboxylic acids such as ketoglutarate and citrate can also accumulate. Various strains of *Candida lipolytica* are used to produce these carboxylic acids [3]. Extensive work in Japanese laboratories has led to strains of this yeast which can accumulate citrate to concentrations close to 200 g/liter [45]. Even with such high yields the preferred biotechnological production of citric acid uses *Aspergillus niger* and carbohydrate feedstocks. Since this fungus can

TABLE 1. Illustrations of Biotechnologically Feasible Product Formation from Hydrocarbon-Utilizing Yeasts and Filamentous Fungi Based on Significant Industrial Funding

Species	Product(s)
Yeasts	
Candida cloacae	Long-chain dicarboxylic acids
Candida lipolytica	SCP, alkane-solubilizing surfactants, citrate, methyl citrate, etc.
Candida petrophilum	Surfactin-like surfactant
Torulopsis bombicola	
Torulopsis gropengiesseri	Sophorose lipids
Torulopsis magnoliae	
Filamentous fungi	
Fusarium sp.	Alkaline protease, ergosterol
Paecilomyces carneus	β-Lactam antibiotics

also metabolize alkanes, a study of citric acid production comparing the two feedstocks might prove enlightening. If odd-chain alkanes are used as carbon substrate, methyl citrate can be produced [46] due to the operation of the so-called methyl citrate cycle of propionyl-CoA oxidation. This pathway is apparently unique to hydrocarbon metabolism and therefore alternative biological production of such methylcarboxylic acids are unlikely. Unlike the case with citrate, which is widely used in both the food and pharmaceutical industries, the market for the methylated analogs is not yet well defined. Other carboxylic acids such as fumarate malate, and succinate can also be overproduced by *Candida* yeasts but levels are considerably lower than for citrate. For all these metabolites, overproduction necessitates a deregulation of the carbon flow through the tricarboxylic acid cycle. Such an effect is relatively simply brought about and capitalizes on the high concentrations of acetyl-CoA derived from the β-oxidation pathway.

Long-chain carboxylic acids, and in particular dicarboxylic acids, are valuable raw materials for the perfume industry. Such dicarboxylic acids result from diterminal oxidation of the parent alkane but do not normally accumulate because they are further oxidized during normal balanced metabolism. In order to produce strains which have suitably high accumulation of these products, the genetic engineers have deleted the enzymes responsible for the further degradation of dicarboxylic acids. A mutant of *Candida cloacae* produces high levels of these acids either using resting cells or during growth with acetate as cosubstrate [47].

2. Lipids

The lipid fraction of the cell can itself be a valuable product. Cell lipids, either as phospholipids or triglycerides, are highly susceptible to overproduction induced by hydrocarbons [48]. Furthermore, the fatty acid component of these lipids reflects the chain length of the alkane substrate. Bearing in mind the current increase in research activity regarding lipid-accumulating microorganisms, it might be thought that alkanes ought to be a suitable feedstock. However, the fatty acid content of lipids from hydrocarbon-grown microorganisms is predominantly of a saturated nature, whereas current research interest is in producing polyunsaturated fats, which have a much higher market price. Such lipids are associated with various yeasts and filamentous fungi, but to date alkane-utilizing species capable of synthesizing polyunsaturated fats have not been described.

3. Carbohydrates

Energetic calculations of the thermodynamics of hydrocarbon degradation suggest that alkanes might be a high-potential feedstock for polysaccharide or other "energy sink" polymers. Some promising results have been achieved with bacteria, but other than surfactants no such products have been developed with fungi. On the other hand, some yeasts, if grown at carefully controlled phosphate concentrations and acidic pH, can be included to accumulate various polyols, e.g., mannitol, erythritol, arabitol, etc.

4. Vitamins

Some advantages to the use of alkanes can be seen if various low-market-size, high-value-added products such as vitamins and antibiotics are examined. Both vitamin B_2 and vitamin B_6 can be produced from alkanes using yeasts, while respiratory chain components such as cytochrome C and coenzyme Q can also be stimulated to higher levels than normal during hydrocarbon metabolism [3]. The yields of coenzyme Q were particularly encouraging until it was discovered that the form of the product made it unsuitable for pharmaceutical usage (the major market for coenzyme Q is as a medicine against anemia, pyorrhea, and dystrophy).

5. Steroids

A number of yeasts show enhanced production of ergosterol during growth on alkanes, which is not surprising considering the high levels of acetyl-CoA and the suitably lipophilic environment created during hydrocarbon fermentation. Higher yields of ergosterol were produced with filamentous fungi; yields in excess of 260 mg/liter have been reported for a *Fusarium* species [49]. For many molds, production was not necessarily better than for growth on glucose, while cultivation times were often significantly longer. This same *Fusarium* species used for ergosterol production was also developed industrially for the production of extracellular alkaline protease with an optimized yield of 46,000 U/ml in a 2000-liter fermentor. No details are available as to whether or not coproduction of protease and steroids is feasible.

6. Antibiotics

The alkane-utilizing fungus *Paecilomyces carneus* is used to produce cepharosporins, members of the β-lactam family of antibiotics. Product concentrations of 450 mg/liter have been achieved but three forms of the antibiotic were present in similar amounts [50], thus increasing downstream recovery costs.

7. Alicyclic and Aromatic Biotransformations

Virtually all biotechnologically interesting applications of hydrocarbon-degrading fungi refer to alkanes, with the exception of depollution strategies which normally use bacteria. The use of alicyclic or aromatic hydrocarbons shows little if any demonstrated utility despite their obvious potential for biotransformation (controlled oxidation). Filamentous fungi are well known to have the capacity to transform certain steroids, but such activity also enables them to modify aromatic ring substituents and bring about ring hydroxylation [51]. The usual products are hydroxyphenyls and similar reactions can also occur with terpenoids. As yet, however, little biotechnological interest has been shown for such reactions.

While the scope for metabolite overproduction is high during alkane metabolism very few products can realistically compete with those synthesized either from alternative feedstocks or by the chemical industry. However, for a limited number of products the potential for industrial development is high. This is particularly true for those compounds uniquely associated with hydrocarbon metabolism or induced to significantly higher concentrations during growth on hydrocarbons.

V. FUTURE DIRECTIONS FOR HYDROCARBON-BASED BIOTECHNOLOGY

To predict the future direction of such a complex industry as biotechnology is dangerous and patently subjective. Having made this statement, some predictive conclusion to this chapter is necessary and for this the author will rely on contacts made over the last few years with industrial biotechnologists. The most telling comments are always those which answer demands for financial support. From such it is clear, at least in Europe, that large-scale financing of ambitious hydrocarbon-based research projects is now a relic of the past. What is attracting interest is the potential of microorganisms following suitable genetic transformation to undertake specific biotransformations with a high degree of compound selectivity and high yields. Such conversions might open the market to an extended variety of plastics having a biodegradable nature as well as a number of pharmaceutical applications. Quite often what is likely to occur is that the desired enzyme will be used either in pure or partially purified form rather than in the form of whole organism, thus eliminating unwanted secondary reactions.

Alternatively, an enzyme associated with hydrocarbon metabolism is destined to be transfered into more easily manipulated host organisms. It would seem that the variety of microorganisms used by industry is decreasing each year, while the diverse properties of the chosen microorganisms are showing a concomitant increase. As a result it is difficult to assign hydrocarbon labels to future processes since it is unlikely that the organism used will be capable of growth on hydrocarbons. One of the directions regularly discussed is the possible use of hydrocarbon-utilizing yeasts or fungi to upgrade the low-cost stocks of animal fat. Ideas vary enormously from simple desaturation procedures to more complex conversions involving chemical radicals as intermediates.

The future direction is unclear if long-term strategies are being considered. There is a growing awareness that the reservoir of hydrocarbon-degrading microorganisms has been largely ignored, but the financial situation is not ideally suited to high-risk screening programs at the moment. Ideally the discovery of such new species ought to come from academic laboratories, but with tendency toward low-risk process optimization studies to attract short-term industrial backing such projects are unlikely to receive enthusiastic support. This chapter has shown that in countries such as Japan where appropriate studies are in progress a great many potential biotechnological processes based on hydrocarbons have been developed. Given the correct research environment it is therefore possible to find novel transformation reactions of industrial interest. Perhaps the expensive SCP research programs which still appear in red within company files dampen enthusiasm. If so, it is incumbent on research scientists to produce results that cannot be ignored. The wide range of skills now available to the biotechnologist ought to enable a dynamic approach to hydrocarbon-utilizing fungi to be fruitful, particularly if the lessons learned during the 1970s are remembered and realistic market appraisal studies pursued at an early stage of new research and development projects.

VI. SUMMARY

Hydrocarbons are a diverse group of compounds which can be degraded by numerous microorganisms, including many species of yeasts and molds. Despite certain problems related to the insoluble nature of hydrocarbons, fungi are capable of rapidly metabolizing some if not all hydrocarbons, provided the specialized biochemical machinery for emulsification and oxidative degradation is present. The characteristics of growth are such that a variety of biotechnological applications can be envisaged, although often similar processes using nonhydrocarbon feedstocks are preferred. Hydrocarbon metabolism is only understood for a limited number of yeasts: others and various filamentous fungi have only been examined superficially. This lack of fundamental knowledge makes industrial applications prone to risk. Armed with a better understanding of the biochemistry of a wider range of fungi, many more applications of the unique features of hydrocarbon-degrading fungi would be feasible.

REFERENCES

1. Hacking, A. J. *Economic Aspects of Biotechnology*. Cambridge University Press, Cambridge, UK, 1986, pp. 74-92.
2. Crawford, R. L., and Crawford, D. L. Recent advances in studies of the mechanism of microbial degradation of lignins. *Enz. Microb. Technol.*, 434-442, 1984.
3. Fukui, S., and Tanaka, A. Production of useful compounds from alkane media in Japan. In *Advances in Biochemical Engineering*, Vol. 17 (A. Fiechter, ed.), Springer-Verlag, Berlin, 1980, pp. 1-36.
4. Rogers, M. R., and Kaplan, A. M. Role of microbial and nonmicrobial contaminants in diesel-fueled vehicle malfunctions. *Dev. Ind. Microbiol.*, *23*:147-165, 1982.
5. Rehm, H. J., and Reiff, J. Mechanisms and occurrence of microbial oxidation of long-chain alkanes. In *Advances in Biochemical Engineering*, Vol. 19 (A. Fiechter, ed.), Springer-Verlag, Berlin, 1981, pp. 175-215.

6. Gallo, M., Bertrand, J. C., and Azoulay, E. Participation du cytochrome P-450 dans l'oxydation des alcanes chez Candida tropicalis. *FEBS Lett.*, *19*:45-49, 1971.
7. Boulton, C. A., and Ratledge, C. The physiology of hydrocarbon-utilizing microorganisms. *Topics Enz. Ferment. Biotechnol.*, *9*:11-77, 1984.
8. Perry, J. J. Microbial co-oxidations involving hydrocarbons. *Microbiol. Rev.*, *43*:59-72, 1979.
9. Anderson, J. J., and Dagley, S. Catabolism of aromatic acids in Trichosporon cutaneum. *J. Bacteriol.*, *141*:534-543, 1980.
10. Ribbons, D. W. Co-oxidation and co-metabolism: Useful terms? and the benefits available. In *Mixed and Multiple Substrates and Feedstocks: Proceedings of EFB Discussion Meeting, Weinfelden* (G. Hamer and T. Egli, eds.) (in press).
11. Fukui, S., and Tanaka, A. Peroxisomes of alkane- and methanol-grown yeasts. *J. Appl. Biochem.*, *1*:171-201, 1979.
12. Tolbert, N. E. Metabolic pathways in peroxisomes and glyoxysomes. *Annu. Rev. Biochem.*, *50*:133-157, 1981.
13. Cooney, J. J., Siporin, C., and Smucker, R. A. Physiological and cytological responses to hydrocarbons by the hydrocarbon-using fungus Cladosporium resinae. *Bot. Mar.*, *23*:227-232, 1980.
14. Meissel, M. N., Medvedeva, G. A., Kozlova, T. M., Pomoshnikova, N. A., Zaikina, A. I., and Fedoseeva, G. E. Regularities of penetration into yeast cells of higher fatty acids and hydrocarbons. *Proc. 3rd Int. Symp. Yeasts*, Otaniemi, Helsinki, 1973, pp. 146-168.
15. Lindley, N. D., and Heydeman, M. T. Alkane oxidation by Cladosporium resinae: The importance of extended lag phases when assessing substrate optima. *FEMS Microbiol. Ecol.*, *31*:307-310, 1985.
16. Rols, J. L., and Goma, G. Enhancement of oxygen transfer rates in fermentation using oxygen-vectors. *Biotechnol. Adv.* (in press).
17. Fonade, C., Goma, G., and Lafforgue, C. A new approach of momentum and mass transfer in high cell concentration biological processes. In *2nd Int. Conf. Bioreactor Fluid Dynamics* (R. King, ed.), Elsevier, London, 1988, pp. 309-318.
18. Goma, G., Pareilleux, A., and Durand, G. Specific hydrocarbon solubilization during growth of Candida lipolytica. *J. Ferment. Technol.*, *51*:616-618, 1973.
19. Lindley, N. D., Pedley, J. F., Kay, S. P., and Heydeman, M. T. The metabolism of yeasts and filamentous fungi which degrade hydrocarbon fuels. *Int. Biodet.*, *22*:281-287, 1986.
20. Siporin, C., and Cooney, J. J. Extracellular lipids of Cladosporium resinae grown on glucose or on n-alkanes. *Appl. Microbiol.*, *29*:604-609, 1975.
21. Miura, Y., Okazaki, M., Hamada, S. I., Murakawa, S. I., and Yugen, R. Assimilation of liquid hydrocarbons by micro-organisms. *Biotechnol. Bioeng.*, *19*:701-704, 1977.
22. Kappeli, O., and Fiechter, A. Component from the cell surface of the hydrocarbon-utilizing yeast Candida tropicalis with possible relation to hydrocarbon transport. *J. Bacteriol.*, *131*:917-921, 1977.
23. Cooper, D. G., and Zajic, J. E. Surface-active compounds from microorganisms. In *Advances in Applied Microbiology*, Vol. 26 (D. Perlman, ed.), Academic Press, New York, 1980, pp. 229-254.

24. Kappeli, O., and Fiechter, A. Properties of hexadecane uptake by Candida tropicalis. *Curr. Microbiol.*, *6*:21-26, 1981.
25. Lindley, N. D., and Heydeman, M. T. Mechanism of dodecane uptake by whole cells of Cladosporium resinae. *J. Gen. Microbiol.*, *132*:751-756, 1986.
26. Lindley, N. D., and Heydeman, M. T. The uptake of n-alkanes from alkane mixtures during growth of the hydrocarbon-utilizing fungus Cladosporium resinae. *Appl. Microbiol. Biotechnol.*, *22*:384-388, 1986.
27. Jones, R. P., and Greenfield, P. F. Effect of carbon dioxide on yeast growth and fermentation. *Enz. Microb. Technol.*, *4*:210-223, 1982.
28. Levy, P. F., Sanderson, J. E., Kispert, R. G., and Wise, D. L. Biorefining of biomass to liquid fuels and organic chemicals. *Enz. Microb. Technol.*, *3*:207-215, 1981.
29. Scrimshaw, N. S. Acceptance of single-cell protein for human food applications. In *Comprehensive Biotechnology*, Vol. 4 (C. W. Robinson and J. A. Howell, eds.), Pergamon Press, Oxford, 1984, pp. 673-684.
30. Johnson, J. J. Growth of yeasts on hydrocarbons. In *Yeasts for Food and Other Purposes*. Noyes Data, New Jersey, 1977, pp. 44-96.
31. Levi, J. D., Shennan, J. L., and Ebbon, G. P. Biomass from liquid n-alkanes. In *Economic Microbiology*, Vol. 4 (A. H. Rose, ed.), Academic Press, London, 1979, pp. 361-419.
32. Shacklady, C. A. Response of livestock and poultry to SCP. In *Single Cell Protein* (P. Davis, ed.), Academic Press, New York, 1974, pp. 115-128.
33. Tanaka, A., Hagihara, T., Kamiryo, T., Mishina, M., Tashiro, S., Numa, S., and Fukui, S. Candida lipolytica mutants defective in acyl-CoA synthetase. *Eur. J. Appl. Microbiol. Biotechnol.*, *5*:79-86, 1978.
34. Champagnat, A., and Adrian, J. *Petrole et Proteines*, Doins, Paris, 1974.
35. Gutnick, D. L., and Rosenberg, E. Oil tankers and pollution: A microbiological approach. *Annu. Rev. Microbiol.*, *31*:379-396, 1977.
36. Atlas, R. M. Microbial degradation of petroleum hydrocarbons: An environmental perspective. *Microbiol. Rev.*, *45*:180-209, 1981.
37. Forbes, A. D. Micro-organisms in oil recovery. In *Hydrocarbons in Biotechnology* (D. E. F. Harrison, I. J. Higgins, and R. Watkinson, eds.), Heyden & Sons Ltd., London, 1980, pp. 169-180.
38. Kan, C. M. L., and Cooney, J. J. Phospholipids of Cladosporium resinae cultured on glucose and on n-alkanes. *Can. J. Microbiol.*, *21*:1205-1210, 1975.
39. Inoue, S., and Ito, S. Sophorose lipids from Torulopsis bombicola as microbial surfactants in alkane fermentations. *Biotechnol. Lett.*, *4*:3-8, 1982.
40. Tullock, A. P., Spencer, J. F. T., and Gorin, P. A. J. The fermentation of long-chain compounds by Torulopsis magnoliae. *Can. J. Chem.*, *10*:1326-1338, 1962.
41. Pareilleux, A. Hydrocarbon assimilation by Candida lipolytica: Formation of a biosurfactant; effects on respiratory activity and growth. *Eur. J. Appl. Microbiol. Biotechnol.*, *8*:91-101, 1979.
42. Iguchi, T., Takeda, I., and Oshawa, H. Bioemulsification during growth of Candida petrophilum on hydrocarbons. *Agr. Biol. Chem.*, *33*:1657-1658, 1969.
43. Parberry, D. G. Biological problems in jet aviation fuel and the biology of Amorphotheca resinae. *Materials and Organisms*, *6*:161-208, 1971.

44. Neihof, R. A., Klemme, D. E., Patouillet, C. E., and Hannan, P. J. Microbial contamination of ships fuels. In *Distillate Fuel Stability and Cleanliness* (L. L. Stavinoha and C. P. Henry, eds.), ASTM, New York, 1981, pp. 126-135.
45. Ikeno, Y., Masuda, M., Tanno, K., Oomori, I., and Takahashi, N. Citric acid production from various raw materials by yeasts. *J. Ferment. Technol.*, *53*:752-756, 1975.
46. Tabuchi, T., Serizawa, N., and Ohmono, S. Isolation and identification of (-)-methylcitric acid and 2-methyl-cis-aconitic acid from the culture of Candida lipolytica. *Agr. Biol. Chem.*, *38*:2565-2566, 1974.
47. Uchio, R., and Shiio, I. Tetradecane 1,14-dicarboxylic acid production from n-hexadecane by Candida cloacae. *Agr. Biol. Chem.*, *36*:1389-1397, 1972.
48. Boulton, C. A., and Ratledge, C. Biosynthesis of fatty acids and lipids. In *Comprehensive Biotechnology*, Vol. 1 (A. T. Bull and H. Dalton, eds.), Pergamon Press, Oxford, 1984, pp. 459-482.
49. Nakao, Y., Suzuki, M., Kuno, M., and Maejima, K. Production of alkaline protease from n-alkanes by a kabicidin resistant mutant strain of Fusarium sp. *Agr. Biol. Chem.*, *37*:1223-1224, 1973.
50. Kitano, K., Kintaka, K., Suzuki, S., Katamoto, K., Nara, K., and Nakao, Y. Screening of microorganisms capable of producing β-lactam antibiotics from n-paraffins. *J. Ferment. Technol.*, *54*:683-695, 1976.
51. Cain, R. B. Transformation of aromatic hydrocarbons. In *Hydrocarbons in Biotechnology* (D. E. F. Harrison, I. J. Higgins, and R. Watkinson, eds.), Heyden & Sons Ltd., London, 1980, pp. 99-132.

30

FUNGAL DEGRADATION OF INDUSTRIAL WASTES AND WASTEWATER

RATNA NANDAN and S. RAISUDDIN *Industrial Toxicology Research Centre, Lucknow, India*

I. INTRODUCTION

The increase in population density with concomitant industrial development is resulting in the discharge of many synthetic chemicals in the environment. The majority of these xenobiotics (foreign agents to biological systems) after deliberate or inadvertent release enter the soil and water, where they either are broken down to simpler forms or remain unaltered for a long time because of their persistent nature, posing a great threat to the ecosystem [1]. Some of the xenobiotics undergo bioaccumulation/biomagnification and cause more danger to the environment. The direct effect of pollutants present in the industrial waste and wastewater is to deplete through the excessive organic load the dissolved oxygen (DO) content of receiving water. In most cases, this depletion occurs to the extent that the water stream becomes incapable of exercising self-purification processes, e.g., dilution, sedimentation, oxidation, reduction, and photodestruction [2-4]. The deoxygenation may be sufficiently high as to destroy aquatic life. The polluted water also causes sporadic outbreak of waterborne disease because it supports the growth of pathogenic organisms [2,3].

Autooxidation, degradation by sunlight, and microbial action are three major forces responsible for the breakdown of chemicals in the environment. The majority of xenobiotics after release go to soil and aquatic sediments. These habitats are rich in microorganisms and microbial activity occurs even in those niches which are inaccessible to sunlight. Hence, microbial degradation is usually the only means for complete mineralization of organic molecules in water and soil [5]. A relatively enormous catalytic power, large surface volume action, and rapid rates of reproduction of microorganisms contribute to the major role they play in the chemical transformation occurring on the earth's surface [5,6].

There are certain metabolic processes, e.g., fermentation, anabolic and chemolithotrophic metabolism, exoenzyme action, and adaptability of microbes to adverse environments through mutation and selection, which make microbes unique tools for biodegradation purposes [1,5,7]. The microbial metabolism for alternation of chemicals may be nonenzymatic or enzymatic. There also exists analog-induced metabolism or cometabolism. In cometabolism, microorganisms are induced by chemicals that structurally resemble the molecule of the chemical meant for degradation [5]. The microorganisms can grow on the given chemical (analog) but not on the chemical which is to be biodegraded even though they can partially metabolize the latter. The microorganisms transform compounds by certain reactions, e.g., oxidation, reduction, hydrolysis, and condensation [1,5].

Both bacteria and fungi have been used in biodegradation of a number of industrial wastes [4-9]. Fungi rank next to bacteria in saprophytic behavior [10]. However, their metabolic rates are higher than that of bacteria. While 33 cal/kg/h is the metabolic rate of bacteria, that of fungi is 50-250 cal/kg/h [11]. Fungi have evolved very specialized morphological and nutritional habits for enduring harsh environmental conditions [12]. Some fungi adopt a dimorphic course in their life cycle under varied nutritional stresses, e.g., *Mucor rouxii* has the capability of existing in two alternate forms depending on the environment and can switch back and forth between a mycelial (under aerobic conditions) and a nonmycelial phase (under elevated CO_2 levels).

Fungi also possess mammalian as well as bacterial metabolic characteristics. Like mammals, some filamentous fungi can detoxify potentially harmful compounds and then reject end products, e.g., hydroxylation of polycyclic hydrocarbons by *Cunninghamella elegans*. Thus, the mechanism of detoxification (a physiological function) is not confined to mammals [13]. In contrast to bacteria, the detoxification reaction is not a prelude to catabolism in fungi. Certain fungal metabolic activities can be compared to bacterial activities. For example, the widely distributed soil yeast *Trichosporon cutaneum* shows versatility comparable to any *Pseudomonas* but has an advantage in having a wider range of metabolic flexibility exhibited in utilization of substrates and cosubstrates. Fungi can utilize several cosubstrates for energy but in the case of bacteria such a diversity in utilization of secondary substrates is not witnessed [7,13]. Due to such a wide spectrum of activities and metabolic flexibility, fungi can have an advantage if employed for degradation in diverse habitats with a number of related or unrelated substrates [13]. Hence, fungi can effectively degrade several high-strength industrial wastes to a better extent than bacteria. These properties of fungi have been exploited in the degradation of different industrial wastes.

In this chapter some aspects related to fungal degradation of industrial wastes and wastewater will be discussed. The existing information on industrial wastes and wastewater degradation by microbes shows that bacteria have been used in most of the studies. However, in view of the above described wide range of metabolic specialization and flexibility, fungi have also been shown to be promising biodegradation tools.

II. PESTICIDES AND PESTICIDE INDUSTRIAL WASTES

Pesticides are necessary for control of pests for better crop production. These compounds before and beyond the area of their deliberate application pose danger to the terrestrial and aquatic ecosystems and represent a major toxicant load in the environment in most developing countries [14,15]. Generally the contamination problem is related to the persistent nature of pesticides and their susceptibility to biomagnification and toxicity to higher animals. An important factor deciding the degree of persistence is the chemical characteristics of the pesticidal compound itself [5]. Pesticides of organophosphorus and carbamate groups are usually degraded to harmless byproducts by natural forces such as sunlight, heat, and moisture whereas polychlorinated pesticides and their residues remain in the biosphere for long periods of time due to their persistent nature.

Some polychlorinated pesticides like DDT,* endrin, and BHC have a low vapor pressure and when sprayed their vapors are aerosolized. These vapors can condense with colloidal or solid particles and finally coalesce to form aerosols which are transported far and wide. Because of such a wide range of transportation, pesticide residues have been detected in the bodies of Eskimos and polar plants, especially in lichens [15]. Pesticides reach water through industrial effluents, rains, and dust in addition to being washed away from spray operations or runoff from land. Pesticide residues in water are either distributed in a truly dissolved form or are incorporated into sediments [5,16].

High dissolved solids and wide pH fluctuations in pesticide wastewater pose difficulty in designing appropriate treatment and disposal systems [16]. Apart from neutralization, pH adjustment, coagulation, and adsorption, the wastewater of pesticide industries can be treated by biodegradation.

A. Microbial Degradation of Pesticides

Pesticides, like other environmental pollutants, can be degraded by chemical, physical, and biological means. Pesticides discharged as industrial effluents can be degraded by various chemical means including mother salt, combustion, microwave plasma destruction, and alkaline hydrolysis to nontoxic products. The ultraviolet region of sunlight is the primary source of physical means of pesticide degradation. Also, certain synthetic photosensitizers are available which in combination with respective pesticides can enhance their photodegradation [17].

Several fungi, bacteria, and actinomycetes have been found to degrade the pesticides with varied potentials [5,6,16,17]. The pesticides are metabolized by the following reactions in microbes [17]:

1. *Hydrolysis*. One of the moieties most commonly present in organophosphates, carbamate, and pyrethroides is ester linkage. Hydrolysis of ester, halide ether, and amide bonds gives rise to nontoxic products in most cases, e.g., the formation of phenol from mexacarbate in *Trichoderma viride*.
2. *Reductive systems*. In chlorinated hydrocarbons, this reaction takes place by replacing a chlorine atom on a nonaromatic carbon with a hydrogen atom, e.g., conversion of DDT to DDD. Reduction processes take place in other groups of pesticides also.
3. *Oxidative reactions*. (a) Epoxidation: In this reaction, an epoxy ring is formed from an unsaturated C-C bond on unchlorinated ring. (b) Oxidative dealkylation of alkylamines. (c) Aromatic ring opening, performed by a series of oxidative ring hydroxylation reactions. (d) Decarboxylation: In the yeast *Rhodotorula grasilis*, DBP is converted to DBP by this reaction.
4. *Dehydrochlorination*. This reaction involves elimination of HCl, e.g., conversion of DDT to DDE.

---

*Pesticide names and abbreviations are defined in the Appendix at the end of this chapter.

5. *Isomerization*. Cyclodiene compounds go through isomerization reactions to yield more stable products.

All groups of pesticides are reported to be degraded by fungi. The extent of degradation varies mainly with the chemical nature of pesticides.

1. Chlorinated Hydrocarbons

Pesticides of this group, e.g., DDT, lindane, heptachlor and chlordane pose serious problems of environmental persistence. They remain unaltered for many years after their application in both soils and aquatic systems.

DDT: This is perhaps the most extensively studied pesticide. The microbial conversion of DDT usually occurs via two pathways: (a) reductive dechlorination, a process favored under anaerobic environments, and (b) dehydrochlorination, which occurs under aerobic conditions (Fig. 1). The iron redox system plays an important role in degradation of DDT under reduced conditions [5,14]. Under anaerobic conditions, DDT forms a complex with iron that is more susceptible to degradation. Under aerobic conditions, the complex is not formed making DDT extremely recalcitrant and inaccessible to microbial attack [14]. The reductive dechlorination of DDT to form DDD or TDE was first described in a yeast [17]. Later this kind of conversion was observed in several other microorganisms [5,17,18].

Many strains of *Trichoderma viride* metabolize DDT. In one study of 18 isolates of *T. viride*, eight formed DDD and dicofol (later identified as DDNS), three formed DDD, and one produced DDE [19,20]. In another study, *T. viride* converted DDT to DDA, DDD, and DDNS [21]. *Fusarium oxysporum* formed DDD from DDT through DDE [22]. In shake cultures of *Mucor alterans*, DDT was partially metabolized to one unidentified polar and two unidentified nonpolar compounds. The R_f values of these metabolites differed from those of DDD, DDE, DDMU, DDA, DDP, dicofol, and DDNS [21]. Cocultures of *Fusarium* sp. and *Hydrogenomonas* degraded DDM to H_2O, CO_2, and HCl [23]. By using 1,1-diphenylethylene as an enrichment source, a strain of *Cladosporium* and three bacterial strains were isolated by Focht and Joseph [24]. The substrate was first hydrated by *Cladosporium* to form 2,2-diphenylethanol, which was then degraded through fission of one benzene ring. Nandan and Polasa [25] reported conversion of DDT to DDE by *Aspergillus flavus* and *A. parasiticus* in the presence of another carbon source. The conversion was rapid in nonsclerotial strains of these two species. A strain of *Saccharomyces cereviseae* could accumulate 253 μg/g of DDT and metabolize it to DDD and DDE [26]. Bumpus and Aust [27] reported mineralization of [^{14}C]DDT by *Pleurotus ostreatus*, *Phallinus weirii*, *Polyporus versicolor*, and *Phanerochaete chrysosporium*.

BHC: The biodegradation of BHC usually occurs at a faster pace in submerged soils under anaerobic conditions. Little information is available on the biodegradation of BHC or its γ isomer lindane by fungi [28,29]. In fungi, lindane is metabolized to chlorinated benzenes including PCB

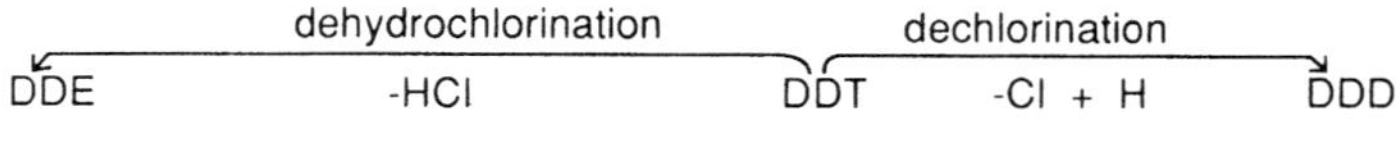

FIGURE 1. Metabolism of DDT under aerobic and anaerobic conditions.

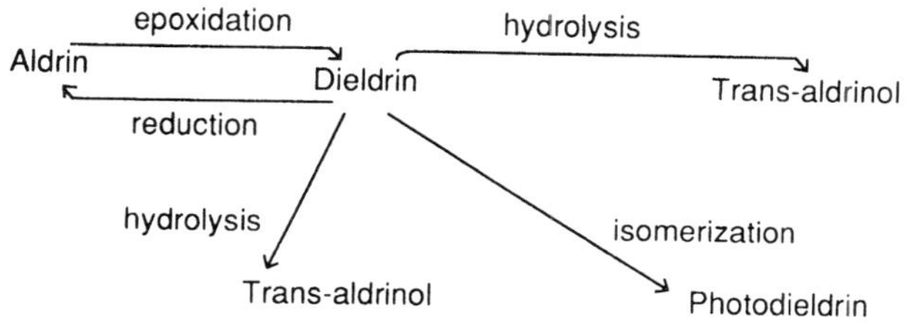

FIGURE 2. Transformation of aldrin and dieldrin by microorganisms.

and phenols [29]. Tu [30] studied the interaction between soil microbes (fungi, actinomycetes, and bacteria) and the formation of four metabolites, i.e., γ-PCCH, α-TCCH, BTCCH, and PCB. He concluded that formation of chloride in lindane-treated soil is due to the conversion of lindane to water-soluble products by microorganisms.

2. Chlorinated Cyclodienes

Aldrin and dieldrin: The epoxidation of aldrin to form dieldrin is a very common phenomenon in microorganisms (Fig. 2). There is a linear relationship between time and metabolization of aldrin to dieldrin in microorganisms, though in some cases an adaptation period is necessary [17,31]. Some species of *Aspergillus* and *Penicillium*, e.g., *A. flavus*, *P. notatum*, and *P. chrysogenum* converted aldrin to dieldrin [31,32] and four other metabolites [32]. Species of *Fusarium*, *Penicillium*, *Aspergillus*, and *Micromonospora* also converted aldrin to dieldrin but in some reports aldrin disappeared without the formation of dieldrin [31]. *Trichoderma viride* in association with bacteria degraded aldrin to several metabolites including one identified as *trans*-aldrinol [31].

Endrin: A few reports are available on the microbial degradation of endrin. *Trichoderma viride* and some other microorganisms were studied for their potential to degrade endrin. These microorganisms could convert endrin to ketoendrin and some other unidentified metabolites [32]. In another study, seven metabolites of endrin including ketoendrin, ketones, and aldehydes with five to six chlorine atoms were reported to be formed by microorganisms [17].

Heptachlor and chlordane: Microorganisms often oxidize heptachlor to its epoxide. Some fungi, e.g., *Aspergillus niger* and *Penicillium urticae*, can slowly degrade dihydroheptachlor. Microbial dechlorination of heptachlor leads to the formation of chlordene, which after epoxidation forms the corresponding chlordene epoxide [17]. Both fungi and bacteria oxidized heptachlor, chlordene, and 1-*exo*-hydroxychlordene to respective epoxides. Miles et al. [33] reported degradation of heptachlor and its epoxide to chlordene and 1-*exo*-hydroxychlordene, respectively, by mixed cultures of soil microorganisms while 1-*exo*-hydroxychlordene and its epoxide were not degraded. Singh et al. [34] studied the utilization of heptachlor by fungi, namely, *A. niger*, *A. flavus*, *A. fumigatus*, *Helminthosporium*, and *Trichoderma viride*, and reported its epoxidation by *A. flavus* in the presence of sucrose. In this study fungi could not use heptachlor as sole source of carbon.

There are a few reports on the metabolism of chlordane by microorganisms in soil or aquatic environments. Iyengar and Rao [35] reported removal of chlordane and heptachlor by *A. niger*. This fungus could remove 12.5 μg/ml of chlordane and heptachlor in 48 h. However, Beeman and Matsumura [36] found *A. niger* to be ineffective in the degradation of chlordane.

Endosulfan: The degradation products of this pesticide are different in fungi when compared to bacteria. The former convert endosulfan to endosulfate while the latter form endosulfandiol [37]. The microbial degradation of endosulfan is significant below pH 7.0 while chemical degradation is more effective above this value. In contrast, El-Zorgani and Omer [38] found the major metabolic product to be endosulfandiol in *A. niger* grown on α- and β-endosulfan. They found that 30% of both isomers of endosulfan were converted to this metabolite.

2,4-D: 2,4-D is a chlorinated herbicide. The hydroxylation of this compound by *Aspergillus niger* results in the formation of 2,4-dichloro-5-hydroxyphenoxyacetic acid and 2,5-dichloro-4-hydroxylphenoxyacetic acid [39].

3. Organophosphates

Unlike organochlorines, the organophosphates are short-lived in soil and aquatic ecosystems. Microbial rather than chemical degradation is the principal means of detoxification of organophosphates in natural ecosystems as well as in simulated laboratory conditions. Microorganisms detoxify this group of compounds through hydrolysis of *p,o*-alkyl and *p,o*-aryl bonds. Under low oxygen tension, some organophosphates undergo reductive transformation also forming their amino analogs, which are easily biodegraded [1].

Parathion: Parathion is easily degraded by biotic or abiotic means and hence do not pose residue problems. Hydrolysis is the major route for detoxification of this pesticide. Both chemical and microbial hydrolysis of parathion occurs in natural environments [14].

Under reduced conditions or under conditions of low oxygen tension, parathion is degraded to aminoparathion due to nitro group reduction. This process is mediated exclusively by microorganisms [14,40]. Aminoparathion is a short-lived, less toxic compound than parathion and can be easily hydrolyzed to *p*-aminophenol for subsequent metabolism (Fig. 3). Several bacteria, fungi, and yeasts have been isolated which carry out parathion degradation in pure cultures [13]. Soil fungi, including strains of *Trichoderma viride* and *Penicillium waksmanii*, can convert parathion to aminoparathion [19,41]. Miller et al. [42] showed that fungi, bacteria, and actinomycetes degraded parathion to aminoparathion and two unidentified metabolites in flooded soils. Reduction of parathion to various amino analogs in river water, lake sediments, and flooded soil by fungi and other microbes has also been reported [40,41]. Formation of aminoparathion from parathion in nonflooded soil was reported by Lichtenstein and Schulz [43]. The nitro group reduction of parathion was associated with yeast proliferation, indicating the possibility of these microorganisms in such conversions [43,44].

Malathion: Malathion belongs to the group of dithioate organophosphate. Matsumura and Boush [45] isolated a strain of *Trichoderma viride* along with *Pseudomonas* spp. which could degrade malathion forming carboxylic acid derivatives as a major metabolite. Both organisms possess soluble

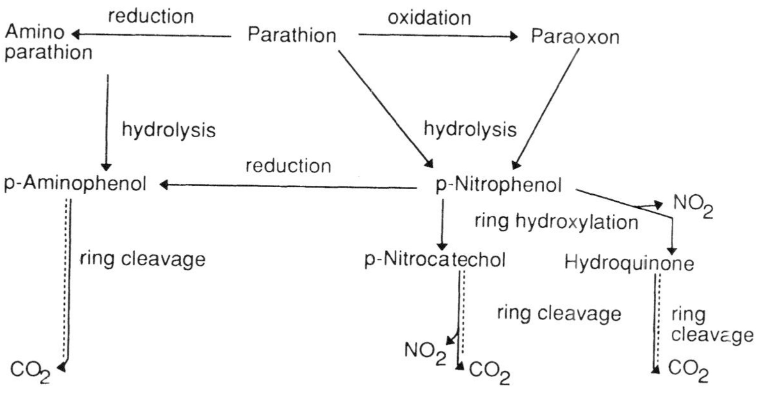

FIGURE 3. Metabolism of parathion.

carboxylesterase activity, the most likely mechanism involved in malathion hydrolysis. These workers isolated the enzyme by an acetone powdering technique and demonstrated its activity in vitro. *Penicillium notatum* metabolized nearly 76% of the applied malathion to a hydrolytic metabolite [46]. *Aspergillus oryzae* isolated from a freshwater pond utilized malathion as source of energy and degraded it to malathion monoacid and malathion dicarboxylic acid [47].

Other organophosphates: *Torulopsis utilis* can degrade phorate and dialkylphenyl phosphates [44]. Flashinski and Lichtenstein [48] reported metabolism of ethoxy ^{14}C and ring ^{14}C fonos (as Dyfonate) by fungi. In this study, *Mucor plumbeus* and *Rhizopus arrhizus* showed the best degradative potentials. The metabolites formed were dyfoxon, ethylethoxyphosphonothioic acid, ethylethoxyphosphonic acid, methylphenyl sulfoxide, and methylphenyl sulfone.

4. Carbamates

These pesticides are ester derivatives of N-substituted carbamic acid. Carbamates do not pose a serious residue problem in nature and can be easily degraded by microbial action. Hydrolysis is the major route of inactivation.

Carbaryl: Several species of fungi have been isolated from carbaryl-treated soils which degraded the insecticide at a slow rate in a mineral medium [44]. *Trichoderma viride* degraded carbaryl through ring hydroxylation to several metabolites [19]. Liu and Bollag [49] isolated the fungus *Gliocladium roseum* from carbaryl-treated soil, which metabolized carbaryl to 1-naphthyl-*N*-hydroxymethylcarbamate and 4- and 5-hydroxyl-1-naphthylmethylcarbamate. These workers [50] further demonstrated that *N*-desmethylcarbaryl and *N*-hydroxymethylcarbaryl were formed by *Aspergillus terreus* and to a lesser extent by *A. flavus* and *A. terreus*. They degraded carbaryl at low concentrations but above the 200 ppm level carbaryl was not degraded, probably because of the toxicity of pesticide to the fungus. Fungi such as *Aspergillus*, *Mucor*, *Penicillium*, *Fusarium*, and *Trichoderma* could degrade carbaryl either by side chain or ring hydroxylation producing water-soluble metabolites [19,51].

Carbofuran: Carbofuran is a broad-spectrum systemic insecticide belonging to *N*-methylcarbamate family. Degradation of carbofuran is either via hydrolysis of carbamate ester to form a phenol or by the oxidation of the methylene groups of the furanyl ring forming alcohols and ketones [44]. Fungi including *Helminthosporium* sp., *A. niger*, and *T. viride* degrade carbofuran to 3-hydroxylcarbofuran [17].

1-Naphthol: This pesticide is formed during the biological and chemical breakdown of carbaryl. The soil fungus *Fusarium solani* used 1-naphthol as an energy source and degraded it in a mineral medium. By using spore suspensions of this fungus as inoculum, 50% of 1-naphthol was hydrolyzed from the medium within 12 h [14]. The enzyme involved in the degradation was of a constitutive nature [52]. Another soil fungus, *Rhizoctonia praticola*, when grown on modified Czapeck-Dox medium transformed 1-naphthol to other polymerized products [53]. The culture filterate of this fungus and another species of *Rhizoctonia* produced a high molecular weight compound from 1-naphthol which was not extractable with benzene or with diethyl ether under anaerobic conditions [54]. At elevated temperature this compound was not produced by the above-mentioned fungus. The extracellular enzymes produced by this fungus polymerized 1-naphthol to a high molecular weight polymer through oxidative coupling [54].

5. Herbicides, Fungicides, and Nematocides

Soil fungi can hydrolyze many acylanilide herbicides with the formation of corresponding anilines. *Fusarium solani*, *Penicillium*, and *Pullularia* degrade acylanilide herbicides by enzymatic action [55]. These enzymes are known as acylamidase and have been isolated from *F. solani* and a *Penicillium* spp. which accelerate the degradation process. Due to acylamidase activity the substrate was cleaved and an aniline and an aliphatic moiety was formed [55]. *Rhizopus japonicus*, *P. jensenii*, *Talaromyces wortmani*, *Rhizoctonia solani*, *F. oxysporum*, and *Geotrichum candidum* can degrade different herbicides, fungicides, and nematocides [55,56]. The mechanisms involved in the degradation have been described [55,56].

Phenoxyalkanoic acid herbicides are metabolized by fungi with an oxidative attack on the ether bond that links the aliphatic acid portion of the molecules to chlorinated aromatic moieties of organic acids. The substituted phenol metabolites are degraded and used as carbon and energy source by microorganisms [57].

B. Fungal Enzymes Involved in the Degradation of Pesticides

Involvement of fungal enzymes for the degradation of DDT by strains of *Trichoderma viride* has been reported by Patil et al. [32]. Enzymatic degradation of DDT to DDD, DDE, DDMU, DDA, DDDH, and DBP by strains of *F. oxysporum* was studied in detail by Engst and Kujawa [22]. Molozhanova et al. [58] found that degradation of trichlorophan was due to culture supernatants of *A. niger*, *Penicillium cyclopium*, *Mucor mucedo*, *T. viride*, and *Actinomyces griseus* indicating exoenzymatic origin of the degradation enzymes. The laccase enzyme system which incites polymerization of aromatic substances into water-insoluble polyaromatics was described for removing xenobiotic compounds from environment [59].

C. Pesticide Waste Treatment

The safe and economical disposal of excess pesticide wastes which are generated from factory effluents and from washings of pesticide containers and application equipments is a problem of considerable magnitude. To avoid groundwater pollution, detoxification of excess pesticides is very important. Degradation of pesticides in pesticide disposal pits [59] and breakdown of organophosphates (e.g., diazinon, ethyl parathion, and malathion) in sewage sludges [60] indicates involvement of microorganisms in the breakdown process. These microorganisms can be potentially utilized for development of low-cost technology for disposal of aqueous wastes or for cleanup of heavily contaminated soils. The presence of high salt concentrations and other micropollutants in sewage sludges and effluents is unfavorable to microbial growth. Therefore, instead of whole cells, cell-free extracts or enzymes have been recommended for detoxification purposes [61]. Enzymatic decontamination (by soluble or immobilized enzymes) of pesticides from wastewater, highly contaminated soils, and soil contaminated by pesticide effluent was reported by Munnecke [62,63] and other workers [61,64,65]. Munnecke used parathion hydrolase for detoxification of pesticide containers and other contaminated sources [62,63].

Enzymatic treatment of high concentrations of parathion and diazinon in soil has been successful [61,62]. In one study using modeled spill situations, 100 ppm diazinon was degraded by 97% in 24 h [62]. The immobilized preparations of parathion hydrolase reduced parathion concentration in industrial effluents below 500 ppb [62,63]. Enzymatic treatment followed by ozonation and alternate aerobiosis and anaerobiosis appear to help in the degradation of many pesticides [15,65].

III. TANNERY EFFLUENT

Tannery effluent is one of the most complex industrial wastes. It contains vegetable tannins in addition to soluble organic matter, suspended solids, chromium, high chloride and sulfide concentrations and high pH (Table 1). Skin or hides that are received in a tannery either in wet-salted or dry-salted conditions are subjected to processes including soaking, liming, dehairing, fleshing, deliming, vegetable tanning, chrome tanning, dying, etc. [66]. There is release of a characteristic waste byproduct with each of these processes. These wastes pose a great threat to the environment and to aquatic life.

A. Fungal Degradation of Tannery Effluent

Tannins have antimicrobial properties but certain fungi and bacteria can grow on tannery wastes [67,68]. Most of the fungal species that have been used for biodegradation of tannery effluent constituents belong to the genera *Aspergillus* and *Penicillium*. Knudson [69] for the first time reported that several *Aspergillus* species can degrade various tannins. Other fungi, including *Chaetomium*, *Fusarium*, *Rhizoctonia*, *Cylindrocarpon*, *Trichoderma*, and *Candida*, are capable of degrading tannery waste constituents [68]. *Psalliota campestris* can oxidize catechin and *A. niger* can degrade gallic acid [70]. The intermediates of this degradation were *cis*-aconitic acid, α-ketoglutaric acid, and citric acid. *A. fumigatus* decomposed

TABLE 1. Chemical Composition of Tannery Waste

Color	Yellow to brown
Total solid	479,000 ppm
Suspended solid	12,166 ppm
Dissolved solid	35,733.33 ppm
Oil and grease	5.15 ppm
pH	8.00
COD	6113.3 ppm
BOD	2502 ppm
Total nitrogen as N	590.98 ppm
Cl^-	11,258.7 ppm
SO_4^{2-}	1499.2 ppm
NO_3^-	5.34 ppm
Phenol	7.5 μg/liter (ppb)
F^-	1.04 ppm
PO_4^{3-}	0.77 ppm
Cr(VI)	Nil
Cr(III)	0.65 ppm

Source: Ref. 66.

gallotannin into gallic acid. Catechin at 0.3% level was degraded by *A. fumigatus* without intermediate formation in 6-8 days. Other fungi including *Fusarium* and *Cylindrocarpon* also transformed tannins but at a slower rate [71]. Chandra et al. [72] reported that catechin was degraded by *A. fumigatus*, *A. niger*, *A. flavus*, and *Penicillium* sp. In another study, species of *Aspergillus* and *Penicillium* were found to utilize gallotannin and gallic acid as carbon sources [73].

Some efforts have been made to isolate and characterize various enzymes involved in the degradation of tannins by fungi [74,75]. Tannase (tannin acyl hydrolase: EC 3.1.1.20), produced by many fungi, is capable of cleaving tannin [70]. This enzyme has been isolated from *A. niger*, *A. carbonarius*, *A. flavus*, *Penicillium* spp., and *Candida* spp. [70,73]. Tannase from *A. niger* is inducible but in some species does not require an inducer. Addition of substrates including glucose, sucrose, urea, and sodium nitrate enhances the growth of fungi and rapid utilization of tannins due to cometabolism [69]. However, this approach needs more validation for use in development of biodegradation models for tannery wastes.

IV. CELLULOSIC AND LIGNOCELLULOSIC WASTES

A. Paper and Paper Pulp Mill Waste

The raw material needed for the manufacture of paper consists of bamboo pulp, eucalyptus hardwoods, grasses, hemps, etc. Paper and paper pulp industries release huge amounts of wastewater into the environment. The characteristics of wastewater from pulp and paper pulp mills depend on the kind of wood used and efficiency of black liquor recovery. The black liquor is the liquid produced after dissolving wood organics in a Kraft digester. The black liquor and wastewater arising from other operations consist of an array of parent compounds and toxic byproducts [76,77] as shown in Tables 2 and 3.

1. Pollution and Toxicity

The discharge of untreated wastewater into water bodies imparts colors to the receiving water and results in oxygen depletion and strong emission of foul odors. The color which is formed due to the presence of lignin and its derivatives persists for a long distance in waste streams. Several constituents of pulp and paper industry wastewater are known to be toxic to fish and other aquatic organisms [76]. Some lignin derivatives have even been described to possess mutagenic properties [78].

2. Fungal Degradation of Pulp and Paper Mill Wastewater

Both bacteria and fungi have capabilities to degrade pulp and paper mill wastewater constituents. However, it is reported that fungi are more efficient in degrading this waste than bacteria [79]. Most of the research on microbial degradation of lignocellulosic waste has been carried out on the different models for the degradation of lignin. Lignin is one of the most abundantly found constituents of the raw materials used for paper production. Lignin and cellulose together with hemicelluloses occur as the structural components of the vascular tissues of higher plants. It is also found abundantly in wastewater of paper pulp mills in different forms.

Fungi and filamentous bacteria are responsible primarily for lignin biodegradation. In the case of lignin, it has been found that bacteria are probably of lesser importance and fungi play the major role [76]. A broad range of fungi have been described that degrade lignin in varying degrees. They include soil fungi (*Fusarium*), soft rot fungi (*Papulospora, Chaetomium*), pseudosoft rot fungi (*Hypoxylon, Xylaria*), litter-degrading fungi (*Collybia, Mycena*), white rot fungi (*Coriolus, Phanerochaete, Polaria*), brown rot fungi (*Gleophyllum, Poria*), and certain other fungi [79]. These fungi can also degrade modified lignins and their derivatives found in paper pulp mill wastewater in the presence of other sources of carbon. The lignolytic system is not induced by lignin but is synthesized in response to nitrogen depletion as a result of primary growth, and shows that lignin degradation is due to secondary metabolism. Kirk et al. [80] reported that a strain of *Phanerochaete chrysosporium* when grown on peach plant waste (pH 4.5) at 39°C resulted in 60% color reduction in 2-4 days in the presence of glucose or cellulose.

The mechanism of lignin degradation is not well understood. One barrier to further understanding of the process is the lack of suitable models. The subunits of lignin, unlike most other natural macromolecules, are bound to one another by nonhydrolyzable carbon-carbon-oxygen bonds. Lignin is comparable in this respect to highly polymerized phenol plastics. Thus the hydrolytic depolymerization that characterizes the degradation of most macromolecules does not apply [79]. In order to understand the events involved in the degradation of lignin by microorganisms, use of radiolabeled lignin has been employed [79].

TABLE 2. Characteristics of Wastewater from Some Paper Mills[a]

Source of wastewater	pH	Total solids	Suspended solids	BOD	COD
		From Bleached Kraft Pulp and Paper Mill			
Chipper house	6.4-8.0	540-900	240-520	30-50	175-450
Digester house	9.0-10.0	1000-2500	140-120	300-360	1850-2200
Pulp washing	8.5-9.5	1400-2500	350-1000	230-480	900-1700
Pulp bleaching	6.0-9.5	2100-2900	140-2200	125-155	550-790
Paper machine	5.3-8.1	850-1250	490-900	100-160	520-780
Combined wastewater	6.5-8.2	1200-2000	350-450	100-235	600-750
		From Paper- and Strawboard and Specialty Paper Mills			
Straw- and paperboard	7.8-10.4	4745	2490	883	2940
Pulp and paper products	7.2	2500	860	330	1122
Mechanical pulping wastewater	4.7-4.9	—	306-390	560-580	1579-1616
Combined wastewater	7.2-7.3	1439	261-375	165-220	508-695

[a]All values except pH are in mg/liter.
Source: Ref. 76.

TABLE 3. Some Toxic Components of Paper and Pulp Mill Wastewaters

Major toxic compounds		Minor toxic compounds
Abietic acids	Trichloroguaicol	Sodium sulfide
Dehydroabietic acid	Tetrachloroguaicol	Methylmercaptan
Isopimaric acid	Monochlorodehydro-abietic acid	Hydrogen sulfide
Palustric acid	Dichlorodehydroabietic acid	Sodium thiosulfate
		Formaldehyde
Pimaric acid	Epoxystearic acid	
Sandaraco pimaric acid	Dichlorostearic acid	Sodium hydroxide
		Sodium chloride

Source: Ref. 76.

3. Biochemistry of Lignin Degradation

The experimental evidences suggest that the lignin polymer-degrading system is oxidative and nonspecific. It is also believed that this system is of extracellular nature. There have been suggestions that the key oxidative reactions in lignin polymer degradation might not be direct enzyme catalysis. Some workers believe that active oxygen species ($\dot{O}_2$, H_2O_2, ^-OH, and O^1) are involved in lignin degradation [81-83]. Evidence that these active oxygen species are produced by fungi has been reported. Such studies have mainly been conducted using *Coriolus versicolor* and *Phanerochaete chrysosporium* [82-84]. The first enzyme for lignin degradation has been reported and H_2O_2 is involved [79]. It was first detected in extracellular culture fluid of lignolytic culture as the agent responsible in the presence of added H_2O_2 for the initial $C\alpha$ and $C\beta$ cleavage reaction in β-1 substructure model compounds such as 1,2-bis(3-methoxy-4-ethoxyphenyl)propane-1,3-diol. Its action on the lignins results in their partial depolymerization. Activity resides in a single protein (MW = 42,000) as evidenced by both gel permeation chromatography and polyacrylamide gel electrophoresis. It is inhibited by azide, suggesting metal involvement.

Primary investigations on enzymes system involved in lignin degradation were focused mainly on phenol-oxidizing enzymes (laccase, peroxidase, tyrosinase), which catalyze the vinyl electron oxidation of phenols. However, these enzymes result in only limited structural changes in lignin.

Another enzyme, cellobiose:quinone oxidoreductase, produced by white rot fungi catalyzes the oxidation of cellobiose to cellobiono-δ-lactone, with transfer of two electrons to any of the various quinone or phenoxy radicals, including those produced in lignin by phenol-oxidizing enzymes. Some workers have also implicated this enzyme in lignin degradation [85].

Enzymes including phenol oxidase, laccase, and peroxidases act as catalysts in lignin-utilizing basidiomycetes. The extracellular enzymes isolated from *P. versicolor*, *Phaenerochaete chrysosporium*, and *Pleurotus ostreatus* were found to degrade lignin [86]. An understanding of how ligninase, the lignin-degrading enzyme of *P. chrysosporium*, acts will lead to insight into the functioning of peroxidase enzymes [86]. Ligninase can also oxidize phenols which are common water pollutants, hence the enzyme will be helpful in reclaiming wastewater. In view of the involvement of various enzymes in lignin degradation, extensive studies are being conducted on improving the enzyme-producing capacity of *P. chrysosporium*, which is the most important producer of lignin-degrading enzymes [87-97].

In recent years, efforts have been made to improve the lignin degradation capacity of white rot fungi by genetic recombination [97]. Attempts are also being made to modify the lignin molecule by photooxidation using UV radiation (210-250 nm) and reactive molecular oxygen. This treatment helps in the opening of phenylpropane rings, thereby rendering the modified lignin molecule biodegradable [98].

B. Degradation of Other Cellulosic and Lignocellulosic Wastes and Production of Fungal Biomass

Several industrial effluents in addition to paper and pulp mill wastes, e.g., agricultural produce wastes, farm wastes, food industry effluents, sugarcane bagasse, etc., contain cellulosic and lignocellulosic matter. Fungi can propagate on these substrates and degrade them, thereby decreasing the pollution load of the effluent. Genera of *Alternaria*, *Aspergillus*,

Penicillium, Fusarium, Chaetomium, Curvularia, Helminthosporium, Trichoderma, Pleurotus sajor-cajci, marine fungi, *Bergenerula spartinae, Phaeospharia typharum*, and *Leptosthaeria oboines* isolated from different sources showed high cellulolytic and lignocellulolytic activities [76,99,100]. Qu et al. [101] isolated a high-cellulase-producing strain of *Pleurotus decumbens* from soil using double-layered agar plates with holocellulose instead of acid-swollen cellulose as the substrate.

Chaetomium cellulolyticum and *C. crispatum* can degrade different types of cellulosic wastes [102]. In *C. crispatum* the cellulolytic enzyme complex displayed optimal activity at pH 5.0 and 25°C. A study of the cellulase complex formation during the growth cycle of this fungus revealed that β-glucosidase was produced mainly intercellularly in early exponential phase, while the overall *exo*-1,4-glucosidase and *endo*-1,4-gluconase formation gradually increased during the fermentation cycle. These workers advocated the use of *C. crispatum* for fungal biomass production as the amino acids of the fungus were found also to be desirable. In *Geotrichum candidum*, the enzyme system consisted of two fractions [103]. One fraction, *endo*-1,4-β-gluconase (Cxt), was able to catabolize cotton fibers in the presence of cellobiohydrolases. Fraction 2 (a different form of fraction 1) was unable to hydrolyze cotton fibers. Species of *Trichoderma* show hypercellulosic activities. A mutant of *Trichoderma reesei*, Rut C30, showed localization of endoglucanase in microsomal fractions [104]. The microsome (light-vesicles) fraction contained a significant proportion of cell-bound CMCase activity while the heavy-vesicle fraction contained amorphous or crystalline electron-dense material and showed low CMCase activity. The majority of ATPase was concentrated in the organelle free cytosol. The residual particle band ATPase was concentrated in the heavy-vesicle fraction in this fungus. In *T. ochrasporum* cellulolytic activity was demonstrated by its ability to produce cell-associated and extracellular cellulolytic enzymes [105,106]. The exoglucanase activity was mostly in the aqueous broth while endoglucanase was associated with the mycelium. The optimum activity of the enzyme was at 50°C. Both monoculture and cocultures of various fungi, e.g., *Aspergillus niger* and *Sporotrichum pulverulentum*, have been used for enhancing biodegradation and biomass production [107]. Cocultures of *T. reesei* and *Aspergillus phoenicis* exerted better cellulase hydrolytic activity compared to the individual fungi [108]. *Scytalidium lignicola* and *Trichoderma longibrachiatum*, when grown together on cellulosic and lignocellulosic wastes including sugarcane bagasse, rice straw, rice bran, wheat straw, newspapers, etc., produced hydrolyzate which was rich in glucose and supported the growth of *Candida lipolytica* [109]. Pretreatment of wastes generally enhanced the cellulase enzyme activity of *T. reesei* [110]. The cultural factors for better cellulase activity and biomass production by *Aspergillus terreus* on bagasse were evaluated and maximum biomass protein recovery was found at pH 4.0 with 1:5 (v/v) culture medium [111]. Photochemical pretreatment of α-cellulose of λ = 254 nm increases cellulase production by 100% using the fungus *Chrysonilia sitophila* [112].

V. BIODEGRADATION OF PLASTIC AND POLYMER WASTES

Plastics and related polymers mainly result in aerial contamination. The wastes formed in the plastic industries are mostly of a solid nature and therefore generally do not enter systems. Therefore, the plastic manufacture does not pose oxygen depletion problems. Vinyl chloride and acrylonitrile are classified as environmental carcinogens. While some whiteners are toxicants, most of the constituents of plastics and film plastics can be microbially degraded. The variety of plastics that are readily degradable do not have important practical applications. Various aspects of biodegradation of plastics and polymers have been reviewed by Krishnamurti [113].

Acrylonitrile fibers can be attacked by species of *Aspergillus*, *Penicillium*, *Stachybotrys*, and *Nigrospora*; Polycarpolactone is degraded by *A. niger*, *A. flavus*, *P. funiculosum*, *Chaetomium globosum*, and *Pullularia pullulans* [113].

Fungi including *Aspergillus fumigatus*, *Chaetomium thermophile*, *Humicola lanuginosa*, *Mucor pusillus*, and *Torula thermophila* can degrade plastic polymers at temperature of 48° [113]. *Trichoderma viride* isolated by soil burial tests using polymerized vinyl acetate/ethylene copolymer dissolved the substrate by depolymerization [114]. *N*-Alkanes, alkenes, and other aliphatic hydrocarbons used in the synthesis of plastic polymers are readily utilized by several filamentous fungi and yeasts. Some of the important fungi which are capable of degrading these compounds are *Acremonium*, *Aspergillus*, *Penicillium*, *Chaetomium*, *Cladosporium*, *Gliocladium*, *Graphium*, *Helicostylum*, *Helminthosporium*, *Monilia*, *Mucor*, *Paecilomyces*, *Rhizopus*, *Candida*, *Hansula*, *Rhodotorula*, *Saccharomyces*, *Torulopsis*, and *Trichosporon* [113].

VI. HEAVY METALS

Among the inorganic pollutants of the environment, heavy metals are the most obnoxious toxic contaminants. They enter the aquatic environment and soil primarily through waste products/water of electroplating units, sewage sludges, tanneries, the paint and pigment industry, the mining and refinery industry, or due to extensive use of heavy metals in fungicides, disinfectants, and as contaminants in fertilizers. Some of the important toxic metals of environmental concern are Hg^{2+}, As^{3+}, Cu^{3+}, Zn^{2+}, Ni^{2+}, Cr^{3+}, Pb^{2+}, Cd^{2+}, and radioactive metals such as uranium, thorium, etc.

The main concern regarding heavy-metal pollution is the possible toxic effects rather than oxygen depletion of water [115]. Some heavy metals are reported to have comprehensive toxic effects on cells, mainly as a result of their ability to interact with biological polymers and make these molecules physiologically ineffective [115]. The presence of heavy metals even in trace quantity can cause toxicity to fishes, crabs, aquatic plants, mammals, and aerobic and anaerobic microorganisms [116].

A. Removal of Metals

Toxic metals can be removed by physicochemical methods such as ion exchange, chemical precipitation, activated carbon adsorption, and electrolysis. But these methods have certain drawbacks which limit their application to industrial situations. The use of microbial biomass to remove heavy metals is a convenient alternative to physicochemical techniques [117]. Over the past three decades, extensive research has been carried out to determine the capabilities of various microorganisms for the accumulation of nonessential or essential heavy metals [118]. Microorganisms can concentrate, adsorb, or absorb heavy metals in the cell interior or on cell walls. These capabilities include not only uptake per se but also solubilization of metals and their subsequent release. In spite of their ubiquitous distribution and essential role as bioconverters, the fungi apart from lichen mycobionts have rarely been accorded a significant role in the biogeochemistry of heavy metals [118]. The term *geomycology* applies to fungal capabilities of attacking and accumulating metals and minerals [118]. The ability of microfungi to grow along a heavy-metals gradient may be due to their positive response to metals [119].

The application of fungal biomass to decontaminate metal-polluted water or to recover economically valuable metals from natural or industrial water is an attractive proposal which is being evaluated by various workers [119-121]. The sensitivity of fungi to metal is influenced by pH. The uptake and binding of nickel, zinc, cadmium, and lead by mycelium of *Penicillium digitatum* was impaired by acidic pH [122]. Siegel et al. [123], while studying biosorption of lead and chromium by a *Penicillium* preparation, reported that *Penicillium* sp. (0.14 g of dry mycelium incubated in lead nitrate, 50 ppm, 2 h) contained 850 μg of Pb and subsequently accumulated 120 μg of Cr in 5 min from 20 ml of 20 ppm dichromate solution. The above authors suggested that the binding of divalent Pb cations to negative fungal wall sites reverses their charge enabling them to accumulate Cr anions. The removal of Cr (VI) from solutions of potassium dichromate by Pb-loaded mycelium of *Penicillium* sp. shows that fungal absorbability for both wastewater treatment and metal recovery may be greatly increased by such modifications. *Rhizopus arrhizus* biomass absorbed different metal ions or cations but could not absorb alkaline metal ions [124]. In this study the uptake of all cations was consistent with absorption of metals at the sites containing phosphate, carbohydrate, and other functional groups in the biomass. The uptake of molybdenum and vandate anions was pH-dependent and the uptake mechanism involved electrostatic attraction to positively charged functional groups. The uptake of cations was related to ionic radii of La^{3+}, Mn^{2+}, Cu^{2+}, Zn^{2+}, Cd^{2+}, Ba^{2+}, Hg^{2+}, Pb^{2+}, and Ag^{2+} [124]. Another species of *Rhizopus* adsorbed a range of mono-, di-, and trivalent metals [125]. Mixed cultures of *Aspergillus niger* and *Trichoderma harzianum* were found to oxidize heavy-metal sulfides [126]. *A. niger* oxidized the sulfides of copper, lead, and zinc to sulfates. With the exception of cadmium sulfide, the sulfide particles in the medium were adsorbed onto the surface of mycelium. The oxidation of sulfides by *T. harzianum* was less. *Penicillium* sp. has been used successfully for removal of metals [127]. *Candida humicola, Saccharomyces cerevisiae, P. digitatum, A. niger, Trichosporon cutaneum, Poxillus filamentous, Paecilomyces varioti*, and some mushrooms can remove cations including toxic and radioactive metal ions from aqueous solutions by bioabsorption [128,129]. Galun et al. [118] demonstrated adsorption, retention, and recovery of uranium (in the form of UO_2Cl) by *Penicillium* biomass. In these studies, many common species of *Penicillium* adsorbed uranium efficiently (about 90% removal from 1 ppm solution), accommodate a heavy load of U about 5000 ppm or more), and operate under relatively severe conditions (40-50°C, pH 2.5-9.5). The fungus *Diploidia natalensis* excels in the number of anastomoses and thus may have more exposed reactive groups in the hyphal surface, which could increase its metal uptake capacity. A high area-to-mass ratio favors absorption processes alike.

These studies add significantly to a growing body of geomycology and also demonstrate the feasibility of using fungal biomass to decontaminate uraniferous wastes. The potential of using fungal biomass, especially that of *Rhizopus*, looks promising in terms of practical exploitation for the removal of metal ions [129].

VII. MISCELLANEOUS INDUSTRIAL WASTES

There are many chemicals found in industrial effluents that are not specific for a particular liquid waste. Phenols are one of these compounds that enter the environment from many sources including partial degradation of phenoxy herbicides, use of wood preservatives, and generation of wastes by the petroleum industry. Yeasts and filamentous fungi can degrade phenols under different cultural conditions [130,131]. As mentioned earlier, production of laccase by the white rot fungus *Phanerochaete chrysosporium*, *Neurospora crassa*, and some other fungi have been extensively studied for removal of pigments and detoxification of phenols and other compounds from liquid wastes [132-135]. Immobilized mycelium of *Coriolus versicolor* is being used for removing colors/pigments from Kraft mill wastes [136]. Several aromatic compounds, alcohols, fatty acids, etc., are easily metabolized by fungi and these properties have been utilized in food and fermentation industries for fungal biomass production and for reduction of COD of wastewater [137,138].

VIII. SEWAGE WASTE

Domestic or municipal sewage is mainly composed of organic matter either in soluble or colloidal form or as suspended solids and inorganic compounds. The contaminants of sewage are classified in three major categories: (a) chemical, (b) physical, and (c) biological. The chemical contaminants are both inorganic and organic. A significant proportion of organic contaminants carried by sewers is derived from runoff water, industrial effluents, and insecticide sprayings. Common organic compounds in sewage are aliphatic, alicyclic and aromatic hydrocarbons, phenols, organochlorines, phthalic acid esters, petroleum hydrocarbons, oils and greases. The inorganic chemicals are sulfides, nitrites, and heavy-metal ions. The physical contaminants are color, turbidity, foams, thermal pollution, and radioactivity. The biological contaminants are pathogenic microorganisms like fungi (*Aspergillus fumigatus*), bacteria, and viruses which cause cholera, typhoid, etc. [139,140].

Cook [141] reviewed the status of fungi in various sewage treatment plants and advocated sewage treatment by fungi. A large number of fungi (about 1000) have been isolated from sewage, sewage treatment plants, and sewage sludges. These include genera of aquatic fungi, soil fungi, and yeasts like species of *Achlya*, *Saprolegnia*, *Phoma*, *Penicillium*, *Aspergillus*, *Fusarium*, *Gliocladium*, *Candida*, *Rhodotorula*, etc.

Fungi have also been utilized for biological oxygen demand (BOD) reduction of municipal wastes [142]. *Aspergillus flavus*, *A. niger*, *A. niveus*, *A. terreus*, *Curvularia lunata*, *Penicillium oxalicum*, and *Trichosporon* sp. were used by Hiremath et al. [143] for domestic sewage treatment and biomass production. Reduction in BOD (53-72%), phosphates (34-77%), and ammonia-N (49-76%) along with biomass production (95-240 mg/liter) was reported by these workers. The enzymatic saccharification of municipal waste by some fungi for production of fermentable sugars is an attractive procedure for sewage treatment [142]. Since fungi have a high pollution reduction capacity, the performance of sewage treatment processes could conceivably be improved by the addition of specific enzymes that augment the ability of natural fungal flora to degrade substances including protein, starch, polysaccharide cellulose, and to enhance adsorption of heavy metals [144,145].

IX. APPLICATION OF BIOTECHNOLOGY AND GENETIC ENGINEERING FOR POLLUTION CONTROL

In recent years the application of biotechnology and genetic engineering in the fields of agriculture, medicine, and environmental improvement programs for human welfare has gained considerable importance. Since microorganisms can readily adapt to adverse and changing environments, researchers have explored the possibilities of using them as tools to help control environmental pollution through biotechnology and genetic engineering [146]. These studies are directed toward two major goals: (a) improvement of the microbial performance through appropriate technology including development of biosensors and immobilization of cells/enzymes and (b) strain improvement for enhancement of overall microbial reactions [145,146].

A. Improvement of Technology

1. Immobilized Cells/Enzymes

The study of microbial processes with the help of immobilized enzymes and cell enzymes/protoplasts is one of the major branches of modern biotechnology. This concept dates back to an era when man learned to make bread dough, which is perhaps the oldest bioreactor in which microbial cells are entrapped and immobilized for fermentation. Now immobilized enzymes and cells are used for the transformation and biosynthesis of organic compounds; degradation of wastes; isolation of heavy metals from sewage, marine, and freshwaters; and in a variety of fermentation processes. The cells or enzymes are immobilized by several methods of entrapment in various gels, adsorption, and covalent binding. Polyacrylamide gel (PAG) is one of the most common carriers for cell entrapment [146]. *Curvularia lunata* is a filamentous fungus of biotechnological relevance used to transform precursor molecules into biologically active steroids [147]. One of the first transformations studied with the help of PAG-entrapped cells was IIβ hydroxylation of cortisol by *Curvularia lunata* mycelium [146]. Now immobilized microorganisms are being used extensively for analysis and removal of contaminants in water purification practices.

Bryers and Homer [148] reported that the production of extracellular enzymes by artificially entrapped cells of fungi, e.g., *Trichosporon cutaneum, Candida tropicalis, Phanerochaete chrysosporium*, is possible using gel entrapment and adherence to a polymer matrix. These findings have resulted in new applications for wastewater treatment. Immobilization of *P. chrysosporium* in polyurethane foam facilitates increased yields of ligninase (the enzyme active in lignin degradation) in batch culture [149].

The entrapped spores of *Fusarium flocciferum* on celite particles or whole cells of *Candida tropicalis* entrapped in a polymeric ionic net [150,151] degraded phenols with high efficiency. The entrapped cell system of *C. tropicalis* remained active for more than 2 months.

2. Biosensors

The development of biosensors is also based on the immobilization of microorganisms. In environmental control studies, monitoring of organic pollutants with the aid of biosensors is an economical and accurate procedure compared to several conventional methods. Karbue [152] constructed captured microbial probes for analysis of BOD of wastewater. The standard procedure of BOD estimation requires a 5-day incubation period and values of test results depend on the skill of operators. For rapid and reproducible test procedures, the yeast *Trichosporon cutaneum* (which is employed in wastewater treatment) was used as sensor in this study. The microorganisms immobilized on a porous acetyl cellulose were pressed between an oxygen-permeable Teflon membrane and a porous membrane. The membrane was directly fixed on the surface of the platinum cathode of an oxygen probe. A continuous-flow system using this new microbial sensor was developed for automatic estimation of 5-day BOD [152].

B. Strain Improvement

Strain improvement is an important technology for achieving higher efficiency in microbial transformations. The microbial strains isolated by natural selection and enrichment techniques can be further improved by additional genetic procedures.

It has often been observed in traditional biological waste systems that microbial populations adjust themselves to the presence of toxic compounds and eventually achieve efficiency in decomposition. This phenomenon, termed *acclimatization*, represents the selection of mutant microorganisms that can tolerate and/or degrade the concerned compounds. This natural acclimatization is the genesis of modern chemostat techniques, which is widely applied in waste treatment studies [145]. Strain improvement can also be achieved by classical genetics/protoplast fusion or by genetic engineering. Classical genetics, based on mutation and recombination of whole genome, has been successfully used for strain improvement for over five decades [146].

Man-made organics, unlike natural products, are highly resistant to microbial action in the environment. Using genetic engineering and recombinant DNA technology (wherein genes can be manipulated toward specific goals), these recalcitrant compounds can be easily degraded [153]. With the help of genetic manipulations such as gene rearrangement, gene amplification, alteration in regulatory genes, and site-specific mutagenesis, the biodegradation capabilities of an organism can be enhanced [153]. Transfer of catabolic enzymes from plants, fungi, and higher animals to bacterial cells (e.g., transfer of metallothioneins from higher animals to bacterial cells for elimination of heavy metals) is also used for enhancing degradation processes. Chakrabarty [154] demonstrated that enzymes of hydrocarbon degradation pathways are plasmid-specific, e.g., the plasmids involved in salicylate catabolism in *Pseudomonas putida*. Since then plasmids have been implicated in degradation of camphor [155] and several other hydrocarbon fractions [156]. Recently, mixed cultures composed of heterotrophic and photosynthetic bacteria developed through genetic engineering have proved to be effective for treating selected industrial wastes in Japan [157].

Most of the studies in genetic engineering for biodegradation application have been conducted on bacteria because of their simple genetic structure. Some information is now available on molecular genetic studies in fungi [146].

Some imperfect fungi undergo parasexual recombination wherein the fusion of two hyphae of equal or different polarity results in a mycelium with nuclei of both parent strains. This heterokaryon is normally stable with the nuclei mingling but not interacting. However, in rare cases, e.g., in *Penicillium chrysogenum*, nuclear fusion occurs and a diploid nucleus is formed. The heterozygotic diploids from parent strains with different cell lines have been used in attempts to produce strains with increased efficiency [146]. The product formation by diploid strains and their segregants is greater than that of the parent strain segregants.

C. Recombination by Protoplast Fusion

Protoplast fusion is one of the most important developments in applied genetics in recent years. Protoplast fusion is normally rare because of the strong negative charges of the protoplast surface, but in the presence of polyethylene glycol (PEG), the protoplasts aggregate and fusion occurs accompanied by DNA exchange. Protoplast fusion for interspecific recombination, interspecific hybridization, and protoplastic transformation or transfection is used for production of superior strains. An enzyme produced by *Trichoderma harzianum* was used for regeneration and protoplast formation from various fungi [158].

Yeasts, particularly strains of *Saccharomyces cerevisiae* and *Candida* spp., are amenable to most of the cultural techniques and genetic manipulations used with laboratory bacteria [159]. It is possible to isolate any yeast gene in the presence of a suitable recipient yeast strain [159]. Some antibiotic resistance genes of bacterial origin are expressed in yeasts [160]. Genetic mapping of *Neurospora*, *Podospora*, *Dictyostelium* (both nuclear and mitochondrial DNA), and *Aspergillus nidulans* is known [161]. The possibility of gene cloning in fungi like *Colchobolus heterostrophus*, *Neurospora crassa*, *A. nidulans*, and *Podospora* spp. to create genetic engineered fungi for weed control has also been suggested [162].

The basidiomycete *Phanerochaete chrysosporium* and other white rot fungi have potential application in lignin degradation. Heterokaryons have been made from autotrophic strains of this fungus [163]. The cloning and sequencing in DNA for ligninase genes of *P. chrysosporium* have been done by several workers. Genetically improved strains of *Trichoderma reesei*, *Penicillium* spp., and *Talaromyces* spp. were developed in the laboratory for fermentable sugar production from the organic fraction of municipal wastes [164-166]. Loper [167] recently studied cytochrome P-450 genes of *S. cerevisiae* and *C. tropicalis* using genetic engineering techniques in order to employ this approach in the degradation of hazardous waste materials.

These studies indicate that fungi also undergo several genetic transformation processes. In recent years, even anerobic rumen fungi, e.g., *Neocallimastic frontalis* and *Sphearomonas cummunis*, are being employed in genetic studies [168]. With the opening of new vistas in biotechnology and genetic engineering, the use of fungi for pollution control is also becoming more feasible.

X. CONCLUSION

The environmental threat posed by several industrial pollutants is well recognized. Most of these pollutants in the form of industrial wastes enter our ecosystem through water. The extent of pollution in most cases becomes such that natural self-purification processes are not sufficient to degrade the industrial wastes. In order to reduce the pollution load, various physical, chemical, and biological means are applied.

The microbial detoxification/degradation of the pollutants is the most effective biological treatment method for combatting industrial pollution. Both bacteria and fungi have been employed for biodegradation. Several fungi have certain advantages including higher metabolic rates than bacteria, efficient enzyme systems, and the ability to form biomass. Fungi also possess some metabolic properties similar to those of mammals. They have very wide metabolic flexibilities. In view of all these salient features, several fungal species have been successfully employed in the degradation/removal of industrial wastes or their constituents from tanneries, paper and paper pulp mills, and sewage. Fungal biomass is also very effective in adsorption and desorption of metals from aqueous solutions.

Applications of biotechnology and genetic engineering represent promising areas for the degradation of industrial effluents. With the aid of modern genetic techniques such as strain improvement by protoplast fusion and genetic engineering, development of biosensors, and immobilization of cells/enzymes cell-free extracts, several fungi are being used for treatment of industrial wastes and wastewater. Further research in these areas will open newer vistas in the field of biotechnology and genetic engineering for biodegradation of industrial wastes and pollution control by the use of fungi.

ACKNOWLEDGMENTS

We are grateful to Prof. P. K. Ray, Director of this center, and to Prof. A. V. S. Prabhakara Rao and Dr. L. Iyengar, of the Indian Institute of Technology, Kanpur, for encouragement. The computer assistance of Mr. Umesh Prasad is acknowledged.

APPENDIX

Chemical Names of Pesticides Used in Text

Aldrin	1,2,3,4,10,10-hexachloro-1,4,4a,5,8,8a-hexahydro-1,4-*endo,exo*-5,8-dimethanonaphthalene
BHC	1,2,3,4,5,6-hexachlorocyclohexane
Carbaryl	1-naphthylmethylcarbamate
Carbofuran	2,3-dihydro-2,2-dimethyl-7-benzofuranylmethylcarbamate
Chlordane	1,2,3,4,5,6,7,8-octachloro-2,3,3a,4,7,7a-hexahydro-4,7-methanoindene
2-4-D	(2,4-dichlorophenoxy)acetic acid
DBP	4,4'-dichlorobenzophenone

DDA	2,2-bis(*p*-chlorophenyl)acetic acid
DDD	2,2'-bis(*p*-chlorophenyl)-1,1-dichloroethane
DDE	2,2'-bis(*p*-chlorophenyl)-1,1-dichloroethylene
DDM	bis(*p*-chlorophenyl)methane
DDMU	1-chloro-2,2-bis(*p*-chlorophenyl)ethylene
DDNS	2,2-bis(*p*-chlorophenyl)ethane
DDOH	2,2'-bis(*p*-chlorophenyl)ethanol
DDT	1,1,1-trichloro-2,2-bis(*p*-chlorophenyl)ethane
Dieldrin	1,2,3,4,10,10-hexachloro-6,6-epoxy-1,4,4a,5,6,7,8,8a-octahydro-1,4-*endo,exo*-5,8-dimethanonaphthalene
Endosulfan	6,7,8,9,10,10-hexachloro-1,5,5a,6,9,9a-hexahydro-6,9-methano-2,4,3-benzodioxathiepin-3-oxide
Endrin	1,2,3,4,10,10-hexachloro-6,7-epoxy-1,4,4a,5,6,7,8,8a-octahydro-*exo*-1,4-*exo*-5,8-dimethanonaphthalene
Heptachlor	1,4,5,6,7,8,8-heptachloro-3a,5,7,7a-tetrahydro-4,7-methanoindene
Lindane	γ-1,2,3,4,5,6-hexachlorocyclohexane
Malathion	*o,o*-dimethyl-*S*(1,2-dicarbethoxyethyl)phosphorodiothioate
1-Naphthol	1-naphthalenol
γ-PCCH	γ-2,3,4,5,6-pentachloro-1-cyclohexane
Parathion	*o,o*-diethyl-*o,p*-nitraphenylphosphorothioate
γ-TCCH	γ-3,4,5,6-tetrachloro-1-cyclohexane
Trichlorphon	dimethyl-(2,2,2-trichloro-1-hydroxyethyl)phosphonate

REFERENCES

1. Alexander, M. Biodegradation of chemicals of environmental concern. *Science, 211*:132-138, 1981.
2. Ramalho, R. S. *Introduction to Waste Water Treatment Process*, 2nd ed. Academic Press, New York, 1983, p. 580.
3. Pickford, J., ed. *Developing World Water*, WEDC. Grosvenor Press International, Hong Kong, 1988, p. 420.
4. Anderson, A. C. Environmental toxicology and biodegradation of xenobiotics. *J. Environ. Health, 48*:196-199, 1986.
5. Matsumura, F., and Krishnamurti, C. R. Degradation of pesticides in the environment by microorganisms and sunlight. In *Biodegradation of Pesticides in the Environment* (F. Matsumura and C. R. Krishnamurti, eds.), Plenum Press, New York, 1982, pp. 67-89.
6. Alexander, M. Microbial degradation and biological effects of pesticides in soil. In *Soil Biology: Reviews of Research*, UNESCO, Paris, 1969, pp. 209-240.
7. Gibson, T. *Microbial Degradation of Organic Compounds*, Marcel Dekker, New York, 1984, p. 535.

8. Chakrabarty, A. M. *Biodegradation and Detoxification of Environmental Pollutants*. CRC Press, Boca Raton, 1984, p. 147.
9. Howe, R. H. L. Toxic waste degradation and disposal. *Process Biochem.*, *4*:3-7, 1969.
10. Cooke, W. B. The usefulness of degradative fungi. *Mycopathol. Mycol. Appl.*, *15*:199-205, 1973.
11. Issac, P. C. G., and James, A. The bacterial ecology of trickling filters. *Vir. Int. Ver. Limnol.*, *15*:620-630, 1965.
12. Garraway, M. O., and Evans, R. C. *Fungal Nutrition and Physiology*. John Wiley and Sons, New York, 1984, p. 401.
13. Anderson, J. J., and Dagley, S. Heat evaluation of microbial metabolism: Effects of monoxygenases. *J. Bacteriol.*, *143*:525-528, 1980.
14. Lal, R. *Insecticide Microbiology*. Springer-Verlag, Berlin, 1984, p. 268.
15. Ray, P. K., Prasad, A. K., and Nandan, R. Pesticides: Major environmental problem. *Sci. Cult.*, *51*:363-371, 1985.
16. Swanminathan, T., and Shivaraman, N. Microbial degradation of waste waters generated in the manufacture and applications of pesticides. In *Proc. National Workshop on Microbial Degradation of Industrial Wastes* (B. B. Sundaresan, ed.), NEERI, Nagpur, India, 1982, pp. 259-290.
17. Matsumura, F., and Benzet, H. J. Microbial degradation of insecticides. In *Pesticide Microbiology* (I. R. Hill and S. J. L. Wright, eds.), Academic Press, London, 1978, pp. 623-667.
18. Nandan, R., and Polasa, H. Impact of pesticides on some microbial metabolism and biodegradation of pesticides. In *Persp. Mycol. Res.*, *1*:265-281, 1987.
19. Matsumura, F., and Boush, G. M. Degradation of insecticides by a soil fungus, Trichoderma viride. *J. Econ. Entomol.*, *61*:610-612, 1968.
20. Matsumura, F., Patil, K. C., and Boush, G. M. DDT metabolism by microorganisms from Lake Michigan. *Nature*, *230*:325-326, 1971.
21. Patil, K. C., Matsumura, F., and Boush, G. M. Metabolic transformation of DDT, dieldrin, aldrin and endrin by marine microorganisms. *Environ. Sci. Technol.*, *6*:629-632, 1972.
22. Engst, R., and Kujawa, M. Enzymatische Abbau des DDT durch Schimm melpilze. Reaktionsverlauf des enzymatischen DDT. Abbaues. *Nahrung.*, *11*:751-760, 1967.
23. Focht, D. D. Microbial degradation of DDT metabolites to carbon dioxide, water and chloride. *Bull. Environ. Contam. Toxicol.*, *7*:52-56, 1972.
24. Focht, D. D., and Joseph, H. Degradation of 1,1-diphenylethylene by mixed cultures. *Can. J. Microbiol.*, *20*:631-635, 1974.
25. Nandan, R., and Polasa, H. Effect of chlorinated hydrocarbons on enzymes associated with kojic acid synthesis in Aspergillus flavus. *Indian J. Microbiol.*, *25*:21-25, 1985.
26. Lal, S., and Lal, R. Bioaccumulation, metabolism and effect of DDT, fenitrothion and chloropyrifos on Saccharomyces cereviseae. *Arch. Environ. Contam. Toxicol.*, *15*:753-757, 1987.
27. Bumpus, J. A., and Aust, S. D. Biodegradation of DDT by the white rot fungus Phaenerochaete chrysosporium. *Appl. Environ. Microbiol.*, *53*:8001-8008, 1987.
28. Engst, R., Macholz, R. M., and Kujawa, M. Identifizierung von metaboliten und der Abbauweg des Hexachlorobenzols in einer Schimmelpulzukultur. *Nahrung.*, *19*:603-608, 1975.
29. Macholz, R. M., and Kujawa, M. Recent state of lindane metabolism III. *Residue Rev.*, *94*:119-149, 1985.

30. Tu, C. M. Interaction between lindane and microbes in soils. *Arch. Mikrobiol.*, *105*:131-134, 1975.
31. Tu, C. M., Miles, J. R. W., and Harris, C. R. Soil microbial degradation of aldrin. *Life Sci.*, *7*:311-322, 1968.
32. Patil, K. C., Matsumura, F., and Boush, G. M. Degradation of endrin, aldrin and DDT by soil microorganisms. *Appl. Microbiol.*, *19*:879-881, 1970.
33. Miles, J. R. W., Tu, C. M., and Harris, C. R. Metabolism of heptachlor and its degradation products by soil microorganisms. *J. Econ. Entomol.*, *62*:1334-1337, 1969.
34. Singh, R., Iyengar, L., and Rao, A. V. S. P. Utilization of heptachlor by fungi. *Indian J. Exp. Biol.*, *14*:466-468, 1976.
35. Iyengar, L., and Rao, A. V. S. P. Metabolism of chlordane and heptachlor by Aspergillus niger. *J. Gen. Appl. Microbiol.*, *19*:321-324, 1973.
36. Beeman, R. W., and Matsumura, F. Metabolism of cis and trans chlordane by a soil microorganism. *J. Agr. Fd. Chem.*, *29*:84-89, 1981.
37. Martens, PR. Der Abbau von Endosulfan durch Mikroorganismen des Bondens. Schriftenr. *Ver. Wasser-Boden-Lifthyg. Berlin Dahlem, 37*: 167-173, 1972.
38. El-Zorgani, G. A., and Omer, M. E. H. Metabolism of endosulfan isomers by Aspergillus niger. *Bull. Environ. Contam. Toxicol.*, *12*:182-185, 1974.
39. Bollen, W. B., Morrison, H. E., and Crowell, H. E. Effect of field treatments of insecticides on members of bacteria, streptomyces and molds in soil. *J. Econ. Entomol.*, *47*:302-306, 1974.
40. Sethunathan, N., Siddaramappa, R., Rajaram, K. P., Barik, S., and Wahid, P. A. Parathion residues in soil and water. *Residue. Rev.*, *68*:91-122, 1977.
41. Rao, A. V., and Sethunathan, N. Degradation of parathion by Penicillium waksamani Zaleski isolated from flooded acid sulphate soil. *Arch. Mikrobiol.*, *97*:203-208, 1974.
42. Miller, C. W., Tomlinson, W. E., and Nargren, R. L. Persistence and movement of parathion in irrigation waters. *Pest. Monit. J.*, *1*:47-48, 1967.
43. Lichtenstein, E. P., and Schulz, K. R. The effect of moisture and microorganisms on persistence and metabolism of some organophosphorus insecticides in soils, with special emphasis on parathion. *J. Econom. Entomol.*, *57*:618-627, 1964.
44. Barik, S. Metabolism of insecticides by microorganisms. In *Insecticide Microbiology* (R. Lal, ed.), Springer-Verlag, Berlin, 1984, pp. 87-130.
45. Matsumura, F., and Boush, G. M. Malathion degradation by Trichoderma viride and a Pseudomonas spp. *Science*, *153*:1278-1280, 1966.
46. Mostafa, I. Y., Bahig, M. R. E., Fakhr, I. M. L., and Adam, Y. Malathion break down by soil fungi. *Z. Naturforsch.*, *27*:1115-1116, 1972.
47. Lewis, D. L., Paris, D. F., and Baughman, G. L. Transformation of malathion by a fungus Aspergillus oryzae isolated from a fresh water pond. *Bull. Environ. Contam. Toxicol.*, *13*:596-601, 1975.
48. Flashinski, S. J., and Lichtenstein, E. P. Metabolism of Dyfonate by soil fungi. *Can. J. Microbiol.*, *20*:399-411, 1974.
49. Liu, S. Y. and Bollag, J. M. Metabolism of carbaryl by a soil fungus. *J. Agr. Fd. Chem.*, *19*:487-490, 1971.

50. Liu, S. Y., and Bollag, J. M. Carbaryl decomposition to 1-naphthyl carbamate by Aspergillus terreus. *Pestic. Biochem. Physiol., 1*:366-372, 1971.
51. Bollag, J. M., and Liu, S. Y. Hydroxylation of carbaryl by soil fungi. *Nature, 236*:177-178, 1972.
52. Bollag, J. M., and Liu, S. Y. Fungal degradation of 1-naphthol. *Can. J. Microbiol., 18*:113-117, 1972.
53. Sjobald, R. D., and Bollag, J. M. Oxidative coupling of aromatic pesticide intermediates by a fungal phenol oxidase. *Appl. Environ. Microbiol., 33*:906-910, 1977.
54. Sjobald, R. D., and Bollag, J. M. Oxidative coupling of aromatic compounds by enzymes from soil microorganism. In *Soil Biochemistry*, Vol. 5 (E. A. Paul and J. N. Ladd, eds.), Marcel Dekker, New York, 1981, pp. 113-115.
55. Cripps, R. E., and Roberts, T. R. Microbial degradation of herbicides. In *Pesticide Microbiology* (I. R. Hill and S. J. L. Wright, eds.), Academic Press, London, 1978, p. 669.
56. Saxena, A., Zhang, R., and Bollag, J. M. Microorganisms capable of metabolizing the herbicide metolachlor. *Appl. Environ. Microbiol., 53*:390-396, 1987.
57. Kilbane, J. J., Chatterjee, B. K., and Chakrabarty, A. M. Detoxification of 2,4-5-trichlorophenoxy acetic acid from contaminated soil with Pseudomonas cipacia. *Appl. Environ. Microbiol., 45*:1697-1700, 1983.
58. Molozhanova, Y. G., Ramizova, L. B., and Brantsevich, L. G. Detoksikatsiya kylarofosa pochvennyml mikroorganizamami. *Khim. Sel. Khoz., 11*:761-762, 1973.
59. Johnson, L. M. Enzymes associated with the microbial metabolism of insecticides. In *Insecticide Microbiology* (R. Lal, ed.), Springer-Verlag, Berlin, 1984, pp. 131-141.
60. McIntyre, A. E., Lester, J. N., and Perry, R. Persistence of organophosphorous insecticides in sewage sludges. *Environ. Technol. Lett., 2*:111-118, 1981.
61. Barik, S., and Munnecke, D. M. Enzymatic hydrolysis of concentrated diazinon in soil. *Bull. Environ. Contam. Toxicol., 29*:235-239, 1982.
62. Munnecke, D. M. Hydrolysis of organophosphate insecticides by an immobilized enzyme system. *Biotechnol. Bioeng., 21*:2247-2261, 1979.
63. Munnecke, D. M., Johnson, L. M., Talbot, H. W., and Barik, S. Microbial metabolism and enzymology of selected pesticides. In *Biodegradation and Detoxification of Environmental Pollutants* (A. M. Chakrabarty, ed.), CRC Press, Boca Raton, 1982, pp. 1-32.
64. Fischer, H. F., Munnecke, D. M., and Domsch, K. H. Final Report: Mikrobielle bzw enzymatische splatung von Parathion BMFT-FB-T80-046, Bonn, 1980, p. 54.
65. Kearney, P. C., Karns, J. S., and Mulbry, W. W. Engineering soil microorganisms for pesticide degradation. In *Pesticide Sciences and Biotechnology* (R. Greeolgh and T. R. Roberts, eds.), Blackwell Scientific, 1987, pp. 591-596.
66. Sastry, C. A. Microbial degradation of waste from slaughter house and tanneries. In *Proc. National Workshop on Microbial Degradation of Industrial Wastes* (B. B. Sundaresan, ed.), NEERI, Nagpur, India, 1982, pp. 122-144.

67. Jacob, F. M. P., and Marie, C. Yeast tannin interaction. Study in tannin medium of some tannic acid hydrolysing yeasts. *Mycopathologia, 57*:139-148, 1975.
68. Mahadevan, A., and Muthukumar, G. Aquatic microbiology with reference to tannin degradation. *Hydrobiologia, 72*:73-79, 1980.
69. Knudson, L. Tannic acid fermentation. I and II. *J. Biol. Chem., 14*: 159-206, 1913.
70. Mahadevan, A., and Sivaswamy, S. N. Tannins and microorganisms. In *Frontiers in Applied Microbiology*, Vol. 1 (K. G. Mukerji, N. C. Pathak, and V. P. Singh, eds.), Print House, Lucknow, 1985, pp. 327-347.
71. Lewis, J. A., and Starkey, R. A. Decomposition of plant tannins by some soil microorganisms. *Soil Sci., 107*:235-241, 1969.
72. Chandra, T., Madhavkrishna, T. W., and Nayduamma, Y. Astringency in fruits. I. Microbial degradation of catechin. *Can. J. Microbiol., 15*:303-306, 1969.
73. Dalvesco, G., Fiusello, N., and Vei Hiranus, M. Utilization of gallotannin and gallic acid by some Aspergillus and Penicillium sp. *Allonis, 17*:25-50, 1972.
74. Hathway, D. E., and Seakins, J. W. T. Enzymic oxidation of catechin to a polymer structurally related to some pholbatannins. *Biochem. J., 67*:239, 1959.
75. Nishira, H. Some aspects on the metabolic fate of tannic acid: Studies on the tannic acid decomposing enzymes of molds. *Sci. Rep. Hyogo. Univ. Agr., 5*:117-123, 1962.
76. Subrahmanyam, P. V. R. Microbial degradation of waste waters from pulp and paper industry. In *National Workshop on Microbial Degradation of Industrial Wastes* (B. B. Sundaresan, ed.), NEERI, Nagpur, India, 1982, pp. 53-77.
77. Veermani, H., and Idress, M. Recovery of pulping chemicals in kraft paper mills. *Chem. Ind. Dev., 21*:16-20, 1975.
78. Bjorseth, A., Corlberg, G. E., and Moller, M. Determination of halogenated compounds and mutagenicity of spent bleaching liquids. *Sci. Tot. Environ., 11*:197, 1979.
79. Kirk, T. K. Degradation of lignin. In *Microbial Degradation of Organic Compounds*, Vol. 13 (D. T. Gibson, ed.), Marcel Dekker, New York, 1984, pp. 399-437.
80. Kirk, T. K., Chang, H. M., and Eaton, D. Fungal decolorization of kraft bleach effluent. *TAPPI, 63*:103, 1980.
81. Nakatsubo, F., Reid, I. D., and Kirk, T. K. Involvement of singlet oxygen in the fungal degradation of lignin. *Biochem. Biophys. Res. Commun., 102*:484-491, 1981.
82. Forney, L. I., Reddy, A. C., Tien, M., and Aust, S. D. The involvement of hydroxyl radical derived from hydrogen peroxide in lignin degradation by white rot fungus, Phanerochaete chrysosporium. *J. Biol. Chem., 257*:11455-11462, 1982.
83. Faison, B. D., and Kirk, T. K. Relationship between lignin degradation and production of reduced oxygen species by Phanerochaete chrysosporium. *Appl. Environ. Microbiol., 46*:1140-1145, 1983.
84. Kutsuki, H., and Gold, M. H. Generation of hydroxyl radical and its involvement in lignin degradation by Phanerochaete chrysosporium. *Biochem. Biophys. Res. Commun., 109*:320-327, 1982.

85. Westermark, U., and Erikson, K. E. Cellobiose:quinone oxidoreductase, a new wood-degrading enzyme from white rot fungi. *Acta. Chem. Scand. B28*:209-214, 1974.
86. Hera, A., Barnett, S. M., Shech, C. H., and Matecalvo (Jr.). An extracellular ligniase: A key to enhanced cellulose utilization. *Al Ch. E. Symp. Series, 84*:17-20, 1978.
87. Milgram, L. Ligniase, biotechnology's new money spinner? *New Scientist, 1456*:16-17, 1985.
88. Asther, M., Capdevila, C., and Corrieu, G. Control of lignin peroxidase production by Phanerochaete chrysosporium. INA 12 by temperature shifting. *Appl. Environ. Microbiol., 54*:3194-3196, 1988.
89. Tien, M., and Kirk, T. K. Lignin degrading enzymes from Phanerochaete chrysosporium. Purification characterization and catalytic properties of a unique H_2O_2 requiring oxygenase. *Proc. Natl. Acad. Sci. USA, 81*:2280-2284, 1983.
90. Buswell, T. A., Mollet, B., and Odier, E. Lignolytic enzyme production by Phanerochaete chrysosporium under conditions of nitrogen sufficiency. *FEMS Microbiol. Lett., 25*:295-299, 1984.
91. Gold, M. H., Kuwahara, M., Chiu, A. A., and Glenn, T. A. Purification and characterization of an extracellular H_2O_2 requiring diarylpropane oxygenase from the white rot basidiomycete Phanerochaete chrysosporium. *Arch. Biochem. Biophys., 234*:353-356, 1984.
92. Faison, B. D., and Kirk, T. K. Factors involved in the regulation of a ligninase activity in Phanerochaete chrysosporium. *Appl. Environ. Microbiol., 49*:294-304, 1985.
93. Leisola, M. S. A., Thaeiwyss, D., and Fiechter, A. Strategies for production of high ligninase activities by Phanerochaete chrysosporium. *J. Biotechnol., 3*:97-107, 1988.
94. Bumpus, J. A. M., Tein, D. W., and Aust, S. A. Oxidation of persisted environmental pollutants by a white rot fungus. *Science, 228*: 1434-1436, 1985.
95. Asther, M., Lesage, L., Drapron, R., Corrieu, G., and Odier, E. Phospholipid and fatty acid enrichment of Phanerochaete chrysosporium INA-12 in relation to ligninase production. *Appl. Microbiol. Biotechnol., 27*:393-398, 1988.
96. Jager, A., Croan, S., and Kirk, T. K. Production of ligninase and degradation of lignin in agitated submerged cultures of Phanerochaete chrysosporium. *Appl. Environ. Microbiol., 50*:1274-1278, 1988.
97. Alic, M., and Gold, M. H. Genetic recombination in the lignin degrading basidiomycete Phanerochaete chrysosporium. *Appl. Environ. Microbiol., 50*:27-30, 1985.
98. Klien, D. A. Improving microbial availability of recalcitrant polymers by photomodification. Studies with lignin sulphonate and straw substances. EPA-R-EC-00278-03, 1970.
99. Rao, M., Deshpande, V., Seeta, R., and Srinivasan, M. C. Hydrolysis of sugar cane bagasse for mycelial biomass by Penicillium funiculosum. *Biotechnol. Bioeng., 24*:1070-1072, 1985.
100. Hang, J. S., Uhm, T. B., Tung, G. T., and Lee, K. B. Studies on the enzymes produced by Pleurotus Sajor-Cajci. I. The production of cellulolytic enzymes. *Korean J. Appl. Microbiol., 12*:59-64, 1984.
101. Qu, Y. B., Gao, P. J., and Wang, Z. M. Screening of catabolite repression resistant mutants of cellulase producing Penicillium sp. *Acta Mycologica Sinica, 3*:238-240, 1989.

102. Tautorus, T. E., and Charmers, W. T. Pilot plant production of SCP utilizing Chaetomium cellulyticum. *Dev. Ind. Mycol.*, *25*:621-631, 1984.
103. Tiunova, N. A., Kobzeva, H. Y. A., Zagustina, N. A., Maximov, V. I., and Benzborodov, A. M. Cellulolytic enzymes of Geotricum candidum. *Mikrobiologia*, *53*:237-241, 1984.
104. Glenn, M., Ghosh, A., and Ghosh, B. K. Subcellular fraction of a hypercellulolytic mutant, Trichoderma reesei Rut C30: Localization of endoglucanase in microsomal fraction. *Appl. Environ. Microbiol.*, *50*:1137-1143, 1985.
105. MacDonald, M. J., Hartley, D. L., and Speedie, M. K. Location of cellulolytic enzyme activity in the marine fungus, Trichocladium aschrasporum. *Can. J. Microbiol.*, *31*:145-148, 1980.
106. Beldan, G., Vragen, A. G. J., Robots, F. M., and Pilnik, W. Synergism in cellulose hydrolysis by endogluconases and exogluconases purified from Trichoderma viride. *Biotechnol. Bioeng.*, *31*:173-178, 1988.
107. Clementi, F., Moresi, M., and Rossi, J. Effect of medium composition of microbial utilisation of citrus waste by mixed fungal culture. *Appl. Microbiol. Biotechnol.*, *8*:305-308, 1986.
108. Duff, B. J. S., Cooper, D. G. and Mynard, F. O. Evaluation of the hydrolytic potential of a crude cellulase from mixed cultivation of Trichoderma reesei and Aspergillus phoenicis. *Enz. Microbiol. Technol.*, *8*:205-208, 1986.
109. Trivedi, S. M., and Ray, R. M. Saccharification of cellulosic wastes by high strength cellulase system mix shake cultivated Scytalidium ligniocola and Trichoderma longibrachiatum. *J. Ferment. Technol.*, *63*:299-301, 1985.
110. Kassim, E. A., and Nagieb, Z. A. Effects of pretreatment of cellulosic wastes on the production of cellulase enzymes by Trichoderma reesei. *J. Ferment. Technol.*, *63*:291-297, 1985.
111. Garg, S. K., and Neelakantan, S. Production of SCP and cellulose by Aspergillus terreus from bagasse substrate. *Biotechnol. Bioeng.*, *24*:2407-2418, 1982.
112. Duran, N., Rodrigues, J., Gomez, E., Campos, E., and Baeza, J. Photochemical pretreatment of cellulose and its fungal degradation. *Biotechnol. Bioeng.*, *31*:215-219, 1988.
113. Krishnamurti, C. R. Biodegradation of waste generated in plastic and polymer industry. In *Proc. National Workshop on Microbial Degradation of Industrial Wastes* (B. B. Sundaresan, ed.), NEERI, Nagpur, India, 1982, pp. 168-196.
114. Griffin, J. I. G. and Mivetchi, H. Biodegradation of ethylene/vinylacetate co-polymers. In *Proceeding of the 3rd International Biodegradation Symposium* (J. M. Sharpely and A. M. Kapalan, eds.), Applied Science, London, 1976, pp. 807-813.
115. Cappola, S. Effect of heavy metals on soil microorganisms. In *Environmental Effects of Organic and Inorganic Contaminants in Sewage Sludges* (R. D. Davis, G. Hucker, and P. L. Hermit, eds.), D. Reidel, London, 1983, pp. 233-243.
116. Gadd, G. M., Mowell, J. L., White, C., and Newby, P. J. Methods of assessment of heavy metal toxicity toward fungi and yeast. In *Toxicity Testing Using Microorganisms*, Vol. 2 (D. L. Lin and B. J. Dutka, eds.), CRC Press, Boca Raton, 1986, pp. 169-185.

117. deRome, L., and Gadd, G. M. Copper adsorption by Rhizopus arrhizus, Cladosporium resinae and Penicillium itallicum. *Appl. Microbiol. Biotechnol.*, *26*:84-90, 1987.
118. Galun, M., Keller, P., Feldstein, H., Galun, E., Siegel, S., and Siegel, B. Recovery of uranium(VI) from solution using fungi. Release from uranium loaded Penicillium biomass. *Water Air Soil Pollut.*, *20*: 277-280, 1983.
119. Nordgren, A., Baath, E., and Soderstrom, B. Soil microfungi in an area polluted by heavy metals. *Can. J. Biol.*, *63*:448-455, 1985.
120. Galun, M., Keller, P., Malki, D., Feldstein, H., Galun, E., Siegel, S., and Siegel, B. Recovery of uranium(VI) from solution using precultured Penicillium biomass. *Water Air Soil Pollut.*, *20*:221-232, 1983.
121. Shanlakadze, T. L., Kosharidze, M. I., Rosishvili, V. N., Tsarsidze, M. A., and Lomsadze, B. A. Mechanism of biodegradation of copper by fungi. *Seobshch. Akad. Nauk Gruz. SSR.*, *124*:633-636, 1988.
122. Babich, H., and Stotzky, G. Sensitivity of various bacteria including actinomycetes and fungi to cadmium and the influence of pH on sensitivity. *Appl. Environ. Microbiol.*, *33*:681-695, 1977.
123. Siegel, S., Keller, P., Galun, M., Lehr, H., Siegel, B., and Galun, E. Biosorption of lead and chromium by Penicillium preparation. *Water Air Soil Pollut.*, *27*:69-75, 1983.
124. Tobin, J. M., Cooper, D. G., and Neufeld, R. J. Uptake of metal ions by Rhizopus arrhizus biomass. *Appl. Environ. Microbiol.*, *47*: 821-824, 1984.
125. Garcia-Toledo, A., Babich, H., and Stotzky, G. Trainup of Rhizopus stolonifer and Cunninghamella blakesleeana to copper: Co-tolerance to Cd, Co, Ni and Pb. *Can. J. Microbiol.*, *31*:485-492, 1985.
126. Wainwright, M., and Graylson, S. J. Oxidation of heavy metal sulphides by Aspergillus niger and Trichoderma harzianum. *Trans. Br. Mycol. Soc.*, *86*:269-272, 1986.
127. Galun, M., Galun, E., Siegel, B. Z., Keller, P., Lehr, H., and Siegel, S. M. Removal of metal ions from aqueous solutions by Penicillium biomass. Kinetic and uptake parameters. *Water Air Soil Pollut.*, *33*:359-371, 1987.
128. Cullen, W. R., McBride, B. C., and Reimer, M. Induction of the aerobic methylation of arsenic by Candida humicola. *Bull. Environ. Contam. Toxicol.*, *21*:157-161, 1979.
129. Volesky, B. *Biosorbent Materials*. Biotechnology and Bioengineering Sym. No. 16. John Wiley and Sons, New York, 1986, pp. 121-126.
130. Tatsuo, S., Kazuyasu, A., Manabu, F., and Ichikawa, K. Basic decomposition parameters of phenols by Candida tropicalis. *J. Ferment. Technol.*, *51*:803-806, 1973.
131. Venkatrao, B., and Bhat, J. V. Characteristics of yeast isolated from phenol and catechol adapted activated sludges. *Antonie Van Leeuvenhoek*, *37*:303-305, 1971.
132. Blaich, R., and Esser, K. Function of enzyme in wood destroying fungi II. Multiple forms of laccase in white rot fungi. *Arch. Microbiol.*, *103*:271-277, 1975.
133. Frochner, S. C., and Erickson, K. E. Purification and properties of Neurospora crassa laccase. *J. Bacteriol.*, *120*:458-465, 1974.
134. Mileski, G. J., Bumpus, J. A., Jurek, M. A., and Aust, S. D. Degradation of pentachlorophenol by a white rot fungus Phanerochaete chrysosporium. *Appl. Environ. Microbiol.*, *54*:2885-2889, 1988.

135. Bollag, J. M., Shuttleworth, K. L., and Anderson, D. H. Laccase mediated detoxification of phenolic compounds. *Appl. Environ. Microbiol.*, *54*:3086-3091, 1988.
136. Livernoche, D., Jurasek, L., Desrochers, M., Dorica, J., and Veliky, I. A. Removal of colour from kraft mill waste waters with cultures of white rot fungi and with immobilized mycelium of Coriolus versicolor. *Biotechnol. Bioeng.*, *25*:2055-2066, 1983.
137. Maheshwari, R., and Kamalam, P. T. Isolation and culture of a thermophilic fungus Melanocarpus albomyces and factors influencing the production and activity of xylanase. *J. Gen. Microbiol.*, *131*: 3017-3027, 1985.
138. Nandan, R., Chandra, H., Bhargava, S. K., and Ray, P. K. Microbial conversion of distillery waste to bioenergy: Effects of media enrichment with low chain fatty acids and Candida sp. *J. Ferment. Technol.*, *65*:49-60, 1987.
139. Cooke, W. B. Potential plant pathogenic fungi in sewage and polluted water. *Plant Dis. Reptr.*, *40*:681-687, 1956.
140. Milliner, P. D. Occurrence of Aspergillus fumigatus during compositing of sewage sludge. *Appl. Environ. Microbiol.*, *34*:765-772, 1977.
141. Cooke, W. B. Fungi in sewage. In *Recent Advances in Aquatic Mycology* (E. B. Gareth Jones, ed.), Paul Elek, London, 1976, pp. 389-434.
142. Cland, M., Durand, H., and Tiraby, G. Enzymatic saccharification of municipal wastes. *Biotechnol. Bioeng.*, *32*:930-934, 1988.
143. Hiremath, A. B., Prabhakar, M. N., and Jayaraj, Y. M. Domestic sewage treatment by fungi and biomass production. *Environ. Ecol.*, *3*:568-571, 1985.
144. Genex, O. Impact of biotechnology on pollution control and toxic waste treatment. Contract report prepared for the Office of Technology Assessment, U. S. Congress, Washington, D.C., April 1983.
145. *Commercial Biotechnology: An International Analysis*. Report by U.S. Government Agency on Commercial Biotechnology. Elsevier, Amsterdam, 1984, p. 612.
146. Crueger, W., and Crueger, A. *Biotechnology: A Text Book of Industrial Microbiology*. Science Tech., Madison, WI, 1984, p. 308.
147. Osiewacz, H. P., and Weker, A. DNA mediated transformation of the filamentous fungus Curvularia lunata using a dominant selectable marker. *Appl. Microbiol. Biotechnol.*, *30*:375-380, 1989.
148. Bryers, J. D., and Homer, G. Application of immobilized captured microorganism in water purification: An overview. In *Meth. Enzymol.*, *137*(D):697-711, 1987.
149. Kirkpatrick, N., and Palmer, J. M. Semicontinuous lignease production using foam immobilized Phanerochaete chrysosporium. *Microbiol. Biotechnol.*, *27*:129-133, 1987.
150. Anselo, A. M., Cabral, J. M. S., and Novais, J. M. The adsorption of Fusarium flocciferum spores on celite particles and their use in the degradation of phenol. *Appl. Microbiol. Biotechnol.*, *31*:200-203, 1989.
151. Klein, J., Hachel, V., and Wagner, F. Phenol degradation by Candida tropicalis whole cells entrapped in polymeric ionic networks. *ACS Symp. Ser.*, *106*:101-118, 1979.
152. Karbue, I. Biosensors in fermentation and environmental control. In *Biotechnology: Application and Research* (P. N. Cheremisinoff and R. P. Quellette, eds.), Technomic, Pennsylvania, 1985, pp. 135-155.

153. Pemberten, J. M. Genetic engineering and biological detoxification of environmental pollutants. *Residue Rev.*, *78*:1-11, 1981.
154. Chakrabarty, A. M. Genetic basis of the biodegradation of salicylate in Pseudomonas. *J. Bacteriol.*, *112*:815-823, 1972.
155. Rheinwald, J. G., Chakrabarty, A. M., and Gunsalus, I. C. A transmissible plasmid controlling camphor oxidation in Pseudomonas putida. *Proc. Natl. Acad. Sci. USA*, *70*:885-889, 1973.
156. Friello, D. A., Mylorie, J. R., and Chakrabarty, A. M. Use of genetically engineered multiplasmid microorganisms for rapid degradation of fuel hydrocarbons. In *Proceedings of the 3rd International Biodegradation Symposium* (J. M. Sharpely and A. M. Kapalan, eds.), Applied Science, London, 1976, pp. 205-214.
157. Qasim, S. Z. Application of biotechnology to marine sciences. In *Recent Trends in Biotechnology and Bioscience* (S. V. S. Rana, ed.), Pragati Press, Muzzafarpur, India, 1986, pp. 65-69.
158. Kitamoto, Y., Mori, N., Yamamoto, N., Ohiwa, T., and Tchikawa, Y. A simple method for protoplast formation and improvement of protoplast regeneration from various fungi using an enzyme from Trichoderma harzianum. *Appl. Microbiol. Biotechnol.*, *28*:445-450, 1988.
159. Morris, D. W., and Roth, B. A. Genetic engineering in yeast: Methods and applications. In *Biotechnology: Application and Research* (P. M. Cheremisinoff and R. O. Quellette, eds.), Technomic, Pennsylvania, 1985, pp. 172-187.
160. Cherallier, M. R., and Aigle, M. Qualitative detection of penicillinase produced by yeast strains carrying chimeric yeast coli plasmids. *FEBS Lett.*, *180*:179-180, 1980.
161. Khachatourians, G. G., and Haffie, T. L. Biotechnology: Applications to genetics. In *Biotechnology: Application and Research* (P. N. Cheremisinoff and R. O. Quellette, eds.), Technomic, Pennsylvania, 1985, pp. 243-250.
162. Turgeon, G., and Yoder, O. C. Genetically engineered fungi for weed control. In *Biotechnology: Application and Research* (P. N. Cheremisinoff and R. O. Quellette, eds.), Technomic, Pennsylvania, 1985, p. 221.
163. Spencer, J. F. T., and Spencer, D. M. Genetic improvement of industrial yeasts. *Annu. Rev. Microbiol.*, *37*:121-142, 1983.
164. Tien, M., and Tu, C. P. D. Cloning and sequencing of DNA for ligninase from Phanerochaete chrysosporium. *Nature*, *362*:520-523, 1987.
165. Smith, T. L., Schalch, H., Gaskell, J., Covert, S., and Cullen, D. Nucleotide sequence of a ligninase gene from Phanerochaete chrysosporium. *Nucleic Acid Res.*, *16*:219-212, 1988.
166. Asada, Y., Kimura, Y., Kuwahara, M., Tsukamotu, A., Koide, K., Oka, A., and Takamani, M. Cloning and sequencing of a ligninase gene from a lignin degrading basidiomycetes Phanerochaete chrysosporium. *Appl. Microbiol. Biotechnol.*, *29*:469-473, 1988.
167. Loper, J. C. Gene engineering of yeasts for degradation of hazardous wastes. *J. Hazard. Mater.*, *19*:323-326, 1988.
168. Guillard, B., Bretan, A., and Bernalier, A. Study of the nuclear cycle of four species of anaerobic rumen fungi by fluorescence microscopy. *Curr. Microbiol.*, *19*:103-107, 1989.

31

USE OF FUNGI FOR BIOCONVERSION OF DISTILLERY WASTE

JOŽEFA FRIEDRICH and ALEKSA CIMERMAN *Boris Kidrič Institute of Chemistry, Ljubljana, Slovenia, Yugoslavia*

ANTON PERDIH *University of Ljubljana, Ljubljana, Slovenia, Yugoslavia*

I. INTRODUCTION

A. Distillery Wastes

Distillery wastes dealt with herein are effluents occurring in the production of alcohol or alcoholic beverages from plant raw materials. These can be produced by fermentation of any naturally occurring sugar, e.g., glucose, fructose, maltose, saccharose; starches or cellulose-containing materials. Sugars are fermented directly, whereas in the case of starch and cellulose, acid or enzymatic hydrolysis is necessary prior to fermentation.

The usual raw materials for alcohol production are sugarcane, sugar-beet, and their molasses; corn, wheat, or barley grain; cassava; potato; different fruits or their juices. In the case of wine, the feedstock already contains alcohol and only distillation is needed. Other raw materials have to be prepared for alcoholic fermentation by dilution or mashing including eventual hydrolysis and/or addition of nutrients.

After fermentation, the alcohol-containing mash is forwarded to distillation. Depending on the technology, some fermentation byproducts like esters, higher alcohols, and aldehydes remain either in the main distillate, are removed as separate streams, or are partly included in wastestreams.

The wastewaters arising in the production of alcohol and alcoholic beverages are washwater, steam water, cleaning and cooling waters, and stillages. They represent a nuisance for the distilleries. The most troublesome are stillages and most attention is devoted to them. *Stillage* (syn. Eng. spent wash, slop, vinasse; Fr. vinasse; Ger. Schlempe; Ital. vinacce; Span. vinaza; Port. vinhoto, vinhaca; Russ. barda; Belorus. braga; Serbo-Croatian dzibra; Slovenian drozga; Bulg. vinasa; Turk. vinas) remains after distilling off the alcohol. Its sugar and alcohol content is largely exhausted. It is either a clear fluid containing spontaneously settling particles or a pulpy material.

1. Composition of Stillages

The composition of stillages depends mostly on the original raw material and therefore a number of different types can occur. Their amounts represent about 10-15 times the volume of the alcohol produced and they contain a high content of organic substances [1]. For each m^3 of wine, 300-400 liters of vinasse is obtained after distillation. In the case of lees, 150-200 liters of the effluent is produced per m^3 of wine. The composition of vinasse of pressed grape husks, lees, or wine reflects the composition of wine with the exception of the alcohol content [2]. When using beet molasses for alcoholic fermentation about 3.6 m^3 of stillage with chemical oxygen demand (COD) of 60,000 mg/liter is produced from 1 ton of molasses [3].

The composition of stillages from different raw materials is presented in Table 1. The stillages contain also 5-18 g/liter of total acids [20,22,25], 0.07-1.6 g/liter of phosphates [13,16-18,20,22,47], 4.4-26.2 mg/liter of glycerol [12,20,22] and some vitamins [25].

Besides the pulpy cell wall debris suspended in the liquid phase, the fruit slops also contain considerable amounts of solid particles such as kernels, stems, stones, and peels. Besides the data in Table 1, additional data for apple slops [29,30] have been reported: in two cases dry matter was 3.4 and 5.9%, acid detergent fibers 37.0 and 47.4% including cellulose 20.4 and 33.0%, and lignin 14.7 and 15.2%; hemicellulose 14.1 and 9.5%, pectin 1.0 and 3.0%, reducing sugars expressed as glucose 12.1 and 9.4%, crude protein 5.2 and 3.5%, crude lipids 6.0%, total ash 5.8%; starch was not found. Also 0.3% ash and 6.1 g/liter of tartaric acid were reported in vinasse [25].

The liquid part can contain substantial amounts of polyols such as glycerol and 2,3-butanediol, of tartaric, malic, citric, succinic, lactic, acetic, isocitric, parasorbic, benzoic, quinic, and other organic acids, as well as amino acids, small amounts of sugars (hexoses, pentoses), flavonoids, etc. [2,23,24].

2. Properties of Stillages

The following characteristics of stillage are the most significant for the technical evaluation of wastewater [24]:

Very high biochemical oxygen demand (BOD)
Very high chemical oxygen demand (COD)
Low pH value

The acidity of the stillages with a pH of 3.3-4.5 precludes their disposal into natural waterways as well as simple treatment in the municipal sewage treatment plants. In the latter case, they must be diluted with a more than 50 times greater quantity of water [23] or domestic sewage [24,28]. Before dilution the slops corrode concrete and carbonaceous materials of the sewage transport system [24].

The high contents of organic substances in stillages need at least a 100,000 times greater natural waterway flux to be disposed of or a 50-100 times greater quantity of low-loaded sewage to be treated in the municipal sewage treatment plants [23,28]. Their actual concentration of organic matter is on the lower side of the nutrient concentration in industrial fermentation substrates. Thus they are promising as complex biotechnological substrates, although they usually lack greater quantities of a specific carbon source. The relationship of BOD5 to N to P of 100:5:1 is usually recommended for sewage treatment [31]. In stillages from sugarcane molasses, this relationship is nearly 250:14:1 and in cane juice stillages it is close to 38,000:70:1. We calculate that in stillage from cassava it is approximately 200:5:1 [32]. In fruit slops it amounts to around 10,000:200:1 [27].

TABLE 1. Composition of Stillages

Origin	BOD, g/liter	COD, g/liter	pH	Solids, g/liter	Nitrogen, g/liter	Protein, g/liter	Red. sug., g/liter	Tot. sug., g/liter	Ash, g/liter
Molasses									
Cane	16-70	32-90	4.2-5.2	31-119	0.2-5.4	3.7-6.2	2-18	6-30	10-40
BS	50-60	74		62-85	0.8	8		9	25
Beet	103	45-158	4.6		3.3-6.0		18.3	26	
Grain	10-340	15	3.2-7.5	20-75	0.2-1.9	5-23	11-30	2.9[a]-5	4-15
Malt	24-43	35-58	2.9-3.8			15-31		6-21[a]	
Cassava	19	23	5.0	24	0.4	2.5	6.8		2.3
SWL	40-50			100		1		8	21
Wine	12-18	26	3.6-4.5	23-125	0.4-1.0			0.8-1.1	
Fruit		38-156	3.3-4.5	30-440	0.4-1.0				

[a]Carbohydrate.

BS, blackstrap molasses; SWL, sulfite waste liquor.

Sources: cane molasses stillage [4-18], blackstrap molasses stillage [1], beet molasses stillage [19,20], grain stillage [21,22], malt stillage [22], wine stillage [2,21], cassava stillage [12], sulfite waste liquor stillage [1], fruit slops [21,23-28].

When disposed into water streams or seas, the high BOD5 of stillages leads to oxygen shortage and seriously jeopardizes water organisms, even though most of the stillages are not toxic in themselves [3]. Overloading occurs if due care is not taken in inadequate treatment facilities. Stillages decay anaerobically and the development of H_2S [24] and lower fatty acids leads to unpleasant odors.

Stillages containing small amounts of insoluble matter already give rise to many difficulties. The fruit distillery slops are viscous liquids containing tiny fibrous and other solid particles and slimy swollen materials. The materials don't settle; their flocculation or centrifugation is inefficient and their filtration is nearly impossible [29,33] except in thin layers because they clog the filters. The filtrate amounts to at most 34% in plum slops and 67% in pear slops and contains 3-6% of dry matter [27].

3. Disposal of Stillages

In a more ecologically conscious world there are substantial pressures on producers of such effluents to develop proper treatment for making wastes ecologically acceptable, and preferably even to profit from them. Many potential solutions of these problems are known. Among differing solutions are simple disposal, sewage treatment, and recovery of useful ingredients. An important point also to be considered is the use of stillages for the production of commercially valuable products.

Perdih et al. [26] suggested the following order of conversion of food industrial wastes:

1. For food
2. For energy or raw materials
3. For feedstuffs
4. As fertilizer
5. Destruction or disposal of wastes that are useless at this time

Holdom and Winstrom-Olsen [34] report a possible plan to pyrolyze wastes with less than 14% moisture. Liquids with a BOD of less than 8000 mg/liter can be aerobically processed in conventional sewage plants, and solids containing high percentages of water could be digested anaerobically.

A text on organic waste conversion [35] as well as several review articles discuss the problem of distillery wastewater. Sheehan and Greenfield [21] presented a comprehensive review of the methods for handling them, citing a great number of references from all over the world. They proposed land applications of stillages with low contents of suspended solids or their anaerobic treatment. They warned that in some countries, such as Australia, fodder yeast production may not be attractive. Weller [24] and Bronn [3,36] reported on the distillery wastewaters in the Federal Republic of Germany. Costa Ribeiro and Castello Branco [32] discussed highlights of technology, economics, market penetration, and opportunities in the recovery and utilization of stillage and its products in Brazil. Kishore et al. [37] and Bhandari et al. [14] reported on the conditions and possibilities in India. Current industrial and experimental utilization and disposal of effluents and properties of stillage of different origins are reviewed and compared by Maiorella et al. [1]. Wastewater treatment in distilling in the United Kingdom is described by Campbell [38] with special attention to aerobic biological treatment.

These and other authors studied a variety of stillages such as vinasses, i.e., stillages remaining after distillation of wine (this name is often used to designate other stillages also) [2,21,24,39], stillages from molasses [1,3,14,21,31,32,37,40-42], from sulfite waste liquor and/or wood hydrolyzates [1], from whisky production [38], from grains, cassava, and potato [1,31,32,38,41,43], as well as from fruit distillery slops [23,28].

The following means of disposal or treatment were discussed: dumping into rivers, sea, landfills, or subsoil [2,21,24,31,39], land applications of the solid and/or liquid phase [2,14,21,36,39], spontaneous separation of liquid and solid phase and their subsequent separate treatment and disposal [21,23], forced mechanical separation and subsequent treatment [21,43,44], evaporation [1,39,41,44], combined thermal treatment, e.g., evaporation and incineration [1,14,40], lime treatment [21,23,24], composting of the solid phase [23,39], aerobic treatment with activated sludge or in lagoons, etc. [2,23,28,31,36-38], anaerobic biological treatment [1,2,14,21,24,32,36,37], combined biological treatment [36,42], single-cell protein (SCP) production [1,2,14,32,37], use as food ingredients [45], use as feed [1,32,43,46], isolation of specific substances, e.g., tartrate [2,39] or potassium salts for use as fertilizer [32,37,40].

There are several methods suggested for solving a problem, and no method is decidedly better than another. Thus, it is not surprising that stillages were treated or upgraded by a variety of yeasts, filamentous fungi, including mushrooms, and several mixed cultures or serial ones as the main process or a part of it.

The role of fungi in treatment of distillery wastes is mainly in microbial biomass production. Experiments in the utilization of these microorganisms for bioconversion are very heterogeneous and seemingly more popular in developing than in developed countries. Consequently, the literature is not easily available or comprehensible as many reports have been published in local journals and languages. Despite these obstacles this chapter should give the reader an overview of the research and development dealing with the use of fungi for distillery waste bioconversion.

The microbiological utilization of stillages shows some similarities with traditional biological sewage treatments. However, there are two major differences. First, in sewage treatment a mixed culture of naturally found microorganisms is overgrown by the most appropriate and strongest ones. However, for the utilization of the waste a specific microorganism or group of microorganisms has to be chosen which must remain uncontaminated during fermentation. Second, the sewage treatment leans toward minimal biomass production, which must be disposed of as sludge, while the utilization process regards biomass as a product to be sold as a protein-rich animal feed [47].

The overall functions of the fungal processes can be defined as follows [47]:

1. The process should tend to produce a maximum amount of a marketable product while it should achieve a maximum degree of treatment.
2. The process should be able to treat wastes at lower cost than traditional methods.
3. It must require a fairly low investment, since companies are rarely prone to invest large capital sums into a plant which does not contribute to their productivity.
4. The process must be easy to operate and should not demand much skilled labor.

B. Historical Overview

The first work on use of fungi for bioconversion of distillery waste was claimed to have been carried out in Brasil in 1943 [48], whereas the first publication was in 1955. This was the paper of the French investigator Lefrançois [49] dealing with the growth of yeast in beet molasses stillage. It was claimed that at that time already seven plants were equipped for the production of yeast from stillage. The aim of the process was to produce protein of high nutritional value for human and animal nutrition. The main reason for the use of stillage was its low price, while the ecological problem was not recognized as such. In 1959 Wonnenberger [50] confirmed the importance of the Lefrançois system for the production of fodder yeast. In France at that time 12 factories producing yeast from stillage were built, and elsewhere 13.

In the 1960s much research was done in Japan [4-7,51-53] and started in Taiwan [9] with yeasts grown in cane molasses stillage. Higher fungi were also proposed as production microorganisms [55]. Wine vinasses were used in France [54].

Since the mid-1970s a number of publications have appeared every year. But it was noticed that in the 1980s their number decreased. Does this mean that the use of fungi in stillage conversion is losing its impetus? Possibly. In the 1970s it was fashionable to investigate wastewater treatment and SCP production. Today in the scientific world other fields have become more popular, so that research on treatment and feed production from waste remains only where it is really necessary.

C. Geographic Overview

The most suitable raw material for alcoholic fermentation is determined by climatic conditions and so the types of stillages change with their geographic origin.

In France, sugarbeet stillage [20,49] and vinasse from winery distilleries [2,54,56] were treated with microorganisms to produce biomass rich in protein. In Northern Ireland, considerable research has been dedicated to the conversion of malt whisky distillery stillage by means of fungi [22,57-59]. In West Germany and Austria, beet molasses stillage has been used for yeast production [19,50]. In Yugoslavia, beet molasses stillage was treated by yeasts [60,61], whereas fruit slops, especially those of the apple, were treated with cellulolytic fungi [29,30,62]. In Bulgaria wine vinasse [70] and in Turkey stillage from raisins was used [25,63]. In the USSR, beet molasses [64-66], grain [67], and potato [68,69] were raw materials for alcohol production and stillages remaining after distillation were treated with fungi.

In Japan and in Taiwan, batch and continuous experiments on laboratory and pilot [4-7,51-53] as well as on an industrial scale [8,9,71-73] were conducted. Besides yeasts, *Aspergillus niger* was cultivated as well [74-77]. In South Korea [10,78] and the Philippines [79] similar kinds of research were performed as well. Indian investigators did not advocate the aerobic biomass production from stillage [14]. However, the use of yeast can substantially improve biogas production in anaerobic digestion [80].

In the United States, little attention was made to these kinds of stillages. However, the only reports on the use of hemicellulose hydrolyzate stillage [81], sulfite waste liquor stillage and corn stillage [1] came from here.

In the Caribbean and South America considerable attention was payed to stillage bioconversion by means of fungi. Brazil, Argentina, Guatemala, and Puerto Rico were involved. In contrast to all other countries, in this region mostly filamentous fungi were chosen. The only reports of mushrooms growing in such waste material originated in Brazil [55]. A combined treatment with yeasts and molds seemed promising in Argentina [18,82].

II. PROCESS

A. Substrate Preparation

Immediately after distillation the stillages are free of vegetative microbial cells and following cooling and settling only the supernatant is often used for bioconversion [22,59]. Wine vinasses are usually clarified [54,56], but this is not necessary in the case of cane molasses stillage [8,51]. Fruit slops must be sterilized [29,62].

Effluents are often deficient in nitrogen and phosphorus; therefore ammonium salts and phosphates are added to obtain a well-balanced medium. Mostly $(NH_4)_2SO_4$ [4-7,10,12,13,16-18,25,51,53-55,72,82], urea [12,13,22,69], $(NH_4)_2HPO_4$ [4,10,11,13,16,54,56,72], and potassium phosphates [10,16,18, 22,55,82] were utilized. In apple slops, the addition of phosphate is not needed whereas ammonium salts and urea have a positive effect [62]. Experiments with unsupplemented stillage have also been performed [13,16, 18,22,57-59,82]. Increased ammonium sulfate addition exerted a beneficial effect on the biomass production in yeast [16] and higher fungi [55]. The effect of nitrogen was noted only when a sufficient amount of phosphorus was present [8]. When growing *Candida tropicalis* in spent potato mash after alcohol distillation, no significant enhancement of growth rates or of protein content were observed by the addition of ammonium sulfate or urea [69]. Presumably in this case the waste was not deficient in nitrogen. Similarly, supplementation with nutrients was not necessary in *Kluyveromyces marxianus* cultivation in beet molasses stillage [19]. As observed by Quinn and Marchant [22], the addition of nutrient salts appeared to increase the biomass production when a larger amount of C source was available in the stillage. When the overall nutrient status of the stillage was low, the supplementation had only a negligible effect. The investigators decided to omit nutrient addition to malt whisky stillage used for growth of *Geotrichum candidum*.

The proportions of elements C/N/P of 50:5:1 were used in growing filamentous fungi in cane stillage [12,13]. For *Acremonium fusidioides* in the same substrate the optimal relation was 40:1:0.1 [17]. Media without N supplementation could possibly be used when *Azotobacter* strains able to fix nitrogen are first grown in the effluent [18].

In some experiments molasses [16,79,83-85] or sucrose [52] additions were successful in improving growth and substrate utilization. Stillage pretreatment with acids [67,84] or hydrolytic enzymes like cellulases [83] or xylanase and arabinase [67] is beneficial.

The effect of stillage dilution was studied as well [8,20,22,57]. Highly loaded vinasses [20] and fruit slops [29,62] require some dilution, e.g., 1:1 to 1:3 [29,62]. In the case of malt whisky stillage, however, it proved to lower the bioconversion yields [22,57].

The pH of the substrate was or was not adjusted to a special value, depending on experimental conditions. Usually pH values between 3.0 and

5.0 were used as the initial pH of the substrate. During fermentation the pH was allowed to change spontaneously or it was kept at a predetermined value. The latter often proved to be beneficial to the microorganism [7,13,16, 18,20,82].

B. Fermentation

As inoculum, cells from agar slants can be used directly or they can be precultured in a smaller volume of the substrate for a short time. Of such cultures, 10% are usually used for inoculation [17,54]. In many cases, the productive strain was propagated on agar slants containing stillage to adapt the fungus to the substrate [16,57]. Special attention was given to inoculum preparation to provide large quantities of cells adapted to the effluent [86]. An increase of the inoculum quantity reduced the fermentation time. The same was observed in *Penicillium* grown on white wine vinasse [56].

When cultivating Basidiomycetes in molasses stillage, the inoculum was cultivated in a malt extract broth for 10 days in the dark, washed, and broken up by strong shaking in flasks with glass beads. When the inoculum of *Agaricus campestris* was cultured for 12 days in stillage medium the conversion was accelerated [55].

For distillery stillage conversion by means of fungi, submerged processes are used with laboratory experiments carried out in shaken flasks and fermentors. Mostly stirred-tank bioreactors were applied. Column fermentors were also used [16,22,57,59]. Culture volumes in laboratory fermentors ranged from 320 ml [18,82] to 14 liters [12].

Pilot and large-scale biomass production from distillery waste was reported as well. Pilot scale volumes from 30 liters [13] to 600 liters [7] were used. In large-scale experiments, volumes from 9 m^3 [16] to 80 m^3 [73] were reported.

Batch or continuous modes of operation were used. A two-stage continuous cultivation with one [22,57] or two different microorganisms [18,82] as well as batch cultivation with serial growth of two microorganisms [47] were described. Continuous cultivation using a cell-recycling technique was reported [68,73]. Tomlinson compared batch and continuous processes [47]. He claimed that a single-step continuous fermentation was not suitable in the case of distillery stillage, where a high biomass productivity and a good substrate consumption had to be reached at the same time. In continuous cultivation dilution rates had to be higher for maximal cell productivity than for maximal cell concentration and full use of the effluent.

In some cases sterile conditions were necessary [11,16]. In others, there was no need for sterility [19,20]. Continuous cultivation in a 2-liter fermentor proved to be difficult for aseptic conditions [47]. Batch cultivation with suitable pH control was proposed to enable the desirable microorganism to overgrow contaminants [86]. Resistance to contamination in continuous cultures can be improved by the use of mixed cultures of different yeast strains [59]. To overcome this problem, the pH of the medium should be kept low [7,13,16]. The cell-recycling technique in 80 m^3 was used in continuous cultivation of *Candida utilis* and no contamination was detected [73]. In *K. marxianus* cultures in a laboratory fermentor in nonsterile conditions, no contamination problems arose [19]. In batch cultivation of *Aspergillus oryzae* stillage contamination by yeast and bacteria occasionally occurred [12] Previous treatment with a clay absorbent followed by settling in a clarifier was recommended for the solution of this problem.

Cultivation temperatures were kept between 22 and 38°C, the temperatures suitable for mesophilic microorganisms. The growth of *G. candidum* [57] and mushrooms [55] in distillery stillages occurred at 22°C. In most cases, 30°C was the chosen cultivation temperature. When *C. tropicalis* was grown in potato mash stillage [69] and *Gliocladium deliquescens* in rum stillage [11], a temperature of 35°C was used. The same temperature was optimal for the cultivation of *C. brumptii* [78]. In large-scale experiments with yeast, the growing temperature was maintained between 34 and 38°C [73] because it was difficult to control the fermentation process [9].

Chemical antifoam agents or mechanical devices [7] are used for foam disruption during fermentation and different aeration rates are employed. It can be stated that in smaller volumes, higher aeration rates are used than in larger ones, presumably for economic reasons. Karaki et al. [51] reported that the use of one volume of air/volume of broth during 24 h was satisfactory for continuous cultivation of yeast in cane molasses stillage. Wang [73] used aeration rates of approximately 1 vvh in large-scale experiments. Mostly aeration values of 0.5-1.0 vvm are employed [8,11,13,20,63,74,76,83]. Exceptions are very high aeration rates in small bioreactors, such as 2.5 vvm [18,82] or even 6 vvm in a column fermentor [57] and agitation rates range from 200 rpm [75] to 1000 rpm [8,20].

In continuous cultivation, dilution rates from 0.025 h^{-1} [57] to 0.383 h^{-1} [73] were tested. At very low dilution rates contamination could occur [22]. With different dilution rates, changes in morphology occurred in the case of *G. candidum* [57]. In two-stage continuous processes, different dilution rates were applied in each stage to optimize the productivity in the first stage and the substrate utilization in the second one [18,22,82]. Matsuo et al. [7] observed that the mean retention time was markedly affected by aeration. The dilution rate was much increased by the use of a cell-recycling technique [73].

The fermentation time is mostly dependent on the microorganism used. When yeasts are applied, the process lasts from several hours [69] to 42 h [8]. In experiments carried out with molds several days are needed to achieve maximal bioconversion, from 72 h [13] to 8 days [15] and with mushrooms 8-12 days [55].

After fermentation the yeast biomass is generally separated from the broth by centrifugation, whereas with filamentous fungi the biomass can usually be separated by filtration or screening. In fruit slops, the swollen slimy substances generally are degraded or physically bound to the fungal mycelium, thus enhancing filtration [29,30,33,62,93,95,96].

III. MICROORGANISMS

The selection of the proper organisms to be used is of the utmost importance. The organism has to be able to utilize the substrate effectively and at the same time its biomass has to have a high nutritive value, especially in its protein content. Bacteria grow more quickly and can utilize a wider range of substrates than fungi, but their small size makes separation more costly and their nutritive value has not yet been fully determined. For growing algae large areas of land as well as high intensities of light are required. Both lower and higher fungi are the most suitable microorganisms [47].

A. Yeasts

Yeasts have been chosen for cultivation in distillery stillages of different origin in the great majority of cases. They have commonly been used in food preparation by mankind for thousands of years so there are no prejudices regarding their use for food or feed as in the case of bacteria or molds. The quick growth of yeasts provides the advantages of time and low cost, as well as limited possibility of contamination by other microorganisms. The fermentation process is easy to handle and continuous cultivation is feasible. The use of yeast is interesting because of its low price and its high protein and vitamin content. Its quality is more important than its quantity. Considering its phosphoproteins, amino acids, and vitamin complexes, yeast can be compared to proteins of animal origin [49]. However, some facts must be taken into account [20]:

1. The yeast should not contain high levels of contaminants if it is to be used for feed.
2. Energy is needed for separation of the yeast from the substrate.
3. Effective aeration is necessary.
4. The remaining effluent still has a high COD and needs to be treated in a sewage treatment facility.

Candida (*Torula*) has been mainly used among yeasts. Cane and beet molasses distillery effluents and other kind of stillages have been treated to a minor degree. SCP production and organic substance reduction in the effluent are the aims of most bioconversion processes. However, yeast has proven to have a positive effect even on the anaerobic digestion of stillages for biogas production [60,61,80].

For cane molasses stillage the *Candida* species *C. utilis*, *C. arborea*, and *C. tropicalis* are most successful for cell mass production and sugar consumption [4]. *C. utilis* is a better protein producer than *C. arborea*, so that the former was chosen for further experiments in pilot scale investigations in a 600-liter open tank [7]. Akaki [87] tested 203 yeast strains and selected *Hansenula*, *Debaromyces*, and *Rhodotorula* for the conversion of cane molasses stillage. Another 224 strains were isolated from soil and sewage. Two were selected and identified as *Candida ciferrii* and *C. brumptii* [78]. The latter was used in experiments for biomass production in cane stillage supernatant. It was concluded that dilution was not necessary but the addition of N, P, and K sources had a positive effect on growth [10].

In addition to *Candida*, *Torulopsis* was also convenient for sugarcane stillage conversion [85,88]. Biomass and protein production can be improved by previous treatment of stillage with HNO_3 and the addition of molasses [84]. Yeasts can be adapted to stillage by growth in molasses media with increasing stillage additions [85]. The following *Candida* species were used: *C. utilis*, *C. intermedia*, *C. tropicalis*, *C. membranaefaciens*, *C. guilliermondii*, *C. vartiovaarai*, and *C. javanica*. The effect of $MgSO_4$ addition to the medium was studied by Tauk and Gambale [89]. It depended on the yeast strain and salt concentration used. Further studies showed that *C. utilis* gave the best results when stillage was supplemented with urea and H_3PO_4 [90].

A new yeast strain, *Pichia tainania*, was isolated from soil and utilized for fodder yeast production from cane molasses stillage. In laboratory experiments very good results were obtained [8], but on an industrial scale it was not as successful as *C. utilis* [9]. When growing *C. utilis* in cane stillage the addition of molasses enhanced the utilization of organic matter with low molar mass [83]. *C. utilis* var. *major* was cultivated in small- and large-scale experiments in airlift fermentors. The biomass yield was improved by addition of nitrogen sources [16]. Several microorganisms were applied in pure and mixed cultures for the conversion of stillage. Among yeasts, *C. utilis* showed better results than *Saccharomycopsis lipolytica* [18]. In wine vinasses, *C. tropicalis* grew well [70]. The same yeast was cultivated on spent granular potato mash from alcohol fermentation [69]. There was no lag phase and maximal growth was observed after 6 h. Grain distillery stillages from wheat and barley were fermented by means of *C. tropicalis* [67]. The yeast consumed pentose sugars in the stillage and thus the production of biomass was enhanced when the waste was first treated with xylanase to hydrolyze pentosans present in the substrate.

Several yeast strains were tested in wine vinasse which was supplemented with ammonium salts. *C. utilis* var. *major* was studied in detail. During the exponential growth phase the following substrate components were metabolized successively: acetic acid, lactic acid, and malic acid. Glycerol was progressively metabolized during the whole growth phase. Tartaric acid was not metabolized at all and the same was supposed for phenolic substances. COD of vinasse was slightly diminished after the cultivation of *C. utilis*. As tartaric acid was not metabolized by *C. utilis*, *Rhodotorula glutinis*, which is capable of using it, was chosen. This yeast had other advantages: it was rich in lipids and formed floccules convenient for separation [54]. *Candida steatolytica* was able to convert stillage from bagasse hemicellulose hydrolyzate containing L-arabinose and xylitol. Besides biomass production, BOD reduction was also achieved [81].

Torula yeasts were grown in cane molasses stillage. The addition of cane molasses stillage to the molasses medium reduced the cost of *Torula* yeast production [79]. *Torula utilis* and *R. glutinis* were cultivated in vinasse from raisins to give biomass for fodder rich in vitamin B [25,63]. *K. marxianus* isolated from beet molasses stillage could effectively utilize this substrate in continuous cultivation. The addition of nutrient salts was not necessary [19]. Besides *C. utilis*, *Torulopsis pinus* and *Trichosporon cutaneum* have been proposed for fodder yeast production from stillage. The composition and qualitative characteristics of the yeasts were emphasized [66]. *Endomycopsis fibuliger*, *C. utilis*, and *R. glutinis* were grown together in a stillage of unknown origin [47]. The use of *Candida* sp. in the coculture for biogas production from distillery spent wash caused reduction of the fermentation time [80]. Aerobic pretreatment of molasses stillage by *Trichosporon fermentans* for 4-6 h enabled high COD reduction during the next anaerobic step [61].

B. Filamentous Fungi

1. Ascomycetes

Compared to yeasts, far fewer reports have been published about the bioconversion of distillery wastes by filamentous fungi. The reasons might be that they are slower growing and are more susceptible to infection. However, fungi have some advantages as they produce a series of extracellular hydrolytic enzymes and therefore they are able to exploit complex carbohydrates without prior hydrolysis. This property enables them to grow in starch and cellulose substrates. The second advantage is in their filamentous morphology, which makes separation of the biomass very easy by simple filtration. As compared to yeasts, filamentous fungi have a lower sensibility to variations in temperature, pH, nutrients, and aeration and they also have a lower nucleic acids content in the biomass [12].

G. candidum, having morphological characteristics of a typical yeast but also producing distinct mycelia [91], was selected for malt whisky stillage conversion. A variety of microorganisms isolated from soil and activated sludge were tested and *G. candidum* was predominant [22]. The undiluted supernatant after settling of stillage was used as the substrate. In batch culture, a diauxic growth curve was observed. This could be a consequence of the inhibition of polyol catabolism by glucose as the complete utilization of glucose approximately coincided with the onset of glycerol utilization. Lactic acid, short-chain polysaccharides, and glucose were consumed during the first growth phase. During the second phase protein and glycerol seemed to be used as carbon sources [57]. Morphological variations were observed with dilution rate changes in continuous culture of *G. candidum*. Spores were always present but their relative amounts in respect to filamentous forms varied considerably [57]. Broad substrate specificity, high yield coefficients, and tolerance to pH values as low as 2.6 were special qualities of *G. candidum*. It is thus a very promising microorganism for conversion of liquid food wastes [58].

Gliocladium deliquescens, *Paecilomyces elegans*, *Myrothecium verrucaria*, *Trichoderma viride*, and *Penicillium* spp. were tested in rum distillery stillage. *G. deliquescens* was selected for further experiments of continuous cultivation [11]. Filamentous fungi were tested for biomass production in cane molasses stillage in Brazil. *Aspergillus oryzae*, *Trichoderma viride*, *Paecilomyces elegans*, *Gliocladium deliquescens*, and *Aspergillus niger* were selected and *A. oryzae* was chosen for biomass production. Feeding experiments for the evaluation of the fungal biomass were made on chicks, which showed weight gain, no toxicity, and excellent animal acceptability [12].

De Lamo and De Menezes [13] studied bioconversion of cane stillage by different filamentous fungi and compared them to *C. utilis* and the natural microflora of the stillage. Best results were obtained with *A. niger* and *A. oryzae*, which exhibited quick filtration, high COD reduction, highest biomass production, and best yields of biomass on COD. *G. deliquescens* and *T. viride* could not compete with both aspergilli. In Brazil, soil was treated with stillage and the amount of soil fungi was much increased. Several fungi were isolated and identified, and some were used for growth in stillage. Four were used in fermentation experiments: *Acremonium fusidioides*, *Penicillium notatum*, *Aspergillus flavus*, and *A. niger*. *Acremonium fusidioides* clarified and decolorized the stillage medium and also produced the highest biomass. Both aspergilli were less effective while *P. notatum* was not successful [17]. Rum distillery stillage was treated with molds to establish its potential use as an animal feed supplement [92]. Filamentous fungi were the preferred microorganisms because of their texture and protein profile as well as their easy and cheap harvesting. Some fungi were isolated from contaminated stillage and identified. Growth curves of 11 strains of *A. niger*, *A. phoenicis*, *A. flavus*, *Penicillium jensenii*, and *Penicillium crustosum* were determined. *A. phoenicis* strains were most promising [15]. *Paecilomyces varioti* showed poor results when grown in unsterilized sugarcane stillage [16]. When the sterilized substrate was supplemented with molasses and salts, the biomass production was much improved. Recently, Nudel et al. [18] reported an aerobic treatment of cane stillage by using several yeasts, molds, and bacteria. *P. varioti, A. niger, A. oryzae, T. viride,* and *Rhizopus* sp. were used as filamentous fungal organisms. *T. viride* and *P. varioti* proved to be the most effective.

Vinasse from Turkish raisin brandy manufacture was treated with yeasts and molds [25,63]. The following molds were tested: *A. niger, Fusarium moniliforme,* and *Penicillium oxalicum*. The latter produced the highest amount of dry mass with a high protein content. *F. moniliforme,* however, had the highest lipid production. Regarding the amino acid pattern of the biomass produced in vinasse, *A. niger* was comparable to FAO standards. French investigators have patented a process for white wine vinasse utilization by means of different *Penicillium* strains. These fungi were chosen because of their good adaptation and utilization of the medium, and satisfactory development at a low pH. A detailed description of the isolated *Penicillium* strains is given in the patent by Magny et al. [56].

A number of different ascomycetous and basidiomycetous fungi were tested on fruit distillery slops [29,93]. From the perspective of rapid growth, consumption of dissolved organic substances, improvement of filterability, and rate and extent of COD reduction, *Aspergillus* spp. and *Myrothecium* sp. perform best. From the point of view of fast processing they would be organisms of choice as well [29,33,62,94-97]. Among them the more cellulolytic strains gave better results [97]. Regarding fiber degradation and protein content of filter cake as well as of cellulolytic enzymes in the filtrate, *Trichoderma* spp. and *Phanerochaete* sp. performed better [29]. An attempt to improve the performance of an *A. niger* strain was made in a mutation series [98-100]. A 50- to 90-fold increase in cellulolytic activities and a 3- to 10-fold increase in amyloglucosidase, pectolytic, and aryl-β-glucosidase activity was achieved after four alternating mutation steps, but the greatly improved enzymatic activities did not improve the technological characteristics such as the decrease of filtration time or of residual COD [99]. A review on fungal treatment of these slops has been published [101].

2. Basidiomycetes

The development of basidiomycetous fungi (mushrooms) in submerged culture is interesting because of their probable suitability as cheap substitutes for mushroom fruiting bodies in certain kinds of food [55]. The suitability of stillage as a substrate for the growth of mushroom mycelia as a source of protein and fat was investigated by Falanghe [55]. Ten strains of *Morchella, Boletus, Xilaria, Agaricus, Tricholoma, Cantharellus,* and *Collybia* were cultivated in submerged conditions. *Agaricus campestris* reached maximal growth after 12 days of fermentation. The protein content of the biomass was high, the fat content was low. *Boletus indecisus* and *Tricholoma nudum* showed higher mycelial weights than *Agaricus campestris,* but lower protein contents. *A. campestris* could adapt to the stillage, which markedly improved the results. *A. campestris* and *B. indecisus* were chosen as the two most suitable fungi. *A. campestris* was more effective concerning mycelial protein contents. *B. indecisus,* however, exhibited a greater ability for mycelial production. The dried mycelia had an agreeable slight flavor and they seemed to have adequate characteristics for improving foods. Their growth in pellet form enabled easy separation from the medium.

C. Mixed Cultures

Some research was done on the use of mixed cultures of different microorganisms to improve the utilization of nutrients in stillage and to obtain better yields of biomass. Combinations of different yeasts [20], yeasts and bacteria [18], yeasts and molds [18,59], or different molds [16,30,82] were applied. An association of 15 yeast strains was grown in beet molasses stillage. The biomass production was near 0.5 kg/kg of converted substrate. This procedure seemed to be an interesting alternative to anaerobic digestion [20]. Inmixed cultures of two different yeasts and a mold [18], *C. utilis* biomass increased rapidly during the early stages of growth and it became the dominant organism at the end of the process. This was presumably due to the higher growth rate of *Candida* or to an antagonism between the organisms. More successful were the experiments using mixed cultures of *C. utilis* with *Corynebacterium acetoacidophilum* and of *C. utilis* with *Brevibacterium flavum*, which were performed to increase the vitamin and amino acid content of the biomass. The biomass yield of the mixed cultures was substantially greater than that of pure cultures. *C. utilis* with *Azotobacter* proved to be a successful combination. This bacterium was chosen to circumvent nitrogen supplementation of the stillage. The increased biomass yield was attributed to nitrogen fixation by *Azotobacter*. Most interesting, however, proved to be a two-step continuous fermentation: *C. utilis* was used during the first step to produce high amounts of biomass and *A. niger* during the second step to markedly reduce the COD of the effluent [18]. Similarly, a two-step aerobic process was employed with another combination of microorganisms [82]. Microbial biomass was first produced by *C. utilis*; *P. varioti* finished the bioconversion by its consumption or organic substances from the supernatant of the first step.

It was found that in addition to *G. candidum* two yeasts could normally be isolated from malt whisky distillery spent wash [59]. They were identified as *Hansenula anomala* and *Candida krusei*. The persistence of all three organisms in the enrichment culture suggested that the use of a three-membered mixed culture might be promising. The microorganisms when grown in pure culture utilized the same substrate constituents. In mixed culture different constituents were used by each fungus, permitting a stable population. The relative proportions of the three organisms were dependent on the dilution rate during continuous cultivation. *C. krusei* used protein as both carbon and nitrogen source. It was the dominant organism in the early stages of batch growth and also at high dilution rates in continuous cultivation. *H. anomala* and *G. candidum* appeared to be utilizing carbohydrates, glycerol, and lactate. *H. anomala* was dominant at low dilution rates.

Mixed cultures of two ascomycetous fungi, *Aspergillus awamori* and *Trichoderma reesei*, were tested in apple slops. They expressed the positive properties of both genera: *Aspergillus* improved the filtration time and COD reduction whereas *Trichoderma* decreased fiber and increased protein content in the biomass [30].

IV. RESULTS

A. Products and Effects

Bioconversion of distillery waste by means of fungi brings a double benefit: a fungal biomass rich in protein is created and a substantially purified effluent. In most experiments *C. utilis* was used as the chosen microorganism. As shown in Table 2, cane molasses stillage was applied in nearly all cases. Biomass concentrations over 10 g/liter and even up to 25 g/liter could be obtained. Crude protein contents accounted for about 28-57.6% of the cell mass. The effluent remaining after the process contained 40-52% of the initial COD and up to 50% of the initial BOD.

Among other yeasts shown in Table 3, *Pichia tainania* grown in supplemented molasses stillage produced more than 20 g/liter of dry yeast. *C. brumptii* biomass contained a high protein amount of 52.6%. COD and BOD reduction, however, were not as high as with *C. utilis*. *Kluyveromyces marxianus* seemed to be a very efficient microorganism considering the COD reduction of 60-70%, the highest among all yeasts tested including *C. utilis*. Biomass concentrations amounted to 10 g/liter, which is relatively low in comparison with *C. utilis*. The quantity of cells produced varied significantly with the concentration of the substrate as well as the addition of N sources. Cabib [16] claimed that with an increasing nitrogen supplementation a concentration of biomass of up to 25 g/liter could be obtained with *C. utilis* in cane molasses stillage. The high quantity of *C. utilis* biomass obtained by Wang et al. [73] was a result of cell recycling in the continuous fermentation, but it should be noted that all cell material could not be harvested since a part of it had to be used for recycling. Similarly, one should keep in mind that the results of COD and BOD reduction present only relative values and that the highest percentage of reduction is obtained in effluents in which initial organic substances are low. The absolute value of pollution reduction is usually much higher in more concentrated effluents, whereas the percentage is lower because of a very high initial value. Thus the treatment of the stillage in terms of COD reduction was most effective with *K. marxianus* growing continuously in beet molasses stillage [19]. The initial value of COD was 45-50 g/liter, and thus a 60-70% reduction means 27-35 g/liter if no dilution or sedimentation was carried out prior to fermentation. However, *C. utilis* grown in cane molasses stillage lowered the COD to only 46% [82]. This means in absolute numbers a decrease of COD from 77 g/liter to 35.4 g/liter, i.e., by more than 40 g/liter. The higher relative reductions of biological or chemical oxygen demand in more diluted effluents occurred in a certain range of pollution load. In stillages of very low initial amounts of reducing organic substances, the COD reduction was small as some nutrients could be lacking [20].

As can be concluded from the available references, filamentous fungi were cultivated mostly in batch processes (Table 4). When compared with yeasts, filamentous fungi were more effective in consumption of polluting substances, as COD and BOD reductions of over 80 and 90% have been recorded. *A. oryzae* and *Gliocladium deliquescens* in cane stillage, *Penicillium* strains in wine vinasse, and *G. candidum* in malt whisky stillage proved to be the most promising in this respect. In a five times diluted effluent the amount of *G. candidum* biomass was 3.5 g/liter. In *P. varioti*, *T. viride*, and *G. candidum* in undiluted waste, the biomass attained to 25 g/liter. The latter grew better in a continuous two-step fermentation process reaching as much as 34 g/liter cell mass with both steps included [22]. It was observed that on average the protein content in the biomass of filamentous fungi was lower than that of yeasts. *G. candidum*, however, was claimed to produce the highest protein contents, being 45.5% of "true" protein in the biomass as determined by the biuret method [22]. Using

TABLE 2. Results of Bioconversion of Stillage of *Candida utilis* Strains

Source of waste	Supplements	Biomass, g/liter	Protein, %	COD red., %	BOD red., %	Cultiv. mode	Ref.
Molasses (cane)	$(NH_4)_2SO_4$	14	35	21	53	Batch	6
Supernat.	Same	6-9	45-55	20-25	51-51.6	"	5,6
"	Urea	8		16	55	"	5
"	$(NH_4)_2SO_4$	10-12	40-46			Cont.	7
"	"	4-7	44-50	30-40		"	51
"	None	10	42			Batch	8
"	N, P	13-16	46-58			"	8
"	"	18-24	42-49			Cont.	9
"	N, P	22	52		37	Cont. c.r.[b]	73
"	$(NH_4)_2SO_4$ + $(NH_4)_2HPO_4$	14-25	41	40-47		Batch	16
"	"	12	41			Cont.	16
"	"	15	41			Semicont.	16
"	$(NH_4)_2SO_4$ + K_2HPO_4	15-22		46-52		Batch	18
"	None	9-13	40-50			"	18
"	None	18		46		"	82
"	None	13	28			"	85
"	Molasses 75%	20	30			"	85
"	Urea + H_3PO_4	7	52			"	90
Molasses + sweet potatoes	$(NH_4)_2SO_4$	6.5	40	19[a]		"	53
Wine	$(NH_4)_2HPO_4$	6.6		48.5		Cont.	54

[a]Total organic matter.
[b]c.r., cell recycling.

TABLE 3. Results of Bioconversion of Stillages by Various Yeasts

Yeast	Stillage origin	Biomass, g/liter	Protein, %	COD red., %	BOD red., %	Cultiv. mode	Ref.
C. arborea	CMS	7-13	33-43	12-22	44-60	Batch	5
P. tainania	CMS	14	33.5			"	8
"	CMS suppl.	19-20	42-49			"	8
Hansenula + Debaromyces + Rhodotorula	CMS	9		>30		"	87
Candida brumptii	CMS suppl.	10	53	20	30	"	10
C. membranaefaciens	"	3.6	45			"	89
C. krusei	"	3-8	31-48			"	89,90
C. stellatoidea	"	3.5	48			"	89
Kluyveromyces marxianus	BMS	10-11		60-70		Cont.	19
Candida tropicalis	WV		48				70
C. steatolytica	HHS	12					81

Abbreviations: CMS, cane molasses stillage; suppl., supplemented; BMS, beet molasses stillage; WV, wine vinasse; HHS, hemicellulose hydrolyzate stillage.

TABLE 4. Results of Bioconversion of Stillages by Filamentous Fungi

Fungus	Stillage origin	Biomass, g/liter	Protein, %	COD red., %	BOD red., %	Ref.
Aspergillus oryzae	CMS	14-17	35-50	61-88	79-83	12
"	"	12-15	39	48-72	78	13
Aspergillus niger	"	8-13	30-40	46-78		13,17
Aspergillus phoenicis	" (rum)	20	21		58	15
Paecilomyces varioti	"	13-25	40	43-70		16,18,82
Paecilomyces elegans	" (rum)	11-14		62-66		11
Trichoderma viride	"	12-28		59-79		11,18
Acremonium fusidioides	"	11	40			17
Myrothecium verrucaria	" (rum)	11-12		79-82		11
Gliocladium deliquescens	" (rum)	11-19		60-85		11
Agaricus campestris	"	13	45			55
Penicillium sp.	" (rum)	12-16		65		11
Penicillium spp.	WV	13	41		91	56
Penicillium oxalicum	Raisins	12	34			25,63
Geotrichum candidum	Whisky[a]	3.5		80.6	92	22
"	Whisky	11-27		32-78	63-91	22
"	Whisky[b]	34	46		87	22
Aspergillus niger	Fruit	4-20	12-36	50-70		29,33,62 93,95-97

[a]Diluted five times.
[b]Continuous, two-stage.
Abbreviations: CMS cane molasses stillage; WV, wine vinasse.

cellulolytic fungi for bioconversion of fruit slops from the filtrate hydrolytic enzymes such as cellulases, glucosidases, pectinases, etc., could be recovered as an additional product [29,30,99,100].

Results obtained with mixed cultures of different microorganisms are shown in Table 5. Coculture of *G. candidum*, *C. krusei*, and *H. anomala* in malt stillage showed the lowest productivity values when compared to the pure cultures of each individual strain. Substrate utilization regarding protein and COD were most effective in mixed cultures. The continuous cultivation process led to a higher protein content in the biomass and enhanced the stability of the culture as well as its resistance to contamination [59]. A coculture of 15 yeast strains in a high-loaded beet molasses stillage of 74.5 g/liter COD, produced the highest amount of biomass, i.e., 28.9 g/liter when grown in batch culture. When diluted medium was applied, the biomass yield was much lower [20]. A successful approach seemed to be that of Bottaro Castilla et al. [82]. A two-step separated cultivation of *C. utilis* and *P. varioti* in batch and continuous processes led to very promising results. In batch experiments a total amount of 22 g/liter cell material and a COD reduction of as much as 92% were obtained. During the first step, *C. utilis* was used mainly for SCP production, whereas *P. varioti* in the second step consumed the reducing organic substances in the liquid phase.

In batch experiments, simultaneous growth of *P. varioti* and *T. viride* in cane molasses stillage showed no perceivable improvement over the pure cultures (Tables 4 and 5). More promising appeared to be the combinations of *C. utilis* with *Corynebacterium acetoacidophilum* or *Brevibacterium flavum*. The yield of biomass as well as the COD reduction substantially increased, particularly in regard to the pure bacterial cultures. An interesting approach was the cultivation of *Azotobacter* strains capable of fixing gaseous nitrogen. After 72 h of growth, *C. utilis* was inoculated and it was shown that the cell concentration increased from 10 g/liter in the control to 16 g/liter in the test culture [18].

A two-step continuous cultivation of *C. utilis* and *A. niger* growing separately [18] proved to be as promising as the process using *C. utilis* and *P. varioti* [82]. Summarizing the results of conversion of both steps, 17g/liter biomass and 89% COD reduction were obtained.

The results of continuous cultivation of fungi in stillages (Table 6) showed that dilution rates of over 0.36 h^{-1} resulted in a high productivity of over 4 g/liter/h and a steady-state biomass concentration of 18-22 g/liter. It is obvious that such an outstanding productivity must be attributed to cell recycling. However, *C. utilis* showed higher productivity than other fungi also in the absence of recycling. *K. marxianus* was also able to lower the COD substantially. This was attributed to the choice of a proper yeast strain [19]. Subsequent culture of a yeast and a filamentous fungus in two steps gave most promising results regarding high biomass and most efficient COD reduction. Two-step cultivation of *G. candidum* also seemed to be very suitable, with the highest steady-state biomass being 31-34 g/liter. Mixed cultures of the same fungus with two yeasts were not as effective. *G. deliquescens* cultivated continuously showed good steady-state biomass concentration and relatively high COD reduction, but only at the lowest dilution rate of all (Table 6).

TABLE 5. Results of Bioconversion of Stillages Using Mixed Cultures

Microorganisms	Stillage origin	Biomass, g/liter	COD red., %	Cultiv. mode	Ref.
Paecilomyces varioti *Trichoderma viride*	CMS	12-17	50-64	Batch	18
Trichoderma viride *Aspergillus oryzae*	"	10-13	50-54	Batch	18
Candida utilis *Paecilomyces varioti*	"	8-12	45-50	Batch	18
Candida utilis *Corynebacterium acetoacidophilum*	"	16-17	65	Batch	18
Candida utilis *Brevibacterium flavum*	"	16	65	Batch	18
Azotobacter *C. utilis*	"	16		Batch	18

C. utilis *A. niger*	"	16-17	89	Cont. serial	18
C. utilis *Paecilomyces varioti*	"	22	92	Batch	82
C. utilis *P. varioti*	"	14-16	85	Cont. serial	82
15 yeasts	BMS (dil)	7	79	Batch	20
"	BMS (dil)	9	72	Cont.	20
"	BMS	29	74	Batch	20
Geotrichum candidum *Candida krusei* *Hansenula anomala*	MWS	13	55	Batch	59
"	"	5	32	Cont.	59
A. awamori *T. reesei*	Apple (dil)	5	31	Batch	30

Abbreviations: CMS, cane molasses stillage; BMS, beet molasses stillage; (dil), diluted; MWS, malt whisky stillage.

TABLE 6. Results of Continuous Cultivation of Fungi in Stillages

Fungus	Dilution rate, h^{-1}	Productivity, g/liter/h	Biomass, g/liter	COD red., %	BOD red., %	Ref.
C. utilis	0.383	4.24	17.9		34.6	73
"	0.365	4.06	21.9		37.1	73
"	0.22	2.64	12			16
"	0.27	3.24	12			16
"	0.2	1.8-2.0	9-10	35		18,82
C. utilis	0.2					
P. varioti[a]	0.1	2.3-2.6	19-21	85		82
C. utilis	0.2					
A. niger[a]	0.1	2.7-3.2	15-17	89		18
K. marxianus	0.3	3	10-11	60-70		19
G. candidum	0.125	2.24		30	50	22
" I. step	0.125					
II. step	0.10		31		50	22
" I. step	0.125					
II. step	0.085	1.72	34		87	22
G. candidum						
C. krusei						
H. anomala	0.10	0.48	4.8	31.5		59
"	0.20	0.36		15.9		59
"	0.35	0.42		2.0		59
G. deliquescens	0.03		17-22	61-50		11
Diff. yeast association	0.134		11	68		20

[a]Two step process.

B. Biomass Evaluation

Microbial biomass rich in protein is the main product of fungal bioconversion of stillages. Some information is available about cell material composition. It appears that *C. utilis* biomass composition is similar to that of other yeasts grown in carbohydrate media. Microbial protein has a good balance of amino acids with the exception of those containing sulfur, such as cysteine and methionine, as is normal for microbial biomass [16,25,58], but lysine is in excess when compared to the recommended level of the FAO standards [58]. Yeast biomass from vinasse is rich in nitrogen, vitamins, and other biologically active substances. The vitamin B complex of yeast is very efficient and could not be adequately substituted by a mixture of analytically pure vitamins [16,25,49]. Nutritional evaluation of yeast biomass grown in malt whisky distillery slop showed that it was suitable for nonruminants with a net protein utilization value of 0.40 and a digestibility of 0.67 [59].

Biomass from distillery stillages is not toxic as confirmed by toxicological tests [59,95]. When molasses stillages were used, K and Mg contents could have laxative effects [12]. Copper accumulation in the biomass of *G. candidum* represented a considerable problem: metal ions were leached from the distillery fittings during distillation when the pH was low [59]. The nucleic acid amounts in cell material should be as low as possible. Filamentous fungi have an advantage over yeasts in this respect [12]. It has been observed that levels of DNA and RNA were lower in batch than in continuous cultures [58].

Some feeding experiments have been carried out. When the *Aspergillus* biomass was used in a diet for chicks, excellent acceptability, good weight gain, and no toxicity could be observed [12]. A 1% methionine addition to *Candida* yeast biomass increased the biological value, which was not much lower than that of casein [16]. *A. oryzae* biomass had a very good protein efficiency ratio being comparable to meat and soya meal [12].

C. Economic Considerations

Little information is available concerning the economics of the bioconversion process. Only rough estimates of costs were given by Maiorella et al. [1] and Tomlinson [86]. As no recent data are available, we cannot give any serious estimates.

Like all other processes, bioconversion of stillage involves advantages and drawbacks. The main costs are high capital investment [14,86], energy needed for aeration, agitation, cooling [1], and biomass separation [86], nutrient supplementation, sterility problems, limited production, and irregularities in raw material composition. However, capital cost may not be too high as was shown in the cost evaluation for molasses stillage treatment methods where aerobic yeast growth was estimated to require the lowest capital investment [1]. Separation of cell materials from the broth would be easy with the application of filamentous fungi. The addition of nutrients is not always required. Sterility problems could be managed or at least reduced by maintaining a low pH [7,16,20], the use of continuous operations [49], and the use of mixed cultures [59]. The problem with large-scale plants is the need for a continuous and relatively homogeneous supply of stillage [86]. A self-sufficient factory would not be feasible under average conditions. It should be connected to an already existing plant which could provide steam, electricity, and water. Special precautions would be needed to decrease irregularities of raw materials [49].

Cost evaluation studies show that problems are being overestimated and are not insurmountable. Cost comparisons with those for conventional biological treatment have shown that potential annual savings could be increased by applying fungal bioconversion of stillages [86]. Similarly, rough estimates of capital and operating costs for molasses stillage treatment by fungi show that savings could be gained even though wastewater would still have to be purified. Methods compared were evaporation to obtain cattle feed, incineration to recover potash, anaerobic digestion to methane, and aerobic *Candida* yeast growth. The last was considered to be the most suitable for molasses waste. The same method was proposed for sulfite waste liquor and wood hydrolysis stillages [1].

The economy of stillage bioconversion with fungi is dependent on many different factors and can vary from country to country. It may not be attractive to those in economic surplus but may be promising in countries with serious malnutrition problems. We conclude that fungi have a great potential in helping mankind to solve deficient nutrition and pollution. Why not continue to use fungi advantageously?

V. SUMMARY

A review of current research dealing with the use of fungi for bioconversion of distillery waste into microbial biomass was presented. The origin of stillage during alcohol production was briefly described including its characteristics and disposal. Based on different raw materials, stillages vary in composition. Their common characteristics of harm to the environment are high contents of organic substances and low pH; however, these properties facilitate their use as substrates for fungal growth. Research work started about 40 years ago, reached its culmination in the 1970s, and started to decline in the 1980s. Beside European countries, the Far East and South America are working on these problems. Sugarcane molasses stillages have been used for the most part. Among fungi, yeasts, especially *Candida utilis*, predominate. To a minor extent filamentous fungi, including Basidiomycetes, as well as mixed fungal or mixed fungal and bacterial cultures have been used. Laboratory, pilot scale, and industrial scale experiments have been performed, and batch or continuous, single- or two-step fermentations introduced. Results of bioconversion are fungal biomass and substantially purified effluents. A well-balanced amino acid composition makes the biomass suitable as an animal feed supplement.

REFERENCES

1. Maiorella, B. L., Blanch, H. W., and Wilke, C. R. Distillery effluent treatment and by-product recovery. *Process Biochem.* (*Aug. 1983*):5-8, 12.
2. Mourgues, J., and Maugenet, J. L'épuration des eaux résiduaires des distilleries vinicoles. *Tribune CEBEDEAU*, *23*:76-86, 1970.
3. Bronn, W. K. Die Abwasserbelastung bei Hefefabriken und Melassebrennereien. *Branntweinwirtschaft*, *115*:388-390, 395-396, 1975.

4. Matsuo, T., Ishikawa, F., Yamanaka, M., and Konishi, K. Yeast production from alcohol distillation slops of cane molasses. 1. *Hakko Kyokaishi, 23*:320-324, 1965.
5. Matsuo, T., Ishikawa, F., Yamanaka, M., Nishioka, S., and Konishi, K. Yeast production from alcohol distillation slops of cane molasses. 2. *Hakko Kyokaishi, 23*:472-478, 1965.
6. Matsuo, T., Ishikawa, F., Yamanaka, M., Nishioka, S., and Konishi, K. Yeast production from alcohol distillation slops of cane molasses. 3. *Hakko Kyokaishi, 24*:67-70, 1966.
7. Matsuo, T., Ishikawa, F., Yamanaka, M., Nishioka, S., Karaki, I., and Konishi, K. Yeast production from alcohol distillation slops of cane molasses. 4. *Hakko Kyokaishi, 24*:223-231, 1966.
8. Lin, S. F., Tsou, C. S., and Chen, W. P. Manufacture of fodder yeast from molasses alcohol slop. *Taiwan Sugar* (*Sept./Oct. 1973*):185-188.
9. Chang, C. T., and Yang, W. L. Study on feed yeast production from molasses distillery stillage. *Taiwan Sugar* (*Sept./Oct. 1973*):200-205.
10. Kim, C. J., Oh, M. J., and Park, S. U. Studies on the production of the single cell protein from alcohol distillation slops of cane molasses. II. On the conditions for cell mass production by Candida brumptii. *Nongop Kisul Yongu Pogo, 3*:244-250, 1976.
11. Rolz, C., De Cabrera, S., Espinosa, R., Maldonado, O., and Menchu, J. F. The growth of filamentous fungi on rum distilling slops. *Ann. Technol. Agr., 24*:445-451, 1975.
12. Araújo, N. Q., Visconti, A. S., De Castro, H. F., Barroso da Silva, H. G., Ferraz, M. H. A., and Salles Filho, M. Produção de biomassa fungica do vinhoto. *Inf. INT., 10*:12-19, 1977.
13. De Lamo, P. R., and De Menezes, T. J. B. Bioconversão da vinhaça para produção de biomassa fúngica. *Coletânea Instituto Tecnologia Alimentos, 9*:281-312, 1978.
14. Bhandari, H. C., Mitra, A. K., and Malik, V. K. Treatment of distillery effluent. *Proc. 43rd Annu. Conf. Sugar Technol. Assoc. India* G65-G72, 1979.
15. De González, I. M., and De Murphy, N. F. Potential use of rum distillery slops as animal feed supplement. II. Nitrogen content of mycelial growth in slops. *J. Agr. Univ. Puerto Rico, 63*:330-335, 1979.
16. Cabib, G., Silva, H. J., Giulietti, A., and Ertola, R. The use of sugar cane stillage for single cell protein production. *J. Chem. Tech. Biotechnol., 33B*:21-28, 1983.
17. Rosalem, P. L., Tauk, S. M., and Nuñes dos Santos, M. C. Efeito da temperatura, pH, tempo de cultivo e nutrientes no crescimento de fungos imperfeitos em vinhaça. *Rev. Microbiol., Sao Paulo 16*:299-304, 1985.
18. Nudel, B. C., Waehner, R. S., Fraile, E. R., and Giulietti, A. M. The use of single and mixed cultures for aerobic treatment of cane sugar stillage and SCP production. *Biological Wastes, 22*:67-73, 1987.
19. Braun, R., and Meyrath, J. Verwertung von Melassebrennereischlempe. *Branntweinwirtschaft, 121*:102-104, 1981.
20. Huyard, A., Malnou, D., and Faup, G. M. Epuration des eaux résiduaires d'industries agro-alimentaires par utilisation de levures. *Eau Ind., 98*:29-32, 1986.
21. Sheehan, G. J., and Greenfield, P. F. Utilization, treatment and disposal of distillery wastewater. *Water Res., 14*:257-277, 1980.

22. Quinn, J. P., and Marchant, R. The treatment of malt whiskey distillery waste using the fungus Geotrichum candidum. *Water Res.*, *14*:545-551, 1980.
23. Denne, A. Die Abwässer von Obstbrennereien. *Wasser Luft Betrieb* (*1964*):607-612.
24. Weller, G. Abwasserbeseitigung in Brennereien und Winzerbetrieben. *Münch. Beitr. Abwasser-Fisch-Flussbiol.*, *31*:107-137, 1979.
25. Yazicioğlu, T., Çelikkol, E., Öcal, Ş., Aran, N., and Ömeroğlu, S. Some trials on the utilization of whey, black water of olive and vinasse for production of SCP in Turkey. *TÜBITAK*, Marmara Scientific and Industrial Research Institut, Publication No. 42:1-23, 1980.
26. Perdih, A., Cimerman, A., Škafar, S., and Kafol, P. Odpadki slovenske zivilske industrije in možnosti za njihovo uporabo. *Nova proizvodnja* (*Ljubljana, YU*), *28*:13-18, 1977.
27. Perdih, A., Škafar, S., Cimerman, A., and Ekar, I. Odpadne vode iz predelave sadja in zelenjave. *Nova proizvodnja (Ljubljana, YU)*, *31*:29-30, 1980.
28. Corajoud, J. M. Essais d'épuration mixte des vinasses de distillerie d'eaux-de-vie de fruits. *Eau Ind.*, *52*:89-94, 1981.
29. Friedrich, J., Cimerman, A., and Perdih, A. Comparison of different cellulolytic fungi for bioconversion of apple distilery waste. *Appl. Microbiol. Biotechnol.*, *24*:432-434, 1986.
30. Friedrich, J., Cimerman, A., and Perdih, A. Mixed culture of Aspergillus awamori and Trichoderma reesei for bioconversion of apple distillery waste. *Appl. Microbiol. Biotechnol.*, *26*:299-303, 1987.
31. Brown, D., McKay, R., and Weir, W. Some problems associated with the treatment of effluents from malt whisky distilleries. *Prog. Water Technol.*, *8*:291-300, 1976.
32. Costa Ribeiro, C., and Castello Branco, J. R. Stillage: A resource disguised as a nuisance. *Effluent Treatm. Biochem. Ind. Conf. Pap., Process Biochem. Int. Conf., 3rd (Watford, Eng., Wheatland)*, Paper 23, 1980.
33. Škafar, S., Perdih, A., and Cimerman, A. Mogućnost koriščenja nekih otpadnih voda plesnima i upotreba micela tih plesni za stočnu hranu. *Hem. Ind.* (*Beograd, YU*), *31*:120-122, 1977.
34. Holdom, R. S., and Winstrom-Olsen, B. Fertilizers and energy from anaerobic fermentation of solid wastes. In *Handbook of Organic Waste Conversion* (M. W. M. Bewick, ed.), Van Nostrand Reinhold, New York, 1980, pp. 383-408.
35. Bewick, M. W. M. *Handbook of Organic Waste Conversion*. Van Nostrand Reinhold, New York, 1980.
36. Bronn, W. K. Die technologischen Möglichkeiten der Abwasserreinigung oder Ablaufverwertung bei Hefefabriken und Melassebrennereien. *Branntweinwirtschaft*, *117*:409-414, 1977.
37. Kishore, K., Jain, A. K., Shanker, G., and Shukla, S. D. Utilization of distillery waste. *Chem. Age India*, *30*:823-826, 1979.
38. Campbell, W. Wastewater treatment in brewing and distilling. *Process Biochem.* (*June 1977*):6-8, 32.
39. Juste, C. Nouvelle possibilite d'utilisation des vinasses de distillerie en agriculture. *Ann. Agron.*, *25*:339-347, 1974.
40. Axelsson, H. Two systems for effluent treatment in the food industry. *Effluent Treatm. Biochem. Ind. Conf. Pap., Process Biochem. Int. Conf., 3rd (Watford, Eng., Wheatland)*, Paper 17, 1980.

41. Eder, K. Evaporation of distillery wash: Energy consideration. *Effluent Treatm. Biochem. Ind. Conf. Pap., Process Biochem. Int. Conf., 3rd (Watford, Eng., Wheatland)*, Paper 3, 1980.
42. Nasr, F. A., and Nawar, S. S. Treatment of distillery waste water. *Int. Sugar J., 89*:150-153, 1987.
43. Michaelsen, Th., Sonnenberg, H., Schuchardt, F., and Baader, W. Aufbereitung organischer Schlämme zu Schüttgut für die Verwertung als Dünger, Futter oder Brennstoff. *Wissenschaft Umwelt, 1*:9-14, 1985.
44. Wilson, D. N. Trattamento per effluenti ad alto carico inquinante. *Impianti, 15*:166-168, 1982.
45. Koivurinta, J., Junnila, M., and Koivistoinen, P. Functional properties of brewer's grain, brewer's yeast and distiller's stillage in food systems. *Lebensm. Wiss. Technol., 13*:118-122, 1980.
46. Garscadden, B. A. By-products from malting, brewing, and distilling. In *Handbook of Organic Waste Conversion* (M. W. M. Bewick, ed.), Van Nostrand Reinhold, New York, 1980, pp. 135-164.
47. Tomlinson, E. J. The production of single cell protein from strong organic waste waters from the food and drink processing industries. 1. Laboratory cultures. *Water Res., 10*:367-371, 1976.
48. Gonçalves de Lima, O. Publicaçao No. 8 da Catedra Microbiologia Industrial, Universidade do Recife, Brasil, 1962; Ref. 16.
49. Lefrançois, L. Production de protéines alimentaires, Levure de vinasse. *Ind. Agr. Alim., 72*:187-191, 1955.
50. Wonnenberger, W. Die Hefezüchtungsanlage System Lefrançois. *Branntweinwirtschaft, 81*:281-284, 1959.
51. Karaki, I., Ishikawa, F., Yamanaka, M., Nishioka, S., Matsuo, T., and Konishi, K. Yeast production from alcohol distillation slops of cane molasses. 5. *Hakko Kyokaishi, 24*:457-471, 1966.
52. Karaki, I., Nishioka, S., and Konishi, K. Yeast production of alcohol distillation slops of cane molasses. 6. *Hakko Kyokaishi, 26*:173-186, 1968.
53. Karaki, I., Nishioka, S., and Konishi, K. Yeast production from alcohol distillation slops of cane molasses. 7. *Hakko Kyokaishi, 26*:263-269, 1968.
54. Flanzy, M., Maugenet, J., Benard, P., and Mourgues, J. Contribution a l'étude du traitement des vinasses des distilleries vinicoles. *C. R. Acad. Agr. Fr.*, 1088-1096, 1967.
55. Falanghe, H. Production of mushroom mycelium as a protein and fat source in submerged culture in medium of vinasse. *Appl. Microbiol., 10*:572-576, 1962.
56. Magny, J., Montant, C., Raynaud, P., Gontier, C., and Dardenne, J. Treating the distillation residues from white wine production. *Ger. Offen.*, 2,630, 680, 1977.
57. Quinn, J. P., and Marchant, R. The growth of Geotrichum candidum on whiskey distillery spent wash. *Eur. J. Appl. Microbiol. Biotechnol., 6*:251-261, 1979.
58. Quinn, J. P., and Marchant, R. The macromolecular composition of Geotrichum candidum grown on whiskey distillery spent wash. *Eur. J. Appl. Microbiol. Biotechnol., 6*:263-270, 1979.
59. Barker, T. W., Quinn, J. P., and Marchant, R. The use of a mixed culture of Geotrichum candidum, Candida krusei and Hansenula anomala for microbial protein production from whiskey distillery spent wash. *Eur. J. Appl. Microbiol. Biotechnol., 14*:247-253, 1982.

60. Ban, S., and Glanser, M. Ekonomska ocjena i mogućnost primjene tehnoloških procesa za istovremenu preradu otpadnih tvari u hranu i energiju. In *Otpadni materijali prehrambene industrije kao sekundarne sirovine* (L. Knežić, ed.), Savez hemičara i tehnologa Srbije, Kikinda, 1983, pp. 12-34.
61. Glanser, M., Dvoraček, L., and Ban, S. N. Fast fermentation of wastewaters to methane. *Wiss. Umwelt* (*1985*):42-47.
62. Friedrich, J., Cimerman, A., and Perdih, A. The use of Aspergillus niger for bioconversion of apple distillery waste. *Eur. J. Appl. Microbiol. Biotechnol.*, *17*:243-247, 1983.
63. Aran, N. A study in the production of single cell protein from vinasse. *T.B.T.A.K.* Marmara Scientific and Industrial Research Institute, Publication No. 25, 1977, pp. 1-16.
64. Shcherback, N. S. Practical ways of using distillery wastes of molasses processing. *Nauchno-Tekh. Ref. Sb.: Spirt. Likero-Vodochn. Prom-st.* *1978*:3-10; Ref. *C. A.*, *89*:195399e, 1978.
65. Skirstymonskii, A. S. Use of wastes in the alcohol industry. *Pishch. Prom-st.* *1977*:24-27; Ref. *C. A.*, *88*:87593j, 1978.
66. Zabrodskii, A. G., and Ilina, L. D. Protein-vitamin food additives produced from wastes of molasses-alcohol plants. *Nauk. Pr.-Ukr. Silskogospod. Akad.*, *191*:101-104, 1976; Ref. *C. A.*, *86*:104375t, 1977.
67. Rodzevich, V. I., Mazur, N. S., and Mikhailova, V. V. Ways of increasing the yield of nutrient yeast. *Ferment. Spirt. Prom-st.* (*1978*): 38-40, 1978.
68. Staheeu, I. V., Kalamiec, E. I., and Ramanouskaya, T. V. Besperapynnae kultyvavanne Candida curvata D 66 na gidralizace sumesi bulbyanoi myazgi i paslyaspirtavoi bragi. *Vyesti Akad. Nauk BSSR, Syer. Biyol. Nevuk*, *1*:54-56, 1985.
69. Ilina, L. D., and Evseichik, B. I. Expediency of growing food yeasts on spent mash from alcoholic fermentation. *Nauk. Pr.-Ukr. Silskogospod. Akad.*, *191*:129-133, 1976; Ref. *C. A.*, *86*:153917e, 1977.
70. Karova, E., Rashkova, Z., Marinov, M., and Koseva, N. Possibilities of using vinasse in the production of feed yeast and L-sorbose. *Priroda* (*Sofia*), *25*:58-59, 1976.
71. Lin, S. F. Studies on feed yeast production on molasses alcohol slop. *Taiwan Sugar Exp. Stn.*, *51*:1-7, 1969.
72. Wang, L. H., Kuo, Y. C., and Chang, C. Y. Studies on the utilization of molasses alcohol slops I. Production of feed yeast by continuous cultivation. *J. Chin. Agr. Chem. Soc.*, *18*:25-33, 1980.
73. Wang, L. H., Kuo, Y. C., Chang, C. Y., and Sang, S. L. Studies on the utilization of molasses alcohol slops II. Factory trial of continuous production of feed yeast with cell recycle technique. *Rep. Taiwan Sugar Res. Inst.*, *90*:45-54, 1980.
74. Chuang, Y. T., Hwang, P. T., Wang, L. H., and Sang, S. L. Studies on the rapid treatment of the yeast waste stream I. The optimal conditions of the mold cultivation. *Rep. Taiwan Sugar Res. Inst.*, *81*:35-45, 1978.
75. Hwang, P. T., and Chuang, Y. T. Studies on the rapid treatment of the yeast waste stream. II. A bench scale study. *Rep. Taiwan Sugar Res. Inst.*, *86*:63-72, 1979.
76. Hwang, P. T., and Chuang, Y. T. Studies on the rapid treatment of the yeast waste stream. IV. Continuous cultivation of Aspergillus sp. 188. *Rep. Taiwan Sugar Res. Inst.*, *88*:65-73, 1980.

77. Chuang, Y. T., and Hwang, P. T. Studies on the rapid treatment of the yeast waste stream. V. Treatment of the fungal waste. *Rep. Taiwan Sugar Res. Inst.*, *89*:45-56, 1980.
78. Kim, C. J., and Oh, M. J. Studies on the production of the single cell protein from alcohol distillation slops of cane molasses I. On the selection and identification of the yeast. *Nongop Kisul Yongu Pogo*, *2*:463-468, 1975.
79. Aguinaldo, J. T. Utilization of the distillery slops for continuous propagation of torula yeast. *Sugar News*, *51*:354, 356, 1975; Ref. *C.A.*, *84*:149182c, 1976.
80. Nandan, R., Chandra, H., Bhargava, S. K., and Ray, P. K. Microbial conversion of distillery waste to bioenergy: Effects of media enrichment with low chain fatty acids and Candida sp. *J. Ferment. Technol.*, *65*:49-60, 1987.
81. Zhang, K., Chiang, L. C., Ueng, P., and Tsao, G. T. Production of yeast single cell protein from bagasse hemicellulose hydrolysate and its alcohol fermented waste liquor. *Abst. 82nd Ann. Meet. Am. Soc. Microbiol.*, Atlanta, 1982, p. 198.
82. Bottaro Castilla, R., Waehner, R. S., and Giulietti, A. M. Aerobic microbial treatment of sugar cane stillage by Candida utilis and Paecilomyces varioti in two step continuous cultures. *Biotechnol. Lett.*, *6*:195-198, 1984.
83. Wang, L. H., Kuo, Y. C., and Chang, C. Y. Distribution of organic matters during molasses alcohol fermentation and feed yeast cultivation on the slop. *Rep. Taiwan Sugar Res. Inst.*, *78*:71-80, 1977.
84. Tauk, S. M. Effect of mineral acid treatments in vinasse on the growth of Candida utilis. *Cienc. Cult.* (*Sao Paulo*), *30*:350-353, 1978; Ref. *C. A.*, *88*:188109m, 1978.
85. Tauk, S. M. Adaptação de leveduras a vinhaça e vinhaça suplementada com melaço. *Cienc. Cult.* (*Sao Paulo*), *31*:522-530, 1979.
86. Tomlinson, E. J. The production of single cell protein from strong organic waste waters from the food and drink processing industries 2. The practical and economic feasibility of a non-aseptic batch culture. *Water Res.*, *10*:372-376, 1976.
87. Akaki, M., Takahashi, T., and Ishiguro, K. Studies on microbiological treatment and utilization of cane molasses distillery wastes. I. Screening of useful yeast strains. *Mie Daigaku Nogakubu Gakujutsu Hokoku*, *62*:155-161, 1981; Ref. *C. A.*, *95*:167101g, 1981.
88. Tauk, S. M. Preliminary study of vinasse as a substrate for yeast growth. *Rev. Microbiol.*, *7*:92-97, 1976; Ref. *C. A.*, *87*:37345p, 1977.
89. Tauk, S. M., and Gambale, V. Efeito da adição de $MgSO_4$ no desenvolvimento de levedura em vinhaça. *Naturalia* (*Sao Paulo*), *6*:5-14, 1981.
90. Tauk, S. M. Culture of Candida in vinasse and molasses: effect of acid and salt addition on biomass and raw protein production. *Eur. J. Appl. Microbiol. Biotechnol.*, *16*:165-170, 1982.
91. Samson, R. A., and van Reenen-Hoekstra, E. S. *Introduction to Food-borne Fungi*. Centraalbureau voor Schimmelcultures, Baarn, 1988, p. 188.
92. Gonzalez, I. M., and Murphy, N. F. Potential use of rum distillery slops as an animal feed supplement I. Mold growth in slops. *J. Agr. Univ. Puerto Rico*, *63*:325-329, 1979.

93. Perdih, A., Cimerman, A., and Friedrich, J. Biotehnološka prerada otpadnih voda prehrambene industrije sa celulolitičkim gljivama. In *Otpadni materijali prehrambene industrije kao sekundarne sirovine* (L. Knežić, ed.), Savez hemičara i tehnologa Srbije, Kikinda, 1983, pp. 77-85.
94. Cimerman, A., Perdih, A., and Škafar, S. Proučavanje rasta i aktivnosti celulolitskih gljiva na nekim industrijskim otpacima. *Mikrobiol. (Beograd), 13*:43-53, 1976.
95. Perdih, A., Cimerman, A., Klemenc, N., Škafar, S., and Friedrich, J. Mogućnost iskorištavanja hranljivih sastojaka iz otpadnih voda industrije za preradu voća. In *Zaštita '78*. Jugoslovensko društvo za zaštitu voda, Beograd, 1978, pp. 91-93.
96. Cimerman, A., Perdih, A., Friedrich, J., and Skafar, S. Mycelentwicklung einiger cellulolytischen Pilze auf Fruchtindustrieabfällen. *Abh. Akad. Wiss. DDR 1979*(3):23-28.
97. Perdih, A., Cimerman, A., and Friedrich, J. Prečiščavanje jako opterečenih otpadnih voda primenom mikroorganizama iz grupe fungi. *Vodoprivreda (Beograd), 13*:129-134, 1981.
98. Friedrich, J., Legiša, M., Cimerman, A., and Perdih, A. Prerada voćnih otpadaka u stočnu hranu pomoću mutanata plijesni Aspergillus niger. *Prehrambeno-tehnološka revija (Zagreb), 20*:173-175, 1982.
99. Gunde-Cimerman, N., Cimerman, A., and Perdih, A. Aspergillus niger mutants for bioconversion of apple distillery wastes. *Enz. Microb. Technol., 8*:166-170, 1986.
100. Cimerman, A., Gunde-Cimerman, N., and Perdih, A. Gamma and MNNG mutants for bioconversion of apple distillery waste. In *Improved Biomass Utilization through the Use of Nuclear Techniques*. IAEA-TECDOC-475, IAEA, Vienna, 1988, pp. 53-69.
101. Perdih, A., Cimerman, A., and Friedrich, J. Fungal bioconversion of fruit distillery wastes. *Enz. Microb. Technol.* (in press).

Part IV

CULTURE COLLECTIONS, LEGAL ASPECTS, AND BIOSAFETY

32

CURRENT STATUS OF FUNGAL CULTURE COLLECTIONS AND THEIR ROLE IN BIOTECHNOLOGY

SHUNG-CHANG JONG and JEANNETTE M. BIRMINGHAM *American Type Culture Collection, Rockville, Maryland*

I. INTRODUCTION

The biology of fungi cannot be adequately studied or assessed without growing them in pure culture. Pure culture techniques are used in the isolation, identification, and propagation of fungi. Changes in the genetic population often take place in pure culture, and no cell population is genetically identical to the parent culture. The increase in any mutants in the total population is dependent on culture conditions and the number of generations of growth permitted. Therefore, once pure cultures are established, they must be maintained in a genetically stable condition. Culture collections have been and are being founded to conserve valuable fungus germplasm important for economic, agricultural, medical, and scientific endeavors [1].

II. MYCOLOGICAL COLLECTIONS AND THEIR ORGANIZATION

For at least 3000 years various microorganisms have been selected and maintained as "starters" to produce fermented foods and drinks, but fungi were not intentionally cultured on plant materials until the 1700s. When sterile technique was developed and solid media introduced in the mid-19th century, it was possible to grow and maintain individual strains in pure culture. As pure cultures came into general use for teaching and experiments, they were amassed on a systematic basis, and culture collections evolved. The first collections were founded by scientists for their own use but were later expanded to include cultures of mutual interest among collaborators. Eventually the collections became so large that a curator was needed to manage them.

Fungus isolates from the German Technical University in Prague were offered commercially by Kral in the late 19th century. It was not until 1904, however, that the first institution to preserve and distribute a wide variety of fungus cultures was founded at the Centraalbureau voor Schimmelcultures (CBS) in the Netherlands. In 1925 the American Type Culture Collection (ATCC) was started in the United States, and in 1940 the USDA

Agricultural Research Services Culture Collection was formally established with the opening of the Northern Regional Research Laboratory (NRRL). The Commonwealth Mycological Institute (CMI), founded as the Imperial Mycological Bureau in 1919, maintains a selection of fungi received as living cultures from its origin and, when the UK National Collection of Type Cultures at the Lester Institute was dispersed in 1947, almost all the filamentous fungi were transferred to CMI.

The discovery of penicillin in the 1930s first drew attention to collections where fungus strains were housed. The 1980s brought a renewed interest in them as major resources for biotechnology [2-4].

In the late 1960s V. B. D. Skerman of Australia began compiling a register of existing culture collections maintained by various institutions throughout the world. The first *World Directory of Collections of Cultures of Microorganisms* was prepared by the World Data Center of Microorganisms (WDC) in 1972. The most recent edition, published in 1986, listed 327 collections in 55 countries, with the names and addresses of the collections registered, their interests and functions, personnel, availability of cultures, and catalog holdings [5].

The total number of culture collections in the world which include fungi is unknown, but at least 200 of the 327 collections contributing data to the World Data Center contain some strains of fungi. There can be little doubt that many existing collections are not included. For example, the Yeast Genetic Stock Center (YGSC) and the USDA Center for Forest Mycology Research (CFMR) are not registered with WDC [1].

Culture collections may maintain a large number of strains of a few species, or a few strains of each species, or a large number of strains of all known species in culture. They may specialize in material relevant to particular fields, such as agriculture, medicine, industry, or forestry, or serve educational needs as well as research and development needs, and be national or international in operation. They may be housed in an educational institution, a research organization, a taxonomic institute, an independent laboratory, or a commercial firm. Any of these collections can be considered as a germplasm resource in that they conserve the diversity of microorganisms that are of past, present, or future interest to scientists [1].

In general, all of these culture collections may be grouped in three major categories: service collections, institutional collections, and private collections. Service collections are usually well established as permanent resource centers which are prime sources of reference and type cultures and actively solicit strains cited in literature. They are concerned with long-term preservation methods and many act as international depository authorities (IDAs) for patent strains. These collections publish catalogs and charge a fee for cultures available on demand to the scientific community. The size of the service collection may vary from a comprehensive range of all types of microorganisms, such as the ATCC, to a limited range of a specific type, such as the Fungal Genetic Stock Center (FGSC), which maintains primarily the genetic stocks of *Neurospora*.

Institutional collections are usually mission-oriented and are established primarily to facilitate ongoing research or teaching programs of the organizations of which they form a part, such as universities, research institutes, government or industrial laboratories, or hospitals. These collections range in size from being even more diverse than some service collections to those specialized in medical, veterinary, agricultural, or industrial microorganisms.

Their supply of cultures to others is usually a secondary function. Generally, no catalogs are available, and the use of such collections by others outside the immediate working unit is somewhat limited. The USDA Agriculture Research Services Culture Collection (NRRL) is an example of an institutional collection. It contains cultures of agricultural and industrial significance.

Private collections are usually built up by individual research scientists from their research work in a university or research institute. These collections are often highly specialized, containing strains with unusual attributes either cited in publications or kept under strict security as a proprietary collection. No catalog of holdings is usually published. However, some researchers do make cultures available from time to time to other scientists free of charge or on an exchange basis when it will not jeopardize their research confidentiality. The continuity of these collections is often at risk of loss due to inadequate storage methods or the termination of research projects or of a scientist's career. Research scientists with private collections should regularly deposit representative strains with institutional or service collections to insure against the loss of important germplasm [1].

The worldwide collection of fungi contains about 17,202 species or varieties including lichens and yeasts. Although some strains, such as those used as standard references for comparative purposes, are represented in a number of collections, duplication is not as great as might be expected—about one-fourth according to a superficial survey of catalogs [2]. Any of these collections could offer new sources of biosynthetic products, genes for hybridization, or even characteristics for genetic engineering. An extensive list of culture collections which include fungi is shown in Table 1.

All culture collections are represented at the international level by the World Federation for Culture Collections (WFCC), which was founded in Paris in 1966 under the auspices of UNESCO. It is accepted by both the International Union of Microbiological Societies (IUMS) and the International Union of Biological Sciences (IUBS). The Federation provides a forum for discussion of all matters relating to culture collections and holds conferences. The European Culture Collection Curators' Organization (ECCCO), established in 1982, provides a forum similar to WFCC for collections in Europe [2]. Organizations for culture collections on a national level are listed in Table 2.

The need for conservation of the genetic uniqueness and diversity of microorganisms was particularly emphasized in the principles and recommendations of the United Nations Conference on the Human Environment in Stockholm in 1972. As a result, the concept of a Global Network of Microbiological Resources Centers (MIRCENs) was proposed in 1974. Today, the global MIRCEN network program, which embodies activities that are carried out within the framework of UNESCO's regular program activities in cooperation with the IUMS, the International Cell Research Organization (ICRO), the WFCC, and other bodies, comprises 18 centers. These centers manage and distribute microbial cultures locally, provide training and advice, and carry on research. Thirteen of the 18 MIRCENs are concerned with the biotechnological aspects of mycology.

TABLE 1. Culture Collections That Include Fungi

Country	Culture collection
Argentina	Collection Catedra Microbiologia Agricola (CCMA); Facultad de Farmacia y Bioquimica de la Universidad de Buenos Aires (CCM-A)
Australia	Australian National Reference Laboratory in Medical Mycology (AMMRL); Australian Wine Research Institute Culture (AWRI); CSIRO Starter Culture Collection (DBR); Microbiol. Dept., Univ. Melbourne (DMCCUM); Dept. Microbiol., Univ. Western Australia (DMPMC); Forestry Commission of New South Wales (DWT); CSIRO Food Research (FRR); Australian National Reference Laboratory in Medical Mycology (IMR-RNSH); King Edward Memorial Hospital for Women (KEMH); Materials Research Laboratories Collection of Microfungi (MRL); Plant Pathology Culture Collection, Dept. Primary Industries (PPCC); Univ. Queensland (UQM): Plant Research Division Culture Collection, Dept. Agriculture (WA); State Health Laboratories, Medical Mycology (WACC)
Belgium	Mycotheque de L'Université Catholique de Louvain (MUCL)
Brazil	Instituto Zimotecnico Culture Collection (ESAP); Tropical Culture Collection (FTPT); Av. Dr. Arnaldo (IAL); UFPE, Dept. Antibiotics (UFPEDA)
Bulgaria	Bulgarian Type Culture Collection (BTCC); National Bank for Industrial Microorganisms and Cellular Cultures (NBIMCC)
Canada	Canadian Collection of Fungus Cultures, Biosystematics Research Inst. (CCFC); Pacific Forest Research Centre (DFF); Dept. Microbiol., Univ. Guelph (CMG); Dept. Microbiol., Univ. Manitoba (DMUM); Forest Pest Management Inst. (EFFPMI); FORINTEK Culture Collection (FORINTEK); Fredericton Stock Culture Collection (FSC); National Research Council Culture Collection, Division of Biological Sciences (NRC); Ontario Ministry of Health (TPHL); Univ. Alberta Microfungus Collection and Herbaria (UAMH); Univ. Western Ontario (UWO); Culture Collection of Wood-inhabiting Fungi (WFPL)
Chile	Catholic Univ. Santiago (CU)
China (Beijing)	Center for Agricultural Culture Collection, Inst. Soil and Fertilizers (CACC); Center for Antibiotic Culture Collection, Inst. Antibiotics (CACC); Center for Collection of General Microbiological Cultures, Inst. Microbiol., Academia Sinica (CCGMC); Center for Forestry Culture Collection, Research Inst. Forestry (CFCC); Center for Industrial Culture Collection, Scientific Research Inst. Food and Fermentation Industry (CICC);

Country	Culture collection
	Center for Medical Culture Collection, Inst. Dermatology (CMCC); China Type Culture Collection, Univ. Wuhan (CTCC); Center for Veterinary Culture Collection, Control Inst. Veterinary Biological Products and Pharmaceutics (CVCC)
China (Taipei)	Culture Collection and Research Center, Food Industry Research and Development Inst. (CCRC)
Czechoslovakia	Culture Collection of Basidiomycetes, Czechoslovak Academy of Sciences (CCBAS); Culture Collection of Fungi, Dept. Botany, Faculty of Sciences (CCF); Czechoslovak Collection of Microorganisms (CCM); Czechoslovak Collection of Yeasts (CCY); The Czechoslovak National Collection of Type Cultures (CNCTC)
Finland	VTT Collection of Industrial Microorganisms (VTT)
France	Laboratoire de Cryptogamie du Museum National d'Histoire Naturelle (LCP); Roussel UCLAF (UCLAF); Unité de Mycologie, Institut Pasteur (UMIP)
Germany	Institut für Pflanzenschutz im Forst (BBLF); Bayerische Landesanstalt für Weinbau und Gartenbau (BLWG); Institut für Mikrobiologie and Landeskultur Mikrobiologie (CCDAM); Deutsche Sammlung von Mikroorganismen (DSM); Academie der Landwirtschaftswissenschaften der DDR (FIE); Friedrich-Schiller-Universitat Jena, Sektion Biologie, Pilzkulturensammlung (MW)
Guatemala	Colleccion de Microorganisms del Instituto Centroamericano (ICAITI)
Hungary	Technical Univ. Budapest (DACT); Microbial Gene Bank, University of Horticulture (HMGB)
Hong Kong	Chinese University of Hong Kong (CUHK)
India	Bose Inst. Dept. Microbiol. (CCDMBI); DMSRDE Culture Collection (DMSRDE); Mycological Herbarium, Delhi Univ. (DUM); Indian Type Culture Collection, Indian Agricultural Research Inst. (ITCC); National Collection of Industrial Microorganisms (NCIM); Culture Collection, Microbiology and Cell Biology Laboratory (NTCCI); Vallabhbhai Patel Chest Inst. (VPCI)
Indonesia	Laboratory for Microbiol. and Fermentation Technology (ITB); Univ. Indonesia Culture Collection (UICC)
Iran	Persian Type Culture Collection (PTCCI)
Ireland	Guinness (Dublin) Culture Collection (GCC)
Italy	Centro di Studio dei Microorganismi Autotrofi-CNR (CSMA); Wine Yeast Collection (DBV-PG); Yeast

Table 1. (Continued)

Country	Culture collection
	Collection, Universita' degli Studi di Perugia (IMAT); Instituto Patologia Vegetale (IPV)
Japan	Faculty of Agriculture, Hokkaido Univ. (AHU); Fermentation Research Inst., Agency of Industrial, Sciences and Technology (FERM); Government Forest Experiment Station (FES); HUT, Dept. Fermentation Technology (HUT); Inst. Applied Microbiol., Univ. Tokyo (IAM); Research Inst. for Chemobiodynamics (IFM); Inst. for Fermentation (IFO); Inst. Public Health (IPH); Japan Collection of Microorganisms, Inst. Physical and Chemical Research (JCM); MR Culture Collection of Microorganisms (MR); National Inst. Hygienic Sciences (NHL); National Inst. Agro-Environmental Sciences (NIAS); Dept. Fermentation Technology Osaka University (OUT); National Research Inst. Brewing (RIB)
Malaysia	Simpanan Kultur Universiti Kebangsaan (SKUK); Univ. Kegangsaan Kultur Perubatan (UKKP)
Mexico	Coleccion del Depto. de Biotecnologia y Bioingenieria (CDBB); Cepario de la Facultad de Quimica, UNAM (CFQ); Coleccion de Cultivos de Hongos del INIF (INIF); Instituto Technoloqico de Durango (ITD); Pathogenic Fungi and Actinomycetes Collection (PFAC); Industrial Culture Collection, Instituto de Investigaciones Biomedicas (UNAM)
Netherlands	Centraalbureau voor Schimmelcultures (CBS); IPO-Collection of Fungal Pathotypes (RIPO)
New Zealand	Victoria Univ. Wellington, Botany Dept. (DBVU); National Health Inst. (NHI); Forest Research Inst. Culture Collection (NZFS); DSIR Plant Disease Division (PDD); DSIR, Mt. Albert Research Centre (PDDCC)
Nigeria	International Inst. Tropical Agriculture (IITA)
Philippines	BIOTECH, UP at Los Banos College (BIOTECH); Fungal Culture Collection, Univ. Philippines at Los Banos College (FCUP); Museum of National History Los Banos MCC): Culture Collection Strain Improvement Project, National Inst. Science and Technology (NIST); Univ. Philippines Culture Collection, National Sciences Research Inst. (UPCC)
Poland	Inst. Fermentation Industry (IPF); Chaire de Microbiologie Technique (LCC); Centre of Industrial Microorganisms Collection (LOCK)
Rumania	Collection of Industrial Microorganisms (ICCF)

Country	Culture collection
Singapore	Dept. Microbiol. National Univ. Singapore (NUSDM); Botany Dept. Culture Collection, National Univ. Singapore (USDB)
Spain	Culture Collection of Microorganisms, Dept. Microbiol., Ciudad Universitaria (CCMCU); Coleccion Espanola de Cultivos Tipo, Valencia (CECT) Sri Lanka Dept. Botany, Univ. Kelaniya (DBUK)
Switzerland	Centre de Collection de Types Microbiens (ICIDTC)
Thailand	Microbiol. Dept. Chulalonghorn Univ. (CHULA); Khon Kaen Univ. (KKU); Microbiol. Section, Dept. Science Service (MDMU); Ramathibodi Hospital, Mahidal Univ. (RHMU)
Turkey	Culture Collections and Industrial Microbiol. (KUKEM)
UK	Aquatic Phycomycete Culture Collection (AP); National Collection of Wood Rotting Fungi (FPRL); Culture Collection, Commonwealth Mycological Inst. IMI (CMI); Kemp *Coprinus* Collection (KCCB); National Collection of Yeast Cultures (NCYC); Culture Collection, Portsmouth Polytechnic (PPCCE)
USA	ARS Collection of Entomopathogenic Fungi (ARSEF); American Type Culture Collection (ATCC); Centers for Disease Control (CDC); Center for Forest Mycology Research (CFMR); Fungal Genetics Stock Center (FGSC); *Fusarium* Research Center (FRC); International Culture Collection of VA Mycorrhizal Fungi (INVAM); Carolina Biological Supply Co. (LMS); Eli Lilly Fermentation Products Research (LRL); National Institutes of Health (NIH); Agricultural Research Culture Collection (NRRL); Upjohn Culture Collection (UCR); Yeast Genetics Stock Center (YGSC)
USSR	Cryobank of Microorganisms (LIA); USSR Research Inst. for Antibiotics (RIA); All-Union Collection of Microorganisms (VKM)
Zimbabwe	Univ. Zimbabwe (BDUZ)

TABLE 2. National and Regional Federations and Committees

1. Australian Collection of Microorganisms Committee
2. Australian Federation of Culture Collections
3. Canadian Permanent Committee on Culture Collections
4. China Committee for Culture Collections of Microorganisms
5. Czechoslovakian Federation of Culture Collections
6. European Culture Collections' Organization
7. Japan Federation for Culture Collections
8. Korean Federation of Culture Collections of Microorganisms
9. KUKEM Society for Culture Collections and Industrial Microbiology
10. Latin American Federation of Culture Collections
11. New Zealand Committee on Culture Collections of Microorganisms
12. United Kingdom Federation for Culture Collections
13. United States Federation for Culture Collections

III. USES OF FUNGAL CULTURES IN BIOINDUSTRY

Fungi, with unlimited synthetic and degradative potential, can bring about chemical changes in almost all types of natural and man-made organic molecules. Their rapid growth rate allows immense quantities to be produced in a short period of time. The study of fungus metabolism and the utilization of its diversity have made possible the industrial production of primary and secondary metabolic products. Fungi are now being used to carry out synthetic steps with great precision [6] and are also being employed in the biodegradation and detoxification of wastes and the extraction of metals from dilute solutions [7].

Over 3000 secondary metabolites have been characterized for fungi, but the actual number is far greater, for the physiological and biochemical activities of most fungi have not as yet been described [8]. Table 3 illustrates some of the many products of fungus fermentation in which strains from the ATCC have been employed.

In addition to the products of fungus fermentation which are popular foods and drinks in many cultures, fungi themselves are used as food. Larger fleshy fungi, particularly mushrooms, found in pastures and forests have been eaten since earliest times. While the number of edible species probably exceeds 500, it is surprising that only about 20 species are grown commercially. The mushroom industry is growing, and new techniques are being devised to increase the number of cultivated species. Naturally, pure culture is important for making spawn for the industry [9]. In addition, filamentous fungi and yeasts are increasingly being employed as protein sources or, after processing, as synthetic meat substitutes. Strains of edible fungi and those which are sources of protein offered by ATCC are also shown in Table 3. In the future fungi could become a valuable food source in underdeveloped countries [6].

TABLE 3. Products of Fungus Fermentation and ATCC Strains Used

Product	Microorganisms	ATCC strains
Beverages		
Beer	*Saccharomyces cerevisiae*	2704, 26292, 32634
Kaffir beer	*Saccharomyces cerevisiae*	48372
Kefir	*Saccharomyces cerevisiae*	2360
	Saccharomyces unisporus	48553, 48555
Tea fungus	*Candida* sp.	46247
	Saccharomyces cerevisiae	46246
Wines		
Ansmushausen	*Saccharomyces cerevisiae*	4113
Fermented apple cider	*Saccharomyces cerevisiae*	4938
Arize	*Saccharomyces cerevisiae*	42926
	Saccharomyces uvarum	32634
Bordeaux	*Saccharomyces cerevisiae*	42928
Bouzy	*Saccharomyces cerevisiae*	42929
Burgundy	*Saccharomyces cerevisiae*	4123, 36024, 42930
Cabernet	*Saccharomyces cerevisiae*	42944
California	*Saccharomyces cerevisiae*	4105
Champagne	*Saccharomyces bayanus*	36023
	Saccharomyces cerevisiae	834, 4097, 42945
Cognac	*Saccharomyces cerevisiae*	42931
Domestic	*Saccharomyces bailii*	8766
French	*Saccharomyces cerevisiae*	4921
Geisenheim	*Saccharomyces cerevisiae*	42932
German, white	*Saccharomyces cerevisiae*	4098
Hoang chiu	*Saccharomyces cerevisiae*	2080
Hsao-hsing	*Saccharomyces cerevisiae*	26607-26611
Ivory Coast	*Saccharomyces chevalieri*	9804
Jerez	*Saccharomyces cerevisiae*	42932, 42933, 42946
Kao-liang	*Monascus araneous*	16361
	Monascus pilosus	16363
	Monascus pubigerus	16364

Table 3. (Continued)

Product	Microorganisms	ATCC strains
[Kao-liang]	*Monascus rubropunctatus*	16368
	Monascus sp.	26264
Laviero	*Saccharomyces cerevisiae*	4114
Malaga	*Saccharomyces cerevisiae*	8258, 42934
Menes	*Saccharomyces cerevisiae*	4117
Montrachat	*Saccharomyces cerevisiae*	26497, 36026
Moselle	*Saccharomyces cerevisiae*	42936
Muscatel	*Saccharomyces cerevisiae*	8256, 42937
Palm	*Saccharomyces rosei*	26813
	Saccharomyces vini	26811
	Torulaspora florentina	26812
	Torulaspora vafer	26814
Port	*Saccharomyces cerevisiae*	42938, 42947
Red	Yeast	38639
Reisling	*Saccharomyces* sp.	10736, 42940
Rhine	*Saccharomyces cerevisiae*	42939
Rice	*Monascus* sp.	16435
Romanee	*Saccharomyces cerevisiae*	42949
Sake	*Aspergillus oryzae*	44193
	Saccharomyces sake	4134, 26421, 26422
	Rhizopus arrhizus	32621
Sauterne	*Saccharomyces cerevisiae*	8255, 8257, 42948
Sherry	*Saccharomyces bayanus*	13055
Spanish, red	*Saccharomyces aceti*	14084
Steinberg, cider	*Saccharomyces* sp.	4938
Steinberg, wine	*Saccharomyces cerevisiae*	11777, 14824, 14825 42941
Sweet	*Saccharomyces bayanus*	36022
	Saccharomyces cerevisiae	36026
Tokay	*Saccharomyces cerevisiae*	4108, 36027
	Saccharomyces sp.	560, 561, 562
Unican House	*Saccharomyces bayanus*	34032, 34033
	Saccharomyces diastaticus	34031, 34034, 34035

Product	Microorganisms	ATCC strains
Valais	*Saccharomyces cerevisiae*	2359, 2364-2366, 2375
Waldenberg	*Saccharomyces cerevisiae*	42942
White	*Saccharomyces cerevisiae*	38636
Winninger	*Saccharomyces capensis*	42943
Miscellaneous	*Saccharomyces cerevisiae*	34081
	Saccharomyces cerevisiae	34082, 38637, 38638 42927
	Saccharomyces chevalieri	34083
	Saccharomyces italicus	34084
	Saccharomyces rosei	34086
	Saccharomyces sp.	10736
Foods		
Blue cheese	*Penicillium camembertii*	4845, 42009, 42293
	Penicillium roquefortii	6989, 9295, 10110, 34905
Bread, sourdough	*Candida holmii*	60586
Lookpang (similar to ragi)	*Saccharomycopsis fibuligera*	32693
Miso	*Aspergillus oryzae*	14895, 26666
	Aspergillus sojae	46244, 46245
	Candida mogii (fuchu miso)	18634
	Saccharomyces rouxii	14679, 14680, 42981 42982, 46261, 46494
	Torulopsis etchellsii	46254
	Torulopsis mogii (red miso)	36590
	Torulopsis versatilis	42980
Ontjom (oncom)	*Neurospora crassa*	24914
	Neurospora intermedia	14151
Ragi (Luk-paeng)	*Rhizopus chiensis* var. *liquefaciens*	22960
	Saccharomycopsis fibuligera	24945, 36213
	Saccharomycopsis malanga	36237
Red rice (anka)	*Monascus albidus* var. *globe*	16358
	Monascus anka	16360

Table 3. (Continued)

Product	Microorganisms	ATCC strains
[Red rice]	*Monascus major*	16362
	Monascus purpurescens	36113, 36114
	Monascus purpureus	16365, 34892, 16427 16436
	Monascus sp.	16437, 34570
Shimozuke	*Torulopsis etchellsii*	28289
	Torulopsis glabrata	28290
Soy sauce (shoyu or sho)	*Aspergillus awamori* (black kuro koji)	14331
	Aspergillus flavus var. *columnaris*	44310
	Aspergillus oryzae	9362, 11494, 14895 20386, 26666, 46249, 44250
	Aspergillus sojae	20387, 20388, 42249 42250
	Aspergillus tamarii (red awamori koji)	16865 38854
	Rhizopus javanicus var. *kauasakiensis*	22961
	Rhizopus oryzae	8466
	Saccharomyces rouxii	46252
	Torulopsis etchellsii	20190, 46254, 46263
	Torulopsis halonitratophila	22989
	Torulopsis mannitofaciens	36632, 36633
	Torulopsis noda	20189
	Torulopsis versatilis	20191
Tape	*Amylomyces rouxii*	42543
	Rhizopus microsporus	22958
	Rhizopus microsporus var. *liquefaciens*	22960
	Saccharomycopsis fibuligera	24945
	Saccharomycopsis malanga	36237
Tempeh (soybean-kedelee)	*Rhizopus cohnii*	46436
	Rhizopus oligosporus	22959
	Rhizopus oryzae	46437

Product	Microorganisms	ATCC strains
Tian toohu (fermented bean curd)	*Monascus albidus*	16357
Tien-chiu-niang (lao chao; rice pudding)	*Amylomyces rouxii*	46243
	Rhizopus arrhizus	46429, 46430
	Rhizopus microsporus	22958
	Rhizopus formesaensis	46431
	Rhizopus javanicus	46432
	Rhizopus oryzae	46242
	Saccharomyces aceti	46433
	Saccharomyces fibuligera	46252
	Saccharomyces sp.	44945
	Torulopsis sp.	44944
Organic Acids		
Citric	*Aspergillus niger*	9142, 11414
	Debaryomyces polymorphus	20280
	Yarrowia lipolytica	20346
Fumaric	*Rhizopus oryzae*	10260, 13310
Gallic	*Aspergillus niger*	16888
Gluconic	*Aspergillus carbonarius*	8740
	Penicillium chrysogenum	10107
	Penicillium puberulum	36363
Itaconic	*Aspergillus terreus*	10020, 36364
Kojic	*Aspergillus oryzae*	7252
	Penicillium jensenii	18227
Lactic	*Rhizopus oryzae*	9363
Industrial Solvents		
Ethanol	*Candida acidothermophilum*	20381
Glycerol	*Pichia farinosa*	20210
Enzymes		
Amylase	*Aspergillus niger*	15475
	Cladosporium nesinae	20495
	Paecilomyces subglobosus	16492, 16493

Table 3. (Continued)

Product	Microorganisms	ATCC strains
Amyloglucosidase	*Aspergillus foetidus*	10254, 14916
	Aspergillus phoenicis	13156, 13157
Cellulase	*Pellicularia filamentosa*	20366
	Sporotrichum cellulophilum	20493, 20494
	Thielavia terrestris	38088
	Trichoderma reesei	13631, 24449
Dextranase	*Fusarium moniliforme*	32426
	Penicillium aculeatum	10409, 20242
	Penicillium funiculosum	10446
Fructosyltransferase	*Aureobasidium pullulans*	9348, 12535
β-Fructofuranosidase	*Candida utilis*	42402
β-Galactosidase	*Absidia griseola*	20430
	Aspergillus oryzae	20423
	Aspergillus niger	13496, 13497
	Cryptococcus albidus	20293, 20294
	Kluyveromyces marxianus	56498
Glucoamylase	*Aspergillus awamori*	22342
	Aspergillus niger	22343
	Amylomyces rouxii	22962, 42543, 46243
Glucose oxidase	*Aspergillus niger*	9029
	Phanerochaete chrysosporium	34541
α-Glucosidase	*Aspergillus niger*	42418
β-Glucosidase	*Dekkera intermedia*	52904
Lactase	*Kluyveromyces marxianus*	8554
Lipase	*Candida rugosa*	14830
	Myriococcum sp.	20374
	Rhizopus oryzae	34612
Maltase	*Saccharomyces* sp.	20488
Acid protease	*Aspergillus niger*	16513
	Polyporus palastris	14624
	Rhizopus rhizopodiformis	48608
Alkaline protease	*Fusarium* sp.	20192
	Gibberella saubinetti	20193

Product	Microorganisms	ATCC strains
Proteases	*Acremonium chrysogenum*	11550, 22571
	Aspergillus awamori	14331, 14333, 14335
	Aspergillus foetidus	14334
	Aspergillus phoenicis	14332
	Talaromyces thermophilus	20186
	Yarrowia lipolytica	16617
Rennin	*Saccharomyces cerevisiae*	20623
Vitamins		
β-Carotene	*Blakeslea trispora*	14059, 14060, 14271
Riboflavin	*Ashbya gosstpii*	10895
Antibiotics		
Penicillin	*Penicillium chrysogenum*	10002, 20444, 58611
Cephalosporin	*Acremonium chrysogenum*	20427, 46117
Griseofulvin	*Khuskia oryzae*	28132
Miscellaneous		
Alkaloids	*Claviceps purpurea*	20103
Gibberellin	*Gibberella fugikuroi*	14164, 58109
Nucleotides	*Candida valida*	14927
Steroid transformations	*Phytophthora infestans*	16981
Edible Mushrooms	*Agaricus bisporus*	56604, 60431, 62494
	Agaricus bitorquis	32675, 62490
	Auricularia auricula	11991, 36636
	Auricularia polytricha	22079
	Boletus edulis	34620
	Cantharellus cibarius	13228, 38974
	Coprinus comatus	12640, 22314
	Flammulina velutipes	9418, 42150
	Grifola frondosa	11936
	Kluehneromyces mutabilis	34481, 44307
	Lentinus edodus	24352, 24462
	Lepista nuda	28844, 38074
	Macrolepiota procera	46378, 46379

Table 3. (Continued)

Product	Microorganisms	ATCC strains
	Morchella angusticeps	14069, 38590
	Morchella crassipes	13227, 18408
	Morchella deliciosa	46778
	Morchella esculenta	10968, 32785
	Morchella rotunda	14071
	Pholiota nameko	42261, 42262
	Pleurotus cystidiosus	28597, 28598
	Pleurotus ostreatus	9415, 9424
	Pleurotus sajor-caju	32078, 32771
	Stropharia aeruginosa	46335, 46380
	Stropharia rugosoannulata	46337, 46338
	Tremella fuciformis	58852, 58859
	Tremella messenterica	24438, 42220
	Volvariella bambycina	44205
	Volvariella esculenta	18105
	Volvariella volvacea	22377, 46888
Edible Protein	*Candida guilliermondii*	20316
	Candida tropicalis	20318, 20021
	Fusarium graminearum	20330, 20334
	Fusarium oxysporium	20328
	Fusarium solani	20327
	Yarrowia lipolytica	20315

Thus, biotechnology is not new but rather represents the coming of age of technology that began decades ago. Although traditional fermentation industries will always occupy a pivotal role in biotechnology, the development of enzyme technology for industry and medicine and genetic engineering will play an important part in many future industrial processes [10].

IV. CONTRIBUTIONS OF CULTURE COLLECTIONS

Various methodologies have resulted in the selection of improved strains of fungi from the natural gene pool, their modification by mutation, and, more recently, the manipulation of their germplasm through genetic engineering. These carefully selected organisms, which represent an enormous genetic resource, must be conserved in a substantially unchanged form and, hence, the value of culture collections, particularly service collections.

Services culture collections perform four basic functions: acquisition, preservation, authentication, and distribution of cultures and associated information. In addition, they offer other services that vary depending on the interests and expertise found in the collection, the role they provide to the parent institution, and the requirements of national authorities [3].

It is obvious that the various functions of collections are not mutually exclusive, and many institutional and private collections provide services as well by supplying cultures and information to other institutions and research scientists.

A. Acquisition

The 170,000 fungus isolates representing more than 7000 species maintained in the world's collections represent only about 11% of the 64,200 accepted species of fungi. The number of species existing in the world has been estimated at more than 250,000, and new species are being described at the rate of 1500 per year. Fungi known up until now only on a specific host of natural substratum continue to be encountered elsewhere or cultured for the first time [2]. With appropriate techniques it is clear that many more species can also be cultured. This is of great interest to the biotechnologist in search of strains with new and unusual properties.

The value of a collection as a resource for biotechnology depends on the range of isolates it is able to supply. To maintain representative isolates of all the fungi is beyond the bounds of all but the largest collections, and even then they are unable to build up holdings of strains of single species, genera, or other small groups [2]. NRRL currently maintains over 80,000 strains of fungi including yeasts, and CBS in the Netherlands has over 38,000 strains. As of December 31, 1988, the total number of fungus strains housed in the ATCC was 22,939 [1]. Some fungus culture collections originated as adjuncts to brewing, pharmaceutical, plant pathology, or medical mycology services, and their holdings reflect this and are limited when compared to service collections.

Collections are always interested in acquiring strains to expand their field of interest, and service collections strive to increase the number of strains important for biotechnology or other aspects of applied biology. To do this, however, a rigorous acquisition policy is necessary. Acceptance of a culture for permanent preservation is a long-term commitment, and consideration must be given to the cost for preservation, storage, and regular purity and identity checks.

Once the parameters for the development of a collection are set, curators identify potential donors and contact them for deposits. This involves extensive literature searches for desirable strains followed by requests for the cultures [11]. Taxonomists are certainly aware of the importance of depositing newly identified species in culture collections, but it is rare for investigators to recognize the need to deposit strains from their research projects or those which may have educational or industrial significance [3]. Researchers should submit subcultures of the strains on which their publications are based, for the significance of the results will diminish if the work cannot be reproduced. If the strain is preserved, their work will continue to be of value even if a taxonomic name change should later occur [2]. It is often the case, especially in biotechnology, that the properties of a culture used are initially more important than the identity [3].

Since there is little hope of preserving even a slight majority of the known fungi, national service collections, such as the ATCC, are very selective in the strains they accession. Each strain chosen exhibits at least one property important to science. The primary significance at the time of deposit of each of the 22,939 strains maintained by the ATCC is as follows: plant pathogens (26%), type and authentic cultures (12%), tools for genetic studies (12%), producers of vitamins, enzymes, and other biochemicals (11%), soil and water fungi (9%), human and animal pathogens (7%), wood fungi (5%), edible mushrooms and food yeasts (5%), producers of antibiotics and mycotoxins (5%), cultures used for fermentation and flavor production (3%), mycorrhizae, cultures used in bioassays, entomogenous fungi, hosts of mycoviruses, and nematode-destroying fungi (each 1%), mycoparasites, fish pathogens, and plant rusts (each <1%).

Accessioning procedures for new cultures have been clearly established by collections. Upon receipt, a new culture is given a unique accession number and tests are carried out to establish its purity and authenticity. All information provided by the depositor must be entered on an accession form [11]. Information requested includes the name of the organism, name of the person who isolated the culture, depositor and previous holders, source, method of isolation, growth requirements, preservation method, known uses, and literature citations. In addition, the ATCC requires information on pathogenicity because of the USDA and Public Health Service (PHS) regulations.

B. Conservation

In keeping microbial cultures, conservation implies both maintenance and preservation. Maintenance means keeping the culture alive and pure, whereas preservation denotes stabilizing the biological properties for a prolonged period. Therefore, in addition to maintaining viability, a culture collection must keep the organisms pure and unchanged in character and appearance and preserve their physiological and genetic properties [12]. For the purposes of conservation, culture collections try to retain samples of the original material deposited with them and to reduce the number of preservation cycles to which samples are subjected.

There is no universal technique for storing microbial cultures. In the early days pure cultures were maintained by serial transfer. This method maintains viability, but it is frequently ineffective in maintaining genetic integrity. Properties of fungi in culture may be unstable through loss of plasmids, spontaneous mutations, or genetic recombination due to the presence of heterokaryons, the parasexual cycle, or normal sexual events. In general, the method used aims to minimize the risk of changes and eliminate frequent subculture by extending the periods between subculture or to bring cellular activity to a halt. Selection of a method by a collection is based on future use of the culture and the advantages or disadvantages of each method and is determined by the equipment, personnel, and financial resources available. Most major collection employ at least two methods in the preservation of their cultures.

1. Subculturing

In a small collection the special requirements (media, pH, light, temperature, frequency of transfer) needed to preserve the characteristics desired in the individual strains can be accommodated with periodic subcultures. This seemingly simple system has disadvantages. With every transfer there is the danger of introducing contamination, of selecting cells with certain characteristics that may eliminate the "normal" character of the strain, and of the gradual weakening and eventual death of the strain through continued growth under artificial conditions.

In larger collections the problems associated with subculturing are multiplied significantly by the increased staff workload and the possible decrease in time available for individual attention to specific details.

The intervals between transfer of cultures for maintenance depend on the requirements of the organism as well as on storage conditions. At room temperature the periods will vary from days to a few months. To avoid rapid drying of cultures on agar medium and to reduce the metabolic rate of the organism, fresh mature cultures, which are not sensitive to cold, can be stored in a refrigerator at -10 to -20°C [13,14]. Under these low-temperature conditions, the viability period of most fungi is lengthened, and the interval between transfers can often be longer than at room temperature.

2. Immersion in Mineral Oil

The storage life of a fungal culture can also be extended by covering a fresh, well-grown agar slant with sterile mineral oil, which reduces dehydration of the medium and retards metabolic activity of the culture [15]. Oil-covered slants are stored at room temperature or under refrigeration depending on the culture [16]. Cultures preserved under oil grow at a reduced rate and will survive for years. As with most methods of preservation, much depends on the condition and vitality of the original isolate.

The oil overlay method is especially applicable to strictly mycelial or nonsporulating fungal forms that cannot be freeze-dried successfully or to small collections in which the installation of lyophilization or liquid nitrogen freezing procedures would not be economical. However, transfer of cultures under oil is messy, and regrowth is sometimes slow.

3. Immersion in Water

Storage in sterile distilled water is an alternative method for extending the shelf life of some cultures. Microbial suspensions, agar blocks, or disks cut from plates of actively growing cultures are placed into test tubes half filled with sterile distilled water and capped to prevent evaporation. The tubes can be stored at room temperature or refrigerated, depending on the thermal sensitivity of the organism. This method is simpler and less messy than the oil overlay method; however, cultures generally will survive longer under oil (reportedly 1-32 years) than in water storage (2-7 years) [17,18].

4. Drying

Microorganisms stored at lowered temperatures, under oil, or in distilled water are in a state of reduced metabolism, which slows the production of metabolites and the aging of the medium. Those stored under dry conditions become dormant and will remain viable for many years. For drying and ease of handling, spores have been stored in soil, sand, silica gel, and gelatin, or on filter paper or on porcelain or glass beads.

Since 1918 scientists have stored fungi in sterilized soil for months or even years with little or no variability in comparison with the same cultures maintained by serial subcultures. This method is used in many industrial collections to preserve viability and the characteristics of commercially important microorganisms [19-22]. In this procedure moistened soil or sand is inoculated with a small amount of inoculum and incubated and grown before drying, or dried soil is inoculated with a large amount of cell or spore suspension and dried immediately. The advantages of the soil culture method are increased longevity of the culture, reduction or elimination of morphological changes, and the availability of a uniform inoculum for many years.

The FGSC dries *Neurospora* spores and mycelia on anhydrous silica gel [23,24]. Ninety percent of 700 strains of genetic stock cultures, including nutritional and morphological mutants, survived for up to 5 years using this method. The viability of nonconidiating strains is greater with this method (drying without freezing) than with lyophilization [23].

5. Freeze Drying (Lyophilization)

Freeze drying or lyophilization is an effective and economical means for the long-term preservation of microbial cultures. It is a process of simultaneously freezing and drying microbes under vacuum, or the freezing part of the process may be carried out separately from the drying. Drying a culture will reduce metabolic activity to a minimum by removal of water necessary for functioning. Freezing a culture will accomplish the same purpose, since water in the frozen state cannot be utilized for metabolism. Both of the methods, freezing and drying, are used for preservation, but both have disadvantages. Lyophilization was developed to try to solve these problems. However, it can be disruptive to the cell, and mycelial cultures or cultures with fragile or elaborate spores usually do not survive the process.

In practice there are limiting factors to the success of freeze-drying for long-term preservation. Fresh isolates are more likely to have longer survival times than cultures maintained for years in subculture before being freeze-dried. Cell suspensions must be sufficiently heavy to ensure the survival of adequate numbers of cells from any freezing injury that may occur. A high percentage of viable cells is also necessary to minimize the selection of mutants.

Living cells will not survive freeze-drying without a protective agent. Many suspending media have been tested for their protective qualities, namely, to prevent total drying and to buffer the effects of electrolyte concentration during dehydration. Bovine serum was the earliest additive used. Others are sodium glutamate, peptone, various sugars or mixtures of sugars, and skim milk. Once a culture has survived the freezing shock, it will probably retain its viability for years.

In the double-vial batch-type method currently used at the ATCC, the spore suspensions "slow-freeze" before primary drying at low temperatures, secondary drying at room temperature, and sealing off under a vacuum. Freeze-dried preparations may be stored at room temperatures, but greater longevity is achieved by storage at 4°C.

Freeze-drying extends shelf life, and isolates generally retain their morphological, biochemical, and physiological characteristics. The rehydration procedure has been found to be an important factor in the viability of freeze-dried cultures. Freeze-dried cells must be allowed sufficient time to reabsorb moisture before transfer to a solid medium [25,26].

Liquid drying in vacuo, referred to as L drying, is similar to freeze-drying insofar as the operation takes place in ampoules attached to a manifold and immersed in water at 20°C. The vacuum level is adjusted to allow evaporation without freezing [27-29]. This method has been successfully used in preservation of various microbial cultures at the Institute for Fermentation in Osaka, Japan (IFO) [30-37]. L drying has the following advantages over freeze-drying: organisms are dried without freezing, nonabsorbent cotton wool plugs are used to prevent contamination, and the cells dry faster.

6. Cryogenic Storage

Since the rates of mitotic recombination and mutation are likely to correspond to those of cell division and metabolic activity, an ideal system for microbial germplasm conservation would be to store living cultures in such a manner as to achieve a complete cessation of cell division and total arrest of metabolism while the cultures still remain viable. This may be accomplished by storage below -139°C, where there is no growth of ice crystals and the rates of other biophysical processes are too slow to affect cell survival. Successful storage of a broad range of living materials has been achieved through the application of cryogenics at below -100°C.

The critical periods in cryopreservation are during the freezing and thawing processes. In general, best results in the survival and recovery of most microbial cultures have been obtained using a slow cooling rate (e.g., 1°C per minute) and rapid thawing.

To reduce injury upon freezing and thawing, one feature common to most successful protocols for cryopreservation of living cells is the application of chemical cryoprotectants. These are of two types: penetrating agents such as glycerol and dimethylsulfoxide (DMSO), which readily pass through the cell membrane and exercise their protective effect within intracellular and extracellular conditions, and nonpenetrating agents such as sucrose, lactose, glucose, mannitol, sorbitol, dextran, polyvinylpyrrolidone, and hydroxyethyl starch, which exert their protective effect external to the cell membrane. By far, glycerol and DMSO have been proven most effective for fungi.

The most satisfactory broad-spectrum, long-term storage technique now in use is storage in liquid nitrogen at -196°C or in liquid nitrogen vapor at a temperature range of -150 to -180°C, which may give almost infinite preservation. This technique causes little or no genetic change in the cells [4,12,38,39]. Cost is the main reason it is not employed exclusively in some collections. For shipping purposes one disadvantage of cryopreservation for a service collection is that strains preserved in liquid nitrogen must first be grown on agar or in liquid medium to avoid the expense of shipping frozen materials.

C. Supply of Strains

Service culture collections differ from other departmental, institutional, or industrial collections in that they supply strains for research, teaching, and industry. Because of the nature of their holdings, different collections distribute more cultures to certain categories of user than others. For example, the ATCC distributes more than half of its cultures to industry. The activities of a service collection take into account the requirements of its users, and adequate stocks are maintained to meet the demand. It is not easy to predict future needs, but attempts are made to ensure that sufficient supplies of the most desirable strains are available. In cases where biotechnologists are searching for particular properties, the most cost-effective method is to screen a large number of isolates already maintained in a culture collection. This allows for the testing of a broad spectrum of taxonomic groups from various substrata and regions of the world and saves the time and expense required for companies to collect their own material [2].

Not all cultures held in collections are available without restriction. In addition to patent strains, where special conditions apply, or safe deposit strains held on a confidential basis for the depositor, strains isolated by the staff or on which there is current research may be held in reserve. They are not included in the catalogs or generally made available to the public but can be supplied on special terms on a case-by-case basis [40].

Very few organisms used in the food, feed, or fermentation industries present any danger to public health or agriculture and are immune from legal restrictions on their movement in the United States and presumably in many other countries. The need for licenses and permits is mentioned in the catalogs of at least three of the major service collections: ATCC, CMI, and CBS. If restrictions are not included in the catalogs, they are either not required or not enforced. Although these restrictions seem formidable, curators ordinarily have little or no difficulty obtaining or sending cultures anywhere in the world [41].

D. Identification

With such a large number of fungus species and a dispersed systematic literature growing at the rate of around 1200 titles each year, the extent to which a nonspecialist can expect to identify fungus isolates with confidence is limited [42]. Historically, many service collections were established for studying or supporting the study of taxonomy, and this function still underlies much of a service collection's activity [43]. It is an area in which the collection staff generally has a great deal of expertise. As culture collections often hold large numbers of isolates of particular genera or species, they are ideal centers from which taxonomists can work to improve the delimitation of species of fungi [2].

The methods used to identify fungi differ widely and may range from microscopic examination to lengthy and sophisticated experiments. The direct comparison of one culture or specimen with another is a critical part of the identification process. Taxonomies are continuously being improved as more isolates of species become available for comparative studies and as new data on their characteristics are obtained. Identification of most groups of filamentous fungi continues to be based almost exclusively on their outward appearance. This is the result of both the relatively large numbers of morphological characters exhibited and the considerable number of species known only on their natural host or substratum [42].

Some groups of organisms, such as yeasts and, more recently, some other fungi, are identified with the help of biochemical tests. Taxonomic clarification using biochemical criteria has provided metabolic and other physiological data to users of the collection. This has been particularly beneficial for biotechnologists seeking organisms on the basis of activity rather than name, and is a real asset to those concerned with differentiating pathogenic from nonpathogenic or toxic from nontoxic strains of the same species of fungi [44].

Computers have been used for the identification of some organisms. A probability matrix has been prepared with which the behavior of unknown strains can be compared. The program gives an identification score to the taxon most like the unidentified microorganism. The success of this kind of system depends on the quality of the data from which the probability matrix has been derived. The advantage is that it enables a large number of phenotypic characters to be used simultaneously, and no taxon is eliminated by virtue of its failure to compare a single character [3].

Nomenclature changes do occur from time to time and affect almost everyone who works with fungi. For examples, the preferred name of the perfect form of *Candida lipolytica* is now *Yarrowia lipolytica* rather than *Saccharomycopsis lipolytica*. The agent of chestnut blight *Endothia parasitica* is now called by the earlier name *Cryphonectria parasitica*. Between the 16th and 17th editions of the ATCC catalog most *Torulopsis* species were redefined as species of *Candida*, and many *Saccharomyces* species became *S. cerevisiae*. To provide proper identification and keep abreast of the nomenclature changes, a collection requires not only specialized personnel but also considerable reference material. The most effective identification services are consequently found in collections where skilled personnel have substantial library facilities as well as large reference herbaria at their disposal [2].

E. Information

Documentation, evaluation, and information retrieval are necessary if a culture collection is to be of any value as a genetic resource. These three processes are separate but so closely related that limitations in any one will reduce the usefulness of any sizable collection, particularly for biotechnology.

The traditional classes of data used to describe strains in culture collections include morphological, physiological, biochemical, genetic, and historical data. Biotechnology places new demands on the informational service of collections. Besides requesting taxa with specific properties, technologists also desire information on the potential use of strains in biotechnological processes and the actual or potential hazards arising from their use. Service collections are now collecting data on additional attributes of interest to biotechnology such as temperature tolerances, behavior in fermentors, toxicity of products, and pathogenicity for unusual hosts [43].

The ultimate source of data is the laboratory records of the collections. Since the quality of each culture held by the collection depends on accurate culture data, a vast amount of paperwork can be generated in routine management activities. Information on a strain includes what comes with the accession form plus any additional data on growth conditions, media requirements, physiology, biochemistry, chemical products, pathogenicity, and known uses [45]. Data may be stored in books, on cards, or computerized. As the volume of information has increased, computers have come to play an increasingly greater role in data management [3]. Because of their economic importance in brewing and baking and the use of biochemical tests in their identification and evaluation, the yeasts have the most complete resource information among the fungi.

The extent to which collections are able to make available information on their holdings will also vary. Most of the major collections periodically publish catalogs. Information on the strains held is as comprehensive as possible, citing growth requirements, biochemical activities, metabolic products, temperature, pH optima, and literature citations. The ATCC catalog, for example, provides an essential resource for research and teaching and is also the primary means for the dissemination of information among those who use ATCC fungi.

In addition to catalogs, some larger culture collections also publish newsletters, technique manuals on culture collection practices and management, and articles on special applications of specific cultures, preservation techniques, and the results of research activities. Many of the smaller collections do not publish catalogs but contribute data to the World Data Bank, which will give an indication of the species held in the *World Directory of Collections of Cultures of Microorganisms*.

In recent years the increased demands of researchers and new or changing regulations, such as Good Laboratory Practices (GLP), have created a need for accessible and up-to-date documentation files relating to the cultures. With the large number of strains available at major resource collections, the recovery of any desired information in a readily usable form must be based on a computerized retrieval system. With the development of computer databases and strain data networks, use of this information service is becoming more frequent and more efficient. The use of computerized retrieval systems is of major interest to biotechnologists and genetic engineers, and has led to proposals to establish biotechnological strain databases with access online by industry.

Development of the RKC microbiological coding system in 1971 was fundamental to the formation of a computerized microbial database. The system uses a six-digit number associated with each test feature and covers morphology, physiology, biochemistry, genetics, serology, ecology, pathogenicity, nomenclature, and strain history. At present it allows encoding data primarily for bacteria and protozoa. In collaboration with M. I. Krichevsky at the National Institutes of Health (NIH), the Mycology Department at ATCC is expanding this system to include fungi and yeasts. A system for the coding of strain features for yeasts has been developed [46].

Dispersed microbial databases represent an untapped wealth of information which can be greatly enhanced if their activities are effectively coordinated. Examples of data banks for regional access include the Tropical Data Base in Brazil, the Microbial Information Network in Europe (MINE), and the Nordic Register, initially developed to provide information on the holdings and services of the collections themselves. Others, such as the Microbial Culture Information Service (MiCIS), have been set up in areas with well-established culture collection systems to provide an online database for searching. Proliferation of microbial data centers worldwide underscores the need for the establishment of international systems such as the Microbial Strain Data Network (MSDN) to direct those seeking information to the appropriate centers. The MIRCEN network will continue to encourage the establishment and development of culture collection activities in the developing world and link them to those in industrial nations [47].

Other types of information associated with service collections include the documentation accompanying patent and safe deposit cultures. This is essential, sometimes statutory, and must often be treated confidentially. Records of the result of identification tests is a type of information which might be used at a later time for taxonomic studies. Even sales records of cultures, which can be extensive, are of value. Analysis of sales and income can show to what extent the collection is being used and, therefore, serving a useful function [3].

Culture collections can give information and advice on broader subjects than their holdings. Those which are located within or in association with mycological research centers can draw on the expertise of the staff in recommending genera or species with particular attributes. Curators are usually familiar with the catalogs and holdings of other collections. Some collections have built up extensive reprint collections concerning the strains they hold which can be very useful for potential users [11].

F. Patent Depositories

Biotechnological inventions may include products, compositions, processes, and uses or methods of use. In biotechnological inventions which involve the use of new microorganisms, those not generally known and available to the public, or those not readily isolated from known and available sources, a written description is not usually considered sufficient for the purposes of disclosure. No matter how carefully the description may be worded, if the microorganism is unavailable, the invention cannot be reproduced. In such a case the strain itself forms an essential part of the disclosure [48].

In general, microbial processes and products can obtain patent protection as long as the basic criteria for patentability are fulfilled. The landmark Chakrabarty decision by the U.S. Supreme Court in 1980 allows the patenting of new life forms such as genetically engineered microorganisms [49].

The patenting procedures of many countries require that a sample of the microorganism used in the patent be deposited so that a ready supply of the microorganism is available at some stage of the patent procedure. In 1949 the U.S. Patent and Trademark Office (USPTO) implemented the requirement that cultures be deposited in conjunction with patent applications of microbiological inventions. The Patent Office asked for a culture collection to serve as depository for patent strains. The ATCC became the first U.S. collection and apparently the first in the world to accession a patent strain. Not only do culture collections act as patent depositories for strains referred to in patentable processes, but they also test for culture viability and can provide information in patent procedures.

To eliminate the problem of multiple deposits to cover each country where patent protection for inventions involving the use of microorganisms was sought, the Budapest Treaty on the International Recognition of the Deposit of Microorganisms for the Purposes of Patent Procedure was signed on April 28, 1977 and came into force on August 19, 1980. Under this treaty a single microorganism deposit with a recognized depository authority satisfies the requirement of all the countries selected in multicountry filings under the Patent Cooperation Treaty (PCT) or the European Patent Convention (EPC). Twenty-one countries and the European Patent Office (EPO) are now party to the Budapest Treaty.

Any culture collection can become an international depository authority (IDA), provided it has been nominated by a contracting member which must assure that the collection can comply with the requirements of the treaty. Qualification for IDA status is based on administrative and technical competence and agreement of the depository with the terms and conditions applicable to deposits for patent purposes. Criteria include the facilities themselves and ability to accept, maintain, and preserve the organisms, professional competency and objectivity, and financial stability. These conditions are important since the patent culture must be maintained under the terms of the treaty for 30 years [49]. Depository institutions with the status of IDAs that accept fungi are shown in Table 4.

All countries party to the Budapest Treaty must recognize a deposit made in an IDA, but not all require deposits to be made in IDAs and recognize other collections that comply with their requirements. Countries not party to the treaty accept deposits made in an acceptable and internationally known culture collection. Deposits made under the treaty are permanent, and the depositor cannot ask for it to be cancelled, regardless of whether a patent is eventually granted.

Most patent offices require the culture of the organism to be deposited not later than the filing date of the application for the patent. Under U.S. patent laws the culture must be deposited before the patent application is filed with the Patent Office because the application must bear the IDA accession number.

The IDAs also have certain procedural conditions which must be met when the microorganisms are submitted. Some will accept cultures on agar slants. Others require lyophilized pellets or both. Often a specific number of cultures must be provided. All must be accompanied by a deposit form with information about proper handling and storage of the culture, viability testing, and the assurance that the sample is a true representation of what is disclosed in the patent. The treaty requires the depository to provide its services to all depositors under the same conditions and with confidentiality. Upon acceptance by an IDA, an official receipt is issued and the culture is tested for viability. The results are forwarded to the depositor along with the accession number to be used on the patent application.

Depending on its policy and the kind of material being deposited, an IDA may or may not prepare subcultures for eventual distribution. For yeasts and fungi it is usual for the IDA to distribute its own preparations. In this case many IDAs will ask the depositor to check the authenticity of their preparations. The depositor is not obliged to do so, but this will ensure that the cultures distributed by the IDA will do what is claimed in the patent application.

When the patent strain becomes available varies in different countries. Since the invention must be reproducible through the life of the patent, the deposited microorganism must remain available for this time, generally 30 years. The depositor is allowed to set restrictions on the availability of the deposit. It is wise to restrict access to the deposit while the patent is pending. Upon issuance, however, the culture must be made available to the public, but only for experimental purposes. The depository will furnish the name of the requestor to the depositor. If it is to be used for commercial purposes, the requestor must get permission of the inventor in the event royalties are due [49].

In the last 15 years biotechnology has grown from a knowledge-building science to a new industry with a major influence on society. Protection of research and development results has become extremely important, and culture collections have played a significant role. Since the Chakrabarty decision the patent activity in the biotechnology industry has escalated rapidly [49].

TABLE 4. Depository Institutions with the Status of International Depository Authority That Accept Fungal Strains

Agricultural Research Culture Collection (NRRL), USA
American Type Culture Collection (ATCC), USA
Australian Government Analytical Laboratories (AGAL), Australia
Centraalbureau voor Schimmelcultures (CBS), the Netherlands
Collection Nationale de Culture de Micro-organismes (CNCM), France
Culture Collection of the Commonwealth Mycological Institute (CMI), UK
Deutsche Sammlung von Mikroorganismen (DSM), FRG
Fermentation Research Institute (FERM), Japan
In Vitro International, Inc. (IVI), USA
National Bank for Industrial Microorganisms and Cell Cultures (NBIMCC), Bulgaria
National Collection of Agriculture and Industrial Microorganisms (NCAIM), Hungary
National Collection of Yeast Cultures (NCYC), UK

G. Other Depository Services

Collections can also offer a safe deposit service for cultures not listed in their catalogs. These might include proprietary cultures of which the depositor retains ownership and confidentiality. Users of this service are generally from industry, where a successful production strain may represent a high investment and there is concern for the continuing availability of the culture for further use [3].

Research projects often extend over a period of years. Laboratories without adequate culture preservation facilities run the risk of losing strains through contamination or accident before an investigation is completed. If isolates are deposited in a culture collection, not only are they recoverable if lost, but they also can be supplied to other investigators to verify or develop work on the same strains [2]. This type of service is intermediate between an open collection deposit and a deposit for patent purposes. Most collections charge to cover the long-range storage and quality control procedures that are required [44].

Collections not only offer backup for culture preservation but can also provide a second location for storage of cultures as insurance against loss through accident at the laboratory of origin [3]. Most major collections preserving important isolates themselves house backup cultures at another facility.

H. Consultancy and Contract Work

Many collections have staff with expertise in specialized areas who are available for the consultancy work which is often necessary before detailed investigations or research programs can begin [44]. Companies involved in biotechnology using fungi can be assisted by the major fungus collections in several ways. In many cases the consultancy must be kept confidential. Collections can recommend selection of strains for specific uses and purposes; techniques on preservation, packaging, and shipping; Good Laboratory Practice requirements regarding the accuracy and traceability of measuring instruments; quality control procedures, record keeping, management, and the administration of nonclinical experiments. They can also provide specialized bibliographies to the investigators.

The most important area in which industry might consider contract research with culture collections is probably the screening of isolates for particular attributes. Investigative work on mycological biodegradation and biodeterioration is offered by the CMI and some other culture collections. Troubleshooting services could handle such issues as the spoilage of food and manufactured goods, fuel contamination, conservation of library and museum materials, damage to building materials and stored agricultural materials, and identification of plant diseases. Moreover, collections carry out studies of growth requirements of cultures, develop educational and diagnostic kits, and establish and maintain computerized data resources [2].

Many modern standards and specifications require the use of microorganisms in the testing of products for toxicity, fungal resistance, sterility, preservative effectiveness, biocompatibility, and mutagenicity. These tests could be carried out in any suitably equipped laboratory, but culture collections are in a particularly good position to do this using their own strains normally supplied to outsiders for the same work. This can be very cost-effective for a company which would have to set up and staff a special laboratory to carry on the testing. Collections themselves may be identified as the source of material for testing standards.

Culture collections are sometimes asked to provide bulk amounts of an organism to commercial organizations or research institutes with limited facilities. This is also quite cost-effective because the collection has the experience of growing the organism and access to the media. Special requests may also come from government agencies for cultures to be used for laboratory certification and proficiency testing in clinical laboratories. Some manufacturers of diagnostic instruments may request cultures for calibrating purposes [44].

I. Training

Many culture collections are involved in educational activities either because of the scientific background of the staff or because of their detailed knowledge of culture collection practices. Larger collections are able to undertake training that is more relevant to biotechnologists. They can arrange programs in direct response to a need identified either by the collection itself or by an outside source. They provide workshops and training courses related to collection activities, such as preservation and maintenance techniques, biotechnology patent conferences, strain data management, identification and quality control of microbial cultures, as well as on a variety of industrial and scientific topics.

J. Research

It is essential for culture collections to carry on research pertinent to their interests and services. All collections want to increase the information on their holdings, and unless service collections keep abreast of the new developments in associated areas of science, they lose sight of the changing requirements of their users.

In association with taxonomic studies new identification methods are devised. Preservation techniques are also developed and improved. Major culture collections are ideal places to undertake research of primarily biotechnological importance. The increased use of genetically engineered strains and strains of industrial importance has emphasized the need for a much greater understanding of the molecular processes that occur during drying or freezing and thawing [2]. New methods that ensure stability of strains need to be developed, e.g., those dealing with plasmid-containing fungi.

Additional research topics of biotechnological importance might include the following: the screening of large numbers of isolates for particular uses or desired properties; comparative studies of strains of the same or closely related species in making a particular enzymic conversion or producing particular metabolites; development and evaluation of detection procedures for particular compounds or activities requiring testing of a large range of fungi; studies on the use of fungi as biocontrol agents against insects, weeds, or other microorganisms; selection of test organisms used in the evaluation of materials; development and testing of short-term preservation and cultivation methods suitable for use in other laboratories; studies on the growth requirements of particular groups or species; testing of types of fermentors for their effectiveness using different groups of fungi; and evaluation of media, culture containers, diagnostic reagents, and procedures.

The curators of most collections are scientists involved to varying degrees in personal research programs. However, most collections do not have the funding or the staff to undertake additional research activities without outside financial support. The potential contribution of culture collections to biotechnological research is likely to be realized only through research grants from the government or industry or the undertaking of research on specific topics on a consultancy basis.

V. CONCLUSION

Culture collections are more than places for the deposit and supply of cultures. They are now being recognized as a unique resource of genetic material and information, and service collections in particular carry on a large number of activities for the basic support of biotechnology. Many collections have evolved from being useful adjuncts to research institutes or teaching institutions to being essential components for a rapidly expanding branch of science. The resources offered by culture collections have been developed to only a fraction of their potential. The microbiological requirements of biotechnology are stimulating efforts to initiate or extend their services. The cost for their support is modest in comparison to the benefits for industry.

The developments in biotechnology have coincided with the advances in computer technology. Data held in the culture collections are becoming increasingly computerized and accessible. The more culture collections can achieve this, the better able they will be to serve biotechnology.

It is difficult to predict which cultures will possess properties of importance to biotechnology in years to come. It is important for the future of biotechnology that culture collections be maintained, supported, and further developed to meet the challenges of the future.

REFERENCES

1. Jong, S. C. *Microbial Germplasm. Proceedings of Beltsville Symposium XIII, Biotic Diversity and Germplasm Preservation—Global Imperatives* (L. Knutson and A. K. Stoner, eds.), Kluwer Academic, Hingham, MA, 1989, pp. 241-273.
2. Hawksworth, D. L. Fungus culture collections as a biotechnological resource. *Biotechnol. Genet. Eng. Rev.*, *3*:417-453, 1985.
3. Kirsop, B. The current status of culture collections and their contribution to biotechnology. *CRC Crit. Rev. Biotechnol.*, *2*:287-314, 1985.
4. Jong, S. C., and Davis, E. E. Germplasm preservation of edible fungi in culture through cryogenic storage. In *Cultivating Edible Fungi* (P. J. Wuest, D. J. Royse, and R. B. Beelman, eds.), Elsevier, Amsterdam, 1987, pp. 213-225.
5. Staines, J. E., McGowan, V. F., and Skerman, V. B. D. *World Directory of Collections of Cultures of Microorganisms*. Brisbane, World Data Center, University of Queensland, 1986.
6. Burnett, J. H. 1975. *Mycogenetics*. John Wiley and Sons, London, 1975, pp. 235-236.
7. Hawksworth, D. L. Resource centres. In *Living Resources for Biotechnology: Filamentous Fungi* (D. H. Hawksworth and B. E. Kirsop, eds.), Cambridge University Press, Cambridge, 1988, pp. 1-30.
8. Hawksworth, D. L. Preface. In *Living Resources for Biotechnology: Filamentous Fungi* (D. H. Hawksworth and B. E. Kirsop, eds.), Cambridge University Press, Cambridge, 1988, p. xi.
9. Jong, S. C. Production and management of spawn for mushroom industry. In *Proceedings of the 5th International Congress of Culture Collections* (L. I. Sly and P. Attasampunna, eds.), World Federation for Culture Collections, Cambridge, 1988, pp. 156-171.
10. Smith, J. E. *Biotechnology* (The Institute of Biology's Studies in Biology no. 136.) Edward Arnold, London, 1981.
11. Jong, S. C. Data management at the American Type Culture Collection. In *Critical Problems of Culture Collections* (L. R. Batra and T. Iijima, eds.), Institute for Fermentation, Osaka, 1984, pp. 234-40.
12. Jong, S. C., and Atkins, W. F. Conservation, collection and distribution of cultures. In *Fungi Pathogenic for Humans and Animals* (D. H. Howard, ed.), Marcel Dekker, New York, 1985, pp. 153-194.
13. Haynes, W. C., Wickerham, L. J., and Hesseltine, C. W. Maintenance of cultures of industrially important microorganisms. *Appl. Microbiol.*, *3*:361-368, 1955.
14. Carmichael, J. W. Viability of mold cultures stored at -20°C. *Mycologia*, *54*:432-436, 1962.
15. Sherf, A. F. A method for maintaining Phytomonas sepedonia in culture for long periods without transfer. *Phytopathology*, *33*:330-332, 1943.
16. Buell, C. B., and Weston, W. H. Application of the mineral oil conservation method to maintaining collections of fungus cultures. *Am. J. Bot.*, *34*:555-561, 1947.
17. Smith, D., and Onions, A. H. S. *The Preservation and Maintenance of Living Fungi*. Commonwealth Mycological Institute, Kews, 1983.
18. Boesewinkel, H. J. Storage of fungal cultures in water. *Trans. Br. Mycol. Soc.*, *66*:183-185, 1976.
19. Greene, H. C., and Fred, E. B. Maintenance of vigorous mold stock cultures. *Ind. Eng. Chem.*, *26*:1297-1298, 1934.

20. Fennell, D. I. Conservation of fungus cultures. *Bot. Rev.*, *26*:79-141, 1960.
21. Atkinson, R. G. Quantitative studies on the survival of fungi in five-year old dried soil cultures. *Can. J. Bot.*, *32*:673-768, 1954.
22. Weiss, F. A. Maintenance and preservation of cultures. In *Manual of Microbiological Methods* (Society of American Bacteriologists, ed.), McGraw-Hill, New York, 1957, pp. 99-119.
23. Perkins, D. D. Preservation of Neurospora stock cultures with anhydrous silica gel. *Can. J. Microbiol.*, *8*:591-594, 1962.
24. Perkins, D. D. Details for preparing silica gel stocks. *Neurospora Newslett.*, *24*:16-17, 1977.
25. Haskins, R. H. Factors affecting survival of lyophilized fungal spores and cells. *Can. J. Microbiol.*, *3*:477-485, 1957.
26. Onions, A. H. S. Preservation of fungi. In *Methods in Microbiology*, Vol. 4 (C. Booth, ed.), Academic Press, New York, 1971, pp. 113-151.
27. Annear, D. I. Preservation of bacteria. *Nature*, *174*:359, 1954.
28. Annear, D. I. The preservation of bacteria by drying in peptone plug. *J. Hyg.*, *54*:487, 1956.
29. Annear, D. I. Recoveries of bacteria after drying on cellulose fibers: A method for the routine preservation of bacteria. *Aust. J. Exp. Biol. Med. Sci.*, *40*:1-8, 1962.
30. Iijima, T., and Sakane, T. A method for preservation of bacteria and bacteriophages by drying *in vacuo*. *Cryobiology*, *10*:379-385, 1973.
31. Iijima, T., and Sakane, T. Method for preservation of bacteria and bacteriophages by drying *in vacuo*. *IFO Res. Commun.*, *6*:4-17, 1973.
32. Banno, I., Mikata, K., and Sakane, T. Viability of various yeasts after L-drying. *IFO Res. Commun.*, *9*:27-34, 1979.
33. Banno, I., and Sakane, T. Viability of various bacteria after L-drying. *IFO Res. Commun.*, *9*:33-45, 1979.
34. Mikata, K., Yamauch, S., and Banno, I. Preservation of yeast cultures by L-drying: Variabilities of 1710 yeasts after drying and storage. *IFO Res. Commun.*, *11*:25-46, 1983.
35. Yokoyama, T., and Asano, I. Preservation of ISP strains of Actinomycetes by L-drying. *IFO Res. Commun.*, *11*:47-59, 1983.
36. Kusaka, T., Sato, K., and Asano, I. Preservation of antimicrobial activity in L-dried and agar-transplanted Actinomycetes. *IFO Res. Commun.*, *12*:90-100, 1985.
37. Kaneko, Y., Mikata, K., and Banno, I. Maintenance of recombinant plasmids in Saccharomyces cerevisiae after L-drying. *IFO Res. Commun.*, *12*:78-82, 1985.
38. Jong, S. C. Conservation of the cultures. In *The Biology and Cultivation of Edible Mushrooms* (S. T. Chang and W. A. Hayes, eds.), Academic Press, New York, 1978, pp. 119-135.
39. Jong, S. C. Isolation, cultivation, and maintenance of conidial fungi. In *Biology of Conidial Fungi*, Vol. 2 (G. T. Cole and B. K. Kendrick, eds.), Academic Press, New York, 1981, pp. 551-575.
40. Hawksworth, D. L., and Allner, K. Administration and safety. In *Living Resources for Biotechnology: Filamentous Fungi* (D. L. Hawksworth and B. E. Kirsop, eds.), Cambridge University Press, Cambridge University Press, Cambridge, 1988, pp. 54-74.
41. Hesseltine, C. W., and Haynes, W. C. Sources and management of micro-organisms for the development of a fermentation industry. *Agriculture Handbook No. 440*. Agricultural Research Service, U.S. Department of Agriculture, Washington, D.C., 1974, pp. 1-38.

42. Hawksworth, D. H. Identification. In *Living Resources for Biotechnology: Filamentous Fungi* (D. L. Hawksworth and B. E. Kirsop, eds.), Cambridge University Press, Cambridge, 1988, pp. 100-114.
43. Krichevsky, M. I., Fabricius, B.-O., and Sugawara, H. Information resources. In *Living Resources for Biotechnology: Filamentous Fungi* (D. L. Hawksworth and B. E. Kirsop, eds.), Cambridge University Press, Cambridge, 1988, pp. 31-53.
44. Allsopp, D., and Simione, F. P. Culture collection services. In *Living Resources for Biotechnology: Filamentous Fungi* (D. L. Hawksworth and B. E. Kirsop, eds.), Cambridge University Press, Cambridge, 1988, pp. 162-172.
45. Jong, S. C., and Gantt, M. J. *ATCC Catalogue of Fungi/Yeasts*, 17th ed.: with an index to industrial applications. American Type Culture Collection, Rockville, MD, 1987.
46. Jong, S. C., Holloway, L., McManus, C., Krichevsky, M. I., and Rogosa, M. Coding of strain features for computer-aided identification of yeasts. *Mycotaxon, 31*:207-219, 1988.
47. Kirsop, B. E., and DaSilva, E. J. Organization of resource centres. In *Living Resources for Biotechnology: Filamentous Fungi* (D. L. Hawksworth and B. E. Kirsop, eds.), Cambridge University Press, Cambridge, 1988, pp. 173-187.
48. Bousfield, I. J. Patent protection for biotechnological inventions. In *Living Resources for Biotechnology: Filamentous Fungi* (D. L. Hawksworth and B. E. Kirsop, eds.), Cambridge University Press, Cambridge, 1988, pp. 115-161.
49. Jong, S. C., and Birmingham, J. M. Disclosure requirements for biological materials in patent law. *Adv. Appl. Microbiol., 35*:255-293, 1990.

33

LEGAL PROTECTION FOR BIOTECHNOLOGY INVENTIONS

ROMAN SALIWANCHIK and DAVID R. SALIWANCHIK *Saliwanchik & Saliwanchik, Biotechnology Patent Law, Gainesville, Florida*

I. INTELLECTUAL PROPERTY

The first step toward effective protection of intellectual property rights is to recognize what intellectual property is. Unlike real estate, a car, a diamond ring, and most other forms of property, intellectual property is often intangible; it is the product of creative human thought.

Some examples of intellectual property include accounting methods, songs, processes for making beer, creative names for products, better mousetraps, and better mice. Clearly, the term "intellectual property" embraces a wide variety of subject matter. The common denominator of this subject matter is that these ideas have value.

The primary purpose of this chapter is to provide information that will enable researchers to derive their fair economic gain from the fruits of their valuable work. This is not intended to be a treatise on patent law. Patent law is an intricate area of legal practice. In order to practice patent law, lawyers must pass a specialized examination in addition to the bar examination. It would not be possible or constructive to delve into the complexities of patent law here. This chapter is not designed to enable scientists to dispense with the need for patent counsel; rather, it is intended to prevent researchers from jeopardizing their property rights before an attorney is consulted. Also, the information provided here will help the researcher work with the patent attorney to obtain the best possible legal protection.

Another reason for avoiding overemphasis on patent law is that patents are not the only means by which research results can be protected. Trade secrets, which will be discussed in detail below, often provide an attractive alternative to patent protection. Also, copyrights and trademarks can be used to protect some forms of intellectual property.

This chapter is intended primarily for researchers in the biological or chemical sciences. Although the same general laws govern inventions in all fields of science, each scientific discipline has evolved its own patent law customs, language, and requirements. This is one reason why it is important to select a patent counsel who is familiar with and experienced in your particular scientific discipline. Choosing a patent attorney who is trained in your scientific specialty increases the chances that the attorney will be able to fully understand your invention and procure the best possible legal protection for you.

II. BIOTECHNOLOGY INVENTIONS

As discussed above, intellectual property can take many forms. The legal means by which this property can be protected is largely dependent on the nature of the property which is to be protected. For example, in order to prevent somebody from copying a poem or book without permission, the author must resort to copyright laws. Trademarks are used to protect property rights in distinctive names or slogans associated with particular consumer goods or services. Although these forms of legal protection can be very valuable and important, the laboratory researcher is less likely to be concerned with these forms of intellectual property than with the laws pertaining to inventions. Therefore, the discussions which follow will focus on the available methods for protecting rights in inventions.

An invention can be defined as "a discovery; something originated; something created by thought or experiment; something devised for the first time." It is probable that an invention has been made when something new and useful has been conceived or developed, or when unusual, unexpected, or unobvious results have been obtained and can be exploited. Obviously, the term "invention" is a very broad concept. There are many different aspects of biotechnology research which can be termed inventive.

It is critical for the researcher to be able to recognize inventions in order to properly patent potential rights to these inventions. Table 1 is a list of some biotechnology inventions. The list is intended to impress on the reader the great diversity in the things which fall under the general category of "biotechnology inventions." We often find that researchers are surprised at the types of things which are considered inventive. Frequently, researchers acknowledge that they have had many inventions go unprotected because their concept of "invention" was too narrow.

Another common misconception regarding the nature of inventions relates to the existence, or nonexistence, of a working model. Often researchers are under the mistaken impression that an invention does not exist until it has been "reduced to practice," i.e., when the experiments have all been carried out or a final result is obtained. This, again, is an overly narrow view of invention. An invention has been made when a person conceives of an idea in sufficient detail that the idea can be realized by simply following the directions of the inventor. Thus, when a researcher has an idea which the researcher knows will work but which will take a lab technician 6 months to implement, it is important for that researcher to take immediate action to protect property rights. The researchers should not wait 6 months until the idea has been finally, and predictably, reduced to practice.

Similarly, if an inventor discovers a particular useful biological control agent, steps should be taken to protect not only the exact agent which was discovered but also any related agent which would be expected to produce the same results. In this regard, the scientist and patent attorney should work together in a common effort to ensure that the coverage of the inventor's legal protection is commensurate with the scope of all that has been invented by the scientist.

III. THE TWO PRIMARY FORMS OF LEGAL PROTECTION FOR BIOTECHNOLOGY INVENTIONS: PATENTS AND TRADE SECRETS

People often associate inventions with patents. Although many inventions are in fact patented, many are not. To understand why an invention may

TABLE 1. Some Biotechnology Inventions

Patent no.	Subject matter
4,689,231	A method for protecting plants from nematodes comprising administering a nematocidally effective amount of *Meria oniospora* spores
4,225,381	A method for removing the odor from a fluid comprising inoculating and cultivating at least one fungus selected from the group consisting of Cephalosporium Ferm-P No. 3876 and Cephalosporium Ferm-P No. 3977
3,999,973	A composition for agricultural application for controlling undesirable vegetation comprising *Colletotrichum malvarum* spores in association with an agricultural carrier
3,849,104	A process for controlling northern jointvetch weeds comprising inoculating the weeds with an amount of the fungus *Colletotrichum gloeosporiodes* f. sp. *aeschynomene* sufficient to infect and kill the weed
3,672,862	A process for killing pigweed by applying a phytotoxin to the leaves of the pigweed
4,331,659	A process for recovering antibiotic U-62,162 from a fermentation beer *Streptomyces verdensis* Dietz and Li
4,736,866	A transgenic nonhuman mammal, all of whose germ cells and somatic cells contain a recombinant activated oncogene sequence
4,735,896	A peptide which is specifically immunoreactive with antibodies to HTLV-III
4,735,896	An immunoassay method for the detection of antibodies to HTLV-III and diagnosis of AIDS, ARC, or pre-AIDS condition
4,082,613	A process for the production of insulin by genetically transformed fungal cells
4,713,336	A biologically pure culture of a lignin-degrading microorganism of the genus *Erwinia* having all the identifying characteristics of ATCC 39873
4,713,336	A vector comprising a DNA segment which codes for an aryletherase protein
4,616,099	A chrysanthemum plant of the *Morifolium ramat* species having a rapid growth rate
4,686,319	A process for synthesizing true-breeding female lines of squash
4,710,377	Monoclonal antibodies produced by hybridomas formed by fusion from a mouse myeloma line and spleen cells from a mouse previously immunized with *Eimeria tenella* sporozoites
4,710,377	A proteinaceous vaccine which comprises anticoccicidally effective amounts of *Eimeria tenella* sporozoite antigen

not be patented, it is first necessary to have some understanding of what a patent is, as well as of the requirements and procedures for obtaining one.

A. Patents

In general, a patent represents a set of legal rights which are bestowed on an inventor by the government when an invention satisfies certain specific requirements. Each country has its own patent laws. In the United States, the recipient of a patent has the exclusive right to make, use, or sell the invention in the United States for a period of 17 years commencing on the day when the patent issues. The patent laws of other developed countries grant similar rights. The text of a patent for a biological invention is provided in Appendix A.

1. Purpose of Patents

The primary purposes of awarding these exclusive rights to an inventor are to (a) encourage the investment of time, money, and effort into the pursuit of new products for society, and (b) encourage and facilitate the dissemination of new knowledge throughout the scientific community and the public at large.

The first of these objectives is achieved by the granting of the limited monopoly to inventors. The rights to financial proceeds is a particularly appropriate and fair reward because it is generally proportional to the value of the idea to society. The second objective—public dissemination of new ideas—is achieved by publishing the full text of issued patents. Thus, when a patent issues, the complete details of that invention become immediately available to the public so that interested parties can build on this knowledge, thus accelerating the rate of progress.

Although public dissemination of the details of patented inventions may be of great benefit to society, it can pose a dilemma for the person seeking to protect property rights. Obviously, if the invention is of a nature where it might be difficult to identify unscrupulous competitors who choose to infringe the patent, then patent protection (and the inevitable public disclosure) may not be an attractive option. This situation, which will be discussed in detail below, is one of the primary reasons why not all inventions are patented.

Another important reason why some inventions are not patented is that not all inventions meet the strict requirements for patentability. Recognizing that the grant of a 17-year monopoly may affect society's ability to commercialize and develop products, the government has enacted strict standards which are designed to ensure that only deserving inventions will be granted patent protection.

2. Patentability Requirements

In the United States there are three primary requirements which an invention must meet in order to be patentable. These are novelty, utility, and nonobviousness [1]. Most foreign countries have analogous requirements.

Although these requirements may sound relatively straightforward, over the years a tremendous amount of case law interpreting them has evolved. A complete treatment of these standards of patentability is well beyond the scope of this chapter. However, a brief introduction to these requirements is in order.

Novelty: A patent applicant must demonstrate that his or her invention is new. To be "new" under the patent laws, an invention must not have been available to the public prior to the time when the applicant for patent "invented" it. Thus, if an uninformed researcher were to independently "discover" penicillin today, a patent would not be awarded because, obviously, penicillin is already in the public domain. Therefore, the *date of invention* is a critical factor in determining if an invention meets the patent law standard of novelty.

Another factor in novelty determinations is the *application date*, i.e., the date on which a patent application is filed. In the United States, even if a researcher is the first to invent something, the invention does not meet the patent law novelty test if the invention has been available to the public, e.g., in a printed publication, more than a year prior to the filing of a patent application.

For example, suppose that on September 3, 1988, a molecular biologist identifies for the first time a gene which produces a useful protein. On June 1, 1989, the researcher publishes the results. If this researcher does not file a patent application for the new gene before June 1, 1990, the researcher will be forever prevented from obtaining patent protection on this invention. The 1-year deadline is frequently referred to as a "statutory bar" because this bar to patentability originates from the patent statutes.

Although this law may seem harsh, most countries do not allow *any* grace period for the filing of an application. For example, Japan, Canada, and most European countries follow an "absolute novelty" system where patentability is barred as soon as any public dissemination occurs. The United States is unique in allowing the 1-year period to file an application.

As the reader can readily see, publications and other public disseminations can have a profound effect on patent rights. Several important aspects of the law regarding publications are discussed in greater detail in Section V below.

Utility: Another requirement for patentability is that the invention be useful. Often this is the easiest requirement to fulfill. The law does not require that the invention be better than existing products or even that the invention work well. Rather, the applicant must merely establish some legitimate use for the product. This is not to say that the patent office never rejects applications on the grounds of lack of utility. This type of rejection is most frequently encountered in situations where the applicant claims what is considered to be an "incredible utility." For example, if an applicant claims that a compound will return people to their youth, that applicant can expect a rejection based on lack of utility. In this situation, the lack-of-utility rejection does not mean that a "fountain-of-youth" compound would not be useful. Rather, the rejection is based on the premise that the claim is so incredible that it cannot be true, and thus the compound lacks legitimate utility.

An applicant faced with a utility rejection can overcome the rejection by providing evidence which supports his or her claim. This can occur, for example, in an application which claims a method for curing cancer. Although such a claim may be incredible, if the applicant can support the claim with convincing data, then the utility rejection will be overcome.

Nonobviousness: The requirement that an invention be "nonobvious" in order to be patentable is the most subjective of the three requirements and is often the most problematic requirement for the applicant to meet. The U.S. patent statutes express this requirement as follows:

> A patent may not be obtained though the invention [satisfies the novelty requirements] if the differences between the subject matter sought to be patented and the prior art are such that the subject matter as a whole would have been obvious at the time the invention was made to a person having ordinary skill in the art to which said subject matter pertains [2].

Stated more simply, if a person trained in your scientific field would have considered your invention to be an obvious improvement over what already is known to the public, then you will not be awarded a patent. One of the critical aspects of this test is that the hypothetical person having ordinary skill in the art is assumed to have knowledge of *all* relevant publications and public information. Also, this hypothetical person is allowed, at least in some instances, to combine the teachings contained in publications which are widely separated geographically, chronologically, and possibly even in subject matter.

The purpose of the nonobvious criterion is to prevent the granting of patents to inventions which are merely predictable and/or small advances over the available art. Thus, it is incumbent on the applicant to demonstrate that the invention was unexpected, highly advantageous, or otherwise more than the next logical step in the course of research. Factors which favor an applicant's assertion of nonobviousness include commercial success, solving a long-standing problem, failure by others to achieve the same results, and the existence of publications which suggest that the claimed invention would not work.

3. Patent Procedures

The Patent Examiner: In addition to the three primary requirements of patentability, there are many other rules and regulations which must be satisfied before a patent will be issued. Many of these rules and regulations are rather technical and often pertain to, for example, the format of the patent application and the payment of appropriate fees.

The Patent Office bears the responsibility of determining, for each and every application, whether the invention is patentable and whether the formalities of the law have been followed. If and when all the requirements of the law are deemed to have been satisfied, then a patent will be issued.

The Patent Office employees given the responsibility of making these determinations are known as patent examiners. There are hundreds of examiners in the U.S. Patent Office. Each examiner has at least a bachelor's degree in some scientific field; many examiners have doctorates, are lawyers, and/or have significant work experience.

Each patent application received by the Patent Office is assigned to an examiner who is trained in the scientific field to which the invention pertains. Because of the tremendous number of patent applications received by the Patent Office, it is frequently 8-24 months before an examiner will have an opportunity to review an application. Applications are reviewed in the order in which they are received by an examiner.

The application: When an examiner begins reviewing an application, the examiner first makes sure that the application includes the necessary parts. A patent application must consist of a complete description of the invention and at least one concise one-sentence statement of what the inventor considers to be the invention. The concise statement of the invention is known as a "claim." The claim, or claims, appears after the description of the invention. The description is commonly referred to as the "specification." There is no limit to the number of claims which an application may have. Also, claims may be redundant. Claims must always be written with great care because the coverage of the patent is determined by what is claimed.

Enabling disclosure: An important rule pertaining to claims is that a claim is valid only if it is supported by what has been disclosed in the specification [3]. Thus, it is not permissible to claim subject matter which is not described within the patent application. For a disclosure to be adequate to support a claim, the disclosure must be of sufficient detail, clarity, and precision so as to enable a person skilled in that field to practice the invention without undue experimentation. Such a full, detailed description is known as an "enabling" disclosure.

Deposit requirements: If a crucial element of an invention is not disclosed or otherwise known in the art, then the disclosure is not considered to be enabling. This situation can occur when an invention requires the use of novel materials. If a source of the novel materials is not disclosed, then the disclosure is not enabling. For example, if a novel process for producing an antibiotic requires the use of a newly isolated fungus, a description of the process would not be enabling if it did not disclose a publicly accessible source of the new fungus [4]. For this reason, patent applications for biotechnology inventions often contain references to deposited cultures. The cultures have been placed on deposit at a depository which is recognized by the Patent Office. The conditions of the deposit must ensure that the biological material will be viable and available to the public for a period of at least 30 years commencing when the patent issues. A list of recognized depositories as well as an example of the form used to transmit biological materials to two of these depositories is provided in Appendix B.

The rules followed in the United States with respect to the deposit of biological materials are derived from, and consistent with, international law followed by many of the developed countries of the world. The primary source of this international law is the Budapest Treaty on the International Recognition of the Deposit of Microorganisms for the Purposes of Patent Procedure [5]. The Budapest Treaty establishes regulations for the deposit of biological materials for patent purposes. A deposit made according to the treaty's provisions will satisfy the patent law requirements with respect to biological deposits in each of the signatory countries.

Examination of applications: Once the examiner is satisfied that an application includes the essential parts, the examiner will then determine whether the invention as described and claimed meets the standards of patentability. If these requirements are satisfied and the proper fees are paid, a patent will be issued.

Generally, however, the examiner will find one or more reasons why the invention as claimed is not patentable. The examiner then puts the reasons for these rejections in a written communication which is sent to the applicant or the applicant's attorney. This written communication is known as an office action.

The applicant is then given the opportunity to respond to the examiner's rejection with appropriate arguments. If the applicant's response successfully overcomes the examiner's rejections, then the claims will be "allowed" and a patent will be issued when the proper fee is paid.

If the examiner does not find the applicant's arguments to be persuasive, then the examiner may renew the rejections and/or issue new rejections. This second set of rejections is often made "final." When a final rejection is issued, the applicant will be forced to appeal or drop the case if the examiner's mind is not changed by another response by the applicant.

In addition to making written arguments to overcome rejections, other means for winning allowances of claims include personal interviews between the applicant and/or the applicant's attorney and the examiner.

If interviews are not conducted or are unsuccessful, the applicant may also amend the claims so that they more clearly satisfy the requirements for patentability. In some circumstances, the applicant may also wish to amend the specification. However, if the amendment of the specification involves the addition of substantively new material, then the filing date for the new material is considered to be the date of the filing of the amended application.

When all else fails, there are appellate procedures which can be followed in order to have the examiner's decision reviewed.

Patent procedures outside the United States: Although an in-depth analysis of worldwide patent procedures is beyond the scope of this chapter, it should be noted that most patent application requirements are not so easily satisfied on the international level as are the requirements for biological deposits.

With a few exceptions, a patent application must be filed in each country where protection is sought. These patent applications and the invention which they describe must meet the substantive and procedural requirements of the particular country where protection is sought. Thus, it is possible that an invention may be patentable in one country but not in another.

Fortunately, although patent requirements are often worded differently in the statutes of various countries, in practice the requirements turn out to be quite similar. There are notable exceptions, however. As already discussed, unlike the United States, most countries do not allow a 1-year grace period between the time of publication and the date of filing a patent application. Also of significance is the law which exists in most European countries which prohibits claims directed toward the treatment of the human body. Thus, in Europe it would not be possible to obtain patent protection for a means of treating AIDS.

There are certain international treaties and agreements which can simplify and expedite international filings of applications. For example, 14 countries belong to the European Patent Office (EPO) [6]. Although each of these countries has its own patent laws, prosecution of an application can be initiated by filing a single application with the EPO.

A recent development in international patent law is the increased use of the filings made according to the Patent Cooperation Treaty (PCT). Over 40 countries are signatory to the PCT, and a single PCT application can designate any or all of these countries as places where patent protection may be sought. Although it is still necessary to prosecute the applications according to the laws of the individual countries, the PCT provisions streamline the initial filing procedures and enable the applicant to postpone a final decision regarding whether to pursue patent action in particular countries until 20 or even 30 months after the original filing of an application [7].

The ability to rapidly and efficiently file a single application which provides an international filing date for many countries can be of great importance. One reason why this procedure is important relates to international laws regarding filing dates. By international agreement, a patent applicant has 1 year from the date when an application is originally filed to file foreign applications and receive the benefit of the initial filing date. For example, if a patent application is filed on June 1, 1989, and if the applicant files corresponding foreign applications by June 1, 1990, then the effective filing date of the foreign applications will be June 1, 1989. The June 1, 1989 date is known as the application's "priority date."

Obtaining the benefit of an early priority date can be very important during international patent prosecution. Because most countries award patents to the first applicant to file an application (rather than to the first to invent the idea), the filing date determines who receives a patent if more than one application is filed for the same invention.

Also, most countries follow an absolute novelty rule whereby any publication of the invention before filing a patent application will bar patentability; no 1-year grace period is allowed. Therefore, in the above example, a publication on August 10, 1989 would bar patentability of any foreign applications filed after that date *unless* the applications were filed before June 1, 1990 and claimed the June 1, 1989 priority date.

It is permissible to file foreign applications more than a year after the original filing of a domestic application. However, if the foreign application is filed after the end of the 1-year priority period, the applicant is not allowed to claim the benefit of the earlier filing date. Hence, any intervening publications will be a bar to patentability.

4. Attributes of Patent Protection

When a patent is issued by the U.S. Patent Office, it grants to the inventor, or the inventor's assignee, for a period of 17 years, the exclusive right to make, use, or sell the invention within the United States, its territories, and its possessions [8]. As a result of the thorough examination given to each application, every patent carries with it a presumption of validity [9]. Despite the presumption of validity accorded to each patent, a patent may be held to be invalid after it has been issued. In the United States, there are two primary means by which a patent may be found to be invalid. One way involves patent office proceedings; the other occurs in the federal courts.

In the patent office, an issued patent, or any of its claims, may be declared invalid as the result of either a reissue proceeding [10] or a reexamination [11]. A reissue proceeding may only be instigated by the patent holder and is intended to correct defects of substance in an issued patent. The defects may be the result of newly discovered prior art which, if known at the time of the filing of the patent application, would have been relevant in the examination of the application. If the newly discovered prior art destroys the novelty of claims in the patent or makes claims obvious, then those claims will be invalidated.

Another patent office proceeding, this one known as a reexamination, also examines the validity of claims in view of newly discovered prior art.

One significant difference between reexamination and reissue proceedings is that reissue is only instigated by the patent holder; reexamination can be requested by any member of the public.

Patent claims can also be declared invalid by federal courts. Often when a patent holder tries to enforce a patent by suing for infringement, the accused infringer will assert, as part of the defense, that the patent is invalid. During the court proceeding, each party will introduce evidence bearing on the validity of the patent. This evidence may be, for example, proof that the inventor had described the invention in a publication more than a year before filing a patent application. If this is found to be true, then the patent will be invalidated because the statutory novelty requirement was not satisfied.

Any claim of a patent which is found to be invalid cannot be enforced against infringers.

B. Trade Secrets

The primary alternative to patents is trade secret protection. Trade secrets are quite different from patent protection both in the manner by which they are obtained and in the protection they provide. Just as patents cannot be obtained for all inventions, trade secret protection is not always available.

1. Subject Matter of Trade Secrets

A trade secret can be defined as information, ideas, or results which are known only by a particular person or business entity and which give that person or business entity an advantage over competitors [12]. Obviously, the subject matter of a trade secret can be virtually anything. Any of the inventions listed in Table 1 could be a trade secret if the invention was known only by a particular business and if it gave that business an advantage.

In addition to things which are normally considered to be inventions, trade secrets can also be things such as customer lists, methods of doing business, marketing plans, and research strategies. Thus, the subject matter which can potentially be covered by trade secret protection is very broad, while the subject matter referred to as inventions and patentable inventions is successively smaller subsets.

2. Obtaining Trade Secret Protection

Unlike patents, it is not necessary to apply for trade secret protection. There are no statutory requirements which must be met and no need for the inventor to demonstrate to the federal government that the invention is new, useful, and unobvious. Therefore, in terms of the procedural requirements for obtaining protection, trade secret protection is much easier to obtain than patent protection. This is not to say, however, that there are never any affirmative acts which may be required of a trade secret holder in order to maintain trade secret protection.

3. Requirements for Maintaining Trade Secret Protection

Not all things which are proper subject matter for trade secret protection can actually be protected as a trade secret. The primary requirements for trade secret protection are that the idea, invention, or information must actually be secret and the holder of the secret must take appropriate steps to maintain secrecy. Thus, a marketing strategy which is intentionally made public is no longer a trade secret. Similarly, publication of a new process for preparing a pesticide eliminates trade secret protection for that process.

The careful reader should immediately recognize that patent protection and trade secret protection are mutually exclusive. When a patent issues, the disclosure of the patent application is published. The publication of a patent disclosure immediately destroys the secrecy, and hence the trade secret protection, of the subject invention. In this regard it is important to note that U.S. patent applications are held in secrecy until a patent actually issues. Therefore, an invention can be held as a trade secret during the pendency of the U.S. application. Also, the patent applicant has the option of abandoning the application at any time until the patent actually issues. Thus, if the applicant decides that trade secret protection is better than patent protection for a particular invention, a patent application can be abandoned with no loss of secrecy.

Many countries do not maintain patent applications in secrecy during the entire pendency of the application. Instead, countries such as Japan and the EPO publish the patent application 18 months after the first filing of the application. For example, if a U.S. patent application is filed on January 1, 1989, and a European application is filed on December 29, 1989 and claims the U.S. filing date as the priority date, then the European patent application will be published on July 1, 1990, which is 18 months after the priority date of January 1, 1989. Once a foreign patent application is published, trade secret protection is lost.

Trade secret protection is also lost if a product is sold and the trade secret is ascertainable from the product itself. If a manufacturer sells a new growth medium which advantageously results in faster growing cultures, trade secret protection will be lost if a competitor can analyze the medium and determine exactly what is in it.

Trade secret protection may also be lost as the result of insufficient efforts to keep the secret confidential. For example, if a researcher goes to a public library to work on a report and carelessly leaves behind valuable data, courts may not be willing to enforce the trade secret against a person who happens to see these data. Also, if research results are transmitted outside a company, and the results are not clearly marked confidential or otherwise transmitted pursuant to a confidential relationship, then trade secret protection may be lost. If a researcher transmits data to an old friend from graduate school and that old friend happens to work for a competitor, then it may not be possible for the researcher to enforce trade secret protection against the competitor when the competitor uses the data to produce a new product.

Trade secret holders are expected to take reasonable precautions to keep their information secret. Reasonable precautions can include limited access to labs so that visitors do not observe confidential experiments. Also, laboratory notebooks should not be left unattended or be accessible to anyone who does not need to have access. Confidential communications should always be clearly marked as "CONFIDENTIAL." Information should never be transmitted without first establishing a confidential relationship between the discloser and the person receiving the information.

One means by which a confidential relationship can be established is an agreement known as a confidential disclosure agreement [13]. These agreements, also known as secrecy agreements, are contracts which are frequently entered into when one company (or person) wants or needs to disclose information to a second party without making the information available to the public. Confidential disclosure agreements typically require the disclosee to keep the disclosed information secret for a specified period of time. These agreements usually do not require the disclosee to keep the information secret if the disclosee can prove that it was already aware of the information prior to the disclosure. For example, the disclosee may have research records showing that internal company research had produced the same information prior to discussions with the discloser.

An example of a simple confidential disclosure agreement is provided in Appendix C. These agreements must usually be edited and styled to suit the particular circumstances involved. One example of particular circumstances is the case involving transfer of biological materials. In this case, the agreement may contain clauses which explicitly prohibit the recipient from reculturing the materials and/or transferring the materials to a culture depository or third party.

One situation which frequently results in disputes over trade secret ownership occurs when an employee leaves and goes to work for a competitor. There is no absolute means for preventing this type of employee movement; however, it can be discouraged by providing employees with favorable working conditions and long-term incentives to make them want to stay with the company. The damage done by having an employee leave can be minimized by employment contracts which explicitly prohibit misappropriation of trade secrets. Also, it is good business practice to limit the dissemination of trade secrets so that only those people who actually need particular information are privy to it.

4. Attributes of Trade Secret Protection

The owner of a trade secret can seek legal redress against anyone who misappropriates the trade secret. Misappropriation refers to obtaining the secret through improper means. Because there is no federal law of trade secrets, the individual states have developed their own standards for determining what conduct is considered to be misappropriation. Thus, it is possible that conduct which would constitute misappropriation in one state may not constitute misappropriation in another.

Despite this possibility for differences in trade secret protection between states, there is general agreement about certain types of activities which will surely constitute misappropriation. For example, a company which pays an employee of a competitor to divulge confidential information or pass on copies of carefully guarded research results would almost certainly be guilty of trade secret misappropriation. Similarly, misappropriation may also occur when a confidential disclosure agreement is broken. If the party receiving the disclosure violates the confidential disclosure agreement by, for example, selling the information to another company, then a misappropriation of the trade secret has occurred and the trade secret holder can sue the misappropriator.

Unlike patent protection, which lasts for 17 years, trade secret protection can theoretically last forever. The only requirements for trade secret protection to continue is that the information remain secret and that the trade secret holder take reasonable steps to maintain secrecy. One of the most famous trade secrets is the formula for Coca-Cola, which has now been held in secrecy for over 50 years.

Although trade secret protection can theoretically last forever, holding a patentable invention as a trade secret can be risky because if the secret somehow becomes known to the public then the trade secret is lost. Of course, if the public disclosure happens as a result of a misappropriation, then the trade secret holder will have a legal cause of action against the misappropriator.

To enforce a trade secret, the trade secret holder can sue the alleged misappropriator pursuant to state laws pertaining to trade secrets. Generally, the trade secret holder in such a lawsuit will seek damages from the misappropriator. Theoretically, the damages will compensate the trade secret holder for the economic injury caused by the misappropriation. An additional remedy often sought in these cases is an injunction to prevent the misappropriator from further use or dissemination of the information. Unfortunately, however, where the misappropriation resulted in a public disclosure of the information, it may not be realistically possible to fully recover the damages caused by the loss of the trade secret protection.

In addition to inadvertent losses of secrecy, another danger of utilizing trade secret protection is that a competitor may develop the same technology and patent it. There is nothing in the patent laws which prohibits the patenting of an invention which has been held as a trade secret by another. In fact, §102(g) of the U.S. patent code explicitly states that the novelty requirement is satisfied even where the invention was previously invented by another, *if* the previous inventor "abandoned, suppressed, or concealed the invention" [14]. Thus, where an original inventor has used trade secret protection to conceal the invention from the public, a subsequent inventor can obtain patent protection for the idea. In the United States, the subsequent inventor who obtains the patent would be able to prevent the original inventor from making, using, or selling the invention for the 17 years of the patent monopoly. Some countries allow the original inventor to continue using the invention after the patent has issued to the subsequent inventor, but the United States does not allow this.

Obviously, there are advantages and disadvantages to both patents and trade secrets. The relative merits of each type of protection will now be discussed in detail.

IV. CHOOSING THE PROPER FORM OF PROTECTION

Patent protection and trade secret protection are mutually exclusive: once a patent is issued, the disclosure is published and, consequently, secrecy and trade secret protection are lost. Therefore, decisions must be made regarding which form of protection is most appropriate to protect valuable ideas and information.

A. Who Should Make the Decision?

The decision as to which form of legal protection is most appropriate for a given invention is best made by a patent attorney with experience working with the particular subject matter in question. Because both scientific and legal considerations will have a bearing on the decision, it is important to utilize an attorney who not only is knowledgeable in the law but who also understands the invention and is able and willing to readily communicate with the scientists. Although patent law, as a whole, is a specialized area of the law, biotechnology patent law is particularly specialized and it is important that the lawyer be familiar with the legal customs and most recent decisions pertaining to the technology.

B. The Timing of the Decision

It is important to consider legal protection for inventions early in the course of a research project. Scientists should be aware that decisions they make regarding publishing and giving talks at scientific meetings can have important ramifications years later when property rights are determined. Even procedures used to keep laboratory notebooks and research records can have important legal implications.

Despite this attention which should be given to legal considerations from the outset of a research project, it is not always necessary to make an early decision regarding the ultimate form of protection which will be relied on. Until information actually becomes public, trade secret protection can be maintained. Therefore, it is possible, and often desirable, to maintain the secrecy of an idea after a patent application is filed.

By maintaining secrecy during the pendency of a patent application, the applicant retains the option of abandoning the application and relying on trade secret protection. This option can be retained until a patent issues or until a foreign application is published, if foreign applications have been filed. Thus, if the applicant runs into a great deal of difficulty obtaining a patent, the applicant may wish to simply abandon the application and retain trade secrecy rather than spend excessive time and money in the pursuit of patent protection.

Therefore, when a patent application is filed and secrecy is maintained, it is not necessary to make a final decision between patent and trade secret protection until a patent is about to be issued or a foreign application is about to be published.

Often, however, the luxury of filing a patent application and maintaining secrecy may not be available. At universities, for example, researchers often desire to publish their results as soon as possible. It is not realistic to expect university researchers to hold up publishing during the 2-5 years of a pending patent application. Under these circumstances, trade secret protection does not present a viable legal alternative, and patent action should be taken as soon as a patentable invention is identified and the appropriate university approval obtained. In this regard, it is crucial for universities to recognize the importance of taking expedient patent action. Once a patent application is filed, the researcher is free to publish without fear or jeopardizing patent rights.

Unfortunately, in the past there was a misconception held by some university researchers and administrators, who were not well informed about patent law, that pursuit of patent protection could prevent or slow down publishing by researchers. Some of these people thought patent protection was antithetical to the role of a state-supported university as a place for the free exchange of ideas and the development of ideas and products to benefit the public at large. In reality, efficient university patent policies and procedures not only expedite the publishing process but ensure that valuable information developed as a result of public-funded research accrues to the benefit of the people of the state, the state-supported institution, and the researcher who did the work. Some companies which have in the past been able to obtain free information by exploiting university researchers have now expressed dismay at recent prudent patent procedures implemented by many universities. The dismay expressed by these companies emanates more from a loss of free valuable information than from any legitimate concern for the researchers, the universities, or the public at large.

In universities where publication is desirable and occurs rapidly, expedient patent action is crucial. There are other situations where patent protection may be the only legitimate alternative. For example, when a researcher sees in a publication or on sale an invention which is the same or similar to the researcher's invention, then trade secret protection is clearly not possible. If the researchers believe that he or she may have been the first inventor, then the filing of a patent application is advised. So long as a patent application is filed within a year of the publication or sale of the invention, then the applicant can still obtain a patent. Of course, in order to obtain the patent, the applicant must establish that he or she was the first to come up with the invention.

If more than one applicant files an application in the U.S. Patent Office for the same invention, the Patent Office institutes an interference proceeding which determines who actually was the first person to invent the subject matter. During an interference proceeding, the parties present evidence in order to establish the dates when they conceived of their invention and when it was reduced to practice. Generally, the first party who conceived of the invention is awarded the patent, assuming that he or she exercised due diligence in reducing the idea to practice once conceived.

Evidence used to establish a date of conception usually takes the form of laboratory notebooks and other verifiable research records. Because laboratory notebooks can be of critical importance in determining patent rights, it is essential that proper procedures be followed for keeping these notebooks. Notebook procedures are discussed in detail in Section VI below.

One important aspect of interference proceedings worth noting here is that the person who files a patent application first has certain procedural advantages during the interference. In particular, the first to file is initially presumed to be the first inventor. It is the burden of the person who filed second to produce evidence establishing an earlier date of invention. Thus, in a case where records are missing or inconclusive, the first to file will be awarded the patent by virtue of having the earliest filing date. Also, in most countries other than the United States, the patent is automatically awarded to the first to file the patent application; dates of conception or reduction to practice are not taken into account. This highlights the importance of taking rapid patent action. Delays in making decisions regarding patent action can delay publication, result in the loss of patent rights due to intervening publications, and have adverse consequences with respect to interference proceedings and critical international filing dates.

C. When Trade Secret and Patent Protection Are Both Available

The previous discussion focused primarily on situations where either patent protection or trade secret was not available or did not present a realistic option. Of course, the decision as to which type of protection to pursue is more difficult when each form of protection presents a viable alternative, i.e., when the invention is still secret and patent action could be taken.

In these situations, there are a number of factors which should be taken into consideration. Among the more important factors are whether the invention can be reverse-engineered, whether patent claims could be enforced effectively, the anticipated commercial longevity of the invention, the criticality of the invention to the business, and the cost. Each of these factors will now be discussed in detail.

1. Whether the Invention Can Be Reverse-Engineered

As mentioned previously, if the constituents of an advantageous new growth medium could be readily ascertained by standard chemical analysis, it would not be prudent to forgo patent action in favor of trade secret protection. In the absence of patent protection, a competitor could determine the constituents of the medium and start producing and selling it. There are many products which are easily reverse-engineered and thus not well suited for trade secret protection.

Processes are frequently better suited for trade secret protection than products. Often it is not possible, by simply examining a product, to determine how that product was made. Therefore, a new process for producing a better tasting beer or a new process which gives enhanced yields of an antibiotic may well be excellent candidates for trade secret protection.

Similarly, a novel ingredient used in a process but not incorporated in the product can be a good candidate for trade secret protection. For example, an enzyme which facilitates a chemical reaction at lowered temperatures or a novel microorganism which gives high yields of product are examples of compositions which might be protectable by trade secret. Competitors would not be able to ascertain the character of these compositions (or even their existence) by merely examining the product which is sold.

2. Whether Patent Claims Could Be Enforced Effectively

In theory, a U.S. patent gives the patent holder the right to exclude others from making, using, or selling the invention for a period of 17 years. In reality, however, it is very difficult to enforce some patents. For example, penicillin can be made using any one of a number of known procedures. If a company were to develop and patent a new, highly efficient method for making penicillin, it might be very difficult for that company to determine which competitors were infringing the new process and which were still utilizing the older processes. It would probably not be possible to determine which process was used simply by examining the penicillin being sold by the competitors. Therefore, in theory, a patent on the process provides 17 years of protection, but in reality the patent merely discloses the process to competitors and gives unscrupulous competitors the opportunity to infringe the patented process with little risk of being caught.

Similarly, patent claims to a novel microbe used in a process may also be hard to enforce because of the difficulty in determining if that particular microorganism has been used to produce a product, unless, of course, the novel microbe was the only known microbe for making the product, or the novel microbe produced significantly larger amounts of the desired product. As previously discussed, the holder of a patent claiming a novel microbe or a process utilizing a novel microbe must make that microbe unconditionally available to the public upon issuance of the patent.

Patent claims may also be hard to enforce if they are easily circumvented. Claims which are relatively narrow in scope are generally the easiest to circumvent. Competitors may be able to make minor changes in order to bring their product outside the coverage of the claims. Of course, this problem is best avoided by drafting claims which provide broad coverage of the inventive concept. Unfortunately, however, gaining allowance in the U.S. Patent Office of broad claims is not always possible. In scientific fields where there is a great deal of research, publishing, and patenting taking place, it is often impossible to obtain broad claims. Instead, the applicant must settle for claims to a relatively narrow niche in a crowded field.

Broad claims may also be hard to obtain in technical disciplines considered to be unpredictable. If an examiner believes that a scientific field is unpredictable, the examiner will not allow the applicant to extrapolate the claims to embodiments which have yet to be demonstrated in the lab. Thus, in the field of biotechnology, which is both crowded and unpredictable, narrow claims are much more common than broad ones, and the possibility of circumvention of these claims by a competitor presents a legitimate concern.

3. Anticipated Commercial Longevity of the Invention

With regard to this factor, it is once again important to focus on the realities of the patent system. In the biotechnology field, it frequently takes 2-4 years from the date of filing to obtain patent protection. If the invention which is being considered for patent action is in an extremely fast-paced area of research and development, and its commercial value will be minimal after 5 years, it may not be worthwhile to proceed with a patent application.

The commercial longevity of a product may in some instances be affected by the issuance of a patent. When a patent issues, the invention is disclosed in full detail to the public at large, including competitors. If the patent disclosure is such that it points competitors in a research direction which they had not previously thought of, then the obsolescence of the patented invention may be hastened. A patent disclosure which reveals an entirely new approach or a new way of thinking may allow others to come up with competitive products which build on the patented idea without infringing it. This danger is again particularly prevalent in areas of research where broad claims are difficult to obtain.

When taking this factor into consideration, it should be borne in mind that future commercial utility is often very hard to predict. Often, a new use will emerge for a product which was thought to be without value. Therefore, commercial determinations should be made keeping in mind the great uncertainties of science as well as the marketplace.

4. Criticality of the Invention to the Business

If a product or process is considered to be essential to the continued survival of a business, it makes sense to try to obtain patent protection rather than relying on trade secrets. The primary danger of relying on trade secret protection for a critical process or product is that a competitor may come along and patent the process or product, thus destroying the business. If there is any possibility of obtaining patent protection for such an invention, it is worth the money and effort to apply for the patent.

As mentioned earlier, if the application for patent is unsuccessful, the application can be abandoned, and the invention can still be kept as a trade secret. Under this scenario, the trade secret holder's anxiety about another person obtaining patent protection should be lessened because the trade secret holder's own unsuccessful efforts to obtain a patent indicate that the invention is not patentable.

5. Cost

In order to obtain a patent, it is necessary to apply to the federal government. Patent applications are frequently long and complicated documents which must always be carefully drafted in order to secure optimal patent coverage. Patent applications are prepared by legal specialists who can charge a premium for their services. The actual cost of a patent application will depend on many factors including the length and complexity of the application, and the expertise of the attorney preparing the application. U.S. patent applications for biotechnology inventions can range in cost from $3000 to $10,000 or more.

The costs associated with filing patent applications outside of the United States vary widely depending, for example, on the countries where applications are filed, the size of the application, and monetary exchange rates. Assuming that an application has already been prepared, international applicants must still pay filing fees, patent agent fees, and, when necessary, translation costs. Often, the larger an application is, the higher each of these fees will be. The cost of filing foreign applications extensively throughout the world can easily exceed $30,000.

In addition to the initial cost of filing a patent application, the applicant can expect to incur additional costs during domestic and foreign prosecution of the case. In many instances, the prosecution costs may be less than the initial costs of preparing an application. However, where long and complex correspondence with the Patent Office becomes necessary, or if an appeal is needed, the cost of prosecution can be significant.

As discussed previously, there is no formal application needed to secure trade secret protection. The primary costs associated with trade secret protection are associated with efforts to keep the information secret. If sophisticated security systems or enclosures are needed, then this cost can be substantial. On the other hand, some inventions can be kept secret at a minimal cost by simply limiting access to the information and making sure that procedures are in place to prevent inadvertent breaches of secrecy.

Unfortunately, the cost of enforcing either trade secrets or patents can be quite high. Litigation of this type frequently involves a great deal of pretrial discovery, which is not only expensive, but can be very time consuming for researchers and management. Often, the best option, at least for small companies faced with intellectual property litigation, is to reach a favorable settlement and avoid the large expenses of discovery and trial.

V. PUBLICATIONS

From the previous discussion of patents and trade secrets, it should be readily apparent that publications can have a profound effect on property fights to inventions. Therefore, a detailed examination of the legal definition and significance of publications may help researchers avoid inadvertent losses of property rights.

A. Effects of Publication

In most countries, publication of the details of an invention before a patent application has been filed results in the irrevocable loss of all possibility of obtaining patent protection for that invention. The United States is the primary exception to that general rule. In the United States, inventors are given a 1-year grace period in which to file an application once the invention has been described in a publication. When an invention claimed in a patent application has been described in a publication more than a year before the filing date of the application, the invention is said to be "anticipated" by the prior art and patentability is precluded.

B. What Constitutes a "Publication?"

Although the word "publication" has a well-known meaning in everyday usage, its legal definition may at times deviate from this common definition. There are three primary factors which must be taken into consideration when determining whether a particular reference anticipates claims in a subsequently filed patent application. These factors are (a) whether the purported publication provided a full and complete description of the invention; (b) whether this description was sufficiently available to the public to be categorized as a publication; and (c) the date on which the reference is deemed available to the public.

A brief discussion about each of these factors will help bring them into focus.

1. Sufficiency of Disclosure

In order to be an anticipating publication, a reference must contain a description which is of sufficient detail to enable a person skilled in that science to practice the invention, i.e., the reference must contain an enabling disclosure. If this requirement sounds familiar, it is because it is the same standard which is required of a patent application. As previously discussed, a patent application is required to contain an enabling disclosure in order for a valid patent to be issued.

In general, the same considerations which are taken into account in evaluating the sufficiency of a patent application's disclosure are also used when assessing whether a publication contains an enabling disclosure. For example, if the publication in question describes a process for making ethanol, and that process requires the use of a novel microbe, then the publication does not constitute an enabling disclosure if a source of the essential novel microbe is not disclosed [15].

Similarly, if a reference describes a method for making a novel product but the procedures outlined in the publication do not actually produce the product, then the reference may not constitute an anticipating publication for either the product or the process. For a description to be enabling, persons skilled in that art must not have to resort to undue experimentation or inventive acts of their own in order to produce the invention.

Therefore, when presented with a prior art reference which appears to bar patent protection, the researcher should carefully examine the reference in order to determine whether it contains an enabling disclosure of the invention. If the reference does not contain an enabling disclosure, then the filing of a patent application may be justified.

2. Extent of Distribution

For a reference to be considered an anticipatory publication, it is not necessary that the reference be in writing or that it be broadly disseminated. Rather, the crucial factor to be considered relates to whether the information was conveyed in such a way that the interested segment of the public actually received the information, or the information was made available to the public if the public wished to have this knowledge. To illustrate these requirements, consider two scenarios which might not be thought of as "publications" according to everyday usage of that term, but which most definitely constitute publications under U.S. patent law.

The first scenario involves the presentation of a poster session at a meeting attended by the preeminent researchers in a highly specialized scientific discipline. Although this poster session may be observed by only a few researchers, these researchers constitute the segment of the public which is most likely to avail itself of the information and to proceed with related research and development. Thus, if the poster session contains an enabling disclosure of the invention, then this invention has entered the public domain and a publication under the terms of the patent law has occurred.

Publication can also occur even if no one has ever actually received the information. The scenario which demonstrates this principle involves the publication of a doctoral dissertation. Most universities follow a procedure whereby dissertations are shelved in the university library and theoretically are available for perusal by any member of the public. It has been established that this availability to the general public constitutes publication for purposes of patent law [16].

In summary, publication occurs any time information is actually conveyed to the interested segment of the public or the information is available to the public at large. The medium through which the information is conveyed is not critical; journals, books, photographs, recordings, and other forms of communication have all been deemed to be publications for purposes of patent law.

3. Date of Publication

The date on which a reference becomes an anticipating publication is often of great importance in determining whether that reference will bar patentability of an invention. This effective date of publication is the date on which the reference first becomes available to the public. For a scientific journal, the effective date of publication is the day on which the journal first appeared on the newsstand or the day it was first received by the subscribers, whichever occurred first. The date printed on the journal is of no consequence to this legal determination.

The date of publication for a dissertation shelved in a library is the date on which the dissertation was shelved and catalogued so that a member of the public could actually have access to the publication.

In each case, the dispositive consideration is the determination of when the publication was actually available to the public.

VI. LABORATORY NOTEBOOKS

The proper recordation of ideas and research results is of great importance in protecting intellectual property rights. Especially in the United States, where patents are awarded to the first to invent (rather than the first to file), laboratory notebooks often provide the evidence on which patents are awarded. Laboratory notebooks are also used to establish ownership of trade secret rights.

In order for laboratory notebooks and research records to have any probative value in judicial or Patent Office proceedings, they must be legible, accurate, and authentic. To achieve these purposes, the researcher should record all laboratory results and ideas in pen in a bound notebook. Each page of the notebook should be consecutively numbered and entries should be made as the experiments are conducted. Incorrect entries should have a single line drawn through them; there should be no attempt to obliterate or erase any entry. All results should be recorded, even negative results.

Immediately after each page of the notebook is completed, it should be initialed and dated by the person doing the work. The notebook should also be reviewed and witnessed on a regular basis by a person who understands the work but who would not be a coinventor of any invention which arose from the work. Unused pages or portions of pages should have a diagonal line drawn through them. The recording of the conclusion of an experiment should be followed immediately by the signature of the experimenter and at least one witness.

By following these simple rules the researcher will be able to produce accurate and verifiable records if a property rights dispute should arise.

APPENDIX A

United States Patent [19]   [11] **Patent Number: 4,643,756**

Cardina et al.   [45] **Date of Patent: Feb. 17, 1987**

[54] **BIOHERBICIDE FOR FLORIDA BEGGARWEED**

[75] Inventors: **John Cardina; Robert H. Littrell,** both of Tifton, Ga.; **Larry J. Stowell,** San Diego, Calif.

[73] Assignees: **Mycogen Corporation,** San Diego, Calif.; **Univ. of Georgia Research Foundation, Inc.,** Athens, Ga.; **The United States of America as represented by the Secretary of Agriculture,** Washington, D.C.

[21] Appl. No.: **715,229**

[22] Filed: **Mar. 22, 1985**

[51] **Int. Cl.**[4] **A01N 63/00**
[52] **U.S. Cl.** **71/79**; 435/911
[58] **Field of Search** 71/79; 435/911; 47/DIG. 11

[56] **References Cited**

U.S. PATENT DOCUMENTS

3,849,104	11/1974	Daniel et al.	71/79
3,999,973	12/1976	Templeton	71/79
4,162,912	7/1979	Charudatan	71/79
4,263,036	4/1981	Charudattan	71/79
4,390,360	6/1983	Walker	71/79
4,419,120	12/1983	Walker	71/79

OTHER PUBLICATIONS

Staples et al, "Appressorium Formation, etc.," (1976), Archives of Microbiology, vol. 109, pp. 75–84 (1976).
Almeida et al, "Curative and Preventative, etc.," (1981) CA 96: 29819s (1982).
Malhotra et al, "Utilization of Oligosaccharides, etc.," Pantnagar J. Res., (1977) 2(2), pp. 119–122.

Primary Examiner—Glennon H. Hollrah
Assistant Examiner—Robert Lelkes
Attorney, Agent, or Firm—Roman Saliwanchik; David R. Saliwanchik

[57] **ABSTRACT**

The subject invention concerns a novel bioherbicide and its use to control a major weed found in many fields in the Southeastern United States where peanuts and soybeans are grown. Specifically, *Colletotrichum truncatum* (Schw.) Andrus & Moore, in an agricultural composition, can be used to effectively control Florida beggarweed without adversely affecting field crops, e.g., peanuts and soybeans. Further, *C. truncatum* (Schw.) Andrus & Moore in a mixture with *Alternaria cassiae* can be used to control Florida beggarweed and other undesired vegetation, such as sicklepod, showy crotalaria and coffee senna.

6 Claims, No Drawings

4,643,756

1

BIOHERBICIDE FOR FLORIDA BEGGARWEED

BACKGROUND OF THE INVENTION

Florida beggarweed (*Desmodium tortuosum* (SW.)DC) is a major weed problem in many fields in the Southeastern United States where peanuts and soybeans are grown. It is ranked as the most troublesome weed in peanuts in the three largest peanut producing states, and its spread is causing increasing concern to farmers throughout the Southeast. This weed occurs primarily in field crops, but is found also in fallowed areas and along roadsides and waste areas. Florida beggarweed is an annual herbaceous legume native to Florida and the Gulf States, extending to the subtropics and tropics of the Western Hemisphere. Although it is currently considered one of the most damaging weeds, particularly in leguminous crops, Florida beggarweed was a popular forage crop prior to 1950. As a forage crop, heavy seeding on land of moderate to high fertility was recommended so that when plants matured 70–80 days later, the crop could be expected to yield 4 to 6 tons of dry matter and return to the soil up to 5000 seeds per plant. As a result, Florida beggarweed, no longer considered an acceptable forage crop, plagues farmers throughout the Southeast on soils that are most suitable for peanuts and soybeans. The damage caused by Florida beggarweed is primarily a result of competition with the field crop for light, water, and nutrients. In addition, this weed, which may reach 10 feet at maturity, overtops the crop canopy and interferes with pesticidal spray programs. Particularly serious is the decreased coverage of biweekly foliar fungicide applications in peanuts caused by infestations of Florida beggarweed. In peanut production, where the crop must be inverted at the end of the season, Florida beggarweed interferes with the digging, turning and curing processes, and thereby causes further reductions in crop yield and quality.

Florida beggarweed harbors several pests of field crops. It is a common late-season food source for *Heliothis virescens* larvae, a pest of numerous crops. This weed is also a host of northern rootknot, lesion and sting nematodes that severely damage many crops. Recent evidence indicates that Florida beggarweed is a host of peanut mottle virus and peanut stripe virus.

Florida beggarweed is a leguminous weed that is not controlled by herbicide programs recommended for use in peanuts and soybeans. Development of chemical control for Florida beggarweed is complicated by its close botanical relationship to peanuts, soybeans, and other leguminous crops. This puts unusually stringent selectivity requirements on development of herbicides.

Chemical weed control programs are seriously inadequate for control of this weed. Florida beggarweed is resistant to many herbicides used in peanuts and soybeans. Frequently the weed germinates below the treated zone and avoids herbicide injury. Although many herbicides have been developed and tested in the last three decades, farmers still rely heavily on dinoseb, a herbicide developed in the 1950's. Dinoseb is a contact herbicide that causes injury to peanuts and soybeans; farmers tolerate this injury out of the necessity to control Florida beggarweed. Alternative approaches include MSMA (monosodium methanearsonate) paraquat, toxaphene and triazine herbicides, but these chemicals are not registered for use on these crops for reasons of toxicology and/or crop safety. Other inadequacies of chemical controls include lack of residual control, in-

2

jury to nontarget organisms, undesirable residues in harvested products and carryover in subsequent crops.

The use of bioherbicides is becoming an increasingly important alternative to chemical herbicides for a variety of reasons, some of which are similar to those experienced in the chemical control of Florida beggarweed, as disclosed above. This importance is accompanied by several patents which have issued for bioherbicides and their use. Some of these patents, by way of illustration, are as follows: U.S. Pat. No. 3,849,104 (control of northern jointvetch with *Colletotrichum gloeosporioides* Penz. aeschynomene); U.S. Pat. No. 3,999,973 (control of prickly sida [teaweed] and other weeds with *Colletotrichum malvarum*); U.S. Pat. No. 4,162,912 (control of milkweed vine with *Araujia mosaic* virus); U.S. Pat. No. 4,263,036 (control of *Hydrilla verticillata* with *Fusarium roseum* Culmorum); U.S. Pat. No. 4,390,360 (control of sicklepod, showy crotalaria, and coffee senna with *Alternaria cassiae*); and U.S. Pat. No. 4,419,120 (control of prickly sida, velvetleaf, and spurred anoda with fungal pathogens.

BRIEF SUMMARY OF THE INVENTION

The subject invention concerns a novel bioherbicide which is effective, when used according to the process disclosed herein, in controlling Florida beggarweed without adversely affecting the growth and yield of the desired field crop. Preferably, the subject invention relates to the use of a composition comprising *Colletotrichum truncatum* (Schw.) Andrus & Moore spores in association with an agricultural carrier wherein said spores are in a concentration of from about 2×10^4 spores/ml of carrier to about 2×10^6 spores/ml of carrier, to control Florida beggarweed infestation. The concentrated spore formulation can be adapted for distribution over geographical locales or a situs where the spores germinate and infect Florida beggarweed.

DETAILED DESCRIPTION OF THE INVENTION

The fungus *Colletotrichum truncatum* (Schw.) Andrus & Moore is the first fungal pathogen that has been used to control Florida beggarweed. This fungus produces relatively insignificant leaf and stem lesions on mature plants, but causes a devastating disease on seedlings up to the 3–4 leaf stage. The initial symptoms of the disease are a twisting of stems and discoloration of leaf midrib and veins. Leaves eventually become mottled brown-yellow and die.

A subculture of *C. truncatum* (Schw.) Andrus & Moore has been deposited in the permanent collection of the Northern Regional Research Laboratory, U.S. Department of Agriculture, Peoria, Ill., USA, on Feb. 19, 1985. The culture was assigned the accession number NRRL 15933 by the repository. This deposit is available to the public upon the grant of a patent disclosing it. The deposit is also available as required by foreign patent laws in countries wherein counterparts of the subject application, or its progeny, are filed. However, it should be understood that the availability of a deposit does not constitute a license to practice the subject invention in derogation of patent rights granted by governmental action.

Taxonomic characteristics of *Colletotrichum truncatum* (Schw.) Andrus & Moore are as follows:

Colonies on V-8 and malt extract agars are composed of grayish-brown mycelium, becoming dark due to

abundant formation of black sclerotia. The sclerotia are irregular in shape, composed of thick-walled, dark brown cells. Acervulate conidiomata erumpent through wheat stem epidermis contain a dense layer of conidiophores. Conidiophores are branched, consisting of dark brown hyphae bearing hyaline phialides at the apex. Numerous stiff, erect, tapered, dark brown setae are also produced in the conidiomata. Setae are variable in length, up to 235 μm long, with an average length of 165 μm. Conidia are hyaline, falcate, with narrowly rounded apices, guttulate, one-celled, but sometimes becoming septate prior to germination. The average size of 50 conidia was 30 μm long$\times$4 μm wide (25–35$\times$3.3–4.4 μm). Appresoria are dark, clavate to irregular in shape, averaging 9.2 μm long$\times$7 μm wide (7.7–10.7$\times$5.4–9.2 μm).

The conidial size is somewhat larger than that recorded by Sutton (Sutton, B. [1980] The Coelomycetes, Fungi imperfecti with pycnidia acervuli and stromata, Commonwealth Mycological Institute, Surrey, England) for this species, but it agrees better with the data from legumes reported by Tiffany and Gilman (Tiffany, L. H. and Gilman, J. C. [1954] Species of Colletotrichum from legumes, *Mycologia,* 46:52–75.

C. truncatum can be grown on solid or in liquid media. Solid media that can be used include water agar, potato dextrose agar, V-8 agar and string bean agar (strained extract of macerated string beans solidified in agar). Spores are produced on solid V-8 medium exposed to florescent light. Specifically, solid media can be, for example, (1) water agar, (2) potato dextrose agar (Difco), (3) lima bean agar (Difco), (4) corn meal agar (Difco), (5) potato-carrot agar (Tuite 19), and (6) Desmodium agar (blend 10 gm Desmodium plant parts or plant extracts in 1000 ml water and solidify with 20 gm agar).

For large scale production in fermentation tanks, liquid media is used, for example:

Formula I - Modified Richard's Solution - V-8*	
Sucrose	50 gm
Potassium nitrate	10 gm
Potassium phosphate, monobasic	5.0 gm
Magnesium sulfate.7 H_2O	2.50 gm
Ferric chloride	0.02 gm
V-8 juice	15 ml
Distilled water to make	1000 ml

*Trademark, The Campbell Soup Company for mixed vegetable juices

Formula II—Modified Richard's Solution—Distillers Solubles—Same as Formula I but substitute 15 gm Distillers solubles for V-8 juice.

Formula III—Modified Richard's Solution—Brewers yeast—Same as Formula I above but substitute 15 gm brewers yeast for V-8 juice.

Formula IV—Modifed Richard's Solution—Torula Yeast—Same as Formula I above but substitute 16 gm torula yeast for V-8 juice.

Formula V—Oatmeal Solution—2%—20 gm oatmeal blended in 1000 ml distilled water.

Formula VI—Oatmeal Solution—4%+2% sugar—
40 gm oatmeal
20 gm sucrose
1000 ml distilled water

Formula VII—Beggarweed Broth—
20 gm beggarweed leaves and petioles
10 gm potassium nitrate
2.5 gm magnesium sulfate 7H_2O

10 gm potassium phosphate dibasic
1000 ml distilled water
Adjust final pH to 6.8

The preparation of spores is commenced in preseed liter flasks containing about 300 ml of liquid medium which have been inoculated with spores. The medium is incubated for 1–3 days with agitation at a temperature of about 26° to about 30° C.

The preseed is then transferred aseptically to 20-liter seed tanks with additional sterile medium as described above. The tanks are provided with sterile air and agitation. The cycle is continued at a temperature of about 26° to about 30° C. for 1 to 3 days.

Larger fermentors (250 liter) are aseptically inoculated with the seed tanks (entire contents), described above. Additional sterile medium, as used above, is added and the pH is adjusted to about 6.0. The fermentor is supplied with sterile air and agitation, and is maintained at a temperature of about 26° to about 30° C. for from 1 to 3 days. The fermentor is then harvested by filtering the contents to remove insoluble solids and mycelia growth. The filtered beer is then centrifuged, the supernatant is discarded, and the remaining spore concentrate is resuspended in water and centrifuged again. The supernate is again discarded and the spore concentrate is collected, placed in plastic bags and stored in ice. The concentrate so stored maintains an 80% germination for up to 21 days.

The spore concentrate is mixed with an agriculturally acceptable diluent or carrier for application to the undesired host vegetation or a situs. By the term "situs" is meant those areas infested with the undesired vegetation or potential infestation sites.

The preferred carrier is water, and the spore concentrate is dispersed to make a concentration of from about 2×10^4 to 2×10^6 spores/ml. The formulation is then sprayed on the undesired vegetation or situs by conventional spraying equipment in an amount of from about 50 to 500 liters per hectare. Applications have been made to Florida beggarweed in the cotyledon to 6-leaf stage at temperatures from 20° to 30° C. Applications are most effective on seedlings with 1–3 leaves to 25° to 30° C.

The fungus is not pathogenic to "Florunner" peanuts or "Centennial" soybeans with spore concentrations up to 2×10^5.

Spores of *C. truncatum* (Schw.) Andrus & Moore can be mixed with those of *Alternaria cassiae* to enlarge the scope of control of undesired vegetation. For example, this mixture can be used to control both Florida beggarweed and sicklepod (*Cassia obtusifolia*), the two most troublesome weeds in soybeans and peanuts in the Southeast. Further, spores of *C. truncatum* (Schw.) Andrus & Moore can be mixed with those of *A. cassiae* to control Florida beggarweed adn showy crotalaria or coffee senna. The use of *A. cassiae* to control sicklepod, showy crotalaria and coffee senna is disclosed in U.S. Pat. No. 4,390,360, which is incorporated herein by reference thereto. The culture, means of growing, and application to these weeds disclosed in U.S. Pat. No. 4,390,360 can be used herein. Mixtures of *C. truncatum* and *A. cassiae,* for example, *A. cassiae* NRRL 12533, can be made by methods well known in the art, utilizing the disclosure of U.S. Pat. No. 4,390,360 and that contained herein.

Though spores are the preferred form of the fungi, the fungi also can be used in their vegetative form. For example, fragmented mycelia can be formulated and applied to Florida beggarweed and/or sicklepod, or showy crotalaria, or coffee senna in much the same manner as described above for the spore form.

Following are tabulated results showing the effectiveness of *C. truncatum* in controlling Florida beggarweed:

TABLE I

Effect of Inoculum Level on Anthracnose of Florida Beggarweed

Thousand spores per ml[1]	Proportion of plants dead in one week[2] %	Visual control rating[3] %	Plant survival after 12 days[4] %
0	0	0	100
25	49	62	35
50	57	65	30
100	57	74	30
250	60	81	20
500	74	97	6
1000	86	96	7

[1]Two-leaf stage plants sprayed to runoff 10-12-84, five replications.
[2]Based on actual counts.
[3]Visual estimate of % diseased tissue.
[4]Proportion of plants with at least one normal leaf.

TABLE II

Temperature C.	Dead %	Necrotic Tissue %
20	17.32	6.26
25	24.70	7.15
30	19.12	5.85

Data are based on plants grown in three growth chambers at the temperatures indicated. Plants were sprayed to runoff with 5×10^4 spores per ml, and covered with plastic bags for 48 hr. Ratings were made 7 days after treatment. Data are means of 3 replications.

TABLE III

Effect of Duration of Moist Period on Disease

Hours	Dead %	Necrotic Tissue %
4	0	0.275
8	0	1.988
12	3.365	4.859
24	24.401	9.174
48	45.068	12.159

Data are means of four replications of growth chamber-grown plants treated in the 3–4 leaf stage with 5×10^4 spores per ml to runoff and covered with plastic bags for the time indicated.

We claim:

1. A composition for agricultural application for controlling Florida beggarweed comprising an isolate of the fungus *Colletotrichum truncatum* (Schw.) Andrus & Moore having the identifying characteristics of culture deposit NRRL 15933, which isolate controls Florida beggarweed without adversely affecting peanuts or soybeans, in association with an inert agricultural carrier.

2. A composition, according to claim **1**, wherein said isolate is in the spore form at a spore concentration of from about 2×10^4 to about 2×10^6 spores/ml of carrier.

3. A process for controlling Florida beggarweed which comprises applying an effective amount of an isolate of the fungus *Colletotrichum truncatum* (Schw.) Andrus & Moore having the identifying characteristics of culture deposit NRRL 15933, which isolate controls Florida beggarweed without adversely affecting peanuts or soybeans, onto said Florida beggarweed or unto the situs of said Florida beggarweed.

4. A composition for agricultural application for controlling Florida beggarweed and sicklepod or showy crotalaria or coffee senna comprising a mixture of the fungi *Colletotrichum truncatum* (Schw.) Andrus & Moore and *Alternaria cassiae* having the identifying characteristics of culture deposit NRRL 12553 in association with an inert agricultural carrier.

5. A process for controlling Florida beggarweed and sicklepod, or showy crotalaria, or coffee senna which comprises applying an effective amount of a mixture of fungi consisting of an isolate of *Colletotrichum truncatum* (Schw.) Andrus & Moore having the identifying characteristics of culture deposit NRRL 15933, which isolate controls Florida beggarweed without adversely affecting peanuts or soybeans, and *Alternaria cassaie* having the identifying characteristics of culture deposit NRRL 12553 onto said Florida beggarweed and sicklepod, or showy crotalaria, or coffee senna or unto the situs of said Florida beggarweed and sicklepod, or showy crotalaria, or coffee senna.

6. A novel *Colletotrichum truncatum* characterized as having the property of controlling Florida beggarweed without adversely affecting peanuts and soybeans, and, when in its essentially biologically pure form, has the taxonomic characterization of *C. truncatum* (Schw.) Andrus & Moore culture deposit NRRL 15933.

* * * * *

APPENDIX B: ACCEPTED CULTURE DEPOSITORIES

Culture depositories which have been accepted as suitable for depositing using the Budapest Treaty provisions are listed below in alphabetical order by country.

Australia

Australian Government Analytical Laboratories (AGAL)
The New South Wales Regional Laboratory
1, Suakin Street
Pymble, N.S.W. 2073
Australia

Bulgaria

National Bank for Industrial Microorganisms and Cell Cultures (NBIMCC)
125, Lenin Blvd.
Block 2
Sofia
Bulgaria

France

Collection Nationale de Cultures de Micro-Organismes (CNCM)
Institut Pasteur
28 Rue du Dr Roux
75724 Paris Cedex 15
France

Germany

Deutsche Sammlung von Mikroorganismen (DSM)
Gesellschaft für Biotechnologische
Forschung mbH
Grisebachstr. 8
3400 Göttingen
Germany

IMET—National Sammlung von Mikroorganismen
IMET—Hinterlegungsstella
Beutanbergstrasse 11
6900 Jena
Germany

Hungary

National Collection of Agricultural and Industrial Microorganisms (NCAIM)
Department of Microbiology
University of Horticulture
Somlói ut 14-16
H-1118 Budapest
Hungary

Japan

Fermentation Research Institute (FRI)
1-3, Higashi 1-chome
Yatabe-machi
Tsukuba-gun, Ibaraki-ken 305
Japan

Republic of Korea

Korean Collection for Type Cultures (KCTC)
Genetic Engineering Center
Korean Institute of Science and Technology
39-1, Hawolgokdong
Sungbuk-gu
Seoul
Republic of Korea

Korean Culture Center of Microorganisms (KCCM)
College of Engineering
Yonsei University
Sodaemun-gu
Seoul
120-749 Republic of Korea

Netherlands

Centraalbureau voor Schimmelcultures (CBS)
Oosterstraat 1
Postbus 273
NL-3740 AG Baarn
Netherlands

United Kingdom

Culture Collection of Algae and Protozoa (CCAP)
Freshwater Biological Association
Windermere Laboratory
The Ferry House
Far Sawrey
Ambleside, Cumbria LA22 OLP
United Kingdom

Scottish Marine Biological Association
Dunstaffnage Marine Research Laboratory
P. O. Box 3
Oban, Argyll PA34 4AD
United Kingdom

Culture Collection of the Commonwealth Mycological Institute
(CMI CC)
Ferry Lane
Kew, Surrey TW9 3AF
United Kingdom

European Collection of Animal Cell Cultures (ECACC)
Vaccine Research and Production Laboratory
Public Health Laboratory Service
Centre for Applied Microbiology and Research
Porton Down
Salisbury, Wiltshire SP4 OJG
United Kingdom

National Collection of Food Bacteria (NCFB)
AFRC Institute of Food Research
Reading Laboratory
Shinfield
Reading RG2 9AT
United Kingdom

National Collection of Industrial Bacteria (NCIB)
c/o The National Collection of Industrial and Marine Bacteria Ltd.
Torry Research Station
P.O. Box 31
135 Abbey Road
Aberdeen AB9 8DG
United Kingdom

National Collection of Type Cultures (NCTC)
Central Public Health Laboratory
61 Colindale Avenue
London NW9 5HT
United Kingdom

National Collection of Yeast Cultures (NCYC)
Food Research Institute
Colney Lane
Norwich, Norfolk NR4 7UA
United Kingdom

United States

American Type Culture Collection (ATCC)
12301 Parklawn Drive
Rockville, Maryland 20852 USA

ARS Patent Collection (NRRL)
Culture Collection Research-Fermentation Laboratory
Northern Regional Research Center
U.S. Department of Agriculture
1815 North University Street
Peoria, Illinois 61604 USA

USSR

USSR Research Institute for Genetics and Industrial Microorganism Breeding
Dorozhnaya Street, No. 8
113545, Moscow
USSR

Institute of Microorganism Biochemistry and Physiology of the USSR Academy of Science
Pushchino-na-Oke
142292, Moscow Region
USSR

USSR Research Institute for Antibiotics
Nagatinskaya Street 3-a
113105, Moscow
USSR

Note: The help of Mr. A. J. Lyons, Curator of the Agricultural Research Culture Collection, in compiling the above information concerning the depositories is gratefully acknowledged.

SAMPLE LETTER TO ACCOMPANY DEPOSIT OF BIOLOGICAL MATERIALS

Mr. A. J. Lyons, Curator
ARS Patent Collection
Culture Collection Research-
Fermentation Laboratory
Northern Regional Research Center
Peoria, Illinois 61604

Dear Mr. Lyons:

Under separate cover, we are sending to you the following (check one):

[ ] two agar slants
[ ] thirty lyophilized preparations
[ ] two lyophilized preparations

of the following viable organism:

[organism being deposited]

This organism is being deposited under the Budapest Treaty on the International Recognition of the Deposit of Microorganisms for the Purposes of Patent Procedure. It is to be added to your permanent collection, i.e., to be maintained for at least thirty (30) years. Should the culture be lost, replacement thereof will be made with additional agar slant transfers or lyophilized preparations. However, it should be known that the culture was viable when it was sent to you.

Please withhold distribution of the culture, except under the conditions hereafter specified. Access to this culture is hereby granted to:

1. the United States Commissioner of Patents and Trademarks;
2. such individuals as the United States Commissioner of Patents and Trademarks shall designate;
3. such individuals whom we designate to you in writing;
4. the public upon the issuance of a United States Patent to [name of applicant or assignee], which patent identifies this culture by the NRRL number assigned to it;

5. the public upon the publication of any patent application which application identifies this culture by the NRRL number assigned to it.

All restrictions on the availability to the public of the above culture deposit will be irrevocably removed upon the granting of a United States patent disclosing the deposit.

Please sign the accompanying form on the designated line, return the original to us, and keep a copy for your files.

Enclosed is a check made out to the Agricultural Research Service, USDA for $500.00 (five hundred dollars). Also enclosed are the executed Budapest Treaty forms.

Very truly yours,

[name]

ACKNOWLEDGMENT OF RECEIPT AND
DESIGNATION OF DEPOSIT NUMBER

[ ] Two agar slants
[ ] thirty lyophilized preparations
[ ] two lyophilized preparations of

[deposited culture],

and the $500.00 check, have been received, and we agree to the above-mentioned conditions.

The culture slants were received on: ______________________.
The NRRL number is: _________________.

Signed: ____
A. J. Lyons Date

Please return the signed original; keep a copy for your files.

APPENDIX C: CONFIDENTIAL DISCLOSURE AGREEMENT

____, (hereinafter referred to as DISCLOSEE), whose mailing address is ____________________ ____, is willing to accept a disclosure from:

(hereinafter referred to as DISCLOSER) relating to:

DISCLOSEE will hold the novel aspects of DISCLOSER'S disclosure in confidence for five (5) years from the date of receipt. The novel aspects of DISCLOSER'S disclosure are those which:

1. Are not already known to DISCLOSEE from their own research or from a third party who has a right to disclose the information to DISCLOSEE; or

2. Are not or do not become subject of a patent issued to a third party; or
3. Are not or do not become open to the public by any means including disclosure in patents or publications, or public use or sales.

DISCLOSEE hereby agrees not to make or sell the disclosed idea without the express written permission of DISCLOSER. Further, DISCLOSEE hereby agrees not to pass the idea to any third party without the express written permission of DISCLOSER.

DISCLOSEE agrees to review the disclosure and notify DISCLOSER within thirty (30) days if DISCLOSEE wishes to obtain an interest in the subject matter of the disclosure.

DISCLOSEE and DISCLOSER have no further obligations with respect to the disclosure.

_______________	_______________
DISCLOSEE	DISCLOSER
Date: _______________	Date: _______________

REFERENCES

1. 35 U.S.C. §101 et seq. (1982).
2. 35 U.S.C. §103 (1982).
3. 35 U.S.C. §112 (1982).
4. Ex parte Kropp, 143 *United States Patent Quarterly* 146 (Patent Office Board of Appeals 1959).
5. Budapest Treaty on the International Recognition of the Deposit of Microorganisms for the Purposes of Patent Procedure (April 28, 1977)., reprinted in 961 *Official Gazette of the United States Patent and Trademark Office, 21* (August 23, 1977).
6. The EPO includes Austria, Belgium, Denmark, France, Germany, Greece, The Netherlands, Italy, Liechtenstein, Luxembourg, Spain, Sweden, Switzerland, and the United Kingdom.
7. *PCT Applicant's Guide*, Vols. 1 and 2.
8. 35 U.S.C. §271 (1982).
9. 35 U.S.C. §282 (1982).
10. 35 U.S.C. §251 (1982).
11. 35 U.S.C. §302 (1982).
12. Saliwanchik, Roman. *Legal Protection for Microbiological and Genetic Engineering Inventions*. Addison-Wesley, Reading, MA, 1982.
13. Saliwanchik, Roman. *Protecting Biotechnology Inventions*. Science Tech/Springer-Verlag, Madison, WI, 1988.
14. 35 U.S.C. §102 (1982).
15. In re Argoudelis, 157 *United States Patent Quarterly* 437 (Patent Office Board of Appeals 1967).
16. Ex parte Hershberger, 96 *United States Patent Quarterly* 54 (Patent Office Board of Appeals 1953).

34

BIOHAZARD AND BIOSAFETY IN BIOTECHNOLOGICAL INDUSTRY

EITAN ISRAELI and JANINA GITELMAN *Israel Institute for Biological Research, Ness-Ziona, Israel*

DANIEL F. LIBERMAN *Massachusetts Institute of Technology, Cambridge, Massachusetts*

I. INTRODUCTION

The impact of a laboratory infection can stretch far beyond the immediate surroundings when industrial processes are employed, and it can have serious effects on the whole plant, on the marketing of the products, on innocent bystanders, and on the environment.

Nowadays scientists from different backgrounds end up in biotechnological laboratories: biochemists, physicists, physicians, chemists, etc. They do not have experience in sterile microbiological techniques—techniques that demand precautions against infections and involve some knowledge on the causes and routes of infection as well as on ways to prevent them. It is especially important to inform and train those scientists on biosafety subjects.

Investment of resources in this area will produce a considerable return by preventing infections, minimizing loss of working hours due to sickness, compensating affected parties, and improving the reliability of the laboratory.

The two stages in biological industrial processes—research and development on the one hand, and production on the other—have different sources of hazard. In the research and development stages the risk mainly lies in exploration of new techniques, new biological agents or products—namely the unknown. On the other hand, in production the techniques are well established but the large scale, the heavy equipment, the high concentrations of biologicals, and even the different level of employee training present hazards to the worker's health from possible accidents or infections, and to the environment from escaped pathogens of animals or plants.

Every few years, a new agent appears in the biomedical field that shakes the current habits, concepts, and information, sometimes to the degree of mass hysteria. Such an event presents new challenges to science and to medical research, until a solution is found by a way of a medicine, vaccination, or international agreement. Examples of such phenomena in the past include polio, cancer, hepatitis B, Marburg virus, and now AIDS. A unique example occurred in the 1970s—the genetic engineering technique.

All of the above phenomena have in common their inspiring a fear of the unknown that urges us to practice extra caution when new and uncommon subjects are involved. But fear of danger, without knowledge of the ways to avoid it, can lead to wrong decisions and faulty execution, resulting in accidents and infections. Conversely, overconfidence, negligence, or ignorance of the danger can lead to underestimation of the importance of safety regulations, which develop into dangerous situations, accidents, and infections. Genetic engineering has made a major contribution to biosafety. Since it was very hard to predict how far such experiments will go or what will be the impact of the new products on the environment once they escaped the laboratory, extreme safety measures were applied and strict safety regulations formulated. Expert committees were formed in order to research, write, and publish regulation for recombinant DNA experiments. They classified all etiological agents on the basis of hazard, classified biomedical laboratories in accordance with biosafety level, and gave a strong initiative to develop safety equipment and procedures.

In the biotechnological industry, safety usually means product safety. This chapter discusses mainly the safety of the worker in this industry as well as that of the environment. It presents an overview of the incidence and severity of laboratory infections with an analysis of their main sources and causes. The main part deals with hazards presented by industrial processes and how to estimate the risk both to the worker and to the environment, as well as suggesting several ways to minimize these hazards in order to prevent infections.

II. LABORATORY INFECTIONS IN THE PAST

The first paper at the beginning of the century described laboratory infection of typhoid, cholera, brucellosis, and tetanus [1]. In 1941 a survey of 74 laboratory-associated brucellosis infections in the United States was published [2]. The survey concluded that "handling of cultures or specimens or the inhalation of dust containing brucella organisms is eminently dangerous to laboratory workers." In 1941 Sulkin and Pike published the first in a series [3] of surveys about laboratory infections that summed up 222 cases of viral infections, 21 of which were fatal. In 1951 the two published a second survey [4] based on a questionnaire sent to 5000 laboratories. Out of 1342 cases mentioned, only about one-third were published in the literature.

The number of brucellosis cases was much bigger than that of other agents and together with tuberculosis, tularemia, typhoid, and streptococcal infections accounted for 72% of all bacterial infections. In only 16% of all cases was an accident blamed for possible infection. Most of these cases were associated with mouth pipeting and syringes.

In 1976 the above survey included 3921 laboratory infection cases [5]. The most commonly reported were brucellosis, typhoid, tularemia, tuberculosis, hepatitis, and Venezuelan equine encephalitis (VEE). Of the cases, 59% were in research laboratories and only 3.4% in laboratories dealing with the production of biologicals.

In 1967 Hanson et al. reported 428 laboratory infections with arboviruses [6]. The common infection source was attributed to infectious aerosols.

In a survey of clinical chemistry laboratories in Denmark about hepatitis, the incidence was 2.3 per year per 1000 employees—seven times the incidence in the general population [7].

In another study the incidence of tuberculosis was found to be five times higher than that of the general population [8].

A. Who Is At Risk?

As was mentioned above, in most of the laboratory infections lab workers were the victims. But are they the only ones at risk? A demonstration of this point is the outbreak of brucellosis at Michigan State College [9-11]. A suspension of *Brucella melitensis* was sedimented in a centrifuge, which created an aerosol that disseminated throughout the building. Of the 94 people infected, 84 were students; 45 were hospitalized, and one died. Of the 45 people hospitalized, 41 were students, one was a laboratory stockroom attendant, one a plumber, one a stenographer, and one a salesman!

One can learn a lot from this accident:

The infection rate in this case, a common biotechnological process, was 27%. The lower statistical value, therefore, is no comfort for the infected people.

Anyone in a laboratory building can be infected, even people remote from the site of the process.

When a large number of infectious microorganisms are inhaled, medical care can prove ineffective.

The severity of the disease can vary according to the specific sensitivity of the individual.

From a survey of 1286 cases of laboratory-acquired infections, 78% occurred in trained scientific personnel, 5% in students, 6% in office clerical staff, and about 10% in maintenance people. This picture emphasizes that indeed most of the people involved with laboratory infection are those working directly with the agent, and that despite their training and scientific knowledge they could not avoid infection. However, even innocent bystanders, such as service people, are victimized. Especially for those it is important to concentrate efforts in biosafety because they do not understand the risks imposed on them by the scientific staff. Even among the scientific staff there are "subpopulations" who have different incidences of infection associated with age, sex, and other points that will be discussed when dealing with the human factor in laboratory infections.

B. Frequency of Laboratory Infections

Statistics carried out on frequency rates of infections in different laboratories in the United States and Europe give an expected number of one to five infections per million working hours [12]. The probability that a certain worker will get infected in a specific year is about 10^{-2}-10^{-3}. This means that in a given institute, university, or biological plant in which there are 500 laboratory workers, one can expect one to five laboratory infections per year. In the study cited, all infections were proved to be laboratory-acquired [12]. But not all cases are reported, nor is there always proof that the disease was contracted from working in the

TABLE 1. Death Rate from Infections with Different Biological Agents

Biological agent	% Death
Viruses	7.3
Bacteria	4.0
Rickettesiae	2.6
Fungi	2.3
Parasites	0

laboratory. Additionally, there are subclinical infections, so the actual frequency rate of infection is much higher.

Infections occur in all types of laboratories: research, diagnostic, production, teaching, and others. From a summary of 3921 infections identified until 1976 [5], 59% were in research laboratories and only 3.4% in the manufacture of biological products. *This significantly lower figure for production laboratories can be attributed to the uniformity of production processes and the well-established and engineered safety procedures and standard operation procedures.*

The frequency of laboratory infections shows a downward trend according to Pike [13]. This positive trend is a result of the momentum that biosafety acquired during the last decades, which was accompanied by suitable safety equipment and laboratories, vaccinations of personnel, and safety procedures.

C. Severity of Laboratory Infections

Based on case reports and communications concerning mortality and morbidity, it is clear that the case fatality rate was 4%. This translates to four lethal infections per 100 laboratory-acquired infections. Of the remaining cases, 70% showed complete recovery and 26% only partial recovery. For comparison, the case fatality rate for motor vehicle accidents was 2.7% [12].

The death rate from viral infections was higher than the rest (see Table 1). In view of the increase in viral-associated laboratory-acquired infections we should be concerned with the higher percentage of deaths in viral infections.

III. HOW LABORATORY INFECTIONS OCCUR

In order to know how to avoid laboratory infections, one must understand the ways in which they occur. Infections in the laboratory, just as in nature, occur by the invasion of the body by an infectious agent by one of four routes:

1. Through the respiratory tract, by inhalation of droplets or dry particles as an aerosol carrying infectious agents.
2. Through the intestinal tract, by swallowing infectious liquids or solids. Because of the close proximity of the esophagus to the trachea, infec-

tious material that was inhaled through the nose can reach the esophagus and vice versa.
3. Through the skin, by pricking or cutting by a sharp contaminated object; a bite of an animal or insect carrying the infectious agent; or by contamination of bruised skin or mucous membrane by infectious materials.
4. Ocular or conjunctival exposure.

After invasion of the body, developing an illness depends on a few factors, among which are the characteristic of the agent, the quantity that penetrated, the immunological state of the person, and the treatment given.

A. Accidents as a Cause of Laboratory Infections

A review of the available literature indicates that 80% of laboratory-acquired infections are due to unknown or unrecognized causes. This means that only about 20% of laboratory-acquired infections can be related to some causative or defined event (or accident) [4,5,13]. The majority of these accidents were caused by human error (about 80%), while only about 20% were caused by equipment failure. Accidents occur when the sequence of actions in the processes or the timing is wrong, when activities are executed at an abnormal speed, when there is a mistake in the quantities used, or when a combination of the above and other factors occurs.

The common types of accidents that cause infections were reviewed by Collins [14] and are shown in Table 2. In an earlier survey of 3700 infections [12] oral aspiration through pipettes was one of the prime causes of laboratory-acquired infections. The decrease in the percentage of accidents involving mouth pipetting can be attributed to the variety of pipetting aids developed and now commonly used by laboratory workers.

The authors would like to stress that using mechanical pipetting devices does not interfere with the accuracy or the efficiency of the work. It improves sterile techniques and is a great help when dealing with larger

TABLE 2. Accidents That Cause Infections

Cause no.	Type of accident	No. of infected persons	%
1	Spillage and splashes	188	26.7
2	Needle and syringe	177	25.2
3	Sharp object, broken glass	112	15.9
4	Bite or scratch of animals or ectoparasites	95	13.5
5	Aspiration through pipette	92	13.1
6	Others	39	5.5
	Total	703	100.0

volumes or many repetitive pipetations in that it does not tire the worker. Those devices are a must when working in biological safety cabinets.

B. Laboratory Processes Creating Aerosols

As indicated previously, the actual causative event could not be identified in the majority (80%) of laboratory infections. Most experts believe that exposure to infectious aerosols or coming into contact with surfaces (benches, equipment, etc.) contaminated by aerosols were at fault. Any time someone disturbs an air-liquid interface, an aerosol is created. As the energy of disturbance increases, the aerosol increases. All laboratory procedures can be shown to produce aerosols; simple laboratory techniques such as pipetation, centrifugation, homogenization, or blending produce aerosols. If the aerosol particle size is in the range of 1-10 μm, a fraction of the particles inhaled will be retained in the respiratory tract (especially the alveoli). Infection is most inevitable if the particles contain viable infective agents.

In some biotechnological processes, the end product is a dry substance that is further processed or formulated. The hazard of dealing with biological agents in the dry form is greater than the hazard of dealing with liquids (dependent on the agent) [15].

Powders tend to hover for a long time in the air and can be carried away to remote places from the site of the accident [16].

C. The Human Factor

The main cause of laboratory accidents is still the worker. The worker is the one who operates the equipment, who handles animals, and unfortunately is the one who becomes infected. As in road accidents, the major cause of laboratory accidents is human failure.

In a study of the human factor in microbiological laboratory accidents carried out at Becton Dickinson Research Center, two groups, each containing 33 people, were selected according to their involvement in accidents and were asked to fill out questionnaires concerning habits and attitudes [17]. The study showed some differences between the accident-involved group and the accident-free one. More people in the accident-free group realized the importance of proper attitudes in safety endeavors, had respect for the infectious agents, and were aware of the hazards. This group placed prime importance on understanding and following safety regulations. The accident-involved group took excessive risks by violating known safety regulations. More nonsmokers and nondrinkers were in the accident-free group, and more members of the accident-involved group were divorced.

Behavioral factors can result in accident, and one way to minimize if not eliminate this cause is to modify unsafe behavior patterns. The subject of behavior patterns was reviewed comprehensively by Martin [18] in a report that discussed the principles of behavioral analysis, modeling and changing behavior step by step, as well as schedules of reinforcement.

Men were involved in seven times more accidents than women (normalized to working hours) [19,20]. There was a difference in accident involvement between age groups as well [20]. The rate in the 17- to 24-year-old group was about twice that of the 45- to 64-year-old group. This finding can be accounted for by the changing of character or attitude with age, and also by the experience and know-how that accumulates with the years.

D. Infections from Laboratory Animals and Zoonosis

Working with laboratory animals involves certain hazards, especially with primates. Sometimes newly purchased (or collected) animals can carry infectious agents asymptomatically, some common and some exotic, including oncogenic or slow viruses. Animals can transmit infectious agents to workers by scratching, biting, or shedding from fur. Disease can be contracted by unsafe handling of tissue samples or clinical specimens.

The most frequent accident in animal research procedure is self-inoculation [5,21] or inoculation of the worker handling the animal. Clearly, suitable restraining devices would prevent this [22]. Autopsies also generate aerosols as the tissues are manipulated and sliced. Autopsy procedures also introduce the possibility of accidental penetrating wounds from sharp objects. The conventional procedures for disposing of animal bedding also generate infectious aerosols; fecal material may contain more than 1 billion microorganisms per gram!

More than 30 of the 150 zoonotic diseases have been described to be transmitted from animals to laboratory workers [23]. These include lymphocytic choriomeningitis, Marburg, herpes B encephalitis, hepatitis, rabies malaria, leptospirosis, tuberculosis, and others. In some cases, animals subclinically carry the infectious agent and there is no way to predict the consequences.

Transmission of disease from man to animal can cause troubles in maintaining SPF animals and can distort test results and evaluation. Such a case was reported in New Zealand [24] in an SPF animal facility, keeping mice, rats, guinea pigs, rabbits, and cats. After 5 years operation without infection, all the various species were found to be infected with *Pasteurella pneumotropica* and staphylococci strains. Phage typing proved that the animals were infected by workers carrying those strains.

IV. BIOINDUSTRY AND OLD-NEW HAZARDS

During the last two decades the balance between diagnostic and research laboratories and industry-oriented laboratories tipped to the side of the latters. Biotechnological laboratories deal with vast variety of processes—most of them multistage ones. Some of them begin on a small scale of research and development (R&D) and are upscaled to processes on an industrial scale that involve large amounts of biological agents and/or their products. Industrial biological processes include techniques and systems from various domains of biology—molecular biology, cell biology, and biochemistry—that interact with chemical and engineered ones.

The materials that start the process can vary—bacteria fungi, yeasts animal and plant cells, animal tissues, or various body fluids. The industrial processes applied also vary—cell growth, cell breakage, genetic engineering, protein extraction and purification, all in different scales.

However, some stages in the processes are common to all [25]:

1. Isolation and preservation of microorganisms or cells
2. Preparations of starter cultures
3. Scale-up and growth
4. Separation of the product from the cells
5. Product purification
6. Waste treatment

Each stage carries its own hazards, some of which render the whole process hazardous. The intrinsic hazards in biological industry involve the nature of the biological agents used, summarized as follows:

1. The potential of some strains to cause disease
2. The possibility of genotypic or phenotypic change to influence certified processes
3. Possible contamination of the system
4. Presence of biological agents in the media

Hazards in some specific bioindustrial processes are discussed below while prevention will be discussed later (Sections VI and VII).

A. Cell Cultures, Hybridomas, and Viruses

Viruses, including oncogenic ones, can be present in any cell culture, even a culture prepared from apparently healty animals [26]. Monkey cell cultures harbor a variety of simian viruses, and herpes viruses were detected in cell cultures derived from horses, rabbits, and guinea pigs. A complete copy of the endogenous type C virus—the virogen—is present in the normal cells of chickens, hamsters, mice, rats, and cats [27]. When working with cell cultures, viruses, and hybridomas, there is at least one common hazard to all three: tumorigenicity. All three are interrelated. Viruses are produced in cell cultures, viruses transform cells, transformed cells can be hybridized with antibody-producing cells, etc. The following are the common hazards with the above-mentioned processes.

Certain human cell strains and lines are susceptible to infection in vitro with chicken and mouse type C leukemia and sarcoma viruses. No evidence of in vivo infection or tumorigenic activity of chicken type C virus for humans has been found [28]. It has also been established that long-term lymphocyte lines contain Epstein-Barr virus (EBV), and laboratory workers who do not have antibodies to EBV may risk getting infectious mononucleosis, which can be devastating [29]. Human lymphocytes can be transformed in vitro and may be oncogenic to humans, imposing another hazard in cell cultures. Some of these cells have been used to make somatic cell hybrids with human cells, and the mouse/human hybrid cells continue to produce type C leukemia viruses. The viruses can alter their host range, either by adaptation or selection. Thus the human hybrid cells constitute a potential hazard.

Several cases of human cancer after kidney transplantation from tumor-bearing donors have been reported. Based on these observations, it is essential that laboratory workers *do not* manipulate cell cultures derived from their own tissues or those of close relatives [30]. The cells can be infected and transformed by a virus in vitro and then infect the worker/donor. After adaptation in vitro, the immune system will be inefficient in protecting the body, and cancer may develop.

In recent years (1980-1986), a series of cancer cases in laboratory workers occurred at the Pasteur Institute. Those laboratories dealt with oncogenes, mutagenic substances, and radioactive isotopes. After four diagnosed cases of cancer, an expert committee was formed to decide if there was a direct connection between the cases and working conditions in the laboratories, and if pressure to meet deadlines of commercial contracts caused deterioration in safety precautions. In June 1986 the illness of Yves Malpiece (a technician who worked only 6 months in that laboratory including working with radioactive isotopes) was classified as an occupational disease by the French authorities. Yves Malpiece died of bone cancer [31].

Lymphocytic choriomeningitis (LCM) deserves special attention. LCM infection in man can be inapparent or can take one of several clinical forms: from a mild influenza-like disorder to fatal meningeal encephalitis. An outbreak of LCM among medical center personnel was reported by Hinman et al. [32]. Forty-eight cases were traced to infected hamsters in which tumor cell lines were being passed. LCM virus was isolated from the hamsters as well as from the tumor cell lines. The study provided evidence that LCM can be transmitted not only by direct contact but also by aerosols.

B. Vaccine Industry

Much has been said about safety to the user of vaccines, but the production of vaccines has received much less attention in terms of safety to workers who produced them.

Three types of formal analyses are used to evaluate viral vaccines: benefit-risk, benefit-cost, and benefit-effectiveness [33]. No consideration is included for safety to the producers of these vaccines. Primary cells in culture that are routeinly used to prepare vaccines include chick, duck, rabbit, dog, and monkey. The problems with indigenous viruses in caught wild animals practically eliminate the monkey kidney cell for future vaccine preparation. Neoplastic and transformed cells are clearly forbidden, though such cultures would provide the means for mass production of virus in submerged cultures. Eliminating those possibilities for the sake of product safety serves the worker quite well.

Many viral vaccines, such as rabies, influenza, some arbovirus vaccines, as well as bacterial vaccines such as plague, typhoid, or cholera, are classified as inactivated vaccines, i.e., they are inactivated by heat treatment, phenol, formalin, or β-propiolacton [34]. The infectious agent must be grown, concentrated, or purified before inactivation. The process involves large quantities and concentrations as well as aerosol-producing equipment, such as centrifuges, and they pose a considerable hazard to the worker. Live attenuated vaccines also post some hazard. If large quantities are inhaled, minor symptoms may occur, as well as side effects such as hypersensitivity reactions. Live attenuated strains for human or veterinary use can cause the disease, same as the wild-type strain. This is the case with *Brucella abortus* strain 19 [14,35], and the case of infection with a laboratory strain of rabies due to aerosol exposure of a worker who was immunized against rabies. This case will be discussed later in a chapter dealing with vaccination as a protective measure for laboratory workers. Subunit vaccines may be the ultimate answer for human vaccination from the point of both product and production safety.

Immunogenic protein can be produced by genetic engineering methods. Subunit viral vaccines derived from the whole virus, however, again have the disadvantage of first needing to grow large quantities of infectious agents. With hepatitis B virus (H-B-VAX), where the subunit vaccine is derived from surface antigen that is purified from the plasma of human carriers of hepatitis B virus, there are hazards indigenous to all processes of the blood product industry.

C. Blood Products Industry and AIDS

The blood product industry includes the separation and fractionation of blood and plasma in large quantities. The products are lymphokines, interferons, coagulation factors, antibodies, antigens, etc. The major hazard in these processes is infection with hepatitis B or acquired immunodeficiency syndrome (AIDS). Laboratory infections with hepatitis B result from contact with infected material, usually blood, or by injection or ingestion [36]. The cited study tends to minimize the importance of aerosol transmission of blood, but not during manipulation with aqueous suspensions. Infection usually results from manual contact with infected material. Such infection can be minimized by good laboratory procedures and a barrier between the biological agent and the worker. Also it is advisable for people in high-risk groups to receive vaccination, including the surface antigen of hepatitis virus, harvested and purified again from the plasma of human carriers [37].

Similar problems arise with AIDS, though on a smaller scale now. The suspected etiological agent for AIDS, human T-lymphocytotrophic virus (HTLV-III, designated today human immunodeficiency virus, HIV), may be present in any plasma or blood bag. By selecting donor panels for certain attributes as, for example, in the production of hyperimmune globulin, we can select for exposure to other antigens as well. Such a case was the commercial antihepatitis B immunoglobulin, which was subsequently found to contain, besides antihepatitis antibodies, high levels of antibodies to HIV [38].

The health authorities in the United States [39] summed up nine cases of infections with AIDS virus (HIV) as a result of laboratory accidents. In four cases the infection was a result of needle pricking that contained blood from AIDS patients (the infections were proved by EIA and Western blot test). In other cases infection occurred due to contact with blood or body fluids of AIDS patients, or by splatter of blood on the faces of the victims. In the mentioned article 1700 cases of laboratory accidents and exposures to body fluids of AIDS patients were cited. Only in one case was seroconversion reported. Adding the other nine cases we arrive at the probability of infection of 1:170. Taking in account the severity of the disease, one cannot ignore the possibility of laboratory infection with AIDS, which is not much different from the overall probability of laboratory infection (10^{-2}-10^{-3}). An alarming fact in this study was the large number of employees that did not follow basic safety rules, and did not cover wounds or scratches while working with potentially infective biological material.

In a study of 265 laboratory and affiliated workers, one individual was found to be seropositive for HIV-1 [40]. Two HIV-1 isolates were derived from a blood sample of this worker. They were indistinguishable from a genotype form of HIV-1 present in the cell line, with which the laboratory was working.

Exposure to this strain of virus most probably occurred during work with concentrated virus or culture fluids from a virus-producing cell line under standard biosafety level 3 containment (see Section VI.F). The infection was attributed to skin contact with virus culture supernatant. The calculated incidence rate was 0.48 per 100 person-years. Exposure to concentrated virus is associated with some risk of HIV-1 infection, which is comparable to the risk for health care workers experiencing a needle stick exposure.

D. Hazard in Recombinant DNA Techniques

Until the early 1970s there was very little interest in safety, but then, as described by Williams [41], "it took off." This phenomenon was attributed to growing concern about laboratory infections from hepatitis, tuberculosis, and typhoid; but for public interest and serious handling of the subject, we have to thank the big boom in recombinant DNA research. This new area of activity raised many questions that had no immediate answer. Research had to be carried out in order to answer them. Because the hazards and consequences of this kind of gene manipulation were initially unknown, health and safety authorities were cautious in their estimation of the hazards and chose the maximum safety procedures and containment available as guidelines for recombinant DNA research. The first official version of the guidelines was published in the *Federal Register* on July 7, 1976 [42].

The experiments were divided into many classes based mainly on taxonomic criteria and were assigned to specific classes of containment. In 1977, a revised proposal was published [43] somewhat relaxing the guidelines, and another in 1982 [44]. In 1978, in an article under the insinuating title "Much Ado About Recombinant DNA Regulations" [45], Szybalski discussed the possible hazards in gene manipulation and stated: "For the bulk of this research (recombinant DNA techniques), the risk is nonexistent from the practical point of view." But this is not true when research is carried out with a host who is in the high-risk class on the basis of hazard. Here safety procedures and containment must be applied as for the original host or vector. The present thinking is that a recombinant cell will not provide more hazard than the modified parent organism [46]. In most cases, the modified cells will be less viable and less likely to become established in an undesirable environment. The addition of foreign DNA is almost always detrimental to the host cell, contributing to benefit but consuming energy and possibly inhibiting the organism directly [47,48]. However, cloning of a gene whose dissemination would be particularly undesirable—e.g., toxin gene—should be carried out in a plasmid or a virus unlikely to be selected for in nature, or to be readily transmitted to another organism. Another measure of safety can be taken by producing the bioengineered desired protein in an inactive state that can be made active only by further chemical treatment.

The only *reported* accident which resulted in an infection with a genetically engineered strain was reported by a group of scientists in which one of them was the infected person. The accident occurred during the study of antigens of vaccinia virus (a vaccine strain WR) harboring genes coding for G and N proteins of vesicular stomatitis virus (VSV).

During injection of mice, 0.05 ml (containing 3×10^4 plaque-forming units) was spilled on a cut in a finger of one of the scientists [49]. The worker experienced a moderate infection of vaccinia virus and produced antibodies to protein N of VSV, the same as in the serum of the infected mice. The group published this incidence as proof that vaccinia as a vehicle for immunization against different proteins will work. However, this case is another example of scientific staff being exposed to infection in the laboratory, and with genetic engineering, depending on the ability of the host to cause disease, there can be problems.

In order to minimize the risk, less virulent strains can be chosen and developed for replacing the virulent agent. This could be done by selection methods, with or without mutagens, or again to isolate the gene from a virulent strain and insert it into nonvirulent microorganisms such as yeasts, *Bacillus*, or *E. coli*.

E. Cell Product Safety

The World Health Organization (WHO) study group on biologicals had published an issue on cell products safety [50]. The main concern was the use of continuous cell lines (CCLs) for the production of biologicals.

However, despite past actions, some concern still exists and this has impeded more general acceptance of CCLs as a substrate for biologicals administered to humans. It was in fact because of concerns regarding product safety that a group of consultants, who met in September 1985 to consider the development of WHO requirements for hepatitis B vaccine produced by recombinant DNA techniques in CCLs, recommended that WHO convene a special group in order to provide an independent international evaluation of the safety issues as well as actions already taken, those in progress, and those which are contemplated.

The probability of the risk that residual cellular DNA contaminant in a biological product will cause a neoplastic transformation event was estimated based on a number of assumptions.

The assemptions were about oncogene size, frequency, biological integrity, and in vitro and in vivo transformation efficiency. The probabilities were calculated to be in the order of 10^{-6}-10^{-9} at the most [51].

In another paper, the risk of gene transfer using CCLs expressing known oncogenes or cells containing oncogenic virus-recombinant DNA vectors was described as "minimal" [52].

V. ASSESSING THE RISKS ASSOCIATED WITH BIOINDUSTRY

When discussing the risk associated with bioindustrial processes, four main points must be considered:

1. The hazards imposed by the *microorganism* or by *the process* on the people performing it—what risk are they taking in getting infected or contracting a disease?
2. The risk imposed by the *industrial process on the environment*; the possibility of an unwanted leak from any stage of the process, resulting in uncontrolled release to the environment of a pathogen or a toxic product.
3. The risk imposed on the *biosystem or the product by the environment*—what are the chances that the whole industrial process will get contaminated and will change its established course, ending up with no product or, worse yet, with an unwanted product?
4. The risks imposed on the *environment* by deliberate release of *bioindustrial products*, the impact of which is very difficult to forsee.

Some aspects to consider in assessing the risk arising from the interrelations among these four entities are the process, the agent, the human factor, and the environment as discussed below.

A. Factors Affecting Risk to Laboratory Personnel

The first stage of safe handling of industrial process and prevention of laboratory infections is the assessment of the risks imposed on the *personnel* by the biological *agent* and by the *process* itself. The risk rises with certain human factors and depends on human judgment, the complexity and variety of the processes, and the nature of the agent. For each of the three entities there are some variables that make it almost impossible to define precisely the risk of infection to laboratory personnel. Some of the factors that are considered in assessing the risk involved with these three entities are discussed below.

1. The Agent

Wedum and Kruse [53] developed an approach for estimating risk of infection as a working guide for laboratories. They considered certain variables dependent on the agent, the occurrence of recorded laboratory infections, the minimal human infectious dose, the recovery of microorganisms from urine and feces of infected animals, and frequency of cage mate infection. Other considerations associated with the agent are the severity of the disease (mild, disabling, fatal), whether there is person-to-person transmission, and the geographic distribution of the agent. All these variables were considered by several institutions and international organizations, and etiological agents were classified on the basis of hazard into four groups, with recommendations for the conditions under which each class should be handled [54,55].

2. The Worker

The variables to be considered when assessing the risk of infection to be the worker are factors affecting susceptibility and resistance: whether the worker is immune to the agent or if a vaccine is available; whether therapy is available and applicable, including medication, antiserum, and interferon; whether the worker is in a state of immunosuppression due to disease, steroids, or cytotoxic medications.

Other human factors to consider in assessing risk are age and sex of the workers, and the level of training and skill of the workers.

The hazard to the human fetus from certain microorganisms is established in many cases (rubella, etc.) but is still unknown for others. Thus, it is most advisable that pregnant women not work with infections viruses or tissue cultures at least during the first trimester.

3. The Process

Usually experiments that involve laboratory animals impose higher hazards than those performed in vitro. A number of variables beyond the control of the worker are present when working with animals whereas in vitro work facilitates control and calculation of every step.

In general, after the research and development stage, biotechnological processes involve large volumes of microorganisms in high concentrations. These two variables (volume and concentration) when taken into consideration in risk assessment dictate higher levels of biosafety than the ones used for the same agents in small-scale research [56].

Freeze-drying (lyophilization) is a common process in biotechnology, and lyophilized cultures may be more hazardous than liquid ones because powder tends to disseminate freely for long distances by air movement. Any procedure that can generate aerosols must be assessed for risk—not only the obvious procedures like blending and centrifugation, but also the very simple ones like pipetting and pouring.

B. Risk of Contamination of the Industrial Process

The risk to people and the environment, associated with properties of the microorganisms and the nature of the industrial process, were discussed above.

Usually, but not always, protecting the system from leaking to the environment also serves the safety of the system from external contamination.

If a contaminant entered the process and could maintain itself, it could disrupt the system in one of several ways. The contaminant could interfere with the biocatalyst function; it could destroy the biocatalyst or the product by using them as an energy source. A contaminant could also introduce substances different from the expected product but inseparable from it, thereby affecting product safety and even rendering the product unusable.

Since the environment inside an industrial fermentor is very different from the natural habitat of a pathogen, it is unlikely that the industrial process will become contaminated with highly pathogenic microorganisms. However, the probability of such an event is far from zero and one must take precautions to keep the process isolated from the environment.

On the other hand, no matter how isolated the system might be, there is still a possibility of contamination from an "inside job" due to a phenotypic or genotypic change.

C. Risk to Environment from Deliberate Release of Microorganisms

The question of risk and safety in connection with bioindustrial products and the environment is not limited to genetic engineering. Similar problems arose when *Bacillus turingiensis* (*B.t.*) was suggested for deliberate release as an insecticide.

For biocontrol of insects only those microbials that are harmless to man, livestock, and crops should be used. Therefore, health authorities insist on risk-benefit analysis as an essential part of the guidelines for the official registration of bioinsecticides.

Hazard from preparations based on living bacteria may include (a) infection or intoxication of nontarget organisms and/or (b) autoreproduction and autodissemination in the environment.

In this case tests with *B.t.* preparations have shown the following: Sporulated cultures have no infectious potency or toxicity on nontarget organisms including man, other animals, and plants. The crystal toxin of sporulated cultures in particular is highly selective; its activity is restricted to a special group of insects. Also studies concerning exotoxins of vegetative cells of *B.t.* have shown no significant risks for vertebrates after application per os. *B.t.* does not have any epizootic potency based on the activity of its spores. Germination and reproduction of *B.t.* have never been found in water, in soil, on plants, or in dry-stored products [57].

In principle, the outcomes associated with the introduction into the environment of organisms modified by rDNA techniques are likely to be the same in kind as those associated with introduction of organisms modified by other methods. Therefore, identification and assessment of the risk of possible adverse outcomes should be based on the nature of the organism and of the environment into which it is introduced, and not on the method (if any) of genetic modification. By the same token, the possible risks associated with the process that produces the organism (e.g., unintended releases from containment) should be distinguished from possible risks arising from the environmental introduction of the organism.

D. Case-by-Case Approach

The approach to evaluation of environmental use of bioindustrial products is to study each separately and individually.

The case-by-case approach to safety assurance has been defined by the OECD as follows: "Case-by-case means an individual review of a proposal against assessment criteria which are relevant to the particular proposal; this is not intended to imply that every case will require review by a national or other authority as various classes of proposals may be excluded" [58]. This case-by-case approach, as currently practiced by both governmental and nongovernmental bodies, is compatible with the development of generic guidelines and methods for determining when and against what standards an application should be screened.

A useful descriptive model for organizing methods for risk assessment of biotechnology applications was presented in a report by the Office of Technology Assessment (OTA) [59]. The model suggests a series of five events, or stages, that need to be considered in assessment:

1. *Formation*. The creation of a genetically altered microorganism through deliberate or accidental means.
2. *Release*. The deliberate release or accidental escape of some of these microorganisms into the environment.
3. *Proliferation*. The subsequent multiplication, genetic reconstruction, growth, transport, modification, and die-off of these microorganisms in the environment, including possible transfer of genetic material to other microorganisms.
4. *Establishment*. The establishment of the microorganisms within an ecosystem niche, including possible colonization of humans or other biota.
5. *Effect*. The subsequent occurrence of human or ecological effects due to interaction of the organism with some host or environmental factor.

The suitability and applicability of risk assessment methods, as tools to answer the above, was reviewed by Fiskel and Covello [60].

VI. PREVENTION OF INFECTIONS IN BIOINDUSTRIAL PLANTS

The law holds the employer responsible for the workers' safety. The employer must explain the hazards to which workers are exposed in the workplace, and is obligated to instruct workers on how to perform safely and to provide them with safety equipment including guidelines for use.

However, a management of an industrial plant, which is usually a profit organization, can interpret and execute the law and safety rules differently from those of a university, a research institute, or a government organization. This raises the next question: Is safety a matter of policy?

A. Safety: A Question of Policy?

The first step in preventing infection in the workplace and in any safety plan is risk assessment. Afterward the action on the part of the management will depend on its policy pertaining to the standard of prevention it wishes to adopt.

Management should decide its strategy (policy) by answering the following questions:

Is it necessary to prevent all infections, including those that can be detected only by serological tests?
Is it enough to prevent only those infections that can cause death or disablement?
What kind of health plan must be provided?
To what extent is it necessary to protect persons outside the plant?
Is it imperative to avoid escape of the microorganism to the environment?

According to decisions for the above questions together with the risk assessment criteria, the management will plan a course of action in the prevention of infections. The strategy will include factors as discussed in the following paragraphs.

1. Suitability of Employees to Their Jobs

The points to consider under this heading should include the character of the person, his or her tendency to be involved in accidents, sense of responsibility, reliability, attitude toward safety, etc. An unreliable person should not be employed in high-risk jobs or in critical positions in production lines.

Physical and health characteristics should be taken into consideration as well. Persons in poor physical condition should not be employed in arduous jobs, and allergic persons should not be assigned to work with laboratory animals or in conditions that will expose them to allergens. Of course, this type of information is rarely known at the beginning of the employment, and when it is established, it is sometimes impossible to reassign employees to other posts. Nevertheless these considerations must be taken into account and can help in preventing accidents and infections.

2. Instruction and Training Program in Safety

Management should develop and execute a program for instructing and training in general and specific subjects of safety pertaining to the activity in the laboratory or plant. Each new employee should be instructed before starting work and periodically thereafter. The program should include lectures, exercises, and exams to test the perception and retention of the material taught.

3. Health Plan

Management will determine whether a health care team should be employed on the premises or to get medical services from the outside. The health plan should include an epidemiological survey for those employed with biohazardous agent, namely, taking blood samples at the beginning of the employment and periodically thereafter, and checking for seroconversion with respect to the agents to which they might be exposed.

The plan should include an appraisal of the worker's health status while looking for high risk factors, such as disorders connected with the immune system, and/or prescribed cytotoxic medications or steroids that could affect one's resistance to infection. The possibility of inoculating employees in high-risk groups with known and approved vaccines should also be considered.

4. Supervision, Report, and Investigation of Accidents

Management should apply methods for supervising the correct execution of safety procedures. This could be carried out by using checklists supplied to the staff or by routine checkups of laboratories by safety officers. Since one always learns best from experience, including mistakes, procedures for reporting and investigating accidents should be applied.

5. Safety Budget

Sometimes at this point the policy of safety is decided. A plant can spend on safety only a certain percentage of its budget because too much expense for safety measures could put it out of business. Still, moderate investment can improve safety tremendously. But high levels of biosafety, namely, containment laboratories, demand high expenses. A budget for these expenses is not always available, and usually priorities dictate the investments in a plant. However, investing in safety equipment will have high benefit by saving on compensations, work day loss due to illness, and even shutoff of production lines by the authorities due to accidents.

B. Defensive Working

After risk assessment and adopting a policy in prevention of laboratory infections comes the "practical" stage, i.e., executing the theoretical approach and actions for "defensive working." The tools for the prevention of laboratory infection can be categorized into four groups:

Safety procedures and techniques
Safety equipment
Containment laboratories
Biological containment

A combination of the four groups provides the worker with the necessary protection for the prevention of laboratory infection.

Another two concepts in safety are primary and secondary barriers. *Primary barriers* are all the physical safety devices that prevent direct contact between the agent and the worker. It includes such personal protective equipment as gloves, coats, respirators, etc., and biological safety cabinets or stainless steel sealed containers. *The secondary barrier* provides a second defense line to block the agent escaping to the environment if the primary barrier collapses. Secondary barriers consist of a variety of safety devices such as an air filtration system, a decontamination and sterilization system, impermeable walls, etc.

1. Instruction in Safety

As was mentioned already, the human factor is predominant in causing accidents and laboratory infections. In order to minimize the risk from the human factor, efforts must be concentrated on instructing all employees involved with bioindustry. Instruction should begin at the universities and technical schools, and should continue in the workplace in research institutes, the industrial plants, and so on.

Courses in safety are organized by CDC, NIH, WHO, and other agencies [61,62]. Still the responsibility is management's, which must write the appropriate and specific safety procedures and practices.

2. Good Laboratory Practices (GLP)

GLP aims to minimize the occurrence of most common accidents caused by human error, poor laboratory practice, and misuse of equipment. It consists of a set of instructions about the use of various laboratory techniques, such as pipettes, bacterial loops, hypodermic needles, centrifuges, homogenizers, and other laboratory equipment [62,63].

These practices are based on sterile techniques, which are learned in every bacteriological course. They intend to protect the experiment from contamination on the one hand, and the worker and the environment on the other. Basically they prevent the production and dissemination of aerosols, and some of them are cited here:

Bacteriological loops should be heated from the far end of the loop.
Loops must be let to cool before the inoculum is touched.
Mixing, blending, and centrifuging will be done in airtight tubes.
The last drop from the pipette should not be blown out.
Lyophilized cultures should be opened wrapped in cloth soaked with decontaminant, preferably in a biological safety cabinet.
Pouring of cultures should be done slowly.
Needles should be wrapped before being withdrawn from bottles.

GLP is based on years of experience and study of accidents and laboratory-acquired infections, and appears in the form of regulations. These regulations must be obeyed, preferably after instruction and explanation of the rationale behind each regulation. GLP for nonclinical laboratories is published in the *Federal Register* [64].

3. Good Manufacturing Practices (GMP)

GMP is also published in the *Federal Register* [65]; it describes the manufacturing, processing, packing, or holding of drugs and of finished pharmaceuticals. GMP deals with organizations and personnel, buildings and facilities, equipment, written procedures, laboratory controls, records, and reports.

GMP is essential not only for product uniformity, quality control, and assurance; it is also good for safety. Written standard operating procedures (SOP) do not give much leeway for improvisation and shortcuts, which usually lead to unforeseen conditions and accidents. SOP is intended to ensure the quality and integrity of the data generated in the course of a study or production process. Deviations in study or production process from standard operating procedures must be authorized by the study director and must be documented in the raw data. Significant changes in SOP have to be properly authorized in writing by the management.

4. Code of Practice (COP)

COP is a listing of the most essential laboratory procedures that are basic to safe laboratory practice. In many laboratory programs, the code may be given the status of "rules" for laboratory operation.

Each worker should read, sign, and even be tested on knowledge of this code. There is a code for basic laboratories and a special one for containment laboratories. Again, these rules, as in the case of GLP and GMP, are derived from accumulated experience and are based on sources and causes of accidents and infection. In order to work safely as well as to produce a good product, one must follow those rules to the letter.

Examples of some rules are the following:

Mouth pipetting is prohibited.
Eating, drinking, and smoking are prohibited in the laboratory work area.
Gloves and laboratory gowns must be worn for procedures that may involve direct contact with infectious materials.
All contaminated materials should be decontaminated before disposal.
Work surfaces should be decontaminated at least once a day and after every spill of infective material.

A detailed list of COP and safety guidelines can be found in Refs. 61-63, and 66.

C. Safety Equipment and Containment

Safety practices are necessary for the prevention of laboratory infections but they are not sufficient; they have to be combined with use of safety equipment which constitutes a primary barrier between the worker and the biological agent. With regard to the operating principles, there are six distinct classes:

Physical barriers, e.g., sealed stainless steel containers
Air barriers, e.g., air locks and biological cabinet
Filtration barriers, e.g., HEPA (high-efficiency particulate air)* and other filters
Extinction barriers, e.g., incinerators
UV barriers, sometimes in combination with air locks
Flexible barriers, e.g., plastic bags and gloves

The containment principles are chosen by the planners of the laboratory or the plant, and the equipment is chosen in accordance. Later the correct way of using safety equipment will be discussed.

---

*"Absolute" filter with about 99.97% efficiency for particles of 0.3 μm in diameter.

1. Personal Safety Equipment

The closest protective equipment to the worker is the personal one, which protects the most vulnerable areas of the body, or the whole body when the hazard level rises. This is the basic protection to the worker and it must be used in accordance with the regulations of the various laboratories. Following is a list of personal protective equipment classified according to the body area that needs protection:

1. *Body*. Laboratory coats; jumpsuits; two-piece laboratory suit; rubber aprons
2. *Head*. Laboratory cap; buffant cap; hooded cap
3. *Hand*. Surgical gloves; heat-resistant gloves; long-sleeve gloves; leather and stainless steel gloves for animal handling
4. *Feet*. Hard safety shoes; water-resistant boots; disposable shoe covers
5. *Eyes*. Safety glasses; goggles; full-face shield
6. *Respiratory protection*. Two kinds: respirators that filter the air, and devices that supply clean and filtered air

   - Respirators: A full-face mask that covers the nose, mouth, and eyes. Contains both biological and charcoal filters; half-face mask, for nose and mouth only; paper dust mask which provides only partial and low-efficiency protection
   - Respiration devices: full-face mask with air supply; continuous flow pressure or demand; regulated suit with air supply; closed-circuit respiration system

One must always bear in mind that these protective equipment devices do not provide complete protection from penetration of contaminated air, but they have certain "protection factors" (PF). For example, the PF for full-face mask is 1:50; a positive pressure suit 1:2000; and closed-circuit system 1:10,000. These devices have to be checked periodically for performance and leaks, and to verify the right way of connection and operation after putting them on.

A summary of safety equipment, the hazards corrected by them, and their safety features is given elsewhere [67].

2. Biological Safety Cabinets

Biological safety cabinets constitute primary barriers of different levels, and reduce the level of contamination in the room and the hazard to the worker. The cabinets do not provide an ultimate barrier and protection; one must choose the right kind of cabinet for the job, to follow the technical certification test, and to use them correctly; only then is maximal protection achieved.

Classes of cabinets: Biological safety cabinets are classified in three ways:

Class I: There is a front opening and the air is flowing from the room into the cabinet. This cabinet does not provide sterility for the process.

Class II: There is a front opening and vertical laminar flow on the work area. The air is filtered before flowing on the work area or from the cabinet to the outside. It provides protection both to the process and to the worker.

Class III: Gas-tight cabinets or "glove boxes." Air flowing in and out is HEPA-filtered. It provides maximum protection to both worker and experiment.

It is important to note that positive horizontal laminar flow cabinets *are not safety equipment*. They provide sterility to the process and are suitable for preparing media. They are definitely not for cell cultures or biological agents. The following are a few examples of the necessity to use biological safety cabinets:

When operating aerosol-producing equipment like blenders, mills, freeze-dryers, sonicators, centrifuges, etc.
When growing microorganisms by shaking or fermentation, which belong to class 2 or higher or contain recombinant DNA fragments
When infecting cells with viruses the characteristics of which are not clear

Certification of safety cabinets: In order to rely on the safety that cabinets provide for the worker, they must pass certain certification test. In routine tests of biological cabinets only 35% passed. The rest posed a safety hazard for the users. Cabinets must pass certification after production, after relocation, and periodically (every 6-12 months) thereafter. When the tests are performed by a skilled technical crew, adequate protection will be provided for the user and the environment.

D. Biological Containment

Physical containment keeps the hazardous agent from the worker and the environment. The biological barrier keeps it from infecting the worker. This concept is achieved either by choosing biological agents with limited *survival* outside the original laboratory host, or by choosing an agent with very narrow host range, so that its *transmission* to other nonlaboratory hosts is not possible.

The guidelines for recombinant DNA research [43,44] define two levels of biological containment for prokaryotes: HV1 and HV2 based on the host-vector system.

HV1 is a host-vector system that provides a moderate level of containment. For example, EK1 system in which the host is *E. coli* K-12 (or derivates) and the vectors include nonconjugated plasmids (pSC101, Col E_1, or λ-phage variant).

HV2 are host-vector systems shown to provide a high level of biological containment. "Escape" of recombinant DNA is less than $1:10^8$ under specific conditions. For example, in the EK2 host-vector systems, in which the vector is a plasmid, no more than one in 10^8 host cells should be able to perpetuate a cloned DNA fragment under nonpermissive laboratory conditions designed to represent the natural environment.

Biological containment must not be limited to recombinant DNA research. Whereas physical barriers may be expected to reduce the probability of infection by 3-6 log, biological barriers may offer much greater degrees of protection. For example, because chicken cells lack receptors for poliovirus, it is impossible to initiate infection in this species with poliovirus stocks containing over 10^{10} plaque-forming units per milliliter!

E. Vaccination

Another approach to the subject of prevention of laboratory infections that may be categorized under "biological containment" is vaccination. Vaccinations provide larger margins of safety for the people working with hazardous agents. When considering the question of immunizing employees, the people in charge should adopt a policy that will define groups of employees according to the risk level to which they are exposed. This consideration should include a risk-benefit calculation and distinguish between mandatory and optional recommended vaccinations.

Recommendations should be directed at the actual agents with which work is being conducted. Vaccines for which the benefit (antibodies titer produced is acceptable as sufficient for protection) is much higher than the risk (local or systemic reaction) should be used for all high-risk groups of workers. Examples for such vaccines are yellow fever, tetanus, rabies, polio, and hepatitis B—diseases that can cause death or permanent disability. On the other hand, vaccinations that involve harsh side effects, local or systemic, or that may cause hypersensitivity upon reinjection should be administered with caution. In this group are cholera, typhoid, and tularemia—diseases that can take severe forms but that can be treated and cured.

It should be stressed that vaccination, even if it produces acceptable neutralizing antibodies titer, should not be regarded as 100% protection against infection in industrial and laboratory process. Infection might result if the dose penetrating the body is very high, and more so if the exposure is respiratory—cases which might occur in the bioindustry. The efficiency of serum antibodies in preventing infection through the respiratory route is not established and the following case illustrates this point. In 1977, a 32-year-old worker at the Department of Health in New York was seeding baits with a live attenuated strain of rabies. The machine performing the task was not protected enough, and produced aerosols, to which he was exposed for a long time. The worker had been vaccinated against rabies, and the neutralizing antibodies titer in his serum was acceptable as protective [68].

At a certain stage the worker came down with rabies and was hospitalized. In 1988, 11 years after the infection, he was paralyzed and 95% crippled (personal communication). As a result of this case, health authorities revised the regulations concerning rabies: mandatory use of biological cabinets with every aerosol-producing process and/or with large quantities; donning of gloves with every activity involving the agent; and a mandatory prerequisite for 1:16 neutralizing antibodies titer before starting work [68].

F. Containment and Safety Laboratories: Biosafety Levels

Safety laboratories, together with safety equipment and safety procedures, contribute to the protection of the worker and the environment from spreading infections. The concept of safety laboratories includes also procedures, techniques, and safety equipment, but usually means the facility itself. As a building, containment laboratories constitute a secondary barrier that prevents the escape of biological agents to the environment. The level of containment rises with the hazard class of the agent.

For classification of containment laboratories, the health authorities use different terms. For example, NIH [61] classifies laboratories in four

TABLE 3. Comparison Between Hazard and Containment Levels as Classified by Different Authorities

WHO		NIH	CDC	NCI	CDC/NIH	England
Laboratories	Hazard class	r-DNA	class of agents	oncogenic viruses	labs	agent categories
Basic	I	P_1	1	—	BL 1	—
Basic	II	P_2	2	Low	BL 2	C
Containment	III	P_3	3	Moderate	BL 3	B
Maximum containment	IV	P_4	4	High	BL 4	A

levels with the letter P, from the lowest (P_1) to the highest (P_4). By a different classification (WHO), the first two safety levels (P_1 and P_2) are designated "basic laboratories," the P_3 "containment laboratories," and the P_4 "maximum containment laboratories. Today the CDC-NIH terminology is BL 1-4 for "biosafety level." This term includes facility design, safety equipment, and safety procedures suitable for each level. This terminology was published jointly by the CDC and NIH in 1984 [66]. Table 3 clarifies the various terminologies used by different authorities in relation to containment laboratories and classification of agents on the basis of hazard. As can be seen from the table, matching the containment level with the assessment of risk is not very complicated. It should be done for rDNA and non-rDNA organisms, and for large-scale industrial processes as well.

Industrial processes are carried out in different biosafety levels. The provisions of the NIH (Appendix K, Physical Containment for Large-Scale Uses of Viable Organisms Containing Recombinant DNA Molecules) supply a list of items which should be considered in the design and operation of culture vessels for various levels of physical containment. Three levels of containment are established and correspond to the levels of physical containment required at laboratory scale [69].

VII. PREVENTION OF THE HAZARDS IN BIOINDUSTRY

Biological hazards of biotechnology are associated with three properties of microorganisms: (a) the potential of few strains to cause disease [70]; (b) the potential for undetected genotypic or phenotypic changes to alter a tested and approved process; and (c) the ubiquity of organisms which can contaminate the system.

While it is unlikely that dangerous human and animal pathogens will be used in very large-scale fermentations, they are used in the laboratory and pilot plant scale fermentation [71,72]. In addition, there is a fundamental lack of information concerning plant diseases and the agents which cause them [73]. The aerosol dissemination of fungal spores [74], bacteria [75], and viruses [66] which are pathogenic to plants have been described. The economic consequences of the inadvertent release of a pathogen in an area where there are susceptible hosts is potentially catastrophic [73].

Therefore it is crucial that efforts be made to prevent the escape of pathogenic organisms to the environment.

Because of the rapid transfer of the new genetic techniques into pharmaceutical R&D programs, the pharmaceutical sector is currently the most active in the commercialization of biotechnology. For this reason, it will form one of the major focuses of this chapter. While the technique may not achieve as rapid an acceptance in other industrial sectors because of a variety of difficulties (such as recovering R&D investments or converting physical plants to accommodate biological methods of production), biotechnology will play an important role in any industrial biological process or any process in which a biocatalyst can replace a chemical one [31,71,72]. This is clearly illustrated in the OTA report, which examined current industrial applications in the pharmaceutical, animal and plant agricultural, specialty chemical, food additive, commodity chemical, energy, bioelectronic, and environmental sectors [76].

A. Risk Assessment

The scientists who originally raised questions in 1975 felt that "few, if any, believe that [genetic engineering] is free from risk" [70,73].

Initially, concern focused on hazard in the laboratory. Now, with the advance of industrial applications, the focus of concern has shifted to include risks associated with industrial use of engineered strains, as well as risks associated with the products obtained from new strains.

Unless the organisms used in an industrial process are thoroughly characterized, the public perception of risk (their ability to cause disease) will continue. Even with a recombinant DNA sequence that is well defined, a breach in containment would leave its behavior in the environment questionable. Sharples states that it was not possible to predict the likelihood of any single organism having a major impact in a foreign ecosystem. The house sparrow, gypsy moth, kudzu vine, chestnut blight fungus, and Dutch elm disease fungus were cited as examples of failures in predicting outcomes [77].

What should be kept in mind is that engineered organisms are not entirely new organisms. The question, then, is whether the newly acquired properties give engineered organisms a selective advantage over unaltered members of the species. The fact that there was a lack of data to support or deny concerns about engineered organisms in the open environment contributes significantly to the uncertainty over the magnitude of the potential harm [78].

Biotechnology encompasses a rather wide range of activity. While each activity is associated with its own starting material (bacterial, fungal, yeast, plant or animal cells or enzymes), technical manipulation (cell fusion, cell or enzyme immobilization, genetic engineering, etc.), processes (continuous or batch fermentation), and scale (small to very large) which are unique to the activity, there are certain common features or stages which lend themselves to a generic treatment. Regardless of where one's interests lie, biotechnological activity will involve (a) the isolation and preservation of the organism or cell; (b) preparation of biocatalyst; (c) process scale-up or large-scale growth of the organism; (d) separation of the desired product from the biocatalyst; (e) purification of the product; and (f) waste treatment.

B. Culture Preservation

The objective of preserving microorganisms, cells, enzymes, or proteins is to maintain the desired properties of a viable population in an unaltered form for long periods of time. Even though the volume of media or the number of micrograms is comparatively small, the biology of the organism (pathogen or saprophyte, mechanism of pathogenicity, identification of suitable hosts at risk, type of illness, etc.) must be understood prior to initiation of activity. Appropriate work practices in conjunction with an appropriate biological safety cabinet should provide the primary containment barriers necessary to protect both the worker and the biological system from cross contamination.

When a lyophilizer is used, care should be taken to prevent the vacuum lines from becoming contaminated. The insertion of a suitable filter or trap is usually sufficient. All lines, tubing, or glass connectors which come in contact with the primary vessel or tube should be decontaminated in an appropriate fashion (chemical disinfection followed by extensive washing to remove residual disinfectant, or steam sterilization) prior to reuse.

C. Preparation and Isolation of the Biocatalyst

The purpose of this stage is to provide the required number and quality of organisms or enzymes to initiate a larger scale cultivation or process.

1. Cells

A tube of preserved materials is opened and its contents transferred into a flask or bottle containing an appropriate volume (on the order of 100 ml) of a sterile nutrient solution, medium, or suspension. The culture is capped or closed with a filter or plug which permits the passage of gases but not of organisms and is then maintained at a suitable growth support temperature until satisfactory growth is achieved. This temperature varies and is dependent on the organism involved. This culture is then used to seed a similar but large vessel which may contain up to a liter of sterile nutrient medium. This flask is maintained at the appropriate temperature and is often placed on an automatic shaker, roller, or stirrer to increase gaseous interchange across the liquid-air interface [79]. After a suitable growth period, the culture is transferred to a sterile container designed to facilitate the further transfer of its contents into a large culture vessel.

2. Enzymes

If the biocatalyst is a purified enzyme, then sufficient enzyme must be isolated from the producing organisms. The preparation of the starting inoculum would be similar to that described above.

When a sufficient number of organisms is attained, the enzyme is isolated and attached to a suitable carrier. By definition, immobilization is the conversion of enzymes from a water-soluble mobile state to a water-insoluble immobile state [80].

All plastic and glassware which comes in contact with the organism should be decontaminated or disposed of promptly after use and not allowed to accumulate. Unattended bottles, flask, plates, etc., which contain media (or even a residual film of media) can support the growth of saprophytes and therefore result in unnecessary contamination of the system. Many of these environmental organisms are spore formers and therefore are capable of extensive contamination.

If pathogens are to be used, precautions to prevent accidental breakage of flasks during shaking are necessary. All glassware should be examined before use and any item which is chipped or cracked should be discarded.

D. Scale-Up of Process

In order to maximize the yield of a desired product, microorganisms are grown in fermentors (the terms "reactor vessel" and "bioreactor" are frequently used) under conditions which are usually rigorously controlled.

Although fermentation systems are usually designed to keep contaminants out, it is possible, when necessary, to retain organisms within the vessel. This containment is important when one is dealing with pathogens, e.g., vaccine production, or organisms which possess an unknown or immeasurable risk or potential risk, such as with genetically engineered organisms [25,66].

The principal biosafety problems associated with bioreactors are:

1. The formation of aerosols of the culture organism (which could endanger the immediate user)
2. Failure to sterilize/decontaminate the liquid and exhaust air from the reactor which may transmit hazardous material to both the surrounding population and the environment
3. Failure to maintain aseptic conditions in the bioreactor

At present, there are no universal codes for the construction of pressure vessels. In the United States, the American Society for Mechanical Engineering (ASME) code is generally applied. Although it is assumed that all pressure vessels are built under the guidance of a code, this is not an enforced requirement. Therefore it may not apply to vessels built outside the United States.

There are three basic fermentor/bioreactor designs for laboratory or pilot plant use [71,79,81]: (a) small portable units, which are filled with medium and sterilized in an autoclave; (b) portable units, which are sterilized in an autoclave and filled aseptically with sterile medium; and (c) fixed units, which are sterilized in place.

Almost all reactor vessels are constructed from glass or stainless steel. Glass is usually limited to vessels less than 30 liters in size [79]. These jars can be flat- or round-bottomed and fitted with a flanged top (with a plastic or stainless steel head plate), or they may be fitted at both ends with head plates.

Specific considerations that relate to vessels and their corrosion, subsequent leakage, and material retention (residual contamination) include the following:

1. Welding procedures and material selection to avoid carbide precipitation
2. Contamination caused by poor welding procedures which go undetected due to the lack of X-ray scanning (analysis) of welds
3. No grinding of external welds which may obscure the true quality of the weld and its potential for failure
4. Passivation of all welds to enhance corrosion resistance

Apart from a massive leak of liquid from the main vessel, which is usually the result of inappropriate operation or worker error, the main avenues for leakage of viable material into the surroundings are as follows:

1. Leaks from drive shaft seals, valve glands, and instrument probe ports
2. Spillage during the inoculation process and sampling procedures
3. The effluent air stream, which is contaminated with seed organism

4. Spillage during harvest operations
5. Incomplete disinfection of a fermentor at the end of a run
6. Collection and treatment of waste accumulated during operation
7. Inadequate cleaning of ventilation air purged from the work area

In either mode (batch or continuous), the design of fermentation equipment and facilities is central to the containment strategy for biological control. The term "biological control" is used because we are required to protect the worker and at the same time protect the integrity of the culture. The reason for this is that most products of biotechnology are synthesized by pure cultures [81,82]. In order to avoid contamination of the culture, all materials introduced to support growth are sterilized. Foreign organisms are filtered from the air by a deep bed of glass wool or by special filters (HEPA) designed to remove particles from air with high efficiency. Steam is employed to sterilize reactor vessels, pipelines, and other surfaces with which the medium comes in contact. The mechanical system must be designed and operated so that the opportunities for contaminants to enter are minimized. Maintaining the integrity of various entry and exit points in the system is both crucial and frequently difficult. The probability for human error or mechanical failure is high and culture contamination does occur.

E. Transfer of Process Fluids and Sampling

The addition of acids, alkalis, and other fluids (nutrients) by peristaltic pump is frequently used in small fermentors for cultures maintained under low containment near atmospheric pressure. These pumps are equipped with flexible tubing, which, unfortunately, is subject to failure. If higher containment is necessary, then these reservoirs should be pressure vessel and fluids should be piped directly to the culture vessel through an electropneumatic valve using air pressure as the driving force.

The pipelines which transport the vessel's contents as well as the air supply, seed culture, sampling lines, etc., required for aseptic use should be chemically or, preferably, steam sterilizable. (The entire system should be designed to facilitate steam or steam/formaldehyde sterilization.) It is important that pipelines be free from steam condensate after sterilization. These lines should be constructed as simply as possible, avoiding repetitive changes in elevation. Condensate in piping which has contact with nonsterile environments can and will become contaminated with environmental organisms.

It is usually necessary to install valves in the pipeline. If a high order of sterility is needed, diaphragm or ball valves should be considered. A diaphragm valve has no packing material and is less prone to contamination. Ball valves can be doubly sealed to minimize chances of contamination and designed to include a steam block. When the valve is open the broth flows through. When the valve is closed, then steam flows through the valve and, if applicable, purges the line as well.

Care must be taken to ensure that all piping and fittings are welded properly and that the use of connecting fittings is minimized. Dead spaces, crevices, and nondraining portions of lines are to be avoided. All connections between sterile and nonsterile segments should be separated by steam blocks. This care is necessary not only to protect the "environment" but to ensure that the product of one batch does not serve as a contamination source for the next.

Recommendations from the National Institutes of Health [69] indicate that "for each level of containment sample collection from a closed system be done in a manner which prevents the release of aerosols or contamination of surfaces." This requirement, while specific for recombinant DNA systems, would be applicable to the use of pathogens as well. Unfortunately, these absolute requirements of preventing the release of aerosols or contamination of exposed surfaces cannot be met when using most types of sampling devices.

A traditional sampler has at least two disadvantages, namely:

1. Aerosol formation in the head space of the sample container, which may escape when the bottle is removed
2. Surface contamination due to droplets of culture which fall from the sampler when the filled container is removed

Flush-mounted valves which eliminate dead spaces should be used where possible. Piston and ball valves can be designed to include a steam block or butterfly valves where steam blows across the valve surface can be used [25].

To avoid the massive release of aerosols on breakage of a sample container, it is essential to place the sample in a bottle or container which is leakproof and will protect against mechanical damage prior to removal from the sampler location [72].

F. Separation of Material of Interest

Separation is the first step in the purification of the desired product, which (if not the organism itself) is normally present, either inside the cell, dissolved in the medium, or both. Yeast and molds are usually harvested by continuous filtration whereas bacteria, viruses, and cells are harvested by centrifugation. It is possible to recover on the order of 100 kg of wet cell material from 1000 liters of culture [71,72]. If a sterile culture fluid is required for further processing, viable organisms can be removed by filtration, inactivated by passage through a heat exchanger, or sterilized in the fermentation tank if the tank is so equipped.

The specific mode is dependent on the hazard associated with the organism, the location of the product (intracellular, dissolved in medium, or both), and the stability of the product with regard to heat, chemical disinfectants, and reagents used in the extraction and purification of the product.

If a hazard to the worker exists, then the organism should be inactivated prior to processing. This can be accomplished by "in-place" sterilization within the reactor vessel (batch sterilization). Alternatively, the reactor contents can be passed through a heat exchanger and sterilized as the material flows through (continuous flow sterilization). Either mode is acceptable if the product of interest is stable under the conditions of sterilization.

If the product is heat-labile, then it is necessary to explore the possibility of chemical inactivation of the organism. This requires that the product be unaffected by the disinfectant selected.

In the event that the product is not stable, then the product must be separated from the cells prior to further processing. Such bulk processing must be performed under closed conditions if personnel or environmental exposures are to be minimized [81]. The vessel contents must be delivered to the filtration columns or centrifuge in a series of lines, the integrity of which should be examined in the same fashion as that of the vessel itself. The columns or centrifuge should be closed-system or placed in a chamber or room specifically designed to control the spread of aerosols, as in containment laboratories.

Centrifugation is the method commonly used to remove organisms from the broth. Three types of machines are widely used for the recovery of cells or cell debris. These are the tubular bowl and multichamber which are batch-operated and the disk centrifuge in which the solids are discharged continuously as concentrated slurry or periodically as a paste.

With the tubular bowl design, the cell slurry after fermentation enters the rotor at its base. The solids are sedimented to the wall and the clarified liquor leaves from the top under centrifugal force to be captured in drain lines. When operating in the continuous mode it is necessary to use intermittent discharge disk centrifuges to handle the large volume of solids.

Cultures are fed to the centrifuge from holding vessels either by air pressure or by pump. Liquid is discharged via a centripetal pump and solids by a hydraulic bowl-opening mechanism. If the desired product is in the broth, then the emerging supernatant fluid can be passed through a filter press containing clarifying and sterilizing cellulose filter mats and subsequently into sterile containers. These centrifuges rely on the introduction of plates or baffles to the rotor for their efficiency. This processing step is associated with the generation of a massive aerosol. Centrifuges must be designed to contain this aerosol or the equipment must be placed in a containment chamber.

G. Processing Material of Interest

Subsequent processing varies widely and is dependent on the purpose of the fermentation. In some cases, the whole culture may be used without separation of cells. Separated microorganisms may be used without further processing beyond drying. Yeast, molds, and bacteria used as single-cell protein fall into this category [75]. In yet other instances the cells may be subjected to some form of physical or chemical disruption to promote the release of a desired product which may subsequently be purified [25].

1. High-Pressure Homogenizer

This device breaks cells by subjecting them to a pressure followed by rapid return to atmospheric conditions. The equipment (Manton Gauling) is quite reliable and mechanical failure should not occur under proper working conditions. Either the unit requires modification to ensure that aerosols are not released or it must be placed inside a containment chamber which will control aerosol spread.

2. High-Speed Agitator Mills

Ten years ago these were basically paint mills with an exposed sieve to retain the accelerated beads which break the cells. Aerosol problems could only be resolved by the use of isolators.

There are units now available where disruption occurs beneath a liquid surface in a contained reactor vessel [83]. The claim is made that this unit contains the aerosol generated. The drawback is that there is a long residence time in the reactor which could lead to protein damage. This system must be tested for aerosol retention capacity before application.

H. Treatment of Waste

Although some of the activities associated with biotechnology are comparatively new (e.g., genetic engineering, enzyme and cell immobilization, and cell fusion), applied microbiology (fermentation), which is the cornerstone, is not [80,82]. The practice of industrial fermentation since the Second World War has provided us with sufficient background to anticipate the kinds of waste that will be generated and, or equal import, the methods one has available to treat these wastes. Clearly, the scale of operation is the major determinant by which treatment methods are chosen.

The important point to remember is that if pathogens are used, they must be rendered inactive before being introduced to the environment. This also applies to genetically engineered organisms which are regulated by the NIH guidelines [69].

In these instances, cells can be inactivated by steam sterilization or chemical treatment. Clearly the scale of operation and, as indicated previously, the stability of the product are essential parameters in the method(s) used.

More often than not, the product is not heat-stable and the use of a disinfectant is not feasible or possible (product instability). In such cases the biocatalyst is separated from the product and the waste that is generated is processed by the appropriate method. Frequently, the product is intracellular and the cells are disrupted to release the product into the milieu. This disruption should render the organism nonviable. Since this is not a certainty, then care should be exerted whenever pathogens are used to avoid the introduction of viable pathogens to the environment. This will necessitate the treatment of all biowaste that is generated in any step which has not undergone previous treatment.

In principle, the disposal of waste is fairly straightforward whether the waste is subsequently processed on site or off site.

In sewage treatment, mixed microbial populations are used to reduce the amount of oxidizable organic material. The main components of biological waste are organic chemicals, nitrogen, phosphorus, and carbon, and inorganic chemicals which include solvents, buffers, acids, and alkalis. Since there are numerous texts and articles on this topic, no further discussion will be attempted.

VIII. CONCLUSION

Biotechnology is at this time a rapidly developing field. Its potential is so vast and diverse that it is impossible to predict all the areas of academic or industrial activity that will benefit from its application. However, it is certain that biotechnology will play an increasingly important role in future industrial processes.

There is a wide variety of bioindustrial processes that include every field in microbiology. Some of them pose hazards to the worker and to the environment. The hazards are not obvious in each case and usually people are unaware of them. Simple and regular laboratory processes create infectious aerosols that can be very concentrated due to the large quantities and concentrations of biomaterials. This phenomenon creates higher risk than in R&D laboratories.

Because human factors are the main cause of laboratory accidents, efforts must be concentrated on developing awareness to hazards and safety, through education and safety courses starting at college and continuing in the research institute and the biotechnological plant.

As a result of the development of safety equipment and laboratories, every process can be carried out successfully while protecting the workers and the environment.

When constructing their large-scale practices, NIH drew on the experiences of the U.S. pharmaceutical industry (vaccine and antibiotic production) and developed a cautious philosophy to control not only the perceptible hazards, e.g., when potential pathogens or organisms which were unclassified were used.

The hazards of concern were associated with exposure of personnel to a viable organism or a biologically active molecule.

The basic principles of industrial or occupational hygiene provide a reasonably clear course to protect personnel from biologically active molecules, namely, isolate the problem (containment) or isolate the personnel (protective equipment).

On the other hand, if the viability of the organism as a contamination is the measure of the threat that it poses, and given that viability is often unknown because it is usually an adventitious contaminant in an unnatural environment, it becomes difficult not only to measure the actual level of the contamination but also to assess the risk involved.

The guidelines define the containment levels to be achieved, but unfortunately they do not provide the specifics on how to achieve it or, once achieved, how to maintain it. In practice, the responsibility lies with the institution. This means that each institution has considerable latitude in determining what constitutes compliance. While this is convenient for the institution, it causes problems for safety professionals to assess the hazards and recommend appropriate criteria to the engineer, who must design the apparatus and operational protocols.

NIH developed a conservative set of criteria, e.g., closed system/no exposure. What was omitted was a set of standards for equipment and procedures; what was and is needed is a policy that avoids excessive regulation but precludes irresponsible action on the part of institutions.

This requires that engineers and safety specialists reach an agreement or consensus about what constitutes adequate containment and compliance. It should be evident that this consensus would be applicable to all fermentation processes and not restricted to recombinant DNA systems.

Realistic limits on the exposure of personnel or the environment to process streams must be established. This requires that a certain degree of common sense be mixed with safety and science to avoid stifling the development of biotechnology under the weight of unnecessary engineering or over-engineering.

There must be a continuous dialogue between the various parties associated with the project to ensure that the safety of personnel and the environment is not compromised.

REFERENCES

1. Wedum, A. G. History of microbiological safety. *18th Biological Safety Conference*. Lexington, Kentucky. 1975.
2. Meyer, K. F., and Eddie, B. Laboratory infections due to Brucella. *J. Infect. Dis.*, *68*:24-32, 1941.
3. Sulkin, S. E., and Pike, R. M. Viral infections contracted in the laboratory. *New Engl. J. Med.*, *241*:205-213, 1949.
4. Sulkin, S. E., and Pike, R. M. Survey of laboratory-acquired infections. *Am. J. Publ. Health*, *41*:769-781, 1951.
5. Pike, R. M. Laboratory associated infections. Summary and analysis of 3921 cases. *Health Lab. Sci.*, *13*:105-114, 1976.
6. Hanson, R. P., Sulkin, S. E., Buescher, E. L., Hammond, W. Mc., McKinney, R. W., and Work, T. E. Arbovirus infections of laboratory workers. *Science*, *158*:1283-1286, 1967.
7. Skinholj, P. Occupational risks in Danish clinical chemical laboratories. II. Infections. *Scand. J. Clin. Lab. Invest.*, *33*:27-29, 1974.
8. Harrington, J. M., and Shannon, H. S. Incidence of tuberculosis, hepatitis, brucellosis and shigellosis in British medical laboratory workers. *Br. Med. J.*, *1*:759-762, 1976.
9. Huddleson, I. F., and Munger, M. A study of an epidemic of Brucellosis due to Brucella melitensis. *Am. J. Publ. Health*, *30*:944-954, 1940.
10. Newitt, A. W., Koppa, T. M., and Gudakunst, D. W. Water-borne outbreak of Brucella melitensis infection. *Am. J. Publ. Health*, *29*:739-743, 1939.
11. Phillips, G. B. Microbiological Safety in US and Foreign Laboratories. Technical Report No 35. U.S. Army Chemical Corps, Biological Laboratories, Fort Detrick, Frederick, MD, 1961.
12. Phillips, G. B. Microbiological hazards in the laboratory. In *Safety in the Chemical Laboratory* (N. W. Steere, ed.), Division of Chemical Education of the American Chemical Society, Easton, 1968, pp. 98-104.
13. Pike, R. M. Laboratory-associated infections: Incidence, fatalities, causes, and prevention. *Ann. Rev. Microbiol.*, *33*:41-66, 1979.
14. Collins, C. H. How laboratory infections are acquired. In *Laboratory-Acquired Infections*. Butterworths, London, 1983, pp. 27-35.
15. Reitmann, M., Moss, M. L., Harstad, J. B., Alg, R. L., and Gross, N. H. Potential infectious hazards of laboratory techniques. I. Lyophilization. *J. Bacteriol.*, *68*:541-544, 1954.
16. Slepushkin, A. N. An epidemic study of laboratory infections with Venezuelan equine encephalomyelitis. *Prob. Virol.*, *4*:311-314, 1959.
17. Phillips, G. B. Human factors in microbiological laboratory accidents. Becton Dickinson Research Center, Research Triangle Park, North Carolina. 1973.
18. Martin, J. C. Behavioral factors in laboratory safety: Personnel characteristics and modification of unsafe acts. In *Laboratory Safety, Theory and Practice* (A. A. Fuscaldo, B. J. Erlich, and B. Hindman, eds.), Academic Press, New York, 1980, pp. 321-342.
19. Phillips, G. B. Casual factors in microbiological laboratory accidents and infections. U.S. Army Biol. Labs. No. 2. Fort Detrick, Frederick, MD, 1965.
20. Vital and Health Statistics. Persons injured and disability days by detailed type and class of accident: United States 1971-1972, DHEW Publ. (HRA) 76-1532. U.S. Department of Health, Education and Welfare, Washington, D.C., 1976.

21. Stark, A. Policy and procedural guidelines for the health and safety in virus laboratories. *Am. Ind. Hyg. Assoc. J.*, *36*:234-240, 1975.
22. Chatigny, M. A. Protection against infection in the microbiological laboratory: Devices and procedures. *Adv. Appl. Microbiol.*, *3*:131-192, 1961.
23. Gerone, P. J. Biohazard associated with laboratory animals. See Ref. 18, pp. 225-248.
24. Blackmore, D. K., and Francis, R. A. The apparent transmission of Staphylococci of human origin to laboratory animals. *J. Comp. Phatol.*, *80*:645-650, 1970.
25. Liberman, D. L., Fink, R., and Schaffer, F. Biosafety and biotechnology. In *Manual for Biotechnology* (A. Demain, ed.), ASM Press, Washington, D.C. 1986, pp. 402-409.
26. Hsiung, G. C., and Swack, N. S. Endogenous viruses in cell cultures prepared from apparently healthy animals. In *Biohazards in Biological Research* (A. Hellman, M. N. Oxman, and R. Pollack, eds.), Cold Spring Harbor Laboratory, New York, 1973, pp. 74-78.
27. Todaro, G. J. Endogenous type C viruses in cell cultures. See Ref. 26, pp. 114-130.
28. Gardner, M. B. Avian and murine RNA tumor viruses: Modes of transmission. See Ref. 26, pp. 143-165.
29. Schneider, J. A. EBV and other potential problems in the handling of long-term human lymphocyte cultures. See Ref. 26, pp. 191-201.
30. West, D. L., Twardzik, D. R., Mckinney, R. W., Barkley, W. E., and Hellman, A. Identification, analysis and control of biohazards in viral cancer research. See Ref. 18, pp. 167-223.
31. Dickson, D. Cancer deaths probed at Pasteur Institute. *News and Comment*, June 27, 1986, p. 1597.
32. Hinman, A. R., Fraser, D. W., Douglas, R. G., Bowman, G. S., Kraus, A. L., Winkler, W. G., and Rhodes, W. W. Outbreak of lymphocytic choriomeningitis virus infection in medical center personnel. *Am. J. Epidemiol.*, *101*:103-110, 1975.
33. Koplan, J. P., and Axnick, N. W. Benefits, risk and cost of viral vaccines. In *Progress in Medical Virology*, Vol. 28 (J. L. Melnick, ed.), Karger, Basel, 1982, pp. 180-191.
34. Centers for Disease Control. Laboratory safety at the Centers for Disease Control. U.S. Dept. Health, Education and Welfare, Public Health Service. No. CDC 76-8118, 1975.
35. Pivnick, H., Worton, H., Smith, D. L. T., and Barnum, D. Infection of veterinarians in Ontario by Brucella abortus strain 19. *Can. J. Publ. Health*, *57*:225-231, 1966.
36. Lauer, J. L., VanDrunen, N. A., Washburn, J. W., and Balfour, H. H. Transmission of hepatitis B virus in clinical laboratory area. *J. Infect. Dis.*, *140*:513-516, 1979.
37. Szmuness, W., Stevens, C. E., Zang, E. A., Harley, E. J., and Kellner, A. A controlled clinical trial of the efficacy of the hepatitis B vaccine (HEPTAVAX-B). A final report. *Hepathology*, *1*:377-385, 1981.
38. Tedder, R. S., Uttley, A., and Cheingsong-Popov, R. Safety of immunoglobulin preparation containing anti HTLV III. *Lancet*, *1*:815, 1985.
39. Centers for Disease Control. Update: Human immunodeficiency in health care workers exposed to blood of infected patients. *MMWR*, *36*:285-289, 1987.

40. Weiss, S. H., Goedert, J. J., Gartner, S., Popovic, M., Waters, D., Markham, P., di Marzo Veronese, F., Gail, M. H., Barkley, W. E., Gibbsons, J., Gill, F. A., Leuther, M., Shaw, G. M., Gallo, R. C., and Blattner, W. A. Risk of human immunodeficiency virus (HIV-1) infection among laboratory workers. *Science, 239*:68-71, 1988.
41. Williams, R. E. O. In pursuit of safety. *J. Clin. Pathol., 34*:232-239, 1981.
42. Recombinant DNA research guidelines. *Federal Register, 41*:27902-27943, 1976.
43. Recombinant DNA research proposed revised guidelines. *Federal Register, 42*:49596-49609, 1977.
44. Guidelines for research involving recombinant DNA molecules. *Federal Register, 47*:38047-38068, 1982.
45. Szybalski, W. Much ado about recombinant DNA regulations. In *Biomedical Scientists and Public Policy* (H. H. Fudenberg and V. L. Melnick, eds.), Plenum Press, New York, 1978, pp. 97-142.
46. Helling, R. B. Recombinant DNA research. See Ref. 18, pp. 145-165.
47. Helling, R. B. and Lomax, M. J. The molecular cloning of genes: General procedures. In *Genetic Engineering* (A. M. Chakrabarty, ed.), CRC Press, West Palm Beach, 1978, pp. 1-30.
48. Adams, J., Kinney, T., Thompson, S., Rubin, L., and Helling, R. B. Frequency dependent selection of plasmid-containing cells of E. coli. *Genetics, 91*:627-637, 1979.
49. Jones, L., Ristow, S., Yilma, T., and Moss, B. Accidental human vaccination with vaccinia virus expressing nucleoprotein gene. *Nature, 319*:543, 1986.
50. World Health Organization. Cell products safety. *Developments in Biological Standardization*, Vol. 68. Karger, Basel, 1987, pp. 1-81.
51. Petricciani, J. C., and Regan, P. J. Risk of neoplastic transformation from cellular DNA: Calculations using the oncogen model. *Science, 219*:43-49, 1987.
52. Moyer, M. P., and Moyer, R. C. Potential risk of tumor virus subgenomes in the production of biologicals. See Ref. 50, pp. 51-52.
53. Wedum, A. G., and Kruse, R. H. Assessment of risk of human infection in microbiological laboratory. Miscellaneous Publication No. 30. Industrial Health and Safety Office, Fort Detrick, Frederick, MD. Dept. Army, 1969, pp. 1-88.
54. DHSS. Code of practice for the prevention of infection in clinical laboratories and post mortem rooms. Dept. Health and Social Services, HMSO, London, 1978.
55. Classification of etiological agents on the basis of hazard. *Federal Register, 46*:59379-59380, 1981.
56. Centers for Disease Control. Proposed biosafety guidelines for microbiological and biomedical laboratories. U.S. Dept. Health and Human Services, 1980.
57. Krieg, A., and Miltenburger, A. G. Bioinsecticides. I. Bacillus thuringiensis. In *Advances in Biotechnological Processes*, Vol. 3 (A. Mizrahi and A. L. Van Wezel, eds.), Alan R. Liss, New York, 1984, pp. 273-290.
58. Church, B. D., and Halvorson, H. Dependence of the heat resistance of bacterial endospores on their dipicolinic acid content. *Nature, 183*: 114-125, 1959.

59. Office of Technology Assessment. Impacts of applied genetics: Microorganisms, plants and animals. (OTA-BA-218). U.S. Congress, 1981.
60. Fiskel, J. R., and Covello, V. T. The suitability and applicability of risk assessment methods for environmental applications of biotechnology. In *Biotechnology Risk Assessment* (J. R. Fiskel and V. T. Covello, eds.), Pergamon Press, New York, 1986, pp. 1-34.
61. National Institute of Health. Laboratory Safety Monograph. A supplement to the NIH guidelines for recombinant DNA research. U.S. Dept. Health, Education and Welfare, Public Health Service, 1978.
62. World Health Organization. *Laboratory Safety Manual*. Geneva, 1983.
63. National Cancer Institute. *Good Laboratory Practice Manual*. Bethesda, 1968.
64. Good laboratory practices for nonclinical laboratory studies. *Federal Register, 43*:60013-60020, 1978.
65. Current good manufacturing practice for finished pharmaceuticals. *Code of Federal Regulations*, Vol. 21, Part 211, 1982, pp. 7-58.
66. Centers for Disease Control and National Institute of Health. Biosafety in microbiological and biomedical laboratories. U.S. Dept. of Health and Human Services. HHS Publication No. (CDC)84-8395, 1984.
67. Israeli, E. Biosafety in biotechnological processes. In *Advances in Biotechnological Processes*, Vol. 6 (A. Mizrahi, ed.), Alan R. Liss, New York, 1986, pp. 1-30.
68. Tillotson, J. R., Axelrod, D., and Lyman, D. O. Follow-up on Rabies—New York. *MMWR, 26*:249-250, 1977.
69. Guidelines for research involving recombinant DNA activity. *Federal Register, 48*:24555-24581, 1983.
70. Berg, P., Baltimore, D., Brenner, S., Roblin, R. O., and Singer, M. M. F. Summary statement of the Asilomar conference on recombinant DNA molecules. *Science, 188*:991-994, 1975.
71. Evans, C. G. T., Preece, T. F., and Sargeant, K. Microbial plant pathogens: Natural spread and possible risk in their industrial use. Publication of the Commission of the European Communities. ECI. 724. United Kingdom, 1981.
72. Sargeant, K., and Evans, C. G. T. Hazards involved in the industrial use of micro-organisms. Publication of the Commission of the European Communities. Brussels, 1979.
73. Gorbach, S. L. Recombinant DNA: An infectious disease perspective. *J. Infect. Dis., 137*:615-626, 1978.
74. Banttari, E., and Venette, J. R. Aerosol spread of plant viruses; potential role in disease outbreaks. *Ann. N.Y. Acad. Sci., 353*:167-173, 1980.
75. Litchfield, J. H. Single-cell proteins. See Ref. 51, pp. 740-746.
76. Office of Technology Assessment. *Commercial Biotechnology: An International Analysis*. U.S. Congress, 1984.
77. Sharples, F. Spread of organisms with novel genotypes: Thoughts from an ecological perspective. *Recomb. DNA Tech. Bull., 6*:43-56, 1983.
78. Krimsky, S. Genetic alchemy: The social history of the recombinant DNA controversy. MIT Press, Cambridge, 1982.
79. Blakebrough, M. Design of laboratory fermenters. *Meth. Microbiol., 1*:473-504, 1969.
80. Cooney, C. L. Bioreactors: Design and operation. *Science, 29*:728-733, 1983.

81. Brown, C. M., Ellwood, D. C., and Hunter, J. R. Growth of bacteria at surfaces: Influence of nutrient limitation. *FEMS Microbiol. Lett.*, *1*:163-166, 1977.
82. Demain, A. L. New applications of microbial products. *Science*, *29*:709-714, 1983.
83. Limon-Lason, J., Hoare, M., Orsborn, C. B., Doyle, D. J., and Dunnill, P. Reactor properties of a high-speed bead mill for microbial cell rupture. *Biotechnol. Bioeng.*, *21*:745-774, 1979.

INDEX

Abiotic, 936
Abortioporus biennis, 801
Abscisic acid, 565-567,569
Absidia blakesleeana, 709
A. corymbifera, 732
A. glauca, 156,163,446,448,455,687
A. griseola, 697,1006
A. ramosa, 588,709,737
Absidia sp., 714
Acanthamoeba sp., 215
Acclimatization, 949
Acervuli, 1047
Acetobacter (Gluconobacter) suboxydans, 363
Achlya ambisexualis, 156,217,446, 448,455
A. bisexualis, 446,447,450
A. heterosexualis, 217
A. klebsiana, 217
A. radiosa, 39
Achlya sp., 38,49,219,221,449, 947
Acidulant, 21,360,362,365,368, 369
Acremonium alabamensis, 710,730, 731,736-738,742
A. cellulophilum, 738
A. chrysogenum, 8-11,24,25,139, 197,520,686,1007
A. fusidioides, 969,974,980
A. kiliense, 519,686
Acremonium sp., 9,11
Acriflavine, 115
Actinomucor elegans 20,686
Actinomucor sp., 743
Actinomyces griseus, 938
Acyltransferase, 10
Adaptability, 931
Adaptation, 604,975
Adsorbents, 405,572
Aerobiosis, 614,939
Aerosol, 692,933,1056,1057,1060, 1063,1064,1068,1072,1075-1077,1082-1085
Aflatoxigenic, 892,893
Aflatoxins, 27,46,82,691,884,891-893,896
Agaricus bisporus, 20,81,283,290, 292,293,295,296,383,449,456, 574,802,890,1007
A. bitorquis, 44,47,216,218,219, 249,250,253,451,1007
A. brunnescens, 38,39,47,61,63, 218,219,221,446,456,868
A. campestris, 61-63,970,975,980
Agaricus sp., 58,457,868,975
Agglutination, 108
Agrobacterium sp., 866
A. tumefaciens, 108
Agroclavine, 476,487-491
Agropyron scabrum, 62
Agropyron sp., 861,863
Aliphatic, 905,907,909,911,912
Alleles, 57,58,62,63,67,68,70,84, 86-90,106,159,163,166,167, 169,173,177,425,625-628,666
Allelism, 67
Allergens, 1070
Allergic reactions, 691
Allergy, 716
Allescheria terrestris, 709,713
Allomyces arbuscula, 39,44,217,221, 446

A. macrogynus, 39,44,217,221,282, 292
Allomyces sp., 37,39,44,220,416, 446
Allosteric, 283,284,315,323,328, 495
Allozymes, 67
Alnus sp., 857
α-Amylase, 3
Alternaria alternata, 143,687, 892
A. cassiae, 28, 1045-1048
A. citri, 134
A. solani, 459
Alternaria sp., 578,943
A. tenuis, 686,891
Alternariol, 891
Amanita forstiana, 578
A. muscaria, 27
A. rubescens, 574
Amanita sp., 37
Amatoxins, 27
American Type Culture Collection (ATCC), 993,999,1019,1051
Ampholytes, 66
Amplification, 245,246,250,251
Amylomyces rouxii, 39,1004-1006
Anabolic, 915
Anabolism, 915
Anaerobiosis, 614,615,939
Anamorph, 48,767
Anastomosis, 457,860,864,865,946
Androgens, 32
Anguillospora pseudolongissima, 858
Anixiopsis sp., 536
Antagonism, 26,856-858,976
Antagonistic, 286,505,839,857
Antiauxin, 577
Antibacterial, 22,23,370,744,745
Antibiotic(s), 2,8,10,14,16,21,22,26, 29,81,86,91,105,106,151,157, 158,174,197,200,202,205,206, 208,284,285,370,445,487,491, 517,520,521,525,532,534-538, 540-543,554,694,701,729,730, 744,745,851,906,924,950,1010, 1027,1031,1040,1085
Antibodies, 287,291,430,455,456,741, 867,1027,1062,1064,1065,1076
Antibody, 289,451,771,867,962
Antibody-antigen interaction, 867
Anticoagulant, 21,362
Anticodon, 226,227
Antifoam, 13,14,17,267,358,539, 540,575,971
Antifungal, 22,26,370,744,745,851
Antigens, 691,741,1027,1063,1064
Antiinsect, 865
Antimicrobial, 589,744,939
Antioxidant(s), 21,360,362,404
Antipathic, 603
Antiserum, 1067
Antitumor, 22,272,273,370,742
Antiviral, 22,370
Aphanomyces euteiches, 221
Aplanaopsis terrestris, 217
Apodachlya brachynema, 217
Apodachlya pyrifera, 217,449
Apodachlya sp., 219
Appressoria, 1047
Appressorium, 1045
Arachnomyces sp., 536
Araujia mosaic virus, 1046
Arbuscules, 824
Armillaria mellea, 47,48,868
Armillaria sp., 58,801
Artemia salina, 415
Arthrobacter luteus, 133,271
Arthrographis kalrai, 686
Arthrospores, 481,482,525
Ascaris sp., 215
Aschersonia sp., 27
Ascobolus immersus, 249,253,868
Aseptic, 1081
Ashbya gossypii, 3,22,1007
Aspergillosis, 716
Aspergillus amstelodami, 111,244, 250,252,451,452
A. awamori, 10,45,81,93,113,140, 156,168,369,686,695,983, 1004,1006,1007
A. awamori var. *kawachi*, 111,113
A. candidus, 714
A. carbonarius, 687,940,1005
A. clavatus, 687
A. echinulatus, 45
A. ficuum, 156,159,168
A. flavipes, 519
A. flavus, 27,41,46,111,113,137, 156,159,168,282,283,366,368, 369,519,579,686,687,690,714, 884,892,893,934,935,937,940, 945,947,974
A. flavus-oryzae, 370
A. flavus var. *columnaris*, 686,1004
A. foetidus, 686,885,886,1006,1007

A. foetidus var. *acidus*, 686
A. fumigatus, 104,111,114,370,686, 708,710,716,718-720,731-733, 737,738,741,743,745,895,935, 939,940,945,947
A. giganteus, 417-421,431,519
A. glaucus, 714
A. griseus, 370
A. heterothallicus, 111
A. japonicus, 111
A. nidulans, 6-11,45,46,81,83-93, 103-107,110-114,116-118, 134,135,137-140,151-156, 158-177,179,180,198,200, 201,203,204,206,213,218, 220-228,282,283,301,313, 316-318,327,328,330,333, 446,447,449,450,456,460, 519,587,687,708,950
A. nidulans var. *echinulatus*, 105,111,113,114,117,218
A. nidulans var. *guadrilineatus*, 218
A. niger, 3,10,17-19,21,45,81,82, 85-88,90,92,93,111,113,114, 135,138,139,151,155,156, 159,161,164,165,167-170, 172,174-177,179,316,334, 357-360,362,364,370,393, 519,587,588,682,687-693, 695-698,700,884-886,892, 894,922,935,936,938-940, 944-947,968,974-976,980, 981,983,984,1005,1006
A. niger var. *macrosporus*, 686
A. niveus, 974
A. nomius, 41
A. ochraceous, 139,687,690,884, 886,893,894
A. oryzae, 3,17-20,30,41,45,46, 81,85,111,113,151,156,168, 339,519,651,686-690,692, 693,695,696,698,699,702, 937,970,974,975,977,980, 982,985,1002-1004,1006
A. oryzae var. *oryzae*, 686
A. oryzae var. *viridis*, 686
A. parasiticus, 41,139,519,882, 884,891,895,896,934
A. phoenicis, 686,688,695,697,944, 974,980,1006,1007
A. quadrilineatus, 105,111,113, 114
A. restrictus, 714
A. rugulosus, 103,106,113,114,116-118,139
A. saitoi, 19,689,699
A. sojae, 41,82,686,688,1003,1004
Aspergillus sp., 3,9-11,18,19,36, 42,45,46,58,84,90,91,102,108, 110,160,161,174,216,327,366, 370,389,391,393,417,449,578, 682,688-691,693,698,699,702, 713,714,782,853,937,939,940, 943,945,947,975,976,985
A. stellatus, 111
A. sydowi, 687
A. tamarii, 45,370,687,1004
A. terreus, 3,10,21,27,28,45,156,168 365,366,587,687,937,944,947, 1005
A. terricola, 687
A. unguis, 45,111
A. ustus, 687
A. versicolor, 708,920
A. wentii, 45,687
Aspiration, 1059
ATCC, 994,1000-1003,1005-1010, 1012-1014,1016,1017,1027
Atherosclerosis, 27
Atriplex gardneri, 854
Aureobasidium pullulans, 3,263,264, 266-269,1006
Auricularia auricula, 1007
A. polytricha, 1007
Autocatalytic, 15
Autoecious, 60
Autolysis, 4,542
Autooxidation, 931
Autotrophic, 950
Auxin(s), 565,567,569,577,578,580
Auxotrophic 8,21,82,84,101,105-108, 110,116,118,132,155,159,169, 175,178,206,623-625,661,730
Auxotrophs, 9,88,108,116,132,327, 331,491,499,521,531
a_w (water activity), 712
Axenic, 841
Azotobacter sp., 857,969,976,981,982

Bacillus circulans, 139
B. megaterium, 108
Bacillus sp., 656,858,1066
B. subtilis, 243,458,459
Bacillus thuringiensis, 865,1068

Bacteriophage, 160,220,252,253
Bacteriostatic, 920
Bacterium lactis aerogenes, 613
Balansia epichloe, 577,578
Band counting, 57
Beauveria bassiana, 27, 102, 113,115,391,686
B. brongniartii, 102,113,115
Beauveria sp., 102,886
B. sulfurescens, 885
Beneckea natriegens, 608
Benomyl, 855,856
Bergenerula spartinae, 944
β-Carotene, 3
β-Glucanase, 3
Beverages, 368,963
Bioassays, 8,577,1010
Biobleaching, 794,795
Biocides, 850,855,856
Biocompatibility, 1020
Biocontrol agents, 856,1021
Bioconversion(s), 577,656,730, 742,746,765,801,967-969, 971,973,974,976,978-982, 985,986
Biodegradation, 263,718,746, 763,764,767,779,784,787, 791,806,807,895,932-934, 939-941,944,945,949-951, 1000,1020
Biodeterioration, 720,746,1020
Bioemulsifiers, 394
Bioengineering, 1
Biogas, 968,972,973
Biogenic, 888
Biohazardous agent, 1071
Bioherbicides, 1045,1046
Bioinsecticides, 1068
Biological bleaching, 794
Biological control, 59,62,102, 445,1026,1081
Biological oxygen demand (BOD), 20,795,947,964,966,973,977
Biomass, 264-268,274,386,389, 391,394,497,503,520,530, 541,544,571,587,604-606, 647,648,650,654,656,659, 664,665,667,732,737,738,743, 763,785,788,791,796,797,906, 911-915,918,944-948,951,967-974,976,977,981,985,986
Biosensors, 948,949,951
Biosphere, 763,933
Biosurfactant, 914,919,920,921
Biotic, 936
Biotransformation, 799,886,893,895, 896,925
Bipolaris sorokiniana, 584
Bjerkandra adusta, 772
Blakeslea sp., 422
B. trispora, 22,409,410,416-418, 420,884,1007
Blastocladiella emersonii, 221,282,289, 290,292,293,296
Blastocladiella sp. 37,39
Boletus cothurnatus, 574
B. edulis, 1007
B. edulis var. *pinicolus*, 574
B. felleus, 578
B. indecisus, 975
Boletus sp., 578,975
Botryodiplodea theobromae, 454,455
Botrytis anthophyla, 574
B. cinerea, 581,582,584
Botrytis sp., 19,58
Bouteloua sp., 851
Bovista plumbea, 701
Bovista sp., 19
Brassica sp., 215
Brassicasterol, 381,382
Bremia lactucae, 156
Brethanomyces sp., 655
Brettanomyces anomalus, 217
B. custersianus, 217
B. custersii, 217,221
B. lambricus, 378
B. naardenensis, 217
Brevibacterium, 366
B. flavum, 976,981,982
Broth, 262,270,273,338,501,530,538, 543,545,586,588,649,694,695, 911,920,971,985,1081,1083
Brown rot, 767,794
Brucella abortus, 1063
B. melitensis, 1057
Byssochlamys sp., 709

Caerulenin, 27,28
Calcarisporiella thermophilia, 710
Caloscypha fulgens, 409,414
Candida acidothermophilum, 1005
C. albicans, 101,111-114,116,217, 264,266,287,290-292,297,299, 447,894
C. arborea, 972,979
C. blankii, 368

C. bogoriensis, 381,383
C. brumptii, 366,370,971,972,977, 979
C. ciferrii, 972
C. cloacae, 923
C. curvata, 386
C. guillermondii, 115,339,884, 972,1008
C. holmii, 1003
C. humicola, 338,946
C. hydrocarbofumarica, 368
C. intermedia, 972
C. javanica, 972
C. krusei, 976,979,981,983,984
C. lipolytica, 361,370,392,459,912, 913,916,917,920,922,923,944, 1015
C. lusitaniae, 622
C. maltosa, 217,221,887
C. membranaefaciens, 972,979
C. mogii, 1003
C. paralypolitica, 392
C. parapsilosis, 217
C. pelliculosa, 339
C. petrophilum, 920,923
C. pseudotropicalis, 105,606,614, 615
C. rhagii, 216,217,221
C. rugosa, 366,391,393,1006
C. shehatae, 631,657,659
Candida sp., 101,300,361,366,391, 456,457,607,621,622,630,655, 912,919,920,923,939,940,945 947,950,972,976,985,986,1001, 1015
C. steatolytica, 973,979
C. stellatoidea, 979
C. tenuis, 656,912
C. tropicalis, 101,104,111,114,116, 339,882,884,910,920,948-950, 969,971-973,979,1008
C. utilis, 101,111,114,135,970,972 973,976-978,981,982-986,1006
C. utilis var. *major*, 973
C. valida, 1007
C. vartiovaarai, 972
C. wickerhamii, 618,621
Cantharellus cibarius, 1007
C. ianthinoxanthus, 409
Cantharellus sp., 408,975
Capsicum sp., 417
Captan, 855
Carcinogen(s), 27,46,885,945
Carcinogenesis, 885
Carcinogenic, 28,794,803,888,918
Carotenogenesis, 401,415-417,419, 420,422,423,425,429,431
Carotenoids, 401-410,412,414-417, 419-423,425,428-430
Carpophore, 855
Cassia obtusifolia, 1047
Cellulase, 3
Cellulomonas sp., 655
Cenococcum geophilum, 850,854
Centraalbureau voor Schimmel-cultures (CBS), 993,998,1009, 1014,1019,1050
Cephalosporin(s), 3,8,9,11,19,24, 25,29,91,174,197-200,205-207, 250,517-520,523,525,532-534, 536-538,540,541,553,555
Cephalosporium acremonium, 3,8,81, 89,91,111,112,114,136,151, 156,158,170,197,198,200,201, 203-207,218,221,226-228,250, 254,255,317,318,449,519,520, 525,536-538,553
C. caerulens, 27,28
Cephalosporium sp., 8,91,107,114, 160,200,206,518,519,532-537, 539,555,687,1027
Cephem, 518
Ceratocystis fimbriata, 865
C. minor, 38
Ceratocystis sp., 38
C. ulmi, 268
C. virescens, 3
Cercospora cruenta, 581
C. kikuchii, 156
C. rosicola, 581,582
Chaetomium cellulclyticum, 731,737, 738,740,743-745,944
C. crispatum, 944
C. fusisporale, 709
C. globosum, 945
Chaetomium sp., 892,939,941,944, 945
C. thermophile, 708,709,719,720, 731,732,734-736,738-740,742, 945
C. thermophile var. *coprophile*, 709,733
C. thermophile var. *dissitum*, 709, 733
Chaotropic, 639
Chaenophora cucurbitarum, 282

Chelating agent, 360,362
Chemical oxygen demand (COD), 964,974-977,981
Chemoinduction, 422
Chemolithotrophic, 931
Chemostat, 525,529,647,662, 949
Chemotherapeutic, 517
Chenopodium quinona, 856
Chitin, 281-291,293-302
Chitosan, 299,300
Chlamydomonas sp., 42,215
Choanophora sp., 887
Cholesterol, 27,28,381,382
Chromophore, 269
Chrysonilia sitophila, 944
Chrysosporium pruinosum, 713, 719,772
Chrysosporium sp., 19,710
Chymosin, 10,699,702
Chymotrypsin, 293
Circinella muscae, 697
Cisternae, 298
Cistrons, 448
Citric acid, 3,20,21
Citrinin, 27
Cladosporium cladosporioides, 587
C. fulvum, 156,168
C. resinae, 696,699,914,916,920, 921,1005
Cladosporium sp., 389,587,686, 934,945
Claviceps fusiformis, 481,482,495, 502,504,506
C. paspali, 27,330,480,481,485,493-496,500,502,504,506
C. purpurea, 3,26,132,138,156, 168,218,249,252,253,338, 339,380,383,453,478-480, 482,483,485,487,489,491-500, 502,503-506,864,882,884,895, 1007
Claviceps sp., 138,381,475,478-482, 485-487,489,493,494,495-502, 504,506,865
C. sulcata, 380,381
Clavine, 476,478,482,489,500,506
Cloning, 88,89,91,151,160,162,170, 172-174,197,198,200,205,206, 453,498,500,779,781,784,807, 1065
Clostridium saccharolyticum, 659
Coagulants, 736
Coagulation, 933
Cochliobolus carbonum, 113
C. heterostrophus, 47,85,86,111,140, 143,156,159,168,170,172,178, 218,219,250,251,864,865,950
C. miyabeanus, 282
C. sativus, 113
Codon, 224-228,782
Coelomomyces sp., 28
Colletotrichum gloeosporioides, 1046
C. gloeosporiodes f. sp. *aeschynomene*, 1027
C. lagenarium, 587
C. lindemuthianum, 156,168,177
C. malvarum, 1027,1046
Colletotrichum sp., 85,1047
C. trifolii, 156,866
C. truncatum, 1045,1046,1047,1048
Collybia dryophila, 40,41
Collybia sp., 941,975
Colonization, 712-716,721,831-841, 851-858,860,861,863-865,867
Cometabolism, 767,909,931
Commensalism, 909
Commonwealth Mycological Institute (CMI), 994,999,1014,1019,1050
Compactin(s), 27,28
Compatibility, 103,849,859,862
Compatible, 655,658,839-841,853,855, 859,865,1069
Competition, 668,707,831,832,838, 851,856,857,922,1046
Complementation, 9,11,84,88,89,110, 116,117,155,160,169,171,172, 200,287,316,423,456,500,604, 866
Compost, 718,719,729,839,840
Composting, 716,718,719,721,802,839
Concanavalin A, 291,298
Concomitant, 11,271,422,573,589,639, 695,697,925,931
Conidiobolus coronatus, 454
C. heterosporus, 686
C. khandalensis, 686
C. lamprauges, 686
Conidiogenesis, 493,494,496
Conidiomata, 1047
Conjugates, 580
Conservation, 170
Contaminant, 82,216,720,887,945, 947,948,970,1066,1068,1081
Contamination, 14,20,27,45,259,266, 282,454,538,542,571,666,734,

[Contamination]
743,906,921,922,932,944,
970-972,981,1010,1013,
1019,1020,1062,1068,1072,
1074,1079,1080-1082,
1085
Contraceptive, 362
Cooxidation, 910
Coprinus atramentarius, 449
C. cinereus, 47,48,63,84,85,87,
137,141,142,156,165,174,
178,218,283,284,287,315,
448,449,687,864
C. commatus, 449,1007
C. lagopus, 446
C. macrorhizus, 142
C. micaceous, 449,574
C. pellucidus, 142
C. radians, 687
C. radiatus, 330
Coprinus sp., 174,219,221,224,
710
C. stercorarious, 218,219
Coriolus sp., 941
C. versicolor, 778,798,943,946
Cortinarius sp., 857
Corynascus thermophilus, 709
Corynebacterium acetoacidophilum,
976,981,982
C. equi, 886
Cosmid, 160
Cronartium fusiforme, 409
Crossing over, 5-8,63,110,163,
176
Cross-pathway control, 336
Cryogenics, 1013
Cryopreservation, 1013
Cryoprotectants, 1013
Cryphonectria cubensis, 58,59,61
C. parasitica, 156,458,1015
Cryptococcus albidus, 385,386,
1006
C. laurentii, 338
Cryptococcus sp., 457
C. terricolus, 385
Culcinomyces sp., 28
Cunninghamella bainieri, 882,884-
886,893,895
C. blakesleeana, 23,884,886,893
C. elegans, 141,282,885,889,890,
891,892,932
Cunninghamella sp., 578
Curdlan, 270,271
Curing, 699
Curling factor, 26
Curvularia falcata, 885
C. lunata, 23,156,168,884,947,948
C. pallescens, 249
Curvularia sp., 944
Cyclase, 109
Cycloids, 291
Cyclosporin, 3
Cyclotella cryptica, 298
Cylindrocarpon radiciola, 23
Cylindrocarpon sp., 939,940
Cytochrome, 489,882,883,884,885,
887,895,896,907,915,924,950
Cytokinin, 565,569,574,576
Cytosol, 368
Cytospora sp., 574

Dactylaria gallopava, 720
Dactylomyces crustaceus, 709
Dealkylation, 885,933
Deamination, 318
Debaryomyces condertii, 370
D. formicarius, 113,116
D. nilssonii, 379
D. polymorphus, 1005
Debaryomyces sp., 116,457,972,979
Decomposition, 2,578,719,720,721,
729,766,949
Decontaminant, 1072
Decontamination, 805,939,1072
Degradation, 17,271,718,736,739,763,
764,766-768,771,774,775,777-
779,784,786,787,790,792,793,
795,800-807,882,884,888,892,
895,896,906,908,909,922,924,
926,931-936,938-941,943,947-
949,951,975
Dekkera bruxellensis, 217
Dekkera intermedia, 217, 1006
Dekkera sp., 221
Denaturation, 37,64
Depollution, 910,919,925
Dermocystidum sp., 379
Desizing, 696
Desmodium tortuosum, 1046
Detergents, 21,691,732,734,736,739
Deterioration, 714
Detoxification, 852,885-888,890,891,
894,932,936,939,947,951,1000
Detoxify, 881
Dextranase, 3
Dichomitus squalens, 778

Dictyostelium discoideum, 574
Dictyostelium sp., 950
Diffusion, 615,619
Diflubenzuron, 285
Digitonin, 284,287,295
Dimorphic, 932
Diplocarpon rosae, 581,582
Diploidia natalensis, 946
Disinfectants, 945,1082
Disomic, 623
Dissemination, 1072
Dopaminergic, 478
Double-stranded RNA, 3
Drosophila sp., 170,215
D. yakuba, 224
Drought, 826

Echinocandin, 23
Ecological niches, 281
Ecosystem, 1078
Ectoparasites, 1059
Eeniella nana, 217,221
Effluents, 933,939,943,947,951, 963,966,969,972,976,977
Eimeria tenella, 1027
Electron-dense inclusions, 910
Electrophiles, 892
Electrophilic, 892
Electroporation, 866
Electrostatic, 946
Elymoclavine, 488,489,490,494
Emericellopsis glabra, 112,114
E. salmosynnemata, 114,519
Emericellopsis sp., 3,107,519, 536
Emulsification, 920,926
Emulsifiers, 360,370,390,698, 920
Emulsifying agents, 913,915, 919
Encapsulation, 274
Endomyces monospora, 452
Endomycopsis fibuliger, 973
E. lipolytica, 884
E. vernalis, 386
Endoplasmic reticulum, 296
Endothia eugeniae, 58
E. parasitica, 690,692,699,736, 1015
Endothia sp., 458
Entomogenous fungi, 1010
Entomophaga aulicae, 446
Entomophaga sp., 447
Entomophordes sp., 28
Entomophthora sp., 132
Entropy, 642
Environmental Protection Agency, 865
Epicoccum nigrum, 386,409,416,421
Epidemics, 60
Epidemiological, 1071
Epidemiology, 83
Epidermophyton floccosum, 519
Epidermophyton sp., 38
Epoxidation, 885,933,935
Eremothecium ashbyii, 22
Ergocornine, 477,491,492,496,500, 506
Ergocristine, 477,492,496,499,500, 506
Ergocryptine, 477,491,492,496,500, 506
Ergoline, 476,478,485-488,490,500
Ergometrine, 476,506
Ergonovine, 487
Ergopeptine, 477
Ergosine, 447
Ergosterol, 21-23,370,382,394,632, 639,640,645,649,650,653,924
Ergostine, 477,491
Ergotamine, 477,480,489-492,496, 500,502,506
Ergotoxine, 480,491,492,498,502
Ergot(s), 27,380,475-478,482,485, 488-495,497,498,500-502,505, 506
Erwinia sp., 1027
Erysiphe graminis f. sp. *hordei*, 59,66,574
Erysiphe sp., 1
Escherichia coli, 10,11,84,88,158, 161,162,168-172,200,201,204, 205,207,223,243,325,328,445, 631,633,639,650,657,691,866, 1066,1075
Estrogenism, 27
Estrogens, 23
Ethanol, 3
Eufolliculina uhligi, 297,298
Euglena sp., 42
Eutrophication, 919
Excision, 245,246,251
Excretion, 459,582
Exobasidium myrtilli, 574,579
Exobasidium sp., 578,587,588
E. uvae-ursi, 574

Exons, 9
Exothermic, 712
Exotic, 1061
Exotoxins, 1068
Expandase, 10
Extracellular, 2,4,9,14,16,17,154, 161,267,270,274,298,378,380, 382-384,390,391,394,492,494, 497,506,542,582,614,616,633, 656,694,695,730-732,735,736, 739-742,745,746,766,768-770, 774-779,784,786-790,796,800, 805,807,895,909,912,913,920, 924,938,943,944,948,973,1013
Extrachromosomal, 862,864,865,866
Exudate(s), 827,832,838,851
Exudation, 833

Fatality, 1058
Flammulina velutipes, 449,1007
Flatulent, 697
Flavinogenic strains, 22
Flocculating agents, 17,263
Flocculation, 626,663,664,966
Flocculence, 624
Flocculent, 647-649,652,661,663, 665,667
Fluidization, 632
Fomes annosus, 142
Fomitopsis castanea, 687
Fossil fuel, 797
Freeze-drying, 1012,1013,1067
Fucosterol, 381,382
Fulvia fulvia, 85
Fumaric acid, 3,368,370
Fumigants, 825
Fungal Genetic Stock Center (FGSC), 994,1012
Fungicide(s), 86,499,500,714, 839,841,842,855,856,866, 921,922,938,945,1046
Fungistatic, 839
Fusants, 655
Fusarium acuminatum, 302
F. aquaeductuum, 421,422,430
F. coeruleum, 687
F. culmorum, 101,134,135,137
F. flocciferum, 949
F. fusarioides, 699
F. graminearum, 20,27,446,1008
F. javanicum, 888
F. lateritium, 28
F. merismoides, 249,253
F. moniliforme, 3,26,156,571,687,975, 1006
F. orthoceras, 687
F. oxysporum, 45,59,60,61,141, 142,156,168,248,253,687,689, 701,851,866,868,882,893,934, 938,1008
F. oxysporum f. sp. *conglutinans*, 248,254,865
F. oxysporum f. sp. *dianthi*, 44,45
F. oxysporum f. sp. *vasinfectum*, 584
F. roseum-culmorum, 1046
F. sambucinum, 156
F. solani, 63,85,249,253,687,892,894, 895,938,1008
F. solani f. sp. *cucurbitae*, 248,865
F. solani-phaseoli, 154,156
Fusarium sp., 3,19,23,39,44,46,248, 249,370,389,421,578,686,694, 698,839,895,923,924,934,935, 937,939,940,941,944,947,1006
F. trictum, 142
Fusicoccum amygdali, 587
Fusidic acid, 3
Fusidium coccineum, 3
Fusionogenic, 153
Fusogen, 108
Fusogenic agent, 108

Gaeumannomyces graminis, 58,142, 156,446,447,450,865
Gaeumannomyces sp., 58
Ganoderma applanatum, 801
G. colossum, 801
G. lucidum, 801
G. oregonense, 801
Ganoderma sp., 801
Gene amplification, 29
Gene expression, 322,324,330
Gene manipulation, 12, 151
Gene pool, 1008
Gene regulation, 790
Genetic code, 213,223,226,228
Genetic engineering, 17,29,180,208, 243,254,655-667,682,691,701, 746,841,849,859,861,865,948-951,995,1008,1055,1056,1061, 1063,1065,1068,1078,1084
Genetic information, 445,493
Genetic loci, 319,324-326,329,331, 332

Genetic manipulations, 2,9,22,82, 151,176,274,491,494,498, 681,730,745,746,851,852, 949,950
Genetic mapping, 950
Genetic reconstruction, 1069
Genetic transformation, 623,661, 666,950
Gene transfer, 151,154
Genome, 45,49,62,82-84,86,88, 89,118,151,160-164,176, 180,213-217,219,222-224, 226-228,243,244-246,250, 251,253-255,445,446-451, 457,479,604,605,622-624, 635,636,640,644,661,666, 667,783,864,866,949
Genotype, 30,36,68-70,82,84, 87,90,91,103,107,110, 826,829,830,833,859,1064
Geomycology, 946,947
Geotrichum candidum, 101,111, 113,135,137,386,392,393, 938,944,969,971,974,976, 977,980,981,983-985
G. lactis, 135,136
Geotrichum sp., 38,391,393
Germplasm, 993,994,995,1008, 1013
Gibberella fujikuroi, 26,140 386,409,413,418,419,421, 423,425,430,571,586,587, 1007
G. saubinetii, 686,1006
Gibberella sp., 417,430,578
G. zeae, 112,140
Gibberellic acid, 566,567
Gibberellins, 3,26,565,569
Gigaspora calospora, 863
G. decipiens, 856
Gigaspora sp., 861,863
Gilbertella persicaria, 292
Gleophyllum sp., 941
Gliocladium deliquescens, 884, 971,974,977,980,981,984
G. roseum, 937
Gliocladium sp., 945,947
Gloeoporus dichorus, 794
Gloeosporium, 578
Glomerella cingulata, 159
G. cingulata f. sp. *phaseoli*, 866
Glomus ambisporum, 854
G. deserticola, 858
G. epigaeum, 861
G. etunicatum, 861
G. fasciculatum, 852,862,863
G. intraradices, 835,854,861
G. margarita, 854
G. mosseae, 852,858,862
G. pecrocarpum, 861
Glomus sp., 58,856,858,861,862, 863,867
Glucoamylase, 3,10,17
Gluconic acid, 21,363,365
Glucose oxidase, 3
Glycolysis, 359,603,619,885
Glycosylation, 741
Graphium sp., 945
Grifola frondosa, 1007
Griseofulvin, 3,26,107
Gyrodon lividus, 218
G. merulioides, 218
Gyrodon sp., 221

Hallucinogen, 27
Hallucinogenic, 478
Haloprus cystisinus, 686
Hanseniospora valbyensis, 339,378
H. vinea, 218
Hansenula anomala, 105,113,338,339, 976,981,983,984
H. bimundalis, 113
H. capsulata, 105,114
H. mrakii, 216,218,221,451
H. petersonii, 218,221
H. polymorpha, 111,112,338,608
H. saturnus, 225,227
Hansenula sp., 457,621,945,972,979
H. wingei, 105,218
Hazardous agents, 1076
Hebeloma cylindrosporum, 860,866
Hebeloma sp., 574,860,864,866
Hedysarum boreale, 851
Helianthemum, 857
Helianthus, 857
Helicostylum piriforme, 884,885,886
Helicostylum sp., 887,945
Heliothis virescens, 1046
Helix pomatia, 133-136,141,144
Helminthosporium avenae, 574
H. gramineum, 134
H. oryzae, 579
H. sativum, 587
Helminthosporium sp., 578,886,935, 938,944,945
H. turcicum, 574
H. victoriae, 138,587

Hemileia vastatrix, 1
Herbicides, 28,365,431,855,886, 888,894,936,938,947,1046
Heterobasidion annosum, 57,61, 62,794,864,867
Heterokaryon(s), 6,7,47,71, 102,103,107,108,110-112, 114-117,176,179,299,357, 425,428,531,623,950,1010
Heterokaryosis, 6,7,176,179,498
Heterokaryotic, 6,7,8,47,62,110
Heterologous, 159,655,656
Heteroplasmons, 637,638,640, 644,662
Heterothallic, 5,6,22,39,40,46, 102,625-629
Heterothallism, 40,625
Heterotrophic, 2,494,949
Heterozygosis, 624,625,627
Heterozygotes, 7,67,68,71
Heterozygotic, 950
Heterozygous, 6,7,8,62,63,67, 158,176,178,622,625
Histoplasma capsulatum, 447
Homokaryons, 108,428,860,864
Homokaryotic, 6,7,47,63,860
Homologous, 159
Homothallic, 6,40,624-627,666
Homothallism, 40,622,625
Homozygosis, 621
Homozygotes, 7,67
Homozygous, 63,67,71,625
Hormone, 580,583
Humicola grisea, 720,732
H. grisea var. *thermoidea*, 710,733,735,737,740
H. insolens, 708,710,719,720 730-733,738,740-742
H. lanuginosa, 393,707,708, 710,713,715,720,730-739 742,745,945
H. stellata, 710
Humicola sp., 733,734,743
Humus, 853
Hybridomas, 1027,1062
Hydrilla verticillata, 1046
Hydrogenomonas sp., 934
Hydrolysis, 655
Hydrophobic, 93,283-285,296, 631,633,639,640
Hydrophobicity, 634,913
Hygromycin, 86
Hymenochaete tabacina, 801
Hypercholesterolemia, 27
Hypersensitivity, 1063
Hypertolerance, 657
Hypochytrium catenoides, 217
Hypoderma roseum, 141
Hypomyces ipomoeae, 249
H. ochraceus, 145
Hypovirulence, 458
Hypoxylon sp., 941
Hysterangium, 957

Idiophase, 489,505,536,538,540, 542,572,786
Immobilization, 259,389,499,506, 524,555,647,655,667,746,791, 795,796,913,948,949,951,1078, 1079,1084
Immune, 1014,1067
Immunization, 1065
Immunoassay, 1027
Immunogenic, 1063
Immunoglobulin, 1064
Immunoreactive, 1027
Imperial Mycological Bureau, 994
Incompatibility, 45,101-103,179,859, 860,864,867
Incompatible, 82,103,839-841,859,864
Inoculant(s), 823,825-827,832,833, 835,838-842
Inoculum, 363,492,493,501-504,538, 540,571,573,575,582,586,694, 718,721,791,825,826,828,831, 835,836,841,842,849-859,861-864,867,938,970,1012,1072
Inoculum potential, 840,861
Inonotus radiatus, 798
Insecticidal, 370
Insecticides, 855,881,885-888,937, 938,947,1068
Insulin, 3,1027
Interferon(s), 3,10,29,451,1067
Intergeneric, 38,40,103,104,107,113, 117,624
International Cell Research Organization (ICRO), 995
International Union of Biological Sciences (IUBS), 995
International Union of Microbiological Societies (IUMS), 995
Interspecific, 102,103,105,107,113, 116-118,145,179,499,500,624, 950
Intoxication, 1068

Intracellular, 14,16,17,250,298, 315,336,384,392,417,420, 422,459,496,497,537,542, 585,589
Intraspecific, 37,40,42,44-48, 58,102-104,107,111,116 145,499,607,614-616,655, 695,700,730-732,736,738, 740,742,778,827,910,914, 1013,1082,1084
Introns, 9,88,169,171,172,200, 205,215,216,220,222-225, 244,252,455,782,783,864
Inversions, 215
Irpex lacteus, 686
Isomerization, 934
Isospora sp., 215
Isozyme, 57-60,62-65,67,69-71
Itaconic acid, 3,21,365,366,367
IUMS, 995

Khuskia oryzae, 1007
Kinetin, 567
Kloeckera africana, 218,221,223
K. apriculata, 379,620
Kloeckera sp., 615
Kluyveromyces fragilis, 115, 116,606,608,618,642,645, 646,650,655,661,662
K. lactis, 104,106,114-116, 216,218,221,222,225,227, 453,457,655
K. marxianus, 613,643,647,654-656,659,969,970,973,977, 979,981,983,1006
K. mutabilis, 1007
K. polysporus, 379
Kluyveromyces sp., 457,655,698
K. thermotolerans, 227
Koji, 20,21,46,358,681,693,735, 1004
Kojic acid, 3,370
Krebs cycle, 20

Laccaria bicolor, 860,866,868
L. laccata, 850,851,859,868
Laccaria sp., 864
Lactase, 3
Lactobacillus sp., 651
Laetisaria arvalis, 380
Lagenidium giganteum, 221
Lagenidium sp., 28
Landfills, 967
Lentinula edodes, 61,63
Lentinus edodes, 143,768,772,794, 801,1007
Lentinus sp., 20
Lenzites styracina, 686
Leptinotarsa decemlineata, 27
Leptographium wageneri, 58,61
Leptomitus lacteus, 217,221
Leptomitus sp., 219
Leptosphaeria maculans, 157,168
Leptosthaeria oboines, 944
Lethal genes, 624,666
Leuconostoc sp., 698
Lipase, 3
Lipophilic, 881,924
Liquefaction, 656
Liquichimica, 916
Liquipron, 916,917
Litter, 718,720,857
Lodderomyces elongisporus, 116
Longevity, 1012
Lovastatin, 27,28
Lyophilization, 17,64,418,1011,1012, 1067
Lyophilized, 489,502,538
Lyophyllum shimeji, 141
Lysergic acid, 476-478,480,485-488, 493,496,500,504,506
Lysergic acid diethylamide, 478

Maceration, 697
Macerozyme, 129,140,143
Macrocyclic, 60
Macrolepiota procera, 1007
Macrophages, 273
Magnaporthe grisea, 85,157,168
Mahanarus posticata, 28
Malbranchea pulchella, 720,731,735, 742,744,745
M. pulchella var. *sulfurea*, 708,710, 713,715,733,736,738,741
Malbranchea sp., 744,746
Manipulation(s), 8,11,478
Mannoprotein, 273
Mantle, 851
Markers, 93,104,105,107,108,116-118, 132,154,155,158-161,165-167, 175,177,179,205,206,296,497, 499,521,622,659,865
Mating, 623,625,626,661,666
Maturation, 303,322,455
Melanin, 269,270
Melanocarpus albomyces, 711,731,733, 737,738

Meria oniospora, 1027
Merulius tremellosus, 772
Mesophiles, 707,715,729,733,734, 736,742
Mesophilic, 713,715,732,733,738, 741,742,971
Metarrhizium anisopliae, 3,27, 686
Methylation, 447,448,454,486, 779,795,891,892
Mevalonic acid, 497
Mevalonin, 27
Microbiological Resources Centers (MIRCENs), 995
Microfibrils, 270,281,289,290, 293-302,763
Micromonospora chalcea, 137
Micromonospora sp., 135,935
Micropollutants, 939
Microsomes, 883-886,907,944
Microsporum sp., 38
Microtubules, 301,302
Mineralization, 777,803,855, 931,934
MIRCEN, 995,1016
Monacolins, 27,28
Monascus albidus, 1004
M. albidus var. *globe*, 1003
M. anka, 1003
M. araneous, 1001
M. major, 1004
M. pilosus, 1001
M. pubigerus, 1001
M. purpurescens, 1004
M. purpureus, 1004
M. ruber, 27,28
M. rubropunctatus, 1002
Monascus sp., 1002,1004
Monilia sp., 945
Monilinia fructicola, 574
M. fructigena, 574
M. laxa, 574
Monoclonal, 1027
Monoculture, 856,859
Monooxygenase(s), 882,884-886, 907,910,915
Monosporium apiospermum, 687
Morbidity, 1058
Morchella angustinceps, 1008
M. conica, 250,253
M. crassipes, 1008
M. deliciosa, 1008
M. esculenta, 1008
M. rotunda, 1008
Morchella sp., 975
Morifolium ramat, 1027
Morphogen, 264
Mortality, 644,1058
Mortierella alpina, 389,390
M. elongata, 389
M. isabellina, 386,886
Mortierella sp., 389,390
M. vinacea, 135,282,283,386,692
Mosaic gene, 224
Mucor alterans, 934
M. ambiguus, 389
M. azygospora, 446
M. circinelloides, 10,84,85,87,157, 386
M. hiemalis, 391,687
M. javanicus, 389,892,893,895
M. miehei, 10,114,140,174,393,689, 690,692,699,700,709,713,730, 731,733,735,736,739,741,744
M. mucedo, 687,938
M. parasiticus, 885
M. plumbeus, 937
M. pusillus, 19,114,139,690,692,698, 699,707-709,713,714,716,718, 720,731,733,734-739,744,945
M. racemosus, 111,116,448,450,687
M. rouxii, 282,283,286,287,290-295, 297,298,301,890,932
Mucor sp., 3,19,20,163,293,391,393, 449,699,743,745,887,937,945
Muscarine, 27
Mutagen, 82,177,499,531,604,745,746
Mutagenesis, 21,23,91,132,178,201, 204,205,207,357,499,502,661, 692,784,788,949
Mutagenic, 482,502,520,531,605,637, 640,659,794,881,895,941,1062
Mutant(s), 8,9,11,13,21,22,24,26, 81,84,87-90,92,105-108,110, 116,132,133,155,158,159,162, 166,167,169,170,173,174,177, 197,200,206,223,244,246,249, 250,286,288,317,318,323,325, 327,360,370,410,411,419,421, 423,425,426,428-430,451,452, 456,459,487,491,494,498,499, 504,519-521,524,531,537,538, 604-606,609,613-615,622-624, 627-630,632,633,635-638,640, 641,644,657,659,660,665,667, 668,740,743,745,746,766,767,

[Mutant(s)]
774,776,777,778,784,787-790,792,796,797,806,866,917,923,944,993
Mutation, 4,6,7,8,12,24,43,44,46,60,69,82,84,85,89,90,104,106,108,155,162,172,175,178,197,204,245,246,266,318,322,337,422,424,425,447,498,499,502,520,531,532,554,605,623-625,629,635,637,640,642,643,659,666,667,693,694,745,894,931,949,975,1008,1010,1013
Mycelia sterilia, 686
Myceliophthora thermophila, 711,731,738,740
Mycena sp., 941
Mycobionts, 946
Mycogen, 28
Mycoherbicides, 28
Mycophenolic acid, 3,370
Mycoprotein, 20
Mycorrhizosphere, 589
Mycoses, 716
Mycoparasite(s), 858,1010
Mycosphere, 857,858
Mycotoxin, 39,691,891,1010
Mycoviral, 457
Mycoviruses, 1010
Myriococcum albomyces, 708, 709,744
M. solens, 720
Myriococcum sp., 1006
Myrothecium sp., 975
M. verrucaria, 974,980

Nadsonia fulvescens, 379
Necrosis, 450
Nectria galligena, 574
N. haematococca, 88,248
Nematicides, 855,938
Neocallimastic frontalis, 950
Neocosmophora vasinfecta, 587
Neoplastic, 1066
Neurospora africana, 40
N. crassa, 3,5,9,10,39,40,46,47,81,83-87,90,134,151-155,157-159,161-173,176,177,179,180,206,213,218,220-228,245,246,251,252,254,282,283,286,287,290,
[N. crassa]
292,293,301,313,315-320,322,323,325,327-330,332-334,336-338,393,408,410-413,416-419,421,423-425,429,430,446-449,451,453,457,459,460,686,742,947,950,1003
N. dodgei, 40
N. galapagosensis, 40
N. intermedia, 40,46,162,247,254,453,457,864,1003
N. lineolata, 40
N. sitophila, 20,40,46,408,686
N. terricola, 40
N. tetrasperma, 40,46,59,248,254
Neurospora sp., 5,6,11,38-40,42,46,160,162,164,174,216,246-248,254,255,318,323,324,328,336,340,413,418,421,430,446,449,453,459,782,950,994
Neutralization, 933
Niche, 707,826,931,1041
Nicotiana knightiana, 118
N. sylvestris, 118
N. tabacum, 109
Nigrospora oryzae, 26
Nigrospora sp., 945,1012
Nikkomycin, 284,285,288
Nocardia sp., 894
Nonaflatoxigenic, 892
Nonphytotoxic, 573
Northern Regional Research Laboratory (NRRL), 994,995,999,1009,1019,1051-1053
Novozym, 83
Nucleophiles, 892
Nucleosides, 568
Nucleotides, 445,453,455,458,459,568
Nystatin, 32,107

Ochratoxin, 27
Oenothera sp., 388
Oerskovia sp., 138
O. xanthineolytica, 152
Oleaginous fungi, 377,384-386,393,394
Oncogene, 1027,1062,1066
Oncogenic, 1061,1062
Oospora laetis, 392
Ophiobolus graminis, 132,138
Ophiostoma ulmi, 157
Organopollutants, 803,804,895
Osmolarity regulators, 852

Osmophilic, 607,660,734
Osmotic shock, 292,299
Osmotic stabilant, 83
Osmotolerance, 607,624,666
Ostrina nubialis, 27
Oudemansiella mucida, 111

Pachytichospora transvaalensis, 218
Paecilomyces carneus, 114,923, 924
P. elegans, 974,980
P. farinosus, 686
P. lilacinus, 687
P. parasiticus, 519
P. persicinus, 519,536
Paecilomyces sp., 370,389,536, 711,886,945
P. subglobosus, 1005
P. varioti, 3,20,366,946,975, 976,977,980-984,
Panus tigrinus, 772
Papulospora sp., 941
Papulospora thermophila, 711, 731
Paragyrodon sphaerosporus, 218,221
Paramecium sp., 215
Parasexual, 6,7,26,81,82,102, 103,110,176,178,357,499, 531,950,1010
Parasexuality, 6,110,864
Partition coefficients, 634
Paspalic acid, 487,504
Pasterella pneumotropica, 1061
Pasteurization, 700
Pathogenesis, 577
Paxillus filamentous, 946
P. involutus, 851
Peanut mottle virus, 1046
Peanut stripe virus, 1046
Pectinase, 3
Pellet(s), 358,694,791,915, 975,1018
Pellicularia filamentosa, 1006
Penicillin acylase, 3
Penicillium aculeatum, 1006
P. album, 11,114
P. amagasakaiense, 700
P. avellaneum, 518
P. baarnense, 114
P. bacculatum, 518
P. camembertii, 1003
P. canescens, 687
P. caseicolum, 111,114
P. chrysogenum, 3,6-10,21,24,81,85, 86,91,103,104,111-114,116-118, 136,137,139,151,155,157-160, 165,168,170,197,198,200,201, 203-206,218,318,323-325,330, 334,339,517-519,521,524-531, 534-536,538,543-554,584,884, 894,935,950,1005,1007
P. citreonigrum, 686
P. citreoroseum, 518
P. citrinum, 27,28,113,114
P. crustosum, 587,974
P. cyaneo-fulvum, 103,111,114
P. cyaneum, 389
P. cyclopium, 113,885,938
P. digitatum, 585,892,946
P. dupontii, 330,697,733,735,736,742
P. emersonii, 688,689,697
P. expansum, 687,696
P. fluorescens, 518
P. frequentans, 111,113
P. funiculosum, 687,688,696,945,1006
P. glabrum, 686,687
P. glaucum, 715
P. granulatum, 687
P. griseofulvum, 26
P. griseum, 687
P. jensenii, 938,974,1005
P. lanosum, 113
P. lilacinum, 386,699
P. nalgiovense, 157,159
P. nigricans, 113
P. notatum, 19,111,113,370,518, 700,935,937,974
P. oxalicum, 947,975,980
P. patulum, 3,26,111,114,117
P. puberulum, 1005
P. purpurescens, 687
P. purpurogenum, 687
P. purpurogenum var. *rubri-sclerotium*, 21
P. raistrichii, 587,687
P. ramigena, 111,113
P. roquefortii, 3,31,103,104,111,113, 114,117,118,686,1003
P. rubens, 518
Penicillium sp., 3,8,9,19,23,26,58, 102,108,160,174,206,323,357, 363,366,370,389,391,393,517, 518,532,533,535-538,555,578, 688,689,698,708,711,886,888,

[*Penicillium* sp.]
935,937-940,944-947,950,
970,974,975,979,980
P. spinulosum, 386,686,687
P. stoloniferum, 3,21,113,
114,117,370
P. thomii, 687
P. turbatum, 518
P. urticae, 935
P. variabile, 587
P. velutinum, 687
P. verrucosum var. *cyclopium*,
114
P. verruculosum, 112
P. waksmani, 687,936
Peniophora gigantea, 793,794
Peniophora sp., 408
Permeability, 785,827
Peronosclerospora philippinensis,
58
P. sacchari, 58
P. sorghi, 61
Peroxisomes, 910
Pestalotina cryptomariaecola,
587
Pestalozzia annulata, 249
Pesticide, 565,825,832,840,842,
856,861,892,894,932-934,
936-939,1035
Phaeospharia typharum, 944
Phaffia rhodozyme, 408,409,415
Phaffia sp., 415
Phakopsora pachyrhizi, 59,60
Phallinus weirii, 934
Phallotoxins, 27
Phanerochaete chrysosporium,
48,111,157,446,686,767-
772,774-780,782-800,802-
807,895,909,934,941,943,
947,948,950,1006
Phanerochaete sp., 941,975
Phascolomyces articulosus, 282
Phenobarbitone, 884,893,896
Pheromone, 450
Phlebia brevispora, 793
P. radiata, 772,782,783,790,807
P. subserialis, 793
P. tremellosa, 793,801
Pholiota nameko, 1008
Phoma sp., 947
Phosphorylation, 4,894
Photodegradation, 933
Photodermatoses, 401
Photodestruction, 931
Photoinduction, 422,429,430
Photooxidation, 943
Photoprotection, 401
Photoreceptor, 420,421
Photoresponse, 430
Photosensitive, 401
Phototropism, 428,429
Phycomyces blakesleeanus, 3,22,42,
111,137,157,158,163,170,217,
286,288,292,295,296,299,301,
410,411,412,416-419,421-426,
428,431,446
Phycomyces sp., 158,172,283,286,
292,298,299,409,410,416-422,
428,430,447
Phyllotopsis nidulans, 59,61
Phymatotrichum omnivorum, 459
Physarum polycephalum, 216,217
Physarum sp., 220
Phytin, 853
Phytoalexins, 833
Phytoene, 410,412-419,421,423-427,
431
Phytohormones, 26,565
Phytoinhibitors, 565
Phytopathogenic, 581
Phytopathogens, 175
Phytophthora cactorum, 44
P. cinnamomi, 61,62
P. cryptogea, 217,449
P. infestans, 1,61,62,135,217,451,
587,1007
P. megasperma, 44,58,217,221,227
P. megasperma f. sp. *glycinea*, 112
P. megasperma var. *medicaginus*, 58
P. megasperma var. *sojae*, 861
P. oligandrum, 44
P. parasitica, 217,221
Phytophthora sp., 58,219,225,226,
228,379,451,578,839,851,867
Phytotoxin, 1027
Picea sp., 852
Pichia farinosa, 1005
Pichia sp., 655
P. stipitis, 656
P. tainania, 972,977,979
Pinocytosis, 910,914
Pinus pinaster, 860
P. radiata, 858
P. resinosa, 851
Pinus sp., 852
Pisolithus tinctorius, 852-854,859
Placopecten sp., 215

Plasmid(s), 865,866,1065,1075
Plasmodiophora brassicae, 574
Plasticizers, 21,888
Plectania coccinea, 409
Pleurochrysis scherffelii, 298
Pleurotus cystidiosus, 1008
P. decumbens, 944
P. eryngii, 801
P. nebrodensis, 59
P. ostreatus, 20,59,143,446,584, 701,768,772,801,890,934, 943,1008
P. sajor caju, 574,801,944,1008
P. salmoneostramineus, 143
P. serotinus, 801
Pleurotus sp., 58,801,802
Podocarpus nagi, 587
Podospora anserina, 10,85,87, 152,155,157,160,161,163, 165,171,178,213,218,221, 223-228,244-246,250-252, 451,452
P. curvicolla, 245,246,252
Podospora sp., 160,163,178, 216,225,246,950
Polaria, 941
Pollutants, 719,764,804,806,807, 850,852,854,865,881,894 895,933,943,945,949,951
Pollution, 741,792,855,918,919, 943,945,947,948,950,951, 977,986
Polycyclic, 905,917,918,932
Polymorphic, 60-62
Polymorphism, 36,43-45,47, 48,60,224,447,450,623
Polyols, 3
Polyoxin, 284,285,287,300
Polyploidization, 23
Polyporus arculais, 801
P. dichomus, 798
P. galactinus, 801
P. ostreiformis, 686
P. palustris, 686,1006
P. versicolor, 795,800,934,943
Polysphondylium pallidum, 298
Population genetics, 57
Poria expansa, 801
P. monticoloa, 800
Poria sp., 941
Poterioochromonas stipitata, 297
Precursor(s), 8,10,15,25,26,91, 197,223,270,297,298,322,
[Precursor(s)] 359,370,388,389,394,401,414, 422,423,454,456,486,489,490, 497,518-521,533,536,537,540-542,544,571,572,580,581,585, 701,764,907,915,948
Priority date, 1033,1055
Prochymosin, 92
Proctoids, 291
Progesterone, 23,742,884
Proliferation, 644,1069
Promoter, 10,161,166,171-173,175, 207,866,867
Propagation, 493,691
Propagules, 720,729
Protease, 3
Proteolysis, 772
Proteus vulgaris, 366
Protomyces sp., 381
Protoperithecia, 5
Protoperithecium, 5
Protoplast fusion, 82,101-103,106-111, 113,116-118,145,197,499,500, 623,624,626,655,657,659,661, 665,666,667,701,746,949,950, 951
Prototroph, 9
Prototrophic, 6,8,104,117,661
Prototrophy, 660
Psalliota compestris, 939
Pseudocercosporella herpetrichoides, 143,157
Pseudohomothallic, 40,46
Pseudohomothallism, 40
Pseudomonas aeruginosa, 522
Pseudomonas putida, 858,949
Pseudomonas sp., 800,857,932,936
Psilocin, 27
Psilocybe mexicana
Psoriasis, 22
Psychrophiles, 707,729
Puccinia graminis, 1,59,62
P. graminis f. sp. *secalis*, 62
P. graminis f. sp. *tritici*, 59-62,65
P. hordei, 61
P. recondita, 59
P. recondita f. sp. *tritici*, 61,62
P. sorghi, 59
Puccinia sp., 58,60
P. striiformis, 60
P. striiformis f. sp. *hordei*, 61
P. striiformis f. sp. *tritici*, 61
Pullulan, 3,259-270

Pullularia pullulans, 383,945
Pullularia sp., 366,938
Pycnidia, 1047
Pycnoporus cinnabarinus, 686
P. sangguineus, 686
Pyricularia oryzae, 60,61,113, 138,139,284
Pyrodictium brackii, 729
P. furiosum, 729
Pyrus betulaefolia, 829
Pyruvic acid, 370
Pythium diclinum, 217
P. irregulare, 217
P. paddicum, 217
Pythium sp., 44,135,136,379, 451,839,851,856,892
P. torulosum, 217,451
P. ultimum, 217,221,386,389, 390,584

Recalcitrant, 445,881,906,910, 919,934,949
Reclamation, 825,826,852,853
Recombinant(s), 7,8,11,26,44, 47,82,90,102,103,105,110, 112,114,155,200,205,207, 243,445,460,498,499,500, 521,622,655,784,790,807, 949,1027,1056,1065,1066, 1075,1077,1078,1082,1085
Recombination, 6,7,9,12,62,87, 88,90,102,104,110,116,163, 164,166,169,176,177-180, 205,215,219,222,357,452, 498-500,622-624,701,702, 746,788,864,943,949,950, 1010,1013
Redox potential, 774
Reforestation, 859
Regeneration, 950
Regiospecific, 885
Renin, 699,735,736
Rennet, 3,692,699,735
Rennin, 10
Replication, 11,162,163,246,250, 253,255
Replicon, 159,163
Repression, 25
Repressor, 207
Resistance, 8,10,11,24,86,87, 104-107,132,157-159,165, 168,169,172,206,248,274, 338,339,499,500,521,537,
[Resistance] 620,621,631,640,659,663,736, 823,851,852,861,865,866,911, 950,970,981,1020,1067,1071, 1080
Resistant, 8,9,21,24,87,93,104-106, 108,117,118,158,159,169,197, 206,288,290,297,302,338,370, 499,531,609,621,640,643,657, 734,745,866,949
Restriction enzymes, 43-46
Retinoids, 401
Retinol, 422,425,428,429
Retrotransposons, 450
Reverse genetics, 200,204
Rhizidiomyces apophysatus, 217
Rhizobium sp., 851,858,861,862,866
Rhizoctonia bataticola, 938
R. solani, 113,129,132,140,144,248, 249,253,254,285,584,865,892, 938
Rhizoctonia sp., 40,839,857,938,939
Rhizomucor miehei, 394,702,736
R. puaillua, 686,735,736
Rhizomucor sp., 714
Rhizophlyctis rosea, 409
Rhizoplane, 857
Rhizopogon luteolus, 574
R. occidentales, 859
R. ochracearubens, 574
R. roseolus, 574
Rhizopogon sp., 58,59,867
R. vinicolor, 850,859
Rhizopus arrhizus, 38,366,368,393, 689,885,937,946,1002,1005
R. chiensis, 41
R. chiensis var. *liquefaciens*, 41, 1003
R. cohnii, 41,1004
R. delemar, 39,688,689,695,700
R. formesaensis, 1005
R. japonicus, 368,938
R. javanicus, 1005
R. javanicus var. *kauasakiensis*, 1004
R. microsporus, 41,391,730,1004, 1005
R. microsporus var. *liquefaciens*, 1004
R. nigricans, 3,20,23,135,368,715, 882,884,895
R. niveus, 688,692
R. oligosporus, 20,41,391,686,1004

R. oryzae, 39,368,687,692, 698,1004-1006
R. pseudochinensis, 41
R. pygmaeus, 41
R. rhizopodiformis, 41,699,732, 739,1006
R. roseolus, 574
Rhizopus sp., 3,17,18,23,41 368,370,391,393,578,693, 700,709,887,892,945,946, 975
R. stolonifer, 455,884,886,893, 894
R. suinus, 577,579,580,588
Rhizosphere, 574,577,590, 719,828,857,862
Rhodosporidium toruloides, 103,112,116
Rhodotorula aureum, 379
R. glutinis, 339,383-385, 409,416,973
R. gracilis, 386,416,933
R. graminis, 383,386
R. minuta, 421
R. mucilaginosa, 416
R. rubra, 416
Rhodotorula sp., 380,416,417, 945,947,972,979
Rhynchosporium secalis, 59
Ribes sp., 388
Riboflavin, 3,22
Ricinoleic acid, 497
Ripening, 700
Root knot, 1046
Rumen, 800

Saccharification, 656,741,948
Saccharomyces aceti, 1002,1005
S. aviformis, 620
S. bayanus, 608,1001,1002
S. bailei, 612,660,1001
S. capensis, 1003
S. carlsbergensis, 216,282,283, 286,453,454,612,614,660, 661,884
S. cerevisiae, 3,5,11,29,83,93, 102-107,109,112-116,118, 152,160,161,170,180,205, 213,216,218,220-228,266, 282,286,287,290-292,295, 299,313,316-318,320,323, 324,327,328,330,332,333, 336,339,378,379,381-383,
[*S. cerevisiae*] 392,394,445-450,452-455,457, 458,610-616,618-621,623,627, 631,634-636,639-644,646-656, 660,661,782,884,894,934,946, 950,1001-1003,1007,1015
S. cerevisiae var. *ellipsoideus*, 610,653
S. chevalieri, 1001,1003
S. clavuligerus, 11,204
S. cordubensis, 626
S. diastaticus, 106,113,114,612,623, 624,656,660,661,665,666,1002
S. exiguus, 218,221,224
S. formosensis, 653
S. gaditensis, 626
S. hispanica, 626
S. italicus, 1003
S. kluyveri, 453
S. malanga, 380
S. mellis, 624
S. montanus, 108
S. oxidans, 626
S. rosei, 105,1002,1003
S. rouxii, 20,612,660,665,1003,1004
S. sake, 610,614,651,653,654,1002
Saccharomyces sp., 3,5,42,115,288, 293,298,327,386,449,452,453, 456,457,603-605,607,609-612, 615,618,620-622,625,627,629-633,635,643,645,649,650,655-657,659,661,663,666,945,1002, 1003,1005,1006,1015
S. tilluris, 218
S. unisporus, 112,218,1001
S. uvarum, 113-115,118,607,610,613, 615,617,620,624,627,635,643, 651,653,654,658,659,665,666, 1001
S. uvarum var. *inulyticus*, 652
S. vini, 1002
Saccharomycopsis fibuligera, 104, 114,1003,1004
S. italicus, 1003
S. lipolytica, 103,112,114,116,218, 221,323,339,361,391,392,973, 1015
S. malanga, 1003,1004
Sake, 603,658
Saprolegnia ferax, 37,39,217,221
S. parasitica, 390
Saprolegnia sp., 38,49,217,446,447, 947

Sapromyces elongatus, 217
Schizochytrium aggregatum, 379
Schizophyllan, 3,259,270,272,273
Schizophyllum commune, 3,37,38, 47,48,63,84,85,157,165, 170,171,178,218,225,227, 270,273,283,284,287,290, 297,299,366,370,446,455, 738,798,866
Schizophyllum sp., 163,228
Schizosaccharomyces octosporous, 114
S. pombe, 103-105,112,114-116, 213,216,218,221-228,330, 447,449,453,456,459,652, 884
Schizosaccharomyces sp., 449
Schwanniomyces alluvius, 112, 113,116
S. castellii, 114,621
Scleroderma flavidum, 852
Scleroglucan, 259,260,270-273
Sclerotia, 380,475,479,481,485, 494,497,498,500,501,1047
Sclerotinia fircariae, 47
S. sclerotiorum, 47,316,587
Sclerotinia sp., 46,47,270,273
S. trifoliorum, 47
Sclerotium, 478
Sclerotium glucanicum, 270,273
S. rolfsii, 270,273
Scopulariopsis brevicaulis, 687
Scytalidium flavo-brunneum, 453
S. indonesiacum, 711
S. lignicolum, 687,944
S. thermophilum, 711
Scytinostroma galactinum, 58,62
S. protrusum, 62
S. protrusum subsp. *protrusum*, 62
S. protrusum subsp. *septentrionale*, 62
Scytinostroma sp., 58
Sedimentation, 931,977
Sedimentation coefficients, 287, 291,295,296,454
Segregation, 57,63,81
Senescence, 244-246,249,251,252, 255,451,452,864
Sensitivity, 106,107
Septoria musiva, 584,585,640,840
S. nodorum, 85,157,168,178,866
Sequestering agent, 21
Sex hormone, 401
Shelf life, 1011,1012
Shiitake, 20
Shoyu, 20
Siderophore, 800
Siredon sp., 215
Sitosterol, 742
Smittium culicis, 454
Soft rot, 767
Soil borne, 825
Solubilization, 299,800,828,855,946
Somatic fusion, 623
Sophorosides, 381,390,394
Specialization, 60
Specification, 1031
Specificity, 336,394,855,859,865,867
Sphacelotheca sorgii, 339
Sphaerobolus stellatus, 408
Sphearomonas communis, 950
Spheroplasts, 152,153,624,661
Spherosomes, 410
Spiroidium sp., 536
Sporopollenin, 401
Sporothrix schenckii, 38,41
Sporotrichum cellulophilum, 696,739, 741,746,1006
S. dimorphosporum, 448
S. pulverulentum, 48,719,731,740, 741,743,745,767,793,801,944
S. thermophile, 711,720,731-733, 735,737-740,742,743,745,746
Sporotrichum sp., 732,743
Sporozoite, 1027
Stachybotrys sp., 389,945
Staphylococcus sp., 291
Stemphylium macrosporoideum, 687
Sterigmatocystin, 891
Stigmasterol, 742
Stilbella thermophila, 711
Stimulants, 197
Stimulate, 858
Stimulation, 854,857
Stimulator, 417,420
Streptococcus lactis, 368
Streptomyces cacaio var. *asoensis*, 284
S. clavuligerus, 520,537,553
S. graminofaciens, 136
S. griseus, 133,137,886
S. lipmanii, 520,537
Streptomyces sp., 134,135,137,139, 174,200,204,205,555,737,800, 884

S. tendae, 284
S. verdensis, 1027
Stropharia aeruginosa, 1008
S. rugosoannulata, 1008
Stropharia spp., 446
Succession, 853,857
Succinic acid, 370
Sufu, 20
Suillus americanus, 218
S. bellini, 866
S. brevipes, 218
S. cavipes, 218
S. granulatus, 854,860
S. grevillei, 218
S. grisellus, 218
S. luteus, 218,852,864
S. neoalbidipes, 218
S. ochraceoroseus, 218
S. placidus, 218
S. punctipes, 574
S. sinuspaulianus, 218
Suillus sp., 216,219,221,225,451, 864
S. spectabilis, 219
S. spraguei, 219
S. subalutaceous, 218
S. tomentosus, 219
S. tridentinus, 219
S. variegatus, 219
S. viscidus, 219
Suppression, 840,850,851,856, 858
Suppressor, 89
Surfactants, 368,377,387,389-391,394,738,784,911,914, 915,919,920-922,924
Surfactin, 921
Susceptibility, 263,290,323,336, 607,797,932,1067
Susceptible, 272,297,865,888, 923,1062
Symbionts, 445,828,841,855, 858,859
Symbiosis, 858-860
Symbiotic, 2,823,840
Syncephalastrum sp., 392
Synergistic, 144,604,615,616, 639,644,645,666,852
Syntrophism, 7

Talaromyces byssochlamydoides, 709,738
T. dupontii, 708,710,720,730,731, 733,742
T. emersonii, 646,701,710,713,730-733,736-742,745,746
Talaromyces sp., 739,745,950
T. spiculisporus, 687
T. thermophilus, 687,710,713,730, 731,733,735-737,1007
T. wortmani, 938
Taphrina amentorum, 574
T. betulina, 574
T. cerasi, 574
T. deformans, 574
Taphrina sp., 381,574,578,588
Teleomorph(s), 39,42,48,249
Tempeh, 681,735
Terfezia sp., 381
Tetrahymena sp., 42,163,215,220,252
Thalassiosira fluviatilis, 298
Thamnidium elegans, 379
Thamnidium sp., 887
Thelophora terrestris, 854
Therapeutic, 175,475,478
Thermoascus aurantiacus, 701,708, 710,720,731-733,737-740,742, 743
T. crustaceus, 710,736,737
Thermoascus sp., 742,744
Thermoidium sulphureum, 711
Thermogenesis, 715,721
Thermoidium sulphureum, 711
Thermolabile, 734
Thermomucor indicae-seudaticae, 730,731,745
Thermomyces ibadanensis, 711
T. lanuginosus, 711,714,719,720, 731-733,742,746
Thermophiles, 701,707,714,715,720, 721,729,730,732,733,735,740, 742,743,745
Thermophilic, 2,654,691,699,707-709,712-716,718-721,729-737, 740-746,801
Thermophilism, 711
Thermophilous, 707,715,720
Thermosensitive, 629,642
Thermostable, 696,698,699,732-734, 736-738,742
Thermotolerance, 624,643,663,665
Thermotolerant, 104,665,701,708,716, 718,720,721,732,743
Thermus aquaticus, 696
Thielavia australiensis, 710
T. minor, 732,733
T. terrestris, 696,737-740,1006

T. thermophila, 710,733
Thielaviopsis sp., 839
Thraustochytrium roseum, 379
Tilletia barclayana, 59
Tilletia controversa, 865
T. indica, 59,61-63
T. tritici, 1
Tintoporia borbonica, 795
Tofu, 20,681
Tolypocladium inflatum, 3
Torula sp., 339,972,973
T. thermophila, 711,730-733,735, 736,945
T. utilis, 973
Torula sp., 339,972,973
Torulaspora florentina, 1002
T. vafer, 1002
Torulopsis bombicola, 382,390,391, 923
T. colliculosa, 379
T. dattila, 379
T. etchellsii, 1003,1004
T. fujisanesis, 380
T. glabrata, 44,104,112,216,218, 221-224,227,451,453,1004
T. gropengiesseri, 920,923
T. halonitratophila, 1004
T. magnoliae, 381,920,923
T. mannitofaciens, 1004
T. noda, 1004
T. pinus, 973
T. petrophilum, 390
T. utilis, 937
T. versatilis, 1003,1004
Torulopsis sp., 382,383,391, 448,449,457,920,945, 972,1005,1015
Toxic, 475,521,531,542,605, 639,647,657,690,764, 794,795,802,803,805,839, 850,855,881,887,892,894-896,936,941,942,945,946, 949,985,1014,1066
Toxicant(s), 932,945
Toxicity, 8,614,917,932,945, 974,985,1020,1068
Toxin, 86,457,656,666,865, 1065
Trade secret, 1034-1042,1045
Trametes hirsuta, 801
T. sanguineara, 19
T. versicolor, 772,783,794,795, 798,807
Transaminase, 319
Transamination(s), 4,316,317,322, 330
Transcription, 29,88,92,170,173, 245,247,337,421,449,450,454, 782
Transduction, 421
Transfection, 950
Transferases, 533
Transformation(s), 9,10,11,23,25, 83,84,86-90,93,102,118,129, 132,144,145,151-155,158-167, 171-173,175-177,179,180,200, 205,206,288,388,394,430,448, 487,603,624,659,719,784,798, 806,807,865,866,882,885,886, 906,925,926,931,936,948,950, 1066
Transgenic, 1027
Translation, 170,171,173-175,790
Translocation, 2,450
Transmethylation, 486
Transmission, 1064,1067,1075
Transplantation, 850,853
Transpositions, 215,450,451
Transposons, 243,245,249,252,253, 500,864
Transsulfuration, 534,535
Tremella fusiformis, 801,1008
T. longibrachiatum, 944
T. messenterica, 1008
T. ochrasporum, 944
T. reesei, 655,696,702,739,790, 944,950,976,983,1006
T. viride, 688,743,933-938,945,974, 975,977,980-982
Trichoderma aureoviride, 334
T. harzianum, 133,137,143,144,152, 206,946,950
T. koningii, 115
Trichoderma sp., 133,136,138,141, 174,389,937,939,944,975,976
T. reesei, 3,108,112,115,141,157, 159,168,175
T. roseum, 134
T. viride, 19,111,115,132,137,138, 140,141,574
Tricholoma matsutake, 140
T. muscarium, 27
T. nudum, 975
Tricholoma sp., 975
Tricholomic acid, 27
Trichomonas vaginalis, 584

Trichophyton mentagrophytes, 519
Trichophyton sp., 38
Trichosporon cutaneum, 882,909, 932,946,948,949,973
T. fermentans, 973
T. pullulans, 386
Trichosporon sp., 945,947
Trichosporum cutaneum, 338,339
Trichothecium roseum, 885
Trifolium subterraneum, 834,836
Trigonopsis sp., 3,19
T. variabilis, 379
Triple helix, 270-273
Trophophase, 489,536,538,541,572
Trypsin, 286,292,293
Tryptophan, 314,328,330,331,334, 338,486-490,495,497,570, 577-579,860
Tumorigenic, 1062
Tumorigenicity, 1062
Tyromyces unicolor, 48

Urogastrone, 29
Uromyces appendiculatus, 60,61
U. fabae, 574
U. phaseoli, 574
Ustilago bullata, 61-63
U. cynodontis, 219
U. esculenta, 574
U. maydis, 85,137,157,158,165, 178,339,383,391
U. nuda, 383
Ustilago sp., 158,160,408,423, 453
U. spinificis, 61
U. violacea, 154,423,424
U. zeae, 1,382,383

VA mycorrhizae, 827,829,839, 854,856,858
VA mycorrhizal association, 824,835,838,842,861
VA mycorrhizal fungi, 823, 825,828,832-834,839-841,842,856-859,861, 866
Variability, 57,58,60,62,83, 110,501,627,693,858-861, 864,868,1011
Variable, 626
Variation, 449
Variotin, 3

Vector, 11,29,154,155,158,160-167, 170,172,174,176,180,200,201, 204-206,254,659,862,863,865, 866,1027,1065,1066,1075
Velum, 636
Verticillium agaricinum, 416
V. albo-atrum, 103,112,114,141,142, 416
V. dahaliae, 103,112,114,141,584, 585
V. lecanii, 28,103,112,687
Verticillium sp., 103,420,886
Vesicles, 824
Vesicular-arbuscular (VA) mycorrhizal fungi, 823
Viability, 24,153,502,607-609,614, 623,625,629,630,638-640,644, 645,646,649,650,651,653,658, 663,665,667,716,831,1010-1012, 1018,1085
Viable, 609,630,657,659,660,1031, 1060,1065,1084,1085
Vinca rosea, 109
Virulence, 59,60,175,865,866
Virulent, 1066
Volvariella bambycina, 1008
V. esculenta, 1008
V. volvaceae, 20,61,63,801,1008
Vulnerable, 1074

Water activity (a_w), 712,713
White rot, 767,768,778,779,784,785, 792,794,795,796,801,806,807, 909,950
Wild-type, 84,107,108,158,159,166, 174,175,201,246,249,336,411, 421-424,428,456,627,629,630, 633,635,745,784,788,790,860, 1063
Wilting, 851
World Data Center of Microorganisms (WDC), 994
World Federation of Culture Collections (WFCC), 995
Wort, 697

Xanthan, 263,272
Xanthomonas campestris, 272,866
Xenobiotics, 764,802-804,806,807, 881,882,884,886,890,891,894, 895,931,938
Xenopus sp., 215
Xerocomus chrysenterum, 219,221

Xerophilic, 734
Xylaria sp., 941,975

Yarrowia lipolytica, 114,449, 450,1005,1007,1008,1015
Yarrowia sp., 361,449
Yeast Genetic Stock Center (YGSC), 994,999
Zygosaccharomyces sp., 655
Zymomonas mobilis, 606,648,653
Zymomonas sp., 649